ARCHITECTURAL, ENERGY AND INFORMATION ENGINEERING

PROCEEDINGS OF THE 2015 INTERNATIONAL CONFERENCE ON ARCHITECTURAL, ENERGY AND INFORMATION ENGINEERING (AEIE 2015), XIAMEN, CHINA, 19–20 MAY 2015

Architectural, Energy and Information Engineering

Editors

Wen-Pei Sung
National Chin-Yi University of Technology, Taiwan

Ran Chen
Chongqing University of Technology, Chongqing, China

CRC Press
Taylor & Francis Group
Boca Raton London New York Leiden

CRC Press is an imprint of the
Taylor & Francis Group, an **informa** business

A BALKEMA BOOK

Description of cover illustration: Architectural and Energy
Copyright 2012: Wenli Yao

CRC Press/Balkema is an imprint of the Taylor & Francis Group, an informa business

© 2016 Taylor & Francis Group, London, UK

Typeset by V Publishing Solutions Pvt Ltd., Chennai, India

Published by: CRC Press/Balkema
      P.O. Box 11320, 2301 EH Leiden, The Netherlands
      e-mail: Pub.NL@taylorandfrancis.com
      www.crcpress.com – www.taylorandfrancis.com

ISBN: 978-1-138-02791-6 (Hbk)
ISBN: 978-1-315-68700-1 (eBook PDF)

Architectural, Energy and Information Engineering – Sung & Chen (Eds)
© 2016 Taylor & Francis Group, London, ISBN 978-1-138-02791-6

Table of contents

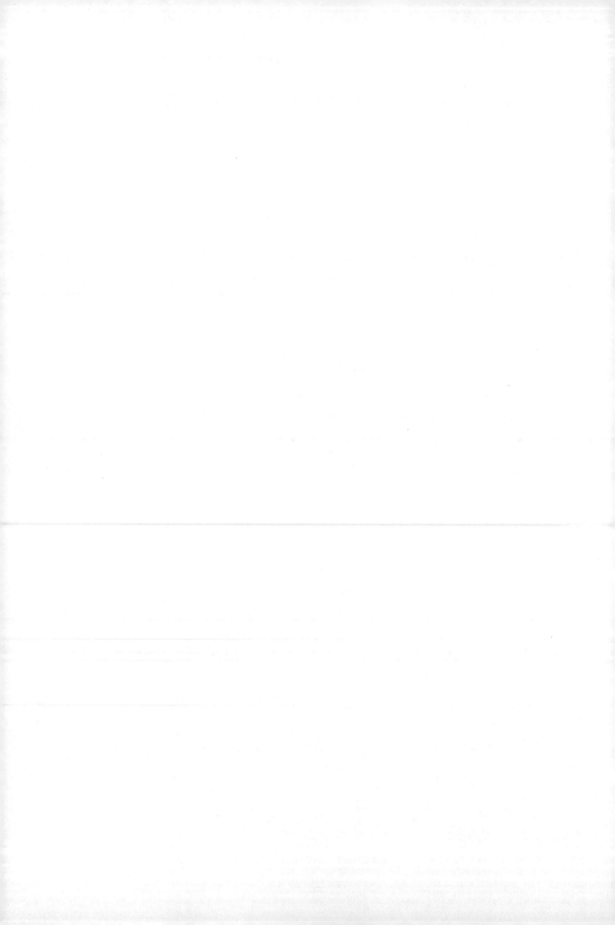

Preface

This book contains select best Architectural, Energy and Information Engineering related papers from the 2015 International Conference on Architectural, Energy and Information Engineering (AEIE 2015) which was held in Hong Kong, during July 15–16, 2015.

Along with the rapid development of society, demand of energy source increases continually. The continued strengthening of requirements for construction energy-saving and the continuous improvement of the living environment encourage energy-saving technologies of the construction to flourish. There are some 228 Architectural, Energy, Environmental Engineering and Information Engineering related papers in these proceedings. The volume is divided into two parts. The first part discusses Architectural, Energy and Environmental Engineering. The second part presents Information Engineering and Its Application. I think the proceedings will promote the development of Architectural, Energy, Environmental Engineering and Information Engineering, strengthening international academic cooperation and communications, and the exchange of research ideas.

I am very grateful to the conference chairs, organization staff, the authors and the members of International Technological Committees for their hard work. We look forward to seeing all of you next year at AEIE 2016.

June, 2015

Wen-Pei Sung
National Chin-Yi University of Technology, Taiwan

Architectural, Energy and Information Engineering – Sung & Chen (Eds)
© 2016 Taylor & Francis Group, London, ISBN 978-1-138-02791-6

AEIE 2015 committee

CONFERENCE CHAIRMAN

Prof. Wen-Pei Sung, *National Chin-Yi University of Technology, Taiwan*

PROGRAM COMMITTEE

Yan Wang, *The University of Nottingham, UK*
Yu-Kuang Zhao, *National Chin-Yi University of Technology, Taiwan*
Yi-Ying Chang, *National Chin-Yi University of Technology, Taiwan*
Darius Bacinskas, *Vilnius Gediminas Technical University, Lithuania*
Viranjay M. Srivastava, *Jaypee University of Information Technology, Solan, H.P., India*
Ming-Ju Wu, *Taichung Veterans General Hospital, Taiwan*
Wang Liying, *Institute of Water Conservancy and Hydroelectric Power, China*
Chenggui Zhao, *Yunnan University of Finance and Economics, China*
Rahim Jamian, *Universiti Kuala Lumpur Malaysian Spanish Institute, Malaysia*
Li-Xin Guo, *Northeastern University, China*
Mostafa Shokshok, *National University of Malaysia, Malaysia*
Ramezan Ali Mahdavinejad, *University of Tehran, Iran*
Anita Kovač Kralj, *University of Maribor, Slovenia*
Tjamme Wiegers, *Delft University of Technology, The Netherlands*
Gang Shi, *Inha University, South Korea*
Bhagavathi Tarigoppula, *Bradley University, USA*
Viranjay M. Srivastava, *Jaypee University of Information Technology, Solan, H.P., India*
Shyr-Shen Yu, *National Chung Hsing University, Taiwan*
Yen-Chieh Ouyang, *National Chung Hsing University, Taiwan*
Shen-Chuan Tai, *National Cheng Kung University, Taiwan*
Jzau-Sheng Lin, *National Chin-Yi University of Technology, Taiwan*
Chi-Jen Huang, *Kun Shan University, Taiwan*
Yean-Der Kuan, *National Chin-Yi University of Technology, Taiwan*
Qing He, *University of North China Electric Power, China*
JianHui Yang, *Henan University of Technology, China*
JiQing Tan, *Zhejiang University, China*
MeiYan Hang, *Inner Mongolia University of Science and Technology, China*
XingFang Jiang, *Nanjing University, China*
Yi Wang, *Guizhou Normal University, China*
ZhenYing Zhang, *Zhejiang Sci-Tech University, China*
LiXin Guo, *Northeastern University, China*
Zhong Li, *Zhejiang Sci-Tech University, China*
QingLong Zhan, *Tianjin Vocational Technology Normal University, China*
Xin Wang, *Henan Polytechnic University, China*
JingCheng Liu, *The Institute of Chongqing Science and Technology, China*
YanHong Qin, *Chongqing Jiaotong University, China*
LiQuan Chen, *Southeast University, China*
Wang Chun Huy, *Nan Jeon Institute of Technology, Taiwan*
JiuHe Wang, *Beijing Information Science and Technology University, China*
Chi-Hua Chen, *Chiao Tung University, Taiwan*

Fu Yang Chen, *Nanjing University of Aeronautics, China*
Huan Song Yang, *Hangzhou Normal University, China*
Ching-Yen ho, *Hwa Hsia College of Technology and Commerce, Taiwan*
LiMin Wang, *Jilin University, China*
Zhang Li Lan, *Chongqing Jiaotong University, China*
Xu Yang Gong, *National Pingtung University of Science and Technology, Taiwan*
Yi Min Tian, *Beijing Printing College, China*
Ke Gao Liu, *Shandong Jianzhu University, China*
Qing Li Meng, *China Seismological Bureau, China*
Wei Fan, *Hunan Normal University, China*
Zi Qiang Wang, *Henan University of Technology, China*
Ai Jun Li, *Huazhong University of Science and Technology, China*
Wen-I Liao, *Taipei University of Science and Technology, Taiwan*
BaiLin Yang, *Zhejiang University of Industry and Commerce, China*
Juan Fang, *Beijing University of Technology, China*
LiYing Yang, *Xian University of Electronic Science and Technology, China*
NengMin Wang, *Xi'an Jiaotong University, China*
Yin Liu, *Zhongyuan University of Technology, China*
MingHui Deng, *Northeast China Agricultural University, China*
GuangYuan Li, *Guangxi Normal University, China*
YiHua Liu, *Ningbo Polytechnic Institute, Zhejiang University, China*
HongQuan Sun, *Heilongjiang University, China*

CO-SPONSOR

International Frontiers of Science and Technology Research Association
Control Engineering and Information Science Research Association

Architectural, energy and environmental engineering

Architectural, Energy and Information Engineering – Sung & Chen (Eds)
© 2016 Taylor & Francis Group, London, ISBN 978-1-138-02791-6

Research status and development analysis on contemporary concrete

Deng Yang, Tao Lin Yang & Hang Chen
School of Civil Engineering, Sichuan Agricultural University, Du Jiangyan, China

Lan Xiu Sun
School of Economics and Management, Sichuan Agricultural University, Chengdu, China

ABSTRACT: In this paper, the history of the development of contemporary concrete is reviewed first. Then, the current development and applications of concrete have been systematically introduced and summarized, which emphasizing on green high-performance concrete including recycled aggregate concrete, self-compacting concrete, etc. Finally, the development trends and future applications of concrete materials have been presented based on the demands of the developing human society.

Keywords: concrete; recycled aggregate concrete; self-compacting concrete; green high-performance concrete

1 INTRODUCTION

1.1 *History of concrete*

Concrete is the mixture of cementing material, aggregate and water (with or without water) according to appropriate proportions, and after a certain time it is hardened into a solid material. In fact, the concrete is an artificial stone[1]. Broadly speaking, the history of the concrete can be traced back to ancient times, Egypt as early as 3000 BC has built pyramids with plaster mortar, China as early as the Qin Dynasty (220 BC) has built a great wall with lime mortar. Subsequently, in 1756 the British John Smeaton invention hydraulic binders and rebuild Eddyston Lighthouse, 1796 British James Parker obtained patent natural hydraulic cement, in 1813 the French Louis Vicat artificial hydraulic cement production by sintering limestone and clay, 1824 British Joseph Aspdin invention Portland cement. When using the Portland cement as cementing material into concrete, in 1850 and 1928 there has been reinforced and prestressed concrete, concrete before they get a wide range of applications, it is currently the most widely used and the use of the world's largest range building materials. To facilitate the description, the following text in the concrete of the term refers to Portland cement concrete.

Concrete with many advantages, such as a rich source of raw materials, low cost, mixing with good plasticity, flexible configuration, room temperature condensation curing, high compressive strength, high temperature performance and low energy consumption, etc. Similarly, concrete also has a self-important, smaller than the strength, thermal conductivity coefficient, hardening slow, long production cycle and other shortcomings.

With the continuous development of social economy, and in order to adapt to the high-rise buildings in the future, high-rise, large-span development, and the development of underground rivers and oceans, the future development direction of the concrete are: rapid hardening high-strength, light, high durability, versatility, and energy saving. For example, the American Concrete Institute AC12000 Commission had envisaged that the future strength of the concrete used for the United States will reach 135 MPa, if necessary, we can technically achieve the concrete strength of 400 MPa. Until then, we will be able to build a high-rise building height of 600~900 m, and the span of 500~600 m bridge.

High-Performance Concrete (HPC) is a new concept raised in the 1990s. It is made by using modern concrete technology, choosing high-quality raw materials, adding the necessary admixtures, and controlling quality strictly[2]. Therefore, the future development of concrete technical route mainly follows the compounding of function, highly fortified, high performance, and ecological.

2 NEW PROGRESS OF CONCRETE TECHNOLOGY

Humans have entered the 21st century, with the rapid development of science and technology, various new kinds of concrete have emerged such as

Table1. Comparison of the typical ultra high-strength concrete.

Component	Mass ratio
Type V Portland cement	1
Fine sand (150–400), >99% quartz	1.1
Silica fume (18m²/g)	0.25
Water	0.15
Superplasticizer	0.044
Micro fibers (essential)	0.05

high-performance concrete, fiber reinforced concrete, self-compacting concrete, recycled concrete, reactive powder concrete and translucent concrete, etc. Whether the concrete can be used as long-term building construction materials lies in its high performance of strength, resistance, durability as well as its ability to become the green material. Therefore, a green high-performance concrete is the inevitable result of the development of modern concrete technology as well as the development direction of the concrete. Table 1 shows the comparison of the typical ultra high-strength concrete. Through this mixture, we can get a high density concrete, it has a high compressive strength, and good durability. It can be seen from the table, high-strength concrete has the following characteristics: no coarse aggregate, mixed with a lot of silica fume, large doses of superplasticizer, high volume fiber, and water-cement ratio is small. In the subsequent sections, we will focus on the recent research status and development of two green high performance concrete—recycled aggregate concrete and self-compacting concrete aggregate concrete.

3 RECYCLED AGGREGATE CONCRETE

3.1 Research status

Recycled aggregate concrete is called recycle concrete in short was formed by the concrete block, which was abandoned during the production process or the using stage after the process of crushing, heat treatment, cleaning, grading, processing and mixing according to a certain proportion or gradation. The extensive use of recycled concrete significantly alleviates the growing shortage situation of building aggregate and deals with the problems of waste concrete from the fundamental solution as well; moreover, it can even achieve harmless and resource. The utilization of recycled concrete has become the important issues discussed by the developed countries, some countries even through legislation to protect the development of the study.

Scholars around the world focused on the study of recycled aggregate concrete and basic material aspects of regeneration, which have been used successfully in the construction of rigid pavement structure examples. Current research on recycled aggregate concrete in the world mainly has the following aspects: (1) Recycled aggregate production process lifting technique; (2) The research of recycle aggregate's using performance; (3) The research of recycled concrete's mix design; (4) The research of recycled concrete's early strength; (5) The research of recycled concrete's deformation performance; (6) The research of recycled concrete's durability; and (7) The research of recycled concrete's structural performance.

3.2 Development analysis

The research time of the recycle aggregate concrete is not very long, thus we have to always carry out comparative experiments that site the comparison between recycled concrete with ordinary concrete and end up to a lot of qualitative conclusion rather than quantitative conclusion. Therefore, as for recycle concrete, we have so many problems to be solved in both research and experimental study part in order to truly achieve a large area of application.

1. Using the recycled concrete could truly solve the problem of waste concrete in technique, but its research has not yet reached a depth of ordinary concrete, its use has not been vigorously promoted. If it is not widely used, it loses its advantages of resolving waste concrete. If it is not widely used, it loses its resolve recycled concrete waste concrete advantages. Therefore, if we can study how to improve its strength, materials durability abrasion resistance, mechanical properties and structural properties and then make it develop to high-performance direction, recycled concrete will be used in all types of building and exert its advantages.
2. The biggest reason for recycled aggregate concrete and ordinary concrete makes the difference is: When recycled aggregate crushed will produce small cracks and voids, thus leading to difference on microstructures, if we can start our research from the micro-structure of the recycled concrete and improve its crushing process, we may improve its various properties.

4 SELF-COMPACTING CONCRETE

4.1 Research status

Japanese scholar Okamura first proposed the concept of self-compacting concrete[3] in 1986. Subsequently, Ozawa and other scholars from the University of Tokyo[4] carried out a study of self-compacting concrete. In 1988, the first use of a commercially available material successfully developed self-compacting concrete with satisfactory performance, including appropriate hydration heat, good compactness, and other properties. Compared with ordinary concrete, self-compacting concrete has the following characteristics[4]: (1) in the fresh stage, it does not require artificial vibration compacting, filling its own weight and density. (2) The stage of early age, it can avoid the original defects. (3) After hardening, it has sufficient resistance to external environmental erosion. For the past 20 years, due to the superiority of self-compacting concrete, its research and application of practice has been widely expanded in the world[5-10]. Currently, the main research aspects of countries in the world are: polypropylene fiber reinforced self-compacting concrete, low self-compacting concrete, mixed fiber strength self-compacting concrete, rubber aggregate self-compacting concrete, etc.

4.2 Development analysis

1. Workability is the key performance of self-compacting concrete, how to quantify and guarantee the performance of self-compacting concrete mixture has been, and will also be the focus of self-compacting concrete research; Developing more standardized, scientific and practical methods to test the performance of self-compacting concrete mixture is the goal of this research field. Table 2 shows the properties

Table 2. Properties test and method evaluating for self-compacting concrete.[11-12]

| Method | Testing property characteristic | Evaluation | |
		Testing index	Typical value
Slump flow	Flowability or filling ability	Total flow spread/mm	~650
Tsoo	Viscosity/ flowability	Flow times	2~5
V-funnel	Viscosity/ flowability	Flow time/s	~8
Oriment	Viscosity/ flowability	Flow time/s	~8
J-ring	Passing ability	Step height/mm	~15
L-box	Passing ability	Passing ratio	≥0.75
U-pox	Passing ability	Height difference/mm	≤30
Wet-sieving segregation	Segregation resistance	Outflow paste percent%	≤3
Penetration test		Depth/mm	~17
Settlement column		Segregation ratio/%	~1

test and method evaluating of self-compacting concrete.
2. To further enhance the research on the design methods and formulation technology of the technical, practical and economic benefits self-compacting concrete, especially for the concrete of common strength grade in engineering; To further study the design methods and preparation techniques of the self-compacting concrete of general (moderate) strength grade i.e., develop the self-compacting concrete of moderate levels, will expand the engineering applications of self-compacting concrete.
3. Strengthening the research about the performance volatility, construction quality assurance system and other aspects in construction process of self-compacting concrete. Surveying the entity structure performance of self-compacting concrete, studying its relationship between the performance of concrete specimens in the same conditions. These will provide a strong guarantee for self-compacting concrete in the construction quality control and engineering applications.

5 CONCLUSION

1. For the past 20 years, concrete has been developed in the direction of compounding function, high strength, high performance, and ecological materials. The high-performance concrete in the world will obtain a relatively wide range of applications. The green high-performance concrete is the development direction of the concrete.
2. Numerous studies have shown that the recycled concrete and self-compacting concrete is a direction of development of high-performance concrete. Adding a large number of fly ash and slag to develop the green high-performance concrete is feasible. The incorporation of a large number of fly ash and slag in the concrete can not only improve the quality of concrete, but also can effectively reduce production costs, which is conducive to the sustainable development of concrete industry.
3. Government should encourage the development and vigorously promote the use of green high performance concrete. We must continue to study and resolve the problems about the application of green high-performance concreteas soon as possible. The development of green high-performance concrete, will raise the level of production and the production of concrete, optimize the product structure of concrete and energy saving, and achieve the important sustainable development of concrete material.

5

REFERENCES

[1] Mehta P Kumar, Monteiro Paulo JM. Concrete, Micro-structure, Properties and Materials [M]. USA: McGraw-Hill 2006.

[2] Grutzeckm, Benesi A, Fanning B. Silicon 229 Magic Angle Spinning Nuclear Magnetic Resonance Study of Calcium Silicate Hydrates [J]. JA m Ceram Soc, 1989, 72(4): 665–668.

[3] Cong Xian dong, Kirkpatrick RJ. 29SiMAS NMR Study of the Structure of Calcium Silicate Hydrate [J]. Adv Ceram Based Mater, 1996, 3(3): 144–156.

[4] Gartner EM, Young EM, Damidot DA et al. Hydration of Portland Cement, Structure and Performance of Cements [M]. London: Spon Press, 2002: 83–84.

[5] Taylor H F W. Cement Chemistry [M]. London: T Telford, 1997.

[6] Spiratos N, Page M, Mailvaganam NP et al. Super plasticizers for Concrete, Supplementary Cementing Materials for Sustainable Development [M]. Ottowa Canada: ACI, 2003: 322.

[7] Tashiro C, Ikeda K and Inome Y. Evaluation of Pozzolanic Activity by the Electric Resistance Measurement Method [J]. Cement and Concrete Research, 1994, 24(6), 1133–1139.

[8] McCarter WJ, Whittington HW, Forde MC. The Conduction of Electricity Through Concrete [J]. Magazine of Concrete and Research, 1981, 33(114): 48–60.

[9] Gu P, Xie P, Fu Y. Micro-structural Characterization of Cementitious Materials: Conductivity and Impedance Methods, Materials Science of Concrete IV [M]. Cincinnati OH: American Ceramics Society, 1985: 94–124.

[10] McCarter WJ, Chrisp TM, Starrs G, et al. Characterization and Monitoring of Cement-Based Systems Using Intrinsic Electrical Property Measurements [J]. Cement and Concrete Research, 2003, 33(2): 197–206.

[11] Bartos PJM. Testing—SCC: towards new European Standards for fresh SCC [A]. In: Yu Zhiwu, Shi Caijun, Khayat KH, et al eds. Proceedings of 1st International Symposium on Design, Performance and Use Self-Consolidating Concrete [C]. Paris: RILEM Publication SARL, 2005. 25–46.

[12] Sonebi M, Rooney M, Bartos PJM. Evaluation of the segregation resistance of fresh self-compacting concrete using different test methods [A]. In: Yu Zhiwu, Shi Caijun, Khayat KH, et al eds. Proceedings of 1st International Symposium on Design, Performance and Use of Self-Consolidating Concrete [C]. Paris: RILEM Publication SARL, 2005. 301–308.

Architectural, Energy and Information Engineering – Sung & Chen (Eds)
© 2016 Taylor & Francis Group, London, ISBN 978-1-138-02791-6

A natural ventilation acoustic noise purification window

X.B. Tuo, Y.K. Ji, H.J. Li, Z.Y. Chen, D. Liu, K. Dong, C.B. Zhou, Z.Y. Yan & L.K. Zhang
School of Energy and Power Engineering

ABSTRACT: Air pollution is getting worse without the existing insulation and ventilation method. According to the theory and computing of Micro-perforated put forward by Ma Dayou academician, we designed a window that can achieve function of natural ventilation, acoustic noise and purification. This design not only ensures natural ventilation but also meets the requirements of acoustic noise, and purifies air entering indoor, and solves the problem of ventilation soundproof window that meets human's requirements for noise and air. In the future, the window will have wide application market prospects.

Keywords: natural ventilation; noise; purifying; micro-perforated plate

1 INTRODUCTION

With the continuous development of city, urban traffic noise and various stores' sound have been the sources of modern city's environment noise pollution. It has caused a serious negative influence on people's physical and mental health, and even cause the social contradictions. The window is the main way for noise to enter indoor. Therefore, we must solve the noise from the window.

Sound insulation window is not only a mechanical ventilation but also a natural ventilation and sound insulation. Mechanical ventilation not only produces large energy consumption, but also produces secondary noise and health risks, and it needs regular maintenance, which an adding additional labor. Existing natural ventilation of sound insulation window cannot completely block out sound and wind at the same time. The contradiction of the ventilation cannot meet the requirements of people in both ventilation and noise reduction.

With the development of the city, air pollution is becoming more and more serious. According to the <<74 city air quality report "in the first half of 2013>>, has nearly half of the city in the first half of 2013 have different air pollution problems. The serious air pollution forces people to shut the window. Human would breath air pollution indoor, rather than the polluted air outdoor.

2 THE DESIGN OF THE WINDOW THAT HAS TWO FUNCTIONS OF NATURAL VENTILATION PURIFICATION AND ACOUSTIC NOISE REDUCTION AT THE SAME TIME

Natural ventilation acoustic noise reduction to purify the whole window is divided into two parts. As shown in Figure 1, the upper part has noise elimination channels and the lower part is push-pull type window day-lighting. The wind outside with noise blows through the window that has been suspended air into the channel. A tunnel entrance is coarse efficiency filter for filtering dust and larger particles. Then, it was posted on noise elimination channels in the porous sound-absorbing material, sound absorption noise reduction of micro-perforated panel. There has active, static in a composite filter on the air in addition to flavor and small particles of solid material purification in indoor outlet. As shown in figure the arrows in the direction of the wind.

3 ACOUSTIC NOISE REDUCTION

Acoustic noise reduction can be divided into porous sound-absorbing material and sound absorption of micro-perforated panel. The porous sound-absorbing material surface has many holes and many internal tiny holes connected to each other, sound through porous sound-absorbing

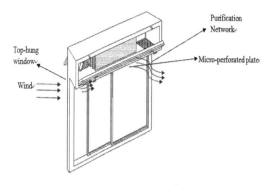

Figure 1. An overall configuration of the windows.

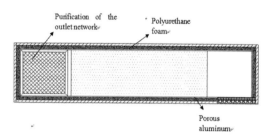

Figure 2. The plan view of the muffler channel.

material was due to the viscous effect between them, the sound energy gets converted into heat energy and it gets consumed, to achieve the effect of noise reduction. And the principle of micro-perforation plate is similar to the porous materials. Considering the demands of window such as day-lighting rate and noise reduction effect, the combination of two kinds of schemes is adopted.

3.1 Porous sound-absorbing material

3.1.1 The choice of porous sound-absorbing material

Within the existing sound insulation window using fiber sound-absorbing material not only can produce secondary pollution, but also is harmful to people's respiratory tract and skin. The effect of acoustic noise reduction would decrease a lot after long-term use. To solve the above problem, we adopt flexible polyurethane foam plastics absorption. Flexible polyurethane products cause no pollution. It is non-toxic, tasteless, no noise, long service life, and high and low temperature resistance, within the range of –20°C–120°C stable performance.

3.1.2 Factors affecting the sound absorption properties of porous materials

Sound absorption properties of porous sound-absorbing material is mainly related to the porosity, pore size, thickness, structure factor, and so

on. According to the most commonly used porous sound-absorbing material that selected porosity of 90%, a pore size of 0.5 mm, the material thickness of 0.5 m, the structure factor for the porous sound-absorbing material 4, its mainly aimed at 500~1000 Hz absorption effect is best.

3.2 Micro-perforated plate material

3.2.1 The select of micro-perforated plate material

Micro-perforation plate material is roughly divided into metal and nonmetal, considering the factor of the window day-lighting rate, this experiment adopts the PETG board with good performance. PETG board has outstanding toughness and high impact strength. It has a wide range of processing, high mechanical strength and excellent flexible. Compared with PVC, it has high transparency, good gloss, easy to printing and environmental protection advantages.

3.2.2 The influencing factors of sound absorption performance of micro-perforated panel abstract frame

The sound absorption performance of the micro-perforated plate mainly relates to perforation rate, diameter, thickness, back cavity thickness and other factors. Micro-perforated theory and theoretical calculations that made on the basis of Ma Dayou academician, respectively, are all factors to simulate and solve the optimal value. Based on the micro-perforated theory and theory calculation that put forward by Ma Dayou academician, now respectively to simulate various influencing factors, and to solve the optimal value.

1. The sound in the Micro-perforated plate resonance structure can be equivalent to the circuit shown in Figure 3.

Micro-perforated panel absorber system consists of thin plate with plenty of Simi level holes and cavity where is after plate. The continuity equation can be derived in the tube resonance structure and equivalent circuit.

The continuity equation can be derived in the tube as follows:

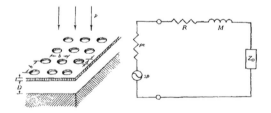

Figure 3. Micro-perforated plate.

$$jw\rho_0 u - \frac{\eta}{r_1} \times \frac{\partial}{\partial r_1}\left(r_1 \times \frac{\partial}{\partial r_1} \times u\right) = \frac{\Delta p}{t} \qquad (1)$$

Relative acoustic resistivity:

$$Z = r + j\omega m - j\cot\frac{\omega D}{C} \qquad (2)$$

$$\omega = 2\pi f; \qquad (3)$$

$$r = 0.147 \times \frac{tK_r}{(d^2 \times p)} \qquad (4)$$

$$m = 0.294 \times 10^{-3} \times \frac{tK_m}{p} \qquad (5)$$

$$K_r = \sqrt{1 + \frac{K^2}{3}} + \frac{\sqrt{2}Kd}{8t} \qquad (6)$$

$$K_m = 1 + \frac{1}{\sqrt{9 + \frac{K^2}{2}}} + \frac{0.85d}{t} \qquad (7)$$

$$K = \sqrt{\frac{f \times d^2}{10}} \qquad (8)$$

$$p = \frac{\pi}{4} \times \frac{d^2}{b^2} \qquad (9)$$

where K_r and K_m are the functions of K, in fundamental constants, and K_r is the K value increases with the increase, particularly in the perforated plate constant K is large, K_r more obvious changes, shown in Figure 4.

Among them:

c—speed of sound (m/s);
D—the thickness of the back chamber (mm);
b—adjacent central moment of micro-pores (mm);
r—relative acoustic resistivity;
ω—angular frequency (Hz);
f—frequency (Hz);

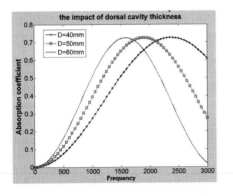

Figure 4.   Dorsal cavity thickness.

m—relative sound quality;
d—diameter of the hole (mm);
t—thickness of plate (mm);
P—perforation rate (%);
K_r—acoustic resistance coefficient;
K_m—sound quality factor.

Absorption coefficient:

$$\alpha = \frac{4r}{(1+r)^2 + \left(\omega m - \cot\frac{\omega D}{C}\right)^2} \qquad (10)$$

When the absorption coefficient comes to maximum, achieving a resonance frequency f_0, the absorption coefficient satisfies the following formula.

$$\alpha_0 = \frac{4r}{(1+r)^2} \qquad (11)$$

$$\omega_0 m - \cot\frac{D\omega_0}{c} = 0 \qquad (12)$$

While the absorption coefficient is equivalent to half of the maximum, the half sound frequency meets the following:

$$\omega_{1,2} m - \cot\frac{Dw_{1,2}}{c} = \mp(1+r) \qquad (13)$$

Approximation depends on the size of the sound absorption bands.

$$\frac{r}{m} = \frac{L}{d^2} \times \frac{K_r}{K_m} \qquad (14)$$

L—Structure constant, wherein the metal plate L = 1140 and non-metallic plate L = 500.

2. The Simulation:
We can get the curve of absorption coefficient of single micro-perforated plate absorber by programming calculation. The calculated parameters are: perforation rate e = 3%, aperture d = 0. 8 mm, the thickness of the back chamber D = 50 mm, Micro-perforated thickness b = 0. 8 mm. And making the corresponding experimental study, we chalk up the simulation results followed.

- To ensure that other parameters remain unchanged, the following results were obtained while changing the thickness of the back chamber as shown in Figure 4.
  If the back cavity thickness reduces, the low frequency sound absorption performance has decreased, and high frequency sound absorption coefficient has increased; If the back cavity

9

thickness increases, the peak will move the low frequency, and low frequency sound absorption performance of the absorption coefficient will improve with increase of the thickness of the back cavity. The thickness of the back chamber is restricted by space constraints and cannot increase indefinitely, thus in the micro perforation system, the appropriate thickness of the back cavity is very important. Although the back cavity thickness D = 80 mm, the low-frequency sound absorption is better, but the bandwidth is narrow. Considering the size of the sound absorption bandwidth, we decide to use the back cavity thickness D = 50 mm.

- To ensure that other parameters remain unchanged, the following results were obtained while changing the aperture as shown in Figure 5.

Aperture has a very significant effect on the maximum value of the absorption coefficient, the absorption coefficient increases rapidly with decreasing pore size; while reducing the aperture, can broaden the sound absorption band. Mechanism is that due to an increase in aperture, the friction loss increases and improving the sound barrier. The problem is that machining process of a very small hole diameter will create difficulties and increased costs. In order to make more cost effective, we select the aperture d = 0.8 mm. Although the ratio of d = 0.7 mm effect is worse, the sheet metal processing cost and precision of d = 0.8 mm is a lot more than d = 0.7 mm.

- To ensure that other parameters remain unchanged, the following results were obtained while changing the plate thickness as in Figure 6.

Absorption coefficient of Micro-perforated plate increases with thickness increasing, the causes of increase short tube radius, resulting in acoustic increase of impedance. It is also found that while the thickness increases, the absorption coefficient peak shifts to lower frequency. Taking into account the size and bandwidth practical application of a fixed and reliable, we select the thickness t = 0.8 mm.

Considering the impact of the perforation rate, the thickness of the back cavity, perforation diam-

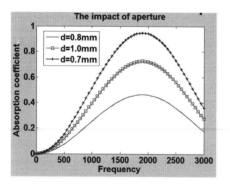

Figure 5.   The impact aperture.

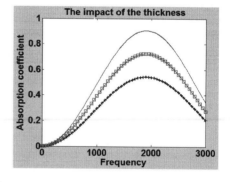

Figure 6.   The impact of the thickness.

	Adsorption method	Filtration and purification method	Plasma purification method	Catalytic method
Purifying contaminant species	Toxic and harmful gases	Particulate pollutants, some of the harmful gases	Organic molecules	Harmful gases, part of particulate matter
Comprehensive purifying effect	Poor	Well	Well	Poor
Corresponding equipment needed for space	Small	Small	Large	Large
Technology maturity	Mature	Mature	Immature	Immature
The ventilation effect	Well	Poor	Well	Well
Price	Low	Not high	High	High
Whether the Energy dissipation	No	No	Yes	Yes

eter, thickness and other design parameters, we make the scheme of parameters as muffler channel noise of micro-perforated plate: perforation rate $e = 3\%$, aperture $d = 0.8$ mm, Micro-perforated thickness $b = 0.8$ mm, and the thickness of the back chamber $D = 50$ mm.

It is shown that the poisonous gas adsorption method has a good effect, but both filtration and purification method are also better integrated in which purification technology maturity is high, occupies little space and excellent economy and environmental characteristics, but there are two kinds of purification purification effect which is not comprehensive enough shortcomings, this design will have adsorption and filtration and purification combination to achieve comprehensive air purification.

4 CONCLUSION

The design of integrated noise reduction, purification in one of the natural ventilation acoustic noise purification window not only to ensure natural ventilation, but also to meet acoustic noise, into the indoor air purification requirements, to make up for the shortcomings of the existing acoustic windows. The overall design to meet the aesthetic and practical premise, also makes the noise performance up to a certain height, must be favored in the future market.

REFERENCES

[1] Maa DY. Microperforated—panel wideband absorbers. Noise Control Eng J 1985; 29(3): 77–87.
[2] Crandall JB. Theory of vibrating systems and sound. New York: Van Nostrand; 1926.
[3] Maa DY. Potential of microperforated panel absorber. J Acoust Soc Am 1988; 104(5): 2861–2866.
[4] Cremer L, Müller H. Principles and applications of room acoustics, vol. 2. Applied Science Publishers Ltd; 1982.
[5] Fahy F. Foundations of engineering acoustics. New York: Academic Press; 2001.
[6] Ingard U. Notes on sound absorption. Maine: KitteryPoint; 1999.
[7] Kang J, Fuchs HV. Predicting the absorption of open weave textiles and micro-perforated membranesbacked by an air space. J Sound Vib 1999; 220(5): 905–920.
[8] Dai HX, Purification of nine kinds of residential air purifier. J Environmental and Occupational Medicine; 2011.7.
[9] Zhang YH, Han SJ, Characteristics of urban environmental noise pollution and Prevention Measures. J Hailun Environmental Protection Agency 2007(7): 92–93+102.
[10] Yu WZ, Wang ZM, A new and efficient natural ventilation window insulation design and test. J Environmental Engineering. 2008(3), 96–99.

Architectural, Energy and Information Engineering – Sung & Chen (Eds)
© 2016 Taylor & Francis Group, London, ISBN 978-1-138-02791-6

Research on influence factors of Ce-HZSM-5 molecular sieve material prepared by cation exchange

Chao Qin Yang & Ya Ming Wang
Faculty of Chemical Engineering, Kunming University of Science and Technology, Kunming, Yunnan, China

Guang Sen Lin & Le Gang Wu
Kunming Sino-Platinum Metals Catalyst Co. Ltd., Kunming, China

ABSTRACT: Molecular sieve is a kind of the HydroCarbon Absorbent (HCA), and it is used extensively in the material of natural gas catalysts, thus the performance of various molecular sieves is different. Before the preparation of Ce-HZSM-5 molecular sieve materials by cationic exchange, the influence factors of molecular sieve materials need to be determined. With decay rate of specific surface area as index, the five factors are considered, such as exchange times (factor A), calcination temperature (factor B), calcination time (factor C), PH (factor D) and deionized water consumption (factors E), while the calcination temperature is the main factor by an orthogonal experiment and a range analysis. It provides the basis of theoretical data for the preparation of modified Ce-HZSM-5 molecular sieve materials.

Keywords: cation exchange method; molecular sieve; influence factors; specific surface area

1 INTRODUCTION

Gas emissions of a vehicle have long cold start time, high exhaust temperature, the exhaust gas purifying catalyst must be able to live on low temperature with methane catalytic oxidation, and have a good anti-aging performance. Through the research literature about methane combustion catalyst, there is a low light-off active fault in traditional methane oxidation catalyst at low temperature of Pd/Al$_2$O$_3$[1]. Although there are other literatures had reported Pd-loaded catalyst[2-5] with oxide carrier, their methane low temperature catalytic oxidation activity has improved than Pd/Al$_2$O$_3$, but still cannot meet requirements of catalyst at low light-off temperature in the gas purification of vehicle. Because of molecular sieve Pd-loaded catalyst has good low temperature catalytic oxidation; therefore, it can effectively solve the problem of CNG catalyst with low light-off temperature methane in the application of methane purification of natural gas vehicle exhaust. An anti-aging performance of molecular sieve needs to be improved at the same time, so the research of modified molecular sieve has the vital significance. Ce^{3+} is added into the molecular sieve by cation exchange, metal Pd can be better distributed in the modified molecular sieve, which lead to more uniform pore structure.

2 EXPERIMENT

2.1 Preparation of Ce-HZSM-5 by cation exchange

Ce (NO$_3$)$_3$ · 6H$_2$O was taken, and the amount of Ce^{3+} oxide was 5 wt % of HZSM-5. It dissolved in deionized water, then added in HZSM-5 100 g, and adjusted the pH value of solution with 30% of the ammonia, and then put the samples in water bath of 80°C, stirring for 4 h, washing until undetectable Ce^{3+}, drying for 4 h in the hot blast oven of 120°C, finally calcination in the box-type resistance furnace. They were labeled as E-Ce-HZSM (1–16) respectively.

2.2 Research of influential factors and the orthogonal experiment table

The better level of five factors was determined, such as the switching frequency (factor A), calcination temperature (factor B), calcination time (factor C), PH (factor D), and dosage of deionized water (E). Four levels of switching frequency (factor A) were 1, 2, 3, and 4; Four levels of calcination temperature (factor B) were 475°C, 550°C, 625°C, and 700°C; Four levels of calcination time (factor C) were 4 h, 5 h, 6 h, and 7 h; Four levels of PH (factor D) were 6, 7, 8, and 9; Four levels

Table 1. Orthogonal experiment table of influence factors.

Level no.	Influence factors				
	A	B	C	D	E
1	1	1	1	1	1
2	1	2	2	2	2
3	1	3	3	3	3
4	1	4	4	4	4
5	2	1	2	3	4
6	2	2	1	4	3
7	2	3	4	1	2
8	2	4	3	2	1
9	3	1	3	4	2
10	3	2	4	3	1
11	3	3	1	2	4
12	3	4	2	1	3
13	4	1	4	2	3
14	4	2	3	1	4
15	4	3	2	4	1
16	4	4	1	3	2

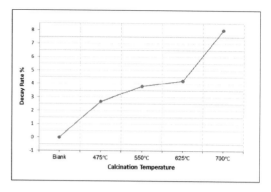

Figure 1. Influence of switching frequency on decay rate of specific surface area.

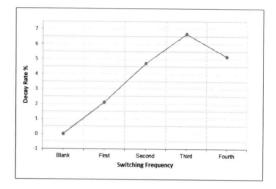

Figure 2. Influence of calcination temperature on decay rate.

of deionized water usage (factor E) were 200 ml, 225 ml, 250 ml, and 275 ml. The conditions were fixed, such as load, drying temperature, drying time, exchange temperature. Five key affecting factors were studied by the orthogonal table L_{16} (4^5), such as PH, dosage of deionized water, exchange times, calcining temperature, and calcining time. The orthogonal experiment table of influence factor is shown in Table 1.

3 RESULTS AND DISCUSSION

3.1 *Exchange times on the influence of specific surface area*

With the switching frequency changed, while other things being equal, different samples were prepared, and their decay rate of specific surface area was measured. The data are shown in Figure 1.

HZSM-5 molecular sieve belongs to the microporous material, with the increase of switching frequency, specific surface area of sample drops. It may be due to cause of congestion through the Ce elements into the molecular sieve hole; Second the Ce ionic radius of exchange is larger, and it will be formed Ce-O with the oxygen in air, which make the hole smaller, affect the specific surface area; Finally, the more exchange times, cleaning Ce is more difficult. Therefore, during drying and calcination Ce of the surface can react with the oxygen in air to generate CeO_2, which attaches on the surface of samples, block the hole, and lead to drop of specific surface area[6].

3.2 *Calcining temperature on influence of specific surface area*

With the calcination temperature changed, while other things being equal, different samples were prepared, and their decay rate of specific surface area was measured. The data are shown in Figure 2.

Figure 2 shows that with the increase of calcination temperature, specific surface area fells quickly. The reasons are as follows, on the one hand, the sample is in the air for burning, Ce was formed with O in the air by the union Ce-O keys, either attached on the surface of the Ce element or have already exchanged Ce into the inside of the molecular sieve. With the increase of groups of atoms, the molecular sieve channel has changed, and the surface will be covered by CeO_2, which make specific surface area decreases; on the other hand, calcination temperature of 700°C of decrease slope is very big, that calcination temperature is too high to collapse and damage in the molecular sieve causing hole.

The molecular sieve internal is caused by certain lattice defects because of large ionic radius of Ce elements[7], which lead to the activity of molecular sieve material is also improved a certain extent.

3.3 Other factors on influence of specific surface area

The influence on specific surface area is shown in Figures 3–5 respectively, such as calcination time, pH, and deionized water consumption of solution concentration.

Figure 3 shows that the decay rate of specific surface area is the lowest when calcination time is 6 h, while specific surface area decreases sharply when calcination time is 7 h. It shows that calcination time more than 7 h is disadvantage for molecular sieve, and some influence of the channel structure is caused because calcining time is too long.

Figure 4 shows that the decay rate of specific surface area is lowest when pH is 7.

Figure 5 shows solution dosage appears a peak in 250 ml.

The specific surface area and decay rate of different samples are shown in Table 2.

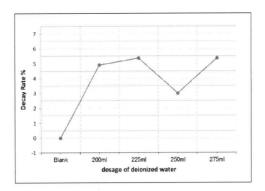

Figure 5. Influence of dosage of deionized water on decay rate of specific surface area.

Table 2. Decay rate of specific surface area of samples after ion exchange.

Sample	BET (m^2/g)	Decay rate (%)
Lgs-kongbai	316.712	
Lgs-01	319.144	−0.77
Lgs-02	309.086	2.41
Lgs-03	321.855	−1.62
Lgs-04	290.128	8.39
Lgs-05	308.114	2.73
Lgs-06	307.768	2.82
Lgs-07	299.112	5.33
Lgs-08	291.625	7.92
Lgs-09	298.147	5.86
Lgs-10	297.238	6.15
Lgs-11	295.046	6.84
Lgs-12	292.028	7.79
Lgs-13	307.409	2.94
Lgs-14	305.532	3.53
Lgs-15	296.927	6.28
Lgs-16	291.815	7.86

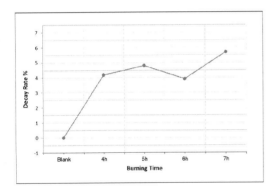

Figure 3. Influence of burning time on decay rate of specific surface area.

Table 3. Range analysis table.

K1	8.41	10.75	16.75	15.88	19.58
K2	18.80	14.91	19.21	20.10	21.46
K3	26.64	16.83	15.69	15.12	11.91
K4	20.60	31.96	22.80	23.35	21.49
k1	2.10	2.688	4.875	3.970	4.895
k2	4.70	3.728	4.803	5.025	5.365
k3	6.66	4.208	3.923	3.780	2.978
k4	5.15	7.990	5.700	5.838	5.373
R	4.56	5.302	1.778	2.058	2.395

With decay rate of specific surface area as index, the less decay rate, the better sample is the better. Therefore, primary and secondary orders of influence factors are:

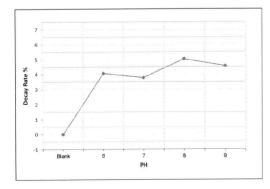

Figure 4. Influence of pH on decay rate of specific surface area.

Factor B > factor A > factor E > factor D > factor C

Factor A: $K_1 < K_2 < K_4 < K_3$ Factor B: $K_1 < K_2 < K_3 < K_4$ Factor C: $K_3 < K_1 < K_2 < K_4$

Factor D: $K_3 < K_1 < K_2 < K_4$ Factor E: $K_3 < K_1 < K_2 < K_4$

Therefore, the optimal program is $A_1B_1C_3D_3E_3$. That is exchange times is one time, calcination temperature is 475°C, calcination time is 6 h, pH is 8, and deionized water consumption is 250 ml.

4 CONCLUSION

With decay rate of specific surface area as index, the five factors are considered, such as exchange times (factor A), calcination temperature (factor B), calcination time (factor C), PH (factor D), and deionized water consumption (factors E), while the calcination temperature is the main factor by an orthogonal experiment and a range analysis. It provides the basis of theoretical data for the preparation of modified Ce-HZSM-5 molecular sieve materials.

ACKNOWLEDGEMENTS

This work was financially supported by the national key new product project of China (2011EG115008). Corresponding Author: Wang Ya-ming Professor.

REFERENCES

[1] D.J. Fullerton, A.V.K. Westwood, R. Brydson, et al. Deactivation and regeneration of Pt/γ-alumina and Pt/ceria–alumina catalysts for methane combustion in the presence of H_2S [J]. Catal. Today 2003, 81, 659–671.

[2] R. Kikuchi, S. Maeda, K. Sasaki, et al. Low-temperature methane oxidation over oxide-supported Pd catalysts: inhibitory effect of water vapor [J]. Appl. Catal. A: General, 2002, 232, 23–28.

[3] W. Lin, L. Lin, Y.X. Zhu, et al. Pd/TiO_2-ZrO_2 catalyst for methane total oxidation at low temperature and their ^{18}O-isotope exchange behavior [J]. J. Mol. Catal. A: Chem., 2005, 226, 263–268.

[4] X.H. Li, K.P. Sun, X.L. Xu, et al. Low-temperature catalytic combustion of methane over Pd/CeO_2 prepared by deposition—precipitation method [J]. Catal. Commun., 2005, 6, 800.

[5] Z.H. Li, G.B. Hoflund. Catalytic Oxidation of Methane over Pd/Co_3O_4 [J]. Catal. Today, 1999, 66: 2–367.

[6] Zhu X.X, Lin S, Song Y.Q, et al. Catalytic cracking of 1-butene to propene and ethane on MCM-22 Zeolite [J]. Appl Catal A, 2005, 2(9): 191–199.

[7] Dehertog W.J.H, Froment G.F. Production of light alkenes from methanol on ZSM-5 catalysis [J]. Appl Catal, 1991, 71(1): 153.

Architectural, Energy and Information Engineering – Sung & Chen (Eds)
© 2016 Taylor & Francis Group, London, ISBN 978-1-138-02791-6

Application of Infiltration Model into urban planning

Xiang Dong Hu
Xi'an Qujiang Daming Palace Ruins Region Preservation and Reconstruction Office Planning Bureau, Xi'an, China

Ming Guo Liu
Shanxi Small Towns Research Center, Taiyuan, China

Wen Hui Wang
School of Architecture, Chang'an University, Xi'an, China

ABSTRACT: With the acceleration of urbanization, urban planning appears more and more important. During the process of the urbanization, the planning of rainwater system is always neglected, because of which there are always many drainage problems. This paper focuses on the rainwater system planning and comes up with Infiltration Model (IM). The theoretical derivation and numerical solution are presented in this paper. What is more, IM is also applied into practical cases which show excellent agreement between the calculated and experimental values. The results are analyzed in this paper and indicate that IM is feasible for describing the infiltration process of water in pavements.

Keywords: Infiltration Model, urban planning; numerical solution

1 INTRODUCTION

The urban planning is more and more important in the process of urbanization, while the planning of rainwater system which plays an important role in urban planning has always been neglected or roughly applied. The porous asphalt pavement has attracted more and more attention in rainwater system planning by its outstanding surface performance of infiltration. For example, this kind of pavements can help reduce the amount of moisture film, especially in rainstorm weather. However, the permeability should not be always enhanced in terms of durability. Therefore, it is necessary to conduct a detailed research into the drainage performance of the porous asphalt pavements. The infiltration performance is related to many factors, including permeability and the slope of the pavements, rainfall intensity and so on. Thus to provide suggestions for the design of drainage pavements.

The filtration includes three phases, which are surface infiltration, void filling and transfusion, respectively. In addition, the transfusion progress has two stages (Figure 1). The first stage is governed by the water supply, in which stage the water supply rate is not beyond the pavement permeability.

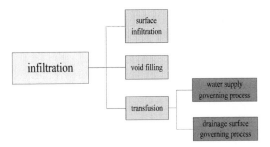

Figure 1. Stages made up of infiltration process.

The second stage is governed by the drainage surface. In this stage, the water supply rate is out of the limit of the pavement permeability.

This paper focuses on the second stage of the transfusion process, corresponding to the saturation state.

2 INTRODUCTION

To analyze the infiltration property, this paper applies infiltration theory to model the infiltration process of the pavements.

2.1 Governing equation

To describe the water state, first we need to set the boundary conditions. In this paper, we assume that there is no water flow on the right side of the calculated length, and the left side is the drainage ditch, which corresponds to a constant height of underground flow in the equation. Therefore, the boundary conditions are as the following:

$$\frac{dh}{dx} = 0 \cdots \cdots x = L$$
$$H = H_0 \cdots \cdots x = 0 \tag{1}$$

where h is the water height;

x is the calculated position;

L is the total length of the drainage road;

H is the thickness of the underground flow;

H_0 is the thickness of the underground flow on the left side of the calculated area, which is constant in the equation.

According to the Boussinesq equation, the underground flow is described by the following differential equation, which is shown in Eq. (2).

$$\frac{d}{dx}\left(Hk \frac{dh}{dx} \right) = -I \tag{2}$$

where k is the infiltration coefficient of the drainage layer in the horizontal direction;

I is the intensity of the rainfall.

To simplify Eq. (2), it is integrated on both sides, thus Eq. (3) is obtained.

$$Hk \frac{dh}{dx} = -Ix + c \tag{3}$$

In the equation above, c is obtained by the boundary conditions, which is as follows.

$$c = IL \tag{4}$$

To get the relationship between H and h, this paper assumes that h is linear with H, as shown in Eq. (5).

$$\frac{dh}{dx} = \frac{dH}{dx} + i \tag{5}$$

where i is the slope of the road.

Applying Eq. (4) and (5) into Eq. (3), the governing equation is obtained.

$$\frac{dH}{dx} = \frac{(L-x)I}{Hk} - i \tag{6}$$

2.2 Numerical solutions

To solve Eq. (6), this paper applies the fourth order Runge-Kutta method into the process of obtaining numerical solutions.

$$y_{i+1} = y_i + \frac{\Delta h}{6}(k_1 + 2k_2 + 2k_3 + k_4) \tag{7}$$

where Δh is the length of the interval; k_1, k_2, k_3, k_4 correspond to the following equation set, respectively.

$$k_1 = f(x_i, y_i)$$
$$k_2 = f(x_i + \frac{1}{2}\Delta h, y_i + \frac{1}{2}\Delta h k_1)$$
$$k_3 = f(x_i + \frac{1}{2}\Delta h, y_i + \frac{1}{2}\Delta h k_2) \tag{8}$$
$$k_4 = f(x_i + \Delta h, y_i + \Delta h k_3)$$

3 COMPUTATIONAL RESULTS AND DATA ANALYSIS

With the numerical solution mentioned above, Figure 2 is obtained assuming $H_0 = 0.001$ m, $L = 5$ m, $i = 2\%$ and $I/k = 5 \times 10^{-5}$.

Generally speaking, what we concentrate on is the maximum thickness of the calculated length. Therefore, this paper calculated the maximum thickness under different conditions.

From Eq. (6), it is obvious that H is controlled by L, i, I and k. In addition, H is not changed if I/k is fixed. Therefore, to simplify calculation, this paper considers I/k as one single unity.

The common road usually contains two lane road, three lane road and four lane road with

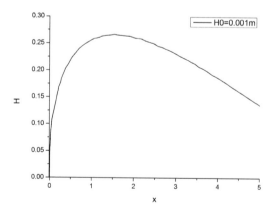

Figure 2. Thickness of underground flow VS distance from drainage ditches.

different width. To be specific, the two lane road width is 11.25 m, three lane road widths is 15.00 m and four lane road width is 18.75 m. This paper conducts a numerical research on the three roads mentioned with $i = 1.5\%$ and 2.0% respectively.

Figure 3 and Fig. 4 indicate an excellent agreement between the computational and experimental data, which shows that the model proposed in this paper is feasible to describe the infiltration performance of the pavements.

From the two graphs mentioned above, we can obviously see that with the higher pavement width, the maximum thickness of the underground flow is also higher. This is caused by the fact that the infiltration speed is slower than the rainfall intensity. In addition, we can obtain that with higher I/k, which means rainfall is fiercer and infiltration coefficient is smaller, the maximum thickness

of the underground flow is surely higher. What is more, when i gets bigger, namely the slope of the road is bigger, the thickness of the underground flow is lower, from the comparison between Fig. 3 and 4. This is caused by the fact that, with bigger slope, the water flows faster because of gravity.

4 MODEL VALIDATION

To ensure the validation of the model, this model was applied into the simulation of the working conditions of a certain highway in Jiangsu Province.

The top layer of this road uses the drainage asphalt which is PAC-13 and whose porosity is 20%. The thickness of the layer is 4.5 cm. The width of the three road lane is 15 m, and the slope is 2.0%.

According to [8], the relationship of porosity (which is V) and transverse permeability coefficient (which is k) of PAC-13 is listed as follows:

$$k = 0.0791V - 0.4979 \tag{9}$$

Then the transverse permeability coefficient was obtained by Eq. (9), which is $k = 1.084$ cm/s.

Using numerical regression, H_{max} was obtained when the slope is 1.5% and shown below:

$$H_{max} = (-1592L + 49891)\left(\frac{I}{k}\right)^{0.8872} \tag{10}$$

To meet the requirement that there is no surface runoff, the maximum thickness of the underground flow should be less than the thickness of the infiltration road (H_{PAC}), which is:

$$H_{PAC} > H_{max} \tag{11}$$

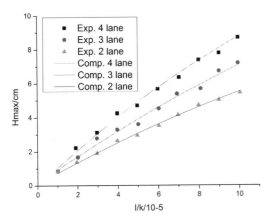

Figure 3. Computational and experimental data when $i = 1.5\%$.

Applying the variables above into the infiltration model proposed in this paper, particularly Eq. (10), the critical rainfall intensity which meets the need of no surface runoff is $I = 2.83$ mm/h.

Table 1 contains the criterion of different levels of rainfall intensity. Compared what we got to the criterion in table 2, it is clear that there is no

Table 1. Criteria for rainfall intensity.

Level	Intensity (mm · h⁻¹)
1	0~0.41
2	0.42~1.04
3	1.05~2.08
4	2.08~4.16
5	4.17~10.42
6	>10.43

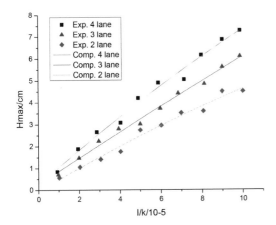

Figure 4. Computational and experimental data when $i = 2.0\%$.

obvious surface runoff in rainstorm days on the objective road, which matches the design and practical performance.

From the analysis above, we can see that, the theoretical calculation have a good agreement with the practical conditions. Therefore, the model this paper proposed is effective and feasible.

5 CONCLUSION AND SUGGESTION

To analyze the process of the infiltration of water in pavements, an infiltration model was proposed in the former part of this paper. This paper applies the Runge-Kutta method to get the numerical solution of the governing differential equation. In addition, an experiment was done to verify this model. The results show that this model has a perfect agreement with experimental data, which indicates that the model proposed in this paper is feasible to describe the infiltration process.

From the analysis above, we could see that, to get a good drainage performance, a bigger slope and higher infiltration coefficient are needed. In addition, narrow pavement could also reduce the thickness of underground flow.

REFERENCES

[1] Xie, Y. Y. & Li, D. M., Li, P. Y. & Shen, S. Q. & Yin, J. M. & Han, S. Q. & Zeng, M. J. & Gu, X. Q. 2005, Research and application of the mathematical model for urban rainstorm water logging, *Advances in Water Science* 16(3): 384–390.
[2] Xie Y. Y. 2007, The Method and Application of Urban Sewer System Simulation, *Master graduation thesis from Tongji University*, Shanghai.
[3] Ma, X. & Ni, F. J. & Li, Q. 2014. Seepage model and parameters of porous pavement layer, *Journal of Southeast University (Natural Science Edition)* 44(2): 381–385.
[4] Ji, Q. K. 2005. Design of road infiltration based on infiltration coefficient of pavement, *Chinese Journal of Geotechnical Engineering* 27(9): 1045–1049.
[5] Chen, B. & Feng, G. L. 2008. A Study on Simulating System of Rain Storm Water-logging in Wuhan City, *Torrential Rain and Disasters* 27(4): 330–333.
[6] Xu, X. Y. & Liu J. & Hao, Q. Q., Ding G. C. 2003. Simulation of urban storm water-logging, *Advances in Water Science* 14(2): 193–196.
[7] Wang, L. & Wang, Z. H. 2014. Research on Problems of Planning of Rainwater Zoology, *Drainage System of Urban road* 4(1): 110–112.
[8] Ma, X. & Ni, F. J. & Wang, Y. & Chen, R. S. 2009. Test and Analysis on Permeability of Porous Asphalt Mixture, *Journal of building materials* 12(2):168–172.

Architectural, Energy and Information Engineering – Sung & Chen (Eds)
© 2016 Taylor & Francis Group, London, ISBN 978-1-138-02791-6

Evaluation on rock bedded slope stability for Huangshan surface mine

H. Ran & W.G. Li
College of Technology, Sichuan Normal University, Chengdu, China

ABSTRACT: Huangshan mine is a surface mine that mining bedded limestone, sometimes, landslide accident happens during the production operation. Affecting factors caused by the landslide accidents are analyzed on the basis of a comprehensive investigation on the mine circumstances, a mechanical model to evaluate rock bedded slope stability is established, and safety coefficient of rock bedded slope stability is appraised under different working states by a limit equilibrium theory. The results show that the slope is stable under natural state, and it is unstable under continuous rainfall or water-saturated state. Technical measures, for example reasonable blasting parameters and open-pit advance direction, timely drainage, and so on, are proposed to effectively prevent landslide accidents.

Keywords: rock bedded slope; limit equilibrium theory; stability evaluation; rainfall

1 INTRODUCTION

Pit mine or mountain surface mine is gradually formed with the continuous development of open-pit mine technology, which resulting in the formation of the slope and consequent occurrence of landslide accidents. As the correctness of adopted stability evaluation of slope not only affects the safety of open-pit mine homework personnel and equipment, but also seriously affects normal production and operating efficiency of open-pit mine. All the 21 rock bedded landslide accidents occurred in Huangshan mountain surface mine are rock bedded landslide along the potential sliding surface I, and caused large casualties and major property damage. This paper takes the mine as an engineering background, and focuses on the stability evaluation of rock bedded slope. It establishes a mechanical model to evaluate rock bedded slope stability, evaluates slope stability in different conditions with a limit equilibrium theory, and proposes some prevention measures such as reasonable blasting parameters and timely drainage. The results have guiding significance for the mine to realize production safety, and provide reference for slope stability evaluation and scientific prevention under similar conditions.

2 AFFECTING FACTORS OF ROCK BEDDED SLOPE STABILITY

Huangshan mine is located at the north slope of Mount Huangshan in Emeishan city. Topography and landform of mining area appear as a mountain formation that is higher in the south and lower in the north, with the basic earthquake intensity of VII, developmental gullies, melting pits and corrosion depressions, and so on. The mine belongs to the subtropical moist climate zone of Sichuan Basin, with an average annual rainfall of 1555 mm, mainly during the months from June to September. The outcropped stratum has Lower Ordovician Dachengsi Formation and Lower Permian Yangxin Formation, the former is the top with a thickness of 2~13 m of powder and close-grained quartz sandstone, and the bottom is with shale containing thin layer, as well as medium-bedded argillaceous and crystalline limestone; the latter is with medium-bedded and thick-bedded limestone of compact lump, which is mainly ore deposits mined by the mine. There are gray black and dark-gray thin-bedded and medium-bedded argillaceous scraps of limestone, which are weak interlayer rock among ore deposits, forming potential sliding surfaces I and II from the top to the bottom successively. Heading-and-bench blasting is adopted in the open pit, 300~600 m from north to south, and 800~1200 m from east to west, forming +670~+720 m level four benches. Bench height is 15 m, bench angel 70°~75°, hole depth 16.5 m, hole aperture 170 mm, hole spacing 6 m, row spacing 5.5 m, and front row burden 6 m. The 2nd rock explosive and millisecond plastic nonel detonator is used. Physical and mechanical parameters of rock mass and potential sliding plane rock for stability evaluation of the slope are shown in Table 1.

Rock bedded slope is formed when limestone is mined in the mine because of two layers of weak interlayer rocks I and II which usually becomes

Table 1. Physic and mechanical parameters of rocks mass.

Rock type	Natural state			Water-saturated state		
	ρ/kN·m^{-3}	C/kPa	φ/(°)	ρ/kN·m^{-3}	C/kPa	φ/(°)
Limestone	26.8	1500	43.5	26.9	1000	36.5
Slip plane I	20.2	90	23.5	27	0	25
Slip plane II	26.7	800	35.5	26.8	400	36.5

potential sliding plane of rock bedded slope, bench angle is greater than dip angle of rock stratum. When limestone is being mined by heading-and-bench blasting, under the influence of production blasting operations, mechanical perforation operation, earthquake, rainfall and other external factors, mechanics parameters like rock cohesive of weak interlayer plane and internal friction angle of rock are reduced, causing the anti-slide force of potential sliding plane decreased. When anti-slide force is reduced to a certain extent and not enough to maintain rock bedded slope stable, the slope will be cut out along the weak sliding plane, and landslide accident happens.

The widespread stratum in the mine is mainly composed of thick-bedded biological limestone, with tectonic fracture development. Mined limestone stratum has generally two groups of tectonic fracture, one is steep fracture oriented south and north of 34°∠82°; another is oriented east and west of 22°∠30°. In general, fracture spacing is 100~500 mm, width 1~30 mm, and the vast majority of joint presents open. Due to stratum with tectonic fissure being in line with the slope dip, weak structure planes are combined with each other, rock mass is roughly cut into prismatic blocks which sizes not equal to orthogonal in strata surface each other, tectonic fissure becomes splay fracture which cuts off the contact between sliding mass and the trailing-edge rock mass.

Because Huangshan mine is an open-pit mine, the dip angle of rock stratum is 20~28°, rock stratum inclination identifies with bench slope direction, and the dip angle of rock stratum is less than the bench angle. From early 1970s to the early 1990s, bench blasting was adopted all the time. The biggest control charge quantity is as high as 121 tons. Blasting operation, on the one hand, destroys the integrity of rock mass slope, on the other hand, closed original cracks are expanded and shape new splay fractures, rock cohesive of weak interlayer plane and internal friction angle of rock are greatly reduced.

Hydrology geological condition in the mine is simple. Because of the average annual rainfall of 1555 mm for many years, rainfall mainly discharges

from the mining area in the form of runoff, and a small amount of rainfall infiltrates into the underground along rock fractures, fractures and caves along the bedding, joints, faults, and so on. Especially under continuous rainfall and heavy rain in the rainy season, the water storage of rock bedded mass is increased, hydrostatic pressure is increased, rock mass density is increased at the same time, and rock cohesive of weak interlayer plane and internal friction angle of rock are greatly reduced, thereby landslide accidents happen occasionally.

3 MECHANICAL MODEL OF ROCK BEDDED SLOPE STABILITY EVALUATION

As previously mentioned, there are many factors affecting the stability evaluation of rock bedded slope, the main factors include production blasting operation, earthquake, and rainfall. As shown in Figure 1, the forces on the rock bedded slope are: the vertical downward gravity W, the production blasting horizontal force kW in bad condition, the hydrostatic pressure V perpendicular to rear crack, the hydrostatic pressure U perpendicular to the potential sliding plane and the slabbing crack, and the anti-slide force is upward and along potential sliding plane. Here the blasting pseudo-static coefficient is k.

For the rock bedded slope shown in Figure 1, the unit width slope is regarded as the research object, according to the limit equilibrium theory, in terms of Mohr-Coulomb criterion, under the most unfavorable conditions, the rock bedded slope stability safety coefficient Fs is

$$F_s = \frac{C(L - L_0) + [W(\cos\beta - k\sin\beta) - U - V\sin\beta]\tan\phi}{W(\sin\beta + k\cos\beta) + V\cos\beta}$$

(1)

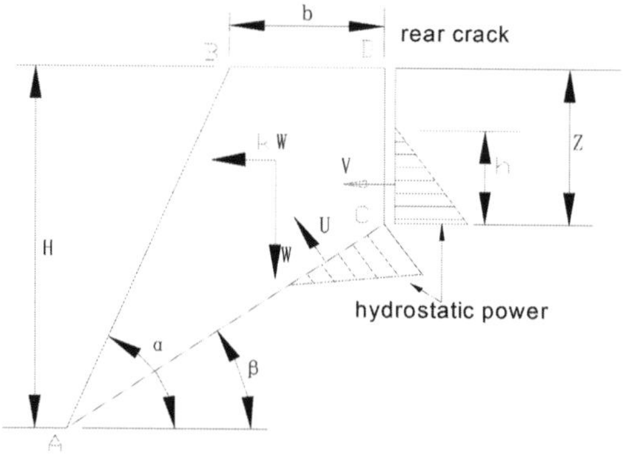

Figure 1. Mechanical model of rock bedded slope.

where α is the slope angle, β is the potential sliding plane angle, L is the potential sliding plane length, h is the rear crack height and L_0 is the slabbing crack water filling height, C is the rock cohesive of weak interlayer plane, ϕ is the internal friction angle of rock, ρ is the density of water, $U = 0.5\ \rho h L_0$ and $V = 0.5\ \rho h^2$.

According to Bishop's definition on safety coefficient of slope stability, when $F < 1$, the slope is in a state of instability destruction; $F > 1$, the slope is in a safe state; $F = 1$, the slope is in the limit equilibrium state.

4 EVALUATION ON ROCK BEDDED SLOPE STABILITY FOR HUANGSHAN SURFACE MINE

According to slope parameters, blasting parameters and characteristic parameters of landslide accidents which along the potential sliding plane I in the mine, when safety coefficients of rock bedded slope stability are evaluated, in such a condition the slope height is 15 m, side slope angel 55°, the potential sliding plane angle 24.5°, rear crack height 2 m, and the distance between rear crack and bench edge 6 m, based on the parameters listed in Table 1. The four different working conditions among the natural state, production blasting under natural state, continuous rainfall or water-saturated state, and production blasting under continuous rainfall or water-saturated state are considered, and the evaluation condition and rock bedded slope stability safety coefficients are shown as follows.

Under the natural state, when rock cohesive of weak interlayer plane I is 30 kPa, internal friction angle of rock is 33°, and the safety coefficient is 2.58.

The production blasting is conducted under natural state, when rock cohesive of weak interlayer plane I is 30 kPa, internal friction angle of rock is 33°, and blasting pseudo-static coefficient of 0.1, and the safety coefficient is 2.13.

Under the continuous rainfall or water-saturated state, when rock cohesive of weak interlayer plane I is 0 kPa, internal friction angle of rock is 25°, height of rear crack and slabbing filling water length are 2 m, and the safety coefficient is 0.99.

When the production blasting is conducted under continuous rainfall or water-saturated state, when rock cohesive of weak interlayer plane I is 0 kPa, internal friction angle of rock is 25°, blasting pseudo-static coefficient of 0.1, height of rear crack and slabbing filling water length are 2 m, and the safety coefficient is 0.84.

The evaluation results show that, when the rock bedded slope is under natural state and the potential sliding plane I is exposed, rock bedded slope stability coefficients have large difference under various working conditions, that is, the rock bedded slope stability coefficient falls to 2.13 in production blasting from 2.58 under natural state, decreasing to 17.5%; it falls to 0.99 under continuous rainfall or water-saturated state, decreasing to 61.6%; it falls to 0.84 in production blasting under continuous rainfall or water-saturated state, decreasing to 67.5%.

Based on the above analysis, when the potential sliding plane I of the rock bedded slope is exposed by production blasting operations in the mine, the slope is stable under natural state; but under continuous rainfall state or water-saturated state, the slope stability coefficients drop sharply, and the slope is unstable.

5 COUNTERMEASURES

When the included angle between the spread direction of seismic wave and the rock stratum strike increases, the attenuation effect in the process of seismic wave spread will increase, and the blasting seismic effect will be weakened. So, it is essential to ensure the advance direction of the working slope parallel or oblique to the rock strata.

Strengthen dewater operation, so as to reduce the amount of rainfall infiltrating into potential sliding planes along rock fractures. First, surface drain must be excavated inside and outside of the surface mine field to intercept rainfall and timely drain away water from the open pit. Second, the berm must have a certain gravity grade to reduce rainfall infiltrating into potential sliding planes to some extent.

Bench blasting parameters must be reasonably determined to reduce the vibration of blasting to potential sliding rock. First, the blasting row-number is 2~3 rows, hole spacing is 4.5~5.5 m, row spacing is 4~5 m, front row burden is 5~6 m, control charge quantity in rear row increase to 10~15%. Second, drilling interval charging is adopted in a borehole, air interval is 1~2 m, millisecond-blasting technique is sequentially adopted between the holes and the rows, delay time of up-and-down grain is 25 ms, charging delay time between front row and rear row is 110 ms. Third, subdrilling in front row is 2.5~3 m, and subdrilling in rear rows is 0.5 m deeper than that in front row.

6 CONCLUSIONS

The rock bedded slope in Huangshan surface mine belongs to pull-type slope. There are potential sliding planes between weak-rock and hard-rock, the anti-slide force of the rock bedded slide mass and its internal friction angle is reduced by many factors, among which are mountain slope advance direction, bench parameters, production blasting parameters and continuous rainfall, leading to cause landslide accident.

The rock bedded slope is stable under natural state. However, under water-saturated state, the slope will be in a limit state, and unstable to slide down. That is, under the same or similar conditions, water content of the slope is the main factor affecting its stability. So, great importance must be paid to the influence of water on rock bedded slope stability.

Reasonable bench blasting parameters, the slope advance direction and timely drainage, and so on, are the effective countermeasures to prevent landslide accidents in Huangshan mine.

ACKNOWLEDGEMENTS

This work is financially supported by The Education Department of Sichuan Province (12ZB115).

REFERENCES

Duncan, C. Wyllie & Christopher, W. Mah. (4th ed.) 2004. *Rock Slope Engineering: Civil and Mining*. New York: Spon Press.

Hoek, E. & Bray, J.W. (3rd ed.) 1981. *Rock Slope Engineering*. London: the Institution of Mining and Metallurgy.

Hu, Yu-wen et al. 2004. Sliding mechanism and stability evaluation for mining slope at a mining area of Sichuan. *Journal of mountain science* 22: 224–229.

Li, Wei guang & Lou, Jian guo 2006. Analysis of affecting factors of along-bed landslide of rock mass and its stability under a rainfall condition. *Journal of Mining and Safety Engineering* 23: 498–501.

Li, Wei guang & Zhang, Ji chun 2007. Study on rock mass bedding slope stability under blast seism. *Explosion and Shock Waves* 20: 426–430.

Luan, Ting ting et al. 2013. Stability analysis and landslide early warning technology for high and steep slope of open-pit mine. *Journal of Safety Science and Technology* 9(4): 11–15.

Lysandros, P. 2009. Rock slope stability assessment through rock mass classification system. *International Journal of Rock Mechanics and Mining Sciences* 46: 315–325.

Mei, Qi yue 2001. Forming conditions and sliding mechanism of switch yard slope at Tianhuangping power station. *Chinese Journal of Rock Mechanics and Engineering* 20: 25–28.

Sun, Yu ke et al.1998. Review and forecast for research of stability of slope on the open tip in China. *Journal of Engineering Geology* 6: 305–311.

Yang, Tian hong et al. 2011. Research situation of open-pit mining high and steep slope stability and its developing trend. *Rock and Soil Mechanics* 32: 1438–1439.

Zhang, Dong ming et al. 2012. *Analysis on slope stability for open-pit mine of Multi-level hillside*. Beijing: Science Press.

Architectural, Energy and Information Engineering – Sung & Chen (Eds)
© 2016 Taylor & Francis Group, London, ISBN 978-1-138-02791-6

Preparation of CuInS$_2$ thin films by chemical bath deposition

H. Liu, J. Li & K.G. Liu
School of Materials Science and Engineering, Shandong Jianzhu University, Jinan, China

ABSTRACT: Copper Indium Sulfide (CuInS$_2$) thin film has become one of the most potential and promising photovoltaic material owing to its high optical absorption coefficient and high conversion efficiency. Chemical bath deposition has the advantages of low cost and is convenient to operate. Therefore, it has become a research hotspot to prepare CuInS$_2$ thin films by chemical bath deposition. In this article, CuInS$_2$ thin films have been prepared using the precursor solution containing copper chloride (CuCl$_2$), indium chloride (InCl$_3$), thiourea (CN$_2$H$_4$S), sodium hyposulfite (Na$_2$S$_2$O$_3$), acetic acid (C$_2$H$_4$O$_2$) and citric acid (C$_6$H$_8$O$_7$) at room temperature. In the experiment design, we are aiming to obtain the uniform and compact thin films by changing the parameters such as pH, deposition temperature, deposition time, stirring speed through one-step and multi-step chemical bath deposition method.

Keywords: CuInS$_2$; chemical bath deposition; thin film

1 INTRODUCTION

There are three main pillars of science and technology in modern times: material, energy and information. In order to solve the energy crisis and protect the environment, it is an urgent task to prepare a material which can reasonably use the clean energy. As a novel green and renewable energy, the solar energy is clean, abundant, and economic. Solar cells can convert solar energy into electricity. Compounds such as CuInS$_2$ can be used as the absorption layer of solar cells [1]. Ternary compound CuInS$_2$ is chalcopyrite crystal structure which is suitable for solar cells [2]. It has good physical properties such as photoelectric, thermoelectric performance and high absorption coefficient (10^5 cm^{-1}). In addition, the ideal conversion efficiency is up to 28%, even 32%. It has a low energy band-gap which is about 1.50 eV and is close to the ideal bandwidth [3,4]. Compared with CuInSe$_2$, CuInS$_2$ thin films have a better band gap. Considering environmental protection, CuInS$_2$ is more appropriate than CuInSe$_2$, because sulfur is non-toxic compared with selenium. Therefore, CuInS$_2$ thin films have become one of the optimum absorption materials for thin film solar cells. As a result, it has become the focus of the present study to choose an appropriate method for the preparation of the CuInS$_2$ thin films.

Thin films for solar cell can be deposited by various methods such as, evaporation [5], sputtering [6], spray [7], hydrothermal synthesis [8], electrodeposition [9], chemical bath deposition [10–13], molecular beam epitaxy [14] etc. Among them, evaporation and sputtering method is the most successful, mature technology. Thin films deposited by these techniques have high photoelectric conversion efficiency, and these methods have realized industrialization operation. However, the two methods both need vacuum equipment, which lead to the high cost for preparation, and can not deposit large area thin film; otherwise, they also have a low utilization ratio of raw materials.

Chemical bath deposition techniques have placed substrates which are treated by surface activation in the solution, under the condition of no applied electric field or other energy at the atmospheric pressure and low temperature, controlling the complexation, dissociation and chemical reaction of the reactants, and deposit films on the substrate ultimately. Chemical bath deposition can use non-conductive substrates such as glasses and ceramics. Thin films can be deposited on the substrate with complex shape, even the inner surface of the substrate. Meanwhile, the deposition process is in solution, so it is not necessary to have a vacuum system and consume more energy. The uniformity of the solution can be ensured, therefore, it is suitable to prepare thin films for large areas. It is widely used to prepare thin films for solar energy materials by chemical bath deposition [15–18].

Chemical bath deposition can be divided into one-step and multi-step. One-step chemical bath deposition technique deposits amorphous CuInS$_2$ thin films on substrates firstly, and then crystallizes by heat treatment in a certain atmosphere. Multi-step chemical bath deposition technique deposits In$_2$S$_3$ and CuS thin films [19] on substrates separately, then crystallize by heat treatment in a certain atmosphere to obtain the CuInS$_2$ thin film.

2   EXPERIMENT

2.1   *Substrate cleaning*

Clean the glass substrates with detergent solution then boil them in sulfuric acid for 20 minutes. Wash the substrates by ultrasonic cleaning in the deionized water for 20 minutes, the substrates were finally stored in the deionized water before use.

2.2   *Preparation of solutions*

All the reagents used for the preparation of CuInS$_2$ thin films were AR grade.

2.2.1   *Preparation of solutions for one-step chemical bath deposition*

The reactive solution for the preparation of CuInS$_2$ thin films was obtained by adding citric acid (30 mM), copper chloride (12.5 mM), indium chloride (12.5 mM) and sodium hyposulfite (25 mM) into 100 mL deionized water respectively.

2.2.2   *Preparation of solutions for multi-step chemical bath deposition*

The reactive solution for the preparation of In$_2$S$_3$ thin films was obtained by adding acetic acid (100 mM), indium chloride (10 mM) and thiourea (100 mM) into 100 mL deionized water respectively.

The reactive solution for the preparation of CuS thin films was obtained by adding copper chloride (25 mM), sodium hyposulfite (90 mM) and thiourea (50 mM) into 100 mL deionized water respectively.

2.3   *Deposition of thin films*

The container storing the reactive solution was placed at room temperature, and the glass substrates were positioned vertically in the solution.

2.3.1   *Deposition of CuInS$_2$ thin films in one-step chemical bath deposition*

The CuInS$_2$ thin film was deposited with a stirring speed of 50 r/min, and the substrate was removed from the solution an hour later.

2.3.2   *Deposition of thin films in multi-step chemical bath deposition*

The In$_2$S$_3$ thin film was deposited with a stirring speed of 50 r/min, and the substrate was removed from the solution an hour later.

The CuS thin film was deposited on the substrate where In$_2$S$_3$ thin film was deposited, with a stirring speed of 50 r/min, and the substrate was removed from the solution an hour later.

All the thin films were washed after deposited with deionized water, dried in air.

2.4   *Heat treatment*

The thin films prepared were crystallized to obtain the phase of CuInS$_2$ under 300°C in argon atmosphere for 30 minutes.

2.5   *Characterization of CuInS$_2$ thin films*

First test the crystallization of films by X-Ray Diffractometer (XRD), if the film was a single-phase and well-crystallized, continuous, uniform, then characterize the morphology of the film by Scanning Electron Microscopy (SEM), as well as further analysis and performance testing.

2.6   *Analysis and comparation*

Analyze and compare CuInS$_2$ thin films prepared by one-step and multi-steps chemical bath deposition method.

3   EXPERIMENT PROCESS

The experiment process is as shown in the figure below:

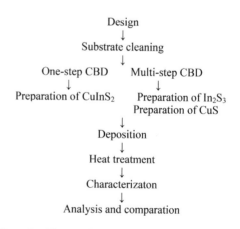

Figure 1.   The experiment process.

4 INFLUENCE

The principle of chemical bath deposition is that the solubility product of ions in the solution is a constant (K_S), when the concentration is approaching to saturation, the ion product is equal to the solubility product (K_S); when the ion product is greater than the solubility product (K_S), the solution is oversaturated, and then the solution will give a precipitation.

The three main factors which influence the solubility product are temperature, solvent and particle size.

4.1 Pretreatment

Pretreatment is the key of a successful experiment. If substrates are not handled properly, the target product won't deposit to the substrate, or the target product will fall off on the substrate.

4.2 Mole ratio of reaction reagent

The mole ratio of reaction reagent has great influence on the reaction process. It will directly affect the solubility product of ions in the solution. Therefore, comparative experiments among different mole ratio of reaction reagent should be done.

4.3 Complex agent

Complex agent: can with metal ions or containing metal ions compounds formed the material with complex. The appropriate complex agent can make the metal ions release slowly. The mechanism of complexation reaction is related to the types of complex agent. The amount of complex agent also affects the degree of reactivity.

4.4 Temperature

The change of temperature will directly determine the reaction direction.

4.5 pH value

The pH of solution has influence on the existing form of ions. For example, some reagents will undergo hydrolysis in a certain pH range. Some metal ions will form hydroxide in certain pH range. What's more, pH will affect the release speed of ions which form complex with complex agent. If the release speed is too fast, the formation rate of films is also too fast, which leads to the weak binding force with substrates. If the release speed is too slow, ions cannot be released, therefore, thin films won't be obtained.

4.6 Stirring rate

The stirring rate impacts thickness of the film. When the speed is higher, the film tends to be thinner and compact. While at a lower stirring rate, a thicker film is formed.

4.7 Heat-treating process

If the film obtained is amorphous state, it cannot get the diffraction peaks in the X-ray Diffractometer (XRD). Through the heat treatment, the amorphous component will be transformed into crystallization after proper annealing process.

4.8 Assisted synthesis

Microwave-assisted synthesis [20,21] can shorten the reaction time and accelerate the heat transfer. Temperature controlling is easy to approaching; avoid local overheating.

In order to improve the reaction rate, other assisted synthesis methods are applied to chemical bath preparation process, such as photo-assisted [22], ultrasonic-assisted [23] chemical bath deposition technique.

All in all, it is a major process to adjust suitable parameters to get the thin film.

5 PROSPECTS

CuInS$_2$ is one of the most promising absorber materials for solar cells, owing to its suitable energy band-gap and high theoretical light conversion efficiency. However, compared with the theory efficiency, the actual conversion efficiency is still quite low. It is still a challenge to prepare CuInS$_2$ thin films with high conversion efficiency. Last but not least, the preparation of CuInS$_2$ thin films with good performance has a long way to go by chemical bath deposition.

ACKNOWLEDGEMENTS

This work was financially supported by the National Natural Science Foundation of China (No. 51272140).

REFERENCES

[1] Mellikov, E. 2008. Research in solar cell technologies at Tallinn University of Technology. *Thin Solid Films* 516(20): 7125–7134.
[2] Angus, A. 2010. Current status and opportunities in chalcopyrite solar cells. *Current Opinion in Solid State and Materials Science* 14(6): 143–148.

[3] Konovalov, I. 2004. Material requirements for CIS solar cells. *Thin Solid Films* 451–452: 413–419.

[4] Hou, X.H. 2005. Synthesis and characteristics of $CuInS_2$ films for photovoltaic application. *Thin Solid Films* 480–481: 13–18.

[5] Abdelkader, D. 2013. The effect of annealing on the physical properties of thermally evaporated $CuIn_{2n}+1S_{3n}+2$ thin films (n = 0, 1, 2 and 3). *Materials Science in Semiconductor Processing* 16(6): 1997–2004.

[6] Liu, X.P. 2007. Reactive sputtering preparation of $CuInS_2$ thin films and their optical and electrical characteristics. *Surface and Coatings Technology* 201(9–11): 5340–5343.

[7] Baneto, M. 2015. Effects of the growth temperature on the properties of spray deposited $CuInS_2$ thin films for photovoltaic applications. *Ceramics International* 41(3): 4742–4749.

[8] Mehdi, M.K. 2015. Facile hydrothermal synthesis, formation mechanism and solar cell application of $CuInS_2$ nanoparticles using novel starting reagents. *Materials Letters* 142: 145–149.

[9] Xu, X.H. 2011. A novel one-step electrodeposition to prepare single-phase $CuInS_2$ thin films for solar cells. *Solar Energy Materials and Solar Cells* 95(2): 791–796.

[10] Lugo, S. 2014. Characterization of $CuInS_2$ thin films prepared by chemical bath deposition and their implementation in a solar cell. *Thin Solid Films* 569: 76–80.

[11] Pan, G.T. 2010. The preparation and characterization of Ga-doped $CuInS_2$ films with chemical bath deposition. *Solar Energy Materials and Solar Cells* 94(10): 1790–1796.

[12] Pena, Y. 2011. $CuInS_2$ thin films obtained through the annealing of chemically deposited In_2S_3-CuS thin films. *Applied Surface Science* 257(6): 2193–2196.

[13] Sharma, R. 2009. Optimization of growth of ternary $CuInS_2$ thin films by ionic reactions in alkaline chemical bath as n-type photo absorber layer. *Materials Chemistry and Physics* 116(1): 28–33.

[14] Calvet, W. 2003. Surface vs. volume stoichiometry of MBE grown $CuInS_2$ films on Si. *Thin Solid Films* 431–432: 317–320.

[15] Savadogo, O. 1998. Chemically and electrochemically deposited thin films for solar energy materials. *Solar Energy Materials and Solar Cells* 52(3–4): 361–388.

[16] Nair, P.K. 1998. Semiconductor thin films by chemical bath deposition for solar energy related applications. *Solar Energy Materials and Solar Cells* 52(3–4): 313–344.

[17] Goudarzi, A. 2008. Ammonia-free chemical bath deposition of nanocrystalline ZnS thin film buffer layer for solar cells. *Thin Solid Films* 516(15): 4953–4957.

[18] Asenjo, B. 2008. Study of $CuInS_2$/ZnS/ZnO solar cells, with chemically deposited ZnS buffer layers from acidic solutions. *Solar Energy Materials and Solar Cells* 92(3): 302–306.

[19] Mukherjee, N. 2011. A study on the structural and mechanical properties of nanocrystalline CuS thin films grown by chemical bath deposition technique. *Materials Research Bulletin* 46(1): 6–11.

[20] Mostafa, S. 2013. $CuInS_2$ nanoparticles: Microwave-assisted synthesis, characterization, and photovoltaic measurements. *Materials Science in Semiconductor Processing* 16(2): 390–402.

[21] Obaid, A.S. 2013. Fabrication and characterisations of n-CdS/p-PbS heterojunction solar cells using microwave-assisted chemical bath deposition. *Solar Energy* 89: 143–151.

[22] Dhanya, A.C. 2013. Effect of deposition time on optical and luminescence properties of ZnS thin films prepared by photo assisted chemical deposition technique. *Materials Science in Semiconductor Processing* 16(3): 955–962.

[23] Gao, X.D. 2004. Morphology and optical properties of amorphous ZnS films deposited by ultrasonic-assisted successive ionic layer adsorption and reaction method. *Thin Solid Films* 468(1–2): 43–47.

Architectural, Energy and Information Engineering – Sung & Chen (Eds)
© 2016 Taylor & Francis Group, London, ISBN 978-1-138-02791-6

Enrichment law of low-rank coalbed methane of Hunchun Basin

Y.Z. Wang
Exploration and Development Research Institute, Daqing Oilfield Company Ltd., Daqing, China

ABSTRACT: This paper systematically describes the main controlling factors of accumulation conditions of Coalbed Methane (CBM). Research on structure setting and sedimentary facies was carried out, and analyzed the enrichment law of CBM in Hunchun Basin. The result shows the CBM Enrichment effect of structure, deposition, and magmatic rock. The anticline core will be local of secondary reservoir; Groundwater perfusion to the basin through fault forms enrichment pattern of "continuous fold", not only supply bio-methane for anticline but also sealing CBM in syncline core. From east to west, the environment of coal-forming, thickness and stability of seams becomes better. The mudstone and tight sand as adjacent rock of seams directly, by swamp and natural levee vertical overriding, will be better for sealing CBM; The local diabase intrusion could increase CBM generating and change physical property of seams.

Keywords: Hunchun basin; Low rank; Accumulation conditions; Enrichment law

1 INTRODUCTION

In China, the low-rank coalbed methane (vitrinite reflectance is less than 0.65%) is 16×10^{12} m³, accounting for 43.47% of total quantity at a large resource scale. The successful development of low-rank coalbed basins in the USA, Canada and Australia has proved the exploration potential of low-rank coalbed methane in China. Compared with typical USA basins, the low-rank basins are mainly continental ones, with complex structure settings. Owing to multi-phases of structure transformation, the coal seam is in strong anisotropy, so as to form the CBM accumulation with Chinese characteristics of "low gas content, saturation and permeability". China's low-rank CBM accumulation is so distinctive that there are different accumulation conditions in basins, not all external exploration theories are applicable for them. The exploration of low-rank coalbed methane needs to be based on geological features to make clear accumulation regularities. Only by this way, the low-rank coalbed methane can be explored rapidly. The industrial breakthrough of coalbed methane has been achieved in Hunchun Basin. It is significance for large-scale development to study its type of accumulation

2 OVERVIEW OF HUNCHUN BASIN

The Hunchun Basin is located in Hunchun city of Yanbian Korean Autonomous Prefecture in the east of Jilin province. It is neighbored with Tumen River to the west, Hulubie to the east, Songshu Village to the north and the south of China-Russia border, which covering an area of 630 km². The Hunchun River goes through the Basin from east to west, then southwest, and finally flows into the Tumen River. In west Hunchun Basin, Baliancheng, Chengxi, Sandaoling and Ying'an mining area, Banshi detailed exploratory areas I, II and III are in its south, and Luotuohezi, Wujiazi general survey area and Miaoling mining are located in its east (Figure 1). It is estimated that the Baliancheng—Banshi area I has a CBM resources of 19.2×10^8 m³.

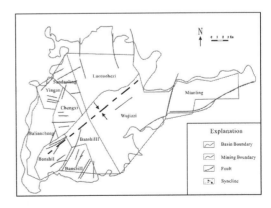

Figure 1. Geological sketch map of Hunchun Basin.

3 CONDITIONS OF CBM ACCUMULATION

3.1 Structural feature

The Hunchun Basin belongs to the Dongning Hunchun fold system of Xing'an Mongolian Variscide fold belt. It is entirely distributed in northeast, with a relieved syncline structure pitching to the west roughly. Small folds are developed in its local areas. There are mainly of normal faults in the Basin, mostly associate with folds, and shown as a series of the NE, NNE and NS faults in distribution.

3.2 Coal formation

The coal formation in Hunchun Basin is the Hunchun Group of the Eogene, with a maximum thickness of 960 meters. There are 30–70 coal seams in the Basin, where there are generally 3–8 minable seams and partial minable seams. The middle part of coal bearing section is the best coal bearing property (Figure 2). The extensive developed coalbeds are 19#, 20#, 21#, 22#, 23#, 26# and 30# in the basin. The coalbeds were characterized by excessive. The coal accumulation center is located in the eastern Baliancheng, middle of Chengxi and northern Banshi area I, where there are cumulative coalbed thickness is more than 20 meters with numerous coal seams and stable distribution. The coal seam becomes thinner and thinner to the east in a poor succession. The coalbed buried depth is deeper in the west than east. However, the burial depth for primary mineable coalbeds is less than 600 meters, which is quite suitable for development of coalbed methane.

3.3 Coal features

In general, the R_o of coal ranges from 0.42% to 0.67% in the Hunchun Basin, so it belongs to the low-rank coal. From west to east see kennel coal, kennel coal—lignitic coal, and lignitic coal. The maceral of coal is made up of vitrinites, at an average content of 81.76% and an ash content at 21–45%. The coal is black with pitchy luster. The cleap within 5 cm contains around 10–20. The coal is classified by half-light to light type, and the half-dark type takes the second place. The fracture is flat, irregular or ladder-like, as well as conchoids. The ash content goes up gradually and the coal structure becomes poor from west to east in the Basin.

So far, there is a small amount of gas content in Banshi Area I and Baliancheng mining area is analyzed. Gas content of Banshi Area I is 0.83–4.11 m^3/t, while that of Baliancheng 19# Coalbed is 4.32–8.18 m^3/t. The gas content of coalbed methane, which is similar to regularities of coal-rank distribution, takes on a high gas content in the west and low in the east, and is somewhat relative with its depth and width.

4 ENRICHMENT LAW OF CBM

The CBM enrichment is controlled by more geological factors such as thickness of coal, burial depth, metamorphic grade, gas content, reservoir pressure, tectonic setting, which affect CBM accumulation. In practical exploration, the decisive factors for CBM enrichment and accumulation are not its primitive conditions but preservation conditions.

4.1 Effect of structure on CBM enrichment

Structure not only controls CBM generating conditions, but also effects CBM preservation conditions.

First, a certain thickness of coal seam is available for the material base of hydrocarbon generation for low-rank coalbed methane. The development of folds has distinctive control effect on the CBM accumulation and distribution in Hunchun Basin. In the effect of Xing'an—Inner Mongolia fold, a series of small-size folds at a proximate grade comes into being in the west of the Basin. Accordingly, a synsedimentary normal fault forms in a zonal distribution in a north-east direction, and it is perpendicular to the trend of the Basin. The center of sedimentation in the west of the Basin guarantees sufficient and stable provenance. The development of folds and faults is to the benefit of gathering and stagnation of water in the wide and gentle tectonic setting. These provide a sedimentary environment for swamp development, so that this area becomes the coal accumulation center of the Basin. With the inclination of the Basin to the

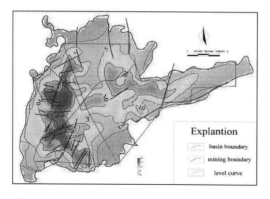

Figure 2. The distribution of coalbeds thickness in Hunchun Basin.

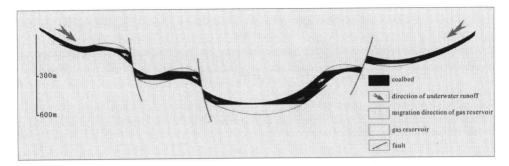

Figure 3. Accumulation model of CBM in Hunchun Basin

east, the coal seam becomes thinner and thinner even wedges out with poor coal-forming environment. Meanwhile, the coal seams are distributed symmetrically nearly along the two sides of syncline structure in Hunchun Basin, indicating the control effect of the tentonic features of the Basin on coalbed development.

Second, dense folds in syncline wing of Hunchun Basin provide the advantageous space for CBM enrichment, forming a "successive-fold" CBM pattern (Figure 3). The fracture near anticlinal axis develops well, which causes partial CBM loss in the same depositional stage. Even so, the anticlinal fracture-developing area enables CBM to be a secondary accumulation with burial depth of coal seams and top overlying strata compacting gradually. The faults develop better in coal enrichment area in the west of the Basin, and they play an active role in underwater runoff; the fault development offers necessary hydrogeological conditions for perfusion of underwater. The stratum water flows to the center of the Basin slowly, not only promoting generation of biological methane, but also increasing degree of mineralization in syncline core in water. This is to the benefit of CBM accumulation because the sealing effect of CBM is available with dissolving capacity decreasing in water. Thus, the CBM enrichment area formed in the NW–SE direction in the west of the Basin is the first target area for CBM exploration and development in the Basin.

4.2 Effect of sedimentary characteristics on CBM enrichment

The CBM reservoir is being in need of formation sealing of available overlying strata in its generative process or after its generation. This sealing mainly depends on lithologic characters of coal roof. When the lithologic character of roof reaches enough densification, the CBM is not lost. Thus, the sedimentary environment of coalbed formation plays an important control role in CBM enrichment.

The coal accumulation in the basin mainly happens in the mid term of deposition of the Hunchun Group. It develops ten sets of coal seams with a maximum cumulative thickness of as much as 40 meters. In the deposition of coal, the basin subsides continually, and the velocity of source supply is less than that of subsidence t of the Basin, indicating a sedimentary characteristic of water subsidence. The main sedimentary environment in the basin shows the large scale of delta depositional system, while deposit of clastic deposits such as alluvial fan only develops in the margin of the basin. Some small-scale lacustrines in the center of the basin are distributed in the finger sand bodies. Main coal accumulation environments in this stage are swapiness in delta plain, relatively low-lying terrains in the west. Plants grow luxuriantly due to sufficient underwater supplementation, where the coal seam comes into being in a stable thickness. The coal accumulation environment becomes poor and coal seam becomes thinner to the west. The CBM well reveals the swamp microfacies are the main developed lithofacies. With riverway moving and discarding, the natural levee and swamps overlay repeatedly, so that the coal roof consists mostly of mudstone, siltstone and fine sandstone. The mudstone is quite to the benefit of CBM preservation for the roof or baseboard, while the tight sandstone plays the same sealing role. In addition, the sandstone near the coal seams is also regarded as a supplementary gas source.

4.3 Effect of magmatic activity on CMB enrichment

Magmatic activity is significance for CBM accumulation especially its invasion activity. The invasion of magmatic rock occurs in the region of Baliancheng mine, forming dyke and sill. The area of dyke is around 0.1 km² between Boreholes 0411 and 0602 in a south-north inclination of around 152 degrees. The distribution area of sill is over

31

1 km² in Borehole 74–21 nearby, presenting an irregular shape. It is mainly consisting of dolerites at tens of meters thick at a dip angle of 15–30 degrees. The intrusive body of magmatic rock is located in the bottom of the mid coal-bearing section, inclining to the southwest. On the one hand, magmatic rock has distinct hot gas field effect, and coal seams are mainly baked but corroded less. The maturity of coal seam rises in a short term with a great deal of secondary gas generation. To some extent, this increases the gas saturation in coal seams, which is no doubt a tremendous advantage for low-rank coalbeds. On the other hand, the cooling of magmatic rock forms a shrinkage fracture system, which improves physical property of coal to some extent. Wells BLCX-1004 and BLCX-1005 are located in the invasion zone of magmatic rock, where cumulative gas is over $20 \times 10^4 \, m^3$, ten times of gas output than the other wells.

5 CONCLUSIONS

The Hunchun Basin presents its syncline structure pitching to the west. Its coal seams mainly develop in the mid coal-bearing seam section in the Hunchun Group, with the "low rank, thin and numerous coalbeds". Its advantages such as coal seam cumulative thickness, stable distribution and shallow bury prove good prospect of CBM exploration and development in Hunchun Basin.

Through analysis on CBM accumulation conditions of Hunchun Basin, it is believed there is the available CBM enrichment in its west part. The fold and fault accompanying area in the west has a distinctive control to accumulation of coal seams and CBM. The anticline axis provides an encirclement trap for secondary CBM accumulation. Underwater exerts a sealing effect on the synclinal CBM, which forms a "successive fold" accumulation model. Coal accumulating environment, and thickness and stability of coal seam are in priority to that of the west of the Basin. The vertical superposition of swamp and natural levee microfacies offers a better cap condition for coal seams. Mudstone and tight sandstone make up baseboard and roof of coal seam, which exerts a good sealing effect on the CBM. Regional dolerite invasion body can promote the secondary gas generation in coal seams of Baliancheng mining. This improves physical property of coal to some extent.

An industrial breakthrough is achieved in the low-rank CBM exploration in Hunchun Basin. Depending on the understanding of enrichment law, we wish this industrial breakthrough would be a certain reference value for other low-rank basins.

REFERENCES

Y. H. Cui & SH. H, Wang & Y. Q. Ma, et al. Analyses on Preservative Feature and Exploitation Prospect of Coal Bed Gas in Hunchun Basins. Coal Technology, 2005, 24(6): 102–104.

L. M. Fan & SH. G Li & Z Sun. 2004. Study the regularity of the igneous rocks in Tiefa coalfield. Coal Technology, 23(3): 107–108.

J. L. Li & Z. SH, Guo, Wu Jingdong, et al. Applied research on reservoir alteration process engineering in Hunchun coalfield CBM exploitation. Coal Geology of China, 2012, 24(11): 22–27.

Q. M. Li & R. Y. Wang & H. L. Chen. 2004. Study on coalbed gas exploration in low rank coal basin Tuha Oil & Gas, 9(3): 267–272.

J. Liu & Y. L, Wang. 2009. Intrusive rock characteristics and intruding mechanism in the Tiefa mining area. Coal Geology of China, 21(9): 29–31.

S. P. Peng & B. Wang & F.J. Sun, et al. 2009. Research on reservoir patterns of low-rank coal-bed methane in China. Acta Petrolei Sinica, 30(5): 648–653.

A. R. Scott & W. R. Kaiser & W. B. Ayers, et al. Thermogenic and secondary biogenic gases, San Juan Basin, Colorado and New Mexico: Implications for coalbed gas producibility. AAPG Bulletin, 1994, 78(8): 1186–1209.

B. Wang & J. M. Li & Y. Zhang, et al. 2009. Geological characteristics of low rank coalbed methane. Petroleum Exploration and Development, 36(1): 30–33.

Y. Q. Wang & T. H. Yang, T. L. Li, et al. 2012. Study on gas geological law of no.19 seam in Baliancheng mine. Coal Science and Technology, 40(1):115–117.

J. P. Ye & Q. Wu & Z. H. Wang. 2006. Controlled characteristics of hydrogeological conditions on the coalbed methane migration and accumulation. Journal of China Coal Society, 26(5): 459–462.

G. ZH, Li & H. Y. Wang & L. X. Wu. 2005. Theory of syncline controlled coalbed methane. Natural Gas Industry, Journal of China Coal Society, 25(1): 26–28.

J. B. Zhang & H. Y. Wang & Q. B. Zhao. 2000. Coalbed gas geology in China. Beijing: Geological Publishing House: 15–30.

Architectural, Energy and Information Engineering – Sung & Chen (Eds)
© 2016 Taylor & Francis Group, London, ISBN 978-1-138-02791-6

Experimental study on frozen-heave influence factors for graded gravel in surface layer of passenger dedicated line

Kai Yi Han, Yuan Liu, Yang Xie & Tan Jiao
School of Civil Engineering, Central South University, Changsha, China

ABSTRACT: Laboratory frozen-heave experiments were performed to study the moisture content, porosity and fine content which influence the frozen-heave's influence on graded gravel, and a partial correlation method is applied to analyze the correlation between influence factors and frozen-heave ratio. The results indicate: the frozen-heave ratio of graded gravel increased with increasing of the moisture content and the fine content; however, with the increasing of the porosity, the graded gravel frozen-heave ratio decreases. Although we actually analyzed the frozen-heave ratio of graded gravel, we should take all factors into consideration.

Keywords: passenger dedicated line; graded gravel; frozen-heave; correlation

1 INTRODUCTION

In the cold northern area of China, roadbed frozen-heave is one of the main reasons that cause stability loss of roadbed engineering. Subgrade in seasonal frozen soil area will change with the phenomenon of frost heaving and thaw collapsing repeatedly, this phenomenon can damage the smoothness of track and influence the safe operation of the railway [1–2]. However, there is no mature experience can be learned to prevent the current frozen-heave, so it has an important significance to study the frozen-heave of passenger dedicated lines subgrade.

Based on the field observation, the 90% of railway subgrade frozen-heave was produced in the subgrade surface range. The graded gravel is the main part of the subgrade surface, the frozen-heave problems will seriously affect the safety, efficient and fast operation of the dedicated passenger line. At present, domestic and foreign scholars have less research on the frozen-heave problem of the coarse materials. In this paper, the Harbin-Dalian passenger dedicated line was selected as the research background to study the influencing factors on the problem of roadbed graded gravel frozen-heave, and the sensitivity of the factors is analyzed, which provided certain reference significance to master the variation law of graded gravel.

2 FROZEN-HEAVE INFLUENCE FACTORS ANALYSIS FOR GRADED GRAVEL

The graded gravel selected for the tests is fresh and slightly weathered hard sandstone, which is massive and sharp, and physical properties of graded gravel

Table 1. Physical properties of graded gravel in natural condition.

Maximum particle size/mm	Uniformity coefficient	Coefficient of curvature
45	15.9	2

in natural condition is shown in Table 1. Ratio of water content, fines content and porosity in the test required is suitable, according to the "code for soil test of Railway Engineering" (TB10102-2010) [3], the specimens (Φ156 mm × 110 mm) were frozen at –5°C. The formula of frozen-heave ratio for specimens is:

$$\Delta h/H \times 100\% \tag{1}$$

In the formula: H is the original height of specimens (110 mm);

$$\Delta h—Frozen-heave\ degree; \tag{2}$$

3 WATER CONTENT TESTS

The fines content and porosity of graded gravel are the controlled constant, the water content of 4%, 8%, 10% and 12% specimens is used to perform frozen-heave tests, the relationship between water content and frozen-heave ratio is shown in Figure 1.

From Figure 1 it can be seen that frozen-heave factor increases with water content. Along with the increase of water content of the sample, fro-

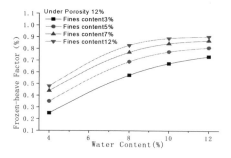

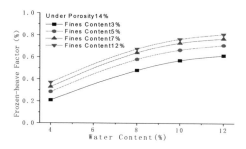

Figure 1. The relationship between frozen-heave and moisture content.

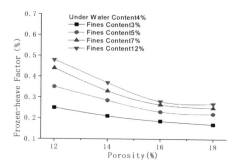

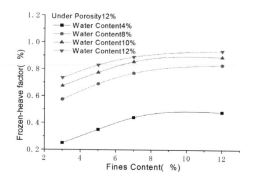

Figure 2. The relationship between frozen-heave and porosity.

zen-heave factor will decrease gradually. When the moisture content is over 10%, the change ratio of frozen-heave will be smaller (less than 0.1%). If the water content and fine content are the same, the frozen-heave factor in the porosity 12% is bigger than that in the porosity 14%.

4 POROSITY TESTS

The samples porosity and fine content are invariable, water content of 4%, 8%, 10% and 12% specimens are used to perform frozen-heave tests, the results are shown in Figure 2.

From the Figure 2 it can be seen that the samples frozen-heave factor decreases with the increase of porosity, and the change rate of frozen-heave factor decreases with the increasing of porosity. When the moisture content is 4%, if the porosity is more than 16%, the change ratio of frozen-heave will be smaller, basically remains unchanged. If the fines content and porosity are constant, the frozen-heave factor in moisture content 8% is bigger than that in moisture content 4%.

5 FINES CONTENT TESTS

The samples porosity and water content are invariable, frozen-heave tests are performed in fines content 3%, 5%, 7% and 12%, the test results are shown in Figure 3.

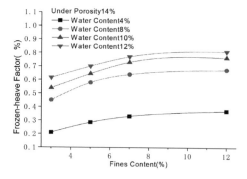

Figure 3. The relationship between frozen-heave and fines content.

From the Figure 3 it can be seen that the frozen-heave factor increases with the increasing of fines content. If the water content and porosity are constant, the frozen-heave factor growth rate is decreasing. When the fines content is more than 7%, the frozen-heave factor is less than 0.1%, and frozen-heave ratio tends to be stable.

6 SENSIBILITY ANALYSIS OF GRADED GRAVEL FROZEN-HEAVE FACTOR

Through the partial correlation method, the water content, porosity and fines' influence on graded gravel were explored based on graph 1, 2, and 3. The specific of this method is: keeps one of the influence factors unchanged and analysis the correlation of the other factor and the creep characteristics. The results are shown in Table 2.

The value used partial correlation method for moisture content, porosity and fines content is 0.948, 0.815, 0.78 respectively, these three influencing factors on frozen-heave are all in 0.01 (bilateral) level significantly related.

7 SUMMARY

1. The graded gravel frozen-heave factor increases with moisture content and fines content, however, the frozen-heave factor growth

Table 2. Sensibility analysis the frozen-heaven factor of graded gravel.

Unchanged factor	Water content (%)	Frozen-heave factor (%)
Correlation	1	0.948
Significant (bilateral)	0.000	0.000
DF (free degree)	0	44
Unchanged factor	Porosity (%)	Frozen-heave factor (%)
Correlation	1	−0.815
Significant (bilateral)	0.000	0.000
DF (free degree)	0	44
Unchanged factor	Fines content (%)	Frozen-heave factor (%)
Correlation	1	0.780
Significant (bilateral)	0.000	0.000
DF (free degree)	0	44

rate will gradually decrease. When the moisture content is over 10%, the frozen-heave factor changes small (less than 0.1%); if the fine content is over 7%, the frozen-heave factor variation is less than 0.1%, and it will be gradually stable. The frozen-heave factor decreases with the increase of porosity; when the moisture content is 4% and the porosity is over 16%, the frozen-heave factor is basically unchanged.

2. The sensitivity analysis of graded gravel frozen-heave factor shows that: the influence of three factors on the graded gravel frozen-heave factor is moisture content > porosity> fine porosity content. The dominant factor is moisture content, and the measures to control the water content <4% of graded gravel are put forward in this paper, whose frozen-heave amount meets the requirements.

3. Moisture content, porosity and fines content's influence on frozen-heave factor are all in 0.01 (bilateral) level significantly related. Therefore, in the analysis of practical frozen-heave problem, the effect of moisture content, porosity and fines content of graded gravel frozen-heave should be considered comprehensively.

REFERENCES

[1] Hong-sheng LI, Yue-dong WANG, Zeng-li LIU. Calculation model of roadbed frozen frost on the cold zone [J]. *Journal of Railway Science and Engineering.* 2011, 8(1): 34–38.
[2] Hong-jiang MA. The search of subgrade of Lanzhou-Xinjiang Railway about the affecting factor sand renovate measure [D] *Master's Degree Thesis.* Lanzhou: Lanzhou Jiao Tong University, 2011.
[3] Code for Soil Test of Railway Engineering (TB10102-2004) [S]. *Beijing: China Railway Publishing House.*
[4] Yi-chen ZHANG, Xin LI, Xi-fa ZHANG et al. Research on frost heave susceptibility and classification of coarse grained soil of highway subgrade in seasonally frozen ground region [J]. *Chinese Journal of Geotechnical Engineering.* 2007, 29(10): 1522–1526.
[5] Vinson, TS, Ahmad, F, Rieke, R. Factors important to the development of frost heave susceptibility criteria for coarse-grained soils [M]. *Transportation Research Board Business Office,* 1986: 124–131.
[6] Ming ZHU, Hui SONG, Wei ZHONG. Mechanism of water transport and treatment measures of subgrade [J]. *Subgrade Engineering.* 2001, 1: 63–64.

Architectural, Energy and Information Engineering – Sung & Chen (Eds)
© 2016 Taylor & Francis Group, London, ISBN 978-1-138-02791-6

The rise of China and the decline of the US in Central Asia: Energy and economic cooperation in the new geopolitical context

Yi Nan Ding
International Business School, Graduate School, Beijing Foreign Studies University, Beijing, China

Long Peng
International Business School, Beijing Foreign Studies University, Beijing, China

Guo Yuan Wang
School of English and International Studies, Beijing Foreign Studies University, Beijing, China

ABSTRACT: Central Asia, with its strategic position and rich oil reserves, has been the center of geopolitical game of world powers. Shifts of strategy of big players in the region has led to changes in its geopolitical landscape. In light of the changing world strategy of the US and China, this paper argues that as the US withdrew from the Middle East, its prominence in Central Asia is on a significant decline, with Central Asia turning into a backup energy safety net. China's heightened presence in Central Asia is driven mostly by economic development, coupled with strategic choice of avoiding conflict with the US in East Asia. Despite China's rise in Central Asia, it should not be interpreted as a gesture of geopolitical assertiveness or a political threat.

Keywords: Central Asia; Sino-US Relations; energy; "The Belt and Road Initiative"

1   INTRODUCTION

Year 2014 has been a transition point for the geopolitical situation in Central Asia. China launched its ambitious "Silk Road Economic Belt" project, making historical progress in economic cooperation with Central Asian countries. The United States' prominence is declining, with Hilary Clinton's "The New Silk Road" initiative withering and the US focus turning to rebalancing East Asia. Russia's strong position on the Ukraine Crisis is also changing the geopolitical landscape of Eurasia. As a region rich in energy and significant in geographical position, Central Asia has been the center of economic and strategic struggle for major world powers. Economically, estimated to have the size of energy reserve as the North Sea, the region is a key energy supplier which is critical to maintaining energy security for its partners. Geographically, it stands at the heart of the grand chessboard, boarding countries including Russia, China, India and the Middle East, making it an important factor for world security. Moreover, given its landlocked position, oil and gas are to be transported through pipelines to countries including Russia, Turkey, Pakistan and China, which also enhanced its strategic importance. With the chang-

ing global strategy of the two countries, the presence of the US in Central Asia has declined and China's influence risen greatly.

This paper argues that as the US withdrew from the Middle East, its prominence in Central Asia is on a significant decline, with Central Asia turning into a backup energy safety net. And China's heightened presence in Central Asia is driven mostly by economic needs, coupled with strategic choice of avoiding conflict with the US in East Asia. The argument is analyzed under the framework that first discusses the Central Asia strategy of the US and China, followed by the analysis of the position of energy and related economic issues under the overall strategic plan.

2   THE UNITED STATES IN CENTRAL ASIA: GOING, GOING…

Central Asia was of strategic importance to the US in the 1990s and early 20th century due to its geographic proximity to Russia and the Middle East, as well as its oil reserve. However, Barack Obama's Central Asia strategy diverted from that of Clinton and Bush Jr's because of the shift of focus of US foreign policy. Since 2009, US strategy in the region include encouraging Central Asia's assistance in stabilizing Afghanistan, promoting democracy and human rights, combating human trafficking and supporting balanced trade policies (Cohen; Feigenbaum). As the US leaves the Middle East and returns to East Asia, its presence in Central Asia is also reduced, and the region has taken more of a supporting, rather than strategic role, in stabilizing post-conflict Afghanistan (Imas).

2.1 US strategy in Central Asia: Power balance

During Clinton and Bush Jr.'s administrations, Central Asia is of strategic importance to the US in order to protect US energy security and deter Russian power. As Zbigniew Brzezinski proposed, the primary interest of the US in Central Asia is to make sure that no single power controls this geopolitical space, and that it remains financially and economically accessible to other countries. Only when oil and gas pipelines connect Central Asia with major world economies can the diversification of geopolitics be realized (122–123). Politically, the US promotes democracy and seeks to ensure the independence of Central Asian countries from Russia. Through sponsoring "colored revolutions", the US aims at developing domestic environment favorable to building the market economy, which would translate in the end to close ties with the US. Moreover, under the grand mission of combating terrorism and protecting energy transportation, the US also enjoys military presence in the region.

However, it should also be noted that although the US can play a role in the region, it may not be a dominant role. The region will continue to be more influenced by Russia, China, and the Muslim world, because of geography, history, and longstanding cultural ties (Cohen). Therefore, despite its strategic position, Central Asia is not at the core of US foreign policy, especially when it leaves the Middle East.

2.2 Energy under US strategy: Economic means to strategic end

The participation of the US in energy issue in Central Asia is not for energy per se. Energy is the economic means to the strategy of maintaining world primacy and power balance in the region. It is not driven by business profit but government strategy (Xu 104). First, the US does not rely on Central Asia as energy supplier because geographically the region is inaccessible to the US soil. Instead, it serve as an alternative energy source that protects its energy security. The "western" route of Baku-Tbilisi-Ceyhan (BTC) pipeline which runs through Azerbaijan, Kazakhstan to

Ceyhan port of Turkey ensures supply to US partners near the Mediterranean and Europe as well as the diversification of sources for itself (Qian 232). The 2.9-billion-dollar worth of pipeline is also viewed as a form of economic penetration—such huge project plays an important role in promoting economic development by adding to more jobs, economic expansion and energy independence (ibid 237–238). Apart from vigorously supporting the building of the BTC pipeline, large US firms also take a hand in Central Asian energy through transferring investment, technology know-how and human capital. Unocal, Exxon Mobil, Pennzoil, Chevron Texaco and Delta Hess are among the transnational energy conglomerates exploiting oil and gas in the Caspian Sea (Xu 114). All this has strengthened US position in the energy competition in Central Asia.

2.3 The decline of the US in Central Asia

During the Obama Administration, US influence in Central Asia is on a significant decline. With US withdrawing troops from the Middle East, its air force leaving Kyrgyzstan and its foreign policy shifting to East Asian, there has been few progress in Central Asian affairs ("The United States in Central Asia"). While still taking a strategic presence, Central Asia is treated more as an economic support to stabilize Afghanistan. Hilary Clinton's "The New Silk Road" project has not come about well. The crux of the initiative, CASA-1000, a 1-billion-dollar electricity transmission project from Kyrgyzstan and Tajikistan to Afghanistan and Pakistan, is faced with strategic failure because of poor infrastructure, problems of cooperation among Central Asian countries, corruption and vulnerability to terrorist attacks and Russian involvement (Imas).

The decrease of US prominence in Central Asia is also related to the growing productivity of oil and liquefied natural gas. The US continues to reduce its reliance on overseas energy and is becoming more powerful in the world energy market. According to the Annual Energy Outlook of 2014 by the US Energy Information Administration, US domestic production of crude oil will witness an increase of 800,000 barrels per day in 2015, reaching 9,500,000 barrels per day in 2016. Production of natural gas will also support domestic consumption, and it is estimated that by 2029, the US would export 3.5 trillion cubic feet of liquefied natural gas, with at least 3 large scale plants built (Annual Energy Outlook: Executive Summary). Increased production of oil and gas and decreased consumption by a growing number of energy-efficient cars together reduce US reliance on overseas energy ("The US to Reduce Reliance on Overseas

Energy"; Yang). Under such background, it is no surprise that the US did not make major progress in energy policy in Central Asia. Energy status quo in the region alone seems suffice to maintain the strategic plan.

3 CHINA IN CENTRAL ASIA: A RISING PARTNER

In contrast to US decline, China is a rising partner for Central Asian countries. Different from the "missionary approach" of the US to spread democracy and system of government, China's presence in Central Asia is a lot more pragmatic—propelled by need to secure energy, metals and strategic minerals in order to support rising living standards of the immense population (Kaplan).

3.1 China's Central Asia strategy: Seeking peaceful development

China's "Silk Road Economic Belt" is by nature a regional integration mechanism under a weak US presence. As the US is actively pushing through its rebalancing strategy in East Asia, China has become the target of pressure and deterrence. Military alliances with Japan, the Philippines were strengthened, closer political ties with China's neighboring countries were formed, and China was excluded from the US-led Trans-Pacific Partnership (TPP). US-China relations in East Asia is becoming increasingly contentious and zero-sum (Sun). Therefore, moving east would add tension to the already unstable situation. Looking west, due to the decline of Russian power in Central Asia and US withdraw from the Middle East, China would have an increasing influence in this less contentious region due to its geographical proximity and strong investment capacity.

3.2 Energy: An important part of economic cooperation

Energy cooperation with Central Asian countries is one of the key aspects of China's Silk Road Economic Belt. Central Asia energy not only serve to meet the growing need of China's economic development, but also reduce its reliance on sources from the Middle East. Through pipelines from Kazakhstan and Uzbekistan to China's Xinjiang, the landline reduces risks in maritime transportation of oil via the Malacca Strait. China's increasing demand for energy, which grows at an annual rate of 2.23% (1.9% higher than the International Energy Agency standard), combined with pressure from the East, make energy cooperation between China and Central Asia all the more necessary.

President Xi Jinping's speech in Nazarbayev University, Kazakhstan marks the official kickoff of the Silk Road Economic Belt. Xi proposed 5 areas of communication—setting up policy communication, improving road connectivity, promoting unimpeded trade, enhancing monetary circulation and increasing understanding between the people ("Promote Friendship between Our People and Work Together to Build a Bright Future"). Under the spirit of Xi's proposal, the Energy Development Strategy Action Plan (2014–2020) released by the State Council has put energy cooperation with Central Asia at a strategic position and strengthen energy cooperation through the Silk Road Economic Belt and the 21st Century Maritime Silk Road.

Kazakhstan, the largest oil producer in Central Asia, is the first and most important partner of China in the new project. Energy cooperation between the two countries dates back to 1997 when China National Petroleum Corporation (CNPC) purchased 60.3% stock rights of Aktobemunaigaz, a Kazak oil company. In 2005, CNPC purchased PetroKazakhstan, Inc. at 4.18 billion dollars and in 2006, China finished the construction of China-Kazakhstan Oil Pipeline, a 1200-kilometer-long pipeline from Aksu, Kazakhstan to Alashankou, China, with the annual transportation capacity of 10 million tons (Xu 84–86). China's exploitation of oil in Kazakhstan has also increased steadily. CNPC runs 6 upstream projects in Kazakhstan and keeps its productivity at 30 million ton level for three consecutive years. In 2013, production of crude oil reached 23.79 million tons, and the oil and gas equivalent amounted to 30.15 million tons. In October 2013, China bought Kashagan Field, Kazakhstan's largest offshore oil field and the world's 5th largest oil field, for 5 billion dollars, becoming the 6th biggest shareholder ("China and Central Asian Energy Cooperation: A National Strategy"). Moreover, in order to ensure security, China also worked with Central Asian countries on anti-terrorism through platforms such as Conference on Interaction and Confidence-Building Measures in Asia (CICA) and Shanghai Cooperation Organization (SCO) ("New Asian Security Concept").

3.3 Beyond energy: The growing chinese capital

The rise of China in the cooperation with Central Asia is not restricted to energy. Advantage in capital is critical for China in the Silk Road Economic Belt. Economic cooperation comes in full scale and with huge amount of capital. On December 14th, Premier Li Keqiang exchanged in-depth views on bilateral relations with President and Prime Minister Kazakhstan, and both sides agreed to enhance mutually beneficial cooperation, especially carry out bilateral cooperation in production capacity. During the trip, Premier Li signed 18 billion dollar's deal including oil, infrastructure and transportation. Different from previous cooperation, this deal is centered not on energy ("China and Kazakhstan Reach Consensus"). Yao Peisheng, Chinese Ambassador to Kazakhstan said "The deal is centered on production capacity cooperation. China and Kazakhstan have achieved many results in energy cooperation, but we can't stay in this one area. In future cooperation, more efforts will be made in the field of infrastructure, agriculture and new energy" ("National Strategy behind the $18 Billion Deal").

Capital is an important advantage China has over the United States in Central Asia. While US influence is fading away along with its failed "New Silk Road" initiative, Chinese capital is vigorous and long-lasting. Scholars believe that China is already a net capital exporter in 2014 although the actual use of foreign capital (106.24 billion dollars) still exceeded non-financial foreign investment (89.8 billion dollars). Until September 2013, China has 6.29-trillion-dollar worth of overseas financial assets, and 4.49 trillion dollars' foreign financial liabilities, with the net asset amounting to 1.8 trillion dollars. As the Silk Road Economic Belt and the 21st Century Maritime Silk Road pursues, the globalization of Chinese capital will significantly accelerate, reaching the annual capital export capacity of 300 billion dollars, and becoming an important competitor against the US ("China to Become a Capital Power with 'The Belt and Road Initiative'").

Therefore, China's Silk Road Economic Belt project is both a strategic and a pragmatic one. Strategically, it seeks peaceful development with Central Asia where cooperation benefits both parties. Turning West is also a geopolitical strategy to avoid conflict with the US in East Asia while targeting the region with declining US presence. Pragmatically, continued energy cooperation, excess production capacity transference and capital export are all efforts on the part of China to maintain its economic growth. Unlike the US which keeps the export of ideology and system of government as its important missions in the region, China's cooperation in Central Asia is more economic than political.

4 CONCLUSIONS

Central Asia, due to its strategic position and rich oil reserves, has been the center of geopolitical game of world powers. Shifts of strategy of big players in the region has led to changes in the geopolitical landscape. As the US withdrew troops from the Middle East and diverted its foreign policy focus to East Asia, its presence in Central Asia is on a significant decline. US energy policy in Central Asia also withered along with the fading of US power in the region. Increased production of oil and LNG add to US energy security, which also reduces the strategic importance of Central Asia. By contrast, China's strategy of Silk Road Economic Belt marks its heightened cooperation with Central Asian countries. The project is driven by the need for economic growth and the pressure to meet increasing demands of the immense population, and also the strategy to avoid conflict with the US in East Asia. Energy cooperation with Central Asian countries, Kazakhstan in particular, plays an important role in ensuring China's energy demand and security, as it diversifies the source of energy and takes a safer transportation route. Economic cooperation between China and Central Asian countries also go beyond energy, reaching to a wider range of areas including infrastructure, agriculture, etc. Capital is a strong advantage for China's cooperation with its neighbors in the west. Despite China's rise in Central Asia, it should not be interpreted as a gesture of geopolitical assertiveness or a political threat. Seen from its policies, China's goal is peaceful economic development, rather than political and military influence.

ACKNOWLEDGEMENTS

The Studies of Chinese folk Financial Development motives and Governance,
Project of National Natural Science Foundation of China(No.71373029).

REFERENCES

"Annual Energy Outlook: Executive Summary." *U.S. Energy Information Administration*. 7 May, 2014. Web. 7 Feb. 2015.

Atal, Subodh. 2003. Central Asia Geopolitics and US Policy in the Region: The Post-11 September Era. *Mediterranean Quarterly* 14(2): 95–109.

Blake, Robert O. Jr. 2012. U.S. Policy in Central Asia. Web. 4 Feb. 2015.

Brzezinski, Zbigniew. 1997. *The Grand Chessboard: American Primacy and its Geostrategic Imperatives*. New York, NY. Basic Books.

China, Kazakhstan to sign $10b deals during Li's Visit. 2014. *Xinhua*. Web. 7 Feb. 2014.

China and the US to Confront Extensively with Low Intensity. (in Chinese) 2015. *Zaobao*. Web. 7 Feb. 2015.

Cohen, Ariel. 2006. U.S. Interests and Central Asia Energy Security. Web. 4 Feb. 2015.

Energy Development Strategy Action Plan (2014–2020). 2014. *State Council of the People's Republic of China*. Web. 6 Feb. 2015.

Feigenbaum, Evan A. 2007. Remarks to the U.S.-Kazakhstan Business Association and the American-Uzbekistan Chamber of Commerce. Web. 6 Feb. 2015.

Imas, Eugene. The New Silk Road to Nowhere. *The Diplomat*. 18 Dec. 2013. Web. 5 Feb. 2015.

Kaplan, Robert D. The Geography of Chinese Power: How Far Can Beijing Reach on Land and at Sea." May/June 2010. Web. 6 Feb. 2015.

Qian, Xuming. *The International Energy Strategy of the U.S.* (in Chinese). Shanghai: Fudan University Press. 2013. Print.

Sun, Yun. 2013. March West: China's Response to the U.S. Rebalancing. *Brookings*. Web. 6 Feb. 2015.

The Belt and Road Initiative' Promotes Capital Export. 2014. *Industrial Securities.* Web. 5 Feb. 2015.

The United States in Central Asia: Going, going... 2013. *The Economist*. Web. 7 Feb. 2015.

Xi, Jinping. 2013. Promote Friendship Between Our People and Work Together to Build a Bright Future. Web. 1 Feb. 2015.

Yergin, Daniel. 2001. "Energy Security and Markets." *Energy and Security: Toward a New Foreign Policy Strategy*. Ed. Kalicki, Jan H., Goldwyn, David L. Washington D.C.: Woodrow Wilson Center Press, 51–65.

Architectural, Energy and Information Engineering – Sung & Chen (Eds)
© 2016 Taylor & Francis Group, London, ISBN 978-1-138-02791-6

The myth or reality of U.S. energy independence: A literature review of Obama's energy security policy

Yi Nan Ding
Graduate School, International Business School, Beijing Foreign Studies University, Beijing, China

Long Peng
International Business School, Beijing Foreign Studies University, Beijing, China

Jing Shi
School of English and International Studies, Beijing Foreign Studies University, Beijing, China

ABSTRACT: In 2011, President Barack Obama defined his version of American energy security, which immediately attracted widespread interest and attention from academy, businesses and governments. To explore the research status and give a clear direction for the future study, the paper adopts comparative approach to review the American and Chinese scholarship and finds that Obama's energy policy is established out of economic and political motivations and brings about great domestic and international influences. Chinese scholars prefer to employ qualitative method with a macro perspective while Americans can combine qualitative and quantitative methods to analyze the issue from two levels. Additionally, Chinese scholars actively learn lessons from American energy policy in order to improve their own ability. In the future, both sides of scholars need to pay heed to specific initiatives so as to follow the latest results in the field.

Keywords: energy independence; Obama administration; economic benefits; international impacts

1 INTRODUCTION

Energy independence has become the centerpiece of American energy security policy ever since the concept first burst out after the Arab oil embargo in 1973. In 2011, President Barack Obama defined his version of American energy security by "reducing oil imports" and "expanding cleaner sources of electricity" (America's Energy Security 2011). This fresh policy immediately attracted widespread interest and attention from scholars and policymakers to researchers and analysts of energy and oil multinational corporations. To explore the research status and give a clear direction for the future study of American energy independence, the paper, through comparative approach, reviews the American and Chinese scholarship in terms of research agenda, characteristics and deficiency.

2 RESEARCH AGENDA

American and Chinese scholars produce much scholarship in their different perspectives but share some topics in common. In research agenda, scholars study five major aspects, namely: motivation of Obama's energy independence, feasibility of the policy, domestic and international influence, initiatives and lessons and inspirations for China.

2.1 Motivation of Obama's energy independence

First of all, energy independence policy helps the United States to upgrade its own energy supply and reduce the imports of foreign oil. The United States was the largest oil consumer with 5% of world population depleting 42% of world energy. This policy can help America fundamentally change the situation (Zhang 2011). Moreover, this policy can integrate the regional cooperation and reduce oil imports from unstable and high-risk countries and regions (Kong 2011; Ge 2012; Wang 2013). For example, the US can increase imports from the North America and West Africa while reducing imports from the Middle East.

Secondly, energy independence can shake up American economy, especially during the gloomy periods (Zhang 2011; Kong 2011; Cui 2012). On the one hand, the investment in the clean energy industry can create millions of jobs; on the other hand, the innovation in clean energy technology can be another engine driving American economy by raising energy efficiency (Zhang 2011).

Thirdly, President Obama's energy independence can improve American environmental image and lay a solid foundation for the US to pursue dominance and leadership in the future negotiations on climate change (Zhang 2011; Kong 2011). Compared to President Bush's setback on the climate change and clean energy, President Obama's policy presents American attention to the environment, projecting a positive image of America. American advancing clean energy technology can get extra bonus points when contesting for leadership in the climate change negotiations.

Fourthly, the policy is conducive to maintaining American hegemony in the long term (Cui 2012; Kong 2011). The clean energy technology innovation helps the United States to occupy the high position in the industry and even control the world energy market in the future. Emerging markets, like China, India, Russia, Brazil and South Africa, successively enters the industrialized period. Oil is the economic blood for them, and whoever controls the energy controls the world.

2.2 Assessment of energy independence

Although the International Energy Agency (IEA) in its 2012 World Energy Outlook (IEA 2012) noted that the United States would become the world's largest oil producer in 2017 and a net oil exporter in 2030, some scholars still question the feasibility and necessity of this policy.

Considering that American energy independence only means the decline of dependence on foreign energy, Zhao (2012) concludes that the absolute energy independence is unrealistic for the United States: first of all, the decline mainly results from economic recession instead of growing domestic oil production and increasing energy efficiency; secondly, it's hard for natural gas to completely replace oil in terms of efficiency and usage and natural gas is just an alternative to coal in many circumstance; thirdly, given the relative advantage and global impacts on North American suppliers, the United States cannot be cut off from the world.

Furthermore, American scholars do not trust IEA completely and establish their own models to test the ability of specific policies to achieve oil independence in an uncertain future (Institute for 21st Century Energy 2012; Greene 2009). *Index of US Energy Security Risk* (2012) employs 37 different measure units, like global fuels, fuel imports, energy expenditure and energy use intensity, uses data from 1970 to 2040 and concludes that the risk will increase slowly after a brief dip. This report

also divides risks into four categories: geopolitical, economic, reliability and environmental, which gives more details about the trend of American energy security (Institute for 21st Century Energy 2012).

2.3 Domestic and international influences

Considering American position in the international relations and global energy consumption market, American energy independence policy will bring about great domestic and international influences to itself and the whole world.

2.3.1 Domestic influence

Domestic influence can be divided into economic and environmental parts. On the one hand, energy independence will produce great economic benefits. First of all, according to the Recovery Plan, large quantity of money will be devoted into clean energy industry, producing millions of jobs (Yergin 2013; Zhang 2011; Zhang 2014; Chen 2013). Next, low costs brought by cheap energy will make American manufacturing industry more competitive (Zhang 2014; Zhang 2012; Zhang 2011; Wang 2013; Yergin 2013). Thirdly, export of oil and natural gas will increase revenues and reduce annual American trade deficit by $ 85 billion at current prices (Yergin 2013, Wang 2013; Zhang 2012; Chen 2013).

However, scholars split in the environmental impacts brought by energy independence. For one thing, as the energy resource of electricity, shale gas will produce much less carbon than coal (Yergin 2013); for another, fracking, the method to produce shale gas, consumes large quantity of water, leading to the waste of farm water. Furthermore, compared to fossil fuel, clean energy, like solar and wind, produces less carbon but higher cost (Yergin 2013; Wang 2013).

2.3.2 International influence

As for international influence, American energy independence has a geopolitical impact on American foreign policy and the rest of the world. First of all, scholars debate on whether energy independence matters in American foreign policy. For one thing, the United States reduces the imports of foreign energy, especially oil, giving Washington foreign policy more room to maneuver (Yergin 2013; Ge 2012). The United States can maintain a tough position when sanctioning Iran as well as shaking oil-exporting countries with low oil price by exporting quantities of shale gas, like Venezuela, Iran, Russia and the Middle East (O'Sullivan 2013; Yergin 2013; Yang 2012; Cambanis 2013). For another, many scholars hold that the United States will not change its Middle East policy in spite of reducing oil imports from that region because (a) the dependence of world economy on Gulf oil is growing and many countries, like BRICS, build their economy on Gulf oil so as an indirect consumer of Gulf oil, the United States is still bound with the Middle East by importing manufactured goods from China (Cordesman 2013); (b) as a gas and oil exporter, the United States needs a stable global market, and U.S.-controlled Middle East is a requisite for a stable global energy system (Sage 2014; Ge 2012; Cambanis 2013); c) besides, terrorism, nuclear proliferation and the security of Israel in this region cling to the United States (O'Sullivian 2013; Cambanis 2013; Cordesman 2013). In addition, this policy will give the United States more cards in the future negotiations on climate change (Zhang 2014).

Secondly, the rest of world is impacted by American energy independence policy, especially China, the largest oil importer. As allies of the United States, Japan and the EU will become the beneficiary of American energy independence by importing cheap oil and shale gas (O'Sullivian 2013); by contrast, China, Russia, the Middle East and the Middle Asia could lose their interest due to American energy independence policy (Zhang 2011; Zhang 2012; Cui 2013; O'Sullivian 2013; Cambanis 2013). For Russia, the Middle East and the Middle Asia, the low price of oil brought by American oil import reduction could destabilize natural resource powers since these countries and regions make a living by exporting oil (Cambanis 2013). For China, the situation is more complicated. As the largest oil consumer, China will pay high prices for national security, carbon trading, goods export, industrial structure and even negotiations on the climate change (Zhang 2011; Cui 2013; Zhang 2014); however, as the largest oil importer, China will gain more influence over oil-producing regions, like the Middle Asia and meanwhile, China will invest more in navy to shoulder the responsibility of policing shipping lanes in the Persian Gulf and other places (Cambanis 2013; O'Sullivian 2013).

2.4 Initiatives

In order to achieve energy independence, both Chinese and American scholars give some suggestions in different sectors. Kong Xiangyong (2011) believes that technology innovation is the key point and proposes to put great efforts into four sectors, namely, alternative resource to transportation fuel, renewable energy, energy efficiency, and smart power grid. American scholars share the same idea and add solar, fusion, batteries, wind, biofuel and small reactors to key fields of research (Brinkman 2012; Lippke et al 2011). Additionally, the Congressional Budget Office (CBO) also advices

to increase government's involvement in enhancing energy security (2012). To specific, the government can dampen the popular needs by increasing prices in electricity and enhancing refinery capability in transportation (CBO 2012).

2.5 Lessons and inspirations for China

In addition to the shared topics by both sides, Chinese scholars also devote themselves to studying the inspirations and lessons from energy independence policy, for example, China should create a good environment for developing new energy, improve the related regulations and laws, speed the change of energy upgrading, enhance the international cooperation with energy-exporting countries and find out the current situation of Chinese shale gas (Cui 2012; Ge 2012; Wang 2013).

3 CHARACTERISTICS AND DEFICIENCY

In the study of energy independence, Chinese and American scholars have present their own advantages and distinctions. Their disparity mainly stem from three aspects: research agenda, research methods and research unit.

In research agenda, both sides share some topics but their emphasis is different. American scholars pay more attention to the impacts on its foreign policy, like American Middle East policy, while Chinese scholars tend to focus on its impact on China and intentionally neglect the potential benefits brought to China; compared to Chinese counterparts, more American scholars follow how to broaden sources of energy and reduce expenditure and meanwhile, the number of their research sectors or initiatives is much larger than Chinese. In addition, Chinese scholars pay a close eye to the motivation of Obama's energy policy and lessons for China, which, I think, has a lot to do with their different positions.

As for research methods, Chinese scholars only use qualitative method and look at the issue from a macro perspective while American scholars adopt both qualitative and quantitative methods to analyze the question and even establish a new model to explain the issue. The combine of two methods can give a more clear result when two voices exist, such as the contradiction in environmental impacts of energy independence.

In research unit, American scholars can solve the problem from both macro and micro perspectives and are good at digging in a certain sector, like biofuel; by comparison, Chinese are adept in dealing with questions from a macro perspective although they also mention the development in a certain sector, like GRID.

Chinese and American scholars have made great contribution to energy security study, but there are still some room for them to improve. First of all, both sides should put more efforts into specific sector and observe the latest results or changes in that field. Next, Chinese scholars can balance their methodology and learn to employ the quantitative method, which can give a clear result when two situations exist. Thirdly, it's possible that American scholars regard the energy independence as a fact so they pay no attention to the motivation behind the policy.

4 CONCLUSIONS

In 2011, President Obama put forward his American energy security policy, which immediately aroused commercial and academic interest in the context of economic recession. Chinese and American scholars have put great energy to the policy from their own perspectives and they have studied motivations, feasibility, influences, and initiatives of energy independence policy and lessons and inspirations for China. In research agenda, they tend to pay more attention to the influences brought to themselves, and additionally, Chinese scholars are more willing to learn lessons to improve their own ability; in research unit, American scholars have a better balance between micro and macro levels, in contrast, their Chinese counterparts prefer the macro level; as for research methods, Chinese scholar need to employ quantitative method more, compared to American scholars. Four years have passed since American energy independence was first published in 2011. With the deepening and developing of this policy, scholars need to pay more attention to initiatives and programs in the future.

ACKNOWLEDGEMENTS

The Studies of Chinese folk Financial Development motives and Governance.

Project of National Natural Science Foundation of China(No. 71373029).

REFERENCES

Brinkman, W.F. 2012. Energy independence with sustainability. US Department of Energy. For *American Geophysical Union Fall Meeting*.

Cambanis, Thanassis. (2013, May 26). American energy independence: the great shake-up. *Boston Globe*. Retrieved from http://www.bostonglobe.com/ideas/2013/05/25/american-energy-independence-great-shake/pO9 Lsad4cVQvjdpyxMI1DO/story.html.

Chen Haibo. 2013. Shale gas revolution & North America's energy independence and how they inspire China. *Natural Gas Technology and Economy* 7(5): 7–12. (in Chinese).

Congressional Budget Office (CBO). 2012. Energy security in the United States. *The United States Congressional Budget Office*.

Cordesman, Anthony H. 2013. The myth or reality of US energy independence. *CSIS*. Working paper.

Cui Nannan. 2013. The "energy independence" strategy of Obama administration. *Journal of Theoretical Reference* 1: 44–47. (in Chinese).

Cui Nanan. 2012. The Obama administration's "energy independence" strategy and China's countermeasures. *Hongqi* 13: 33–38. (in Chinese).

Ge Xubo. 2012. Thoughts on American energy independence. *State Grid* 5: 52–55. (in Chinese).

Institute for 21st Century Energy. 2014. Index of US energy security risk: assessing America's vulnerabilities in a global energy market. US Chamber of Commerce.

Kong Xiangyong. 2011. On the new energy policy of Obama administration. *Forum of World Economics & Politics* Sep. (5): 28–42. (in Chinese).

Lippke, Bruce et al. 2011. Sustainable biofuel contributions to carbon mitigation and energy independence. *Forecast* 2: 861–874.

Liu Yue. 2013. Two sides of US "energy independence". *International Petroleum Economics* 22(5): 10–20. (in Chinese).

O'Sullivian Meghan L. (2013, Feb. 14). "Energy independence" alone won't boost US power. *Belfer Center*. Retrieved form http://belfercenter.hks.harvard.edu/publication/22768/energy_independence_alone_wont_boost_us_power.html.

Sage, Christopher S. 2014. The myth of US energy independence: why the rapidly changing global energy landscape will increase US military engagement around the world. Working paper prepared by Weatherhead Center for International Affairs, Harvard University.

The White House. (2011, March 30). American energy security. *FACT SHEET*. Retrieved from https://www.whitehouse.gov/the-press-office/2011/03/30/fact-sheet-americas-energy-security.

Wang Wei. 2013. An effectiveness analysis of US independent energy strategy and its inspirations. *Peace and Development* 2:96-107. (in Chinese).

Zhang Hongfan. 2011. An analysis on the strategy of new energy of Obama administration. *Journal of China University of Petroleum* 27(5): 13–18. (in Chinese).

Zhang Maorong. 2014. The prospect of US "energy independence" and geo-economics impacts. *Contemporary International Relations* 7: 52–60. (in Chinese).

Zhang Monan. 2012. Will American energy independence reshape the global pattern? *Development and Research* 5: 34–37. (in Chinese).

Zhao Hongtu. 2012. An analysis of US "energy independence". *CIR* 22(4): 59–74.

Architectural, Energy and Information Engineering – Sung & Chen (Eds)
© 2016 Taylor & Francis Group, London, ISBN 978-1-138-02791-6

Western debate on China's quest for foreign energy: Mercantilism or Liberalism?

Yi Nan Ding
Graduate School, International Business School, Beijing Foreign Studies University, Beijing, China

Long Peng
International Business School, Beijing Foreign Studies University, Beijing, China

Bo Wang
School of English and International Studies, Beijing Foreign Studies University, Beijing, China

ABSTRACT: This article intends to shed some light on how western scholars and researchers view China's quest for foreign energy—the impact of China's energy diplomacy on energy exporting countries, the global energy market as well as on western countries, by reviewing scholarly writings dating from 2000. From this heated debate, this article identifies three schools of thoughts—the Mercantilism School, the Liberalism School and the Middle Ground School. The Mercantilism School argues that China's quest for foreign energy is mercantilist in nature and has generated negative effect, while the Liberalism School holds otherwise and celebrates liberal factors in China's energy diplomacy. However, the Middle Ground School believes that China's energy diplomacy reflects both mercantilism and liberalism under certain circumstances. Despite different even conflicting opinions, these three schools offer identical policy recommendation to their officials—cooperation with, rather than confrontation against China on energy issues.

Keywords: China's quest for foreign energy; mercantilism; liberalism

1 INTRODUCTION

In the past three decades, China has achieved economic takeoff, becoming the second largest economy in the world. However, sustaining high rate economic growth requires compatible energy supply. Since the 1990s, China has actively established and furthered energy relations with the Middle East, African, Latin American and Asian countries to solve its energy shortage caused by inadequate indigenous supply and continuing rising demand. Different from the western countries which are market-driven economies in energy, China strives to ensure its energy security through promotion of its National Oil Companies (NOCs). From the perspective of western scholars, it is to "control" energy supply instead of relying on market. Besides, as the second largest economy and the largest energy consumer in the world, China's energy policy now has great impact not only on itself, but also on the world economy. Against this backdrop, western scholars, researchers and pundits dedicate themselves to investigating the strategic nature of China's quest for foreign energy and its potential implications so as to offer sound

policy recommendations to their respective officials. Three schools of thoughts, therefore, has emerged from this debate—Mercantilism, Liberalism and the Middle Ground school. This article aims to review this heated scholarly debate with an attempt to see how western scholars and pundits view China's quest for foreign energy, or China's energy diplomacy.

2 CHINA'S QUEST FOR FOREIGN ENERGY

2.1 *China's energy security*

Since the beginning of the reform and opening up, China's economy has expanded at a spectacular average annual growth rate and so has its energy consumption. Gradually, its domestic energy supply was no longer sufficient to meet the demand. In 1993, the oil demand exceeded supply. The discrepancy between energy demand and domestic supply for the first time in history emerged as a problem for the nation to solve. But during the 1990s, the problem imposed little threat to China's energy security because it imported comparatively small quantities of oil and the global oil market during that time was abundant in oil supply with relatively low price (Christie 2009).

Things, however, began to change from the very beginning of the 21st century. The 9/11 terrorist attack, the Iraq War and the Afghanistan War rendered the Middle East, the most important area of oil importing for China, insecure. At the same time, domestic demand for energy increased dramatically due to its rapid economic development and urbanization. Energy shortage thus threatens to disrupt economic growth and social stability. In such situation, the Chinese government has taken three steps to meet its growing demand for energy to power its economic growth and social activities: 1) expanding foreign oil from the Middle East; 2): diversifying energy sources by reaching out to Africa, Russia, Central Asia and the Americas; 3) securing oil transport routes (Lai 2007).

2.2 China's energy diplomacy

The early stage of China's quest for foreign energy started in 1993 when it became a net oil importer. As a new player in the international oil market, China had to buy oil on spot market. During this period, the Middle East was the major oil origin, which has accounted for the largest share of China's imported oil since then. Major exporting countries in this area were Sudan, Iran and Iraq. Since early 21st century, motivated by growing domestic energy demand and volatile international environment, the Chinese government has adopted a more comprehensive energy diplomacy. First, China has been diversifying its sources for oil with an emphasis on the Middle East, Africa, Russia, Central Asia and Latin America. Second, China has been diversifying its modes to acquiring oil from international market, such as long-term supply contracts and equity investments. Third, China has been offering packages deals to oil suppliers for their energy, for example, oil-for-infrastructure and oil-for-loan. Fourth, in order to reduce transit risks, China has been diversifying and securing its energy transport routes, for example, constructing gas and oil pipelines liking Myanmar and China. Fifth, to improve its NOCs competitiveness, China has been providing them with financially and politically support (Christie 2009).

These measures indicate China's strong resolve and momentum to secure its energy supply, which has invited academic discussions about the target or strategic nature of China's energy diplomacy. Viewpoints of these scholars or pundits fall into three categories: the Mercantilism school, the Liberalism School and the Middle Ground school.

3 THE STRATEGIC NATURE OF CHINA'S QUEST FOR FOREIGN ENERGY

3.1 The mercantilism school

Mercantilism refers to the kind of economic practice that advocates government control of national economy with an aim to enhance state strength at the cost of competing states. Another remarkable feature of mercantilism is its promotion of governmental protection of national industries through various means, tariffs and subsidies, for instance (Goldstein 2003). The Mercantilism School in this debate believes that China's energy diplomacy is mercantilist in nature because the main Chinese actors in oil market are state-owned companies, not private enterprises and the Chinese government supports them economically and politically to "hunt and nationalize" foreign oil (Holslag 2006; Leverett 2007; Herberg 2011). According to this school, China's mercantilist approach to energy supply has generated negative impact on the international oil market and energy exporting countries and has imposed threats to the interests of the US-led western countries.

To begin with, the Mercantilism School argues that China's mercantilist approach helps it to "lock up" oil resources by formulating unilateral relations with energy exporting countries, which might choose to trade off their oil reserves for loans or infrastructure. The supply of oil from these countries will thus contract and then cause a rise in oil price in international oil market (Leverett 2007). Second, China's growing interest in foreign oil makes the energy exporting countries increasingly dependent on oil as main commodity, which renders them more vulnerable to negative price fluctuations and precludes them from diversifying their economies. In other words, the "neo-colonial" feature of China's energy policy will bring about negative influence in the long term that could override the short term economic profits (Taylor 2007). Third, China's quest for foreign energy is believed to have imposed threats to the US-led western countries in the following ways. In terms of energy issue, China's search for foreign energy in the Middle East, Africa, Latin America and Russia has made it a major competitor to the United States for oil reserves. Politically, China has maintained energy relations with states that the US lists as "pariah states" or "rouge states", such as Sudan, Iran and Iraq, which hinders the western efforts to address the issue of proliferation of nuclear weapons and human rights violation. With regard to security matters, China has been, in recent years, improving its naval capacities to secure its major oil transport route, the Strait of Malacca, which, according to this school, will challenge the

US dominance in this area (Leverett 2007). Last but not least, growing competition among China, Japan, India and Korea to promote their NOCs and control foreign energy supplies is weakening public confidence in fair access to energy supply in the future and deepening strategic distrust (Giljum 2009; Collins 2011).

3.2 The liberalism school

Liberal theories of international relations propose that deepening economic interdependence among states will raising the prospects for cooperation and moderate those of conflict. Given the crucial significance of energy to the economic and national security and the desire for a peaceful environment, China's energy dependence will encourage it to secure its energy supply through diplomacy, international market and multilateral institutions rather than through military tools that result in conflict and confrontation. As demonstrated in China's energy activities, China has been actively cooperating with exporting countries, as well as other importing countries through various forums and international institutions, such as the WTO, the International Energy Agency (IEA) and the Shanghai Cooperation Organization (SCO) (Ziegler 2006). Instead of competing in a "zero-sum" game depicted by the Mercantilism School, China is working cooperatively in a "win-win" situation. Another misinterpretation from the Mercantilism school, according to the Liberalism School, is that the Chinese government plays a major role in its NOCs oversea activities. For the last two decades, China's energy sector has been greatly liberalized and decentralized. It is the NOCs's stakeholders (shareholders and banks), not the government, that play the major role in pursuing profits (Downs 2007).

As for the impact of China's quest for foreign energy, the Liberalism still holds different even conflicting opinions. First of all, China does not "lock up" oil reserves for itself. Two thirds of oil bought by China's NOCs in 2006 were sold on the international market. China increases rather than contracts oil supply to the global market, which can help reduce price fluctuations and supply disruptions caused by unexpected incidents (Downs 2007; Leung 2010). Therefore, China's energy policy strengthens international energy security rather than harms it. Second, China's NOCs helps energy exporting countries construct infrastructure, such as railroads, schools, hospitals, or offer them loan in return for their energy, which, with no doubt, will help them develop and thrive (Hanauer & Morris 2014). Third, China's energy activities does not fundamentally threatens US economic and political interests. On the contrary, they are, to some extent, helpful. Infrastructure constructed by China in Africa, for instance, helps reduce transaction costs and expand regional oil markets for possible American investors (Yergin 2006; Hanauer & Morris 2014).

3.3 The middle ground school & policy recommendations

The Middle Ground School, as its designation suggests, stands in between. This school contends that China's energy diplomacy reflects both mercantilism and liberalism under certain conditions. In energy countries where China has nurtured unilateral agreements such as Sudan and Iran, policies or strategies lean to mercantilism. One the other hand, when it comes to securing the maritime transport routes, China relies on other countries, such as the United States. In such circumstances, China looks for multilateral cooperation or free riding. According to the Middle Ground School, China has been quite pragmatic and flexible in employing energy policies to fit certain situations and maximize its energy interests (Lieberthal & Herberg 2006; Christie 2009; Lee 2010; McCarthy 2013).

Although these schools hold different opinions towards the strategic nature of China's energy diplomacy, they converge on what their governments should do about China, which is to cooperate with, rather than confront against it. Take the United States for an example. As the first two largest economies and energy consumers in the world, energy policy adopted by these two countries will have tremendous impact on each other and more importantly on the world. Therefore, the US and China have vital common interests on energy issues. Both countries share an interest in maintaining stability in energy exporting regions, securing energy transits, avoiding global oil price fluctuations and supply disruptions as well as accelerating the research and development of clean energy. It is undeniable that strategic distrust between the US and China still exists and will do in the years to come. But the common interests will impel the two to work out some solutions.

4 CONCLUSIONS

This article has thus presented the main arguments of each school on the western debate about China's quest for foreign energy respectively—whether China's energy diplomacy reflects mercantilism or liberalism or both. Apparently, these three schools interpret Chinese energy activities form different perspectives. The Mercantilism School, to a great extent, follows the rationale of realism that emphasizes states' instinct and resolve to pursue national interests and power at the expense of other rivals, which usually leads to conflicts. The liberalism School, on the other hand, is in favor of the power of economic interdependence and multilateral institutions to facilitate cooperation among states. The Middle Ground School synthesizes the theoretical rationales of the former two schools and analyses China's search for foreign energy on a case by case basis.

Mercantilism has gradually become a word used to describe China's energy activities that go against the US, or the western interests. Some use it to criticize Chinese government for directing its national oil companies to acquire foreign energy assets, especially oil. But it should not be neglected that mercantilism was the dominant economic theory and practice in the west between the 16th and 18th century, although they now have replaced it with reliance on global oil market. China is not the only one that subordinates its foreign policy to energy interests. The United States, for example, has done some political intervention and even fought wars to serve its oil interests (Iran and Iraq). Both China and the western countries has employed certain strategies to secure energy supplies. As Mr. Zweig once commented, it is the responsibility of Chinese leaders to secure the nation's energy supply and China has the right to do that through market strategies, which the west countries should recognize (Zweig 2006).

"Energy is a politicized commodity." Without electricity and gasoline, life goes dark and slow in a modern society. Without energy, an economy dies. The vital importance of energy to economic and national security has made it preponderant over other commodities in foreign trade. Neither can China, nor the west countries, afford to leave energy to market forces alone.

ACKNOWLEDGEMENTS

The Studies of Chinese folk Financial Development motives and Governance.

Project of National Natural Science Foundation of China (No.71373029).

REFERENCES

Christie, E.H. (Ed.) 2009. China's foreign oil policy: genesis, deployment and selected effects. FIW Research Reports (03): 1–83.

Collins, G. 2011. Asia's rising energy and resource nationalism Implications for the United States, China, and the Asia-Pacific Region. The National Bureau of Asian Research. (31): 1–80.

Downs, Erica S. 2007. The fact and fiction of Sino-African energy relations. China Security 3(3): 42–68.

Downs, Erica S. 2007. China's quest for overseas oil. Far Eastern Economic Review. 52–56.

Giljum, J. P. 2009. The future of China's energy security. The Journal of International Policy Solutions. 11: 12–24.

Goldstein, J. S. 2003. *International Relations*. New York: Longman.

Hanauer, Larry & Lyle J. Morris. 2014. Chinese engagement in Africa-drivers, reactions and implications for US policy. RAND Corporation Reports. 1–173.

Herberg, Mikkal. 2011. China's energy rise and the future of US-China energy relations. New America Foundation. 1–16.

Holslag, Jonathan. 2006. China's new mercantilism in Central Africa. African and Asian Studies 5(2): 134–169.

Lai, H. H. 2007. China's oil diplomacy: is it a global security threat? Third World Quarterly 28(3): 519–537.

Leung, G.C.K. 2011. China's energy security-perception and reality. Energy Policy 39:1330–1337.

Leverett, Flynt. 2007. The geopolitics of oil and America's international standing. Committee on Energy and Natural Resources. 1–9.

Lee, M.S.L. 2010. China's energy security: the grand "hedging" strategy. United States Army Command and General Staff College. 1–61.

Lieberthal, Kenneth & Mikkal Herberg. 2006. China's search for energy security: implications for US policy. The National Bureau of Asian Research. 17(1): 1–52.

McCarthy, Joseph. 2013. Crude "oil mercantilism"? Chinese oil engagement in Kazakhstan. Pacific Affairs 86(2): 257–278.

Taylor, Ian. 2007. Unpacking China's resource diplomacy in Africa. Center on China's Transnational Relations. (19): 1–34.

Yergin, Daniel. 2006. Ensuring energy security. Foreign affairs 85(2): 69–82.

Ziegler, Charles. 2006. The energy factor in China's foreign policy. Journal of Chinese Political Science. 1–26.

Architectural, Energy and Information Engineering – Sung & Chen (Eds)
© 2016 Taylor & Francis Group, London, ISBN 978-1-138-02791-6

Application of the distributed energy system in a data center in Beijing

Y. Zhao
CNPC (Beijing) Technology Development Co. Ltd., Beijing, China

ABSTRACT: The Combined Cooling, Heating and Power (CCHP) is one of the advanced technologies of the distributed energy, which has the properties such as energy conservation, security, and flexibility and is more suitable for the development in China. This paper calculated the load of the CCHP system in a data center in Beijing. The system configuration and operating strategy of the CCHP system were analyzed, which reflected a development trend of the energy supply. Results show that by comparing the power generation efficiency, economy and security, the Gas Internal-Combustion Generator (GICG) sets have more advantages and are adopted to work for the data center. The CCHP system in this paper has demonstration significance for the development of the distributed energy.

Keywords: Combined cooling; Heating and power; Power generation efficiency; Gas internal-combustion generator; Distributed energy

1   INTRODUCTION

With the rapid development of economy in China, the energy demand is increasing gradually, which makes it urgent to adjust the energy structure and improve the energy utilization efficiency. The distributed energy overcomes shortcoming of the conventional energy, such as the simplicity of the system and the limitation to improve the energy utilization efficiency, and opens a new avenue of the comprehensive utilization of energy as a result of its properties such as energy conservation, security, and flexibility.

The Combined Cooling, Heating and Power (CCHP) system is one of the advanced technologies of the distributed energy, which is based on energy cascade utilization and produces cold, heat and electricity using a natural gas as the primary energy. CCHP system has many advantages such as high energy efficiency, less negative impact on the environment, security to local power grid and load shifting to natural gas and power grid. Taking the economy into consideration, CCHP system is suitable for the project in which the heating and power load is relatively stable and cooling or heating is needed all year round.

The data center in this paper has a Class A construction standard with a total construction area of about 55,000 square meters and 4500 surface data set cabinet, including three functional areas such as offices, data services, and energy supply. This paper analyzes the system configuration and operating strategy of the CCHP system in the data center, which reflects a development trend of the energy supply.

2   CALCULATION OF THE COLD, HEAT AND ELECTRICITY LOADS

2.1   *Calculation of the electricity loads*

This data center was designed in accordance with GB-50174-2008 Code for Design of Electronic Information System Room. The heat generation of the cabinet is 3.0 kW per set, which can be improved to 7.0 kW per set in the future. The electricity loads include: IT, lighting, self-electricity consumption, precision air conditioner, and UPS losses. The calculation results of the hourly electricity loads are shown in Figure 1.

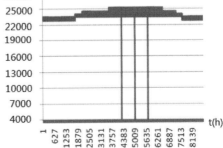

Figure 1.   The hourly electricity loads of the data center.

Table 1. Annual energy consumption of the data center.		
Electricity consumption (kWh)	Cold consumption (kWh)	Heat consumption (kWh)
207240817.5	147862385.3	219480.05

2.2 Calculation of the cold and heat loads

This data center was designed in accordance with GB50019-2003 Design Code for Heating Ventilation and Air Conditioning of Civil Buildings. The cooling loads include: heat dissipation of the equipment room, heat transfer of the maintenance, the introduction of new wind, lighting and heat dissipation of the human body. The heat loads mainly include that of the structure maintenance and that of the fresh air.

The installation method of hot and cold aisles is adopted by the computer equipment and the racks in the data room, which can further improve the cooling effect. According to the latest ASHRAE standard, the cold/hot aisle temperatures are set to be 18/29°C. The software EQUEST is used to calculate and analyze the load.

2.3 Summary of the energy consumption

According to the analysis of the cold, heat and electricity loads, the annual energy consumption of the data center is obtained as shown in Table 1.

The balance among the cold, heat and electricity is the most serious problem affecting the configuration of the CCHP system in the distributed energy. Currently, the "Power Determined by Heat/Cold" and "Heat/Cold Determined by Power" are the two most typical configuration methods in the CCHP system.

3 CONFIGURATION OF THE CCHP SYSTEM

3.1 Configuration principles

The safety and reliability of the power and cooling supply must be ensured in the configuration of the CCHP system in the data center. The specific principles are as follows:

1. Gas generators work as the main power source and the number of generators is set to be N + X. When a single generator fails to work, there is a standby generator that can start to work at any time.
2. The parallel operation between the gas generator and the commercial power is utilized. The commercial power is prepared as the alternative energy source.

3. There is supplemental combustion in the residual heat unit, which is set to be 100%. When the generator fails to work, the hot and cold supply is ensured by supplemental combustion.
4. Prepare a spare electric refrigerator, which can provide the cooling loads in case if there is a problem with the gas supply.
5. Set the natural cooling system and supplemental combustion boiler system in order to improve the safety and economy of the heat and cold supply in winter.

3.2 Selection of the gas generator

Currently, there are two main gas generator sets used in the CCHP system: small Gas Turbine Generator (GTG) sets and Gas Internal-Combustion Generator (GICG) sets. For the data center, the GICG has the following advantages:

1. It can be seen from Table 1 that in the data center, the energy consumption and cold load both need to be supplied, while the cooling to power ratio is less than 1. Typically, the cooling to power ratio of the GTG set is greater than 1.5, and that of the GICG is less than 1. So, the GICG has better adaptability to the load.
2. The output of the GTG changes greatly with the ambient temperature and the variation of the generating capacity is up to 20%. But the output of the GICG changes slightly with the ambient temperature. Since the change of the annual load in the data center is small, the GICG is more suitable.
3. It can be seen from Figure 2 that the indicators of the GICG are superior to those of the GTG according to the power generation efficiency.
4. Taken the cost into account, the cost of the GICG is 30% lower than that of the GTG.

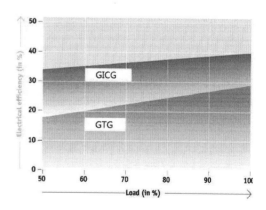

Figure 2. Power generation efficiency of GTG and GICG.

3.3 Waste heat utilization

The waste heat utilization technology of the gas turbine generator is:

1. Directly docked with the "flue gas" or "flue gas-hot water" waste heat absorption air conditioning units. (Hereinafter referred to as DCT).
2. Connected with the vapor or hot water waste heat absorption air conditioning units through the waste heat boiler. (Hereinafter referred to as ICT).

In recent years, the DCT develops quickly with simple process and small land occupancy area. Since the heat transfer process is reduced, the thermal efficiency becomes higher. At the same time, taking into account that there is an annual cooling need, no heating or domestic hot water in the data center room, the DCP is chose.

3.4 Parallel operation mode

Because of the high electricity load and high levels of electrical safety, in order to ensure the security and economy of the system, a parallel operation mode between the gas generators and the commercial power is adopted. The generators only provide the base load, leaving the insufficient part provided by the commercial power. The output power of the gas generator can be adjusted according to the load demand. In this case, the overall efficiency of the gas turbine is high and the economy of CCHP system is good. When there are fluctuations in the electricity load, less impact could happen on the generators, which can produce the best power quality and reliability.

4 OPERATION MODES OF CCHP

Figure 3 describes the flow process of the CCHP system. The operation modes are introduced in detail in this section.

4.1 Operation mode of the power supply system

The electricity of the data center is supplied by GICG sets, with the commercial power supplying the insufficient part.

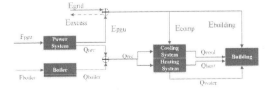

Figure 3. The flow process of the CCHP system.

4.2 Operation mode of the heat supply system

Natural heating season is from November 15 to March 15 the next year. During this time, the heating system supplies heat to the outside. The residual heat lithium bromide units are the primary heat source while the gas-fired hot water boiler works for the peak regulation.

4.3 Operation mode of the cold supply system

Natural cooling season is from May to October. During this time, the residual heat lithium bromide units are the primary cold source and the electric cooling system works as the spare cold source for the data center. The electric refrigeration cooling system is the primary cold source for the external and can be switched with the residual heat lithium bromide units when the latter cannot supply cold.

During the natural heating quarter, the natural cooling system is the primary cooling source for the data room. When the outdoor wet bulb temperature is greater than 4°C from 10:00 to 22:00 for three consecutive days, the natural cooling system cannot meet the cooling needs of the data center, so the electric cooling system needs to work for the cooling supplement. The residual heat lithium bromide units are the spare cooling source for the data room and do not supply cold for the external.

At other time, the residual heat lithium bromide units are the primary cold source for the data center and the electric refrigeration system is the spare cold source. The system does not supply cold for the external.

5 CONCLUSION

By reasonable design and argumentation, the CCHP system can meet the power and cooling needs of the data center and ensure the reliability of the electricity and cold supply. The utilization of CCHP in the data center will be beneficial to cost savings, energy conservation and environmental protection, and it will become a development trend for the energy supply in the data center.

REFERENCES

Arsalis A. & Alexandrou A. 2015. Design and modeling of 1–10 MWe liquefied natural gas-fueled combined cooling, heating and power plants for building applications [J]. *Energy and Buildings* 86 (2015): 257–267.

Choa H., Smith A.D. & Mago P. 2014. Combined cooling, heating and power: A review of performance improvement and optimization [J]. *Applied Energy* 136 (2014): 168–185.

Casisi M. et al. 2015. Effect of different economic support policies on the optimal synthesis and operation of a distributed energy supply system with renewable energy sources for an industrial area [J]. *Energy Conversion and Management* 95 (2015): 131–139.

Govardhan M. & Roy R. 2015. Economic analysis of unit commitment with distributed energy resources [J]. *Electrical Power and Energy Systems* 71 (2015): 1–14.

Havelesy V. 1999. Energetic efficiency of cogeneration systems for combined heat, cold and power production [J]. *International Journal of Refrigeration* 22 (1999): 479–485.

Kong X.Q. & Wang R.Z. 2004. Energy efficiency and economic feasibility of CCHP driven by stirling engine [J]. *Energy Conversion and Management* 45 (2004): 1433–1442.

Liu M.X., Shi Y. & Fang F. 2014. Combined cooling, heating and power systems: A survey [J]. *Renewable and Sustainable Energy Reviews* 35 (2014): 1–22.

Lazzarin R.M. 1996. A new HVAC system based on cogeneration by an I.C. engine [J]. *Applied Thermal Engineering* 16(7): 551–559.

Sun Z.G. 2004. Energetic efficiency of a gas-engine-driven cooling and heating system [J]. *Applied Thermal Engineering* 24 (2004): 941–947.

Architectural, Energy and Information Engineering – Sung & Chen (Eds)
© *2016 Taylor & Francis Group, London, ISBN 978-1-138-02791-6*

Experiment study for preparation of soluble potassium through thermal reaction of potassium feldspar, calcium sulfate & calcium oxide

Chao Qin Yang & Ju Pei Xia
Chemical Engineering Faculty, Kunming University of Science and Technology, Kunming, Yunnan, China

ABSTRACT: The experimental study is carried out for the preparation of soluble potassium by using the raw materials potassium feldspar, calcium sulfate and calcium oxide. In this study, mole ratio of n(potassium feldspar): n($CaSO_4 \cdot 2H_2O$): n(CaO) is 1:1 (2~16) was selected. The results show that the optimum ratio is 1:1:10. The XRD analysis results show that calcium silicate is CS, C_3S and C_2S in the selected ingredients ratio calcined products. Composition and proportion of calcium silicate are associated with the reaction temperature of system. Only a sialic acid calcium salt C_2AS is existed in the form of soluble potassium $K_2S_2O_8$, while not K_2SO_4 and C_3A ingredients. The experimental results of TG and DSC show that initial temperature of system displacement reaction is about 1000°C. The reaction is violent when temperature is more than 1100°C. It is consistent with the XRD spectra at different temperatures.

Keywords: Potassium feldspar; Thermal response; Soluble potassium

1 INTRODUCTION

Potassium feldspar belongs to feldspar minerals, which is a kind of frame structure with silicate, and its formula is $KAlSi_3O_8$, the theoretical elements[1]: K_2O is 16.9%, SiO_2 is 64.70%, and Al_2O_3 is 18.40%. Its crystal structure is made of the silicon oxygen tetrahedron $[SiO_4]_4$—and alumina tetrahedral AlO_4 5—as skeleton together. K^+ ions are filled in the gaps of skeleton to balance the extra negative charge and form a solid tetrahedral network structure. Its chemical stability is good. In addition to the high concentration of sulfuric acid and hydrofluoric acid, it can avoid being corroded from other acid, alkali[2].

Many experts and scholars had studied thermal decomposition behavior of potash feldspar since the nineteen fifties last century at domestic. The studies were carried out from the early simple choice of technological condition gradually deep into the study of mechanism of extraction potassium. Qiu Long-hui et al.[3] have used thermodynamic data and system phase diagram, and systematically studied the six different potassium feldspar system behavior of thermal decomposition process and kinetics. They pointed out that the thermal decomposition process of the system all conforms to g-gold—cloth dynamics equations and is controlled by the solid membrane diffusion. Han Xiao-zhao et al.[4] have studied the different additives on the effect of potassium ion exchange degree of potassium feldspar. They had put forward the mechanism of extraction potassium was ion exchange reaction, and made relevant ionic reaction kinetics and thermodynamic studies. Qi Long-shui et al. have studied decomposition of potassium feldspar by roasting to use potassium carbonate as additive, and the mechanism of decomposition reaction was preliminary analyzed. Shi Lin[5] has mainly studied thermal decomposition process in the system of potash feldspar, calcium sulfate & calcium carbonate. Through thermodynamic calculation and experimental verification, the optimum ratio of the system is potassium feldspar: calcium sulfate: calcium carbonate of 1:1:14, and main phases of products are K_2SO_4, $3CaO \cdot A_{12}O_3$ and $2CaO \cdot SiO_2$.

On the basis of summarizing predecessors' research results, this paper proposes potassium feldspar, calcium sulphate & calcium oxide system as the research object. Potassium dissolution rate, XRD and TG-DSC of calcined products are tested, and the reaction mechanism of system is proven.

2 EXPERIMENT

Potassium feldspar is from a mine in Yunnan, it is gray after crushing and grinding, after 200 target quasi sieve to chemical analysis, the main components of the mineral data are provided by Yunnan metallurgy research institute and the results are shown in Table 1.

Table 1. Chemical component analysis of potassium feldspar.

K₂O	Na₂O	Al₂O₃	T_{Fe}	CaO	MgO	SiO₂
13.51	0.97	17.09	1.15	0.19	0.60	62.91

Table 2. Leaching test results of roasting sample.

No.	n(K-feldspar): n(CaSO₄): n(CaO)	K₂O dissolution rate (%)	Soluble K₂O content of roasting sample (%)
1	1:1:2	18.57	2.03
2	1:1:4	33.17	3.18
3	1:1:6	38.61	3.34
4	1:1:8	62.93	4.96
5	1:1:10	88.85	6.53
6	1:1:12	88.90	6.08
7	1:1:14	89.52	5.75
8	1:1:16	90.91	5.67

The experiment data in the Table 1 show that the composition of ore used mainly are SiO₂, Al₂O₃, K₂O, while other oxide content is few. Iron content is low, to a certain extent, which reduces the effects of impurities on the reaction system. Chemical composition of raw material meets the requirements for preparation of potash in the industrial production by potash feldspar [6]: K₂O > 9%, Na₂O < 3%, and MgO + CaO < 2%. In order to facilitate the analysis of the reaction products, calcium sulfate and calcium oxide are all chemical pure.

3 RESULTS AND DISCUSSION

3.1 Determination of ingredients

Ingredients of test raw materials conform to n(potassium feldspar): n(CaSO₄ · 2H₂O): n(CaO) = 1: 1 (2~16)[7], and is mixed after adding water, the same quality of the mixture is respectively weighed and is put in porcelain crucible, then put into the muffle furnace, finally it is calcined for 120 minutes at 1200°C, and the calcined products are used for analysis after ground.

Calcined samples are sieved through 200 mesh standard screen after ground. Under the same condition, water-soluble potassium is leached by water. Effective K₂O content in filtrate is calculated, soluble potassium dissolution rate is used as index. The test results are shown in Table 2.

Table 1 shows that, with the increase of CaO content in the ingredients, the soluble potassium increases as a whole. When n(potassium feldspar): n(CaSO₄ · 2H₂O): n(CaO) exceeds 1:1:10, growth decreases significantly. Compared with the ingredients ratio of 1:1:16, only lower about 2%. On the other hand, with the increase of CaO content in ingredients, soluble potassium oxide content in the roasting sample first increases and then decreases. When the ingredients ratio is 1:1:10, maximum of soluble K₂O content is 6.53%. The above two aspects results are taken into account comprehensively, the ratio of experiment materials is chosen as n(potassium feldspar): n(CaSO₄ · 2H2O): n(CaO) = 1:1:10. Due to the dosage of calcium oxide reduces greatly, it is not only beneficial to improve the concentration of soluble potassium dissolution, but also conducive to improve load and reduce the cost.

3.2 Roasting process research of K-feldspar and mixed additives of CaSO₄·2H₂O and CaO

The dicating of research system is a complex solid—solid multiphase reaction, at the same time the main process is the component diffusion and chemical reaction between components, and the diffusion includes surface diffusion, external diffusion and internal diffusion. In addition, the roasting process also includes generation of lattice defects of solid phase, formation and decomposition of solid solution.

In a perfect ideal of lattice solid, atoms or ions vibrate around their equilibrium position, while particle migration does not occur. When particle vibrate in the thermal form, adjacent atoms can arbitrarily change position, the displacement can lead to moment distortion of lattice. Heating can cause lattice atom vibration amplitude to increase and the number of lattice defects to increase. The more defects of crystal exist, the easier the diffusion carries on. Therefore, it can improve interaction ability between solid phases of the study system.

The ingredients of test raw materials conform n(potassium feldspar): n(CaSO₄ · 2H₂O) of 1:1:10, first they are mixed and 10 g are taken to put in the porcelain crucible, second they are put into the muffle furnace, finally the samples are roasted for 2 h at 900°C, 1000°C, 1100°C, 1200°C and 1300°C respectively. The roasted samples are quickly taken out to put in the air for cooling, and they are ground for the XRD diffraction analysis, the result is shown in Figure 1. Analytical instruments are the Japan's neo-confucianism company D/max-3 B type X-ray diffractometer, at the same time test conditions: rated current of Cu target is 2~50 mA, rated voltage is 20~60 kV, scanning speed is 10°/min, and 2 theta scope of scanning angle is 10°~90°.

Figure 1 shows that KAlSi₃O₈ diffraction peaks in the XRD pattern is very clear at 900°C.

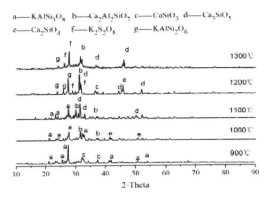

Figure 1. XRD diagram when ingredients of n (potassium feldspar): n(CaSO₄ · 2H₂O): n(CaO) of 1:1:10 at different temperature.

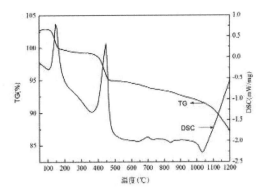

Figure 2. TG-DSC curves of K-feldspar with $CaSO_4 \cdot 2H_2O$ and CaO.

It instructs that $CaSO_4 \cdot 2H_2O$ and CaO of system fail to destroy the structure of potash feldspar. SiO_2 characteristic peak of raw material disappeared, while $CaSiO_3$ characteristic peak appears. The results show SiO_2 of raw material has all the transformation with a plentiful supply of CaO under the experiment condition. $KAlSi_3O_8$ characteristic peaks intensity is abate in the XRD spectrum at 1000°C, while $CaSiO_3$ characteristic peaks disappear and $Ca_2Al_2SiO_7$, while Ca_2SiO_4 characteristic peaks appear. $KAlSi_3O_8$ characteristic peaks are more obvious in the XRD pattern at 1100°C, at this time a stronger characteristic diffraction peaks of Ca_3SiO_5 appear, while Ca_2SiO_4 characteristic peaks disappear. When temperature rises to 1200°C, $KAlSi_3O_8$ diffraction peaks disappear, while breakdown products $KAlSi_2O_6$ characteristic peaks appear. A new phase $K_2S_2O_8$ appears, at the same time $CaSiO_3$ and Ca_2SiO_4 characteristic peaks appear; When the reaction temperature reaches 1300°C, $KAlSi_2O_6$ characteristic peak is abate, while $K_2S_2O_8$ diffraction peaks is the main peaks, $CaSiO_3$ characteristics peak disappear again.

Above XRD pattern analysis shows that, when n(K-feldspar): n(CaSO₄ · 2H₂O): n(CaO) is 1:1:10, reaction of CaO and SiO_2 is a continuous process, and product can be CS, C_2S, C_3S, which mainly depends on reaction temperature. Sialic acid calcium salt is existed only in the form of C_2AS. Soluble potassium is existed in the form of $K_2S_2O_8$ while not K_2SO_4.

3.3 DSC-TG analysis of K-feldspar and additives of CaSO₄ · 2H₂O, CaO

K-feldspar, $CaSO_4 \cdot 2H_2O$ and CaO conform to molar ratio of 1:1:10 are first mixed, then the samples weighed 15 mg are tested. DSC-TG analysis

test equipment is STA499F3 thermogravimetric analyzer produced by Germany NETZSCH companies. Its heating rate is 10°C/min, while the atmosphere is N_2, which flow rate is 50 ml/min. The heating temperature range is 25~1300°C.

Figure 2 shows that three times weight loss process appears in the course of the entire test process. Weightlessness for the first time appears at 138~180°C, accompanied by a strong endothermic peak, at this time lost quality should be because of crystallization water of $CaSO_4 \cdot 2H_2O$ lost. The second weightlessness appears at 356~426°C, which is caused by decomposition of Ca (OH)₂ formed because of adding water into materials, at the same time accompanied by a strong endothermic peak. Weightlessness for the third time appears at 1000~1200°C, which is mainly because K_2O and SiO_2 are system substitution reaction and decomposition reaction product, which are partly volatiled, at the same time system has low melting material to generate and the new phase appears. The overall phenomenon reflects the strong absorption of heat.

DSC-TG analysis shows that initial temperature of displacement reaction is about 1000°C, then strong reactions are occurred after 1100°C, it fit for the analysis of XRD spectra of the roasting sample.

4 CONCLUSION

1. The experimental results show that potassium dissolving rate increases with the increase of the content of calcium oxide in the system of potassium feldspar, calcium sulfate & calcium oxide at different ratio. If n(potassium feldspar): n(CaSO₄ · 2H₂O): n(CaO) exceeds 1:1:10, the growth decreased. Combined with the analysis result of effective potassium oxide content in

the calcined samples, the system optimum ingredient ratio is n (potassium feldspar): n(CaSO$_4$ · 2H$_2$O): n(CaO) = 1:1:10.

2. When the ratio of n(potassium feldspar): n(CaSO$_4$ · 2H$_2$O): n(CaO) is 1:1:10, water-soluble potassium content of roasting product is as high as 6.53%, at this time the potassium oxide dissolution rate is 88.85% in potassium feldspar.

3. The XRD analysis indicates that, when n (potassium feldspar): n(CaSO$_4$ · 2H$_2$O): n (CaO) is 1:1:10, CaO first reacts with SiO$_2$ in the raw material to produce CS. With the increase of temperature, CaO and CS further react to generate C$_2$S and C$_3$S. Only a kind of sialic acid calcium salt C$_2$AS is existed in the replacement reaction products, while soluble potassium is existed in the form of K$_2$S$_2$O$_8$. In the experiment ingredients system of n(K-feldspar): n(CaSO$_4$ 2H$_2$O): n(CaO), the K$_2$SO$_4$, C$_3$A ingredients of literature reports are not found.

4. The DSC-TG analysis results show that initial temperature of displacement reaction is about 1000°C, then strong reactions are occurred after 1100°C, it fit for the analysis of XRD spectra of the roasting sample at different temperature.

ACKNOWLEDGEMENTS

This work was financially supported by the Natural Science Foundation of China (21166012).

REFERENCES

[1] Sanda Lupan. Phosphates and Fertilizers for the Future [J]. Proc. Indian Natl. Sci. Acad, Vol, 58 (2008), p. 119.

[2] M.Y. Bakr, A.A. Zatout, M.A. Mohamd. Orthoclase, gypsum and limestone for production of aluminum salt and potassium salt [J]. Interceram, Vol, 28 (2009), p. 34–35 (in China).

[3] Long-hui Qiu, Lisheng Wang, Zuomei Jin. The preparation method of potash fertilizer by using potassium feldspar [J]. Sichuan chemical and corrosion control, Vol, 2 (1999), p. 17–19 (in China).

[4] Xiaozhao Han, Weitang Yao, Bo Hu, et al: Chinese Journal of Applied Chemistry. Vol, 20 (2009), p. 373–374 (in China).

[5] Lin Shi, Dingsheng Chen: Journal of South China University of Technology (Natural Science Edition). Vol, 35 (2008), p. 94–99 (in China).

[6] Hongwen Ma. Industrial Minerals and Rocks, Geological Publishing House, Beijing (2010), p. 48 (in China).

[7] Xue-jiao Ren, Ju-pei Xia, Zhao-shu Zhang. Thermodynamic analysis of phosphogypsum reductive decomposition reaction [J]. Environmental engineering, Vol, 7 (2013), p. 1128–1132 (in China).

Architectural, Energy and Information Engineering – Sung & Chen (Eds)
© 2016 Taylor & Francis Group, London, ISBN 978-1-138-02791-6

An exploration of the water poverty index in Kinmen

Tung Tsan Chen
Department of Civil Engineering and Engineering Management, National Quemoy University, Taiwan, ROC

Chien Lu
Center for General Education, National Quemoy University, Taiwan, ROC

ABSTRACT: Kinmen has a subtropical ocean climate. Because of poor natural conditions, Kinmen has a dry climate with low rain and high evaporation. Constrained by this geographical environment, small lakes and dams, the overlapping of living areas and lakes and dams, rapid land development, and pollution of lakes and dams have caused eutrophication. After the investment of considerable resources and manpower, the water supply remains unstable; therefore, more than half of the water supply is extracted by the Kinmen Water Factory to satisfy the overall demand. The liberalized tourism and three mini links have caused severe over-extraction of underground water in Kinmen. In 2002, the Water Resources Agency (Ministry of Economic Affairs), with the assistance of the Centre for Ecology and Hydrology, constructed a regional Water Poverty Index (WPI) for the purpose of analyzing data on water sources in Taiwan. However, this research excluded Ponghu, Kinmen, and Lianjiang. This study is the first attempt to perform WPI calculations in Kinmen County. Furthermore, this study selected primary indices from the WPI based on time order to estimate the water resources that will be available in the future. The findings could be a useful reference for establishing and moderating future policies.

Keywords: Eutrophication; Water poverty index; Water resources; Sustainable development

1 INTRODUCTION

1.1 *Research background*

A lack of water supply is a problem for every outer island, including Kinmen, the climate of which is influenced by north-eastern winds, and the average annual temperature is 21.3°C; furthermore, the winter and summer temperatures vary considerably.

Generally, typhoons strike Kinmen from May to October and supply water for 12 lakes and dams. The average annual rainfall is 1,047 mm, which is 41.63% of the average total rainfall in Taiwan (2,515 mm). Because the total average annual evaporation of 1,653 mm exceeds the total of average annual rainfall, hence Kinmen has a dry climate. The Water Poverty Index (WPI) can provide decision makers with critical and concise information regarding water resources. These indices can be used to ascertain whether a geographical area can provide the water resources and sanitary services required by rapid economic development and population growth (Feitelson & Chenoweth, 2002).

1.2 *Research purpose*

The lack of water supply is a severe problem in Kinmen. This research adopted a domestic scale to examine Kinmen's water resource problem. The findings provide an effective reference for those regulating and moderating relevant policies.

Although Kinmen's government has aggressively attempted to develop various methods of reserving water, the demand for water has increased as a result of three mini links. The water supply is crucial to Kinmen's tourism industry. Low rainfall, high evaporation, narrow drainage areas, limited dam depth, and population increases have aggravated the lack of water supply. The water supply crisis occurs during the dry season. Therefore, the purpose of this research was to perform WPI calculations in Kinmen County for the year 2002 and to predict the changes in the results of these calculations for the year 2025.

This study was conducted on the basis of the following hypotheses: 1) Every dimension, subdimension, and indicator in the WPI can be used to determine the water resource problem in Kinmen; 2) The calculated data retrieved from every indicator are sufficiently accurate; 3) The measuring scale of the indices accurately reflects the conditions in Kinmen; 4) Large-scale changes in the basic characteristics of Kinmen will occur.

1.3 Research limitations

The WPI, which ranges from 0 to 1, was used to evaluate the degree of water poverty in Kinmen. A literature review was conducted to establish a primary calculation method for Kinmen. The alignment of the index structure with the water resource problem requires long-term modifications and further feedback from experts regarding the representativeness of the index calculation and the selection of scales.

2 LITERATURE REVIEW

2.1 Development of the WPI

Three-fourths of the globe is covered by ocean. However, the water resources that can be directly used or developed are restricted. According to data from the fourth Water Resources Forum (UNESCO), the world contains 14 billion square meters of water resources, but only 2.5% is usable.

With population growth and expanding industrial and agricultural development, the demand for fresh water has increased rapidly worldwide; in the twentieth century, demand for water increased 7-fold and the industrial demand increased 20-fold. According to statistics from the United Nations Environment Program, more than 35 billion people and more than 40 countries will experience a shortage of water by 2025.

Underground water is a primary source of drinking water. Worldwide, more than half of the drinking water is underground water. In some areas, underground water extraction has reached critical levels. More than 90% of natural disasters involve water, and the frequency and intensity of these disasters are also increasing, which will eventually affect the economic development. Water supply crises cause ecological deterioration and biological destruction, which threaten the existence of mankind. In addition, these crises destabilize international relationship because countries must compete for water resources (The United Nations World Water Development Report 4, 2012).

2.2 Origin of the WPI

In 2002, the World Commission on Environment and Development shortlisted water, energy, health, agriculture, and biodiversity as the five primary issues in the twenty-first century. Among them, the water resources were ranked as the top priority.

Based on the previous discussion, the Center for Ecology and Hydrology (CEH), supported by the Department for International Development, developed the WPI in 2001. The purpose of the WPI was to construct a consolidated compound indexing system to identify water resource issues and provide an accurate water resource management system.

The main factors that are analyzed using this index include water usability, accessibility, and retrievability, as well as the background and quality of water conservation. The index enables nations and communities to be ranked and compared. It also accounts for physical and socioeconomic factors (Sullivan, 2002).

2.3 The structure and application of the WPI

Initially, the WPI was applied regionally, and it was later applied internationally. Nations view water resources from various perspectives. For example, Tanzania, South Africa, and Sri Lanka focus on the manner of water usage, water quality and changes thereof, water resources for food production, water management ability, and environmental and spatial scales (Sullivan et al., 2003).

After referencing various factors, the CEH originally used communities as measuring scales to establish the WPI, which was designed for all communities nationwide. Finally, the WPI was applied nationally. The WPI was developed into a cross-national indexing system that applied the human development index to measure a nation's position (1 is the highest, and 0 is the lowest). The WPI includes five components with subcomponents or indicators, namely, resources, access, capacity, use, and environment (Sullivan & Meigh, 2007).

2.4 Methodology

The previously mentioned dimensions have 2 to 6 sub-components. The basic framework is listed in Table 1. The indicators in every dimension have specific meanings.

$$Y_j = (X_j - X_{min})/(X_{max} - X_{min}) \qquad (1)$$

The variable x_j is a nation; j is the actual score; and x_{max} and x_{min} are the highest and lowest scores of one variable. The variable y_j is the index corresponding to the WPI.

$$C_i \sum_{k=1}^{n} Y_k/n \qquad (2)$$

$$WPI = 0.2 * \sum C_t \qquad (3)$$

The total score of each nation was calculated and geographical position system results are shown in Figure 1, which indicates that the top five countries are Finland, Canada, Iceland, Norway,

Table 1. WPI basic architecture (International).

Component	Sub-component or (Indicator)
Resources	Internal volume of fresh water sources, Foreign water, Population.
Access	Percentage of the population can get clean water, Percentage of the population can use sanitary equipment, Percentage of the population for irrigation acceptable.
Capacity	GDP Per Capita, Under-five infant mortality, The proportion of education, Gini index. (income distribution)
Use	Daily water consumption, The proportion of industrial water and agricultural water.
Environment	Z value of the water quality, water pressure, Environmental laws and management, biodiversity, IT capacity indicators in the ESI architecture.

Source: Yeh Shin-cheng, et al., 2004, research and application of water poverty index of the Taiwan region.

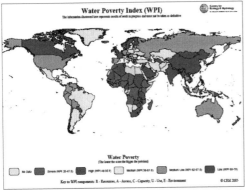

Source: Water Management and Policy Group - Oxford Centre for Water Research - School of Geography and the Environment-2004- The Water Poverty Index.

Figure 1. Total score of WPI of countries.

and Guyana. The lowest five countries are Haiti, Niger, Ethiopia, Eritrea, and Malawi (Lawrence et al., 2002).

1. The formula for the WPI first identifies the index for a particular nation and ascertains its position in relation to other countries (1 is the highest, and 0 is the lowest).
2. Next, the score of the index is averaged and multiplied by 100; the total score of the dimension is then obtained. The highest score is 100.

The variable i is the score of one specific dimension, and this dimension is assumed to have n components. With 5 dimensions, 100 is the highest score. The total score of each dimension is averaged, and the total score of the WPI is obtained.

3. The WPI is the nation's total score, with 100 being the highest score. A higher score indicates superior water resources.

2.5 The development of water resources in Taiwan

Taiwan is a subtropical island, and two-thirds of its terrain consists of hills and mountains. It has a rainy season in May and June and a typhoon season between July and October. The distribution of precipitation during the rainy season is uneven, which contributes to natural disasters and detrimentally affects people in Taiwan. Taiwan's total rainfall is three times the worldwide average. However, the average water resource per person is 4,290 m^3 (1/4.6 of the worldwide average). Recently, draughts have been common in Taiwan; rainfall is decreasing annually. Overall, the annual changes in rainfall differ depending on the region of Taiwan (e.g., the rainfall is decreasing in Kaohsiung and increasing in Hualien).

2.6 Current status of domestic water resources

To construct the WPI for Taiwan, the Water Resources Agency (Ministry of Economic Affairs) invited prominent WPI researchers from Britain (i.e., the CEH) to discuss WPI calculations and the methods of constructing a regional WPI. Next, the Taiwanese WPI team collected the necessary data for each index. Each index was calculated mutually by Taiwanese and British researchers, and the total scores and rankings were calculated by the CEH (Yeh et al., 2004). The Taiwanese WPI team referred to the regional structures of other countries and compared WPI structures worldwide with draft from the WPI of Taiwan. The Delphi method was applied and interviews were conducted to confirm the index structure. After the consistency verification, the selected index items and their individual weightings were determined, and a calculation of the WPI of Taiwan was developed. Table 2 shows the index items of Taiwan's WPI (Yeh et al., 2005).

2.7 Development of the regional WPI

A lack of water resources has always been an issue in Kinmen, particularly after the development of tourism was prioritized. Long dry seasons and limited watersheds render the quality and quantity of water difficult to control. Moreover, low rainfall, high evaporation, and uneven rainfall periods amplify

Table 2. WPI indicators in Taiwan.

Component	Sub-component
Resources	Surface water (R1)
	Groundwater (R2)
	Storage capacity (R3)
Access	Living water (A1)
	Industrial water (A2)
	Agricultural water (A3)
Capacity	Health (C1)
	Education (C2)
	Consumption capacity (C3)
	Water investment (C4)
Use	Living water usage (U1)
	Water efficiency (U2)
	Cost (U3)
	Leakage (U4)
Environment	Mudslide (E1)
	Flood (E2)
	Subsidence (E3)
	Biodiversity (E4)
	Pollution (E5)

Source: Environmental Quality Culture and Education Foundation, 2005, the application of water poverty index of Taiwan and international collaborative research.

the problem of water shortage. Kinmen has developed multiple sources to satisfy the need for water resources, including surface water, underground water, and desalinized sea water. The geology of Kinmen's eastern island is granite, and the fresh water is mainly derived from dams; however, the soil in Kinmen's western island is permeable, with an abundance of high quality underground water. To conclude, underground water is the main water source in Kinmen, whereas Lieyu depends on surface and underground water. The number of tourists who visit Kinmen has reached 600,000 annually and continues to increase. Most hotels in Kinmen dig wells to retrieve underground water for consumption; therefore, the phenomena cannot reflect the actual supply and demand for the tourism industry.

3 FINDINGS AND DISCUSSIONS

3.1 *Comparisons with other counties and cities*

The Total Scores of the WPI of Kinmen and Comparisons With Other Counties and Cities.

The WPI structure in Taiwan has five dimensions (including resources, access, capacity, use, and environment) and 30 subdimensions. After gathering the data in Kinmen, we found the following:

1. The resources score: The total score of the Kinmen region was 0.621, which ranked Kinmen 13th among all counties and cities;

2. The access score: The total score of the Kinmen region was 0.828, which ranked Kinmen 11th among all counties and cities;

3. The capacity score: The total score of the Kinmen region was 0.523, which ranked Kinmen 17th among all counties and cities;

4. The use score: The total score of the Kinmen region was 0.532, which ranked Kinmen 17th among all counties and cities;

5. The environment score: The total score of the Kinmen region was 0.809, which ranked Kinmen first among all counties and cities;

6. An analysis of the overall scores and county rankings shows that the total score of Kinmen is 0.664, which ranks Kinmen ninth among the 23 counties and cities.

3.2 *The future development of the WPI of Kinmen*

Since 2002, Taiwanese governments have developed the Taiwanese regional WPI and published the trial results of every city and county. However, the data have been updated since the initial WPI development. Most of the data (e.g., water reservation development, industrial and service development, agricultural water usage, total industry output, forestry coverage, and extinct fish species) are changing constantly.

4 CONCLUSIONS

According to the indicators established in this research, the conclusions are as follows. First, the WPI scores for the Kinmen region were 0.621, which ranked Kinmen 13th among the 23 cities and counties in the resource dimension; 0.828, which ranked Kinmen 11th among the 23 cities and counties in the access dimension; 0.523, which ranked Kinmen 17th among the 23 cities and counties in the capacity dimension; 0.532, which ranked Kinmen 17th among the 23 cities and counties in the use dimension; and 0.809, which ranked Kinmen first among the 23 cities and counties in the environment dimension. Compared with other cities and counties, Kinmen was above average. However, Kinmen was ranked last when compared by region. The results show that Kinmen has a lack of water resources compared with other regions. Therefore, if Kinmen's government is incapable of increasing water access and improving the economic viability of water resources, the entire Kinmen region will face a severe water shortage. This research shows that Kinmen's scores regarding access to water and use of water are comparatively low and therefore strongly recommends that corresponding policies should be implemented to solve the problem of water poverty in Kinmen.

REFERENCES

[1] Feitelson, E. & Chenoweth, J. (2002) Water poverty: towards a meaningful indicator. *Water policy*, 4, pp. 263–281.

[2] Lawrence P., Meigh J.R., Sullivan C.A., (2002) The Water Poverty Index: an International Comparison, *Keele Economics Research Papers*, pp. 5~16.

[3] Sullivan C.A. (2002) Calculating a water poverty index. *World Development* Vol. 30 (7): pp 1195–1210.

[4] Sullivan C.A., Meigh J.R., Giacomello A.M., (2003) The Water Poverty Index: Development and application at the community scale. *Natural Resources Forum* 27, pp 189–199.

[5] Sullivan C.A., Meigh J.R. (2007) Integration of the biophysical and social sciences using an indicator approach: Addressing water problems at different scales. *Water Resource Manage*, pp 114–125.

[6] United Nations Educational, Scientific and Cultural Organization, (2012) Managing Water under Uncertainty and Risk, *The United Nations World Water Development* Report 4.

[7] Water Management and Policy Group, (2004) Oxford Centre for Water Research—School of Geography and the Environment-2004-The Water Poverty Index.

[8] Yeh S.C., Liu M., Cai Y.L., (2004) Research and application of water poverty index of the Taiwan region, Ministry of Economic Affairs Water Resources Agency Environmental Quality Culture and Education Foundation commissioned research project report, 3–91, Taipei.

[9] Yeh S.C., Liu M., Cai Y.L., (2005) The application of water poverty index of Taiwan and international collaborative research. *Environmental Quality Culture and Education Foundation*, Taipei.

Architectural, Energy and Information Engineering – Sung & Chen (Eds)
© 2016 Taylor & Francis Group, London, ISBN 978-1-138-02791-6

Emission characteristics of waste cooking oil biodiesel and butanol blends with diesel from a diesel engine exhaust

P.M. Yang, Y.C. Lin, S.R. Jhang & S.C. Chen
Institute of Environmental Engineering, National Sun Yat-Sen University, Kaohsiung, Taiwan

Y.C. Lin
College of Pharmacy, Kaohsiung Medical University, Kaohsiung, Taiwan

K.C. Lin
Mechanical and Electromechanical Engineering, National Sun Yat-Sen University, Kaohsiung, Taiwan

T.Y. Wu
Department of Chemical Engineering and Materials Engineering, National Yunlin University of Science and Technology, Yunlin, Taiwan

ABSTRACT: The purpose of this work is to investigate the suitability of butanol-diesel fuel with waste cooking oil biodiesel blends as an alternative fuel for the diesel engine operated at a steady-state condition, and determine their effect on the engine emission of regulated harmful matters and $PM_{2.5}$. The three test fuels are Diesel (100 vol% of pure diesel), WCOB10 (10 vol% of biodiesel made from waste cooking oil), and Bu10 (10 vol% of butanol). Experimental results indicate that when using the 10 vol% of biodiesel, the CO can be reduced to about 1.56%. Using Bu10 instead of diesel that decreases NOx by 13.1%, and $PM_{2.5}$ by 16.5%. During engine performance tests, WCOB10, and Bu10 blends show 1.50%–5.26% higher brake specific fuel consumption and 1.99%–4.92% lower brake thermal efficiency compared with that of diesel. In conclusion, 10 vol% of biodiesel and 10 vol% of butanol blends with diesel can be used in diesel engines without modifications.

Keywords: Alternative energy; Butanol; Nitrogen oxides; $PM_{2.5}$; Diesel engine

1 INTRODUCTION

In recent years, the increasing depletion of petroleum resources from environment and the worsening pollution problems have led to concerns regarding alternatives to petroleum fuels. It is by now well known that the EU has set a target of replacing 10% of conventional fuels with biofuels by 2020 (Georgios et al. 2010). As renewable, biodegradable, and nontoxic fuel research has continued to the present, biodiesel has attracted considerable amount of attention over the past decade. Among biodiesel feedstock, the biodiesel made from waste cooking oil can be used to effectively reduce the raw material cost as well as solve the problem of waste oil disposal (Kumaran et al. 2011).

Butanol is preferable to methanol and ethanol for blends in diesel engines owing to its good solubility in diesel, greater heating value, higher cetane number and miscibility, and lower vapor pressures. Butanol is produced by the fermentation of biomass, resulting in lower cost of production and less corrosive. Furthermore, the carbon chain of butanol is twice than that of methanol and ethanol, which leads that the combustion of butanol has higher heating values and efficiency (Ballesteros et al. 2012).

Most of researches have indicated that biodiesel manufactured from waste cooking oil can reduce emissions of PM, HC, CO, and PAHs from engines (Lin et al. 2011). However, previous engine studies have showed there was an increase in NOx emissions from biodiesel combustion (Sun et al. 2010). In order to reduce the NOx emissions from biodiesel-diesel three methods have been proposed: (a) using low-temperature combustion, (b) employing reformulated biodiesel, selective catalytic reduction, and (c) utilizing exhaust gas recirculation (Fang et al. 2008; Muncrief et al. 2008). The other superior potential solution for reducing NOx emissions from biodiesel is to adopt the bioalcohol, which has a relatively greater vaporization heat as an additive.

The objective is to study gaseous pollutants produced by burning diesel fuels blended with butanol and biodiesel, which is made from waste cooking oil, in diesel engines. The work is to evaluate the potential of blended fuels that are able to decrease traditional emissions from diesel engines. In addition, the feasibility of biodiesel blends of biodiesel in fuel blends are assessed.

2 MATERIALS AND METHODS

2.1 Test fuels preparation

The premium diesel fuel produced by Chinese Petroleum Corporation, Taiwan, is used as a base fuel in the current study. The biodiesel from WCOB is produced by Taiwan Greatec Green Energy Corporation and the butanol is obtained from J. Baker (>99.5% purity). The properties of the tested fuels are listed in Table 1. Fuels with three different blends are used. Two of them are blended fuels WCOB10 (containing 10 vol% of waste cooking oil biodiesel and 90 vol% of diesel); and Bu10 (containing 10 vol% of butanol zand 90 vol% of diesel). A magnetic stirring plate is used to blend butanol in biodiesel. Complete mixing of butanol and biodiesel is achieved to avoid the emission uncertainty caused by blends not perfectly mixed.

2.2 Diesel engine testing

The diesel engine (Robin SDG 2200, manufactured by Subaru Co. Ltd.) has a displacement of 230 cm³ and a rated output of 2.8 kW. The fuel injection pressure is approximately 50 MPa. This engine is four-cycle, air-cooled, overhead valve and single-cylinder. The combustion system is direct injection and no further modification is needed. All tests were operated under a constant load of 2.8 kW at a constant speed of 3000 rpm (steady-state condition)

Table 1. Properties of fossil diesel, waste cooking oil biodiesel and butanol.

Property	Diesel	Biodiesel	Butanol
Chemical formula	C_xH_y	–	$C_4H_{10}O$
Heating value (Cal/g)	10,930	9,800	8,009
Stoichiometric air/fuel ratio	14.3	12.5	11.2
Lower heating value (MJ/kg)	42.7	37.5	32.2
Flash point (°C)	70.0	174	37.0
Density @ 20°C (kg/m³)	835	881	810
Viscosity @ 40°C (cSt)	2.66	4.42	2.24
Cetane number	50.2	54.9	25.6
Oxygen content (%)	0	11.2	21.6

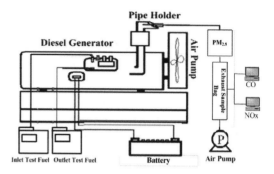

Figure 1. Test bench and the sampling equipment diagram.

throughout the experiment. Figure 1 shows the test bench and sampling equipment diagram.

2.3 Sample collection

The gaseous pollutant emissions in the tail pipe exhaust are monitored online using a portable gaseous pollutant analyzer (Telegan Sprint V4). Specifically, nitrogen oxide for each sample is analyzed using a ChemiLuminescent Detection (CLD) (model 404, Rosemount, UK). Carbon monoxide from each sample is determined using a Non-Dispersive Infrared Detector (NDIR) (model 880 A, Rosemount, UK). Particulate Matter (PM) from each filter sample is weighed again using an electronic analytical balance with fully automatic calibration technology (AT200, Mettler, Switzerland) to determine the net mass of collected $PM_{2.5}$.

3 RESULTS AND DISCUSSION

3.1 Break Specific Fuel Consumption (BSFC)

BSFC refers to the ratio between the fuel mass flow rate and engine power. Figure 2 shows the brake specific fuel consumption of three test fuels. It is observed that at a constant speed of 3000 rpm, the BSFC values of biodiesel and butanol were generally higher compared with diesel. The trends increase with the increase of butanol and biodiesel concentrations. In the fuels tested, diesel fuel has the lowest BSFC as compared with butanol and biodiesel blends since diesel has a relatively high heating value that is used for energy production in the engine. We obtain mean BSFC values for blends, which are 135 and 140 gkWhr⁻¹ for WCOB10 and Bu10, respectively. The BSFC values for diesel are 133 gkWhr⁻¹. The average BSFC values for WCOB10 and Bu10 were found to be 1.50% and 5.26% higher than the BSFC values of diesel, respectively. The tendency of BSFC values observed in this study is consistent with other studies (Behcet 2011; Nabi et al. 2009). Comparatively low calorific values, high viscosity and high density in biodiesel-diesel blends may be responsible for the minor increase in BSFC values (Behcet. 2011).

3.2 Break Thermal Efficiency (BTE)

Brake thermal efficiency is defined as break power of a heat engine as a function of the thermal input from the fuel. It is used to evaluate how well an engine converts the heat from a fuel to mechanical energy. The variation in the engine BTE output with different mixture ratio at a steady-state condition is also presented in Figure 2. From the figure, the BTE of pure diesel was highest at a constant speed of 3000 rpm while that of 10% butanol blends (Bu10) was lowest. The primary reason for the decrease in the BTE of biodiesels is the higher BSFC due to biodiesel having a lower calorific value, which is also supported by other literature (Sayin et al. 2011).

3.3 Carbon monoxide (CO) emissions

Figure 3 shows the emission factors of carbon monoxide with different butanol and biodiesel blends in the exhaust of the diesel engine. When butanol is used with diesel blended fuels, the CO emissions increase by 19.3% from Bu10. When butanol is not added in blends, CO formation is inversely proportional to the biodiesel and butanol used in the blends, the CO emissions decreases by 1.56% in WCOB10. The increase of CO in the combustion of butanol in biodiesel blends is because of the relatively low reaction temperatures that would slow down the oxidation reactions in forming CO_2 from CO. The increase of CO may be ascribed to relatively low cheating value for the alcohol fuels, which increase ignition delays that lead to an

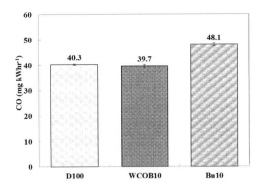

Figure 3. Emission factor of CO from the exhaust of the diesel engine fueled with various butanol/biodiesel blends.

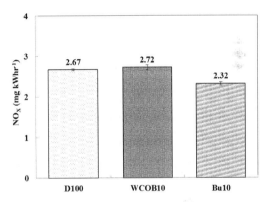

Figure 4. Emission factor of NO_X from the exhaust of the diesel engine fueled with various butanol/biodiesel blends.

incomplete combustion and further increases CO emissions (Nadir. 2012; Nadir et al. 2014).

3.4 NO_X emissions

The variation of NO_X emissions factor for various blends is shown in Figure 4. NO_X emissions are reduced by mixing butanol concentrations in blends. Compared with the pure diesel, NO_X emissions decrease by 13.1% from Bu10. On the other hand, when biodiesel is used in blends without butanol, the NO_X emissions increase by 1.98% from WCOB10. The NO_X contents increasing in blends were also reported by other studies (Muralidharan et al. 2011; Varuvel et al. 2012). It is seen that the addition of butanol to diesel is able to lower NOx emissions. As reported by Nadir et al. (2014), the addition of butanol in blends creates a cooling effect and hence decreases a combustion temperature to reduce the NOx formation. And

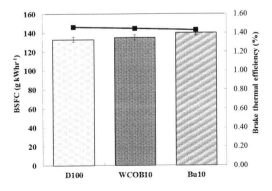

Figure 2. BSFC and BTE for diesel engine running on blended fuels of biodiesel-butanol-diesel.

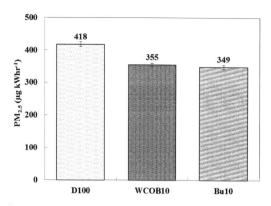

Figure 5. Emission factor of $PM_{2.5}$ from the exhaust of the diesel engine fueled with various butanol/biodiesel blends.

the observation is explained by blended fuels that have relatively low heating value and lower cetane number is able to decrease the temperature of combustion.

3.5 *$PM_{2.5}$ emissions*

Figure 5 shows the $PM_{2.5}$ collected on a filter paper with different butanol and biodiesel blends in the exhaust of the diesel engine. The result shows that $PM_{2.5}$ formation is lowered when addition of biodiesel and butanol in blends is increased. Compared with that of pure diesel, $PM_{2.5}$ formation decreases by 15.0% (WCOB10) without butanol, and by 16.5% without biodiesel in blends. The experiment results confirm that addition of biodiesel and butanol for petroleum diesel decreases the $PM_{2.5}$ emissions of the diesel engine regardless of addition percentage. This finding is explained by the fuel-bonded oxygen atoms in butanol and WCO remove carbons from the pool of hydrocarbon products that are potentially precursors of PM (Lapuerta et al. 2009). In addition, the addition of butanol inhibits the formation of sulfuric acids that may contribute to accumulation and condensation of soot or metallic ash in exhaust gases (Tsai et al. 2010).

4 CONCLUSIONS

The regulated harmful matters of premium diesel and blends with biodiesel and butanol are measured and analyzed in this paper. The results indicate different trends on regulated harmful matter emissions from different fuels tested. In this study, we identified that by blending 10 (v/v)% of butanol and 10 (v/v)% of biodiesel in diesel, NO_x and $PM_{2.5}$ emissions are able to be reduced significantly. The $PM_{2.5}$ reduction rate is 15.0% for WCOB10, and 16.5% for Bu10, respectively. NO_x is able to be reduced approximately by 13.1% when 10% vol. of butanol is blended in diesel fuel. Furthermore, as compared with the premium diesel, diesel/biodiesel and diesel/butanol blends increase the BSFC while significantly decreasing the BTE. The outcome of using diesel/biodiesel and diesel/butanol blends as alternative fuels is encouraging. In general, the variation of regulated harmful matters emissions of biodiesel and butanol in engines can be affected by several factors, such as engine load, biodiesel components, and driving cycle. Further research can be carried out to analyze the effect of combustion characteristics with the increase of different proportions of butanol and biodiesel blend.

REFERENCES

Ballesteros R, Hernandez JJ, Guillen-Flores J. 2012. Fuel; 95: 136–145.

Behcet R. 2011. Fuel Process Technol; 92: 1187–1194.

Chen Z, Liu J, Han Z, Du B, Liu Y, Lee C. 2013. Energy; 55: 638–646.

Fang T, Lin YC, Foong TM, Lee CF. 2008. Environ Sci Technol; 42: 8865–8870.

Georgios F, Georgios K, Marina K, Leonidas N, Evangelos B, Stamoulis S, Zissis S. 2010. Environ Pollut; 158: 2496–2503.

Kumaran P, Mazlini N, Hussein I, Nazrain M, Khairul M. 2011. Energy; 36: 1386–1393.

Lapuerta M, Armas O, Garcia-Contreras R. 2009. Energy Fuels; 23: 4343–4354.

Lee WJ, Liu YC, Mwangi FK, Chen WH, Lin SL, Fukushima Y, Liao CN, Wang LC. 2011. Energy; 36: 5591–5599.

Lin YC, Hsu KH, Chen CB. 2011. Energy; 36: 241–248.

Muncrief RL, Rooks CW, Cruz M, Harold MP. 2008. Energy Fuels; 22: 1285–1296.

Muralidharan K, Vasudevan D, Sheeba KN. 2011. Energy; 36: 5385–5393.

Nabi MN, Rahman MM, Akhter MS. 2009. Appl Therm Eng; 29: 2265–2270.

Nadir Y, Francisco MV, Kyle B, Stephen MD, Antonio C. 2014. Fuel; 135: 46–50.

Nadir Y.2012. Energy; 40: 210–213.

Sayin C, and Gumus M. 2011. Appl. Therm. Eng; 31, 3182–3188.

Sun J, Caton JA, Jacobs TJ. 2010. Prog Energy Combust Sci; 36: 677–695.

Tsai JH, Chen SJ, Huang KL, Lin YC, Lee WJ, Lin CC, et al. 2010. J Hazard Mater; 179: 237–243.

Varuvel EG, Mrad N, Tazerout M, Aloui F. 2012. Appl Energy; 94: 224–231.

Architectural, Energy and Information Engineering – Sung & Chen (Eds)
© 2016 Taylor & Francis Group, London, ISBN 978-1-138-02791-6

Development of wind power in Taiwan and the communication for control and monitoring of offshore wind turbine

Yun Wei Lin & Yung Hsiang Wu
Measurement/Calibration Technology Department, Electronics Testing Center, Taiwan

Cheng Chang Chen & Jian Li Dong
Bureau of Standards, Metrology and Inspection, Ministry of Economic Affairs, Taiwan

ABSTRACT: In this paper, we will introduce the development of wind power in Taiwan and the policies or subsidies issued by government of Taiwan for development of offshore farm, which includes Formosa Wind Power (FWP), Fuhai Wind Farm (FWF) and Taiwan Power Company offshore demonstration sites. Then, we will focus on the IEC 61400-25 which is the communication standard for control and monitoring. The information models and information exchange model are used to describe the services provided by the wind power plant. The mapping from the models to the communication profiles and the condition in Taiwan are then illustrated.

Keywords: wind power plants; IEC 61400-25; wind turbine; communication protocol; information model

1 INTRODUCTION

1.1 Background

With the growth of global economy, developed and developing countries consume more and more energy. The use of fossil fuels even increases the risk of climate change. In order to improve the problem of energy crises and climate change, lots of countries are active in developing renewable energy, such as wind power, solar energy, hydropower, geothermal energy.

1.2 Wind power

Among various renewable energy, wind power is a well-developed technology to produce energy by wind. It can be divided into two categories, onshore and offshore. An offshore wind farm, such as the London Array, is even with a capacity of 630 MW which, by yearly average, is sufficient to supply approximately more than 480,000 homes. In order to deal with such a large amount of capacity connecting to the main power system, the control and monitoring becomes more and more important in this system, especially the communication protocol. There are some standards have profiled the communication between power system and wind turbines, but it cannot be sure that the interoperability between these standards. In order to resolve the problems that the wind turbines from different vendors may not be able to communicate with each other, and the SCADA system may not interoperate with different wind turbines. The IEC 61400-25 series standards are promoted. It increases the interoperability between wind turbines from different vendors.

2 DEVELOPMENT OF WIND POWER IN TAIWAN

Taiwan is surrounded by the sea. The geographic environment and wind source is perfect to develop wind energy. Especially in the west coast of Taiwan and Penghu area, the winter northeast monsoon and the summer southwest monsoon is a great potential for developing wind power.

2.1 Onshore wind power

In 1980, due to the crisis of energy, renewable energy become important. Industrial Technology Research Institute (ITRI) had been entrusted by the government of Taiwan to assess the potential of wind power for Taiwan and develop the technology of wind power in Taiwan. In the end of 1991, they complete the 4 kW, 40 kW, and 150 kW wind power electricity generator.

In 2000, the government of Taiwan issued the "Wind Power Demonstration Incentive Program". The government provides subsidy for both equipment and developing processes. Due to these successful experiences in onshore wind power, more and more manufacturers are willing to invest in wind power in Taiwan.

2.2 *Offshore wind power*

The advantage of the offshore wind power is that the wind is much stronger than the wind over the continent. Table 1 is the rank of average wind speed in 23 years which is evaluated by 4C Offshore. Taiwan has the great potential to develop offshore wind power. But the development of offshore wind power needs lots of technologies, such as remote monitoring and control system and the offshore foundations. It made this expensive energy generating technology keep on hold in Taiwan.

However, after the Fukushima Nuclear Power Plant accidents, people are seeking new energy to substitute the nuclear power. The large scale deployment of offshore wind power becomes a solution to re-solve the problem.

On 3rd July, 2012, Taiwan government officially announced the "Offshore Demonstration Incentive Program" and provided subsidies for both equipment and developing processes for offshore wind power. Taiwan's Ministry of Economic Affairs (MOEA) had signed deals with Formosa Wind Power (FWP), Fuhai Wind Farm (FWF) and Taiwan Power Company to build three offshore demonstration sites on the west coast of Taiwan.

As shown in Table 2, The government of Taiwan had also set the "Thousand Wind Turbines Project". The goal is to build 600 MW of offshore wind energy by 2020 and 3 GW by 2030. By the end of 2030, there would be a thousand of offshore and onshore wind turbines in Taiwan. The Bureau of Energy (BOE) has estimated the gross value of industrial output will achieve 500 billion TWD in 2030.

At the beginning, the offshore will combine the technologies from advanced countries and the ones from Taiwan. The government of Taiwan will continuously support industries in Taiwan to build the key technology of offshore wind energy.

Even though there are lots of benefits for development of offshore, we still have to conquer many grave challenges in Taiwan. For example, we locate the subtropical zone, so there are much tougher conditions for building key technology of the offshore wind power. The turbines must bear the

Table 1. Rank of average wind speed in 23 years (4Coffshore).

Rank	Project	Sea	City	Speed (m/s)
1	Fujian Pingtan Development Zone Changjiangao 200 MW offshore Wind Farm	South China Sea	China	12.11
1	Putian Pinghai bay offshore project phase D-E	Taiwan Strait	China	12.11
1	Pingtan experimental zone 300 MW offshore wind power project	South China Sea	China	12.11
4	Fujian Putian Shicheng Yugang Offshore Wind Farm	Taiwan Strait	China	12.06
5	Putian Pinghai bay offshore Wind Farm demonstration project phase 2–250 MW	Taiwan Strait	China	12.04
5	Longyuan Putian Nanri Island 400 MW project—Phase 3 350 MW Commercial	Taiwan Strait	China	12.04
5	Longyuan Putian Nanri Island 400 MW project—Phase 2–34 MW Commercial	Taiwan Strait	China	12.04
5	Longyuan Putian Nanri Island 400 MW project—Phase 1–16 MW Commercial	Taiwan Strait	China	12.04
5	Putian Pinghai bay offshore demonstration project phase A	Taiwan Strait	China	12.04
10	Xidao	Taiwan Strait	Taiwan	12.02
10	Zangfang—TGC	Taiwan Strait	Taiwan	12.02
10	Changhua—North - Taipower	Taiwan Strait	Taiwan	12.02
13	13 Changhua—South - Taipower	Taiwan Strait	Taiwan	11.94
13	Changhua—Pilot - Taipower	Taiwan Strait	Taiwan	11.94
13	Hanbao—TGC	Taiwan Strait	Taiwan	11.94
13	Changhua Offshore Pilot (COPP)—TGC	Taiwan Strait	Taiwan	11.94
13	Fuhai Offshore Windfarm—TGC	Taiwan Strait	Taiwan	11.94
13	Fuhai Phase III—TGC	Taiwan Strait	Taiwan	11.94
19	NW1 Medium Term Option for Offshore Wind Development	Scottish Continental Shelf	Scottish	11.91
20	NW2 Medium Term Option for Offshore Wind Development	Scottish Continental Shelf	Scottish	11.86

Table 2. Roadmap of the thousand wind turbines project in Taiwan.

Year	2014	2015	2020	2025	2030
Onshore (No. of turbines)	632.6 MW (319)	866 MW (350)	1,200 MW (450)	1,200 MW (450)	1,200 MW (450)
Offshore (No. of turbines)	0	15 MW (4)	600 MW (120)	1,800 MW (360)	3,000 MW (600)
Total	632.6 (319)	881 (354)	1,800 (570)	3,000 (810)	4,200 (1,050)

wind with more salt than Europe or America and the earthquake usually happened in Taiwan.

Because it is in the initial phase of developing smart grid in Taiwan, we have to solve the problems of interoperability of communication which might happen in the future. It is a significant issue we must figure out and research the standard in interoperability. Thus, each part of smart grid could communicate with each other, and the main backbone of power system can manage the power easily.

3 COMMUNICATION OF WIND TURBINE

3.1 Communication standard for wind power plant—IEC 61400-25

IEC 61400-25 is a communication standard for control and monitoring in wind power plant. The standard is based on IEC 61850, which is originally a communication standard for substation automation. Recently, it is extended as a communication standard for power utility automation.

The IEC 61400-25 standards defines the information models in the wind power plant (IEC 61400-25-2), which is used to describe the real world component of the wind power plant, so that there is a common method to describe many sub-components in the wind power plant. The information exchange model is al-so defined in the series of standards (IEC 61400-25-3). The functions and data of each information model (component) are illustrated in detail. Finally, the models are mapping to the existing communication profiles, so that the wind turbines can be controlled and monitored by the users.

3.2 Information model

As shown in Fig. 1, the real world wind power plant physical devices are modeled as logical devices in the virtual world. The information of each component (logical node class) of the wind power plant logical devices, such as WTUR-wind turbine general information, WROT-wind turbine rotor information, WGEN-wind turbine generator information, is de-fined in this standard. There are also some logical node classes, for example, XCBR-circuit breaker, MMXU-measuring for operative

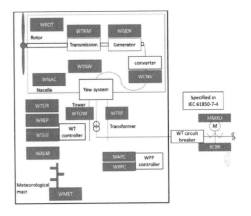

Figure 1. The example of logical device model of physical device.

purpose, have been defined in the IEC 61850 series standards. The IEC 61400-25 and the IEC 61850 series standards de-scribe the wind power plant well, and let the external clients and users have the common language to access these logical devices and logical nodes.

3.3 Information exchange model

In order to access the structure and content of the information models, the information exchange model is provided in IEC 61400-25-3.

As shown in Fig. 2, the services and interfaces are defined for the external clients to interact with the logical devices and logical nodes in the wind power plant. The logical nodes contain many data, which can be grouped into data sets.

The Get/Set and Control are the functions to write and read data in the data set of logical nodes. The data sets are also important for the Report and Log functions. For example, some devices could record the data periodically. If the external clients want to retrieve some of the logged data, it can send the request to the correspondent device. Then, the rules of logging are defined in the Log Control Block (LCB). The logged data can be collected into data sets and retrieved by user. Similarly, the rules of reporting are defined in the Report Control Block (RCB). The clients can subscribe and retrieve the data.

71

3.4 *Mapping to communication profile*

The information model and information exchange model are defined in IEC 61400-25-2 and IEC 61400-25-3. In order to exchange the data over the network, the models are mapping to the specific communication profile. There are 5 communication profiles defined in the IEC 61400-25-4, which are web services, OPC XML-DA, IEC 61850-8-1 MMS, IEC 60870-5-104, and DNP3. Fig. 3 shows the conceptual architecture of the mapping to communication profiles.

Different communication profiles have their own characteristics. For example, DNP3 and IEC 60870–5–104 are the TCP/IP based communication protocol. However, IEC 61850 even provides metadata in the devices, real-time applications (GOOSE, SMV), and system configuration language. In Taiwan, the main backbone of power system uses the DNP3 as the communication protocol, and the RTU of Taiwan power system also provide the DNP3 communication port. In order to let the offshore wind turbine interoperates well with the original power system, the communication profile of offshore wind turbines should be able to integrate with the RTU.

4 CONCLUSIONS

In this paper, the development of wind power in Taiwan is illustrated, which includes the background of onshore wind power and the demonstration project of the offshore wind power.

Recently, Taiwan official BSMI will set up a test site in Taichung. The offshore wind turbine should meet the highest standard of IEC Type Certification class 1 A. The turbine under test will connect to the power system. The communication for control and monitoring would be important. The integration of communication protocols will be considered in the phase of testing to ensure the success of interoperability in the offshore wind plant.

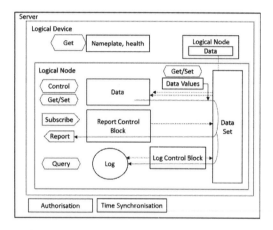

Figure 2. Information exchange model of wind power plant.

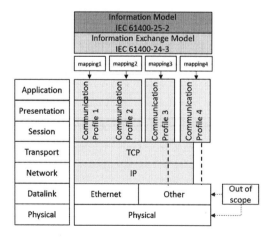

Figure 3. Concept of mapping the information model and information exchange model to the communication profile.

ACKNOWLEDGEMENTS

This paper is based on the project "Study of Standard and Certification Program for Communication of Offshore wind turbine" which is supported by Bureau of Standards, Metrology and Inspection, Ministry of Economic Affairs (BSMI, MOEA), R.O.C.

REFERENCES

Ahmed, M. & Kim, Y.-C. 2014. Communication Network Architectures for Smart-Wind Power Farms. *Energies 7, no. 6*: 3900–3921.

Ahmed, M. & Kim, Y.-C. 2014. Hierarchical Communication Network Architectures for Offshore Wind Power Farms. *Energies 7, no. 5*: 3420–3437.

Global Energy Research Council, 2013. Global wind report-annual market update 2013. *Tech. Rep.* Global Energy Research Council.

International Electrotechnical Commission, 2006. IEC 61400–25: Communications for monitoring and control of wind power plants.

Johnsson, A. & Højholt, L. 2007. Use of IEC 61400-25 to secure access to key O&M data. *European Offshore Wind conference.* Berlin, Germany, 2007.

Quecedo, E. T., Canales, I., Villate, J., Robles, E. & Apñanz, S. 2007. The use of IEC 61400-25 standard to integrate wind power plants into the control of power system stability. *European Wind Energy Conference & Exhibition.* Milan, 7–10 May 2007.

Architectural, Energy and Information Engineering – Sung & Chen (Eds)
© 2016 Taylor & Francis Group, London, ISBN 978-1-138-02791-6

Processing behavior of electrodes made with addition of nanopowder

S.V. Makarov
Yurga Technological Institute of National Research Tomsk Polytechnic University, Yurga, Russia

ABSTRACT: The study analyzes the processing behavior of welding electrodes made with addition of nanopowder, and compares them to standard welding electrodes (without addition of nanopowder). The effect of nanopowder of complex composition on the electrode metal transfer process is analyzed as well.

Keywords: fusion welding, welding electrode, nanopowder, oscillogram, droplet transfer, spatter.

1   INTRODUCTION

In the early 21st century it has been essential to raise the scientific and technical level of the economy throughout the world. To resolve this issue, it is essential to conduct extensive scientific research and to insure the large-scale application of cutting edge technologies in production processes. According to the forecasts of many relevant organizations, priority must be given to research on nanomaterials development [1].

As mentioned in studies [2–5], the essence of an experiment in adding nanopowder to the welding electrode is as follows: a complex-composition nanopowder (Al_2O_3, SiO_2, Ni, TiO_2, W) with the nanoparticle size 90 nm and purity 99.85% is added to a water glass (Table 1). The nanopowder is added to water glass in the mechanical cavitation activator type facility at 30°C–35°C for 2 minutes.

A series of experiments was performed to define the processing behavior of the electrodes under study. During these experiments pins welded on the surface of the plates. During the welding process, the changes in energy response (amperage, voltage) were recorded by the digital oscillograph.

Steel plates were used for welding: SS type St3, 300 mm long, 50 mm wide, and 4 mm thick. On the surface of the plates a bead of the electrode metal was welded. For welding, electrodes 4 mm in diameter were used, types MP-3, OK-46.00, УОНИ 13/55. The ВД 306-У3 rectifier was used as a source of supply. Welding was performed in single-pass under following conditions: amperage—140–160A, voltage 24–26 V.

Along with the study of the processing behavior of electrodes, an experiment was performed to loss of metal during burning and spatter.

On the basis of the obtained results, the following values were determined:

Mass of the molten electrode:

$$m_{me} = m_e - m_{e\,gen}$$

Mass of the weld electrode material:

$$m_{we} = m_{p\,sv+s} - m_{p\,before}$$

Mass of the weld metal material:

$$m_{we} = m_{p\,w} - m_{p\,before}$$

Mass of electrode loss from burning and spatter:

$$p = \frac{\left(m_{me\,r} - m_{we}\right)}{m_{me}} \cdot 100\%$$

where $m_e = mass$ of the electrode; $m_{p\,before} = mass$ of plate before welding; $m_{p\,sv+s} = mass$ of plate after welding with dross and spatter; $m_{p\,w} = mass$ of plate after welding without dross and spatter; $m_{e\,r} = mass$ of the rest of the electrode after welding; $m_{e\,gen} = mass$ of the rest of the electrode without dross.

The results of the experiment on welding a bead on the steel plate surface are shown in Figures 1–6. The digital oscillograph was used.

As is obvious, the process is characterized by high stability of the transfer of droplets of the electrode metal into the weld pool (in this situation stability is understood as the frequency of

Table 1.   Performance index of double water glass.

Name of index	Serial	Experimental
Module	3.130	3,200
Viscosity, Pa·s	0.604	0.292
Density, g/cm³	1.433	1.430

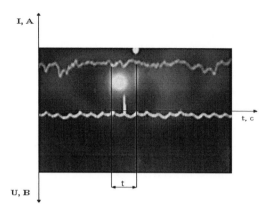

Figure 1. First specimen: oscillogram of welding process.

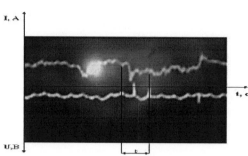

Figure 4. Third specimen: oscillogram of welding process.

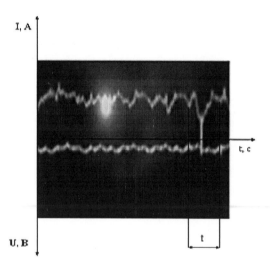

Figure 2. Fourth specimen: oscillogram of welding process.

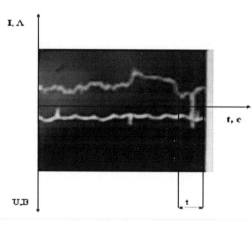

Figure 5. Fifth specimen: oscillogram of welding process.

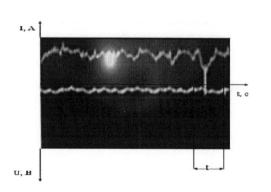

Figure 3. Second specimen: oscillogram of welding process.

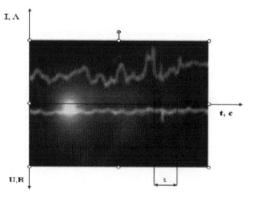

Figure 6. Sixth specimen: oscillogram of welding process.

Table 2. Results of studies on burning and spatter loss.

	MR-3		OK-46.00		UONI 13/55	
	1	2	3	4	5	6
Mass of plate before welding, g	558.0	565.2	558.2	568.3	526.4	534.5
Mass of electrode before welding, g	59.0	59.1	57.3	57.4	58.5	58.4
Mass of plate after welding, with skimming, g	597.4	608.1	592.3	604.5	563.8	564.3
Mass of electrode after welding, g. Stubs	18.2	15	20.1	17.8	25.6	19.2
Electrode loss, %	3.4	2.7	8.5	8.3	12	10.6

the microcycle transfer). The period of the droplet transfer amounts to 45–50 ms, prompting suggestions that the droplets transferred to the weld pool are small.

The processes in Figures 5 and 6 are less stable compared to the previous specimens and the frequency of the electrode metal droplet transfer is not evident. The period of droplet transfer is 30–60 ms. The amplitude values are different, which is a sign of a minor destabilization of the welding process.

As seen from the pictures obtained, the welding process of specimens 1 and 2 is characterized by low stability of process, recurrence of short circuits, and a lower perturbation level.

A series of calculations was performed to determine the effect of the complex-composition nanopowder on the electrode metal loss during welding. Experimental results are shown in Table 2.

As is seen from the derived experimental results, the lowest spattering was during welding of specimens 1 and 2, which demonstrably augments the data previously obtained, because stability of the transfer of droplets substantially affects both the stability of the welding process in general and the electrode losses in spattering.

The complex-composition nanopowder has an integrated effect on the processing behavior of the welding electrodes. As seen from the data and oscillograms, the nanopowder makes it possible to improve the stability of the transfer of droplets of the electrode metal.

REFERENCES

[1] Makarov, S.V., Zorina, T.J., Kremneva, M.A. 2014. The nature of cavitation as applied to manufacturing of welding electrodes. Life Science Journal 11(8 s): 99–102.

[2] Makarov, S.V. & Sapozhkov, S.B. 2013. Use of Complex Nanopowder (Al$_2$O$_3$, Si, Ni, Ti, W) in Production of Electrodes for Manual Arc Welding. World Applied Sciences Journal 22: 87–90.

[3] Makarov, S.V. & Sapozhkov, S.B. 2014. Production of Electrodes for Manual Arc Welding Using Nanodisperse Materials. World Applied Sciences Journal 29(6): 720–723.

[4] Makarov, S.V. & Sapozhkov, S.B. 2012. Production of Electrodes with Nanosized Powder of Complex Structure (Zr, Si, Ni, Ti, Cr). In the Proceedings of VIII International Scientific and Practical Conference 2012 Development of the Contemporary Science, Prague: Education and Science Publishing House: 88–91.

[5] Makarov, S.V., Gnedash, E.V. & Ostanin, V.V. 2014. Comparative characteristics of standard welding electrodes and welding electrodes with the addition of nanopowders. Life Science Journal 11(8s): 414–417.

Architectural, Energy and Information Engineering – Sung & Chen (Eds)
© 2016 Taylor & Francis Group, London, ISBN 978-1-138-02791-6

Cosmogenic mechanisms of earthquakes

M.S. Hlystunov & Zh.G. Mogiljuk
Moscow State University of Civil Engineering, National Research University, Moscow, Russia

ABSTRACT: This article is devoted to one of the main problems of earthquake engineering: verification of the dominant mechanisms and causality of the earthquake's intensity dangerous evolution. It discusses the results of the comparative analysis of the Earth's gravitational interaction energy variation amplitudes with the Sun, the Moon and the planets in the solar system. It also presents the results of the comparative evaluations of the Earth's geosphere gravitational perturbation amplitudes with the solar radiation energy and with the energy of its own heat. It is shown that the energy of its own heat and Sun exposure of the Earth is much less than gravitational perturbations in the near-Earth space. This article presents the results of the spectral analysis of earthquake's global daily energy on the Earth before and after the Shoemaker-Levy comet explosion on Jupiter. It is shown that the number of seismic events on Earth with a magnitude greater than 2.5 on the Richter scale after the comet explosion increased 10 times. The earthquake's global daily energy spectrum shows the spectral manifestations of the solar system planets' gravitational resonances.

Keywords: construction; earthquakes; cosmo-terrestrial relations; planets; gravidynamic resonances

1 INTRODUCTION

Currently, constructions in seismically active areas are faced with a serious problem of powerful earthquake risk reliable assessment at an urban planning depth (up to 100 years and more). One of the key tasks of this problem is the need for the verification of a poorly known causality of the earthquake's intensity dangerous evolution. As a first step in solving this problem, the authors examine the analysis and identification of dominant power sources and mechanisms of seismic activity activation on Earth. Among these sources are undoubtedly gravitational perturbations in the near-Earth space. To date, however, the world's scientific schools focused on endogenous causes and mechanisms of earthquakes. This attitude to the problem by Geophysics explains the global capacity of the Earth's own heat, the processes of heat and mass transfer, and thermodeformative in the lithosphere. However, authors making a comparative assessment of energy from these processes with cosmogenic gravitational perturbation energy of the Earth have proved the fallacy of this approach. First, from the modern science's viewpoint, the concept of the space (mainly solar activity) influence on the Earth's natural processes was formulated in the late 19th and early 20th centuries by a number ofscientists from different countries (Lehrbuch, 1903; Das Werden, 1907; First Report, 1926). A comprehensive approach

and interdisciplinary formulation of natural disasters and catastrophes cosmogenic concept were formulated by the honorary member of the USA Sciences Academy and the Russian scientist Czyzewskiy A. L. (Chizhevskiy, 1976). Despite the scientist's assumptions in the early 20th century about the gravitational effects of the solar system planets' movement on solar activity, and other evolutionary processes in all the Earth's spheres (atmosphere, hydrosphere, lithosphere, biosphere, society and other spheres of the Earth), this mechanism still remains one of the little-studied problems in the field of Cosmo-terrestrial relations. This problem stems from the fact that unexplained fluctuations in the cycles of the climatic and geophysical processes intensity variations on Earth does not correspond to the stable orbital periods of gravitational perturbations in the solar system space. The special interest of the Hlystunov's scientific school to this problem has arisen in connection with the unique result of this school's research in the orbital microgravity field and the theoretical opening in 1992, the radial gravidynamic resonances orbital space objects (Khlystunov, 1994). This discovery helped to explain the cause of gravitational perturbations' instability cycles and the corresponding oscillation periods of the geosphere processes intensity variations. Individual research results of the author's discoveries with colleagues on this topic have been published in a series of articles dedicated to the total influence of orbital and resonant gravitational disturbances in the near-Earth space environment on the global meteorological and seismic processes evolution.

2 RADIAL GRAVIDYNAMIC RESONANCE OF SPACE OBJECTS

After centuries of oblivion, the first step of the cosmogenic gravitational concept's further development has resulted in the opening by the Russian scientists [professor Hlystunov M.S. and professor Nikitskiy V.P.] the phenomena of orbital objects radial gravitational resonance and its impact on the natural disasters and natural catastrophes intensity evolution (Khlystunov, 1994). According to their formula, the excitation of the orbital object radial gravitational resonance on its orbital motion will impose radial resonant oscillations with an amplitude of A_{gr}. This movement for a perfect circular orbit, without considering the effect of the angular momentum conservation law (assuming that the mass of the central object M_g is significantly less than the mass m_s of the satellite, $M_g \gg m_s$, and the satellite radius r_s is significantly less than the radius R_g of the central object, $r_s \ll R_g <_{orb}$), will appear, as shown in Figure 1.

The opening formula of these resonances was first published by the authors in the International Aerospace Congress IAC'94 theses in August 1994 as follows: "Orbital object revolving around the center of gravitational attraction, has a radial gravidynamic resonance whose period is approximately equal to 2/3 of the period of his conversion." In the future, the hypothesis and the radial gravidynamic resonance theory will be verified by the authors of the discovery from data on the orbital flight trajectory height variations of the geodetic satellite «Topex-Poceidon» and satellite «GEO_IK». Along with this, the interesting fact that the radial

gravidynamic resonance period, for example, at the sea level height should be 1 hour ws established. Attempts to find historical sources introduction in practice of such units (1 hour) have been unsuccessful. It should be noted that the satellites' radial vibrations relative to the classical orbit was an imposing classical orbital motion, radial resonance frequency, and this movement phase modulation results from the action of the angular momentum conservation law when changing the orbit radius by these resonant oscillations.

3 THE PARAMETERS OF THE ORBITAL AND RESONANT GRAVITATIONAL PERTURBATIONS

The amplitude evaluation results of the planets and Sun orbital gravitational perturbations energy E_{orb} and the solar system planets' radial gravidynamic resonances mid-frequency calculations (in accordance with the opening formula) are summarized in Table 1.

Considering the most powerful catastrophic Geosphere processes (endogenous and exogenous-cosmogenic), the authors focused their attention on a very important fact. All major endogenous and exogenous processes of heat and mass transfer are realized due to the presence of the gravitational field or the direct action result of its perturbations. A comparison of the cosmogenic gravitational perturbations energy distribution average density in the Earth's geospheres confirms their dominant role compared with the endogenous component

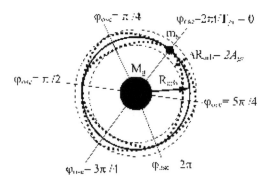

Figure 1. Schematic representation of the resonant oscillations of the satellite (the planet) relative path in the form of the unperturbed circular orbit.

Table 1. The special parameters of the orbital and resonance gravitational perturbations.

Source of disturbance	The parameters of the perturbation*				
	F_{orb}, Hz	T_{orb}, day	F_{gr}, Hz	T_{gr}, day	E_{orb}, J/kg
Moon	4.25 E-7	27.32	6.01 E-7	19.32	–
Mercury	1.32 E-7	87.97	1.87 E-7	62.2	–
Venus	5.165 E-8	224.7	7.3 E-8	158.9	6.1E33
Earth	3.177 E-8	365.2	4.49 E-8	258.3	5.55E33
Mars	1.69 E-8	686.7	2.39 E-8	485.6	–
Jupiter	2.68E-9	4332	3.79E-9	3063	3.3E35
Saturn	1.079E-9	10760	1.53 E-9	7609	5.68E34
Uranium	3.782E-10	30689	5.35 E-10	21700	–
Neptune	1.928 E-10	60187	2.73 E-10	42558	–
Pluto	1.281 E-10	90616	1.81 E-10	64075	–

*F_{orb}, T_{orb}, F_{gr}, T_{gr} and E_{orb} are, respectively, the frequency and the period of the orbital motion, the frequency and the period of the radial gravidynamic resonance, and the specific orbital energy of the Sun's gravitational perturbations.

Table 2. The special parameters of the orbital and resonance gravitational perturbations.

Source of disturbance	The parameters of the perturbation*		
	E_g/E_{sr}, r.u./year	E_g/E_{gt}, r.u./year	Q_g/Q_{sr}, r.u./year
Sun	1.0E+09	8.3E+12	1.81E+07
M	1.85E+05	1.52E+09	1.03E+04
Mercury	2.36E+02	1.96E+06	1.22E+02
Venus	9.02E+03	7.48E+07	7.56E+03
Mars	6.18E+02	5.13E+06	5.21E+02
Jupiter	2.27E+05	1.89E+09	7.38E+04
Saturn	3.27E+04	2.71E+08	6.41E+03
Uranium	2.36E+03	1.96E+07	2.39E+02
Neptune	1.78E+03	1.48E+07	1.09E+02
Pluto	1.31E-02	1.09E+02	6.41E-03

*E_g/E_{sr}, E_g/E_{gt}, Q_g/Q_{sr} are, respectively, the amplitude variation ratio of the geosphere's perturbation gravitational energy to the solar irradiation, the ratio of the geosphere's perturbation gravitational energy amplitude variations to Earth's own energy dissipation, the ratio of the geosphere's gravitational energy density perturbation variations amplitude to the specific energy of the solar radiation (on a 1 kg mass of the Earth during the variation period).

(see Table 2). The gravitational interaction energy density of the Earth with the objects in the solar system and the Moon is several orders of magnitude higher than the solar radiation average energy density and the Earth's own heat.

The calculations in this table are made by taking into account only the eccentricity of the solar system planets around the Sun. In this regard, the authos' particular interest was the planets' resonant vibration influence on the geosphere's processes under the action of gravitational perturbations, for example, caused by the fall of the Shoemaker-Levy comet on Jupiter in July 1994. 'According to preliminary estimates made by the "father" of the hydrogen bomb Edward teller, when the comet collided with Jupiter, the released energy was equivalent to an energy blast of 10 billion megatons of TNT, or a hundred million explosion energy of the Tunguska meteorite ($2{,}77 * 10^{22}$ to $4.61 * 10^{25}$ J). The resulting energy of Jupiter's gravitational perturbations can be much more than the solar irradiation of the Earth for a year. The self-heat generation energy of the Earth for the year will be over five orders of magnitude lower than Jupiter's gravitational perturbations energy. In this regard, it can be assumed that the comet explosion on Jupiter may have a significant impact on the intensity of seismic, volcanic and meteorological processes on Earth. In this article, we will consider the global analysis results of the seismic

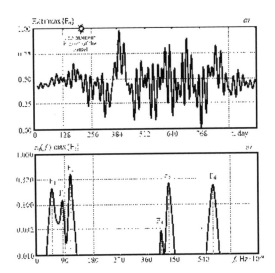

Figure 2. The normalized amplitude and the spectrum of earthquakes daily energy variation in the period from 01.01.1994 to 19.09.1996.

activity evolution on Earth in the period before and after the collision of the Shoemaker-Levy comet on Jupiter. For example, data on global seismic activity on Earth with an array length of 1024 days are used. The original data were represented by the earthquake magnitudes more than 2.5 points on the Richter scale. The array primary analysis showed that the number of such seismic events and their daily energy increased considerably after the fall of the Shoemaker-Levy comet on Jupiter. Figure 2 shows the amplitude (a) and spectrum (b) variations of earthquake daily energy on Earth for a period of 2.8 years. The normalization of daily energy earthquake curve was performed by the maximum value in the array.

4 SPECTRUM ANALYSIS OF SEISMIC ACTIVITY DAILY VARIATIONS ON EARTH

As the array length is only 1024 days, thespectrum is clearly observed only for the gravitational perturbation manifestations of Mercury, Venus, Earth and Mars, as well as their combined harmonics. The background spectrum below 1% of the maximum amplitude in Figure 3b is not shown, since it has no resonance or cyclic manifestations and does not present opportunities for making correct analytical conclusions. A comparative analysis of the spectrum and fundamental frequencies and combined harmonics of the gravitational perturbations in the near-Earth space according to the data from Table 1 reveals the following coincidence frequencies of the dominant resonant manifestations in the spectrum:

- F1 = 5.16 * 10^{-8} equal to the frequency of Venus circulation around the Sun;
- F2 = 8.04 * 10^{-8} equal to the difference in frequency of Venus and Mercury circulations;
- F3 = 1.08 * 10^{-7} equal to the difference in the frequency of Mars' radial gravidynamic resonance and the orbital circulation frequency of Mercury. In the frequency band of this peak, there is also the frequency difference harmonic of radial gravidynamic resonances of Venus and Mercury, the frequency sums harmonic of the Earth circulation and radial gravidynamic resonance of Venus, and the second harmonic of the Venus orbital circulation around the Sun;
- F4 = 3.90 * 10^{-7} equal to the difference in the frequency of radial gravidynamic resonance of the Earth and the Moon circulation around the Earth, but in the frequency band peak of this resonance, there is also the frequency difference harmonic of the Moon circulation around the Earth and the Earth circulation around the Sun;
- F5 = 4.50 * 10^{-7} equal to the frequency sum of Mars' radial gravidynamic resonance and the Moon circulation around the Earth, the speed of the layers of the Sun near the equator;
- F6 = 5.50 * 10^{-7} equal to the synodic period of the Moon circulation around the Earth, the difference in the frequency of the Moon's radial gravidynamic resonance and Venus circulation around the Sun.

5 CONCLUSION

The results of the study on the radial gravidynamic resonances and the solar system planets around the Sun on the evolution of the geophysical processes intensity on Earth allow drawing the following conclusions:

1. A comparative power analysis of endogenous and cosmogenic sources of the geosphere's excitation showed that the Earth's cosmogenic gravitational perturbations power is several orders of magnitude higher than the total capacity of its own heat and Sun exposure of the Earth, and the technosphere power, including the military.
2. The variation analysis of the earthquakes daily energy on the Earth before and after the Shoemaker-Levy comet explosion on Jupiter showed that the number of seismic events on Earth with a magnitude of more than 2.5 points after the comet explosion increased 10 times. This fact confirms the presence of the dominant gravitational influence of the Cosmo-terrestrial relations on the Earth's seismic activity.
3. A reliable intensity evolution prediction of natural disasters and natural catastrophes on Earth cannot be implemented without taking into account the effect of the orbital gravitational perturbations and perturbations caused by the radial gravidynamic resonance excitation of the solar system planets, the Earth and the Moon.

The results of this study can be useful for the natural risk assessment on the long-term development of urbanized territories (Managing the Risks, 2012) in an era of global climate change (Climate Change, 2013) and growth activity of geophysical processes.

REFERENCES

Chizhevskiy A.L. Zemnoe eho solnechnyih bur. Izdatelstvo «Myisl». ISBN 978-5-517-92874-0, 1976. 367 pp.
Climate Change 2013. The Physical Science Basis/Working Group I Contribution to the Fifth Assessment Report of the Intergovernmental Panel on Climate Change. UK: Cambridge University Press, 2013, 1536 pp.
Das Werden der Welten. Ueb. von L. Bamberger L., 1907, p. 50.
First Report of the Commission appointed to further the study of solar and terrestrial relationships. Paris, 1926.
Khlystunov M.S., Nikitskiy V.P. Task of the control of microgravitation level on board of space vehicles and international problem of global catastrophes. Abstracts. International Aerospace Congress. August 16–19, 1994, Moscow, Russia, p. 225
Lehrbuch der kosmischen Physik. Leipzig, 1903.
Managing the Risks of Extreme Events and Disasters to Advance Climate Change Adaptation Special Report of the Intergovernmental Panel on Climate Change. UK: Cambridge University Press, 2012, 582 pp.

Architectural, Energy and Information Engineering – Sung & Chen (Eds)
© 2016 Taylor & Francis Group, London, ISBN 978-1-138-02791-6

Previously unknown regularities of extreme wind load formation

Zh.G. Mogiljuk, M.S. Hlystunov & V.I. Prokopiev
Moscow State University of Civil Engineering, National Research University, Moscow, Russia

ABSTRACT: This article presents previously unknown regularities of extreme wind load formation on buildings and structures. It discusses the hypothesis verification results of the presence of wind gust formation quantum regularities. It also explains the statistical analysis methodology of the wind gust implementation frequency in high definition based on the meteorological observation data from Tokyo. This paper presents graphical materials and calculations made on the basis of official weather information stored in two arrays with a time interval of 5355 days each. The interval between array observations is more than 22 years. In addition, it presents the dependence graph of the wind gust amount from wind speed in Tokyo as well as the diagram of wind gust formation velocity distribution in Tokyo over the peaks' ordinal numbers on the curves of this dependence.

Keywords: buildings and structures; wind loads; wind gusts; formation; quantum regularities

1 INTRODUCTION

According to the latest report of the UN Inter-governmental Panel on Climate Change [IPCC], almost all the nations of the world are not prepared for precautionary measures on vitally important sectors of economy and population adaptation (Managing the Risks, 2012) to global warming (Climate Change, 2013). The lack of progress and a common position for the world's leading scientific schools on the global climate change issue poses a particularly difficult situation for the construction industry in all states in practice, including countries with highly developed economies. The relationship of this problem with the construction activity and its acuteness follows from the direct dependence of design decisions from the global climate change risk assessment on the urban planning depth (100 years and more). Construction activity in the world, connected with many trillions of dollars of long-term investments cannot stop because of differences in meteorological science schools. This is due to the daily necessity of such design solution selection that will provide the necessary security and stability of building structures to all forms of climatic and meteorological loads and impacts for a long period in the life cycle of industrial and civil construction objects. Among such loads, special place has been given to wind load, which is characterized as average daily wind speed and the maximum speed of the shock wind effects on building structures. In our opinion, mainly, the period of

statistical modeling in theoretical meteorology was delayed. Undoubtedly, the use of modern digital technologies and supercomputers has significantly increased the level, volume and efficiency of full-scale meteorological information processing. However, this was insufficient for long-term prediction of the evolution of climatic and meteorological processes in the life cycle of object construction up to 100 years and more. In this regard, the authors conducted a comprehensive study on poorly known regularities of the dangerous natural processes intensity evolution and changes, including climatic and geophysical characteristics. In addition, the authors' attention was drawn by the results of unique research, such as Fundamental nonlinearity weather theory by Edward Lorenz and arising from his theory risks of causing hurricanes remote local aerodynamic micro-processes (Hilborn, 2004). The research results on meteorological risks were published in a series of articles by Hlystunov's scientific school (Hlystunov, 2013; Hlystunov, 2014; Hlystunov, 2015). This paper presents the fundamental research results of the micro-processes role on the gale-force wind gust formation. In our view, there is a new stage in the theoretical meteorology development, based on the quantum laws and phenomena along with the classic. On the one hand, this article is fundamental; on the other, new knowledge about the quantum character of shock wind load formation opens up new opportunities for the protection methods development of buildings and constructions from such loads.

2 FORMULATION OF THE HYPOTHESIS

According to the fundamental law of quantum mechanics to change the parameters of micro-particle motion (in this case, the molecules of the atmosphere), it is necessary to set the angular momentum equal to Planck's constant $\hbar = 1.05457E-34$ J*s. Then, for a wind gust excitation of each air molecule, the parameter concerned in the wind gust formation, it is necessary to additionally impart angular momentum $h_m = \Delta H_m$, equal to Planck's constant, that is,

$$h_m = \Delta H_m = m_m \Delta V_m r_{mm} = \hbar \tag{1}$$

where $m_m, \Delta V_m, r_{mm}$, respectively, are the mass of the molecule the molecule's increment speed, necessary for the next wind gust formation, and the distance between the molecules.

In addition, growth velocity and angular momentum wind gust should also be distinguished from impulse at a slower rate by Planck's constant value.

3 THE DISCOVERY THEORETICAL EVIDENCE OF THE WIND GUST FORMATION QUANTUM LAWS

According to the hypothesis formulation, the statistical-mechanical distribution of wind velocity for dry air must be "comb-like" in nature, that is,

$$nh_m = n\Delta H_m = nm_m \Delta V_m r_{mm} = n\hbar \tag{2}$$

or

$$n\Delta V_m = \frac{n\hbar}{m_m r_{mm}}, \tag{3}$$

where n is the burst or crest number of the statistical frequency dependence graph of wind speed implementation.

It should also be noted that the number of molecules per atmospheric unit volume in the general case depends on the pressure increment in the wind gust, temperature, humidity and aerosol concentrations. Furthermore, wind gust formation in the air increases the solid and liquid aerosol concentrations (mineral and organic dust, fog and precipitation in the form of liquid water and ice crystals). For example, with regard to aerosol impurities, the formula is given by

$$n\Delta V_m + n\Delta V_{dust} = \frac{n\hbar}{m_m r_{mm}} + \frac{n\hbar}{m_{dust} r_{dm}}, \tag{4}$$

where $\Delta V_{dust} \approx \Delta V_m, m_d, r_{dm}$, respectively, are the speed treatment of aerosols captured by the wind gusts, the average mass of particles in aerosols and precipitation, and the distance between the particles and the air molecules.

In summary, the hypothesis is a basic tenet that considers the meteorological observation data analysis methods on the example of a quantum regularities theoretical statistical model of wind gust formation, for example, for the atmosphere volume unit.

4 STATISTICAL ANALYSIS METHODOLOGY OF WIND GUST FORMATION REGULARITIES

The sole criterion of the truth-value of the hypothesis, in this case, is the statistical distribution ridge nature presence of velocity pores wind gust, obeying the quantum mechanics laws, that corresponds to the results of the observations on a variety of the planet's geographical locations. The hypothesis is proved by statistical methodology based on the assumption that the wind speed increment is sufficient to generate a wind gust, for example, in a dry air volume unit necessary for the increment of angular momentum:

$$\bar{K} = \Delta V_s \rho_o r_{mmo}. \tag{5}$$

As the number of molecules per air unit, for example, under normal conditions is equal to N, then for each molecule, it is necessary to give a momentum:

$$k = \frac{\bar{K}}{N}. \tag{6}$$

The influence of the preliminary assessment of humidity, temperature and aerosol concentrations, including dust, showed that the maximum total degree of their influence on the value of k does not exceed 10%. In this regard, we will use the derived expressions for the evidence of quantum hypothesis. For this purpose and to justify the fundamental findings, the authors conducted an analysis of real meteorological data on velocities of wind gusts formation in the cities on different continents. These materials are published by the authors in a special series of articles and presentations at international conferences. The individual study results for wind gust formation quantum mechanics laws have been previously published by the authors (Hlystunov, 2013; Hlystunov, 2014; Hlystunov, 2015). This paper considers the generalized detailed results of similar studies in Tokyo.

5 THE QUANTUM LAWS VERIFICATION OF THE WIND GUST FORMATION IN TOKYO

The underlying data array uses the meteorological observation data from Tokyo for the period from 01.01.1973 to 31.08.1987 (5355 days or 14.66 years), and the array of data about the current state of Research Institute of Meteorological Processes data on analogical meteorological observations for the period from 01.01.1995 to 31.08.2009 (5355 days or 14.66 years). The time interval between the arrays is 22 years. The wide spacing between the observation arrays allows to eliminate the possibility of accidental correlation manifestations and mutual influence of these arrays at each other. Figure 1 shows the dependence graph of the wind gust amount $s(V)$ in Tokyo from wind speed (m/s) for the period from 01.01.1973 to 31.08.1987 and from 01.01.1995 to 31.08.2009 with resolution $dV = 0.5 \ m/s$.

The curves, as shown in Figure 1, for both periods of observation are comb-like in nature. However, watches by significant expansion of the peaks of the curves may affect the accuracy of subsequent calculations. In this regard, the authors have made a more detailed analysis of the data with a resolution $dV = 0.1 \ m/s$. The result reveals the lined nature of the wind velocity statistical distribution for the observation period from 01.01.1973 to 31.08.1987, as shown in Figure 2, and for the observation

period from 01.01.1995 to 31.08.2009, as shown in Figure 3. The analysis results of the wind velocity statistical distribution nature for both observation periods allow to establish the wind gust formation speed dependence $V(n)$ on ordinal numbers n of the peaks of the curves in Figure 2 and Figure 3. The number of the peaks in this case was chosen conditionally, i.e. starting with the most sharp peak expressed by the curves in Figure 2 and Figure 3.

In accordance with Figure 4, the velocity increment between the wind gust formation peaks is (on average)

$$\Delta V = \frac{V_{24} - V_1}{23} = \frac{49.8 - 7.8}{23} \ m/s = 1.826 \ m/s. \quad (7)$$

We then calculate the angular momentum required for the wind gust formation as in an air volume unit and for one molecule:

$$\bar{K} = \Delta V_s \rho_o r_{mmo} = 1.826 \frac{m}{s} \times 1.228 \frac{kg}{m^3} \times 1.025 \times 10^{-9} m$$
$$= 2.298 \times 10^{-9} J \times s, \quad (8)$$

where the average angular momentum for one molecule will be

$$\bar{K} = \frac{\bar{K}}{N} = \frac{2.298 \times 10^{-9}}{2.07689 \times 10^{25}} N \times s \times m$$
$$= 1.107 \times 10^{-34} J \times s. \quad (9)$$

Thus, the resulting angular momentum exceeds Planck's constant value, which is less than 4.7%:

$$\frac{\Delta \bar{k}}{k} = 100\% = \frac{\bar{k} - \hbar}{\bar{k}} \times 100\%$$
$$= \frac{1.107 \times 10^{-34} - 1.054571726 \times 10^{-34}}{1.107 \times 10^{-34}} \times 100\%$$
$$= +4.7\% \quad (10)$$

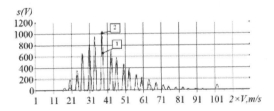

$s(V)$

Figure 1. The dependence graph of the wind gust amount from wind speed in Tokyo (with resolution $dV = 0.5 \ m/s$): curve (1) represnts the period from 01.01.1973 to 31.08.1987; curve (2) represents the period from 01.01.1995 to 31.08.2009.

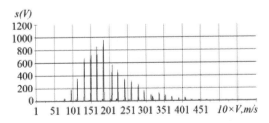

$s(V)$

Figure 2. The dependence graph of the wind gust amount in Tokyo from wind speed for the observation period from 01.01.1973 to 31.08.1987 (resolution $dV = 0.1 \ m/s$).

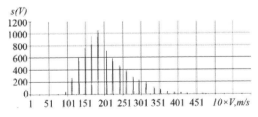

$s(V)$

Figure 3. The dependence graph of the wind gust amount in Tokyo from wind speed (m/s) for the observation period from 01.01.1995 to 31.08.2009 (resolution $dV = 0.1 \ m/s$). The exact values of peak velocity (m/s) are indicated above the relevant column of the diagram in Figure 4.

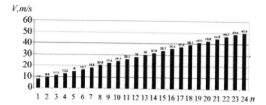

Figure 4. Diagram of wind gust formation velocity distribution $V(n)$ in Tokyo over ordinal numbers n of the peaks of the curves in Figure 2 and Figure 3.

This greater value over Planck's constant is understandable, which is the variation result of temperature, humidity and aerosol concentrations. Also for Tokyo, it is characterized by fog and precipitation in the liquid form. Only the temperature change within $\pm 10^{\circ}C$ may lead to changes in the density more than 11%.

6 CONCLUSION

The rising problem of risks related to building construction accidents, both in Russia and abroad, greatly exacerbated not earlier forecast of aerodynamic processes fluctuations power amplitude growth. A special place among the phenomena is associated with aerodynamic processes in high-rise buildings encountered by sudden gusts of wind, storms, hurricanes and tornadoes. The generalized statistical analysis results of aerodynamic manifestations of global climate change show that the greatest threat to various human activity spheres, including the building industry, is not only global warming, but also the extreme states risks of wind processes. The wind processes extreme states realization risk in urban areas, both stagnant and hurricane, also poses a threat to forestry and parks, marine and air transport. Preserving this trend for the current period of urban planning (for example, 100 years) is a very real risk of the increasing intensity of these fluctuations. This is quite probable that development of the global climate change process cannot be ignored by investors, self-regulatory organizations, owners of construction objects control and energy companies, insurers and, importantly, municipal, regional and Federal

services regulation and urban development planning, as well as aviation and marine transport companies. This problem in both Russia and abroad greatly exacerbated not earlier forecast of the intensity growth and non-dangerous man-made and natural climatic and geological-geophysical processes and factors, which are responsible for the implementation of new and, consequently, irregular comprehensive excess loads and impacts on the technosphere objects.

Fundamental quantum laws of the squall wind formation in Tokyo are global in nature and were also verified by the authors in other cities across the globe, including London, Moscow, Niamey and Anchorage.

REFERENCES

Climate Change 2013. The Physical Science Basis/Working Group I Contribution to the Fifth Assessment Report of the Intergovernmental Panel on Climate Change. UK: Cambridge University Press, 2013, 1536 pp.

Hilborn, Robert C. Sea gulls, butterflies, and grasshoppers: A brief history of the butterfly effect in nonlinear dynamics./American Journal of Physics 72 (4): pp. 425–427.// DOI: 10.1119/1.1636492. Bibcode: 2004 Am J Ph.72.425H.

Hlystunov M.S., Prokopjev V.I. and Mogiljuk Zh.G. Quantum Regularities of Shock Wind Processes Formation. World Applied Sciences Journal, ISSN/E-ISSN: 1818-4952/1991-6426, 2013, No 26(9) pp. 1219–1223.

Hlystunov M.S., Prokopiev V.I., Mogiljuk Z.G. Verification of the squall wind formation quantum laws in Anchorage. Life Sci. J. 2014; 11(12s): pp. 767–771, (ISSN: 1097-8135). http://www.lifesciencesite.com. 165.

Hlystunov M.S., Prokopiev V.I. & Mogiljuk Zh.G. Verification of the Squall Wind Formation Quantum Laws in New York. Modern Applied Science; Vol.9, No.1; 2015, pp. 96–102, ISSN 1913-1844 E-ISSN 1913-1852, Published by Canadian Center of Science and Education. Received: September 30, 2014 Accepted: October 6,2014 Online Published: December xx, 2014, doi: 10.5539/mas.v9n1pxx, URL: http://dx.doi.org/10.5539/mas.v9n1pxx.

Managing the Risks of Extreme Events and Disasters to Advance Climate Change Adaptation Special Report of the Intergovernmental Panel on Climate Change. UK: Cambridge University Press, 2012, 582 pp.

Architectural, Energy and Information Engineering – Sung & Chen (Eds)
© 2016 Taylor & Francis Group, London, ISBN 978-1-138-02791-6

A QoS collecting model based on an automated reporting method

J. Li & G.Sh. Wu
School of Software Engineering, Beijing University of Posts and Telecommunications, Beijing, China

X.Y. Zou
Agricultural Bank of China Limited, Beijing, China

ABSTRACT: With the widespread deploying and applying of web services, service users not only want to satisfy their functional requirements, but also pay attention to the needs of non-functional requirements, such as Quality-of-Service (QoS). Researchers have intensively explored many QoS-related problems, such as QoS attribute model, QoS evaluation model, QoS-based service composition. Although existing approaches emphasize on how to custom the QoS attribute model, evaluate the QoS of web services and make use of QoS information, the technologies on QoS information collecting are ignored. These fundamental technologies are also important for QoS management, evaluation and application. From the perspective of these problems, the paper proposes a QoS collecting model based on an automated reporting method, analyzes the shortcoming of the existing QoS information collecting approaches, and introduces the technologies used in this model. Finally, this paper demonstrates the feasibility of this model.

Keywords: QoS attributes; Web Service; Management Model

1   INTRODUCTION

Spurred on by the recent application environment and the new business requirement, SOA, which is based on web services, has developed rapidly in different fields including enterprise application integration field and distribution computing field. The combinational web services, which meet complex application requirements of different customers by composing multiple simple services, are inevitable demands of SOA development. Since the amount of web service has increased and thus many can satisfy the functional requirements, thus service users pay more attention to the quality-of-service (QoS) of web services. However, due to the dynamic and unpredictable characteristics of the web services, it is not an easy task to provide the desired QoS for web service users [1].

At present, the research on the QoS of web services focuses on the following aspects: QoS-based web service information collection model research (Yueyi Luo et al. 2013)[2], QoS-based web service choice and composition method research (S. Wang et al. 2012[3], Sachan, D. et al. 2013)[4] and QoS-based web service recommendation research (Liang Chen et al. 2014[5], Z. Zheng et al. 2011)[6]. From these aspects of research, we find that QoS-based web service information collection, as an infrastructure work, determines the accuracy of web service of the QoS evaluation and the availa-

bility of QoS-based applications; therefore, a good web service QoS information collecting model is necessary, in order to achieve high availability. This model must meet following requirements: (1) it should adapt to the change of the QoS attributes; (2) the information collecting work should not interfere with the work of the web service and should not take up too much network resources.

Many studies have been conducted on QoS information collecting. For instance, Sun Su-yun[7] presented two methods to collect the QoS information: one is voluntary monitoring of the web service; the other is processing of customer feedback information. All of the QoS information collecting work is fulfilled by the QoS certification center, so the QoS information's objectivity can be guaranteed, but the real-time property is far from satisfactory. Zheng Yi [8] presented a QoS information collecting and measurement mechanism method based on the API Hook technology. This method is able to intercept callings to Sockets APIs made by web service applications when SOAP messages are being transferred back and forth through a network, and will record QoS-related information about the observed operation of a web service. This method can collect information about different QoS properties from both the client and server sides objectively, automatically and light weightly, but the QoS information's transmission frequencies in the network may be very high, so it takes up too much network resources. In addition, all of these methods cannot be applied to the changing QoS attribute circumstances.

In accordance with these problems, we propose corresponding solutions in this paper, and based on these solutions, a QoS collecting model based on an automated reporting method is presented. It realizes that QoS information can be collected even when the QoS attributes change dynamically, and can ensure the QoS information's objectivity and real-time property without interfering with the network environment and the web service work.

2 SOLUTIONS TO THE PROBLEMS

The QoS information collection work process mainly involves the following steps:

1. Collecting the original QoS information.
   The original QoS information cannot be used directly. In this step, first, we must identify QoS information collecting locations, and take the original information collection of the web services' response time for example, the location of information collection should be the web services' deployment server. In this case, we can monitor the request and response message free from outside distraction, and the objectivity of original QoS information can be assured. Second, the methodologies of original QoS information collection should be established, and then the original information collecting model can be realized based on those methods. The locations and methodologies of QoS information collection vary with different QoS attributes and web services.
2. Measuring the QoS information.
   The QoS information is calculated by processing the original QoS information, and different QoS attributes have different measurement methods. During the measurement of QoS information, we need to analyze and process the original information, and then calculate the value of the QoS attribute. As we know, inputs of measuring the QoS information model are the original information, which is provided by the QoS original information collection model, so the format of inputs is determined by the corresponding QoS attribute's definition, and outputs are used by QoS evaluation models, in order to ensure the comparability of the QoS attributes' value, we should use the same measurement methods on the QoS attributes of different web services, so for every QoS attributes, the corresponding measurement methods are defined. In other words, a QoS information module can process the original information of a QoS attribute.

3. Sending the QoS information to the QoS information management center.
   The QoS information management center should define APIs to publicize the QoS information, so that clients can send the SOAP request message with QoS information.
4. Storing the QoS information to the database.
   By analyzing the QoS information collection process, we can find that collecting the original QoS information and measuring the QoS information are related to the QoS attributes' definition. So, if we want to realize a QoS information collection model that can adapt to the change of QoS attributes, the original QoS information collection model and the QoS information measuring model must be fully extensible. In addition, the third step involving the control logic of sending the QoS information can be added to decide that about sending the QoS information. As a result, we can find the solutions to the problems of the existing QoS information collection model.

First, the QoS information collection system adopts the C/S model and creates the QoS information management center in the server, and the QoS information management center provides inquiry APIs and publication APIs. The clients of the system can be deployed in anywhere on the Internet and exchange information by calling the APIs. The exchange between the server and the client is based on SOAP, which is the standard communication protocol of SOA. So, the QoS information collection system can be integrated in systems based on SOA.

Second, the procedure module of the original QoS information collection can be implemented based on thread technology, and the information collection thread runs automatically and collects the QoS information continuously when we start it. In this way, the real-time property of QoS information can be ensured.

Third, the original QoS information collection model can be designed as a container. It contains zero or more original QoS information collection sub-modules, and these sub-modules can be added and removed dynamically. The container maintains the life circle of these sub-modules. Moreover, the information of these module invocations, which is expressed using XML, is stored in configuration files. The container manages these modules based on the configuration files. Similarly, the QoS information measurement model can also be designed as a container that can load the sub-modules dynamically and manage them by using configuration files. Using reflection mechanisms, the original QoS information model invokes the QoS information measurement model easily. Thus, the original QoS information collection model and the QoS information measurement model can well adapt to the change of QoS attributes.

Fourth, during a certain time, QoS information may be constant or change according to certain rules. So, we can judge whether the QoS information changes before the QoS information sends to the server, and then determine whether to send this QoS information. The variations judgment module can add to the QoS information sending model. The details about the changing rules are stored in the configuration files, and the variations judgment module judges the variation of QoS information based on the rules recorded in the configuration files. Different QoS attributes have different changing rules, thus the changing rules express the mathematical relationships between the QoS attributes' variation quantity Δv and the collection time's variation quantity Δt. For example, if the value of QoS attributes is constant, the changing rules can be expressed as: $\Delta v = 0 * \Delta t$.

3 A QoS INFORMATION COLLECTION MODEL DESIGN

Based on the solutions to the problems of the existing QoS information collection model, an improved QoS information collection model is proposed in this paper. The model architecture is shown in Figure 1.

In this model, there are three configuration files, and they express information using the standard XML: (1) collection.xml, a configuration file where the information of the original QoS collection modules' invocation is recorded; (2) measure.xml, a file where the QoS information measurement rules are recorded. (3) variation.xml: the QoS attributes' changing rules are recorded in this file.

The descriptions of this information collection model process are as follows.

First, the collection services management module constantly monitors the change of collection.xml. If a new original QoS information collection module is added to the collection.xml, the management module starts this collection module automatically. When the collection module completes the collection of the original QoS information, this module reads the QoS information measurement rules from the measure.xml by the attribute's identifier which is recorded in the collection.xml and calls the corresponding QoS information measurement module, and the QoS information can be obtained after processing, and then the QoS information is sent to the variations judgment module. The variations judgment module finds the changing rule by reading the variation.xml and judges the variation of the QoS information according to the changing rules. If the QoS information has changed, the variations judgment module will call the communication module to send this QoS information to the QoS information management center by calling publication APIs. Then, the QoS evaluation model can get the QoS information by calling inquiry APIs.

This QoS information collection model reports the QoS information automatically according to the variation of the QoS information, so it called the QoS information collection model based on an automated reporting method.

4 AN APPLICATION CASE

Let us consider the response time attribute, for example. The original QoS information is the time when the web service receives a SOAP request message and the web service sends a corresponding SOAP response message, and the QoS information can be calculated by subtracting the response time by the request time. So, we can implement a QoS information collection module to collect the original information by monitoring the SOAP message, and a QoS information measurement module to measure the original QoS information and add these modules to the QoS information collection system.

Next, we should modify the configuration files. In the collection.xml, the following content is added:

<collections>
...

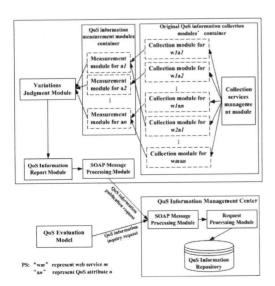

Figure 1. A QoS collecting model based on an automated reporting method.

```
<module wsid="ws-01" attributeId="attribute-01">
<class>cn.edu.bupt.collection.original.
ResponseTimeCollection</class>
</module>
</collections>
```

And in measure.xml, the new content is:

```
< measurements>
…
<rule attributeId="attribute-01">
<class> cn.edu.bupt.collection.measure.
ResponseTimeMeasure </class>
<function>measure</function>
<params>
<param name="startTime" type="date" />
<param name="endTime" type="date" />
</params>
</rule>
</measurements>
```

Similarly, the new changing rule is added to variation.xml:

```
<variations>
…
<rule attributeId="attribute-01">
<metric> Δv=0*Δt</metric>
</rule>
</variations>
```

Subsequently, we collect the response time information automatically.

5 CONCLUSION

This paper analyzes the problems and shortcomings of the existing QoS information collection approaches, and proposes solutions to those problems and shortcomings. Then a QoS information collection model based on an automated reporting method is presented to implement these solutions. Finally, this paper gives an application case to show the validity and feasibility of the model.

REFERENCES

[1] KangChan Lee, JongHong Jeon. QoS for Web Services: Requirements and Possible Approaches, http://www.w3c.or.kr/kr-office/TR/2003/ws-qos/.
[2] Yueyi Luo, Jun Long, Jingjiao Wen. QoS collection for web services based on WS-monitor model. *Proceedings of the 5th Asia-Pacific Symposium on Internetware*, 2013.
[3] S. Wang, Q. Sun, and F. Yang, Quality of service measure approach of web service for service selection, *IET Software*, 2012, pp. 148–154.
[4] Sachan, D., Dixit, S. K., Kumar, S. A system for Web Service selection based on QoS, Information Systems and Computer Networks (ISCON), 2013 International Conference, pp. 139–144.
[5] Liang Chen, Jian Wu, Hengyi Jian, Hongbo Deng, Zhaohui Wu. Instant Recommendation for Web Services Composition. Services Computing, *IEEE Transactions*, 2014(7), pp. 586–598.
[6] Z. Zheng, H. Ma, M. R. Lyu, and I. King, QoS-Aware Web Service Recommendation by Collaborative Filtering, *IEEE T. Services Computing*, 2011, pp. 140–152.
[7] SUN Su-yun. QoS Model of Web Service Based on Extending UDDI. *Computer engineering*, 2008(34), 14.
[8] Zheng, Y. Research on Web service QoS model and measurement, Master thesis, Shanghai: Fudan University, 2005.

Architectural, Energy and Information Engineering – Sung & Chen (Eds)
© 2016 Taylor & Francis Group, London, ISBN 978-1-138-02791-6

Experimental investigation of combustion characteristic of a hybrid hydrogen–gasoline engine under the idle driving condition

S.R. Jhang, Y.C. Lin, K.S. Chen, P.M. Yang & S.C. Chen
Institute of Environmental Engineering, National Sun Yat-Sen University, Kaohsiung, Taiwan

Y.C. Lin
College of Pharmacy, Kaohsiung Medical University, Kaohsiung, Taiwan

K.C. Lin
Mechanical and Electromechanical Engineering, National Sun Yat-Sen University, Kaohsiung, Taiwan

C.B. Chen
Heavy Duty Diesel Engine Emission Group, Refining and Manufacturing Research Center, CPC Corporation, Chia-Yi, Taiwan

ABSTRACT: As the global fossil fuel depletion and the petroleum price are increasing, developing new alternative fuels has become a pressing issue. Improving combustion performance and gas emit characteristics of the Spark-Ignited (SI) engines has attracted increasing attention during the past few years. As a result, this research investigates a new strategy of alternative fuel for the Spark-Ignited (SI) engines, using the original gasoline fuel and hydrogen–enriched gasoline engines, with the pure hydrogen at the idle condition to produce rarely emissions. In this paper, hydrogen at flow rates of 0, 4 and 8 L/min was produced by a water electrolysis system and pumped into the intake manifold. The addition of hydrogen at the idle condition effectively reduced the CO_2, CO, HC and NOx emissions. The experiment also showed that the consumption of fuel reduced 23.3~33.4% at the idle condition with various H_2 flow rate inputs. Consequently, this study partially solved the CO-NOx trade-off problem, especially the petro-energy could be saved at the idle operation.

Keywords: Spark-Ignited (SI) engine; Hydrogen/Gasoline; Traditional Pollutant Emission; Fuel Consumption

1 INTRODUCTION

Concerning the fossil fuel depletion and increasing global price, developing new alternative fuels has become a pressing issue. Combustion performance improvement and gas emit characteristics of the Spark-Ignited (SI) engines have attracted enormous attention during the past few years. More researchers hunger for more sustainable energies for vehicles to both effectively relax the fossil fuel demands and greenhouse gas emissions. Hydrogen (H_2) is one of the "cleaner" fuels, which has high energy density and no carbon emission, and can also save energy due to improved combustion efficiency (Wang et al., 2012). Additionally, hydrogen has not only the advantage of lower toxic emits and high efficiency for these applications in the engine, but also long-time availability, which is recyclable and renewable, provides improved air quality and is a potential clean fuel, with a wide range of flammability limits, high mass and thermal diffusivity, low minimum ignition energy, and high stoichiometric air-to-fuel ratio. Therefore, the addition of hydrogen can effectively improve the thermal efficiency of the engine (Ma et al., 2008). The study by Ji et al (2012) examined the hydrogen–gasoline blends found that the addition of hydrogen could decrease HC, CO and NOx emissions at the idle and stoichiometric conditions. At the lean condition, compared with gasoline, hydrogen has higher flame speeds, which could lower ignition delay and increase the thermal efficiency of the engine. The other advantage of hydrogen is a wider flammability, which enables the hydrogen-fueled engine to run smoothly (Kahraman et al., 2007; Ma et al., 2011; Ji et al 2013). With respect to engine performance, the results of the study by Ganesh et al (2008) showed that the addition of hydrogen decreased the power output by about 20%. The traditional emissions of hydrogen carbon and carbon monoxide were negligible. On the contrary, nitrogen oxide emissions from the hydrogen engine were about four times higher than those from the gasoline engine. Lucas et al (1982) ran an SI engine with the pure hydrogen gasoline blends at the idle condition, and found that thethermal efficiency of the engine was improved by at least 10% after using this strategy with increasing load. The results of Ji et al (2012) indicated that the addition of hydrogen reduced fuel energy consumption. Although many studies have investigating the pollutant emissions characteristics, fueled by the hydrogen–gasoline blends, investigations on the passenger car under various testing conditions have not yet been reported. Therefore, this paper added three different hydrogen flow rates (0, 4 and 8 L/min), and then purged into the cylinder within the intake stroke, and investigated the effects of the addition of hydrogen on traditional pollutants, such as carbon monoxide (CO), Total HydroCarbon (THC), and nitrogen oxides (NOx), and fuel consumption monitored by Non-Dispersive Infrared (NDIR) and Electro-Chemical (ELC) sensor, respectively.

2 EXPERIMENTAL SETUP AND PROCEDURE

2.1 Experimental setup

A schematic diagram of the experimental setup is shown in Figure 1. The experiments are performed on a 1.997 L gasoline engine manufactured by France Peugeot 406. The detailed specifications of the test engine are described in Table 1. The mixture of H_2 was generated by electrolyzing

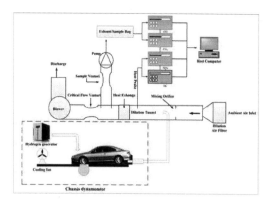

Figure 1.   Experimental setup.

Table 1.   Engine specifications.

Engine specifications	
Engine type	EW10 JF
Cylinders/valves	4/16
Bore (mm)	85
Stroke (mm)	88
Displacement (cm³)	1997
Maximum torque (Nm)	19.8 kg-m
Maximum power (kW)	137 hp/6000 rpm
Compression ratio	10.8:1

Table 2.   Properties of hydrogen (Senthil et al., 2003).

Properties	Hydrogen
Density (kg/m³)	0.082
Calorific value (MJ/kg)	119.8
Flame velocity (m/s)	2.70
Auto ignition temp. (°C)	585
Carbon residue (%)	0.0

Table 3.   Properties of the fuel.

Properties	E3
Density (kg/m³)	0.7458
Octane value	95.6
Vapor pressure (at 37.8 °C, kPa)	56.4
Lead content (g/L)	<0.0025
Final boiling point	210.9
Net heating value (cal/g)	10196
Carbon content (wt.%)	89.45
Hydrogen content (wt.%)	10.34

water using a hydrogen generator machine, GOC HGMS-1000. The H_2 gas was then directly pumped into the combustion chamber of the test engine as a supplemented fuel. A gas flow meter was used to measure the flow rate of H_2. The flame arrestor was installed in the H_2 line to suppress explosions before the mixture was transported to the engine via the air inlet manifold. The properties of hydrogen are given in Table 2. An unleaded blend with 3% ethanol (E3) was tested in this study, whose properties are given in Table 3.

The equipment of Horiba MEXA-7400 LE was used to analyze the engine emissions and applied to detect traditional pollutants, such as carbon monoxide (CO), HydroCarbon (HC), and Nitrogen Oxides (NOx), which were monitored by Non-Dispersive Infrared (NDIR) and Electro-Chemical (ELC) sensor, respectively. The emissions analyzer measures the instantaneous emissions through a direct sampling line from the constant volume sampling system (CVS-7200s) during the test and gas concentration measurement in the CVS bag.

3 RESULTS AND DISCUSSION

In this study, the exhaust gas and fuel consumption of the gasoline engine are studied using H_2 fuel blends in the engine, which was operated at the idle condition. Hydrogen was added at three different flow rates: 0, 4 and 8 L/min, respectively.

3.1 Carbon monoxide emissions

Emission factors of carbon monoxide for the test of the passenger car using the gasoline and hydrogen mixture are shown in Figure 2. We can find that hydrogen added is also effective in reducing carbon monoxide. At the idle condition, the carbon monoxide emissions dropped from 5.20 g/km to 1.09 g/km and from 5.20 g/km to 0.01 g/km with 4 L/min and 8 L/min of hydrogen, respectively. The average decreases in CO emissions were79.1% and 96.8% in comparison with the original gasoline engine. The addition of hydrogen to gasoline fuel increases the H/C ratio, reduces combustion duration and increases the diffusivity of hydrogen, compared with other fuels, providing a homogeneous combustion mixture and producing more oxygen to enhance the combustion (Bari et al., 2010; Jia et al., 2005; Lin et al., 2011). The results of specific CO emissions in this study are consistent with those of Jia et al (2005).

3.2 Hydrocarbon emissions

At the idle condition, the variation of hydrocarbon emissions factors with the flow rate of H_2 is shown in Figure 3. Hydrogen carbon emissions from the hybrid hydrogen–gasoline engine are obviously lower than those of the original gasoline engine. The hydrocarbon emissions dropped from 1.12 g/km to 0.18 g/km and from 1.12 g/km to

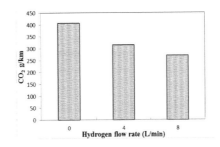

Figure 3.   HC emission factor with H_2 flow rate.

0.09 g/km with 4 L/min and 8 L/min of hydrogen, respectively. The average decreases in hydrocarbon emissions were 83.9% and 92.1%, respectively. The emission of hydrocarbon was probably related to the amount of hydrogen and some crucial radicals (OH and H) with the addition of hydrogen at 4 and 8 L/min. In addition, the addition of hydrogen prevents the carbon atom entry to the engine, and the wide flammability of hydrogen enables the hydrogen–gasoline blends to be burnt more completely at lean conditions (Lin et al., 2014). The results of specific HC emissions in this study are consistent with those of Ji et al (2012).

3.3 Oxides of nitrogen emissions

The effect of a small amount hydrogen added on the specific NOx emissions at the idle condition are shown in Figure 4. When the fuel is burned in a hydrogen gasoline engine, a significant reduction is observed in the specific NOx emissions, but slightly increased at the hydrogen flow rate of 8 L/min. The NOx emissions factor dropped from 0.17 g/km to 0.11 g/km and dropped from 0.17 g/km to 0.13 g/km with 4 L/min and 8 L/min of hydrogen, respectively. The average decreases in NOx emission at different hydrogen flow rates are 37.6% and 25.3%, respectively. The addition of hydrogen–gasoline engine tends to exhaust less NOx emissions than those of the original gasoline engine at the idle condition. The NOx emissions for both the hydrogen and gasoline mixtures with gasoline engines are related to combustion temperature. This is because the addition of hydrogen reduces the combustion duration and decreases the cylinder temperature. The increase in thethermal efficiency of the engine is due to less fuel consumption than the amount of the gasoline engine. Thus, the addition of hydrogen in the gasoline engine not only causes the lower cylinder temperature than that of the original engine, but also reduces the fuel energy flow rate, which helps reduce the NOx emissions at the idle condition. The results for the NOx emission factor in this study are consistent with the those of Ma et al. (2011).

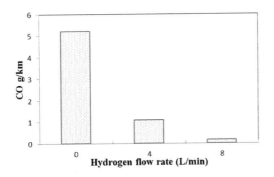

Figure 2.   CO emission factor with H_2 flow rate.

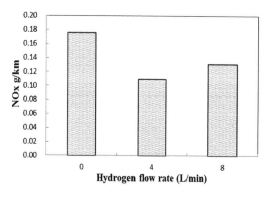

Figure 4. NOx emission factor with H_2 flow rate.

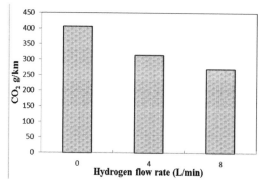

Figure 5. CO_2 emission factor with H_2 flow rate.

3.4 Carbon dioxide emissions

The effect of hydrogen addition on the specific CO_2 emissions at the idle condition is shown in Figure 5. When the fuel is burned in a hydrogen gasoline fuel engine, a significant reduction is observed in the specific CO_2 emissions. With the addition of hydrogen at flow rates of 0, 4 and 8 L/min, the emission factors of carbon dioxide decrease from 407.1 g/km to 315.0 g/km and 271.5 g/km, respectively (original gasoline fuel). The average decreases in the CO_2 emission at different amounts of hydrogen are 22.7% and 33.3%, respectively. At the idle condition, the addition of hydrogen results in a rich mixture, due to the formation of CO, and the low cylinder temperature reduces the oxidation of CO into CO_2. Additionally, the reduction in specific CO_2 and CO emissions is due to the carbon-free property of the hydrogen fuel, as it is substituted by gasoline and thus the mixture produces less carbon dioxide emissions.

3.5 Fuel consumption

The effect of fuel consumption at the different flow rates of hydrogen is shown in Figure 6. The decrease in fuel consumption is observed with the increase in the hydrogen pumped into the cylinder. At the hydrogen flow rates of 4 and 8 L/min, fuel consumption decreases from the baseline value of 0.16 L/km (original gasoline fuel) to 0.12 L/km and 0.11 L/km, respectively. The average decreases in fuel consumption at the different amount additions of hydrogen are 23.3% and 33.4%, respectively. The effect of hydrogen–gasoline mixtures occur with the wide flammability of combustion than that of the original gasoline fuel. Thus, during the addition of hydrogen at the flow rate of 8 L/min, the burning speed of the fuel is the highest and the combustion amounts of the fuel are much lower during the idle condition.

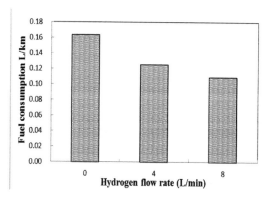

Figure 6. Fuel consumption with H_2 flow rate.

4 CONCLUSION

The H_2 was produced by a water electrolysis system and pumped into the intake manifold with the flow rate of 0, 4 and 8 L/min. The premixed air-H_2 was then purged into the cylinder within the intake stroke. The addition of hydrogen at the idle condition effectively reduced the CO_2, CO, HC and NOx emissions, and its addition to gasoline fuel increased the H/C ratio, reduced the combustion duration and increased the diffusivity of hydrogen, compared with other fuels, providing a homogeneous combustion mixture and producing more oxygen to enhance the combustion. The results also indicated that the fuel consumption reduced from 23.3 to 33.4% at the idle condition with the various H_2 flow rate inputs.

ACKNOWLEDGMENTS

This work was supported in part by the National Science Council and Environmental Protection

Administration in Taiwan under NSC102- EPA-F-009-001. The authors also thank Dr. C.B. Chen and Mr. S.H. Gua at the Chinese Petroleum Corporation for their help in the dynamometer testing of the heavy-duty diesel engine and the Company of General Optics Corporation for the support of Hydrogen Generators.

REFERENCES

Bari S, Esmaeil MM. Fuel 2010; 89: 378–383.

Ganesh RH, Subramanian V, Balasubramanian V, Mallikarjuna JM, Ramesh A, Sharma RP. Renew Energy 2008; 33: 1324–1333.

Jia LW, Shen MQ, Wang J, Lin MQ. J Hazard Mater 2005; 123: 29–34.

Ji C, Wang S. Applied Energy 2012; 97: 584–589.

Ji C, Wang S, Zhang B, Liu X. Fuel 2013; 106: 873–875.

Kahraman E, Ozcanl SC, Ozerdem B. Int J Hydrogen Energy 2007; 32: 2066–2072.

Lucas GG, Richards WL. SAE Paper No. 1982; 820315

Lin YC, Hsu KH, Chen CB. Energy 2011; 36: 241–248.

Lin YC, Wu TY, Jhang SR, Yang PM, Hsiao YH. Bioresource Technology 2014; 161: 304–309.

Ma F, Liu H, Wang Y, Li Y, Wang J, Zhao S. Int J Hydrogen Energy 2008; 33: 816–822.

Ma F, He Y, Deng J, Jiang L, Naeve N, Wang M, et al. Int J Hydrogen Energy 2011; 36: 4454–4460.

Senthil KM, Ramesh A, Nagalingam B. Use of hydrogen to enhance the performance of a vegetable oil fuelled compression ignition engine. Int J Hydrogen Energy 2003; 10: 1143–1154.

Wang HK, Cheng CY, Chen KS, Lin YC, Chen CB. Fuel 2012; 93: 524–527.

Architectural, Energy and Information Engineering – Sung & Chen (Eds)
© *2016 Taylor & Francis Group, London, ISBN 978-1-138-02791-6*

Study on cluster supply chain mode in the mechanical manufacturing industry

Chuang Li & Jun Li
School of Economics and Management, Henan Polytechnic University, Jiaozuo, China

ABSTRACT: Combined with the industry cluster characteristics in the field of mechanical manufacturing, this paper discusses the leading enterprise-oriented vertical industry chain collaboration and cluster-oriented horizontal industry chain collaboration, and describes various collaboration modes in the above two cases from the supply, production, sales, logistics and service nodes.

Keywords: Cluster supply chain; Coordination; Logistics

1 INTRODUCTION

Cluster supply chain is a complex logistics value chain system. After being optimized from the perspective of value chain of cluster supply chain, we can establish the goals for optimal investment proportion of the fuzzy linear planning model to establish the supply chain logistics activities of the value contribution by using value analysis tool for enterprises cluster supply chain logistics activity (procurement logistics, production logistics, sales logistics, reverse logistics) to measure the value, in order to reconstruct a more optimized logistics value chain to realize added logistics value (reduce the logistics cost) and enhance the competitive advantage of cluster supply chain logistics integration [1].

In the current machinery manufacturing, regional agglomeration and industry supply chain has a strong coupling. A single business collaboration model is unable to meet the business needs of the cluster supply chain synergies. In order to meet different types of inter-enterprise business collaboration needs, two basic industrial clusters are introduced: a single company as the leading vertical industry chain collaboration and industry cluster as the leading horizontal industry chain collaboration.

2 ENTERPRISE-ORIENTED VERTICAL INDUSTRY CHAIN COLLABORATION

For modern industrial cluster, a single company as the leading vertical industry chain collaboration model includes almost all forms of a single coordinated approach. It is an integrated test bed; the collaborative model is mainly manifested in the areas described below.

2.1 *Procurement coordination*

A single enterprise's procurement is through the professional division of labor within the enterprise and horizontal integration, as well as between enterprises in the supply chain vertical coordination, and gradually establishes a virtual network of suppliers. The ultimate goal of procurement must meet the following four aspects: first, introduction of category management to create sustainable value; second, development of supplier relationship management collaborative planning, low investment and good quality deep suppliers; third, optimization of cost structure and procurement of outsourcing operations; last but not the least, more strategic and value-based contract to promote the implementation of the overall cost of ownership approach. In order to achieve coordinated procurement, manufacturing companies need to make improvements in the areas mentioned below.

First, enterprises must strengthen the management of suppliers to establish long-term, comprehensive supplier management mechanisms. Second, companies need to continuously improve the procurement performance, behavior patterns and inventory levels of information transparency to reduce "non-contract purchases." To this end, companies should consider the formation of centralized procurement organization, the implementation of centralized vendor management and development and a major supplier of the main contract. Third, purchasing managers should improve effectiveness in the aspects of contract management and expenditure analysis technology and the implementation of processes. In addition, enterprises should develop performance appraisal system that is based on the extending of supply chain to meet the needs of integrated supply networks, global purchasing organization and outsourced process management [2].

2.2 Design coordination

A single enterprise can work closely with customers, corporate with customers and supply chain service providers in the design process, and use component-based and appropriate standards in order to design a variety of products at a low cost. To transfer the innovation into profitable results, companies need to recognize the importance of the new product development in the market competition and make a rational assessment. It relates to the business rational allocation of resources and time, and can ensure new product development process operation. Second, companies need to significantly improve interoperability with customers and to achieve the further integration with suppliers, and to discuss possible measures with suppliers to meet the customer's needs. They also need to achieve integrated design and procurement, and introduce more competitive new products. Third, companies must expand the current practice of new product development. For example, they need to use standardized components for designing and developing custom product configuration and technical specifications, and a specific product or service development process, thereby the company can introduce more new products to the market more quickly and efficiently.

2.3 Manufacturing coordination

Advanced product design and analysis can ensure that the premise of reducing the design cost of materials becomes possible. Through continuous process optimization and validation, we can implement lean manufacturing and continuous improvement and collaborate with customers, suppliers and service providers to improve manufacturing competitiveness force, and use advanced manufacturing capabilities to respond quickly to market changes. In this collaborative manufacturing mode, the manufacturers transfer the quality requirements into the production and profit performance. So, quality management plan should be widely implemented, such as lean manufacturing, continuous improvement and Six Sigma management system. Meanwhile, many successful cases that companies achieve successfully by outsourcing have provided good example for manufacturers to improve production flexibility and optimize processing cost structure.

2.4 Logistics coordination

Enterprises establish a single highly integrated logistics network in order to meet the changing needs of customers flexibly, thereby the structure and the cost can be flexibly adjusted through the network. Non-core logistics functions will be outsourced to leading third-party logistics service providers, and major service providers, and other supply chain partners can integrate end-to-end process. In addition, the real-time visibility and event monitoring of the entire supply chain customers, products, and supply and logistics information can be used to manage the logistics networks by monitoring event and exception conditions.

In order to carry out logistics activities more effectively, the current model of logistics outsourcing of the manufacturing should be expanded to improve supply chain flexibility and reduce costs. Meanwhile, we must again reduce the order cycle by re-organizing the logistics processes and improve order fill rates and on-time delivery. Warehouse management systems or transportation management systems and other advanced technology are used to help improve logistics management capabilities and performance indicators. Second, the technology has become important initiatives to achieve end-to-end supply chain integration and synchronization, as well as significantly improve the visibility and reality of the inter-department inventory and shipping information. Supply chain technology can provide customers with online and EDI tool to submit orders and receive delivery. In this way, it allows buyers, sellers, financial institutions and other payments between related parties to automate the process of paying, related documents and information [3].

2.5 Sales coordination

By close collaboration through the entire network, manufacturing enterprises can respond quickly to market changes, coordinate with the customer in demand planning and make forecasts, improve dealer inventory planning and deployment plan, effectively develop cross-functional sales and operation planning of the enterprise and extended supply chain network, and develop targeted differentiated supply chain strategy according to the customer's actual situation on the market.

In order to achieve cross-selling model, first, manufacturers must be integrated within the organization, and then this integration must be extended to customers and suppliers in order to achieve collaborative supply chain planning, and improve the transparency of real-time information and market demand response. Then, companies must formulate performance evaluation indicators to track progress and benefits on the basis of established goals. Enterprises need to strengthen the full use of technology and systems, such as technology in sales and operation planning, demand planning, production and inventory planning process, and make better performance of the supply chain performance management. Meanwhile, enterprises should expand the corporate network risk-sharing capacity and reduce supply chain risk. They also need to take further steps to improve the ability to predict with sellers, and to improve supply chain performance levels through the establishment of a weekly planning process and the system-generated supply and demand plan.

2.6 *Service coordination*

By forming further close customer relationships through vertical coordination in the supply chain, single enterprise can ultimately integrate the business processes with other major service providers and supply chain partners from end to end, and can provide differentiated products and bundled services according to customer segmentation. Meanwhile, it can provide value-added features for comprehensive and real-time service, such as capacity reserves and/or implementation of co-order configuration that is based on the design. Next, internal and external supply chain partners can share product design specifications, demand information, capacity and supply constraints. Lastly, it can meet the customer's demand on the condition of paying attention to market fluctuations.

Therefore, the manufacturers first need to identify each customer or group of customers' profitability. To this end, companies first need to know how to clarify the breakdown and management of customers, and how to organize customer service management structure to support the business strategy and to improve the cost-effectiveness of services. Subsequently, companies should implement Customer Relationship Management (CRM) practices more widely, such as customer focus groups and automatic cross-service, and improved customer satisfaction. Enterprises also need to establish key performance indicators to track and control the effective management mechanisms to ensure that customer service management and business strategy is consistent, and constantly improve the service performance levels.

3 CLUSTER-ORIENTED HORIZONTAL INDUSTRY CHAIN COLLABORATION

3.1 *Supply coordination*

Manufacturers at the same level of the supply chain have many common parts needs, and bulk material procurement requirements under the bulk of economic demand make the gathering area of collaborative procurement and supply more important.

Most parts production can be done by the procurement and outsourcing. In the mode of collaborative supply chain business-led horizontal collaboration, the cluster number of leading enterprises is in accordance with market plan component material procurement plan, and consolidation of the classification can be done by a third-party vendor, and then multiple clusters and third-party suppliers negotiate the supply prices, and the third-party vendors will deliver the goods directly, or purchase separate third-party suppliers. Procurement of spare parts can be optimized to determine whether the logistics can be delivered directly to the manufacturer. Collaborative logistics supply has an important supporting role: it helps companies reduce production costs in the international market for commodities and raw materials pricing.

3.2 *Logistics coordination*

In the machinery manufacturing industry, there is a close contact between collaborative supply logistics, collaborative marketing and collaborative services, which form the bases. Multi-enterprise business collaboration provides a full range of collaborative logistics services for the cluster member companies, such as supplying, manufacturing, sales and after-sales service. On the one hand, a leading logistics company business can provide with more than one product logistics business collaboration services; On the other hand, the third-party logistics company from the cluster center provides logistics services for multi-product collaboration. In this cooperation mode, logistics, collaborative supply, sales and collaboration services together will produce the synergistic interaction. In the collaborative logistics business, supporting suppliers of leading companies can provide products directly to supporting parts, and directly supply vendor directly to the two leading companies to provide a common or standard parts. The third-party supplying vendors purchase from parts suppliers, and then sell directly to the leading business [4]. In the collaborative logistics sales business, leading manufacturer provide products to sales agents, and the sales agents can provide professional sales and service for all leading companies.

3.3 *Service coordination*

After the delivery of finished products to customers, the collaborative service management processes begin. This process involves the parts suppliers, leading enterprises of direct service providers and agency service providers working together. Remote fault diagnosis of collaborative business processes should be started when customer service requests or abnormal monitoring of product information are received, and then parts suppliers, direct service providers will assist clients via remote fault diagnosis and exclusion. If equipment failure cannot be ruled out through the remote, but on-site equipment maintenance service is needed, then service dispatching collaborative business processes need to be started. If failure occurs in the parts suppliers, workers should be sent to the suppliers to ask for on-site maintenance services. If failure occurs in the leading manufacturer of home-made parts, workers should be sent to the direct service provider to ask the frontline service engineers for on-site maintenance. If the product service agent has been licensed to service providers, workers should be sent to the service provider to ask service engineers for on-site repair [5].

3.4 *Building a network of overseas service*

The international trend of manufacturing industry has gradually become clear; many large manufacturing enterprises regarded an international business as the future development of the strategic objectives. To meet the development trend of the internationalization of enterprises and to make the companies within the industrial cluster share the international service system, we need to build a network of overseas service as soon as possible. When being driven by the demand of the leading manufacturers' international sales and service parts, industry clusters can provide machinery parts suppliers through a global service center network to the global storage accessories; leading global storage center will focus on storage and sorting according to the needs of manufacturing companies, and send to the overseas regional storage centers. Regional warehouse will distribute the parts to the international sales agents, international agency service providers and clients to meet their sales or service needs.

4 CONCLUSION

In short, in the modern market, competition is always present and the cooperation is an inevitable trend. All the manufacturing enterprises, be it the vertical industry chain cluster leading by a single company or the horizontal business chain collaboration dominated by cluster companies, need to take a different combination of collaboration according to the actual situation to improve their competitiveness, and ultimately achieve a win-win and harmonious situation.

ACKNOWLEDGMENT

This work was supported by the Youth Project of the National Social Science Fund (13CJY045); the Henan Province Education Department Science and Technology Research Project (14A790010); the Henan Province College Humanities and Social Science Research Project (2014-ZD-067); the Henan Polytechnic University Young Teacher Support Project (72105002).

REFERENCES

[1] Zhou Xingjian, Li Jizi, Xie Shaoan. The optimization of cluster supply chain logistics value chain. China's Circulation Economy. 2011(4): 1.
[2] Huo Jiazhen, Wu Qun, Chen Long. Cluster supply chain network connection mode and co governance framework. China Industrial Economy, 2007(10): 64–67.
[3] Zhang Peiliang, Cluster supply chain organization mode. Dalian: Dongbei University of Finance and Economics, 2006.
[4] Lu Shan, Gao Yang. Supply chain coordination theory system research. The Commercial Times, 2007(30): 15–23.
[5] Li Jizi, Cluster supply chain and management. Wuhan: Huazhong Agricultural University, 2006: 21–22.

Architectural, Energy and Information Engineering – Sung & Chen (Eds)
© 2016 Taylor & Francis Group, London, ISBN 978-1-138-02791-6

The simulation calculation of micro hybrid vehicle based on refine

Hui Wang, Lv He Gao, Mei Yang & Yong Hou
Beijing Polytechnic College, Beijing, China

Jun Wang
Jiang Huai Automobile Corp. Technology Center, Power Research Institute, Anhui Province, Hefei City, China

ABSTRACT: In this paper, using Cruise software, simulation analysis of a BSG micro hybrid power car is carried out, to build a hybrid model for fuel saving and emission characteristics index to predict the vehicle. It is a good reference for the later prototype development.

Keywords: Hybrid; BSG; Motor; Generator

1 FOREWORD

Micro hybrid dynamic BSG technology can be applied to a variety of models, including cars, multi-function cars and trucks, and can provide users with good oil-saving characteristic and meet emissions targets of regulatory requirements. The system is also adaptive to different engines and avoids the risk of changing the engine or transmission system.

Micro hybrid power system includes complex dynamical control systems, with precise motor and generator, and also has the braking energy regeneration and effective charging function. In the idling and engine-off condition, the NEDC heat engine cycle CO_2 emissions reduction is 6%. A super capacitor can supply about 12 volt electricity.

2 INTRODUCTION

The advantage of fuel economy of the hybrid is very attractive, but the high cost of the system compared with conventional vehicles cannot be ignored. It can save fuel economy by about 4–6%, for engine-off during the idling condition. That is, the vehicle usually shuts down the engine in the idle condition. If the driver needs torque, such as stepping on the accelerator or getting a command of restarting engine based on the standard hybrid control unit, the engine restarts.

This paper studies the hybrid system installed in the Refine business car of JAC, the main power source using the 2.4 L naturally aspirated engine, and auxiliary power source adopts the design of motor/generator Mana Austria generator to replace the original engine. The feasibility of its design, fuel economy and emission can be evaluated through the simulation analysis.

The micro hybrid function mainly includes:

Idling engine-off, quick start.
Optimization of battery charging.
Use of renewable braking capability.

2.1 System structure

The overall scheme consists of the belt-driven starter/generator system, brake energy regeneration system and control system. The belt-driven starter motor or generator system is composed of a Belt Driven Starter/Generator System (BSG) to replace the standard starter motor, alternator and the new belt. A generator and a motor complete the starter functions and provide electrical energy when the engine is running. The actual connection between the generator and the crankshaft is accomplished through a belt-driven system. When starting, the starter and the generator get electricity from the super capacitor, and then the motor quickly runs and starts the engine by a belt. In the road, when a traffic stop shows the red light, the control system will automatically shut down the engine.

The braking energy regeneration system consists of the super capacitor group, AC/DC, starter/generator and other important parts. Deceleration from the vehicle brake, gear driving, the engine is motored continuously to rotate, driving the starter/generator to generate electricity through a belt drive, charging the super capacitor, the realization of regenerative braking energy function, the belt drives the starter/generator system, are referred to as BSG, as shown in Figure 1; it replaces the original starter and generator. This system is useful in referring to the start and stop function, and the price is reasonable. The micro hybrid is to reduce fuel consumption and CO_2 emissions.

Overall, the system provides three different functions:

Starter/generator can reverse function, aiming at the start/stop function and application of the generator function.

Braking energy recovery function, in addition to acid battery, through using super capacitor energy storage.

Assisting torque, by storing dynamic force in a super capacitor module, to help the engine running and stop protection.

2.2 Main electronic components

As shown in Figure 1, the micro dynamic power vehicle electronic system is mainly composed of different electronic components:

Reverse starter—generator device, provides adjusting mode of voltage and the power, which is mainly composed of a power converter and a motor.

Reversible DC/DC converter, allowing the energy conversion in the super capacitor and the network.

A super capacitor module allows energy exchange between the motor and the network (charge—discharge).

The dynamic control unit, as the main control module in the BSG system.

All the micro dynamic system modules communicate through the CAN bus.

3 SIMULATION ANALYSIS

In the selection phase of the technical scheme, system selection, hybrid systems and components are modeled and simulated using computer technology. Through the analysis of the performance and characteristics of the system, the evaluation is an indispensable link. Simulation by alternating use of alternative sub-system finds the best solution. Computer models provide detailed specifications and design parameters for each alternative sub-system, which is convenient to the work of designers, but also help to develop the project goals and plans for the design and fabrication of the prototype. In the research and software development of simulation technology, the target is accurate, which makes the power transmission system comparison between different structures significant.

In this study, the simulation analysis of vehicle evaluates the fuel-saving potential of the micro dynamic system (in the standard operating conditions). It uses a vehicle company AVL analysis software—Cruise hybrid module to simulate, and can predict the fuel economy, emission and driving

performance. As shown in Figure 2, it is a micro hybrid power vehicle model.

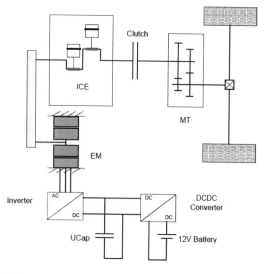

Figure 1. Micro hybrid system structure diagram.

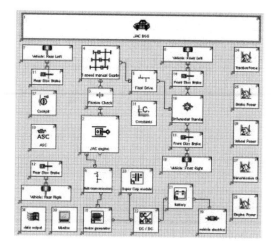

Figure 2. The vehicle simulation model.

3.1 Vehicle simulation boundary

Table 1. Simulation boundary input parameters.

The vehicle data1	kg
Kerb mass/net mass	1890
The inertia weight	2040
Maximum gross weight	2800
Axle	3080

(Continued)

Table 1. (*Continued*).

The vehicle data2	m²	mm
Windward area	3.1	
Slide damping	Experimental measurements	
Slide damping coefficient	0.014	
Static rolling radius	302	
Dynamic rolling radius	324	

Engine data1	L	rpm
Displacement	2.35	
Idle speed		750
Maximum speed		5500

Engine data2	kg/h	KJ/kg
Idle fuel consumption	0.8	
Fuel ratio of energy		42700
Engine data3	kg/l	RON
Fuel density	0.74	
The type of fuel		91

The transmission data1		
Type	MT	
1 gear ratio	3.986	
2 gear ratio	2.155	
3 gear ratio	1.414	
4 gear ratio	1.000	
5 gear ratio	0.813	
Reverse gear ratio	3.814	
The transmission data2	Nm	
Maximum clutch torque	250	

3.2 Simulation results

There is a causal relationship between the driving cycle mode and the potential of reducing fuel economy in the hybrid vehicle. Obviously, for example, the idle engine-off function is appropriate only in a frequent stop condition. NEDC cycle analysis shows that the engine idling time accounted for 30% (Figure 3), and can continue to improve fuel economy when turning off the engine during the idling condition.

Figure 4 shows the fuel-saving potential for different driving cycles of the Refine micro hybrid car due to the idle engine shutdown function. There are obvious differences between the NEDC cool engine (4.9% fuel-saving) and the NYCC hot engine (15.5% fuel-saving).

Regenerative braking energy can further reduce fuel consumption, and this potential depends on the power of the motor. In the NEDC loop, 144 model 4kw motor of three axles can produce about 60% of the available energy (Figure 5).

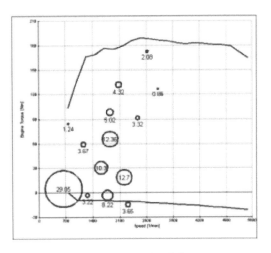

Figure 3. The map condition (NEDC).

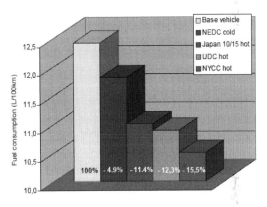

Figure 4. Fuel consumption reduction due to the idle engine shutdown.

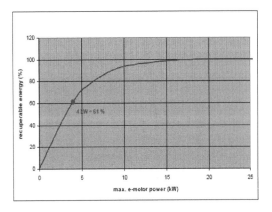

Figure 5. Motor power and renewable energy.

101

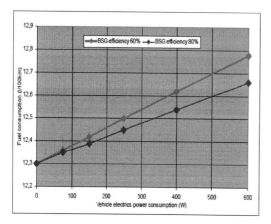

Figure 6. Efficiency of fuel consumption.

In the micro hybrid power vehicle, regenerative energy is for electronic devices, especially in the engine shutdown. Therefore, the engine used as the generator (under normal driving conditions) can be weakened to reduce 2% fuel consumption.

Finally, the increased efficiency of the BSG (relative to a conventional generator) has a positive effect on fuel economy improvement, mainly depending on power consumption of the electronic components. Figure 6 shows the two different efficient driving motors, where extra fuel consumption is caused by the electronic device.

4 CONCLUSION

The BSG hybrid vehicle uses a super capacitor to supply the energy to a motor. In the control system, the start/stop switch mode and system state conversion is realized by the hybrid control unit. Models of the micro hybrid vehicle are established through a software. The predicted results for micro hybrid models on fuel saving and emission performance have achieved the expected goal.

REFERENCES

Chen Jian et al. The logic control strategy research and Simulation of [J]. automotive engineering fuzzy hybrid electric vehicle.

Chen Qingquan et al, a modern electric car technology, Beijing Institute of Technology press.

Fangwu, vehicle aerodynamics, Mechanical Industry Press.

Feng Guosheng & Yang Shaopu, the modern design method of vehicle, Science Press.

Lv Shengli, left the dawn. The parallel hybrid electric vehicle control strategy of the comprehensive analysis of [J]. in Shanghai.

Ni Youmin, a car coasting test method and data processing, Automotive Engineering.

Qian Lijun et al. Hybrid electric vehicle control design and simulation [J]. Journal of system simulation strategy. 2005 (03).

Yu Zhisheng, the theory of automobile, machinery industry press.

Zhang Xin et al. A parallel hybrid electric vehicle powertrain control strategy simulation of [J]. automotive engineering.

Architectural, Energy and Information Engineering – Sung & Chen (Eds)
© 2016 Taylor & Francis Group, London, ISBN 978-1-138-02791-6

Key technology of offshore wind power and its development tendency

Shan Lin Zhong
North China Electric Power University, Baoding, China

ABSTRACT: With the rapid development of wind power technology, offshore wind power has become a major concern of the world's renewable energy development. Based on the present status of offshore wind power, this paper focused on the key technology of offshore wind power, including resource assessment, wind turbine and construction. Combined with the latest research, the development tendency of offshore wind power was put forward at the end of the paper.

Keywords: Offshore wind power; Present status; Key technology; Development tendency

1 INTRODUCTION

Wind power generation, as a sustainable and environmental energy utilization method, has great significance in saving the use of resource, preserving the ecological environment and promoting sustainable development.

Compared with onshore wind, offshore wind has plenty of advantages. It does not involve the issue of land acquisition, so countries with a small land area, but near the ocean can rely on it. Wind speed at sea is higher and more constant than on the land, and there is less limitation of noise and landscape, which allows the possibility of using large wind turbines [1].

The aim of this paper is to introduce the present status of offshore wind farm development, and analysis of the key technology of offshore wind power and its development tendency.

2 PRESENT STATUS

Europe has leadership position in global offshore wind development. In 2013, offshore wind in Europe added 1,567 MW installed capacity, with 34% of year-on-year growth and a cumulative installed capacity of 6,562 MW. Currently, Britain captures 47% of Europe's new additions in offshore wind capacity, with 733 MW, and the second is Denmark, with 350 MW and 22% of market shares. Next is Germany (240 MW) and (192 MW), with 15% and 12% of market shares respectively. Table 1 presents the global cumulative installed capacity of offshore wind power.

Britain has the most developed offshore wind farm in the world, capturing over a half of cumulative installed capacity all over the world, with 3,681 MW of installed capacity. In 2013, four projects (London Array, Lincs, Teesside & Gunfleet Snds), with a total of 733 MW, have been connected to the grid. Britain will fulfill 18 GW installed capacity by 2020, which will provide 18%~20% of the country's electricity needs.

The USA started late on offshore wind. The fastest developing projects are Cape Wind (468 MW) and Block Island (30 MW). In addition, other districts in the USA have a long project list of offshore wind power, and most of them are located in the northeast sea area and Great Lakes.

In Asia, a 390 MW offshore wind farm has been built in China, and the approval of the establishment is 1830 MW. Japan has a strong maritime industry and the sixth largest Marine Special

Table 1. Global cumulative installed capacity of offshore wind power (MW).

State	2012	Cumulant	2013	Cumulant
Britain	854	2947.9	733.1	3681
Denmark	46.8	921	350	1271
Belgium	185	379.5	191.5	571
Germany	80	280.3	239.7	520
China	127	389.6	0.4	390
Holland	0	246.8	0.2	247
Sweden	0	163.7	48.3	212
Japan	0.1	25.3	24.3	49.6
Finland	0	26.3	0	26.3
Ireland	0	25.2	0	25.2
Korea	3	5	0	5
Norway	0	2.3	0	2.3
Portugal	0	2	0	2
Total	1295.9	5414.9	1587.5	7002.4

Economic Zone in the world. So far, Japan has 49.6 MW installed capacity, including 4 MW floating wind power [2, 3].

3 KEY TECHNOLOGY OF OFFSHORE WIND POWER

Offshore wind power is a high-technical content field in the wind power industry. It could not have developed so fast without the support of sophisticated techniques. The key technology of offshore wind power includes offshore wind resource assessment and short-term prediction, design of wind turbine generator, construction of wind farm (including infrastructure, installation and transportation, operation monitoring and controlling) and grid-connection (high-voltage system on the sea, submarine cable, access device on the land) [4].

3.1 *Offshore wind resource assessment and short-term prediction*

Marine meteorological observation is an extremely difficult problem. Utilizing the observation data from meteorological station, satellite, oil drilling platform and ships near the wind farm can make a preliminary assessment of the wind resource. However, these data have great uncertainty, which makes it difficult to estimate the generating capacity. For this reason, building anemometer tower for field measurement is necessary.

In recent years, with the popularization of computer technology, some numerical simulation software, such as POWER, have been applied for measurement and prediction. By building the numerical model of the offshore wind filed, it can easily acquire the distribution of wind filed, and

solve the lack and discontinuity of the marine meteorological observation data [5].

3.2 *Offshore wind turbine generator*

Offshore wind turbine generator is comprised of four modules: blade, driving system, control system and power generation system. Compared with onshore wind turbines, the design of the offshore wind turbine refers to the experience of the traditional maritime construction technology, such as lighthouse, bridge and offshore drilling platform, and pays more attention to the performance of the wind turbine in order to improve utilization and reliability, reduce the rate of repair.

3.3 *The construction of wind farm*

The construction of wind farm is the most expensive and difficult part in offshore wind project; it can capture 75% of the total cost. Limited by the complex weather conditions, offshore wind farm has to withstand gale load, seawater corrosion and wave impact.

According to the environment wind field, depth of water and seabed condition, the infrastructure can be divided into four categories: strut, gravity, tubbiness and floating.

There are two main construction methods: overall installation and split installation. The former is to assemble the wind turbine on the land, then transport the turbine to the farm by ship, and install it to the infrastructure with the help of a crane. On the contrary, the latter is to assemble the wind turbine on the sea [8].

3.4 *Grid-connection*

The characteristic of offshore wind power is different from the conventional power generation

Table 2. Comparison of different infrastructures.

Type	Characteristic	Depth	Advantage	Disadvantage
Strut	Steel pile	Single <20 m, tripod >30 m	Simple production process, good corrosion protection	High construction cost
Gravity	Reinforced concrete structure	Less than 10 m	Simple structure, low cost, high stability	Seabed must be flat
Tubbiness	Fixed on the seabed with the help of water pressure	Deeper than 30 m	Low material cost, convenient transportation	Seabed must be flat
Floating	Including floating type and semi-submersible type	Deeper than 75 m	Low cost, suitable for deep water	Unstable, affected by wave easily

method, which makes it to face new technology challenge in grid-connection. On the one hand, power transmission mainly uses the submarine cable. When the capacity of the wind farm is larger than 100 MW and the distance to coast is over 15 km, an offshore substation is needed. On the other hand, with the rapid development of wind turbine control technology (variable propeller control, automatic control and automatic halt), and the use of short-term weather forecast and centralized control strategy, the stability of wind power grid-connection is improved vastly.

4 DEVELOPMENT TENDENCY OF OFFSHORE WIND

4.1 *Larger generator capacity*

At present, mainstream generator capacity is 2~3.5 MW. However, in the 1990s, the generator capacity was 500~600 kW. ENERCON, the most famous turbine manufacturer, has developed a 6 MW direct-driven generator successfully. Meanwhile, many manufacturers have been researching a 10 MW generator. All of these reports indicate that the development of generator capacity is towards large scale [9].

4.2 *Essentiality of carbon fiber blade*

Blade technique is a core of large wind generator development. Offshore wind generator has a large capacity, and the generating capacity is in direct proportion to the squared length of the blade, which requires a better strength and stiffness of blade materials. Glass fiber in the manufacture of a large composite blade appears to be insufficient performance. Therefore, carbon fiber will be essential for blade manufacture.

4.3 *Higher tip speed*

Noise reduction is the factor that needs to be considered in the project of land wind farm. However, offshore wind farm is far away from human settlement, so it can take full advantage of aerodynamic efficiency. The higher tip speed and the smaller area of the blade are good factors for the structure and transmission of the wind turbine.

4.4 *Mainstreaming of variable pitch turbine*

Variable pitch turbine can work in the highest speed around the rated speed, make the most of wind energy at high wind speed, and improve the reliability of the wind turbine. The traditional fixed pitch turbine has a low utilization of wind;

therefore, it is inevitable for the variable pitch turbine to replace it.

4.5 *Direct drive permanent-magnet synchronism generator as the main force*

The direct drive permanent-magnet technology can drastically reduce the size of the synchronism generator, and it is easier than the double-fed induction generator to satisfy low-voltage ride through after connected to the grid, which can be of benefit to the stability of the grid.

5 CONCLUSION

The offshore wind is the development trend of the wind power industry due to its abundant resources, stable wind speed and less negative effects on the environment. Compared with the land-based wind power development, the offshore wind power development is being faced with some new challenges. However, with the development of new technology and the market expansion, offshore wind power will be the mainstream of wind power utilization in the future.

REFERENCES

[1] Mikel de Prada Gil, Oriol Gomis-Bellmunt, Andreas Sumper. Technical and economic assessment of offshore wind power plants based on variable frequency operation of clusters with a single power converter [J]. Applied Energy. 2014.

[2] 2013 Annual Review and Outlook on China Wind Power.

[3] 2014 Annual Review and Outlook on China Wind Power.

[4] Chen Xiao-ming, Li Lei, Wang Hong-mei, et al. Development Trend of Offshore Wind Power [J]. Design & Development. 2010.

[5] Huang Dong-feng. The survey of off-shore wind power development in Europe [J]. New Energy and Technology. 2008.

[6] Zhang Xian-liang. Research on development trend and key technology of offshore wind power [J]. New Energy and Technology. 2013.

[7] Yao Xing-jia, Sui Hong-xia, Liu Ying-ming, et al. The development and current status of offshore wind power technology [J]. Shanghai Electric Power. 2007.

[8] Liu Lin, Ge Xu-bo, Zhang Yi-bin, et al. Analysis on Status of China's Offshore Wind Power Development [J]. Energy Technology and Economy. 2012.

[9] Sun Zi-mo, Zhang Jian, Shu Xin-xin, et al. Analysis on construction technology of offshore wind farm [C]. Collected Papers of the 14th China Ocean Engineering Academic Seminar.

[10] Li Ze-chun, Zhu Rong, He Xiao-feng, et al. Research on technology of wind source assessment [J]. Acta Meteorologica Sinica. 2007.

[11] Rehana Perveen, Nand Kishor, Soumya R. Mohanty Off-shore wind farm development: Present status and challenges [J]. Renewable and Sustainable Energy Reviews. 2014.

[12] Brian Snyder, Mark J. Kaiser. Ecological and economic cost-benefit analysis of offshore wind energy [J]. Renewable Energy. 2009.

[13] Brian Snyder, Mark J. Kaiser. Biodiversity offsets for offshore wind farm projects: The current situation in Europe [J]. 2014

[14] I. Spiropoulou, D. Karamanis, G. Kehayias. Offshore wind farms development in relation to environmental protected areas [J]. Sustainable Cities and Society. 2014.

[15] Bergström, L., Kautsky, L., Malm, T., Rosenberg, R., Wahlberg, M., Åstrand Capetillo, N., et al. (2014). Effects of offshore wind farms on marine wildlife—A generalized impact assessment. Environmental Research Letters, 9, 034012–34024.

Architectural, Energy and Information Engineering – Sung & Chen (Eds)
© *2016 Taylor & Francis Group, London, ISBN 978-1-138-02791-6*

National sustainability evaluation model

Shan Lin Zhong
North China Electric Power University, Baoding, China

ABSTRACT: Sustainable development is the topic of the 21st century. It is important for a country to evaluate the sustainability accurately. This paper introduces an evaluation model of sustainability. The model contains five support systems: Living, Development, Environment, Society and Intelligence. Each support system has a number of indices, respectively. The evaluation score is calculated by means of Entropy Method, Analytic Hierarchy Process (AHP) and the weighted average method. Based on fuzzy set theory, we finally put forward the criterion (evaluation score higher than 0.45 is sustainable) to define the sustainability of a country.

Keywords: Evaluation model; Sustainability; Entropy Method; AHP

1 INTRODUCTION

Since the publication of the report "Our Common Future" (Brundtland Report) by the World Commission on Environment and Development in 1987 and the accomplishment of the United Nations Conference on Environment and Development (UNCED) in 1992, the concept of "sustainability" has been adopted as a key political principle by most governments worldwide [1].

In order to put forward a sustainable development plan, this paper provides an efficient mathematical model to evaluate the sustainability of countries.

2 INDICATORS SYSTEM

A reasonable indicators system of sustainable development should contain the indices that can describe the relationship between humans and nature from different aspects, and it should be as concise as possible at the same time. Based on this principle, we construct our indicators system that contains five support systems: Living, Development, Environment, Society and Intelligence (LDESI). Each support system has a number of indices.

- *Living support system (L)*. It represents the basic conditions for sustainable development. In other words, it is the condition that humans and the ecosystem need to survive. The indices in this system include Average precipitation in depth, Renewable internal freshwater resources per capita, Energy use, Arable land, Cereal production, and Agriculture value added.

- *Development support system (D)*. It represents the driving force for sustainable development. The indices include GDP per capita, GDP per unit of energy use, Internet users, Industry value added, External balance on goods and services, Access to electricity, and Road density.

- *Environment support system (E)*. It represents the restricted condition for sustainable development. In other words, the state of the ecosystem can be reflected via this system. The indices include Forest area, CO_2 emissions, Terrestrial and marine protected areas, Nitrous oxide emissions, and Terrestrial and marine protected areas.

- *Society support system (S)*. It represents the guarantee condition for sustainable development. It reflects the state of the society. The indices include Population, GINI index, Public health expenditure, Life expectancy at birth, Unemployment, and Age dependency ratio.

- *Intelligence support system (I)*. It represents the persistence condition for sustainable development. The indices include Public spending on education, Primary school enrollment, Researchers in R&D, Hightechnology exports, and Teachers of primary education.

3 PREPROCESSING OF THE DATA

3.1 *Interpolation*

When we collect data from the Internet, we notice that some data are missing. One of the reasons for this is that these data are not surveyed every year. Given this fact, we fill in the data mainly based on interpolation, and we can obtain the data in the nearby years.

3.2 Negative index

Some indices in the indicators system have a negative correlation with the sustainability. In order to reflect the characteristics of these indices correctly, we arrive at an approximate solution by replacing the value of the index with its reciprocal.

3.3 Data normalization

There are 30 indices with different units and magnitudes in our evaluation model, which might produce incorrect results when we use the data to evaluate directly. Taking this factor into consideration, we have to transfer the original data to a comparable sequence. For this purpose, the data are normalized in the range between zero and one:

$$v_i^*(k) = \frac{v_i(k) - minv_i}{maxv_i - minv_i}, \tag{1}$$

where k is the object k; $v_i(k)$ is the value of the index i of the object; $maxv_i$ is the largest value of v_i; $minv_i$ is the smallest value of v_i; and $v_i^*(k)$ is the value after normalization.

3.4 Growth rate

In order to describe the dynamic process of sustainable development, we define $r_{i,year}$ to denote the development tendency of each index. It is defined as follows:

$$r_{i,year} = \frac{v_{i,year} - v_{i,(year-1)}}{v_{i,year}}. \tag{2}$$

As defined in the previous section, $v_{i,year}$ represents the value of the index i in years (e.g. $v_{L1,2015}$ is the value of the index L1 (Arable land) in 2015) and $r_{i,year}$ is the growth rate of the index i in years.

From the above discussion, we build our evaluation model based on two kinds of indicators, namely static and dynamic, which represent the current state of development and the tendency of development, respectively.

4 THE WEIGHTING PROCESS

4.1 Entropy method

There are 5 to 7 indices in each support system. Considering the amount of date, for each support system, we use the entropy method to obtain the weight of its indices.

Step 1. Calculate the ratio of the country k with index i:

$$p_{ik} = \frac{v_i^*(k)}{\sum_{i=1}^{n} v_i^*(k)}. \tag{3}$$

Step 2. Gain the entropy of index i:

$$e_{ik} = -\frac{1}{\ln n}\sum_{i=1}^{n} p_{ik} \ln p_{ik}. \tag{4}$$

Step 3. Calculate the difference coefficient of index i. For index i, the bigger the difference of the data, the greater the weight of the index and the smaller the entropy. We define the difference coefficient as follows:

$$g_k = \frac{1 - e_k}{m - E_e}, \tag{5}$$

$$E_e = \sum_{k=1}^{m} e_k, \tag{6}$$

where $\sum_{k=1}^{m} g_k = 1$.

Step 4. Calculate the weight:

$$w_k = \frac{g_k}{\sum_{k=1}^{m} g_k}. \tag{7}$$

Since $\sum_{k=1}^{m} g_k = 1$, the difference coefficient is equal to the weight.

We next consider the Living support system for example, and the weight of its indices, as given in Table 1.

The evaluation score of static state indices V_i should be

$$V_i = \sum w_i v_i^*, \tag{8}$$

where v_i^* is the value after normalization.

Since we assumed the state and tendency of development have the same importance as the sustainability of a country, so the weight of the index is equal to its growth rate. The evaluation score of the dynamic tendency index R_i is

Table 1. The weight of the indices in the Living support system.

Index	L1	L2	L3	L4	L5	L6
Weight	0.11	0.24	0.08	0.16	0.11	0.30

$$R_i = \sum w_i r_i^*. \tag{9}$$

To comprehensively consider the effect of the static and dynamic characteristics of sustainable development, we adopt a linear weighted method:

$$y_i = \omega V_i + (1 - \omega)R_i, \tag{10}$$

where y_i is the evaluation score of the support system i and ω and $(1 - \omega)$ are weights that add to 1. From the above assumption, we can know that $\omega = 0.5$.

By applying this method, we can calculate the evaluation score of the five systems, namely y_L, y_D, y_E, y_S, y_I.

4.2 Analytic Hierarchy Process (AHP)

Based on the above calculation, we get the evaluation grade of the five support systems. We then use AHP to distribute weight for the systems in order to fulfill the evaluation.

Determine the judging matrix. We use the pairwise-comparison method and 1–9 method of AHP to construct the judging matrix $A = (a_{ij})$:

$$a_{ij} = a_{ik}a_{kj}, \tag{11}$$

where a_{ij} is set according to the 1–9 method. The judging matrix that we built is as follows:

$$A = \begin{bmatrix} 1 & 7 & 5 & 4 & 3 \\ 1/7 & 1 & 2/3 & 1/2 & 1/2 \\ 1/5 & 3/2 & 1 & 1 & 1 \\ 1/4 & 2 & 1 & 1 & 1 \\ 1/3 & 2 & 1 & 1 & 1 \end{bmatrix}. \tag{12}$$

Calculate the eigenvalues and eigenvectors. The greatest eigenvalue λ_{max} of matrix A has the corresponding eigenvector $l = (l_1, \ldots, l_n)$.

Perform a consistency check. The indicator of consistency is

$$CI = \frac{\lambda_{max} - n}{n - 1}, \tag{13}$$

where n is the dimension of the matrix and CI is the indicator of the consistency check.

The expression of the consistency ratio is

$$CR = \frac{CI}{RI}, \tag{14}$$

where CR is the consistency ratio and RI is the radom consistency index.

Table 2. The weight of the support system.

Support system	L	D	E	S	I
Weight	0.05	0.36	0.22	0.20	0.17

Table 3. Evaluation of the 16 countries (2012).

Rank	Country	Y	Rank	Country	Y
1	USA	0.56	9	Djibouti	0.40
2	Singapore	0.51	10	Gambia	0.39
3	China	0.51	11	India	0.39
4	Britain	0.47	12	Angola	0.39
5	Sweden	0.47	13	Brazil	0.39
6	Japan	0.45	14	Benin	0.36
7	Finland	0.45	15	Zambia	0.32
8	Germany	0.45	16	Senegal	0.31

We then calculate the value of $CR = 0.0054$, and find that it satisfies the criterion. The weight of the support system is given in Table 2.

The final evaluation score Y should be

$$Y = \Sigma W_i y_i. \tag{15}$$

Here, W_i is the weight of y_i, and y is the state of sustainable development.

5 MODEL APPLICATION

To verify our evaluation model, we choose 16 countries to evaluate their sustainability. These countries include developed country, developing country and the least developed country, which can make the verification more representative. The result of the evaluation is summarized in Table 3.

As shown in Table 8, the United States has the highest score in the evaluation, while Senegal has the lowest score. Developed countries occupy four positions on the top 5, because they possess a number of resources and advanced technologies, which make them to be able to focus more on the problem between humans and the ecosystem. Developing countries and the least developed countries have a lower rank due to low finance rates and resources.

6 CRITERION OF SUSTAINABILITY BASED ON FUZZY SET THEORY

To define whether a country is sustainable or unsustainable is an extremely difficult problem. As there are many factors associated with it, we implement fuzzy set theory to describe sustainability and confirm its criterion.

- *Identify alternative and attributes*

The alternative is the evaluation object, namely the country we want to evaluate. The attributes are the ten indices, five state indices and five tendency indices. We derive the ideal alternative from the evaluation result of the 16 countries:

$$u = (u_1, \ldots, u_5) = (0.67, 0.88, 0.80, 0.62, 0.78), \quad (16)$$

where u_1 to u_5 represents the ideal value of y_L, y_D, y_E, y_S, y_I.

- *Determine membership functions*

A fuzzy set is defined in terms of a membership function that maps the domain of interest onto the interval [0, 1]. The value of the membership represents the degree that the domain item belongs to the set.

The membership function is represented by the following equation:

$$m = \frac{x_i^*}{u_i}, \quad (17)$$

where x_i^* is the value of y_i. If $m > 0.6$, we define that the country is qualified in index i. Therefore, we obtain the criterion of eligibility as follows:

$$u_a = 0.6(u_1 \ldots - u_5) =$$
$$(0.40, 0.53, 0.48, 0.37, 0.47), \quad (18)–(19)$$

- *Criterion of sustainability*

Based on equation (15), we can get the criterion of sustainability as follows:

$$Y_a = \sum_{i=1}^{5} W_i u_{ai}. \quad (20)$$

When $Y > Y_a$, the country is sustainable, and vice versa. The value of Y_a in our model is 0.45.

Figure 1 illustrates the sustainability of the 16 countries. The line represents the criterion of sustainability.

7 CONCLUSION

The model that we developed in this paper could distinguish more sustainable countries and policies

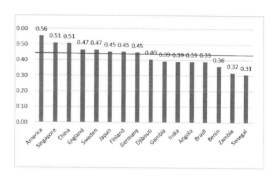

Figure 1. The sustainability of the 16 countries.

from less sustainable ones, and the country with an evaluation score higher than 0.45 is sustainable.

From the above discussion, we can infer that our model is feasible, as it can reflect the real situation correctly.

REFERENCES

[1] Ewald, Rametsteiner. Sustainability indicator development—Science or political negotiation. Vienna, Austria: Ewald Rametsteiner, 2011.
[2] http: //en.wikipedia.org/wiki/Developing_country.
[3] United Nations. The future we want. Resolution adopted by the General Assembly. 66th Session of the General Assembly, 123rd plenary meeting; 2012 July 27. New York: UN; 2012 Sep 11 (Resolution A/RES/66/288) [cited 2013 Jul 23].
[4] http://en.wikipedia.org/wiki/Analytic_hierarchy_process.
[5] http: //en.wikipedia.org/wiki/Fuzzy_set.
[6] Qingmin Wan, Dongjun Cui. The evaluation of the sustainable development in Tai Jin based on DPSIR Model. Business studies, 2013, 03: 27–32.
[7] Xiaoqing Chai. The existing problem in the evaluation of sustainable development and solving problems. Business studies, 2012, 02: 8–13.
[8] Wei Bai, Ying Yi, Changzhi Cai. The revelation sustainable communities in Sweden give to China. Urban planning, 2013, 09, 60–66.
[9] Ruixin Liu, Xuelin Li, Qingxiao Wang. Compare with Four nonlinear fitting methods of Weibull's dissolution curve. Chinese Journal of Hospital Pharmacy, 2009, 15: 1315–1316.
[10] Donnelly, A., Jones, M., O' Mahony, T., Byrne, G. 2007. Selecting environmental indicator for use in strategic environmental assessment. Environmental Impact. Assessment 27, 161–175.

Architectural, Energy and Information Engineering – Sung & Chen (Eds)
© 2016 Taylor & Francis Group, London, ISBN 978-1-138-02791-6

Comparative analysis of transcritical CO_2 two-stage cycles

Ying Fu Liu
Baoding Electric Power Voc. & Tech. College, Baoding, China

Ying Zheng Rong
The Affiliated School of Hebei Baoding Normal, Baoding, China

Guang Ya Jin
North China Electric Power University, Baoding, China

ABSTRACT: According to throttling type, a complete inter-cooling transcritical CO_2 two-stage compression cycle can be divided into a Single-Throttling (STDC) cycle and a Double-Throttling (DTDC) cycle. These two cycles are investigated, and are compared with the single Stage Compression (STSC) cycle. The results indicate that at a given condition, the DTDC cycle has maximum COP, while the STDC cycle has the optimum high pressure. For these three cycles, increasing gas cooler outlet temperature results in decreasing COP and increasing optimum high pressure. The optimum inter-cooling pressure for the STDC cycle and the DTDC cycle deviates from the geometric mean of high pressure and low pressure, but this has a little effect on system performance.

Keywords: transcritical CO_2 two-stage cycle; COP; optimal intermediate pressure

1 INTRODUCTION

The application of the natural refrigerant CO_2 is of great significance to reduce the greenhouse effect and ozone depletion. CO_2 as a refrigerant has its unique advantages, but the COP of the transcritical system is not high because of their lower critical temperature (31°C). Therefore, the key to promote the transcritical CO_2 cycle is to improve the efficiency of the system [1–3]. Two-stage compression, which could reduce the compressor work, is an effective method to improve the COP [4–5]. Researchers have shown that there is an optimum heat rejection pressure $P_{c,opt}$, where the two-stage cycle has the best COP, the cooling effect of the intercooler has a greater impact on COP [6], the COP of the complete inter-cooling mode is higher than that of the incomplete mode [7].

According to throttling type, a complete inter-cooling CO_2 transcritical two-stage compression throttling cycle can be divided into a Single-Throttling (STDC) cycle and a Double-Throttling (DTDC) cycle. The effects of the outlet temperature of gas-cooler and heat rejection pressure on the COP of these two cycles are investigated, and are compared with that of the single Stage Compression (STSC) cycle, which could provide a basis for the optimization of CO_2 transcritical two-stage cycle.

2 TRANSCRITICAL CO_2 TWO-STAGE CYCLES

Figure 1 shows the flow description and pressure-enthalpy diagram of the Single-Throttling, Double-Compression cycle (STDC). High pressure CO_2 vapor coming out of the gas cooler can be divided into two ways: one is throttle to the intermediate pressure P_m through the throttle valve A and into the intercooler; another is passing through the intercooler, which is cooled and then throttled to the evaporation pressure Te through the throttle valve B.

Figure 2 shows the flow description and pressure-enthalpy diagram of the Double-Throttling, Double-Compression cycle (DTDC). Compared with the STDC cycle, the entire high pressure CO_2 vapor coming out of the gas cooler is passing into the intercooler through the throttle valve A. Saturated CO_2 vapor is passing into the HP compressor and saturated CO_2 liquid is passing into the evaporator through the throttle valve B, which produces the refrigeration effect.

3 THERMODYNAMIC ANALYSIS

The following assumptions have been made for the analysis: (1) CO_2 liquid out of the intercooler and CO_2 vapor out of the evaporator is in the

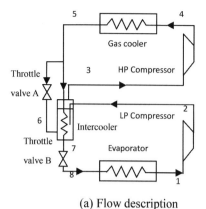

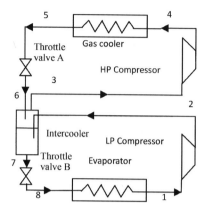

(a) Flow description

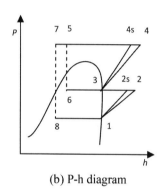

(b) P-h diagram

Figure 1.   STDC cycle.

(a) Flow description

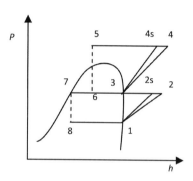

(b) P-h diagram

Figure 2.   DTDC cycle.

saturated state. (2) Compression process is adiabatic but non-isentropic. (3) Heat transfer with the ambient is negligible. (4) Pressure drop in the heat exchanger and piping is negligible.

Mass and energy conservation equations for the STDC cycle are as follows:

$$m_2 + m_6 = m_3, \tag{1}$$

$$m_6 + m_7 = m_5, \tag{2}$$

$$m_1 = m_2 = m_7 = m_8, \tag{3}$$

$$m_3 = m_4 = m_5, \tag{4}$$

$$m_2 h_2 + m_5 h_5 = m_3 h_3 + m_7 h_7. \tag{5}$$

Mass and energy conservation equations for the DTDC cycle are as follows:

$$m_2 + m_6 = m_3 + m_7, \tag{6}$$

$$m_1 = m_2 = m_7 = m_8, \tag{7}$$

$$m_3 = m_4 = m_5 = m_6, \tag{8}$$

$$m_6 h_6 + m_2 h_2 = m_3 h_3 + m_7 h_7. \tag{9}$$

Coefficient of performance of the STDC cycle and the DTDC cycle can be determined by

$$COP = \frac{Q}{W}, \tag{10}$$

$$Q = m_1(h_1 - h_8), \tag{11}$$

$$W = W_1 + W_2 = m_1(h_2 - h_1) + m_3(h_4 - h_3), \tag{12}$$

where Q is the Cooling capacity, and W_1, W_2 is the work of the LP Compressor and the HP Compressor:

$$R = \frac{r_2}{r_1} = \frac{p_4 / p_2}{p_2 / p_1} = \frac{p_1 p_4}{p_2^2}, \tag{13}$$

where r_1, r_2 is the pressure ratio of the LP Compressor and the HP Compressor, and R is the pressure ratio of the HP Compressor and the LP Compressor.

4 ANALYSIS OF THE RESULTS

Figure 3 shows the variation in the COP with the heat rejection pressure (P_c) of the STDC cycle, the DTDC cycle and the STSC cycle (Single-Throttling, Single-Compression) when the evaporation temperature T_e is 0°C, the gas cooler outlet temperature T_c is 35°C, and R = 1. It shows that, first, the COP of the three cycles increases and then decreases with increasing P_c, and the system performance of two-stage compression is better than that of single-stage compression. There is a maximum COP when P_c is about 9 MPa, and the pressure is the optimum high pressure ($P_{c,opt}$). Therefore, the practical system should be run at the optimum heat rejection pressure as far as possible, and then the DTDC cycle will have the best COP. The COP of the STDC cycle and the DTDC cycle is the same when P_c is about 11 MPa.

Figure 4 shows the variation in the maximum COP (COP_{max}) of the three cycles with T_c. It shows that the DTDC cycle has the best COP_{max} under the same temperature conditions. The gas cooler outlet temperature has a great impact on system performance, when T_c increased, COP_{max} decreased

rapidly; the higher the T_e, the better the COP_{max} and the larger the COP_{max} drop.

Figure 5 shows the gas cooler outlet temperature (T_c) at which the best COP is achieved. The $P_{c,opt}$ is basically the same when T_c is 35°C. With the increase of T_c, and $T_e = 0$°C, the $P_{c,opt}$ of the STDC cycle has the maximum $P_{c,opt}$, followed by the STSC cycle and the DTDC cycle. When $T_e = 0$°C, the $P_{c,opt}$ of the STDC cycle has the maximum $P_{c,opt}$ and the STSC cycle has the minimum $P_{c,opt}$.

Figure 6 shows the variation in the COP with the pressure ratio of the HP Compressor and the LP Compressor (R) of the STDC cycle and the DTDC cycle for $T_e = 0$°C and $T_c = 35$°C. It indicates that the optimum inter-cooling pressure for the STDC cycle and the DTDC cycle deviates from the geo-

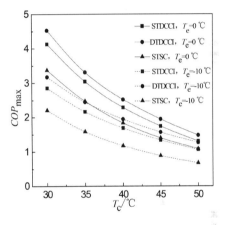

Figure 4. Effect of the gas cooler outlet temperature on COP_{max}.

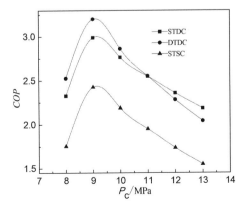

Figure 3. Variation in COP with heat rejection pressure.

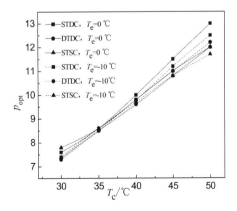

Figure 5. Variation in $p_{c,opt}$ with gas cooler outlet temperature

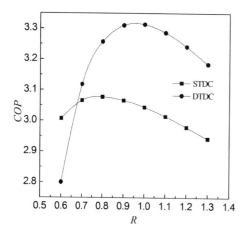

Figure 6.   Variation in COP with the value of R.

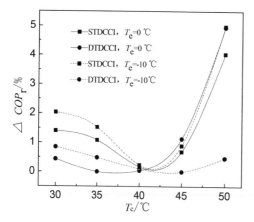

Figure 7.   Variation in ΔCOPr with gas cooler outlet temperature.

metric mean of high pressure and low pressure. The COP of the STDC cycle and the DTDC cycle first increases and then decreases with increasing R. The STDC cycle has a maximum coefficient of performance when R = 0.8, while for the DTDC cycle, the optimum value of R (R$_{opt}$) is about 0.9.

Under different T$_c$ and corresponding optimal high pressure, the maximum coefficient of performance when R = Ropt (COP$_{Ropt}$) is compared with the maximum coefficient of performance when R = 1 (COP$_{max}$). The relative COP improvement (ΔCOP$_r$) can be calculated as follows:

$$\Delta COP_r = \frac{COP_{Ropt} - COP_{max}}{COP_{max}} \qquad (14)$$

The relative COP improvement is shown in Figure 7. The COP improvement first decreases

and then increases with increasing gas cooler outlet temperature T$_c$. When T$_e$ is 0°C, and T$_c$ is less than 40°C, ΔCOP$_r$ of the STDC cycle is greater than the ΔCOP$_r$ of the DTDC cycle. When T$_c$ is higher than 40°C, the opposite is true. When T$_c$ = 50°C, ΔCOP$_r$ of the STDC cycle and the DTDC cycle is about 4% and 5%. When T$_e$ is –10°C, ΔCOP$_r$ of the DTDC cycle is less than 1%, in the whole range of T$_c$.

As shown in Figure 4, in order to ensure a greater coefficient of performance, the system should be operated under lower gas cooler outlet temperature conditions, with T$_c$ = 35°C for example, When T$_e$ is 0°C and –10°C, ΔCOP$_r$ of the STDC cycle is 1.1% and 1.5%, and ΔCOP$_r$ of the DTDC cycle is 0.1 % and 0.5 %, respectively. Due to the lower ΔCOP$_r$ value, the use of the geometric mean of high pressure and low pressure in the system design has a little effect on the system for optimum performance.

5   CONCLUSIONS

Thermodynamic analysis of the two-stage transcritical CO_2 cycle for the STDC cycle and the DTDC cycle has been presented in this paper. The effects of the outlet temperature of the gas cooler and heat rejection pressure on the COP of these two cycles have been investigated, and compared with that of the STSC cycle.

1. The results indicate that at a given condition, the DTDC cycle has the maximum COP, while the STDC cycle has the optimum high pressure, and the gas cooler outlet temperature and evaporation temperature have a great impact on system performance.
2. Optimum inter-cooling pressure for the STDC cycle and the DTDC cycle deviates from the geometric mean of high pressure and low pressure, but this has a little effect on system performance. The use of the geometric mean of high pressure and low pressure in the system design has a little effect on the system for optimum performance.

REFERENCES

[1] Jahar Sarkar. Review on Cycle Modifications of Transcritical CO_2 Refrigeration and Heat Pump Systems. Journal of Advanced Research in Mechanical Engineering, Vol. 1, 22–29, 2010.
[2] Ramon Cabelloa, Daniel Sancheza, Jorge Patinoa, Rodrigo Llopisa, Enrique Torrellab. Experimental analysis of energy performance of modified single-stage CO_2 transcritical vapour compression cycles based on vapour injection in the suction line. Applied Thermal Engineering, Vol. 47, 86–94, 2012.

[3] Mehdi Aminyavaria, Behzad Najafib, Ali Shirazic, Fabio Rinaldib. Exergetic, economic and environmental (3E) analyses, and multi-objective optimization of a CO_2/NH_3 cascade refrigeration system, Applied Thermal Engineering, Vol. 65, 42–50, 2014.

[4] Eunsung Shina, Chasik Parkb, Honghyun Choc. Theoretical analysis of performance of a two-stage compression CO_2 cycle with two different evaporating temperatures. International Journal of Refrigeration, Vol. 47, 164–175, 2014.

[5] Alberto Cavallini1, Luca Cecchinato, Marco Corradi, et al. Two-stage transcritical carbon dioxide cycle optimization: A theoretical and experimental analysis. International Journal of Refrigeration, Vol. 28, 1274–1283, 2005.

[6] Wei Hua Li. Optimal analysis of gas cooler and intercooler for two-stage CO_2 trans-critical refrigeration system. Energy Conversion and Management, Vol. 71, 1–11, 2013.

[7] Xie Yingbai, Sun Ganglei, Liu Chuntao, Liu Yingfu. Thermodynamic analysis of CO_2 trans-critical two-stage compression refrigeration cycle. Journal of Chemical Industry and Engineering, Vol. 59, 2985–2989, 2008.

Architectural, Energy and Information Engineering – Sung & Chen (Eds)
© 2016 Taylor & Francis Group, London, ISBN 978-1-138-02791-6

Thermodynamic analysis of different CO_2 cascade refrigeration cycles

Ying Fu Liu & Ming Xing Zhao
Baoding Electric Power Voc. & Tech. College, Baoding, China

Ying Zheng Rong
The Affiliated School of Hebei Baoding Normal, Baoding, China

Guang Ya Jin
North China Electric Power University, Baoding, China

ABSTRACT: CO_2 as a refrigerant has unique advantages such as environmental safety, favorable thermodynamic properties, and non-flammability. This paper analyzes the performance of five different (NH_3/CO_2, $R22/CO_2$, $R32/CO_2$, $R290/CO_2$, $R404a/CO_2$) cascade refrigeration cycles. The results showed that, at a given condition, the NH_3/CO_2 cycle had the maximum COP, the COP of the $R32/CO_2$ cycle and the $R290/CO_2$ cycle were almost the same. The COP of five different cascade refrigeration cycles increased with increasing evaporating temperatures t_1, while it decreased with the heat-transfer temperature difference of condenser-evaporator Δt and increasing condensing temperature of LTC t_3. In addition, different cascade refrigeration cycles, except for $R404a/CO_2$ cycle, had the best COP with increasing condensing temperature t_7. The $R404a/CO_2$ cycle had the minimum COP when the condensing temperature t_7 was less than –10°C, but it almost had the same COP as the NH_3/CO_2 cycle when condensing temperature t_7 was 0°C.

Keywords: cascade refrigeration cycle; CO_2; thermodynamic analysis; coefficient of performance

1   INTRODUCTION

In refrigeration applications for temperatures between –40 and –80°C, the cascade system is often a convenient option, usually using Freon. Natural refrigerants are increasingly becoming the refrigerant of choice to replace the environmentally harmful CFCs and HCFCs. The former IIR chairman G. Lorentzen [1] was the first to advance the CO_2 transcritical cycle, and suggested that CO_2 is "the most prospects refrigerant of the 21st century." CO_2 as a refrigerant has unique advantages [2] such as environmental safety, favorable thermodynamic properties, and non-flammability.

The results indicated that the Low-Temperature Circuit (LTC) of the refrigeration system may use CO_2, whereas the High-Temperature Circuit (HTC) of a cascade refrigeration system can normally be charged with NH_3, R32, N_2O, R134a, and R1270 [3–7]. However, they have disadvantages that limit their application.

In the paper, thermodynamic analysis is carried out with five different refrigeration cycles (NH_3, R22, R32, R290, R404a) as the high temperature

(HTC) fluid and carbon dioxide as the low temperature (LTC) fluid in the cascaded system. The effects of evaporation temperature, condensing temperature, heat-transfer temperature difference of condenser-evaporator and condensing temperature of LTC are investigated, which could provide a basis for the optimization of the CO_2 cascade refrigeration cycle.

2   THE INVESTIGATED CYCLES

Figure 1(a) schematically depicts different CO_2 cascade refrigeration cycles. There are two different refrigeration cycles: the High-Temperature Cycle (HTC) with five different refrigeration cycles (NH_3, R22, R32, R290, R404a) as the fluid and the Low-Temperature Circuit (LTC) with carbon dioxide as the fluid. The cycles are thermally connected to each other through a condenser-evaporator, which acts as an evaporator for the HTC and a condenser for the LTC. Figure 1(b) schematically presents the corresponding temperature-entropy.

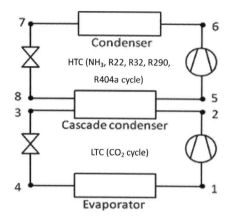

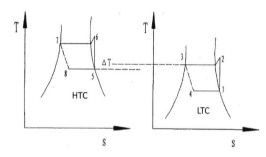

(a) Flow description

(b) Temperature-Entropy diagram

Figure 1. CO_2 cascade refrigeration cycle.

3 THERMODYNAMIC ANALYSIS

According to the quality and energy conservation, a theoretical model of the CO_2 cascade refrigeration cycle was established, and the following assumptions were made to simplify the thermodynamic analysis:

1. All components are assumed to be a steady-state and steady-flow process. The changes in the potential and the kinetic energy of the components are negligible.
2. The compression processes have been assumed to be adiabatic but not isentropic. An isentropic efficiency of 0.8 has been assumed for both compressors.
3. The heat loss and pressure drops in the piping and components are negligible.
4. All throttling devices are isenthalpic.
5. The outlet of the condenser and the cascade-condenser are at saturated liquid states and that of the evaporator is at the saturated vapor state.

In view of the schematic and state points of Figure 1, the following equations were applied for the analysis.

The mass flow rate of LTC can be expressed as

$$m_L = \frac{Q_e}{h_1 - h_5}. \tag{1}$$

Power input for the LTC compressor can be expressed as

$$W_L = \frac{m_L(h_{2s} - h_1)}{\eta_L}$$
$$= \frac{m_L(h_{2s} - h_1)}{\eta_m \eta_{is,L} \eta_{el}} = \frac{m_L(h_2 - h_1)}{\eta_m \eta_{el}}. \tag{2}$$

The coefficient of performance of LTC can be formulated as:

$$COP_L = \frac{Q_e}{W_L}. \tag{3}$$

The rate of heat transfer in the cascade-condenser can be determined from the following equation:

$$Q_{cas} = \frac{Q_e(h_2 - h_4)}{h_1 - h_4}. \tag{4}$$

The mass flow rate of HTC can be expressed as

$$m_H = \frac{Q_{cas}}{h_6 - h_{10}} = \frac{Q_e}{h_6 - h_{10}} \cdot \frac{h_2 - h_4}{h_1 - h_4}. \tag{5}$$

The work input for the HTC compressor is represented by

$$W_H = \frac{m_H(h_{7s} - h_6)}{\eta_H}$$
$$= \frac{m_H(h_{7s} - h_6)}{\eta_m \eta_{is,H} \eta_{el}} = \frac{m_H(h_7 - h_6)}{\eta_m \eta_{el}}. \tag{6}$$

Heat transfer from the HTC condenser is estimated from the following equation:

$$Q_H = m_H(h_7 - h_9). \tag{7}$$

The coefficient of performance of HTC can be formulated as

$$COP_H = \frac{Q_{cas}}{W_H}. \tag{8}$$

The COP of the system can be determined from the following equation:

$$COP = \frac{Q_e}{W} = \frac{COP_L \cdot COP_H}{1 + COP_L + COP_H}. \qquad (9)$$

4 ANALYSIS OF THE RESULTS

The thermodynamic analysis evaluation revealed that the evaporating temperature t_1 ranged from $-55°C$ to $-45°C$, the condensing temperature t_7 ranged from $30°C$ to $40°C$, the heat-transfer temperature difference of condenser-evaporator Δt ranged from $5°C$ to $9°C$, and the condensing temperature of LTC t_3 ranged from $-30°C$ to $0°C$.

Figure 2 shows the variation in COP with the evaporating temperature t_1 of the five different cascade refrigeration cycles (NH_3/CO_2, $R22/CO_2$, $R32/CO_2$, $R290/CO_2$, $R404a/CO_2$) at $t_3 = -10°C$, $\Delta t = 5°C$, and $t_7 = 40°C$. It showed that, at a given condition, the COP of the five different cascade refrigeration cycles increased with increasing evaporating temperature t_1. The NH_3/CO_2 cycle had the maximum COP, while the $R404a/CO_2$ cycle had the minimum COP, and the COP of the $R32/CO_2$ cycle and the $R290/CO_2$ cycle were almost identical.

Figure 3 shows the variation in COP with condensing temperature t_7 of the five different cascade refrigeration cycles (NH_3/CO_2, $R22/CO_2$, $R32/CO_2$, $R290/CO_2$, $R404a/CO_2$) at $t_3 = -10°C$, $\Delta t = 5°C$, and $t_1 = -50°C$. It showed that, at a given condition, the different cascade refrigeration cycles, except for the $R404a/CO_2$ cycle, had the best COP with increasing condensing temperature t_7. The NH_3/CO_2 cycle had the maximum COP, and the COP of the $R32/CO_2$ cycle and the $R290/CO_2$ cycle

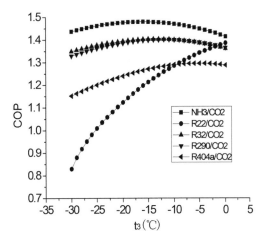

Figure 3. Variation in COP with t_7.

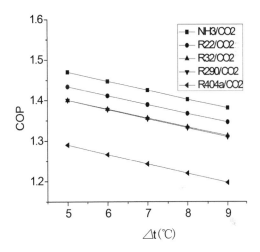

Figure 4. Variation in COP with Δt.

were almost identical. The COP of the $R404a/CO_2$ cycle increased rapidly with increasing condensing temperature t_7, and had the minimum COP when the condensing temperature t_7 was less than $-10°C$, but it almost had the same COP as the NH_3/CO_2 cycle when condensing temperature t_7 was $0°C$.

Figure 4 shows the variation in COP with heat-transfer temperature difference of condenser-evaporator Δt of the five different cascade refrigeration cycles (NH_3/CO_2, $R22/CO_2$, $R32/CO_2$, $R290/CO_2$, $R404a/CO_2$) at $t_3 = -10°C$, $t_1 = -50°C$, and $t_7 = 40°C$. It showed that, at a given condition, the COP of the five different cascade refrigeration cycles decreased with the heat-transfer temperature difference of increasing condenser-evaporator Δt. The NH_3/CO_2 cycle had the maximum COP,

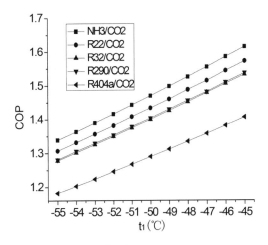

Figure 2. Variation in COP with t_3.

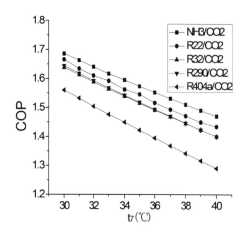

Figure 5. Variation in COP with t_3.

denser-evaporator Δt and increasing condensing temperature of LTC t_3. Then, the NH_3/CO_2 cycle had the maximum COP, while the R404a/CO_2 cycle had the minimum COP, and the COP of the R32/CO_2 cycle and the R290/CO_2 cycle were almost identical.

2. It showed that, at a given condition, the different cascade refrigeration cycles, except the R404a/CO_2 cycle, had the best COP with increasing condensing temperature t_7. The NH_3/CO_2 cycle had the maximum COP, but the COP of the R404a/CO_2 cycle increased rapidly with increasing condensing temperature t_7. It had the minimum COP when the condensing temperature t_7 was less than $-10°C$, but it almost had the same COP as that of the NH_3/CO_2 cycle when the condensing temperature t_7 was $0°C$.

while the R404a/CO_2 cycle had the minimum COP, and the COP of the R32/CO_2 cycle and the R290/CO_2 cycle were almost identical.

Figure 5 shows the variation in COP with condensing temperature of LTC t_3 of the five different cascade refrigeration cycles (NH_3/CO_2, R22/CO_2, R32/CO_2, R290/CO_2, R404a/CO_2) at $\Delta t = 5°C$, $t_1 = -50°C$, and $t_7 = 40°C$. It showed that, at a given condition, the COP of the five different cascade refrigeration cycles decreased with increasing condensing temperature of LTC t_3. The NH_3/CO_2 cycle had the maximum COP, the COP of the R32/CO_2 cycle and the R290/CO_2 cycle were almost identical, while the R404a/CO_2 cycle had the minimum COP.

5 CONCLUSIONS

Thermodynamic analysis of five different cascade refrigeration cycles (NH_3/CO_2, R22/CO_2, R32/CO_2, R290/CO_2, R404a/CO_2) have been present here.

1. The COP of five different cascade refrigeration cycles increased with increasing evaporating temperature t_1, while it decreased with the heat-transfer temperature difference of con-

REFERENCES

[1] G. Lorentzen. Revival of carbon dioxide as a refrigerant. International Journal of Refrigeration, 17(5): 292–301, 1994.

[2] Andy Pearson. Carbon dioxide-new uses for an old refrigerant. International Journal of Refrigeration, 28, 1140–1148, 2005.

[3] Mehdi Aminyavaria, Behzad Najafib, Ali Shirazic, Fabio Rinaldib. Exergetic, economic and environmental (3E) analyses, and multi-objective optimization of a CO_2/NH_3 cascade refrigeration system, Applied Thermal Engineering, Vol. 65, 42–50, 2014.

[4] Jian Xiao, Yingfu Liu. Thermodynamic analysis of a R32/CO_2 cascade refrigeration cycle. Advanced Materials Research, Vol. 732–733, 527–530, 2013.

[5] Souvik Bhattacharyya, Anirban Garai, Jahar Sarkar. Thermodynamic analysis and optimization of a novel N_2O-CO_2 cascade system for refrigeration and heating International Journal of Refrigeration, Vol. 1, 1–8, 2008.

[6] Carlos Sanz-Kocka, Rodrigo Llopisa, Daniel Sáncheza, Ramón Cabelloa, Enrique Torrellab. experimental evaluation of a R134a/CO_2 cascade refrigeration plant, Applied Thermal Engineering, Vol. 73, 41–50, 2014.

[7] Alok Manas Dubey, Suresh Kumar, Ghanshyam Das Agrawal. Thermodynamic analysis of a transcritical CO_2/propylene (R744–R1270) cascade system for cooling and heating applications, Energy Conversion and Management, Vol. 86, 774–783, 2014.

Architectural, Energy and Information Engineering – Sung & Chen (Eds)
© 2016 Taylor & Francis Group, London, ISBN 978-1-138-02791-6

A low-carbon city power energy optimization configuration structure

Yi Peng Li
State Grid Jiangxi Electric Power Research Institute, Nanchang, China

Wei Xu
Nanrui Technology Co. Ltd., Nanjing, China

ABSTRACT: We propose a low-carbon city power energy optimization configuration data center structure. In the construction of an energy optimization configuration platform including six professional application modules (distribution network dispatch, operation and maintenance, marketing, planning, distributed power supply and low-carbon evaluation and validation), the data requirement of all the modules was realized through an unified data center.

Keywords: Low carbon; energy; optimization

1 INTRODUCTION

Low-carbon power grid comprehensive demonstration project is a composition of distributed generation, energy storage, active load and the associated control equipment [1–6]. Gongqing city is located in the north of Jiangxi Province, China, the middle of the Nanchang-Jiujiang industrial corridor. In June 2011, the Japanese NEDO launched the Gongqing "smart community" demonstration project with Jiangxi Province officially in Tokyo. The core idea of the project is to construct an intelligent technology demonstration area by considering both economic and low-carbon development.

The data center of the constructed low-carbon city power energy optimization configuration platform is mainly based on the data of generation, transmission, transform, distribution and usage systems, including the following: GIS system, dispatching automation system, distribution automation system, PMS system, power energy quality management system, electricity usage information acquisition system, voltage monitoring system, photo-voltaic subsystems, distribution network dispatch management system, operation and maintenance management system, marketing and service management system, distributed generation and micro grid system, and planning and design management system.

2 IMPLEMENTATION OF HIERARCHICAL DATA CENTER

According to the subject, data scattered in different functional subsystems in various business departments are extracted and stored through the regularized data model organization in the data warehouse, to form the statistical information data source. The six professional application modules—distribution network dispatch, operation and maintenance, marketing, planning, distributed power supply, and low-carbon evaluation and validation—obtain relevant data directly from the data center and store back the computation results in it.

The system application of the data center is divided into two levels, namely management system and business system, maintained by the data center uninterruptedly. Information services are provided according to different application levels, as discussed below.

- Management system level: it provides comprehensive monitoring of each business system information, and provides comprehensive statistics, correlation analysis and assistant decision-making functions, according to basic data from various business systems.
- Business system level: it provides business statistics and data supporting functions for business systems, and improves the data subject according to the expansion of business systems.

The platform can be divided into the user level and the function level, as illustrated in Figure 1.

Business data source includes both the business system processing data source and the external acquisition data source. The data center system is an expandable warehouse to both realize the comprehensive management of the data and meet the expanding information requirement. The system interface can be functionally divided into the following factors: data extraction, conversion and

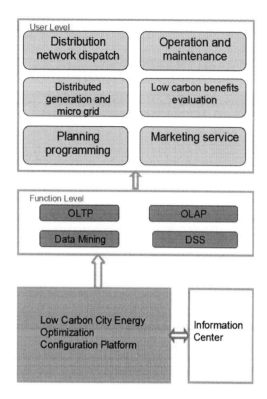

Figure 1. Hierarchical data center structure.

storage interface, information transmission interface, and interface management. Data cleaning, warping and final loading of various professional systems and other data sources can be realized through data extraction, conversion and storage interface, to realize the building of the data warehouse.

At the same time, on the basis of the data warehouse, supporting functions such as OLAP, DSS, and Data Mining are provided, which include public/internal information network interface and OA system interface. The management interface is the interactive interface of the system, including system management interface and metadata management interface. The system management interface provides system management functions (such as authentication function), and the metadata management interface realizes the maintenance and management of the various rules, and the definition of the metadata.

According to the subject, data scattered in different functional subsystems in various business departments are extracted and stored through the regularized data model organization in the data warehouse, to form the statistical information data source. The data model remains relatively stable, but the analysis theme can have flexible expansion based on the extended business system.

The degree of use of various statistical information is growing, so it is impossible to follow the traditional manual statistical analysis model. In order not to affect the normal operation of the business system, statistical analysis jobs that were originally rooted in the business systems are collected, the data center is built independently, and statistical information is separated from business operation data to form independent statistical information data source. At the same time, considering fully the design of the data warehouse metadata, the fact table is maintained expandable with respect to stable business, while the dimension table can be expandable according to the business expansion.

System data is extracted regularly from various business support systems, which guarantees the truth of the data source, the data analysis process is correct, and the statistical information has certain effectiveness, which also effectively avoids the non-uniform caliber situation.

3 SUPPORTING FUNCTIONS OF THE STRUCTURE

3.1 Query

- Clear formatting of the preview of query result, separable query format and data information.
- Stacked and covered type of OLAP function.
- Interactive information between graph and table, multi-table and multi-graph simultaneous query drill function.
- Providing OLAP extended functions such as custom indicators and custom dimension member, supporting complex industrial function and user expandable requirements.
- Integration of the OLAP and OLTP query.

3.2 Report forms

- Export of report forms and storage of multiple report elements.
- Contents of report forms being printable and exportable, effective integration with external systems.

3.3 Alarm monitoring

- Quickly and accurately locating the key data location based on OLAP alarm rules.
- Providing a variety of analysis tools to assist the customization of alarm rules, including the forecast, the breakeven point, and providing the guidance of association rules.

- Providing many kinds of analysis means to find deeper information behind alarm data.
- Providing customization of warning rule set to meet the complex requirements of the scheduling module.
- Providing various relation analysis functions, and continuation of users' hypothetical reasoning.
- Providing correlation analysis of report form levels, and flexible customization of relational path.
- Providing correlation analysis of cross analysis topic to assist jumping relational analysis.

3.4 *Data mining*

- Providing a variety of expert evaluation methods and construction of comprehensive scoring model based on multiple indices for the use of business analysts or industry experts.

3.5 *Access control*

- Based on the access control mechanism of user-role-function-resource, providing users' resource integration strategy across the role to meet the actual needs of the customers.
- Realization of access control and management inside the platform.

3.6 *Application closed loop*

- Realization of the application closed loop on the platform level: problem definition; problem identifying; problem analysis, analytical thinking closer to the user, first hypothesis; analysis; re-hypothesis; and re-analysis.
- Mutual use of information based on core resource of warning and association rules during the closed-loop application.

4 CONCLUSION

With the integration of various kinds of adjustable resources including distributed power supply, reactive power compensation devices, and energy storage devices, the traditional passive distribution network is evolving into an active distribution network, and the low-carbon power grid comprehensive demonstration project is a composition of distributed generation, energy storage, active load and the associated control equipment. A low-carbon city power energy optimization configuration data center structure was proposed to meet the requirement of all the sub-modules.

REFERENCES

[1] Zhou Tianrui, Kang Chongqing, et al. Analysis on distribution characteristics and mechanisms of carbon emission flow in electric power network [J]. Automation of Electric Power Systems, 2012, 36(15): 39–44.
[2] Xiao Xiangning, Chen Zheng, et al. Integrated mode and key issues of renewable energy sources and electric vehicles' charging and discharging facilities in microgrid [J]. Transactions of China Electrotechnical Society, 2013, 28(2): 1–14.
[3] IPCC. Intergovernmental panel for climate change: the fourth assessment report [M], Cambridge, U.K.: Cambridge University Press, 2007.
[4] Grubb M, Jamasb T, et al, Delivering a low carbon electricity system [M], Cambridge, U.K.: Cambridge University Press, 2008.
[5] Jia Wenzhao, Kang Chongqing, et al, Capability of smart grid to promote low carbon development and its benefits evaluation model [J], Automation of Electric Power Systems, 2011, 35(1), 7–12.
[6] Chen Xiaoke, Zhou Tianrui, et al, Structure identification of CO_2 emission for power system and analysis of its low-carbon contribution [J], Automation of Electric Power Systems, 2012, 36(2), 18–25.

Architectural, Energy and Information Engineering – Sung & Chen (Eds)
© 2016 Taylor & Francis Group, London, ISBN 978-1-138-02791-6

The preliminary exploration for city entrance landscape planning based on semiotics: An example of east entrance planning of Weinan City

J.Z. Zhou & Q.H. Hou
Chang'an University, Xi'an Shanxi, China

Y. Hu
University College London, London, UK

ABSTRACT: Along with rapid urbanization, the entrance space of cites is usually in chaos because of urban expansion. Although government advocates distinctive city culture, most of the cities' entrances are struggling with the absence of the consciousness that shapes itself as a window to display city characteristics. This article addresses the city entrance issue, using the landscape semiotics expression method. It contains the means of pragmatics, syntactics and semantics. The semiotics method is able to integrate the entrance space and shape landscape characteristics. Finally, this paper briefly introduces the east entrance planning of Weinan City to explain that the semiotics design method has the potential to solve the chaotic entrance issue as well as create a distinctive urban landscape.

Keywords: city entrance; Weinan city; landscape express; semiotics; landscape symbol

1 INTRODUCTION

City entrance is a connection between the inner city and the external environment. It contains the airport, railway stations, high speed railway stations, highway entrance, and wharf. Among these categories, this article only discusses highway entrance. Along with the development of the city, the public and the government pay more attention to the characteristics of entrance. High way entrance should not only meet the needs of the urban external traffic, but also show the characteristic of the city. However, rapid urbanization results in the spontaneous urban edge extension, which leads this area fail to be constructed as qualified as the urban development required. The entrance areas not only have a lot of construction sites, industrial workshops, old residences, old facilities, but also have some new public service buildings and residential buildings. There are difficulties coordinating them with landscapes. Moreover, as the window of the city, most of the city entrances contain little performance to display local characteristics and urban landscapes. At present, these two aspects are usually valued by the government. If taking the former planning of city entrance as references, the main spatial design techniques include "point-line-domain" (Zhang et al., 2005) aspect and "planning integration-space design-operating management"

(Cheng et al., 2008) aspect. By using these techniques, city entrances can meet the functional requirements as well as the design of the basic landscape. However, when concerning the aspects of reflecting the local characteristics and urban landscape, there is a necessity to imbed relevant measures into design process. In order to provide a reference for future city entrance landscape planning, this paper proposes a method that can express landscape by the semiology method. Through the means of pragmatics, designers are able to convert present-situation landscape elements to language symbols, and then use the means of syntactics and semantics to integrate the entrance space as well as shape the characteristics of city entrance.

2 SEMIOTICS RESEARCH FOR CITY ENTRANCE LANDSCAPE EXPRESSION

2.1 *Theoretical research*

Semiotic theory originated in the early 20th century, which was first proposed by a Swiss philosopher, linguist Saussure. During the same period, the American philosopher, logician Pearce also proposed semiotic theory. The two scholars played an important role in the establishment of the semiology subject. During its development, semiotics gradually penetrated into other disciplines, and emerged into a diverse semiotic theory system. The most prominent is the trichotomy theory of semiotics, which was proposed by Morris in 1946. He divided semiotics into three parts, namely pragmatics, syntactics and semantics. He also associated all symbols and symbolic phenomena with human life, using the method of semiotics. It is a reasonable approach to study social and cultural issues including all kinds of material, spirit and behavior.

Landscape is also a kind of symbol. Thus, landscape can be read as the literature. In the modern landscape planning, it can be expressed in various ways. Aiqing Wang summarized its methodology as direct, reproduce, simulate, freehand brushwork, abstract, associate, and reference (Wang, 2009). Min Chen summarized its methodology as reference, retention, transformation, recurrence, symbol and metaphor (Chen, 2008). Some relevant professional refined as direction, simulation, abstract, and metaphor (Liu et al., 2005). Tongyu Li performed research on the semantic communication mechanism in landscape design (Li, 2013). Mojv Xu considered that symbol is an element of visual expression that can express urban culture (Xu, 2010). In conclusion, semiotics and expression of landscape planning has its inner link. Semiotic theory based on language rhetoric can be applied to landscape expression.

2.2 The application of semiotics in entrance landscape expression

In the entrance design, we can apply trichotomy theory of semiotics, which was proposed by Morris, to build the research framework of city entrance landscape planning. Among them, pragmatics can be used to explore local characteristics as well as interpretation and transformation of the cultural connotation. Syntactics can be used to integrate landscape morphology, which is based on material space, while semantics can be used to express non-material cultural elements.

2.2.1 Interpretation and transformation based on pragmatics

According to certain combination (grammar), landscape symbols (words) are able to form a particular landscape (coded into the text). Then, visitors receive and interpret the landscape symbols and obtain internal information though their senses. In landscape design, excavation, transformation and interpretation are equally important (Figure 1).

Before the entrance landscape planning, it is necessary to pay more attention to understand the characteristic of the city where the entrance

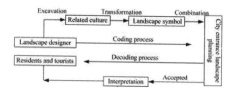

Figure 1. Framework for landscape planning based on pragmatics.

is located in, such as architecture, customs and culture. City scape of the region, for example vegetation and hydrological topography, is also important. Only by converted culture resources to landscape symbols followed by material space construction and cultural atmosphere creation can landscape planning express the connotation appropriately. In order to allow the audience to accept the ideas better, the methods of syntactics and semantics should be used to arrange the converted landscape symbols.

2.2.2 Interpretation and transformation based on pragmatics

Similar to the arrangement of texts and symbols, the principles of syntactics can be followed through different ways of coding method of syntactics to form landscape with cultural connotations. In the city entrance landscape design, these landscape symbols can be recombined and arranged into a new environment. There are five encoding methods, namely retention, restoration, citation, reconstruction and replacement. In fact, these methods can also be used synthetically (Table 1).

By using the methods of syntactics, we can integrate landscape space with material space. It can not only be used to avoid the entrance chaotic, but also provide reasonable protection to urban historical cultural relics.

2.2.3 Interpretation and transformation based on pragmatics

Similarly, the immaterial form of landscape is a symbol of semantic. It includes traditional history culture, regional spirit culture, regional characteristic culture, science culture, and philosophy culture. Its expression methods include simile, metonymy, metaphor and inscription (Table 2).

In fact, the technique of semantics is mainly used to show the characteristics of the city, as well as to express city spirits and develop city context. The use of the semantics technique flexibly can solve the problem of lacking non-material cultural in entrance. Furthermore, it can solve the issues of the absence of landscape characteristics and the need for historical relic protection.

Table 1. The design technique application of syntactics.

Design technique	Design connotation	Application object
Retention	Retain original historical and cultural landscape in the original position, and combine them with new landscape symbols, so that the historical and cultural aspects can continue in the new landscape	The cultural relics of the city; urban architecture with the regional characteristics; historic districts; villages with regional characteristics, etc.
Restoration	Restore it through repairing	Old buildings; broken road; difficulty used space; muddy riverbank; patches of green space
Citation	Introduce the ready-made statements to illustrate the problem, and strengthen the persuasion	The color and detail of the building and structure, sculpture, and landscape sketch
Reconstruction	Dissolve the original landscape symbol structure, and rearrange it by new design rules	Hybrid function and the discord between the old and the new buildings make the entrance environment to lack coordination
Replacement	Replace the original symbols with similar meaning symbols, so as to keep the meaning unchanged	Old village; more damage buildings and structures; old sculpture landscape

Table 2. The expression application of semantics.

Design technique	Design connotation	Application object
Simile	Express the cultural connotation of landscape symbols by a realistic technique	Urban characteristic cultures include architectures, customs, cultures, celebrities and famous things
Metonymy	Through another landscape symbols to express the original cultural connotation, but its cultural connotation does not change	The substance once owned but no longer exists now, but its cultural connotation is still preserved
Metaphor	Through the metaphor technique to form a symbolic sign	The color of the urban architectures, and its characteristic culture
Inscription	Write text messages directly on the landscape to understand easily	Signs, inscriptions, horizontals and some record on stones

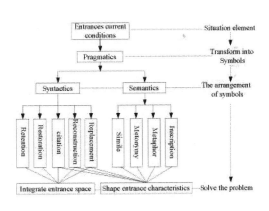

Figure 2. Construction of the method system.

Although the semiology method can be divided into pragmatics, syntactics and semantics, these three parts need to be mutually merged in practical applications. The paragraph text of the landscape is often complex and delicate, which needs a variety of methods to arrange paragraphs and sentences (Figure 2).

3 APPLICATION RESEARCH ON ENTRANCE PLANNING OF WEINAN CITY

3.1 *Present situation analysis*

The planning area is located in the east entrance of Weinan City. However the image of status quo of entrance is poor. The buildings on both sides of the roads are dilapidated and unrepaired. The various space elements and urban public service facilities are the lack of systematic design. So, they did not form a unique urban landscape (Figure 3).

3.2 Excavation and expression of cultural connotation based on pragmatics

Weinan City has a profound cultural background. It is located in the east of Guanzhong plain, next to the confluence of three rivers. It is generally considered as the birthplace of the Chinese nation whose ancestors are Yan Emperor and Yellow Emperor. The old city zone in the north of the planning area is rich in historical and cultural resources, such as Drum tower, Confucian temple, City temple, Wall site, Huidui site, and Liugu spring site. (Table 3).

3.3 Arrangement of landscape symbols based on syntactics

Syntactics can be used to prevent the chaos of the entrance space. We use the technique of reconstruction to remove all the illegal structures such as huts, buildings, as well as shops and stalls that obstruct the traffic. In addition, various poles, overhead lines, and disorder advertisements and signs need to be cleaned up, and repairing and painting the outside walls that have been damaged or very old are also beneficial. The main color of the buildings on the street is gray, ochre, and mineral yellow. We use the technique of citation to apply distinctive symbols to architecture components, door heads, windows, advertisements and lamps. By using the method of simile and citation, we can show the characteristics of the city in a more explicit way, for example, sculptures, sketches, relief walls, road pavements, and service facilities.

Syntactics can also be used to shape the landscape characteristics of city entrance. Let us take the landscape design of historical and cultural park as an example, which is a showcase for Weinan culture. It can be used for retaining the original city wall site, City temple, Confucian temple and other historical monuments, repairing some damaged walls and the moats before the walls, and then the moat reconstruction for landscape water corridor, decorated with fountains,

Table 3. Characteristic excavation and expression of Weinan City.

City characteristic resources		Symbol conversion
Culture of "Three Saint"	"Word Saint" named Cangjie Chinese character culture	Chinese characters culture square; character element expression on sculpture, landscape sketch and street lamp
	"Wine Saint" named Dukang Dukang wine	Wine culture park and square; wine culture museum and block; landscape sketch
	"History Saint" named Sima Qian Book of Shiji	Square; sculpture; culture wall
Building remains	Drum tower, Confucian temple, City temple, Wall site, Huidui site, Liugu spring site	Combined with modern landscape to shape some important nodes in Weinan City
	Guanzhong residential, three-in-one dwelling and four-in-one dwelling	Traditional folk blocks
Waterfront landscape	Youhe eight sceneries, Ancient bridge, Youhe park and Bridgehead square	To build harmonious space of hydrophilic by plank way and water platform
Marketplace life	Clothing, Diet, Music residential and Entertainment	Sculpture square, folk blocks and cultural park
Urban color	Gray, Ochre, Khaki	The main colors of the new city area are khaki, white and ochre, but gray and ochre are the colors of the old city area

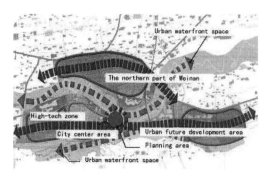

Figure 3. The location of the planning area in Weinan.

columns and sketches. The technique of citation can be used to arrange hard floor tiles, with soft floor tiles forming some characteristic patterns of Weinan. There is a landscape green ladder inside the wall, by using the technique of reconstruction to make the green ladder distinct levels and abundant changes.

3.4 *Performance of city connotation based on semantics*

In fact, the technique of semantics is mainly used to show the characteristics of the city, as well as to express the spirit of Weinan City and continue its city context. In order to show the spirit of "Sima Qian" which is the core spirit of Weinan, there is a culture square build in the old city area, which is located in the northwest of entrance. Its square refers to the technique of extraction, extracting the element of Han dynasty's stele. It is also set up eight landscape columns standing on both sides of the square flatter with dark red pavement to highlight the grand momentum of the entrance square. Sima Qian's statue is placed in the middle of square (Simile). There are two relief walls beside the statue, carved with Sima Qian's life story (Inscription). The technique of metaphor is used to set the spot 'Historical picture', whose style is primitive and simple. However, the special floor is designed with a historical atmosphere. Behind the square is the spot 'Ink pond of Chunqiu'. The landscape is dominated with black stone decorated with small waterfall.

The east entrance planning of Weinan also include other squares and blocks, the details of which is not described in this paper owing to lack of space.

4 CONCLUDING REMARKS

Under the background of rapid expansion of the city, the city entrances face many problems. In order to seek some new methods for the entrance landscape design, this paper preliminarily discusses the application of semiotics in the entrance landscape planning. The research and the method remains to be further explored. However, the original intent of this research is a starting point, which will be helpful for conducting further study.

REFERENCES

Chen, M. 2005. Exploration and Expression of History & Culture in Urban Landscape Design. *Master degree thesis of Beijing Forestry University.*

Cheng, D. & Xiao, M. 2008. Streetscape shaped in city portal area. *Science & Technology Information* (17): 104.

Li, T.Y. & Zou, G.T. 2013 Semiotic Interpretation Expressed in the Cultural Connotation of Landscape Design. *Special academic papers* (10): 89–92.

Liu, Z.H. 2005. Harbin city parks regional culture expression of research. *Master degree thesis of Northeast Forestry University.*

Wang, A.Q. 2009. Study on exploration and expression of cultural elements in urban public space—Take Xiang monarch's hometown detailed planning and Songshan cultural court planning as examples. *Master degree thesis of Xi'an University of Architecture & Technology.*

X, M.J. 2010 Symbol—Visual Expression of Urban Culture. *Journal of HuBei Vocational College of Ecological Engineering* 8(2): 43–46.

Zhang, L.Z. & Xiang, Z.Y. 2005. Study on the Method of Planning for Road Landscape Improvement of City Entry and Exit—With the West City Entry-Exit of Wenzhou as an Example. *Planners* 7(21): 43–44.

Architectural, Energy and Information Engineering – Sung & Chen (Eds)
© *2016 Taylor & Francis Group, London, ISBN 978-1-138-02791-6*

Multiple scales trend analysis of marine water quality of recent ten years in Tianjin nearshore district

M.C. Li, Q. Si & Y. Wang
Laboratory of Environmental Protection in Water Transport Engineering, Tianjin Research Institute of Water Transport Engineering, Tianjin, China

ABSTRACT: Trend analysis of marine water quality by the multiple scales is an important means to obtain the marine environmental situation and the developed trend of marine water quality for guiding the environmental management. In this paper, the max, min and average field data of Dissolved Inorganic Nitrogen (DIN) and petroleum in Tianjin nearshore marine district, Bohai Bay is analyzed. The whole water quality situation of DIN and petroleum could be grasped through the multiple scale trend analysis. The research results show that the pollution level of research marine district is on the rise. So more management means must be implement to decline the level of environmental pollution.

Keywords: marine; water quality; trend analysis; nearshore; multiple scales

1 INTRODUCTION

The multiple scale trend analysis of marine pollution time series is an important method for obtaining the marine environmental status. It is very useful for guiding the environmental management and the ecological protection. The environmental status of various pollutants has been studied to guide the marine environmental management and nearshore engineering construction, such as the water quality [1], phytoplankton biomass [2], the petroleum hydrocarbon [3], etc. Assessment and analysis methods are applied for marine environment status analysis problem, such as the set pair analysis method for water quality and carrying capacity [4], the neural network model for carrying capacity [5], etc.

In this paper, the max, min and average of Dissolved Inorganic Nitrogen (DIN) and petroleum in the Tianjin nearshore district, Bohai Bay is analyzed. The whole water quality situation could be grasped through the trend analysis of recent ten years.

2 RESEARCH MARINE DISTRICT

2.1 *Basic information*

In this paper, Tianjin nearshore marine district is studied. The total area of Tianjin is about 2270 km^2. The coastline is about 153 km (Fig. 1A). Coastal area is 3000 km^2. It is very shallow and its averaged water depth is less than 20 m. The sea bottom is very flat and its mean slope is less than 2%. With the development of the economy exploitation, marine pollution and ecological hazards are aggravated continuously [6]. So the environmental quality of Tianjin nearshore marine district must be studied to guide the marine management, engineering construction and ecological protection.

2.2 *Research marine position*

Ten years' field data in Tianjin nearshore marine district near the Tianjin port are selected for the trend analysis. More stations are selected for obtaining the max, min and average value. The main Tianjin nearshore marine district named Tianjin port is shown in Figure 1B.

3 MULTIPLE SCALE TREND ANALYSIS

The least squares fitting method is applied to construct the trends of two indexes with ten years' field data. The trends of DIN and petroleum are shown in Figure 2–4 respectively.

3.1 *Max value of research district*

The max value time series of research marine district and its trend is shown in the Figure 2.

Figure 2 shows the max value time series in the research marine district. The DIN belongs to the third water quality standard. The phosphate belongs to the third water quality standard. The

A

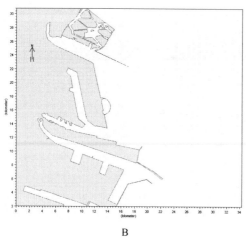

B

Figure 1.　The position of research marine district.

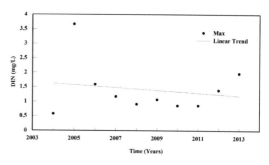

DIN

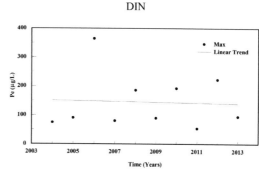

Petroleum

Figure 2.　The max value in the research marine district.

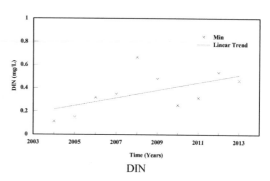

DIN

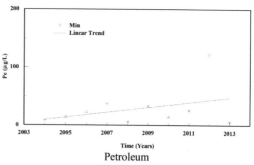

Petroleum

Figure 3.　The min value in the research marine district.

trend of DIN and petroleum show the downtrend in the research ten years.

3.2　Min value of research district

The min value time series of research marine district and its trend is shown in the Figure 3.

Figure 3 shows the min value time series in the research marine district. The DIN belongs to the first water quality standard. The phosphate belongs to the second water quality standard. The trend of DIN and petroleum is on the rise in the research ten years.

3.3 *Average value of research district*

The average value time series of research marine district and its trend is shown in the Figure 4.

Figure 4 shows the average value time series in the research marine district. The DIN belongs to the first water quality standard. The phosphate belongs to the second water quality standard. The trend of petroleum is on the rise in the research ten years. But the trend line of DIN is horizontal.

4 CONCLUSIONS

In this paper, the Dissolved Inorganic Nitrogen (DIN) and petroleum in Tianjin nearshore marine district, Bohai Bay is analyzed through the max, min and average value field data. The max value in the research marine district belongs to the third water quality standard. The others are better than the max. The analysis results show that the distribution of pollution is disequilibrium. The high value district of pollution exists in research marine district. So muchmore research need be used to analyze the cause of the high value district. Although the max value of DIN and petroleum show the downtrend, the min and average value is on the rise, which shows the pollution level of research marine district is improved. So more management means must be implement to decline the level of environmental pollution.

ACKNOWLEDGMENTS

This work was supported by the National Natural Science Foundation of China (No.51209110), the project of Science and Technology for Development of Ocean in Tianjin (KJXH2011-17) and the National Nonprofit Institute Research Grants of TIWTE (TKS130215 and KJFZJJ2011-01).

REFERENCES

[1] Q. Si, M.C. Li, G.Y. Zhang, S.X. Liang, Z.C. Sun, Set pair analysis method for water quality evaluation based on nonlinear power function. Advanced Materials Research, vol. 573–574, pp. 497–500, 2012.

[2] H. Wei, L. Zhao, Z.G. Yu, J. Sun, Z. Liu, S.Z. Feng, Variation of the phytoplankton biomass in the Bohai Sea, Journal of Ocean University of Qingdao. Vol. 33, pp. 173–179, 2003.

[3] X. Wang, L.R. Xu, L.Y. Li, W.Q. Li, Relationship between the distribution of petroleum hydrocarbon and environmental factor in Dapengao, Daya Bay, Journal of Oceanography in Taiwan Strait. Vol. 21, pp. 167–171, 2002.

[4] M.C. Li, G.Y. Zhang, S.X. Liang, Z.C. Sun, Marine water quality assessment's nonlinear set pair analysis method, Advanced Materials Research. Vol. 113–116, pp. 185–190, 2010.

[5] M.C. Li, Q. Si, S.X. Liang, Z.C. Sun, Carrying capacity assessment of Tianjin Binhai New District by artificial neural network, The 3rd International IEEE Workshop on Intelligent Systems and Applications. pp. 792–794, 2011.

[6] X.H. Liu, Analysis on environment situation of coastal areas of Bohai Sea, Environmental Protection Science, vol. 36, pp. 14–18, 2010.

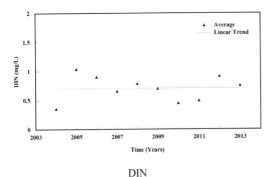

DIN

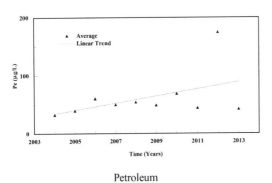

Petroleum

Figure 4. The average value in the research marine district.

Architectural, Energy and Information Engineering – Sung & Chen (Eds)
© *2016 Taylor & Francis Group, London, ISBN 978-1-138-02791-6*

Synthesis of SiO$_2$ aerogel using coal gangue at an ambient pressure

Jin Meng Zhu & Hui Wu Cai
Xi'an University of Science and Technology, Shaanxi, China

ABSTRACT: The hydrophobic SiO$_2$ aerogel was prepared by using coal gangue as the starting material—which is viewed as a waste in most coal mining and preparation plant—via ambient pressure drying through the sol-gel method. We achieved SiO$_2$ aerogel with a low density of 0.2599 g/cm^3, a high porosity of 88.97%, a strong hydrophobicity contact angle with water of 140.61, and a high temperature stability of 360°C~400°C after surface modification using the TMCS/ethanol/n-hexane mixture solution, which demonstrated that the final product exhibited an excellent property. Differences in SiO$_2$ aerogel properties are discussed based on industrial sodium silicate (A) as a precursor, sodium silicate derived from coal gangue (B) as a precursor and coal gangue-derived sodium silicate through exchange resin (C) as a precursor. The results revealed that C had the best properties compared with the others. The SEM image of B showed that the sponge-like silica structure and pore are uniformly distributed along with a small amount of NaCl, contributing to the increased density and reduced contact angel with water. Removal of impurities will be helpful in the improvement of properties. The SiO$_2$ aerogel synthesized using coal gangue has better properties than that obtained using industrial sodium silicate whensubjected to the same procedure.

Keywords: Coal gangue; SiO$_2$ aerogel; Surface modification

1   INTRODUCTION

Aerogels are low-density nano-porous non-crystalline solids whose structures can be controlled through preparation [1]. In recent years, they have attracted significant applications in many aspects such as catalysis, adsorption, window insulating systems, and drug delivery systems due to its high porosity (>90%), high specific surface area (500~1000 m^2/g), low bulk density (~0.03 g/cm^3), extremely low thermal conductivity (0.005 W/mk), and low sound velocity (100 m/s) [2–8]. However, its high cost and toxicity have restricted its production, which is now produced using costly silica gel and alumina gel or metal organics as the starting material. Chen et al. successfully synthesized SiO$_2$-Al$_2$O$_3$ binary aerogel from fly ash at an ambient pressure, which overcame the above-mentioned drawbacks, giving a good example for making good use of waste material and reducing material costs [9]. Shi et al. prepared silica aerogels using fly ash via a cost-effective route at normal pressure [10].

As with the composition of coal ash, coal gangue consists of a large amount of valuable mineral resources such as SiO$_2$ and Al$_2$O$_3$, which can be reused to prepare other high-cost products. As it is well known, coal gangue is the main residue in the process of coal mining and washing. Although its comprehensive use has been studied for many years, the utilization efficiency of coal gangue is still low and mostly made as brick, cement, and constructing materials, especially in China. In the literature, the SiO$_2$ aerogel synthesized using coal gangue as the raw material has not yet been reported. In this paper, we demonstrate the production of the SiO$_2$ aerogel employing coal gangue via the sol-gel and ambient-pressure drying methods.

2   MATERIALS AND METHODS

The coal gangue sample used in this experiment was obtained from Shenmu Shanxi. The chemical composition was examined according to GB/T 212-2008, and found that SiO$_2$ accounts for 66.31%. The mineralogical phase of coal gangue was examined by XRD, as shown in Figure 1. The XRD pattern illustrates that coal gangue mainly consists of α-quartz, which demonstrated that it is possible to synthesis SiO$_2$ aerogel utilizing coal gangue if an appropriate method is adopted. In addition, analytically, pure Hydrochloric acid (HCL) and sodium hydroxide (NaOH) were purchased from Xi'an Shanxi. Activation methods of coal gangue and optimal parameters of leaching rate of SiO$_2$ from coal gangue have been described in our previous work [11].

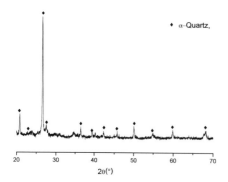

Figure 1.   XRD pattern of raw coal gangue.

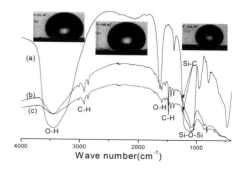

Figure 2.   FTIR spectra of SiO₂ aerogel based on (a) A, (b) B and (c) C.

Sodium silicate of industrial grade was purchased from Xian, China. The chemical content of sodium oxide (Na₂O) was 2.6 wt%, with the molar ratio of silica (SiO₂) to Na₂O being 1.08. The 732 cation exchange resin, trimethyl-chlorosilane (TMCS ≥98.0%), anhydrous ethanol (≥99.7%), and n-hexane (≥97.0%) were purchased from Xian, China. The solution of sodium silicate prepared from coal gangue was adjusted to a pH of 8–9 with H₂SO₄ (5 mol/L). Then, the solution system was carried out by keeping motionless to generate hydrogel. In order to eliminate Na+ and SO42- as much as possible, the hydrogel was positioned in deionized water for 72 h, changing the water every 24 h. Then, the achieved silica aqueous gel was immersed in anhydrous ethanol for 24 hours. To induce hydrophobicity, the aged gel was soaked in a mixed solution of TMCS/ethanol/hexane, with the volume ratio of ethanol to TMCS of 2:3, and then used for solvent exchange/surface modification of the silica aqueous gel [12]. Thereafter, the alcogel was subjected to drying at ambient temperature for 24 h, and then heat-treated at 50°C for 1 h and at 150°C for 1 h in air. The temperature during ageing and surface modification stage was kept constant at 50°C. Industrial sodium silicate as the precursor was employed at the same condition for comparison. On the other hand, a certain amount of water glass extracted from coal gangue was passed through an ion exchange column filled with strong acidic styrene type cation exchange resin. Then, silicic acid was collected and adjusted to a pH of 5–6 with 1.0 mol/l NaOH solution in order to obtain the silica aqueous gel. Meanwhile, after gelation, the wet gel was soaked in anhydrous ethanol for 24 h. The conditions of drying and surface modification are the same as described previously. Density, porosity and pore volume were calculated as described in a previous report [13]. Features including thermal stability and hydrophobicity were confirmed by thermogravimetric and differential thermal analysis (TG-DTA), Fourier transform infrared spectroscopy (FTIR), and contact angle measurements, respectively.

3   RESULTS AND DISCUSSION

SiO₂ aerogel was prepared using different precursors. It can be seen that for the three precursors, the peaks around 3447 cm⁻¹ and 1607 cm⁻¹ were attributed to the O-H stretching band. However, the broad absorption peaks of (a) at the two points indicated that the hydrophobicity of (a) was low compared with that of (b) and (c), which can be confirmed by the contact angle with water of (a) 131.01, (b) 140.61, and (c) 143.62. For (b) and (c), the existence peaks centered at 1259 cm⁻¹ were related to Si-C bonding, which demonstrated that the SiO₂ aerogel was successfully modified by surface modification, while those at 2924 cm⁻¹ and 1495 cm⁻¹ corresponded to the terminal–CH₃ group. The strong absorption band near 1093 cm⁻¹ was attributed to Si-O-Si asymmetric stretching vibrations for all of them. The complete surface modification of (b) and (c) caused strong hydrophobicity, which showed that it was feasible to produce the SiO₂ aerogel utilizing coal gangue as the raw material and not the optimal parameters for (c) to prepare the SiO₂ aerogel if the same conditions were applied to (b) and (c).

The mineralogical phase and structure of SiO₂ aerogel based on B were estimated by X-ray diffraction (XRD) and scanning electron microscope (SEM). Steamed bun shape and weak diffraction intensity peak shown in the XRD pattern (Figure 3) demonstrated that the SiO₂ aerogel produced based on B was identical to the disordered amorphous silicon oxide [14]. The sharp and strong peaks of NaCl crystal were also observed, which indicated that impurities, such as NaCl, were not eliminated entirely in the process of the experi-

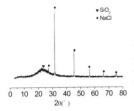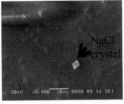

Figure 3. XRD pattern and SEM image of SiO₂ aerogel based on B.

Table 1. Physical properties of different samples.

Sample	Taping density (g/cm³)	Porosity (%)	Pore volume (cm³/g)
Based A	0.2559	87.08	3.40
Based B	0.3962	79.99	2.02
Based C	0.2030	89.75	4.42

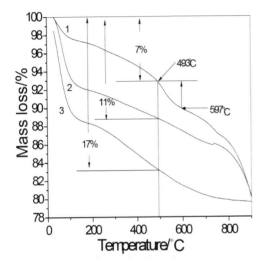

Figure 4. TG curve of the hydrophobic SiO₂ aerogels based on 1 C, 2 B, and 3 C.

ment. The SEM image of the SiO₂ aerogel based on B, as shown in Figure 3, suggested a structure of silica aerogel synthesized by the ambient-pressure drying method. We can see that the larger pores were distributed uniformly. The highly porous network structure with continuous porosity and coagulated particle was maintained, which was caused by the spring-back during ambient pressure drying. However, a distinct cube NaCl crystal present in the particle of the SiO₂ aerogel was observed. It can be inferred that NaCl caused an increase in the density of the SiO₂ aerogel based on B and a decrease in its contact angle (Table 1), and further elimination during preparation will enhance the significant properties of the SiO₂ aerogel based on B.

Figure 4 shows the TG curve of the three kinds of hydrophobic SiO₂ aerogels in oxygen at temperatures up to 900°C. Compound 1 experienced minor mass changes (7%) before 493°C, and from 493°C to 597°C, a rapid weight loss was observed, which is attributed to the oxidation of surface

methyl ($-CH_3$) groups. The thermal stability (of compound 1) can be maintained up to 493°C, and it will be converted into hydrophilic groups beyond that temperature. For compounds 2 and 3, this phenomenon (rapid weight loss) does not appear in the range of 0-900°C. The decrease in their weight (11%, 17%) at 493°C indicated that the $-CH_3$ groups attached to the silica aerogel surface were oxidized, that is, their thermal stability was less than 493°C. Those mentioned above demonstrated that it was favourable for the SiO₂ aerogel produced by the precursor prepared from coal gangue to undergo ion exchange resin method. According to compound 1, experiencing the same mass changes for compound 2 needs the temperature raise up to 360°C~400°C; hence, it can be speculated that the thermal stability of compound 2 will reach up to 360°C~400°C. Similarly, we know that the temperature for the conversion of compound 3 to hydrophilic groups is 250°C~300°C.

4 CONCLUSIONS

Silica aerogel can be synthesized by using coal gangue as the starting material via the ambient pressure drying process. The FTIR spectra of SiO₂ aerogel based on coal gangue show that the aerogel achieved hydrophobicity with the surface modification of TMCS. SiO₂ aerogels that we synthesized using coal gangue possess a high pore volume (3.40 cm³/g), larger contact angle (140.61), low density (0.2559 g/cm³), high porosity (87.07%) and thermal stability (360°C~400°C). Particularly, the prepared silica aerogels by the resin-exchange method show a higher pore volume (4.42 cm³/g), porosity (89.75%), thermal stability (493°C), greater contact angle (143.62), and lower density (0.2030 g/cm³) than those obtained without exchanging with resin. Furthermore, impurities such as NaCl existing in the final product have adverse effects on the entire properties of SiO₂ aerogels. By comparison with industrial sodium silicate as the precursor, synthesizing SiO₂ aerogels based on coal gangue possess better properties in the same parameters. Our work provided a potential way to synthetize a high value-added product from the wasteby-product coal gangue.

REFERENCES

A. Parvathy Rao, Venkateswara Rao. J., Mater. 2010. Sci 45: 51–63.

Bond, G.C., Flamerz, S. 1987. Appl Catal 33: 219.

Chen, N., Yan, Y., Hu, Z.H., 2011. J Wuhan University of Technology 33: 37.

Gao, G.M., Miao, L.N., Ji, G.J., Zou, HF & Gan, SC. 2009. Mater Lett 63: 2721–4.

Hrubesh, L.W., 1990. Chem Ind. 17: 824.

Lee, C.J., Kim, G.S., Hyun, S.H., & Mater, J. 2002. Sci 37: 2237–2241.

Lin, G.T., Zhang, D., & Cent, B.J., 2006. South Univ 37: 1117.

Rao, A.V., Kulkarni, M.M., Amalnerkar, D.P., & Seth, T. 2003. Appl Surf Sci 206: 262–70.

Rao, A.V., Pajonk, G.M., Bhagat, SD & Barboux, P. J. 2004. Non-Cryst Solids 350: 216–23.

Shi, F., Liu, J.X., Song, K., & Wang. Z.Y., 2010. J Non-Cryst Solids 356: 2241.

Shi, F., Wang, L.J., Liu, J.X., 2006. Mater Lett 60: 3718.

Wang, J., Uma, S. & Klabunde, K.J., 2004. Micropor. Mesopor. Mater. 75: 143.

Yuan, B., Ding, S.Q., Wang, D.D., Wang, G., & Li, HX. 2012. Mater Lett 75: 204–6.

Zhu, J. 2014. m. 3rd International Conference on Chemical Engineering, Metallurgical Engineering and Metallic Materials, Gui Lin, China.

Architectural, Energy and Information Engineering – Sung & Chen (Eds)
© 2016 Taylor & Francis Group, London, ISBN 978-1-138-02791-6

The study of acute toxicity of perfluorooctanoic acid on Zebra fish

Wen Chao Zhao
Guizhou University, Guiyang, Guizhou, China
Guizhou Academy of Testing and Analysis, Guiyang, Guizhou, China

Qing Li, Fei Xiao, Guang Lin Song, Yang Lu, Hong Bo Yang & Chao Xuan Liao
Guizhou Academy of Testing and Analysis, Guiyang, Guizhou, China

ABSTRACT: To study the toxic effect of Perfluorooctanoic Acid (PFOA) on Zebra fish, this research adopts static experiment method to study the acute toxicity and conducts ecotoxicology harmfulness evaluation. It is found that 24h-LC_{50} of PFOA on the sample fish is above 150.00 mg · L^{-1}; 48h-LC_{50} of PFOA on sample fish is 145.52 mg · L^{-1}, with 95% confidence limit up to 134.52~174.90 mg · L^{-1}; 72h-LC_{50} of PFOA on sample fish is 130.29 mg · L^{-1}, with 95% confidence limit up to 119.22~146.64 mg · L^{-1}; 96h-LC_{50} of PFOA on sample fish is 118.82 mg · L^{-1}; with 95% confidence limit up to 107.46~134.33 mg · L^{-1}. Thus, it is proved that PFOA is a chemical substance with low toxicity.

Keywords: PFOA; Zebra fish; acute toxicity

1 INTRODUCTION

Perfluorochemicals (PFCs) are a kind of organic pollutants that can persistently exist in the environment. Perfluorooctane Sulfonates (PFOS) and Perfluorooctanoic Acid (PFOA) are the commonly-seen PFCs[1]. Due to its stable chemical structure and water and oil repellency, PFOA are widely used in many industry fields such as textile, paper-making, packaging, pesticides, carpet, leather, floor sanding, shampoo and fire-extinguishing foam etc[2–4]. PFOS and PFOA are the two typical PFCs in the environment. The ecotoxicological study has found that PFCs can cause hepatotoxicity, developmental toxicity, immune toxicity and internal secretion disturbance and have potential carcinogenicity[5–7]. The human-beings exposure to PFOA includes drinking water, fish, seafood, meat, eggs, indoor dusts etc. It is found that fish and seafood have the highest exposure risk to human beings. However, the study of acute toxicity of PFOA on fish is rarely seen in the literature. In order to prevent the environmental damage of using PFOA, there is a need to conduct the ecological risk evaluation over the PFOA. Among the eco-risk evaluation of the chemicals, USA Environmental Protection Agency (USEPA), *Code of Organization for Economic Cooperation and Development* (OECD, 1992), *Code of Chemical Pesticide Safety Evaluation Experiment* revised by State Environmental Protection Administration in 1989 all contain the safety evaluation over the fish products. Using zebra fish as the subjects, this research studies the acute toxicity of PFOA on fish and evaluates the harmfulness of ecotoxicology.

2 MATERIAL AND METHOD

2.1 Samples

Zebra fish: The samples of the research are provided by Wuhan Institute of Hydrobiology, Chinese Academy of Science. The sample are 3-month old with the total length up to 2.5–3.0 cm.

2.2 Testing material

Perfluorooctanoic Acid (PFOA). Purity: 98%, J&K Scientific. 1.3 Instruments Static Testing System (Beijing Aisheng Technology), Dissolved Oxygen Meter (USA HACH Company), Portable PH Meter (METTLER TOLEDO), High Performance Liquid Chromatograph (Agilent), Glass fish tank.

2.3 Testing procedure

2.3.1 The determination of PFOA stability in water
The study uses High Performance Liquid Chromatograph to determine the PFOA solution concentration of 50 mg · L^{-1}, 100 mg · L^{-1} and 200 mg · L^{-1} under different time period so that the stability of PFOA in testing water can be reflected.

Table 1. Classification guidelines for the ecotoxicology harmfulness.

Harmfulness ranking			
Extremely High	High	Middle	Low
≤1 mg L^{-1}	>1~10 mg · L^{-1}	>10~100 mg · L^{-1}	>100 mg · L^{-1}

2.3.2 The selection of testing method

By reference to OECD Guidelines for the Testing of Chemicals, "203, Toxicity Test", "Chemical Test Guideline" and "Chemical Test Method", on the basis of the results from stability test, this study adopts static water testing system.

2.3.3 Test operation

(1) Testing device parameter adjustment
Temperature: 23 ± 2°C
Glass fish tank: 5 L
Testing water: 3 L
(2) Exposure Concentration
1$^{#}$: 0.0 mg/L; 2$^{#}$: 72.3 mg/L; 3$^{#}$: 86.8 mg/L; 4$^{#}$: 104.2 mg/L; 5$^{#}$: 125.0 mg/L;
6$^{#}$: 150.0 mg/L
(3) 10 zebra fish are put in each treatment group randomly. During the experiment, the toxic symptoms and mortality rate are collected every 24 hours.

2.3.4 LC$_{50}$ and 95% confidence limit statistic analysis and method

Adopting EPA PROBIT ANALYSIS PROGRAM USED FOR CALCULATING LC/EC VALUES Version 1.5, this study conducts the statistical analysis over the pre-determined concentration and the corresponding mortality rate. In the following, the 96h-LC$_{50}$ value and 95% confidence limit are calculated and the harmfulness ranking is evaluated.

2.3.5 Evaluation code

We adopt *New Chemical Harmfulness Evaluation Guideline* (HJ/T154-2004) to analyze the acute toxicity of 96h-LC$_{50}$ and evaluated the harmfulness of ecotoxicology. The evaluation code is listed in Table 1.

3 RESULTS AND ANALYSIS

3.1 The stability of PFOA in the testing water

Through the determination of the actual PFOA solution concentration of 10 mg · L^{-1}, 100 mg · L^{-1} and 200 mg · L^{-1} under different time period, it is found that PFOA is rather stable in the water. (The real-time concentration within the original concentration 80%~120% is defined as stable.)

Table 2. Mortality rate of Fish during the test.

Treatment group	0.0 mg·L^{-1}	72.3 mg·L^{-1}	86.8 mg·L^{-1}	104.2 mg·L^{-1}	125.0 mg·L^{-1}	150.0 mg·L^{-1}
Mortality rate (%)						
24 h	N	N	N	N	N	30
48 h	N	N	N	N	10	60
72 h	N	N	N	10	40	80
96 h	N	N	10	30	50	90

Table 3. LC$_{50}$ and 95% confidence limit.

Item	Time			
	24 h	48 h	72 h	96 h
LC$_{50}$	>150.00 mg·L^{-1}	145.52 mg·L^{-1}	130.29 mg·L^{-1}	118.82 mg·L^{-1}
95% Confidence Limit	>150.00 mg·L^{-1}	134.52~ 174.90 mg·L^{-1}	119.22~ 146.64 mg·L^{-1}	107.46~ 134.33 mg·L^{-1}

3.2 The statistic result of the mortality rate

The fish mortality rate is listed in Table 2. Judging from the table above, under various PFOA solution groups, there is no fish found dead in 24 h 0.0, 72.3, 86.8, 104.2 and 125.0 mg · L^{-1} concentration groups. While the mortality rate in 150.0 mg · L^{-1} concentration group is 30%; there is no fish found dead in 48 h 0.0, 72.3, 86.8 and 104.2 mg · L^{-1} concentration groups while the mortality rate in 125.0 and 150.0 mg · L^{-1} is 10% and 60% respectively; there is no fish found dead in 72 h 0.0, 72.3 and 86.8 mg · L^{-1} concentration groups while the mortality rate in 104.2, 125.0 and 150.0 mg · L^{-1} is 10%, 40% and 80% respectively; there is no fish found dead in 96 h 0.0 and 72.3 mg · L^{-1} concentration groups while the mortality rate in 86.8, 104.2, 125.0 and 150.0 mg · L^{-1} is 10%, 30%, 50% and 90% respectively.

3.3 The results of sample fish LC$_{50}$ and confidence limit

The results of sample fish LC$_{50}$ and confidence limit are listed in Table 3. We adopt EPA PROBIT ANALYSIS PROGRAM USED FOR CALCULATING LC/EC VALUES Version 1.5 to conduct statistical analysis over the pre-determined concentration and the mortality rate. It is found that 24h-LC$_{50}$ of PFOA on the sample fish is above 150.00 mg · L^{-1}; 48h-LC$_{50}$ of PFOA on sample fish is 145.52 mg · L^{-1}, with 95% confidence limit up to 134.52~174.90 mg · L^{-1}; 72h-LC$_{50}$ of PFOA on sample fish is 130.29 mg · L^{-1}, with 95% confidence limit up to 119.22~146.64 mg · L^{-1}; 96h-LC$_{50}$ of

PFOA on sample fish is 118.82 mg · L^{-1}; with 95% confidence limit up to 107.46~134.33 mg · L^{-1}.

4 CONCLUSIONS

It is found that PFOA in water is rather stable and the research adopts the static testing method to study the acute toxicity of PFOA on Zebra fish. Under the various exposure concentrations, there is no fish found dead in 96 h 0.0 and 72.3 mg · L^{-1} concentration group, while the mortality rate in 86.8, 104.2, 125.0 and 150.0 mg · L^{-1} is 10%, 30%, 50% and 90% respectively, with LC$_{50}$ up to 118.82 mg · L^{-1} and 95% confidence limit up to 107.46~134.33 mg · L^{-1}. Adopting *New Chemical Harmfulness Evaluation Guidelines* (HJ/T154-2004) to conduct the analysis and ranking over its ecotoxicology, it is concluded in this study that PFOA is a chemical substance with low toxicity on Zebra fish.

REFERENCES

Bhhatarai B & Gramatica P. 2010. Per-and polyfluoro toxicity (LC50 inhalation) study in rat and mouse using QSAR modeling. *Chemical Research in Toxicology* 23(3): 528–539.

Bots J & De Bruyn L & Snijkers T, et al. 2010. Exposure to perfluorooctane sulfonic acid (PFOS) adversely affects the life-cycle of the damselfly Enallagma cyathigerum. *Environmental Pollution* 158(3): 901–905.

D'Hollander W & Roosens L & Covaci A, et al. 2010. Brominated flame retardants and perfluorinated compounds in indoor dust from homes and offices in Flanders, Belgium. *Chemosphere* 81(4): 478–487.

Lau C & Anitole K & Hodes C, et al. 2007. Perfluoroalkyl acids: A review of monitoring and toxicological findings. *Toxicological Sciences* 99(2): 366–394.

Pan Y Y & SHi Y L & Chai Y Q. 2008. Determination of the Perfluorinated Compounds in Aquatic Products Fish and Shell. *Chinese Journal of Analytical Chemistry* 36(12): 1619–1623.

Sun H & Gerecke A C & Giger W, et al. 2011. Long-chain perfluorinated chemicals in digested sewage sludges in Switzerland. *Environmental Pollution* 159(2): 654–662.

Wang L & Sun H & Yang L, et al. 2010. Liquid chromatography/mass spectrometry analysis of perfluoroalkyl carboxylic acids and perfluorooctan esulfonate in bivalve shells: Extraction method optimization. *Journal of Chromatography A* 1217(4): 436–442.

Architectural, Energy and Information Engineering – Sung & Chen (Eds)
© 2016 Taylor & Francis Group, London, ISBN 978-1-138-02791-6

Attitude and heading reference of UAVs with adaptive SO (3) complementary filter based on fuzzy logic

J.Y. Zhai & X.J. Du
School of Aerospace Engineering, Beijing Institute of Technology, Beijing, China

ABSTRACT: Algorithms are crucial for the precision and reliability of a navigation system based on MEMS (Micro-electromechanical Systems) technology. It is the key issue to be discussed in this paper. The definition of fuzzy logic and the structure of SO (3) complementary filter are introduced. By means of compensating sensor errors and analyzing influential factors of the UAV attitude algorithms accuracy, the adaptive complementary filtering algorithm based on the fuzzy logic is proposed. The actual flight test results show that the algorithm can estimate the UAV attitude well and satisfy the accuracy requirements of small UAV attitude and heading reference systems.

Keywords: MEMS; sensor modeling; fuzzy logic; complementary filter

1 INTRODUCTION

UAV derives from reconnaissance platforms program proposed by the Weapon and Equipment Research Department of the United States. It can be applied in aerial reconnaissance, target localization, electronic countermeasures, and is widely used in military and civilian application domains[1]. The navigation and positioning system is of great importance in its autonomous flight. Limited in the size, weight and cost of navigation device requirements, an attitude and heading reference system with Micro Electromechanical Systems (MEMS) that satisfies the requirements is essential.

In general, when the system hardware configuration is determined, the navigation algorithm becomes a key factor in the precision and reliability of the system. In this paper, the adaptive complementary filtering algorithm based on the fuzzy logic is proposed. With the corresponding fuzzy logic rules, filtering parameters can be determined on the basis of flight conditions to improve accuracy and anti-jamming capability of the whole system.

2 MEMS SENSOR ERROR MODELING AND ANALYSIS

According to the spectral characteristics[2], MEMS sensor errors can be divided into deterministic errors and random errors. The former mainly includes constant errors, scale factor errors and non-orthogonal errors. The latter mainly refers to the random drift caused by uncertainties, whose error model parameters can be obtained by Allan variance[3].

2.1 Error modeling

The relationship between the actual and theoretical values of the MEMS sensor is

$$A^m = KA^e + A_0 \tag{1}$$

where A_0 is the zero bias, K is the error coefficient matrix, $K = K_1 K_2$, K_1 is the scale factor error coefficient matrix, K_2 the non-orthogonal error coefficient matrix with α, β, γ small angle errors normally. Specific relationships are as follows:

$$\begin{bmatrix} A_x^m \\ A_y^m \\ A_z^m \end{bmatrix} = \begin{bmatrix} k_x & 0 & \alpha k_x \\ \beta k_y & k_y & \gamma k_y \\ 0 & 0 & k_z \end{bmatrix} \begin{bmatrix} A_x^e \\ A_y^e \\ A_z^e \end{bmatrix} + \begin{bmatrix} A_{0x} \\ A_{0y} \\ A_{0z} \end{bmatrix} \tag{2}$$

2.2 Allan analysis of variance

The inertial system is an integral system; all the noises the IMU outputs will possibly interfere the system up to divergence over time[4]. Therefore, it is necessary to analyze the system noise and build error models. Allan variance analysis is currently the most widely used analyzing method. Taking the gyroscope for example, a variety of error model parameters can be obtained by analyzing the measurement data over a certain period of time.

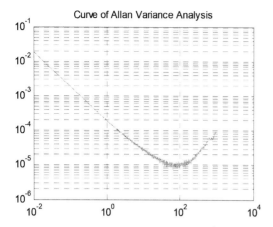

Figure 1. Allan variance analysis of crossbow.

Collecting the gyro output data with the sampling interval T_s and the acquisition number N, sample points can be divided into M groups to obtain the mean angular velocity of each segment of time. we assume that the k-th gyro point is $\omega(k)$, and the mean values of the i-th and the (i+1)-th groups are:

$$\bar{\omega}_i(\tau) = \frac{\tau_s}{\tau} \sum_{k=mi}^{m(i+1)-1} \omega(k), \bar{\omega}_{i+1}(\tau) = \frac{\tau_s}{\tau} \sum_{k=m(i+1)}^{m(i+2)-1} \omega(k) \quad (3)$$

Allan variance is defined as:

$$\sigma^2(\tau) = \frac{1}{2(M-1)} \sum_{i=0}^{M-1} [\bar{\omega}_{i+1}(\tau) - \bar{\omega}_i(\tau)]^2 \quad (4)$$

When the collected sample is large enough, the Allan variance curve can reflect some of the characteristics of the sample data. Random noise errors in gyros mainly include the quantization noise errors, angle random walk errors, zero impartial instability errors, angular rate random walk errors, and angular rate random slope errors[6]. Allan variance can be represented by the gyro error term:

$$\sigma^2(\tau) = \sigma_{\omega,Q}^2(\tau) + \sigma_{\omega,N}^2(\tau) + \sigma_{\omega,B}^2(\tau) + \sigma_{\omega,K}^2(\tau)$$
$$+ \sigma_{\omega,R}^2(\tau)$$
$$= \frac{3Q^2}{\tau^2} + \frac{N^2}{\tau} + \frac{2B^2}{\pi} \ln 2 + \frac{K^2\tau}{3} + \frac{R^2\tau^2}{2} \quad (5)$$

The lg τ-lg σ curve can be drawn with the data, and the error coefficient is directly obtained from different slopes of the curve. Take the Allan variance analysis curve of 6 hours data collected by crossbow for example, which is shown in Figure 1.

3 ADAPTIVE COMPLEMENTARY FILTERING ALGORITHM BASED ON FUZZY LOGIC

3.1 Fuzzy logic

Fuzzy logic is a precise method that can resolve inaccurate and incomplete information by means of imitating the cognitive process to analyze uncertain information. Fuzzy set F in the given domain U is represented by a membership function taking values in a closed interval [0, 1]:

$$\mu_F : U \to [0,1] \quad (6)$$

In the fuzzy inference process, first the input should be obtained and obscured by the membership function; then fuzzy reasoning is done by applying the fuzzy operator and the results should be clustered as the output. Finally a single-valued output can be obtained by means of the defuzziness of clustered output.

The general fuzzy logic diagram is shown in Figure 2.

3.2 SO (3) complementary filter

SO (3) represents the special orthogonal group, its application of a complementary filter in the special orthogonal group is called SO (3) complementary filter. Let R represents the actual direction cosine matrix of the aircraft, $\hat{R}$ the matrix calculated by the complementary filter, R_0 the observed direction cosine matrix by the accelerometer and digital compass, μ_H the high-frequency measurement noise of R_0, $R_0 = R + \mu_H$, R_c the attitude calculated by the gyro, and μ_L the low-frequency accumulated error in Rc; then $R_c = R + \mu_L$.

Take $G_L(s) = \frac{C(s)}{s+C(s)}$, $G_H(s) = 1 - G_L(s) = \frac{s}{s+C(s)}$, and if C(s) has the all-pass characteristic, then $G_L(s)$ has the low-pass filter characteristic, and $G_H(s)$ has the high-pass filter characteristic.

$$\hat{R}(s) = G_L(s)R_o(s) + G_H(s)R_c(s)$$
$$= R(s) + G_L(s)\mu_H(s) + G_H(s)\mu_L(s) \approx R(s) \quad (7)$$

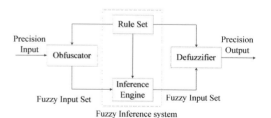

Figure 2. Diagram of the fuzzy logic.

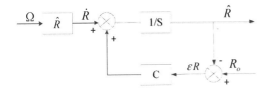

Figure 3. Model of the complementary filter.

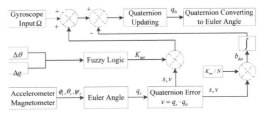

Figure 4. Flow diagram of the filtering algorithm.

From the attitude update equation $\dot{R}_b^n = R_b^n \Omega_{nb}^b$, we obtain the formula (13):

$$s\hat{R}(s) = C(s)(R_o(s) - \hat{R}(s)) + \hat{R}(s)\Omega(s) \qquad (8)$$

Then the complementary filter model can be obtained as is shown in Figure 3.

If $C(s)$ takes a fixed constant k_p, then the cut-off frequency between the low-pass filter and the high-pass filter is designed as $f_T = k_p / 2\pi$. At a frequency higher than f_T, the gyro plays a major role in the attitude estimation; at a low frequency below f_T, the digital compass and the accelerometer play a major role.

3.3 Error analysis of UAV attitude algorithm

3.3.1 Harmful acceleration and magnetic interference

In general, observations of the pitch angle and the roll angle are obtained by the three-axis accelerometer as well as the heading angle obtained by three-axis magnetometer in the attitude and heading reference system.

However, in the actual flight, UAVs often need to perform climbing, acceleration, cornering and other maneuvers during the flight, which will bring harmful acceleration to the accelerometer, resulting in a large observational attitude error that affects the output accuracy of the system.

Due to the wide spectral range of the magnetic field, the magnetic measurement is easy to be interfered from the vector itself, the magnetic field generated by the electronic equipment and the electromagnetic environment of space. If the airspace of the UAV has a high-intensity magnetic field, the measurement data will be anomalous, even leading to the output of the entire system divergent.

3.3.2 The body vibration

Compared to other models, small UAVs are small and flexible; however, their volume restricts the attitude and the heading reference system equipped to work in a very severe environment. Not much damping measures can be done and therefore the output of the inertial sensor is seriously affected.

3.4 UAV attitude determination program

Taking the low hardware computing power of the embedded systems and the poor accuracy of sensor into account, we will adopt SO (3) complementary filter based on quaternion as the main filtering algorithm, while the fuzzy logic control is also added to establish a strong robust adaptive filtering algorithm.

3.4.1 Complementary adaptive filter model

For the observed input of the error disturbance, the fuzzy logic link is increased in the feedback input of the complementary filter, and the gain value can be adjusted adaptively according to the relevant characteristics of observation. The flow diagram of the filtering algorithm is shown in Figure 4.

Reference attitudes ϕ_r, θ_r, ψ_r are obtained through the data from the accelerometer and magnetometer in the complementary filter, and are then converted into the reference quaternion q_r by Euler angles to get the quaternion error. The fuzzy logic controller can dynamically adjust the value of the filter gain K_{est} by monitoring the difference values $\Delta\theta$ and Δg between the reference value and the output value of the filter.

The attitude update equation based on the SO (3) complementary filter is[8]

$$\dot{\hat{q}} = \frac{1}{2} \cdot \hat{q} \otimes p(\Omega - \hat{b} + \omega) \qquad (9)$$

where $\hat{b} = -k_b(t)\dot{s}\dot{v}$ is the gyroscope bias value, $\omega = k_{est}(t)\dot{s}\dot{v}$ is the correction amount of the angular rate, $k_{est}(t), k_b(t)$ are the variable gain output by the fuzzy logic, and v is the scalar and vector part of the quaternion errors.

The correction term of the three-axis gyroscope can be compensated by formula (10):

$$\omega_x = k_x(t)\dot{s}\dot{v}_x, \omega_y = k_y(t)\dot{s}\dot{v}_y, \omega_z = k_z(t)\dot{s}\dot{v}_z \qquad (10)$$

Based on the index-related bias model $\dot{b}_1(t) = -\frac{1}{\tau}b_1(t) + w_{b_1}$, the modified zero partial update model is:

$$\hat{b} = \frac{-1}{\tau}\hat{b} + \frac{k_{est}(t)}{N}\dot{s}\hat{v} \qquad (11)$$

where N is a proportionality constant, and τ is a time constant of the first-order Markov process. Gains k_b and k_{est} are associated to decouple the gyro bias from the attitude error. Under dynamic conditions, the fuzzy logic controller can ensure the estimation of zero values in a stable state by reducing the proportion of constant k_b.

3.4.2 Design of fuzzy logic controller

According to the interference analysis, the fuzzy logic controller designed is divided into two kinds, one completing the loop self-adaptive gain output of the pitch angle and the roll angle based on the interference strength of the harmful acceleration, and the other completing the loop self-adaptive gain output of the heading angle based on the magnetic disturbance intensity. In this paper, three fuzzy logic controllers are designed, corresponding to $k_x(t)$ (the roll angle), $k_y(t)$ (the pitch angle) and $k_z(t)$ (the heading angle), respectively. Here, we only take the pitch angle loop controller for example.

The fuzzy logic controller has two inputs, namely Δg and $\Delta \theta$, and the output is the adaptive gain adapt K, and satisfies

$$\Delta g = | g - \sqrt{a_x^2 + a_y^2 + a_z^2} |, \Delta \theta = | \theta_r - \tilde{\theta} | \qquad (12)$$

where, a_x, a_y, a_z are the outputs of the three-axis accelerometer, θ_r is the observed value of the pitch angle, and $\tilde{\theta}$ is the pitch angle estimated previously.

The input and output diagram of the fuzzy logic as well as the corresponding membership function are shown in Figure 5. Input Δg is divided into four fuzzy levels which are small, mid, big and vbig, input $\Delta \theta$ into three fuzzy levels which are small, mid and big, and output adapt K into four fuzzy levels which are zero, small, mid and big.

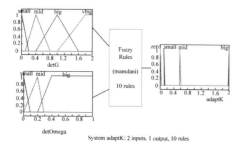

System adaptK: 2 inputs. 1 output, 10 rules

Figure 5. Input and output diagram of fuzzy logic.

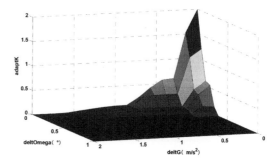

Figure 6. Output table of adaptive gain adapt K.

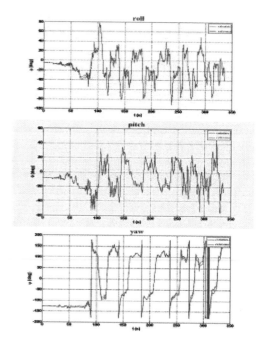

Figure 7. Calculation of the attitude angle.

According to the established fuzzy rule table, adaptive gain output adapt K is shown in Figure 6. The results show that the fuzzy logic controller can automatically increase the gain to enhance the dynamic response capability of the system if the external perturbation is small; when the external disturbance is large, the controller will reduce the gain or even shut down the filter to weaken the interference outside.

4 ATTITUDE ACCURACY TEST ANALYSIS

To verify the performance of the filtering algorithm proposed in the real flight environment, a

146

Table 1. Calculation of the heading error.

	Roll error	Pitch error	Heading error
Standard deviation (°)	−1.75	1.60	7.59
Mean (°)	−1.07	−0.15	−2.04

test was conducted with an AHRS-400CC inertial measurement unit used to simultaneously output the information of heading, attitude, angular rate, acceleration. A variety of aircraft maneuvers were operated to verify the estimation ability in cases of motion. The roll angle ranges between ±80° and the pitch angle between ±40°. After the flight, the data collected by the AHRS-400CC was calculated with the algorithm proposed to obtain the attitude and make a compared analysis with the AHRS-400CC attitude angle.

Statistics of the attitude angle error are shown in Table 1.

Figure 7 shows that the algorithm proposed can estimate the UAV attitude well.

5 CONCLUSIONS

In this paper, the sensor modeling, attitude filtering algorithm design and a flight test are completed for the UAV attitude and heading reference system.

Based on the compensation of the MEMS sensor modeling, the fuzzy logic controller is designed with its input and output as well as membership function, and the adaptive complementary filter algorithm is finally proposed based on the fuzzy logic. Actual flight test shows that the algorithm proposed is both practical and effective. The attitude and heading reference system can accomplish the task of attitude determination and satisfy the accuracy requirements of small UAVs.

REFERENCES

[1] Meng, Li. 2010. *Development of Modeling and Control Technologies for Small UAV System*. Beijing: Beijing Institute of Technology.
[2] Weston J L & Titterton D H. 2000. *Modern inertial navigation technology and its application*. Electronics & Communication Engineering Journal, 12(2): 49–64.
[3] Haiying, Hou. 2004. *Modeling Inertial Sensors Errors with Allan Variance*. University of Calgary: Department of Geomatics Engineering.
[4] Feng. Liang. 2011. *Design of small-scale attitude measurement system based on MEMS inertial devices*. Harbin: Harbin Engineering University.
[5] Haitao, Liu. 2007. *Allan analysis of MEMS gyroscope random error*. Remote Telemetry, 11 (28): 158–162.
[6] Yande Liang, et al. 2011. *Attitude* estimation *of a quad-rotor* aircraft *based on complementary filter*. Sensors and Micro-systems, 30 (11): 56–61.

Architectural, Energy and Information Engineering – Sung & Chen (Eds)
© *2016 Taylor & Francis Group, London, ISBN 978-1-138-02791-6*

Study on the analysis method of Imazamox in water

Qing Li, Fei Xiao, Guang Lin Song, Xi Jun Xu & Wei He
Guizhou Academy of Testing and Analysis, Guiyang, China

Wen Chao Zhao
Guizhou University, Guiyang, China

ABSTRACT: The detection method for imidazolinone in water was established by using high-performance liquid chromatography. Under the optimized conditions, quantitative analysis of the Imazamox in water was determined by the ZORBAX SB-C18 column at the wavelength of 203 nm. The result suggested that the linear equation was $y = 10.5509x - 3.85536$, the correlation index was 0.9999 and the limits of detection for Imazamox was 0.005 mg $\cdot$ kg^{-1}. Water sample recovery was founded to be in the range of 99.13%–100.77% and the varying factor was 0.63%. The method is simple, with features of high accuracy and precision, good linear relationship and low output limit, making it an ideal method for product quality analysis.

Keywords: Water; Imazamox; Analysis

1 INTRODUCTION

Water security is directly related to people's health. According to the EU regulations, individual pesticide residues in drinking water should not be more than 0.1 μg/L, and the total amount of pesticide residues should not be greater than 0.5 μg/L[1]. Imidazolinone is a kind of efficient herbicide, which is used mainly for postemergence weeding of soybean and peanut fields. It can effectively control most annual grass and broadleaf weeds. Imazamox is an efficient herbicide belonging to imidazolinone, Imazamox is its generic name, and its trade name is Raptor, Sweepe, Odyseey and gold beans[2]. Its chemical name is (RS)-2-(4-isopropyl-4-methyl-5oxo-2-imidazolin-2-yl)-5-methoxy-methylnicotinic acid. Its molecular formula is $C_{15}H_{19}N_3O_4$ and the CAS number is 114311-32-9. Pure Imazamox is a colorless solid, whose melting point and vapor pressure are 166~166.7°C and 1.3×10^{-5} Pa respectively, and distribution coefficient is 5.36.

The chemical structure of Imazamox is as follows.

The detection methods used for imidazo-linone herbicides are mainly LC-MS and HPLC. LC-MS is too expensive to spread; therefore, HPLC is the most commonly used detection methods[3–5]. The analytical methods of Imazamox in water have not yet been reported. Thus, in this paper, high-performance liquid chromatography is used to detect Imazamox in water. The method is simple, accurate and fast. It also has a precise and accurate separation that can meet the requirements of quantitative analysis. It also can be used as testing organizations and business reference methods of production analysis.

2 MATERIALS AND METHODS

2.1 *Instruments and reagents*

The instruments used were as follows: Agilent 1220 obtained from Agilent chromatography workstation; Column: ZORBAX SB-C18 (4.6 × 250 mm × 5 μm); Electronic balance (AL204, Exact to 0.1 mg, LHY016-1); Centrifuge; PH meter.

The reagents used were as follows: methyl alcohol, chromatographically pure, Honey Well Burdick & Jackson Ulsan, 680–160 Korea; acetonitrile, chromatographically pure, Honeywell Burdick & Jackson Ulsan, 680–160 Korea; ultrapure water, with resistivity of 18.25 MΩ · cm; Imazamox standard substance (98.5%),

Dr Ehrenstorfer GmbH; phosphoric acid, a guaranteed reagent.

2.2 *Chromatographic conditions*

The conditions set were as follows: column, ZORBAXSB-C18 (4.6 × 250 mm × 5 μm); Mobile phase, methyl alcohol + water (90:10 (v/v), containing 0.2% phosphoric acid); flow rate, 1 mL/min; detection wave length, 203nm; column temperature, 20°C; run time, 6 min; injection volume, 5 μL.

2.3 *Preparation of standard solution*

Briefly, a 0.0250 g imazamox standard was added into a 25 ml volumetric flask, made a constant volume in methanol solution to get a stock solution of imazamox methanol to get a stock solution of imazamox standard with a concentration of 1000 mg/L, and was shaken well for use. The stock solution of the above concentration was diluted to concentrations of 25, 50, 100, 200, and 400 mg/L, and was shaken well to serve as standard solutions.

2.4 *Sample processing*

The experiment sample was tap water from the laboratory, pond water, river water, and farmland water. The 4 water samples were filtered, respectively, and then the pH value was adjusted by phosphoric acid. The C18 column was activated by methyl alcohol and pure water, Then, the pH value of the water samples were adjusted. The column was drained and eluted with acetonitrile 5 times, and the eluate was collected, controlling the extraction rate and elution rate of the operation process. The eluate was dried under nitrogen at room temperature, and then transferred to a 0.5 mL volumetric flask and diluted with acetonitrile and prepared for subsequent chromatographic analysis.

2.5 *Calculation*

The content of the effective constituent was calculated using the following formula:

$$W = \frac{C \times P \times V \times D \times 10^{-6}}{M},\qquad(1)$$

where W is the content of the effective constituent in the test samples (%); C is the concentration (mg/L); P is the standard sample purity (%); V is the sample volume (mL); D is the sample dilution ratio; and M is the sample quality (g).

3 RESULTS AND ANALYSIS

3.1 *Optimization of chromatographic conditions*

3.1.1 *Column selection*

The 4 column liquid phases of KB-5 capillary column (30 m × 0.32 mm × 0.25 μm), Thermo HYPERSIL GOLD C18 (250 × 4.6 mm × 5 μm), ZORBAX SB-C18 (4.6 × 250 mm × 5 μm) and Cloversil ODS-U C18 (150 × 4.6 mm × 5 μm) were compared. The results showed that the four kinds of columns all had a good response to Imazamox. Of these, ZORBAX SB-C18 was more suitable, which had a high proportion of organic mobile phase. It was characterized by good peek shape, low pressure, high signal response and ideal retention time. The chromatogram of Imazamox is shown in Figure 1.

3.1.2 *The selection of detection wavelength*

Most parent compounds of imidazolinone herbicides had a maximum absorption within the range of 200–260 nm. The UV wavelength scan image was aquired by the pectral data acquisition capability of Agilent 1220 HPLC, as shown in Figure 2. As can be seen from the figure, imazamox had a greater absorption at 190–205 nm as well as at 203 nm. The detection wavelength was determined

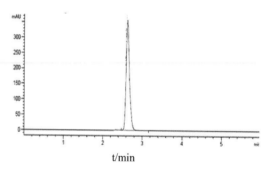

Figure 1. The standard chromatogram of Imazamox.

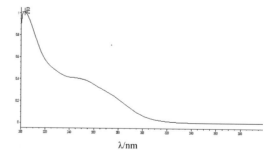

Figure 2. UV spectra.

at 203 nm in order to minimize the interference of the solvent.

3.1.3 *The selection of the mobile phase*

The experiment was carried out under two kinds of mobile phase: methanol and acetonitrile. It showed that compared with acetonitrile, methanol had low absorbance values at the wavelength of 203 nm, its viscosity was low and both the separation and peek shape were good. So, methanol was chosen as the mobile phase. When the mobile phase was acidic, it can inhibit the dissociation of analytes and extend the retention time of the components on the stationary phase based on the weak acid of imidazolinone herbicides. The pH of the aqueous mobile phase was adjusted at 2.5–4.0 in order to separated completely. At the same time, the pH of the mobile phase was adjusted by adding hydrochloric acid, phosphoric acid and formic acid, respectively. It showed that it had a minimum baseline drift and best sensitivity when adding phosphoric acid. So, 1 moL/L of phosphoric acid was added to adjust the pH.

3.2 *Linear relationship and detection limit test*

A standard solution of 5 concentrations prepared in line with 1.3 was tested under the test condition of 1.2, and showed a good linear relationship between the peak area (y) and concentration (x), with the linear equation given as follows: y = 10.5509x – 3.85536. The correlation index was 0.9999.

The minimum detection limit of Imazamox was 0.005 mg · kg[-1]. For detailed data, see Table 1. The calibration curve is shown in Figure 3.

3.3 *Degree of the precision test*

A total of six measurements were taken for samples of the same concentration under the same chroma-

tography conditions, and calculated the concentration of Imazamox. The weight of the sample was 0.0101 g, 100 mL constant volume in methanol solution was obtained, and made into as solution of 101 μg/mL in accordance with the operating conditions of 1.2, and repeated the measurment 6 times. The results are presented in Table 2. The relative standard deviation of this method was 0.20%, proving its feasibility.

3.4 *The recovery rate test*

To evaluate the accuracy of this method, the recovery test must be conducted. Briefly, a certain amount of the standard constituent to be tested was added pricisely into a known sample containing the constituent to be tested, and then measured and calculated its recovery ratio. A 0.0098 g sample was added to methylalcohol to make a 100 ml sample solution with aconcentration of 99 μg/ml for analysis. A 100 ug/ml Imazamox standard solution, in a 1:1 volume ratio, was added to the above-mentioned sample solution to make a to-be-tested solution with the concentration of 99 μg/ml. The aforementioned procedure is repeated to make 5 to-be-tested solutions that is of the same concentration as the to-be-tested constituent. The measurement was repeated 5 times in terms of 1.2. The results are presented in Table 3. Imazamox was not detected in the tap water from the laboratory, the urban river water, a lotus pond water and farmland water, as revealed by the method proposed in this paper. The average recovery ratio was 100.10%, and the coefficient of variation was 0.63%. It proved the flexibility of this method.

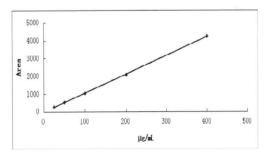

Figure 3. Calibration curve of Imazamox standard.

Table 1. Linear equation data.

Concentration (μg/mL)	25	50	100	200	400
Peak area	278.01	533.93	1037.32	2079.36	4243.95

Table 2. Test results of the precision of Imazamox.

Serial number	1	2	3	4	5	6	Average value	RSD (%)
Peak area	1061.96	1057.02	1060.56	1056.82	1061.40	1059.00	1059.46	0.21
Test contents (μg/mL)	100.78	100.32	100.65	100.30	100.73	100.51	100.55	0.20

Table 3. Results of the sample recovery rate.

Sample	Blank μg/mL	Concentration μg/mL	Test content μg/mL	Recovery %	RSD (%)
Lab water	ND	99	99.76	100.77	0.63
The urban River water	ND	99	99.51	100.52	
Suburb River	ND	99	98.98	99.98	
Pond water	ND	99	99.11	100.11	
Farmland water	ND	99	98.14	99.13	

4 CONCLUSION

The method proposed in this paper was carried out to determine the presence of Imazamox in the tap water from the laboratory, downtown river water, a lotus pond water and farmland water. The results revealed that Imazamox was not detected in all the water samples analysed. The data analysis showed that the average recovery was 99.63% and varying factors was 0.63%, which proved the feasibility of the method.

REFERENCES

[1] Chengyan HE, Yuanqian LI & Shen jiao WANG, 2010. Determination of Biphenyl Ether Herbicides in Water Using HPLC with Cloud point Extraction. *J Sichuan Univ (Med Sci Edi)* 41(1): 148–152.

[2] Qiang BI, Wei WANG & Zhiming CHENG, 2007. Review on Synthesis of Imazamox. *Modern Agrochemicals* 6(2): 10–14.

[3] Chengyan HE, Yuanqian LI & Hexing WANG, 2008. Determination on Five Sulfonylurea Herbicides in Water by High Performance Liquid Ceromatography with Solid Phase Extraction. *Modern Preventive Medicine* 35(3): 538–540.

[4] Haiying ZUO, Lin ZHANG & Fei LIU, 2014. Determination of Herbicide Residues in Groundwaters Using Liquid/Liquid Extraction and off-line Purification with Liquid Chromatography-Mass Spectrometry. *Rock and Mineral*: 96–10.

[5] Shasha BAI, Zhi LI & Xiaohuan ZANG, 2013. Graphene-based Magnetic Solid Phase Extraction-Dispersive Liquid Liquid Microextraction Combined with Gas Chromatographic Method for Determination of Five Acetanilide Herbicides in Water and Green Tea Samples. *Chinese Journal of Analytical Chemistry* 41(8): 1177–1182.

Architectural, Energy and Information Engineering – Sung & Chen (Eds)
© 2016 Taylor & Francis Group, London, ISBN 978-1-138-02791-6

Study on optimization design for the diaphragm wall of a subway station

Xun Chen Liu, Xiu Zhi Sui & Jun Jun Li

Shijiazhuang Institute of Railway Technology, Shijiazhuang, Hebei, China

ABSTRACT: By considering the influence of the 3D effect, ring-beam effect and angle effect, a new design method for diaphragm wall was put forward. The results of numerical calculation based on the elastic foundation beam theory and the new method are discussed in this paper. In a subway station, the maximum displacement and internal force were obtained by analyzing the monitoring data. The results of numerical calculation based on the new method compared well with the monitoring data, suggesting the feasibility of this new method for optimization design of diaphragm wall.

Keywords: Deep excavation; Diaphragm wall; Monitoring

1 INTRODUCTION

With the development of high-rise building and urban mass transit, deep excavation has become increasingly extensive in civil engineering. Diaphragm walls are commonly used as retaining structures in many deep foundation pits, especially when the environment protection is severe (Cong, 2001; Shen, et al., 2008; Yang, et al., 1998; Liu, et al., 2014; Jiang, et al., 1999; Feng, et al., 2009; Zhang, et al., 2012). Furthermore, diaphragm walls can also be used as part of the basement of a high-rise building or subway station structure.

Because of complexity of soil property, load, and construction environment, the design and construction scheme based only on engineering geological prospecting materials and soil test parameters has many uncertain factors. The actual stress and deformation condition can be obtained by studying monitoring data, and also analyzing the shortcomings of current calculation methods to validate the advantage of the new method.

2 PROJECT PROFILE

2.1 *Introduction*

The design depth of deep excavation of the subway station is 27.3 m. The supporting structure is the rowing pile and diaphragm wall with interior bracing (see Figure 1). The first bracing consists of the enclosing structure with the reinforced concrete beam and interior bracing, and other four bracings consist of the steel pipe bracing and steel beam. The diameter of the steel pipe is 600 mm and the thickness is 12 mm; the steel beam is 2I45C I-steel.

2.2 *Monitoring scheme*

The layout of measurement points is as follows (see in Figure 1):

1. Settlement and displacement of the top of the diaphragm wall: around the top beam, each has 16 points;
2. Stress of the diaphragm wall and horizontal bracing: in the middle and end of bracing, each layer has 8~12 points;
3. Settlement and incline of adjacent building: the building within 3H (H is the depth of excavation) from excavation, each building with at least 2 settlement and incline points;
4. Ground settlement and displacement of roads around, underground pipeline: ground settlement, horizontal displacement and settlement of underground pipeline use the same point that is located within 3H from excavation and along the road pavement underground pipeline; the total number of points is 10 and the distance between each point is 15 m.

The initial values should be measured at least twice before the construction of the supporting structure. The measurement intervals are based on the construction process, being less than 1 day during excavation and fast unloading.

3 THEORETICAL ANALYSIS

The results of theoretical calculation are based on the elastic foundation beam theory and optimization method that consider the influence of the 3D effect, ring-beam effect and angle effect. The values during construction and service time are calculated sepa-

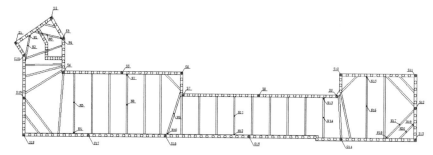

Figure 1. Layout of support and measurement points.

rately. According to the internal force calculation during construction using an increment method, the pre-existing displacement of the structure is first considered and then the deformation of bracing.

3.1 Optimized calculation of displacement

With the popularization of computer technology, the numerical solution method has become the effective design and the calculation method of the deep foundation pit soil retaining structure such as the underground diaphragm wall (Sui and Liu 2015). The theory of beams on elastic foundation is a widely used calculation method, but there are certain defects in this theory. For example, it does not reflect the various factors in the process of deep foundation pit excavation on the internal forces of the earth pressure distribution on the wall under the interaction between the support and the wall, not does it consider the impact on the construction process of the wall.

In order to make the design of the underground diaphragm wall closer to the actual working condition, the impacts that the 3D effect of underground diaphragm wall has on the mechanical characteristics of building envelopes are considered under the interaction between the support and the wall. With the introduction of correlation coefficients, based on the theory of beams on elastic foundation, the actual underground diaphragm wall design internal force and the displacement formula are proposed.

Based on the solution of Winkler foundation beam (Zhang's method), the formula of displacement is given by

$$y = \frac{e - \beta x}{2EI}[S_o cos\beta x + M_o \beta(cos\beta x - sin\beta x)] \quad (1)$$

$$\beta = \sqrt[4]{K_n B / 4EI}, \quad (2)$$

in which x is the depth from an excavation face; E (unit: kg/cm⁴) is the elasticity modulus of the underground diaphragm wall; K_n (unit: kg/cm³) is the modulus of foundation support along the horizontal direction of the soil; B is the width of the under-

Table 1. Value of C1.

Underground diaphragm wall and support conditions	C_1
1 Set-up ring-beams and the first bracing is reinforced concrete brace, within the scope of 5 m below ground surface	0.90
2 Set-up ring-beams, length-width ratio greater than 1.5 m and the space between the bracings less than 3.5 m, the scope of (L-30) m in the long side	0.95
3 Inside corners within 10 m set-up bracings, in the corresponding range	0.9
4 Depth of the foundation pit greater than 15 m, excavation face below 15 m	1.05–1.1
5 Length less than 50 m and length-width ratio greater than 1.5, all set-up bracings, within the scope of (L-20) m in the middle	1.1
6 Horizontal space between the bracings greater than 3.5 m, the middle of the two bracings	1.05

ground diaphragm wall; So is the external force acting on the underground diaphragm wall; and Mo is the primary bending moment acting on the underground diaphragm wall and the wall excavation face.

By considering the impact of the 3D effect, ring-beam effect and angle effect on internal force, we can get the optimizing formula of displacement through introducing the coefficient C_1:

$$y = C_1\left\{\frac{e - \beta x}{2EI}[S_o cos\beta x + M_o \beta(cos\beta x - sin\beta x)]\right\}, \quad (3)$$

in which C_1 is the adjustment coefficient related to the 3D effect, ring-beam effect and angle effect (See Table 1).

3.2 Results of axial force

The axial force of the bracing based on the elastic foundation beam theory and the new method is presented in Tables 2 and 3.

154

Table 2. Axial force based on the elasticity beam method.

No.	Axial force (KN)	No.	Axial force (KN)
1st bracing	993.6	4th bracing	2167.8
2nd bracing	1925.8	5th bracing	1288.3
3rd bracing	2195.6		

Table 3. Axial force based on the new method.

No.	Axial force (KN)	No.	Axial force (KN)
1st bracing	1075.3	4th bracing	1400.1
2nd bracing	1555.2	5th bracing	1313.5
3rd bracing	1409.7		

Table 4. Displacement based on the elasticity beam method.

No.	Displacement (mm)	No.	Displacement (mm)
1	12.79	12	13.16
2	4.48	13	21.86
3	19.12	14	18.32
4	11.15	15	14.46
5	10.06	16	14.46
6	12.79	17	14.45
7	12.80	18	18.67
8	12.80	19	10.68
9	13.16	20	6.83
10	12.49	21	11.35
11	12.87	22	7.25

Table 5. Displacement based on the newmethod.

No.	Displacement (mm)	No.	Displacement (mm)
1	27.27	12	23.44
2	14.38	13	21.33
3	19.61	14	6.11
4	28.05	15	13.66
5	13.50	16	17.35
6	32.11	17	26.14
7	10.56	18	17.92
8	13.71	19	29.88
9	6.41	20	25.83
10	19.78	21	29.57
11	42.87	22	33.15

3.3 Results of displacement

The displacement of the wall based on the elastic foundation beam theory and the new method is presented in Tables 4 and 5.

4 THE DIFFERENCES BETWEEN THEORETICAL VALUE AND MONITORING DATA

4.1 Comparison of calculating values by elastic foundation beam theory with monitoring data

Comparing the displacement of the wall based on the elastic foundation beam theory with monitoring data, as shown in Table 6, most of the monitoring data are less than the calculated values in the long side. The result provides a safer value than practical figures, but the maximum error is up to 84%. The monitoring data are more than the calculated values in the corner of excavation presented haphazardly; the result provides more dangerous values than practical figures, in which the average error is more than 100% and the maximum error is up to 332%.

Table 6. Comparison of displacement based on the measuring and elasticity beam methods.

No.	Calculated value (mm)	Measured value (%)	Errors (%)
1	12.79	36.77	−187.5
2	4.48	15.66	−249.6
3	19.12	19.99	−4.6
4	11.15	31.53	−182.8
5	10.06	8.61	14.4
6	12.79	32.13	−151.2
7	12.80	7.49	41.5
8	12.80	9.61	24.9
9	13.16	4.15	68.5
10	12.49	22.63	−81.2
11	12.87	55.72	−332.9
12	13.16	32.25	−145.1
13	21.86	14.87	32.0
14	18.32	2.93	84.0
15	14.46	10.33	28.6
16	14.46	20.65	−42.8
17	14.45	32.78	−126.9
18	18.67	20.44	−9.5
19	10.68	33.49	−213.6
20	6.83	27.60	−304.1
21	11.35	36.28	−219.6
22	7.25	31.32	−332.0

Table 7. Comparison of axial force based on the measuring and elasticity beam methods.

No.	Calculated value (KN)	Measured value (KN)	Errors (%)
1st bracing	993.6	1170.0	−17.8
2nd bracing	1925.8	1350.0	29.9
3rd bracing	2195.6	1330.0	39.4
4th bracing	2167.8	1177.0	45.7
5th bracing	1288.3	1053.0	18.3

155

Table 8. Comparison of displacement based on the measuring and improving methods.

No.	Calculated value (mm)	Measured value (%)	Error (%)
1	27.27	36.77	−34.8
2	14.38	15.66	−8.9
3	19.61	19.99	−1.9
4	28.05	31.53	−12.4
5	13.50	8.61	36.2
6	32.11	32.13	−0.1
7	10.56	7.49	29.1
8	13.71	9.61	29.9
9	6.41	4.15	35.3
10	19.78	22.63	−14.4
11	42.87	55.72	−30.0
12	23.44	32.25	−37.6
13	21.33	14.87	30.3
14	6.11	2.93	52.0
15	13.66	10.33	24.4
16	17.35	20.65	−19.0
17	26.14	32.78	−25.4
18	17.92	20.44	−14.1
19	29.88	33.49	−12.1
20	25.83	27.60	−6.9
21	29.57	36.28	−22.7
22	33.15	31.32	5.5

Table 9. Comparison of axial force based on the measuring and improving methods.

No.	Calculated value (KN)	Measured value (KN)	Errors (%)
1st bracing	1075.3	1170.0	−8.8
2nd bracing	1555.2	1350.0	13.2
3rd bracing	1409.7	1330.0	5.7
4th bracing	1400.1	1177.0	15.9
5th bracing	1313.5	1053.0	19.8

Comparing the axial force of the bracing based on the elastic foundation beam theory with monitoring data, as shown in Table 7, the difference between the monitoring data and the calculated values is large, and the maximum error is up to 45%.

4.2 Comparison of the calculating values by optimization method with monitoring data

Comparing the displacement of the wall based on the optimization method with monitoring data, as shown in Table 8, most of the monitoring data are identical with the calculated values. Although several negative errors are up to 38%, the result provides a safer value than practical figures and the maximum positive error is less than 52%.

Comparing the axial force of the bracing based on the optimization method with monitoring data, as shown in Table 9, the difference between the monitoring data and the calculated values isd very small, and the maximum error is only 19.8%.

5 CONCLUSIONS

The elastic foundation beam theory was optimized by introducing the coefficient C_{1that} considers the 3D effect, ring-beam effect and angle effect. The axial force of the bracing and the displacement of the diaphragm wall can be obtained from field monitoring, and also by calculating using the elastic foundation beam theory and the new method. The comparison of the calculating value and monitoring data shows that the new method can obtain smaller error and the result can become closer to actual conditions. It is also more reliable and accurate than the elastic foundation beam theory.

ACKNOWLEDGMENTS

The authors would like to acknowledge the funding support by the Department of Education of Hebei Province (Grant No. ZH2012084).

REFERENCES

Cong. A.S., 2001. The design, construction and application of underground continuous wall, 2–3. Beijing, China Water Power Press.

Feng S., X.X., Chen & G.Y. Gao. 2009. Analysis of underground diaphragm wall by iterative incremental method. Rock and Soil Mechanics 30 (1): 226–230.

Jiang P.M., Z.X., Hu & J.H., Liu. 1999. Analysis of space-time effect on the stability for slurry-trench of diaphragm wall. Chinese Journal of Geotechnical Engineering 21 (3): 338–342.

Liu N.W., X.N., Gong & C.H., Lou. 2014. Deformation characteristics analysis of diaphragm wall for foundation pit support in soft soil. Chinese Journal of Geotechnical Engineering 33 (s1): 2707–2712.

Shen J., W.D., Wang & Q.P., Weng. 2008. Analytic method of diaphragm walls of circular foundation pits. Chinese Journal of Geotechnical Engineering 30 (Supp): 280–285.

Sui X.Z., & X.C., Liu. 2015. Research on optimum design of underground diaphragm wall with interior bracing. Applied Mechanics and Materials 716: 227–231.

Yang M., Z.Y., Ai & Y.Q., Feng. 1998. Research on diaphragm wall's spatial effect by finite-element of beam on elastic subgrade. Chinese Journal of Rock Mechanics and Engineering 17 (4): 440–445.

Zhang Y.X., M. Ding & H., Wang. 2012. Analysis method for diaphragm wall structure based on Melan's solution. Rock and Soil Mechanics 33 (10): 2890–2896.

Architectural, Energy and Information Engineering – Sung & Chen (Eds)
© 2016 Taylor & Francis Group, London, ISBN 978-1-138-02791-6

Environmental assessment of biochar for security applications

F. Yang, H.Y. Guo, D.L. Su, Z.Q. Cheng & S.H. Liao
Faculty of Environmental Science and Engineering, Kunming University of Science and Technology, Yunnan, China

ABSTRACT: Many studies have reported on the general properties and applications of biochar, such as an adsorbent, carbon sequestration, and alleviation of greenhouse gas emissions. However, only a few studies have reported that there are heavy metals and Polycyclic Aromatic Hydrocarbons (PAHs) in the biochar. Heavy metals and PAHs pose a series of risks for the environment and humans. In addition, the contents of heavy metals and PAHs may connect with feedstock and pyrolysis temperature. This research mainly compares the normal characteristics of biochar and the risks of heavy metals and PAHs during biomass pyrolysis. The assessment of the risks of biochar has an important significant for security applications.

Keywords: soil amendment; biochar; pH; heavy metal; PAHs

1 INTRODUCTION

Biochar is a kind of carbon that has a rich content of carbon. It refers to black carbon formed by the pyrolysis of biomass, and the product of thermal decomposition of organic materials in the absence of oxygen (pyrolysis) (Roberts, K.G. 2009) or low oxygen environment. Feedstock mostly include wood waste (Clough, T.J. 2010, Spokas, K. 2009), crop residue (Chun, Y. 2004), animal waste (Cao, X. 2009), carbide sediment, papermaking sludge, and all kinds of biomass resources (Cao, X. 2009). Biochar is mainly composed of carbon, hydrogen, and oxygen, and has a high carbon content (about 70%~80%) (Goldberg, E.D. 1985).

Many studies have reported on the general properties and applications of biochar, such as an adsorbent (Tsai, W.T. 2012), carbon sequestration (Sparrevik, M. 2013), and alleviationg of greenhouse gas emissions (Sparrevik, M. 2013). Biochar properties vary greatly, depending on the biomass used to produce biochar and the production conditions (such as pyrolysis temperature and retention time). A high pH can be a key feature of biochar in improving soil acidity, but may not be required in naturally basic or even sodic soil (Antal Jr., M.J. 1985). However, only a few studies have reported that there are heavy metals and Polycyclic Aromatic Hydrocarbons (PAHs) in the biochar (Brady, N.C. 2002, Bridle, T. 2004, Kloss, S. 2012). Heavy metal and residual PAHs in biochar are harmful for soil and water. For instance, phenolic compounds are strongly adsorbed by biochar and are unlikely to leach and be bioavailable (Keech, O. 2005). Recently, the Ministry of Environmental Protection (MEP) and the Ministry of Land and Resources (MLR) of China issued a joint report on the current status of soil contamination in China. Chinese soil is divided into three levels. Class II, which could be applied to agriculture, orchard and pasture land, ensures agricultural production and human health via the food chain (Zhao, F.J. 2014). Overall, the risks of heavy metals and PAHs in the biochar during pyrolysis and the thermal degradation reaction should be taken into account before it is used safely.

2 RESULTS AND DISCUSSION

Interestingly, as shown in Table 1, biochar with higher ash content has higher pH values except for poultry litter. The pH values of biochar may change depending on the types of feedstock. For all samples, pH values increase with increasing pyrolysis temperature. The pH values increase with the temperature increasing from 60°C to 800°C, showing that the resulting biochar exhibits more alkaline nature because the minerals separate from the organic matrix in the form of ash (Cao, X. 2010). The biochar of high pH values may be used as soil amendments to reduce soil acidity. The pyrolysis temperature also affects the ability of carbon sequestration and chemical composition (Manyà, J.J. 2012, Tsai, W.T. 2012).

The yields of biochar are observed to decrease at a higher pyrolysis temperature, which can be attributed to the volatilization of tar products

Table 1. Characteristics of biochar from literatures (Manyà, J.J. 2012, Tsai, W.T. 2012).

Feedstock	T (°C)	Yield(%)	C (%)	H (%)	N (%)	pH	Ash (%)
Rice straw	RT	n. m.	39.7	5.20	0.59	n. m.	n. m.
	300	61	62.2	3.64	0.91	n. m.	n. m.
	500	23	66.6	2.27	0.79	n. m.	n. m.
Sewage sludge	400	n. m.	28.2	2.0	3.8	7.7	n. m.
	600	n. m.	27.1	1.1	3.1	11.5	n. m.
	800	n. m.	26.4	0.4	1.6	11.0	n. m.
Swine manure	400	n. m.	41.8	1.0	3.2	7.5	43.5
	600	n. m.	41.1	0.8	2.5	10.7	47.5
	800	n. m.	42.1	1.1	1.6	11.4	51.8
Oak wood	60	n. m.	47.1	5.81	0.11	3.73	0.3
	350	n. m.	74.9	3.43	0.16	4.80	1.1
	600	n. m.	87.5	2.41	0.18	6.38	1.3
Corn stover	60	n. m.	42.6	5.54	0.51	6.70	8.8
	350	n. m.	60.4	3.78	1.18	9.39	11.4
	600	n. m.	70.6	2.29	1.07	9.42	16.7
Poultry litter	60	n. m.	24.6	3.09	1.89	7.53	36.4
	350	n. m.	29.3	1.39	1.95	9.65	51.2
	600	n. m.	23.6	0.35	0.94	10.3	55.8

"n. m." NO MENTIONED;
"T" TEMPERATURE.

Table 2. Heavy metals and PAHs of biochar (Freddo, A. 2012, Kloss, S. 2012, Williams, S.M. 2004).

Feedstock	T (°C)	Metal variation range (mg · kg^{-1})	PAHs (mg · kg^{-1})	Exceeding the limit PAHs (%)
Rice straw	400	B (0.4–1.25),	5.2±2.9	35
	460	Cr (0.01–0.02),	10.7±5.6	172
	525	Mo (0.20–0.62),	33±22.1	93–830
		Ni (0.019–0.026),		
		As, Cd, Pb (n. d.)		
Spruce	400	B (1.24–2.34),	30±9.0	262–562
	460	Cr (0.025–0.15),	5.8±2.0	30
	525	Mo (0.0029–0.0078),	1.8±0.6	
		Ni (0.0016–0.0099),		
Poplar	400	B (1.37–4.49),	4.3±1.8	2
	460	Cr (0.05–0.025),	17.9±3.6	138–258
	525	Mo (n.m.), Ni (n. m.),	2.0±0.8	
		As,Pb (n. d.)		

"n. m." NO MENTIONED;
"n. d." NO DETECTED
"T" TEMPERATURE

derived from the lignocellulosic components of separated feedstock (Cao, X. 2010). There is a clear phenomenon that the H/C ratios are lower with increasing pyrolysis temperatures due to the lower H/C ratios that demonstrate an increased aromaticity of the biochar (Manyà, J.J. 2012). The pH values, ash and carbon contents of all biochar are found to increase with increasing temperature, while the yields of biochar and hydrogen contents decrease as a result of pyrolytic volatilization (Tsai, W.T. 2012).

Many studies havemainly reported on the conventional applications of biochar and their advantages, but only a few data are available that deal with the content of trace elements. In addition, high contents of heavy metals and PAHs have been reported in the biochar produced from some feedstock, such as sewage sludge and tannery wastes (Brady, N.C. 2002, Bridle, T. 2004, Kloss, S. 2012). The heavy metal and PAH contents of biochar are listed in Table 2. Despite the fact that the organic contamination such as PAHs can lead to severe public health problems (Manyà, J.J. 2012), relatively little attention has been focused on this issue. This situation may imply potential safety concerns with regard to the application of biochar to the soil (Fernandes, M.B. 2003, Manyà, J.J. 2012). However, somereports have indicated that the trace element can be used to improve the necessary microelement (Fernandes, M.B. 2003). Therefore, further studies are required for this argument. Pyrolysis temperature not only has an effect on the contents of Cu, B, Mo, and Cr in biochar, but also dominates the concentrations of PAHs (Kloss, S. 2012). In addition, although the US Environmental Protection Agency (USEPA) has set the limit value for PAHs in the biochar for soil application as 6 mg kg^{-1}, about 78% have exceeded the limit for these samples. The concentrations of PAHs are increasing with temperature for rice straw, and it is opposite with spruce because it probably contains oily substances and could convert into small-molecular-weight PAHs during pyrolysis, and these PAHs could in turn dissipate with increasing temperature. So, the biochar from spruce at a high temperature with less amounts of PAHs could be used in the soil. Zhao et al. reported that Class II soil limit of Cr is 150 mg · kg^{-1}–350 mg · kg^{-1} (Zhao, F.J. 2014), and the limit of Ni is 40 mg · kg^{-1} –60 mg · kg^{-1} (Epstein, E. 2002, Williams, P.N. 2009). Although the contents of heavy metals in the biochar are low, the continuous application of biochar into the soil will lead to the accumualtion of heavy metals, so that the heavy metals could exceed the limit in the soil.

3 CONCLUSION

Biochar properties vary greatly, depending on the kind of feedstock and the production conditions, which could affect its applications. Based on the data of chemical elements, heavy metals and PAHs, first, the biochar at a high temperature could be used to increase soil acidity, but it may not be suitable at low temperatures. In addition, the biochar frombiomass such as rice straw at a low temperature and spruce at a high temperature may also be applied to the soil in order to achieve the maximum benefit and be harmless to the soil environment.

REFERENCES

Antal Jr., M.J. 1985. Biomass Pyrolysis: a Review of the Literature Part 2—Lignocellulose Pyrolysis. *Advances in solar energy*, Springer: 175–255.

Brady, N.C. 2002. Nitrogen and sulfur economy of soils. *The nature and properties of soils:* 524–575.

Bridle, T. 2004. Energy and nutrient recovery from sewage sludge via pyrolysis. *Water Sci. Technol* 50(9): 169–175.

Cao, X. 2010. Properties of dairy-manure-derived biochar pertinent to its potential use in remediation. *Bioresour. Technol* 101(14): 5222–5228.

Cao, X. 2009. Dairy-manure derived biochar effectively sorbs lead and atrazine. *Environ. Sci. Technol.* 43(9): 3285–3291.

Chun, Y. 2004. Compositions and sorptive properties of crop residue-derived chars. *Environ. Sci. Technol.* 38(17): 4649–4655.

Clough, T.J. 2010. Unweathered wood biochar impact on nitrous oxide emissions from a bovine-urine-amended pasture soil. *Soil Sci. Soc. Am. J* 74(3): 852–860.

Epstein, E. 2002. Land application of sewage sludge and biosolids, *CRC Press.*

Fernandes, M.B. 2003. Characterization of carbonaceous combustion residues: II. Nonpolar organic compounds. *Chemosphere* 53(5): 447–458.

Freddo, A. 2012. Environmental contextualisation of potential toxic elements and polycyclic aromatic hydrocarbons in biochar. *Environ. Pollut* 171: 18–24.

Goldberg, E.D. 1985. Black carbon in the environment: properties and distribution. *Environ. Sci. Technol.*

Keech, O. 2005. Adsorption of allelopathic compounds by wood-derived charcoal: the role of wood porosity. *Plant Soil* 272(1–2): 291–300.

Kloss, S. 2012. Characterization of slow pyrolysis biochars: effects of feedstocks and pyrolysis temperature on biochar properties. *J. Environ. Qual* 41(4): 990–1000.

Manyà, J.J. 2012. Pyrolysis for biochar purposes: a review to establish current knowledge gaps and research needs. *Environ. Sci. Technol* 46(15): 7939–7954.

Roberts, K.G. 2009. Life cycle assessment of biochar systems: Estimating the energetic, economic, and climate change potential. *Environ. Sci. Technol* 44(2): 827–833.

Sparrevik, M. 2013. Life cycle assessment to evaluate the environmental impact of biochar implementation in conservation agriculture in Zambia. *Environ. Sci. Technol* 47(3): 1206–1215.

Spokas, K. 2009. Impacts of woodchip biochar additions on greenhouse gas production and sorption/degradation of two herbicides in a Minnesota soil. *Chemosphere* 77(4): 574–581.

Tsai, W.T. 2012. Textural and chemical properties of swine-manure-derived biochar pertinent to its potential use as a soil amendment. *Chemosphere* 89(2): 198–203.

Williams, P.N. 2009. Occurrence and partitioning of cadmium, arsenic and lead in mine impacted paddy rice: Hunan, China. *Environ. Sci. Technol* 43(3): 637–642.

Williams, S.M. 2004. Crop cover root channels may alleviate soil compaction effects on soybean crop. *Soil Sci. Soc. Am. J* 68(4): 1403–1409.

Zhao, F.J. 2014. Soil Contamination in China: Current Status and Mitigation Strategies. *Environ. Sci. Technol.*

Architectural, Energy and Information Engineering – Sung & Chen (Eds)
© 2016 Taylor & Francis Group, London, ISBN 978-1-138-02791-6

Effects of operating conditions on the performance of the Proton Exchange Membrane Fuel Cell

Y. Zhou
State Key Laboratory of Engines, Tianjin University, Tianjin, China
China Automotive Technology and Research Center, Tianjin, China

Y. Guan, J. Wang & D. Yu
China Automotive Technology and Research Center, Tianjin, China

ABSTRACT: The experimental method in this study is used to test the performance of the inhouse-assembled Proton Exchange Membrane Fuel Cell (PEMFC) by changing the inlet pressure, inlet temperature, inlet relative humidity and assembly force. The results indicate that the performance of the PEMFC can be better when the inlet pressure and inlet temperature increase within a certain range, while the inlet relative humidity has little effect on the performance of the PEMFC. It is shown that the performance of the PEMFC is not fitted with the increase in assembly force; especially for the high current density, the performance was even worse.

Keywords: Proton Exchange Membrane Fuel Cell; Performance; Experiment; Operating Conditions

1 INTRODUCTION

Proton Exchange Membrane Fuel Cell (PEMFC) is widely considered as one of the most promising clean energy conversion devices because of its high energy-conversion efficiency, high power density, capability of quick start-up, and clean and quiet operating characteristic (Yi 1998, Li 2006). However, during the working process of the PEMFC, different operating conditions would lead to a big change in its performance (Alaefour I.E. et al. 2011, Dong Q. et al. 2005). This study will build the PEMFC's performance test systems and use experimental methods to, respectively, complete the PEMFC's performance testing under different operating conditions by varying the PEMFC's inlet pressure, inlet temperature, inlet relative humidity and assembly force. It not only provides a practical basis for the numerical model analysis, but also determines the optimum operating parameters for future work.

2 EXPERIMENTAL METHODS

The test system used in this experiment is the G60 system produced by Greenlight Company, which mainly includes the gas supply system, the automatic control system of anode and cathode, the deionized water humidification system and the cooling system. The single cells used in this study are assembled through the procurement of different components, as shown in Figure 1. The flow channel is serpentine, proton exchange membrane is Nafion 211 obtained from DuPont, gas diffusion layer is from Toray's TGP-H-060 carbon paper, and the density of the cathode and anode catalyst platinum loading is 0.2 mg cm^{-2} and 0.5 mg cm^{-2}, respectively.

The reaction gas used in this experiment is made up of 99.99% hydrogen and compressed air. In addition, pure nitrogen is used to establish the backpressure and to complete the scavenging work before and after the reactions. The three kinds of gas are stored in high-pressure gas cylinders, and put into test devices after converting into low-pressure hydrogen, air and nitrogen through the decompression valve. The gas flow and working pressure in the cathode and anode is controlled, respectively, by using the mass flow meter at the test bench and the back pressure valve. Besides, the test system controls the temperature, pressure, humidity and other operating conditions of the reaction gas by making use of the temperature sensor, pressure sensor and heating belt through a control unit, and simultaneously detects and display operating parameters in real time by the monitor.

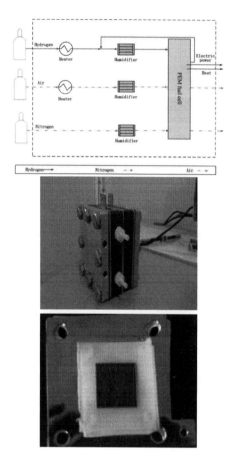

Figure 1. The experiment system and the single cell.

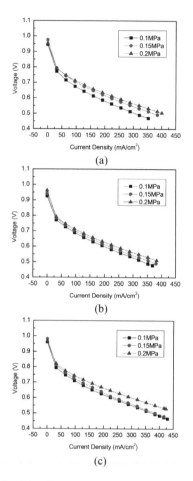

(a)

(b)

(c)

Figure 2. The effect of the inlet pressure on the performance of the PEMFC under different operating conditions: (a) inlet temperature, 50°C; (b) inlet temperature, 60°C; (c) inlet temperature, 70°C.

3 RESULTS AND DISCUSSION

3.1 *The effect of the inlet pressure on the performance of the PEMFC*

In order to investigate the effect of inlet pressure on the performance of the PEMFC, the inlet temperature is set at 50°C, 60°C, and 70°C, and the inlet humidity is 100%. From Figure 2, it can be seen that the performance of the PEMFC improves when the inlet pressure increases from 0.1 MPa to 0.2 MPa. The polarization curve of 0.2 MPa is much smoother, which means the voltage drops more slowly as the current density increases. This is because the density of the inlet gas increases and the gas is more likely to reach the surface of the catalyst at a higher inlet pressure. The electrochemical reaction is completed and thereby a relatively high current density is obtained. In addition, the increasing inlet pressure also increases the water vapor partial pressure, so that the water content of the membrane rises, then the conductivity of the membrane increases,

and finally a better performance of the PEMFC is accomplished. However, it should be emphasized that the inlet pressure cannot always be increased especially when considering the mechanical strength of the membrane. The high inlet pressure will damage the membrane and thereby the performance and life of the PEMFC would decrease. Therefore, the inlet pressure should be improved to obtain a higher performance of the PEMFC on the basis of membrane material tolerance in the process of the experiment.

3.2 *The effect of the inlet temperature on the performance of the PEMFC*

The effect of inlet temperature on the performance of the PEMFC is shown in Figure 3. The inlet pressure is set at 0.1, 0.15, 0.2 MPa and the inlet humidity is 100%. It is shown that the inlet temperature has a little impact on the performance of the PEMFC when the current density is low. However, with the increased current density, the performance of the PEMFC can be improved at a relatively higher inlet temperature. The rate of the electrochemical reaction increases due to the rising of inlet temperature, which improved the activity of the catalyst. Besides, the relatively higher inlet temperature can further enhance the diffusion rate of the inlet gas and effectively improve the transfer condition inside the PEMFC. The rising of the inlet temperature can also increase the saturated vapor pressure of water and the content of water vapor in the inlet gas, which accelerates the migration of the

proton in the membrane by reducing the ohmic loss and improving the conductivity. This is why the performance of the PEMFC is almost the same when the inlet temperature is 50 and 60°C. In addition, the effect of the inlet temperature is gradually weakened with the increasing inlet pressure. Therefore, the inlet temperature should be kept above 70°C to ensure the proper performance of the PEMFC.

3.3 The effect of the inlet relative humidity on the performance of the PEMFC

On the basis of the above experiment, the effect of the inlet relative humidity on the performance of the PEMFC is investigated when the inlet pressure is set at 0.2 MPa and the inlet temperature is set at 70°C, as shown in Figure 4. It can be seen from Figure 4 that the performance of the PEMFC increased with the increasing inlet relative humidity. This phenomenon is more obvious when the inlet relative humidity is less than 80%. The reason for this is that the inlet relative humidity can directly affect the diffusion of water in the PEMFC. The increasing water content of the membrane can decrease the resistance of proton transport and affect the conductivity of the membrane, which makes the current density distribution more uniform, and then the performance is better. When the inlet relative humidity is over 80%, the water content of the membrane basically

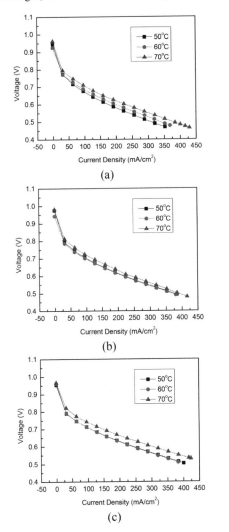

(a)

(b)

(c)

Figure 3. The effect of the inlet temperature on the performance of the PEMFC under different operating conditions: (a) inlet pressure, 0.1 MPa; (b) inlet pressure, 0.15 MPa; (c) inlet pressure, 0.2 MPa.

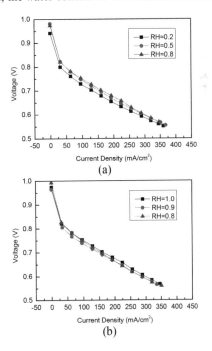

(a)

(b)

Figure 4. The effect of the inlet relative humidity on the performance of PEMFC at the inlet pressure of 0.2 MPa and the inlet temperature of 70°C.

163

meets the requirements of the electrochemical reaction. Also, the electrochemical reaction can generate liquid water. However, the water content of the membrane needs to reach a certain balance to prevent flooding. Thus, a reasonable inlet relative humidity is also a very important factor for the performance of the PEMFC.

3.4 The effect of the assembly force on the performance of the PEMFC

Based on the previous experimental analysis, the performance of the PEMFC can be changed under different operating conditions. When the inlet pressure is 0.2 MPa, the inlet temperature is 70°C and the relative humidity is 100%, the performance of the PEMFC is better than under other conditions. Therefore, the effect of the assembly force on the performance of the PEMFC is studied under this operating condition. It should be noted that it cannot accurately measure the assembly force due to the experimental equipment and conditions. The assembly force can only be adjusted according to the fastening bolt.

It can be seen from Figure 5 that the performance of the PEMFC improves with the increasing assembly force while the other operating conditions remain unchanged. The effect of the assembly force on the performance is more obvious under the medium-pressure and low-pressure conditions. When the assembly force continues to reach a high level, the performance only improves slightly at low current density, but begins to deteriorate at high current density. This result is not consistent with previous research. In general, with the increasing assembly force, the contact resistance between different components is reduced and the current density is increased. On the other hand, the increasing assembly force will cause the deformation of the

component in the PEMFC, especially in the porous media GDL. The deformation can affect the porosity, permeability and other transport parameters, and then the performance of the PEMFC. (Ge J.B. et al. 2006, Yim S.D. et al. 2010). However, these transport parameters cannot be accurately measured due to the limitation of experimental equipment and conditions. It should be investigated by establishing the three-dimensional multi-flow model, which will be studiedin the future.

4 CONCLUSIONS

In this study, the experimental method is used to test the performance of the inhouse-assembled Polymer Electrolyte Membrane Fuel Cell (PEMFC) by changing the inlet pressure, inlet temperature, inlet relative humidity and assembly force. The results show that the performance of the PEMFC increases while the inlet pressure increases from 0.1 MPa to 0.2 MPa. The inlet temperature has a little impact on the performance of the PEMFC when the current density is low. However, with the increasing current density, the performance of the PEMFC can be improved at a relatively higher inlet temperature. The performance of the PEMFC increases with the increasing inlet relative humidity when the inlet relative humidity is less than 80%. It is also demonstrated that the effect of the assembly force on the performance of the PEMFC is not always increased, which is different from that reported in previous research. (Lee W.K. et al. 1999) Therefore, the effect of the assembly force on the performance of the PEMFC should be investigated in the future.

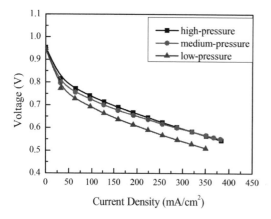

Figure 5. The effect of the assembly force on the performance of the PEMFC at the inlet pressure of 0.2 MPa, the inlet temperature of 70°C and the relative humidity of 100%.

REFERENCES

Alaefour I.E. et al. 2011. Experimental study on the effect of reactant flow arrangements on the current distribution in proton exchange membrane fuel cells. Electrochimica Acta. 56(5): 2591–2598.

Dong Q. et al. 2005. Distributed performance of polymer electrolyte fuel cells under low-humidity conditions, Journal of Electrochemical Society. 152(11): A2114–A2122.

Ge J.B. et al. 2006. Effect of gas diffusion layer compression on PEM fuel cell performance, Journal of Power Sources. 159(2): 922–927.

Lee W.K. et al. 1999. The effects of compression and gas diffusion layers on the performance of a PEM fuel cell, Journal of Power Sources. 84(1): 45–51.

Li X. 2006. Principles of fuel cells. New York: Taylor & Francis;

Yi B. 1998. Present and future situation for fuel cells. Battery Technology. 22(5): 216–221.

Yim S.D. et al. 2010. The influence of stack clamping pressure on the performance of PEM fuel cell stack, Current Applied Physics. 10(2): S59–S61.

Architectural, Energy and Information Engineering – Sung & Chen (Eds)
© *2016 Taylor & Francis Group, London, ISBN 978-1-138-02791-6*

The method of building digital core based on adaptive model

Hui Fen Xia
Key Laboratory of Enhance Oil and Gas Recovery of Education Ministry, Northeast Petroleum University, Daqing, Heilongjiang, China

Hong Yu Qiao, Xu Lai & Miao Xu
Northeast Petroleum University, Daqing, Heilongjiang, China

ABSTRACT: According to the distribution rules of the size of pore from the results of mercury penetration experiment, three-dimensional digital core, in which microscopic parameters such as wettability, shape factor, coordination number and pore throat ratio are considered, is built on the basis of axle wire principle of pore and adaptability principle of rock core. The value of three-dimensional digital core's permeability can be calculated by simulating seepage flow of single-phase with the constant speed. The pressure of pores can be calculated through the algorithm of symmetric successive overrelaxation in the simulation. It is indicated that the value of core permeability calculated by building the model is consistent with that obtained through performing the experiment. Three-dimensional digital core can be applied to simulate the process of water flooding and polymer flooding.

Keywords: Three-dimensional network; adaptive; constant speed flooding

1 INTRODUCTION

To explore the study of fluid flow in the porous media from microscopic perspective, the capillary bundle simulation method is proposed firstly, but due to the actual situation of porous media pore space is very complex, simple parallel capillary bundle is difficult to reflect the interaction between the pore connectivity and to simulate the real flow in pore. With the development of computer technology, three-dimensional digital core is used to simulate all kinds of fluid percolation process.

If the microscopic pore structure of three-dimensional digital core is fine described quantitatively, and the mechanism of multi-phase flow is to be understand right enough, oil displacement efficiency can be calculated through the simulation of fluid flow in the digital core. Because of avoiding the physical simulation experiments of repeatability, and the simulation of fluid transport properties in porous media is in conformity with the seepage mechanism, digital core has become a hot research field of many research scholars.

The network model is put forward by Fatt [1-3] for the first time, Levine and Chandler first used network model to simulate the fluid flow in the pore throat. Hui-fen xia et al. [4-6] used network model to study the mechanism of the polymer solution on oil. Blunt and Jackson [7] and others studied the effect of wettability on oil displacement efficiency by using network model. Network model was used to study the influence on two phase relative permeability by Wang Jin xun, Ping-ping shen[8-10], Hou Jian[11-13] studied the microscopic remaining oil distribution, Hui-mei gao et al.[14] introduced the present situation of the application of pore network model through a lot of research.

Analysis shows that, while seepage law under the network model is described and the corresponding results is obtained, but due to the use of three-dimensional network model for study is regular three-dimensional network model, the distance between nodes is constant and the network do not represent the real core pore structure. Then, the result has a certain deviation with real physical simulation experiment results because of the fixed injection side pressure. In this paper, on the basis of others' study, we built the digital core conform to the actual pore throat structure using the principles of Central Axis and adaptive. At the same time, the constant speed displacement algorithm is used to calculate permeability of the constructed three-dimensional digital core, and the correctness of the model is verified.

2 CONSTRUCTION OF THREE-DIMENSIONAL DIGITAL CORE

Currently, there are two kinds of method of building 3D digital core. The first method, image is obtained by thin section, CT scans or scanning electron microscopy, holographic imaging, etc, then the digital core is obtained by 2D and 3D reconstruction, the reconstruction method is relatively long and the core have limitations; the second method, statistical properties and the topological parameters of pore structure are obtained through the experiment, so as to generate three-dimensional network. But finding the method of the constructing digital core conforming to the real core topological properties is very important.

2.1 Radius of throat and pore

The constructed three-dimensional digital core pore throat structure is based on the natural core mercury injection experiment data, the distribution range of pore and throat and distribution of pore throat frequency is stored in the original TXT file. At the beginning of the building of 3D digital core, the original file is called, random radius is generated according to the distribution and distribution frequency. Pore radius is obtained by the product of the average pore-throat ratio and average throat radius connected to the pore, and considering the pore radius is greater than or equal to its adjacent throat radius value, the formula (1) is used to determine the pore radius of the digital core:

$$r_p = \alpha * \left[\sum_{i=1}^{n} r_{th} / n, \max(r_{th}) \right] \quad (1)$$

where r_{th} is the throat radius μm; rp is the pore radius, μm; α is the average pore-throat ratio; n is the coordination number.

2.2 Coordination number of the digital core

The principle of three line controlled by one point is used in the process of building digital core, the throat is the connection between nodes. Random numbers from 0 to 1 are generated, z is the average coordination number, if $r \leq \frac{6-z}{6}$ then throat does not exist between nodes.

2.3 Throat shape of digital core

Real core pore shape is very complex, in order to effective describe the irregularity of porous media, the section of throat is abstracted to circular, triangle, and square. Three angle of triangle is obtained from random assignment, shape factor characterize the irregularity of porous media, and the shape factor is calculated by formula (2): the biggest shape factor of circular is 0.07958, a square shape factor is 0.0625, and the shape factor of the triangle is range of 0.01 ~ 0.01.

$$G = \frac{A}{P^2} \quad (2)$$

where A is the throat area, μm^2; P is the throat section circumference, μm.

2.4 Physical nodes coordinates and the throat length

Because the rule of three-dimensional network model is used to study by researchers, the distance between two adjacent nodes is a fixed value, the model cannot be characterization of the real core of complex features. CT scan slice image analysis shows that the complexity of the pore can be dealt with by using the principle of the central axis, and the throat length distribute in line with Weibull distribution. In the process of building the digital core the X axis length between adjacent pore is given according to the distribution pattern, pore carried out random drift on both sides of central axis in accordance with the Weibull distribution formula according to the principle of central axis. Displacement of two adjacent nodes along the X direction is given by formula (3), node displacement deviation is given by the formula (4).

$$d_x = (d_{max} - d_{min}) \times \left[-\alpha \ln(z(1 - e^{-1/\alpha}) + e^{-1/\alpha})^{1/r} \right] + d_{min} \quad (3)$$

$$\Delta y = (\Delta y_{max} - \Delta y_{min}) \times \left[-\alpha \ln(z(1 - e^{-1/\alpha}) + e^{-1/\alpha})^{1/r} \right] + \Delta y_{min} \quad (4)$$

where dx-X is the direction distance between nodes, μm; d_{max}, d_{min} is the vertical distance X direction node, μm; Δy is the offset value of Nodes in the Y direction; α is the β-characteristic parameters $\alpha = 0.8$, $\beta = 1.6$; -Random number between 0 ~ 1.

How can represent pore throat existence state of the real core is vital, X direction length, and offset parameter values based on the X axis directly influences whether the construction of digital core can be applied to seepage simulation. Porosity is the important parameters of network model and macro core, the author using the adaptive principle to build digital core based on Daqing oilfield natural core porosity, constructing the core program automatically adjust parameters, d_{max}, d_{min}, Δy_{max}, Δy_{min}, to match the porosity. Then the distance between two nodes and the corresponding length of throat can be determined. The porosity of digital core is the ratio of the sum of all the pore and throat volume and total volume and the digital core volume. Program flow chart is shown in Figure 1.

Three-dimensional digital core rendering is obtained through the visual process compilation as shown in Figures 2 and 3, the throat section for the triangle and square is expressed with the cylinder.

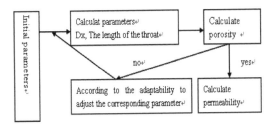

Figure 1. Program flow chart of determining the length of throat adaptive.

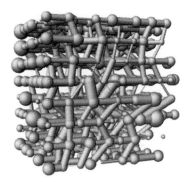

Figure 2. Stereogram of digital core.

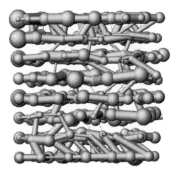

Figure 3. Vertical view of digital core.

3 SIMULATE WATER SINGLE-PHASE FLOW AND CALCULATE PERMEABILITY

3.1 Drag coefficient calculation method

Because the flow type is single-phase flow, fluid exists in the whole section of the throat, the unit length diversion coefficient can be calculated approximately by formula (5) based on Poiseuille's law.

$$g = \frac{\pi}{128} \left(\sqrt{\frac{A_t}{\pi}} + r_t \right)^4 \qquad (5)$$

After calculating the diversion coefficient, Drag coefficient is calculated by formula (6):

$$R_{whole} = \frac{\mu_f * 128}{\pi} \left(\frac{1}{\left(\frac{C}{\sqrt{\pi}} + 1 \right)^4} \right) \frac{x}{r_t^4} \qquad (6)$$

where μ_f is the viscosity, mpa·s; C is the area coefficient; X is the length of fluid in the throat, μm.

3.2 Method of pore pressure is calculated

Permeability is calculated based on the constant speed displacement method. Flow rate per unit time between the first row of the pore and the second row pore of inlet end of the digital core is considered as the flow rate of the model.

$$V = \sum_{z=1}^{n} q_{ij} \qquad (7)$$

where q_{ij}, flow rate per unit time between the first row of the pore i and the second row pore j. n, the number of pores connected.

Because capillary force does not exist, formula (8) can be used to calculate flow rate of the digital core.

$$q_{ij} = \frac{p_i - p_j}{R_{whole}} \qquad (8)$$

The sum of injection flow and traffic flow of the pore should be zero based on the law of conservation of mass flow, $\sum_{t=1}^{n} q_{ij} = 0$. Every pore pressure value can be calculated.

3.3 Permeability calculation and results analysis

The digital core, in which microscopic parameters such as a comprehensive wettability, shape factor, coordination number and pore throat ratio is considered, is built. Darcy percolation formula (9) can be used to calculate permeability.

$$k = \frac{Q * \mu * L}{A * \Delta P} \qquad (9)$$

where Q is the flow rate, cm³/s; μ is the viscosity, mpa·s; L is the length of digital core, cm; A is the injection side cross-sectional area, cm²; ΔP is the differential pressure, 0.1 MPaand K is the permeability, md

The pore throat distribution frequency of natural core of the different areas of Daqing is shown

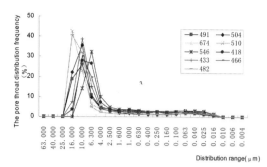

Figure 4. Pore-throat distribution frequency diagram.

Table 1. Real permeability value and calculated permeability value of 3D digital core.

Core number	510	504	674	491	337	482
Porosity (%)	33.2	31.6	29.9	29.4	29.3	30.9
Real value (md)	2067.8	1262.7	552.3	609.1	244	3145.2
Calculated value (md)	2096.6	1311.9	524.6	607.3	220.1	3139.8

in Figure 4, digital core is built where circular throat and square throat is 20%, triangle throat is 60%, nodes is $8 \times 8 \times 8$. Its permeability calculation results are shown in Table 1, the graph shows that water permeability value and real physical simulation experiment value have good match.

4 CONCLUSION

1. According to the distribution rules of the size of pore from the results of mercury penetration experiment, the method on the basis of axle wire principle of pore and adaptability principle of rock core of building three-dimensional digital core is obtained.
2. The value of three-dimensional digital core's permeability is calculated by simulating seepage flow of single-phase with the constant speed. It is indicated that the value of core permeability calculated by building the model is consistent with that obtained through performing the experiment. Three-dimensional digital core

can be applied to simulate the process of water flooding and polymer flooding.

REFERENCES

[1] Fatt, I. 1956. The Network Model of Porous Media, I. Capillary Pressure Characteristics. Trans. AIMM, 207: 144–159.
[2] Fatt, I. 1956. The Network Model of Porous Media, II. Dynamic Properties of a Single Size Tube Network. Trans. AIMM, 207: 160–163.
[3] LENORMAND. 1983. Mechanism of the displacement of one fluid by another in a network of capillary ducts [J]. J Fluid Mechanics, 135:3372353.
[4] Hui-fen xia zhai, et al. 2013. Based on the CT scan technology core pore coordination number of experiment research. [J]. Journal of experimental technology and management.
[5] Hui-fen xia zhai, et al. 2013. Based on the CT scan technology core pore coordination number of experiment research. [J]. Journal of experimental technology and management.
[6] Hui-fen xia, r, etc. 2001. The displacement efficiency of viscoelastic polymer solution to improve the micro mechanism research [J]. Journal of oil.
[7] Blunt M J, Jackson M D, 2002, Piri M et al. Detailed physics, predictive capabilities and macroscopic consequences for pore-network models of multiphase flow. Adv Water Res, 25: 1069–1089.
[8] Wang Jinxun. 2001. Application of pore network model study of two phase percolation rule [D]. Beijing: China petroleum exploration and development research institute.
[9] Ping-ping shen. 2003. Oil and water movement in the porous media theory and practice [M]. Beijing: petroleum industry press. 1–30.
[10] Wang Jinxun qing-jie liu, Yang Puhua etc. 2001 More than seepage theory are used to calculate unsteady method [J]. Journal of oil-water relative permeability curve of oil exploration and development, 28 (2): 79–82.
[11] Hou Jian zhen-quan li, shut the teng, etc. 2005 Based on the three-dimensional network model of water flooding microscopic seepage mechanism research [J]. Journal of mechanics, 37 (6): 783–787.
[12] Hu Xuetao Li Yun. 2000. Stochastic network simulation study microscopic remaining oil distribution [J]. Journal of oil, 21 (4): 46–51.
[13] TaoJun, Charles lai etc. 2007. 3 d random network model was used to study oil/water two-phase flow [J]. Journal of oil, 28 (2): 94–101.
[14] Hui-mei gao, jianghan bridge, min-feng Chen. 2007. The current status of application of the porous media pore network model [J]. Journal of daqing petroleum geology and development.

Architectural, Energy and Information Engineering – Sung & Chen (Eds)
© 2016 Taylor & Francis Group, London, ISBN 978-1-138-02791-6

An experimental study on the mechanical performance of an overhead floor system

C.G. Wang, Y.L. Zheng, G.C. Li & D.C. Wang
Shenyang Jianzhu University, Shenyang, Liaoning, China

ABSTRACT: Based on the functional requirements of existing beams, a new cross section of the cold-formed cap steel beam was designed. It effectively increases the use-height above the floor. The test studied the mechanical properties and parameters of the cold-formed cap steel beam in the serviceability limit state. Three different cross sections of cold-formed cap steel beams and one square steel beam were tested under a homogeneously distributed load. It is shown that, in the serviceability limit state, the bending stiffness of the square steel beam is not obviously better than that of the cap steel beam. For cold-formed cap steel beams, the bending stiffness is influenced by the flange width and the thickness. From the deformation and the load-displacement curves, it is found that the bending stiffness of specimen ML-3 is better than that of specimen ML-2.

Keywords: cross section; cold-formed cap steel beam; square steel beam; mechanical properties; bending stiffness

1 INTRODUCTION

As a branch of steel structure integration decoration, overhead floor is applied to realize the separation of decoration and structure. All pipes and lines of equipment are not required to be embedded in the main structure. To meet the requirements of the architectural decoration design and reasonable construction process, on the overhead floor edge link, there should be matching fittings, and functions such as ramp, step and others are supposed to have a special product design. In 1995, the American T.C. Hutchinson and other scholars [1–2] conducted a test on the connection mode. The beam was connected to the bracket by screws and exhibited a good effect. Gradually, this technique became widespread globally. In 2002, an experimental study was carried out by the American Mehdi Setareh and others [3–4] on seismic performance, and some suggestions on the overall stiffness and strength of the structure were proposed. In 2010, the American Alashker and others [5–6] conducted a study to compare the test for different support component sizes, using the finite element analysis method under different load conditions, and introduced the advantages and disadvantages of different support component sizes.

Overhead floor system is composed of main parts of floor, beam, and bracket. In order to study the mechanical performance and maneuverability of the overhead floor system, the mechanical properties of the existing beam and the function requirement are considered. A new cross section of the cold-formed cap steel beam is designed. Cold-formed cap steel is endowed with good resistance capacity to bending and torsion, and the use-height above the floor is effectively increased. However, for the selected cold-formed cap steel beam, its mechanical performance and the influence factors of mechanical performance are not very clear. To overcome these problems, an experimental study of a new type of cold-formed cap steel beam and square steel beam is conducted. Finally, the force mechanism of two kinds of specimen, namely the influence parameters and load-displacement curve, is obtained.

2 TEST PROGRAM

2.1 Specimen design

Three different section sizes of cold-formed cap steel beams and a square steel beam are designed, and these four beams are made of Q235 steel. A square steel beam specimen is numbered for FL, whose cross section size is $30 \times 20 \times 1.2$ mm and specimen length is 570 mm. The section of the cold-formed cap steel beam and the forms of the lap joint are shown in Figure 1 and Figure 2, and specimen size parameters are shown in Table 1.

The standard pipe bracket is used for the test, whose diameter is 22 mm. Cold-formed cap steel

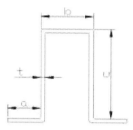

Figure 1. The cross section of the cold-formed cap steel beam.

Figure 2. Cold-formed cap steel beam and lap joint.

Table 1. Specimen parameters of the cold-formed cap steel beam.

Specimen number	Steel	Length/ mm	Section size a×b×c/mm	Thickness t/mm
ML-1		570	10×15×30	1.2
ML-2	Q235	570	10×20×30	1.2
ML-3		570	10×20×30	1.5

beams used a straight edge, which is based on the lap joint size of square steel beam. The design of the lap joint size is shown in Figure 3 and Figure 4.

Specimens are made according to the section size of the experimental need with a rolling bending machine in the Shenyang xingming steel mill. Specimens are shown in Figure 5.

2.2 Material test

The test is continued until the specimens are snapped, and the material test data is obtained, which is presented in Table 2.

2.3 Test point arrangement

2.3.1 Strain gauge arrangement

Two resistance strain gauges are placed on every surface of cold-formed cap steel beam and square steel beam. All specifications are 3 × 2 mm. In order to determine the deformation phenomenon, a longitudinal strain gauge (M1) is placed across

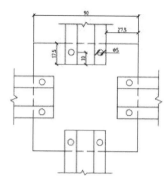

Figure 3. Specimen ML-1 and ML-3.

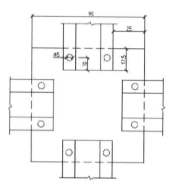

Figure 4. Specimen ML-2.

(a) Integral form of cold-formed cap steel beam

(b) Integral form of square steel beam

Figure 5. Specimens.

the lateral middle position of the web plate on cold-formed cap steel beam. The strain gauge (M2) on the middle of the midspan lower flange is used to determine the strain distribution of the specimen

Table 2. Material test results.

Material property					
Test number	Yield load (kN)	Yield strength (MPa)	Ultimate load (kN)	Ultimate strength (MPa)	Elasticity modulus (MPa)
1	9.13	228	15.5	387	1.85×10^5
2	9.24	231	15	375	2.14×10^5
3	9.01	225	15.8	395	1.98×10^5

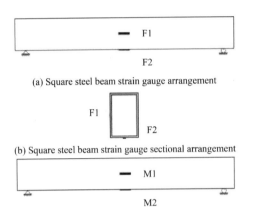

(a) Square steel beam strain gauge arrangement

(b) Square steel beam strain gauge sectional arrangement

(c) Cap steel beam strain gauge arrangement

(d) Cap steel beam strain gauge sectional arrangement

Figure 6. Strain gauge arrangement.

by combining the web plate strain distribution. To compare the analysis of the mechanical property of the cap steel beam, strain gauges are placed on the same position of the square steel beam, namelyfor F1 and F2. The strain gauge arrangement of these two specimens is shown in Figure 6.

2.3.2 Displacement meter arrangement

Displacement meters are arranged on the middle of the midspan lower flange for the cap steel beam and the square steel beam. During the process of the heaped load test, the displacement meter is used to monitor the deflection change with the increase in load. The arrangement of the displacement meter is shown in Figure 7.

2.4 Test device and loading system

This experiment was conducted in the laboratory of Shenyang Jianzhu University. The *in situ* sandbag heaped load was used to make a uniform

Figure 7. Displacement meter arrangement.

load distribution on the specimen. Using the static strain gauge, all data were collected, and made a good record of the experimental phenomenon.

Preloading: first, the specimen was geometrically centred to ensure that the joint component size is correct. Then, it was physically centred to make the parts in good contact with the solid to ensure the normal working condition of the test device and acquisition instrument. All specimens were controlled within the elastic range, and there was no deformation. The overhead floor was checked for its stabilization of standing.

Formal loading: according to the load code for overhead floor within the scope of normal use, under uniformly distributed load of $q = 7.35 \, \text{kN/m}^2$, the board deflection amount should be less than 3 mm, and there was no permanent deformation. After calculation, $P = 7.35 \times 0.6 \times 0.6 = 2.646 \, \text{kN}$, so the load of two overhead floor was 2.646 kN. Within the scope of the normal use, the experimental load was divided into 11 levels, with each level of the load being 0.245 kN. During the loading process, each one was supposed to stay for 3 minutes after every loading, and collected data, and then moved to the next level.

3 TEST ANALYSIS

3.1 Deformation shapes analysis

In the serviceability limit state, all specimens are still in the elastic state, and their appearance deformation are not obvious. With the increase in load, the specimens show some local deformation. The deformation process of the four specimens are generally similar, as shown in Figure 8.

3.2 Test results analysis

Through the process and analysis to record data during the test, the load-displacement curve and the load-strain curve are generated on the middle of the midspan lower flange for the cold-formed cap steel beam and the square steel beam, as shown in Figure 9 and Figure 10.

From Figure 9(a), it can be concluded that the load-displacement curves of the square steel beam and the cold-formed cap steel beam ML-1 are basically coincidental, and so is the curve slope.

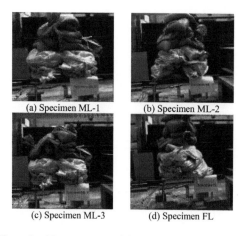

(a) Specimen ML-1   (b) Specimen ML-2

(c) Specimen ML-3   (d) Specimen FL

Figure 8.   The appearance deformation of the specimens.

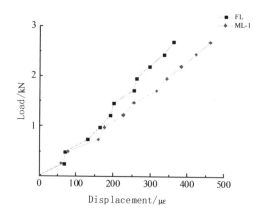

Figure 10.   Load-strain curve of the middle of the mid-span lower flange.

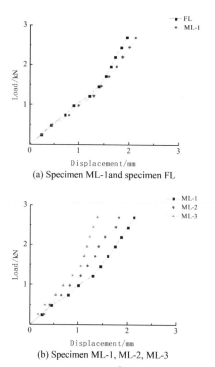

(a) Specimen ML-1and specimen FL

(b) Specimen ML-1, ML-2, ML-3

Figure 9.   Load-displacement curve of the middle of the midspan lower flange.

Compared with the load-strain curve of the square steel beam from Figure 10, the strain increase in beam ML-1 is not very obvious.

From Figure 9(b), it can be seen that the slopes of the load-displacement curve for specimen ML-2 and specimen ML-3 are bigger than that of specimen ML-1. So, their bending stiffness are bigger, and the ability to resist bending deformation is stronger. For specimen ML-2 and specimen ML-3,

the bending stiffness is different. Once the flange of the specimen is widen, the flexural stiffness becomes smaller, and the ability to resist deformation becomes weaker.

4   CONCLUSION

In the serviceability limit state, the mechanical property of the square steel beam is not obviously superior. Moreover, the cold-formed cap steel beam can effectively increase the use-height above the floor. After comprehensive consideration, the cold-formed cap steel beam is feasible in the overhead floor system, and thus considered as the alternative of the square steel beam.

In the serviceability limit state, the bending stiffness of the beam is closely related to the flange width and the thickness. From the deformation degree and load-displacement curve of each beam, it can be concluded that the larger the thickness of the specimen is, the more significant advantage the bending stiffness has.

ACKNOWLEDGEMENTS

This paper was supported by the National Science and Technology Supporting Plan (serial number: 2012BAJ13B05). The authors gratefully acknowledge its support.

REFERENCES3

[1] T.C. Hutchinson. 2012. Shake Table Testing and Numerical Simulation of Raised Access Floor-Computer Rack Systems in a Full-Scale Five-Story Building [J]. Bridges, (3): 1385–1396.
[2] Wang F, Chen G, Li D. 2008. The formation and operation of modular organization: A case study on

Haier's "market-chain" reform [J]. Frontiers of Business Research in China, 2(4): 621–654.

[3] Mehdi Setareh. 1984. Structural serviceability: floor vibrations [J]. Journal of Structural Engineering, 110(2): 401–418.

[4] Fredriksson P. 2006. Mechanisms and rationales for the coordination of a modular assembly system: the case of Volvo cars [J]. International Journal of Operations & Production Management, 26(4): 350–370.

[5] Alashke. 2010. Progressive collapse resistance of steel-concrete composite floors [J]. Journal of Structural Engineering, 136(10): 1187–1196.

[6] Frigant V, Talbot D. 2005. Technological determinism and modularity: lessons from a comparison between aircraft and auto industries in Europe [J]. Industry and Innovation, 2(3): 337–355.

Architectural, Energy and Information Engineering – Sung & Chen (Eds)
© 2016 Taylor & Francis Group, London, ISBN 978-1-138-02791-6

Study on the performance of purlins with new connections in steel sloping roof

C.G. Wang, Y. Bai, G.C. Li & X. Wang
Shenyang Jianzhu University, Shenyang, Liaoning, China

ABSTRACT: This paper presents detailed results on the performance of new connected purlins. It is necessary to study the influence of changing purlin connection on enhancing the bearing capacity of purlins. A test was conducted to compare new connected purlins with original connected purlins in many aspects, such as bearing capacity, stiffness, distribution of internal force and failure mode. In addition, it made a research about the influence of the bearing capacity and stiffness of each new wave purlin or new isolation wave purlin. It is shown that the new type of purlin connection can improve the bearing capacity, stability and stiffness of purlins. The promotion effect of each new wave purlin connection is better than that of new isolation wave purlin connection. Each new wave of purlin connection compared with the original connection can improve the bearing capacity of purlins by 50%, and the new isolation wave of purlin connection can improve the bearing capacity by 25% compared with the original connection.

Keywords: light steel; purlins; new connection; bearing capacity; stiffness

1 INTRODUCTION

With the rapid and continuous development of economy in China, the light steel residential structure has obtained fast development because of its advantages, such as the light self-weight, material uniformity, accurate and reliable stress calculation, simple manufacture, high degree of industrialization, convenient transportation and installation, materials province, and convenient sampling. There are three ways to study the performance of purlins with new connections in steel sloping roof, namely the test method, finite element numerical simulation and theoretical simulation[1]. Test method is the most direct way to study the performance of purlins and is the most accurate and reliable than the other ways[2]. So, this research uses the test method to study the performance of purlins with new connections in steel sloping roof.

2 TEST SPECIMEN

In current practical engineering applications, purlins are put on the steel beams and connected with beams by purlin holders[3–4]. However, under this connection, purlins' stiffness is small, they are just connected on the edge, and this connection is not beneficial to purlins' lateral restraint and global

stability[5]. If making the best use of the function of restraint by covering the effect, it could improve the bearing capacity of purlins and reduce the steel consumption.

In this article, a new connection is proposed (Figure 1). On the basis of the former function of purlin holder and self-tapping screws[6–7], a new upper connection is proposed to connect the purlins with roof boarding. It strengthens the connection of purlins and roof boarding, and helps to take full advantage of the function of restraint by covering the effect of boarding. So, the use of the upper connection could improve the global stability of purlins, bearing capacity and stiffness.

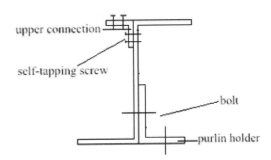

Figure 1. Diagram of the new connection.

3 TEST PROGRAMS

3.1 The objective of the test

The objective of this test is to study the stress performance and failure mode of a new type of Z purlin connection. The test is conducted to make a comparison between the new connected purlins and the original connected purlins in many aspects, such as bearing capacity, stiffness, distribution of internal force and failure mode. In addition, it makes research on the influence of the bearing capacity and stiffness of each new wave purlins or new isolation wave purlins.

3.2 Test setup

The test condition is as follows: the purlin span is 4 m, its spaces is 1.5 m, and its section is Z100-40-20-2. A roof sheathing system consisting of three-span purlins and steel sheet is chosen, as shown in Figure 2.

The type of the pressure steel sheet used in the test is type 970 produced by Shenyang Hongyuan color steel tile corporation. The size is shown in Figure 3.

The upper connection uses equilateral angle steel L50-50-4, which is cut according to the size of the upper connection. The thickness of the upper connection is bigger than that of purlins, in order to prevent the damage of the upper connection. Three self-tapping screws are used to connect purlins with the upper connection and two are used to connect purlins with the steel sheet. The number of specimens is three, namely the original connected purlins, each new waveconnected purlins and new

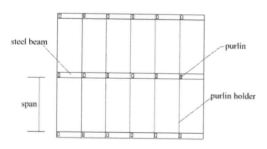

Figure 2. Schematic of the sloping roof.

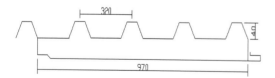

Figure 3. Schematic of Type 970 pressure steel sheet size.

Table 1. Specimen selection table.

Serial number	Remark	Section type (mm)	Span (m)	Space (m)
SJ-1	Each new wave connection	Z100-40-20-2	4	1.5
SJ-2	New isolation wave connection	Z100-40-20-2	4	1.5
SJ-3	Original connection	Z100-40-20-2	4	1.5

isolation wave connected purlins. The size of the specimens is listed in Table 1.

According to purlin space, the position of the purlin holder on the steel beam is ensured. The purlin holder is welded to the steel beam, and the position of the self-tapping screw and the connecting position of upper connection are ensured. Then, the purlins and the upper connection are installed. Finally, the steel sheet is fastened.

The measurement items of this test are mid-span deflection of the side span and the middle span; side deflection of the side span and the middle span; and mid-span web strain and mid-span bottom flange strain of the side span and the middle span. The test instruments used in the test are displacement transduction and strain gauge.

3.3 Loading

This test loading method is evenly distributed. The position and amount of sandbags is changed to achieve loading distribution. In order to ensure a uniform increase in loading, the crane and the raising pole are used to make symmetrical loading[8].

3.4 Test phenomena

The destruction form of the new connection purlins is different from that of the original connection ones. When the roof system composed of the new connection purlin profiled steel sheet reaches the limit load, buckling occurs. Subsequently, in the the midspan, purlin is twisted towards one side. The top flange of purlin is obviously bended and has a concave trend, as shown in Figure 4.

When the roof system composed of the original connection purlins profiled steel sheet reaches the limit load, buckling occurs. During this time, in the midspan, purlin is twisted towards one side. The degree of torsion is significantly larger than that for the new connection at the same position. However, the bending of the original connection in the top flange is not obvious, nor is the concave trend, as shown in Figure 5.

Figure 4. The deformation picture of the new connection purlins in the midspan.

Figure 5. The deformation picture of the original connection purlins in the midspan.

The destruction form of the new connecting purlins is different from the original connection ones. When the isolation wave or each new wave connection purlins reaches the limit bearing capacity, the purlin is crushed suddenly. When the original connection purlins reaches the limit load, purlins' displacement in the midspan suddenly increases and continuously rises in the process, but it does not crush suddenly.

4 TEST RESULTS AND ANALYSIS

The results of the test are summarized in Table 2. From this table, it can be seen that a new connection has an obvious effect to enhance the bearing capacity of the purlins. The bearing capacity of each new wave connected purlins is significantly higher than the isolation wave connected purlins. The spacing distance of the upper connection has an obvious influence on the new connection purlins' bearing capacity. The smaller distance of the connection more obviously increases the bearing capacity.

Table 2. The bearing capacity of the new connection purlins.

Serial number	Bearing capacity (kN/m²)	Increasing ratio with original connection
SJ-1	3.83	50.79%
SJ-2	3.18	25.5%
SJ-3	2.54	–

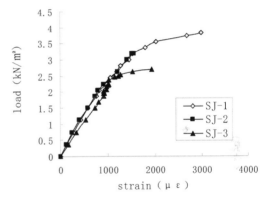

Figure 6. The load-strain curves of the bottom flange of the mid span purlin.

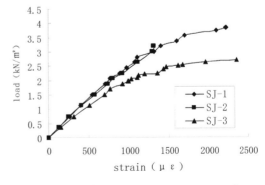

Figure 7. The load-strain curves of the bottom flange of the side span purlin.

Both the new isolation waveconnection purlin and each wave connection purlin promote the stiffness of purlins. However, in the midspan, compared with the new isolation wave connection, the improvement of stiffness of each new wave connection purlin is apparently higher. In the side span, both are similar.

From the load-strain curve of purlins at the bottom flange in the midspan, it can be seen that the effect of the new connection on the strain is obvious, as shown in Figure 6 and Figure 7. The stress of the new connection purlin at the bottom flange compared with the original one under the same load is much less.

Compared with the one in the side span, the decline in the strain of the new connection purlins at the middle flange relative to the original connection purlins in the midspan is more serious. The stress of each new wave connection purlins at the bottom flange is similar to the isolation wave one under the same load. The main reason for this phenomenon is that the new connections of both are added to the midspan position. In this position, there is a little difference between them.

When loading up to the 80 percent of ultimate load, the strain of the original connection purlin at the bottom flange in the midspan begins to have non-uniform changes. However, the load-strain curve of the new connection purlin is relatively homogeneous.

5 SUMMARY

This article presented the roof system design and test of purlins and profiled steel sheet. The comparison of three kinds of connections with the roof system, namely the original connection, new isolation wave connection and each new wave connection, were made with respect to the bearing capacity and stiffness of purlins in this test. On the basis of the above results, the following conclusions were drawn:

1. In the test condition, where purlins span is 4 m, purlins spaces is 1.5 m, and purlins section is Z100-40-20-2, each new wave compared with the original connection can improve the bearing capacity of purlins by 50%.
2. In the test condition, the new isolation wave of purlins connection can improve the bearing capacity by 25% compared with the original connection.
3. Limit bearing capacity of the new connection purlin is higher than that of the original one.

4. The stiffness of the new connection purlin is stronger than that of the original one. Concerning the stiffness order, the stiffness of each new wave connection purlin is maximum. The second one is the new isolation wave connection purlin, and the last one is the original connection purlin.
5. The destruction form of new connection purlins is different from that of the original one. Compared with the roof system consisting of the original connection purlin, the new one is greatly improved in the limit bearing capacity. The new connection purlin, but not the original one, crushes suddenly.
6. When the new connection purlin is destroyed, compared with the original connection purlin, the buckling at the top flange becomes more obvious and the bending becomes smaller. In contrast, when the original connection purlin is destroyed, the crankle in the midspan becomes more obvious.

ACKNOWLEDGEMENTS

This work was supported by the National Science and Technology Supporting Plan (serial number: 2012BAJ13B05). The authors gratefully acknowledge its support.

REFERENCES

[1] Michael C. 2000. Neubert. Estimation of Required Restraint Forces: Z-Purlin Supported, Sloped Roofs Under Gravity Loads [J]. American Society of Civil Engineers, 18(2): 253–259.
[2] M. Mahendran. 2004. Splitting Failures in Trapezoidal Steel Roof Cladding [J]. American Society of Civil Engineers, 22(7): 21–25.
[3] Davies JM, Leach P. 1994. First-order generalized beam theory [J]. Constructional Steel Research, 31(3): 187–220.
[4] D.a. Heinz. 1994. Application of generalized beam theory to the design of thin-walled purlins [J]. Thin-Walled Structures, 19(3): 311–335.
[5] Davies JM, Leach P, Heinz D. 1994. Second-order generalized beam theory [J], Constructural Steel Research, 31(3): 187–220.
[6] Catherine J. Rousch, Gregory J, Hancock. 1997. Comparison of tests of bridged and unbridged purlins with a non-linear analysis model [J]. Journal of Constructional Steel Research, 13(41): 197–220.
[7] B. Beshara, R.A. Laboube. 2001. Pilot study: lateral braced C-sections under bending [J]. Thin-Walled Structures, 39(2): 827–839.
[8] Dimos Polyzois. 1987, Sag rods as lateral supports for girts and purlins [J]. Journal of Structural Engineering, 113(1): 1521–1531.

Architectural, Energy and Information Engineering – Sung & Chen (Eds)
© *2016 Taylor & Francis Group, London, ISBN 978-1-138-02791-6*

The finite element analysis of cold-formed steel stud walls under compression

C.G. Wang & Q.Q. Liu
Shenyang Jianzhu University, Shenyang, Liaoning, China

ABSTRACT: Based on the experimental results of five cold-formed steel stud walls under axial compression, the static behavior of walls was stimulated by the finite elements analysis using the program ANSYS. The ultimate bearing capacity of the specimens matched the experimental data well, and the accuracy of the analysis was proved by the experiments. The parameters of the test specimens including sheathing, the spacing of screws and the spacing of studs that affect the bearing capacity of wall studs under axial loads were measured using the finite element analysis. Furthermore, the behaviors of the cold-formed steel wall system with web-stiffened studs were analyzed. The results indicate that the stud with web stiffener increases the ultimate compressive strength obviously.

Keywords: cold-formed steel walls; compressed under axial loads; finite element analysis; parameter analysis; stud web stiffener

1 INTRODUCTION

Cold-formed steel housing system [1] is a kind of new residence, which is based on cold-formed thin-wall steel components and light plates such as the Gypsum Board (Gyp) and the Oriented Strand Board (OSB) as the bearing and maintenance of the structure. It has been widely used in North America, Japan, Australia and other developed countries. In recent years, it has also been widely used in China.

Wall is the main bearing component of the cold-formed steel housing system. The sheathing of the wall is used not only as retaining structures, but also to provide effective lateral support and torsional restraint, so that it will increase the ultimate compressive strength [2–5]. The finite element method has a high precision of calculation, and as with a numerical calculation method, it is effective and economic. With the further promotion of cold-formed steel housing systems, the section form of cold-formed steel has become increasingly complicated. Research has shown that it effectively improves the axial bearing capacity when the stiffening rib is set in the web of edge channel steel components. At present, there are only a few studies on the performance of the wall of edge channel steel with a complicated cross-section form. Therefore, based on the experimental results of the existing litera-

ture, this article analyzes the axial compressive properties of cold-formed steel framing and stud walls that cover different boards using ANSYS finite element software. It verifies the correctness of the simulation method. In addition, it carries out the finite element analysis of the behaviors of the cold-formed steel wall-stud system with web-stiffened studs under axial load.

2 TEST INTRODUCTION

Based on the test results of reference [6], this article establishes the finite element model of specimens, and verifies the correctness of the simulation method. The C-section and U-section were used for studs and tracks, respectively, whose cross-sectional dimension notations are shown in Figure 1. Five specimens were selected from the literature [6]. The specimens were stimulated by using the finite elements analysis of its axial compression properties; of these five specimens, one is cold-formed steel framing and the other four are sheathed stud walls. The section size of the studs is $89 \times 41 \times 13 \times 1.0$ mm (H × B × d × t), the section size of the tracks is $92 \times 40 \times 1.0$ mm (H$_1$ × B$_1$ × t$_1$), and the height of the walls is 2700 mm. However, the studs are always attached to the sheathing of Oriented Strand Boards (OSB) and gypsum boards: the thickness of the gypsum board is

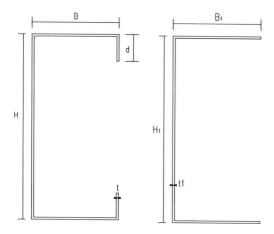

Figure 1. Cross-sectional dimension notations of the studs and tracks.

Table 1. Parameters of the test specimens.

Reference	Specimen number	Sheathing	Spacing of screws mm	Spacing of studs mm
[6]	WS1	bare		
	WSS2	OSB-OSB	300	600
	WSS4	OSB-OSB	150	600
	WSS3	OSB-Gyp	300	600
	WSS15	OSB-Gyp	300	300

12 mm and that of the OSB is 14.5 mm. In addition, each of the specimens is composed of a series of studs and tracks. It only takes the middle stud and the half length of two studs spacing, which not only simplifies the calculation, but also simulates the experiment condition effectively. The specimen parameters are listed in Table 1.

3 FINITE ELEMENT SIMULATION

3.1 Element type and material properties

ANSYS is used to simulate the test specimens used in this article, which selects the shell element 181 to simulate the sheathing and the cold-formed steel components. The steel material properties of the specimens are set as follows: Young's modulus $E = 20600$ MPa, Yield strength $fv = 370.4$ MPa, and Poisson's ratio $v = 0.3$. The OSB and gypsum boards are determined as orthotropic plates by the material test; therefore, the material properties are different between the vertical and horizontal directions, as presented in Table 2.

Table 2. Mechanical properties of sheathing.

Sheathing	Thickness mm	Sample position	E MPa	v
OSB	14.5	Vertical	3506.67	0.3
		Horizontal	1966.67	0.3
Gyp	12	Vertical	3036.67	0.23
		Horizontal	1943.33	0.23

3.2 Geometric model building and meshing

The geometric modeling was established from the bottom to the top. First, each key point is entered according to the modeling size and cross-section formation; second, the key point is connected to form a straight line; the end to form the surface. Besides, each surface of the model represents the actual board of the wall.

This article adopts the method of free-meshing to mesh. According to the importance of different components in the force analysis, it will deal with the different components as different levels of details. As the main components, the stud should have detailed meshing in the finite elements analysis. Considering the condition of the location of the screws, the contact part of the stud flange and track flange should be syncopated and meshed alone, with the grid size of 5×5 mm and the grid size of the rest stud webs and flanges of 15×5 mm (along the stud vertical direction of 15 mm and along the stud horizontal direction of 5 mm). The edge in the direction of cross section is divided into two, and along the vertical direction, it is further divided into 15 mm. As the less important component in the structure analysis, the track should be meshed in large size. The grid size of the flange is 5×20 mm (along the track vertical direction is 20 mm and along the track horizontal direction is 5 mm); the web grid size of 20 mm × 10 mm (along the track vertical direction is 20 mm and along the track horizontal direction is 10 mm); and sheathing of the grid size of 20 mm × 30 mm (along the stud vertical direction is 30 mm and along the stud horizontal direction is 20 mm).

3.3 Tapping screw simulation and boundary conditions

With regard to the tapping screw connecting the stud and track, this article adopts the method of node coupling to simulate the tapping screw. It couples the translational and rotational displacement of the node in tapping screw location. In order to simplify the unloading process, it builds the rigid surface on both ends of the stud in finite element simulation (see Figure 2). The translational degree of freedom is restricted in the plane (X-axis) and out of the plane (Y-axis) on both ends

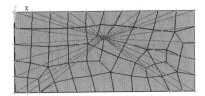

Figure 2.   Rigid region formation at the reaction end.

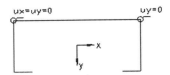

Figure 3.   Boundary condition at the top of the stud.

of the stud. Besides, the translational degree of freedom is restricted in the Z-axis on the bottom of the stud, as shown in Figure 3.

It only takes the middle stud and the half length of two studs spacing in the finite element simulation, due to the symmetry, symmetrical boundary conditions applied to the end of the track and sheathing on both sides of the wall. The bottom guide web and the translational degree of freedom are constrained along the X, Y, Z directions as well as a free rotation around the Z axis. The node is coupled at the contact stud and track; at the same time, the node is coupled at the contact stud and sheathing. The constraint conditions are shown in Figure 4.

3.4 *Residual stress and initial imperfection*

For cold-formed steel materials, the increasing yield strength and residual stress have the opposite effect to the cold-formed steel axial compression bearing capacity characteristics. When the influence of the residual stress is not considered in the model, the bearing capacity of the specimen has a little effect on the limit, so this article does not consider the influence of the residual stress.

In this article, the initial imperfection is selected according to the relevant light steel specification of USA [7]. When components are bent buckling, the defect value is L/1000, where L is the stud length; when components are flexural and torsional buckling, the defect value is L/(d × 1000), where d is the height of the stud section; when local buckling occurs, the defect value is d/200.

3.5 *Solving and post-processing*

The first step involves the solving of the eigen value buckling analysis, regardless of whether the component of elastic materials is considered; therefore,

(a)   Boundary condition of up tracks.

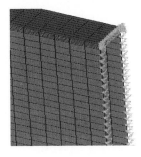

(b)   Boundary condition of sheathing.

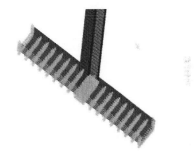

(c) Boundary condition of bottom tracks.

(d) Boundary condition of screws.

Figure 4.   Boundary condition of the finite element.

Table 3. The comparison between the literature results and the simulation.

Reference	Specimen number	P_t kN	P_f kN	Pt//Pf
[6]	WS1	19.68	21.29	1.08
	WSS2	50.95	54.62	1.07
	WSS4	53.33	55.02	1.03
	WSS3	45.51	47.98	1.05
	WSS15	52.71	55.32	1.05

a size unit load of 1 is applied on the structure. It should be applied to the initial defect according to the first-order buckling mode after analysis. Then, it involves the solving of the nonlinear buckling by the arc-length method. In the simulation process, the parameters are set as follows: the maximum step number is 200, the maximum arc-radius is 20 mm, and the minimum arc-length radius is 0.00000001 mm.

3.6 The result analysis

The ultimate loads of the tests in the literature are compared with the results of the finite element values, as presented in Table 3 (P_t denotes the test value; P_f denotes the finite element value).

Comparing the finite element values with the test results shows the correctness of the finite element analysis process. At the same time, from the above results, it can be concluded that the presence of the sheathing can enhance the axial bearing capacity of cold-formed steel stud walls; if the stud spacing is changed from 600 mm to 300 mm, the axial bearing capacity of the wall that covered the gypsum board and the OSB can improve by 15.8%; however, if the middle stud screw pitch is changed from 300 mm to 150 mm of the wall covering the OSB, it will not improve the ultimate axial bearing capacity.

4 WALL SYSTEM FINITE ELEMENT SIMULATION WITH WEB-STIFFENED STUDS

This article uses the finite element analysis process to set up the finite element model of a cold-formed steel framing with stud web stiffener and the other two walls to cover the sheathing on both sides to enhance the axial compression bearing capacity. Material properties and component forms are as described in the literature [6]. From Table 2, it can be seen that the place of changing is just built 14.14 mm in length of stiffening rib at the center of the stud web plate, stiffening rib and web

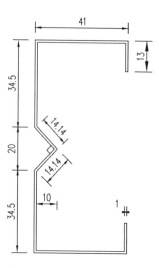

Figure 5. Dimensions of the studs.

Table 4. The comparison between the simulation results and the test results.

Specimen number	Sheathing	P_f kN	P_s kN
WS1	Bare	21.29	30.74
WSS2	OSB-OSB	54.62	81.25
WSS3	OSB-Gyp	47.98	78.45

connection at a 45 degree angle, and the size of cross-section of the stud is shown in Figure 5. The comparison of the finite element results and test results are presented in Table 4 (Pf denotes the finite element value; Ps denotes the finite element value of the stud web stiffener).

From the above information, it can be seen that the axial bearing capacity test value increased by 44.38% from the cold-formed steel framing stud web stiffening to the web non-stiffened steel framing; the value increased by 48.75% compared with the test value of bearing capacity of the stud stiffening wall with OSB board on both sides; the axial bearing capacity increased by 63.49% compared with the tested value from two sides covered gypsum board and OSB, respectively. So, further study must be conducted on the wall of the stud with web stiffener.

5 SUMMARY

1. Sheathing can effectively improve the bearing capacity axial of cold-formed steel wall, and OSB can increase the axial bearing capacity better than gypsum boards for cold-formed steel wall connected with the same sheathing on both sides;

2. Reduction in the stud spacing will increase the axial bearing capacity of cold-formed steel wall that, respectively, covered gypsum boards and OSB on each side;
3. For the cold-formed steel wall covering the OSB on both sides, the axial bearing capacity is not improved when the screw spacing is decreased;
4. The existence of the stud web stiffener can improve the axial bearing capacity of the cold-formed steel wall system effectively.

REFERENCES

[1] North American Steel Framing Alliance. Prescriptive Method for Residential Cold-Formed Steel Framing [S]. Year 2000 Edition.
[2] Yaip Telue, Mahen Mahendran. 2004. Behavior and Design of Cold-Formed Steel Wall Frames Lined with Plasterboard on Both Sides [J]. Engineering Structures, 26: 567–5791.
[3] Baokang, Zhou Tianhua. 2004. Experiment of load caring capacity for wall studs in low-rise cold-formed steel house [J]. Steel Construction, 19(4): 26–29.
[4] Fiorino L., Della Corte, G., 2007. Landolfo R. Experimental tests on typical screw connections for cold-formed steel housing [J]. Journal of Structural Engineering. 29(8): 1761–1773.
[5] Vieira, L. C. M., Jr., Shifferaw, Y., Schafer, B. W. 2011. Experiments on sheathed cold-formed steel studs in compression [J]. Journal of Constructional Steel Research. 67(10), 1554–1566.
[6] Qin Yafei. 2006. Theoretical and Experimental Research of Cold-formed Steel Residential Building Wall Stud system Subject to Centric Axial loads [D]. Shanghai: Tongji University.
[7] American Iron and Steel Institute. North American Specification for the Design of Cold-Formed Steel Structural Members [S], 2001, Washington, DC, USA.

Architectural, Energy and Information Engineering – Sung & Chen (Eds)
© 2016 Taylor & Francis Group, London, ISBN 978-1-138-02791-6

Research on the simulation analysis of rib concrete pouring sequence of V-type rigid continuous beam combination arch bridge

Y.Y. Fan & Z.Q. Li
School of Civil Engineering, Lanzhou Jiaotong University, Lanzhou, Gansu, China

ABSTRACT: The V-type rigid continuous beam combination arch bridge is a new form of bridge structure that has the combined advantages of the concrete-filled steel tube arch bridge and V-type rigid bridge in terms of shape, appearance, and stress. Different rib concrete lifting and pressed pouring construction sequences will produce different effects on the stress and deformation of the main arch rib steel pipe. In this paper, Nianchu Bridge's main bridge upper structure using a (60 + 148 + 60) m V-type rigid plus arch structure and the rib concrete using the two lifting and pressed pouring construction sequences are simulated for analysis. The main purpose of this paper is to study the impact of the two rib concrete lifting and pressed pouring construction sequences on the main arch rib stress and deformation of V-type rigid continuous beam combination.

Keywords: V-type rigid combination arch, lifting and pressed pouring, construction sequences, simulation, stress, deformation

1 INTRODUCTION

Owing to its advantages such as high strength, light weight, short duration, handsome appearance, and large span ability, the concrete-filled steel tube arch bridge is more widely used in bridge construction. Compared with the same span continuous beam, the V-type rigid continuous beam bridge has the advantages of small span calculation, low-height beam cross-section, less-weight load main beam, less-negative moment beam of pier, and positive moment span central point. At the same time, the V-support of V-type rigid bridges also makes the bridge most beautiful linearly. So, the V-type rigid continuous beam bridge becomes one of the more popular bridge types of the city bridge and landscape bridge constructions.

The V-type rigid continuous beam combination arch bridge is a new form of bridge structure that has the combined advantages of the the concrete filled steel tube arch bridge and the V-type rigid bridge in terms of shape, appearance, and stress. The V-type rigid continuous beam combination arch bridge has the characteristics of vivid, beautiful, dynamic and streamlined appearance, and a good effect of bridge landscape, and its architecture has the characteristics of beam bridges and arch bridges. The beam arch combination structure has become a more popular bridge design, which greatly improves the possibility of the bridge span[1].

The arch structure of the V-type rigid continuous beam combination arch bridge uses the steel tube concrete bowstring arch, and its upper structure is made of three parts, namely the main arch ribs of the concrete filled steel tube, the prestressed concrete continuous beam and the hanger. A part of constant load and live load of the structure is borne by the prestressed concrete girder of V-type rigid continuous beams; another part is borne by the hanger of the main beam passing through the main arch rib; the main arch ribs are subjected to axial pressure, resulting in outward horizontal thrust. The horizontal thrust of the main rib can be balanced by a prestressed concrete girder, forming a self-balanced system[2].

The arch structure construction sequence of the V-type rigid continuous beam arch bridge is as follows[3,4]: ① construct the scaffolding system of V-type rigid support and prestressed concrete continuous beams, and bury the arch foot of the steel tube; ② after the continuous beam construction is completed, erect the arch rib bench on the beam, and assemble and weld the steel tube arch ribs; ③ lift and press the pouring rib concrete; ④ install and tension the hanger; and ⑤ demolish the stents.

There are two design ideas of lifting and pressed pouring construction sequences of the rib concrete: ① when the assembly and welding of the rib is completed, the first step involves pouring concrete; the concrete in the steel tube reaches a certain intensity, and then demolish the rib stents; ② when the assembly and welding of the rib is completed, the first step involves demolishing of the rib stents, and then pouring concrete. These two design ideas have their own advantages and disadvantages. In this paper, Nianchu Bridge's main bridge upper structure using a (60 + 148 + 60) m V-type rigid plus arch structure and the rib concrete using the two lifting and pressed pouring construction sequences are simulated for analysis. The main purpose of this paper is to study the impact of the two rib concrete lifting and pressed pouring construction sequences on the main arch rib stress and deformation of V-type rigid continuous beam combination.

2 ENGINEERING SITUATION

Nianchu Bridge's main bridge upper structure uses the (60 + 148 + 60) m V-type rigid plus arch structure[5], the arch rib uses 1.8 times parabolic, with the calculated span of 134 m, the bilge of 39.3 m, the rise-span ratio of 1/3.4, the rib transverse space of 7.6 m, and the rib with a variable height dumbbell-shaped cross-section. It sets 7 crossbars between the two rib specimens, one is l-shaped in the top and the remaining six are K-shaped, and the crossbars are empty steel with a truss structure.

The full bridge has 13 pairs of hangers: the hangers use the double suspender form, the center distance is 8 m, each of the hanger is composed of parallel strand wires of double root 73 wires and the diameter of φ7 mm, and the vertical spacing between the hangers is 60 cm. To protect from human destruction, at a distance of top beam into the 3 m range and in hangers PE sheath, a 0.8 mm-thick flat stainless steel pipe is added. The tensioning end of hangers is located in the rib upper.

The V-pier lateral oblique leg and the vertical plane's angle is 54 degrees, the inside oblique leg and the vertical plane's angle is 42.8 degrees, the cross-section uses a rectangular solid cross-section, the wide of transverse is 9.7 m, the thickness of each is 2 m, roots of oblique legs and soffit are provided with a circular arc chamfering connection. The bottom of the V configuration to beam ceiling height is 11 m, the bearings are stalled on the bottom of the V-pier.

The beams use C55 concrete, its section is the straight web section of the single cell and the single box, the beam height is at the range of 80 m from the main span middle and 24.85 m of the boundary span end is 3 m, the beam height on the support of the V pier two oblique leg is 4.6 m. In addition, the beam height changes according to the 1.8 times parabolic, the width of the box beam's top is 10.7 m, the width of the bottom is 6.8 m, the thickness of the box beam roof is 40~60 cm, and the thickness of web is 50~120 cm. The height of the main beam inside V configuration changes according to the arc, and the minimum height of the beam is 3.2 m.

3 CALCULATION MODEL

Using MIDAS Civil 2012 software, the two rib concrete lifting and pressed pouring construction sequences of Nianchu Bridge are simulated[6,7]. Condition 1 involves the following steps: first, pumping concrete, and then removing the rib bench after the concrete in the pipe reaches a certain intensity. Condition 2 involves the following steps: first, removing the rib bench, and then pumping the concrete. In Condition 1, the lifting and pressed pouring concrete is completed in the rib bench. When removing the rib bench, the steel tube arch rib and the concrete in the steel tube arch are stressed collaboratively, and jointly provide the rib structure flexural rigidity and pressure section, and simulated via the steel-concrete joint section. In Condition 2, the process of lifting and pressed pouring concrete does not have bench support, and during the process of the injection of concrete, the concrete does not have any strength, nor does it have flexural rigidity and pressure section, and the rib pipe continuously deforms with concrete pumping. At this time, most of the stiffness of the rib structure is borne by the steel, and simulated by using the form of applying uniformly distributed load of concrete in the upper rib. So, different rib concrete lifting and pressed pouring construction sequences will produce different effects on steel rib stress and deformation. The main beam and arch consists of a discrete beam unit, and the hangers comprise only a discrete tension unit.

The supports of the beam are simulated by the tension unit, and the V structure and the beam use the rigid connection. The rib concrete lifting and pressed pouring finite element model of the Nianchu Bridge's main bridge is shown in Figure 1 and Figure 2.

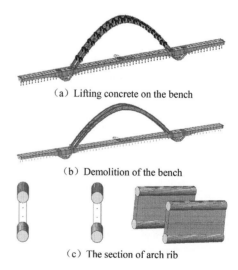

(a) Lifting concrete on the bench

(b) Demolition of the bench

(c) The section of arch rib

Figure 1. The finite element model of Condition 1.

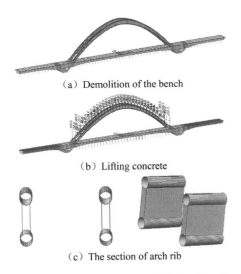

（a）Demolition of the bench

（b）Lifting concrete

（c）The section of arch rib

Figure 2.　The finite element model of Condition 2.

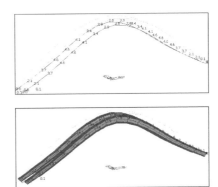

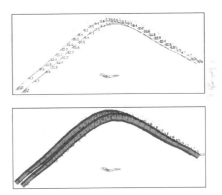

Figure 3.　The curve of rib deformation in Condition 1.

4　THE ANALYSIS OF SIMULATION RESULTS

4.1　*The deformation of the arch rib*

Different rib concrete lifting and pressed pouring construction sequences of Nianchu Bridge's main bridge using the (60 + 148 + 60) m V-type rigid plus arch structure will produce effects on the deformation of the steel rib, as shown in Figure 3, Figure 4 and Table 1.

4.2　*The stress of the arch rib*

Different rib concrete lifting and pressed pouring construction sequences of Nianchu Bridge's main bridge using the (60 + 148 + 60) m V-type rigid plus arch structure will produce effects on the stress of the steel rib, as shown in Figure 5, Figure 6 and Table 2.

The comparison of rib deformation and stress in Condition 1 and Condition 2 is shown in Table 1, Table 2 and Figure 4 to Figure 8.

5　CONCLUSIONS

The result from the numerical simulation calculation shows the rib concrete lifting and pressed pouring construction sequences of Nianchu Bridge's main bridge using the (60 + 148 + 60) m V-type rigid plus arch structure, where the maximum vertical deformation of rib is 4.4 mm and the maximum stress of rib is –18.2 MPa in the construction sequence of Condition 1; the maximum vertical deformation of rib is 12.4 mm, and the maximum stress of rib is –65.7 MPa in the construction sequence of Condition 2. The maximum

Figure 4.　The curve of rib deformation in Condition 2.

Table 1.　The vertical deformation value of the rib in Condition 1 and Condition 2.

No.	Rib node number	Distance from the vault (m)	Vertical deformation (mm) Condition 1	Condition 2
1	212	–63.0	0.0	–0.5
2	211	–56.0	2.1	4.7
3	210	–48.0	3.5	9.9
4	209	–40.0	4.3	12.4
5	208	–32.0	4.4	12.3
6	207	–24.0	4.0	10.5
7	206	–16.0	3.4	8.1
8	205	–8.0	2.9	6.1
9	204	0.0	2.6	5.3
10	213	8.0	2.9	6.1
11	214	16.0	3.4	8.1
12	215	24.0	4.0	10.5
13	216	32.0	4.4	12.3
14	217	40.0	4.3	12.4
15	218	48.0	3.5	9.9
16	219	56.0	2.1	4.7
17	220	63.0	0.0	–0.5

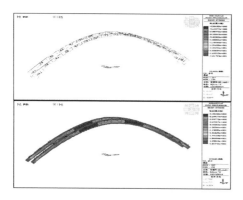

Figure 5. The stress curve of the rib in Condition 1.

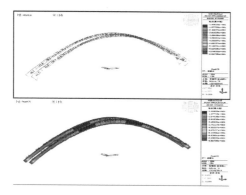

Figure 6. The stress curve of the rib in Condition 2.

Table 2. The stress value of the rib in Condition 1 and Condition 2.

No.	Rib node number	Distance from the vault (m)	Rib stress (MPa) Condition 1	Condition 2
1	212	−63.0	−18.1	−65.7
2	211	−56.0	−7.8	−32.8
3	210	−48.0	−8.6	−30.9
4	209	−40.0	−8.0	−28.9
5	208	−32.0	−6.7	−23.6
6	207	−24.0	−5.1	−16.4
7	206	−16.0	−6.4	−18.7
8	205	−8.0	−7.1	−23.3
9	204	0.0	−7.1	−26.1
10	213	8.0	−7.1	−23.2
11	214	16.0	−6.4	−18.3
12	215	24.0	−5.1	−16.4
13	216	32.0	−6.7	−23.6
14	217	40.0	−8.0	−28.9
15	218	48.0	−8.5	−30.9
16	219	56.0	−7.8	−32.8
17	220	63.0	−18.2	−65.7

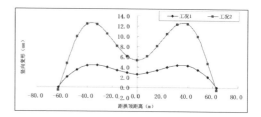

Figure 7. The vertical deformation comparison chart of the rib in Condition 1 and Condition 2.

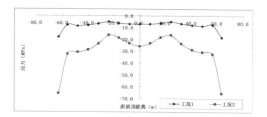

Figure 8. The stress comparison chart of the rib in Condition 1 and Condition 2.

vertical deformation of the rib and the maximum stress of the rib in the construction sequence of Condition 1 are less than that observed in Condition 2. Consequently, comprehensively considering the vertical deformation and stress of the rib, it can be seen that the construction sequence of Condition 1, which is first lifting and pressed pouring concrete on the rib bench and then removing the rib bench, is beneficial in controlling the stress and deformation of the rib.

REFERENCES

Liping Zhang. Span Tied Arch Bridge Concrete perfusion method [J] China Municipal Engineering, 2006, (05): 40–42.

Ming Yin. Concrete Pipe Tied Arch Bridge construction boom tensile force calculation [J] Highway Engineering, 2010, 35 (2): 101–103.

Xiaosen Yang, Weiming Yan. etc. Large-span steel arch bridge concrete pouring of concrete sequential optimization [J] Highway and Transportation Research, 2010, 27 (1): 67–83.

Yanbin Xiao. V-shaped rigid frame continuous beam bridge construction portfolio structure stress analysis [J] Urban Roads Bridges and Flood Control, 2013, (10): 69–71.

Zhaofeng Liu. V-frame Composite Arch grillage analysis [J] Science Technology and Engineering, 2012, 12(31): 8465–8468.

Measurement of trichlopyr residual in soil

Fei Xiao, Qing Li, Guang Lin Song, Tian Ying Nie, Bo Na Wang & Ying Li Chen
Guizhou Academy of Testing and Analysis, Guiyang, China

Yi Zhang
Guizhou Research Institute of Chemical, Guiyang, China

ABSTRACT: In this paper, we established the method of analysing trichlopyr residual in soil, using the High-Performance Liquid Chromatography (HPLC) method on a ZORBAX SB-C18 chromatographic column with methanol + water solution (containing 0.5% ice acetic acid) in a proportion of 80 + 20 (V/V), and made a quantitative analysis of trichlopyr at a flow velocity of 1 mL/min at a wavelength of 291 nm. When the concentration of trichlopyr is within 600 µg/mL, the linear equation is $y = 3.76659x + 0.4397$, the correlation coefficient is 0.99999, the lowest detection limit is 0.001 mg · kg^{-1}, the standard deviation is 0.58, and the average recovery is 100.53%. This method is proved to be accurate, rapid, and suitable for the quantitative analysis of trichlopyr.

Keywords: Soil; trichlopyr; residual

1   INTRODUCTION

Trichlopyr, Common name: Trichlopyr, CAS No.: 55335-06-3, Chemical name: [(3,5,6-trichlorine-2-pyridine) oxygen] acetic acid, belongs to the herbicide group, which is applied after seedling leaf processing of uptake and translocation, where the weeds quickly absorb the drug and transfer it to the whole plant. The top of sensitive weeds is wilted after absorbing the drug for 2 to 3 days. Herbicidal symptoms include typical hormone appearance, plant malformation, warping, and ultimately death. Its main application involves preventing and killing the annual broadleaf weeds, such as cleavers, purslane, nightshade, and alternanthera philoxeroides, in the field of winter wheat and corn. In recent years, the literature on fluroxypyr is scarce. Most studies on fluroxypyr have reported on the safety use, pesticide residual and degradation in the environment of pest control. However, no reports are available with respect to content determination and analysis by RP-HPLC. This article presents the quantitative analysis of trichlopyr in soil and obtained satisfactory results by using RP-HPLC, C18 reverse phase column and detector of variable wavelength ultraviolet. The advantages of the method are easy operation, high accuracy, and excellent reproducibility, providing a reference for production quality control and analytical method on trichlopyr.

2   MATERIALS AND METHODS

2.1   *Instruments and reagents*

The instruments used are as follows: Agilent 1220 HPLC, Diode Array Detector (DAD), Agilent chromatographic work station, electronic scale BD324, accurate to 0.1 mg, ultraviolet spectrophotometer Agilent UV-V is 8453.

The reagents used are as follows: methyl alcohol of chromatographically pure grade from Honeywell Burdick & Jackson Ulsan, 680-160 Korea, ultrapure water, electrical resistivity of 18.25 MΩ · cm, glacial acetic acid of analytically pure grade, Kelong Chemical Reagent Factory in Chendu.

Trichlopyr standard substance: 99.0%, Dr Ehrenstorfer Gmbh.

2.2   *Instrument conditions*

The conditions set are as follows: chromatographic column is ZORBAX SB-C18, 4.6 × 250 mm × 5 µm, mobile phase is methyl alcohol + water (including 0.5% glacial acetic acid) = 80 + 20 (V/V), flow rate is 1.0 mL/min, wavelength is 291 nm, column temperature is 25°C, run time is 7 min, sample injection is 5 mL. Atlas is determined by the above condition for trichlopyr, as shown in Figure 1.

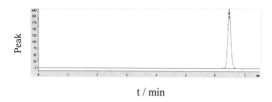

t / min

Figure 1.　Atlas of the standard substance of trichlopyr.

2.3　Preparation of the standard solution

A standard substance of 0.0250 g trichlopyr was placed in a 25 mL volumetric flask, made a constant volume in methanol solution to obtain a stock solution of trichlopy standard with a concentration of 1000 mg/L, and was shaken well for analysis. The stock solution of the above concentration was diluted to concentrations of 25, 50, 100, 200 and 400 mg/L, and was shaken well to server as standard solutions.

2.4　Sample processing

In a 50 ml centrifuge tube, an 8.00 g soil sample was added and filtered through a 1.5 mm sieve and was mixed. Then, 5 mL ultrapure water was added to it, and subjected to volute vibration for 1 min. Thereafter, 10 mL methanol extract of volume fraction 2% of glacial acetic acid was added, and subjected to volute vibration for 5 min. After vibrating 5.00 g anhydrous magnesium sulfate and 2.00 g sodium chloride, the mixture was centrifuged for 5 min. Then, 25 mg PSA and 100 mg anhydrous magnesium sulfate were added to the 1.5 mL supernatant, and then vortex centrifuged and filtered the supernatant through a 0.22 μm organic filter. Finally, the supernatant was measured in accordance with the chromatographic condition in Section 2.2.

2.5　Calculation

The content of the effective constituent was calculated using the following formula:

$$W = \frac{C \times P \times V \times D \times 10^{-6}}{M}, \qquad (1)$$

where W is the content of the effective constituent in test samples (%); C is the concentration (mg/L); P is the standard sample purity (%); V is the sample volume (mL); D is the sample dilution ratio; and M is the sample quality (g).

3　RESULTS AND ANALYSIS

3.1　Optimization of extract and purification conditions

In this study, it is found that, when using acetonitrile as the extraction solvent, the recovery rate of trichlopyr is lower than 50%, which cannot meet the requirments of residual analysis. In view of the faintly acid nature of the trichlopyr, its appropriate addition to the extraction solvent is taken into consideration in order to increase the recovery rate of the target object. By comparing the extraction efficiencies between methanol solutions containing glacial acetic acid of separate quality scores 0.5% and 2%, it is found that a satisfactory recovery rate of trichlopyr can be obtained when using methanol solution containing 2% of glacial acetic acid to conduct the extraction. In the process of sample purification, PSA is mainly adopted to remove the disturbing impurities from the matrix material. However, due to the interference of other impurities in the soil, the purifying effect is not good enough when using PSA only. Therefore, on the basis of PSA purification, appropriate anhydrous magnesium sulfate is added for auxiliary purification, and then an ideal purifying effect is obtained. Meanwhile, better reproducibility is also realized when obtaining a better recovery rate, thus simplifying the operating process and increasing the efficiency.

3.2　The confirmation of detection wavelength

The standard solution of trichlopyr was scanned at the wavelength of 200~400 nm by UV-VIS to obtain the UV spectra, as shown in Figure 2. Absorption peaks of wavelength appeared at 201, 231, and 291 nm, but it was not possible to conduct the microanalysis at 201 and 231 nm due to a narrow linearity range and the appearance of the interference peak. The third absorption peak of trichlopyr was observed at 291 nm, which met the requirements of the quantitative analysis, with a wide linearity range.

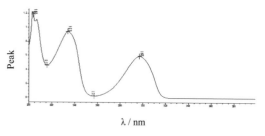

λ / nm

Figure 2.　UV spectra.

3.3 Optimization of the mobile phase

The selection of the chromatographic column and the mobile phase is key to the separating effect by LC, using methanol—water at a volume ratio of 80:20 as the mobile phase, since the stationary phase of the separation of chromatographic column was ZORBAX SB-C18, when using strong ability of elution and methyl alcohol proportion greater than 80% as the mobile phase. The appearance time of trichlopyr was fast. At less than 80% as the mobile phase, the retention time of trichlopyr lasted for a long period and its peak type turned wider. While using methanol—water at a volume ratio of 80:20 and 0.5% glacial acetic acid as the mobile phase, the peak type of chromatography was symmetrical, and the baseline was smooth and steady.

3.4 Methodological linear relation

A standard solution of 5 concentrations prepared in Section 2.3 was measured according to the above operating conditions shown in Section 2.2. Both peak area (y) and concentration (x) had a good linear relation: linear equation was $y = 3.76659x + 0.439785$ and the correlation coefficient was 0.99999. The results showed that the minimum limit of detection was 0.001 mg · kg^{-1} when trichlopyr was in the range of 5~600 μg/mL, suggesting

a good linear correlation between the mass concentration and the peak area. The detailed data are presented in Table 1.

3.5 Precision test

In the same operating conditions of chromatography, 5 samples with the same concentration were prepared, the average value and the relative standard deviation of trichlopyr were calculated and separately weighed 0.0102 g, 0.0099 g, 0.0097 g, 0.0098 g, and 0.0102 g of samples, made constant volume to 100 mL with methyl alcohol, and repeatedly and separately measured 5 concentrations according to the above operation conditions described in Section 2.2. The results are summarized in Table 2. The relative standard deviation of the method was 0.13%, which indicated that the method was acceptable.

3.6 Accuracy of the experiment

To evaluate the accuracy of this method, an experiment of the recovery rate was included, namely, to precisely add a certain amount of the standard sample to the known content of the components tested. To calculate the recovery rate, a 0.0099 g sample was added to methyl alcohol, and made a constant volume to 100 ml, and a 99 ug/mL solution was prepared and added 100 ug/ml standard solution of trichlopyr into the above sample solution at the volume ratio of 1:1 to obtain the solution with a concentration of 99.50 μg/ml for analysis. The above procedure was repeated to prepare 5 solutions having the same concentration with the content of the components tested. The test was repeated for 5 times according to the operation conditions described in Section 2.2. The results are summarized in Table 3. The average recovery rate was 100.53% and the variable coefficient was 0.58%, thus concluding that the method was acceptable.

Table 1. Concentration and peak area of trichlopyr standard.

Concentration (μg/mL)	Peak area
25	92.84
50	191.38
100	379.64
200	749.57
400	1508.32

Table 2. The results of the precision measurement of trichlopyr.

Serial number	1	2	3	4	5	Average value	Relative standard deviation (%)
Sample amount (g)	0.0102	0.0099	0.0097	0.0098	0.0102	–	–
Measurement of concentrations (μg/mL)	102.69	99.68	97.56	98.71	102.40	–	–
Measurement of contents (μg/mL)	99.67	99.68	99.57	99.72	99.39	99.61	0.13

Table 3. Results of the sample recovery rate.

Sample	Blank (µg/L)	Concentration (µg/L)	Test content (µg/L)	Recovery (%)	Relative standard deviation (%)
Soil-1	ND	99.50	100.22	100.72	
Soil-2	ND	99.50	99.10	99.60	
Soil-3	ND	99.50	99.87	100.37	0.58
Soil-4	ND	99.50	99.88	100.38	
Soil-5	ND	99.50	100.21	100.71	

4 CONCLUSIONS

In this paper, the method of residue analysis of trichlopyr in soil was put forward. With the development of liquid chromatography and chromatographic column, high-performance liquid chromatography has become a major method in the quality inspection of pesticide products, environmental pollution monitoring and pesticide residue. This article presented the analytical method of High-Performance Liquid Chromatography (HPLC) that is simple, quick and of accurate result, thus suggesting its suitability for the content analysis of trichlopyr.

REFERENCES

Hai-xia Wan & Xiao-hui Gao, 2014. Determination of Active Ingredient Content in Fluroxypyrmepthyl by HPLC. Agro chemicals: 31–33.

Li-qun Guo, Jun Xu & Feng-shou Dong, 2012. Simultaneous determination of nicosul Furon, atrazine and fluroxypyr in soil and corn by ultraperformance liquid Chromatography-mass spectrometry. Chinese Journal of Pesticide Science: 177–184.

Yang Sun, Dong-mei Qin & Ying-ming Xu, 2011. Residue Detection and Degradation of Fluroxypyr in Wheat and Soil. Environmental Chemistry: 760–765.

Architectural, Energy and Information Engineering – Sung & Chen (Eds)
© 2016 Taylor & Francis Group, London, ISBN 978-1-138-02791-6

Clogging of the subsurface infiltration system

Y.L. Duan, X.C. Ji & Y.Y. Yu
Key Laboratory of Regional Environment and Eco-Remediation, Ministry of Education, Shenyang University, Shenyang, China

Y.H. Li
School of Resource and Civil Engineering, North Eastern University, Shenyang, China

ABSTRACT: The subsurface infiltration system of sewage treatment is based on recycling, harmony and other ecological principles. It has many virtues such as high quality of effluent, low investment, simple management and undamaged surface landscape. However, there are also issues such as low hydraulic loading, large area, clogging and other factors.

Keywords: Subsurface infiltration system; mechanism of clogging; physical blockage; biological clogging; solutions

1 INTRODUCTION

The subsurface infiltration system of sewage treatment is a perennial ecological engineering (Yang et al., 2007). The system uses the self-regulation mechanism of the terrestrial ecosystem usually combined by soil, microorganism and plants, and comprehensive purification of pollutants to treat municipal sewage and some industrial wastewater, improving the quality of water to varying degrees. With the biogeochemical cycles of nutrients and moisture, the system can promote green growth and waste water recycling and is harmless.

Subsurface infiltration system is widely used at home and abroad for its advantages such as high quality of effluent, low investment, simple management, no odor and undamaged surface landscape (Li et al., 2009). However, the technology researched in recent years has revealed that there are still shortcomings in the design and practical applications of subsurface waste water infiltration system, which are described below. ① Low hydraulic load: it is a major problem that exists in the subsurface waste water infiltration system. With the increase in hydraulic load, the speed of sewage through the system becomes faster, thereby reducing the reaction time with microorganism, affecting the quality of effluent. Also, it increases the possibility of clogging. Therefore, it is necessary to maintain the balance between hydraulic load, quality of effluent and clogging (Wang et al., 2006). ② Large area: the matrix of the subsurface infiltration system is filled by soil *in situ*, which has a poor infiltration rate. In order to improve the processing capacity of the system, the area should be expanded. Besides, the large area is the main factor that hampers the development of the technology, which has a close relationship with the strain resources of the land. ③ Clogging: this is the problem that frequently occurs in the subsurface wastewater infiltration system. In addition, it is one of the most important factors that hampers the application of the technology. Clogging not only affects the hydraulic loading, but also affects the life of the system. A good land treatment system can run steadily more than a decade or even 20 years (Zhang et al., 2006). According to reports, 72% of wastewater land treatment facilities need to be excavated to repair or to replace the packing because of the clogging in the USA, and 70% of the project in Australia are paralyzed due to the clogging (Capra et al., 2007). The subsurface infiltration system has been widely applied in the foreign country. It has also been reported that 30% of the sewage made by 115 million households in the USA was treated by the subsurface infiltration system, and about 40% of new homes were equipped with the subsurface infiltration system (Yang et al., 2002). Subsurface infiltration system is suitable for the treatment of decentralized sewage, and has a great use of space in rural area without a perfect municipal pipe network. It is in favor of promotion and application of the subsurface infiltration system to analyze the mechanism of clogging and the solutions.

2 MECHANISM OF CLOGGING AND THE SOLUTIONS

The causes of clogging are complex, but physical blockage and biological clogging are considered as the main cause of system crashes. Specifically, the mechanism of clogging involves fluid dynamics, structural mechanics, fluid mechanics and other aspects. In addition, bubble blockage also occurs in some cases. Matrix blockage will reduce the hydraulic conductivity of the bed and impede ventilation, reducing the purifying effect of the subsurface infiltration system (Li et al., 2008).

2.1 Physical blockage and solutions

Physical blockage is an important cause of clogging of the subsurface infiltration system. The suspended solids in sewage retained by the soil matrix occupy the interconnected pores of the soil matrix (Carroll et al., 2006), and then the saturated permeation rate of the system is gradually reduced, and finally clogging occurs. The physical blockage body of the soil matrix includes suspended solids from the sewage and unstable particles from the soil matrix, and clogging is mainly caused by the retention and adsorption of the clogging body, including the three aspects described as follows. ① Precipitation: the particulate matter with a better property setting is in suspension because of the fluid state of the sewage. This part of the suspension flowing into the infiltration system along with the influent would precipitate quickly due to the gravity. The precipitation collected on the surface of the packing forms a sediment layer with smaller permeability, triggering the clogging of the system. ② Mechanical filtration: the percolation dielectric layer is composed of particles with different sizes, where there are many pores, just like a "sieve." The size of the suspended particles that is greater than the medium pore cannot pass and thus are retained. When it has accumulated to a certain extent, the smaller particles of the suspended solids would be further limited into the system. This phenomenon usually occurs on the surface of the percolation medium. ③ Adsorption: It includes medium particles, biofilm, and biological floc between the particles and the suspended particulates trapped. The three kinds of the above-mentioned effect show that the majority of trapping and adsorption of the suspended matter occurs on a surface layer of the system with a small thickness, and a little on the lower layer, thereby causing the phenomenon of the uneven distribution of suspended solids retained in the filter layer. Therefore, the clogging caused by the retention and adsorption of suspended solids mainly appears in the border between the distribution layer and the

filter layer, especially in a 0~15 cm section of the filter layer. With the increasing depth, the degree of clogging would be reduced greatly.

The process of physical blockage caused by suspended solids is complex and irreversible, so it should be avoided emphatically in the subsurface infiltration system (VanCuyk et al., 2001). Usually, cloggingb can be prevented by using the following 3 aspects: ① Proper filler: the soil with good crumb structure and greater porosity can be used as the filler. ② Proper operating parameters: proper hydraulic load can avoid the erosion of packing in the subsurface infiltration system. ③ Efficient pretreatment measures: precipitation, flotation and other pretreatment measures can reduce the concentration of suspended solids, and the pretreatment can guarantee the normal operation of the system (Zhang et al., 2003).

2.2 Biological clogging and countermeasures

Biological clogging is another important cause of clogging in the subsurface infiltration system. The processing layer of the subsurface infiltration system is a functional layer with a strong degradation to contaminants formed during the progress of sewage treatment, namely biological cushion. Biological cushion is acquired, mainly located in the range of 0~10 cm under the water layer. There are many organisms, including microbial cell body, extracellular polymeric, refractory organisms adsorbed on the packing and extracellular polymers. There are also some inorganic substances, including inorganic salts adsorbed on extracellular polymers, microbial metabolites, nitrogen, and methane (Nie, 2011). Biological clogging body include the cell body, extracellular polymers, gaseous substance adsorbed on the surface of microorganisms, sulfide mineral, and iron bacteria trapped on the surface of microorganisms. Gaseous substances include methane, nitrogen and other gases produced during the degrading of pollutants in sewage. The increasing microbial biomass and the formation of a large number of extracellular polymers reduce the capillary porosity, decreasing the permeability of the soil. Finally, the biological clogging occurs.

There are 4 causes for the biological clogging: ① Clogging caused by the growth of microbes means the blockage formed by the decreasing capillary porosity due to the increasing microbial biomass and the thickness of the biofilm adsorbed on the surface of packing, during the treatment progress in sewage. Because the subsurface infiltration system is located underground and there is no sludge, the amount of the growth of microorganisms equals the amount of its attenuation, when the system is stabilized. The clogging caused by the growth of microorganisms would then stop after the stabilization of the system. Usually, controlling the quantity of the influent and the concentration of the pollutant can release the biological clogging. ② Blockage caused by extracellular polymers of microorganisms means the clogging formed by the decreasing capillary porosity due to the metabolites. These metabolites are produced in the process of growth and reproduction of microorganisms using the nutrients from the sewage. Polysaccharide is the main product of the metabolic processes of soil microorganisms. In addition, the material needs a long time to degrade, about 23 to 50 days. So, it easily collects in the soil gap, causing blockage. Usually, biological or chemical inhibitors are added to the system to prevent the biological clogging. ③ Blockage caused by falling biofilm. The falling biofilm includes microbial cell body, extracellular polymers, refractory organisms and organisms adsorbed on the soil matrix and extracellular polymers. These substances are difficult to degrade, so they easily gather in the soil gap, causing blockage. ④ The blockage caused by gases means that the gases produced from the biochemical reaction in the soil clog the soil pore with the complex mechanical effect. If carbon dioxide, methane, nitrogen and other gases produced during the process of microbial aerobic-anaerobic reaction in the soil cannot be discharged from the soil, they would be divided into a number of segments remaining in the soil pore, forming a gas block like "solid". Or the gases would accumulate in the soil capillary net, resulting in excessive internal pressure and limiting sewage flows, causing the clogging in soil pore (Luanm et al., 2002). When the subsurface infiltration system is used to treat decentralized sewage, there is little clogging caused by gases, because the packing of the system can be alternating between wet and dry conditions as a result of intermittent drainage.

The main causes of biological clogging are the overgrowth of microorganisms and the accumulation of extracellular secretion or the loss of the biofilm. Usually, the biological clogging can be prevented by taking the actions described below (Zhang et al., 2003). ① Intermittent dosing: continuous dosing makes the soil of the system in a reduced state, causing the accumulation of extracellular polymers, then leading to the clogging gradually. The use of intermittent dosing can release the soil and ensure the soil in a certain degree of aerobic state. So, the excessive accumulation of extracellular polymers can be avoided and the clogging can be prevented (Balks et al., 1997). In general, the longer the interval time, the better the ability of treatment recovery, and greater the penetration rate. However, the interval time should not be too long, considering the treatment efficiency and processing load. Moreover, usually, the ratio of dry and dosing is 1~8. ② Aeration: the anaerobic condition is the important cause of the accumulation of extracellular polymers; therefore, the aeration of sewage can play a role in the prevention of clogging. Sewage aeration can improve the value of Dissolved Oxygen (DO), but the value of DO decreases rapidly during the infiltration. Suppling the air around the pipe dispersing the sewage can improve the DO effectively, and maintain the soil in the aerobic state. So, the microbial decomposition can be maintained, and the accumulation of extracellular polymers can be prevented. ③ Application of microbial inhibition or dissolving agents (Magesan et al., 2000): the application of microbial inhibition or dissolving agents can prevent the clogging by limiting the growth of microbes or killing the microbes. However, the removal of pollutants in sewage depends on the metabolic activity of microorganisms, therefore the measures that do not damage the habitat of soil microbes should be adopted to restore the hydraulic conductivity of the soil. So, the measures that limit or kill the microbes to prevent clogging has a little value in the application.

3 CONCLUSION

With the extension of the underground infiltration system uptime, the infiltration rate of soil would gradually decline, and the clogging would appear due to the influence of physical, chemical, biological and other factors. Moderate soil blockage can expend the unsaturated low area in the subsurface infiltration system, and the treatment effect could be improved. However, it would be hard for sewage to flow through the soil layer because of the excessive blockage, so that the ability of the system to treat sewage would be seriously degraded, reducing the life of the system. The clogging of the subsurface infiltration system can be divided into physical and biological clogging. Physical clogging usually appears on the surface of the infiltration system, and it is greatly influenced by SS concentration. Usually, physical clogging can be prevented by selecting proper packing and operation parameters and intensifying the pretreatment. Biological clogging is caused by the overgrowth of microbes, the accumulation of metabolites and the loss of the biofilm. Usually, the biological clogging can be released by selecting the proper operating mode and aeration and adding some inhibitors. Overall, it is of great significant for improving the life of the system and guaranteeing the efficiency of the treatment to enhance the primary treatment.

ACKNOWLEDGMENTS

This work was financially supported by the National Natural Science Foundation of China (no. 51108275), the Program for Liaoning Excellent Talents in Universities (LNET) (no. LJQ2012101), the Program for New Century Excellent Talents in Universities (no. NCET-11-1012), the Science and Technology Program of Liaoning Province (nos. 2011229002 and 2013229012), and the Basic Science Research Fund in Northeastern University (nos. N130501001 and N140105003).

REFERENCES

Balks, M.R., Mclay C.D.A., Harfoot C.G. et al. 1997. Determination of the progression in soil microbial response and changes in soil permeability following application of meat processing effluent to soil [J]. Appl Soil Ecol, 6: 109–116.

Capra, Seicolone, et al. 2007. Recycling of poor quality urban wastewater by drip irrigation systems [J], Journal of Cleaner Production, 15: 1529–1534.

Carroll S., Goonetilleke A., Khalil W.A.S. et al. 2006. Assessment via discriminant analysis of soil suitability for effluent renovation using undisturbed soil columns [J]. Geoderma. 131(1–2), 201–217.

Li Xiaodong, Sun Tieheng, Li Haibo, et al. 2008. Research progress of sewage treatment mode in a housing estate [J], Journal of Ecology, 27(2): 269–272.

Li Yinghua, Sun Tieheng, Li Haibo, et al. 2009. Technical difficulties and solutions of subsurface infiltration system in domestic wastewater treatment [J], Chinese Journal of Ecology, 28(7): 141–1418.

Luanmanee S, Boonsook P, Attanandana T. 2002. Effect of intermittent aeration regulation of am ult-i so i-l laye ring system on domestic wastewater treatment in Thailand [J]. Ecological Engineering, 18: 415–428.

Magesan G.N., Williamson J.C., Yeates G.W. et al. 2000. Wastewater CBN ratio effects on soil hydraulic conductivity and potential mechanisms for recovery [J]. Bioresource Technology, 71: 21–27.

Nie Junying, 2011. Studies on wastewater treatment by fortified subsurface soil infiltration system as well as its mechanism [D], Shanghai Jiao Tong University, Shanghai.

VanCuyk, S., Siegrist, R., Logan A. et al. 2001. Hydraulic and purification behaviors and their interaction during wastewater treatment in soil infiltration systems [J]. Water Research, 35: 953–964.

Wang Jizheng, Qiao Pengshuai, Lu Zhili, et al. 2006. Analysis of several rea men skills in sewage land rea menechnology; [J], Journal of Beijing Technology and Business University (Nature Science Edition), 2006, 24(2): 13–16.

Yang Wentao, Liu Chunping, Wen Hongyan, et al. 2007. Introduction to wastewater land treatment system [J], Soil science, 38(2): 394–398.

Yang Xingyu, Peng Renzhi. 2002. The mechanism and application of trench style land treatment in sewage purification—in Guizhou Province [J], 20(3): 65–68.

Zhang Jian, Shao Changfei, Huang Xia, et al. 2003. Soil problems of wastewater land treatment process [J], Chinese Water & Wastewater, 3: 17–20.

Zhang Zhiao, Lei Zhongfang. 2006. The main problems and mitigation programs in process soil percolation process, Environmental Science and Management [J], 31(5): 41–43.

Architectural, Energy and Information Engineering – Sung & Chen (Eds)
© 2016 Taylor & Francis Group, London, ISBN 978-1-138-02791-6

Micrositing of a container roof mounted wind turbine using the neutral equilibrium atmospheric boundary layer

Q. Wang
College of Energy and Power Engineering, Inner Mongolia University of Technology, Hohhot, Inner Mongolia, China

J.W. Wang & Y.L. Hou
Key Laboratory of Wind Energy and Solar Energy of the Ministry of Education, Hohhot, Inner Mongolia, China

ABSTRACT: In the future program of the city, it is very significant to consider the great renewable energy sources in the city environment. Based on the Task 27 of International Energy Agency project, this study uses the Computational Fluid Dynamic technique to simulate the turbulent wind flow over the building. A container modelled a building and the flow field around the container placed in the Zhangbei wind power construction base. First, more precise inflow boundary conditions are introduced based on the Neutral Equilibrium Atmosphere Boundary Layer (NE-ABL), which is suitable for the local wind resource. Second, the turbulent characteristics are analyzed in the vertical direction above the container in order to find the suitable mounting locations and height at the top of the container to install a HAWT. It was found that 1.17H to 1.50H above the container is the suitable height for wind turbine installation. The frontier point is the appropriate installation location for the selected wind turbine and the suitable installation height is 1.42H to 1.52H. The maximum ideal power increase factor is up to 54.5% at annual average wind speed. This will provide corroboration for further experiment of building mounted wind turbine.

Keywords: Wind turbine; container; micrositing; NE-ABL; CFD

1 INTRODUCTION

With the development of the green buildings in cities, one of the hotspots has been the proper use of wind energy and the study of the micrositing and output of rooftop wind turbines [1–3].

These wind turbines have been studied initially both at home and abroad. Burton noted that the wind resource should be evaluated before installing the wind turbine with more output and better economy efficiency [4]. Bavuudorj proposed a method of micrositing of wind turbine on building's rooftop by using joint probability density function of wind speed and direction to analyze the wind distribution and predict the wind acceleration influenced by the building [5]. Islam presented the effect of different roof shapes on the energy yield and position of roof mounted wind turbines, covering different buildings' heights under different wind directions [6]. Yuan introduced both new concepts: wind speed increment effect and the wind tunnel, providing a new way to make use of the wind energy [7]. Yuan developed a math simulation of the wind environment in Anzhong building, and the results suggested that the more obvious the effect on mass, the faster the wind speed in the

roof. Thus, based on the size of the roof, multi-row wind turbines can be installed [8].

However, there are many studies on the analysis of the characteristic of wind around the buildings; however, none of them reported the aspect in the micrositing and output of rooftop wind turbines. According to the Task 27 of International Energy Agency project, this paper predicted the changing rules of the power's increment factor under the condition of the annual mean wind speed and the different installed positions and heights with different wind directions, providing the theory for future study.

2 MODELED PARAMETERS

The simulation towards the building mounted wind turbine replaced the waste container, developing the model by using the Design Modeler of ANSYS software. The length and breadth of the container is 12.4 m × 2.4 m and the height is 3 m.

To investigate the effect of wind direction, simulations were undertaken under different wind directions (Figure 2), with direction 315° (NW), prevailing wind direction of Zhangbei, and 270°,

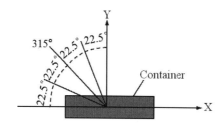

Figure 1. The wind flow direction.

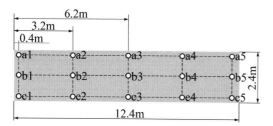

Figure 2. Locations for WT installation (top view).

292.5°, 315°, 335.5° and 360° being chosen for the simulations, respectively. Then, the results are processed and the 15 points' turbulence parameters are analyzed in the vertical direction, to find a suitable position and height, to avoid the region of strong turbulence as far as possible. Finally, based on the above results, the optical position is chosen for a selected wind turbine. The locations of the proposed mounting positions are demonstrated as a grid superimposed onto the roof of the container (Figure 2).

3 NEUTRAL EQUILIBRIUM ATMOSPHERIC BOUNDARY LAYER PROFILE

CFD is the tool used in this paper for predicting the above investigated cases of wind turbine output. The minimum requirements for carrying out a uniform CFD simulation can be summarised in the following points:

1. For flows over the container, the standard k-ε turbulence model is preferred [9].
2. If H is the height of the container, the vertical direction dimension = 5H, the lateral dimension = 5H + container width and flow direction dimension = 15H, while the blockage ratio is 1%.
3. Mesh cells are equidistant while fairing and encrypting the mesh in zones of complex flow using the tool of ICEM-CFD.

4. For the boundary conditions, the ground should be a non-slip wall with standard wall functions (roughness height = 0.20 m), the top and side should be symmetrical, the outlet should be a pressure outlet and inflow should be a log law atmospheric boundary layer profile.
5. Second-order schemes should be used for solving the equations.
6. The scaled residuals should be in the range of $10^{-5} \sim 10^{-6}$.

Surface roughness, obstacles or other factors affect the construction near the ground due to the viscous shear stress that forms the atmospheric boundary layer containing the urban canopy layer, roughness sub-layer and inertial sub-layer. Usually, wind turbine should be installed in the roughness sub-layer. Zhangbei located in the flat country that its weather is cloudy and windy, conforming to the neutral atmosphere boundary standard proposed by Gu [10]. In this paper, the inlet boundary condition was described using a UDF, satisfying Esq. (1)~(3) for the velocity U_Z, turbulent kinetic energy k and turbulent diffusion rate ε, which is modified by the logarithmic law:

$$U_Z = \frac{U_*}{\kappa} \ln\left(\frac{Z - Z_0}{Z_0}\right) \tag{1}$$

$$k = \frac{U_*^2}{\sqrt{C_\mu}} \sqrt{C_1 \ln\left(\frac{Z + Z_0}{Z_0}\right) + C_2} \tag{2}$$

$$\varepsilon = \frac{U_*^3}{\kappa(Z - Z_0)} \sqrt{C_1 \ln\left(\frac{Z + Z_0}{Z_0}\right) + C_2} \tag{3}$$

where U_Z is the average speed at the height Z; U_* is the friction velocity; κ is von Karman's constant; Z_0 is the surface roughness; and C_μ is the turbulence model constant. C_1 and C_2 are constants equal to –0.17 and 1.62, which depend on the suggestion of Yang [10].

4 RESULTS

When the wind flows over the container, it causes two effects: one is an acceleration of wind speed around the container and the other is the increase in the turbulence intensity at the top of the container. Therefore, there is a need to analyze both the turbulence parameters and the optical locations and theheight of the wind turbine at the top of the container under different directions.

The annual average wind speed is 6.2 m/s in Zhangbei. Turbulence intensity is used to estimate the low turbulence zones in the vertical direction, and the wind accelerated factor is used to determine the location where high output is produced at the top [11]. Then, the optical height can be found by average turbulent thickness, which is confirmed by the high turbulence zone and the high velocity gradient zone.

4.1 Turbulence intensity and wind accelerated factor

In this section, the turbulence intensity and wind accelerated factor distributions of the wind at variable elevations above the top are presented and discussed. The turbulence intensity I and wind accelerated factor C_V are defined as:

$$I = \frac{\sigma}{U_Z} \tag{4}$$

$$C_V = \frac{U_Z}{U_{Z0}} - 1 \tag{5}$$

where σ is the fluctuating velocity and U_{Z0} is the undisturbed velocity at height Z. Due to the symmetry of the selected locations, the distribution of I and C_V about raw b at the top of the container is only shown in Figure 3.

According to the International Electrotechnical Commission (IEC) Standard 61400-12, a turbine should not be exposed to wind with a turbulence intensity greater than 0.25 and installed at the zone without a high velocity gradient. Hence, it is vital to estimate I and C_V at the turbine mounting locations. Figure 3 (a) shows that the maximum value of turbulence intensity reaches up to 0.31 at location b1, and the zone where Z/H is greater than 1.17 does not belong to the high turbulence section. As shown in Figure 3 (b), the most pronounced speed-up effect occurs at the frontier sites compared with other locations above the container. This significant phenomenon disappears when the height reach up to 1.60H. Therefore, the wind turbine should be installed at frontier points at the height of 1.17H~1.60H.

4.2 Average turbulent thickness

Statistical analysis of turbulent parameters of 15 locations above the container under 5 wind directions was performed. The turbulent thickness δ is introduced to discuss the influence of the wind horizontal direction on the wind characteristics, and to provide guidance for the height of wind turbine installation. The thickness of turbulence δ is defined as the greater value of height between the

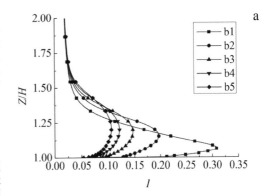

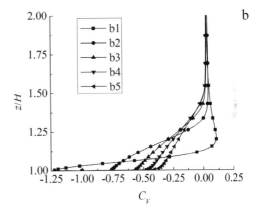

Figure 3. Distribution of I and C_V about raw b at the top of the container.

high turbulence zone and the high velocity gradient zone. The average turbulent thickness of each column δ_i can be calculated through the turbulent thickness of 15 locations δ_{ai}, δ_{bi} and δ_{ci}, where i takes a value from 1 to 5.

The distribution of δ_i for each column above the container at different wind directions is shown in Figure 4. In this case, as the horizontal wind direction θ increases from 270° to 360°, the turbulent thickness rises, while the deviation of the average turbulent thickness of each column at Y direction decreases. The maximum average turbulent thicknesses of the container δ_{imax} are 1.13H, 1.14H, 1.25H, 1.28H and 1.32H, respectively. This phenomenon occurs due to the high length-width ratio of the container, which makes the wind inflow position relative to the container variable under different wind directions. Thus, considering the effects of wind directions, the height of wind turbine installation should be higher than the maximum average turbulent thickness of the container under different horizontal wind directions.

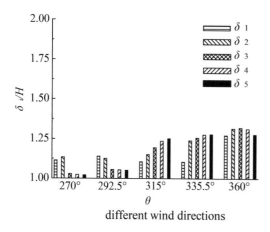

different wind directions

Figure 4. Distribution of δ_i for each column above the container at different wind directions.

Table 1. The turbulent parameters of different locations at the top of the container at different horizontal wind inflow angles.

	270°	292.5°	315°	335.5°	360°
h_0	1.36H	1.37H	1.47H	1.51H	1.52H
L_{op}	b 1	b 1	c 1	a 1	b 1
C_{Vmax}	0.11	0.11	0.15	0.12	0.10
h	1.42H	1.43H	1.47H	1.51H	1.52H
$C_p/\%$	34.9	37.5	54.5	39.7	31.3

4.3 Output of the wind turbine

We hypothesize the installation of a small horizontal axis wind turbine at the top of the container, whose type is HX1400, start-up wind speed is 3 m/s, and wind wheel diameter is D = 1.4 m. To reduce the impact of gradient wind and high turbulence on wind turbine, wind turbine installation height should satisfy the following condition: $h \geq \delta + D/2$, where δ is the largest average turbulent thickness δ_{max} under different θ. Initially, $h_0 = \delta + D/2$ should be determined by δ, then the locations L_{op} and height h should be found to obtain the maximum wind acceleration factor within the height of $Z \geq h_0$. Finally, the output of the selected wind turbine with the power increase factor should be predicted.

Table 1 shows the optimum mounting positions and different heights under different wind directions for the selected wind turbine. It was found that the frontier point is the appropriate installation location for the selected wind turbine, the suitable installation height is 1.42H to 1.52H, and in the condition of annual average wind speed, the maximum ideal power increase

factor is up to 54.5%. The results not only validate the accuracy of the above analytical results, but also provide a theoretical basis for wind turbine installation.

5 CONCLUSION

Applying new ideas for finding the suitable mounting location of roof mounted wind turbines depends on the indicators such as wind acceleration factor, average thickness of turbulence and power increase factor. In this paper, a new inflow condition of the neutral equilibrium atmospheric boundary layer was used to investigate the flow around the container to determine wind turbine installation locations in different directions. Based on the investigation, 1.17H to 1.50H above the container is the suitable height for wind turbine installation. The frontier points are the appropriate installation locations for the selected wind turbine; the suitable installation height is 1.42H to 1.52H, and in the condition of annual average wind speed, the maximum ideal power increase factor is up to 54.5%. This will provide corroboration for further experiment of building mounted wind turbine.

ACKNOWLEDGMENTS

This work was supported by the Open Research Subjects for the Inner Mongolia Autonomous Region (Nos No. 20130902 and 20130905) and the Open Research Subjects for Key Laboratory of Wind Energy and Solar Energy of the Ministry of Education (No. 201412).

REFERENCES

[1] Keith, S., Thomas, W., Small wind turbines in turbulent (urban) environments: A consideration of *normal* and *Weibull* distributions for power prediction. Journal of Wind Engineering and Industrial Aerodynamics, 121(10), 70–81, 2013.

[2] Dursun Ayhan. A technical review of building-mounted wind power systems and a sample simulation model. Renewable and Sustainable Energy Reviews, 16(10): 1040–1049, 2011.

[3] Ishugah, T., Li Y., Wang, R.Z., Advances in wind energy resource exploitation in urban environment: A review. Renewable and Sustainable Energy Reviews, 37(5): 613–626, 2014.

[4] Tony, B., Nick, J., David, S. et al. Wind energy handbook. UK: Wiley, 9–36, 2001.

[5] Ovgor, B., A method of micrositing of wind turbine on building roof-top by using joint distribution of wind speed and direction and computational fluid

dynamics. Journal of Mechanical Science & Technology, 26(12): 3981–3988, 2012.

[6] Abohela, I., Hamza, N., Dudek, S., Effect of roof shape, wind direction, building height and urban configuration on the energy yield and positioning of roof mounted wind turbines. Renewable Energy, 50(2): 1106–1118, 2013.

[7] Yuan Anmin, Tian Sijin. Research on the utilization of the wind energy among city highrises. Energy Technology, 26(10): 154–156, 2005.

[8] Yuan Xingfei, Zhang Ying. Wind environment simulation and wind power utilizability of Anzhong Building. Journal of Zhejiang University, 47(10): 1790–1797, 2013.

[9] Yoshie, R., Mochida, A., Tominaga, Y. et al. Cooperative project for CFD prediction of pedestrian wind environment in the Architectural Institute of Japan. Journal of Wind Engineering and Industrial Aerodynamics, 95(10): 1551–1578, 2007.

[10] Yang, Y., Gu, M., Chen, S. et al. New inflow boundary conditions for modelling the neutral equilibrium atmospheric boundary layer in computational wind engineering. Journal of Wind Engineering and Industrial Aerodynamics, 97(2): 88–95, 2009.

[11] Akira Nishimura, Takuya Ito. Murata1 Wind Turbine Power Output Assessment in Built Environment. Smart Grid and Renewable Energy, 4: 1–10, 2013.

Architectural, Energy and Information Engineering – Sung & Chen (Eds)
© *2016 Taylor & Francis Group, London, ISBN 978-1-138-02791-6*

The application of the matrix in the Subsurface Wastewater Infiltration System

S.M. Wang, X. Wang & H. Wang
Key Laboratory of Regional Environment and Eco-Remediation, Ministry of Education,
Shenyang University, Shenyang, China
College of Environmental Sciences, Shenyang University, Shenyang, China

ABSTRACT: The Subsurface Wastewater Infiltration System (SWIS) is considered to be an efficient, common, economic and ecological process for decentralized domestic wastewater treatment. Matrix is the main part of the SWIS, whose properties have a close relationship with the effect and lifespan. This paper introduces several kinds of commonly used matrices and their functions, the mechanism of decontamination, as well as the choice principles of the matrices. Finally, it puts forward the research direction of the matrix in the future, which can improve the removal of pollutions.

Keywords: subsurface wastewater infiltration system; matrix; decentralized domestic sewage

1 INTRODUCTION

With the constant improvement of the residents' living standard, the quantity of sewage at villages, small towns and scattered residential areas has increased rapidly, but the decentralized wastewater treatment facilities is increasing slowly, which is unable to dispose the increased amount of sewage. The Subsurface Wastewater Infiltration System (SWIS) at a low cost can dispose domestic sewage effectively and improve the aquatic environment quality of these areas on-site (Nie, 2011).

Matrix is the main part of the SWIS. It is not only the medium of pollutant removal, but also the carrier of plant growth and microbial living. Its compositions and properties determine the quality of the output water directly. Therefore, it is important to research the matrix systemically in the SWIS.

2 THE MATRIX COMPOSITION AND FUNCTION

The SWIS is mainly composed of the matrix, the surface plant and the microorganism. The three parts mutually influence and interact, and become as a whole, to achieve the goal of purifying the sewage. There are many approaches to degrade the pollutants in the SWIS, but matrix adsorption, microbial degradation, as well as plant absorption are closely related to the matrix. Currently, the SWIS often adopts the mixture, which contains the natural materials (soil, gravel and sand) as the basic raw material, and then adding suitable industrial by-products or artificial products (coal cinder, zeolite, activated sludge, limestone and slag) (Wang et al., 2010). The different matrices with characteristic functions and applications are listed in Table 1.

3 THE POLLUTION REMOVAL MECHANISM OF THE MATRIX

3.1 *The removal mechanism of the suspended solids*

The removal of suspended solids mainly depends on the adsorption, interception and filtering effect of the matrix in the SWIS.

3.2 *The removal mechanism of organic pollutants*

The removal of organic pollutants in the SWIS mainly relies on the adsorption of matrix and the oxidation of biological compounds. The latter is the main effect: organic pollutants flow into the ground with the current, when passing the upper matrix; the pollutants are removed under the action of microorganisms after the filtration and adsorption of the matrix (Pan et al., 2008). This process generally adopts the intermittent operation mode, which is beneficial to maintain the soil aerobic condition to improve the removal effect of the organic pollutants.

Table 1. The different matrices with its characteristic, function and application.

Species	Name	Characteristic and function	Application
Natural material	Soil	Natural filter material with adsorption and interception; the carrier of plant and microbial life (Cai & Deng 2007a b, Kong et al., 2005); the interface of a variety of physical, chemical and biochemical reactions.	Large numbers of experiments show that the soil physical and chemical properties (ORP, soil particle size, soil specific surface area) can significantly influence pollution removal (Zhang et al. 2002a,b, VanCuy kS. et al., 2001). Yang Jian and Dai Qiang et al, found that the sand can prevent the soil plugging and increase the soil adsorption of phosphorus, respectively (Yang et al., 2010 a, b, Dai et al., 2012). Experiments also show that vermiculite, zeolite, limestone can increase the permeability of matrix significantly, and contribute to the adsorption of NH_4^+ and TP (Guo et al. 2010).
	Sand	Big particle size improves matrix permeability to restore oxygen; particle surface area is large, which is conducive to biofilm to adhere; dispersing water flow and distributing water, preventing the system from physical clogging (Xue et al., 2005a, b, Robin W. E. 2009).	
	Gravel	Zeolite and vermiculite have a certain particle size of the cavity and channel, large and well distributed; large specific surface area; strong ability of adsorption and ion exchange (Chen et al., 2009 a, b, Wu et al 2005); limestone has a certain ability of adsorption; it can buffer pH value of wastewater (Song et al., 2003).	
Industrial by-product	Cinder and fly ash	Porous structure, big particle size; large specific surface area, strong adsorption ability; internal pore is bulky and coarse, which can be used as the microbial reaction bed, containing the metal cation (aluminum, silicon abd magnesium), which is beneficial to the adsorption of phosphorus (Chen 2005 a, b, Zhang et al., 2006);	Experiments prove that cinder will be conducive to the removal of organic matter, NH_4^+-N and TP (Que 2011 a, b, Yang et al., 2011). Many researchers (Westholm L. J. 2006 a, b, Kang et al., 2009) have found that the presence of Fe_{3+} and Al_{3+} can increase the adsorption capacity of phosphorus, and improve the removal efficiency of phosphorus; and can improve the removal rate of ammonia nitrogen by adding the active sludge in the matrix (Zhang Zhiyin et al., 2006).
	Scrap iron	Fe and C will generate galvanic cell (Liu et al., 2009), which is advantageous to the reduction of NO_3–N and NO_2–N, and it can generate the iron hydroxide, which is a good flocculant (Wang, 2000).	
	Sludge	Containing different kinds of microbes that can shorten the time of domestication and maturation, and improve the removal of nitrogen.	
Artificial product	Foam	Big particle size, large specific surface area; carrier for the microbial living.	Nie Junying [(Nie, 2011), Dai Ying (Dai, 2009) and Zou Yi (Zou, 2007) found that the removal rate of TN can be improved by using polyurethane foam materials and bacterial preparation, respectively.
	Bacterial preparation	Increasing the amount of specific microorganisms, which can shorten the time of domestication and maturation, and improve the removal of nitrogen.	

3.3 The removal mechanism of nitrogen

Many denitrogenation mechanisms exist in the SWIS, which include: the volatilization of ammonia nitrogen, the adsorption of plants, the decomposition of microorganisms, and the adsorption of the matrix. The main role is played by the microorganisms. The matrix provides attachment for microbial growth and plays a main role in wastewater treatment: the nitrogen in the sewage generally exists in the form of ammonia nitrogen and organic nitrogen. First, the organic nitrogen is intercepted or adsorbed by the matrix, and then changes into ammonia nitrogen under the action of ammonia-oxidizing bacteria. The soil is negatively charged, so it is easy to adsorb NH_4^+, and then it changes into NO_3^-N under the action of nitrosation bacteria and denitration bacteria. NO_3^-N can be converted to N_2 or N_2O under the action of denitrifying bacteria and then volatile out.

3.4 The removal mechanism of phosphorus

The removal of phosphorus in the SWIS depends on plant absorption, the adsorption of the matrix and the effect of microbes. Among them, the adsorption of the matrix plays an important role. On the one hand, the soil is the repository of phosphorus, which has strong adsorption ability for phosphorus; on the other hand, in the acidic soil, iron, aluminum and calcium can react with orthophosphate, and then generate insoluble phosphate. In the soil, minerals with a granular structure, such as $Fe(OH)_3$, have a good adsorption effect on phosphorus. In addition, part of the inorganic phosphorus in the soil isabsorbed by microorganisms, and then convertsinto organic phosphorus.

4 THE PRINCIPLES OF SELECTING THE MATRIX

The selection of the matrix is always an important part of the SWIS: selecting the matrix appropriately not only is related to the pollutants' removal effect of the whole system, but also has a relationship with the life and the cost of the system. So, the selection of the matrix should consider comprehensively the system own needs and the property of the matrix (Wang, 2007).

4.1 Selecting different matrices according to different pollutants

Different matrices have different removal effects on different pollutants, thus we need to select different matrices according to different pollutants. For high phosphorus concentration in the sewage, we can select slag or shale as the matrix (Ruan et al., 2009), or consider the combined matrix to enhance the purification capacity of the matrix layer. Putting the material with high removal of phosphorus and nitrogen in the matrix has a better treatment effect than a single material (Zhang et al., 2013).

4.2 Selecting the matrix with appropriate physical and chemical properties

A variety of physical and chemical properties of the matrix have an effect on its removal. First, the matrix should have enough intensity, and be susceptible to different intensities of hydraulic shear and the friction collision between the matrix (Wang, 2010). In the SWIS, the matrix will become soft, so the matrix must have a high mechanical intensity. Second, the matrix must have a larger specific surface area and high porosity. The specific surface area of the matrix is related to the particle size and shape. In general, the smaller the particle size, the larger the specific surface area and the stronger the adsorption ability of the matrix. This is conducive to the growth of microorganism and the removal of pollutants. Third, the matrix should have good chemical stability. The matrix itself should not react with other matrices so as to prevent secondary pollution or damage the matrix layer.

4.3 The matrix should be accessible, efficient and cheap

When selecting the matrix, the first step is to choose the local matrix with high removal ability of pollutants, which can not only improve the effect of disposing sewage and reduce the cost of the SWIS, but also prolong the lifespan of the SWIS. The wide source of the matrix is another aspect of the accessibility. Efficient and cheap are also two important factors of the matrix: high efficiency is conducive to improve the removal effect; cheap is helpful to reduce the cost of the system.

Besides, other conditions should be considered: whether the matrix can regenerate and whether the matrix is convenient to collect and transport. Experiments (Lu et al., 2006) have shown that some matrices can regenerate under certain conditions, which can not only avoid replacement matrix frequently, but alsoreduce the cost.

5 RESEARCH PROSPECT

The SWIS is suitable for decentralized wastewater treatment and has a broad application prospect. The matrix is an important part of the SWIS and its property has a close relationship with the effect of the effluent. It has been shown that the matrices restrict the application and spread of the SWIS. So, the research on the matrix should be conducted with respect to the following several aspects:

First, the new type, high efficiency and low price of the matrix material are still the important research direction in the future.

Second, it is necessary to take advantage of the synergism of the different matrices, to research two or more matrices and select the optimum combination.

Third, it is important to find the matrix that can prevent the soil clogging and provide nutrition for the microorganisms.

ACKNOWLEDGMENTS

This work was financially supported by the Program for Distinguished Young Scholar of Liaoning Province (LJQ2012102), the Liaoning Natural Science Foundation (2013020146), the National Natural Science Foundation of China (51008198 and 51108275), and the Major Science and Technology Program for Water Pollution Control and Treatment (No. 2013ZX07202-007).

REFERENCES

Cai Mingkai & Deng Chunguang. 2007. Influential factors and research suggestion of nitrogen transfer of artificial wetland. Journal of Anhui Agricultural Science, 35(22): 6902, 6933.

Chen Mingli et al. 2009. Mechanism of nitrogen removal by adsorption and bio-transformation in constructed wetland systems. Chinese Journal of Environmental Engineering, 3(2): 224–228.

Chen Huaiman. 2005. Environmental Soil Science. Beijing: Science Press.

Dai Qiang et al. 2012. Screening on medium of subsurface filtration system. Journal of Anhiui Agri. Sci, 40(9): 5446–5447, 5458.

Dai Ying et al. 2009. Effect of hydraulic load on efficiency of enhanced biological nitrogen removal in subsurface infiltration system. Journal of Heilongjiang Institute of Technology, 23(2): 71–74.

Guo Zhenyuan et al. 2010. Phosphorus removal in improved constructed rapid infiltration, Technology of Water Treatment, 36(6): 116–118, 135.

Kang Aibin et al. 2009. Pilot study treatment on improved constructed rapid infiltration system for domestic sewages water. Purification Technology. 28(4): 42–45.

Kong Gang et al. 2005. Subsurface soil infiltration systems for sewage treatment, Environmental Science & Technology, 37(3): 251–257.

Liu Shenyang et al. 2009. Study on cocking waste water treatment by using on metallization pellets with high carbon content. China Metallurgy, 19: 42–45.

Lu Shaoyong et al. 2006. Nitrogen adsorption and reactivation of zeolite and soil in constructed wetland. Transactions of the Chinese Society of Agricultural Engineering, 22(11): 64–68.

Nie Junying. 2011. Studies on wastewater treatment by fortified subsurface soil infiltration system as well as its mechanism. Shanghai: Shanghai Jiao Tong University.

Pan Jing et al. 2008. Spatial distribution of microorganisms in subsurface wastewater infiltration and their correlation with purification of wastewater. China Environmental Science, 28(7): 656–660.

Que Yuanlin. 2011. Performances of infiltration system with composite filler for domestic wastewater. Shanghai: Tongji University.

Robin, W.E., 2009. Performance analysis of established advanced on-site wastewater treatment systems in a subarctic environment: Recirculating trickling filters, suspended growth aeration tanks, and intermittent dosing sand filters, Alaska: University of Alaska Anchorage.

Ruan Jingjing et al. 2009. Research development of substrates in constructed wetland. Journal of Capital Normal University, 30(6): 85–90.

Song Tiehong et al. 2003. The Characteristic on surface constructed wetlands treating wastewater. Journal of Jilin Architectural and Civil Engineering Institute., 20(4): 13–14.

VanCuy kS. et al. 2001. Hydraulic and purification behaviors and their interaction during wastewater treatment in soil infiltration systems. Wat Res, 35(4): 953–964.

Wang Ping. 2000. Study on removal of phosphate with sponge iron. Acta Scientiae Cirumstantiae, 20: 798–800.

Wang Shihe. 2007. Theory and technology of artificial wetland wastewater treatment, Beijing: Science Press.

Wang Xin et al. 2010. Progress on matrix in a subsurface wastewater infiltration system. Environmental Science & Technology, 33(12): 86–89.

Westholm, L.J., 2006. Substrates for phosphorus removal-Potential benefits for on-site wastewater treatment. Water Research, 40(1): 23–36.

Wu Xiaofu et al. 2005. Sorption of NH4-N by vermiculite minerals as a function of initial solution concentration and sorbent quantity. Research of Environmental Sciences, 18(1): 64–66.

Xue Shuang et al. 2009. Behavior and characteristics of dissolved organic matter during column studies of soil aquifer treatment. Water Research, 43(2): 499–507.

Yang Jian et al. 2010. Intensified denitrification of compound mixed subsurface wastewater infiltration system. Journal of Tongji University, 38(5): 697–703.

Yang Jinhui et al. 2011. Experimental study of adsorbing ammonia nitrogen and phosphorus in water by cinders. Uranium Mining and Metallurgy, 30(4): 221–224.

Zhang Dan et al. 2013. Progress on substrates selection for constructed wetland research. Academic essays of Chinese environmental sciences association: 3505–3512.

Zhang Fang et al. 2006. Analysis of packings absorption function for domestic wastewater. Technology of Water Treatment, 32(11): 37–40.

Zhang Jian et al, 2002. Nitrogen and phosphorus removal mechanism in subsurface wastewater infiltration system. China Environmental Science, 22(5): 438–441.

Zhang Zhiyin et al. 2006. Relationship between Variation of nitrite-nitrogen concentration and removal efficiency for soil aquifer treatment process. Journal of Fudan University (Natural Science), 45(6): 755–758.

Zou, Yi. 2007. The emulation experiment research that increased the subsurface wastewater infiltration system processing. Shenyang: Shenyang Pharmaceutical University.

Architectural, Energy and Information Engineering – Sung & Chen (Eds)
© 2016 Taylor & Francis Group, London, ISBN 978-1-138-02791-6

The research status of RDB of composite beams in negative moment region

Z.N. Zhang & Z.Z. Zhang
Civil Engineering of Shenyang, Jianzhu University, Shenyang, Liaoning, China

ABSTRACT: Restrained Distortional Buckling (RDB) is a special kind of buckling form of distortional buckling, which usually occurs in the negative moment region of steel-concrete composite beams. Full restraint against lateral displacement and twist along its tension flange make it complex to simulate the mechanism of buckling. It is not mature that the domestic and overseas research of restrained distortional buckling, and needs to be studied further. A literature review is presented in this paper about the research on the distortional buckling of composite beams in hogging moment region among domestic and foreign scholars, and the prospect, which might provide the reference to the research of restrained distortional buckling, is proposed.

Keywords: Steel-concrete composite beams; The negative moment region; Restrained distortional buckling

1 INTRODUCTION

The distortional buckling lateral torsional buckling and local buckling are steel-beam buckling modes. Lateral torsional buckling mode is known as a flexural torsional buckling mode, in which the cross sections of a beam translate in lateral direction and twist as rigid bodies with no distortion in the web (Figure 1 (a)). The Local Buckling (LB) is shown as in Figure 1 (b), web or flange appears wavy convex curved deformation, relative angle does not occur in the intersection of flange and web. As the cross-section parameter keeps constant, LTB and LB are the characteristic buckling modes of the long- and short-span beams, respectively. Distortional buckling is a buckling mode that combined with lateral deflection and twists in the cross-sectional shape of medium span with weak web (Figure 1 (c), (d)). And lateral distortional buckling mode is a failure mode between lateral torsional buckling and local buckling modes. Restrained distortional buckling mode is a special mode of the lateral distortional buckling mode, which usually occurs in the negative moment region of composite beams.

Many scholars have studied the distortional buckling, and found out that there is a great influence on the critical load due to web distortions when composite beams experience restrained distortional buckling. The research of restrained distortional buckling of composite beams in negative

moment region is not yet mature, a lateral bracing is added to ensure stability of composite beams in hogging moment region usually. Therefore, further study is necessary to this problem. A literature

(a) lateral torsional buckling (LTB)

(b) The local buckling(LB)

Figure 1. *(Continued)*.

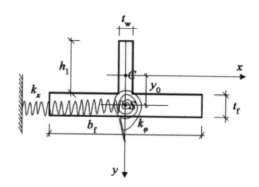

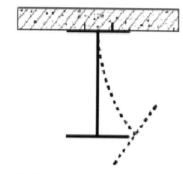

（c）lateral distortional buckling （LDB）

（d）restrained distortional buckling (RDB)

Figure 1.   Different buckling modes of steel I-beams.

Figure 2.   Elastic foundation beam model.

review is presented in this paper about the research on the distortional buckling of composite beams in hogging moment region among domestic and overseas scholars, and the prospect, which might provide the reference to the research of restrained distortional buckling, is proposed.

2   THE RESEARCH STATUS

At present, the numerical method or simplified theoretical method is usually adopted to study the composite beams' RDB at hogging moment region. Numerical method, including finite element method and finite strip method, is normally used for the parameter buckling analysis of component and the inelastic buckling analysis; simplified theoretical method mainly includes the method of elastic foundation beam and energy method, which is usually used for elastic buckling analysis of component. It is of great significance that the simplified theory to explain the mechanism of buckling and puts forward practical design method.

With full restraint against lateral displacement and twist along its tension flange.

2.1   *Method of elastic foundation beam*

The buckling mode of I-section beams with full restraint against lateral displacement and twist along its tension flange, is RDB mode rather than LTB mode. Elastic foundation beam model, used by researchers to simulate the RDB mode, take the compressive flange as the research object, and also regard the restriction in compressive flange due to the web as a continuous elastic support (Figure 2). The derivation and conclusion of this method are simple. In the 1970s, Light steel Standard Specification of Sweden adopted elastic foundation beam model with constant axial force (i.e., constant along the axial length of elastic foundation compressive bar model) to analyze the RDB of composite beams. The specification of Britain's bridge (BS5400) also uses the method in the continuous composite bridge design.

Due to elastic foundation beam method without considering influence of the variety of axial forces on RDB instability, Stevenson (1985) introduced coefficient of the variety of axial forces on the basic of elastic foundation beam model. Later Williams considered the problem of different distribution of axial force problems on the basis of Stevenson, and compared with the result of finite element method. Weston in 1991 puts forward empirical formula of tender ratio of the lateral instability through the analysis of parameter by finite element method, and revised the British standards (BS5400) Bridges about the design method of the stability of the continuous composite bridge, but he only considered uniformly distributed load condition. In 2002, Shiming Chen also analyzed the variety of axial force problem and got an ordinary differential equation of fourth order considering the different distribution of axial force on the basis of Stevenson. Formula (1) shows as follows:

$$\frac{d^4y}{d\zeta^4} + \pi^2\lambda\frac{d}{d\zeta}\left[N(\zeta)\frac{dy}{d\zeta}\right] + (\beta L)^4 y = 0 \qquad (1)$$

where $\zeta = x/L$; c is the elastic foundation restraint stiffness; If only consider web torsional effect $c = \frac{3EI_w}{h_w^3}$.

Since elastic foundation beam model cannot accurately reflect the actual buckling model without considering the impact of web, Wei Chen and Jihong Ye of southeast university in 2012 revised Swenson's model, took the impact of the web into consideration, the deformation of the mode was more close to RDB mode, and they deduced two kinds calculation expression of column stability with variable axial compressive on the basis of the new model. They also analyzed the accuracy of his method through finite element program.

2.2 Energy method

Considering foundation beam stability model is more difficult to consider gradient of flexural moment and the influence of web buckling reasonably, and it is not easy to ensure its accuracy. Many scholars use energy method to study the RDB.

Goltermann et al. were derived by the expression of total potential energy of the buckling of I-section steel, assumed that the lateral deformation of the web is equivalent to flexural deformation of cantilever beam, and the angle of compressive flange and web keeps constant. Finally, the total potential energy expression is relatively simple deduced by Goltermann.

In 2006, Jun Xia under the guidance of professor Genshu Tong, researched the performance of RDB of composite beam. He used energy method deduced critical load expression under pure bending moment condition, assuming that twist angle of the web has nothing to do with the flange, and he verified the assumption with ANSYS, found that twist angle of the web distortion is slightly smaller than the actual web buckling twist angle. So the twist angle cannot be ignored. Wangbao Zhou, Lizhong Jiang and others in 2012, deduced the calculation formula of lateral restraint stiffness and torsion restraint stiffness of web by using energy method, the web of composite I-section beam under the longitudinal linear distribution of stress.

Considering the existing potential energy expression of RDB not considered the transverse distribution of load condition, Wei Chen and Jihong Ye (2012) used the energy principle deduced general expression of potential energy of the RDB of double symmetric I-section beam, as shown in formula (2). This expression, equivalent to classic expression of the lateral torsional buckling without considering the influence of web distortion, can be used for a variety of load conditions considering the influence of web distortion.

$$\begin{aligned}
\Pi = &\frac{1}{2}\int_0^l EI_{zt}\upsilon_{t,xx}^2 dx + \frac{1}{2}\int_0^l GJ_t \cdot \theta_{t,x}^2 dx + \frac{1}{2}\int_0^l EI_{zb}\upsilon_{b,xx}^2 dx \\
&+\frac{1}{2}\int_0^l GJ_b \cdot \theta_{b,x}^2 dx + \frac{1}{2}\cdot\frac{Et_w^3}{12}\iint \upsilon_{w,xz}^2 dzdx \\
&+\frac{1}{2}\cdot\frac{Gt_w^3}{3}\iint \upsilon_{w,xz}^2 dzdx - \frac{1}{2}\int \sigma_{tx}(\upsilon_{t,x}^2 + y^2 \cdot \theta_{t,x}^2)dV \\
&-\frac{1}{2}\int \sigma_{bx}(\upsilon_{b,x}^2 + y^2 \cdot \theta_{b,x}^2)dV \\
&-\frac{1}{2}\int(\sigma_{wx}\upsilon_{w,x}^2 + 2\tau_{wxz}\upsilon_{w,x}\upsilon_{w,z})dV \\
&+\frac{h}{2}\int \sigma_{tx}(\theta_{t,x}\upsilon_{b,x} + \theta_t \upsilon_{t,xx})dV \\
&-\frac{h}{2}\int \sigma_{bx}(\theta_{b,x}\upsilon_{b,x} + \theta_b \upsilon_{b,xx})dV \\
&-\int \sigma_{wx}z(\theta_{w,x}\upsilon_{w,x} + \theta_w \upsilon_{w,xx})dV \\
&-\frac{h^2}{4}\int \sigma_{tx}(\theta_{t,x}^2 + \theta_t \theta_{t,xx})dV \\
&-\frac{h^2}{4}\int \sigma_{bx}(\theta_{b,x}^2 + \theta_b \upsilon_{b,xx})dV \\
&-\int \sigma_{wx}z^2(\theta_{w,x}^2 + \theta_w \upsilon_{w,xx})dV
\end{aligned}$$
$$(2)$$

3 PROSPECT

In the past few decades, the research of RDB of composite beams in hogging moment region has obtained some achievements. The lateral bracing is added to ensure stability of composite beams in hogging moment region. So it is necessary to make further study on this problem. Based on the research of I-section beam's RDB, this paper puts forward following proposals for the research:

1. In view of study about RDB of the continuous composite beams in negative moment region, it is necessary to do some related experimental research and compare the result of experimental research with the current extrapolation expression, so as to apply the expression in engineering.
2. Some current methods, under the complex load condition, is difficult to meet the accuracy of requirements, which is not overestimate or conservative, needs to be further studied.
3. Putting forward a simplified method of equivalent bending moment of continuous composite beams, is a new direction to the study of the

stability of the continuous beam, and hoped to get more in-depth studies on the basis of mentioned above.

ACKNOWLEDGEMENTS

This work described in this paper was supported by National Science Found of China (51108279), their supports are gratefully acknowledged.

REFERENCES

BS5400. 1982. Steel, concrete and composite bridges. Part3: Code of practice for design of steel bridges, London, British Standards Institution.

Goltermann P, Swenson S E. 1988. Lateral distortional buckling: predicting elastic critical stress. Journal of Structural Engineering114(7): 1606–1625.

Jun Xia & Genshu Tong. 2006. The simplified formula for calculating the deflection of steel-concrete composite beams and the distortional stability analysis of its negative moment zone, [D]. Zhejiang: Zhejiang University, 1–91.

Lee D S. 2005. Inelastic lateral-distortional buckling of continuously restrained continuous beams. Steel and Composite Structures 5(4): 305–326.

Shiming Chen. 2002. Stability of steel-concrete continuous composite beams [J]. Industrial Construction 32(9): 1–4.

Svensson S E. 1985. Lateral buckling of beams analyzed as elastically supported columns subjected to a varying axial forcesd. Journal of Constructional Steel Research 5(3): 179–193.

Vrcelj Z, Bradford M A. 2009. Inelastic restrained distortional buckling of continuous composite T-beams. Journal of Constructional Steel Research 65(4): 850–859.

Wangbao Zhou & Lizhong Jiang. 2012. Elastic distortional buckling analysis of steel-concrete composite beams in negative moment region, Journal of Central South University (Science and Technology) 43(6): 2316–2323.

Wei Chen & Jihong Ye. 2011. Elastic restrained distortional buckling of I-steel-concrete composite beams. Journal of Building Structures 32(6): 82–91.

Wei Chen & Jihong Ye. 2012. Elastic lateral buckling of doubly symmetric I-beams based on potential energy method. Engineering Mechanics 29(3): 95–109.

Weston G & Nethercot D A. 1991. Lateral buckling in continuous composite bridges. The Structural Engineer 69(3): 79–278.

Williams F W & Jemah A K. 1987. Buckling curves for elastically supported columns with varying axial force, to predict lateral buckling of beams 7(2): 133–147.

Architectural, Energy and Information Engineering – Sung & Chen (Eds)
© 2016 Taylor & Francis Group, London, ISBN 978-1-138-02791-6

Tenability analysis in a tobacco workshop in case of fire with a 2-layer zone model

J.L. Niu & L. Yi

Institute of Disaster Prevention Science and Safety Technology, Central South University, Changsha, China

ABSTRACT: The FED model was used with a self-developed model to the quick tenability analysis in a large space tobacco workshop in case of fire. Burning characteristics including heat release rate, mass loss rate and generation rate of CO_2, CO and particles of the tobacco combustibles in the workshop are measured by cone calorimeter. Smoke filling in the workshop is simulated and tenability criteria including temperature, CO concentration and visibility are considered. It indicates that the predicted Available Safe Egress Times (ASET) are 747 s in the natural filling case. With mechanical exhaust, available safe egress time in the workshop can be extended effectively, the predicted ASETs are more than 2000 s.

Keywords: Tenability analysis; smoke control; tobacco workshop; zone model

1 INTRODUCTION

Tenability analysis is commonly performed in fire safety assessment in buildings. In this paper, tenability analysis will be carried out in a recently built tobacco workshop. Burning characteristics of the tobacco combustibles used in workshop are tested by a cone calorimeter. Smoke filling in the tobacco workshop both with and without mechanical exhaust is simulated by using a simple two-layer zone model[1]. Available Safe Egress Time (ASET) for occupants in the workshop is obtained, and safety assessment for the occupants in the workshop is performed.

2 BURNING CHARACTERISTICS OF TOBACCO COMBUSTIBLES

In this study, tobacco from northeast of Hunan province of China is supplied to a newly built tobacco workshop. The burning characteristics of the tobacco are measured using a cone calorimeter for the tenability analysis in the tobacco workshop. Cut tobacco, leaf tobacco (loose and compacted, respectively) and stem tobacco with the same size were selected and tested in the cone calorimeter. The test results including effective heat of combustion, heat release rate and CO yield, and so on are listed in Table 1.

Variations of the heat release rate of the samples were plotted and compared with t^2 fire after neglecting the incubation period. It is demonstrated that the growth rate of the heat release rate can be approximately classified as "fast" or "medium" for cut tobacco and leaf tobacco and "slow" for stem tobacco.

Table 1. Conditions and results of the tests in cone calorimeter.

Ambient temperature/°C	25
Ambient pressure/Pa	101325
Relative humidity/%	26
Radiative heat flux/kWm^{-2}	25
Exhaust rate/ls^{-1}	24

No.	1	2	3	4
Sample	Cut tobacco	Leaf tobacco (loose)	Leaf tobacco (compacted)	Stem tobacco
Time to ignition/s	7	6	6	140
Time to flameout/s	21	16	65	350
Mass of sample/g	5.8	3.7	13.0	27.6
Effective heat of combustion / MJkg^{-1}	18.5	11.9	11.3	10.0
Peak heat release rate/kWm^{-2}	55.9	26.6	40.0	96.6
CO yield/kgkg^{-1}	0.06	0.08	0.08	0.04
CO_2 yield/gkg^{-1}	3.74	4.4	2.4	1.7

3 TENABILITY CRITERIA

3.1 Temperature of smoke

NFPA 130[2] and ISO/TS 13571[3] developed an FED assessment model for heat incapacitation. The model gave the correlation between the duration time and heat flux:

$$t_{Irad} = \frac{240}{(q_r^{"})^{1.35}} \qquad (1)$$

where t_{Irad} is the time to incapacitation caused by radiation, s. $q_r^{"}$ is the intensity of thermal radiation, kwm^{-2}.

Assuming the heat flux can be replaced by the smoke temperature. Then, Equation (1) can be changed to the following form:

$$t_{Irad} = \frac{1.63 \times 10^{15}}{(T + 273)^{1.35}} \qquad (2)$$

where T is the temperature of smoke in °C.

The duration time caused by heat convective can be estimated by the following equation:

$$t_{Iconv} = 2.5 \times 10^{10} T^{-3.61} \text{(subjects are fully clothed)} \qquad (3)$$

$$t_{Iconv} = 3 \times 10^{9} T^{-3.4} (subjects\ are\ lightly \qquad (4) \\ clothed\ or\ unclothed)$$

where t_{Iconv} is the time to incapacitation caused by flux convective, s.

Considering FED is the sum of heat effect of the combined influence of radiation and convection. Therefore, the FED model for heat effect can be detailed below.

$$FED = \sum \left(\frac{1}{t_{Irad}} + \frac{1}{t_{Iconv}} \right) \Delta t \qquad (5)$$

In the above equation, Δt is the increment time between two adjacent calculation steps, s.

3.2 Toxicity of smoke

For assessing the synthetical effects of different gases, the following equation is widely used,

$$FED = \sum_{i=1}^{n} \sum_{t_1}^{t_2} \frac{C_i}{(Ct)_i} \frac{\Delta t}{60} \qquad (6)$$

where C_i is the concentration of gas species i, ppm (mL/m^3 for gas). $(Ct)_i$ is the specific exposure dose, ppm·min.

3.3 Obscuration of smoke

The obscuration of smoke reduces the visibility, thus affects the evacuation. The relationship between obscuration and visibility is:

$$S = \frac{S_c}{D} \qquad (7)$$

where S is visibility of smoke, m. S_c is a constant, 3 for light-reflecting sign and 8 for light-emitting sign. D is the extinction coefficient, m^{-1}.

According to NFPA 2003a[4], 2003b[5], the time when FED reaches 0.3 is regarded as the critical time when toxicity or high temperature of smoke is fatal to the occupants. The smoke optical densities typically used as tenability criteria in hazard analysis are 10 m for visibility for large space buildings and 2 m for visibility for small buildings such as dwelling units [6].

4 TENABILITY ANALYSIS WITH A ZONE MODEL

To perform tenability analysis in the workshop in case of fire, an improved single two-layer zone model is also applied[7].

4.1 Zone model

Mass conservation of the smoke layer and air layer:

$$\dot{m}_s = \dot{m}_p - \dot{m}_e \qquad (8)$$

$$\dot{m}_a = \dot{m}_i - \dot{m}_{p.} \qquad (9)$$

In the above equations, m, h and ρ are the mass, depth and density of the gas layer, respectively. Subscript s and a represent smoke layer and air layer respectively.

For the height of the workshop, H,

$$H = h_a + h_s \qquad (10)$$

Considering that a fire is basically a phenomenon at atmospheric pressure, the equation of state of the ideal gas in this model is simplified as follows:

$$\rho_i T_i = const \qquad (11)$$

Marching solution is selected for solving the energy and species equation of the smoke layer. Applying the first law of thermodynamics on the smoke layer, energy conservations during a time step Δt is given by:

$$\dot{Q}_c \Delta t - \dot{Q}_l \Delta t = c_v (m_s^{(t+\Delta t)} T_s^{(t+\Delta t)} - m_s^{(t)} T_s^{(t)})$$
$$+ c_p (\dot{m}_e \cdot \Delta t) T_s^{(t)} - c_p (\dot{m}_p \cdot \Delta t) T_a \quad (12)$$

In the above equation, $m_s^{(t)}$ and $m_a^{(t)}$ are the mass of the smoke layer and air layer at time t respectively. c_p and c_v are the specific heat of gas at constant pressure and at constant volume respectively. $\dot{Q}_c$ is the convective heat release rate of the fire, taking about 70% of the total heat release rate of the fire. $\dot{Q}_l$ is the heat loss rate of the smoke layer to the walls and ceiling.

t^2 fire was used in the calculation of $\dot{Q}$:

$$\dot{Q} = \alpha t^2 \quad (13)$$

where α is fire intensity coefficient, kW/s^2, according to the experiment results of the sample tests, 0.04689 was used in the model.

Assuming one production line was totally burned, a conservative value of 10 MW (100 kW/m^2 * 100 m^2) is set as the maximum stable heat release rate of the fire.

The heat loss rate of the smoke layer to the walls and ceiling which can be calculated using the Newton's law of cooling,

$$\dot{Q}_l = h_w A_w (T_s^{(t)} - T_a) \quad (14)$$

where h_w is the equivalent heat transfer coefficient, kW/m^2 K. 0.025 is applied in the calculations according to the results from Hu et al.[8]. A_w is the comprehensive area of the convective heat transfer.

Combining Equations (8–14), the variables h_s, h_a, T_s and ρ_s in the above equations can be solved numerically with initial conditions, boundary conditions, a plume model and state equation of gas. Then ASETs can be achieved.

4.2 Fire scenarios

The workshop is 144 m long, 36 m wide and 10 m high. Two tobacco manufacture lines of about 100 m long and 1 m wide are equipped in the workshop. 10 openings of 2 m high are located on 4 walls of the workshop. Mechanical exhaust system is installed in the workshop with exhaust rate of 95 m^3/s.

The CO yield, particulate yield and effective heat of combustion are set as 0.08 kgkg^{-1}, 0.02 kgkg^{-1} and 15 MJkg^{-1} for the combustion of the tobacco respectively according to Table 1. The tobacco is evenly placed on the manufacture lines with thickness of 8 cm. An ignition source is placed on the middle of the manufacture line and removed after the ignition of the tobacco combustibles.

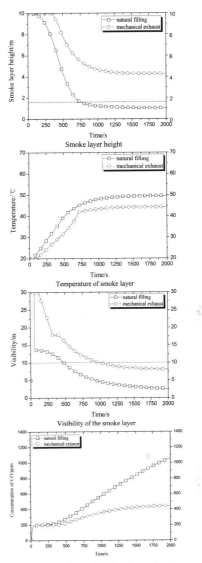

Concentration of CO of smoke layer.

Figure 1. Development of the smoke layer predicted by the zone model.

The two fire scenarios in the tobacco workshop are considered:

Scenario 1: natural smoke filling.

Scenario 2: Smoke filling with mechanical exhaust (exhaust rate of 95 m^3/s and operates at 120 s after ignition).

4.3 Results and discussion

Height of the smoke layer interface, temperature and visibility of the smoke layer in the natural filling and mechanical exhaust scenarios are plotted in

Figure 1. Due to the large space of the workshop, the smoke layer interface descends slowly, even in the natural filling scenario. Mechanical exhaust affects the interface height, temperature, CO concentration and visibility of the smoke layer significantly. Smoke layer descends to the characteristic height at 747 s in the natural filling scenario and keeps above 4 m above the floor in the mechanical exhaust scenario. Due to high particulates yield of combustion of the tobacco combustibles, visibility of the smoke layer reduces rapidly and drops to 10 m just 478 s and 997 s after ignition in the two scenarios respectively.

Figure 2 shows the variation of FEDs based on the predicted temperature and CO concentration of the smoke. Times for FEDs of heat to reach 0.3 are all more than 2000s in both scenarios; for FED of CO, the times are 1452s and 1962s in the 2 scenarios, respectively. For the tobacco workshop, tenability analysis shows that the toxic gas generated by the combustion is more dangerous than the heat under fire. The influence of heat and CO concentration on safe evacuation is weaker than the visibility of the smoke. Therefore, the ASETs are 747 s and greater than 2000s for scenarios 1 and 2, respectively.

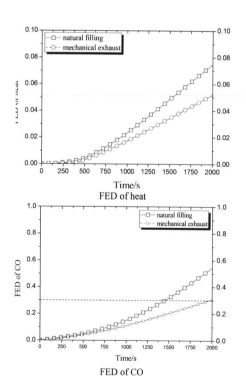

Figure 2. Results of the FED analysis based on the data predicted by the zone model.

5 CONCLUSIONS

Among the tenability criteria in the large space tobacco workshop in case of fire, visibility and CO are more sensitive for evacuation due to large particle and CO yield of the tobacco combustible. In both fire scenarios, FEDs of heat do not reach the threshold due to the large volume of the workshop and heat loss of the smoke. The available safe egress time for the natural filling scenario is more than 10 minutes (747 s predicted by the 2-layer zone model) and more than 30 minutes for the mechanical exhaust scenario.

In modern combine workshop, the number of the employees work in the workshop is relatively small due to the high automation. For the tobacco workshop studied in this paper, required safe egress time for the staff is within 2 minutes under fire based on the result of evacuation exercise, less than the ASETs in natural filling scenario. Therefore, the purpose of installation of mechanical exhaust system in such large space workshop is to help firefighting more than to help evacuation.

ACKNOWLEDGEMENTS

This work is supported by the Freedom Explore Program of Central South University under Grant No. 2013zzts234 and National Natural Science Foundation of China under Grant No. 51406241.

REFERENCES

[1] Quintiere, J.Q., 1989. Fundamentals of enclosure fire "zone model". J. Fire. Prot. Eng. 1, 99–119.
[2] National Fire Protection Association, 2002. NFPA 204: Standard for Smoke and Heat Venting.
[3] ISO/TS 13571, 2002. Life threatening components of fires-Guidance on the estimation of time available for escape using fire data. International Standardisation Organisation (ISO), New York.
[4] National Fire Protection Association, 2003. NFPA. 2003a. NFPA 101.
[5] National Fire Protection Association, 2003. NFPA. 2003b. NFPA 130, Standard for Fixed Guideway Transit and Passenger Rail Systems.
[6] Grampton, G.P., Lougheed, G.D., 2004. Comparison of smoke measurements with standard and non-standard systems. Institute for Research in Construction, National Research Council Canada, Canada.
[7] Yi, L., Chow, W.K., Li, Y.Z., Huo, R., 2005. A simple two-layer zone model on mechanical exhaust in an atrium. Build. Environ. 40, 869–880.
[8] Hu, L.H., Huo, R., Li, Y.Z., Wang, H.B., Chow, W.K., 2005. Full-scale burning tests on studying smoke temperature and velocity along a corridor. Tunn. Undergr. Sp. Tech. 20, 223–229.

Architectural, Energy and Information Engineering – Sung & Chen (Eds)
© 2016 Taylor & Francis Group, London, ISBN 978-1-138-02791-6

Comparison of the reinjection of sandstone reservoir geothermal used water between Xi'an and Xianyang

Z.Y. Ma & Y. Meng
The Environmental Science and Engineering School, Changan University, Xi'an, China

ABSTRACT: The reinjection clogging of sandstone reservoir geothermal used water in Xi'an and Xianyang severely restricts the sustainable development and utilization of geothermal water. Thus, we introduced a study for the first time to investigate the cause for the reinjection clogging of used geothermal water in the Xianyang No. 1 reinjection wells and Xi'an sanqiao reinjection wells in 2011 and 2012, respectively. The coupling block simulation experiment in the laboratory with field investigation on reinjection, and sandstone reservoir used water reinjection in Xi'an and Xianyang have been comparatively studied in this paper with a combination of geological conditions and thermal storage environment characteristics of the two wells. Dynamic simulation in the laboratory with the target reservoir core indicates that all clogging rates in Xi'an reinjection simulation are larger than that of Xianyang No. 1 reinjection wells under the same temperature. The reason for this is as follows: an open-type reinjection manner of Xi'an sanqiao reinjection wells increased the congestion of chemical substances and microorganisms. Thus a closed-type reinjection is suggested in the future.

Keywords: sandstone reservoir; reinjection; comparison

1 INTRODUCTION

Deep-porous geothermal resource is a precious green energy, whose exploration and utilization plays a key role in dealing with energy shortage and environment pollution[1]. However, excessive exploration of geothermal water could cause environment pollution and the problem of decreasing geothermal water level, which could seriously hinder the construction of the harmonious society. Under this circumstance, artificial reinjection of used water could be employed, which is more active and effective than the limited exploration of geothermal water[2]. Nevertheless, many geothermal field investigations conducted both at home and abroad have shown that almost 80% of the sandstone geothermal reinjection wells were blocked and some reinjection wells were even forced to halt operation[3]. Thus, dealing with the reinjection clogging has become the top priority of geothermal water exploration and utilization at the present stage. Zheng Xilai et al. (1996) and Bi Erping (1998) attempted to analyze the chemical precipitation of some related substances in the shallow geothermal system with the help of mineral balance mode[4]. Liu Xueling, et al. discussed the formation of suspended matter in the geothermal water and the clogging mechanism[5]. In the present

study, we attempted to make a deep exploration on the reinjection clogging mechanism: from 2011 to 2012, we investigated the used geothermal water in Xianyang and Xi'an. By comparing the clogging situation under different conditions such as different structural units, different geothermal reservoir properties and different reinjection methods, we analyzed the common points and differences of different clogging mechanisms. We hope that the present study could provide some references for the studies of reinjection clogging mechanism in areas with a similar condition.

2 GEOGRAPHICAL CONDITION COMPARISON

Xi'an reinjection wells are located in the west of the fault block section between Sanqiao and Hanyuandian in the Xi'an geothermal field and in the Weihe fault depression basin. These wells are at the deep sag belt of Xi'an central depression whose geological structural conditions make them convenient channels for the storage and transportation of geothermal water. Xianyang No. 1 reinjection wells are located in the fault zone of north on the gently dipping slope of Xi'an depression with a relatively simple structure. The fault

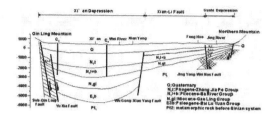

Figure 1. Profile map of the Xi'an depression and Xianyang-Liquanxian fault steps.

structures mainly include the near east-west fault on the north side of Weihe in the south and the north-west Xianyang-Chang'an fault in the west. In fact, the fault on the north side of Weihe is a base lithological boundary with lower Paleozoic carbonate on the north, and Proterozoic metamorphic rocks and Indosinian and Yanshannian granite on the south. The fault plane is inclined to the south with an angle of 67–80° and the fault displacement can reach 700 to 1000 m. The Xianyang-Chang'an fault has a north-west trend and is inclined to the southwest with an angle of about 70°. This fault is about 3 km in the west of the reinjection wells with a not too big fault displacement (Figure 1). Since the reinjection wells of both areas are not at the fault belt, a comparison between them is necessary.

3 A COMPARISON OF THE HEAT STORAGE STRATUM OF THE REINJECTION WELLS

For most of the exploited wells and reinjection wells in the urban area of Xi'an and Xianyang, the Lantian-Bahe group (N_2l+b) is the major water intake layer. Within the reinjection target stratum of Xi'an Sanqiao reinjection wells, there are Sanmen group, Lantian-Bahe group and Gao-ling group with a depth ranging from 1611.90 meters to 2876.60 meters and with a thickness of 1213.9 meters. According to the logging interpretation, there are a total of 133 layers of sandstone, whose thickness is 321.7 meters, and the ratio of sandstone thickness to the thickness of the whole stratum is 26.50%. The sandstone has a porosity ranging from 15.22% to 38.94% and permeability ranging from 3.65 md to 867.53 md. Within the reinjection target stratum of Xianyang No.1 reinjection wells, there is the Lantian-Bahe group with a depth ranging from 1270.7 meters to 2452.15 meters and a thickness of 1181.45 meters. Overall, there are 91 layers of sandstone whose total thickness is 412.3 meters, with the maximum thickness

being 19.3 meters and the minimum thickness being 0.9 meter. The ratio of the sandstone thickness to the thickness of the whole stratum is 34.90%. The sandstone has a porosity ranging from 10.22% to 22.44%, and permeability ranging from 2.06 md to 35.96 md.

4 COMPARISON OF THE RESULTS OF LABORATORY SIMULATION

4.1 Comparison of the results of static compatibility experiments

The results of the static compatibility experiments of Xianyang wells and Xi'an wells indicate that both the used water and raw water quality in Xianyang reinjection wells are incompatible (compatibility standard[6]: 100 mg/L) at 50°C, while there is good compatibility in Xi'an Sanqiao reinjection wells at 50°C. At 30°C, the water quality is still compatible in Xi'an Sanqiao reinjection wells. Thus, it can be concluded that there is better compatibility of water quality in Xi'an Sanqiao reinjection wells at 50°C, and that the precipitation capacity in Xi'an wells is only 15.41% compared with that of Xianyang reinjection wells.

4.2 Comparison of the results of dynamic displacement simulation experiment

The results of the clogging simulation of Xi'an Sanqiao reinjection wells and Xianyang No.1 reinjection wells are shown in Figure 2, respectively. From these two figures, it can be found that at the same temperature, the clogging rate in the simulation of Xi'an reinjection wells is greater than that in the simulation of Xianyang No.1 reinjection wells, and at 50°C, the chemical clogging rate of Xi'an Sanqiao reinjection wells is 30.91% higher than that of Xianyang reinjection wells. According to Figure 3, the leading factor causing a high clogging rate in Xi'an Sanqiao reinjection wells is chemical clogging, and under open-type circumstances, Fe^{2+} can be easily oxidized to Fe^{3+} with mineral deposits such as $Fe(OH)_3$ and Fe_2O_3 being produced, which could aggravate chemical clogging. However, it must be noted that with a water temperature of 30°C, the suspended matter still dominated Xi'an Sanqiao reinjection wells even after 5 μm filtering, which means that under open-type circumstances, clogging caused by the suspended matter cannot be ignored. Thus, in order to guarantee successful reinjection at the primary stage, it is of great necessity to employ closed reinjection to improve the filtering level of the suspended matter and increase the filtering volume.

Table 1. The comparison results of the injection displacement experiment of Xi'an and Xianyang.

Position of the well	Second-level filtering	Grit catcher	Same floor reinjection
Xianyang reinjection wells	√	√	√
Xi'an reinjection wells	√	√	√

Position of the well	Compression	Closed reinjection	Highest reinjection volume
Xianyang reinjection wells	3 MP	√	70 m³/h
Xi'an reinjection wells	0.3 MP	×	24 m³/h

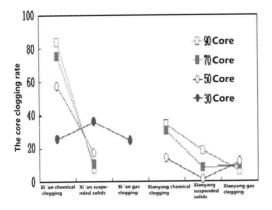

Figure 2. The map of the comparison results of the injection displacement experiment of Xi'an and Xianyang.

5 COMPARISON OF GROUND EQUIPMENTS

Xi'an Sanqiao reinjection wells are opening big pool wells. Under open-type circumstances, microorganisms, water quality and suspended matter would undergo great changes, which constitutes a great hindrance to the reinjection process. Without second-level filtering and other process equipments such as grit catcher, iron removal and aeration, the reinjection wells in Xi'an Sanqiao has a highest reinjection capacity of only 33 m³/h. However, for Xianyang No. 1 reinjection wells, there are second-level filtering and equipments such as grit catcher and closed reinjection wells are also

Table 2. The comparison of field ground equipment between Xi'an and Xianyang.

Clogging rate (%)		Temperature			
		30°C	50°C	70°C	90°C
Xi'an	Chemical clogging	25.5	57.2	75.1	83.6
	Suspended matter clogging	36.0	17.2	9.9	7.2
	Gas clogging	24.6			
Xianyang	Chemical clogging		26.3	33.4	42.4
	Suspended matter clogging		1.4	8.4	18.5
	Gas clogging		11.5	9.4	6.3

employed, which greatly reduced the difficulties met in Xi'an and the highest reinjection capacity could reach 60 m³/h. Thus, in order to reduce reinjection clogging of used geothermal water, closed ground equipments are often adopted to deal with suspended matter clogging and gas clogging. At present, the ground process equipments that are often employed include filter, vent tank, grit catcher, and iron removal. which could relieve clogging and increase reinjection volume effectively. By comparing the ground equipments of the reinjection wells in Xi'an and Xianyang and comparing the reinjection volume (Figure 5), it can be found that though many factors such as thermal reservoir property, degree of pressure and ground equipment lead to the difference in highest reinjection volume between Xi'an Sanqiao reinjection wells and Xianyang No. 1 reinjection wells, the selection of open-type reinjection or closed-type reinjection is another factor that cannot be ignored.

6 CONCLUSION

1. According to the laboratory simulation, it can be found that at the same temperature, the clogging rate in the simulation of Xi'an Sanqiao reinjection wells is greater than that in the simulation of Xianyang No. 1 reinjection wells. Besides, the iron pipe can produce large quantities of iron substances, and under open-type circumstances, Fe^{2+} can be easily oxidized to Fe^{3+}, with mineral deposits such as $Fe(OH)_3$ and Fe_2O_3 being produced, which could aggravate chemical clogging. However, it must be noted that with a water temperature of 30°C, the suspended matter still dominated Xi'an Sanqiao reinjection wells even after 5 μm filtering, which means that under open-type circumstances, clogging caused by the suspended matters cannot be ignored. Thus, in order to guarantee successful reinjection at the primary stage, it

is of great necessity to employ a closed-type rein-jection to improve the filtering level of the sus-pended matter and increase the filtering volume.

2. Under open-type circumstances, saprophytic bacteria and iron bacteria will multiply rapidly and canabsorb the fine suspended particles and dead microbes in the water. Besides, a large num-ber of iron bacteria can produce $Fe(OH)_3$, which further develop into Fe_2O_3 surrounding the bac-teria in the form of brown sticky mud, and these sticky mud accumulates on the surface of the wells and on the pores, which would aggravate suspended matter clogging, chemical clogging and microbe clogging. Thus, it is of great sig-nificance to adopt a closed-type reinjection.

3. The maximum injection quantities (70 m³/h) of Xianyang No. 1 reinjection wells and that of Xi'an Sanqiao (24 m³/h) show that only a closed-type injection can avoid congestion, which then guarantees the smooth work of the reinjection process.

REFERENCES

[1] Gallp, D.L., 1997. Aluminum silicate scale formation and inhibition (1): scale characterization and labora-tory experiments. Geothermics 26, 483–499.

[2] Stefansson V., Geothermal reinjection experience. Geothermics 1997, 26: 99–139.

[3] Liu Jiurong., The development status of geother-mal injection. Hydrogeology & engineering geology. 2003, 3(100).

[4] Zheng Xilai, Guo Jianqing. The mixing effect of geo-thermal system and its application research. Xi'an institute of geological. 1996, 18(04): 53–57.

[5] Liu Xueling, Zhu Jialing. A study of clogging in geo-thermal reinjection wells in the Neogene sandstone aquifer[J]. Hydrogeology & engineering geology. 2009, 138(5).

[6] Bian Chaofeng, Zhu Qijia, Chen Wu. Study on Com-patibility of Injected Water with Reservoir in Oilfield, 2006, 23(7): 48–50.

Architectural, Energy and Information Engineering – Sung & Chen (Eds)
© 2016 Taylor & Francis Group, London, ISBN 978-1-138-02791-6

Effect of Na_2O dosage and SiO_2/Na_2O ratio on the properties of Alkali-Activated Fly Ash geopolymers

M.C. Chi
Department of Fire Science, WuFeng University, Taiwan

R. Huang
Department of Harbor and River, National Taiwan Ocean University, Taiwan

ABSTRACT: This study investigates the effect of Na_2O dosage and SiO_2/Na_2O ratio on the properties of Alkali-Activated Fly Ash (AAFA) geopolymers. Sodium oxide (Na_2O) dosages of 121 and 150 kg/m^3 and liquid sodium silicate with alkaline modulus ratios (mass ratio of SiO_2 to Na_2O) of 1.23 and 0.8 were prepared as activators. The alkaline solution to binder ratio was kept at a constant of 0.5. The compressive strength test, drying shrinkage test, water absorption test, initial surface absorption test, Mercury Intrusion Porosimetry test (MIP), and Scanning Electron Microscopy (SEM) were conducted and their performance is discussed. The test results show that AAFA geopolymers with a higher alkaline modulus ratio and dosage of Na_2O have the superior properties than the others. The SEM analysis demonstrates that the hydration products of AAFA geopolymers are mainly amorphous alkaline aluminosilicate gel, which attributed to their compressive strength.

Keywords: Sodium oxide; Modulus ratio, property; Alkali-Activated Fly Ash (AAFA); Geopolymer

1 INTRODUCTION

Given that alumina and silica are necessary for the alkali-activated reaction, the utilization of fly ash as a raw material for synthesizing the Alkali-Activated Fly Ash (AAFA) geopolymer has attracted the interest of many researchers for the development of geopolymer products (Hu et al. 2009; Komljenović et al. 2010; Zhang & Liu 2013; Torres-Carrasco & Puertas 2015; Ryu et al. 2013). The AAFA geopolymer has a great potential to be an effective alternative to ordinary Portland cement since it possesses excellent mechanical properties and durability in an aggressive environment, extended fire and acid resistance, and environmental benefits (Bakharev 2005; Němeček et al. 2011; Pacheco-Torgal et al. 2008). However, the chemical composition of alkaline activators is still the subject of much debate in the scientific literature, which depends on the physical-chemical characteristic of the raw materials, the characteristic and quantity of the activators, and the curing condition (Collins & Sanjayan, 1999; 2000; Bakharev et al. 1999; Bernal et al. 2010). Chi and Huang (2012) concluded that the Na_2O dosage and alkaline modulus ratio of the alkali-activated solution are two key factors that influence the properties of Alkali-Activated Slag Mortars (AASM). Peng et al. (2015) investigated that Na-activated systems show a higher strength than the K-based ones after thermal activation, with the 8 M NaOH activation being the most efficient. Jun and Oh (2014) studied the micro-structural characteristics of the alkali-activated fly ashes, and found that the reaction products after activation were dissimilar between the activated ashes with not much different chemical compositions in XRD and XRF data. Meanwhile, the activated fly ashes have a higher compressive strength as the reaction products possessed a lower Si/Al ratio of 1.

Although many previous studies have researched on the AAFA geopolymer, there is still a significant need for further investigations. In the alkaline activation of fly ash, the type of alkaline activator, the concentration of alkaline-activated solution, and the dosage of the alkali component have a considerable influence on the mechanical and structural performance of the AAFA geopolymer. This study investigates the effect of Na_2O dosage and SiO_2/Na_2O ratio on the properties of Alkali-Activated Fly Ash (AAFA) geopolymers.

2 EXPERIMENTAL PROGRAM

2.1 *Materials*

Class F Fly Ash (FA) was used as the main aluminum and silicate source for synthesizing the geopolymeric binder. Its specific gravity and Blaine specific surface area were 2.06 and 0.237 m^2/g, respectively. Standard sand conforming to ASTM C778 was used as a fine aggregate in the manufacture of mortars. The specific gravity of the standard sand is 2.65. The most used alkaline activators are a mixture of NaOH with sodium silicate ($Na_2O \cdot \gamma SiO_2$) (Sievert 2005). In this study, the alkaline activation of the FA was carried out using NaOH pellets with a density of 2130 kg/m^3 and the sodium silicate solution ($Na_2O \cdot \gamma SiO_2 \cdot nH_2O$) composed of 29.2% SiO_2, 14.8% Na_2O and 56.0% H_2O by mass. Na_2SiO_3 and NaOH solutions were prepared one day prior to use.

2.2 *Mix design and specimen preparation*

Mixing of Alkali-Activated Fly Ash Geopolymers (AAFAG) with alkaline solution/binder ratios of 0.5 were designed. The sand/binder ratio was kept at a constant of 2.75. Liquid sodium silicate with two modulus ratios (mass ratio of SiO_2 to Na_2O) of 1.23 and 0.8 with symbols "A" and "B" and sodium oxide (Na_2O) with two levels of dosage (121 kg/m^3 and 150 kg/m^3 per cubic meter of mortar) with symbols "L" and "H" were used as alkaline activators. The mortar mix proportions are summarized in Table 1. The mixing of the mortar was conducted in a mechanical mixer. After mixing, the mixture was poured into steel molds. The specimens were cast and kept in steel molds for 24 hours, and then the specimens were demolded and moved into a curing room at a relative humidity of 80% and a temperature of 25°C until testing. The specimens were tested in triplicate sets until the time of testing.

2.3 *Methods*

The compressive strength tests of the specimens were conducted according to ASTM C109.

Table 1. Mix proportions of AAFA geopolymers.

Mix no.	SiO_2/ Na_2O	Water (kg/m³)	FA (kg/m³)	Fine agg. (kg/m³)	Activator (kg/m³) SiO_2	Na_2O
AL	1.23	118	528	1453	148.8	121
AH	1.23	93	528	1453	184.5	150
BL	0.8	128	528	1453	96.8	121
BH	0.8	106	528	1453	120	150

The drying shrinkage test was done in accordance with ASTM C596. Water absorption was made in accordance with ASTM C642. The initial surface absorption test was performed in accordance with BS 1881-201. The Mercury Intrusion Porosimetry (MIP) test was conducted in accordance with ASTM D 4404-10 by injecting mercury into the dried specimens. SEM analyses were performed using a HITACHI S-3400 microscope with an Energy Dispersive Spectroscopy (EDS).

3 RESULTS AND DISCUSSION

3.1 *Compressive strength*

The compressive strength development of AAFA geopolymers at the ages of 7, 14 and 28 days is shown in Figure 1. As expected, the compressive strength of all specimens increases with the increasing ages. The compressive strength of AAFA geopolymers shows an increase with an increasing dosage of Na_2O. Meanwhile, the higher the SiO_2/Na_2O ratio (M_s) of the alkaline solution, the superior the compressive strength of AAFA geopolymers. At the age of 28 days, specimen AH ($M_s = 1.23$, $Na_2O = 150$ kg/m^3) has the highest compressive strength, with a compressive strength of 66.0 MPa, followed by the specimen BH ($M_s = 0.8$, $Na_2O = 150$ kg/m^3) with a compressive strength of 42.0 MPa, and then the specimen AL ($M_s = 1.23$, $Na_2O = 121$ kg/m^3) with a compressive strength of 32.7 MPa. Specimen BL is on the other end of the scale, with a compressive strength of 13.4 MPa. The high compressive strength of alkali-activated fly ash is attributed to the amorphous hydrated alkali-aluminosilicate produced for fly ash.

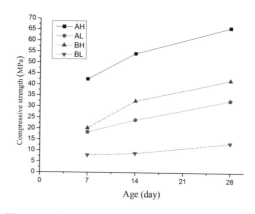

Figure 1. Compressive strength development of AAFA geopolymers vs. age.

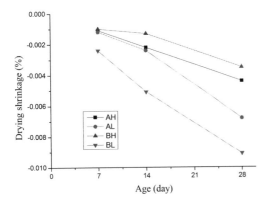

Figure 2. Drying shrinkage of AAFA geopolymers vs. age.

Table 2. Water absorption and porosity of AAFA geopolymers.

Mix no.	Water absorption (%)	Porosity (%)
AH	5.8	11.3
AL	6.4	12.2
BH	6.7	13.4
BL	7.9	14.9

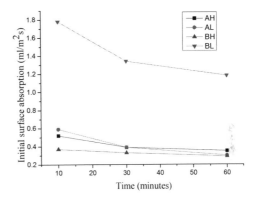

Figure 3. Initial surface absorption of AAFA geopolymers vs. age.

3.2 Drying shrinkage

The drying shrinkage or length change of AAFA geopolymers at the ages of 7, 14 and 28 curing days is shown in Figure 2. The length change of all specimens increases with an increasing ages. At the age of 28 days, the rate of length change of AAFA geopolymers with Na_2O of 121 kg/m^3 (AL and BL) was higher than that of AAFA mortars with Na_2O of 150 kg/m^3 (AH and BH). It indicates that the sodium oxide (Na_2O) concentration has a significant influence on the length change of AAFA geopolymers. Generally, an increase in the sodium silicate level caused a decrease in the total porosity and the increase in mesopore volume, which is directly related to the shrinkage due to self-desiccation (Neto et al. 2008).

3.3 Water absorption and porosity

Table 4 lists the water absorption of AAFA geopolymers at the age of 28 days. Water absorption ranged from 5.8% to 7.9%. The higher dosage of sodium oxide (Na_2O) leads to more effective alkali activation of fly ash and results in decreased water absorption. Thus, the higher the dosage of sodium oxide (Na_2O), the lower the porosity. During the alkali activation process, when the alkali activation of fly ash is completed, the internal structure of the AAFA geopolymer becomes more dense. On the contrary, when the alkali activation of fly ash is not completed, the remaining water is kept inside the AAFA mortars to form pores and causes increased porosity.

3.4 Initial Surface Absorption Test (ISAT)

The variations in initial surface absorption with respect to the testing time are plotted in Figure 3. The initial surface absorption values of the AAFA

geopolymers decreased with the testing time. Initial surface absorption values for AAFA geopolymers with an alkaline modulus ratio of 0.8 and sodium oxide (Na_2O) dosage of 121 kg/m^3 (BL) were obtained, which were higher than those of the others. It indicates that the initial surface absorption value of AAFA mortars is affected by the dosage of sodium oxide. The increasing dosage of sodium oxide for AAFA geopolymers can effectively reduce the initial surface absorption.

3.5 Mercury Intrusion Porosimetry (MIP)

The MIP test was carried out for the specimens at the age of 28 days. By tracking pressure and intrusion volume at each increment, the cumulative intrusion volume, capillary pore intrusion volume, and gel pore intrusion volume were obtained and are summarized in Table 3. The AAFA geopolymer with Na_2O of 150 kg/m^3 at the modulus ratio of 1.23 (AH) has a lower cumulative intrusion volume at 0.0807 ml/g (capillary pore intrusion volume 0.0731 ml/g and gel pore intrusion volume 0.0076 ml/g), followed by the AAFA mortar with the Na_2O dosage of 150 kg/m^3 at the modulus ratio of 0.8 (BH), with a cumulative intrusion volume of 0.0983 ml/g (capillary pore intrusion volume

Table 3. Intrusion volume of mercury of AAFA geopolymers.

Mix no.	Gel pore intrusion volume (ml/g)	Capillary pore intrusion volume (ml/g)	Cumulative pore intrusion volume (ml/g)
AH	0.0076	0.0731	0.0807
AL	0.0092	0.0911	0.1003
BH	0.0088	0.0895	0.0983
BL	0.0099	0.1023	0.1122

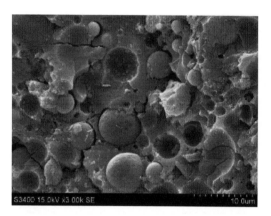

Figure 4. SEM image of AAFA geopolymers with an alkaline modulus ratio of 1.23 and sodium oxide (Na$_2$O) dosage of 150 kg/m^3 (AH) at the age of 28 days at 3K x.

0.0895 ml/g and gel pore intrusion volume 0.0088 ml/g). The cumulative intrusion volume decreases with an increasing dosage of Na$_2$O. The reduction in the total pore volume is due to the high reactivity in the alkaline activation of fly ash and the pore-filling effect of the small particles of fly ash. It indicates that both the modulus ratio and the dosage of the alkali-activated solution have a significant influence on the cumulative intrusion volume of AAFA mortars.

3.6 *Scanning Electron Microscopy (SEM)*

SEM was conducted on geopolymer mortar-based fly ash with the mixture of Na$_2$O and SiO$_2$ as an alkaline activator, in order to make the reactants of fly ash consistent with the alkaline activator and to confirm the internal microstructure. Figure 4 shows the SEM images of AAFA geopolymers with an alkaline modulus ratio of 1.23 and sodium oxide (Na$_2$O) dosage of 150 kg/m^3 (AH) at the age of 28 days at 3K x. Some unreacted fly ash particles still existed in the specimen after it was cured for 28 days. Some fly ash particles that have reacted with the alkali-activated solution were observed to

coexist with some remaining unreacted spheres and even with some other particles partially covered with reaction products. However, there are a number of amorphous alkaline aluminasilicates on top of the fly ash particles. Due to the amorphous aluminasilicates formed in the geopolymerization reaction, which is a consequence of the alkali-activated fly ash, the specimen shows a compact microstructure and a very low porosity.

4 CONCLUSIONS

This study investigates the effect of the Na$_2$O dosage and SiO$_2$/Na$_2$O ratio on the properties of Alkali-Activated Fly Ash (AAFA) geopolymers, based on which the following conclusions can be drawn:

1. Both alkaline modulus ratio and the dosage of Na$_2$O are two significant factors influencing the properties of AAFA geopolymers.
2. The compressive strength of AAFA mortars shows an increase with an increasing dosage of Na$_2$O. Meanwhile, the higher the SiO$_2$/Na$_2$O ratio (M$_s$) of the alkaline solution, the superior the compressive strength of AAFA mortars.
3. The increasing dosage of sodium oxide for AAFA mortars can effectively reduce the drying shrinkage, porosity and initial surface absorption.
4. The hydration products of AAFA mortars are mainly amorphous alkaline aluminosilicate gel, which attributed to the compressive strength.

REFERENCES

Bakharev, T. 2005. Geopolymeric materials prepared using Class F fly ash and elevated temperature curing. *Cement & Concrete Research* 35(6): 1224–32.
Bakharev, T. Sanjayan, J.G. Cheng, Y.B. 1999. Alkali activation of Australian slag cements. Cement & Concrete Research 29: 113–20.
Bernal, S. Gutierrez, R.D. Delvasto, S. Rodriguez E. 2010. Performance of an alkali-activated slag concrete reinforced with steel fibers. Construction & Building Materials 24: 208–14.
Chi, M. Huang, R. 2012. Effects of Dosage and Modulus Ratio of Alkali-Activated Solution on the Properties of Slag Mortars. *Advanced Science Letter* 16(1): 7–12.
Collins, F. Sanjayan, J.G. 2000. Strength and shrinkage properties of alkali-activated slag concrete containing porous coarse aggregate. Cement & Concrete Research 29: 607–10.
Collinsa, F. Sanjayan, J.G. 1999. Cracking tendency of alkali-activated slag concrete subjected to restrained shrinkage. Cement & Concrete Research 30: 791–8.
Hu, M. Zhu, X. Long, F. 2009. Alkali-activated fly ash-based geopolymers with zeolite or bentonite as additives. Cement & Concrete Composites 31: 762–68.

Jun, Y. Oh, J.E. 2014. Mechanical and microstructural dissimilarities in alkali-activation for six Class F Korean fly ashes. *Construction & Building Materials* 52: 396–403.

Komljenović, M. Baščarević, Z. Bradić, V. 2010 Mechanical and microstructural properties of alkali-activated fly ash geopolymers. Journal of Hazardous Materials 181: 35–42.

Němeček, J.I. Šmilauer, V. Kopecký, L. 2011. Nanoindentation characteristics of alkali-activated aluminosilicate materials. Cement & Concrete Composites 33: 163–70.

Neto, A.A.M. Cincotto, M.A. Repette, W. 2008. Drying and autogenous shrinkage of pastes and mortars with activated slag cement. Cement & Concrete Research 38: 565–74.

Oswaldo, B-D. Ivan, E-GJ. Rat, A-A. Alexander, G. 2010 Statistical analysis of strength development as a function of various parameters on activated metakaolin/slag cements. *Journal of American Ceramic Social* 93(2): 541–7.

Pacheco-Torgal, F. Castro-Gomes, J. Jalali, S. 2008. Alkali-activated binders: A review: Part 1. Historical background, terminology, reaction mechanisms and hydration products. Construction & Building Materials 22(7): 1305–14.

Peng, Z. Vance, K. Dakhane, A, Marzke, R. Neithalath, N. 2015. Microstructural and ^{29}Si MAS NMR spectroscopic evaluations of alkali cationic effects on fly ash activation. *Cement & Concrete Composites* 57: 34–43.

Ryu, G.S. Lee, Y.B. Koh, K.T. Chung, Y.S. 2013. The mechanical properties of fly ash-based geopolymer concrete with alkaline activators. Construction & Building Materials 47: 409–18.

Sievert, T. Wolter, A. Singh, N.B. 2005. Hydration of anhydrite of gypsum ($CaSO_4.II$) in a ball mill. Cement & Concrete Research 35: 623–30.

Torres-Carrasco, M. Puertas, F. 2015. Waste glass in the geopolymer preparation. Mechanical and microstructural characterisation. Journal of Cleaning Products 90: 397–408.

Zhang, Y. Liu, L. 2013. Fly ash-based geopolymer as a novel photocatalyst for degradation of dye from wastewater. PARTICUOLOGY 11: 353–8.

Architectural, Energy and Information Engineering – Sung & Chen (Eds)
© *2016 Taylor & Francis Group, London, ISBN 978-1-138-02791-6*

B-spline curve approximation to planar data points based on dominant points

Z.Z. Lin & S.H. Shu
College of Mathematics and Computer Science, Jiangxi Science and Technology Normal University, NanChang, China

Y. Ding
Jiangxi Normal University Science and Technology College, NanChang, China

ABSTRACT: In this paper, an algorithm of curve approximation to planar data points based on dominant points is presented. Three kinds of dominant points are sampled from the original data point set, such as inflect points, curvature extreme value points, and discontinuous curvature points. Dominant points sampling algorithms based on discrete curvature are also described in this paper. Next, a B-spline curve is constructed to interpolate these dominant points and a new interpolation point is inserted at the place where the maximum error is greater than the given accuracy. Finally, numerical examples of this algorithm are presented to demonstrate its effectiveness and validity.

Keywords: In order to; MS Word

1 INTRODUCTION

With the development of laser measurement technology, data points from laser scanner are usually dense and noisy; it is very difficult to construct a curve or a surface to approximate these dense points. B-spline curve approximation to planar data points is the core and basic problem in reverse engineering, but this approximation depends on several factors such as parameters of data points, knot vector and the stability of the approximation system (Farin, G. 1994; Piegl. L et al. 1997; 2001). Unknown shape is often appear in this approximation curve, because this curve can not approximate many dominant points exactly such as inflection point, the crease point and extreme curvature point (LIU, G.H. 2002; Deng, C. 2014; Fang, J.J. et al. 2013), so dominant points play an important role in the curve approximation, smoothing treatment of data point is adopted according to the discrete curvature (LIU, G.H. 2002), this method provides a good way to use the discrete curvature in approximation. RAZDAN (RAZDAN, 1999) has proposed a knot placement method using the arc and curvature information, but this method is applied to smooth data with no noisy. Li (LI, W.

et al. 2005) has designed a low-pass filter to sample data point from original data according to the discrete curvature, this filter only can recognize the inflection points as the dominant points, but the geometric information provided only by inflection points is too little. The dominant points can be identified automatically according to the discrete curvature of each data point in (Park. H 2007), a B-spline curve is constructed to approximate data points by interpolating these dominant points (Ma, W.Y. et al. 1995; Kosinka, J. 2014).

An algorithm of curve approximation to planar data points based on feature points is proposed in this paper. Dominant data points such as inflect points, curvature extreme value points and are sampled from the original data points set. Dominant points and the end points consist of the initial approximation data set K, a B-spline curve is constructed to interpolate these data points in K, and data point is inserted in K to start a new interpolation if the maximum error is less than the given error bound. This process is repeated until the maximum fitting error is met. The numerical examples are presented to show that the dominants of the data points profile well, and both the numbers are reasonable.

2 PRELIMINARIES

2.1 B-spline curve

A B-spline curve is defined by

$$C(u) = \sum_{i=0}^{n} N_{i,p}(u) P_i \qquad (1)$$

where $\{P_i\}$ are the control points, and $\{N_{i,p}(u)\}$ are the pth-degree B-spline basis functions defined on knot vector $U = \{u_0 \leq u \leq L \leq u_p \leq L \leq u_n \leq L \leq u_{n+p+1}\}$. is defined as

$$N_{i,0}(u) = \begin{cases} 1, u_i \leq u < u_{i+1} \\ 0, u < u_i \ or \ t \geq t_{i+1} \end{cases},$$

$$p = 0 \begin{cases} N_{i,p}(u) = \dfrac{u - u_i}{u_{i+p} - u_i} N_{i,p-1}(u) + \\ \dfrac{u_{i+p+1} - u}{u_{i+p+1} - u_{i+1}} N_{i+1,k-1}(u), \quad p \geq 1 \\ \dfrac{0}{0} = 0 \end{cases}$$

2.2 B-spline curve interpolation to point

If data points $\{Qk\}$, their corresponding parameters $\{t_k\}$ $k = 0, L, m$ and knot vector U of the approximation curve are given, a B-spline curve is computed to interpolate $\{Q_k\}$ in this sense

$$Q_k = C(t_k) = \sum_{i=0}^{n} N_{i,p}(t_k) P_i \qquad (2)$$

The control points can be computed by solving a linear equation $N_p = Q$, where $N = [N_{i,j}] = [N_{i,p}(u_j)]$, $P = [p_j]$ and $Q = [Q_i]$. Chord length parameterization is usually used to compute the parameters of data points in this way:

$$t_0 = 0; \quad t_i = \dfrac{\sum_{j=0}^{i} |Q_j - Q_{j-1}|}{\sum_{j=0}^{m} |Q_j - Q_{j-1}|} \quad i = 0 \ldots m \qquad (3)$$

These parameters are assigned to every knot span averagely.
Let

$$d = \dfrac{m+1}{n-p+1} \qquad (4)$$

Then define the internal knots by

$$i = \text{int}(jd) \quad \alpha = jd - i \quad u_{p+j} = (1-\alpha)t_{i-1} + \alpha t_i \qquad (5)$$

$$j = 1 \ldots, n - p.$$

3 CURVE APPROXIMATION TO PLANAR DATA POINTS

In this section, the curve approximation algorithm based on dominant points and three kinds of dominant data sampling algorithms based on discrete curvature are described.

3.1 The curve approximation algorithm

1. Step 1. Add dominant points and end points into the approximation data set K.
2. Step 2. Construct a B-spline curve $C(u)$ to interpolate data point in K.
3. Step 3. Compute the approximation error of each point. If the maximum error is less than the given accuracy, output this B-spline $C(u)$ If not, go to step 4.
4. Step 4. Find the point Q_j with the maximum error, insert Q_j into data set K, turn to step 2.

Details of this algorithm are explained in the subsequent sections. The approximation data points set contains all of the dominant data points and the end points, the curve approximate algorithm only needs few iterations to meet the high accuracy. The dominant points are sampled according to the discrete curvature, so we first introduce the method of computing the discrete curvature.

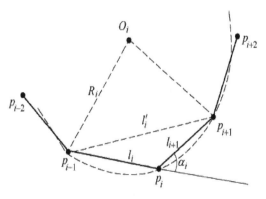

Figure 1. Circle arc interpolating three adjacent points.

3.2 The discrete curvature of each point

The discrete curvature k_i of each point Q_i is computed by

$$k_i = \frac{2p_{i-1}p_ip_{i+1}}{l_il_{i+1}l_i'} = \pm(p_{i-1}p_ip_{i+1})\frac{\sin\alpha_i}{l_i'} \qquad (6)$$

The k_i is the curvature of P_i on the circle arc that interpolates p_{i-1}, p_i, p_{i+1}. The circle arc is shown in Figure 1. α_i is the angle of two line segments $\overline{P_iP_{i+1}}$ and $\overline{P_{i-1}P_i}$ l_i is chord length of line segment $\overline{P_{i-1}P_{i+1}}$. If P_{i-1}, P_i, P_{i+1} are arranged in anti-clockwise direction, k_i is positive, otherwise is negative. We need not compute the curvature of end points, because the approximation data set K contains these two end points.

After computing the discrete curvature k_i of each point, we can determine three kinds of dominant data points based on k_i: inflect point, curvature extreme value points, and discontinuous curvature points. In the next section, we will describe the dominant points determining algorithm.

3.3 Inflect points

If the discrete curvature k_i, k_{i+1} of P_i, P_{i+1} meets these three conditions:

1. $k_ik_{i+1} < 0$

2. Every $k_{i-2}, k_{i-1}.k_i$ is positive and every $k_{i+1}, k_{i+2}.k_{i+3}$ is negative, and vice versa.

3. $\Delta k_i^- = \max(\|\,k_{i-3}\,|-|\,k_i\,\|, \|\,k_{i-2}\,|-|\,k_i\,\|,$
$\qquad \|\,k_{i-1}\,|-|\,k_i\,\|)$

   $\Delta k_i^+ = \max(\|\,k_{i+3}\,|-|\,k_i\,\|, \|\,k_{i+2}\,|-|\,k_i\,\|,$
$\qquad \|\,k_{i+1}\,|-|\,k_i\,\|)$

   $\Delta k_i^- > 3\Delta\overline{k}$ and $\Delta k_i^+ > 3\Delta\overline{k}$. Then the point is identified as the inflection point.

3.4 Curvature extreme value points

If each point P_i and its discrete curvature k_i meet these three conditions at the same time as follows:

1. $k_i > k_{i-1}$

2. $k_i > k_{i+1}$

3. $k_i > k_{avg}/4$

where k_{avg} is the mean curvature of all data points, then this data point is identified as the curvature value points.

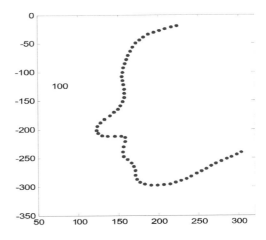

Figure 2.   Data points.

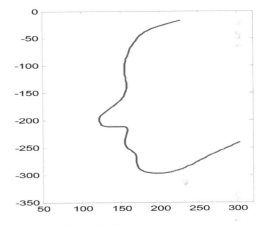

Figure 3.   Approximation curve.

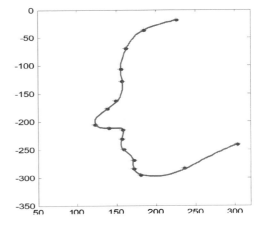

Figure 4.   Dominant points.

3.5 Discontinuous curvature points

For each point P_i, if it meets the following conditions, the points are recognized as the discontinuous curvature points:

1. $k_{i-3}, k_{i-2}, k_{i-1}$ have the same sign and, $|k_{i-j}| < |k_i|$, $j = 1, 2, 3$

2. $k_{i+1}, k_{i+2}, k_{i+3}$ have the same sign and $|k_{i-j}| < |k_i|$, $j = 1, 2, 3$.

4 EXAMPLE

In this section, an experimental example is presented to prove that the proposed algorithm is feasible and efficient. Figures 2 and 5 show the algorithm process in approximating planar data points based on dominant points

Figure 2 shows a set of data points from human profile, and there are 135 data points in this set. The final approximation curve is shown in Figure 3, and only 29 data points are sampled in this approximation within the maximum error bound $\varepsilon = 0.2$ mm. These data points are shown in Figure 5.

The sampling process is an iteration process, about 19 dominant points are first detected by the dominant points sampling algorithm, these dominant points are shown in Figure 4. We insert another 10 dominant points in approximation data set K to meet the accuracy $\varepsilon = 0.2$ mm, the final interpolation points and final approximation curve are shown in Figure 5.

5 CONCLUSION

In this paper, we given a method of B-spline curve approximation to planar data points based on

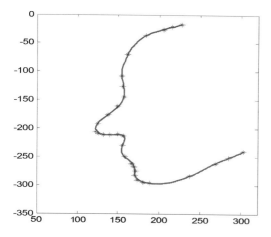

Figure 5. Approximation points.

dominant points. The dominant data points are detected according to the discrete curvature of each point. From the experimental example, we can see that the approximation algorithm is feasible and efficient in data approximation. Most of the dominant points can be sampled by the given algorithm, the B-spline curve perfect well because that the approximation curve interpolates all the dominant points.

But the discrete curvature of each point is the curvature of the circle interpolating three adjacent data points, so this method is limited to the planar data points. How to apply this method on space data is valuable to further research.

ACKNOWLEDGMENT

This work is financially supported by the Natural Science Foundation of the Jiangxi Province of China under Grant No. 20122BAB201004 and the General Natural Science Foundation of the Jiangxi Science & Technology University under Grant No. 2013XJYB002.

REFERENCES

Deng, C.Y. & Lin, H.W. 2014. Progressive and iterative approximation for least squares B-spline curve and surface fitting, Computer-Aided Design, 47(2): 32–44.

Fang, J.J & Hung, C.L. 2013. An improved parameterization method for B-spline curve and surface interpolation, Computer-Aided Design, 45(6): 1005–1028.

Farin, G. 1994. NURBUS curve and surfaces. Projective geometry to practial use. MA, USA.

Kosinka, J. & Malcolm S, Dodgson, N. 2014. Creases and boundary conditions for subdivision curves, Graphical Models, 76(5): 240–251.

Li, W.S. & Xu S.H, Zhao G, et al. 2005. Adaptive knot placement in B-spline curve approximation. Vol. 37. Computer- Aided Design, Aug.: 791–797.

Liu, G.H. & Wong, Y.S. Zhang Y.F, et al. 2002. Adaptive fairing of digitized point data with discrete curvature. [Computer-Aided Design], 34(4): 309–320.

Ma, W.Y. & Kruth. J.P. 1995. Parameterization of randomly measured point for least squares fitting of B-spline curves and surfaces. Computer-Aided Design, 27(1): 663–675.

Park. H & Lee, J.H. 2007. B-spline curve fitting based on adaptive curve refinement using dominant points, Computer-Aided Design, 39(6): 439–451.

Piegl, L. & Tiller, W. 1997. The NURBS book. 2nd ed. New York: Springer Verlag.

Piegl, L. & Tiller, W. 2001. Parametrization for surface fitting in reverse engineering, Computer-Aided Design, 33 (8): 93–603.

Razdan, A. 1999. Knot placement for B-spline curve approximation [R]. Arizona State University.

Architectural, Energy and Information Engineering – Sung & Chen (Eds)
© 2016 Taylor & Francis Group, London, ISBN 978-1-138-02791-6

Safety prediction research on building demolition work

L. Jiao, T.J. Zhang, Q.X. Shi & Y.L. Zheng
Shenyang Jianzhu University, Shenyang, Liaoning, China

ABSTRACT: Aiming at the faults of the safety accident prediction method in building demolition work, based on the Least Squares Support Vector Machine (LS-SVM), a safety prediction model was established. The analytic hierarchy process and the fuzzy comprehensive evaluation were combined to establish a comprehensive evaluation model (AHP-FCE). This model was used to make safety assessment for 12 domestic building blasting demolition projects. The evaluation results were considered as the training sample and verified sample of the least squares support vector machine safety prediction model. The accuracy and feasibility of the model were verified. It is shown that the predicting accuracy of the model is higher, and that it can be used for safety prediction evaluation of building demolition work.

Keywords: Building demolition; safety evaluation; Least Squares Support Vector Machine; safety prediction

1 INTRODUCTION

With the expanding quantity and scale of building demolition, there have been increasing proportions of accidents during demolition work. In addition, the loss is becoming increasingly huge. Among numerous demolition methods, the development of blasting demolition technology is the most rapid and most widely used. Therefore, taking blasting demolition technology as an object, it is necessary to do further research on safety accident prediction in building demolition work.

At present, for safety accident prediction of building demolition work, some safety evaluation methods have been set up such as the fault tree analysis, the fuzzy comprehensive evaluation, and the neural network prediction model. These methods solved some problems and also played an important role, but have shortcomings. These include the following: the fault tree analysis is not applied to the problems of no clear boundaries, the fuzzy comprehensive evaluation has strong subjectivity factors, and the neural network prediction model is based on large sample data.

Suykens J.A.K proposed the Least Square Support Vector Machine (LS-SVM) algorithm [1–6] in 1998. The safety prediction model of building demolition work is established by using the LS-SVM to overcome the above shortcomings. It is not longer limited by sample quantity, and avoids the quadratic programming problem of the Support Vector Machine (SVM). Moreover, it makes the algorithm simple, the convergence speed of the model is faster, and the calculation result is more accurate.

In this paper, the analytic hierarchy process and the fuzzy comprehensive evaluation are combined to establish the comprehensive evaluation model AHP-FCE. Taking blasting demolition technology as an research object, the quantitative evaluation results of the AHP-FCE are used as the training sample and verified sample of the LS-SVM model. Finally, the model is verified for its reliability and accuracy after training.

2 AHP-FCE MODEL

The comprehensive evaluation model AHP-FCE is the comprehensive application of the analytic hierarchy process and the fuzzy comprehensive evaluation method. The analytic hierarchy process is used to calculate all the levels of index weight, and then the total weight is obtained. The evaluation results are quantified by the fuzzy relationship matrix and the maximum membership degree principle of the fuzzy comprehensive evaluation method. From the safety perspective of building demolition work, it is important to find out where it is not safe and how to prevent. The specific establishing process of the comprehensive evaluation model is shown in Figure 1.

A comprehensive evaluation of the model AHP-FCE shows that it has good simulation, and the relevant factors affecting the decision problem are decomposed into the target layer, criterion layer and index layer. Then, each layer isquantitatively and qualitatively analyzed to find the factors of each layer and the relationship influence between

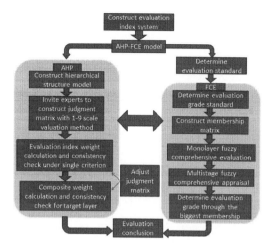

Figure 1.   AHP-FCE model.

the layers. At the same time, based on the fuzzy mathematical theory and fuzzy reasoning using the qualitative and quantitative analysis, with the combined idea of precision and fuzziness, a comprehensive analysis and evaluation is conducted on some ambiguous problems. Especially for a large and complex objective, i.e. a big system with many structure layers, the model can not only objectively describe the subject of evaluation, but also accurately determine the weights of evaluation indices. Particularly aiming at fuzzy comprehensive evaluation, the method is a more ideal method. Therefore, the comprehensive evaluation results of the AHP-FCE model are very close to the actual situation.

3   LS-SVM MODEL

3.1   *The modeling process*

Based on the regression method of the least squares support vector machine, aiming at the relevant data of construction site, the specific steps of the LS-SVM model can be formulated as follows:

1. Input fuzzy comprehensive evaluation results data as learning samples.
2. According to the sample data, set up the appropriate least squares support vector machine model. The model chooses the radial basis function as the kernel function. The least squares error criterion is considered as the loss function. The generalization of the model is guaranteed by controlling the capacity parameter α.
3. Input the prediction sample data.
4. Calculate the error of the LS-SVM model output data.

5. If the permissible error limit is not corresponding, then turn to step 2 to adjust the parameter α and capacity, control regularization parameter γ.
6. Output results and fit prediction graphics.

Traditional safety prediction evaluation models are obtained through the rich experience and professional knowledge of relevant experts, the evaluation standards and specifications, and the raw data in practical constructions.

The specific working processes of the LS-SVM model are as follows. According to the evaluation criteria, the collected data is considered as the input. The expert evaluation results are considered as the output. Then, a pair of sample data set is formed. The LS-SVM model is trained by making use of these samples, and the error limit is ensured. In this way, the model learns the function relation between the input and the output; furthermore, it memorizes that. Therefore, the whole model is equivalent to a series of experts, with the evaluation of their experience and related professional knowledge, so as to evaluate the new input, and the output is needed forecast results.

3.2   *Programming of the model*

The program of the LS-SVM model is complex, the calculated amount and the difficulty involved are both challenging. The realization of the algorithm uses MATLAB software. The sample data is changed into a vector form, and saved into an m file; in this way, the data is acquired through file. First, the training data is compiled into an m_1 file. Then, 8 sets of training sample data is randomly selected from 12 projects. The training data is input to the model through the file. After training, the model can provide the relevant parameters. Next, the evaluation data of 12 projects are randomly disturbed as the model validation sample. Besides, it is compiled into an m_2 file. Finally, the validation sample is input into the model through the file, as shown in Figure 2 to Figure 5.

```
1    %% get the data through file
2 -  [filename pathname] = uigetfile('*.file','select the data file');
3 -  filefullname = [pathname filename];
4
5 -  fid = fopen(filefullname);
6 -  if fid == -1
7 -      error (['open ' filefullname ' file Failed!']);
8 -  end
9
```

Figure 2.   Sample data form an m file.

```
%% data preparation
[M N] = size(data);   %N is the project number, M is the total number of x+y
nyc = 3;              %the number of rows of y
nxc = 6;              %the number of columns of x to calculate a b
signa = 20;           % σ
gamma = 40;           % γ

disp('%-- — — — — — — — — — — — — -%');
disp(['σ = ' num2str(signa)]);
disp(['γ = ' num2str(gamma)]);
disp('');
```

Figure 3. Parameter optimization.

```
%% {{ calculate K
clear K
K = zeros(N,M-nyc);
for i = 1:N
    for j = 1:M-nyc
        K(i,j) = 0;
        for jj = 1:M-nyc
            K(i,j) = K(i,j) + sqrt(abs(data(j,i)-data(jj,i)))
        end
        K(i,j) = K(i,j)^2/(sigma^2);
        K(i,j) = exp(-K(i,j));
        if(i == j)
            K(i,j) = K(i,j) - 1/gamma
        end
    end
end
% complement
K = [ones(N,1) K];
K = [[0 ones(1,M-nyc)]; K]
%}}
```

Figure 4. Kernel function calculation.

```
%%
clear Y
for iy = 1:nyc
%{{ calculate a b
    disp(['%---------------- the line of num2str(iy)......
        '  ----------------------------%']);
    disp('%-- result ------------------------------------%');
    Y = data(M-nyc+iy,:);

%{{ calculate Yc
    Yc = K*a;
    disp('%% test value Y=');
    disp(num2str(Yc(2:N+1)));
     disp('%% real value Y=');
    disp(num2str(Y(2:N+1)'));

%}}
    disp('%---------------- end ----------------%');
    disp('');
end
%% END
```

Figure 5. Training and test of the model.

After the training, the knowledge, experience and methods of the sample are transformed into the properties of itself, and they are memorized down. In other words, the model will have the ability to predict the safety situation of building demolition work. When a new data is input, a reasonable judgment result will be output.

3.3 *Model prediction*

The forecast results are shown in Figure 6 to Figure 8. The curve ordinate is the membership degree value of blasting demolition project safety conditions for safety grade, general safety grade, and hazard grade. The curve abscissa is 12 domestic blasting demolition projects.

From the contrasting curves, it can be observed that the predictive values fluctuate around the true values. However, the error is within the scope of the controllable error limit. The error of project 11 is the biggest. The reason for this may be that there is noise in the sample data. At the same time, in the comprehensive evaluation, the evaluation score section of this level is smaller. This resulted in bigger floating data, so that the predictive data are very unstable compared with the real data.

It can be found from the contrastingcurves that the test values and true values have good stability, basically conforming to the trend of the true

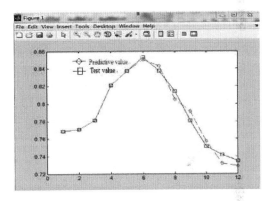

Figure 6. The contrast curve of the test value and predictive value of 12 domestic blasting demolition projects with a safety grade membership degree.

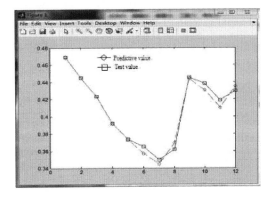

Figure 7. The contrast curve of the test value and predictive value of 12 domestic blasting demolition projects with a general safety grade membership degree.

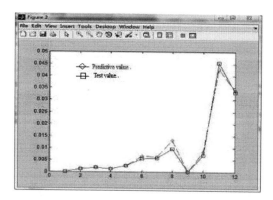

Figure 8. The contrast curve of the test value and predictive value of 12 domestic blasting demolition projects with a hazard grade membership degree.

values. During the comprehensive evaluation, because the evaluation score range is rather large, the numerical change is moderate. Thus, the sample data is provided for decent stability instead of a lot of noise.

Contrasting the two curves, it can be observed that the test values and true values fit well. Since the safety protections of 12 projects are very effective. There is little danger condition with respect to the evaluation index, so the floatability of the data is small. Therefore, the predictive values are stable compared with the test values.

From the above results, it can be concluded that the LS-SVM model is good for fitting the evaluation results of the fuzzy comprehensive evaluation method after training, and the accuracy is better. It is able to reflect the safety situation of 12 blasting demolition projects. Although there is a certain error between the predictive value and the true value, they are within the prescribed error limits. The cause of the error between the predictive value and the true value may be that the training sample size is not big enough. So, the model learns little information and knowledge; moreover, the training sample has a certain noise. On the whole, the LS-SVM model is able to study the experts' experience, knowledge and logic judgment through training, and then memorizes everything. For the new building demolition project, to determine the security situation, only the evaluation indices of the project safety evaluation system are input into the model, and thus the safety condition of the project can be predicted. The model realized the

dynamic evaluation of engineering. It is very convenient; furthermore, the manpower and financial resources are saved.

4 CONCLUSIONS

Based on the basic principle of the Least Squares Support Vector Machine (LS-SVM), the building demolition safety prediction model is established, and the MATLAB programming is used to calculate. The evaluation results of 12 domestic building blasting demolition projects are obtained by using the AHP-FCE comprehensive evaluation method. They are considered as learning samples and test samples to train the LS-SVM model. It was found that the model is able to approximately predict the safety state of the building demolition security system under the influence of many factors. Its predictive effect is more ideal, and it studies the expert's experience, knowledge and logical thinking. Therefore, it is feasible to apply the LS-SVM model in the safety prediction of building demolition work.

ACKNOWLEDGMENTS

This paper was supported by the Ministry of Housing and Urban-Rural Construction. The authors gratefully acknowledge its support.

REFERENCES

[1] Vapnik V.N., 1995. The Nature of Statistical Learning Theory [M]. NY: Springer-Verlag: 66–70.
[2] Vapnik V.N., Levin E., Le C.Y., 1994. Measuring the VC-dimension of a learning machine [J]. Neural Computation, (6): 851–876.
[3] Muller K.-R, Smola A.J., Ratsch G., et al., 1997. Predicting time series with support vectormachines. In: Proc. of ICANN'97 [C]. Springer Lecture Notes in Computer Science: 999–1005.
[4] Zhang H.B., Zhong P., Zhang C.H., 2003. The Newton-PCGA 1 gorithmvia Automatic Differentiation. OR Transaction, 7(1): 28–38.
[5] Scholkopf B., Sung K.-K., Burges C., et al., 1997. Comparing support vector machines with Gaussian kernels to radial basis function classifiers [C]. IEEE Trans. On Signa Processing, 45(11): 2758–2765.
[6] Burges C.J.C., 1998. Atutorial on support vector machines for pattern recognition [J]. Data Mining and Knowledge Disco-very, 2(2): 57–60.

Architectural, Energy and Information Engineering – Sung & Chen (Eds)
© 2016 Taylor & Francis Group, London, ISBN 978-1-138-02791-6

The adiabatic temperature rise and the early strength of concrete mixed with mineral admixture

X.G. Zhang, R.J. Huang, S.J. Lou, D.C. Liu & P. Xu
Guangdong Provincial Key Laboratory of Durability for Marine Civil Engineering, College of Civil Engineering, Shenzhen University, Shenzhen, China

ABSTRACT: In the paper, the adiabatic temperature rise of concrete mixed with different amounts of fly ash and slag powder, together with its early strength, are experimentally tested to find out the proportion at which it can effectively contribute to the improvement of the early cracking performance of mass concrete. Without changing the total amount of these two admixtures, the proportions of fly ash and slag powder are adjusted in 6 groups to measure the optimal mixture ratio. It is shown that when the amount of fly ash and slag powder is equal, there is the best selection both to reduce the adiabatic temperature rise and to keep its essential early strength at the same time. The results of the paper can be used to compare and assess the early cracking performance of mass concrete in different mixture ratios.

Keywords: Fly ash; slag powder; adiabatic temperature rise of mass concrete; early strength; proportion of mixture

1   INTRODUCTION

The control of early thermal cracking induced by the hydration heat of cementitious materials is the most important subject in practical engineering of mass concrete structures. Previous studies [1–5] have found that adding mineral admixture materials such as fly ash or slag powder in concrete is an economic and effective way to avoid the corresponding early temperature cracks. This is because mixing with slag powder or fly ash in mass concrete will reduce the hydration heat and also decrease the peak of the early material's temperature [6–8]. At the same time, slag powder can increase the compressive strength of mass concrete, and fly ash can improve its tensile and compressive strength [9,10]. However, the lower amount of fly ash or slag powder will not largely contribute to the control of the hydration heat, while its higher amount will obviously reduce the early tensile strength [11,12]. Therefore, varying contents of fly ash or slag powder have different influences on the adiabatic temperature rise and the early strength of mass concrete. Thus, the appropriate adding ratio and respective proportions of these two admixtures are very important for both reducing the adiabatic temperature rise and keeping its essential early-age strength. Although the early cracking performance of mass concrete mixing with only slag powder or fly ash has been widely studied [13–15], there is

still a lack of research on mixing both fly ash and slag powder. The effect of the different ratios of slag powder and fly ash on the early strength was researched by Kula [16], but the corresponding rules of material temperature rise were not been mentioned. Therefore, the present paper aims to study the adiabatic temperature rise and the early strength influenced by the different ratios of fly ash and slag powder mixed in concrete. Without changing the total amount of these two admixtures, the proportions of fly ash and slag powder are adjusted in 6 groups to measure the optimal mixture ratio and to find out the best proportion. The results of the paper can be used to compare and assess the early cracking performance of mass concrete in different mixture ratios.

2   THE MATERIAL PROPERTIES AND MIX PROPORTION

The physico-chemical properties of cement, fly ash and slag powder adopted in the present paper are presented in Table 1–3. The proportions of the mixture are divided into groups A and B by adjusting the different ratios of fly ash and slag powder, as shown in Table 4. From Table 4, it can be seen 44.8% cement in group A and 45.5% cement in group B are, respectively, replaced by both fly ash and slag powder. The ratio of water to cementitious materials of the two groups is 0.36.

Table 1. Physico-chemical properties of cement.

Specific surface/m²/kg		365
Standard consistency/%		24.2
Setting times/min	Initial	143
	Final	184
Invariability		Qualified
Insoluble residue/%		0.76
SO₃/%		2.30
MgO/%		1.52
Loss-on-ignition/%		2.90
Chloride ion content/%		0.016
Alkali content/%		0.42
Admixture/%		4
Gypsum/%		Natural:3.0
		Desulfurized:3.0
Grinding aid/%	0.04	
Flexural strength/MPa	3d	6.5
	28d	8.8
Compressive strength/MPa	3d	35.5
	28d	58.6

Table 2. The physico-chemical properties of fly ash.

Particle size (0.045 µm square hole sieve residue)/%	SO₃/%	Free CaO/%
15.8	0.73	0.61

Moisture content/%	Loss on ignition/%	Water demand ratio/%	Stability
0.10	2.42	98	Qualified

Table 3. The physico-chemical properties of slag.

Mobility ratio/%	Moisture content/%	Loss on ignition/%	SO₃/%	Activity index/%	
				7d	28d
101	0.1	1.87	0.60	82	120

Table 4. Proportions of the mixture.

Test specimen	Strength grade	Cement	Slag powder	Fly ash	Sand	Gravel	Water	Superplasticizer	Proportion of slag powder	Proportion of fly ash
A1	C40	232	94	94	757	1030	154	9.2	22%	22%
A2	C40	232	112.8	75.2	757	1030	154	9.2	27%	18%
A3	C40	232	75.2	112.8	757	1030	154	9.2	18%	27%
B1	C40	240	100	100	762	990	159	10.1	23%	23%
B2	C40	240	120	80	762	990	159	10.1	27%	18%
B3	C40	240	80	120	762	990	159	10.1	18%	27%

3 ADIABATIC TEMPERATURE RISE TEST

In order to find out the rules of the early temperature rise of concrete with different mix proportions given in Table 4, the FH-ATRD adiabatic temperature rise detector is used in the experimental study. First, the FH-ATRD detector is checked by using 20 L water before the test to ensure the accuracy of the results. It is shown that the fluctuation in temperature is less than 0.02°C at 20°C, 40°C, 60°C and 80°C during 24 h, which proves that the detector is reliable. The experimental results of the adiabatic temperature rise and its corresponding temperature rise rate affected by different amounts of fly ash and slag are illustrated in Figures 1 and 2.

As can be seen from Figure 1, the specimens containing the minimum amount of fly ash and the maximum amount of slag powder (A2 and B2) reach the maximum value of the adiabatic temperature rise. Also as shown in Figure 2, the specimens of A2 and B2 still have the highest rate of

the adiabatic temperature rise. As a result, it indicates that fly ashcontributes to the decreased rates of the hydration heat and temperature rise better than slag powder. Conversely, the specimens of A1 and B1, which incorporated an equal amount of fly ash and slag powder, obtain the lowest value of the adiabatic temperature and adiabatic temperature rise. Therefore, the equivalent incorporation of fly ash and slag powder is optimal for reducing the adiabatic temperature and adiabatic temperature rise rates.

4 SPLITTING TENSILE STRENGTH AND COMPRESSIVE STRENGTH TEST

Standard specimens of 100 mm³ for measuring the splitting tensile strength and compression strength are tested and placed in a standard curing chamber for 3d, 7d and 28d. According to the Chinese code-GBT 50107-2010, the test is conducted by a digital

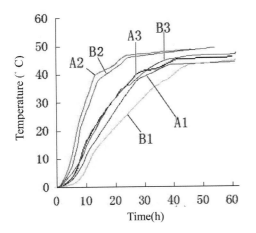

Figure 1. The development curves of the adiabatic temperature rise.

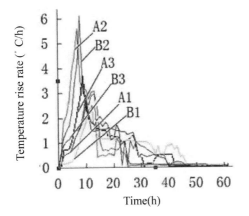

Figure 2. The rate of the adiabatic temperature rise in different mix proportions.

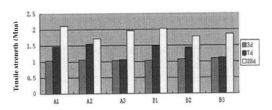

Figure 3. Development process of the splitting tensile strength of concrete.

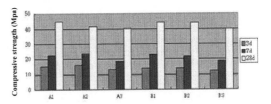

Figure 4. Development process of the axial compressive strength of concrete.

universal testing machine. The results are shown in Figures 3 and 4.

From Figures 3 and 4, it can be observed that specimens A2 and B2 containing the largest amount of slag powder in each group have the largest tensile and compressive strength on day 7. However, its strength becomes the least on day 28. It indicates that the effect of slag powder on improving strength is obvious on day 7, and then the effect decreases significantly.

Specimens A3 and B3 containing the largest amount of fly ash have the lowest strength on day 7. It is indicated that the effect of fly ash on improving strength is not obvious on early age. Specimens A1 and B1 incorporating the same amount of fly ash and slag powder have the largest tensile and compressive strength, which shows that the same

amount of fly ash and slag powder results in the best effect on improving the strength of concrete.

5 CONCLUSION

By studying the adiabatic temperature rise and the early strength of concrete mixed with different amounts of fly ash and slag powder in this paper, the optimal proportion at which it can effectively contribute to the improvement of the early cracking performance of mass concrete has been found. From the results of the present study, the following conclusions can be drawn:

1. For concrete hydration heat control action, fly ash is better than slag powder. The effect is more pronounced within 7d. However, the effect decreases significantly afterwards. Therefore, admixing 20% to 25% of fly ash in concrete can effectively reduce the hydration heat of mass concrete, in order to achieve the purpose of controlling thermal stress in mass concrete.
2. When the incorporation of fly ash in concrete equals that of slag, the adiabatic temperature rise of concrete is minimum and the values of tensile and compressive strength are maximum. Thus, the fly ash:slag ratio of 1:1 is the best mixed ratio.
3. Slag powder has more early activity than fly ash. A larger incorporation of slag powder leads to a higher strength of concrete at an early age. So, by the incorporation of 20% to 25% of the slag can compensate for the loss of strength caused by fly ash replacement in cement.

ACKNOWLEDGMENTS

This research was supported by the Technology Research and Development Program (Basic research project) of Shenzhen, China (JCYJ20130329143859418). The authors gratefully acknowledge the support from the institution.

REFERENCES

[1] Spring en schmid R., Prevention of thermal cracking in concrete at early ages [M]. London: E & FN Spon. 1998: 1–10.

[2] Holt, E., Leivo, M., Cracking risks associated with early age shrinkage. Cement and Concrete Research, 2004, 26: 521–530.

[3] Yuan, Y., Wan, Z.L., Prediction of cracking within early-age concrete due to thermal, drying and creep behavior. Cement and Concrete Research, 2002, 32: 1053–1059.

[4] Isamu Yoshitake, Howe Wong, Takeo Ishid6a, Ayman Y. Nassif. Thermal stress of high volume fly-ash (HVFA) concrete made with limestone aggregate. Construction and Building Materials, 2014, 71: 216–225.

[5] Jeon, D., Jun, Y.B., Jeong, Y., Oh, J., Microstructural and strength improvements through the use of Na_2CO_3 in a cementless $Ca(OH)_2$-activated class F fly ash system. Cement and Concrete Research, 2015, 67: 215–225.

[6] Schutter, G., Hydration and temperature development of concrete made with blast-furnace slag cement. Cement and Concrete Research, 1999, 29: 143–149.

[7] Haque1, M., Kayali. O., Properties of high-strength concrete using a fine fly ash. Cement and Concrete Research, 1998, 28: 1445–1452,

[8] Mo, L.W., Deng. M., Thermal behavior of cement matrix with high-volume mineral admixtures at early hydration age. Cement and Concrete Research, 2006, 36: 1992–1998.

[9] Li, X.Y., Chen, W.Y., Chen, W.M., Multi-powder dam concrete with high performance and low adiabatic temperature rise Trans. Nonferrous Met. Soc. China 2009, 1: 727–733.

[10] Obada Kayali, M. Sharfuddin Ahmed. Assessment of high volume replacement fly ash concrete. Construction and Building Materials, 2013, 39: 71–76.

[11] Johanna Tikkanen, Andrzej Cwirzen, Vesa Penttala. Effects of mineral powders on hydration process and hydration products in normal strength concrete. Construction and Building Materials. 2014, 72: 7–14

[12] Shi C.J., and Day, Robert L., Acceleration of the reactivity of fly ash. Cement and Concrete Research, 1995, 25: 15–21.

[13] Wang, X.Y., Lee, H.S., Evaluation of Properties of Concrete Incorporating Fly Ash or Slag Using a Hydration Model. Journal of Materials in Civil Engineering, 2011, 23: 1113–1123.

[14] Hiroshi Uchikawa, Shunsuke Hanehara, Hiroshi Hirao. Influence of microstructure on the physical properties of concrete prepared by substituting mineral powder for part of fine aggregate. Cement and Concrete Research 1996, 26: 101–111.

[15] Ivindra Panea, Will Hansen. Investigation on key properties controlling early-age stress development of blended cement concrete. Cement and Concrete Research 2008, 38: 1325–1335.

[16] Kulaa, A., Olgun, V., Sevinc, Y., Erdogan. An investigation on the use of tincal ore waste, fly ash, and coal bottom ash as Portland cement replacement materials. Cement and Concrete Research, 2002, 32: 227–232.

Architectural, Energy and Information Engineering – Sung & Chen (Eds)
© 2016 Taylor & Francis Group, London, ISBN 978-1-138-02791-6

Prediction of corrosion induced cracking-time relationship of reinforced concrete by considering initial defects

X.G. Zhang, M.H. Li, F. Xing, P. Xu & L.F. Chen
Guangdong Provincial Key Laboratory of Durability for Marine Civil Engineering, College of Civil Engineering, Shenzhen University, Shenzhen, China

ABSTRACT: Corrosion induced cracking-time relationship is very important for the lifetime assessment of corroded Reinforced Concrete (RC) structures. In this paper, on the basis of the fracture mechanics and the complex variable function method, a theoretic model is proposed in order to predict the time when the cover cracking appears. By considering the influence of initial defects of concrete cover, the corrosion induced cracking-time relationship of reinforced concrete is established. The calculations by means of the proposed method are compared with the help of Finite Element Method (FEM). As the study shows that a reasonably good agreement is observed. This implies that the prediction of the proposed model has a good accuracy.

Keywords: initial defects of concrete cover; corrosion-induced cracking time; stress field of concrete cover; corroded RC structures

1 INTRODUCTION

The steel corrosion of RC structures due to chloride ingress followed by its deterioration is a growing and serious problem throughout the world. Corrosion-induced cover cracking is the most important degradation of the performance of RC structure in terms of safety and serviceability [1]. Up to now, the critical point of time to cover cracking is mostly used to predict the service life of RC structures under corrosive environment. Therefore, it is of great importance for accurately predicting the time to cover cracking in evaluating the service life of RC structures. Considerable researches have been undertaken on corrosion-induced cover-cracking time and different analytical and empirical models [2–19] have been suggested to predict it. However, these models often neglect to consider the actual effects of practical initial defects of concrete cover. In order to accurately predict the time to cover cracking, the internal initial defects of concrete cover with different location and size must be taken into account.

Considering the effects of the surface initial defects of concrete cover, this paper attempts to model the corresponding corrosion-induced cracking time of RC structures. By assumption the internal actual initial defects as fine cracks, formula for predicting the stress field around steel reinforcement is proposed. Comparisons of the predicted values with FEM show that the predictions given

by the present model are reasonably agreement with the FEM results and have good precision.

2 MODEL OF CORROSION-INDUCED CRACKING TIME CONSIDERING SURFACE PRACTICAL INITIAL DEFECTS

As illustrated in Fig. 1, concrete with embedded reinforcing steel bars often can be modeled as a thick-walled cylinder [1–18] for studying its corrosion-induced cover cracking. It is considered that the pore band on the interface of concrete and rebar will be filled with the corrosion products firstly. Then further increasing of corrosion products will inevitably cause an internal pressure on the inner surface of the bar.

According to the elastic mechanics analysis in polar coordinates, the radial stress σ_r and circumferential stress σ_θ of a thick-walled cylinder subjected to an internal radial pressure q can expressed as follows [18]:

$$\sigma_r = \frac{\frac{n^2}{r_0^2} - 1}{\frac{n^2}{m^2} - 1} q \quad \sigma_\theta = \frac{\frac{n^2}{r_0^2} + 1}{\frac{n^2}{m^2} - 1} q \tag{1}$$

where r_0 is the distance from the different position of concrete cover to the centre of reinforcing bar,

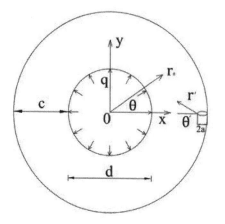

Figure 1. Schematic of the defects on the cover surface of corroded RC utilized for numerical analysis via fracture mechanics..

m is the radius of the inner cylinder (i.e. $m = d/2$), n is the radius of the outer cylinder (i.e. $n = d/2 + c$), and c is the thickness of concrete cover, d is the diameter of reinforcement bar.

By assumption the practical initial defects on the surface of concrete cover as fine cracks with the length of 2a (shown in Fig. 1), the global solutions of the stress field around steel reinforcement can be determined under the corrosion-induced pressure q. The corresponding stress value can be expressed as follows:

$$
\begin{cases}
\sigma_x = \dfrac{\sigma_r + \sigma_\theta}{2} + \dfrac{\sigma_r - \sigma_\theta}{2}\cos 2\theta \\[2mm]
\sigma_y = \dfrac{\sigma_r + \sigma_\theta}{2} - \dfrac{\sigma_r - \sigma_\theta}{2}\cos 2\theta
\end{cases}
\tag{2}
$$

where σ_x is the stress of the x direction and σ_y is the stress of the y direction.

The rust expansion stress field of the surface initial crack of concrete cover can be derived according to the Muskhelishvili and Westergaard complex variable function method:

$$
\sigma_x = \frac{qm^2}{n^2 - m^2} \frac{\sqrt{\frac{d}{2}+c-2a}}{\sqrt{\frac{\frac{d}{2}}{c-2a}}}\left[r'^{-\frac{1}{2}}\cos\left(\frac{\theta}{2}\right) \right.
$$
$$
\left. + \frac{yr'^{-\frac{3}{2}}\sin(\frac{3\theta}{2})}{2} \right] - \frac{q\cos 2\theta m^2 n^2}{r_0^2(n^2-m^2)}
\tag{3}
$$

$$
\sigma_y = \frac{qm^2}{n^2 - m^2} \frac{\sqrt{\frac{d}{2}+c-2a}}{\sqrt{\frac{\frac{d}{2}}{c-2a}}}\left[r'^{-\frac{1}{2}}\cos(\frac{\theta}{2}) - \frac{yr'^{-\frac{3}{2}}\sin(\frac{3\theta}{2})}{2} \right]
$$
$$
+ \frac{q\cos 2\theta m^2 n^2}{r_0^2(n^2 - m^2)}
\tag{4}
$$

The corresponding Stress Intensity Factor (SIF) of concrete cover when containing surface practical initial defects can be also determined by:

$$
K_1 = \sqrt{2\pi}\,\frac{qm^2}{n^2 - m^2}\frac{\sqrt{\frac{d}{2}+c-2a}}{\sqrt{\frac{\frac{d}{2}}{c-2a}}}
\tag{5}
$$

It is generally known that when SIF of the concrete cover exceeds its fracture toughness, the crack may grow quickly and propagate to the reinforcement surface rapidly. Hence, the following fracture criterion can be used if the concrete cover cracking happens:

$$
K_I \geq K_{IC}
\tag{6}
$$

The radial displacement of concrete δ_c at the surface of the reinforcing bars is given as [17, 24]:

$$
\delta_c = \frac{q}{E}\left[\frac{(r_0+c)^2 + r_0^2}{(r_0+c)^2 - r_0^2} + v_c \right] r_0
\tag{7}
$$

in which $r_0 = d/2 + \delta_0$, $E = E_c/(1.0 + \varphi)$, where δ_0 is the thickness of the annular layer of concrete pores at the interface between reinforcing bar and concrete, d is the diameter of reinforcement bar, c is the thickness of concrete cover, E is the effective modulus of elasticity of the concrete cover, E_c and φ are the initial tangent modulus and creep coefficient for the concrete cover, respectively, V_c is Poisson's ratio of concrete.

The corresponding mass of steel (mg/mm) per unit length of the reinforcement being consumed by the corrosion process M_{loss} can be evaluated as follows:

$$
M_{loss} = \frac{\pi\rho_{st}[(r_0+\delta_c)^2 - (d/2)^2]}{n-1}
\tag{8}
$$

where M_{st} is the mass density of reinforcing steel, n is the ratio of the volume of expansive corrosion products to the volume of iron consumed in the corrosion process, which is determined in [11, 16].

Table 1. Comparison between the elastic mechanics model, the model proposed in the paper and the FEM when the direction angle Θ=0,45,90

Θ=0 Θ=45 Θ=90	$r_0=10$ (mm)	$r_0=14$ (mm)	$r_0=18$ (mm)	$r_0=22$ (mm)	$r_0=26$ (mm)
Elastic mechanics	1.77	0.97	0.64	0.47	0.37
	0.13	0.13	0.13	0.13	0.13
	-1.50	-0.70	-0.37	-0.20	-0.11
The present model	2.11	1.36	1.11	1.08	1.30
	0.398	0.372	0.323	0.261	0.205
	-1.29	-0.523	-0.23	-0.09	-0.02
FEM	2.08	1.34	1.09	1.11	1.33
	0.390	0.359	0.316	0.271	0.211
	-1.25	-0.51	-0.22	-0.1	-0.03
Error of elastic mechanics (%)	16.08	28.97	42.74	56.55	71.08
	66.54	64.14	58.68	49.01	35.09
	-16.2	-33.9	-63.5	-122	-349
Error of the present model (%)	1.3	1.7	2.5	-2.8	-2.95
	2.1	3.4	2.2	-3.5	-2.7
	2.9	3.5	4.1	-3.8	-5.5

Finally the time to cover cracking T_{cr} (in years) can be gained from the M_{loss}, which is given by [17]:

$$T_{cr} = \frac{M_{loss}^2}{0.196\pi di_{cor}\alpha} \qquad (9)$$

is the ratio of molecular weight of iron to the molecular weight of corrosion products, r is the radius of the reinforcement bar, i_{cor} is the constant annual mean corrosion rate (A/cm^2).

3 VERIFICATION OF PRESENT MODEL

To assess the accuracy of the proposed model, the FEM (ABAQUS software) is adopted as illustration to simulate the corrosion-induced cracking of concrete cover containing surface initial defects. Comparisons of the corrosive stress field of concrete cover given by the present model with FEM, along with the corresponding values obtained by Elastic mechanics for contrast, are listed in Table 1. In order to present the detailed stress values at

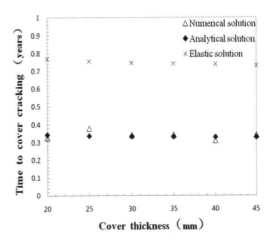

Figure 2. Comparison between the previous solution, the present solution and numerical solution.

different position r_0 along different direction angle Θ, the diameter of the steel bar is assumed to 20 mm, the thickness of the concrete cover is supposed as 25 mm and the length of surface initial crack is deemed with 2 mm considering the conventional case in practical engineering. The other parameters are taken as: $E_c = 27$ GPa, $v_c = 0.18$, and $q = 1.5$ MPa.

From Table 1 it can be seen that the predictions of stress values given by the present model are reasonably agreement with the FEM results and have good precision. One can also find that from Table 1 the practical initial defect of concrete cover indeed have great influences on the corresponding distribution of stress field of concrete cover.

Subsequently, the comparison of time to cover cracking predicted by the proposed model of the paper with the results of numerical solution is shown in Fig. 3. As illustrated in Fig. 3, there is a reasonably agreement between the results of the two methods with lower error of less than 10%. As a result, it is no doubt that the proposed analytical model introduced by this paper can gives a reasonable and accurate prediction for time to cover cracking when containing deferent internal initial defects.

4 CONCLUSIONS

In this paper, a modeling method for forecasting the time to cover cracking initiation from the defects on the cover surface of corroded RC is presented. By assumption the surface initial defects as fine cracks, formulas for predicting time to cover cracking, SIF of concrete, and the stress field around steel reinforcement etc. are proposed based on complex variable function method and fracture mechanics. The comparison calculated by the theory of elasticity and the proposed model show a significant difference prediction in the stress field distributions and the time to cover cracking.

ACKNOWLEDGEMENTS

This research was supported by the following projects: the National Natural Science Foundation of China (51278304), the National Basic Research Program (973 Program) of China (2011-CB013604), the Technology Research and Development Program (Basic research project) of Shenzhen, China (JCYJ20120613174456685). These supports are gratefully acknowledged.

REFERENCES

[1] G. Xu, J. Wei, K.Q. Zhang, X.W. Zhou. A calculation model for corrosion cracking in RC structures [J]. Journal of China University of Geosciences, 2007, 18(1): 85–89.

[2] C.Q. Li, S.T. Yang. Prediction of concrete crack width under combined reinforcement corrosion and applied load. Journal of Engineering Mechanics, ASCE, 2011, 137(11): 722–731.

[3] J. Zhong, P. Gardoni, D. Rosowsky. Stiffness degradation and time to cracking of cover concrete in reinforced concrete structures subject to corrosion. Journal of Engineering Mechanics, ASCE, 2010, 136(2): 209–219.

[4] N. Yüzer, F. Aköz, N. Kabay. Prediction of time to crack initiation in reinforced concrete exposed to chloride. Construction and Building Materials, 2008, 22(6): 1100–1107.

[5] C.Q. Li, R.E. Melchers, J.J. Zheng. Analytical model for corrosion-induced crack width in reinforced concrete structures. ACI Structure Journal, 2006, 103(4): 479–487.

[6] C. Leonid, V. Dimitri. Prediction of corrosion-induced cover cracking in reinforced concrete structures [J]. Construction and Building Materials, 2011, 25: 1854–1869.

[7] C.H. Lu, W.L. Jin, R.G. Liu. Reinforcement corrosion-induced cover cracking and its time prediction for reinforced concrete structures [J]. Corrosion Science, 2011, 53: 1337–1347.

[8] S.C. Li, M.B. Wang, S.C. Li. Model for cover cracking due to corrosion expansion and uniform stresses at infinity [J]. Applied Mathematical Modelling, 2008, 32(7): 1436–1444.

[9] X.G. Zhang, Y.G. Zhao, Z.H. Lu. Dynamic corrosion-induced cracking process of RC considering effect of initial defects [J]. Journal of Asian Architecture and Building Engineering, 2010, 9(2): 439–446.

[10] G. Malumbela, M. Alexander, Moyo P. Model for cover cracking of RC beams due to partial surface steel corrosion. Construction and Building Materials, 2011, 25(2): 987–991.

[11] X.H. Wang, X.L. Liu. Modelling effects of corrosion on cover cracking and bond in reinforced concrete. Magazine of Concrete Research, 2004, 56(4): 191–199.

[12] L. Chernin, D.V. Val, K.Y. Volokh. Analytical modelling of concrete cover cracking caused by corrosion of reinforcement. Material of Structure, 2010, 43(4): 543–556.

[13] S.J. Williamson, L.A. Clark. Pressure required to cause cover cracking of concrete due to reinforcement corrosion. Magazine of Concrete Research, 2000, 52(6): 455–467.

[14] S.J. Pantazopoulou, K.D. Papoulia. Modelling cover-cracking due to reinforcement corrosion in RC structures [J]. Journal of Engineering Mechanics, ASCE, 2001, 127 (4): 342–351.

[15] K. Bhargava, A.K. Ghosh, Y. Mori, S. Ramanujam. Model for cover cracking due to rebar corrosion in RC structures [J]. Engineering Structures, 2006, 28: 1093–1109.

[16] Y. Liu, R.E. Weyers. Modelling the time-to-corrosion cracking in chloride contaminated reinforced concrete structures. ACI Materials Journal, 1998, 95(6), 675–681.

[17] K. Bhargava, A.K. Ghosh, Y. Mori, S. Ramanujam. Modeling of time to corrosion-induced cover cracking in reinforced concrete structures. Cement and Concrete Research, 2005, 35: 2203–2218.

[18] B.S. Jang, B.H. Oh. Effects of non-uniform corrosion on the cracking and service life of reinforced concrete structures [J]. Cement and Concrete Research, 2010, 40(9): 1441–1450.

Architectural, Energy and Information Engineering – Sung & Chen (Eds)
© 2016 Taylor & Francis Group, London, ISBN 978-1-138-02791-6

The early temperature stress and cracking resistance of mixed fly ash and slag powder concrete

X.G. Zhang, D.C. Liu, S.J. Lou, R.J. Huang & P. Xu
Shenzhen Municipal Key Laboratory for Durability of Civil Engineering Structure,
College of Civil Engineering, Shenzhen University, Shenzhen, China

ABSTRACT: This paper aims to investigate the influence of different incorporating proportions of fly ash and slag powder on the early cracking resistance of mass concrete by using the temperature-stress test and slab cracking test. Without changing the total amount of these two admixtures, 6 different mix proportions of fly ash and slag powder are selected to find the influence rules in the present study. The results show that when a higher proportion of slag powder is incorporated, the cracking temperature is lower and the corresponding cracking tensile strength is a larger value in the temperature-stress test, while the cracking number and total cracking area are also less, respectively, in the slab cracking test. As a result, improving the amount of slag powder will be better than incorporating the same amount of fly ash to contribute to the early cracking resistance of restrained concrete.

Keywords: Cracking resistance on early age, temperature-stress test, slab cracking test, fly ash, slag powder

1 INTRODUCTION

The underground super-length concrete slab or wall belongs to the mass concrete structure. The great thermal and shrinkage stress in the construction period caused by the hydration of cementitious materials will result in early random cracks under the restrained conditions [1]. These cracks will obviously accelerate the ingress of corrosive products and finally cause the deterioration of the structure's service life [2–4]. According to the engineering practice, about 80% of early cracks in mass concrete are mainly caused by restrained deformation such as thermal and shrinkage deformation, only 20% of cracks are induced by the external load [5]. As a result, the assessment of early cracking resistance of mass concrete structures has always been one of the most important topics in previous research [6]. A large number of methods to control the random cracks such as using the cementitious material of low hydration heat, reducing the amount of cement, and installing cooling pipes or automatic curing systems have been widely proposed [7,8]. Among these methods, replacing cement with fly ash and (or) slag powder has been the favorite one. It can largely reduce the hydration heat of cementing materials and greatly improve the properties of concrete including the mechanical characteristics, chloride permeability as well as anti-carbonation ability [9–13]. However, to date, the influence of incorporating both fly ash and slag powder on early cracking resistance

of mass concrete, especially its early temperature stress, has been rarely mentioned.

By using the temperature-stress test and the slab cracking test, the effects of different incorporating proportions of mineral admixture on the early cracking resistance of mass concrete are presented and investigated in this paper. Without changing the total amount of these admixtures, 6 different mix proportions of fly ash and slag powder are selected to find the influence rules. The results of the paper can be applied to describe the early cracking performance of mass concrete at different mixture ratios.

2 EXPERIMENT SCHEDULE

2.1 Raw materials and mix proportion

The raw materials such as PII42.5R cement, Grade II fly ash, S95 slag powder and polycarboxylic acid superplasticizers are applied in the present temperature-stress test and slab cracking test. The selected 6 mix proportions with different incorporating amounts of fly ash and slag powder are listed in Table 1.

The mass concrete of actual structures is often confined with different degrees, and this restraint also greatly affects the cracking position of early concrete. The fully restrained state of mass concrete structures in actual engineering can be easily simulated by using the present temperature stress

Table 1. Mix proportion used in the experiment.

Mix proportion	Strength class	Cement	Fly ash	Slag powder	Sand	Gravel	Water	Superplasticizer	Proportion of Fly ash	Proportion of slag powder
A1	C40	232	94	94	757	1030	154	9.2	22.38%	22.38%
A2	C40	232	112.8	75.2	757	1030	154	9.2	26.86%	17.90%
A3	C40	232	75.2	112.8	757	1030	154	9.2	17.90%	26.86%
B1	C40	240	100	100	762	990	159	10.1	22.73%	22.73%
B2	C40	240	120	80	762	990	159	10.1	27.27%	18.18%
B3	C40	240	80	120	762	990	159	10.1	18.18%	27.27%

Figure 1. The specimens of the temperature-stress test.

Figure 2. The specimens of the slab cracking test.

test. Therefore, the early cracking tendency and its cracking resistance with different mix proportions are objectively evaluated under the semi-adiabatic state. In the experiment of temperature stress, the environmental temperature is maintained at 23 ± 2°C and the placing temperature of concrete is controlled at 28 ± 1°C. For the slab cracking test, the statistical data of cracking at 24 ± 0.5h are obtained at an environmental temperature of 20 ± 2°C, a humidity of 60 ± 5% and a wind velocity of 0.6 m/s, according to the Chinese code-GBT500822009.

3 RESULTS AND DISCUSSION

3.1 Temperature-stress test

Considering the influence of restrained degree and temperature history, not only the ability of the crack resistance of concrete can be objectively and precisely depicted, but also the relevant data can be comprehensively recorded by the temperature-stress test method. This is the advantage of the above temperature-stress test method when studying the crack resistance of concrete at early age.

The time-varying temperature and its corresponding stress are shown in Figure 2. During the test process, the temperature history is found

to be divided into 2 stages: temperature rise stage and cooling stage. At the same time, there are also 3 stages of stress history: increasing compressive stress, reducing compressive stress and increasing tensile stress. The results of the experiments are presented in Table 2. The parameters are described below.

The first zero-stress temperature is the temperature at which the compressive stress occurs under a restrained condition, with the development of cement hydration and elastic modulus. It describes the turning point from plastic to viscoelastic of concrete.

Maximum compressive stress is the maximum value of compressive stress throughout the whole experiment. After concrete mixing, shrinkage and hydration results. With a small shrinkage value, the compressive stress of specimens occurs by expansion, which is mainly induced by a rise in temperature.

The second zero-stress temperature is the temperature at which the stress of the specimen changes from the compressive stress to the tensile stress. The expansion of the temperature rise induces the compressed tendency of specimen, while its shrinkage causes the tensile tendency. The second zero-stress temperature is affected by both temperature rise and shrinkage.

Table 2. Results of the temperature-stress test.

Parameters	A1	A2	A3	B1	B2	B3
First zero-stress temperature/°C	27.33	29.38	28.23	28.32	28.45	28.16
Maximum compressive stress/MPa	0.30	1.25	0.55	0.54	0.27	0.34
Second zero-stress temperature/°C	45.32	68.76	57.75	55.93	50.83	50.11
Temperature peak/°C	45.69	68.97	58.49	56.32	50.93	50.33
Cracking stress/MPa	2.15	2.31	2.01	1.98	2.11	1.95
Cracking temperature/°C	40.41	39.63	47.09	38.44	34.08	39.60

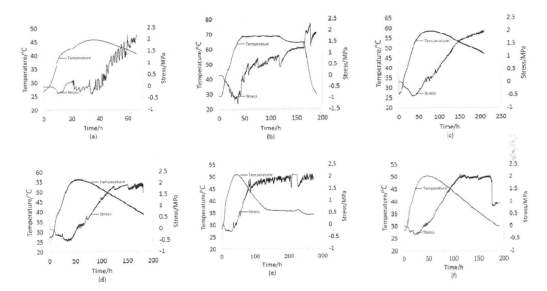

Figure 3. The evolution curves of temperature and stress: (a) A1; (b) A2; (c) A3; (d) B1; (e) B2; and (f) B3.

Temperature peak is the maximum temperature under the semi-adiabatic condition and constant room temperature in the process of hardening.

Cracking stress and cracking temperature are the stress and temperature at which the crack occurs in the specimen. The cracking stress represents the mechanical properties of concrete. Besides, the higher cracking stress means better tensile properties. The mutual influences of factors such as temperature rise of concrete caused by hydration, compressive stress of heating stage, tensile stress of cooling stage, stress-relaxation, elastic modulus, permissible value of tensile strain, tensile strength, linear expansion coefficient, and autogeneous volume deformation are comprehensively denoted by cracking temperature, which can be the evaluation index of cracking sensitivity of concrete [14,15]. The lower cracking temperature represents the lower cracking sensitivity. The parameters of cracking stress and cracking temperature denote that the cracking resistance on the early age of A2 and B2 is the best in each group.

With the increasing amounts of slag powder and the acceleration of the hydration of cementitious materials, the tensile strength is higher under the fully restrained and semi-adiabatic conditions. Being influenced by the heating rise, shrinkage, strength and modulus evolution, the cracking temperature of specimens is lower, and the cracking resistance of concrete on early age is better.

3.2 Slab cracking test

After the pouring of concrete, the shrinkage occurs when the evaporation velocity of the surface water is greater than the flow velocity of the groundwater. At this time, the strength of concrete is very low. Besides, the cracks occur on the surface of concrete and gradually expand when the shrinkage stress is greater than the tensile strength. The cracking number ascends; meantime, the length and width of cracking increase. In addition, the high temperature of the environment, the low humidity and the high rate of air flow may result

Table 3. The results of the slab cracking test.

Mix proportion	A1	A2	A3	B1	B2	B3
Cracking number/strip	18	15	23	16	12	19
Total cracking area/mm²	43.08	19.95	58.5	51.37	27.6	73.6

in the high evaporation velocity of the surface water, obvious shrinkage and severe cracking. The results of the slab cracking test presented in Table 3 indicate that the cracking status of A2 and B2 is better than the others. The results also show that, with the reduction in the incorporating proportion of fly ash and the increasing proportion of slag powder, keeping the total amount of admixture constant, the crack resistance enhanced under the restrained conditions.

4 CONCLUSIONS

The temperature-stress test and slab cracking test have been developed to investigate the early temperature stress and cracking resistance of mixed fly ash and slag powder concrete. Based on the experimental results, it was found that when the total amount of admixture is kept constant the cracking resistance of mixed fly ash and slag powder concrete at early age can be enhanced by increasing the incorporating proportion of slag powder.

ACKNOWLEDGMENTS

This research was supported by the Technology Research and Development Program (Basic research project) of Shenzhen, China (JCYJ20130329143859418). The authors gratefully acknowledge the support from the institution.

REFERENCES

[1] M. Briffaut, F. Benboudjema, J.M. Torrenti, G. Nahas. A thermal active restrained shrinkage ring test to study the early age concrete behaviour of massive structures. Cement and Concrete Research, 2011, 41: 56–63.

[2] Gerard B, Marchand J. Influence of cracking on the diffusion properties of cement-based materials, Part I: influence of continuous cracks on the steady-state regime, Cement and Concrete Research, 2000, 30(1): 37–43.

[3] Y. Yuan, Z.L. Wan, Prediction of cracking within early-age concrete due to thermal, drying and creep behavior. Cement and Concrete Reach, 2002, 32: 1053–1059.

[4] Muhammad Nasir Amin, Jeong-Su Kim, Yun Lee, Jin-Keun Kim. Simulation of the thermal stress in mass concrete using a thermal stress measuring device, Cement and Concrete Research, 2009, 39: 154–164.

[5] M. Briffaut, F. Benboudjema, J. Torrenti, G. Nahas. Effects of early-age thermal behavior on damage risks in massive concrete structures. *European Journal of Environmental and Civil Engineering*, 2012, 16(5): 589–605.

[6] R. Springenschmid. Prevention of Thermal Cracking in Concrete at Early ages. E&FN SPON, 1998.

[7] Mats Emborg, Stig Bernander. Assessment of risk of thermal cracking in hardening concrete. Structural Engineering, 1994, 120(10): 289–329.

[8] Ju-Hyung Ha, Youn su Jung, Yun-gu Cho. Thermal crack control in mass concrete structure using an automated curing system. Automation in Construction, 2014, 45: 16–24.

[9] Isamu Yoshitake, Howe Wong, Takeo Ishida, Ayman Y. Nassif. Thermal stress of high volume fly-ash (HVFA) concrete made with limestone aggregate, Construction and Building Materials, 2014, 71: 216–225.

[10] Aveline Darquennes, Bernard Espion, Stéphanie Staquet. How to assess the hydration of slag cement concretes. Construction and Building Materials, 2013, 40: 1012–1020.

[11] Cengiz Duran Atis. Heat evolution of high-volume fly ash concrete, Cement and Concrete Research, 2002, 32: 751–756.

[12] O. Kayali, M. Sharfuddin Ahmed. Assessment of high volume replacement fly ash concrete. Construction and Building Materials, 2013, 39: 71–76.

[13] Linhua Jiang, Zhenqing Liu, Yiqun Ye. Durability of concrete incorporating large volumes of low-quality fly ash. Cement and Concrete Research, 2004, 34: 1467–1469.

[14] Springenschmid R, Gierlinger E, Kiernozycki W. Thermal stress in mass concrete: a new testing method and the influence of different cement, Proceedings of 15th Congress on Large Dams. Lausanne, Switzerland. 1985: 57–72.

[15] Breitenbuecher R. Investigation of thermal cracking with the cracking-frame. Materials and Structures, 1990, 23(3): 172–177.

Architectural, Energy and Information Engineering – Sung & Chen (Eds)
© *2016 Taylor & Francis Group, London, ISBN 978-1-138-02791-6*

Research and application of clay stabilizer used for water injection in the Jidong oil field

Ling Li
Tangshan Jiyou Ruifeng Chemical Co. Ltd., Tangshan, China

ABSTRACT: Most of the oil deposits with low permeability are characterized by strong water sensitivity. During the water injection process, as water absorption of strata gradually decreases and injection pressure rises, water flooding becomes poor. The result of the evaluation test shows that the anti-swelling rate of the new type clay stabilizer is 80% and the core damage rate is 7.1%. In the field test, the new clay stabilizer performs well in preventing clay expansion, and can effectively reduce the injection pressure or increase the water volume.

Keywords: clay stabilizer; anti-swelling rate; water injection; evaluation

1 INTRODUCTION

In the process of developing oil fields with medium or low permeability, water injection is an important measure to keep the reservoir pressure and realize high and stable yields. Protecting the water injection reservoir from damage plays a crucial role in water injection. For reservoirs with relatively strong water sensitivity, using a suitable clay stabilizer can reduce the damage of water sensitivity in the reservoir. This method is of great significance for sustainable and high efficient development of the entire reservoir (Li, Q.Y. 2008). Most of the oil deposits with low permeability are characterized by strong water sensitivity. During the water injection process, as water absorption of strata gradually decreases and injection pressure rises, water flooding becomes poor. In order to solve this problem and improve water flooding extraction, a new type of clay stabilizer has been synthesized for the purpose of water injection. This product has been widely used in the Jidong oil field, and obtained a good result in reducing pressure and increasing water injection.

2 MECHANISMS

All kinds of clay minerals may cause different degrees of expansion with water, and expansion includes crystalline expansion and osmotic expansion. Crystalline expansion is generated by clay and concentrated brine or an aqueous solution containing a lot of bivalent or multivalent cations, as the water molecules and oxygen among the clay particles can form hydrogen bonds. This expansion is less harmful for strata since it is relatively small. Osmotic expansion is generated by water and diluted solution or solution containing a large amount of $Na+$ ions. It is caused by the electric double layer formed on the surface of clay minerals. This kind of expansion usually occurs after crystallization, and it may cause severe damage. This phenomenon generates a repulsive force to separate chips, and usually occurs in montmorillonite (Lu, H.S. et al., 2012; Luo, T. et al., 2010).

A large amount of positively charged atoms (e.g. N, P and S) are present in the molecular chain of a cationic polymer clay stabilizer. The cation with strong electrostatic forces adsorbs on the surface of clay particles through ion exchange, together with the effects of hydrogen bonds and Van der Waals forces, which make large molecules firmly adsorb on the clay and other particles to form the adsorption film. The adsorption film "neutralizes" the negative charges between the crystal layer and the surface of the clay, which can reduce the electrostatic repulsion between the crystal layer and particles. The crystal layer shrinks with hydrous disintegration, and the long-chain polycation macromolecules simultaneously adsorb on multiple particles, thus the clay and particles can be stablized through the inhibition of clay dispersion and particle migration. There are plenty of adsorption centers in a molecular chain. Once adsorbed on clay, it is not easy to remove the polycation from all adsorption sites at the same time. Thus, the adsorption isdelayed (Fan, Z.Z. et al. 2005; Jin, B.J. et al. 2002).

3 SYNTHESIS AND EVALUATION

3.1 Synthesis

In high permeable reservoirs, the clay-stabilizing agent uses a high molecular weight polymer, while a low molecular weight polymer is needed in low permeable reservoirs. The higher molecular weight clay stabilizing dosage is used in medium-low permeable reservoirs.

The agents used are as follows: chloropropene, methylamine, epoxy chloropropane, sodium hydroxide, and initiator.

The synthetic steps involved are as follows: add a certain amount of chloropropene and methylamine in one 250 ml of three-necked flask under a certain temperature. After a suitable amount of catalyst, add a certain amount of initiator and control the reaction temperature below 50°C. After certain reaction time, add epoxy chloropropane and control the reaction temperature below 45°C, and then obtain the intermediate, and then hydrolyze the intermediate in sodium hydroxide solution to get the clay stabilizer.

In the synthetic process, a series of target products are obtained by changing the ratio of the monomer, the resultant temperature, the reaction time at each stage, the stirring speed and the concentration of the initiator. Then, they are evaluated according to the methods stipulated by the SY/T 5971-94 "Performance Evaluation of the Clay Stabilizer Used for Water Injection", and eventually obtain the new type of the clay stabilizer.

3.2 Loss rate test

The loss rate test is conducted for the new type of clay stabilizer and the existing clay stabilizer at different temperatures. The results obtained at 90°C are presented in Table 1.

It can be seen from Table 1 that the loss rate of the new type of clay stabilizer and the existing clay stabilizer at 90°C is 12.9% and 13.1%, respectively. This suggests that the new type of clay stabilizer performs slightly better than the existing one at 90°C.

The tests on the anti-swelling rate and core damage rate on the new type of clay stabilizer and the existing clay stabilizer (Table 2) are conducted. From the testing results, we can see that the anti-swelling rate of the new type of clay stabilizer and the existing clay stabilizer is 80% and 68%, respectively, while their core damage rate is 7.1% and 62.7%, respectively. This indicates that the anti-swelling rate of the new type of clay stabilizer is higher than that of the existing one, while its core damage rate is far lower than that of the existing one.

3.3 Field test

The clay stabilizer solution is used at a concentration of 2%, and a certain amount of clay stabilizer

Table 1. A comparison of the loss rate between the new type of clay stabilizer and the existing clay stabilizer at 90°C.

Name of the sample	Sieve weight (g)	Core mass (g)	Dried mass (g)	Difference (g)	Loss rate (%)
Water	37.644	1.019	38.236	0.4273	41.905
New type	37.984	1.210	38.964	0.2295	12.967
Existing	37.762	1.019	38.647	0.1339	13.140

Table 2. Testing results of the anti-swelling rate and the core damage rate.

Name of the sample	Anti-swelling rate	Core damage
New type	80%	7.1%
Existing	68%	62.7%

Note: The anti-swelling rate and the core damage rate are tested at 90°C.

is injected into an aqueous solution when oil well is converted or water injection well is put into use. During the formation of water absorption, the disposal radius is between 3 and 5 meters.

After putting Well G59-5 of the Jidong oil field into use, we conducted an eight-month tracking, from which we achieved the desired goal (Figure 1).

It can be seen from the above figure that the casing pressure drops when the injection amount maintains under the same condition. This suggests that clay expansion is curbed.

From the water injection curve of the four wells in different blocks (Figures 2–5), we can see that the injection pressure in most of the wells maintains a steady or downward trend for a long time after adding the clay stabilizer in the injected water, and the expansion, dispersion and migration of clay particles have been effectively inhibited. This is sufficient to prove that the clay stabilizer performs positively in inhibiting the clay from expansion, dispersion and migration, maintaining the permeability of the stratum, protecting the seepage channels. Thus, it is able to lay a good foundation for long-term water injection. In Well NP11, the increased injection pressure is due to a sharp increase in the injection rate; when the injection is increased to 400 m³/d, the injection pressure is stable for a long time. This can also indicate the effectiveness of the clay stabilizer.

Up to now, water injection clay stabilizer has been used for over 142 well times, a total of 746 tons of clay stabilizer has been used, the injection rate has been increased in more than 80% of the wells, while the rest also maintained a steady water injection pressure.

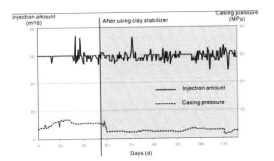

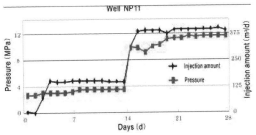

Figure 1. The effect of the new type stabilizer in Well G59-5.

Figure 5. The injection curve of Well NP11.

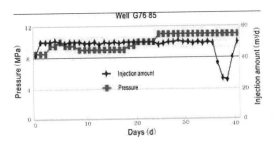

Figure 2. The injection curve of Well G76-85.

4 CONCLUSIONS

An advanced method is used to synthesize the clay stabilizer in this paper. The indoor test shows that the anti-swelling rate of the new clay stabilizer reaches 80%, while the core damage rate is only 7.1%, and the loss rate is only 6.2% and 12.9% at 60°C and 90°C, respectively. All of these performance indicators are better than domestic products in the same industry.

The new clay stabilizer is applied in the field tests of 10 wells situated in different low permeability reservoirs. The experimental results show that the new clay stabilizer performs well in preventing clay expansion, and can effectively reduce the injection pressure or increase the water volume.

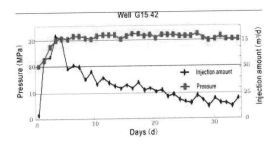

Figure 3. The injection curve of Well G15-42.

ACKNOWLEDGMENTS

This project was supported by Tangshan Jiyou Ruifeng Chemical Co., Ltd and Jidong oil field of Petrochina. We also thank the collaborating research organizations and universities, and our industry collaborators for their support.

REFERENCES

Fan, Z.Z., Wang, B.K., Liu, Q.W., 2005. Application and preparation of clay stabilizer FL-1 used in water injection, *Advances in Fine Petrochemicals*, 6(10): 9–11.

Jin, B.J., Song, G.R., Li, S.G., et al., 2002. Screening clay stabilizers for reservoir anti-swelling treatment in water injection wells, *Oilfield Chemistry*, 19(3): 244–247.

Li, Q.Y., 2008. Experimental selection and evaluation of clay stabilizer for injection water, *Oilfield Chemistry*, 25(2): 151–154.

Lu, H.S., LI, W., Guo, F., 2012. Synthesis and performance evaluation of oligomer quaternary ammonium salt as an anti-swelling chemical, *Speciality Petrochemicals*, 29(5): 11–13.

Luo, T., Ma, X.P., Liu, Y., et al., 2010. Study on synthesis and combination of anti-swelling agent for clay, *Advances in Fine Petrochemicals*, 11(9): 5–7.

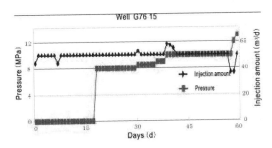

Figure 4. The injection curve of Well G76-15.

Architectural, Energy and Information Engineering – Sung & Chen (Eds)
© 2016 Taylor & Francis Group, London, ISBN 978-1-138-02791-6

Study on foamy oil flow in heavy oil recovery by natural gas huff and puff

J.H. Li & Y. Sun
A.A. Key Laboratory of Exploration Technologies for Oil and Gas Resources, Yangtze University,
Wuhan, Hubei, China
Institute of Petroleum Engineering, Yangtze University, Wuhan, Hubei, China

C. Yang
Tuha Oil Field Company, Petrochina, Shanshan, Xinjjiang, China

ABSTRACT: Natural gas injection has become an important method of improving the heavy oil recovery factor. In in-house physical experiments and field pilots in natural gas injection for heavy oil recovery, foamy oil flows were observed. It is believed that in the mechanism of heavy oil displacement based on natural gas injection, conventional dissolution-induced viscosity reduction is involved, and more importantly, the pseudo-bubble point of the heavy oil formed after natural gas injection enables dispersion-based viscosity reduction. The objective of this paper is to investigate increased recovery mechanism for natural gas huff and puff with foamy fluid through experiments and numerical simulations. The paper presents an advanced study on the oil displacement mechanism of foamy oil flows in the natural gas huff and puff process for heavy oil reservoirs, which was based on in-house physical experiments and numerical simulations.

Keywords: foamy oil; natural gas huff and puff; heavy oil; lab experiment; numerical simulation

1   INTRODUCTION

Two-phase non-Darcy flow of gas oil and dispersion bubbles are very common to be seen in heavy oil reservoir in Canada and Venezuela. Primary production from some heavy-oil reservoirs in a solution gas drive mechanism has unexpectedly high primary recovery with relatively low pressure decline rate and high oil production rates. The process of solution gas drive involves nucleation of gas bubbles as the pressure in the reservoir falls below the bubble point. Gas bubbles tend to be dispersed in heavy crude oil for a relatively long period of time. In order to understand the mechanisms involving in the process, a number of theoretical, experimental, and field studies have been conducted since the end of last century. The manmade foamy oil flow by utilizing natural gas huff and puff has been also investigated in the understanding of mechanisms.

2   EXPERIMENTAL STUDY ON NATURAL GAS HUFF AND PUFF

2.1   *Lab PVT test*

The purpose of laboratory PVT test is to reveal the physical property of heavy oil which saturated by natural gas and to get PVT parameters such as viscosity, volume factor and GOR. This paper also includes several unconventional PVT tests in order to reveal the un-conventional behavior of heavy oil in super-saturation state. The main difference of between conventional (or traditional) PVT and unconventional PVT is that the unconventional PVT is carried out without unintermitted agitation, avoiding the quick gas liberation.

2.2   *Unconventional PVT of quasi saturation pressure test*

The unconventional PVT experiments are carried out to reveal the phenomena and properties of man-made foamy oil by high pressure natural gas injection. These tests might be able to represent the gas trapping ability of certain heavy crude oils in a relative simple way.

2.2.1   *Experiment procedure*
Under the temperature of 100°C and the pressure of 30 MPa, natural gas is injected into PVT container. After 20 hours solution, excess natural gas is released, then make the pressure rise to 50 MPa maintains for 10 hours. Subsequently, depletion begin and make each step of waiting times of tank volumes enlarged respectively in 10 minutes and 60 minutes, without agitation.

In the PVT tube or in porous media, natural gas is in unsaturated state in crude oil.

In unconventional PVT analysis is far less than that of bubble point.

We found that true core model or slim tube model has some disadvantages for huff and puff simulation: the model is too small to simulate the elastic energy of whole formation, and the procedure of gas huff and puff is limited to only 3~5 cycles. For the most part of oil is produced in the first two cycles and after the third cycle the pressure of model is decreased too low to maintain depletion production. At the same time, the small size model do not completely embody the contribution of foamy oil. Consequently, it leads to rather differ from field behavior and fails to be aware of gas intake location and study regular patterns about energy transmission and transfer during gas huff and puff. In order to solve all the problems, we redesigned a pseudo one dimensional model and experiment apparatus of series core tubes with different inner diameters.

3 NUMERICAL SIMULATION STUDY ON NATURAL GAS HUFF AND PUFF

The critical parts of the foamy oil model are bubble nucleation, bubble growth, and coalescence. A model similar to the non-equilibrium foamy oil models and developed by the commercial reservoir simulator STARS was used. The model accounts for the kinetics of physical transition occurring in the process of gas-oil dispersion.

3.1 Simulation and experimental matching studies

3.1.1 Simulation of experimental tests
The fluid and core properties in Table 1 were used in the model. The other PVT data were obtained by using the Standing correlations. The core properties, such as permeability and porosity, seem not to be changed during the history matching. In the history matching, the pressure at the outlet end of the sand-pack was collected as the Bottom-Hole-Pressure (BHP) in the process of the simulation at different time point of the experiment.

3.1.2 History matching results
The history matching were performed to obtain a reasonable match of the observed cumulative oil and gas productions, as well as the observed oil and gas rates. The history data was collected through five cycles' simulation with pressure depletion of 4.3 MPa/cycle in natural gas huff n puff process, from the lab.

Despite the limited experiment data, it was done to match the injection pressure, shut-in pressure,

Table 1. Properties of fluid and core.

Parameters	Value
Initial reservoir pressure, MPa	27
Permeability, $\times 10^{-3}$ μm^2	
Horizontal (Kh)	125
Vertical (Kv)	125
Porosity, %	25
Gas oil ratio, m^3/m^3	0
Reservoir temperature, °C	80
Initial oil saturation, %	79
Formation compressibility factor, 1/MPa	1.0E-5

Table 2. Injection and shut-in pressure.

Cycle Number	Injection pressure MPa	Shut-in pressure MPa
1	31.5	23.5
2	30.8	20
3	29.8	16
4	26.7	14
5	23	6

and cumulative oil at every cycle. The matching results were shown in Table 2.

3.2 Effects of reservoir sensitivity parameters on foamy oil flow

3.2.1 Simulation of grid size
By comparing the two sets of grid size of 5 and 50 vertically in 2D profile model, It can be seen that the produced cumulative oil was increased and the cumulative gas was decreased with an increase of the grid size, as shown in Figure 1. The gas formed in S-bubble or dissolved in oil phase would improve the mobility of oil.

3.2.2 Heterogeneity
Two types of heterogeneity, 0.2 and 0.8 vertically in 2D profile model were compared with 4 diffusion coefficients, including 0.00005, 0.0005, 0.005, 0.05 m^2/d, used for porous media. The effect of heterogeneity on huff and puff recovery was nearly able to be neglected at low gas diffusion, as shown in Figure 2. However, the heterogeneity effect on recovery was obvious at high gas diffusion. When the gas diffusion reached a certain degree, cumulative produced gas would be increased and cumulative produced oil would be decreased in low heterogeneity because of the gas break.

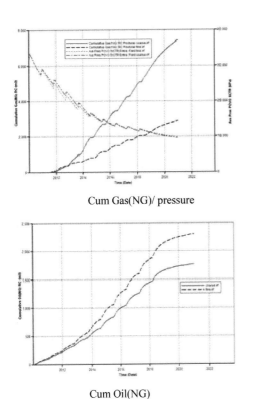

Cum Gas(NG)/ pressure

Cum Oil(NG)

Figure 1. Effect of Grid Size to Cumulative Oil and Gas Produced.

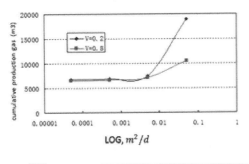

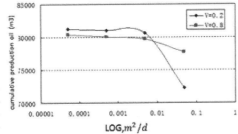

Figure 2. Effect of Heterogeneity on Cumulative Production in Different Diffusion Coefficient.

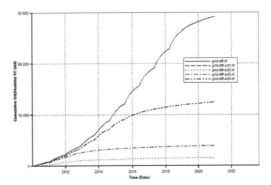

Figure 3. Effect of Dissolution Rate Constant to the S-bubble in Cumulative Oil.

3.2.3 Dissolution rate constant

The original dissolution rate constant was obtained by matching experimental results. Four different dissolution rate constant of 0.1, 0.2, 0.3, and 0.5 were further given by multiplying the original value. It was also found that the lower the dissolution rate was, the higher the distribution of component NG in gas phase was. The rate constant reflects the S-bubble growth rate to form the foamy oil in oil phase. Figure 3 shows that the foamy oil in cumulative produced oil was decreased by a half when the dissolution rate was $0.5*K_0$.

3.2.4 Injection gas molar density

The oil phase volume was decreased and the produced gas was increase when the molar density of injection gas was increased. With increasing the produced cycles, the molar fraction of the produced NG in gas phase was increased. The molar fraction of the produced NG was proportional to the injection gas molar density. The NG molar fraction in oil phase went down at beginning and then rose. The reservoir pressure at the inflection point is pseudo-bubble pressure. At this point, there were a large number of NG in small bubbles in oil phase, causing the molar fraction of the produced NG turning to rise.

4 CONCLUSIONS

The objectives of this work were to identify the parameters in the foamy oil model and clarify their effects on foamy oil flow in porous media by matching the physical tests.

The following conclusions can be drawn:

1. The pressure of pseudo bubble point at unconventional PVT analysis was much less than

that of initial bubble point. The pseudo bubble points of Block L oil reservoir were 9.8 MPa and 13.3 MPa at different dissolvent gas, whereas the initial bubble point was 30 MPa.

2. A numerical simulation model was developed for the physical behavior of the experimental results, in which five components, including Heavy oil, Natural Gas, Small Bubble, Big Bubble, Free Gas, were defined by using reservoir simulator STARS.

3. The sensitivity study was performed based on a vertical heterogeneous model. The results showed that the parameters of diffusion, dispersion, and non-equilibrium effects dominated the recovery process in large grid scale. Low oil recovery efficiency was observed in low heterogeneity formation, while the diffusion effects were appeared in high heterogeneity formation.

4. Displacement instability effects occurred due to different reaction rate and molar density between the injected gas and reservoir oil. The lower the dissolution rate was, the higher the distribution of component NG in gas phase was. The high molar density of injection gas reduced oil phase volume and increased high gas production.

ACKNOWLEDGEMENTS

The authors acknowledge the Innovation Fund of China National Petroleum Corporation for research funding. The valuable contribution of Ms. Zhao jian, in Lab tests is also acknowledged.

REFERENCES

[1] Metwally, M., Solanki, S.C., Heavy oil reservoir mechanisms, Lindbergh and Frog Lake Fields, Alberta: Part I. Field observations and reservoir simulation, 46th Ann. Meeting of the Petroleum Society of the CIM, Banff, Alberta, Canada, 1995.

[2] Ceng Satik, Carlon Roberttson, B. Kalpaki and Deeper G. "A Study of Heavy Oil Solution Drive for Hamaca Field: Depletion Studies and Interpretation" SPE 86967, 2004.

[3] Kumar Rahul, Mahadevan Jagan. "Well performance relationships in heavy foamy oil reservoirs", SPE117447, 2008.

[4] Sheng J.J., Maini, B.B., Hayes, R.E. Critical review of foamy oil flow. Transport in Porous Media 35, 157–187, 1999.

[5] Wang Bojun, Wu Yongbin, Jiang Youwei etc. "Physical simulation experiments on PVT properties of foamy oil". ACTA PETROLEI SINICA, 2012.1.

Architectural, Energy and Information Engineering – Sung & Chen (Eds)
© 2016 Taylor & Francis Group, London, ISBN 978-1-138-02791-6

Research on the solar cell arrangement and its effect on the photovoltaic protected agriculture systems in the Hainan tropical region

Zhi Wu Ge
School of Physics and Electronics Engineering, Hainan Normal University, Hainan, China

Xue Fei Jiang
School of Gardening, Hainan University, Hainan, China

ABSTRACT: To compare the different effects between the solar cell's square format and straight-line arrangement above farmland, we use a solar energy light intensity meter and an infrared thermal imager to detect the two shadow images. The detection results showed that in the square format arrangement, the difference between the maximum and minimum light intensity value is less than 80 w/m^2 and a temperature is less than 3°C, while for the straight-line arrangement, the corresponding value is 205 w/m^2 and 6°C. This suggests that the square format arrangement is better than the other arrangement.

Keywords: Solar cell; arrangement; light intensity; temperature

1 INTRODUCTION

In most places in China, due to rapid industrial development and the rapid advancement of urbanization, energy shortage and environmental pollution have become an increasingly serious problem. Therefore, advantages such as cleanness, security, convenience and efficiency, and solar photovoltaic power generation have been widely concerned and have become the focus for the development of emerging industries in countries worldwide. As the only tropical region in China, Hainan Province has a huge potential of developing clean solar photovoltaic power. However, the energy density distribution of sunshine is small, and solar power needs a large area. Nowadays, most of the solar plates used for solar power are installed on the exclusive field, and they occupy a large amount of land resource, which makes the cost higher and its promotion of using is extremely restricted. This problem seems to be more prominent on this land-scare international tourism island, and it has become the bottleneck problem that restricts the development of solar power in Hainan Island. In Hainan Island, summer is long, the glare of sunlight is intense, and its climate is not suitable for crop growth. Arranging solar cells above farmland to shade the land and to generate electricity at same time are suitable measures to solve the above-mentioned problems. However, the solar cell's arrangements are not the most optimal in the existing photovoltaic green-

houses. So, in this paper, we detect the optimal solar cell's arrangements, and conduct experimental research on its effect.

2 ARRANGEMENT OF SOLAR CELLS

Based on the analysis of almost all theexisting photovoltaic greenhouses, we discover that the solar cell's arrangements are not the most optimal, and can cause a non-uniform distribution of light intensity as well as temperature intensity. The typical arrangements are shown in Figures 1 and 2.

Based on the thermal and optical theory, we put forward a suitable set-up mode of the solar cell for the Hainan tropical photovoltaic power generation and crop cultivation composite agriculture system, as shown in Figure 3.

In order to simulate the effect of the solar cell's different arrangements above farmland, we arrange the cardboard as the required arrangement. In the following experiment, we choose two typical arrangements, and the contrast experiment was carried out to show the corresponding different effects of different arrangements. The cardboard are shown in Figure 4.

The solar cell's arrangement experiment was carried out, as shown in Figure 5. We can adjust the height of the cardboard to the ground. We can obtain the shadow image as shown in Figure 6 to Figure 9.

Figure 1. Solar cell's row arrangement.

Figure 2. Solar cell's interval arrangement.

Figure 3. Solar cell's optimal arrangement.

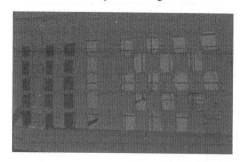

Figure 4. Cardboard simulating the solar cell's arrangement.

Figure 5. Experiment simulating the solar cell's arrangement.

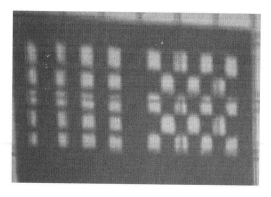

Figure 6. Solar cell's arrangement – experiment I.

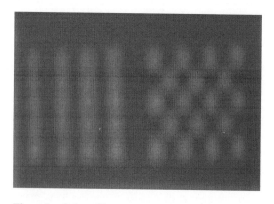

Figure 7. Solar cell's arrangement – experiment II.

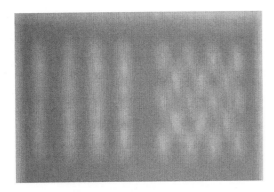

Figure 8. Solar cell's arrangement – experiment III.

Figure 9. Solar cell's arrangement – experiment IV.

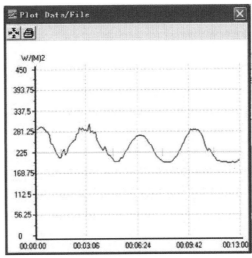

Figure 10. Light intensity of the square format arrangement.

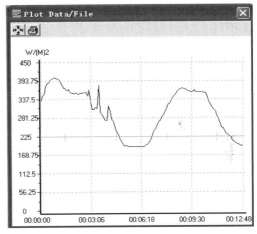

Figure 11. Light intensity of the straight-line arrangement.

As shown in Figure 6, the height of the cardboard to the ground is less than 1.8 m. As shown in Figure 9, the height of the cardboard to the ground can be increased to more than 3 m. We find that the shadow image gets increasingly dim as the height increases, especially in the optimal arrangement.

In order to compare quantitatively the different effects between the solar cell's square format arrangement and straight-line arrangement, we use the solar energy light intensity meter TES-1333R made in Taiwan to have a real-time detection of the two shadow images shown in Figures 5 and 7. The detection results in Figures 10 and 11 clearly show the differences quantitatively: the peak value of the low light intensity is about 200 w/m², and the peak value of light intensity is about 280 w/m². However, in Figure 11, the value of light intensity is about 180 w/m², which is low, and the peak value is 385 w/m². The differences between the trough value and the peak value is

205 w/m², while in the square format arrangement, the differences between the trough value and the peak value is only 80 w/m². As the height of the cardboard to the ground increases, the differences between the trough value and the peak value will be more obvious in the straight-line arrangement and more fuzzy in the square format arrangement. In fact, the differences between the trough value and the peak value will be zero in the square format arrangement as the height of the cardboard to the ground increases to more than 3 m.

The differences will be more obvious in experiment III to experiment IV, as shown in Figures 8 and 9. We can see that in Figure 9, the solar cell's square format arrangement, shadow area and lighting area cannot be distinguished. Besides, the shadow area and the lighting area have become fuzzy as a whole.

Figure 12 shows the thermal images of the right part in Figure 7.

Figure 13 shows the line temperature analysis graph of Figure 12, while Figure 14 shows the histogram of Figure 12.

Figure 15 shows the thermal images of the left part in Figure 7.

Figure 16 shows the line temperature analysis graph of Figure 15, while Figure 17 shows the histogram of Figure 15.

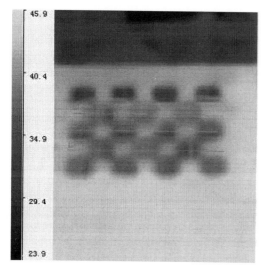

Figure 12.   Thermal images of the right part in Figure 7.

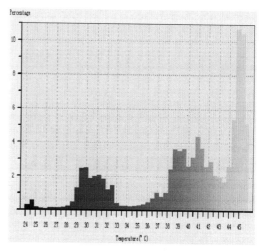

Figure 14.   Histogram of Figure 12.

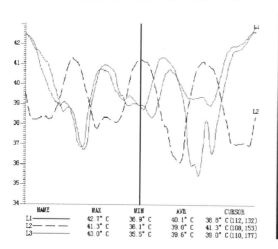

NAME	MAX	MIN	AVE	CURSOR
L1	42.7° C	36.9° C	40.1° C	38.8° C (112, 132)
L2	41.3° C	36.1° C	39.0° C	41.3° C (108, 153)
L3	43.0° C	35.5° C	39.6° C	39.0° C (110, 177)

Figure 13.   Line temperature analysis graph of Figure 12.

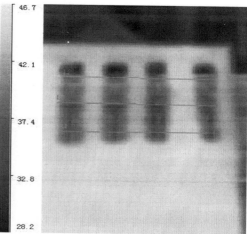

Figure 15.   Thermal images of the left part in Figure 7.

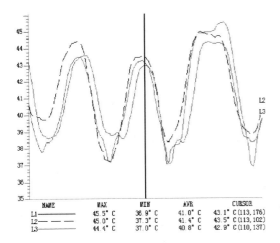

NAME	MAX	MIN	AVR	CURSOR
L1 ——————	45.5° C	36.9° C	41.0° C	43.1° C (113, 176)
L2 — — —	45.0° C	37.3° C	41.4° C	43.5° C (113, 102)
L3 ————	44.4° C	37.0° C	40.8° C	42.9° C (110, 137)

Figure 16.   Line temperature analysis graph of Figure 15.

3   CONCLUSIONS

The square format arrangement is better than the straight-line arrangement for the uniform distribution of light as well as temperature intensity.

REFERENCES

[1] Masayuki Kadowaki, Akira Yano, Fumito Ishizu, Toshihiko Tanaka, Shuji Noda. 2012, Effects of greenhouse photovoltaic array shading on Welsh onion growth, 3rd ed, Amsterdam, Biosystems Engineering. vol. 111: 290–297.

[2] GeZhiwu, China. Patent 201420489433.3.

[3] GeZhiwu, China. Patent 201410430562.X.

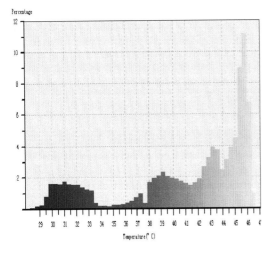

Figure 17.   Histogram of Figure 15.

Architectural, Energy and Information Engineering – Sung & Chen (Eds)
© 2016 Taylor & Francis Group, London, ISBN 978-1-138-02791-6

Effect of MICP on the properties of the cement paste

Xiao Lu Yuan, Yan Zhou Peng, Dong Mei Liu & Qiao Sen Zhu
College of Civil Engineering and Architecture, Three Gorges University, Yichang, China

ABSTRACT: This paper investigates the influence of MICP on the properties of the sulphoaluminate cement paste. Results indicate that the addition of the bacterial culture increases the compressive strength of the cement paste, decreases its water absorption and promotes the formation of carbonate precipitation of calcite and magnesite in the cement paste, leading to the lower porosity, more micro-pores in the microbial cement paste.

Keywords: microorganism; mineralization; pore structure; cement past

1 INTRODUCTION

Concrete of favorable performance and low cost is a versatile and most popular construction material. However, due to flaws within it such as pores and cracks, concrete performances are so decreased that the service security of civil engineering is threatened. Therefore it is necessary to find effective ways to improve the structure and the overall behavior of concrete.

Microbially Induced Carbonate Precipitation (MICP) is resulted from the metabolic activities of some specific microorganisms, forming the precipitation of carbonate. Among the natural pathways to form MICP, the degradation of urea is the easiest to control and the most effective to produce carbonate precipitation [1–6].

Owing to its high early strength, low alkalinity, high frost resistance, high impermeability and high corrosion resistance, sulphoaluminate cement is often used in emergency salvaging, precast elements, GRC, cryogenic engineering, et al. In this paper, MICP was applied to sulphoaluminate cement paste. Effect of MICP on the properties of sulphoaluminate cement paste was discussed with the strength and water absorption experiment and the microstructure analysis (XRD, SEM, and MIP).

2 EXPERIMENTAL

2.1 Microorganism and culture conditions

The microorganism needed in this study should show a high urease activity and a continuous ability to form dense calcium carbonate crystals, as well as have no harmful effect on human health. Thus bacillus sphaericus was used.

Liquid culture medium consisted of 12 g/L peptone, 3.6 g/L beef extract and 6 g/L NaCl. The liquid medium was sterilized by autoclaving. Bacillus sphaericus was grown with the culture at 30°C on a shaker at 200 rpm for 36 h to obtain the stock culture. The concentration of bacterial cells in the suspension was 10^8 cells /L.

To protect bacteria from the cementitous environment, Diatomaceous Earth (DE) [7] was added to the stock culture with the proportion of 1 g diatomaceous earth: 5 ml stock culture. The diatomaceous earth has the particle size distribution ranging from 8~12 μm. The mixture was put on a shaker at 100 rpm for 1 h to make the DE immobilized bacteria.

2.2 Mortar mixture proportion and performance measurement

Sulphoaluminate cement 42.5 (SC) was used to make the cement paste ($40 \times 40 \times 40$ mm). The mix proportion of cement paste is shown in Table 1.

The cement specimens were put in an air-conditioned room with a temperature of 20°C and a relative humidity of more than 90% for 1d. Then the specimens were demolded and immersed into the solutions containing different concentrations of magnesium sulfate. The solution consisted of 30 g/L sodium acetate, 4 g/L monopotassium phosphate, 1 g/L yeast extract, 2 g/L ammonium sulfate, 4 mg/L ferrous sulfate, 0.53 g/L magnesium sulfate, 10 mg/L manganese sulfate, 0.8 mg/L cobalt chloride, 20 g/L urea, 1.4 g/L calcium nitrate.

Table 1. Mix proportions of cement pastes.

Sets	SC /g	DE /g	Water /ml	Stock culture /ml	Yeast extract /g	Urea /g	Calcium nitrate /g
1	400	32	160	0	0.8	8	16
2	400	32	0	160	0.8	8	16

After 28 d of immersion, the specimens were taken out to measure the compressive strength and the water absorption. The experimental samples were taken randomly from the specimens and measured for the microscopic analysis and the pore structure by the method of the Scanning Electron Microscope (SEM), X-Ray Diffraction (XRD) and the Mercury Intrusion Porosimetry (MIP).

3 RESULTS AND DISCUSSION

Fig. 1 shows the compressive strength and the water absorption of cement pastes. It can be seen that the addition of the bacterial culture increases the compressive strength of the cement paste and decreases its water absorption.

Fig. 2 shows the XRD and SEM images of Set 2 cement paste. XRD diagram presents obviously calcite and magnesite diffraction peak, which indicates that the carbonates of calcite and magnesite are formed in the cement paste by the microbial mineralization. In the SEM image, it can be seen that these white prismatic and flocculent minerals fill in the pores of cement paste, making the structure denser.

Pore structure parameters and the pore size distribution of cement pastes are shown in Table 2 and Fig. 3. It can be seen that the addition of the bacterial culture decreases the porosity and the volume intruded of cement paste. Compared with Set 1, Set 2 has larger pore distribution in <20 nm and lower pore distribution in 50~200 nm. These indicate that the cement paste containing the bacterial culture has lower porosity and more micropores than that containing no bacterial culture.

When the cementitious materials containing bacteria are immersed in the culture solution, various nutrients in the culture solution diffuse into the pores of material. Urease produced by microorganisms catalyzes the hydrolysis of urea into carbamate and ammonia. Carbamate spontaneously hydrolyses to form ammonia and carbonic acid, which subsequently produce bicarbonate, ammonium and hydroxide ions, giving rise to a pH increase and the formation of carbonate ions. Carbonate ions subsequently react with the cations which deposit on the cell surface of microorganisms, such as calcium and magnesium ions, lead-

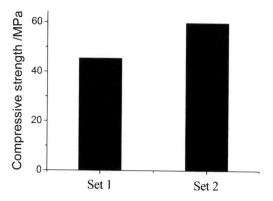

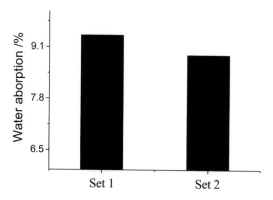

Figure 1.   Compressive strength and water absorption.

Table 2.   Pore structure parameters of cement pastes.

Sets	Porosity (by volume) %	Pore distribution (by volume) %			
		<20 nm	20~50 nm	50~200 nm	>200 nm
1	23.28	28.53	42.57	16.58	12.32
2	22.14	42.19	32.61	11.75	13.45

ing to the precipitation of carbonate. The reaction process is seen in Eqs. (1) and (2)[8].

$$CO(NH_2)_2 + H_2O \rightarrow NH_2COOH + NH_3$$
$$NH_2COOH + H_2O \rightarrow NH_3 + H_2CO_3$$
$$H_2CO_3 \leftrightarrow HCO_3^- + H^+$$
$$2NH_3 + 2H_2O \leftrightarrow 2NH_4^+ + 2OH^-$$
$$HCO_3^- + H^+ + 2NH_4^+ + 2OH^- \leftrightarrow CO_3^-$$
$$+ 2NH_4^+ + 2H_2O \tag{1}$$

$$Cell - Ca^{2+} + CO_3^{2-} \rightarrow Cell - CaCO_3 \downarrow$$
$$Cell - Mg^{2+} + CO_3^{2-} \rightarrow Cell - MgCO_3 \downarrow \tag{2}$$

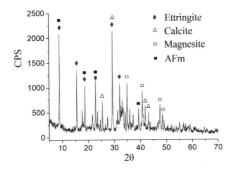

Figure 2. XRD and SEM analysis.

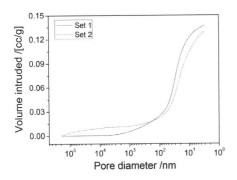

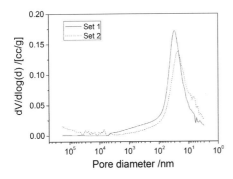

Figure 3. MIP analysis of cement paste.

These carbonates formed in Eq. (2) are prone to grow attaching along the surface of pores and cracks within the cement paste, which makes the microstructure of cementitous materials denser.

4 CONCLUSIONS

This paper investigated the effect of MICP on the properties of the microbial cement paste. The addition of the bacterial culture increases the compressive strength of the cement paste, decreases its water absorption and promotes the formation of carbonate precipitation of calcite and magnesite in the cement paste, leading to the lower porosity, more micro-pores in the microbial cement paste.

ACKNOWLEDGEMENTS

The authors acknowledge support from China Hubei Province Education Department Natural Science Research Item (SN: D20141204).

REFERENCES

[1] Willem De Muynck, Nele De Belie, Willy Verstraete. Microbial carbonate precipitation in construction materials: A review [J]. Ecological Engineering, 2010, 36: 118–136.

[2] Willem De Muynck, Kim Verbeken, et al. Influence of urea and calcium dosage on the effectiveness of bacterially induced carbonate precipitation on limestone [J]. Ecological Engineering, 2010, 36: 99–111.

[3] V. Ramakrishnan, Ramesh K. Panchalan, et al. Improvement of concrete durability by bacterial mineral precipitation [J]. 11th International Conference on Fracture, Turin (italy), March 20–25, 2005.

[4] Willem De Muynck, Dieter Debrouwer, et al. Bacterial carbonate precipitation improves the durability of cementitious materials [J]. Cement and Concrete Research, 2008, 38: 1005–1014.

[5] H.K. Kim, S.J. Park, J.I. Han, H.K. Lee. Microbially mediated calcium carbonate precipitation on normal and lightweight concrete [J]. Construction and Building Materials, 2013, 38: 1073–1082.

[6] Varenyam Achal, Xiangliang Pan, et al. Improved strength and durability of fly ash-amended concrete by microbial calcite precipitation [J]. Ecological Engineering, 2011, 37: 554–559.

[7] J. Wang, N. De Belie, W. Verstraete. Diatomaceous earth as a protective vehicle for bacteria applied for self-healing concrete [J]. J Ind Microbiol Biotechnol, 2012, 39: 567–577.

[8] Kim Van Tittelboom, Nele De Belie, et al. Use of bacteria to repair cracks in concrete [J]. Cement and Concrete Research, 2010, 40: 157–166.

Architectural, Energy and Information Engineering – Sung & Chen (Eds)
© 2016 Taylor & Francis Group, London, ISBN 978-1-138-02791-6

Analysis of the combined capture of CO_2 and SO_2 using aqueous ammonia

Ji Fa Zhang, Fang Qin Li, Huan Liu, Ji Yong Liu & Hai Gang Ji
Shanghai University of Electric Power, Shanghai, China

ABSTRACT: The atmospheric pollutant emissions seriously influence the environmental changes; especially, acid rain and the greenhouse effect have caused great threat to human life. Therefore, measures must be taken to control the emissions of SO_2 and CO_2. Compared with conventional pollutant control method, aqueous ammonia can achieve the desulfurization and decarburization simultaneously. This article mainly describes the reaction mechanism, the research status and the existing problems of the combined capture of CO_2 and SO_2 using aqueous ammonia.

Keywords: SO_2; CO_2; aqueous ammonia

1 INTRODUCTION

In recent years, with the development of industry, a large number of pollutants have been emitted to the atmosphere, which has caused great pollution. This leads to frequent severe weather; especially, acid rain and the greenhouse effect, which are caused by SO_2 and CO_2, respectively, have been the focus of attention of most researchers. So, it is necessary to take effective measures to control and reduce SO_2 and CO_2 emissions.

As one of the important sources of pollutants, coal-fired power plants have committed to the work of reducing the emissions of pollutants. Currently, a vast majority of plants fitted with the flue gas desulfurization device, a carbon capture technology, are also studied. However, the current pollution control method is mainly every kind of pollutants removal alone, such as the Limestone/Lime method to remove SO_2 in flue gas, SCR to remove NOx, and ESP and WESP to remove particulate matter. There is no doubt that this will make the control system of flue gas pollution very complex, and increase the investment of equipment and the operating cost. Research has shown that the combined capture of CO_2 and SO_2 using aqueous ammonia can realize a small reduction in power plant efficiency and an effective control of investment cost[1]. It may be worth to consider the combined removal technology. Related research[2,3] has shown that aqueous ammonia can remove SO_2 and CO_2 in flue gas, and the combined capture of CO_2 and SO_2 using aqueous ammonia has great feasibility.

2 THE REACTION MECHANISM OF THE COMBINED CAPTURE OF CO_2 AND SO_2 USING AQUEOUS AMMONIA

Ammonia absorption of SO_2 and CO_2 is a typical gas-liquid two phase reaction process, which can be explained by the dual mode theory, as shown in Figure 1. Both SO_2 and CO_2 are transferred from the gas phase body to the gas membrane and liquid membrane, and then make a chemical reaction by absorbing to the liquid phase body, but the reaction process is not the same. Danckwerts[4] found that the absorption of SO_2 in ammonia solution is a transient absorption process, and the total mass transfer resistance is focused on the gas film, which means that it is controlled by the gas film. In contrast, the absorption of CO_2 is a rapid reaction, and the total mass transfer resistance is focused on the liquid film, which means that it is mainly controlled by the liquid film.

The reaction equation for CO_2 absorption by aqueous ammonia is as follows:

$$CO_2 + 2NH_3 \rightarrow NH_2COONH_4 \qquad (1)$$

$$NH_2COONH_4 + H_2O \Leftrightarrow NH_3 + NH_4HCO_3 \qquad (2)$$

$$NH_3 + NH_4HCO_3 \Leftrightarrow (NH_4)_2CO_3 \qquad (3)$$

$$CO_2 + H_2O + (NH_4)_2CO_3 \Leftrightarrow 2NH_4HCO_3 \qquad (4)$$

The absorption process of CO_2 is considered as a rapid reaction, which can be explained by the

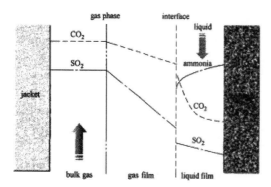

Figure 1. Two-film theory model of SO_2 and CO_2 absorption.

intermediate state ion mechanism that was proposed by Caplow and promoted by Danckwert[5]:

$$CO_2 + NH_3 \Leftrightarrow NH_3^+COO^- \tag{5}$$

$$NH_3^+COO^- + B \rightarrow NH_2COO^- + BH^+ \tag{6}$$

The reaction equation for SO_2 absorption by aqueous ammonia is as follows:

$$SO_2 + NH_3 + H_2O \Leftrightarrow NH_4HSO_3 \tag{7}$$

$$2NH_3 + H_2O + SO_2 \Leftrightarrow (NH_4)_2SO_3 \tag{8}$$

$$(NH_4)_2SO_3 + H_2O + SO_2 \Leftrightarrow 2NH_4HSO_3 \tag{9}$$

$$NH_4HSO_3 + NH_3 \Leftrightarrow (NH_4)_2SO_3 \tag{10}$$

The ammonium sulfite produced by the above reaction is a good absorbent for SO_2. In the application, the absorption process of SO_2 actually chooses the mixture of NH_4HSO_3-$(NH_4)_2SO_3$ as the absorbent for SO_2 removal.

In addition, in the process of CO_2 and SO_2 absorption simultaneously, the absorption of SO_2 will be followed by the desorption of CO_3^{2-} and HCO_3^-[6]. So, the following reaction can be considered:

$$NH_4HCO_3 + SO_2 \Leftrightarrow NH_4HSO_3 + CO_2 \tag{11}$$

$$(NH_4)_2CO_3 + SO_2 \Leftrightarrow (NH_4)_2SO_3 + CO_2 \tag{12}$$

The reaction mechanism of aqueous ammonia reacting with SO_2 and CO_2 indicates the possibility of the combined capture of CO_2 and SO_2 using aqueous ammonia.

3 THE RESEARCH STATUS OF THE COMBINED CAPTURE OF CO_2 AND SO_2 USING AQUEOUS AMMONIA

Although some research on ammonia desulphurization and ammonia decarbonization has been made, as well as some progress and achievements, the study on the combined capture of CO_2 and SO_2 using aqueous ammonia has been relatively scarce. In this context, we list three of the research made as follows.

Qiu Zhongzhu et al.[7] performed an experimental study on the simultaneous absorption of SO_2 and CO_2 in a packed tower and analyzed the influence of the inlet concentration of SO_2 and CO_2, the ratio of liquid and gas and concentration of ammonia on the volume total mass transfer coefficient. The results showed that with the increase in the inlet concentration of SO_2, the volume total mass transfer coefficient of SO_2 and CO_2 decreased gradually; with the increase in the inlet concentration of CO_2, the volume total mass transfer coefficient of SO_2 decreased gradually, but the volume total mass transfer coefficient of CO_2 increased gradually. With the increase in ammonia concentration and the ratio of liquid and gas, the volume total mass transfer coefficient of SO_2 and CO_2 increased gradually, the effect of ammonia concentration and the ratio of liquid and gas on the total volume mass transfer coefficient SO_2 was greater than that on CO_2. Ammonia concentration had the greatest effect on the volume total mass transfer coefficient of SO_2 and CO_2, followed by the ratio of liquid and gas.

Alstom's frozen ammonia method integrated the ammonia solution as the desulfurizer of a wet desulfurization device and decarburization process at a low temperature, as shown in Figure 2[8]. The process used a mixture of ammonium carbonate and ammonium bicarbonate as the absorbent. CO_2 reacted with the poor solution that contained high ammonium carbonate concentration to produce ammonium bicarbonate and crystallize into solid. The crystallization went into the regeneration link to release CO_2 by heating, and the poor solution after regeneration became the new absorbent to absorb CO_2. Alstom set up the demonstration system in the Milwaukee area. This system's removal efficiency of CO_2 was higher than 90% and could also remove a small amount of SO_2, SO_3, particulate matter and other pollutants.

Qi Guojie[9] drew the conclusion that the combined removal of CO_2 and SO_2 in the new process using aqueous ammonia has a certain application value and economy, and can achieve higher CO_2 and SO_2 removal efficiency and has the potential to reduce renewable energy. He also revealed the mechanism and law of combined removal of CO_2

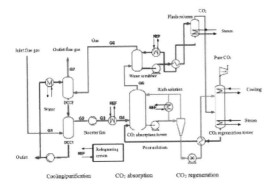

Figure 2. Alstom's frozen ammonia decarbonization process.

and SO_2, with the increased SO_2 concentration, the decreased CO_2 mass transfer coefficient in the SO_2 instantaneous reaction area, and increased mass transfer resistance, and reduced total mass transfer coefficient of CO_2. The packing tower could be divided into the lower CO_2 and SO_2 mixed reaction area and the upper CO_2 reaction zone alone. The absorption heat of SO_2 was higher than that of CO_2, and the join of SO_2 load could significantly reduce the energy consumption of regeneration. When ammonia solution had a certain amount of SO_2 load (less than 0.2 mol SO_2/mol NH_3), it still had good absorption.

4 THE EXISTING PROBLEMS OF THE COMBINED CAPTURE OF CO_2 AND SO_2 USING AQUEOUS AMMONIA

There are mainly two problems: toxicity and explosiveness; ammonia escape and crystallization.

First, ammonia has a certain toxicity and explosiveness, which is the main reason for its limited application in explosives. Ammonia is a colorless gas with strong pungent odor, inflammable, explosive, toxic, corrosive, and easily water-soluble properties. Ammonia belongs to the category of dangerous chemicals, can irritate the skin mucous membrane, and result in corrosion. Ammonia at high concentrations can cause chemical strep throat and pneumonia; when it is inhaled, it can cause reflective breathing stops and cardiac arrest. Its direct contact can cause skin frostbite. However, ammonia is widely used as a fertilizer and a refrigerant, so only by taking the strict enforcement prevention rules and regulations of fire, explosion, poisoning can a safe and orderly production be guaranteed generally.

Next, it is ammonia escape and crystallization, which is usually related. The strong ammonia vola-

tilization characteristics lead to serious ammonia escape problems existing in the process of desulphurization and decarbonization using aqueous ammonia. This will cause the loss of effective ammonia in ammonia solution and is likely to cause pipeline corrosion and plugging, and affect the stability of the system. At present, there are two main kinds of method to control ammonia escape: by mixing with another absorber to reduce the volatilization of ammonia, or by adding washing devices or by optimizing the system to realize the capture of the escaped ammonia and use again.

5 CONCLUSION

In principle, aqueous ammonia has the ability to absorb SO_2 and CO_2 simultaneously, and the combined capture of these pollutants may have more advantages. By taking the strict enforcement prevention rules and regulations of fire, explosion, poisoning, the combined capture of CO_2 and SO_2 using aqueous ammonia has great feasibility. Besides, ammonia escape and crystallization still requires more investigation.

REFERENCES

[1] Ciferno Jared P, DiPietro Philip, Tarka Thomas. An economic scoping study for CO_2 capture using aqueous ammonia [R]. Washington DC: US Department of Energy (DOE). 2005.
[2] Xiu Changxiang, Fu Guoguang. Review on ammonia flue gas desulfurization [J]. *Electric Power Environmental Protection*. 2005 (21) 5: 17–20.
[3] Zeng Qing, Guo Yincheng, Niu Zhenqi et al. Mass transfer performance of CO_2 absorption into aqueous ammonia in a packed column [J]. *CIESC Journal*. 2011(62) S1: 146–150.
[4] Danekwerts P V. Gas-Liquid Reaction [M]. New York: McGraw-Hill 1.1970.
[5] Danckwerts P V. The reaction of CO_2 with ethanolamine [J]. *Chem. Eng. Sci.* 1979, 34: 443–446.
[6] Ebrahimi S, Picioreanu C, Kleerebezem R, Heijnen J J, van Loosdrecht M C M. Rate-based modeling of SO_2 absorption into aqueous $NaHCO_3/Na_2CO_3$ solutions accompanied by the desorption of CO_2 [J]. *Chemical Engineering Science*, 2003, 58: 3589–3600.
[7] Qiu Zhongzhu, Gong Shaolin, Zheng Chaorun et al. Simultaneous Absorption of SO_2 and CO_2 by Aqueous Ammonia in a Packed Column [J]. *Journal of Chinese Society of Power Engineering*, 2011, 31(9): 700–704.
[8] Fred Kozak, Arlyn Petig, Ed Morris, et al. Chilled Ammonia Process for CO_2 Capture [J]. *Energy Procedia*. 2009, 1419–1426.
[9] Qi Guojie. Study on Combined Capture of CO_2 and SO_2 in Aqueous Ammonia [D]. Beijing: Tsinghua University. 2013.

Architectural, Energy and Information Engineering – Sung & Chen (Eds)
© *2016 Taylor & Francis Group, London, ISBN 978-1-138-02791-6*

Integrated topographical and inclinometric procedures for long term monitoring of complex civil structures

C. Balletti, G. Boscato, V. Buttolo, F. Guerra & S. Russo
University Iuav of Venice, Tolentini, Venezia, Italy

ABSTRACT: In this paper, the complementary nature of topographical and continuous inclinometric monitoring procedures for the assessment of the movements of the complex structure being built is analysed. The processes used in the two different monitoring systems and the results obtained are described. These are characterized by a significant homogeneity in terms of the precision of the two different types of monitoring which can be considered interchangeable. It is therefore demonstrated that as the two systems are completely independent from each other and not correlated they can be used to validate each other reciprocally.

Keywords: long term structural monitoring; periodic and continuous monitoring; inclinometric monitoring; topographic survey

1 INTRODUCTION

The monitoring of civil engineering structures is a needed procedure especially in critical scenarios characterized both by historical buildings or by complexity of new structures during the yard phase.

For the heritage structures and historical monuments the assessment of structural health condition, using an appropriate monitoring system is recommended (Fregonese et al. 2013, Hill & Sippel 2002, Fratus et al. 2013, Balletti et al. 2014), in order to help register and understand future developments.

The dynamic monitoring is an important step for model updating techniques, structural health monitoring (Boscato et al. 2012 a,b, Boscato et al. 2014) and damage prognosis. For the structural control and health monitoring long term, in addition to a dynamic, a static system was adopted (Russo 2012a,b, Boscato et al. 2011).

The same procedures have been adopted to identify and control the structural integrity of new buildings (Boscato & Russo 2014).

The integrated procedures of topographical and continuous inclinometric monitoring of the movements of the complex structure being built is proposed (Simeoni & Benciolini 2007).

In particular, during the excavation operation, the monitoring of the retaining walls (diaphragms) was considered particularly critical and a great deal of attention was paid to them. The possible deformation of these elements was therefore monitored using two systems: the first the traditional topographical type, and the second of a geotechnical origin through inclinometric survey. The reliability of both monitoring systems is investigated.

2 GENERAL DESCRIPTION

In Mestre, Venice (Italy), a building site was set up for the construction of a street level car park and an underground multi-storey garage. The area of the site occupies a space between two of the main traffic routes of the city (Fig. 1). In fact, within the site area there are two buildings, each composed of 10 residential floors above ground and a portico with shops underneath of the project area. The presence of these large structures close to the excavation caused the client concerns and therefore the structural monitoring system was installed to check the elevation of the buildings and monitor the structure itself.

In particular the plan for the project was to build a containment basin for the underground levels consisting of a considerably deep structure for which diaphragm walls 80 cm thick were built.

In the three dimensional diagram of the volumes (Fig. 1), one can observe the high elevation from ground level of the two buildings being monitored and their proximity to the site area.

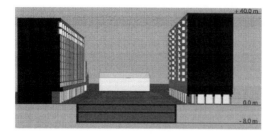

Figure 1. The three dimensional diagram of the volumes.

3 DESCRIPTION OF MONITORING

The monitoring plan devised envisaged a series of measurements being taken, with the aim of validating the planned project model.

The monitoring plan from June 2013 to February 2014 included:

- checking for movements of the facades of the buildings using target detection on the facade;
- checking for vertical movements of the facades using precision levelling of supporting benchmarks on the pillars on the ground floor;
- checking for movement of the diaphragms using topographical measurement;
- inclinometric checks on the diaphragms at the bottom.

3.1 Topographical monitoring

High precision topographical monitoring is based on the study and analysis in time of the variations to the spatial position of sets of points opportunely materialised with the benchmarks on the structure. The structural engineer indicates the quantity and distribution of the benchmarks non so that the deformative model of the structure can be analysed and validated.

The most important and consistent part of topographical monitoring is concentrated directly on the structure, during the construction stage of the structure itself. When building the benchmarks for the topographical detection, supporting elements had to be built with the foundations of the structure; 12 bars, each 25 centimetres long were inserted, so that their possible movement could represent a phenomenon of the structure. 12 prism holders were inserted onto these bars for the detection of the position of the benchmarks materialised by the bars.

The impossibility of materialising the station points with pillars—for the excavation operations, the movement of vehicles on the building site and the deposit of construction materials—led to the

decision not to carry out the comparison between the coordinates of points but to check the distance variation between the points. Indeed, unlike the coordinates, the distance is unchanging compared to the reference system and can be calculated starting with the coordinates of the benchmarks with a slightly different network each time.

The distances calculated are shown in Figure 2.

The changes to the network for each measuring session were controlled as much as possible with respect to the basic scheme planned and simulated. The typical network plot with benchmarks and topographic station was simulated, in order to obtain acceptable precisions as compared with the quantities at play; that is less than ±1 mm, expressed as mean square deviation on coordinates x, y, z of the points. The simulation of the topographical network in fact allows, starting with its geometry (therefore with the arrangement of the benchmarks) and with the instrumental precision, the calculation of the obtainable precisions. The scheme hypothesized for the simulation, which was very similar to that actually carried out is shown in the Figure 2.

As mentioned before, the results of the simulation shows that precisions on the estimation of the coordinates varies in a range between ±0.0002 m and ±0.0007 m.

Using the same simulation the arrangement of the error ellipses was also verified. Analysis of the scheme brings to light the fact that the result is not perfectly homogenous in terms of precision; the error ellipses are different from each other and some are eccentric, (Fig. 2). This is a consequence of the network plot, which however is in turn conditioned by the narrowness of the area accessible for the measurements and by the works on the site. Precisely for this reason the observations have always been compensated so that the minor axis of the ellipse, which illustrates the greatest precision of the measurements, was arranged in the direction of the most significant coordinate in terms

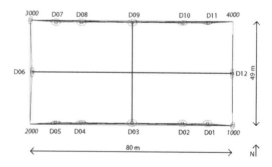

Figure 2. The plan and the distances calculated, with the standard network plot and the simulation.

of the anticipated movement of the diaphragm, therefore the orthogonal direction with respect to main development of the diaphragm wall itself. The possible movement to be monitored, which would be dangerous to the structure, is indeed an out of plumb with respect to the panel plan. The mean square deviation for the estimation of the coordinates is, however, such that it provides significant information for the analysis of possible movements of the structure. The simulation was therefore deemed satisfactory.

3.1.1 Instrumentation and processing of the measurements

The angular and distance observations were acquired using a Leica TCA first order total station and adjusted to the least squares using the software Micro Survey Starnet; using double intersection, (where possible triple) in advance, the three-dimensional coordinates of the prisms inserted onto the structure and therefore the points subjected to monitoring, were calculated. Starting with the coordinates x, y, z of the prisms, the distances between the walls overlooking the underground garage were calculated.

The monitoring was effectively organised with periodical measurement checks, or rather in measurement campaigns in successive periods, always using the same instrument, the Leica TCA 2003 and a similar, but not identical, measurement network. The angular and distance observations were always carried out from at least two stations, thus guaranteeing a redundancy equal to double the observations of the iso determined network. The treatment of the survey data highlighted average precisions that are lower than ±1 mm, which are usually required by monitoring purposes.

3.1.2 The distance variation topographically

The distances between the corresponding prisms were calculated, after the compensation of the azimuth, zenith and distance observations, starting with the three dimensional coordinates of the points, using the formula of the 3d distance:

$$d_{AB} = \sqrt{[(X_B - X_A)^2 + (Y_B - Y_A)^2 + (Z_B - Z_A)^2]} \quad (1)$$

where X_A, X_B, Y_A, Y_B, Z_A, Z_B are the coordinates of the points, determined after the least square adjustment of the observation.

The movements of the points were calculated as a difference between the distances calculated in various measurement sessions, as compared with the distances measured during the first measurement session, which represents step 0, that is the measurement taken as reference.

In order to compare the topographic distances with the inclinometric ones, the mean square deviation of the topographic distances was calculated starting with the mean square deviation of the coordinates x, y, z of the points, by applying the law of the variance propagation.

$$\sigma_f^2 = \left(\frac{\partial f}{\partial x}\right)_m^2 \sigma_x^2 + \left(\frac{\partial f}{\partial y}\right)_m^2 \sigma_y^2 + \cdots + 2\left(\frac{\partial f}{\partial x}\right)_m \left(\frac{\partial f}{\partial y}\right)_m \sigma_{xy}$$

$$(2)$$

For simplicity the coordinates were considered as independent and the equation of the distance between two points was linearised thus:

$$\left(\frac{\partial d_{AB}}{\partial X_A}\right)_0 dX_A + \left(\frac{\partial d_{AB}}{\partial Y_A}\right)_0 dY_A + \left(\frac{\partial d_{AB}}{\partial Z_A}\right)_0 dZ_A$$

$$+ \left(\frac{\partial d_{AB}}{\partial X_B}\right)_0 dX_B + \left(\frac{\partial d_{AB}}{\partial Y_B}\right)_0 dY_B + \left(\frac{\partial d_{AB}}{\partial Z_B}\right)_0 dZ_B$$

$$= \left(d_{AB} - d_{AB}^0\right) + v \quad (3)$$

The development of the derivatives leads to:

$$-\left(\frac{X_A^o - X_B^o}{d_{AB}^o}\right)_0 dX_A - \left(\frac{Y_A^o - Y_B^o}{d_{AB}^o}\right)_0 dY_A$$

$$-\left(\frac{Z_A^o - Z_B^o}{d_{AB}^o}\right)_0 dZ_A + \left(\frac{X_A^o - X_B^o}{d_{AB}^o}\right)_0 dX_B$$

$$+\left(\frac{Y_A^o - Y_B^o}{d_{AB}^o}\right)_0 dY_B + \left(\frac{Z_A^o - Z_B^o}{d_{AB}^o}\right)_0 dZ_B$$

$$= \left(d_{AB} - d_{AB}^o\right) + v \quad (4)$$

Which becomes:

$$\sigma_{d_{AB}}^2 = -\left(\frac{X_A^o - X_B^o}{d_{AB}^o}\right)^2 \sigma_{X_A}^2 - \left(\frac{Y_A^o - Y_B^o}{d_{AB}^o}\right)^2 \sigma_{Y_A}^2$$

$$-\left(\frac{Z_A^o - Z_B^o}{d_{AB}^o}\right)^2 \sigma_{Z_A}^2 + \left(\frac{X_A^o - X_B^o}{d_{AB}^o}\right)^2 \sigma_{X_B}^2$$

$$+\left(\frac{Y_A^o - Y_B^o}{d_{AB}^o}\right)^2 \sigma_{Y_B}^2 + \left(\frac{Z_A^o - Z_B^o}{d_{AB}^o}\right)^2 \sigma_{Z_B}^2 \quad (5)$$

With the aim of automising the calculation work a software was specifically created, which reads a file in input containing the coordinates and the square deviations derived directly from the least square adjustment performed in Starnet and it restores in output the 3d distances calculated and their mean square deviation.

Figure 3 shows the trend of the variation of the differences, divided by the distance itself. It

269

displays the variation in the different measurement campaigns. The distances considered here and in the following study are the distances D03-D09 (short distance, blue line) and D06-D12 (long distance, red line), that is those used, due to proximity, for the comparison with the inclinometers.

3.2 Inclinometric monitoring

Inclinometers are specific instruments which are used to measure horizontal movements along vertical alignments. The inclinometric probe is a transducer which allows the inclination of the instrument with respect to the vertical measured at the time of installation to be recorded continuously.

The continuous monitoring plan using inclinometric probes on the site in included the checking of the diaphragms using 12 fixed biaxial inclinometers, distributed on 4 different columns placed on each diaphragm (Fig. 4), with the aim of monitoring the behaviour in plane and/or out-of-plane.

For every column A, B, C and D the biaxial inclinometers allow the reading of the movements from the vertical with respect to two orthogonal tracks between them which given the shape of the structure being monitored, coincide with the out-of-plane of the diaphragm (reading from channel a) and in the plane (reading from channel b) (Figure 4).

Each of these columns was linked to a data collection unit which automatically took the readings, equested via modem GSM/GPRS, for the entire duration of the monitoring.

The fixed inclinometers were installed in series—the depths of installation are: –4 m, –8 m, –11 m—within an inclinometric tube and connected to each other by a stainless steel cable with a diameter of 2 mm.

The data collection is carried out using an automatic datalogger. The datalogger is composed of a 24- channel relay multiplexer for reading the instruments, a 220 W—6/12 V power supply.

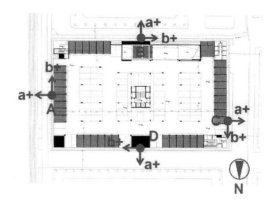

Figure 4. Scheme of the reading of the biaxal inclinometers.

The data processing consists of the conversion of the sizes from electrical to physical units.

The presence of precise peaks which exceed 0.05°, associated with a trend, can be attributable to structural type variations. The values of the alarm threshold were set at +0.07° and –0.07°. In the event of the threshold being exceeded at least three times in a row, which generally represents a trend, an alarm SMS is automatically sent by the system.

Table 1 summarises the technical characteristics of the fixed inclinometers (Model S412HA15, Solid State (MEMS)) used for the monitoring of the walls of the diaphragms during the construction of the car park.

The horizontal movements of the inclinometers were calculated starting with the angular values, using the simplified calculation procedure.

$$x = L^{\cdot} sin\ \alpha \qquad (6)$$

L is the distance between the axes of the wheels of the inclinometric probe; α the measurement of the deviation from the initial vertical, recorded by the inclinometers; x represents the horizontal deviation of the point.

This hypothesis, which introduces simplifications in the processing of the data normally used for the calculation of deviations with inclinometers was implemented in relation to possible monitorable movements of the structure. These prove to be very small, given the insertion of the probe in inclinometric columns anchored to and built in with the diaphragm walls, which represent the structure supporting the garage being built. The anticipated movements are therefore very small, if compared with those which are usually monitored with instruments of this type.

The precision of the calculation of the horizontal movements was therefore calculated starting with the angular deviations, by means of the application of the law of variance propagation.

Figure 3. The variation of the distance for the short (blue line) and the long distance (red line).

Table 1. Technical characteristics of fixed inclinometers.

Technical characteristics	
Range of measurement	±15°
Sensor resolution	0.001% FS
Total precision	Better than ±0.4% FS
Total length	1.17 m
Wheel axis distance	1 m

Therefore it will be:

$$\sigma_x^2 = \left(\frac{\partial x}{\partial L}\right)^2 \sigma_L^2 + \left(\frac{\partial x}{\partial \alpha}\right)^2 \sigma_\alpha^2$$
$$+ 2\left(\frac{\partial x}{\partial L}\right)\left(\frac{\partial x}{\partial \alpha}\right)\sigma_L\sigma_\alpha r_{L\alpha} \qquad (7)$$

The coefficient $r_{L\alpha}$ is logically null, because they are different quantities, therefore also

$$2\left(\frac{\partial x}{\partial L}\right)\left(\frac{\partial x}{\partial \alpha}\right)\sigma_L\sigma_\alpha r_{L\alpha} = 0 \qquad (8)$$

σ_L is zero, because it expresses the mean square deviation of a measurement, which is considered constant and therefore infinitely precise: the distance between the axes of the wheels of the inclinometric probes which measures one metre.

Therefore the formula after the derivation is:

$$\sigma_x^2 = L^2 * cos^2\alpha * \sigma_\alpha^2 \qquad (9)$$

σ_α is the precision of the inclinometer; L is equal to 1 m and is the length between the axes of the wheels of the inclinometer, which is considered to be a fixed length and therefore infinitely precise. The angle α is calculated from the data logged, as a deviation from the initial vertical recorded by the inclinometric sensor.

The precision of the inclinometer was calculated based on the data from the installed instrumentation; the full-scale, in the field of systems of measurement, represents the maximum value which can be measured by a given instrument of measurement. In particular, for this model of inclinometers it is stated that the precision of the instrument should be calculated as ±0.4% of the full scale.

Only the components that were significant with respect to the direction of the anticipated movement were considered, that is orthogonal to the panel (which is equivalent to saying the reading from channel A of the inclinometer). Moreover, only the value of the inclinometer situated at −4 metres with respect to the surface level, that is the one which is closest to the topographical points, was considered.

The precision calculated on the lateral deviation of the inclinometer, expressed as mean square deviation is 1 mm or less.

4 ANALYSIS OF RESULTS

The data related to one of the "short" measurements (the distance between prisms D03-D09 and inclinometers B and D), taken at the midpoint of the excavation and that of the "long" side, (the distance between prisms D06-D12 and inclinometers A and C), also at the midpoint, will be displayed hereunder. These points and these distances were chosen because the points are the closest to the position of the inclinometers.

The topographical movements were calculated, in Table 2 the differences are shown, compared with the measurement calculated the previous time.

The movements of the inclinometers in millimetres were calculated based on the simplified method.

The movement of the inclinometric columns corresponding to the topographical measurements (Tab. 3) was verified, in particular for distances D03-D09; and D06-D12.

4.1 Comparison

The comparison of the resulting horizontal movements calculated using the inclinometric and topographical measurements is shown in the graphs below (Figs. 5 and 6), for the two reference measurements (with respect to the short side of the future garage and one on the long side). The orange line

Table 2. Topographical differences.

Date	D03-D09 (m)	Difference (mm)	D06-D12 (m)	Difference (mm)
18 06 2013	49.10467	−2.688	80.08131	−1.202
25 06 2013	49.10198	0.872	80.08011	−2.259
01 07 2013	49.10285	1.924	80.07785	2.232
09 07 2013	49.10478	−1.461	80.08008	−1.13
16 07 2013	49.10332	−0.762	80.07895	1.712
06 08 2013	49.10256	−3.292	80.08066	−4.218
28 08 2013	49.09926	−2.739	80.07645	−0.027
12 09 2013	49.09652	1.498	80.07642	−7.163
11 10 2013	49.09802	−3.625	80.06926	−8.745
06 11 2013	49.09440	−5.806	80.06051	−8.156
17 12 2013	49.08859	−1.006	80.05235	2.603
16 01 2014	49.08759	2.955	80.05496	−0.296
13 02 2014	49.10467	−2.688	80.05345	−1.202

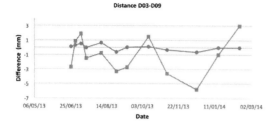

Figure 5. Graph of the inclinometric (blue) and topographic (orange) variation of the distance.

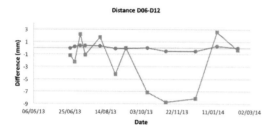

Figure 6. Graph of the inclinometric (blue) and topographic (orange) variation of the distance.

Table 3. Inclinometrical differences.

Date	Inclinometers B-D (mm)	Inclinometers A-C (mm)
18 06 2013	0.124450480	−0.074055787
25 06 2013	0.342821652	0.218554870
01 07 2013	0.532361865	0.355828867
09 07 2013	−0.019754564	0.361246688
16 07 2013	0.645836512	0.290803448
06 08 2013	−0.651162839	−0.110180810
28 08 2013	0.021683139	−0.101148977
12 09 2013	0.117495379	−0.010838238
11 10 2013	−0.306550848	−0.536450885
06 11 2013	−0.629278270	−0.543677496
17 12 2013	0.001917165	0.288997516
16 01 2014	0.009035247	−0.003612571
13 02 2014	0.124450480	−0.074055787

Table 4. The correlation between the two measurement sets.

Measurement	Correlation
Short side D03-D09	0.81325691
Long side D06-D12	0.81271862

Table 5. Total movement of the inclinometers.

Inclinometer	Total movement (mm)
Inclinometer A(D06-D12)	0.444333460
Inclinometer C(D06-D12)	−0.308866835
Inclinometer B(D03-D09)	0.675560586
Inclinometer D(D03-D09)	0.675560586

represents the variation of the differences measured by means of topography, the blue one the inclinometric differences, in both the graphs.

In general, as is clearly visible from the graphs, the topographical measurements record quantatively greater measurements, very probably due to the effects of surface movements, not recorded by the inclinometric columns which are incorporated in the perimeter walls of the diaphragms. The topographical points are certainly also more subject to variations in position due to intervening variations in the environmental parameters of temperature, humidity and pressure. The measurements were taken, in fact, both during the summer season with detection campaigns which were very close together in terms of time, and also during the winter season, with a much longer time between measuring sessions.

The trend of the two graphs is, however, well comparable; one shrinkage of the structure recorded by the inclinometric measurements corresponds to a datum with the same tendency measured using traditional type topographical observations.

The calculation of the correlation coefficient between the two measurement sets, which is equal to around 0.81 for both the distances (Tab. 4), confirms the presence of correlation between the data.

The correlation is slightly greater for the measurements that refer to the short side of the structure and therefore to the inclinometric columns B- D and to the topographical distance D03-D09. In fact, the inclinometric column C, used for the calculation of the difference in the long distance is not situated exactly at the midpoint of the side like the other columns (as one can see from the map which shows the indication of the placement of the columns, Fig. 4), but due to site demands was installed closer to the long wall. This causes a greater structural link of the column, which in fact moves less than the others, because it is closer to the corner of the foundation structure. Point D06 in fact, is not close to the head of the column, as is the case in the other 3 sides of the rectangular structure (Table 5).

5 CONCLUSIONS

A comparison and integration between the horizontal movements obtained using topographical and inclinometric measurements is possible, as demonstrated, because the data obtained have comparable precisions.

The capacity of the inclinometers to record continuously guarantees of continuous checking in real-time of the phenomena that can occur on the structure.

Topographical checks on the other hand, have the advantage of providing statistically robust data and therefore reliable information. This is precisely because of the measurement schemes adopted (with an overabundance of observations) which lead to decisive hyperdeterminant systems. In contrast, the measurement campaigns occur in specific moments on site, chosen based on the work going on at the time, at times in which there could be discontinuity.

From all of this, it is inferred that it is convenient to activate protocols in which two systems coexist, allowing the validation of the respective data and best exploitation of the characteristics of the two systems; on one hand the continuity, on the other the robustness.

Therefore the phenomena identified independently by the two monitoring systems, having similar precision and measurements, can be considered true and adequately described.

REFERENCES

Balletti C., Guerra F., Tsioukas V., Vernier P., 2014. Calibration of Action Cameras for Photogrammetric Purposes. Sensors 2014, 14: 17471–17490.

Boscato, G. Riva, G. Russo, S. Sciarretta, F. 2011. ND tests for a first assessment of mechanical behaviour of the stone-covered façades of Palazzo Ducale in Venice. *12th International Conference on Structural Repairs and Maintenance of Heritage Architecture 2011*, September 2011 Chianciano Terme, Italy. WIT Transactions on the Built Environment 118, pp. 615–625.

Boscato, G. Pizzolato, M. Russo, S. Tralli, A. 2012a. Seismic Behaviour of a Complex Historical Church in L'Aquila, *in INTERNATIONAL JOURNAL OF ARCHITECTURAL*, ISSN: 1558–3066, doi: 10.1080/15583058.2012.736013.

Boscato, G. Rocchi, D. Russo, S. 2012b. Anime Sante Church's Dome After 2009 L'Aquila Earthquake, Monitoring and Strengthening Approach. *Advanced Material Research*, vol. 446–449, p. 3467–3485, ISSN: 1022–6680, doi: 10.4028/www.scientific.net/AMR.446–449.3467.

Boscato, G. Dal Cin, A. Russo, S. Sciarretta, F. 2014. SHM of Historic Damaged Churches. *Advanced Materials Research*, 838–841, pp. 2071–2078.

Boscato, G. & Russo, S. 2014. Dissipative capacity on FRP spatial pultruded structure, Composite Structures (2014), doi: http://dx.doi.org/10.1016/j.compstruct.2014.03.036.

Fratus de Balestrini, E., Ballarin, M., Balletti, C., Buttolo, V., Gottardi, C., Guerra, F., Mander, S., Pilot, L., and Vernier, P. 2013. Survey Methods for Earthquake Damages in the "Camera Degli Sposi" OF MANTEGNA (MANTOVA), *Int. Arch. Photogramm. Remote Sens. Spatial Inf. Sci.*, XL-5/W2, 265–270, doi:10.5194/isprsarchives-XL-5-W2-265-2013, 2013.

Fregonese, L. Barbieri, G. Biolzini, L. Bocciarelli, M. Frigeri, A. Taffurelli, L. 2013. Surveying and monitoring for vulnerability assessment of an ancient building. *Sensors*, 2013, 13.

Hill, C. & Sippel, K. 2002. Modern deformation monitoring: a multi sensor approach, *FIG 22nd International Congress*, Washington D.C., USA, 19–26 April 2002.

Russo, S. 2012a. Testing and modelling of dynamic out-of-plane behaviour of the historic masonry façade of Palazzo Ducale in Venice, Italy. *Engineering Structures*, 46, 2013, 130–139 [DOI 10.1016/j.engstruct.2012.07.032.

Russo, S. 2012b. On the monitoring of historic Anime Sante church damaged by earthquake in L'Aquila. *Structural control and health monitoring*, ISSN: 1545–2263, doi: 10.1002/stc.1531.

Simeoni, L. & Benciolini, G. B. 2007. Complementarity between inclinometric and topographical monitoring for the analysis of the stability of a slope. *23rd National Geotechnical Convention*, Abano Terme, Padua, 16–18 May 2007.

Architectural, Energy and Information Engineering – Sung & Chen (Eds)
© 2016 Taylor & Francis Group, London, ISBN 978-1-138-02791-6

Research on road traffic congestion in Shanghai based on the system dynamics theory

Y.P. Wang
School of Traffic and Transportation, Beijing Jiaotong University, Beijing, China

ABSTRACT: This paper uses the system dynamics theory and method to study the road traffic congestion problem in Shanghai. Based on the analysis of social and economic influential factors, a system dynamics model was built to simulate the trend of Shanghai's traffic condition in the next decade. The simulation result indicated that without proper measures, Shanghai's traffic density will increase further and the traffic congestion will become worse in the next decade. By means of changing the parameters of the model, the influence of traffic policy and traffic construction was simulated. Based on the result, certain measures were proposed to relieve the traffic congestion problem.

Keywords: road traffic congestion; system dynamics

1 INTRODUCTION

With the rapid development of China's economy, the amount of vehicles in cities has increased dramatically in recent years, making road traffic congestion increasingly common, especially in large cities. Traffic congestion not only increases the cost of going-out, but also brings a huge number of social problems such as energy wasting, environment pollution and road accidents, which has become a serious obstacle to the city's sustainable development.

As one of the largest cities in China, Shanghai has been facing much serious road traffic congestion in recent years. According to the Shanghai Comprehensive Traffic Annual Report, compared with the year 2013, Shanghai's gross traffic flow increased by about 7% in 2014, which made the rush hour to extend about 1 hour and traffic congestion became more common in everyday life. As predicted, if certain measures are not taken in time, Shanghai's traffic condition will become worse in future.

Shanghai's problem of road traffic congestion is only an epitome of traffic problems that every China's large city is facing where urbanization is proceeding promptly. Thus, as a typical research model, analyzing Shanghai's traffic congestion problem will contribute to finding the solution of other large cities' traffic problems.

Based on the deep analysis of Shanghai's road traffic condition, the paper tried to take the viewpoint of the system dynamics theory to research this problem. A system dynamics model including a variety of influential factors of traffic congestion was built, and through the simulation, certain traffic policy and construction advices are addressed with respect to the traffic congestion problem of Shanghai.

2 BASIC THEORY

System dynamics was created during the mid-1950s by Professor Jay Forrester of the Massachusetts Institute of Technology. It is a methodology and mathematical modeling technique for framing, understanding, and discussing complex issues and problems.

System dynamics model is able to solve the problem of simultaneity by updating all variables in small time increments with positive and negative feedbacks and time delays structuring the interactions and control. Originally developed to help people improve their understanding of industrial processes, system dynamics is currently being used throughout the public and private sectors for policy analysis and design.

Nowadays, the urban traffic system has become rather complex, which has a close relationship with the social system, economic system and other related systems. Thus, applying the system dynamics theory to the research of road traffic congestion helps to take all these influential factors into consideration, which will be helpful to analyze how these influential factors affect the road traffic congestion.

3 SYSTEM DYNAMICS MODELING

3.1 Causal loop diagram

The basic construction of the system dynamics model is a feedback cycle, which is constructed based on the cause-and-effect relationship. Based on the analysis of all the influential factors of traffic congestion, a causal loop diagram was built, as shown in Figure 1.

The analyses of each feedback cycle are explained below:

1. GDP→ + GDP Per Capita→ + Amount of Personal Vehicles→ + Road Traffic volume Per Kilometer→ − GDP

   It is a negative feedback cycle. With the continuous development of urban economy, GDP Per Capita rises, and the people's demand for private vehicles increases. The increase in private vehicles makes the road become more crowded, which in turn hinders the development of the economy.

2. GDP→ + GDP Per Capita→ + Traffic Volume of Personal Vehicles→ + Road Traffic volume Per Kilometer→ − GDP

   It is a negative feedback cycle. With the economy developing and GDP Per Capita rising, personal traffic demand increases. As a result, roads become more crowded and it hinders the development of the economy.

3. GDP→ + Traffic Facilities Investment→ + Scale of Urban rail transit→ − Traffic Volume of Personal Vehicle→ + Road Traffic Volume Per Kilometer→ − GDP

   It is a positive feedback cycle. With GDP rising, investment of traffic facilities from government increases, and the mileage of the urban rail transit extends. The convenience of the urban rail transit reduces the traffic demand of personal vehicles, which will relieve the traffic congestion.

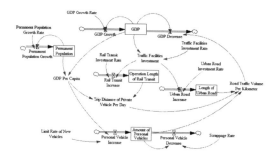

Figure 2. System flow chart.

4. GDP→ + Traffic Facilities Investment→ + Scale of Road Network→ − Road Traffic Volume Per Kilometer→ − GDP

   It is a positive feedback cycle. With GDP rising, investment for traffic facilities from government increases, leading to the increase in investment for road reorganization and expansion. The scale of the urban road network expands, which will relieve the traffic congestion.

3.2 Stock and flow diagram

Based on the analyses above, the stock and flow diagram was constructed, as shown in Figure 2.

Road traffic volume per kilometer was intensively observed through the model simulation, because it was the key measurable indicator of road traffic congestion.

3.3 Mathematical equations

In order to reflect the actual condition of Shanghai's traffic, the 2003 to 2012 data of Shanghai Statistical Yearbook, Shanghai Comprehensive Traffic Annual Report and reports from Shanghai Traffic Office were selected. The regression analysis of the data was conducted to obtain the correct mathematical equations. The mathematical equations are calculated as follows.

1. Level Variables:

$$GDP_{.k} = GDP_j + DT \times (GDPGrowth_{jk} - GDPDecrease_{jk})$$

$$PermanentPopulation_k = PermanentPopulation_j + DT \times PermanentPopulationGrowth_{jk}$$

$$OperationLengthofRailTransit_k = OperationLengthofRailTransit_j + DT \times RailTransitIncrease_{jk}$$

$$LengthofUrbanRoad_k = LengthofUrbanRoad_j + DT \times UrbanRoadIncrease_{jk}$$

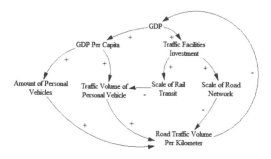

Figure 1. Causal loop diagram.

$$AmountofPersonalVehicles_k$$
$$= AmountofPersonalVehicles_j + DT$$
$$\times (PersonalVehicleIncrease_{jk}$$
$$- PersonalVehicleDecrease_{jk})$$

2. Rate Variables:

$$GDPGrowth = GDP \times GDPGrowthRate$$

$$GDPDecrease$$
$$= (0.024 \times Ln(RoadTrafficVolume$$
$$PerKilometer) + 0.027) \times GDP$$
$$PermanentPopulationGrowth$$
$$= PermanentPopulation$$
$$\times PermanentPopulationGrowthRate$$
$$RailTransitIncrease$$
$$= 263.9 \times Ln(TrafficFacilitiesInvestment$$
$$\times RailTransitInvestmentRate) - 642.7$$
$$UrbanRoadIncrease$$
$$= 279.1 \times Ln(TrafficFacilitiesInvestment$$
$$\times UrbanRoadInvestmentRate) - 930.5$$
$$PersonalVehicleIncrease$$
$$= (1e - 008 \times GDPPerCapita \wedge 2 - 0.001$$
$$\times GDPPerCapita + 39.77)$$
$$\times LimitRateofNewVehicles$$
$$PersonalVehicleDecrease$$
$$= AmountofPersonalVehicles$$
$$\times ScrappageRate$$

3. Auxiliary Variables:

$$GDPGrowthRate = 7.8\%$$
$$PermanentPopulationGrowthRate = 1.4\%$$

$$RailTransitInvestmentRate = 44.21\%$$
$$TrafficFacilitiesInvestment = GDP$$
$$\times TrafficFacilitiesInvestmentRate$$
$$TrafficFacilitiesInvestmentRate = 3.137\%$$

$$UrbanRoadInvestmentRate = 45.31\%$$
$$TripDistanceofPrivateVehiclePerDay = 0.000368$$
$$\times GDPPerCapita - 0.000115$$
$$\times operationLengthofRailTransit + 8.733$$

$$LimitRateofNewVehicles = 1$$
$$ScrappageRate = 10\%$$
$$RoadTrafficVolumePerKilometer$$
$$= AmountofPersonalVehicles$$
$$\times TripDistanceofPrivateVehiclePerDay$$
$$\div LengthofUrbanRoad$$

3.4 Data validation examination

The amount of personal vehicles, the length of the urban road and road traffic volume per kilometer are chosen to examine the validation of the equations (Table 1). Comparing the error value between

Table 1. Data validation result.

	Personal vehicle 10 thousand	Road km	Traffic density 10 thousand/km
Simulation	154.06	17964	0.35315
Reality	161.38	17498	0.36468
Deviation	4.54%	2.66%	3.16%

the simulation data and the reality data, the rate of deviation was below 5%, which meant that the model was reflecting the actual traffic condition of the reality correctly.

4 MODEL ANALYSIS

The simulation result of road traffic volume per kilometer is shown in Figure 3.

Through the system dynamics simulation, the trend of Shanghai's traffic condition was obtained. The simulation showed that the road traffic volume per kilometer will increase from 0.321 in 2012 to 1.027 in 2022, about 3 times throughout this decade. By means of changing the influential factors, the influence of traffic policy and traffic construction was simulated.

4.1 Analysis of the influence on the traffic facility investment rate

Traffic facility investment rate refers to the proportion that the traffic facility investment accounts for GDP. According to the data, it was set as 3.137%. When we set it as 2% and 4%, the road traffic volume per kilometer was changed, as shown in Figure 4. As can be seen, when the rate increased, the road traffic volume per kilometer increased slightly as well.

4.2 Analysis of the influence of component ratio on traffic facility investment

The component ratio of traffic facility investment refers to the proportion that urban rail transit investment and urban road investment account for all the traffic facility investments. According to the data, they were both about 45% in recent 10 years. We reset them in two conditions. In condition 1, the urban rail transit investment rate was 30% and the urban road investment rate was 60%. In condition 2, the two rates reversed. The result is shown in Figure 5. The numerical value showed that condition 1 was the best and condition 2 was the worst, which means that at present, investing more on the urban rail transit has a better effect on traffic congestion.

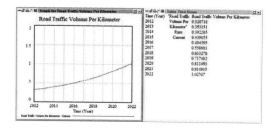

Figure 3.   Simulation result.

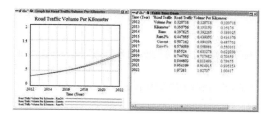

Figure 4.   Influence of traffic facility investment rate.

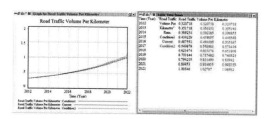

Figure 5.   Influence of the component ratio on traffic facility investment.

4.3 Analysis of the influence of limit rate on new vehicles

The limit rate of new vehicles refers to the ratio that new vehicles can gain license plate under the government vehicle management policy. In the model, the rate was set to 100%, but when it was reset to 60% and 80%, the road traffic density decreased obviously, as shown in Figure 6. As can be seen, when the limit rate of new vehicles get stricter, the traffic density goes down obviously.

4.4 Analysis of the influence on the vehicle scrappage rate

It was reported that according to the Vehicle Mandatory Scrapping Standard released recently, the average age of vehicle scrapping was 10 years, which meant that every year about 10% vehicles were required to be scrapped. We set the scrappage rate at 8.33% and 12.5%, and the road traffic density changed, as shown in Figure 7. As can be seen, when the scrappage rate rises, the traffic density goes down obviously.

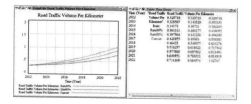

Figure 6.   Influence of the limit rate on new vehicles.

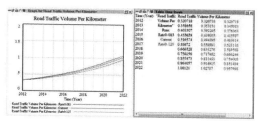

Figure 7.   Influence of the vehicle scrappage rate.

5   CONCLUSION

According to the simulation and analyses of Shanghai's road traffic condition, it is known that without efficient road traffic management, the traffic density of Shanghai will be going up sharply and the traffic congestion problem will be even worse in the next decade. Hence, on the one hand, the government ought to increase the traffic facility investment to improve the convenience of public transit and the scale of urban road network; on the other hand, the government should conduct strict vehicle management by means of limiting the number of new license plates and executing vehicle mandatory scrapping in time.

In addition, combined with the simulation result and development reality, as Shanghai's urbanization process come into a stationary phase, the available urban construction land will be reducing gradually. Thus, massive urban traffic facility construction has to be restricted in the future, which will not play a decisive role at solving the traffic congestion problem. It is obvious that only by strict vehicle management and the limit of the vehicle's number can the road congestion problem be solved fundamentally.

REFERENCES

[1] Richardson, G., & Pugh III, A. I. 1981. *Introduction to system dynamics modeling with DYNAMO.* Productivity Press Inc.
[2] Shanghai Statistical Yearbook. 2013. Shanghai Statistical Yearbook. China Statistical Press.
[3] Shanghai Institute of Urban Comprehensive Transportation. 2013. Shanghai Comprehensive Traffic Annual Report. *Traffic and Transportation,* 24(6).
[4] Stopher, P. 2004. Reducing road congestion: a reality check. *Transport Policy, 11*(2), 117–131.

Architectural, Energy and Information Engineering – Sung & Chen (Eds)
© *2016 Taylor & Francis Group, London, ISBN 978-1-138-02791-6*

Study on prediction model of Ningdong coal chemical gasification coal calorific value and industrial analysis

Gai Rui Qiao, Hui Wu Cai, An Ning Zhou & Fu Sheng Yang
College of Chemistry and Chemical Engineering, Xi'an University of Science and Technology, Xi'an, China

ABSTRACT: Based on Ningdong coal chemical base, calorific value and industrial analysis testing data of Hongliu mine were analyzed. Using the Explore and Correlation Analysis functions of SPSS statistical software, after multiple regression analysis, deduced a mathematical model between calorific value and industrial analysis index, and detected the forecast effect of the prediction model. The results showed that this prediction model has higher forecasting precision and reliability, and the regression model can be used to forecast the calorific value of the NingDong HongLiu coal seam.

Keywords: Ningdong coal chemical industry; Calorific value; Industrial Analysis; SPSS; Mathematical model.

1   INTRODUCTION

Ningxia is rich in coal resources, proven reserves of 31 billion tons. Currently, it has built 250,000 tons/year of coal-based methanol, 210,000 tons/year of DME, 600,000 tons of methanol, 520,000 tons/year of coal-based olefin, etc. One of these has been completed and planning projects' core technology is coal gasification technology. Industrial analysis and calorific value of coal are the most important indicators to study the basic gasification technology. There is a close relationship between these indicators. And from the coal moisture, ash and volatile components accurately calculate the calorific value, has the vital significance to the coal preparation, gasification of coal blending.

Never use letterspacing and never use more than one space after each other.

2   CALORIFIC VALUE AND INDUSTRIAL ANALYSIS INDEX MULTIPLE REGRESSION GETTING STARTED

2.1   *Regression analysis*

Regression analysis is the study of one or several variables change on the influence of the change of another variable method, according to the known information or data, find out the relational expression between them, with the known values of the independent variable to speculate about the value of the dependent variable or scope. Linear statistical principle is as follows: set

Y as the dependent variable, and X as the independent variable

$$Y = \beta_0 + \beta_1 X_1 + \beta_2 X_2 + \ldots + \beta_p X_p \qquad (1)$$

If the dependent variable Y is p independent variables' linear function, $X_1, X_2 \ldots X_p$, under the condition of the independent variables of different values, $X_1, X_2 \ldots X_p$, and N trials were conducted,

$$X_{n1}, X_{n2} \ldots X_{np}, Y_n, n = 1, 2, \ldots, n.$$

The structural model of the N sets of data can be written as:

$$Y_n = A_0 + X_i + C_n \sum_{i=1}^{p} A_i \qquad (2)$$

$C_1, C_2, \ldots, C_n$ are random variables, independent to each other, and subject to the same normal distribution $N(0, \sigma^2)$, where σ^2 is the experimental error variance. The regression equation can be written as:

$$Y = B_0 + X_i \sum_{i=1}^{p} B_i \qquad (3)$$

Undetermined coefficients $B_0, B_1, \ldots B_p$ of P+1 are respectively the estimator of regression coefficients $\beta_0, \beta_1, \beta_2, \ldots \beta_p$ in the equation.

2.2   *SPSS*

SPSS as an important mathematical statistics software, with its powerful functions, simple operation, etc., has been applied to various fields[6]. This paper takes the Ningdong coal chemical industry base in Hongliu coal mining area as the research object. By using SPSS for $Q_{gr, ad}$ and A_{ad}, V_{daf}, Mt the multiple regression analysis between each index, to establish mathematical model between them.

3 THE ESTABLISHMENT OF CALORIC AND INDUSTRIAL ANALYSIS' MATHEMATICAL MODEL LAYOUT OF TEXT

3.1 Coal quality index correlation analysis

The main process in the analysis is: first, Hongliu mine 50 sets of data from the Excel table wrote into SPSS system, establish the data file. Then click on the main menu: Analyze → Correlate → Bivariate, open Bivariate Correlations. Import Qgr, ad, Aad, Vdaf,

Mt, and select the Pearson's correlation coefficient and then "OK". By the Pearson's correlation coefficient can be concluded that Qgr, ad, Aad, and Vdaf has a strong negative correlation. Negative correlation coefficients were –0.925 and –0.609 respectively, but the correlation coefficient of Mt is –0.278, whic is significantly weak. Thus to establish Qgr, ad, Aad, and Vdaf binary regression equation.

3.2 Build binary regression model

Click on the main menu Analysis → Regression → Linear, choose Qgr,ad as the dependent variable, Aad and Vdaf as independent variables for binary linear regression analysis, click the "OK" button to get two elements model. The regression results: coefficients (see Table 1), analysis of variance (see Table 2), and the coefficient table (see Table 3).

It can be seen from Table 1, the correlation coefficient R = 0.933, illustrate the dependent variable Qgr,ad with the independent variable Aad, Vdaf has a strong correlation, degree of fitting R2 = 0.871.

Explain the independent variables can explain the dependent variable 87.1%. The independent variable can be explained by the variability of the 87.1% dependent variables and Sig. = 0.000, below the reference value of 0.05.

As we can see from Table 2 Anova table, the F statistics of significant is Sig. = 0.000, less than 0.05 confidence, the dependent and independent variables significantly are related.

As we can see from Table 3 the Qgr, ad, Aad, Vdaf significant probability of Sig. were Sig. = 0.000 and Sig. = 0.762, in which Vdaf significant probability

Table 1. Model summary.

Model	R	R2	Adjust the R2	Standard error of estimate	R2 change	F change	Sig.F change
					Change statistics		
1	.933a	.871	.863	.24154	.871	103.649	.000

Predict variables: (Constant), Vdaf(%), Mt(%), Ad(%).

Table 2. Anova.

Model		Sum of squares	df	Mean square	F	Sig.
1	Regression	18.141	3	6.047	103.649	.000a
	RSS	2.684	46	.058		
	Total	20.825	49			

a. Predict variables, Vdaf(%), Mt(%), Aad(%). b. dependent variable Qgr, ad.

Sig. = 0.762 larger than the standard range of 0.05, at a two elements regression equation accuracy is not high, therefore need to delete variables Vdaf, Qgr, ad and Aad regression equation of, to proceed the test of significance, to get the optimal model.

3.3 To establish a element regression model

Operation steps according to the two elements regression model, finally got model coefficients and regression model (see Table 4), the table of variance analysis(see Table 5) and the coefficient table (see Table 6).

As we can see from Table 4, Qgr,ad and Aad R value and R square values are 0.925 and 0.856, strong significant, and Sig. was 0.000, precision is high.

As we can see from Table 5 significant probability Sig. = 0.000 F statistics, less than 0.05 confidence, the dependent and independent variables significantly were related.

As we can see from Table 6 Qgr,ad and Aad significant probability Sig. = 0.000, less than the standard 0.05, so you can get the optimal linear regression equation. We can see from Table 7 Qgr,ad and Aad negative correlation, coefficient is –.305, and the constant term is 24.349. Whereby to obtain a regression model is: Qgr, ad = 24.349 – 0.305 Aad

Therefore, the use of the regression model can predict the calorific value in pencil in the bottom margin of each page and number the pages correctly.

Table 3. Coefficient[a].

| Model | | Non standardized coefficient | | Standard coefficient | | | 95.0% confidence interval of B | |
		B	Standard error	Trial version	t	Sig.	Lower limit	Limit
1	constant	24.169	.616		39.214	.000	22.930	25.409
	Aad(%)	−.310	.025	−.940	−12.604	.000	−.360	−.261
	Vdaf(%)	.008	.027	.023	.304	.762	−.046	.062

a. dependent variable: Qgr, ad.

Table 4. Model summary.

| Model | R | R2 | Adjust R2 | Standard error of estimate | Change statistics | | |
					R2 change	F change	Sig. F change
1	.925[a]	.856	.853	.24999	.856	285.23	.000

a. Predict variables: (Constant), Aad(%).

Table 5. Anova[b].

Model		Sum of squares	df	Mean square	F	Sig.
1	Regression	17.825	1	17.825	285.230	.000[a]
	RSS	3.000	48	.062		
	Total	20.825	49			

a. Predict variables: (Constant), Ad(%). b. dependent variable: Qgr, ad.

Table 6. Coefficient[a].

| Model | | Non standardized coefficient | | Standard coefficient | | | 95.0% confidence interval of B | |
		B	Standard error	Trial version	t	Sig.	Lower limit	Limit
1	constant	24.349	.170		143.071	.000	24.007	24.692
	Aad(%)	−.305	.018	−.925	−16.889	.000	−.342	−.269

a. dependent variable: Qgr, ad.

4 CONCLUSION

In this paper, Hongliu coal mine data of Ningdong coal chemical for the study, in detail described the SPSS software to use in coal quality data processing, correlation analysis and regression analysis. In conclusion to the mathematical model of calorific value and coal industry analysis, the model can accurately response the variation between calorific value and ash content, the regression model can be used to forecast the calorific value of the NingDong HongLiu coal seam. It provides an easy method for coal preparation, and gasification of coal blending.

REFERENCES

Beibei Zhang & Yandong Xue. Multivariable linear regression of the correlation between calorific value and moisture and ash content of coal [J]. Coal Geology and Exploration. 2014, 42(4): 9.

Chehreh Chelgani, S., Makaremi, S., Explaining the relationship between common coal analyses and Afghan coal parameters using statistical modeling methods [J]. Fuel Processing Technology, 2012, 110: 79–85.

Hongbo Chen & Xiangfei Bai. Linear regression of the correlation between calorific value and moisture and ash content of coal [J] Coal Technology. 2007, 26(2): 112–113.

Huancheng Wei Calculation of calorific value of coal based on MATLAB [J] Coal Technology. 2007, 26(2): 112–113.

Lingguo Meng A Study on the Characteristics of Coking Coal from KaiLuan Mining Area and the Trend Prediction of Coal Quality. [M] Beijing, Beijing University of Chemical and Technology, 2013.

Mesroghli, Sh., Jorjani, E., Chehreh Chelgani, S., Estimation of gross calorific value based on coal analysis using regression and artificial neural networks [J]. International Journal of Coal Geology, 2009, 79: 49–54.

Xiangbao Gao, Hanqing Dong. Data analysis and application of SPSS [M]. Beijing: Tsinghua University press, 2007.

Xiaopeng Tian. Linear correlation between coal's calorific value and ash content [J]. Sci-Tech Information Development & Economy, 2011, 21(10): 179.

Xiaoren Mei, Peng Chen & Yongsheng Gao. Study on related factors of coal calorific value analysis and regression model based on SPSS [J]. Coal Technology. 2011, 37(7): 29.

Architectural, Energy and Information Engineering – Sung & Chen (Eds)
© 2016 Taylor & Francis Group, London, ISBN 978-1-138-02791-6

Experimental study on the effect of double pipe wall thickness on the instantaneous air source heat pump water heater

Shao You Yin

Heat Pump Engineering and Technology Development, Center of Guangdong Universities,
Shunde Polytechnic, Foshan, China

ABSTRACT: The performance of the instantaneous air source heat pump water heater was studied, and the influence of double pipe wall thickness on heat pump water heater performance was analyzed in this paper. The experimental results show that double pipe wall thickness has a certain impact on the system performance, such as heat capacity, energy efficiency ratio, and other properties. The energy efficiency of heat pump water heater is improved by the appropriate design of double pipe wall thickness, in addition, the heat capacity can be increased, and the power consumption can be reduced.

Keywords: heat pump water heater, double pipe condenser, wall thickness, performance, energy efficiency ratio

1 INTRODUCTION

In order to save energy, to governance haze, to protect environment, to advocate low carbon environmental protection, to eliminate coal burning small boiler, and to vigorously promote the heating with electricity instead of coal. To produce the same amount of hot water conditions, the energy consumption of air source heat pump water heater is only 1/4 of electric water heater, and 1/3 of gas water heater, and 1/2 of electric assisted solar water heater, so it has a great potential in energy saving. Air source heat pump water heater is a new product, the characteristics of high efficiency, energy saving and environmental protection, etc. It is becoming the mainstream products in domestic and foreign water heater industry. In order to study the performance of air source heat pump water heater, some domestic and foreign scholars have made a series of research on the air source heat pump water heater. Minsung Kim etc. [1] have performed the study on the heat transfer characteristics of water source heat pump water heater. J. Sarkar etc. [2] have performed the study on the characteristics of the high temperature heat pump water heater natural working substance. Pei Gang etc. [3] have analyzed the performance of instantaneous air source heat pump water heater and circulating heat air source heat pump water heater. Zhang Taikang etc. [4] have performed the study on the performance of air source heat pump water heater of refrigerant for R134a, R417a and R22, R134a, R417a and R22 as refrigerant. Tian Hua etc. [5] have simulated the system of air source heat pump

water heater by scroll compressor, Hao Jibo etc. [6] have tested and simulated the performance of static heat air source heat pump water heater. The air source heat pump water heater of one horsepower to two horsepower is widely used in domestic air source heat pump water heater. Therefore, it has a great competitive potential market and very good market prospects for double pipe instantaneous air source heat pump water heater. The performance of the two horsepower double pipe instantaneous air source heat pump water heater was tested in this paper, the research results will provide the reference and technical support on the air source heat pump water heater design.

2 EXPERIMENTAL SETUP

This experiment testing principle is shown in Figure 1. The temperature sensor uses the T-type thermocouple. Pressure sensor range is 0 MPa–4 MPa, precision is 0.5%; Flow rate sensor with Dalian KRC, Inc. KRC-1518H time single channel ultrasonic liquid flow sensor, velocity range is 0.01 m/s–30 m/s, flow accuracy is ±0.5%. The experimental tests were performed at the laboratory of heat pump water heater, the environmental working conditions of the laboratory is from −10°C to 50°C. Test nominal working condition is: the air side air dry-bulb temperature is 20°C, the wet-bulb temperature is 15°C; water side entering water temperature is 15°C, the leaving water temperature is 55°C. The products of 5 different double pipe heat exchangers are made

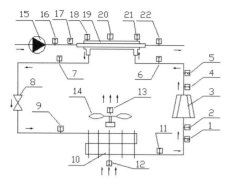

Figure 1. Schematic diagram of air source heat pump water heater test systems. 1, 5-pressure sensor, 2, 4, 6, 7, 9, 11, 12, 13, 16, 18, 20, 21, 22-temperature sensor, 3-compressor, 8-throttle valve, 10-finned evaporator, 14-fan, 15-Variable frequency pump, 17-Flow sensor, 19-double pipe heat exchanger.

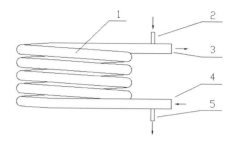

Figure 2. The double pipe condenser of air source heat pump water heater. 1-double pipe heat exchanger, 2-refrigerant entering, 3-heat water leaving, 4-cold water entering, 5-refrigerant leaving.

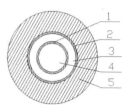

Figure 3. The double pipe condenser section of air source heat pump water heater. 1-inner pipe of double pipe, 2-outer pipe of double pipe, 3-ring channel, 4-inner pipe channel, 5-thermal insulation layer.

of the double pipes that are 1860 mm expansion length of the double pipe, 0.3 mm-1.0 mm thickness of the inner tube wall, and 170 mm height of the double pipe heat exchanger. The products of 5 different double pipe instantaneous air source heat pump water heaters are made of the products of 5 different double pipe heat exchangers respectively. The double pipe instantaneous air source heat pump water heater is tested, and the system parameters for heat capacity, power, power consumption, COP (coefficient of performance), temperature, pressure, flow rate, etc., are recorded. The Guangzhou Panasonic Corp rotary compressor is used, the model is 2V32S225AUA, and the rated cooling capacity is 5375W, and rated power is 1700 W, and refrigerant is R22 in the double pipe instantaneous air source heat pump water heater. The pump rated flow is 3 m³/h. The motor of fan is YDK120–40–6, rated power is 40W, and the fan impeller specification is Φ420 × 148. The double pipe heat exchanger is shown in Figures 2 and 3, the inner pipe and the outer pipe of double pipe are brass, the brand is TP2M; the outer diameter of the outer copper pipe is Φ15.88 mm, the thickness of the outer copper wall is 0.5 mm; the outer diameter of the inner copper pipe is Φ9.52 mm, the thickness of the inner pipe wall is 0.3 mm–1.0 mm respectively. The outer pipe wall of the double pipe heat exchanger is used PE thermal insulation pipe, the insulation thickness is 8 mm. The curvature radius of the double pipe spiral coil is 110 mm. Water flows passage of the inner tube inside, and the refrigerant flows in the ring channel between the outer tube and inner tube. The evaporator is L type and double row aluminum finned. The outer diameter of the brass pipe is Φ9.52 mm, the brand is TP2M. The thickness of the hydrophilic finned

aluminum is 0.095 mm. Emerson throttle valve is used, the model is AAE2HC-30FT, the degree of superheat is adjusted.

3 DATA PROCESSING

The products of 5 different double pipe instantaneous air source heat pump water heaters are tested, and the power, temperature, water flow are tested in the experiment, the double pipe instantaneous air source heat pump water heater energy efficiency ratio can be calculated, the calculation formula is as follows:

$$COP = \frac{Q_c}{W_C + W_F} \qquad (1)$$

where Q_c-heat capacity of heat pump water heater, W; W_C-compressor power consumption, W; W_F-fan power consumption, W.

$$Q_c = KA\Delta t_m \qquad (2)$$

where K-heat transfer coefficient of double pipe heat exchanger, W/(m²·°C); A-heat transfer area of double pipe heat exchanger, m²; Δt_m-calculation average temperature, °C.

$$K = \cfrac{1}{\cfrac{1}{\alpha_w} + \cfrac{d_i}{2\lambda}\ln\cfrac{d_o}{d_i} + \cfrac{d_i}{\alpha_r d_o}} \qquad (3)$$

where d_i-inner diameter of inner copper pipe, m; d_o-outer diameter of outer copper pipe, m; λ-thermal conductivity of inner copper pipe, W/(m.°C); α_w-convection heat transfer coefficient of inner copper pipe inner wall surface water side, W/(m².°C); α_r-convection heat transfer coefficient of inner copper pipe outer wall surface refrigerant side, W/(m².°C).

4 RESULTS AND DISCUSSION

The double pipe heat exchangers are made of the double pipes which are 1860 mm expansion length of the double pipe, 0.7 mm thickness of the inner tube wall, and 170 mm height of the double pipe heat exchanger. The double pipe instantaneous air source heat pump water heater is made of the double pipe heat exchanger. The double pipe instantaneous air source heat pump water heater is tested. Test working condition is: the air side air dry-bulb temperature is from −5°C to 45°C, water side entering water temperature is 15°C, and the leaving water temperature is 55°C. The energy efficiency ratio of COP and the heat capacity of double pipe instantaneous air source heat pump water heater are concerned of the different environment temperatures, as shown in Figures 4 and 5. When the ambient air dry-bulb temperature rises, the energy efficiency ratio of COP and the heat capacity of the double pipe instantaneous air source heat pump water heater are increased gradually. When the ambient air dry-bulb temperature is 20°C, it is the highest energy efficiency ratio of COP and the highest heat capacity. When the ambient air dry-bulb temperature continuously rises, energy

efficiency ratio of COP and heat capacity are gradually reduced.

The products of 5 different double pipe heat exchangers are made of the double pipes that are 1860 mm expansion length of the double pipe, 0.3 mm, 0.5 mm, 0.7 mm, 0.9 mm, 1.0 mm thickness of the inner tube wall, and 170 mm height of the double pipe heat exchanger. The products of 5 different double pipe instantaneous air source heat pump water heaters are made of the products of 5 different double pipe heat exchanger respectively. The double pipe instantaneous air source heat pump water heater is tested. Test results are shown in Figures 6 and 7.

See Figures 6 and 7, the energy efficiency ratio of COP and the heat capacity of the double pipe instantaneous air source heat pump water heater are concerned of the double pipe inner tube wall thickness under a test nominal working condition. In Figures 6 and 7, when the thickness of the double pipe inner pipe wall is increased, the water side heat transfer area of the double pipe inner tube is reduced, the flow velocity is increased, the convection heat transfer coefficient of copper pipe inner wall surface water side is increased in double pipe heat exchanger, heat transfer coefficient of double

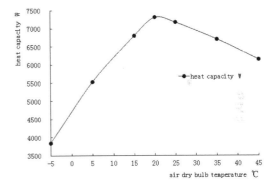

Figure 5. The heat capacity of air source heat pump water heater with different ambient temperatures.

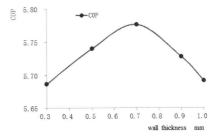

Figure 6. The COP value of air source heat pump water heater with double pipe wall thickness.

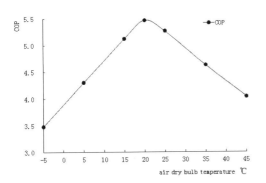

Figure 4. The COP value of air source heat pump water heater with different ambient temperatures.

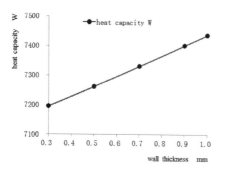

Figure 7. The heat capacity of air source heat pump water heater with double pipe wall thickness.

pipe heat exchanger is increased, energy efficiency ratio of COP and heat capacity of double pipe instantaneous air source heat pump water heater are increased gradually. When the double pipe inner tube wall thickness is 0.7 mm, it is the highest energy efficiency ratio of COP. When the double pipe inner tube wall thickness is continuously increased, energy efficiency ratio of COP is gradually reduced, and heat capacity is gradually increased.

In Figures 6 and 7, when the thickness of the double pipe inner pipe wall is increased, the heat transfer area of the double pipe inner tube water side is reduced, the flow velocity is increased, the resistance of water pipe system is increased, the operation efficiency is reduced, and the cost is increased in the double pipe instantaneous air source heat pump water heater. When the thickness of the double pipe inner pipe wall is reduced, the inner pipe ability to withstand pressure is reduced. Comprehensive results show that the proper of double pipe inner pipe wall thickness not only can save materials, but also improve the efficiency of the system, and improve the system stability and reliability.

5 CONCLUSION

1. The double pipe inner pipe wall thickness is an important parameter of the double pipe instantaneous air source heat pump water heater in the design process. When the thickness of the double pipe inner pipe wall is increased, the energy efficiency ratio and the heat capacity of the heat pump water heater are increased gradually; when the double pipe inner tube wall thickness is 0.7 mm, it is the highest energy efficiency ratio. When the double pipe inner tube wall thickness is continuously increased, the energy efficiency ratio is gradually reduced, and the heat capacity is gradually increased.

2. In the design of air source heat pump water heater, it should choose the proper of double pipe inner pipe wall thickness. When the thickness of the double pipe inner pipe wall is increased, the heat transfer area of the double pipe inner tube water side is reduced, the flow velocity is increased, the resistance of water pipe system is increased, the operation efficiency is reduced, and the cost is increased in the double pipe instantaneous air source heat pump water heater. When the thickness of the double pipe inner pipe wall is reduced, the inner pipe ability to withstand pressure is reduced.

3. When the ambient air dry-bulb temperature rises, the energy efficiency ratio and the heat capacity of the double pipe instantaneous air source heat pump water heater are increased gradually. When the ambient air dry-bulb temperature is 20°C, it is the highest energy efficiency ratio and the highest heat capacity. When the ambient air dry-bulb temperature continues rises, the energy efficiency ratio and the heat capacity are reduced gradually.

ACKNOWLEDGMENTS

This work was supported by the National Science Foundation of China (No. 51246006) and Project of Foshan City Science and Technology Innovation Foundation (No. 2013AG100063).

REFERENCES

[1] Minsung Kim, Min Soo Kim, Jae Dong Chung. Transient thermal behavior of a water heater system driven by a heat pump [J]. International Journal of Refrigeration, 2004, (27): 415–421.

[2] Sarkar, J., Souvik Bhattacharyy, M. Ram Gopal. Natural refrigerant-based subcritical and transcritical cycles for high temperature heating [J]. International Journal of Refrigeration, 2007, (30): 3–10.

[3] Pei Gang, Li Guiqiang, Ji Jie. Comparative study of air-source heat pump water heater systems using instantaneous heating and cyclic heating modes [J]. Applied Thermal Engineering, 2011, (31): 342–347.

[4] Zhang Tai-kang, Weng Wen-bing, Yu Jin. Performance research of air-source heat pump water heater using R134a, R417a and R22 [J]. Fluid Machinery, 2010, 39(5): 72–76.

[5] Tian Hua, Ma Yi- tai, Wen Zi- qiang. Simulation of air- source heat pump water heater [J]. Journal of Thermal Science and Technology, 2008, 7(4): 273–378.

[6] Hao Jibo, Wang Zhihua, Jiang Yuguang, Wang Fenghao. Analysis of system performance of air source heat pump water heater [J]. Refrigeration and Air Conditioning, 2013, 13(1): 59–62.

Architectural, Energy and Information Engineering – Sung & Chen (Eds)
© *2016 Taylor & Francis Group, London, ISBN 978-1-138-02791-6*

Design of the simple household solar heating systems

Li Liu
Shenyang Polytechnic College, Shenyang, China

ABSTRACT: In order to need the energy saving requirements, this paper presents a simple heating system that is comprised of regular heating pipeline, heating system, and heat pump. The hot water from solar thermal collector supplies heating via heating pipeline loop; when the water temperature in water tank is too low to supply heats, the heat pump heating system will be started for heating; the recycle of well water in heating pipelines can also be used to achieve summer refrigeration. The solenoid valve and commutator can be controlled by reading the time shown on the clock clip to support the conversion between underground water and solar hot water, therefore to achieve the switch between refrigeration and heat supply. The innovation of this paper is to realize heating and refrigeration by solar energy.

Keywords: solar energy; refrigeration; heat supply

1   INTRODUCTION

Heating is a must in winter in northern China. So far, domestic heating mainly relies on coal and natural gas, which cost high and therefore become an albatross to residential users. The exhaust gas from coal-fired boiler in northern area also brings a serious pollution to the atmospheric environment[1]. A significant decline of air quality occurs in many northern cities. It is imperative to develop a new environmental friendly, renewable, energy-saving, and economical energy heating means[2].

Solar energy, featuring unlimited and clean reserves, is an ideal energy for heating[3].

2   OVERALL DESIGN ABOUT HEATING SYSTEM

2.1   System components

The solar heating system consists of two solar heating control systems where heat is supplied via heating pipeline and heat pump air-conditioner. When it is in a low-temperature environment, the water in solar heat collectors will be delivered to heating pipelines via water pumps and supply indoor heating via heat sinks.

2.2   Works and process

When the temperature sensor detects that the water temperature in the heat preservation storage water tank cannot cater to the indoor heating requirements, the heat pump air-conditioner heating system works then to supply heats. The heat pump air-conditioner is composed of heat collectors, storage water tank, compressor, and condenser. The hot water from heat preservation storage water tank will enter the evaporator, commutator, compressor and then to condenser to supply indoor heat. The water that is condensed via condenser will be transported to the heat collector for cyclic utilization.

When it is in a high-temperature environment, the water in the solar heat collector is too hot to supply indoor refrigeration. The underground well water will fill this role to cool the indoor environments. The water from underground well can be sent to the heating pipelines for recycling directly. When the well water is cold, it will absorb the heat while going through heating pipelines and therefore to cool the indoor environment. If the well water is not cold enough to supply indoor refrigeration, the heat pump air-conditioner will be started to supply refrigeration. The whole systems process in this way to ensure the system's integrity while controlling the indoor temperature based on consideration of the actual application values.

3   HARDWARE CIRCUIT DESIGN

3.1   The overall design

The hardware part shall be achieved by using AT89S55 single chip. The main circuit consists of temperature detection module, keyboard display module, and drive module. The indoor temperature shall be detected through a three-point test.

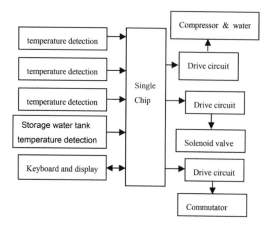

Figure 1. Overall design of the system.

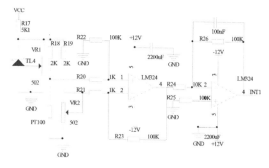

Figure 2. Pt100 temperature measuring circuit.

The single chip will control the compressor according to the temperature it detects.

The temperature of cold water from heated heat collectors will be detected through storage water tank temperature detection. The detected analog quantity will be sent to A/D converter and then the single chip for data process after module/data has been conveyed. The display unit will then take the responsibility to display the temperature of the storage water tank. The room temperature can be set according to requirements via keyboard from the system. The overall structure of the system hardware can be shown as Figure 1. This system applies CH451 interface chip for the keyboard and display circuit. The eight digits on the nixie tube displays respectively the indoor temperature sensed by DS18B20 (three digits), the storage water tank temperature detected by Pt100 (three digits) and the set indoor temperature (the last two digits). The four buttons are mainly used to set the indoor temperature.

3.2 Temperature detection modules

This module covers temperature detections as follows: detection of temperature of storage water tank via Pt100. Pt100 outputs the analog signal of the temperature detection circuit and then conveys these signals to the single chip via A/D for data process. This data is used to provide users with the temperature of water heated by solar energy; detection of indoor temperature via DS18B20. DS18B20 detects the indoor temperature, which conveys the temperature signal into a series digital signal and then is sent to the P1.0 pin for data process.

After settings of the indoor temperature, the single chip will compare the indoor temperature detected by DS18B20 with the set value for a better

control of compressor operation. Water supply switches between well water and solar energy water from summer refrigeration to winter heating. Evaporator and condenser also switch with each other via four-way commutator or software. The Pt100 temperature measurement circuit is shown in Figure 2.

The circuit applies TL431 and potentiometer VR1 to adjust 4.096V of reference power supply; measuring electric bridge is composed of R18, R19, VR2 and Pt100 (R8 = R9 and VR2 is the 100Ω precision resistance). When the resistance value of Pt100 does not equal to that of VR2, the electric bridge will output mV level of pressure difference signals. After going through the operational amplifier LM324 circuit for amplifying, the output voltage signal is conveyed to an AD conversion chip for AD conversion. In differential amplifier, R20 = R21, R20 = R21.

3.3 Drive & control circuit modules

The drive circuit is composed of the optocoupler and relay, as shown in Figure 3, and is used to realize controls of four-way commutator, four-way valve and water pump. If the AT89S52 chip pin P20 is at a low level, both the photoelectric coupler and transistor Q1 will achieve breakover, the relay is actuated and the electrical appliance is broken over as well. The four-way commutator, four-way valve and water pump are all controlled with application of this circuit.

3.4 Compressor control circuit

The system applies the irreversible PWM control circuit to control the compressor, as shown in Figure 4. It is comprised of two power transistors V1 and V2 and two diodes VD1 and VD2. V1 acts as the master tube for modulation and V2 acts as the auxiliary tube. Two impulse voltages with opposite polarity from PWM will work respectively onto the base electrodes of V1 and V2.

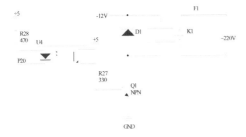

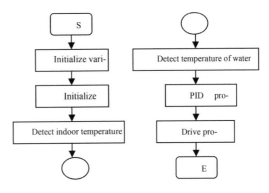

Figure 3. Drive circuit.

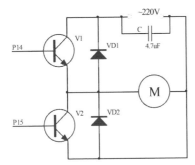

Figure 4. PWM control circuit.

4 DESIGN OF SYSTEM SOFTWARE

The master flowchart is shown in Figure 5. It is used to initialize the system data, self-check the program, and determine whether the device can work regularly or not, set up the interrupt response and distinguish certain functions. In the very beginning of system operation, a self-check and initialization are required. The switch-on self-checks will be performed before a system initialization. If there is no error, the initialization can then be started. Normally, both the hardware initialization and software initialization are included. The hardware initialization is set to clear initialized status for all the hardware in the system, for example, to program various programmable chips, set up the initial status for all the I/O terminals and assign tasks for the hardware resources of the single chip.

Figure 5. Master flowchart.

5 SUMMARY

At present, the high cost and serious pollution of winter heating are the practical problem for people, which need to solve urgently. The clean and low-cost solar heating system is designed and developed for this problem. This system takes full advantage of the solar energy for heating in winter. The system reduces the heating costs, overcomes the high fees resulting from coal heating and instability of indoor temperature during the coal heating. Besides, the system can also be used for summer refrigeration.

REFERENCES

[1] Yu Bi Ju, Zhaohai, Zhu Jian. Archive room temperature and humidity control system based on WSN, The eighth Shenyang Science Conference paper sets (2011).
[2] Chaxiu Guo, Wujun Zhang. Numerical simulation and parametric study on new type of high temperature latent heat thermal energy storage system [J]. Energy Conversion and Management. 2007 (5).
[3] Rakesh Kumar, Marc A. Rosen. A critical review of photovoltaic—thermal solar collectors for air heating [J]. Applied Energy. 2011 (11).

Architectural, Energy and Information Engineering – Sung & Chen (Eds)
© 2016 Taylor & Francis Group, London, ISBN 978-1-138-02791-6

Study on viscoelastic modulus of heat-resistant polymer solution

Tao Ping Chen, Qi Hao Hu & Rui Jie Cui
The Key Laboratory of Enhancing Oil and Gas Recovery Efficiency of Educational Ministry,
Northeast Petroleum University, Daqing, China

ABSTRACT: In order to explore the BH heat-resistant polymer solution viscoelasticity, elastic modulus and viscous modulus are measured at 2 temperatures and 7 concentrations of BH polymer solution with RS6000 type rheometer, and the elastic modulus and viscous modulus are measured with aging 0~90 d at 2 aging temperatures. The results show that: The elastic modulus and viscous modulus of BH polymer solution increase with the increase of the concentration, and decrease with the increase of temperature. Under the same temperature, the angle frequency of elastic modulus and viscous modulus equivalent point decrease with increasing concentration of the polymer solution. Under the same concentration, the angle frequency of elastic modulus and viscous modulus equivalent point increase with increasing temperature. Aging at 50°C and 95°C, BH polymer solution elastic modulus and viscous modulus present reduce after increase firstly, and then decrease slowly with the increasing aging time. After aging 90 d, viscoelastic modulus decreases, but the viscoelastic modulus retention rate is over 70%, what shows that the polymer has good heat resistance.

Keywords: polymer; heat resistance; viscoelasticity; viscoelastic modulus

1 INTRODUCTION

With the wide application of polymer in EOR and other fields, people realize that the polymer solution has high viscosity, which also has elasticity. Polymer solution has viscoelasticity[1–2]. At present the main polymer used in EOR are polyacrylamide and xanthan gum, whose rheological properties and viscoelasticity are researched widely[3–5], and viscoelasticity is related with oil displacement efficiency[6–7]. Core displacement experiment points out that the key factor for improving microscopic oil displacement efficiency of polymer solution is the fluid elasticity[8]. Reference[9] points out that elasticity of polymer solution can improve the microcosmic oil displacement efficiency on residual oil film. Reference[10] gives the mechanism of polymer solution elasticity displacing residual oil in the blind end. The results show that, the viscoelasticity of polymer solution is an important factor to affect the oil displacement efficiency. However, with the increase of temperature, the viscoelasticity of conventional polymer solution weaken rapidly, what restricts its application in high temperature reservoir. Currently some types of heat-resistant polymer for high temperature reservoirs have been developed, which provides a material basis for polymer flooding of high temperature reservoir. In order to provide technical support for polymer flooding of high temperature reservoir, BH heat-resistant polymer solution viscoelastic modulus are tested and analyzed.

2 EXPERIMENTAL MATERIAL AND METHOD

2.1 *Preparation of the polymer solution*

1. Sample preparation. The mother polymer solution is manufactured with oilfield sewage at room temperature, whose concentration is 4500 mg/L. In order to make the polymer molecular chains fully hydrate and extend, the stirring time of the mother polymer solution is 2 h. And then the mother polymer solution is diluted into 7 kinds of concentration with sewage.
2. Filtration. The polymer solution is placed in the intermediate container with filter membrane. Then the polymer solution is pressurized and filtered into a beaker.
3. Deoxidization. The filtered polymer solution is placed into a container and vacated until no air bubble.
4. Aging. The polymer solution after deoxygenation is placed in the incubator, whose temperature is 50°C and 95°C respectively, aging for 10 d~90 d.

2.2 *Experimental equipment and condition*

The viscoelasticity of polymer solution are tested by the RS6000 rheometer at angle frequency ranging 0.01~100 1/s. The testing temperature is 50°C and 95°C respectively. Testing data are recorded by computer automatically.

3 THE CHANGE RULE OF POLYMER SOLUTION VISCOELASTIC MODULUS

The change rule of elastic modulus and viscous modulus are the main parameters of viscoelastic fluid properties. The elastic modulus G' reflects the energy stored in the process of polymer solution elastic deformation, reflects elastic properties of the fluid. The viscous modulus G'' reflects the energy dissipated in the process of polymer solution deformation, reflects viscous properties of the fluid.

According to the curves of elastic modulus G' and viscous modulus G'' with angle frequency are measured at 2 temperatures, 7 concentrations of BH polymer solution, as shown in figure 1~ figure 2.

The figure 1 and figure 2 shows that with the increase of polymer solution concentration and angle frequency, its corresponding elastic modulus G' and viscous modulus G'' all increase, shows that the higher the concentration, the more significant the viscoelasticity of polymer solution. While the viscous modulus and elastic modulus curve under different concentrations of polymer solution has a point of intersection, which is the equivalent point of viscous modulus and elastic modulus. When the angle frequency is less than the equivalent point, the viscous modulus is greater than the elastic modulus, shows that the solution viscous effect is greater than the elastic effect. On the contrary, the elastic modulus is greater than the viscous modulus. This shows that the polymer molecular viscosity is dominant in slow deformation frequency, elasticity is not obvious. When deformation is rapid, the increase deformation energy is absorbed by the intramolecular elastic deformation or intermolecular elastic deformation. The polymer molecular does not have enough time to make the material produces viscous flow, thus elasticity is greater than viscosity. With the increase of polymer solution concentration, the twine between polymer molecules is more serious, the elastic modulus and viscous modulus increase. At low concentrations, molecular chain is not intertwined into instantaneous bond, so the viscous feature is obvious. At medium concentration, on the left of the equivalent point, viscous flow characteristic is obvious. On the right of the equivalent point, the elastic characteristic is obvious. At high concentration, polymer molecular internal chain binds into the ring and molecular chains intertwine into instantaneous bond, the polymer solution shows greater elasticity than the viscosity.

According to the angle frequency ω_c of G' and G'' equivalent point of polymer solution with different concentrations at 50°C and 95°C, angle frequency of G' and G'' equivalent point variation with the concentration is drawn as shown in figure 3. As can be seen from figure 3:

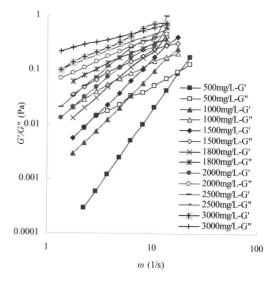

Figure 1. G' and G'' curve of BH polymer solution at 50°C.

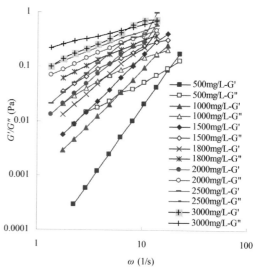

Figure 2. G' and G'' curve of BH polymer solution at 95°C.

292

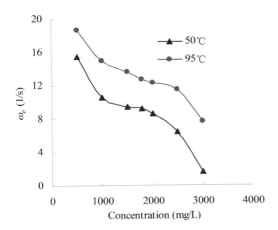

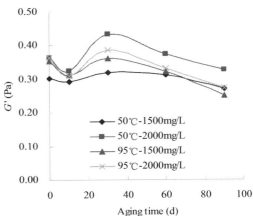

Figure 3. The relationship of G' and G'' equivalent point angle frequency with concentration and temperature.

Figure 4. The relationship of G' with aging temperature and aging time.

1. Under the same temperature, with the increase of the concentration of polymer solution, the angle frequency of G' and G'' equivalent point decreases. This is because the concentration of solution increases, the twine of polymer molecules is more serious. The polymer can exhibit strong elastic effect solution with lower angle frequency.

2. When the concentration of polymer solution is certain, the angle frequency of G' and G'' equivalent point increases with the increase of temperature.

3. For a viscoelastic effect equivalent point angle frequency, the equivalent point is in low concentration at low temperature (50°C), high concentration at high temperature (95°C).

4 THE RELATION OF VISCOELASTIC MODULUS WITH AGING TEMPERATURE AND AGING TIME

In order to study the effect of different aging temperature and aging time on BH polymer solution viscoelasticity, G' and G'' are measured with the concentration of 1500 mg/L and 2000 mg/L after aging polymer solution at 50°C and 95°C. Under the angle frequency of 7.34 1/s, the curve of aging BH polymer solution G' and G'' with the change of aging time at 50°C and 95°C is drawn as shown in figure 4 and figure 5.

As can be seen from the figure 4 and figure 5, within the aging time 30 d days, elastic modulus G' and viscous modulus G'' change ups and downs. After 30 d, the elastic modulus G' and viscous modulus G'' decrease with the increase of aging time, the elastic modulus G' and viscous modulus

G'' at low temperature (50°C) is greater than the value at high temperature (95°C), the elastic modulus G' and viscous modulus G'' of high concentration (2000 mg/L) is greater than the value of low concentration (1500 mg/L).

According to the value of elastic modulus G' and viscous modulus G'' with different aging time measured at 2 temperatures, 2 concentrations of BH polymer solution, the relation curve of G' and G'' retention rate with the aging temperature and aging time is drawn as shown in figure 6 ~ figure 7.

As can be seen from the figure 6, when polymer solution of 1500 mg/L and 2000 mg/L concentration is aged at 50°C, the elastic modulus of polymer solution first increases then decreases with the increase of aging time (based on fresh polymer solution). Calculation shows that, the retention rate of BH polymer solution elastic modulus G' of the concentration 1500 mg/L and 2000 mg/L after aging 90 d are 89% and 90% respectively. The change rule of the elastic modulus increase with the aging time at 50°C is the same as the aging at 95°C.

As can be seen from figure 7, when polymer solution of 1500 mg/L and 2000 mg/L concentration is aged at 50°C and 95°C, viscous modulus change rule and elastic modulus change rule are basically identical with aging time increasing.

As can be seen from figure 6 and figure 7:

1. Aging at 50°C and 95°C, with the increase of aging time, elastic modulus and viscous modulus experience three processes of decrease, increase and decrease slowly. The G' and G'' retention rate of BH polymer solution decrease firstly and then increase and then decrease slowly. The G' and G'' retention rate of BH polymer solution still has a high retention rate after 90 d aging, elastic modulus retention rate is between

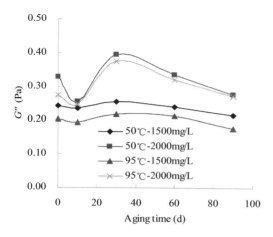

Figure 5. The relationship of G'' with aging temperature and aging time.

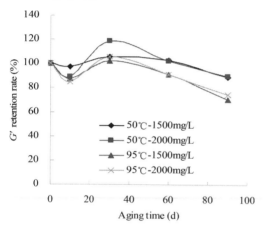

Figure 6. The relationship of G' retention rate with aging temperature and aging time.

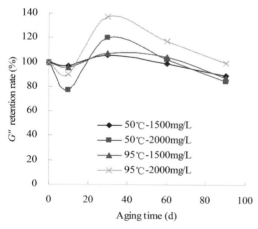

Figure 7. The relationship of G'' retention rate with aging temperature and aging time.

70%~90%, the viscous modulus retention rate is between 84%~100%.

2. After aging 90 d, BH polymer solution viscoelastic parameters retention rate of low concentration is greater than high concentration at 50°C, and the retention rate of low concentration is smaller than high concentration at 95°C.

3. Under the same temperature after aging 90 d, 2000 mg/L BH polymer solution G' and G'' retention rate is smaller than that of 1500 mg/L at 50°C, 2000 mg/L BH polymer solution G' and G'' retention rate is larger than that of 1500 mg/L at 95°C, the difference is around 10%.

5 CONCLUSIONS

1. The elastic modulus and viscous modulus of BH polymer solution increase with the increase of the concentration, and decrease with the increase of temperature.

2. Under the same temperature, the angle frequency of elastic modulus and viscous modulus equivalent point decrease with increasing concentration of the polymer solution. Under the same concentration, the angle frequency of elastic modulus and viscous modulus equivalent point increase with increasing temperature.

3. After 90 d aging, the BH polymer solution viscoelastic parameters retention rate of low concentration is larger than that of high concentration at 50°C, the BH polymer solution viscoelastic parameters retention rate of low concentration is smaller than that of high concentration at 95°C.

4. Aging at 50°C and 95°C, BH polymer solution elastic modulus and viscous modulus present reduce after increase firstly, and then decrease slowly with the increasing aging time. After aging 90 d, viscoelastic modulus decreases, but the viscoelastic modulus retention rate is over 70%, what shows that the polymer has good heat resistance.

ACKNOWLEDGEMENTS

The project was supported by National Science and Technology Major Project of the Ministry of Science and Technology of China (2011ZX05009).

REFERENCES

[1] Manli Tong. The viscoelastic effects of polymer dilute solution in porous media. *Natural Gas Industry*, 7(1), pp. 64–71, 1987 (In Chinese).

[2] Daoshan Li, Wanli Kang, Hongjun Zhu. Studies on Viscoelasticity of Aqueous Hydrolyzed Polyacrylamide Solutions. *Oilfield Chemistry*, 20(4), pp. 347–350, December 2003 (In Chinese).

[3] [3] Hua Liu, Yufu Zhang. Rhelogical Property of the Xanthan Biopolymer Flooding Systems. *Journal of the University of Petroleum*, 19(4), pp. 41–44, August 1995 (In Chinese).

[4] Weiying Wang. Viscoelasticity and Rheological Property of Polymer Solution in Porous Media. *Journal of Jianghan Petroleum Institute*, 16(4), pp. 54–63, December 1994 (In Chinese).

[5] Qinghai Chen, Fulin Yang, Yingjie Liu, et al. Study on Rheology of Partially Hydrolized Polyacrylamide Solution in Low Shear Rate. *Petroleum Geology & Oilfield Development in Daqing*, 25(1), pp. 91–92, February 2006 (In Chinese).

[6] Huifen Xia, Demin Wang, Jirui Hou, et al. Effect of viscoelasticity of polymer solution on oil displacement efficiency. *Journal of Daqing Petroleum Institute*, 26(2), pp. 109–111, June 2002 (In Chinese).

[7] Huifen Xia, Demin Wang, Zhongchun Liu, et al. Study on the Mechanism of Polymer Solution with Viscoelastic Behavior Increasing Microscopic Oil Displacement Efficiency. *Acta Petrolei Sinica*, 22(4), pp. 60–65, July 2001 (In Chinese).

[8] Demin Wang, Jiecheng Cheng, Qingyan Yang. Viscous-Elastic Polymer Can Increase Micro-Scale Displacement Efficiency in Cores. *Acta Petrolei Sinica*, 21(9), pp. 45–51, September 2000 (In Chinese).

[9] Gang Wang, Demin Wang, Huifen Xia. Efffect of viscoelastisity of HPAM solution on residual oil film. *Journal of Daqing Petroleum Institute*, 31(1), pp. 25–30, February 2007 (In Chinese).

[10] Huifen Xia, Demin Wang, Gang Wang, et al. Elastic behavior of polymer solution to residual oil at deadend. *Acta Petrolei Sinica*, 27(2), pp. 60–65, March 2006 (In Chinese).

Architectural, Energy and Information Engineering – Sung & Chen (Eds)
© 2016 Taylor & Francis Group, London, ISBN 978-1-138-02791-6

Research on the factors influencing flow resistance of plate heat exchanger

W. Rao
China Nuclear Power Engineering Co. Ltd., Shenzhen City, Guangdong Province, China

X. Xiao, J.Q. Gao & L. Zhang
School of Energy, Power and Mechanical Engineering, North China Electric Power University, Baoding City, Hebei Province, China

ABSTRACT: Plate heat exchanger is a kind of high efficient and compact heat transfer equipment, and most nuclear power stations are applying it to the closed cooling water system. The plate heat exchanger showed high pressure drop during the operation; the reason is the affection of the flow resistance. The flow resistance relates to the structure and the factors of operation. In this paper, flow resistance from the plate structure and the number of port and passes have been studied. It concludes that the corrugated inclination is the most important geometric parameters of corrugated plate heat exchanger, and the optimal corrugated inclination should be near the 60 degrees.

Keywords: plate heat exchanger; flow resistance; plate structure; the number of port and passes

1   INTRODUCTION

Plate heat exchanger is a kind of high efficient and compact heat transfer equipment, which involves the application of almost all industrial fields, and it is also a wide application prospect. With the application of plate heat exchanger in the industry, enhancing heat transfer effect and reducing the flow resistance become a focus in the study of plate heat exchanger. At the same time, the plate heat exchanger also went wrong more or less because of making and designing during the operation. For example, there are 1000 MW nuclear power units in China. A unit include four plate heat exchangers, which work as the RRI/SEC heat exchangers. But the heat exchangers of the closed cooling water system showed high pressure difference during the operation. The actual pressure difference is 30% higher than the designed pressure difference. The high pressure drop can affect the heat exchange and the work of the pump and fan. Through the analysis and research from the locale, it concluded that the main reason of high pressure drop is that the margin is not enough at the design calculation.

The structure of plate heat exchanger is more complex. So there are many factors that influencing the pressure difference. Such as the plate structure, port number and passes number[1]. All these factors can affect the pressure drop, but how can they affect? And which is the main influencing factors? Therefore, the flow resistance is more worth studying.

2   THE COMPOSITION OF FLOW RESISTANCE

For single-phase fluid, the flow resistance consists of the frictional resistance and the local resistance[2].

2.1   The frictional resistance

In flow channel of fluid flow, the frictional resistance is because of the viscous fluid and solid wall contacting, and making displacement. In general, the higher flow velocity, the greater the viscosity, surface roughness, the longer the process, and the greater the friction resistance.

The basic formula for calculating frictional resistance is:

$$\Delta P = 4f \frac{L}{d_e} \frac{\rho \omega^2}{2} \tag{1}$$

The basic formula of f is:

$$f = CR_e^n \tag{2}$$

where C and n depend on the structure.

2.2 The local resistance

Fluid in the process of flow, the resistances result from the changing of flow direction or speed. The basic form of formula for calculating local resistance is:

$$\Delta P = \xi \frac{\rho \omega^2}{2} \qquad (3)$$

where ξ is the coefficient of local resistance.

The coefficient of local resistance relates to the geometry of the size of the local resistance coefficient and local disturbance flow shape, size, shape and surface roughness.

3 CALCULATION OF FLOW RESISTANCE

3.1 Flow process

Figure 1 shows that the flow process consists of the corner hole and the flow channel. The flow resistance is different in the two processes [3].

3.2 The calculation formula containing the coefficient of friction

When there is no rule of Euler equation, the pressure drop and coefficient of friction correlations to calculate can be used. The pressure drop of plate heat exchanger consists of angle pore and flow pressure drop [4]:

$$\Delta p = \Delta p' + \Delta p'' \qquad (4)$$

(1) The corner hole pressure drop

Corner hole pressure drop is a fluid flows through the corner hole flow channel, in order to overcome its resistance to flow pressure drop

$$\Delta p' = m f_1 \left(\frac{\rho \omega^2}{2} \right) \left(1 + \frac{n}{100} \right) \qquad (5)$$

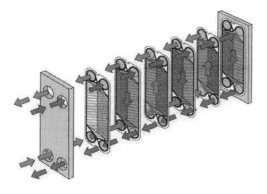

Figure 1.   The diagram of flow process.

where f_1 is the coefficient of friction, n is the channel number in a routine, m is the passes number, and it depends on the different type of plate heat exchanger

(2) The flow pressure drop

The flow pressure drop is that flowing into the angle of hole and coming out of the other hole and overcome its resistance and formation pressure drop, the herringbone:

$$\Delta p'' = 2 f_2 \frac{L}{d_e} \rho \omega^2 m \left(\frac{\mu}{\mu_\omega} \right)^{-0.17} \qquad (6)$$

where f_2 is the coefficient of friction, and L is the length of flow, μ is the Dynamic viscosity of fluid side, and μ_ω the dynamic viscosity of wall lateral.

4 THE DESIGN FACTORS OF FLOW RESISTANCE

Under the same flow rate, different type or Different geometry parameters, the flow resistance is also different [5].

4.1 The design of plate structure

Heat transfer plate is the key element of plate heat exchanger, the ripple forms have an important influence on the performance of the plate heat exchanger. To meet the needs of different occasions, people has developed various kinds of corrugated plate, such as chevron bellows, horizontal flat ripple, the ripple, dimple shaped rippled, and the chevron and horizontal straight corrugated are mostly used in industrials currently. To measure the performance of corrugated plate, its main parameters are heat transfer efficiency, fluid resistance and bearing capacity. In general, herringbone corrugated plate heat transfer efficiency and the fluid resistance are high, and good bearing capacity; relative to the herringbone corrugated plate, the heat transfer efficiency of horizontal straight corrugated plate is lower, fluid resistance small, bearing capacity and slightly lower.

The herringbone corrugated plate as the research object. In the slab structure parameters, there are three main influences for the plate heat exchanger: corrugated inclination, corrugated depth, and ripple spacing. To explore the relationship between the performances of plate heat exchanger, we studied respectively. Its influence on the pressure drop is described below:

4.1.1 Corrugated inclination

When corrugated inclination increased gradually, the flow resistance loss increases, but the pressure

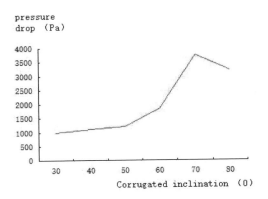

Figure 2. The carve of the pressure drop along with the change of corrugated inclination.

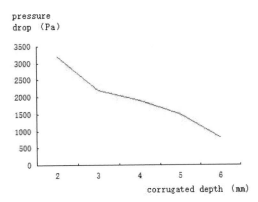

Figure 3. The carve of the pressure drop along with the change of corrugated depth.

drop begins to decrease at the degree of 70. The reason why pressure drop begins to decrease at the degree of 70 is that corrugated inclination increases caused a decline of the contactor between the plates, and when contactor between the plates reduces, turbulence effect is abate, the heat transfer effect and resistance are smaller.

4.1.2 *Corrugated depth*

The corrugated depth increases by 4 mm to 6 mm, when it continues to reduce but change trend is relatively moderate. In the practical work of plate heat exchanger, the smaller the ripple depth, the smaller the scaling tendency.

By the above analysis, we can see that: between heat transfer and resistance problems should be considered, in the meantime to find the most advantage. Nu was similar when at the degree of 70 and 60, but at the degree of 70 the pressure drop is twice at the degree of 60. Based on the above consideration, 60 degrees heat transfer effect is good and not the maximum resistance, so the optimal corrugated inclination of herringbone corrugated plate heat exchanger should be near the 60 degrees.

The use of smaller corrugated depth is useful to high-speed turbulent for cleaning effect into full play and can be particularly effective to reduce the deposition of dirt. In addition, under the condition of the ripple spacing, it is difficult in processing corrugated depth, so from the analysis, it concluded that the reasonable corrugated depth should be around 4–5 mm.

4.1.3 *Ripple spacing*

When the ripple spacing increases, the gradient of the pressure drop is larger. Nu is smaller but not gradient, so we should perform comprehensive analysis on the influence of the pressure drop and heat transfer.

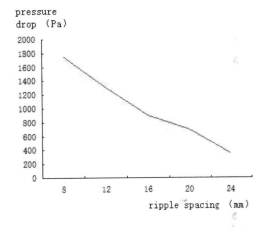

Figure 4. The carve of the pressure drop along with the change of ripple spacing.

The corrugated inclination is one of the most important geometric parameters of corrugated plate heat exchanger in this study.

Corrugated inclination changing leads to the change of flow state, thus affecting the heat transfer and resistance of the plate heat exchanger. At the designing of the plate heat exchanger, the three important factors must be considered, not only considered as one or two elements. Not only to pursuit reducing resistance, but also to improve the heat transfer coefficient at the same time, looking for the matching style of heat transfer and resistance is the key to the work.

4.2 *Design of the port number*

The corner hole consists of sheet bar and corn pore, its length increases with the increase in port number N of the plate. The pressure drop of corner hole passages not only related to the structural type and geometric size of angle hole, but also related to the port number [6]. When flow velocity is constant, the pressure drop of corner hole increases quickly with the increase of the flow channel number N. When N increased to a certain value, the pressure drops of corner hole channel more than that in port. For example, for BR-0.05 plant heat exchanger, when the Re of water is 1000, the pressure drop of corner hole is 1.6 times as the pressure drop of flow. So, the relationship between the pore channel resistance and port number should design reasonably.

4.3 Design of the passes number

It is generally believed that the pressure drop of a multi-pass plate heat exchanger is the one-way pressure drop multiplies the number M, which is the number of passes. Practice has proved that the calculated value is higher than the measured values, and the reason is more complex. For one-way, the total pressure drop includes the equipment import and export pressure drop, and multi-pass exchanger does not exist the pressure drop started from the second path. As another example, in the inlet side corner hole channel of a single heat exchanger, the direction of fluid flow and channel and vertical fin, so fluid disturbance is large, and the pressure drop is also big. We know that fluid is along the inner finned channel spin forward after the second path channel, the fluid disturbance is small, so the pressure drop is small. Selecting reasonably passes number is necessary.

5 CONCLUSIONS

1. The corrugated inclination is the most important geometric parameters of corrugated plate heat exchanger. But at the designing of the plate heat exchanger, the three important factors must be all considered.
2. At the degree of 60, heat transfer effect is good and not the maximum resistance, so the optimal corrugated inclination of herringbone corrugated plate heat exchanger should be near the 60 degrees.

REFERENCES

[1] Zhang J.Z. et al., 2012. Effects of the contact points distribution on heat transfer and resistance performances of plate heat exchangers, *journal of Shandong University (engineering science)*: 121–126.
[2] Yang C.L. 1998. The design manual of Plate heat exchanger,
[3] Yang R.Z. & Gong B., 2009, Effectiveness of Design and Calculation on Plate Heat Exchangers, *petrochemical equipment*: 12–14.
[4] Qiu X.L. 2013. Study of Heat Transfer and Flow Resistance in Chevron Plate Corrugation Geometry Parameters. *South China University of technology*.
[5] Cui L.Q. 2008. Theoretical 3D Numerical Simulation of Plate Heat Exchanger Based on FLUENT, *Zhejiang University*.
[6] Zhao Z.N. 2001, Effects of the corrugated inclination angle on heat transfer and resistance performances of plate heat exchangers, *petro-chemical equipment*: 1–3.

Study on chromium-free iridescent passivator for zinc coating

Lei Shi, Jing Ya Ye & Sheng Nan Wang
School of Materials Science and Engineering, Shandong Jianzhu University, Jinan, China

ABSTRACT: In this paper, using the chromium-free iridescent passivator for zinc coating was studied, a passivation film of bright iridescent appearance and uniform color can be achieved. Single variable method is used to determine the scope of the content of each component in passivation solution, and by the method of optimization contrast experiment, components of chromium-free iridescent passivation solution for zinc coating are optimized. Various process parameters on the influence of passivation film appearance and corrosion resistance have been studied. Electrochemical testing and dropping experiment are carried out to study the corrosion resistance of passivation film, and scanning electron microscopy and X-ray diffraction are used to observe the surface morphology of passivation film and analyze components.

Keywords: Zinc coating; chromium-free; iridescent passivation

1 INTRODUCTION

Iridescent passivation film technologies that commonly used are high chromate passivation, low chromate passivation, ultra low chromate passivation, and chromium-free passivation. Traditional passivation technologies all use chromate passivation, which are high toxicity and cause serious environmental pollution. Chromium-free passivation controls the existence of chromium ions at the source, to make production enterprises achieve the purpose of clean production by using new products, so that the products do not contain any chromium ion after passivation, and also ensures that the final products meet the requirements of environmental protection. The anti-corrosion effect of some chromium-free passivation is similar to hexavalent chromium or chromium-free passivation, and some indexes even exceed hexavalent chromium passivation; while some of the chromium-free passivating technologies are simple, and there is no need to change the existing galvanizing technology, therefore using chromium-free passivation to replace chromium passivation has been proved to be feasible in both effect and production. In this paper, using the chromium-free iridescent passivator for zinc coating studied, a passivation film of bright iridescent appearance and uniform color can be achieved.

2 EXPERIMENT

2.1 Base material

The material used in the experiment is low-carbon steel of size 4 cm × 8 cm. The reagents selected are all analytical reagents.

2.2 Technological process

Matrix degreasing ~ washing ~ acid pickling ~ washing ~ polishing ~ washing ~ activating (diluted hydrochloric acid of concentration 3%~5%) ~ washing ~ electrogalvanizing ~ washing ~ bright dipping (dilute nitric acid of concentration 30–50 g/L) ~ washing ~ passivation ~ washing ~ drying.

This experiment uses alkaline non-cyanide zinc plating.

2.3 Formula and process

2.3.1 Zinc plating

Zinc oxide/ (g/L)	10
Sodium hydroxide / (g/L)	110
DPE-III/(ml/L)	6
WB-3 ml/L	4
Temperature	20–30
Cathode-current density/(A/dm^2)	1.5–2.0
pH	12.5~13
Electroplate time	5~10 min

2.3.2 Passivation

Cerous nitrate (g/L)	50
Sodium citrate (g/L)	15
Cobalt nitrate (g/L)	20
Temperature/°C	20~30
pH	1~2.5
Passivation time	8~20 s

2.4 Test methods

2.4.1 Micro-morphology

The surface micro-morphology of coating was analyzed by type JSM-6380LA Scanning Electron Microscope (SEM) of Japan Electronics Co., Ltd.

2.4.2 Passivation film phase analysis

Passivation film phase is analyzed by XRD diffraction.

2.4.3 Corrosion resistance of coating

Tafel curve of coating was measured by type PARSTAT 2273 electrochemical workstation of Princeton Applied Research Co., Ltd. Solution is the potassium chloride solution of mass fraction 5%; scanning rate is 10 mv/s, reference electrode is the saturated calomel electrode, auxiliary electrode is the platinum electrode, and working electrode is the experiment sample.

3 RESULTS AND DISCUSSION

3.1 Influence of components of passivation film on passivation film

3.1.1 Cerous nitrate

Regarding cerous nitrate as a single variable, the content of cobalt nitrate is 18 g/L, the content of sodium citrate is 20 g/L, and PH value adjusts to 1.5 by the concentrated nitric acid. The influence of the content of cerous nitrate on the appearance of passivation film is shown in Figure 1.

From the Figure 1, it is concluded that when the content of cerous nitrate is 40~60 g/L, passivation film appears to be iridescent film of bright, uniform, and good color and lustre.

3.1.2 Sodium citrate

Regarding sodium citrate as a single variable, the content of cerous nitrate is 48 g/L, the content of cobalt nitrate is 18 g/L, and PH value adjusts to 1.5 by the nitric acid, the influence of the content of sodium citrate on the appearance of passivation film is shown as in Figure 2.

From Figure 2, it is shown that when the concentration range of sodium citrate is 15~25 g/L, passivation film appears to be iridescent film of bright and uniform, so the concentration range of sodium citrate is 15~25 g/L.

3.1.3 Cobalt nitrate

Regarding cobalt nitrate as a single variable, the content of cerous nitrate is 48 g/L, the content of sodium citrate is 20 g/L, and PH value adjusts to 1.5 by the concentrated nitric acid. Because in the formula of passivation solution, cobalt nitrate affects the corrosion resistance of passivating film, so the copper sulfate intravenous drip experiment has been carried out in this experiment to observe the appearance and corrosion resistance of the passivation film. The influence of the content of cobalt nitrate on the appearance of passivation film is shown as in Figures 3–6.

Through the appearance in Figure 3 and the analysis in Figure 4, synthesizing the appearance and corrosion resistance, it can be concluded that the concentration range of cobalt nitrate is 15~25 g/L.

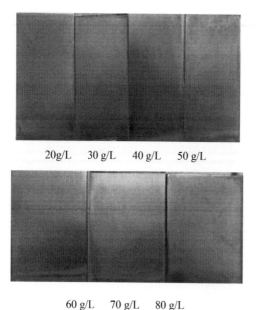

| 20g/L | 30 g/L | 40 g/L | 50 g/L |

| 60 g/L | 70 g/L | 80 g/L |

Figure 1. The influence of the content of cerous nitrate on the appearance of passivation film.

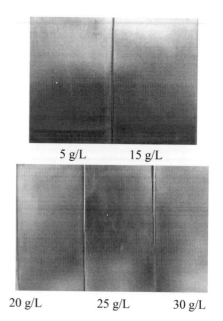

| 5 g/L | 15 g/L |

| 20 g/L | 25 g/L | 30 g/L |

Figure 2. The influence of the concentration of sodium citrate on the appearance of passivation film.

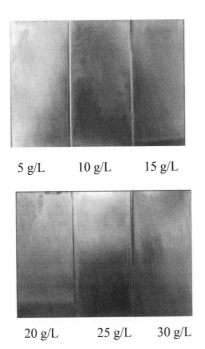

5 g/L   10 g/L   15 g/L

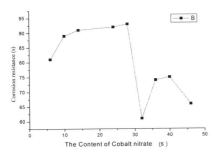

20 g/L   25 g/L   30 g/L

Figure 3.   The influence of the content of cobalt nitrate on the appearance of passivation film.

Figure 4.   The line chart of the relationship between the concentration of cobalt nitrate and drip time.

Figure 5.   The relation diagram of the concentration of cobalt nitrate and self-corrosion current density.

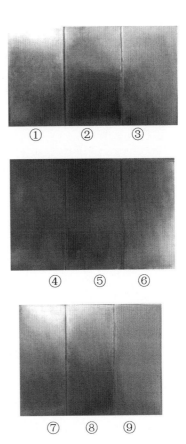

Figure 6.   The morphology of samples.

Since the self-corrosion current density is smaller and the corrosion resistance is higher, it has been verified in Figure 6 that when the content of cobalt nitrate is 15~25 g/L, it has a very good corrosion resistance, small corrosion current density, and good appearance effect. Nitrate ions provided by cobalt nitrate, have the effect of bright dipping. When the content is high, the passivation film achieved has a high brightness, and cobalt ions can increase the corrosion resistance, and self-corrosion current density of the passivation film decreases with the increase of concentration of cobalt nitrate, which explains its corrosion resistance increases with the increase of the content of cobalt nitrate.

3.2   *Optimize the concentration of each component of passivation solution by the orthogonal test*

Through the above single factor experiments, the approximate concentration of each component in passivation solution has initially been

Table 1. The orthogonal formula.

	Cerous nitrate	Sodium citrate	Cobalt nitrate
①	40	15	15
②	40	20	20
③	40	25	25
④	50	15	20
⑤	50	20	25
⑥	50	25	15
⑦	60	15	25
⑧	60	20	15
⑨	60	25	20

Table 2. The comparison table of corrosion resistance of samples passivation.

No.	corrosion resistance(s)
④	92
⑤	79
⑥	85

set at: cerous nitrate 40–60 g/L, sodium citrate 15–25 g/L, cobalt nitrate 15–25 g/L, pH value 1.5, passivation time 15 s, empty stop 8 s, and room temperature. List the orthogonal formula as the following Table 1.

Do nine experiments respectively and the appearance of passivation film achieved has been shown as in Figure 6.

From the figure, the passivation film of ④⑤⑥ appears to be iridescent film of bright, good color and luster, and uniform. The copper sulfate intravenous drip experiment has been carried out to observe its corrosion resistance, and intravenous drip results are shown in Table 2 below.

Analyzing Figure 6 and Table 2 by contrast, the best formula of passivation solution is ④, which the formula of passivation solution is: 50 g/L of cerous nitrate, 15 g/L of sodium citrate, and 20 g/L of cobalt nitrate.

3.3 Performance test

3.3.1 Surface morphology
Under the optimal technological conditions, the surface morphology of samples after passivation is analyzed under the scanning electron microscopy. Test results are shown in Figures 7 and 8.

From the figure, it is shown that the surface of samples after passivation is smooth and dense, which greatly improves the corrosion resistance of zinc coat.

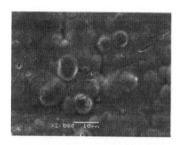

Figure 7. The surface morphology of zinc coating samples is not passivated under the scanning electron microscopy.

Figure 8. The surface morphology of zinc coating samples after iridescent passivation under the scanning electron microscopy.

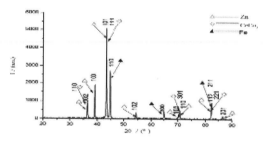

Figure 9. The XRD pattern of the passivation film of samples.

3.3.2 Passivation film phase analysis
The passivation film of samples is observed after passivation under XRD. The phase analysis table is shown in Figure 9.

4 SUMMARY

1. The best formula of chromium-free iridescent passivation solution for zinc coating is, 50 g/L of cerous nitrate, 15 g/L of sodium citrate, 20 g/L of cobalt nitrate, temperature is room temperature, pH = 1.5, and passivation time

is 15s. The passivation film achieved after the optimum process is smooth, uniform, bright, and good corrosion resistance.

2. The formula of passivation solution determined after test has advantages of convenient operation, low cost, high corrosion resistance of passivation film, and beautiful appearance, which has a certain positive significance to replace chromate.

REFERENCES

Anfu Zhang. The new technology of surface treatment based galvanized steel. Surface Technology, 1991, 20(5): 29–33.

Daren Qian, Shijia Li, Jianping He, etc. Preliminary study on chromate-free passivation for zinc. Electroplate and Environmental Protection, 1992, 15(5): 7–10.

Haili Pu, Jianhua Wang, Xiong Jiang. Discuss to chromium-free passivation [J]. Electroplate and Environmental Protection, 2004, 24(2): 25–26.

Hinton, Romanl, Cristescuc. Non-Chromate Conversion Coating Treatments for Electrodeposited Zinc-Nickel Alloys. Electrochem, 1996, 74(5): 171–173.

Huili Hu, Jinning Cheng, Ning Li, etc. The application and prospect of phytic acid in the metal protection. Material Protection, 2005, 38(12): 39–41.

Huili Hu, Ning Li, Jinning Cheng. Study on corrosion resistance of phytic acid passivation film for zinc coating. Electroplate and Environmental Protection, 2005, 25(6): 21–25.

Jianghong Zhang, Yingjie Zhang, Lei Yan, Yuxing Yan. The research and development of chromiumfree passivation technology for galvanized products, Material Protection, 2009, 42(3): 48–50.

Ju Kang, Lihua Han, Yinghua Liang. Research progress of chromate-free passivation for zinc coating. Shanghai Chemical Engineering, 2008, 33(6):18–22.

Li Gong, Lin Lu, Yanping Lu. Study on the corrosion resistance of thin chromium-free organic composite coated steel, Material Protection, 2008, 41(2): 68–71.

Ninglin Zhou. Introduction to organic silicon polymer. Beijing: Science Press, 2000: 168–172.

Plating & Surface Finishing, 2001 (2): 68.27 1 Bellezze T. Royenti G. Fratesi R. Electrochemical Study of on the Corrosion Resistance of Cr(m)- Based Conversion Layers on Zinc Coatings. Surface and Coatings Technology, 2002 (155): 221–23.

Shuangcheng Wu. The application of tannic acid in surface treatment. Surface Technology, 2000, 29(2): 36–37.

Urlton P. The Effect of the Sealers on Increase of Corrosion Resistance of Chromate, Free Passivation on Zinc & Zinc Alloys.

Xunjun Chen. The corrosion inhibition and mechanism of ammonium acetate molybdate. China Corrosion and Protection Journal, 1995, 15(4): 279–284.

Z W Chen, N F Kennon, J B See, et al. Technigalva and Other Developments in Batch Hot-Dip Galvanizing. JOM, 1992, 44(1): 22–26.

Architectural, Energy and Information Engineering – Sung & Chen (Eds)
© 2016 Taylor & Francis Group, London, ISBN 978-1-138-02791-6

Study on the galvanized chromium-free blue-white passivator

Lei Shi, Jing Ya Ye & Wen Ting Song
School of Materials Science and Engineering, Shandong Jianzhu University, Jinan, China

ABSTRACT: Electroplating zinc coating usually needs passivation, as the application of hexavalent chromium and trivalent chromium is limited, a kind of chromium-free passivator is studied in this paper, and a blue-white uniform galvanized passive film with bright appearance can be achieved. The single variable method is used to determine the scope of the each component's content of passivation solution, and using the orthogonal experiment to optimize the composition of chromium-free passivation solution. This paper studies the various process parameters on the influence of passivation film appearance and corrosion resistance. The corrosion resistance of passivation film is studied by electrochemical test and salt spray test and the surface morphology of passivation film is observed and its components are analyzed using scanning electron microscopy and X-ray diffraction.

Keywords: zinc coating; chromium-free; blue-white; passivation

1 INTRODUCTION

In the industrial production, in order to reduce the thickness of zinc coating, and improve the corrosion resistance of zinc coating at the same time, passivation is usually used after galvanization. Passivation is a treatment process that using oxidizers to form a layer of conversion film on zinc coating, to improve the corrosion resistance of zinc and give a beautiful outer conversion film. Currently, passivation in market is mainly the passivation of hexavalent chromium and trivalent chromium. As hexavalent chromium is a highly toxic and carcinogenic substance, and its harm to the human body and environment, it has caused a great attention in various countries. On July 1st, 2011, the European Union issued a new directive ROHS2.0 to replace the old directive. In the ROHS2.0 article 4 (1) and the annex II, it is clearly pointed out that the maximum permissible concentration of Cr^{6+} in the homogeneous material is 0.1%. Hexavalent chromium passivation obviously can not meet the requirement, and trivalent chromium in passivation solution is easily transformed into hexavalent chromium, which is also difficult to achieve the standard. Passivation of chromium-free blue-white can be comparable to that of hexavalent chromium, and has a similar corrosion resistance, and the appearance is also better than that of trivalent chromium, so it has a good protective decorative property. On the basis of the original, this paper further explores chromium-free galvanized blue-white passivation.

2 EXPERIMENT

2.1 Base material

The material used in the experiment is Q235 steel, and the size of test block is 40 mm × 40 mm × 0.8 mm. The reagents selected are all analytical reagents.

2.2 Technological process

Matrix degreasing ~ washing ~ acid pickling ~ washing ~ polishing ~ washing ~ activating (diluted hydrochloric acid of concentration 3%~5%) ~ washing ~ electrogalvanizing ~ washing ~ bright dipping (dilute nitric acid of concentration 30–50 g/L) ~ washing ~ passivation ~ washing ~ drying.

This experiment uses alkaline zincate to galvanize it.

2.3 Formula and process

2.3.1 Galvanization

Zinc oxide/(g/L)	10
Sodium hydroxide/(g/L)	110
DPE-III/(ml/L)	6

WB-3 ml/L	4
Temperature	20–30
Cathode-current density/ (A/dm^2)	1.5–2.0
pH	5~5.5
Electroplate time	20 min

2.3.2 Passivation

Sodium molybdate (g/L)	30
Sodium citrate (g/L)	15
Cobalt nitrate(g/L)	35
Temperature/°C	25~35
pH	1~2
Passivation time	15~25 s

2.4 Test methods

2.4.1 Micro-morphology

The surface morphology of the coating is analyzed by type JSM-6380 LA Scanning Electron Microscope (SEM) of Japan Electronics Co., Ltd.

2.4.2 Passivation film phase analysis

Passivation film phase is analyzed by XRD diffraction.

2.4.3 Corrosion resistance of coating

Tafel curve of coating is measured by type PARSTAT 2273 electrochemical workstation of Princeton Applied Research Co., Ltd. Solution is the potassium chloride solution of mass fraction 5%; scanning rate is 10 mv/s, reference electrode is the saturated calomel electrode, auxiliary electrode is the platinum electrode, and working electrode is the experiment sample.

In the salt spray test, brine is 5% sodium chloride, pH value is 6.5~7.2, temperature of test chamber is required at (35 ± 1)°C, humidity is higher than 85%, the amount of fog spray is 1~2 mL/(h · cm^2), and nozzle pressure is 78.5~137.3 kPa (0.8~1.4 kgf/cm^2).

3 RESULTS AND DISCUSSION

3.1 Influence of components on passivation film

3.1.1 Sodium molybdate

As sodium molybdate is the main salt, so first the range of sodium molybdate dosage is determined. Regarding the content of sodium molybdate as a single variable, PH value adjusts to 1.5, the influence of the sodium molybdate content on the appearance of passivation film and its self-corrosion potential is obtained as in Tables 1 and 2.

As Table 1 shows that, passivation film begins to be in blue-white when molybdate is 25 g/L, and when it is 50 g/L, the appearance of passivation film is dark blue. It is supposed that the molybdate plays a certain role in the appearance of passivation film.

The passivation film is in darker color when the molybdate content is higher. The corrosion resistance is higher when the self-corrosion current density is smaller.

From Table 2, it can be seen that the concentration of sodium molybdate must be 20~40 g/L.

3.1.2 Cobalt nitrate

The main function of cobalt nitrate is to determine the appearance and corrosion resistance of passivation film. Regarding the content of cobalt nitrate as a single variable, PH value adjusts to 1.5, the influence of the cobalt nitrate content on appearance of passivation film and its self-corrosion potential is obtained as in Tables 3 and 4.

As Table 3 shows that the color and lustre of passivation film are very good when cobalt nitrate is 15 g/L~40 g/L, and from Table 4, with the increase of the concentration of cobalt nitric, self-corrosion current density of passivation film decreases, which explains that its corrosion resistance increases with the increase of the content of cobalt nitrate. And when the content of cobalt nitrate is 25~40 g/L, corrosion resistance of passivation film is the best. Therefore, the range of

Table 1. The influence of the content of sodium.

Content of sodium molybdate	Appearance of passivation film
5g/L	Yellow, bright, uniform
10g/L	Yellow, bright, uniform
15g/L	Yellow, bright, uniform
20g/L	Yellow, bright, uniform
25g/L	Blue-white, bright, uniform, good color and lustre
30g/L	Blue-white, somewhat blue, bright, uniform
40g/L	Dark blue, bright, uniform, good color and lustre

Table 2. The relation of the sodium molybdate content on the appearance of passivation film molybdate and self-corrosion current density (Electrode area: 0.8 cm^2).

Content of sodium molybdate (g/L)	Self-corrosion current density (mA/cm^2)
5	2.22E-5
10	5.58E-6
15	4.93E-6
20	1.94E-6
25	1.57E-6
30	1.72E-6
40	2.44E-6

Table 3. The influence of the content of cobalt.

Content of cobalt nitrate	Appearance of passivation film
1 g/L	Blue-white, bright, uniform
5 g/L	Blue-white, bright, uniform
10 g/L	Blue-white, bright, uniform
15 g/L	Blue-white, bright, uniform, good color and lustre
20 g/L	Blue-white, bright, uniform, good color and lustre
25 g/L	Blue-white, bright, uniform, good color and lustre
30 g/L	Blue-white, bright, uniform, good color and lustre
40 g/L	Blue-white, bright, uniform, good color and lustre

Table 4. The relation of the cobalt nitrate content on the appearance of passivation film nitrate and self-corrosion current density (Electrode area: 0.8 cm^2).

Content of cobalt nitrate (g/L)	Self-corrosion current density (mA/cm^2)
20	1.09E-5
25	9.36E-6
30	3.53E-6
35	8.85E-7
40	6.21E-7

Table 5. The influence of the content of sodium (Electrode area: 0.8 cm^2).

Content of sodium nitrate (g/L)	Appearance of passivation film
5	Blue-white, fogged, non-uniform
10	Blue-white, bright, non-uniform
15	Blue-white, bright, uniform
25	Blue-white, bright, uniform
30	Blue-white, bright, uniform

the amount of cobalt nitrate dosage is determined to be 25~40 g/L.

3.1.3 Sodium citrate

Sodium citrate is a kind of common complexing agent, and mainly plays a complexation role in the passivation solution. Regarding the content of sodium citrate as a single variable, pH value adjusts to 1.5, the influence of sodium citrate content on the appearance of passivation film and its self-corrosion potential is obtained as in Tables 5 and 6.

Table 6. The relation of the citrate content on the appearance of passivation film sodium citrate and self-corrosion current density.

Content of sodium citrate (g/L)	Self-corrosion current density (mA/cm^2)
5	7.03E-6
10	5.98E-6
15	1.98E-7
25	2.61E-6
30	5.58E-7

Table 5 shows that when the content of sodium citrate is between 10 and 30 g/L, the appearance of passivation film is good.

Table 6 shows that when the content of sodium citrate is between 15 and 30 g/L, self-corrosion current density is small, and corrosion resistance is high. Therefore, the range of the amount of sodium citrate dosage is 15~30 g/L.

3.2 Optimize the concentration of each component of passivation solution by the orthogonal test

Based on the influence of sodium molybdate, cobalt nitrate, and sodium citrate contents on the appearance and corrosion resistance of passivation film, factors in the orthogonal test are determined. According to the appearance and corrosion resistance of passivation film, the level of each factor in the orthogonal test is determined. Select the amount of sodium molybdate to be 20 g/L, 25 g/L and 30 g/L, select the amount of cobalt nitrate to be 25 g/L, 30 g/L and 35 g/L, select the amount of sodium citrate to be 15 g/L, 22 g/L and 30 g/L. Choose L9 (three factors and three levels) orthogonal table to arrange the experiment. According to the result of orthogonal test, determine the optimal formula of galvanized passivation solution by observing the appearance and corrosion resistance of passivation film.

Table 7 is the result of orthogonal test. It is realized that the appearance of sets 3, 5, and 6 affects the beauty of passivation film, so the electrochemical test (i.e., corrosion resistance test) was not performed with them. Table 8 is the self-corrosion current density of each set, the self-corrosion current density of set 7 has the minimum density, so the corrosion resistance is the best. Synthesizing the appearance and corrosion resistance, finally the test 7 is determined to be the optimal formula.

3.2.1 Surface morphologies

Under the optimum process conditions, the surface morphologies of samples after passivation are analyzed under the scanning electron microscopy. Test results are as in Figures 1 and 2.

Table 7. The result of orthogonal test.

Test factors	Appearance of passivation film
1	Blue-white, bright, uniform
2	Blue-white, bright, uniform
3	Blue-white, somewhat yellow, fogged
4	Blue-white, bright, uniform
5	Blue-white, somewhat blue, bright, uniform
6	Blue-white, somewhat blue, bright, uniform
7	Blue-white, bright, uniform
8	Blue-white, bright, uniform
9	Blue-white, bright, uniform

Table 8. Orthogonal test and self-corrosion current density.

Test	Self-corrosion current density (mA/cm^2)
1	4.44E-6
2	4.98E-6
4	1.50E-5
7	2.07E-6
8	2.17E-6
9	2.27E-6

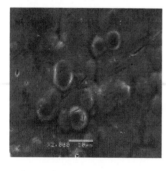

Figure 1. The surface morphologies of samples not passivated under the scanning electron microscopy.

Figure 2. The surface morphologies of samples passivated under the scanning electron microscopy.

3.3 Performance test

The figure shows that the surface of samples passivated is smooth, dense, and seamless, which greatly improves the corrosion resistance of galvanized sheet.

3.3.1 Salt spray test

During the test, on the basis of the GB6458-56 national standards for salt spray test, in the particular test chamber (plating equipment), put samples on the special shelf, and appears 15°~30° in the plane. According to the GB6458-86 standard requirements, neutral salt spray test is done with samples. Spray the brine that contains 5% of sodium chloride, pH value is 6.5~7.2 to samples through a spraying device, and let the salt fog settle onto the test pieces. The temperature of test chamber is required at (35 ± 1)°C, humidity is higher than 85%, the amount of fog spray is 1~2 mL/(h · cm^2), and nozzle pressure is 78.5~137.3 kPa (0.8~1.4 kgf/cm^2).

The test finds that, after 72 hours continuous spray test with 5% neutral salt, it have the phenomena with no rust, no white spot, and no corrosion. According to GB/T9800-1988 "The chromate conversion film of electrogalvanizing and cadmium plating layer" or ISO4520:1981, it meets the requirements of the corrosion resistance of conversion film.

4 SUMMARY

The best formula of galvanized chromium-free blue-white passivation solution is, 30 g/L of sodium molybdate, 15 g/L of sodium citrate, 35 g/L of cobalt nitrate, the temperature is room temperature, pH = 1.5, and passivation time for 15s. The passivation film achieved after the optimum process is smooth, uniform, bright, and good corrosion resistance.

REFERENCES

[1] Yali Liu, Zhiyi Yong, Jing Zhu, et al. A preparation method of magnesium alloy surface molybdate compound conversion film. Patent number: 200710034700.

[2] Ju Kang, Lihua Han, Yinghua Liang. Research progress of zinc coating chromate-free passivation. Shanghai Chemical Engineering, 2008, 33(6): 18–22.

[3] Wilcox G.D., Gabe D.R. Chemical molybate conversion treatments for zinc. Metal Finishing, 1988, 86(9): 71–74.

[4] Jintang Lu, Gang Kong, Jinhong Chen, et al. Hot-dip Zn coating molybdate passivation process. Corrosion Science and Protection Technology. 2001, 13(1): 46–48.

[5] Wilcox G D, Gabe D R. Passivation studies using group VIA anions. V cathodic treatment of zinc. British Corrosion Journal, 1987, 22(4): 254–258.

[6] Keping Han, Jingli Fang. Study color anti-corrosion film of zinc surface by XPS and AES. China Corrosion and Protection Journal, 1997, 7(1): 41–50.

[7] Tang P T, Bech-Nielsen G, Moller P M. Molybdate based alternatives to chromating as passivaiton treatment for zinc. Plating and surface Finishing, 1994, 81(11): 20–23.

[8] Tang P T, Bech-Nielsen G Moller P M. Molybdate based passivation of zinc. Transactions of the Institute of Metal Finishing, 1997, 75(4): 144–148.

[9] Yuanchun Yu, Ning Li, Huili Hu, et al. Research progress of chromium-free passivation and trivalent chromium passivation. Surface Technology. 2005, 34(5): 6–9.

[10] Liqun Zhu, Fei Yang. Study on environmental protection zinc coating blue passivation film corrosion resistance. Corrosion and Protection. 2006, 27(10): 503–507.

[11] Liqun, Zhu, Fei Yang, Huijie Huang. Investigation of Formation Process of the Chromium-free Passivation Film of Electrodeposited Zinc. Chinese Journal of Aeronautics.. 2007, 20: 129–133.

[12] Lin Lu, Xiaogang Li, Li Gong, et al. Current situation and development of zinc coating chromium-free (VI) passivation. Steel Rolling. 2007, 24(5): 41–44.

[13] Keping Han, Xiangrong Ye, Jingli Fang. Study on zinc coating surface silicate anti-corrosion film. Corrosion Science and Protection Technology. 1997, 9(2): 167–70.

[14] Fengjun Shan, Changsheng Liu, Xiaozhong Yu, et al. Research progress of galvanized steel sheet chromium-free passivation technology. Material Protection. 2007, 40(10): 45–47.

[15] Jinming Long, Xiayun Han, Ning Yang, et al. Rare earth surface modification of zinc and galvanized steel. Rare earth. 2003, 24(5): 52–56.

[16] Lei Shi, Ludan Shi, et al. A kind of galvanized trivalent chromium blue-white passivator. Patent number: 201210136629.X.

[17] Lei Shi, Mingxiao Zhang, et al. Galvanized chromium-free blue-white passivator. Patent number: 2013101591413.

Architectural, Energy and Information Engineering – Sung & Chen (Eds)
© 2016 Taylor & Francis Group, London, ISBN 978-1-138-02791-6

Research on the CCHP system based on natural gas engines

Bing Kun Li & Wen Zhuang Zhang
School of Energy and Power Engineering, Wuhan University of Technology, Wuhan, China

ABSTRACT: Considering the YC6MK375 N-30-type natural gas engine as the research objective, this paper discusses the principles and the components of the Combined Cooling, Heating and Power system (CCHP system). This paper calculates the cold, heat and power supply of the system in summer and winter. Based on the result of the calculations, the thermal economic analysis was carried out on some indicators, including the system's primary energy ratio, relative energy saving rate, and equivalent exergy ratio, which confirmed the advantage of using the CCHP system based on natural gas engine, such as energy saving and high economic benefits.

Keywords: natural gas engines; CCHP; thermal economic evaluation

1 INTRODUCTION

Natural gas is a clean energy with a high heat value. The development of the Combined Cooling, Heating and Power system (CCHP system) driven by natural gas is an important way for reasonable utilization of natural gas in the city. It is also the effective method to optimize city energy structure and improve the quality of the city environment. CCHP is a comprehensive system of energy production and use, which on the basis of the energy cascade utilization system can be dispersed at the users' living needs. First, using primary energy drives the engine to provide power, and then recycle the waste heat through a a variety of waste heat utilization equipment, and ultimately achieve a higher energy utilization rate, a lower energy cost, a higher security of energy supply and a better environmental performance and other multi-function targets.

2 CCHP SYSTEM BASED ON NATURAL GAS ENGINES

The CCHP system is the main form of distributed energy system, according to the function of each part, which can be roughly divided into four subsystems, namely the power subsystem, subsystems of heating, cooling, and the corresponding control subsystem. Based on heat emission, the heat recovery subsystem of downstream cascade utilization is designed, including subsystems of heating and cooling.

2.1 Natural gas engines

The system selects inline, six-cylinder, four stroke natural gas engine as power equipment, the engine

model is YC6MK375 N-30, the nominal power is 276 Kw, and the declared speed is 2100 r/min. The natural gas engine with turbocharged inter-cooled and electronic control technology has good performance and fuel economy, and its emission reaches the GB18352.3-2005.

Zhu Xianbiao studied the thermal balance test of this type of engine and obtained the distribution of fuel calorific value when the engine was operating under different conditions, as shown in Figure 1.

Under various operating conditions presented in the graph, the heat was converted into useful work (Q_e) release of 34%~38% of total fuel combustion heat, about 22%~26% of heat released by exhausting (Q_1), 21%~28% of heat released by the cooling water (Q_2), 2%~5% of heat released by the intercooler (Q_3), the residual heat accounts for about 4%~10% (Q_4), and the three account for more than 80%. Obviously, making full use of the energy can greatly improve the efficiency of the system.

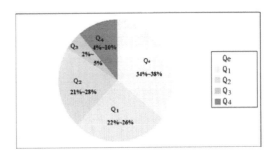

Figure 1. Distribution map of fuel combustion heat.

2.2 Lithium Bromide (LiBr) absorption refrigerator driven by waste heat

Engine waste heat utilization mainly includes heat recovery of waste heat from exhaust gas and cylinder liner cooling water. Flue gas is commonly a high grade heat source between 400°C and 600°C, used in the double-effect LiBr absorption refrigeration unit. In addition, there is about 200°C exhaust gas discharged from it, which can provide hot water through the gas-liquid heat exchanger: the normal jacket water temperature range is 80°C~90°C, which can satisfy the heat source requirement of the single-effect LiBr absorption refrigeration.

Ordinary single-effect LiBr absorption refrigeration unit of COP is about 0.7, and the COP of double-effect LiBr unit can reach 1.2; due to the different requirements of heat source temperature, both at the same time do not meet the better utilization of waste heat of gas engine. This system adopts single- and double-effect complex absorption refrigeration, on the basis of the double-effect LiBr absorption refrigeration unit; furthermore, it adds a single generator, as well as the cooling water high temperature and low temperature condenser at the same time. High temperature flue gas is added to the high voltage generator for double-effect refrigeration, while cylinder liner cooling water flows into the single-effect generator for single-effect refrigeration. This innovative combination achieves full utilization of waste heat, obtains more refrigerating capacity, and also reduces the volume and cost of the equipment.

3 CALCULATION OF CCHP SUPPLY DURING SUMMER AND WINTER

In winter and summer, the rated power of CCHP is $276\,Kw \times 0.95 = 262.2\,Kw$ (0.95: reactive power conversion efficiency). Apart from electricity generation, it is mainly used for refrigeration and provides less amount of hot water in summer, and via heating provides hot water in winter. In spring and autumn, CCHP is used only for electricity, which will not be dicussed in this paper.

3.1 Calculation of refrigerating capacity and hot water supply in summer

3.1.1 Calculation of total heat of engine exhaust Q_1 and the coolant heat Q_2

$$Q_1 = N / \eta_0 \cdot \eta_1 = 236.6\,\text{Kw} \qquad (1)$$
$$Q_2 = N / \eta_0 \cdot \eta_2 = 157.7\,\text{Kw} \qquad (2)$$

where $N = 276$ Kw: the rated power of engine; $\eta_0 = 0.35$: the effective efficiency of engine;

$\eta_1 = 0.3$: the percentage of engine exhaust heat; and $\eta_2 = 0.2$: percentage of engine coolant heat.

3.1.2 The calculation of single- and double-effect complex LiBr absorption refrigeration unit total refrigerating capacity Q_A

The temperature of exhaust gas at the entrance of the refrigerator: $t_1 = 500$°C; the temperature at the exit: $t_2 = 200$°C, the available exhaust heat are given by

$$Q_B = Q_1 \times \frac{t_1 - t_2}{t_1} = 142\,\text{Kw}. \qquad (3)$$

The refrigerating capacity that is produced by the heat used for the double-effect LiBr absorption refrigeration is given by

$$Q_{A1} = Q_B \times COP_{Double} = 170.4\,\text{Kw} \qquad (4)$$

where $COP_{Double} = 1.2$: thermo-coefficient of flue gas double-effect LiBr absorption refrigeration.

The calculation of hot water single-effect LiBr absorption refrigeration capacity is described below.

Let 80°C~90°C cylinder liner cooling fluid be added to the single-effect generator for single-effect LiBr absorption refrigeration, which is given by

$$Q_{A2} = Q_2 \times COP_{Single} = 110.4\,\text{Kw} \qquad (5)$$

where $COP_{Single} = 0.7$: thermo-coefficient of hot water single-effect LiBr absorption refrigeration.

Therefore, total refrigerating capacity of the single- and double-effect complex LiBr absorption refrigeration unit is given by

$$Q_A = Q_{A1} + Q_{A2} = 280.8\,\text{Kw}. \qquad (6)$$

3.1.3 The calculation of the amount of hot water produced by 200°C exhaust gas

Let the 200°C exhaust gas from the refrigeration unit be added into a primary heat exchanger at the rear of the unit for recovery of the residual heat and producing hot water. People use hot water between 60°C to 65°C, according to the temperature of the hot water dosage standard. To prevent corrosion of the exhaust pipe cold end, the exhaust's outlet temperature must be set above 120°C. So, we can get the maximum recoverable heat from 200°C waste gas:

$$Q_{200°C} = Q_1 \times \frac{t_2 - t_3}{t_1} = 37.9\,\text{Kw} \qquad (7)$$

where $t_3 = 120$°C: final temperature of exhaust gas; the amount of hot water provided an hour:

$Q_{Summer} = Q_{200°C} \times (t_2 - t_3) = 722$ kg/h; $c = 4.2$ kJ/(kg · °C): specific heat of water; $t_4 = 65°C$: temperature of living hot water; and $t_5 = 20°C$: average temperature of water in summer.

Based on the above calculation, the maximum refrigerating capacity in summer is 280.8 Kw, and the amount of hot water provided an hour is 722 kg/h.

3.2 Calculation of heating quantity and hot water supply in winter

3.2.1 Calculation of heating area

$$A = \frac{Q_b}{A \cdot c \cdot (t_7 - t_6)} = 2433.6 \text{ m}^2 \quad (8)$$

where $t_6 = 70°C$: inlet temperature of heating water; $t_7 = 90°C$: outlet temperature of heating water; and $A_0 = 2.5$ kg/(h · m²): required amount of hot water per square meter and per hour.

3.2.2 Calculation of hot water

The calculation of the amount of hot water supplied by the coolant is described below.

In practice, the heat coolant taken away only about 50% can be recycled. So, the amount of hot water coolant provided per hour is

$$G_1 = \frac{Q_2 \times 0.5}{c \cdot (t_4 - t_8)} = 1231 \text{ kg/h} \quad (9)$$

where $t_8 = 10°C$: average temperature of water in winter.

The calculation of the amount of hot water 200°C exhaust gas can produce

$$G_2 = \frac{Q_{200°C}}{c \cdot (t_4 - t_8)} = 590 \text{ kg/illegal children in m}_i \quad (10)$$

The amount of total hot water is given by

$$G_{W \text{ int} er} = G_1 + G_2 = 1821 \text{ kg/h} \quad (11)$$

Based on the above calculation, the maximum heating area of the system in winter is 2433.6 m², and the amount of hot water provided an hour is 1821 kg/h.

4 EVALUATION OF THERMAL EFFICIENCY

4.1 Primary energy ratio

Also known as thermal efficiency of system and total energy use efficiency, the primary energy ratio is equal to the ratio of output energy and input energy. The power, heat, and cold can be considered the same and directly added. The higher the primary energy ratio is, the better the thermal performance of the CCHP system is.

The primary energy ratio in summer is given by

$$PER_{CCHP} = \frac{Q_{heating-load} + Q_{cooling-heat} + P_{el}}{Q_{in}}$$

$$= \frac{Q_{200c} + Q_A + N \cdot 0.95}{N} \cdot \eta_0 = 73.66\% \quad (12)$$

The primary energy ratio in winter is given by

$$PER_{CCHP} = \frac{Q_{recovered-heat} + P_{el}}{Q_{in}}$$

$$= \frac{Q_1 \frac{t_1 - t_3}{t_1} + Q_2 \cdot 0.5 + N \cdot 0.95}{N} \cdot \eta_0$$

$$= 66.05\% \quad (13)$$

where $Q_{heating-load}$: heat supply; $Q_{cooling-heat}$: refrigerating capacity; P_{el}: generated energy; Q_{in}: input energy; and $Q_{recovered-heat}$: recoverable heat.

4.2 The relative energy saving rate

In order to reflect the differences in energy use between the CCHP system and conventional individual production systems, we put forward the relative energy saving rate as

$$FESR = (Q_{re} - Q_{cog}) / Q_{re} \times 100\% \quad (14)$$

where Q_{re}: the total energy consumption of individual production system as reference; and Q_{cog}: the total energy consumption of CCHP system.

The fuel input of the system as the reference, corresponding with the output of the CCHP system, is given by

$$Q_{re} = w / \eta_a + Q_A / (\eta_a \cdot COP_{re}) + Q_{hot} / \eta_b \quad (15)$$

where $w = 262.2$ Kw: generated energy; $\eta_a = 33.3\%$: average efficiency of individual production system; $COP_{re} = 4$: electric compression refrigeration coefficient of individual production system; and $\eta_b = 90\%$: thermal efficiency of the boiler.

The total energy consumption of an individual production system is given by $Q_{re-Summer} = 1040.31$ Kw; $Q_{re-Winter} = 1074.89$ Kw. The total energy consumption of the CCHP system is given by $Q_{cog} = 276 \div 0.35 = 788.57$ Kw. The relative energy saving rate is given by $FESR_{Summer} = 24.2\%$, $FESR_{Winter} = 26.64\%$.

4.3 Equivalent Exergy Ratio (EExP)

Equivalent Exergy Ratio is the evaluation of the energy utilization of the CCHP system according to the grade of energy. It can evaluate thermal devices reasonably with a single criteria standard:

$$EExR = \frac{N + a_c Q_1 + a_h Q_{hot}}{Q_{all}} \quad (16)$$

where $a_c = T_d/T_c - 1$: the coefficient grade of cooling capacity; $a_h = 1 - T_d/T_h$: the coefficient grade of cooling capacity; T_a: environment temperature; T_c: temperature of refrigerant water; and T_h: temperature of heating steam (or hot water).

The result obtained is as follows: $a_c = 0.019$; $a_h = 0.019$; $EExR_{Summer} = 36.29\%$; $EExR_{Winter} = 39.20\%$.

The more the energy of the cooling or heating system is, the more smaller the EExP than the primary energy ratio will be.

5 CONCLUSIONS

With the YC6MK375 N-30-type natural gas engine for power equipment, the CCHP system provides the corresponding cold, heat and power supply to meet the users' living needs in the winter and summer to a certain extent. From the thermal economic analysis on some indicators including the system's primary energy ratio, relative energy saving rate and equivalent exergy ratio, we confirm that the CCHP system based on natural gas engine has advantages such as energy saving and high economic benefits, and should be popularized and applied widely.

REFERENCES

Ge Bin. 2011. Principle and technology of cogeneration cooling heating and power [M]. Beijing: China electric power press.

Yue Yongliang. 2003. Single and double effect complex and waste-heat utilization LiBr absorption refrigeration refrigerator [J]. Hefei: Fluid machinery.

Zhang Chunxin. 2012. Energy-saving analysis of combined cooling, heating and power system [D]. Beijing: North China Electric Power University.

Zhang Ke. 2013. Configuration optimization research of office building combined cooling, heating and power system [D]. Harbin: Harbin Institute of Technology.

Zhu Xianbiao. 2013. Thermal equilibrium modeling and experimental research of YC6MK375 N-30-type natural gas engine [D]. Wuhan: Wuhan University of Technology.

Architectural, Energy and Information Engineering – Sung & Chen (Eds)
© 2016 Taylor & Francis Group, London, ISBN 978-1-138-02791-6

A strategy research on low impact development and green infrastructure with purpose of solving the problem of China's urban waterlogging

Yi Luo
School of Architecture, South China University of Technology, Guangzhou, China

ABSTRACT: The implement of green infrastructure and low impact development with basic service function is a systematic approach to solving the urban ecological environment problems, such as rainstorm and water logging. It has an effect of the urban ecological system, which plays a significant role in sustainable development. Thus, to introduce the green infrastructure and low impact development in a good time, will be conducive to the realization of urban ecological and sustainable development in China.

Based on exploration of the fundamental reason for the urban waterlogging issue in China, this article elaborates on the approach of replacing grey infrastructure with green infrastructure, introducing strategies comprising the low impact development so as to tackle the problem and achieve the urban ecological goal.

Keywords: green infrastructure; low impact development; rainstorm and waterlogging

1 INTRODUCTION

Recently, rainstorm and waterlogging is a big problem for most Chinese cities, including the historic centers (e.g. Beijing) and even some emerging cities (e.g. Shenzhen). However, through years of study and exploration, low impact development and green infrastructure may be treated as a future shift in this area. Accordingly, on April 2, 2015, there were 16 Chinese cities becoming the first pilot cities of the "sponge city construction", leading a start of a new phase.

2 THE CONCEPT OF GREEN INFRASTRUCTURE

The Green Infrastructure (GI), which is recently introduced by western countries, is a strategy for open space planning and land conservation. Compared with ecological infrastructure, GI emphasizes the value of continuous green space network and relative vegetation. Hence, it is based on the ecological theory, and aims at the concept of "building facilities", such as the "grey infrastructure" (e.g. roadways and municipal sewage pipe networks) as well as the social infrastructure (e.g. hospitals and schools). It regards urban open spaces, forests, wildlife, parks and other green space networks as essential infrastructure for urban and community development. In words, to take advantage of GI will effectively minimizes dependencies on grey infrastructure, as well as saves investments on public resources, and reduces the sensitivity of natural disasters. It has a close relationship with human health and the urban ecosystem health; hence, it should be treated as an essential infrastructure in order to maintain the natural life process.

3 THE ELEMENTS OF GI SYSTEM

The space formation of GI is made up of three elements, i.e. rainfall flood infrastructure for rehabilitating the city wetland; urban recreational facilities represented by the greenway; and urban habitats especially the wasteland protection and education. They respectively provide solutions for the urban waterlogging and pollution, the crisis of urban residents' recreational facilities as well as the urban environmental problem caused by biodiversity loss and lack of education.

3.1 *Rainfall flood infrastructure for rehabilitating the city wetland*

The urban waterlogging could be seen in the majority of cities, which even caused death in someplace. During the agricultural era, since various farmlands took charge of wetlands regulate function while wastelands were responsible for biota; the wetland had not received enough attention. Accordingly, it was not an essential infrastructure in this period.

However in the urban age, farmland loss and pasture degradation result in insufficient biological habitat. Thus, the wetland is highly required nowadays. In other words, the urban wetland system is the most significant GI system.

3.2 *Urban recreational facilities represented by the greenway*

The statements concerning the function of greenway inside and outside the country generally include four functions: recreation, green commuting, environment protection, and culture experience. But with the limitation of the input, and the high density of the city construction, the option for the function of greenway is not so wide. The specific circumstance of the city in China leads to the fact that the recreation for the citizens should be the major function of the city greenway, with the consideration to green commuting, environment protection and culture experience.

3.3 *Urban habitats especially the wasteland protection and education*

For the agricultural era, wasteland is a critical place. But it becomes a scare resource in the urban age because of its biota and environmental education functions.

In the planning map, planners tend to consider there must be flowers and birdsong in a public park. However, the garden-style will only damage the diversity of space, while the wasteland exactly to be in charge of biological habitat and environmental education duties.

4 THE CONCEPT OF LOW IMPACT DEVELOPMENT

The sponge city construction refers to Low Impact Development (LID), which is a new leading technique to control storm runoff and non-point source pollution. In 2010, it was defined as the 21-century green infrastructure by USEPA.

To be exact, LID capitalizes on ecological engineering method rather than wipe water infrastructure with high consumption. By using scattered small-scale source control mechanisms (e.g. plant absorption, evaporation, soil infiltration and retention), it could reduce and defer the storm runoff and crest value from the source, so that the developed urban hydrologic cycle could return to the past.

5 THE CLASSIFICATIONS BASED ON GI AND LID

5.1 *Rain garden*

In recent years, especially in many developed countries, rain gardens are widely used on stormwater management and runoff pollution control,

which is a very important green water infrastructure. Generally, to select a piece of low-lying land with indigenous crop and the rainfall can infiltrate through the land and other natural hydrological effects, so that rainwater can not only intercept, but also be purified. Pollutants can be reduced by 30% before rainwater into streams.

Rain gardens usually plant indigenous crops to reduce cost, and without deliberate maintenance; And these crops can adapt the local climate, soil, water better; In a meanwhile, rain gardens can be used as landscaping and play an important role in recharging of groundwater, as well as reducing the flood peak volume and storm water runoff.

5.2 *Permeability pavement*

Using permeable pavement of urban roads can effectively reduce runoff to ease the burden on urban drainage system and pollutants in the rain. The drawback of the permeable pavement is the low durability, high cost. However, from the perspective of long-term sustainable development, it should be promoted.

5.3 *Vegetation in shallow groove*

Usually the plants grow in surface ditches, contaminants in rainwater flow through the shallow groove can be removed in osmosis "filtering" precipitation, but rainwater is collected through gravity. The depth and width of vegetation in shallow trenches can change with the amount of displacement in the region. Vegetation in shallow ditches can achieve the purpose of rainwater collection and pollutants reduction in storm water.

5.4 *Landscape storm storage pond*

Landscape storm storage pond is a high-efficient system to combine landscape with storm detention rank, so that a sustainable and green infrastructure can be developed. In general times, it keeps a normal water level, with car park, school or greenway on the ground. While when the storm comes, the large gap between usual level and the ground works, which could store and adjust the storm flow, so as to reduce the damage for around areas and low reaches. After that, stored water level decrease to normal by filtration, evaporation or discharge etc..

6 THE CAUSES OF CHINA'S URBAN RAINSTORM AND WATERLOGGING

Why the waterlogging appeared in China? Reasons given by the irrigation experts are quite simple: the exploration of modern cities has broken the original water circulation, the water permeation is not allowed due to the hard surface of the construction and the road. When rainstorm comes, the rain supposed to permeate or to be stored by the vegetation is impossible but be ejected out. The ejection relays on the underground pipeline web which bears overload pressure during the storm. Therefore, the rain stays at the lower ground of the city, road etc., and then comes the waterlogging.

Obviously, it is inspiring if proper construction can solve this problem. Some experience in western country worthy to draw. However, merely using more galleries does more bad than good: it costs a large number of money and the pollution brought by the rain is uncountable. The amount of water brings the garbage from the road and flows into natural water system, such as rivers and lakes. The water goes through the pipeline web, gathering pollution, phosphorus, nitrogen and heavy metal etc. into the natural water in a short period.

7 STRATEGY TO TACKLE WATERLOGGING WITH LID AND GI

Another approach can be taken to deal with thunderstorms before the pipes are completed, i.e. to "soften" the city. Whereas the original landform cannot be kept intact due to urban development, the current hydrologic cycle will be protected as best as we can via technologies provided by LID.

Sponge City is a more vivid expression, meaning a city with good flexibility can easily adapt to the changeable environment and cope with natural disasters. Whenever it rains, the water will be able to be absorbed, stored, seeped and purified or even be released in case of emergent use.

When it rains, the green roof can hold the rainwater. The vegetation undergone special design will neither be an excessive burden to the building nor seep into the houses. The plants and improved soil can reserve part of the water on the green roof and filter as well as purify it. The excess water will enter the ecological sequestration facilities on the ground through the drainage pipes.

To make it short, the ecological sequestration facilities are improved green facilities with stronger water sequestration ability. Under the vegetation, the aquifer can store water and its role is even more instrumental when confronted by thunderstorms.

Permeable road surface of certain area can not just expedite the water flow on itself but also accelerate the speed of that from the neighboring areas. In this manner waterlogging will be avoidable.

The hardened roads and squares can be made permeable with permeable tiles and concrete.

Grassed swales are channels with vegetation on the ground. It can collect, transit and drain rain water and also can purify it. It is applicable to the connection with other independent facilities, urban rainwater channel system and excessive rainwater drainage system.

Based on the philosophy of rainwater management and use, different countries have conducted various practices. Comparatively speaking, the LID development is more systematic and complete. According to data released by EPA, performance of LID technologies varies under different climate conditions. But on the whole, LID can reduce the thunderstorm runoff by 30–90% and delay its peak for 5–40 minutes. In this case the pressure faced by the municipal drainage system can be greatly reduced.

8 CONCLUSIONS

To sum up, the system of LID and GI is an eco-system for sake of the sustainable development of environment, society and economy. Its significance lies in the urban ecological system, which aims to achieve the goal of natural resource protection and utilization in urban open space.

It provides a macro understanding of sustainable framework to maintain ecosystem's value and functions, to balance human needs with flora and fauna, and to guide the urban development. Therefore, starting from the long-term goal of urban sustainable development, the implement of LID and GI requires long-term and unremitting efforts.

REFERENCES

[1] Wei Wu, Xi'e Fu: The Concept of Green Infrastructure and Review of Its Research Development (Urban Planning International, China, 2009.10) In Chinese.

[2] Ying Jun, Zhang Qing-ping, Wang Mo-shun, Wu Xiao-hua: Urban green infrastructure system and its system construction (Journal of Zhejiang A&F University, China, 2011, 28(5): 805–809) In Chinese.

[3] Zhou Yanni, Yin Haiwei: Foreign Green Infrastructure Planning Theory and Practice (Urban Studies Vol. 17 No. 8 2010) In Chinese.

[4] Zhang Jin-shi: Green Infrastructure: A Systematic Solution to Urban Space and Environmental Issues (Modern Urban Research, 2009.11) In Chinese.

[5] Jia Zhang, Ge Lu: The Ideas and Strategies of Constructing the Green Infrastructure in Hangzhou Urban Fringe Villages (Modern City, Vol. 6, No. 2, 2011) In Chinese.

Architectural, Energy and Information Engineering – Sung & Chen (Eds)
© 2016 Taylor & Francis Group, London, ISBN 978-1-138-02791-6

The reuse of nonmetals recycled from waste printed circuit boards as reinforcing fillers in cement mortar

Wei Jiang, Yu Hua Qiao, Yong Biao Xu & Yan Hong Zheng
Inner Mongolia Shuangxin Environment-Friendly Material Co. Ltd., Ordos, P.R. China

Fei Xiong Zhang
Inner Mongolia Shuangxin Energy and Chemical Industry Co. Ltd., Ordos, P.R. China

ABSTRACT: The feasibility of reusing nonmetals recycled from waste Printed Circuit Boards (PCBs) as reinforcing fillers in cement mortar is studied in this paper. The effects of nonmetals recycled from waste PCBs on cement mortar are investigated with different particle sizes. The bending strength, water absorption, moisture movement of the nonmetals/cement mortar was measured. The results showed that the general properties of the nonmetals/cement mortar can be significantly improved by filling nonmetals recycled from waste PCBs. Cement mortar made with recycled waste PCBs nonmetallic powder is a new type of green building materials. All the above results indicate that the reuse of nonmetals as reinforcing fillers in the PP composites represents a promising way for recycling resources and resolving the environmental pollutions.

Keywords: nonmetals; waste printed circuit boards; reinforcing fillers; cement mortar

1 INTRODUCTION

Printed Circuit Boards (PCBs) are the typical and fundamental component for almost all electronic products and they contain diversiform materials, including metals such as Cu, Al, Fe, Sn, Sb, Pb, etc and nonmetals such as thermoset resins, glass fibers, etc [1–2]. Recycling of PCBs is an important subject not only for the treatment of waste but also for the recovery of valuable materials as the amount of deserted PCBs is dramatically increasing. Air classification is a cleaner separation method that does not use any polluting medium for separation [3–4]. After being mechanical separated, recycled metals such as Cu, Al, Fe, Sn, Sb, Pb, etc, are sent to recovery operations and the processes are already quite mature. However, significant quantities of nonmetals (up to 70%) especially present a huge challenge for reusing. The main components of the nonmetals are thermoset resins and reinforcing materials. Traditionally, these nonmetals are landfilled or incinerated without further disposing or reusing, which will cause resource waste and potential environment problems. Recently, many scientists and engineers are exploring to reuse these nonmetals in a more profitable and environmentally friendly way. The nonmetals could be used as fillers for epoxy resin products, such as paints, adhesives, decorating agents and building materials [5–6]. Recently, in our earlier publications, the nonmetals recycled from waste PCBs can be successfully reused as reinforcing fillers in the PP composites [7].

To our knowledge, there is little published information about using nonmetals recycled from waste PCBs as reinforcing fillers in cement mortar and evaluating bending strength, water absorption, moisture movement of the nonmetals/cement mortar. Cement mortar is a great consumption of construction material. If the nonmetals recycled from waste PCBs can be successfully reused as reinforcing fillers in cement mortar, the environmental problem will be solved effectively.

In this article, the objective of the research is to reuse the nonmetals recycled from waste PCBs in cement mortar, with the aim to recycle the resources in a more profitable and environmentally friendly way. The feasibility of using nonmetals as reinforcing fillers in cement mortar is studied by testing bending strength, water absorption, moisture movement of the nonmetals/cement mortar. All the results show that the reuse of nonmetals as reinforcing fillers in the cement mortar represents a promising way for recycling resources and resolving the environmental pollutions.

2 EXPERIMENT

2.1 Materials

The detailed nonmetals recycling process from waste PCBs had been reported in a previous publication [7]. The nonmetals with particle sizes of 25~80 (coarse), 80~150 (medium) and less than

150 meshes (fine) are selected and compounded in cement mortar. The main components of the non-metals recycled from waste PCBs are thermoset resins and reinforcing materials. Physical characterisation of the nonmetals is complicated by the presence of both fibrous and particulate fractions. The shapes and compositions of the nonmetals vary with different particle sizes through careful observations on SEM as shown in Figure 1.

Standard cement is purchased from China building materials science research institute. The chemical composition of standard cement is listed in Table 1. The sand is purchased locally. The size of the sand is 0.35–0.5 mm medium sand.

2.2 Fabrication procedure

The formulation and curing method of the non-metals/cement mortar are listed in Table 2. All of the raw materials were based on DRY weight in the recipe. Sample preparation was according to Shuangxin Fiber cement lab standard. After sample production, all the samples were packaged in plastic bags and cured in a wooden box 50°C with a bottle of water inside for 24 hours, then each plastic bag was removed and the samples were transferred to an autoclave for curing. Autoclav-

ing temperature and time were setup according to industry values and the autoclaving process was automatic. After autoclaving, the samples were cured in a climate cabinet at 23 ± 2 centigrade and 50% relative humidity for 4 days, after which time the samples were ready for measurement.

2.3 Measurements

The flexural properties of the pure cement mortar (without nonmetals) and nonmetals/cement mortar are measured using automatic cement bending compression testing machine at room temperature (23°C) according to ISO Standards 679: 2009. Dry density, water absorption, moisture movement of the nonmetals/cement mortar is measured according to China standard GB/T 7019-1997.

3 RESULTS AND DISCUSSION

3.1 Bending strength of nonmetals/cement mortar

Figure 2 shows the bending strength of cement mortar with different particle sizes of 25~80 (coarse), 80~150 (medium) and less than 150 meshes (fine). The bending properties results show

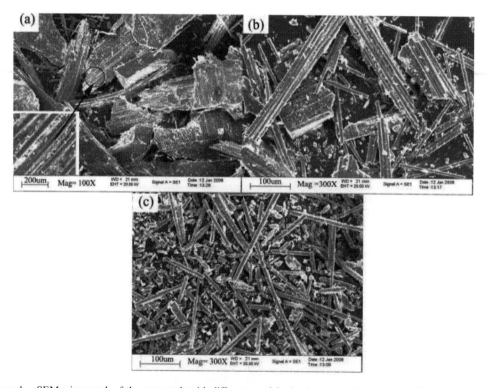

Figure 1. SEM micrograph of the nonmetals with different particle size (a: coarse; b: medium; c: fine).

Table 1. Chemical composition of standard cement (wt%).

SiO$_2$	Al$_2$O$_3$	Fe$_2$O$_3$	CaO	MgO	SO$_3$	Na$_2$Oeq	f-CaO	Loss	Cl$^-$
20.76	4.58	3.27	62.13	3.13	2.8	0.057	0.76	1.86	0.013

Table 2. The formulation and curing method of the cement mortar (Parts by weight).

| Sample | Nonmetals | | Cement | Sand | Water | Maintenance method |
	Size	Content				
1	none	0	200	400	100	Aircured
2	coarse	40	200	800	100	Aircured
3	medium	40	200	400	100	Aircured
4	fine	40	200	400	100	Aircured

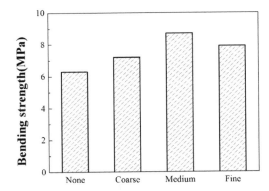

Figure 2. Bending strength of the nonmetals/cement mortar.

that bending strength of the nonmetals/cement mortar are improved significantly by adding the fine, medium and coarse nonmetals particles respectively. That is mainly because the nonmetals recycled from waste PCBs contain 50–70 wt % glass fibers. And these glass fibers possess many excellent characteristics, such as high length diameter ratio (L/D ratio), high elastic modulus and low elongation. Every dispersed fiber triggers effective stress concentrations and leads to mass crazes so that the weak point cannot be formed in the nonmetals/cement mortar. Thus, the nonmetals/cement mortar properties are improved through the interaction of high strength particles and cement matrix. The size of nonmetals also can affect the bending properties of the composites. The bending properties of the nonmetals/cement mortar adding the fine and medium nonmetals are both greater than those of the cement mortar adding the coarse ones. It is evident that the size of the nonmetals is an important factor in affecting the bending properties of the nonmetals/cement

mortar. Decreasing particle size results in dramatic increases in the specific surface area of particles, which leads to an increase in interfacial contact area between the filler and PP matrix. The increase in interfacial contact area would be beneficial to transfer the stress from the matrix to particles, therefore resulting in higher tensile strength of the composite. But there is the optimum particle size of the nonmetals for the best bending properties of the nonmetals/cement mortar. Based on comprehensive consideration of the bending properties, the optimum particle is the medium nonmetals recycled from waste PCBs.

3.2 *Water absorption of nonmetals/cement mortar*

The absorbing water rate of the nonmetals/cement mortar is shown in Figure 3. The water absorption results show that absorbing water rate of the nonmetals/cement mortar decrease by adding the fine, medium and coarse nonmetals particles respectively. That is mainly because the nonmetals recycled from waste PCBs contain 20–30 wt% thermoset resins. These thermoset resins are hydrophobicity. Therefore, absorbing water rate of the cement mortar decreases by adding the nonmetals into them. The smaller the particle size of the nonmetals is, the lower water absorption of them is. But the optimum absorbing water rate of the cement mortar is depended on the properties of the bonding material.

3.3 *Moisture movement of nonmetals/cement mortar*

Moisture movement of the cement mortar adding nonmetals recycled from waste PCBs on are investigated with different particle sizes. The moisture movement of the nonmetals/cement mortar is

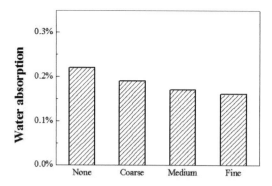

Figure 3. Water absorption of nonmetals/cement mortar.

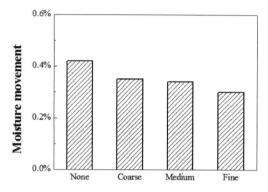

Figure 4. Moisture movement of the nonmetals/cement mortar.

shown in the Figure 4. The results show that the moisture movement decrease by adding the fine, medium and coarse nonmetals particles respectively. That is mainly because the nonmetals recycled from waste PCBs contain 50–70 wt % glass fibers. And these glass fibers possess many excellent characteristics, such as high length diameter ratio (L/D ratio), high elastic modulus and low elongation. The glass fibers of the nonmetals act as the concentration of stress in the cement mortar matrix which undertakes the loading. The single glass fibers possess high elastic modulus, low elongation and can first undertake the loading. Therefore, the moisture movement of the cement mortar is improved by adding the nonmetals particles. No matter using the fine, medium and coarse

nonmetals particles, the moisture movement of the cement mortar is improved.

4 CONCLUSIONS

The nonmetals recycled from waste PCBs can be successfully used in cement mortar. The bending strength, water absorption, moisture movement of nonmetals/cement mortars are improved significantly by adding the nonmetals, especially the bending strength.

The water absorption and moisture movement of the nonmetals/cement mortar decrease by adding the fine, medium and coarse nonmetals particles respectively.

Based on comprehensive consideration of the bending properties, water absorption, moisture movement, the optimum particle is the medium nonmetals. The use of nonmetals recycled from waste PCBs as reinforcing fillers in the cement mortar represents a promising way for resolving the environmental pollutions and health problems.

REFERENCES

[1] E.Y.L. Sum, The recovery of metals from electronic scrap, JOM. 43 (1991) 53–61.
[2] M. Goosey, R. Kellner. Recycling technologies for the treatment of end of life Printed Circuit Boards (PCB), Circuit. World. 29 (2003) 33–37.
[3] R.L. Franz. Optimizing Portable Product Recycling Through Reverse Supply Chain Technology. In: Proceeding of the 2002 IEEE International Symposium on Electronics and the Environment. 2002, pp. 274–279.
[4] C. Eswaraiah, T. Kavitha, S. Vidyasagar, S.S. Narayanan. Classification of metals and plastics from Printed Circuit Boards (PCB) using air classifier. Chem. Eng. Process. 47 (2008) 565–576.
[5] M. Iji. Recycling of epoxy resin compounds for moulding electronic components. J Mater Sci 33 (1998) 45–53.
[6] Z. Gao, J. Li, H.C. Zhang. Printed circuit board recycling: a state-of-art survey. In: Proceeding of the 2002 IEEE International Symposium on Electronics and the Environment. 2002, pp. 234–241.
[7] Y. Zheng, Zhigang Shen, Chujiang Cai, Shulin Ma, Yushan Xing. The reuse of nonmetals recycled from waste printed circuit boards as reinforcing fillers in the polypropylene composites[J]. Journal of hazardous materials, 2009, 163(2–3): 600–606.

Architectural, Energy and Information Engineering – Sung & Chen (Eds)
© 2016 Taylor & Francis Group, London, ISBN 978-1-138-02791-6

Influence of silane coupling agent-treated Polyvinyl Alcohol (PVA) fiber on an air-cured fiber cement product

Yong Biao Xu, Yu Hua Qiao, Wei Jiang & Yan Hong Zheng
Inner Mongolia Shuangxin Environment-Friendly Material Co. Ltd., Ordos, P.R. China

Fei Xiong Zhang
Inner Mongolia Shuangxin Energy and Chemical Industry Co. Ltd., Ordos, P.R. China

ABSTRACT: Over the past years, Polyvinyl Alcohol (PVA) fiber has been used as the main substitute fiber for asbestos in the air-cured non-asbestos fiber cement product. In this paper, the influence of silane coupling agent-treated Polyvinyl Alcohol (PVA) fiber on an air-cured fiber cement product was investigated with different fiber contents. The bending strength, density, water absorption, moisture movement of PVA fiber cement products were measured. The results showed that the general properties of the fiber cement products can be significantly improved by filling modified PVA fibers. The modified PVA fibers have higher contact angles for water and lower moisture adsorption than the unmodified PVA fibers. The use of silane coupling agents can improve the properties of PVA fiber and increase its compatibility with cement.

Keywords: silane coupling agent; PVA fibers; fiber cement product

1 INTRODUCTION

Asbestos fibers can be harmful to human's health. These fibers have been studied and confirmed by many scientists. So, on the basis of these studies, many countries and regions have made laws to prohibit the use of asbestos fibers in industries [1–5]. As a result, the construction industry had to find other materials to replace asbestos to maintain the same quality and productivity in cement production. A wide variety of synthetic fibers, such as Polyvinyl Alcohol (PVA) fiber, PET, PP, and glass fibers, have been investigated academically as a replacement for asbestos fibers. Rafael Farinassi Mendes et al. researched on the modification of eucalyptus pulp fiber using silane coupling agents with aliphatic side chains of different lengths [6]. Wang W et al. researched on the progress of the surface modification of PP fiber used in concrete [7]. However, to our knowledge, there is little information about the use of a silane coupling agent-modified PVA fiber cement product and evaluation of the bending strength, water absorption, moisture movement of the PVA fiber cement product.

In this article, the objective of the research was to use the silane coupling agent-modified PVA fiber in the fiber cement product. The influence of silane coupling agent-treated Polyvinyl Alcohol (PVA) fiber on an air-cured fiber cement product was investigated with different PVA fiber contents.

The bending strength, dry density, water absorption, moisture movement of the PVA fiber cement products were measured. The results showed that the general properties of the fiber cement products can be significantly improved by filling modified PVA fibers. The modified PVA fibers have higher contact angles for water and lower moisture adsorption than the unmodified PVA fibers. The use of silane coupling agents can improve the functional properties of PVA fiber and increase its compatibility with cement.

2 EXPERIMENT

2.1 *Materials and fabrication procedure for the fiber cement product*

Standard cement was purchased from China Building Materials Science Research Institute. The chemical composition of the standard cement is listed in Table 1. Microsilica (GSE) and pulp were purchased locally. PVA fibers were obtained from Inner Mongolia Shuangxin Environment-Friendly Material Co., Ltd. PVA fiber properties are listed in Table 2. The length of the PVA fiber sample was 6 mm. To improve the dispersion of the fiber in the cement matrix and the compatibility between the PVA fibers and the matrix, the PVA fibers were modified with 1.0 wt% content of silane coupling agent 3-Aminopropyltriethoxysilane (KH-550,

Nanjing Shuguang Chemical Group, Nanjing, China) through silanization with a high-speed mixer (SHR-5 A, Zhangjiagang Qiangda Plastics Machinery, Suzhou, China) at 1800 rpm. Before silanization, 40 vol% content of KH-550 was mixed and hydrolyzed in the solvent (ethanol–water, volume ratio 7:3) for 30 min at room temperature (23°C) and 150 rpm with a stirrer. The formulation and curing method of the fiber cement product are listed in Table 3. All of the raw materials were based on dry weight in the recipe. Sample preparation was according to the Shuangxin Fiber cement laboratory standard. After sample production, all the samples were packaged in plastic bags and cured in a wooden box at 50°C with a bottle of water inside for 24 hours, then each plastic bag was removed and the samples were transferred to an autoclave for curing. Autoclaving temperature and time were set up according to the industry values and the autoclaving process was automated. After autoclaving, the samples were cured in a climate cabinet at 23 ± 2°C and 50% relative humidity for 4 days, and then the samples were ready for measurement.

2.2 Measurements

The flexural properties of the pure (without the PVA fiber) and fiber cement products were measured using an automatic cement bending compression testing machine at room temperature (23°C) according to the ISO Standards 679: 2009. The dry density, water absorption, moisture movement of the pure (without the PVA fiber) and fiber cement products were measured according to the China Standard GB/T 7019-1997.

3 RESULTS AND DISCUSSION

3.1 Bending strength of fiber cement products

Figure 1 shows the bending strength of the pure (without the PVA fiber) and fiber cement products using fiber with 6 mm modified and unmodified PVA fibers (1.0 wt%, 1.5 wt% and 2.0 wt%). It shows that regardless of using the modified or unmodified PVA fibers, the bending strength of fiber cement products was improved significantly by adding PVA fibers. The content of the PVA fibers added in the fiber cement product could also affect the bending strength of the fiber cement product. It was evident that bending strength of the fiber cement product by adding PVA fibers increased with increasing fiber contents. Meanwhile, the surface modification of the PVA fibers could significantly improve the bending strength of the fiber cement products. This is mainly due to the fact that the surface modification provided an improved fiber to the cement matrix interfacial bond. The modified PVA fibers had higher contact

Table 1. Chemical composition of the standard cement (wt%).

SiO_2	Al_2O_3	Fe_2O_3	CaO	MgO	SO_3	Na_2Oeq	f-CaO	Loss	Cl^-
20.76	4.58	3.27	62.13	3.13	2.8	0.057	0.76	1.86	0.013

Table 2. PVA fiber properties.

Sample	Linear density (dtex)	Tensile strength (CN/dtex)	Tensile modulus (CN/dtex)	Elongation (%)	Solubility (%)
PVA Fiber	2.13	12.64	314.65	6.48	0.59

Table 3. The formulation and curing method of the fiber cement product.

Sample	PVA fiber		Cement	Pulp	GSE	Limestone	Maintenance method
	Modified	Unmodified					
1	0.0%	0.0%	81.5%	3.5%	5.0%	10.0%	Aircured
2	1.0%	0.0%	80.5%	3.5%	5.0%	10.0%	Aircured
3	1.5%	0.0%	80.0%	3.5%	5.0%	10.0%	Aircured
4	2.0%	0.0%	79.5%	3.5%	5.0%	10.0%	Aircured
5	0.0%	1.0%	80.5%	3.5%	5.0%	10.0%	Aircured
6	0.0%	1.5%	80.0%	3.5%	5.0%	10.0%	Aircured
7	0.0%	2.0%	79.5%	3.5%	5.0%	10.0%	Aircured

angles for water and lower moisture adsorption than the unmodified PVA fibers. The silane coupling agent-treated fibers had the highest moisture resistance. The use of silane coupling agents can improve the functional properties of PVA fiber and increase its compatibility with cement. Based on the comprehensive consideration of bending properties, economy, environment and technology, the optimum PVA fiber was 1.5 wt% content fibers with surface modification.

3.2 *The dry density of fiber cement products*

The dry density of fiber cement products are shown in Figure 2. The dry density of the fiber cement products gradually decreased with the addition of the PVA fibers compared with the fiber cement product without the PVA fiber. This is mainly because the fiber's density is less than the cement and the stuff density. Furthermore, the amount of the hole increased due to the addition of the PVA fiber. Meanwhile, the surface modification of the

PVA fibers could not improve the bending strength of the fiber cement products.

3.3 *Water absorption of fiber cement products*

The absorbing water rates of the PVA fiber cement products are shown in Figure 3. It was evident that the absorbing water rate of the fiber cement products by adding PVA fibers increased with increasing fiber contents. The lower the density of the PVA fiber cement products, the higher their water absorption. Meanwhile, the surface modification of the PVA fibers can decrease the absorbing water rate of the fiber cement product. This is mainly because the silane coupling agent-treated fibers had the highest moisture resistance.

3.4 *Moisture movement of fiber cement products*

The moisture movement of the PVA fiber cement products is shown in Figure 4. The addition of the PVA fiber tended to increase the moisture movement. Meanwhile, the surface modification of the PVA fibers can decrease the moisture movement of the PVA fiber cement products. This is

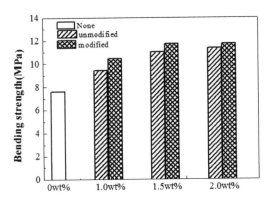

Figure 1.   Bending strength of the fiber cement product.

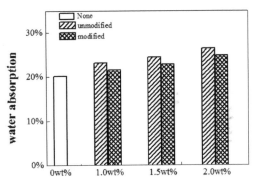

Figure 3.   Water absorption of fiber cement products.

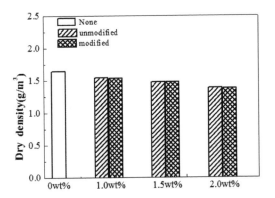

Figure 2.   Dry density of fiber cement products.

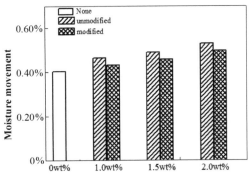

Figure 4.   Moisture movement of fiber cement products.

mainly because the surface modification provided an improved fiber to the cement matrix interfacial bond. The modified PVA fibers had lower moisture adsorption than the unmodified PVA fibers. The silane coupling agent-treated fibers had the highest moisture resistance.

4 CONCLUSIONS

The PV-A fibers can be successfully used in the air-cured non-asbestos fiber cement product. The bending strength, dry density, water absorption, moisture movement of PVA fiber cement products were improved significantly by adding the PVA fibers.

The surface modification of the PVA fibers can significantly improve the general properties of the fiber cement products. The general properties of the fiber cement products can be significantly improved by filling modified PVA fibers.

Based on the comprehensive consideration of the bending properties, dry density, water absorption, moisture movement and economy, environment and technology, the optimum PVA fiber was 1.5 wt% content fibers with surface modification. The use of PVA fibers as reinforcing fillers in the cement products represents a promising way for resolving the environmental pollutions and health problems.

REFERENCES

[1] E.Y.L. Sum, The recovery of metals from electronic scrap [P]. JOM. 1991, 43: 53–61.

[2] Akers, S.A.S., Meier, P., Studinka, J.B., Dodd, M.G., Johnson, D.J., and Hikasi, J. Long term durability of PVA reinforcing fibres in a cement matrix. The international Journal of Cement Composites and Lightweight concrete [P]. 1989, 11(2): 79–91.

[3] Jian Ping Han, Jin Yu Tao. Experimental Study on Flexural Performance of PVA Fiber Reinforced Cement-Based Composites [P]. Applied Mechanics and Materials. 2013, 356–357: 963–967.

[4] Kanda, T. and Li, V.C. Effect of Apparent Fiber Strength and Fiber-Matrix Interface on Crack Bridging in Cement Composites [P]. ASCE J. of Engineering Mechanics, 1999, 125(3): 290–299.

[5] Akkaya, Y., Peled, A., Picka, J.D. and Shah, S.P. Effect of sand addition on properties of fibre reinforced cement composites [P]. ACI materials journal. 2000, 97(3): 393–400.

[6] Rafael Farinassi Mendes, Lourival Marin Mendes, Juliano Elvis de Oliveira, Holmer Savastano Junior, Gregory Glenn, Gustavo Henrique Denzin Tonoli. Modification of eucalyptus pulp fiber using silane coupling agents with aliphatic side chains of different length [P]. Article first published online: 28 MAR 2015.

[7] Wang, W., Wang, L., Shi, Q., Yu, H.J., Chen, T., Wang, C.L., Sun, T.X. Progress of the Surface Modification of PP Fiber Used in Concrete [P]. 2006, 45(1): 29–34.

Architectural, Energy and Information Engineering – Sung & Chen (Eds)
© *2016 Taylor & Francis Group, London, ISBN 978-1-138-02791-6*

Development status and trends of rural renewable energy in China

D. Yang & Y.F. Shi
Key Laboratory of Renewable Energy Utilization Technology in Building of National Education Ministry,
Jinan, China
Shandong Provincial Key Laboratory of Building Energy-Saving Technology, Jinan, China
School of Thermal Energy Engineering, Shongdong Jianzhu University, Jinan, China

T.Y. Luan
School of Thermal Energy Engineering, Shongdong Jianzhu University, Jinan, China

ABSTRACT: China is an agricultural country, whose rural economy has an important position in the whole national economy. On the other hand, energy is an important factor that restricts the development of national economy. Under the background of rapid development of national economy but scarcity of natural resources, encouraging the development of rural renewable energy has important practical significance. This paper introduces the present situation of rural renewable energy in China, and analyzes the trend of development of renewable energy. It is concluded that China must vigorously develop renewable energy in rural areas. According to the social and economic development trend, energy supply and demand situation, domestic and foreign development background, resource situation and technical conditions, a preliminary judgment is made on the prospect of renewable energy development and utilization in the future.

Keywords: renewable energy; development status; rural

1 INTRODUCTION

As an agricultural country, Chinas' rural economy has an important position in the whole national economy. However, the rural energy problem has become the "bottleneck" of China's rural agricultural development, and it is an important factor of social progress and economic development in the rural areas of China[1]. Solving this problem will actively develop and promote the rural renewable energy technology and equipment, to solve the rural renewable energy resource utilization, the difficulty of construction, in order to alleviate the pressure of conventional energy supply, optimize the structure of energy consumption in rural areas, protect the ecological environment, and realize sustainable development of rural areas.

Renewable energy refers to the energy that can be regenerated in the biosphere, circulating along with the nature of the biochemical circulation without depletion, which cannot run out and is clean and pollution-free[2]. In addition, the problem of resource depletion will not exist, nor will there be a threat to environment pollution, as a result renewable energy will be an important part of sustainable energy system for the future social development. The rural renewable energy includes biomass energy, solar energy, wind energy, small hydropower, and geothermal energy.

2 THE DEVELOPMENT AND UTILIZATION OF HYDROENERGY IN RURAL AREAS

Small hydropower resources in China are very rich. The small hydropower installed capacity is about 120 million kW, and its distribution is very wide, which is suitable for the rural areas and remote mountainous areas in the development and utilization, not only developing local economy, but also solving the problems of the local people's electricity difficult, which in turn brings considerable benefit to them[3]. After decades of construction, by the end of 2006, the country built 46989 small hydropower projects whose installed gross capacity was 44934 MW, which accounted for about 37.4% of the capacity development and about 34.9% of the total installed hydropower. In recent years, through a series of national projects based on small hydropower, the small hydropower has become the China's largest high-quality renewable energy, which has formed a social[4], environmental and economic beneficial new industry.

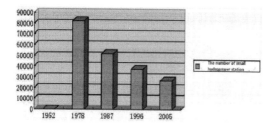

Figure 1. Small hydropower construction situation in rural areas throughout the country.

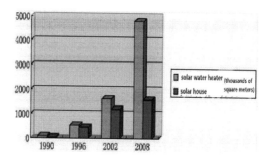

Figure 2. The solar water heater and solar house construction in rural areas.

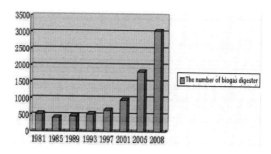

Figure 3. The construction of the biogas digester in rural areas.

Due to the development of some large hydropower stations, rural small hydropower stations have been washed out gradually, which has reduced the number of small hydropower stations, but the total installed capacity and power generation have increased[5]. Because the rural power grid renovation was completed in 1998, the rural electricity consumption ability improved greatly, and large power stations added power gap of rural areas.

3 THE DEVELOPMENT AND UTILIZATION OF SOLAR ENERGY IN RURAL AREAS

Solar energy utilization technology refers to the solar energy conversion and utilization technology directly. It converts solar energy into heat energy and takes advantage of the solar energy heat utilization technology[6]. Using the solar energy photovoltaic effect principle of the semiconductor device can convert it into electricity, which is called the solar photovoltaic technology.

It can be seen from Figure 2, the use of solar energy in each region is not balanced, solar water heaters are used commonly in various areas in the country, and the majority are in North China and East China[7]. Sun rooms are widely used in North China, Northeast China and Northwest China, where the winter season is cold, and in other regions, the number is small. Solar stoves are widely used in the Northwest and the Southwest where there is sufficient sunlight.

4 THE DEVELOPMENT AND UTILIZATION OF BIOMASS ENERGY IN RURAL AREAS

China has abundant biomass resources. According to the estimates, China's theory of biomass resources is about 50 million tons. The currently available resources for development are mainly biomass waste, including straw, firewood, manure, industrial organic waste and municipal solid waste, and other organic material[8]. In the rural economy, the development of rural renewable energy, particularly biomass technologies for sustainable development of the rural economy, is very important.

4.1 Biogas technologies

Among the bioenergy technologies, biogas technology is more mature, especially in rural areas, promoting its use more effectively and more extensively. Methane as a clean, renewable energy, the country's decades of research and development in the rural household biogas technology, is quite mature, has a higher penetration rate, and has produced significant rural ecological environment and economic benefits, while large biogas development and utilization is concerned, at the same time, we also have accumulated some evidence, but the current large biogas is not universal, and mainly UASB, auxiliary heat collection box biogas digesters and tunnel-type digesters are used currently.

Since the 1990s, in rural areas, a large number of biomass digesters have been built. In 2000, some of the large biogas project were put into use in rural areas. Biogas is clean and pollution-free, and the waste can be recycled, so it is widely promoted and

Table 1. Future demand for renewable energy.

Year	The proportion of renewable energy consumption %	The total energy consumption (million tons of standard coal)	Renewable energy consumption (million tons of standard coal)
2010	9.6	3246.93	312
2015	11.5	3900.15	445
2020	15	4519.39	678

applied in the countryside. Due to regional, temperature, and natural resource differences, biogas has been widely applied in central China and southwest, and used less in northwest China and northeast China where there is less vegetation or low temperature, which presents an obvious difference in regional distribution.

Generally, an ordinary household biogas digester supplies more biogas than needed when it functions well or in a period of seasonally higher temperatures, resulting in large amounts of waste and additional atmospheric pollution.

Biomass energy technology in rural renewable energy has an obvious impact on the development of rural economy. The development of biogas technology in rural areas promotion is earliest, which is also the most widely distributed. In addition, the particular form of rural biogas technology in China is the biogas digester, and, currently, the water pressure digester is the main form.

4.2 Methane power generation

Biogas combustion power generation is a new direction with the development of biogas utilization of biogas technology and the emergence of a high-grade use. It is used for biogas engine and equipped with an integrated power plant to generate electricity and heat, which is an important way for the effective use of biogas. Currently, the device that is mainly used for biogas power includes internal combustion engines and turbines. Biogas is mainly a dual fuel engine type with a complete-burning system. Generally, combustion of biogas generation and biogas combustion engine are used to generate electricity in large biogas projects.

In the field of biomass utilization, we have made a significant progress, especially in biogas technology, in which the production of energy per year has reached 1.15 million tons of oil equivalent, accounting for 0.24 percent of rural energy; annual energy savings has reached 52.5 million tons of oil equivalent by usingfirewood stoves[9]. According to the analysis and prediction, using 60% of all kinds of straw in rural China each yearcan produce 74.4 billion cubic meters of biogas, per year livestock manure can produce 153.62 billion

cubic meters, human excrement (urine) can produce 8.84 billion cubic meters of biogas per year in rural areas, biogas production reaches to a total of 236.86 billion cubic meters, and has a 118.43 cubic meters of biogas per household resources, which can solve the energy problems of rural life.

5 THE PROSPECT OF RENEWABLE ENERGY UTILIZATION

China has a wealth of new energy and renewable energy resources, but the development and utilization of new energy and renewable energy development started relatively late. In recent years, because of the increasingly serious environment problem and energy problem, China has began to pay close attention to the development and utilization of new energy and renewable energy, and new energy policy has gradually been promoted and guaranteed of function. The renewable energy will be considered as a source for improving the energy structure, climate change and energy security of alternative energy sources. By promoting the establishment and development of the renewable energy market, the level of the localization of renewable energy technologies will be improved, the cost of renewable energy use will be reduced, and the use of renewable energy will be gradually increased as an alternative to conventional energy.

According to "The Long-term Development of Renewable Energy", China's renewable energy consumption will reach to 15% of the total energy consumption by 2020. According to the energy demand forecasting[10], the total energy demand in 2020 will reach to 4.52 billion tons of standard coal, so we deduce that the renewable energy demand is 678 million tons of standard coal. In 2010, China's renewable energy consumption was equivalent to 312 million tons of standard coal, so in the next decade, China's renewable energy demand gap will be larger, and the total demand will be about 4 billion tons of standard coal.

In China's rural areas, the biogas residue and slurry is used as a fertilizer directly without any treatment. Such practice, on the one hand, restricts its wide application because of the inconvenience of application and transportation. Because of the seasonal nature of agricultural production and limited dissolving and absorbing ability of farmland, a significant amount of biogas residues and slurries are discharged directly into the nearby water bodies or farmland, creating a serious threat to the environment, drinking water sources and agricultural ecological safety[11]. Such practices ignore both sanitary and safety concerns. Biogas fertilizer in the form of anaerobic fermentation residue not only contains nutrients, but also heavy metals, pesticide residues and pathogenic bacteria. The unbalance of nutrients and the excessive manure load per unit area of farmland caused by the continual application of a kind of fertilizer cause secondary pollution, such as soil phosphate, soil phosphorus leachate and heavy metal accumulation. The amount of biogas residue increases considerably with the development of biogas projects. So, with a broad market and great potential, commercial fertilizers should be produced through the deep processing of biogas residues, which is of long shelf-life and good effect, convenient to transport and apply and, above all, pollution-free.

6 CONCLUSIONS

The development of rural renewable energy can contribute to the emergence of new economic growth point. The development of renewable energy in rural areas is mainly based on local natural resources and human resources, and can greatly promote regional economic development using rural renewable energy, which is still in its infancy. Human development diversified rural renewable energy can create new areas of economic growth and new jobs. As the large-scale development of the intensive breeding industry and the advantage of developing straw biogas, the full use of straw as a raw material will be the direction of biogas development in China. At present, dry methane fermentation will become an important method in the large-scale production of biogas from agricultural wastes[12]. The development of renewable energy in rural areas, to meet the energy needs of farmers and new rural constructions, has important significance. This will be conducive to China's transformation of economic development in rural areas, to enhance the level of economic development in rural areas and improve the economic efficiency of rural areas. Moreover, for the development of sustainable use of resources in rural areas, the new socialist countryside ecological civilization blazes an important trail. It has a great significance for rapid, low-carbon energy development in rural areas.

ACKNOWLEDGMENTS

This paper was financially supported by the Independent Innovation Project of Academics in Jinan (No. 201401223).

REFERENCES

[1] Bloch H, Rafiq S, Salim R. Economic growth with coal, oil and renewable energy consumption in China: Prospects for fuel substitution [J]. Economic Modelling, 2015, 44: 104–115.

[2] Ming Z, Ximei L, Yulong L, et al. Review of renewable energy investment and financing in China: Status, mode, issues and countermeasures [J]. Renewable and Sustainable Energy Reviews, 2014, 31: 23–37.

[3] Niu H, He Y, Desideri U, et al. Rural household energy consumption and its implications for eco environments in NW China: A case study [J]. Renewable Energy, 2014, 65: 137–145.

[4] Yao L, Chang Y. Energy security in China: a quantitative analysis and policy implications [J]. Energy Policy, 2014, 67: 595–604.

[5] Ouyang X, Lin B. Levelized cost of electricity (LCOE) of renewable energies and required subsidies in China [J]. Energy Policy, 2014, 70: 64–73.

[6] Li C, Li P, Feng X. Analysis of wind power generation operation management risk in China [J]. Renewable Energy, 2014, 64: 266–275.

[7] Lund H. Renewable energy systems: a smart energy systems approach to the choice and modeling of 100% renewable solutions [M]. Academic Press, 2014.

[8] Chen Y, Hu W, Feng Y, et al. Status and prospects of rural biogas development in China [J]. Renewable and Sustainable Energy Reviews, 2014, 39: 679–685.

[9] Deng Y, Xu J, Liu Y, et al. Biogas as a sustainable energy source in China: Regional development strategy application and decision making [J]. Renewable and Sustainable Energy Reviews, 2014, 35: 294–303.

[10] Sun D, Bai J, Qiu H, et al. Impact of government subsidies on household biogas use in rural China [J]. Energy Policy, 2014, 73: 748–756.

[11] Yang J, Chen B. Extended energy-based sustainability accounting of a household biogas project in rural China [J]. Energy Policy, 2014, 68: 264–272.

[12] Song Z, Zhang C, Yang G, et al. Comparison of biogas development from households and medium and large-scale biogas plants in rural China [J]. Renewable and Sustainable Energy Reviews, 2014, 33: 204–213.

Architectural, Energy and Information Engineering – Sung & Chen (Eds)
© 2016 Taylor & Francis Group, London, ISBN 978-1-138-02791-6

Thermal gravimetric experimental study on the combustion properties of biomass

Dong Yang & Yan Zhang
Key Laboratory of Renewable Energy Utilization Technology in Building of National Education Ministry, Jinan, China
Shandong Provincial Key Laboratory of Building Energy-Saving Technology, Jinan, China
School of Thermal Energy Engineering, Shandong Jianzhu University, Jinan, China

ABSTRACT: With the rapid development of economy, the problems of environmental pollution and energy shortage have become increasingly prominent. Biomass has become an important direction of development and utilization of renewable energy with the features of low content of N, S, near-zero net CO_2 emissions, and inexhaustible. From the current point of view, biomass direct combustion technology is the most widely used and the most extensively used biomass energy utilization technologies. Therefore, the study on the combustion characteristics of biomass fuels is necessary. This paper adopted four typical biomass fuels in Shandong Province as the object of the study, using thermal gravimetric analysis methods to study biomass fuel combustion characteristics. This paper investigated the effects of the heating rate and combustion atmosphere on four biomass fuel combustion characteristics through a series of experiments. The analysis of biomass fuel combustion characteristics and mechanism will have an important theoretical and practical value to provide evidence for efficient combustion power generation of biomass.

Keywords: biomass; fuel; combustion characteristics; thermal gravimetric analysis

1 INTRODUCTION

Rapid economic development requires us to adjust the current form of energy consumption, and to develop and utilize clean renewable energy. In a large number of renewable energy and new energy projects, the large-scale development of biomass energy is undoubtedly a realistic option. Biomass energy is rich in China. The calorific value of biomass is approaching to that of medium bituminous coal, and using them will produce less environmental pollution; moreover, SO_2 and NO_X emissions will be far less than the emissions from coal. Biomass production can make waste to be used as resources and to be harmless, which is conducive to environmental protection and recycling of resources, and is one of the important means to achieve China's development of renewable energy and to control carbon emission targets.

Biomass combustion is the most direct way of biomass utilization. Fuel properties are the main factors that cause defects in the power feeding system, wear of the precipitator, and the abnormal operation of a unit. Therefore, the study of the combustion characteristics of biomass is necessary.

At present, China's biomass combustion power generation has begun to take shape, but there are still problems: the types of biomass are many, different biomass fuels have different combustion characteristics, so we cannot find a uniform combustion, which will not be conducive to resource utilization. Therefore, this paper uses a thermal gravimetric analysis of experimental methods to study the fuel characteristics of four biomass fuels: apple wood, poplar, corn stalks and cotton stalks. It investigates the effects of the heating rate and combustion atmosphere on four biomass fuel combustion characteristics by using TG and DTG curves, providing a scientific basis for the mature use of biomass fuel technology.

2 RAW MATERIALS OF BIOMASS

The test selected four typical biomass fuels, including cotton straw, corn stover, poplar and apple wood. Corn is the main crop in north China, where the planting area of cotton is relatively large and its straw tends to be treated as waste for direct fire in the open air, neither achieving the recycling of resources, nor seriously polluting the environment.

The latter two belong to woody plants. The samples were prepared using a grinder.

3 BIOMASS BURNING TGA EXPERIMENTS

3.1 Experimental instrument

Thermal gravimetric analysis used in this study consists of the balance, the stove, the temperature-controlled system of program, and recording system, which is an instrument used in the TGA method to detect material temperature—mass change relations under the control of a computer program. It adopted TGA/SDTA851 thermal balance produced by Mettler Toledo Company, accuracy of which is 1 microgram at 0.01°C. The protective gas flow rate into the combustion chamber is 60 ml/min, and the flow rate into the balance chamber is 40 ml/min.

3.2 Experimental principle

The experiments selected four typical biomass fuels, including cotton straw, corn stover, poplar and apple wood, using the thermal gravimetric analysis system, analyzing biomass characteristics according to different heating rates (20°C/min, 40°C/min and 60°C/min, respectively). The final temperature rose to 900°C and the amount of the material is about 10 mg. Under the conditions of the programmed temperature, TGA tests were conducted on the four biomasses, respectively, and obtained biomass burning TG curves, with the temperature as the abscissa. During the pyrolysis process, the furnace temperature set according to the rate increases and the quality decreases. The TG curve represents the variation in quality.

Figure 1.   Thermo gravimetric analyzer.

3.3 Experimental results and analysis

Biomass volatile matter process is usually divided into warm-drying stage, volatile precipitation stage, and carbonization stage. After the start of the experiment, the furnace temperature gradually increased, the water evaporated, and the TG curve tended to level off when the moisture content was reduced to a certain extent. Then, volatile matter began to precipitate, the weightlessness rate increased rapidly, the TG curve steepened significantly, about 300°C, and the weight loss rate reached the maximum. Then, due to the precipitation of a large number of volatile matter and not obvious carbonation, the weightlessness rate decreased and the TG curve tended to be constant. With a further increase in the furnace temperature, the carbonization rate continued to accelerate, and the TG curve had a downward trend. At about 830°C, volatile matter ended basically and quality remained unchanged. The quality variation in the pyrolysis process is shown in Figures 2–5:

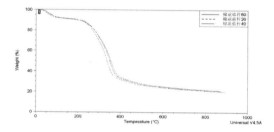

Figure 2.   TG curve of cotton stalks.

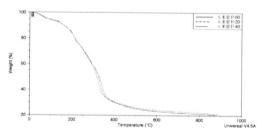

Figure 3.   TG curve of corn stalks.

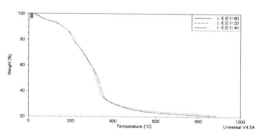

Figure 4.   TG curve of apple wood.

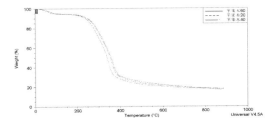

Figure 5.   TG curve of poplar.

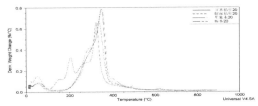

Figure 6.   Pyrolysis weight loss curve of different biomass fuels.

3.3.1   *Analysis of four kinds of biomass pyrolysis characteristics at different heating rates*

Figures 2.2–2.5 show the case of volatile matter of cotton stalks, corn stalks, apple wood and poplar at different heating rates.

At three different heating rates, the combustion curves showed a similar characteristic, but showed different temperature ranges and volatile matter rates with different heating rates. Increasing the heating rates resulted in an increase in all the characteristics of the temperature. The TG weight loss curve moved to the high temperature side. However, the temperature corresponding to the maximum rate of pyrolysis was generally around 300°C. The main reaction space increased slightly. This is because when achieving the same temperature, the higher the heating rate, the shorter the response time experienced by the sample, and the lower the reaction degree. Different heating rates had a little effect on the ultimate weight loss or eventually closed gas rate, but the time required to achieve the same precipitation amount of volatile matter varied widely. For example, when cotton straw was pyrolysed, to achieve a 50% weight loss rate, the required time at a heating rate of 40°C/min was about only half of 20°C/min. At the pyrolysis temperature of 830°C, volatile matter was fully precipitated basically. Volatile matter activation energy at different heating rates was different. The higher the heating rate, the greater the activation energy of the volatile matter reaction. This is mainly because the higher the heating rate, the greater the heat transfer resistance inside the reactant particles. In addition, the maximum volatile matter rate at the heating rate of 40°C/min was about 2 times when the heating rate was 20°C/min; however, the time required for volatile matter was about half of 20°C/min. It shows that the higher the heating rate, the faster the volatile matter rate, and the shorter the time required to achieve the same degree of volatile matter, so in a short time, the maximum amount of volatile can be reached. So, the heating rate increase will affect many aspects of biomass volatile matter, which should all be taken into consideration.

3.3.2   *Analysis of four kinds of biomass pyrolysis characteristics at the same heating rates*

This paper studies the case of volatile matter for four biomass fuels. In order to compare the different combustion characteristics of biomass, DTG curves at the heating rate of 20°C/min are summarized, as shown in Figure 6.

As shown in Figure 6, the temperature region and the peak height of volatile matter show the differences in volatile matter performance for different types of fuel. We can see that the evaporation of water is basically identical in the first stage, and the final temperature is substantially the same. The variation in volatile and fixed carbon content for four kinds of biomass lead to differences in the second stage and the third stage. The volatile content of biomass is much higher than the fixed carbon content. The combustion rate in the second stage was significantly higher than that in the third stage.

4   CONCLUSIONS AND DISCUSSION

Based on the investigation of the current biomass situation and the analysis of TGA experimental data, we draw the following conclusions:

Fuel combustion atmosphere and heating rate have a significant impact on the combustion characteristics of a fuel. It was found that, through the TGA, the combustion loss rate of biomass fuel increases in the full process with the increase in the heating rate, under the same atmosphere, the larger the heating rate, the higher the peak temperature, and the greater the corresponding conversion rate. It will produce the phenomenon of thermal hysteresis.

The first volatile matter stage of four kinds of biomass fuel, which is the process of water evaporation, is basically the same, as well as the final temperature of volatile matter. However, the variation in fixed carbon and volatile matter contents among the four biomass fuels causes differences between the second stage and the third stage. The same kind of biomass showed similar characteristics at different heating rates for TG curves, but exhibited different temperature ranges and volatile matter rates with different heating rates. With the increase in the heating rate, various characteristics of the temperature have increased more or less. However, the temperature corresponding to the maximum volatile matter rate is generally around 300°C. Different heating rates have a little effect on the ultimate weight loss rate or eventually closed gas rate, but the time required to achieve the same precipitation amount of volatiles varies widely.

(2) Biomass is a clean and renewable energy, and is one of the primary energy in the future. In the process of a variety of conversion and utilization for biomass, biomass combustion technology is the best suitable way for China's national conditions, and is one of the most mature and feasible approach in the clean and efficient use of large-scale biomass. Better burning results can be obtained under the premise that the device does not need to make great changes right now. So, promotion and application of biomass combustion technology has an important significance to the development of biomass utilization technologies and protection of ecological environment. However, at present, research on the aspects of biomass burning is relatively scarce. Although moisture in biomass fuel is too large, and is not conducive to processing and transportation, the volatile content of biomass fuel is more than 65%, which can be more easily organized for ignition and burning. So, the continuous improvement of combustion technology of biomass efficiently is an important direction of biomass development.

ACKNOWLEDGMENTS

This paper was financially supported by the Independent Innovation Project of Academics in Jinan (No. 201401223).

REFERENCES

[1] 5E-MACIII infrared Coal Analyzer manual.
[2] Xu De-liang, Sun Jun, Chen Xiao-juan, Liu Xiang. Biomass burning application status of industrial boilers [J]. Forest Products Industry, 2009, 36(6): 3–7.
[3] Yuan Zhen-hong, Wu Chuang-zhi, Ma Long-long. Utilization principles and techniques of biomass energy [M]. Beijing: Chemical Industry Press, 2004. 11.
[4] Wei Xue-feng, Liu Jian-ping. Utilization status and prospects of biomass fuels [J]. Yunnan Environmental Science, 2005, 6, 24(2): 16–19.
[5] Zhu Shi-qing, Yan Li-feng, Guo Qing-xiang. Biomass Clean Energy [M]. Beijing: Chemical Industry Press, 2002.
[6] Jiang Shu-jie, Zhao Wei-ying. Superficial talk on the straw biomass direct combustion power generation technology [J]. Boiler manufacturing, 2009, 7, Fourth (Total 216): 40–42.
[7] Liu Rong-hou, Niu Wei-sheng, Zhang Da-lei. Biomass thermochemical conversion technology [M]. Beijing: Chemical Industry Press, 2005. 5.
[8] Beijing Agricultural Information Network. Current Assessment Situation of Chinese crop straw resources, 2010. 7.
[9] Matti Hiltunen, Vesna Barišić, Edgardo Coda Zabetta. Combustion of Different Types of Biomass in CFB Boilers. Presented at 16th European Biomass Conference Valencia Spain June 2–6, 2008.

Architectural, Energy and Information Engineering – Sung & Chen (Eds)
© 2016 Taylor & Francis Group, London, ISBN 978-1-138-02791-6

Thermal gravimetric experimental study on the co-combustion properties of biomass and coal

Dong Yang & Yan Zhang

Key Laboratory of Renewable Energy Utilization Technology in Building of National Education Ministry, Jinan, China
Shandong Provincial Key Laboratory of Building Energy-Saving Technology, Jinan, China
School of Thermal Energy Engineering, Shongdong Jianzhu University, Jinan, China

ABSTRACT: Over the past decade, biomass's co-firing with coal has become a hot research topic, because of the features of low cost and low risk, and as a renewable resource. This paper introduces the research status for the co-combustion properties of biomass and coal at home or abroad, and conducts an experimental study for the mixed combustion properties of biomass and coal using the thermal gravimetric analysis. At different heating rates (20°C/min, 40°C/min, 60°C/min), a separate test analysis was carried out for coal, sycamore wood, cotton stalks and poplar. At the heating rate of 60°C/min, three types of biomass burned mixed with coal at the proportion of 10%, 20%, and 30% respectively. This paper qualitatively analyzes the impact of two factors, heating rate and mixing ratio, on the mixed fuel combustion characteristics, and proposes that coal combustion characteristics can become better when adding biomass, in order to provide theoretical guidance for taking advantage of heavy-duty boilers and the large-scale use of biomass.

Keywords: biomass; coal; combustion; combustion characteristics

1 INTRODUCTION

At present, China's energy and environmental problems have been highlighted, which is the bottleneck in current socio-economic development. In addition, the use of fossil fuels has caused serious environmental pollution and ecological damage. Biomass energy is the world's fourth largest energy source, with features of renewable, resources-abundant, and environment-friendly. It can replace part of oil, coal and other fossil fuels, having become one of the important ways to solve energy and environmental problems. Thus, domestic and foreign scholars have conducted extensive research on biomass. Western Kentucky University has developed a new technology, in which biomass air gasification produces fuel gas with a high calorific value and low tar. Tar content is very low, and carbon conversion and gasification efficiency are higher[1]. The US National Renewable Energy Laboratory conducted a study of combined gasification of coal and biomass in a fluidized bed under high pressure conditions, obtaining satisfactory results[2], and a series of analyses and evaluations were conducted on a variety of biomass utilization technologies. At domestic, Forest Chemical Industry Research Institute in the Chinese Academy of Forestry achieved results in the fluidized

gasification technology, circulating fluidized cone-enriched biomass gasification technology[3]. Xi'an Jiao tong University in recent years has been committed to basic research in aspects of super-critical biomass catalytic gasification[4]. The China University of Technology conducted a plasma gasification of biomass[5] and biomass gasification synthesis. However, the existing biomass energy utilization methods have many disadvantages. In addition, the problems of collecting difficulty of biomass directly and high cost cannot be solved. So, this paper puts forward the use of biomass for mixed combustion in the existing large coal-fired power plants; in this way, the problem of high efficiency, large-scale use of biomass can be effectively solved, leading to a realistic significance for the use of renewable energy in China.

2 PROMOTING THE COMBUSTION MECHANISM OF BIOMASS TO COAL

Biomass mixed with coal can greatly improve the fuel combustion characteristics, because it contains a large amount of cellulose and lignin, poses a large number of volatile matter, with rapid evaporation combustion in the initial period, which can make it easy to burn at a low temperature, to reduce the ignition temperature of coal and to shorten the time needed for fire, significantly improving the burning of coal properties. At the same time, cellulose volatile analysis can form a loose shape of structure, which is advantageous for the coal to further combine with oxygen, and makes it easier for coal combustion. In addition, biomass burning will form biomass ash content, and the existence of high alkali metal and alkaline earth metal elements in ash content can play a catalytic role in the burning of coal, which improves the ignition and burning of coal.

3 EXPERIMENTAL INTRODUCTION

3.1 *Laboratory instruments*

The test equipments used in this paper are as follows: Thermogravimetric Analyzer, Fast Infrared Coal Analyzer, Dry dish, Blast Oven, Electronic analytical balance, and Small laboratory pulverizer. Experimental samples are Cotton stalks, Corn stover, Apple wood, Poplar, Indus wood and coal.

Thermogravimetric analyzer is an instrument to test the temperature—mass change relations of the substance. The TGA method is used to measure the changes in the relationship of the quality of the substance that varies with the temperature (or time) under the programmed temperature. When the test substance sublimates, vaporizes, decomposes gases or loses water of crystallization during heating, the measured mass of the substance will change. By analyzing the TG curve, we can find the number of degrees when the test substance produces changes, and according to the weight lost, we can calculate how much substance is lost. Experimental TGA helps to study the changes in crystal properties, such as melting, evaporation, sublimation, adsorption and other physical phenomena of the substance; in addition, TGA also helps to study dehydration, dissociation, oxidation, reduction and other chemical phenomena of the substance. Curves obtained by the TGA method are called the thermal gravimetric curve (TG curve). The TG curve uses quality as the ordinate, representing the reduction in the quality from top to bottom, and uses temperature (or time) as the abscissa, from left to right representing the increase in the temperature (or time).

3.2 *Experimental principle and aim*

The thermal balance is used to study the combustion characteristics of coal, which is the combustion characteristic parameter represented according to the analysis of the curve. It includes the determination of a fuel ignition temperature, the maximum burning rate, and burning temperature, which is a common analytical methods to

Figure 1. Thermogravimetric analyzer.

determine the raw coal combustion characteristics. This method began in the 1970s in a foreign country, while the study of China's domestic coal combustion characteristics began in the 1980s. The method using the thermal balance to study the combustion characteristics of coal has not been of uniform standards at home and abroad. American B&S Company determines the ignition characteristics of coal with the combustion distribution curve after testing. Xi'an Thermal Power Research Institute showed that the discriminant index R and burn characteristics distinguish the index Ri with stable ignition combustion characteristics. Although more information is available about the thermal gravimetric analysis of single coals, research on mixed fuel is scarce. For a mixed fuel, due to the interaction between different fuels, tts combustion characteristics, such as burn-out characteristics and fire characteristics, show a very big difference compared with a single fuel. If the difference is greater when the fuel is mixed, this difference becomes more pronounced. By using the thermal gravimetric analysis method, the study of fuel characteristics is conducted on single coal and biomass alone and mixed in different proportions. Through a comparative analysis on the TGA curves of single biomass, coal and fuel mixture, it conducts the analysis of the similarities and differences between a single fuel and a mixed fuel.

3.3 *Biomass and coal combustion characteristic research methods*

We adopted the thermogravimetric analysis method, at different heating rates (e.g. 20°C/min, 60°C/min). A separate test analysis was carried out using coal, wutong wood, cotton stalks and cypress; in addition, three kinds of biomass and coal combustion experiment were carried out, respectively, in different proportions (e.g. 10%, 20%) under the

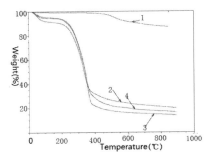

a) Heating rate of 20°C/min, pure coal (1), cotton straw (2), sycamore (3), aspen (4) TG graph

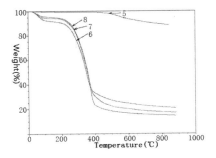

b) Heating rate of 40°C/min, pure coal (5), cotton straw (6), sycamore (7), aspen (8) TG graph

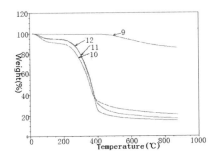

c) Heating rate of 60°C / min, pure coal (9), cotton straw (10), sycamore (11), aspen (12) TG graph

Figure 2. Different heating rates (e.g. 20°C/min, 40°C/min, 60°C/min): coal, sycamore wood, cotton stalks, and poplar burning alone TG curve.

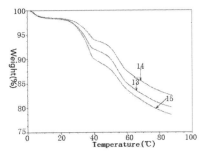

d) Heating rate of 60°C / min, 10% proportion of cotton stalks (13), sycamore (14), Aspen (15) blended coal burning TG graph

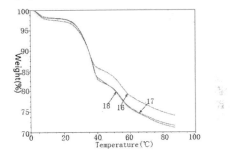

e) Heating rate of 60°C / min, 20% proportion of cotton stalks (16), sycamore (17), Aspen (18) blended coal burning TG graph

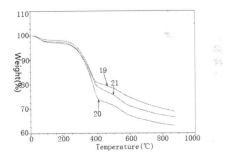

d) Heating rate of 60°C / min, 30% proportion of cotton stalks (19), sycamore (20), Aspen (21) blended coal burning TG graph

Figure 3. Heating rate of 60°C/min: three biomass mixed coal at the proportion of 10%, 20%, and 30%, respectively burning TG curve.

heating rate of 60°C/min. The test mainly uses TG curves, and conducts the research on two factors, heating rate and mixing ratio, and qualitatively analyzes the effect of the two factors on the combustion characteristics of the blended fuel.

3.4 Analysis of the test results

Coal and sycamore wood, cotton stalks, poplar and other biomass burn after mixing. There are many factors that affect the combustion characteristics.

This article studies the two factors heating rate and the blend ratio on the basis of TGA experiments, qualitatively analyzing the impact of the two factors on the mixed fuel combustion characteristics. Based on the above results, the following conclusions can be drawn:

1. For the pure biomass, with the increased heating rate, TG curves move to high temperature side. This is mainly because the biomass burning is an endothermic reaction, and the increase in the heating rate reduces the time required for the combustion environment to reach the same temperature.
2. For pure coal, the heating rate is high, the burnout temperature is higher, the burnout time is less, and thermal hysteresis phenomenon generated is more serious, often making the initial and final temperatures on TG curves to increase. Under different heating rates, the final percent content of coal combustion tends to be the same value, because the final product of coal burning is ash, and the ash content is always constant.
3. For mixed combustion, the peak formed in a low temperature range is similar to the case of single biomass combustion, where the peak is merely low. In addition, for the high temperature segment, with an increasing biomass mixing proportion, the weight loss region gradually moves to the direction of decreasing temperature. With the increase in the biomass mixing ratio, the maximum weight loss rate increases in the low-temperature region and decreases in the high-temperature region.

When coal is mixed with biomass, the maximum rate of combustion has a trend to advancement, the time required is shorten and the temperature drops down. It shows that the addition of biomass in the coal results in the maximum combustion rate.

Biomass and coal combustion makes the ignition temperature lower that that of coal. This is because the biomass contains large amounts of volatile matter that can be rapidly precipitated at a lower temperature, improving the coal ignition performance. With the increase in biomass adding quantity, coal fire performance can be improved to different degrees.

When biomass is mixed with coal, the burnout temperature of the coal decreases, to obtain a better burning efficiency at a lower temperature. With the proportion of the biomass increasing, the burning temperature gradually decreases, the time required to burning is shorter than coal, and burnout characteristics are obviously better than elemental coal. This shows that the addition of biomass is conducive to the complete combustion of the coal, and to improve the utilization of coal.

4 CONCLUSION

Biomass and coal combustion technology is a potential research, where the mixed combustion of biomass and coal is beneficial to improve the efficiency of the combustion of coal and protect the ecological environment. Thus, the development and utilization of biomass energy and the research of mixed combustion of biomass and coal are effective measures for the sustainable development of energy in China. It is helpful to establish a biomass fuel market and promote the development of economy. However, research on the mixed combustion of biomass and coal still indicates many problems: how to transform the existing coal-fired power plants to meet the needs of the biomass fuel is the main challenge of biomass blended with coal combustion technology.

ACKNOWLEDGMENTS

This paper was financially supported by the Independent Innovation Project of Academics in Jinan (No. 201401223).

REFERENCES

[1] Yan, C., Yang, W., John, T.R, et al. A novel biomass air gasification process for producing tar-free higher heating value fuel gas [J]. Fuel Processing Technology, 2006, (87): 343–353.
[2] McLendontr, Lui AP, Pineaultrl, et al. High pre-ssure co-gasification of coal and biomass in a fluidized bed [J]. Biomass and Bioenergy, 2004, (26): 377–388.
[3] Jiang Jian-chun, Ying Hao, Dai Wei-di. Study on the industry technology of catalytic gasification of biomass with a fluidizated bed [J]. Acta Energiae Solaris Sinica, 2004, 25(5): 678–684.
[4] L&U You-jun, Ji Cheng-meng, Guo Lie-jin. Experimental investigation on hydrogen production by agricultural biomass gasification in supercritical water [J]. Journal of Xi'an Jiaotong University, 2005, 39(3): 238–242.
[5] Zhao Zeng-li, Li Hai-bin, Wu Chuang-zhi, et al. The study of the plasma gasification of biomass [J]. Acta Energiae Solaris Sinica, 2005, 26(4): 468–472.
[6] Zhu Xi-feng, Venderbosch R H. Experimental research on gasification of biooil derived from biomass pyrolysis [J]. Journal of Fuel Chemistry and Technology, 2004, 32(4): 510–512.

Architectural, Energy and Information Engineering – Sung & Chen (Eds)
© 2016 Taylor & Francis Group, London, ISBN 978-1-138-02791-6

Research on fast parallel clustering algorithm based on cloud platform in digital archives

X. Wang
Archives, Southwest University, Chongqing, China

Z.P. Li
Information Centre, Southwest University, Chongqing, China

ABSTRACT: In order to improve the performance of traditional K-Means data mining algorithm, this paper proposes a universal optimization strategy based on a new dynamic cloud platform. The new dynamic cloud platform takes advantages of hadoop model and lustre module. The new optimization strategy solves the problem of how to select the initial cluster centers in K-Means. Experimental results show that the optimized algorithm is more efficient than traditional algorithm, which is suitable for cluster analysis in large-scale data, such as digital archives.

Keywords: cloud computing; clustering algorithm; digital archives; K-means

1 INTRODUCTION

During the past decades, vast amounts of data have been generated every day in the world, which are growing exponentially. Important information about enterprises, research institutions, government departments and other areas can be obtained through analyzing these massive data. It is inefficient to analyze data through the traditional manual analysis, which cannot make a strategic decision quickly and accurately. So a lot of data mining algorithms have been proposed by experts and scholars to replace traditional manual analysis [1,2], for example cluster analysis algorithm, but the current clustering algorithms have encountered a bottleneck when we process massive data. It has become a hot issues that experts and scholars concern how to obtain valuable information from large-scale data efficient, fast, accurate and cost-effectively.

Although a lot of effort is being spent on improving these weaknesses, the efficient and effective method has yet to be developed. This paper examines a new measure of clustering analysis based on cloud computing which overcomes the difficulties found in current clustering algorithms measures.

This paper proceeds as follow.

We shall first briefly introduce cloud computing and related research, then details on hadoop and K-means algorithm are discussed, an entirely Platform of dynamic cloud and clustering algorithms are proposed, lastly, the correctness and clustering efficiency and acceleration effect of the algorithm are validated through experiments.

2 BACKGROUND

In this section, we first introduce several models on distributed data processing model based on cloud platform. Then, we present the latest researches on distributed parallel clustering algorithm.

2.1 *The distributed data processing model based on cloud platform*

Kun Wang [3] summarizes several common model of distributed concurrent data processing, there are mainly three kinds of model that is Variable-Sharing Concurrency Message-Passing Concurrency and Dataflow Concurrency. Variable-Sharing Concurrency and Dataflow Concurrency could possibly go wrong of starve to death and conditions of competition, because the principle of them are the shared variable technology. Therefore, the model of Message-Passing Concurrency based on the parallel execution is relatively mature. The representative model are Google Pregel [4], Yahoo Giraph [5] and MapReduce [6]. In this paper, the mapreduce were used to solve massive data clustering problem. The map reduce is efficient distributed programming model which of typical models are the HOP [7] system and

Hadoop system. So far, in order to improve the disk I/O performance and job scheduling model of traditional Hadoop system, experts and scholars put forward the optimization model based on Hadoop, such as HaLoop, PrIter, Twister etc.

2.2 The distributed clustering algorithms

Currently, researchers have designed Distributed clustering algorithms, in 1995, parallel algorithm based on hierarchical clustering were proposed by Clark F. Olson, the time complexity of Hierarchical clustering was optimized from O (N2) to O (nlogn) [8]. References [9] provide a detailed introduction about parallel K-Means clustering algorithm based on MPI, and achieving the process of data based on the resume through this algorithm. Dhillon I.S. and Modha D.S studied the parallel K-Means, which speeds up the clustering speed of K-Means. Jian Wan Realizing the distributed text clustering of K-Means [10] in Hadoop. Yi Feng Liu studied [11] the hierarchical text clustering in hadoop.

By means of the above research results, the main contributions of this paper include the following:

1. A universal optimization strategy is proposed to improve the efficiency of K-means algorithms. The proposed strategy focuses on selection of initial cluster centers, which were based on model of dynamic cloud platform.
2. Systematic experiments have been conducted to compare the clustering efficiency and acceleration effect between the traditional k-means algorithms and optimized algorithms based on dynamic cloud platform.

3 THE PLATFORM OF DYNAMIC CLOUD

3.1 The hadoop

The reading and writing process of hadoop as shown in (Figure 1) [12].

When dealing with the high complexity of the application, Hadoop will cause the following problems:

1. If the intermediate calculation results are relatively large, the intermediate results are written to disk piecewise, in addition to controlling the number of output file from single Map task, there is dedicated thread to merge these files in hadoop, these operations will cause memory consumption and disk blocking.
2. The Map results are stored on the local disk in hadoop, so before the Reduce operation were executed, there is the special thread to acquire the necessary results of Map through HTTP,

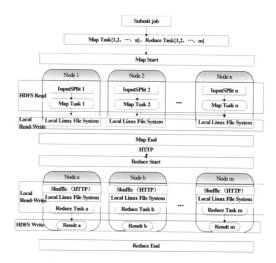

Figure 1. Hadoop read-write flow.

the network traffic will be outbreak at this stage that can result in network resource consumption and network instability.
3. The universal property and scalability of the HDFS were limited because it can not be mounted to the local file system.

3.2 A new dynamic cloud platform based on hadoop and luster

To solve the problems presented in Section 3.1, we design a new dynamic cloud platform, which is based on hadoop and lustre, lustre module design is shown in Figure 2 [13–15], the single task I/O of Map/Reduce were implemented concurrently through this platform, as shown in Figure 3, the JobTracker and TaskTracker of Hadoop are still responsible for the job management and task management. The optimization of the new dynamic cloud platform includes the following:

1. The I/O operation was managed by TaskTracker through the client of lustre, rather than the local linux file system and HDFS. The single task of Map/Reduce can be output to Multiple OST of lustre by the client of lustre, this can reduce the possibility of disk block.
2. At the beginning of lustre, HTTP transmission were replaced by the hard link function in lustre, the actual I/O operation were read when needed by the client of lustre, this technique is called "read delay", can effectively reduce the network storm of reduce task in the beginning, disperse network traffic, reduce the possibility of network bottleneck.

342

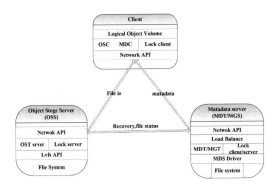

Figure 2.   Lustre module design.

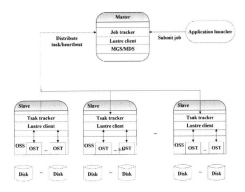

Figure 3.   New dynamic cloud platform design.

3. Scheduling tasks in the JobTracker, using Runtime.getRuntime () shell command is executed at run time through the lustre analysis of each band of the OSS data, thereby reducing unnecessary network overhead.

4   A OPTIMIZATIONAL STRATEGY FOR THE K-MEANS ALGORITHMS

4.1   *Basic idea of the k-means*

The k-means algorithm is an evolutionary algorithm that gains its name from its method of operation. The algorithm clusters observations into k groups, where k is provided as an input parameter. It then assigns each observation to clusters based upon the observation's proximity to the mean of the cluster. The cluster's mean is then recomputed and the process begins again.

The traditional K-Means algorithm (Figure 4) is described as follows:

1. The algorithm arbitrarily selects k points as the initial cluster centers.
2. Each point in the dataset is assigned to the closed cluster, based upon the Euclidean distance between each point and each cluster center.
3. Each cluster center is recomputed as the average of the points in that cluster.
4. Steps 2 and 3 repeat until the clusters converge. Convergence may be defined differently depending upon the implementation, but it normally means that either no observations change clusters when steps 2 and 3 are repeated or that the changes do not make a material difference in the definition of the clusters.

4.2   *Optimizational k-means algorithm*

One of the main disadvantages to k-means is the fact that you must specify the number of clusters as an

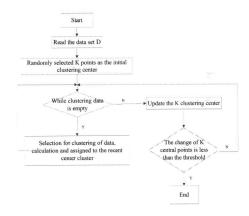

Figure 4.   Traditional k-means algorithm calculation process.

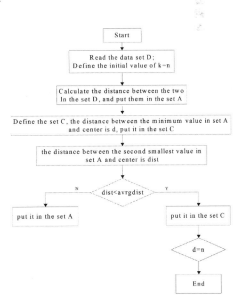

Figure 5.   Optimizational k-means algorithm.

343

input to the algorithm. As designed, the algorithm is not capable of determining the appropriate number of clusters and depends upon the user to identify this in advance. So we propose a universal strategy to select the initial clustering center. The optimizational K-Means algorithm as shown below (Figure 5).

5 EMPIRICAL STUDY

In order to validate the correctness and clustering efficiency and acceleration effect of the optimizational algorithm, we build a platform of dynamic cloud in which the k-means and optimizational k-means are compared. The data of digital archives was chosen as the experimental object.

The first set of experiments verify the time efficiency of traditional k-means algorithm and the optimizational algorithm running on a different scale of database, the size of the tree database: A = 100000, B = 200000, C = 300000, D = 400000, E = 500000, the experimental results as shown below (Figure 6).

From the above chart shows, the optimizational algorithm based on dynamic cloud platform is better than the traditional k-means clustering algorithm, especially in large scale database.

With the increase of data size, the required time of traditional algorithm and the algorithm is increased synchronously , the required time of traditional algorithm show nonlinear growth, while this algorithm show more moderate growth, this indicate that it is a good way to improve the algorithm efficiency through Distributed processing.

In the second experiment, the acceleration effect of algorithm was verified based on dynamic cloud platform. The database Contains 100000 tree using simulated tree generation process. The minimum support degree is 200, 400, 600, 800, 1000, the speedup as shown below (Figure 7).

As can be seen from the figure, the data scale is larger, the accelerated algorithm performance is better. Because the calculation performance of dynamic cloud platform is better in a large database.

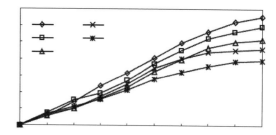

Figure 7. Speedup performance test results of traditional algorithm.

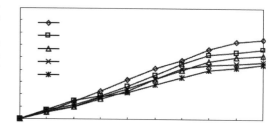

Figure 8. Speedup performance test results of optimizational algorithm.

At the same time, the optimizational algorithm reduces the cost of communication between nodes using a master control program Driver to coordination algorithm implementation process, the minimal tasks with reasonable values in the dynamic cloud platform ensure that the algorithm is efficient and accurate operation.

In summary, the optimizational clustering algorithm based on dynamic cloud platform is effective and feasible, Because of good clustering effect, the algorithm is suitable for clustering analysis on massive data.

6 CONCLUSIONS

This paper puts forward an entirely Platform of dynamic cloud and clustering algorithms. The experimental results show that the proposed algorithm has good clustering effect and feasibility. Clustering algorithm based on cloud platform has become a hot at home and abroad. In the future, we will further improve the algorithm and the platform of application be apply to a broad range of data type.

ACKNOWLEDGMENTS

This work was supported by the Fundamental Research Funds for the Central Universities (no. XDJK2014C009).

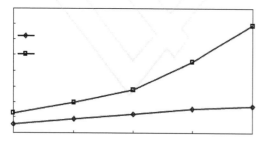

Figure 6. The efficiency of clustering algorithm.

REFERENCES

[1] Zamir O, Etzioni O, Madani O, et al. Fast and Intuitive Clustering of Web Documents [C]// Proceedings of the 3rd International Conference on Knowledge Discovery and Data Mining. New York: AAAI Press, 1997: 287–290.

[2] Osinski S, Weiss D. Conceptual Clustering Using Lingo Algorithm: Evaluation on Open Directory Project Data [J]. Intelligent information processing and Web mining: proceedings of the International IIS. Springer Press. 2004:369–377.

[3] Kun Wang. Parallel Program Design and Optimization Based on Multi core [D]. Nanjing: Nanjing University, 2012.

[4] Grzegorz Malewicz, Matthew H Austern, Aart J.C.Bik et al. Pregel: A system for large-scale graph processing C] //Proceedings of the SIGMOD. Indianapolis, Indiana, USA, 2010: 135–145.

[5] Ching Avery. Giraph: Large-scale graph processing infrastructure on Hadoop [C]//Proceedings of the Hadoop Summit. Santa Clara, 2011.

[6] Dean Jeffrey, Ghemawat Sanjay. MapReduce: Simplified data processing on large clusters [J]. Communications of the ACM, 2008, 51(1): 107–113.

[7] Condie Tyson, Conway Neil, Alvaro Peter et al. MapReduce Online [C] // Proceedings of the NSDI. San, Jose, California, USA, 2010: 33–48.

[8] Clark F. Olson. Parallel algorithms for hierarchical clustering [J]. Parallel Computing, 1995, 21(8):1313–1325.

[9] Lina Feng. The research application of Parallel K-Means clustering method in the Resume data [D]. Yunnan: Yunnan University, 2010.

[10] Jian Wan, Wenming Yu, XiangHua Xu. Design and Implement of Distributed Document Clustering Based on MapReduce [C]// Proceedings of the Second Symposium International Computer Science and Computational Technology (ISCSCT09), 2009: 278–280.

[11] Yifeng Liu, Lucio Gutierrez, Xiaoyu Shi, Xiaodi Ke. Parallel Document Clustering with Hadoop on Amazon Elastic Computing Cloud [R]. CMPUT 681 Project Report, 2009.

[12] Jason Venner. Hadoop build scalable, distributed applications in the cloud [M]. The United States of America: Apress, 2009.

[13] Lustre document [EB/OL]. http://lustre.org/. (2015-4-11).

[14] Jun Liang, Wenjun Xie. Research on Lustre File System [J]. Information Technology, 2014(4).

[15] Britta Wülfing. Sun Assimilates Lustre File system. Linux Magazine, 2007.

Architectural, Energy and Information Engineering – Sung & Chen (Eds)
© *2016 Taylor & Francis Group, London, ISBN 978-1-138-02791-6*

A simulation system of cutter suction dredger-based Quest3D

X.B. Meng
College of Mechanical Engineering, Donghua University, Shanghai, China
School of Art and Design, Changzhou Institute of Technology, Changzhous, China

S.R. Sun
School of Art and Design, Beijing Institute of Fashion Technology, Beijing, China

Y.M. Wang
College of Design, Jiaxing University, Jiaxing, China

ABSTRACT: The simulation of dredging is an advanced training method for people working in the dredging industry, and a new research area in the discipline of dredging. In this paper, we discuss how to establish a simulation system of cutter suction dredger-based Quest3D, which can be used in the training of operating dredger and familiarizing the environment of ports for crews. The system not only cuts the costs of equipment, but also provides a more real effect compared with the systems of physical simulation and semi-physical simulation.

Keywords: simulation of dredging; Quest3D; cutter suction dredger; virtual reality; simulation in three-dimension

1 INTRODUCTION

With the continuous improvement of computer technology in recent years, the technology of virtual reality has made a rapid development and makes profound impacts on the areas of education training, design of mechanical engineering, and manufacturing. Research on education training, design, and assembling in virtual reality has attracted increasing attention. There are many new progresses in the accessory equipment and developing environment in virtual reality. The technologies of force feedback data gloves, multi-channel circular-screen movie, and immersed wearable VR equipments, have improved the data communication between human and computer, increased the channel numbers of human-computer interaction, and made the communication between human and computer more smooth and natural. The appearance of virtual simulation platforms such as Virtools and Quest3D, which are more integrated, helps designers to get rid of detailed and complicated coding and to focus the projects themselves.

The system of dredging engineering is complex, and its equipments are expensive. Cutter suction dredger is one of the most common and typical equipment, and the training of skilled technicians for operating the cutter suction dredger needs a long time. The critical problem is how to train

the dredging technicians quickly and effectively to improve their skills. In the previous research of the center of dredging engineering of Hohai University, we developed a semi-physical simulation system of cutter suction dredger, and built a real console of the dredger and a 160° circular screen, which simulated various construction sites including lake, reservoir, port, waterway, and coastal areas and natural environments such as diurnal variation, sunny, rainy, snowy, and foggy day. Basically, the system simulates all the aspects of cutter suction dredger's operation and the working environments very well, and completes the tasks of crew training, engineering prediction, and experimental study. However, there are still many defects such as the equipment's high cost and the inability to move the system. So, to overcome these main problems, we develop a simulation system of cutter suction dredger based-Quest3D running in PC, which provides training conveniently for the dredging technicians anytime and anywhere.

2 ARCHITECTURE AND DESIGN OF THE SYSTEM

The main target of the system is simulating the working state of cutter suction dredger on PC with 3D virtual technology, simulating the environment of port, lake and shore. The dredger's operation is controlled by the keyboard. The system must complete the task of the primary training work to make the dredging technicians understand the principle, working condition and working environment of the cutter suction dredger, and make a good foundation for the actual operation in the dredger. The system should realize four functions: ① moving control and operation to the dredger; ② the real effect of simulating port, lake and sea shore; ③ demonstration of the cutter working condition under the water; and ④ roaming in the virtual environment and the control of camera lens.

In the previous research, the virtual display of the system used a 160° circular screen, which is divided into four areas. The whole picture is formed by four projectors. In order to ensure the visual integrity of the picture, the overlap regions by the projectors are handled with the technology of edge fusion. In this system design, we initially intend to use the computer monitors to replace the 160° circular screen. Although there are visual obstructions by the borders of computer monitors, the scheme is still viable, considering its lower cost. In this project, we use two monitors to simulate the circular screen. Three or more monitors can be used according to the requirements.

There are three technical difficulties in the system. The first one is how to communicate between Quest3D and the PCL console through the C++ interface in the later expansion of the system. The second is how to realize the effect of the circular screen by combining computer monitors. The third one is the particle system in Quest3D.

For the first difficulty, we found a feasible technical route through the analysis and research. The general idea is that we load the Quest3D SDK in C++, code the dynamic data link library by Quest3D standard interface, and then load it as a plug-in into Quest3D. In Quest3D, the compiled plug-in becomes the module of channel. Dragging it into the scene, Quest3D will package it into the excitable file. When the program is running, the system will receive the dynamic motion signals by the console.

For the second difficulty, it can be solved by graphic card supporting multi-monitors and high-performance PC. Generally, the graphic card will have four interfaces, and the effect of circular screen can be simulated by using three or four monitors. This technology is mature, so we can focus only the specific problems in the equipment settings.

For the third difficulty, the particle system in Quest3D is totally different from 3dMax. Accomplishing the real effect of the dredger cutter's working state, and the ripple effect generated by the ship's moving, we need a deep research on the particle system in Quest3D.

3 THE BASIC PROCESS AND THE KEY STEPS

System development steps involved are as follows: ① complete the 3D solid model and the virtual scenes in 3dMax according to CAD drawings and the real digital pictures. The system's efficiency is improved by the technologies of texture mapping, level of detail and billboard to minimize complexity of the graphics, and at the same time, we must ensure the necessary accuracy and realness. ② The 3D models are controlled and the corresponding scenes are rendered, when the real-time dredging data are received and read on the Quest3D platform. In order to achieve the function of playback and data analyses, the database and network sources will be established.

3.1 Modeling and rendering

We need to make the following models in three-dimensional software: a fine dredger model, simple models used for interaction, cutter models and

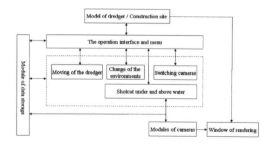

Figure 1. Framework of the system.

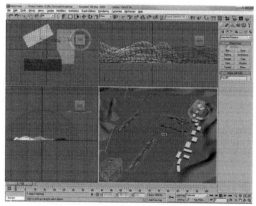

Figure 2. Model of port.

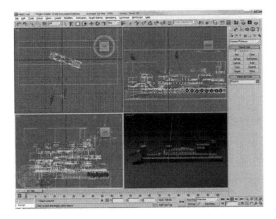

Figure 3. Rendering model.

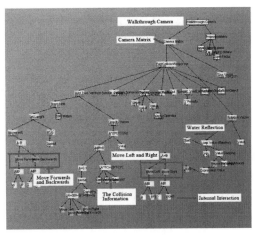

Figure 4. Channel group of dredger.

port models. There are three types of models in 3dMax: wire frame model, surface model and solid model. For making the interaction in Quest3D later, the completed models need to be rendered to textures using VRAY renderer.

3.2 Interaction design in Quest3D

After the modeling and rendering, all the interaction tasks of the system are completed on the Quest3D platform. Quest3D consists of three core parts: ① channel section is very important in Quest3D. Channel section can be shown after the Quest3D starts, whichis a foundation for creating a project. ② The functions of animation section are to fix the position, and control the movement of 3D models, cameras and lights. The part of animation section has a window in which we can preview the project. ③ The part of objection section has spare objects. Quest3D provides a number of options to deal with the object surface properties such as color and texture.

The construction of a program in Quest3D is completed with modules, which are called channels. Channels are the key parts to construct the project. The channel group of dredger contains all the interaction information including the texture of dredger, relationships of interaction and collision. There are manyinteraction modules that can be used directly in Quest3D. After the realization of dredger's controllable interaction, the dredger must connect with the camera module for the switching of cameras and the continuous operation of the dredger in the final display system. Figure 4 shows the channel group of dredger including all the properties of cutter suction dredger and interaction information. Figure 5 shows the settings of cameras in port.

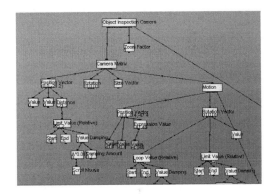

Figure 5. The settings of cameras in port.

Figure 6. The final effect I.

4 SUMMARY

In the future, the technology of virtual reality for training of dredging technicians will become a trend. The system of cutter suction dredger virtual

Figure 7.   The final effect II.

simulation based on Quest3D not only provides the real effect of dredging equipments' working state and the operating environment for training technicians, but also makes the training faster and easier. The system running on PC is easy to deploy, and reduces the costs of the physical simulation and semi-physical simulation systems. The functions involved are as follows. ① the introduction of dredger and animation demo; ② the scenes above and under the sea, working state demonstration of dredger cutter stirring and sucking the sand under the sea; ③ the interaction of the dredger (based the keyboard); ④ changing of the port's environment (the changing of day and night, and foggy weather); ⑤ multi-angle observation and any angle of view to the port's environment and the dredger; and ⑥ the research of display using multi-monitors, which accomplishes the simulation using two monitors.

ACKNOWLEDGMENT

This project was supported by the Fundamental Research Funds for the Central Universities (26120132013B10414). The corresponding author is MENG Xiang-bin.

REFERENCES

[1] NI Fu-sheng, ZHAO Li-juan, GU Lei, et al. Simulation of Dredging of Cutter Suction Dredger [J]. Journal of System Simulation, 2012, 24(11): 2271–2274.
[2] HE Kun-jin, NI Fu-sheng, GU Lei, JIN Yong-xia. A Kind of Virtual Implementation Method of Terrain and Cutterhead [J]. Journal of System Simulation, 2009, 21(20): 6524–6528.
[3] Pan Zhigeng, Check Adrian David. Virtual Reality and Mixed Reality for Virtual Learning Environment [J]. Computer & Graphics, 2006, 30(1): 20–28.
[4] NI Fusheng, ZHAO Lijuan, XU Liqun, W J Vlasblom. A Model Calculation for Flow Resistance in the Hydraulic Transport of Sand [C]// Proceedings of the 18th World Dredging Congress, Orlando, Florida, USA, 2007. Lake Buena Vista, Florida, USA: Newman: Printing Company, 2007: 1377–1383.
[5] LI Chang- guo, ZHU Fu- quan, TAN Liang, YANG Chun. Research on Development Approaches of Virtual Experiment Based on 3D and Virtools Technologies [J]. Computer Engineering and Application, 2006, 31: 84–86.
[6] Sankar Jayaram, Uma Jayaram, Yong Wang, et al. A Virtual Assembly Design Environment [J]. Computer Graphics and Applications, 1999, 19(6): 44–50.
[7] Javaram S, Connacher H, Lyons K. Virtual Assembly Using Virtual Reality Techniques [J]. Computer-Aided Design, 1997, 29(8): 575–584.
[8] ZHENG Yi, NING Ru-xin, LIU Jian-hua, DU Long. Survey on Key Techniques of Virtual Assembly [J]. Journal of System Simulation, 2006, 18(3): 649–654.

Architectural, Energy and Information Engineering – Sung & Chen (Eds)
© 2016 Taylor & Francis Group, London, ISBN 978-1-138-02791-6

Fluid simulation analysis of 800GS80 single-stage double-suction centrifugal pump

Xing Rong Chu, Jun Gao & Lin Yuan Wang
School of Mechanical, Electrical and Information Engineering, Shandong University, Weihai, China

Wang Yao
Research Institute of CSIC, Harbin, China

Jia Bin Wang
Shandong Shuanglun Co. Ltd., Weihai, China

ABSTRACT: In this work, based on the Computational Fluid Dynamics (CFD) knowledge and the actual working condition, the fluid simulation of 800GS80 centrifugal pump was carried out with Fluent using Reynolds time-averaged N-S equation and Multiple Reference Frame (MRF) method. By setting the appropriate fluid parameters, the accuracy of the simulation result was improved and it is revealed that these results can well simulate the actual working state of the centrifugal pump. Based on the simulation model, the pressure and velocity in the main working parts of the centrifugal pump were analyzed and discussed under rated and non-rated condition.

Keywords: centrifugal pump, fluent simulation, fluid parameters, pressure, velocity

1 INTRODUCTION

800GS80 centrifugal pump is a single-stage double-suction pump. It has the advantages of convenient maintenance, less pressure on the bearing, good work stability, large flow, good cavitation erosion performance and so on. Due to these advantages, the single-stage double-suction pump was widely used and has a broad application prospect. Because of the complex spatial structure and the complex flow pattern, the pump internal structure and the flow components both have different effects on the liquid flow. So for the pump flow field design, it's difficult to get a satisfactory result only with the traditional structural mechanics and fluid mechanics. In the past, most producers have to do many prototyping and testing work, and then modify the pump body structure based on the obtained data. Hence, for the design of a new type centrifugal pump, high cost and long design period are needed, which cannot fulfill the market flexible requirements.

The traditional pump design methods have difficult to meet the company design need. Now, more and more centrifugal pump design is carried out with computer aided and analysis software. In this work, the computational fluid dynamics analysis is adopted to analyze the 800GS80 centrifugal pump flow field. The obtained simulation results are used to guide the pump design, which accordingly shortens the design cycle and reduces the design cost.

2 HYDRAULIC MODEL

According to the original related data provided by Shuanglun Company, the hydraulic model of the suction chamber, volute and impeller for 800GS80 pump was built in Soliworks 2013, as shown in Fig. 1. The hydraulic model is complex and has many irregular curved surfaces, so for the model meshing method, the main unstructured mesh mixed with the structured mesh is adopted. The impeller, suction chamber and volute were meshed by tetrahedral meshes and hexahedral meshes, and in order to get an accuracy result, the mesh of the impeller outlet boundary is refined. The mesh generation procedure is the mapping process of the model computing domain to the solving geometric domain, so the generation of the high-quality mesh is the key point for the simulation analysis [1]. In this work, the professional software ICEM CFD is used to mesh the hydraulic model [2]. According to the ICEM CFD meshing result, the meshed model is shown in Fig.2.

Figure 1. Hydraulic model.

Figure 2. Hydraulic mesh model.

3 FLOW FIELD ANALYSIS METHODS

The flow filed analysis was carried out with Fluent. The used fluid is water, for which there is no concentration gradient and no mass diffusion. So the Navier-Stokes equation was adopted as the control equation. The calculated Reynolds number in the 800GS80 pump is much larger than 4000, according to this calculated value and the pump working condition, the Reynolds time-averaged N-S equation, the Multiple Reference Frame method and the Realizable k-ε model were chosen in Fluent to carry out the flow filed analysis.

In this work, the suction chamber and the volute are set as fixed field. The impeller is set as rotating field and the rotating speed is applied according to the pump working condition. To obtain the appropriate solving method and under-relaxation factor, related parameters in Fluent should be set. Based on the pressure-velocity coupled solver, the SIMPLE method is used. Standard control method is selected for pressure control and second-order upwind mode with second order accuracy is selected for momentum, turbulent kinetic energy and dissipation rate control. For under-relaxation factor setting, the pressure entry is set to 0.3, the momentum equation entry is set to 0.7, the turbulent kinetic energy is set to 0.8, and density, volume force term are set to 1.

4 FLOW FIELD SIMULATION RESULTS

With the simulation results obtained by MRF method in Fluent, the related calculated parameters are imported in the Eq. (1), and the 800GS80 pump performance parameters can be obtained [3].

$$H = \frac{\Delta P}{\rho g}, N = \frac{Mn \times 2\pi}{60}, \eta = \frac{\rho g Q H}{3600 \times N} \quad (1)$$

where, H is the pump head, N is shaft power, η is the efficiency, ΔP is the pressure difference between the pump outlet and the pump inlet, M is the torque on the impeller shaft direction, and Q is the pump flow under rated condition. Considering the effects of mechanical loss and the volume loss, based on the above calculated values, the revised 800GS80 pump performance is shown in Table 1. Compared to the data provided by Shuanglun Company, the simulation results well match the design, which proves that the adopted model can be used to simulate the 800GS80 pump working condition.

The pressure and velocity distributions of the pump impeller and the pump volute are presented in Fig. 3. After the fluid flow through the impeller inlet, the pressure increases in the axial and radial direction, as shown in Fig. 3 (a). With the help of impeller rotation, the mechanical energy is gradually transformed into the liquid pressure energy and kinetic energy. The total pressure is the highest near the impeller outlet. With the rotation, the pressure of the impeller working surface is greater than the pressure of the impeller back face. In the impeller inlet area, where the front cover connects with the impeller root, a negative pressure region is found. This is because when the liquid flows into the impeller inlet, due to the lateral impact caused by the inclination front root portion of the impeller blade, an offset was applied to the liquid and brings a secondary flow. When the offset arrives at the maximum, the pressure negative area is formed. These pressure negative regions are the surface parts where the cavitation erosion is most likely to happen. Meanwhile, due to the impact of the liquid, this region is also the area where most of the hydraulic loss exists.

Table 1. 800GS80 pump performance parameters.

Flow (m³/h)	Head (m)	Shaft power (kw)	Efficiency (%)
1440(0.8Q)	83.9	1606.7	85.2
1800(Q)	82.8	1740.5	88.5
2160(1.2Q)	79.9	1854.3	87.9

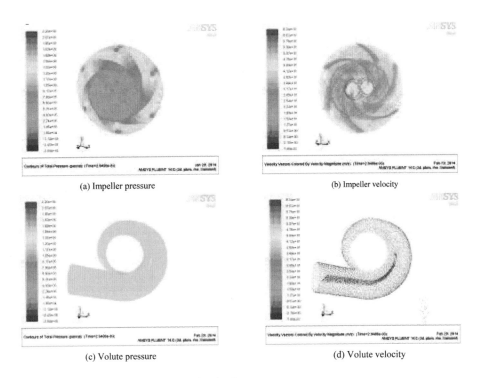

(a) Impeller pressure

(b) Impeller velocity

(c) Volute pressure

(d) Volute velocity

Figure 3.   Simulation results of impeller and volute in Fluent.

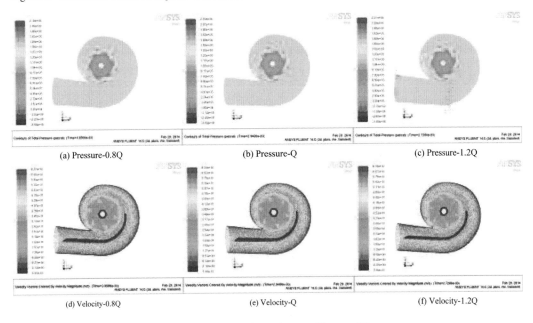

(a) Pressure-0.8Q

(b) Pressure-Q

(c) Pressure-1.2Q

(d) Velocity-0.8Q

(e) Velocity-Q

(f) Velocity-1.2Q

Figure 4.   Total pressure and velocity distribution under various working conditions.

For the impeller velocity analysis, as shown in Fig. 3 (b), the flow rate increases with the impeller rotation. The liquid flow rate of the impeller blade working surface is greater than that of the back surface. The liquid flow rate in the blade working surface maintains at a high value, while the flow rate in the blade back surface is lower and uniform. The maximum flow rate is observed at the outlet of the impeller. But due to the presence of vortex flow near the outlet channel, part of the hydraulic efficiency is lost. And this provides theoretical basis for the impeller optimization. The pressure and velocity distributions in the volute are shown in Fig. 3 (c) and Fig. 3 (d). The fluid pressure is low at the volute inlet and increases gradually, the maximum is found at the volute outlet. The flow velocity is the maximum at the volute inlet, with the diffusion in volute, the liquid decreases and reaches a minimum value at the outlet. The above analysis well coincide with the impeller and volute real working condition, and gives useful information for the pump hydraulic model design and optimization.

Based on the selected model, to evaluate the pump performance, the working state of the 800GS80 centrifugal pump under low flow condition (0.8Q), rated flow condition (Q) and high flow condition (1.2Q) are analyzed. In this work, the stable suction chamber was not considered, only the states of impeller and volute are discussed.

As one can be seen in Fig. 4, for various working conditions, the tendency of pressure and velocity distribution keeps the same. All meet the centrifugal pump working condition design. With the flow rate increases, the total pressure of the pump and the flow rate peak increase. Meanwhile, with the flow rate increases, the impact recirculation region at the impeller outlet is enlarged and the energy lost increases. At high flow condition (1.2Q), in the outlet of the pump, the pressure drop is found. With the increase of the flow rate, the high-speed zone expands gradually and the effect of the flow vortex on the flow rate decreases. Under non-rated working condition, the velocity gradient is higher and the velocity distribution is not uniform,

meanwhile the effects of liquid impact, flow vortex and reflux increase, hence, the centrifugal pump energy loss is greater than the rated condition.

5 CONCLUSION

In this work, the working state and the pressure and velocity distributions in the impeller and volute under rated and non-rated condition were obtained in Fluent. Compared to the pump actual working condition, the Reynolds time-averaged N-S equation and MRF method are proved to be suitable to model the 800GS80 pump field analysis simulation. The Fluent simulation can give a reasonable analysis for centrifugal pump design and optimization.

ACKNOWLEDGEMENTS

Thanks for the technical support provided by Shuanglun Company. The corresponding author is Jun Gao.

REFERENCES

[1] Zhang Wenming, Liu Bin, Xu Gang. Three-dimensional solid mesh adaptive partitioning algorithm [J] Mechanical Engineering, 2009, 45(11): 266–270.
[2] Ji Bingbing, Chen Jinping. ANSYS ICEM CFD meshing technology example explanation [M] Beijing: China Water Power Press, 2012.
[3] Wang Jianhua. Centrifugal impeller design optimization parameters [J] Drainage and Irrigation Machinery, 1994, 2: 13–15.

Architectural, Energy and Information Engineering – Sung & Chen (Eds)

Positioning of university physical culture under the background of students' physical decline

Qing Bo Gong

PE Department, Shandong Institute of Business and Technology, Yantai City, Shandong Province, China

ABSTRACT: Under the current serious situation that college students' physical quality is becoming worse, this paper gives a very comprehensive analysis in terms of the factors such as teaching goals, contents, methods, PE teacher and teaching evaluation in order to provide some references for the teaching reform of college PE teaching. College PE should follow the basic rules of physical exercises in accordance with teaching goals, contents, teaching modes and methods, and exam evaluation. Based on students' practical needs, it should have a short- and long-term teaching goal planning and positioning by means of mastering sports skills and knowledge, by taking physique and certain program test as motivation, by using teaching to promote learning and using evaluation to promote exercising so as to cultivate students' interests and the hard-working spirit and lay a solid foundation for students' life-long exercising habit.

Keywords: physique; teaching goal; teaching contents; teaching evaluation

1 INTRODUCTION

Since the end of the last century and with the college teaching reform, there have been a lot of changes in the PE teaching: earlier it was an optional course; now it becomes a required course. In the past, courses such as track and field, gymnastics, Kungfu and ball sports all took a certain percentage in all their courses. At present, students are only required to choose one or two programs for half term or one school year, which has certain advantages such as helping students learn to concentrate their time on a certain technical skill based on their interests. Teachers can also use their skills well to enhance their teaching effect. However, when we reflect our teaching effect by observing students' one year's PE class, they could only maser two or three sets of basic movements like in some performance classes and they seldom practice after the test. Most students forget all after graduation. It is not practical for them to practice later. In class, there also existed some problems such as continually repeating the same movements at a relatively small place, which is hard to stimulate students in light of the amount of exercise and intensity. Therefore, the effect is not obvious to enhance physical quality. In the end, students cannot learn the skills well and seldom exercise, which is hard to improve their physique situation.

In sports circle, there has been an argument about which is important: quality or technique? However,

in terms of the guiding principle of "health first", the sports circle has come to an agreement. The purpose of PE class is to improve students' health, invigorate health effectively and help them form a good exercise habit. All the teaching of certain sport events is to improve students' health by the mastery of certain technical movements of the sport events. Both the learning of technical movements and the later self-exercise are all for the need of self-interest and being healthy. Therefore, based on this, the learning of sport event is only one tool. It is also a method that can both arouse an interest and improve and invigorate health effectively. Exercisers will have different psychological experiences from the process such as playing a basketball match and finishing the same amount of running for basketball can be done at an atmosphere of fierce competition and psychosomatic cheerfulness, but the long time speed endurance running might be finished by gritting one's teeth and bearing a lot of pain. These two have big different psychological experiences. The reason for students to choose certain sports event is to exercise joyfully, learn the technical movements while exercising happily so as to experience the benefits of sports, and help form the life-long exercise habit. Therefore, it is very necessary to know the sports event features while teaching and know which one is more important. For some technical movements that is hard to improve in a short time, it is not reasonable to use too much of class practice time.

2 THE TARGET POSITIONING OF COLLEGE PE CLASS

The teaching target of PE class is the intermediary element between curriculum objectives and classroom teaching of PE class. The exploration of the problems such as the features, connotation division and writing can be very helpful for the further research of teaching objectives of PE class. It can also help enhance the bridging of different target levels and the scientificity of the writing objectives[1]. The target of college education has a direct influence and can guide the content, standard and methods of the evaluation of college PE education, which plays a role in deciding the application of content and methods in PE education and the final effect of PE education. The basic content of college education target includes the full enhancement of students' physique, the healthy development in both body and mind, health first, the mastery of certain sports skills and hygienic knowledge, the training of sports cultural quality, and the formation of life-long sports habits. It is very hard to improve students' physical quality effectively only by the once- a-week PE class. The improvement of students' physical quality should rely more on the long-term self-exercise. PE class should indicate directions for students' self-exercise, help them solve the various questions while excising, and arrange some self-exercising tasks. It should also play the role of laying foundation and indicating directions for students' self-exercise, and monitor and test students' self-exercise status. However, the main problem for the current situation is that except for a small number of students, many students still do not have a good self-exercise habit. Besides, the number of students who are willing to go to the sports field and exercise is very small, mostly for the girls. Due to the influence from the sports field and self-interest, many students do not have the experience of self-exercise after class.

One of the basic rules for current college PE class is to help students away from computer games and help them walk out of their dormitories, go to sports field, love sports and exercise so as to realize the best optimization grouping of the various aspects such as college PE management, PE teachers, PE courses, students' participating level of PE class, the evaluation method of college PE education and college sports fields, which can ensure to realize the maximum college PE class target. For the college PE education, if there is too much attention on the test results, it deviates not only its fundamental target to develop and improve students' physical quality, but also its final target to realize the full development in both body and mind[2].

3 THE CONTENT POSITIONING OF COLLEGE PE CLASS

The choice and setting of the teaching content of college PE class is one of the important items of the college PE teaching reform. The choice and change of teaching content should meet social needs and the need of education development, and follow strictly the guiding principle of 'health first' advocated in *the Sports and Health Standards*.[3] Centering on the target of enhancing students' physical quality, PE class should choose some sports events with a certain amount of exercise and intensity. It is quite reasonable to control some entertaining sports events with much less amount of exercise and intensity so as to avoid the blind and unpractical pursuit of diversity of some special optional sports events, which can result in the reduction of its competitive nature of sports. Much attention should also be paid to the reasonable grouping in terms of their own features of sports events. The PE class of non-PE major is obviously different from students of the other majors. It is not correct to simply take the teaching of optional courses of non-PE class as the beginning period of the teaching of required courses of PE major students because it is an conclusive teaching process without a higher or follow-up phase. The arrangement of teaching content should be diversified. Their sketchy and extensive feature should be highlighted to meet the needs of students of different levels by combining the use of hierarchical teaching methods. It is quite unreasonable to waste too much of class practice time on the correction of some detail standardability of certain movement. By contrast, the maximum encouragement of students' enthusiasm to practise in class and the fulfillment of the maximum amount of exercise under the atmosphere of joyful experience both in body and mind should be introduced. The arrangement of content should be based on the mastery of skills and full improvement of students' physical qualities in terms of students' features. The adding of practice content of special event and physical quality test is also very important. Teachers should give some time to practice in class and arrange some after class practice assignment, which can be easy to practice and have less influence from negative factors such as the limitation of sports field and time. Therefore, this is a way to help students practice more by testing and supervising their self-exercise after class.

4 THE POSITIONING OF TEACHING METHODOLOGY OF COLLEGE PE CLASS

PE class teaching methodology is an important factor to realize the PE class teaching target process. It is also the path operational program to realize the PE class teaching target process[4]. Due to some college PE class programs such as tennis, yoga and some other programs which cannot be operated in primary and middle schools, the other programs such as basketball, volleyball, and football are almost the same as they did in primary and middle schools. Students' foundation are still not good enough. The teaching difficulty level is low. Students are still treated like beginners in terms of teaching methodology, which exposes the absence of teaching of stroking techniques of sports events at primary and middle schools. This makes college PE class still repeat the most basic stroking techniques methodology that accompany students for many years already. What is worse, many students after one semester or one year's study still cannot master the basis techniques. Therefore, during teaching, teachers should experience the practical situation and understand different students' levels and physical qualities. They should also adopt the methodology of group training and teaching in terms of different levels and groups to avoid the phenomenon "Some don't have enough to eat while some have too much to eat even when they don't like" so as to assist students of different levels to make the self-exercise plan after class. In class, teachers should practice more, talk less and teach better. It is also good for teachers to take the teaching content as center and try the ones that can motivate the students most to realize the perfection of teaching methodology in the process of exploration and summarization.

5 THE POSITIONING OF EVALATION ABOUT UNIVERSITY PHYSICAL EDUCATION

Teaching evaluation is an invisible baton, which has a deep influence on various aspects such as students' intellectual curiosity, learning development and emotions[5]. The evaluation of PE class includes not only the evaluation of the teaching contents, methods, organization and effect, but also includes the evaluation about aspects such as the mastery of the teaching contents, exercise effect, the formation of exercise habits, and physical quality level. Currently, the evaluation of college PE classes are mainly based on the students' evaluation, scores from college teaching supervision team, and partly from the evaluation of the administrators. Final quantitative rating will be made in terms of certain weight of each part. Among these parts, students' evaluation takes the absolute high proportion part. On the one hand, it is to fully respect the scientific importance of students who are the subject of education; on the other hand, some teachers want to avoid some negative effects from students who do

not understand teachers' hard work well, which, to some extent, lead to the phenomenon such as teachers' soft teaching and polite testing. Teachers try to get along well with students and make them happy. In the end, they try to make most of the students to pass their tests. If students like their class, teachers are very happy. However, this can lead to undesired consequences as follows. Students are not serious about their classes and become very lazy about their exercises. The intensity and amount of exercise are not enough. Therefore, the teaching effect cannot be good enough. The self-exercise after class cannot be ensured, which damage both class teaching and health.[6]

Currently, the evaluation on students' PE class test results mainly includes the aspects such as sports specific movements, qualities, sports theory and class behavior. Mostly, many teachers still choose the summative assessment, which is to test students by taking some classes close to the end of semester. This kind of summative assessment test only cares about test results and ignores students' practice process and lacks supervision to their self-exercise process. Students just want to pass the test without thinking to improve their skills. Many of them even just barely pass their tests. The lack of the practice process finally leads to the unsatisfactory teaching effect. Excessive worries about the risks hinder the realization of college sports education targets, let alone the formation of the life-long exercise habit. It is even very hard to improve students' physical qualities, which is a commonly discussed topic. Therefore, it is very hard to realize the effective change of physical qualities. In the process to fully promote college students' quality education and carry out the healthy sports education and life-long sports concept, attitude research on college students' PE class should be an important research project, to improve PE teaching content, methods, and curriculum structure.

6 CONCLUSION

Factors such as social progress, economic development and the improvements of people's life make people ignore the cultivation of the spirits like hard-working. Students prefer to spend more hours on computer games. There are also some problems such as the imbalance between less sports facilities and more students. Some good quality sports stadiums want to earn money. Current sports facilities can hardly meet students' exercise needs. Some boring running makes students away from exercise. All these reasons lead to the deterioration of students' physical qualities. For the third- and fourth-year students who do not have the PE class, without proper supervision, they are more unwilling to do exercises mostly for those events such as endurance exercise programs. Therefore, their physical qualities, especially their endurance qualities, decline quickly.

The rules in the teaching of sports should be strictly obeyed by the change of concepts, the acquirements of information resources, and the innovation of knowledge and technique. It is also very necessary to use good spirits such as responsibility, concept, methodology, and practices, and accurately master rules and freedom and successfully motivate, trigger, sustain and improve the students' PE study[7]. It is important to make short- and long-term PE education target and self-exercise supervision assessment system in terms of students' physical qualities. These can help supervise third- and fourth-year students at regular intervals, and also help to realize the purpose to encourage exercise by assessment and promote health by exercise. The setting of college PE class and methods of teaching should strictly follow the college PE education target system, which can lay a solid foundation for the formation of life-long exercise habit in terms of physical quality development, the improvement of skill and the mastery of PE basics. Therefore, it is very reasonable to raise the ability of PE class administration and service to realize effective teaching results, and match and promote the current PE class and teaching reform and development.

REFERENCES

[1] Shao Dewei, Liu Zhongwu, Li Qidi. PE Teaching Teleology [J]. Beijing Sports University Journal 2012, 35(9): 96–101.

[2] Yao Lei. The Thinking on the Current School PE Education Development and Reform [J]. Nanjing Physical Culture Institute Journal 2011, 25(2): 4–6.

[3] Li Qidi, Zhou Yan PE Education Teaching Method Differences. [J] Sports and Science, 2012, 33(6): 113–115.

[4] Wang Dengfeng. School PE Education Predicaments and Methods-The Report on Tianjian City PE Education Meeting [J]. Tianjian Physical Culture Institute Journal 2013, 28(1): 1–7.

[5] Huang Juyun. The Negative Influence on Students Physical Quality and Health of the Softening of PE Education [J]. Journal of Sports, 2012, 19(6): 5–10.

[6] Lu Yuanzhen. Several Theory and Practice Questions in Current School PE Education [J]. Jilin Physical Culture Institute Journal, 2009, 25(5): 1–6.

[7] Zhang Jingping, Lu Jisi, Wang Hongbing. The Research on the Application of "Development Assessment "in Assessing Students PE Education.

[8] Huang Cong. The Freedom of Rules: The Potential Concept and Application of the "Herding Sheep Style Teaching" [J]. Journal of Sports, 2011, 18(2): 74–77.

NOTES

This is a research program from our Division of Teaching Affairs from Shandong Institute of Business and Technology in 2013.

Activation of the low-temperature non-equilibrium plasma water

Michail Bruyako, Larisa Grigorieva & Angela Orlova
Technical Sciences, The Department of Composite Materials Technology and Applied Chemistry,
Moscow State University of Civil Engineering (MGSU), Moscow, Russian Federation

Vyacheslav Pevgov
Physical and Mathematical Sciences, Moscow Institute of Physics and Technology (State University),
Moscow Region, Russian Federation

ABSTRACT: This paper presents the equations for the calculation of some characteristics of the plasma at a known distribution function of electron energy, the dependence of the constants different excitation degrees of freedom in water steam from the E/N.

Keywords: low-temperature plasma; activation of water; nonequilibrium; function of the electron energy distribution

1 INTRODUCTION

Processing techniques to control the structure and properties of composite materials are different. The activation of source components, including water, affects the chemical and physico-chemical processes that form the matrix structure. The modification of the water is carried out in different ways: physical, chemical, or combined. Plasma treatment is one of the ways of physical activation.

Plasma methods are effective and useful for the oxidation of water and organic substances. They are distinguished by the ability to initiate oxidative processes with the highest oxidative chemical potential (atoms and molecules of oxygen in the electronically excited state). These methods also have a low inertia in the restructuring of the operating modes. The latter fact is crucial for cases that are characterized by significant variations inchemical composition, concentrations and ratios of the various components.

The general scheme of action for "Non-thermal plasma methods" is as follows [1]. The gas stream to be treated are passed through the low-region non-equilibrium temperature plasma (discharge region). In the low-temperature non-equilibrium plasma, there are energetic electrons with an average kinetic energy of 2–5 eV (this energy corresponds to a temperature of 20000–50000 K) at concentrations in the range of 10^{10}–10^{13} cm^{-3}. Electrons efficiently transfer their kinetic energy to the internal degrees of freedom of the molecules, without causing significant heating of the processed gas. The selective excitation of internal degrees of freedom leads to the dissociation of the numerous highly chemically and environmentally harmless particles, ions, atoms and radicals such as N_2^+, O_2^-, O^-, O, O_3, H, OH, HO_2, metastable molecules N_2^*, O_2^*, N^*, and others. Non-equilibrium processes involve the following features: the cold gas creates a very high concentration of active particles, which in equilibrium is achieved only by heating the gas to several thousand degrees.

They are formed under the action of plasma active species that react with the molecules of the component-treated medium.

To solve these problems requires the knowledge of the mechanisms of physical and chemical processes in plasma—ionization, excitation, dissociation, and chemical transformations. In particular, a major issue is the relationship between the contribution of the processes under the influence of electron collisions, the charged particles and radicals, accompanied by various non-equilibrium processes: the excitation of rotational, vibrational and electronic levels of molecules, their dissociation and ionization. The rates of initiation is largely dependent on the function of the electron energy distribution, which is a sufficiently low degree of ionization and is non-equilibrium. The distribution function of the electron energy determines important plasma parameters such as the electron drift velocity and coefficient of diffusion. The distribution function of the electron energy is determined by solving the kinetic equations.

The only reliable criterion for evaluating the reliability of a set of cross sections is the agreement of the calculated kinetic coefficients, drift velocity and diffusion coefficients of electrons with the data obtained in the measurements on the drift tubes [2]. This criterion is used traditionally for the correction of cross sections measured in beam experiments, as suggested earlier by Phelps [3]. In finding the scattering cross sections on the basis of the experimental data on the drift tube, special attention should be given to the correctness of the numerical model used for the experimental conditions. In particular, it is necessary to take into account that the value of the same kinetic coefficient is different for different production processes of the experiment. This effect becomes significant when the motion of electrons is in high fields, or in the presence of strong electron sticking [4]. A set of sections satisfy the procedure, called the self-consistent. Currently, such self-consistent set of sections is known for many atoms and molecules [5].

The known distribution function of the electron energy can define the characteristics of the plasma, such as drift velocity of electrons in an electric field

$$V_{\partial p} = -\frac{1}{3}\left(\frac{2e}{m}\right)^{1/2}\left(\frac{E}{N}\right)\int_0^\infty u \frac{df}{du}\left[\sum_j y_j Q_{mj}\right]^{-1} du, \quad (1)$$

the average electron energy

$$u_{cp} = \int_0^\infty u^{3/2} f(u) du, \quad (2)$$

electron diffusion coefficient

$$D = \frac{1}{3N}\left(\frac{2e}{m}\right)^{1/2}\int_0^\infty u\left[\sum_j y_j Q_{mj}\right]^{-1} f(u)\, du, \quad (3)$$

elastic collision frequency

$$k_{mj} = \left(\frac{2e}{m}\right)^{1/2}\int_0^\infty \frac{u^2}{u_{cp}} Q_{mj}(u) f(u)\, du, \quad (4)$$

frequency of inelastic collisions

$$k_{ij} = \left(\frac{2e}{m}\right)^{1/2}\int_0^\infty u Q_{ij}(u) f(u)\, du, \quad (5)$$

fraction of the energy expended in the inelastic and elastic processes in relation to the total energy deposited into the plasma

$$\eta_{ij} = \frac{y_j u_{ij}}{v_{\partial p}\left(\frac{E}{N}\right)} k_{ij}, \quad (6)$$

$$\eta_{mj} = y_j \frac{2m}{M_j} u_{cp} \frac{k_{mj}}{v_{\partial p}\left(\frac{E}{N}\right)}. \quad (7)$$

must satisfy the condition

$$\sum_{ij}\eta_{ij} + \sum_j \eta_{mj} - \sum_{ij}\tilde{\eta}_{ij} = 1 \quad (8)$$

Here, $\tilde{\eta}_{ij}$ is the fraction of the energy acquired by an electron in the process of deactivation of the excited particles. In the numerical solution of the Boltzmann equation, Condition (1) can be used to control the accuracy of the solution.

A detailed derivation of the formulas has been given in [6].

Figure 1 shows the results of the calculation of the rate constants of elementary reactions that take place with the participation of the electron plasma in the water steam.

Figure 2 shows the results of the calculation of the electron energy balance for a pure water steam, obtained by the numerical solution of the Boltzmann equation for the function of the electron energy distribution in the binomial approximation. E/N denotes the value of the reduced electric field in Townsend (Td), where E is the magnitude of the electric field and N is the total density of neutral particles. 1 Td = $1 \cdot 10^{-17}$ B $\cdot$ cm².

Figure 2 shows the amount of shares to the excitation of different vibrational modes of the water molecule. The maximum proportion of energy that is involved in the dissociation of water molecules co-constitutes 79% at E/N = 125 Td.

Changes in the activity of water in the low-temperature non-equilibrium plasma due to the regulation of plasma characteristics, in particular

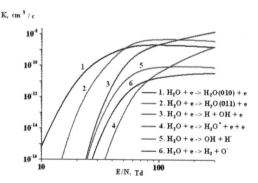

Figure 1. Dependence constant excitation of different degrees of freedom in the water steam from the E/N.

360

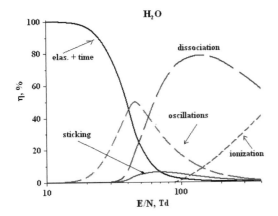

H₂O

Figure 2. Electron energy balance for a pure water steam on the basis of the Boltzmann equation for the function of the electron energy distribution.

the E/N, can provide the varying rates of chemical and physico-chemical processes in the formation of mineral, organic-matrix composite materials.

ACKNOWLEDGMENTS

We acknowledge the financial support from the Russian Ministry of Education (Contract No. 7.2200.2014/K).

REFERENCES

[1] Y. S. Akishev: *Nonequilibrium plasma in dense gases (physics, chemistry, technology and applications in ecology)* (Publishing MEPhI, Moscow 2002).
[2] L. Huxley, R. Crompton: *Diffusion and drift of electrons in gases* (Publishing World, Moscow 1977).
[3] L. S. Frost, A. V. Phelps: *Phys. Rev.,* Vol. 127, (1962), p. 1621.
[4] A. P. Napartovich and I. V. Kochetov: Mater. Plasma Sources Sci. Technol., Vol. 20, No. 2, (2011), 025001.
[5] LXcat community, www.lxcat.laplace.univ-tlse.fr.
[6] Y. S. Akishev, A. A. Deriugin, V. B. Karalnik et al.: submitted to Journal Fizika Plasmy (1994).

Architectural, Energy and Information Engineering – Sung & Chen (Eds)
© 2016 Taylor & Francis Group, London, ISBN 978-1-138-02791-6

Modern methods used in foreign language learning in the 21st century

Klara Rybenska
Faculty of Education, University of Hradec Kralove, Hradec Kralove, Czech Republic

Karel Myska
Institute of Social Work, University of Hradec Kralove, Hradec Kralove, Czech Republic

ABSTRACT: This article deals with individual online instruments determined for foreign language learning via the Internet (eLearning) in most suitable conditions, if possible. The first section of this article deals with the possibilities of individual instruments and determines the criteria that the suitable online instrument in foreign language learning should carry out. The second section of this article deals with the description of selected online instruments and with its individual possibilities for use in foreign language learning.

Keywords: Babbel; duolingo; eLearning; gamification; lingQ; livemocha

1 INTRODUCTION

Nowadays, it is very popular and also entertaining to study via the Internet. If the offered service is of an adequate quality and free, many supporters and keen users appear. One of the requirements to ensure the adequate service being entertaining, but mainly attractive for a potential online student, is to involve gamification elements as the application of game-thinking techniques in a non-game environment, for example websites [1] that orientate on different user categories. The principle of user reward system providing success on Web or platforms is the guaranty of adequate application popularity.

One sphere to use gamification principles is to learn via eLearning, thus online distant learning. One of the possibilities to use the aforementioned aspect is to involve gamification elements into online language learning, which at once poses like an entertaining game. The service is created via a non-forced method that motivates to gradual achievements in learning [2]. This service can be based on the following principles:

- Levels,
- Advancing on higher levels,
- Badge inquiry,
- Virtual currency,
- Indicator of "player's" development on appropriate level,
- Exchange, gifts, rewards,
- Competitiveness between "players", and
- Also "game in game", thus a smaller application within one main aspect [3].

Playfulness spreads at many workplaces, being the motivation instrument in the newspapers, marketing but also in school systems, which can help to make learning more intensive and to develop covered competencies of contemporary inattentive generation being surrounded with computer entertainment [4]. One of the great advantages of online foreign language learning, except for its significant flexibility (user determines himself/herself when and how intensively to learn the appropriate language), is the fact that similarly held learning has unquestionable results. At the University of New York and University of South Carolina, an independent study was performed on Duolingo instrument. It was proved that 34 hours on average spent in online learning Duolingo instrument is identical to one study semester of foreign language at university [5]. Participants studied Spanish and in the end of the study passed the test of Spanish language at one university. From this study, it can be suggested that a person without knowledge of Spanish would need 26–49 hours to explain the subject matter of Spanish language of one university semester. One university semester course takes usually 34 hours, therefore learning in Duolingo is much effective [6].

Internet offers many additional services or private institutions promising foreign language learning via online courses. However, some solutions of the aforementioned type contain a few problems that not all users are willing to accept. First, there are significant costs, very often reaching thousands for a course, or standard learning methods that are not entertaining for many users [2]. Even though top well-known instruments of worldwide response exist (e.g. Rosetta Stone instrument), it is necessary to pay for web browser courses or smart device applications [7], despite the fact it is possible to test the instrument in the free trial version.

However, applications allowing foreign language learning in the entertaining way exist. These applications differ, of course, in the quality of information provided, in price (some of them are free, the others have to be paid, although charges are not high), in language instruction offer but also in the manner of foreign language learning.

The aim of this article is to compare and describe selected online web instruments intended for foreign language learning, and to define the most suitable variant based on the determined criteria.

2 METHODOLOGY

This article is based on the method of comparative analysis. It examines and compares the possibilities of the application of online instruments intended for foreing language learning. Instruments examined were selected in a random sampling manner. These include Babbel, Duolingo, LingQ and Livemocha. The aforementioned instruments have many things in common but differs in certain possibilities of use as well. For examination purposes, the basic criteria that the ideal online learning instruments should carry out were determined.

The criteria determined were as follows:

- Instrument should be free,
- No instrument installation,
- Operating version on web browser,
- Version for smart devices (Android, iOS),
- Instrument uses the gamification method,
- Instrument explains grammer,
- Instrument allows writing, reading, listening, speaking,
- Offer of at least 6 foreign languages intended for learning,
- Possibility of email registration.

3 DESCRIPTION OF THE SELECTED INSTRUMENTS EXAMINED

Babbel. It deals with online language learning. Babbel instrument is paid, although it is possible to test the basic lesson free. Many languages are available, courses are interactive, and it is possible to pass without any installation. Courses for beginners, such as grammer courses, word dictionaries, tongue-twisters, proverbs or song courses, are also available. It is possible to use the microphone (and practice individual words being corrected). It is also possible to use Babbel in mobile devices. Applications are available not only for iPhone or iPad but also for devices affixed with the Android operation system. Version for Windows Phone is also available. Babbel is based on present technologies and popular learning methods. Its purpose is to make language learning entertaining but mainly easy. Babbel puts significant emphasis on pronunciation [8, 9]. Technology identifies language in a real time, and the user is provided immediately with feedback concerning education attainment. Nevertheless, it is possible to cancel the microphone. Afterwards, the instrument operates with the applications not operating with the microphone. Babble also offers possibility to contact students around the world and to communicate in appropriate language. Students can have their own profile pages, and can discuss in a real way or send internal messages [9].

Learning process proceeds in such a way that we are not only tested but also words/sentences appear simultaneously explaining us the terms. First, the test can also be passed without previous registration; afterwards, registration is required if the user wants to proceed ahead, download the selected terms into the word dictionary, and make the course available. The system also contains many clues that are directly available in the lesson (e.g. partly invisible letters providing a clue concerning writing the appropriate letter). The letters are switched, therefore the clue is not univocal and the user has to think concerning writing the word. During word writing, the letter becomes more invisible in order to make evident it was used before (Figure 1).

If the word written is wrong, then a clue is provided to write it properly. It is more comprehensive to fill in the texts. Babbel provides the user with whole paragraphs to fill in individual words or whole sentences. When making a progress at an advanced level, the possibility to write the text is provided, without any clue. However, it is possible to provide the clue—there is the reference to the "*Help*" button, making the clue of the aforementioned mixed letters available. If the word is filled in properly, then the sentence is read in the appropriate language. In the end of the test/lesson, the result is displayed (amount of points) and the possibility to correct mistakes. Each wrongly filled letter represents one mistake. It is better to correct mistakes. It is a necessity from one point of view because it is possible to make a progress at an advanced level only if writing absolutely correct test. The whole course is paid. If having an ambition to make a progress at an advanced level, it is necessary to buy a 1–12 month course. If not, then only basic tests and lessons are provided, which will be exhausted soon. Babbel sometimes provides

Figure 1.　Example of filling out the word in lessons of Italian (Source: Babbel, 2014).

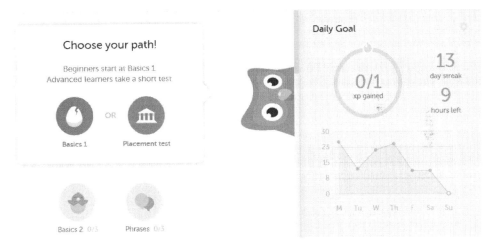

Figure 2.　Example of lesson selection in Italian course in Duolingo (Source: Duolingo, 2014).

discounts and various advantageous campaigns of appropriate language courses.

Duolingo is the combination of language lessons from basics with the elements of traditional games. Through gradual learning, a student is provided with XP (experience) and also with virtual currency, for which bonus lessons can be provided [2].

Duolingo instrument has existed since 2012. Thereafter, the offer of language learning has expanded extensively, even the first Czech—English course is already available. Otherwise, most of the language courses are focused on learning of Czech—foreing language. The instrument is absolutely free (e.g. there are no microtransactions for virtual currency), and virtual platforms arealso very popular. It is another project of reCaptcha's author, providing the description of books into electronic version via the crowdsourcing method. As mentioned above, the main language of the courses designed is still English. However, there already exist courses for foreing-speaking students (e.g. the already aforementioned Czech courses). Currently, many courses are in beta version; certain ones are not available for public. It is considered to be a great advantage that every native speaker or bilingual speaker can contribute and forward its development quickly. In case the student does not observe the discipline concerning language learning, a warning message is sent to email to motivate for learning.

It is possible to be provided with more courses; however, after supplementing additional one, it is necessary to respect the fact concerning the unavailability of all courses together, and it is essential to switch over to native (or foreign) language than to switch over to an appropriate course. Duolingo course structure is similar to traditional lessons—in the beginning, there are primary basics with a few words and expressions branching to another categories. Each category usually contains 2–7 lessons, making a progress in learning. It is possible to complete the summary test verifying a possibility to advance at an advanced learning level or the fact concerning the revision of previous lesson. Primary basics can be omitted via writing the summary test immediately. Thereafter, the student advances to another lesson [2].

It is possible to motivate the users to monitor their friends to compete with or to monitor their friends' learning advance. Through experience, the user passes the test and makes a progress at an advanced level. Therefore, the user's level of progress is evident. Another Duolingo motivation instrument is a graph representing the user's learning attitude and whether being a diligent student. Moreover, the instrument offers the testing of the explained curriculum and verification of extended and forgotten knowledge. The last one and very helpful Duolingo motivation instrument is the virtual currency called *Lingots*. It is provided for passing the lessons, for endurance (if at least one lesson a day will be completed week by week) but also for advancing at a new level. It is possible for *Lingots* to buy various bonus lessons corresponding with season during the year or with a season like Christmas and instruments making learning more pleasant. Another interesting motivation instrument is the mistake tolerance. Duolingo tolerates three wrong answers only. During the first lessons, only four mistakes are tolerated. When wrong answers are made in more questions, it becomes necessary to repeat the whole lesson [2].

After choosing the language, each user passes the first lesson containing primary basics. Learning principles are also determined. It involves usually one sentence or word expression upon which the student works with. The course contains the possibilities of word transcription according to figures, word or sentence writing according to listening, or word or sentence transcription from one language to another. It is possible to use the microphone when testing and to record words in the appropriate language directly into the course (possibly being corrected). However, the users do not have to apply the aforementioned possibilities; it is possible to switch off not only the microphone but also the sound and to complete tests by writing only (without listening or speaking). Sometimes the student has to fill the words in the sentences choosing from 3 possibilities usually. Two answers can be correct. It is necessary to point out that in some languages, certain small mistakes are still present. It is registered immediately and discussed about it in the wide community participating on the development of individual Duolingo courses. One of the many advantages is that this project is based on many volunteers around the world, who not only speak the native language and via this knowledge participate on course development and its revision, but also explain certain sentence constructions intended for keen students.

Of course, Duolingo cannot substitute teachers properly. It is not possible to communicate in such an intensive and flexible manner via this instrument. However, everything written in foreign language can be read randomly—sometimes in usual conversational speed or slowly word by word where the emphasis is put on individual word differentiation or definite and indefinite articles or sentence elements. New words are highlighted and after clicking on it, the translation is provided, and its meaning and characteristics are explained (e.g. declension).

Duolingo instrument is the social website which is not based only on observing learning advance in friends. The greatest advantage is that the community Duolingo is created by and the possibility to discuss learning troubles and to explain subject matter being not clear. These discussions are selected automatically according to the language learnt; discussions are arranged according to its attendance. If advancing appropriate advanced knowledge of language learned, it is possible to deepen the knowledge directly in practice. Some users record documents for translation; these are very often various brochures or Internet articles, Wikipedia page possibly. It is not necessary to translate the whole text directly; it is sufficient to choose the paragraph or sentence and then to translate it. The rest of the translation is translated by other users. Translations are evaluated afterward. The appropriate student—translator is honored for correct translation [2].

The last advantage to be mentioned is the fact that the possibility of learning is not only via websites, but also via mobile applications, such as Android platform [10] and also iOS platform [11]. Everything is optimised for smartphones and tablets, applications provide learning in the same manner as in browser, and it is possible to connect it on the way to work. Moreover, mobile version contains more levels of competition, the so-called duals, in which users compete having the same tasks offered by an ordinary lesson (e.g.sentence translation, filling in the word). It is also necessary to be faster than the competitor [2].

LingQ. This instrument is based on *The Liguist* method that destroys the obstacles preventing people from foreign language learning. Learning principles are based on methods teaching children to speak. Each student is exposed to perceptions and drifts providing the user with knowledge. Users can participate in the conversation, find friends, provide texts for translation and revision download and listen to the text or learn new vocables, and are provided with tutor advice. This instrument is available with the possibility to extend an account and to buy more lessons [12]. During learning, the user can choose from a wide range of various texts according to the topic that the user is interested in. Each text is read and subsequently a dictionary is provided for translating individual languages into the user's native one (Google translator is used; Figure 3).

Sound texts can be downloaded to disk and listened offline. The system offers possibility to speak to the native speaker that can be chosen according to the user's knowledge of appropriate language. Similarly, it is also possible in return to teach somebody the language that the user commands. The instrument's idea of learning foreign language is based on the principle of the user's language learning interest (it is possible to listen to or read real stories, interviews, radio programs, and novels); moreover, vocabulary is advanced a lot. The student reads and listens to foreign language in many different situations. LingQ system je designed to learn the user, for example, writing in the native speaker way. It learns phrases used by native speakers, therefore writings are more native. Similarly, during speaking to teachers, the student learns to use the most common phrases and expressions. Pronunciation is also practised. LingQ also measures everything the user performs so that an advance is relatively evident. It is clear that there is a possibility to determine aims. Instrument is also available in application not only in version for

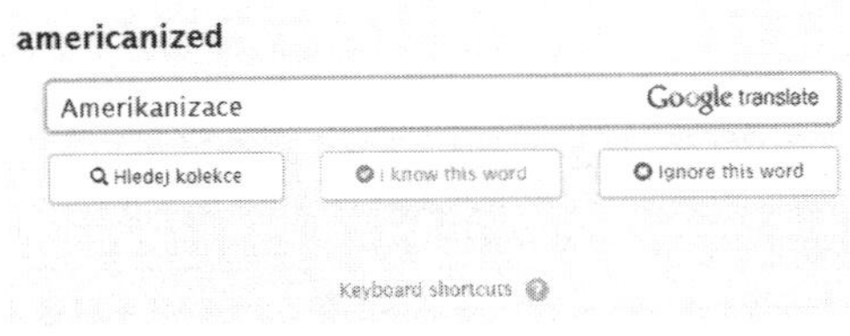

the Android operation system but also for the iOS operation system [13].

Change of language is also in LingQ performed fast via the switch on the left upper page corner. Learning via lessons is solved in LingQ in a very interesting way. In fact, there is no Lesson 1. The user has to select a course corresponding with the level of knowledge extended via setting the modification and when selecting the sphere of interest. After that, it goes through the list of courses to select something corresponding with the user's interest. Afterwards, the course is tested and if considered being an interesting one proceeds with the course. Conversely, it is possible to return back on the list of courses and select a new one. It is possible to select more courses. New lesson is selected on the user's request. Generally, it is possible to declare that if 70–80% of the lesson concerning listening is clear, the user should proceed with learning ahead not losing time. The user learns to keep up his/her motivation for learning. If some lessons are boring but having an ambition to advance, it is possible to do so. To understand words in context or to return back in case not able to understand the subject matter properly is also possible. It is possible to highlight familiar words as familiar ones. However, sometimes the user does not understand the words in different context. It is possible to click on it and the translation is provided again. It is also possible to return back the words from a familiar form to an unfamiliar one. Statistics are modified [12].

A variety of possibilities to the user provided are considered as being great instrument advantage. Student can save the courses, delete, select or print it. It is possible to create dictionary of familiar words and being motivated via the amount of new words extended. The whole course is supplemented by the animal which emerges from an egg and gradually grows alongwith the advance of the user's knowledge. For passing the lesson or for learning new words, the student credits coins (*Avatar Coins*), and this *Avatar* (aforementioned animal) grows gradually which excessively represents gamification again. It is also possible to buy dress for *Avatar Coins* and to dress the animal according to the level of extended knowledge achieved in the course. It is also possible to buy accessories for the animal. The results of the aforementioned process can be published in social networks or put directly on blog website reference with the current condition of "badge", providing the student's progress in learning. When sharing advances, another *Avator Coins* can be earned as well as for inviting friends. The amount of words written or texts listened to is provided in the course. Course provides with the possibility of either online conversation with native speaker or with a person having an advanced knowledge of appropriate language.

americanized

Amerikanizace	Google translate

| Q Hledej kolekce | O I know this word | O Ignore this word |

Keyboard shortcuts @

Figure 3. Example of translation via Google translator in LingQ (Source: LingQ, 2014).

Just examine schedule of tutor and enrol for time. Each user can become a teacher, a tutor in fact if having sufficient advanced knowledge and being capable to spare time for it. Although it is possible to pass a few lessons free in the course, the course is paid mainly. Not all instruments are available in the basic free version. The course is well designed. Although appears to be complicated to orientate in it, it is considered positively its language and instrument accessibility. Although, for example, in Czech language, the translations are not always absolutely accurate, the user with the interest to learn English for example without having any knowledge of language orientates in the course properly and understands assignment easily.

Livemocha. It deals with language community providing education materials in many languages. Registration and access to selected service characteristics is free, nevertheless certain components are paid [14]. Websites are presented in many languages, primary language is of course English, but it is also possible to present websites in Turkish or Chinese. Primary courses are free and one course contains approximately 30–50 hours. Users can participate on individual course development in case of having advanced language knowledge. Paid courses are active. It deals with the courses intended to reach the conversational fluency. An advantage of the course is that texts and written compositions can be revised by a graduated teacher. It is also possible to learn the language via social network, meet new friends and native speakers, communicate online and also provide own work for revision, possibly revise work of other community members [14].

After registration containing also question concerning user native language and language preferred the websites of required language are made available. The system inquires the user about language knowledge. Lessons are bought for virtual currency. The choice of the lesson depends fully on the user. Courses consist of a few steps. The introduction course contains video in alphabet learning sound video with alphabet speaking is used for example. After watching the video, it is possible to use the dictionary. It is possible to skip between individual course sections as well. The course contains instruments for writing, speaking,

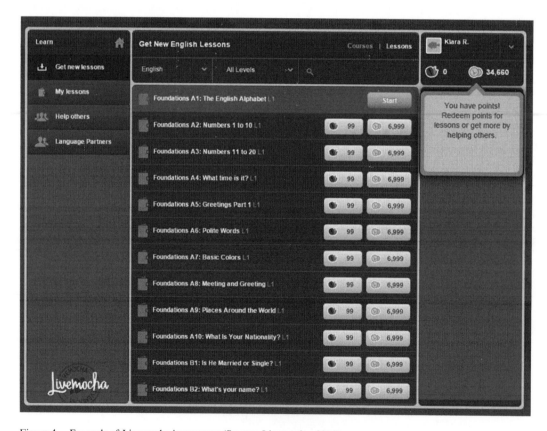

Figure 4.   Example of Livemocha instrument (Source: Livemocha, 2014).

reading and listening. In first lessons, an excessive emphasis is on the alphabet management and also on correct listening to individual words and letters. Nevertheless, certain parts of the course seem to be too complicated for a self study. For correct answers, *Points* are credited. For example, an advantage of English learning is that all sentences (instructions for use to complete the exercise) are in English affixed with Czech (different) translation equivalent. This translation is provided for a short occasion only but long enough to read it. Whole learning is constructed in a more difficult way to understand individual lessons seemingly, but with excellent results. Additional lessons are bought for credit points or for real money.

4 CONCLUSIONS

It is impossible to determine unequivocally that some of the instruments examined are the best ones because each has its advantages and disadvantages (Figure 5).

For example, Babbel instrument emphases on basic conversational abilities construction and grammer explanation [15]. It is not completely free. The basic version is free, but thereafter it is necessary to buy supplemental courses. Conversely, Duolingo has significant potential in entertaining form of gamification, simple and friendly look and not demanding user surrounding. Moreover, it enables relatively fast advance in knowledge inquiry and the users participate actively on its development. Duolingo also explains grammer, but does not emphasis on it so much. Lesson is mainly based on correct translations. LingQ is the instrument whose basic form is free. It is possible to learn foreign language via the lowest (free) program version. However, after that, it is considered that many additional instruments are missing, making foreign language learning more effective and more interesting. Livemocha is also free

in basic version. It offers also a possibility to buy game menu (for real money), for which individual courses can be bought much faster than to save for it via own activities.

As evident from the above graph, Duolingo and LingQ came out as being the best ones concerning fulfilling the requirements. To mix these two instruments together (simple and user smart design and easy Duolingo operation with many possibilities that LingQ brings) and provide this instrument free would bring into the world of modern technologies and foreign language eLearning new competitor. Online foreing language instruments enable effective language learning. It fascinates the user and also makes staying in the learning process only if having much to offer. The instruments motivate via gamification, offer attractive learning courses supplemented with interesting topics or combine standard principles of foreing language learning with practical word expressions in a way the student can learn the language fast. As it is evident, it still does not substitute language stay abroad or private lessons with a native speaker. However, it enables students to learn foreign language or advances the knowledge really effectively, and in many cases fast equally or faster than in ordinary language courses. The users manage and

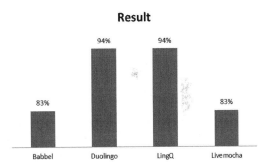

Figure 6.   Examination results (Source: Authors).

	Babbel	Duolingo	LingQ	Livemocha
Tool should be free	Yes/No	Yes	Yes/No	Yes/No
Tool that doesn't require installation	Yes	Yes	Yes	Yes
Working in web browser	Yes	Yes	Yes	Yes
Smart devices compatibile	Yes	Yes	Yes	No
Tool that uses gamification	Yes/No	Yes	Yes	Yes
Tool explaining grammar	Yes/No	Yes/No	Yes	Yes
Tool that enables writting, reading, listening and speaking	Yes	Yes	Yes	Yes
Tool that offers at least 6 languages for learning	Yes	Yes	Yes	Yes
Option to log-in via e-mail	Yes	Yes	Yes	Yes

Figure 5.   Schedule of examination results (Source: Authors).

regulate learning according to their abilities. Many of these eLearning courses offer the possibility of native speaker communication. Each of the testing instruments offers the possibility of communication, corresponding, listening and speaking. Moreover, it is absolutely free in many cases when the student decides to learn the language. It is considered that similar eLearning language instruments bring a huge effort for future not only for students of primary and secondary schools and universities, but also for public, possibly seniors who would like to learn some language.

REFERENCES

[1] Gamifikace. In: Wikipedia: the free encyclopedia, (2014) on: http: //cs.wikipedia.org/wiki/Gamifikace.

[2] S. Janů: Nejzábavnější výuka cizích jazyků. Computer. Zive. Vol. 9/14. (2014), p. 90–91.

[3] R. Holečková, in: *M & B: Marketing and Business Department*: Gamifikace—Hra v reálném životě. (2012), on: http: //www.marketingostrava.cz/ novinky/208-gamifikace-hra-v-realnem-ivot.

[4] P. Kočí, in: *Lupa: Server o českém internetu.* Kdo si hraje, neuteče? Co je gamifikace a jak může pokračovat válka o naši pozornost. (2011), on: http://www.lupa.cz/clanky/kdo-si-hraje-neutece-co-je-gamifikace-a-jak-muze-pokracovat-valka-o-nasi-pozornost/?discussionBox-tabId = top&do = discussionBox-switch.

[5] O nás: Opravdu Duolingo funguje?, in: *Duolingo*, (2014), on https: //www.duolingo.com/ effectiveness-study.

[6] R. Vesselinov, J. Grego. Duolingo Effectiveness Study. (2012), on: http: //static.duolingo.com/s3/ DuolingoReport_Final.pdf.

[7] Rosetta Stone (2014), on http: //www.rosettastone. eu/.

[8] Babbel (2014), on http: //www.babbel.com/ dashboard.

[9] Babbel Wikipedia: the free encyclopedia, (2014) on: http: //en.wikipedia.org/wiki/Babbel.

[10] GooglePlay (2014), on: https: //play.google.com/ store/apps/details?id = com.duolingo.

[11] J. Graham, in: *USA Today.* Duolingo: Apple's choice for App of the Year. (2013), on: http://www.usatoday. com/story/tech/columnist/talkingtech/2013/12/17/ duolingo-apples-iphone-app-of-the-year/4042469/.

[12] LingQ (2014), on: http: //www.lingq.com/.

[13] S. Kaufmann, in: *TheLinguist.* My Method (2013), on: http: //www.thelinguist.com/method/.

[14] Livemocha, in: *Wikipedia: the free encyclopedia,* (2014), on: http: //en.wikipedia.org/wiki/ Livemocha.

[15] Review: Babbel and Duolingo, in: *The Economics,* (2013), on: http: //www.economist.com/blogs/ johnson/2013/06/language-learning-software.

Architectural, Energy and Information Engineering – Sung & Chen (Eds)
© *2016 Taylor & Francis Group, London, ISBN 978-1-138-02791-6*

Numerical simulation and remaining oil distribution in the South Konys oilfield, Kazakhstan

D.X. Wang
China University of Geosciences (Beijing), Beijing, China

T. Jiang & Y. Lei
Research Institute of Exploration and Development, Tuha Oilfield Company, PetroChina, Xinjiang, China

Y.D. Wu
CNPC Kazakhstan Corporation, Almaty, Kazakhstan

ABSTRACT: The South Konys oilfield is characterized by low formation pressure, low permeability of reservoirs and rapid waterflooding. On the basis of seismic, logging and geological data, a 3D geological model with fine grid was built. The numerical simulation was conducted to study the remaining oil distribution in the M-II layer. Reserves calculated by fitting and volume methods have a minor error of 5.27%. Furthermore, through historical production performance fitting, the parameters including production rates, water cut ratio, and cumulative oil production volume fit well with the actual data for the M-II layer. The results of reserve fitting and production performance fitting suggest that the numerical simulation results are reliable. According to the numerical simulation results, the remaining oil in plane is mainly distributed in areas with low-permeability reservoir, with less water injection wells, stagnant zone among oil or water wells, with low permeability resulted from reservoir heterogeneity and structural uplifts. The remaining oil is mainly vertically distributed in the M-II-4 and M-II-5 sub-layers.

Keywords: remaining oil distribution; numerical simulation; South Konys oilfield; Kazakhstan

1 INTRODUCTION

South Konys oilfield, discovered in 1989, is located in the south part of South Turgai basin, Kazakhstan. Although oil production has been rising steadily since 2006, there are three problems with oil production. First, the formation pressure is relatively low because oil is produced by natural energy at the early stage of production and water injection is implemented at a later stage. Second, due to the low permeability of reservoirs, the volume of oil, which is mainly produced by hydraulic fracturing, decreases rapidly. Third, water in plane is injected in one direction. Aiming to lower the decreasing rate of oil production and improve the production performance, the present study numerically simulated the remaining oil distribution in the M-II layer. The results will provide guidelines for the further promotion of development potential of the oilfield.

2 GEOLOGICAL BACKGROUND

The South Konys oilfield consists of several oil-bearing layers, including M-II, J-0-1, J-0-2, and J-0-3, among which M-II of the Cretaceous Lower Daul formation is the main oil-bearing layer. Conglomerates and pebbly sandstones form the reservoir, which is of worse reservoir quality with porosities of 4.1~16.7% (averaging 9.6%) and with a permeability of $0.16 - 1573 \times 10^{-3} \mu m^2$ (averaging $62.6 \times 10^{-3} \mu m^2$). The South Konys oilfield commenced production in June 2001, and water injection was implemented in 2008 with steadily rising production rates by drilling new wells (Figure 1). By the end of November 2013, the total volume of oil produced from the M-II oil-bearing layer was $321.5 \times 10^4 m^3$ with the oil recovery rate, recovery percent of reserves, comprehensive water-cut and accumulative injection to production ratio of 1.34%, 7.81%, 45.2%, and 0.49, respectively.

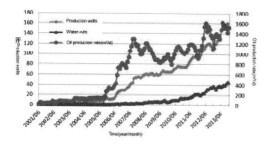

Figure 1. Production rate curve of the South Konys oilfield.

3 3D GEOLOGIC MODEL

On the basis of seismic, logging and geological data, a 3D geological model with fine grid was built by using the Petrel software. The plane grid size is 50 m × 50 m, and the vertical grid step is 1 m.

4 NUMERICAL SIMULATION

Model grid building. The numerical simulation was conducted by using Eclipse software. The upscaling grid of the 3D geological model has steps of 50 m in DX and 50 m in DY. The step of the vertical grid is 5 m on average, and the bulk grid volume is 1337490 (193 × 330 × 21). Figure 2 shows the upscaling attribute model of porosity, permeability, oil saturation and net to gross ratio.

Reserve fitting and production performance fitting. Reserve calculation is mainly influenced by the oil-bearing area, effective thickness of strata, porosity, and oil saturation (Li, 2014). By the methods of reserve fitting, the calculated reserve of the M-II layer in the South Konys oilfield is $4358.95 \times 10^4 \, m^3$. The reserve calculated by volume methods is $4169 \times 10^4 \, m^3$. The reserves calculated by these two methods have a minor error of 5.27%, which proves that the fitting geological reserve is reliable.

The production performance fitting mainly focuses on the pressure and production rates (Hu, 2012). Through the historical production performance fitting, the parameters including production rates, water cut ratio, and cumulative oil production volume fit well with the actual data for the M-II layer in the South Konys oilfield (Figure 3).

5 REMAINING OIL DISTRIBUTION

The understanding of the remaining oil distribution provides fundamental guidelines to produce oil, improve waterflood performance, and increase oil recovery rates (Zhou, 2010; Li et al., 2005; Lu et al., 2001). The remaining oil distribution is con-

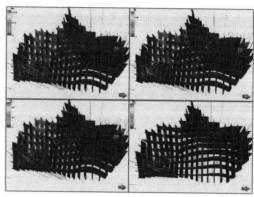

Figure 2. Attribute models of the M-II layer in the South Konys oilfield.

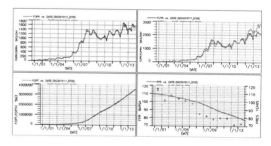

Figure 3. Comparison curves of oil production rate, fluid production rate, cumulative oil production, and pressure.

Figure 4. Oil saturation distribution in the M-II-4 sub-layer.

trolled by many factors, including depositional microfacies of reservoir, structural characteristics of reservoirs, flow unit, reservoir heterogeneity, and well patterns.

372

Table 1. Remaining oil reserve in different sublayers in M-II layer in South Konys oilfield.

Sublayer	Fitting reserve (10⁴ cubic meters)	Cumulative production oil (10⁴ cubic meters)	Remaining geological reserve (10⁴ cubic meters)	Recovery ratio (%)	Averaging oil saturation (%)	Remaing reserve percentage per unit (%)
M-II-1	176.88	3.51	173.37	1.98	49.57	4.51
M-II-2	447.33	68.48	378.85	15.31	41.41	9.85
M-II-3	772.45	54.68	717.77	7.08	46.93	18.66
M-II-4	1587.06	118.80	1468.26	7.49	50.01	38.17
M-II-5	1212.78	104.01	1108.77	8.58	42.65	28.82
Total	4196.50	349.47	3847.03	8.33		

Plane distribution of remaining oil. According to numerical simulation results, the remaining oil in plane is mainly distributed in areas including areas with low-permeability reservoir, areas with less water injection wells, stagnant zone among oil or water wells, low-permeability area resulted from reservoir heterogeneity and structural uplifts. Taking the remaining oil distribution in the M-II-4 sub-layer into consideration, remaining oil is mainly distributed in the SW part in a continuous sheep shape, especially in the neighboring area of wells SK-6 and K-58 (Figure 4).

Vertical distribution of remaining oil. The vertical distribution of the remaining oil is mainly controlled by intra-layer heterogeneity, inter-layer heterogeneity, and well pattern efficiency (Zhang et al., 2001). According to numerical simulation results, the remaining oil is mainly distributed in the M-II-4 and M-II-5 sub-layers. By the end of 31, November 2013, the total volume of the remaining geologic reserve was $3847.03 \times 10^4 \, m^3$. The volumes of the remaining geologic reserves in the M-II-4 and M-II-5 sub-layers were $1468.26 \times 10^4 \, m^3$ and $1108.77 \times 10^4 \, m^3$, respectively, which is 67% of the total volume of the reserve. The averaging oil saturation values were 50.01% and 42.65%, respectively (Table 1).

6 CONCLUSIONS

The present study investigated the 3D numerical simulations of the M-II layer in the South Konys oilfield. The results obtained from the simulation were used to evaluate the remaining oil distribution.

The fitting reserve and that calculated by volume methods have a minor error of 5.27%. Furthermore, through the historical production performance fitting, the parameters including production rates, water cut ratio, and cumulative oil production volume fit well with the actual data for the M-II layer.

The results show that the remaining oil in plane is mainly distributed in areas including areas with low-permeability reservoir, areas with less water injection wells, stagnant zone among oil or water wells, low-permeability area resulted from reservoir heterogeneity and structural uplifts. In addition, the remaining oil is mainly vertically distributed in the M-II-4 and M-II-5 sub-layers.

REFERENCES

Hu, H. 2012. Quality evaluation method on history matching of reservoir numerical simulation. *Fault-block Oil & Gas Field*. 19 (3): 354–358.

Li J., Li X., Gao W. & Jiang T. 2005. Residual oil distribution research of high water-cut stage in Tuha oil field. *Natural Gas Geoscience*. 16 (3): 378–381.

Li, B. 2014. Determination of parameters for volumetric method based on SEC rules. *Petroleum Geology and Experiment*. 36 (3): 381–384.

Lu J., Li G., Fan Z., Li C. & Xue Y. 2001. Residual oil distribution research of high water-cut stage in an oil-field. *Acta Petrolei Sinica*. 16 (5): 48–52.

Zhang Z., Liu J. & Li W. 2001. A quantitative study on vertical distribution of remaining oil through injection profiles. *China Offshore Oil and Gas (Geology)*. 15 (5): 345–349.

Zhou W., Tang Z., Wen J. & Gong Y. 2010. Study on distribution law of remaining oil by using reservoir numerical simulation technology. *Fault-block Oil & Gas Field*. 17 (3): 325–329.

Architectural, Energy and Information Engineering – Sung & Chen (Eds)
© 2016 Taylor & Francis Group, London, ISBN 978-1-138-02791-6

Sedimentary Facies of Cretaceous to Paleocene in C Oilfield, P Basin

Gao Jie Xiao
*Research Institute of Petroleum Exploration and Development, CNPC, China University of Geosciences,
Beijing, China*

Xing Yan Fan, Zhi Yun Yuan & Gui Lin Zhang
Research Institute of Petroleum Exploration and Development, CNPC, Beijing, China

ABSTRACT: Based on integrated analysis of drilling profiles, cores, well logging and seismic data, the sedimentary facies of Cretaceous to Paleocene in C Oilfield were comprehensively studied. The result shows that there are seven types of sedimentary facies in this region, including alluvial fan, fan delta, subaqueous fan, turbidite fan, delta, lacustrine and marine. The structural evolution controls migration process of the depocenter and sedimentary evolution: (1) Affected by back arc extension during the Cretaceous, the study area is a fault depressed lacustrine basin, and mainly developed coarse-grained alluvial fan-fan delta facies. The depocenter located in the east of the study area. During the Paleocene, affected by foreland extruding effect, the depocenter moved to the west of study area and mainly developed fine grained meandering stream-delta sedimentary system.

Keywords: P Basin; Cretaceous to Paleocene; C Oilfield; Sedimentary Facies

1 INTRODUCTION

1.1 *Type area*

Located within the Sub-Andean foreland basin group that is known to be hydrocarbon-rich in the South America continent[1-4], the P basin is the Colombian portion of the P-Oriente-Maranono Basin. It is a typical foreland basin bounded by Cordillera Oriental to the west, the Brazilian Guyana Shield to the east, the LIanos Basin to the north and the Ecuadorian Oriente Basin and the Maranon Basin to the south, covering an area of 80000 km². Exploration activities in the past years have suggested that the basin contains abundant petroleum resources[5-8], with 19 discovered oil fields and ultimately recoverable oil reserves of about 6.3×10^8 t[9].

1.2 *Tectonic and stratigraphic characteristics*

As part of the Sub-Andean foreland basin group, the P Basin's tectonic evolution is considered to be always closely related to the Andean orogenic event and Gondwanal and break-up & convergence. Its tectonic evolution can be divided into five periods [10-11]: Pre-Mesozoic basement evolution stage during Pre-Cambrian to Devonian time: Paleozoic non-marine and shallow-sea continental clastic rocks and limestone strata are preserved in the western portion of the basin which has close proximity to Cordillera Oriental slope belt; rifting—back-arc stretching stage during Middle Jurassic to Lower Cretaceous age: a deep rift comprising synrift terrigenous clastic rocks was formed in the northern and western portions of the basin, and the normal fault-controlled titled fault blocks were developed in the eastern portion of the basin; basin-scale thermal subsidence stage during Late Jurassic to Early Cretaceous: a stable and wide continental is present, and Hollin/Caballos formations uncomfortably overlie the strata formed in rifting period and then are overlain by Villeta/Napo/Conglomerate formations; early foreland basin stage during Late Cretaceous to Paleocene: a foreland basin is formed due to uplifted Andes fold, in which the Rumiyaco and Pepino/Mirador formations are deposited, successively, sourced by Cordillera Central and East Guyana Shield; and late foreland basin stage since Miocene: structures are further squeezed and the western portion of the basin is wholly uplifted and eroded, with typical continental molasse red bed sediments of Orteguaza/Arrayan-Orito/Sexrrania formation deposited.

The C oil field is located within the P Basin, which contains the Mesozoic and Cenozoic strata, and a Pre-Paleozoic basement consisting mainly of Cambrian crystalline rocks and metamorphic rocks (schist, gneiss, granite or migmatite)). The Cretaceous Conglomerate Fm., which consists of conglomerate, pebbly sandstone and sandstone interbedded with argillaceous siltstone and sandy mudstone, can be divided into three units: the Lower Conglomerate (CL), Middle Conglomerate (CM) and Upper Conglomerate (CU) members. The Paleocene Eocene Mirador FM, comprising interbedded grey and greywhite medium-coarse sandstone and mudstone of varying thicknesses, contains two units: the Lower Mirador (ML) and Upper Mirador (MU) members. The ML member is a thick section of mudstone containing an interval of tuffaceous shale that extends across the region and is considered a regional marker bed; and the MU member consists of interbedded medium to coarse sandstone and mudstone of varying thicknesses, with pebbly sandstone locally present. The Oligocene Arrayan Fm ranges from 175 to 1000 ft thick, containing grey and red mudstone interbedded with sandstone and thin carbonate rocks, with oolitic and brownish red granular limestone present

2 SEDIMENTARY FACIES TYPES

Based on analysis of stratigraphic characteristics of the Cretaceous-Paleocene formations in the C oil field by combining core data with geophysical data, the study area is considered to have developed six types of sedimentary facies: alluvial fan, fan delta, delta, lacustrine, subaqueous fan and slope fan facies.

2.1 Alluvial fan

An alluvial fan is a fan-shaped deposit formed where a temporary flood stream flattens, slows, and spreads typically at the exit of a canyon onto a flatter plain[12]. The alluvial fan in the study area contains a variety of lithologies including conglomerate, pebbly sandstone and argillaceous conglomerate. The mudstone is brownish red, purplish red and brown colored, coarse-grained, low in maturities of components and structures, and poorly sorted, with poorly to moderately developed beddings including inconspicuous oblique and cross beddings. The sub-facies of the alluvial fan exhibits differing responses on resistivity log: root-fan is massive high-resistivity shaped, middle-fan presents interbedded low-resistivity and high-resistivity which corresponds to braided channel deposits, and the argillaceous sediments-dominated fan edge shows a flat and straight low-resistivity shape. Seismically, an alluvial fan is characterized by chaotic and wavy reflections with poor to worse continuity.

2.2 Fan delta

A fan delta is the alluvial fan built by clastic sediments moved from mountainous terrain to flood detention basin, and due to its steep gradient and close proximity to provenance, contains a variety of lithologies including thick grey and greywhite conglomerate, sandy conglomerate and sandstone, interbedded with thin-bedded Grey and greysiltstone and mudstone. The coarse-grained sandstone is less thermally mature. This is a typical feature of proximal deposits. The sedimentary structure is dominated by wedge and parallel beddings, with commonly present erosional surfaces. A prograding and aggrading combination that is abnormally high in amplitude is shown by Gamma Ray log. This involves an upward-increasing, funnel-like abnormal amplitude combination at the base, the box-shape feature in the middle, and a bell-like and straight curve combination in the upper part, suggesting a transition from fan delta front to fan delta plain facies.

2.3 Delta

A lacustrine delta is the deposits of terrigenous clast that form at the mouth of a river, where the river flows into a lake [13]. This facies is distributed broadly in the study area, with major lithologies including medium-fine sandstone, siltstone and dark mudstone. It can be planarly divided into three sub-facies, from lakeshore to lake-basin center: delta plain, delta front and pro-delta. The delta plain sub-facies mainly contain distributary channel sand bodies that are reported to be moderately mature. The delta front sub-facies is characterized by the development of river mouth bar sand bodies and is considered the most typical region of alluvial-lacustrine actions. Its lithology is dominated by fine sandstone and siltstone, in which lenticular bedding, ripple bedding, flaser bedding and bioturbation structures are present. It shows a continuous, funnel-like combination that gradually increases upward in amplitude on Gamma Ray log.

2.4 Coastal subaqueous fan

A costal subaqueous fan is a fan-shaped deposit formed where the proximal mountainous flood carrying substantial amounts of terrigenous materials flows into a lake. This facies is common in the downthrown side of a boundary fault that controls a basin (Sun Yongchuan, 1978; Zheng Junmao, 1984) and extends into the semi-deep—deep lacustrine facies.

Its lithology is characterized by grey mudstone interbedded with grey to light grey conglomerate, breccia, pebbly sandstone, medium sandstone and fine sandstone. The conglomerate is poorly sorted and rounded, and mostly sub-angular. Clast-supported conglomerate facies, massive sandstone facies and superimposed & eroded lithologies are present in the medium sandstone. Unlike the fan delta, the slope fan is formed in a deep-water environment, and a dark Grey and grey or dark, pure mudstone was therefore deposited. Massive bedding, parallel bedding, erosional surface and current bedding are present. The coastal sub-aqueous fan from inner to outer portions grades from medium-amplitude box and bell-like to low-amplitude finger-like on well log.

2.5 Turbidite fan

A turbidity fan is a fan-shaped classic rock body of deep-water turbidity origin. The most common lithology of the turbidity fan is dark grey. mudstone interbedded with thin-layered fine sandstone. Superimposed graded bedding, massive bedding, parallel bedding, small-scale ripple & current bedding, and bedding-plane structure representing gravity flow are present. The turbidite fan commonly shows a large set of combination, in which amplitude increases upward, occurred in a flat and straight curve, and its log shape changes from jugged box-like and bell-like to funnel-like.

2.6 Lacustrine facies

A lake is a relatively low-lying terrain that lies on land and is fed by rivers and streams. Lake may block a considerable amounts of sediments carried by rivers, and hence is considered an important site for continental sediments to deposit (Chi Qiue, 2001). Based on the lake water level, the lacustrine facies can be divided into shore-shallow lacustrine and semi-deep-deep lacustrine sub-facies. The shore-shallow lacustrine sub-facies consists of variegated mudstone interbedded with greyish green mudstone, sandy mudstone and silt-stone, in which small-scale cross bedding, ripple bedding, bioturbation and wormhole are present, as shown by cores. The lacustrine facies shows a tooth-like to micro-tooth-like to flat & straight combination on GRY log. The semi-deep—deep lacustrine sub-facies are characterized by a grey and black mud shale lithology and are very rich in organic matter.

2.7 Marine facies

The Paleocene Oligocene strata in the P Basin contain marine deposits due to a transgression. These marine deposits are dominated by coastal swamp facies and neritic facies deposits. The argillaceous-dominated bathyal deposits are present locally in the western portion of the study area.

3 EVOLUTION CHARACTERISTICS OF SEDIMENTARY FACIES

Based on a variety of techniques including core observation & fine characterization, single well facies analysis, cross-well facies analysis, formation thickness, sandstone sand factor mapping and seismic attribute analysis, the planar distribution of Cretaceous-Paleocene sedimentary facies in the C oil field in the P Basin is mapped using a quantitative research methodology of lithofacies paleogeography (Feng Zengzhao) that involves a single-factor analysis and multi-factor mapping.

3.1 Conglomerate Fm

A series of half-grabens developed in the deformed P Basin during Early Mesozoic time in a back-arc setting as a result of the Gondwanaland subducting. The North American plate later was separated from the Gondwanaland. Meanwhile, the Cordillera volcanic arc was formed as a result of the Pacific plate subducting, and the back-arc extension has resulted in the formation of titled fault blocks that are controlled by normal faults in the eastern portion of the basin. A post-rift thermal subsidence stage commenced as a result of the South Atlantic extension during Late Cretaceous time. The transition from fault depression to sag occurred during the deposition of Conglomerate Fm. Fault block movement significantly becomes weak but titling & uplifting is intense. The north-western portion of the study area was uplifted and the southeastern portion was plunged. The depo-center moved southward, forming the terrain that descends from east to west and from north to south but still remains the geomorphic feature of a steep and narrow fault subsidence basin. During the deposition of CM, a depositional system consisting of alluvial fan to fan delta to coastal subaqueous fan to sub-lacustrine fan to lake was formed in the high-lying western portion of the study area from the mountain pass to lake-basin center; and in the relatively low-lying eastern portion rivers emerged from the mountainous terrain and flowed into a lake, forming a fan-delta to sub-lacustrine fan depositional system. The deposition of alluvial fan dominated the northern portion of the study area. During the deposition of CL which shares similar depositional feature with CM, the folded & uplifted West Andes has resulted in an increasingly more elevation difference between the western and eastern portions of the study area. A fan delta dominated deposition system consisting of alluvial fan to fan delta to lake was formed in the high-lying western portion; and in the low-lying eastern portion fan delta deposits were present as water level increases significantly, and a depositional system of alluvial fan to fan delta to coastal subaqueous fan was formed as the provenance to the north extended into the lake basin.

3.2 Mirador Fm

The Caribbean plate converged with the South American plate during Paleocene time, with the Andes Mountains uplifted continuously, and a foreland extension commenced since Late Oligocene time, with some overthrust or reverse faults occurred. As a result, the western portion of the study area is plunged while the eastern portion is uplifted, forming the terrain that descends from east to west and from north to south. The deposits in the western portion grade from Conglomerate Fm sandy/gravelly clastic sediments to argillaceous bathyal sediments, the terrain in the eastern portion is squeezed and uplifted but still remains flat and gentle, and as the deposition system of delta to subaqueous fan to turbidity fan enters into the marine basin from northeast, the deposits in the northern portion of the study area are dominated by delta facies sands. Structures were further squeezed during the deposition of Upper Mirador, and the depocenter continued to move southwestward. The deposits in the eastern portion of the study area remained deep-water argillaceous sediments-dominated. The provenance to the northeast still exists, but sea level drops, forming a depositional system of meandering river to delta to turbidite fan that extends from northeast to southwest.

3.3 Characteristics and evolution laws of sedimentary facies

Tectonism controls the migration of depocenter and exerts a considerable control on sedimentary evolution of the study area. During the Cretaceous time, the separation of the North American plate and the subduction of the Gondwanaland and Pacific plate have resulted in the formation of the Cordillera Central volcanic arc, which leads to the occurrence of a back-arc extension in the P Basin. The study area at this period is within a fault depression stage, developing the depositional system of alluvial fan to fan delta to turbidite fan to lake at both western and eastern portions, and the depocenter is located in the eastern portion of the study area. During the Paleocene time, the Andes Mountains were uplifted due to the extensional collision and strike-slipping of the Caribbean

and South America plates, and the pre-existing foreland extension has resulted in the occurrence of thrust napping in the P Basin. As a result, the western portion of the study area subsided rapidly, developing bathyal argillaceous deposits, while the eastern portion is relatively uplifted, developing a depositional system of meandering river to delta to subaqueous fan to turbidite fan, and the depocenter is located in the western portion of the study area.

4 CONCLUSIONS

1. The Cretaceous to Paleocene strata of the C oil field in the P Basin has developed seven types of sedimentary facies: alluvial fan, fan delta, delta, coastal subaqueous fan, turbidite fan, lacutrine and marine, each of which has different lithologic features, sedimentary structures and log shapes.
2. The study area has experienced an evolution process characterized by the transition from shallow to deep to shallow water level during the Cretaceous to Paleocene time. A depositional system of alluvial fan to fan depth to sublacustrine fan to lake was developed during the deposition of Cretaceous-age Middle Conglomerate Fm., with alluvial fan sand body predominating; and as lake-basin water area expanded, and a depositional system of fan delta to sublacustrine fan to lake was developed during the deposition of Upper Conglomerate Fm., with fan delta sand body predominating. As a high water level remained during the deposition of Paleocene-age Lower Mirador, the depositional system of delta to subaqueous to turbidite fan to bathyal was developed, with delta sand body predominating. As seawater retreated during the deposition of Upper Mirador, a deposition system of meandering river to delta to subaqueous fan to turbidite fan to bathyal was developed, with meandering river to delta sand bodies predominating.
3. Tectonism controls the migration of the depocenter and exerts a considerable control on sedimentary evolution of the study area. During Cretaceous time, the study area presents a fault depression basin feature due to the effect of a back-arc extension, developing coarse-grained alluvial fan to fan delta deposits, with a depocenter located in the eastern portion; and during Paleocene time, the depocenter moved westward due to the effect of a foreland extension, with meandering river to delta deposits predominately developed.

378

ACKNOWLEDGMENTS

This research was supported by National Major Science and Technology Special Project (2011ZX05029-001) and Science Foundation of CNPC (2013D-0902), We also recognize the support of CNPC for the permission to publish the paper.

REFERENCES, SYMBOLS AND UNITS

[1] Zou Caineng, Tao Shizhen. Classification, formation and distribution of giant oil and gas zone [J]. Petroleum Exploration and development, oil 2007, 34 (1): 5–12.

[2] Zou Caineng, Zhang Yaguang, Tao Shizhen, et al. Geological features, major discoveries and unconventional petroleum geology of the global oil and gas exploration field [J]. Petroleum Exploration and Development, 2010, 37 (2): 129–145.

[3] Xie Yinfu, Ji Hancheng, Su Yongdi, et al. Petroleum geology characteristics and exploration potential of Oriente-Maranon basin. Petroleum Exploration and Development [J], 2010, 37 (1): 51–56.

[4] Liu Yaming, Xie Yinfu, Ma Zhongzhen, et al. Analysis of Paleozoic natural gas reservoir characteristics and exploration potential of South foreland basin [J]. Natural Gas Geosciences. 2012, 23 (6): 1045–1053.

[5] Ye Deliao, Xu Wenming, Chen Ronglin. Exploration and development potential of oil and gas resources of South American [J]. Chinese Petroleum Exploration, 2007, 12 (2): 70–75.

[6] Yang Fuzhong, Wei Chunguang, Yin Jiquan, et al. Typical structural characteristics of petroliferous basin in northwestern South America [J]. Geotectonica et Metallogenia, 2009, 33 (2): 230–235.

[7] Bai Guoping, Qin Yangzhen. The petroliferous basins of South America and the distribution of oil and gas [J]. The Modern Geology, 2010, 24 (6): 1102–1111.

[8] He Hui, Fan Tailiang, Lin Lin. Petroleum geology characteristics and exploration potential analysis of Brazil's Amazon basin [J]. Journal of Southwest Petroleum University: Natural Science Edition, 2010, 32 (3): 61–63.

[9] Zhao Qing, Yuan Bingqiang, Zhang Chunguan, Song Lijun. Gryavity field and Cretaceous distribution characteristics of Columbia P basin [J]. Journal of Xi'an Petroleum University, 2013, 28 (2): 35–39.

[10] Zhao Changyu, Tian Lihua, Yang Zhenwu, Tan Fulin. Analysis of oil and gas exploration potential in Caguan-P basin, Southeast of Columbia [J]. ProGryess in Geophysics, 2014, 29 (6): 2843–2850.

[11] Wen Qin, Zhang Xionghua, Liang Yu, et al. Tectonic evolution and petroleum geological characteristics of typical petroliferous basin in western of South America [J]. Petroleum geology and engineering, 2011, 25 (3): 27–32.

[12] Zhao Chenglin. Sedimentary petrology [M]. Beijing: Petroleum Industry Press, 2000: 275–278.

[13] Wang Liangchen, Zhang Jinliang. Sedimentary environment and facies [M]. Beijing: Petroleum Industry Press, 1996: 145–146.

Architectural, Energy and Information Engineering – Sung & Chen (Eds)
© 2016 Taylor & Francis Group, London, ISBN 978-1-138-02791-6

A new SAR and optical image fusion method based on Nonsampled Contourlet Transform

N. Yang, H.H. Li & P.D. Yan

College of Automation, North Western Polytechnical University, Xi'an, China

ABSTRACT: SAR and optical images have large differences in imaging mechanism and spectral characteristics. Moreover, SAR images are always severely contaminated by speckle noise. Consequently, it is very difficult to obtain satisfactory results when we fuse SAR and optical images. Considering the advantage of Nonsampled Contourlet Transform (NSCT) compared with other multiscale decomposition methods, a new SAR and optical image fusion method based on NSCT is proposed. Unlike previous SAR and optical image fusion methods, the proposed algorithm combines denoising and fusing steps in the fusion rule design together rather than separates the two steps. The whole process is more simple and easy to implement. First, the original images are decomposed by NSCT, and then low- and high-frequency subband coefficients can be obtained. Second, as to the SAR image, according to the different properties of noise and signal's NSCT decomposition coefficients, we carry out hard threshold denoising at the finest scale level of high-frequency coefficients. While some faint signals may be mistaken as noise and deleted in this step, the fusion method based on region is introduced to improve the result. Other scale levels of high-frequency coefficients are fused in accordance with the finest scale level. Third, we employ 'choose absolute maximum' to select the low-frequency coefficients of the fused image. Finally, the fusion coefficients are inversed to get the final fusion result. The experimental results show that it not only preserves the salient information of original images well, but also restrains the noise effectively.

Keywords: image fusion; Nonsampled Contourlet Transform; SAR image; optical image

1 INTRODUCTION

Because the SAR image could reflect special information of the target and be obtained in all weather conditions, it is usually fused with other images to provide complementary information, such as SAR and infrared image fusion [1], multiple polarimetric SAR image fusion [2–4], and SAR and MS image fusion [5–7]. There are much differences in imaging mechanism and spectral characteristics of the SAR image and other images. Taking SAR image and optical image as examples, SAR image information lies on the object's geometrical property and dielectric property, and optical image information lies on the resonance of the object's surface molecule [8], and then SAR and optical image fusion can make full use of the complementary information of both. Usually, the Intensity-Hue-Saturation (IHS) [9–10] method is the most common procedure used for the fusion of SAR and optical images. With this fusion method, the I component was directly substituted by the SAR image, which introduces a great spectral distortion in the fusion image. To overcome the drawbacks

of this method, another common method based on the Intensity Modulation (IM) was proposed by Garzelli in 2003 [11]. Afterwards, this method was developed by some researchers [7, 12–13]. This kind of methods obtained better performance in spectral preservation than the IHS method, but it could only inject some isolated points with great illumination of SAR data into the optical data, and the continual urban areas and some important targets are lost in the fusion results. For example, in [12], the underground bunker in the left bottom of the SAR image (Figure 3(a)) could not be preserved in the fusion result (Figure 3(d), (e)). Recently, the fusion methods based on multi-resolution analysis are most often used in remote-sensing image fusion, among them the methods based on multiscale geometric analysis are always the hotspot. Multiscale geometric analysis is more appropriate for the analysis of the signals that have line, plane or hyperplane singularity than wavelet, and it has better approximation precision and sparsity description. When introducing multiscale geometric analysis to image fusion, we can pick up the characteristics of original images well and provide more information for fusion. The ability of noise restraint is also better than wavelet transform. Multiscale geometric analysis involves Ridgelet transform [14–15], Curvelet transfom [16], and Contourlet transform [17]. Li [18] proposed a SAR and optical image fusion method based on Ridgelet transform. Although Ridgelet could express the line characters effectively, it is the same as wavelet when processing curve characters in image. Then, Curvelet transform was used in the fusion process [19]. The experiment showed that better results could be obtained than wavelet. However, in line with [18], because the size of the images, especially high-frequency subband images, are not the same after the application of Ridgelet and Curvelet transforms, the simple 'choosing absolute maximum' fusion is often used. Moreover, it is difficult to design a better fusion strategy according to the relationships of coefficients, such as the relationships between different scales. The Contourlet Transform (CT) allows for different and flexible numbers of directions at each scale. The CT can efficiently capture the intrinsic geometrical structures in natural images such as smooth contour edges, and is a real multiscale and multidirection expansion. The Nonsampled Contourlet Transform (NSCT) [20] was developed on the basis of the CT theory. The NSCT is fully shift-invariant. When the image is decomposed with NSCT, we could obtain the low- and high-frequency subband images with the same size, which is convenient for designing the fusion strategy. Thus, a new SAR and optical image fusion algorithm based on the NSCT is proposed.

In many previous documents, first, SAR images are often denoised, and then the denoised SAR images are fused with optical images. These two steps are carried out separately [7–8, 11, 21]. Our proposed algorithm combines the denoising and fusing steps together, making the implementation of the whole process more simple and easy. In addition, this algorithm is effective for SAR and optical image fusion, and could preserve the original images' salient information adequately and has a good noise suppression.

According to different properties of noise and signal's NSCT decomposition coefficients, we propose a hard threshold denoising and fusion algorithm at the finest scale level. We carry out a hard threshold denoising at the finest scale level of high-frequency coefficients the when SAR image is decomposed by the NSCT. While some faint signals may be mistaken as noise and deleted in this step, the fusion method based on region is introduced to improve the result. Other scale levels of high-frequency coefficients are fused in accordance with the finest scale level. In other words, if the coefficient of the SAR image at the finest scale level

is regarded as an important one and not the noise coefficient, the coefficient with the same location at other scale levels is also picked up and added to the fusion result, or else the coefficient of the optical image is added to the fusion result. As for the low-frequency coefficients, we employ the 'choose absolute maximum' fusion rule. The experimental data show that it not only results in good fusion results, but also restrains the noise effectively.

The rest of this paper is organized as follows. In Section 2, NSCT decomposition is introduced. In Section 3, as for the SAR image, different properties of noise and the signal's NSCT decomposition coefficients are mainly discussed, which serves as a support to the denoising algorithm proposed in Section 4. In Section 4, the proposed algorithm is described in detail. Section 5 presents the experiments and comprehensive discussions on the results. Finally, the main conclusions of this paper are discussed in Section 6.

2 NSCT DECOMPOSITION

NSCT is a kind of multi-scale and multi-direction image decomposition method. It is an improved version of Contourlet Transform. It uses the dual filter group to realize the multi-scale and multi-direction decomposition. The NSCT decomposition structure is shown in Figure 1 [20].

NSFB, including NSPFB (Nonsubsampled Pyramid Filter Bank) and NSDFB (Nonsubsampled Directional Filter Bank), are used when implementing the NSCT. Without down sampling during decomposing and reconstructing, the NSCT is a shift-invariant decomposition and the coefficients' image has the same size with the original image after decomposition. In addition, the NSCT removes the frequency compared with Contourlet Transform.

3 THE PROPERTIES OF NOISE AND SIGNAL'S NSCT DECOMPOSITION COEFFICIENTS IN THE SAR IMAGE

SAR images have a serious speckle noise due to its imaging mechanism. The noise makes it hard to interpret the original image. Sometimes the noise could cover up the salient information of the ground object. Consequently, when we fuse the SAR and optical images, it is very significant to restrain the speckle noise. One kind of popular noise-restraining method is based on the multi-resolution analysis, which is proposed by many former researchers, for example, the thresholding denoising method based on wavelet transform by D L. Donoho [22]. Because wavelet transform have

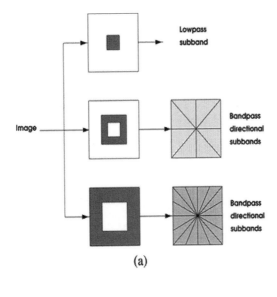

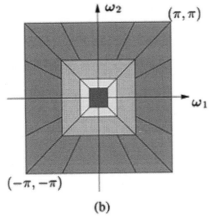

Figure 1. NSCT decomposition structure. (a) NSFB structure that implements the NSCT. (b) Idealized frequency partitioning.

let coefficients into two types. One is from the true signal's transformation, and this kind of coefficients has a large value of amplitude but less numbers. The other is from the noise's transformation, and this kind of coefficients have a small value of amplitude but more numbers. In addition, with the increase in the number of decomposition levels, the values of coefficients from the noise's transformation decrease rapidly because a finer scale could smooth the noise better to a certain degree. Based on this, we can set a threshold to restrain noise according to the difference in the wavelet coefficients' amplitude. In the NSCT domain, because the NSCT also has a multiscale property as wavelet, the NSCT coefficients have similar properties to that of wavelet. Then, with the increase in the number of decomposition levels, the noise's NSCT transformation values decrease rapidly, but that of the true signal's transformation maintain the large value at each scale level. So, the noise and signal at the finest scale level could be easier to be distinguished. We could set a fixed threshold at the finest scale level to classify the noise and the signal, and the coefficients in the same location at other scale levels are determined as the noise or the signal corresponding to the finest scale.

4 IMAGE FUSION BASED ON THE NSCT

4.1 The flow of the image fusion

The flow of the SAR (source image A) and optical image (source image B) fusion based on the NSCT is shown in Figure 2.

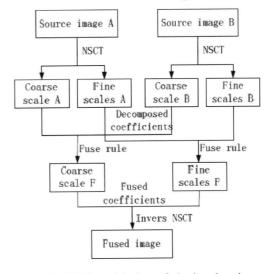

Figure 2. The flow of the image fusion based on the NSCT.

not only good performance of locality in both time and frequency domain, but a multiscale property, it possesses the ability of "centralizing" the signal. In other words, the wavelet coefficients of the natural image centralize on few coefficients with a large amplitude, yet the noise coefficients have a uniform distribution in the whole wavelet space, and the amplitudes do not have great differences with each other. If the energy of the signal is centralized on few coefficients in wavelet transform, then in contrast, the values of these coefficients are inevitably greater than those of noise coefficients whose energy is distributed in numerous wavelet coefficients. Accordingly, we can classify the wave-

1. The source SAR and optical images are decomposed by level NSCT, and then transform coefficients $\left\{H_{j,k}^A, L^A\right\}$ and $\left\{H_{j,k}^B, L^B\right\}$ can be obtained, where L is the low-frequency subband coefficients; H is the high-frequency subband coefficients; j is the current decomposition level; and k is the number of the decomposition direction;
2. Aiming at low- and high-frequency subband coefficients, different fusion algorithms are designed to obtain the fusion coefficients;
3. The fusion coefficients are inversed to get the final fusion result.

4.2 The proposed algorithm

The entire brightness of the SAR image is lower than the optical image, and the regions such as city are expressed as bright points and regions, which are often the salient features needed to add in the fusion result. Then, the fusion objective is not only to preserve the detailed background information of the optical image, but also to add the salient features of the SAR image, and have a good noise suppression.

4.2.1 Fusion of high-frequency subband coefficients

Because high-frequency coefficients include the noise and the detailed information of the original image, the objective of high-frequency coefficients' fusion is to retain the detailed information of the original image and to restrain the noise effectively.

1. High-frequency coefficients' fusion rule at the finest scale level

Based on the properties of noise and signal's NSCT decomposition coefficients in the SAR image mentioned in Section 3, we can distinguish the noise and true signals at the finest scale level easily. The source SAR image is decomposed by level NSCT, and the decomposition coefficients $\left\{H_{j,k}^{SAR}, L^{SAR}\right\}$ are obtained. At the finest scale level (J_{level}), we define the coefficients at the k direction as $H_{J,k}^{SAR}$. A threshold value T is set to determine whether the coefficient is noise or signal. Suppose that $\tilde{H}_{J,k}^{SAR}$ are the results. If the coefficient at any location $H_{J,k}^{SAR}(i,j)$ is less than T, then the coefficient is determined as noise, and we set $H_{J,k}^{SAR}(i,j)$ as zero; otherwise, the coefficient is determined as signal, and the value is retained. The process can be expressed as follows:

$$\tilde{H}_{J,k}^{\text{SAR}}(i,j) = \begin{cases} H_{J,k}^{\text{SAR}}(i,j) & H_{J,k}^{\text{SAR}}(i,j) > T \\ 0 & H_{J,k}^{\text{SAR}}(i,j) < T \end{cases} \quad (1)$$

The threshold value can separate the noise and signal effectively. In this paper, we set $T = \omega \times$ Mean,

where ω is the weighting coefficient and the average of all coefficients at the finest scale level. Generally speaking, the larger the T, the more effective the suppression of the noise. However, some real signals of weak texture in the source image are also suppressed. On the contrary, the signal can remain more completely, but the noise cannot be suppressed effectively. In order to improve the result that is influenced by setting the threshold, we use Spatial Frequency based on region to fuse the images. Because the decomposition coefficient of the real signal is not isolated, it has a strong continuity. In other words, the region near the coefficient of the real signal shows similar values. However, for noise, isolated points are shown. Spatial Frequency could reflect the trend of gray change in the region (the region can be 3-by-3 or 5-by-5 window) [23]. The larger value of Spatial Frequency means some break of gray, for example, the edge or texture. In high-frequency coefficients, it could be used to detect the detailed information. Consequently, we can use Spatial Frequency to determine whether the coefficient is real signal or noise further after the above-mentioned setting T threshold.

The Spatial Frequency of the image reflects active degree in the spatial domain of the image. Its definition is given by:

$$SF = \sqrt{RF^2 + CF^2} \quad (2)$$

where RF is the row frequency and CF is the column frequency, which are defined as follows:

$$RF = \sqrt{\frac{1}{M \times N} \sum_{i=0}^{M-1} [F(i,j) - F(i,j-1)]^2} \quad (3)$$

$$CF = \sqrt{\frac{1}{M \times N} \sum_{i=0}^{M-1} [F(i,j) - F(i-1,j)]^2} \quad (4)$$

where $F(i,j)$ is the coefficient value at location RF.

After the threshold T is utilized to suppress the noise, we set the nonzero coefficients at the finest scale level of the SAR image as the centre of window. We then compare the Spatial Frequency in this window of the SAR and optical image. If the Spatial Frequency of the SAR image is larger, the coefficient before suppressing the noise using T will remain again, taking the value of zero, and then we could add again the real signals of weak texture that are mistakenly deleted by noise suppression. If the Spatial Frequency of the optical image is larger, we use the coefficient of the optical image as the fusion coefficient. In this way, the fusion result could retain the important information coming from the optical image and suppress the noise

384

effectively. The formula can be expressed as follow:

$$H^F_{J,k}(i,j) = \begin{cases} H^{SAR}_{J,k}(i,j) & SF^{SAR}_{window} > SF^{PAN}_{window} \\ H^{PAN}_{J,k}(i,j) & SF^{SAR}_{window} < SF^{PAN}_{window} \end{cases} \quad (5)$$

We consider the Landsat 7 Panchromatic image as the original optical image, where "PAN" refers to Panchromatic.

2. High-frequency coefficients' fusion rule at other scale level

At the finest scale level, we set noise as zero, and the corresponding locations at other scale levels could be equally considered as noise. Therefore, when we fuse the other scales' coefficients, we could use the coefficient of the SAR image as the fusion result if the corresponding coefficient at the finest scale level is nonzero, or else the coefficient of the optical image is selected.

The coefficients at different scale levels have the following relationships: all the subband images are of the same size as that of the original image after NSCT decomposition, thus the coefficients could be found easily in the same location at different scale levels. However, the problem is that there may exist different direction numbers at different scale levels. Figure 3 shows the relationship between two adjacent scales. The number "1" refers to the nonzero value. One coefficient in the coarse scale matches with respect to two coefficients of two directions at the finest scale. In Figure 3, four coefficients are taken as an example. Only both coefficients of two directions at the finest scale are nonzero, and the matching coefficient in the coarse scale is nonzero. This fusion rule's design could be supported by the theory of the coefficients' continuity after NSCT decomposition.

4.2.2 Fusion of low-frequency subband coefficients

Presently, the weighted average is often used to fuse the low-frequency coefficients [24–25]. However, SAR and optical images have large differences in spectral characteristics, so there often exists an opposite pixel intensity in the same location of the SAR and optical image. If we average the low-frequency coefficients to fuse, some salient information may be lost in the fusion result. The regions with high brightness are often reflected by a large absolute value in low-frequency coefficients after NSCT decomposition, and the selection of maximum absolute values is used to preserve regions with high brightness in the SAR image:

$$L^F(i,j) = \begin{cases} L^{SAR}(i,j) & if \; L^{SAR}(i,j) > L^{PAN}(i,j) \\ L^{PAN}(i,j) & if \; L^{SAR}(i,j) < L^{PAN}(i,j) \end{cases} \quad (6)$$

5 EXPERIMENTAL RESULTS AND DISCUSSION

The original images are the ERS-2 SAR image and the Landsat 7 Panchromatic image of Pavia city in Italy [26], as shown in Figure 4.

We mainly compare the result with that of the NSCT algorithm based on pixel and region. In addition, we also compare the result with that of the NSCT algorithm based on region after the SAR image is first filtered by some frequently used denoising algorithm. Figure 5 shows the SAR images filtered by different denoised algorithms. The Median filter, the Gamma MAP filter, the Lee filter and the Wiener filter are used. From Figure 5, we can see that the SAR images with Lee Filtering and Wiener filtering have less noise and the salient information are preserved better, compared with that with Median filtering and Gamma MAP filtering. Consequently, we use the two filtered results as the original SAR images to fuse with the Panchromatic image in order to compare the fusion results with that of the proposed algorithm.

Figure 6 shows the different fusion results. Figure 6a shows the NSCT algorithm based on pixel where the fusion rule of its high-frequency subband coefficients chooses the maximal absolute value. Figure 6b shows the NSCT Algorithm based on region where the rule chooses the maximum Spatial Frequency, which are calculated within the window. The window has a size of 5 * 5.

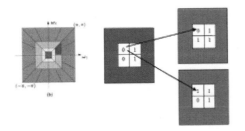

Figure 3. Corresponding relationship between two adjacent scales of NSCT coefficients.

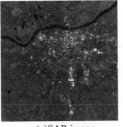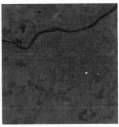

(a)SAR image   (b) Panchromatic image

Figure 4. Original images.

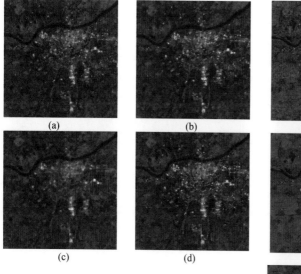

(a)　　　　　(b)

(c)　　　　　(d)

Figure 5. SAR image with different denoised algorithms.

Figure 6c–d shows the results of the NSCT algorithm based on region after the SAR image is filtered by the Lee filter and the Wiener filter. Figure 6e–g shows the different fusion results of the proposed algorithm, as mentioned previously. In view of checking the algorithms' effectivity, we choose same mode of decomposition: the filter for the Laplacian pyramid is set as 'maxflat', the directional filter is set as 'dmaxflat7', the decomposing level is 4 levels, the number direction of each level is 2, 4, 8 and16, respectively. The fusion rules of low-frequency subband coefficients all choose maximum absolute values.

From Figure 6a and 6b, we can see thatdue to the influence of speckle noise in the SAR image, the results of the NSCT algorithm based on pixel and region still have much noise, the luminance of the whole image is close to that of the SAR image, and the information of the Panchromatic image could not be added to the fusion results well. From Figure 6c–d, we can see that if the SAR image is filtered first, the results are improved with respect to noise suppression and information preservation. However, compared with the proposed algorithm, the latter have less noise, and the salient features of the Panchromatic and SAR images are both preserved better, especially $\omega = 3$ and $\omega = 4$ (the differences can be clearly seen in the left bottom of the images). Moreover, the proposed algorithm could obtain flexible fusion results, because ω can control the degree of noise suppression and information preservation. If ω is small, we can get the result with much information preservation, but with rela-

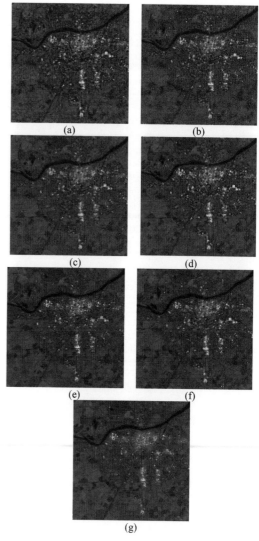

(a)　　　　　(b)

(c)　　　　　(d)

(e)　　　　　(f)

(g)

Figure 6. Different fusion results. a) NSCT algorithm based on pixel; b) NSCT algorithm based on region; c) NSCT algorithm based on region after the SAR image is filtered by the Lee filter; d) NSCT algorithm based on region after the SAR image is filtered by the Wiener filter; e) Proposed algorithm ($\omega = 2$); g) Proposed algorithm ($\omega = 4$).

tively no good noise suppression. If ω is large, we can get the result with good noise suppression, but with reduced weak texture and bright region.

We also evaluate the results with some objective evaluation. Because the fusion objective is not only to preserve the detailed background information of the Panchromatic image, but also to add the salient features of the SAR image, and to have good noise suppression, we select the Correlation

Table 1. Performance evaluation of the image fusion method.

	NSCT based on pixel	NSCT based on region	NSCT based on region (Lee filter)	NSCT based on region (Wiener filter)	Proposed algorithm ($\omega = 2$)	Proposed algorithm ($\omega = 3$)	Proposed algorithm ($\omega = 4$)
CC	0.3005	0.3320	0.4556	0.464	**0.3903**	**0.4774**	**0.5521**
DC	0.2867	0.2747	0.1877	0.1591	**0.2152**	**0.1608**	**0.1361**
ENL	6.1607	7.3146	10.6925	10.4649	**7.7785**	**9.5963**	**11.8467**

Coefficient (CC), the Difference Coefficients (DC) and the Equivalent Number of Looks (ENL) to evaluate the fusion results.

The CC reflects the similarity of the fusion image (A) and the Panchromatic image (B), the value is closer to 1, and the proximity of the two images is better. The definition is given by:

$$CC(A,B) = \frac{\sum_{i,j}\left[(A(i,j) - \overline{A}) \times (B(i,j) - \overline{B})\right]}{\sqrt{\sum_{i,j}\left[(A(i,j) - \overline{A})^2\right]\sum_{i,j}\left[(B(i,j) - \overline{B})^2\right]}}$$

(7)

DC reflects the intensity deviation of the fusion image (A) and the Panchromatic image (B). The smaller value means the less intensity deviation of the fusion image and the Panchromatic image. The definition is:

$$DC = \frac{1}{M \times N}\sum_{i=1}^{M}\sum_{j=1}^{N}\frac{|B(i,j) - A(i,j)|}{B(i,j)}$$

(8)

where $M \times N$ is the size of the image.

ENL could be used to measure the degree of noise suppression and the clarity of edges. The greater value means good noise suppression and the clarity of edges. The definition is given by:

$$ENL = [E(A)]^2 / Var(A)$$

(9)

$E(A)$ is the average value of the fusion image and $Var(A)$ is the variance of the fusion image.

The computing results are summarized in Table 1.

From Table 1, we can see that the proposed algorithm could obtain the results of high relevance for the Panchromatic image compared with the NSCT based on pixel and region, with the values of DC being obviously smaller than those of the two methods. Although the SAR image is filtered first, we obtain the improved results, but the results are similar to that of the proposed algorithm ($\omega = 3$), and not good as that of the proposed algorithm

($\omega = 4$). In fact, with the increase of ω, we could obtain better noise suppression at the cost of some weak texture and bright region of the SAR image, as identical with the visual impression.

6 CONCLUSION

A new SAR and optical image fusion method based on Nonsampled Contourlet Transform (NSCT) was proposed. Because SAR images are always severely contaminated by speckle noise, the proposed algorithm combines denoising and fusing steps together, and designs an effective method that could preserve the original images' salient information adequately and have a good noise suppression, making the implementation of the whole process more simple and easy. The experiment results showed that this SAR and optical image fusion method achieved a satisfactory result, especially for SAR images severely contaminated with speckle noise.

ACKNOWLEDGMENTS

This work was supported by the Natural Science Foundation of Shaanxi Province (2014JQ2-6035), the Fundamental Research Funds for the Central Universities (3102014JCQ01062), and the Aero-Science Fund (20131953022).

REFERENCES

Alparone, L., L. Facheris, S. Baronti et al. Fusion of Multispectral and SAR Images by Intensity Modulation. The 7th International Conference on Information Fusion, 2004: 637–643.

Alparone, L., S. Baronti, A. Garzelli. Landsat ETM+ and SAR image fusion based on generalized intensity modulation, IEEE Transaction on Geoscience and Remote Sensing, 2004, 42(12): 2832–2839.

Candes E J. Ridgelet: theory and applications. Ph.D Thesis. Department of Statistics, Stanford University, 1998.

Candes, E.J., L. Demanet, D. L. Donoho et al. Fast discrete curvelet transforms. Applied and Computational Mathematics, California Institute of Technology, 2005.

Chandrakanth, R., J. Saibaba, and G. Varadan. Fusion of High Resolution Satellite SAR and Optical Images, 2011 International Workshop on Multi-Platform/Multi-Sensor Remote Sensing and Mapping (M2RSM), 2011: 1~6.

Chen, D., B Li, ZK Shen. Research on a Data fusion algorithm of SAR and optical images. Systems Engineering and Electronics. 2000, 22(9): 5–7.

Chen, S.H., S. Zhang. SAR and multispectral image fusion using generalized IHS transform based on à Trous wavelet and EMD decompositions, 2010, 10(3): 737: 745.

Chibani, Y. Additive integration of SAR features into multispectral SPOT images by means of the à trous wavelet decomposition. ISPRS Journal of Photogrammetry & Remote Sensing, 2006, 60: 306–314.

Cunha, A L., JP. Zhou, M N. Do. The nonsubsampled contourlet transform: theory, design and applications. IEEE Transactions on Image Processing, 2006, 15(10): 3089–3101.

Do, M. N., M. Vetterli. The Contourlet transform: an efficient directional multiresolution image representation. IEEE Trans Image Processing, 2005, 14(12): 2091–2106.

Dong, C., Q. Yuan, Q. Wang,. A combined wavelet analysis- fuzzy adaptive algorithm for radar/infrared data fusion. Expert Systems with Applications, 2010, 37: 2563–2570.

Donoho D.L. Fast ridgelet transforms in dimension 2. Department of Statistics., Stanford CA 94305–4065, Tech. Rep. Stanford University, 1997.

Donoho, D.L., I.M. Johnstone, G. Kerkyacharian, and D. Picard. Wavelet shrinkage, Asymptopia Journal of the Royal Statistical Society, Series B, 1995, 57(2): 301–369.

Garzelli, A. Wavelet based fusion of optical and sar image data over urban area. International archives of photogrammetry remote sensing and spatial information sciences. 2002, 34: 59–62.

Garzelli, A., F. Nencini. Integration of Landsat and SAR images based on intensity modulation. Proc. SPIE Image and Signal Processing for Remote Sensing. Barcelona, Spain, 9–12 September, 2003, 5238: 37–344.

Harris, J.R., R. Murray. IHS transform for the integration of radar imagery with other remotely sensed data, Photogrammetric Engineering and Remote Sensing. 1990, 56(12): 1631–1641.

Li, H. H., L. Guo, G, X. Li. Is ridgelet transform better than wavelet transform in SAR and optical image fusion? Journal of Northwestern Polytechnical University, 2006, 24(4): 418–422.

Li, H. H., L. Guo, K. Liu. Remote sensing image fusion based on curvelet transform. Journal of Optoelectronics Laser, 2008, 19(3): 400–403.

Li, S.T., J.T. Kwok, Y N Wang. Combination of images with diverse focuses using the spatial frequency. Information Fusion, 2001, 2: 169–176.

Tang, L., F. Zhao, Z. G. Zhao. The Nonsampled Contourlet Transform for Image Fusion. Proceedings of the 2007 International Conference on Wavelet Analysis and Pattern Recognition, Beijing, China, 2007.

Yang, S., M. Wang, Y. Lu, "Fusion of multiparametric SAR images based on SW-nonsubsampled contourlet and PCNN," Signal Processing, 2009, 89: 2596–2608.

Ye, Y., B. Zhao, L. Tang. SAR and visible image fusion based on local non-negative matrix factorization. ICEMI. 2009: 263–266.

Yésou, H., Y. Besnus, J. Besnus, J.C. Pion. A. Aing. Merging Seasat and SPOT imagery for the study of geological structures in a temperate agricultural region. Remote Sensing of Environment, 1993, 43(3): 265–279.

Zhang, Q., BL. Guo, Research on Image Fusion Based on the Nonsubsampled Contourlet Transform, 2007 IEEE International Conference on Control and Automation. Guangzhou, China. May 30 to June 1, 2007: 3239–3243.

Zhang, X., P. Huang, P. Zhou. Data fusion of multiple Polarimetric SAR images based on combined curvelet and wavelet transform. APSAR, 2007: 225–228.

Zheng, Y., S. She, W.M. Zhou. False color fusion for multi-band SAR images based on contourlet transform. Acta Automatica Sinica, 2007, 33(4): 337–341.

Architectural, Energy and Information Engineering – Sung & Chen (Eds)
© *2016 Taylor & Francis Group, London, ISBN 978-1-138-02791-6*

A modified MBOC modulation for next generation navigation satellite system

W. Liu

Merchant Marine College, Shanghai Maritime University, Shanghai, China

ABSTRACT: This paper presents a new modified Multiplexed Binary Offset Carrier (MBOC) modulation in Global Navigation Satellite System (GNSS) signal design. The general mathematical model of modified MBOC modulations is developed. Simulations show that modified MBOC modulation indeed offer improved performance of the autocorrelation, code tracking, multipath and compatibility. The proposed modulation can provide capacity for compatible with other signals while offering advantages and multipath resistance, and the opportunity for flexible receiver design.

Keywords: signal modulation; signal performance; satellite navigation

1 INTRODUCTION

The increasing number of navigation systems can provide a bright future for navigation applications. New signal modulation design for next-generation of Global Navigation Satellite Systems (GNSS) must offer improved performance and the opportunity for spectrum compatible with existing and planned signals. The first generation of spreading modulation is based on Binary Phase Shift Keying with rectangular pulse shape, denoted BPSK-R (Betz, J.W. 2001) (e.g. GPS C/A code with BPSK-R(1)). Then, Binary Offset Carrier (BOC) modulations that use a square wave sub-carrier to create separate spectra on each side of the transmitted carrier were proposed (Betz, J.W. 2001). It can provide spectral isolation and lead to significant improvements in terms of tracking, interference and multipath mitigation than BPSK-R modulation. Based on the BOC modulations, the Multiplexed Binary Offset Carrier (MBOC) modulation (Avila-Rodriguez, J. A. et al. 2008) is recommended for GPS L1C and Galileo E1 OS signals and the alternative BOC (AltBOC) modulation (Issler, J.L. et al. 2003) is planned for Galileo E5 signal. More recently, the combinations of MSK and BOC modulations were shown to offer opportunities for high spectral confinement while still provide constant envelope (Pasupathy, S. 1979). A generalization of the BPSK-R and BOC modulation, called Binary Coded Symbol (BCS) modulation (Hegarty, C.J. 2005), is proposed for offering additional degrees of freedom for shaping the signal spectrum and thus its correlation function. Therefore, the BPSK and BOC modulations can be understood as a particular case of the BCS modulations.

This paper presents the extension of the Minimum Shift Keying (MSK) theory with MBOC modulations, denoted as MSK-MBOC or M-MBOC in GNSS signal design. Examples are presented to illustrate the construction and performance of MSK-MBOC modulation for GNSS. The MSK-MBOC modulation can provide potential opportunities for Beidou signal design. The MSK-MBOC modulation can provide capacity for compatible with other signals while offering advantages and multipath resistance, and the opportunity for flexible receiver design.

This paper is organized as follows. Section 2 of this paper describes the signal model and spectral characteristics of MSK-MBOC modulations. Section 3 displays the signal performance with regards to noise, interference and multipath. The application of MSK-MBOC modulation in Beidou signal design is also investigated, while Sections 4 draws conclusions.

2 MSK-MBOC MODULATIONS

2.1 Signal model

A direct sequence spread spectrum (DSSS) signal, s (t) can be represented as follows

$$s(t) = \sum_{k=-\infty}^{\infty} a_k q(t - kT_c) \qquad (1)$$

where $\{a_k\}$ represents the spreading sequence, $q(t)$ is the spreading symbol, and T_c is the spreading code period.

For the BCS modulations, the spreading symbol is divided into K segments, each of equal length T_c/K. Then the spreading symbol is given by (Hegarty, C. 2005)

$$q(t) = \sum_{k=0}^{K} s_k p_{T_c/K}(t - kT_c / K) \qquad (2)$$

where

$$p_\Delta(t) = \begin{cases} 1, & 0 \le t \le \Delta \\ 0, & \text{elsewhere.} \end{cases} \qquad (3)$$

The notation, BCS $([s_0, s_1, ..., s_{K-1}], f_c)$, is used to denote a BCS modulations that uses the vector $[s_0, s_1, ..., s_{K-1}]$ for each symbol and a spreading code rate of $f_c \times 1.023$ MHz $= 1/T_c$.

The symbol pulse of MSK is (Pasupathy, S. 1979)

$$p_{MSK\_pulse}(t) = \begin{cases} \cos(\pi t / T_{sc}), & 0 \le t \le T_{sc} \\ 0, & \text{elsewhere.} \end{cases} \qquad (4)$$

If we substitute the symbol pulse waveform (3) of MSK waveform, a new class of particularly attractive modulations with spreading code rate $f_c \times 1.023$ MHz, denoted as MSK-BCS $([s_0, s_1, ..., s_{K-1}], f_c)$, can be obtained, where $T_c/K = T_{sc}$. Typically, we can define BCS-BPSK-R (f_c) and BCS-BOC(f_s, f_c) modulations, where $K = 1$ and $K = 2$ f_s/f_c, respectively.

The PSD for MSK-BCS modulations with perfect spreading code can be written as

$$S(f) = f_c \frac{8(Kf_c)^2}{\pi^2} \frac{\cos^2\left(\dfrac{\pi f}{Kf_c}\right)}{(K^2 f_c^2 - 4f^2)^2} \left| \sum_{k=0}^{K-1} s_k e^{-j2\pi f/Kf_c} \right|^2 \qquad (5)$$

2.2 Spectra characteristic

China is now proposing to move its signal modulation almost entirely into the Binary Offset Carrier (BOC) family. The Beidou signals at L1 will place a MBOC (6,1,1/11) Open Service (OS) and a BOC (14,2) Authorized Service (AS). Note the MBOC is also recommended for the future Galileo L1 OS and GPS L1C signals (Avila-Rodriguez, J. A. et al. 2008). The MBOC design continues the trend in most modernized signal designs to provide improved performance for GNSS. Although very good performance can be obtained with the MBOC signal, it has been recognized that better performance can be obtained using spreading modulations that provide more power at high frequencies away from the center frequency.

Since MSK-BCS modulations present potential benefits for GNSS, it may become a new choice for Beidou B1 OS signal. Based on the MSK-BCS technique, a new MBOC-like modulation can be obtained, denoted MSK-MBOC. The PSD of MSK-MBOC can be expressed as

$$G_{signal}(f) = \frac{10}{11} G_{BOC(1,1)}(f) + \frac{1}{11} G_{MSK-BOC(6,1)}(f) \qquad (6)$$

where $G_{BOC(1,1)}(f)$ is the normalized PSD of BOC (1,1), and $G_{MSK-BOC(6,1)}(f)$ is the normalized PSD of MSK-BOC(6,1).

Figure 1 shows the PSD of MSK-MBOC. Compared to MBOC, the MSK-MBOC spectrum has place a small amount of additional power at higher frequencies. In other words, MSK-MBOC can provide improved tracking performance.

3 SIGNAL PERFORMANCE

3.1 Code tracking performance

The Cramér-Rao Lower Bound (CRLB) is usually employed to assess the performance of the code tracking errors estimation (Betz, J. W. 2001). The CRLB is defined as

$$\sigma_{CRLB} = \frac{1}{2\pi} \sqrt{\frac{B_L}{\dfrac{C}{N_0} \int_{-\beta_r/2}^{\beta_r/2} f^2 G_s(f) df}} \qquad (7)$$

where B_1 is the code tracking loop bandwidth, β_r is the receiver front-end bandwidth, and $G_s(f)$ is the normalized power spectral density of the signal.

Figure 2 shows the Cramér-Rao lower bound for the receiver configuration of most interest. It has a 24 MHz two-sided receiver bandwidth and 1 Hz code tracking loop bandwidth. As shown, the MSK-MBOC signal has better code tracking performance than MBOC signal.

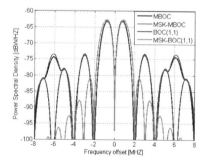

Figure 1. Normalized PSDs of MBOC, MSK-MBOC, BOC(1,1) and MSK- BOC(1,1).

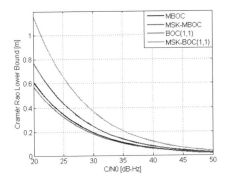

Figure 2. Code tracking errors (24 MHz).

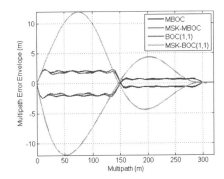

Figure 3. Multipath error envelopes (24 MHz).

3.2 Multipath performance

Besides the code tracking accuracy, code multipath performance is another argument in the signal design. In this paper, the performance of BOC, MBOC and MSK-MBOC examples in terms of multipath using the multipath error envelopes are evaluated. In order to obtain an estimate for the multipath error we will take the distributions of path delays and relative amplitudes into account. The normalized multipath probability density function $D(\tau)$ and running average multipath errors e_{mp} are given by (Hein, G. W. 2005)

$$D(\tau) = \frac{3e^{\frac{3\tau}{2\tau_0}}}{2\tau_0} \qquad (8)$$

$$e_{mp} = \frac{1}{2} \int_0^{\infty} \frac{\|E_{max}(\tau)\| + \|E_{min}(\tau)\|}{2} \cdot D(\tau) d\tau \qquad (9)$$

with E_{max} and E_{min} are maximum and minimum multipath envelopes.

Figures 3 and 4 show the corresponding results for 24 MHz two-sided receiver bandwidth, with a narrower early late spacing 24.4 ns. The multipath to direct path signal power ratio (MDR) is assumed to be -10 dB. Note that MSK-MBOC signal provides error envelopes that are smaller than those for MBOC signal. The average errors for MSK-MBOC are also smaller than those for MBOC. All results show the proposed MSK-MBOC signal yields smaller multipath errors than the other signals.

3.3 RF compatibility

The GNSS radio frequency compatibility has become a matter of great concern for the system providers and user communities. In this section, we will investigate the RF compatibility for MSK-MBOC as a choice for Beidou B1 OS signal. Since MSK-MBOC places more power at higher

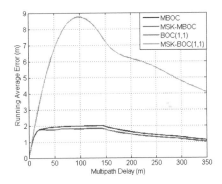

Figure 4. Running average multipath errors (24 MHz).

frequencies, it may also provide some additional benefits in RF compatibility.

According to (Betz, J. W. 2001), the degradation of effective C/N_0, which is the general quantity for the RF compatibility assessment, can be expressed as

$$\Delta(C/N_0)_{eff} = \frac{\dfrac{C}{N_0 + I_{Intra}}}{\dfrac{C}{N_0 + I_{Intra} + I_{Inter}}} = 1 + \frac{I_{Inter}}{N_0 + I_{Intra}} \qquad (10)$$

Therefore, the expression of intrasystem interference in dB as

$$\phi_{Inter} = 10 \cdot \log\left(1 + \frac{I_{Inter}}{N_0 + I_{Intra}}\right) \qquad (11)$$

where C is the received power of the desired signal. I_{Intra} is the equivalent noise power density of interfering signals from satellites belonging to the same system as the desired signal. I_{Inter} is the aggregate equivalent noise power density of interfering signals from satellites belonging to the other

Table 1. GPS, Galileo and Beidou signal parameters in L1 band.

System	Service type	Carrier frequency (MHZ)	Modulation type	Chip rate (Mcps)
GPS	C/A	1575.42	BPSK	1.023
	P(Y)	1575.42	BPSK	10.23
	M	1575.42	BOC(10,5)	5.115
	L1C	1575.42	MBOC	1.023
Galileo	E1OS	1575.42	MBOC	1.023
	E1PRS	1575.42	BOCc(15,2.5)	2.5575
Beidou	B1C	1575.42	Option 1 MBOC	1.023
			Option 2 MSK-MBOC	
	B1A	1575.42	BOC(14,2)	2.046

system. N_0 is the power spectral density of thermal noise. In this paper, we assume N_0 to be -201.5 dBW/Hz.

The following Table 1 summarizes the characteristics of GPS, Galileo and Beidou signals to be transmitted in L1 band. The detailed information about the signal parameters can be found in the GPS and Galileo Interface Control Document, respectively. For Beidou B1 OS signal, we assume two options in the calculations, including: MBOC (as described in (Avila-Rodriguez, J.A. et al. 2008)) and MSK-MBOC.

Ideal spreading codes, normalized (unit area over infinite bandwidth) power spectrum of each signals and two sided receiver bandwidth using 24 MH are assumed in the RF compatibility analysis. Table 2 shows the RF compatibility analysis results.

As shown in Table 2, compared to MBOC for Beidou B1C, the proposed MSK-MBOC for Beidou B1C provides more compatibility with GPS and Galileo civil signals. As we can recognize, MSK-MBOC modulations can provide potential opportunities for Beidou signal design.

Table 2. C/N_0 degradations for different signals.

Case	Option 1	Option 2
Beidou B1C←GPS C/A	0.38	0.39
Beidou B1C←GPS L1C	0.69	0.65
GPS C/A←Beidou B1C	0.52	0.52
GPS L1C←Beidou B1C	1.16	1.08
Beidou B1C←Galileo L1 OS	0.58	0.52
Galileo L1 OS←Beidou B1C	1.48	1.37

ACKNOWLEDGEMENT

This paper is sponsored by Science and Technology Program of Shanghai Maritime University (GN: 20120092).

4 CONCLUSIONS

A new modified Multiplexed Binary Offset Carrier (MBOC) modulation in global navigation satellite system signal design was presented in this paper. The general mathematical model of modified MBOC modulations was developed. Simulations showed that modified MBOC modulation indeed offer improved performance of the autocorrelation, code tracking, multipath and compatibility. The proposed modulation can provide capacity for compatible with other signals while offering advantages and multipath resistance, and the opportunity for flexible receiver design. For GNSS modernization and construction, the spreading symbols of MSK-MBOC must be chosen carefully in signal design.

REFERENCES

Avila-Rodriguez, J. A., Hein, G. W. & Wallner, S. etc. 2008. The MBOC Modulation: The Final Touch to the Galileo Frequency and Signal Plan, *NAVIGATION: Journal of the Institute of Navigation*: 55(1), 15–28.
Betz, J. W. 2001. Binary Offset Carrier Modulations for Radio navigation, *NAVIGATION: Journal of the Institute of Navigation*: 48(4), 1–10.
Hegarty, C. J. 2005. Binary Coded Symbol Modulations for GNSS, *Proceedings of ION NTM 2005, San Diego, USA, 7–9 June 2004*.
Hein, G. W. 2005. Performance of a Galileo PRS/GPS M-Code Combined Service, *Proceedings of ION NTM 2005, San Diego, USA, 24–26 January 2005*.
Issler, J. L., Ries, L. & Lestarquit, L. etc. 2003. Spectral measurements of GNSS Satellite Signals: Need for wide transmitted bands, *Proceedings of ION GNSS 2003, Portland, USA, 9–12 September 2003*.
Pasupathy, S. 1979. Minimal Shift Keying: A Spectrally Efficient Modulation, *IEEE Communications Magazine*: 17(4), 14–22.

Architectural, Energy and Information Engineering – Sung & Chen (Eds)
© 2016 Taylor & Francis Group, London, ISBN 978-1-138-02791-6

Study on cross-line footbridge response analysis based on a multi-scale model

J. Liu

Hubei Key Laboratory of Roadway Bridge and Structure Engineering, Wuhan University of Technology, Wuhan, China

J.X. Lu & L. Jiang

School of Civil Engineering and Architecture, Wuhan University of Technology, Wuhan, China

ABSTRACT: The multi-scale simulation and response analysis of cross-line footbridge was researched by considering the effects of train wind. First, train wind turbulent numerical simulation was carried out according to Reynolds averaged N-S control equations, and train wind pressure time history loading on the cross-line footbridge was obtained. Taking into account the precise scope of the strain, the cross-line footbridge was simulated in multi-scale by the sub-model method, and the multi-scale finite element model was established. Then, the dynamic response of the cross-line footbridge was calculated, and the cross-line footbridge strain time history curve of dangerous spots (the maximum composite deformation) was achieved after the train wind pressure time history was loaded on the finite element model. Finally, the cross-line footbridge response was analyzed. The results show that the cross-line footbridge multi-scale model can well serve for the simulation of the footbridge wind-induced response, and that train wind has a great influence on the response of the cross-line footbridge.

Keywords: cross-line footbridge; multi-scale model; response analysis; train wind

1 INTRODUCTION

Because of the large span and structure flexibility, cross-line footbridge will be affected by a strong train wind when the main line train passes through the station, which will have a serious influence on the comfort and durability of the cross-line footbridge. Therefore, it is very important to simulate the cross-line footbridge in multi-scale by considering the train wind. In recent years, many scholars in China and abroad have conducted a series of studies[1-2] on train wind simulation and its influence on the cross-line footbridge, and have reached a consensus that train wind has a great influence on the cross-line footbridge. For example, Yin Guogao et al. 2014[3] performed a study on the vibration and vibration reduction measures of passenger footbridge in the Xuzhou East Station of Beijing-Shanghai high-speed railway, and found that train wind is the main factor that causes the vibration of passenger footbridge, and proposed feasible vibration reduction measures through the comparative analysis of the measured and calculated results. He Lianhua et al. 2008[4] used the commercial software fluent to simulate train wind induced by

high-speed train passing by Wuhan Station, and the calculation results showed that train wind was concentrated on the front and rear of the train. Song Jie[5] used the cross-line footbridge and station awning as the engineering background to simulate and analyze the train wind pressure distribution and structure responses caused by high-speed train passing by the station, and the results showed that the vibration of structures would increase with the increasing train speed. Song Ruibin[6] used the Finite Volume Method and the dynamic grid technology to establish a dynamic simulation model of high-speed trains, and studied the aerodynamic characteristics of high-speed train running on the bridge. Yang Yijun et al. 2005[7] researched on the cross-line bridge made of steel plate beam and its wind-induced vibration caused by high-speed train passing by the station, and the results showed that it was effective to set small TMD for restraining wind-induced vibration. Although many researches have been made on this aspect, so far, none has reported on the simulation of the cross-line footbridge in multi-scale and accurately analyzed the response of cross-line footbridge by considering the effect of train wind. Therefore, the cross-line footbridge response analysis based on the multi-scale model is studied in this paper.

2 TRAIN WIND SIMULATION

2.1 *Numerical simulation of train wind*

Currently, research on train wind is very mature. Generally, the corresponding flow field of the cross-line footbridge computational fluid dynamics model considering the effect of the main line train is regarded as the three-dimensional incompressible flow field. Train wind turbulent numerical simulation was carried out according to Reynolds averaged N-S control equations[8]. The train wind calculation model built by running a high-speed train was established and meshed using commercial software GAMBIT, and CFD equations were solved by the FLUENT program in this paper. Thus, the entire process of train wind loading on the cross-line footbridge was simulated. Specific contents are available elsewhere[9].

2.2 *Train wind pressure time history curve*

Taking into account the current speed of train passing by the station is about 300–310 km/h, train wind pressure time history loading on the cross-line footbridge was simulated when the main line train is passing by the speed of 310 km/h in this paper, as shown in Figure 1. Figure 1 shows the wind pressure time history curve at the bottom of the cross-line footbridge, which is just above the main line train. It can be concluded from Figure 1 that wind pressure increased rapidly as the train was getting closer to the bridge and wind pressure reached the first negative maximum reduction instantly. This process produced a positive to a negative pressure fluctuation. The wind pressure

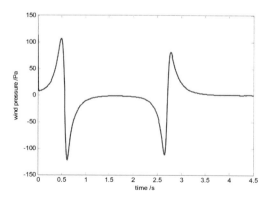

Figure 1. Train wind pressure time history curve in the bottom of the footbridge.

of the footbridge surface decreased rapidly when the train rear passed through the cross-line footbridge. Furthermore, it resulted in a negative after the first positive pressure fluctuation when the entire rear of the train passed by the footbridge entirely with a rapid increase in pressure. The wind pressure of the cross-line footbridge underside also decreased rapidly to zero when the train was getting as far as away from the footbridge. The wind pressure changed greatly when the front and rear of the main line train just passed by the cross-line footbridge, resulting in two pressure pulses. The positive pressure maximum was 240 Pa, and the negative pressure maximum was 245 Pa. The wind pressure was substantially zero when the main line train was just passing by the footbridge.

3 CROSS-LINE FOOTBRIDGE SIMULATION IN MULTI-SCALE

In order to precisely analyze the internal forces of the cross-line footbridge, first, the cross-line footbridge strain time history curve of dangerous spots (composite deformation) should be achieved. Taking into account the precise scope of the strain, the cross-line footbridge was simulated in multi-scale by the sub-model method, because it was relatively coarse to get strain using the housing unit model. The strain at the relevant point would be achieved from the multi-scale finite element solid model.

3.1 *Sub-model theory*

The sub-model method, also known as the cut boundary displacement method or specific boundary displacement method, is the finite element method to get exact solutions of partial regions of the overall model. It is often necessary to consider the detail of the results and locally dense meshing in the finite element analysis. However, a part of the overall structure does not need a very detailed division. In this case, very dense meshing is time-consuming and uneconomical to compute, and very sparse meshing can hardly lead to a good result, whereas the use of the sub-model method can solve this problem perfectly. The sub-model method is based on an analysis of the Saint-Venant principle, and the stress and strain of the structure will change only in the vicinity of the area where the load is applied if the actual distribution load is replaced by the equivalent load. This shows that the stress concentration effect is just in a load centralized area and the sub-model can get a more accurate result if the sub-model boundary is far away from the position of stress concentration. The sub-model method needs to establish the overall crude division and the local meticulous division finite element model of the analysis objectively, and the cutting boundary is divided into the border of the corresponding local detailed model and the overall coarser model. The calculating displacement of the corresponding position of the overall model is the sub-model boundary conditions of the sub-model.

3.2 Cross-line footbridge finite element model in multi-scale

Taking the cross-line footbridge of Zhuzhou West Railway Station in Wuhan-Guangzhou high-speed railway passenger line as the engineering background, the multi-scale footbridge finite element analysis model was established in this paper using ANSYS. The cross-line footbridge of Zhuzhou West Railway Station is the frame structure. The bridge deck is the beam and the slab structure is made of a profiled steel sheet and I-shaped steel beams, and the footbridge pier is the steel tube-reinforced concrete column. The width of the bridge deck is 11.1 m and the span over the main lines is 37.5 m. The cross-line footbridge overall coarser finite element model is shown in Figure 2. The total number of units of the model is 12301. Moreover, the main parts and staircase of the footbridge are established by the SHELL63 housing unit. The steel tube-reinforced concrete column is established by the SOLID45 unit.

By loading the achieved train wind pressure time history on the overall model of the cross-line footbridge, the overall structure was analyzed. The calculation time step is 0.01 s, and the total number of time steps is 850. Then, the dynamic response time history curve of the overall structure in the vertical direction was achieved, as shown in Figure 3.

The housing unit I-beam displacement response time history of the cross-line footbridge at the mid-span corresponding position was extracted as the next sub-model boundary conditions. Then, the

Figure 2. Overall and local model of the cross-line footbridge.

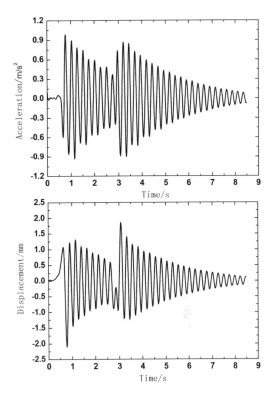

Figure 3. Cross-line footbridge dynamic response of vertical vibration by train wind.

I-beam dangerous area of the cross-line footbridge in the mid-span was selected for a further detailed analysis, and its solid model was established by the SOLID45 unit, as shown in Figure 1.

The total number of units for the solid model was 2080. The I-beam model cross-sectional dimension was consistent with the shell element overall model, and the longitudinal length was 4 m. The I-beam displacement response time history of the cross-line footbridge at the corresponding mid-span position in the first step was applied on both ends of the solid model as sub-model boundary conditions. In addition, applying the same external load on the overall model at the corresponding position, and taking the same time step and the number of load steps of the overall model for analysis, more accurate results of the I-beam dangerous area were obtained.

4 RESPONSE ANALYSIS OF DANGEROUS SPOTS

Based on the multi-scale simulation results of the cross-line footbridge, the footbridge stress-strain state of the danger area was obtained accurately.

In view of the multi-axial fatigue damage suffered by the structure, the composite deformation was used to determine the dangerous spots of fatigue. The maximum composite deformation was obtained using Von-Mises criteria, which was regarded as the dangerous spots of fatigue in this paper. First, the approximate area of the dangerous spots of fatigue was determined by the relatively coarse housing unit model. Figure 4 shows the maximum composite deformation distribution of the finite element model determined by the Von-Mises criterion (maximum strain response time t =1.0 s). After the approximate area of the dangerous spots of fatigue was determined, the accurate sub-model was established for the strain field analysis to obtain the exact value of the strain. From Figure 4, the area of the maximum composite deformation or the approximate area of the dangerous spots of fatigue was in the beam flange and web connection just above the passing trains when main line trains were passing by with a high speed. So, the sub-model at this area can be used for the further analysis of the strain.

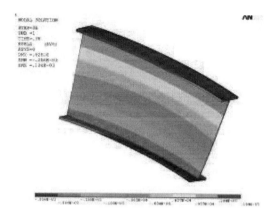

Figure 6.   Normal strain distribution in the Z-direction.

Figure 4.   The Von-Mises composite deformation distribution of the overall model.

Figure 5.   The Von-Mises composite deformation distribution of the I-beam.

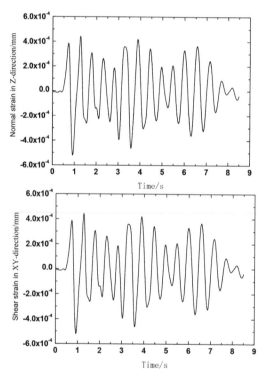

Figure 7.   Strain time history curve of dangerous spots.

Figure 5 shows the strain distribution of the local meticulous division finite element model at this moment. From Figure 5, the composite deformation of flanges and web junction is larger than other areas in this beam. Taking into account the fillet welds of the web and flange of the I-beam, it is the most likely area of brittle fracture. So, the spots in the web and flange connection of the I-beam were chosen as the maximum dangerous stress spots.

Through the analysis of normal strain and shear strain of the I-beam in 3 directions, it was shown that the normal strain in the Z-direction was the main strain, which is the section along the direction of the bridge, as shown in Figure 6.

In order to clearly analyze the stress time history curve of the dangerous spots of fatigue, the strain time history curve of these dangerous spots by considering the effect of train wind was obtained, as shown in Figure 7. It can be shown that the normal strain of the dangerous spots in the Z-direction was obviously larger than the strain in other directions.

5 CONCLUSIONS

Based on the present analysis, the following conclusions can be drawn: a) the area of the maximum composite deformation or the approximate area of the dangerous spots of fatigue was in the beam flange and web connection just above the passing trains when main line trains were passing by with a high speed. b) The composite deformation of flanges and web junction is larger than other areas in this I-beam. Taking into account the fillet welds of the web and flange of the I-beam, it is the most likely area of brittle fracture. c) The vertical vibration of the cross-line footbridge primarily considers the effect of train wind.

ACKNOWLEDGMENTS

This work was supported by the National Natural Science Foundation Program (51108363) and the Natural Science Foundation of Hubei Province of China (2014CFB849).

REFERENCES

He Lianhua, Zhao Pengfei et al. 2008. Study on simulation of train wind induced by high-speed train passing by Wuhan station. *Railway Engineering* 8: 108–110.

Jiang Li, Liu Jia et al. 2014. Analysis on vibration responses of footbridge induced by wind caused of main line train pass by. *Journal of Wuhan University of Technology* 36(6): 79–83.

Song Jie. 2012. Numerical simulation of roof passing by high speed railway train and vibration analysis. *Beijing Jiaotong University*.

Song Ruibin. 2011. The numerical study of aerodynamic characteristics of each other when high-speed trains move on bridge. *Central South University*.

Sun Daoyuan. 2011. Analysis of dynamic characteristics of a light flexible steel footbridge. *Sichuan Building Science* 37(3): 60–64.

Wang Shaoqin, Xia He et al. 2012. Nonlinear coupling vibration analysis of wind load-train-long-span bridge system. *Journal of Beijing Jiaotong University* 36(3): 36–40.

Xia He. 2010. *Traffic induced environmental vibrations and controls*. Beijing: Science Press.

Yang Yijun, Wang Qinyun et al. 2005. Researched of wind-induced vibration of steel plate girder overhead bridge and measures against vibration thereof due to high-speed running train. *World Bridges* 1: 46–49.

Yin Guogao, Zhang Gaoming et al. 2014. Vibration study of passenger footbridge in Xuzhou east station of Beijing-Shanghai high-speed railway. *Building Structure* 44(1): 53–57.

Information engineering and its application

Architectural, Energy and Information Engineering – Sung & Chen (Eds)
© 2016 Taylor & Francis Group, London, ISBN 978-1-138-02791-6

The study on the influence of a computer-aided writing instruction on the aspects of writing performance

Z. Ma
Beijing Wuzi University, Beijing, China

ABSTRACT: The aim of the study is to focus on the effectiveness of a computer-based teaching approach in writing, and examine its influence on different writing aspects. The comparative research was conducted on students who were randomly divided into the Experimental Group and the Control Group. With Pretest result as assurance of reliability of the whole experiment, a new computer-assisted pedagogy via the NHCE online system was employed in teaching EG in language labs, while the conventionally accepted writing mode was dealt with CG subjects. After the operation, the Posttest writing was set, and data on mistakes made in different aspects of writing were collected and manipulated in Spss. The results of statistical analysis shows the better effects of the new approach compared with the traditional one, and at the linguistic level, the lexical and morphological aspects were better improved. Finally, interviews and investigations on the backwardness of discourse aspect were carried, and related discussions are presented.

Keywords: Effectiveness; Computer-aided pedagogy; Aspects of writing

1 INTRODUCTION

In recent years, under the prevailing college English reform, researchers in Chinese educational and study organizations have been trying every possible strategy to optimize learning and teaching approaches, in order to cultivate qualified intellectuals with comprehensive capabilities. Among these approaches, investigation and innovation of college English writing methods have been widely conducted, since writing proficiency is regarded as critical ability for non-English majors to compose professionally. Despite the ability to master all five languages, students' performance on writing is least satisfactory, ranking the lowest in Cet 4 (College English Test, Band 4). The underlying reason is admittedly attributed to the traditionally adopted writing methodology that trapped both instructors and learners. Usually, instructors assign 3–4 writing tasks each term, directing students to utilize and follow newly learned vocabularies, sentences and discourse structures to write. Yet, within a given format of structures, students find certain assignments dull enough, hence losing their interest in writing. They would in turn download and copy online to cope with these tasks. During the time of evaluation, teachers select several well-shaped papers to share with students, which definitely dis-

courage the learners' creating initiative and enthusiasm. Likewise, they feel that the teachers' limited feedback is insufficient to stimulate writing urges. Consequently, the learners' writing competence is gradually deteriorating.

In order to enhance language learners' writing competence, this study attempts to examine the effectiveness of a new computer-assisted writing approach via the NHCE online learning system, and to explore its influence on different aspects of writing. The NHCE online system (New Horizon College English) is developed by Foreign Language Teaching and Research Press in Beijing, China. It is the assisting study system of NHCE Textbooks. Multiple interactive functions of this system can provide the possibility of constructing knowledge in a collaborative way. In language labs where the campus-connected NHCE online system is available, the subjects from the EG are required to follow a new writing instructing procedure, such as the Preparation Stage, the Writing Stage, and the Evaluation Stage. For the same 14-week semester, the subjects from the CG are required to undergo the traditional writing approach. At the end of the "experiment, a Posttest was conducted, and the data collected was analyzed using Spss to compare the detailed performance of the subjects, and corresponding discussions as well as conclusions are presented.

2 LITERATURE REVIEW

Kelly (1991), as Constructivism advocate, proposed that we look at the world through mental constructs or patterns which we create. According to Brown, J.S et al. (1978), the sense of things does not exist separately from us, but depends on our active construction. Under the present foreign languages learning condition, as the operational technological support, the modern highly advanced computer-net technology paves ways for realization of constructivism hypothesis. Collaborative Learning stresses on sociality, openness, diversity, creativity as well as procedural learning. Besides, all-round students-centered communications among teachers and students in multi-level are strongly advocated. CALL (Computer-Aided Language Learning) applications include "guided drill and practice exercises, computer visualization of complex objects, and computer-facilitated communication between students and teachers" (CAI, 1998).

3 RESEARCH METHODOLOGY

This study examines on the effectiveness of a new computer-based teaching method, and subjects' specific writing performances on morphology, lexicon, syntax and discourse, based on the frequency of mistakes made and the general score in each aspect, so as to pose questions as follows: generally speaking, can the computer-assisted new writing instruction prove more effective than the conventional teaching method, as far as the frequency of mistakes is concerned? Where do the high frequencies of mistakes lie? Does each item of Posttest show significant difference? How well and poor do subjects in the CG and the EG perform as far as the 4 items are concerned?

The research first selected randomly students of equal learning level as the CG and the EG, with the same number and writing instructor. Before the experiment, they took a Pretest, and the results indicated that they grasped the equivalent level of writing proficiency, so a conclusion was drawn that any difference found in the Posttest can be explained by the treatment itself. Then, for a whole term, a new English writing technique and a traditional instruction were administered in language labs and classrooms. By the end of the study, a Posttest was conducted to examine the results of the experiment by two raters following the same standard, in which the students from each group were required to write under the same directions. According to *The Classified Statistics of Mistakes in Chinese students' English Writing* (Yu, 2004), mistakes in morphology, lexicon, syntax and discourse were listed separately. Lastly, all the collected statistics were processed and analyzed in response to the research questions, followed by the corresponding discussions and explanations.

4 DATA COLLECTION AND DISCUSSION

In this section, the statistics of the frequency of mistakes in the Pretest and the Posttest, Descriptive Statistics and Multivariate Tests in the Posttest of the EG and the CG are analyzed.

In Table 1, for the EG and the CG in the Pretest, the frequency of mistakes is almost equal, with the overall number of 280 and 276, guaranteeing the same-level writing proficiency of the EG and the CG.

In Table 2, the statistics of the frequency of mistakes in the Posttest are presented, ranging from morphology, lexicon, syntax to discourse.

Generally, it is obvious that after training on computer-based learning mode, the overall mistakes occurring in the Posttest of the EG are apparently less than those occurring in the Posttest of the CG, that is 407:529, with respective aspects bearing the same condition, that is, 97:123 in morphology, 69:113 in lexicon, 120:151 in syntax, 121:142 in discourse, which show the effectiveness of the computer-aided college English writing instruction. Then, as far as the linguistic level is concerned, mistakes made usually gathers more in syntax and discourse, accounting for 29% and 30% of the total mistakes for the EG and 29% and 27% of the total mistakes for the CG. The values tells that for students' feedback and evaluation, more mistakes were detected in morphology and lexicon, while those in syntax and discourse were neglected to some extent.

In Tables 3 and 4, the Descriptive Statistics of the EG and CG are given, and the Multivariate Tests were run to judge whether there exists any significant difference among the four writing aspects within each group.

In Table 3, multivariate tests show that p = 0.008, less than 0.05; therefore, there exists a significant difference among the four aspects of writing competence. Since for the mean score of each item, 22.35 > 21.27 > 20.38 > 20.35, the mastery degree of each aspect is: lexicon > morphology > syntax > discourse, which conveys that the command of lexicon and morphology of the EG is better than that of syntax and discourse. Table 4 provides the data for the CG.

In this table, since p = 0.040 in the multivariate tsts, less than 0.05, a significant difference within the aspects of the CG is also observed. The mastery degree of each aspect is: lexicon > morphology > discourse>syntax, with 20.65>20.27>19.54>19.19, which suggests the students' command of lexicon and morphology is better than that of discourse and syntax.

The statistics in Tables 3 and 4 reconfirm the conclusion in Table 2, that is, the students' performance

Table 1. Statistics of the frequency of mistakes in the Pretest.

Linguistic level	Sublevel	Pretest of the EG		Pretest of the CG	
		Frequency of mistakes	Percentage	Frequency of mistakes	Percentage
Morphology	Article, Case, Tense, Adjective Comparative and Superlative, Derivation	71	25	78	28
Lexicon	Choice and Collocation of words	51	18	42	15
Syntax	Ellipsis, Agreement, Word Order, Avoidance, Repetition, Passive sentence	70	25	75	27
Discourse	Unclear Identification of First Person, Overuse of Second Person, Expressive Indirection, Coherence	88	31	81	29
Overall		280	100	276	100

Table 2. Statistics of the frequency of mistakes in the Posttest.

Linguistic level	Sublevel	Pretest of the EG		Pretest of the CG	
		Frequency of mistakes	Percentage	Frequency of mistakes	Percentage
Morphology	Article, Case, Tense, Adjective Comparative and Superlative, Derivation	97	24	123	23
Lexicon	Choice and Collocation of words	69	17	113	21
Syntax	Ellipsis, Agreement, Word Order, Avoidance, Repetition, Passive sentence	120	29	151	29
Discourse	Unclear Identification of First Person, Overuse of Second Person, Expressive Indirection, Coherence	121	30	142	27
Overall		407	100	529	100

Table 3. Descriptive statistics and Multivariate Tests in Posttest of EG.

	Mean	Std. Deviation	N		
Morphology	21.27	2.631	26		
Lexicon	22.35	1.623	26		
Syntax	20.38	2.714	26		
Discourse	20.35	3.599	26		
	Multivariate Tests				
Effect	Value	F	Hypothesis df	Error df	Sig.
Posttest					
Pillai's Trace	.396	5.027[a]	3.000	23.000	.008
Wilks' Lambda	.604	5.027[a]	3.000	23.000	.008
Hotelling's Trace	.656	5.027[a]	3.000	23.000	.008
Roy's Largest Root	.656	5.027[a]	3.000	23.000	.008

Table 4. Descriptive statistics and Multivariate Tests in Posttest of CG.

	Mean	Std. Deviation	N
Morphology	20.27	2.146	26
Lexicon	20.65	2.134	26
Syntax	19.19	2.638	26
Discourse	19.54	2.596	26

		Multivariate Tests			
Effect	Value	F	Hypothesis df	Error df	Sig.
Posttest					
Pillai's Trace	.299	3.269[a]	3.000	23.000	.040
Wilks' Lambda	.701	3.269[a]	3.000	23.000	.040
Hotelling's Trace	.426	3.269[a]	3.000	23.000	.040
Roy's Largest Root	.426	3.269[a]	3.000	23.000	.040

on syntax and discourse needs to be improved, which will be the future topic of this research. However, from the sums of mistakes collected in the Posttest, the students from the EG made less mistakes than those from the CG, a sign indicating that on the reliable condition of the Pretest, with the original same writing proficiency level, the EG students achieved more scores than those of the CG after the new treatment in language labs was conducted. Therefore the effectiveness of the computer-aided writing mode is observed.

5 CONCLUSIONS

From the research questions raised above, after the evaluation of the whole experiment and data analysis, it can be concluded that the new computer-aided writing instruction can better enhance student's writing ability. For the specific aspects ranging from syntax, discourse, lexicon and morphology, it is shown that students perform better in lexicon and morphology than in discourse and syntax. Even both groups have done better in the lexical aspect, and those from the EG on the new treatment exhibit more progress than the subjects from the CG. This study basically proves the effectiveness of the innovative teaching method. Yet, in terms of different performances in various aspects of writing, more reviews and reflections are necessary. As for the solution to students' poor performance on the content of writing, according to Krashen's Second Language Acquisition Theory (1982), comprehensible input is one of the true elements for promoting second language acquisition. Generally speaking, only by reading extensively and effectively can students accumulate knowl-

edge, absorb information in the writing process, and end up as successful writers. However, with the interviews conducted among EG subjects, students confided their problems in reading online. For enormous materials stored, they were not capable of handling and taking in what they need, which could be attributed to their poor language foundation and limited practice of critical thinking, since reading online "requires the users' best reasoning which draws on pertinent background knowledge and specific information literacy skills for interpreting and evaluating information" (Hancock, 1994). So, for the improvement of students' performance on online reading and writing, a similar practice needs to be carried for later semesters, with more definite and efficient measures and assisted directions.

REFERENCES

Brown, J.S., Collins, A. & Duguid, P. 1978. Situated cognition and the culture of learning. *Educational Researcher* 18(1): 32–42.

CAI. 1998. Encarta Encyclopedia 99 [CD-ROM]. New York: Microsoft.

Hancock. Alternative Assessment and Second Language Study: What and Why? 1994. http: //www.cal.org/resources/digest/hancoc01.html.

Kelly, G.A. 1991. *The psychology of Personal Constructs: Volume one-A theory of personality*. London: Routledge.

Krashen, S.D 1982. *Principles and Practice in Second Language Acquisition*. Oxford: Pergamon Press.

Mingli, Yu. 2004. *Language Transfer and Second Language Acquisition—Review, Reflection and Research*. Shanghai: Shanghai Foreign Language Education Press. (in Chinese).

Architectural, Energy and Information Engineering – Sung & Chen (Eds)
© 2016 Taylor & Francis Group, London, ISBN 978-1-138-02791-6

On learning community based on network applied in distance education

N. Liu
Foreign Languages School, Tianjin University of Technology and Education, Tianjin, China

C.X. Wang
Information School, Tianjin University of Technology and Education, Tianjin, China

ABSTRACT: Modern distance education enables students to learn the lesson with no limitation of time and space. Based on the network, learning community constructed by students provides learners and teachers a new type to organize their learning and teaching. Problems and advantages are analyzed, with the suggestions given in the conclusion.

Keywords: distance education; learning community; network-based

1 INTRODUCTION

In the modern distance education, teachers and students can interact instantly or non-instantly, with no limitation of time and space. This is the most obvious difference between modern distance education and traditional education. In addition, this is how modern distance education can overwhelm others. It is admitted that distance education based on network provides learning community more opportunities. Learning community based on network ensures distance education more interactive. These two aspects promote and reinforce the effects of each other.

2 DISTANCE EDUCATION

Holmberg declared that distance education contains various education levels as well as various teaching and learning methods. The teaching and learning is not under the direct instruction of teachers. In other words, students do not sit in the classroom or live on campus. Therefore, distance education possesses the following features.

First, teaching and learning is not face to face between teachers and students. Second, education companies devote themselves in it. Third, the technological multi-media is adopted in distance education. Information technology connects teachers and students. It carries the education material. Fourth, the communication is between teachers and one single student. Fifth, there are no more chances of communication among students. Sixth, distance education has the tendency to become an industry, which is totally different from traditional education.

3 LEARNING COMMUNITY

Learning community is a new type of organization in which teachers and learners can communicate and cooperate more than that in the traditional class. Members of the community join in the learning process, together with the realization of self-respect, self-reflection, and social being.

Dewey was the first to introduce "community" to the field of learning and teaching. He assumed that the society is a community of thoughts and emotions, which are based on the interpersonal communication [1]. The concept of community is used in education to emphasize the social interaction among learners. Through social interaction and the power of peer models, learners understand the knowledge system, its structure and the development. In this way, learners can acquire some effective cognitive strategies and develop better ability of learning. The concept of "community" in the field of education evolves into a new concept, that is, "learning community".

Learning community consists of learners (students) and learning assistants (teachers). All of them have the same learning task to finish, with the same aim to make progress altogether. In the learning community, to influence and to be influenced among the members is emphasized through the approaches of inter-personal communication, sharing the learning resources [2]

According to the constructivism, acquiring knowledge is not a simple process of receiving or copying, but a constructing process. One has a unique way to understand the world and to construct his own sense [3]. The problem is that the sense constructed by oneself is not ensured to be exactly and totally correct and reasonable. In this sense, it is urgently necessary to share and communicate in a community. That is why the reasonable knowledge needs negotiation. Thus, one's independent thinking and dealing with knowledge is not enough. It is a complicated and long-term process to construct a sensible knowledge system.

Two types of learning community are included: Internet-based virtual type and blended type with virtual and practical features.

To build and manage a learning community, two functions are satisfied.

The first is to consolidate the sense of social being. Constructing the learning community is an important way to satisfy the need of building learners' self-respect and belongingness [4]. In the inner structure of the learning community, members can have a strong sense that they belong to the same group, take the same tasks, obey the same rule and have the same sense of value. This kind of belongingness and the respect gained from other members can attract more members to join and be more active. In this way, members have more interests and supports in learning, which is not a personal affair any more.

The second is to share the knowledge resources. Learners can communicate with both learners and assistants (teachers). In the way of constructing and sharing, learners can get touch with new information as well as different perspectives to deal with problems [5]. They are motivated to have more self-reflection and reorganize one's own knowledge system.

4 ADVANTAGES OF LEARNING COMMUNITY BASED ON NETWORK APPLIED IN MODERN DISTANCE EDUCATION

The advantages of modern distance education based on learning community are as follows.

First, students can share free education resources within the community. With the development of modern multi-media technology and network, distance education possesses great conditions to develop itself. Students have convenient devices, such as cell phone, ipad, and portable computer. It makes it possible that students connect to the Internet, search for information needed, and interact with fellow students and teachers. Sharing with others becomes a key way of learning and achieving knowledge. Learning is not a way of individual behavior, but a method of sharing. It reinforces the relationship among students and distance is not a "distance" any more.

Second, peer work can be realized on the network within the community. The previous distance education lacks communication among students. Students go through their learning individually.

Within the learning community, students form a special type of relation, which is both strong and loosely tied. Peer work is realized in this way. Students can know more about the level of other students and be clear about their own gap from others. Thus, the study is promoted.

Third, teachers can know about students' learning levels in the organization unit of community. Teachers can join in the students' learning community as well. On the network, besides the distance learning environment, the learning community provides a new space for teachers to know about students. Teachers can grasp more information about students' interests, character, motivation, purposes and learning methodology. In this way, teachers can adjust the teaching mode and make it adapted to students. Moreover, teachers can judge and assess the teaching method and material so as to satisfy students' needs and meet the social requirements.

5 CONSTRUCTING A LEARNING COMMUNITY

How to build a learning community is another important problem concerned by the educators. It is proved that it is a gradual procedure to satisfy some regular conditions. Furthermore, it can also be allowed to be adjusted for the common benefits of all the members.

First, possessing a common purpose is the basic condition to ensure the running of the learning community. The common purpose is the starting point and also the final aim to build a community [6]. The individuals should first get it clear why they want to study together, so that it is possible to organize activities and manage the community. This common purpose unites all the separate individuals to be in unity. It makes individual learning behavior to be a team work.

Second, the teacher plays an important role in building and running the learning community. As learners are individuals in the learning process, it is necessary for the teacher to organize a certain amount of learners with a common purpose and share features together to form a community. This is the starting step to build a community.

With the participation of the teacher, the community can be organized routinely, formally and harmoniously. What is more, the teacher can also assist learners to solve problems, orient directions, accelerate and advance the course. With the confidence and understanding system built among members, the community can be mature and developed.

Third, gaining the sense of "social being" is the most important value in running the learning community. In the learning community, the learning process is based on the cooperation and communication within the community among members. Learning is never a separate and individual process. Each member should share the resources, express the ideas, and listen to each other. All the members should respect each other and admit the being of each other as well. The social being identity is magnified to be big enough to achieve self-respect and realize self-value. In this way, the efficiency and effect of learning is promoted.

Fourth, learners devote one's emotion and sensation into the community and accept the counterparts from other members. This is an invisible clue, like a backbone, going through and sustaining the whole community. Learners understand, accept and admit the perspective, expression and emotion of other members. The sense of belongingness is built gradually with the trust on and from the community. Without it, the community cannot develop itself with inner motivation.

6 EVALUATION AND SUGGESTIONS

The application of the learning community in the above perspective enlightens the researchers to consider how an Internet-based learning community can be well constructed and managed in practice. Some suggestions are given in the following as the reference.

First, how the teacher plays the role of guiding and monitoring decides whether the start of the learning community can be successful. Teacher is not a lecturer any more, but a bridge and one of the participants. After the guidance of the teacher for a period of time, the management of the community can be ensured formally and seriously. Furthermore, the teacher can help to develop the role of some key men and allow more responsibilities to run the community done by students themselves.

Second, students should possess a stronger sense to cooperate with partners and to explore new information. Students are not the blind followers and recipients, but are the controllers, as the core of the community. They decide whether the community can be sustained. The benefits of constructing a learning community should be presented to the students. Based on it, students are expected to be motivated enough in the process of managing a learning community. Each member contributes to it, and the role of each one is reflected and acknowledged. It is admitted that some of the students are "invisible" and cannot make their voice heard. Teachers should pay equal attention to this part of students and exert their potential by creating more chances. All the members should respect each other and listen to each other. In this way, all can achieve equal chances of development.

Third, more activities, tasks and principles that accord with the need of the community and unite all the members should be designed. On the one hand, members can pay more attention to the community; on the other hand, the community can serve better for the students.

7 CONCLUSIONS

The research puts the conception and construction of learning community in distance education. It is proved that the network-based learning community influences the effects and improves the ability of learning in distance education.

The correlation of the learning community with distance education can make the education purpose clear and make the activities organized and oriented with cooperation-directed goals. The teacher can take more responsibilities to monitor the whole process of constructing and managing the learning community.

Internet plays an important role, as the supported platform in running the community. The online community overcomes the constraints of time and space, and provides abundant resources of information collection.

However, it cannot be ignored that some problems exist. It is expected that a better solution will be found in future research.

ACKNOWLEDGMENTS

This paper is part of the fruit of the following programs: young teachers' program of the Education Ministry in Educational Science plan "On production and acquisition method of teachers' practical knowledge in vocational colleges" (EJA140376) and the academic research program of Tianjin University of Technology and Education (SK14-17).

REFERENCES

Dewey J. Democracy and Education. An Introduction to the Philosophy of Education, 2nd ed., Beijing: People's Education Press, 2001, p. 32.

Feng Pan. "The difficulty and solution of cooperative learning" in Social Work. 2012 (11): 49–51.

Guangxin Wang, Chengjie Bai. "The formation and development of virtual learning community" in Electronic Education Research. 2005 (1).

Jianwei Zhang. On Virtual Learning Community [DB/OL]. http://www.being.org.cn/sikao/netgtt.html.

Man cur Olson, The Logic of Collective Action, Cambridge, Mass: Cambridge University Press, 1965: p. 23.

Shuhong Liang. "Advantages and strategies of PBL in college education" in China Electronic Power Education. 2012 (35): 26–28.

Architectural, Energy and Information Engineering – Sung & Chen (Eds)
© *2016 Taylor & Francis Group, London, ISBN 978-1-138-02791-6*

An optimization segmentation approach for high resolution remote sensing image

L. Wu & Z.S. Liu
Chongqing Communication Institute, Chongqing, China

ABSTRACT: One of the indispensable prerequisites for remote sensing image processing is image segmentation. The approach presented in this paper aims for an optimization segmentation effective and adaptable to high resolution remote sensing images. This is achieved based on region growing and Particle Swarm Optimization (PSO). The position vector of each particle corresponds to a set of segmentation parameters which can be used for segmentation based on region growing. A new quantitative image segmentation evaluation function which calculates weighted sum of global intra-region homogeneity and global inter-region heterogeneity of segmentation result is constructed and used as a fitness function for PSO. The PSO tries to find the near-optimal segmentation result with the maximum fitness value by evolving the initial particle swarm through iterations. The proposed approach has been applied to ZiYuan-3 (ZY-3) high resolution remote sensing images; the experimental results demonstrate its effectiveness and adaptability in the near-optimal segmentation result selection.

Keywords: high resolution remote sensing image, Image segmentation, Region growing, Particle Swarm Optimization (PSO), Quantitative image segmentation evaluation

1 INTRODUCTION

In recent years, substantial high resolution remote sensing images are increasingly applied in the field of environmental protection, land survey, disaster assessment, military target monitoring, etc. Nowadays, object-oriented image processing techniques have received great attention and demonstrate more advantages in remote sensing change detection, object detection, etc. An indispensable step for object-oriented image processing is image segmentation.

A large variety of image segmentation algorithms were developed during the last 20 years. Due to the complexity of the high resolution remote sensing image scene, two main groups of segmentation algorithms (boundary-based and region-based) were widely used to segment the high resolution remote sensing images. The boundary-based algorithms detect object contours explicitly by using the discontinuity property, and the region-based algorithms locate image regions explicitly according to the similarity property (Zhang 1997). The region-based algorithms are less sensitive to texture and noise which is a significant advantage in high resolution remote sensing image segmentation and the implementation of multi-level segmentation is easier with region-based technique as long as the heterogeneity tolerance is increased,

consequently, the region-based algorithms will be the mainstream in the foreseeable future (Carleer et al. 2005).

A quintessential region-based algorithm is region growing, the procedure starts at each pixel which is called one-pixel region and in numerous subsequent steps, and smaller regions are merged into bigger ones, if the two adjacent regions satisfy the merging criterion (Burnett et al. 2003). The determination of appropriate segmentation parameters is the most arduous task and a parameter optimization problem in essence; it is still a challenge and a hot research topic for researchers.

PSO is a new evolutionary computing technique based on swarm intelligence of bird flocks (Shi 2004). It has some intelligent properties such as adaptation and self-organizing, and has the strong ability to search for the optimal solutions for optimization problems. In image processing fields, PSO has been successfully used to solve the problem of thresholding-based segmentation (Zheng et al. 2009, Mohsen et al. 2011, Gao et al. 2013).

The goal of this paper is to propose an optimization segmentation approach for high resolution remote sensing image based on region growing and PSO. A new quantitative evaluation function will be proposed and used as a fitness function for PSO. The PSO will try to find the near-optimal segmentation parameter values that can help to obtain a near-optimal segmentation result for a given image according to the proposed fitness function.

The remainder of this paper is organized as follows. Section 2 provides a brief review on related work. The new fitness function is given in Section 3. The details of the proposed method are described in Section 4. The experimental results and analysis are shown in Section 5. Finally, the conclusions are stated in Section 6.

2 A BRIEF REVIEW ON RELATED WORK

2.1 Region growing

In this paper, the region growing algorithm proposed by Wu et al. (2014) is selected as the segmentation algorithm, in which the region merging procedure starts with one-pixel regions. And smaller regions are merged into bigger ones according to the increase of heterogeneity in numerous subsequent steps. There are four major steps of this algorithm: segmentation initialization, pixel couples sequence construction, merging criterion definition and region growing. The segmentation results are mainly affected by three parameters: the scale T, the weight of spectral heterogeneity $w_{spectrum}$ and the weight of compactness $w_{compact}$. Scale is a measure of the maximum size of regions in segmented image (i.e. the larger the scale, the bigger the segments), and is the stop criterion for segmentation process. However, it is difficult to get optimal segmentation parameters by using conventional methods.

2.2 Particle swarm optimization

PSO is initialized with a group of random particles and then searches for optima by updating generations. Each particle moves stochastically in the direction of its own best previous position and the whole swarm's best previous position. The performance of each particle, i.e. how close the particle is from the global optimum, is measured using a fitness function which depends on the optimization problem.

Each particle possesses three basic characters: the current position, the personal best position and the current velocity, the personal best position is the best position that the particle has visited so far. For particle j, they are denoted as X_j, P_j and V_j, respectively. And the best position discovered by the whole swarm is denoted as Pg.

Suppose that the search space is M-dimensional, the vector X_j and V_j are constructed such as:

$$X_j = \left(x_{j1}, x_{j2}, ..., x_{jM} \right) \tag{1}$$

$$V_j = \left(v_{j1}, v_{j2}, ..., v_{jM} \right) \tag{2}$$

And they are manipulated according to the following equations:

$$V_j^{k+1} = w \times V_j^k + c_1 \times rand() \times (P_j^k - X_j^k) \\ + c_2 \times rand() \times (P_g^k - X_j^k) \tag{3}$$

$$X_j^{k+1} = X_j^k + V_j^{k+1} \tag{4}$$

where c_1 and c_2 are acceleration coefficients; k and k_{max} are the current iterative time and the maximum iterative time, respectively; w is the inertia weight; w_{max} and w_{min} are the maximum and minimum value of w, respectively; $rand()$ is the random number with uniform distribution $U(0, 1)$.

3 CONSTRUCT A NEW FITNESS FUNCTION

In order to obtain optimal segmentation result of high resolution remote sensing image based on region growing and PSO, we need a quantitative segmentation evaluation criterion, i.e. a fitness function. Zhang et al. (2008) summarized a large variety of segmentation evaluation criteria and described them in more depth. These evaluation criteria are quantitative, objective and do not require segmentation results to be compared against a reference segmentation. This ability not only enables evaluation of any segmented image, but also enables the unique potential for self-tuning. But instead single evaluation criterion is difficult to fully evaluate the segmentation quality; consequently, a combination of two or more evaluation criteria is quite frequently used in practice.

For this study, a new fitness function is constructed. It unifies the global intra-region homogeneity and the global inter-region heterogeneity evaluation criteria and enables them to counteract each other nicely and provide a reliable overall segmentation quality evaluation. Firstly, the standard deviation of regions in each spectral band is calculated and used to represent the global intra-region homogeneity. Secondly, the mean value of regions and the mean value of the full image in each spectral band are calculated and used to represent the global inter-region heterogeneity (Shortridge 2007). The calculation formulas are defined as follows:

$$H = \sum_{j=1}^{B} \left(w_j \left(\sum_{i=1}^{N} a_i \sigma_{ji} \right) \middle/ \sum_{i=1}^{N} a_i \right) \tag{5}$$

$$Y = \sum_{j=1}^{B} \left(\frac{w_j \left(N \sum_{i=1}^{N} \sum_{k=1}^{N} r_{ik} \left(\mu_{ij} - \mu_j \right) \left(\mu_{kj} - \mu_j \right) \right)}{\left(\sum_{i=1}^{N} \left(\mu_{ij} - \mu_j \right)^2 \right) \left(\sum_{i=1}^{N} \sum_{k=1}^{N} {}_{i \neq k} r_{ik} \right)} \right) \tag{6}$$

where H is the global intra-region homogeneity of the full image; Y is the global inter-region heterogeneity of the full image; B is the total number of image spectral bands; N is the total number of regions in the segmented image; w_j is the weight of spectral band j; a_i is the area of region i; σ_{ji} is the standard deviation of region i in spectral band j; μ_{ij} is the mean value of region i in spectral band j; μ_j is the mean value of the full image in spectral band j; r_{ik} is a measure of spatial contiguity of regions i and j, $r_{ik} = 1$ if region i and region k are neighboring, otherwise $r_{ik} = 0$. In Equation 5, combining with area factor that gives more weight to the regions of large area and avoids the instability which caused by the regions of small area. The smaller the H, the higher the global intra-region homogeneity. Likewise, the smaller the Y, the higher the global inter-region heterogeneity.

In order to make the fitness value objectively and accurately reflects the image segmentation quality, we must achieve the balance in the contradiction between the global intra-region homogeneity and the global inter-region heterogeneity. Consequently, they should be merged, and each of them is normalized. The new fitness function is denoted as $Fit(H, Y)$ and defined as:

$$F(H) = (H_{max} - H) / (H_{max} - H_{min}) \quad (7)$$

$$F(Y) = (Y_{max} - Y) / (Y_{max} - Y_{min}) \quad (8)$$

$$Fit(H, Y) = \beta \times F(H) + (1 - \beta) \times F(Y) \quad (9)$$

where $F(H)$ and $F(Y)$ are the normalized global intra-region homogeneity and global inter-region heterogeneity, respectively; $Fit(H, Y)$ is the weighted sum of $F(H)$ and $F(Y)$; $0 < \beta < 1$, β is the weight of $F(H)$. The fitness value of $Fit(H, Y)$ is a measure of the position of each particle. Obviously, the higher the $Fit(H, Y)$, the smaller the H and Y, i.e. the better the segmentation quality, and vice versa.

4 THE PROPOSED METHOD

This section develops an optimization segmentation method of high resolution remote sensing image based on region growing and PSO, the new quantitative evaluation function proposed in section 3 will be used as a fitness function for PSO. The new method will be named as RGPSO. The algorithm of PSO, in RGPSO, tries to find the near-optimal segmentation parameter values that can give us a near-optimal segmentation result.

4.1 Representations of PSO

One of the key issues in designing a successful PSO algorithm is the representation step, i.e. finding a suitable mapping between the problem to be solved and particle swarm. As described in Section 2.1, three segmentation parameters dominate the segmentation results; consequently, a single particle represents three segmentation parameters. That is, the vector X_j and V_j are reconstructed such as:

$$X_j = \left(x_{j1}, x_{j2}, x_{j3} \right) \quad (10)$$

$$V_j = \left(v_{j1}, v_{j2}, v_{j3} \right) \quad (11)$$

where x_{j1}, x_{j2}, x_{j3} refer to the segmentation parameter T, $w_{spectrum}$ and $w_{compact}$ of particle j, respectively.

4.2 RGPSO method

Flow chart of RGPSO is shown in Figure 1 and the procedure of RGPSO is described as following:

Step 1: RGPSO starts by randomly initializing particle swarm; each particle contains three segmentation parameters.

Step 2: The target image is segmented into image regions using the segmentation parameters of each particle separately, and the fitness value is calculated for each particle.

Step 3: Compare the current fitness value of each particle with the fitness value of its personal

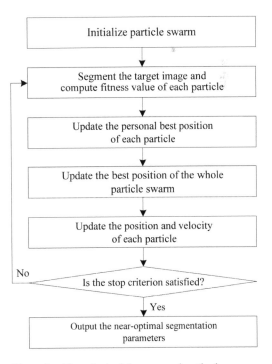

Figure 1. Flow chart of the proposed method.

411

best position. If the former is better, then set the current position of each particle as its personal best position.

Step 4: Compare the current fitness value of each particle with the fitness value of the whole swarm's best position Pg. If the former is better, then set the current position of each particle as Pg.

Step 5: Update the position and velocity of each particle according to Equations 3–4.

Step 6: This procedure is repeated until the number of iterations has been satisfied.

5 EXPERIMENTAL RESULTS AND ANALYSIS

In order to test the effectiveness and adaptability of RGPSO in this paper, experiments are carried out on two ZY-3 remote sensing images. ZY-3 is China's first civilian high resolution cartographic satellite launched in Jan. 9, 2012. The ZY-3 images contain four spectral bands with a 5.8 meters spatial resolution and a panchromatic band with a 2.1 meters spatial resolution. It should be noticed that, the data used in the experiments are pan-sharpened multi-spectral images.

The two data sets shown in Figure 2 are sub-regions of Guiyang China with size of 256×256 pixels. The first image was acquired on Mar. 31, 2012 (see Fig. 2a); the other was acquired on Mar. 30, 2013 (see Fig. 2d). Multi-spectral image pan-sharpening is conducted on ENVI 4.7. The size of the particle swarm and the maximum iterations are 40, 100. The parameter β in Equation 9 is 0.5.

For the test images I_1 and I_2, the near-optimal segmentation parameter values selected by RGPSO are shown in Table 1.

In order to quantitatively verify the validity of the near-optimal segmentation parameter values shown in Table 1, any two of the segmentation parameters are remained unchanged and the last one is modified, subsequently the test images are segmented by RGPSO method. Specific steps are as follows:

Step 1: $w_{spectrum}$ and $w_{compact}$ are remained unchanged, T is modified in the range [500, 9000] and the bin size is 200. Subsequently, segment the test images and the segmentation quality curves are shown in Figure 3.

Step 2: T and $w_{compact}$ are remained unchanged, $w_{spectrum}$ is modified in the range [0, 1] and the bin size is 0.05. Subsequently, segment the test images and the segmentation quality curves are shown in Figure 4.

Step 3: T and $w_{spectrum}$ are remained unchanged, $w_{compact}$ is modified in the range [0, 1] and the bin size is 0.05. Subsequently, segment the test images

and the segmentation quality curves are shown in Figure 5.

As can be seen from Figures 3–5, for images I_1 and I_2, the near-optimal segmentation parameter values are shown in Table 2, they are apparently very close to or in accordance with the corresponding values shown in Table 1. This indicates the effectiveness and adaptability of RGPSO method and the new fitness function.

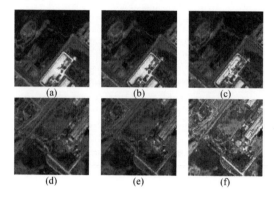

Figure 2. (a) image I_1, (b) reference segmentation of I_1, (c) the near-optimal segmentation result of I_1, (d) image I_2, (e) reference segmentation of I_2, (f) the near-optimal segmentation result of I_2.

Table 1. The near-optimal segmentation parameter values and segmentation quality by RGPSO method.

Parameters and segmentation quality	I_1	I_2
T	2437	4164
$w_{spectrum}$	0.41	0.39
$w_{compact}$	0.65	0.40
Fit(H,Y)	0.7714	0.7195

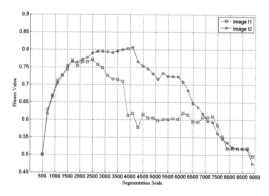

Figure 3. The fitness value (i.e. segmentation quality) of different T (i.e. the segmentation scale).

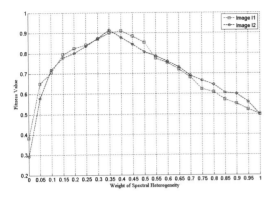

Figure 4. The fitness value (i.e. segmentation quality) of different $w_{spectrum}$ (i.e. the weight of spectral heterogeneity).

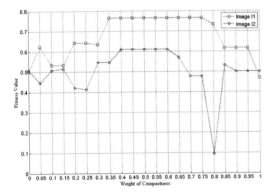

Figure 5. The fitness value (i.e. segmentation quality) of different $w_{compact}$ (i.e. the weight of compactness).

Furthermore, we can draw the following conclusions:

Firstly, T and $w_{spectrum}$ have strong influence on the image segmentation quality, they dominate the segmentation quality, by contrast, $w_{compact}$ has softer influence on image segmentation quality. Consequently, the reasonable choice of T and $w_{spectrum}$ is very crucial.

Secondly, the RGPSO method is effectiveness and adaptability. The conventional methods are heavily reliant on expert experience or repeatable experiments, in which some representative disadvantages such as subjectivity, arbitrariness and inefficiency are avoided in RGPSO method. Consequently, the segmentation quality can be significantly enhanced based on RGPSO and a substantial foundation will be built for the subsequent remote sensing object detection or change detection.

For images I_1 and I_2, the reference segmentation and the near-optimal segmentation results

Table 2. The near-optimal segmentation parameter values from Figures 3–5.

Parameters	I_1	I_2
T	2500	4100
$w_{spectrum}$	0.40	0.35
$w_{compact}$	0.65	0.40

by RGPSO are shown in Figures 2b, c, e and f, respectively. Human eye is still a strong and experienced source for evaluation of image segmentation (Baatz et al. 2000). Compared with the reference segmentation by visual interpretation, there is no evident under-segmentation in the segmentation results; some surface features comprise some over-segmentation phenomena, but the primary surface features, such as roads, forest, bare land and houses were well segmented. Shortridge (2007) pointed out that a desirable segmentation result should comprise reasonable over-segmentation and no evident under-segmentation phenomena; this indicates the near-optimal segmentation results obtained by RGPSO method are reasonable.

6 CONCLUSIONS

This paper combines PSO and region growing algorithm to propose an optimization segmentation approach for high resolution remote sensing image, RGPSO. In RGPSO method, a new evaluation function has been proposed and used as a fitness function for PSO which tries to find a near-optimal segmentation result for a given image. The experimental results have demonstrated that the effectiveness and adaptability of RGPSO method and the new fitness function. More discussion on our work and comparison with other existing approaches will be added in the future work.

REFERENCES

Baatz, M. & Schape, A. 2000. Multiresolution Segmentation: an optimization approach for high quality multi-scale image segmentation. *Angewandte Geographische Informationsve-rarbeitung XII*: 12–23.

Burnett, C. & Blaschke, T. 2003. A multi-scale segmentation/object relationship modelling methodology for landscape analysis. *Ecological Modelling* 168: 233–249.

Carleer, A.P., Debeir O. & Wolff, E. 2005. Assessment of very high spatial resolution satellite image segmentations. *Photogrammetric Engineering and Remote Sensing* 71(11): 1285–1294.

Gao, H., Kwong, S., Yang, J. & Cao, J. 2013. Particle swarm optimization based on intermediate disturbance strategy algorithm and its application in multi-threshold image segmentation. *Information Sciences* 250: 82–112.

Liu, Y., Bian, L., Meng, Y.H., et al. 2012. Discrepancy measures for selecting optimal combination of parameter values in object-based image analysis. *ISPRS Journal of Photogrammetry and Remote Sensing* (68): 144–156.

Mohsen, F.M.A., Hadhoud, M. M. & Amin, K. 2011. A new optimization-based image segmentation method by particle swarm optimization. *International Journal of Advanced Computer Science and Applications, Special Issue on Image Processing and Analysis* 10–18.

Shi, Y. 2004. Particle swarm optimization. *IEEE Connections* 2(1): 8–13.

Shortridge, A. 2007. Practical limits of Moran's autocorrelation index for raster class maps. *Computers, Environment and Urban Systems* 31(3): 362–371.

Wu, L., Zhang, Z., Wang, Y. & Liu, Q. 2014. A Segmentation Based Change Detection Method for High Resolution Remote Sensing Image. *In Pattern Recognition*, Springer Berlin Heidelberg 314–324.

Zhang, Y.J. 1997. Evaluation and comparison of different segmentation algorithms. *Pattern Recognition Letters* 18: 963–974.

Zhang, H., Fritts, J.E. & Goldman, S.A. 2008. Image segmentation evaluation: A survey of unsupervised methods. *Computer Vision and Image Understanding* 110(2): 260–280.

Zheng, L., Pan, Q., Li, G. & Liang, J. 2009. Improvement of grayscale image segmentation based on PSO algorithm. *Fourth International Conference on Computer Sciences and Convergence Information Technology* 442–446.

Architectural, Energy and Information Engineering – Sung & Chen (Eds)
© 2016 Taylor & Francis Group, London, ISBN 978-1-138-02791-6

Based on the controlling process to improve the purity of C72DA

Q.W. Zuo & D.Q. Cang
School of Metallurgical and Ecological Engineering, University of Science and Technology Beijing, Beijing, P.R. China

X. An
Tangshan Iron and Steel Co. Ltd., Tangshan, P.R. China

ABSTRACT: This paper mainly discusses on the technology of converter, refining, and continuous casting processes, which greatly affect the purity of the tire cord steel C72DA coil domestically. It was shown that the cases can guarantee the purity of steel, in which the contents of nitrogen and oxygen reach 30 ppm and 20 ppm, respectively. In addition, the content of carbon in molten steel is no less than 0.45%, the basicity of the refining slag is about 1.0, the time and intensity of Ar bubble soft blow must be kept at a certain time period, the range of overheat is limited to 10°C~15°C strictly, the cast speed is kept stable, and the secondary cooling and intensity of electromagnetic stirring must be proper. The results show that the grade of inclusions A, B, C, and D is no more than 0.5, the rate of sorbite is more than 90% of microstructure, the tensile strength reaches up to 1000 MPa, and the reduction of area is more than 40%.

Keywords: tire cord steel C72DA; metallurgy process; purity; converter; refining; continuous casting

1 INTRODUCTION

Cord steel C72DA is widely used for radial tire, which is manufactured by a Ø5.5 mm rolled coil. It is required that the diameter of the final product is only 0.15 mm–0.2 mm, drawing for many passes including both dry and wet, thus the purity of steel is a key factor. The microstructure of the base is mostly sorbite. Many researchers[1–5] have investigated the generation of inclusions from a certain point of view of thermodynamics or single segment during the production process; however, the fact is that the whole controlling process is even more important to guarantee the purity.

With the development of modern times, there is a huge demand for heavy load trucks for transportation, so tire cord steel C72DA is regarded as the candidate material for its high tensile strength. Tire cord steel whose tensile strength reaches as high as 2800 MPa and 3600 MPa is applied in practice and even a higher grade level is under development.

This paper mainly discusses about the technology of converter, refining, and continuous casting processes, which affect the purity of tire cord steel C72DA coil domestically. The microstructure gets better, the quantities and size of inclusions decrease, and the percentage of fracture in drawing gets lower through optimizing the parameters of the whole process. The diameter of the final product is as smaller as 0.16 mm.

1.1 Flow and composition

The entire process of tire cord steel C72DA, from smelt to drawing, is as follows: Converter → refining (LF + VD) → CC → Φ5.5 mm coil → descaling → straightening → washing → dilute sulphuric acid → electrolytic pickling → washing → boron coated drying → 7 passes drawing → Φ2.8 mm → 7pass drawing → Φ1.4 mm → heating → lead bath quenching → copper plating → 19 wet drawing → Φ0.3 mm → Φ0.15 mm.

The composition and stander of tire cord steel C72DA are provided in Table 1.

1.2 Deoxidation and alloying

During the converter process, the basic parameters are as follows: the temperature of tapping is 1600°C–1620°C, high carbon catching operation, and the content of molten steel is no less than 0.45%. In order to decrease the quantities of inclusions brought in raw materials, ultra-low aluminum alloy and nitrogen carburant are needed.

The added order of alloying elements can be as follows: when the amount of molten steel discharges about one-third, manganese-silicon is added followed by ferrosilicon, Si-Ca-Ba. The quantities of the alloys are 4.0~4.5 kg/t, 1.5~2.2 kg/t, and 0.3~0.5 kg/t, respectively. The amounts of the alloys can also be determined by the other remaining elements in molten steel.

Table 1. Composition and stander.

	Composition (%)											
				P	S	Cr	Ni	Mo	Cu	Al	N (ppm)	
Grade	C	Si	Mn				≤					
C72DA(GB)	0.69–0.75	0.15–0.30	0.40–0.60	0.025	0.020	0.10	0.10	0.05	0.10	0.01	50	
C72DA (stander)	0.70	0.18	0.45	0.015	0.010	0.06	0.06	0.05	0.08	0.01	30	

* batch that has values ranging between 0.1% and 0.15% for Ni, Cr, and Cu needs to be categorized as lower grade.

During tapping, the pressure of Ar bubble blowing should be kept at 0.5~0.7 MPa, and turned down to 0.3~0.4 MPa to the end. Violent blowing is forbidden. If too much slag is poured into the ladle, 200~300 kg lime is needed. Because of the high content of carbon, the slag of the molten slag is of high viscosity. The ladle is critical to the purity of liquid steel, the remaining slag should not be deposited at the bottom and the temperature must not be higher than 1000°C.

2 REFINING PROCESS

2.1 Application of the diagram

In order to form a proper refining slag, which possesses at a low melting point, desulfurization is arranged at pretreatment of hot metal stage. Usually, the percentage of sulfur after pretreatment is 0.002%~0.005%. Although the values may increase during the converter smelting process, the content of sulfur is still in control.

According to the law[6] that the higher the melting point of inclusion, the lower the deformation index, we can confirm that the basicity of the refining slag is most important. Whether the basicity is too high or too low, the quantities and size inclusions must be out of control. The analysis shows that the best range of basicity is 1.0~1.1.

Many research results show that the main inclusions in the steel can be divided into two categories[7,8]: $CaO-SiO_2-Al_2O_3$ and $MnO-SiO_2-Al_2O_3$. Their percentages are 82% and 18%, respectively. According to the diagram, the low melting point inclusions distribute around spessartine ($3MnO \cdot Al_2O_3 \cdot 3SiO_2$), where the range of Al_2O_3 content is 15%~30%. While for $MnO-SiO_2-Al_2O_3$, the low melting point inclusions distribute around anorthite ($CaO \cdot Al_2O_3 \cdot 2SiO_2$) and pseudowollastonite ($CaO \cdot SiO_2$)[9].

2.2 Tire cord steel C72DA refining slag

In order to improve the purity of molten steel, slag forming route was in line with the isotherm and as narrow as possible. During the process, the time and intensity of soft blow must be appropriate.

Table 2. Composition of tire cord steel C72DA at the end of refining.

	Composition (%)				
Grade	C	Si	Mn	P	S
C72DA	0.71	0.20	0.52	≤0.015	≤0.010

The conditions of the refining temperature should be no less than 1525°C and the sampling depth must be about 400 mm, respectively. The next stage is how to adjust the composition of the slag in accord with the results of the sample. Usually, the white basic slag lasts for 20 minutes.

The target of composition at the end of refining is presented in Table 2.

3 CONTINUOUS CASTING PROCESSES

3.1 Control of mold and tundish

Mold fluxes must especially be used dry for tire cord steel. The flow was controlled by the stopper, and the retaining wall was built in tundish. A submerged nozzle was made from Al-C material with seals. The parameters of electromagnetic stirring were 250 A and 5 Hz.

The automatic open casting rate is no less than 95%, and a carbonless cover agent is added to the tundish when the molten steel arrives at the bottom of the long nozzle. The casting process is initiated when the depth of the steel in tundish reaches 350 mm, and then the electromagnetic stirring is started.

During the casting process, the depth of the molten steel should be kept no less than 600 mm, even the depth increases to 650 mm when the ladle is changed. The first furnace is very important, when the depth reaches to 300 mm; the basic cover agent should be input into the tundish. The cover agent used for the second furnace and after can be instituted by carbonized rice husk.

The casting process is protected by Ar, the submerged depth of the long nozzle is 200–250 mm

Table 3. Temperature of casting.

Grade	Temperature of the first furnace	The second furnace	Third furnace and after	Tundish temperature
C72DA	1565–1575°C	1525–1540°C	1520–1535°C	1490–1505°C

Table 4. Parameters of secondary cooling.

Cooling intensity (L/Kg)	Secondary cooling water (m³/h)			
	Spraying water	1#	2#	3#
1.72	10	20.4	9	39.4

Figure 1. Metallograph of C72DA.

and that of the submerged nozzle is 100–150 mm. At the end of casting, the depth of the molten steel in the tundish is no less than 200 mm. The depth of mold fluxes should be kept properly.

3.2 Control of temperature and secondary cooling

Proper casting temperature is a key factor to quality of billets; in fact, the control of temperature is the control of overheat. It is know that crack, segregation, and porosity are affected by overheat. The casting temperature is provided in Table 3.

Casting speed is usually kept as 2.20 m/min; the parameters of secondary cooling are listed in Table 4.

When the casting speed is set at 2.2 m/min, the water flow of the 4# segment is no more than 5 m³/h. While the speed turns down, the flow should be 1.72 L/Kg and the flow of 1#, 2#, 3# should be turned down.

4 QUALITY OF THE PRODUCT

The base and microstructure of tire cord steel C72DA is shown in Figure 1, and the state of inclusion is shown in Figure 2.

The results of the analysis of energy spectrum are shown in Figure 3.

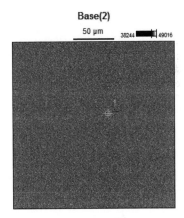

Figure 2. Microstructure and inclusion.

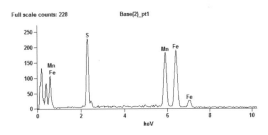

Figure 3. Spectrum analysis of inclusion.

5 CONCLUSIONS

1. The content of carbon in molten steel should be no less than 0.45%, the basicity of the refining slag must be about 1.0, the time of soft blow should be no less than 30 minutes, the range of overheat is 10°C~15°C, the casting speed must be kept at 2.20 m/min, the contents of nitrogen and oxygen are 30 ppm and 20 ppm, respec-

tively, and the grade of inclusions A, B, C, D should be no more than 0.5.

2. The results of the analysis show that the rate of sorbite is more than 90% of microstructure, and the tensile strength can reach up to 1000 Mpa; the reduction of area is more than 40% after the control of the entire process.

3. The purity of steel should be considered with respect to the whole production process.

REFERENCES

Bernsman G P. Inclusions controlling steel for tyre cord [J]. The Minerals, Metals & Materals Society, 1996, 1: 123–133.

Bernard G, Ribound P V, Urbain G. Oxide inclusions plasticity [A]. La Revue de Metallurgie-CCT [C], Mai, 1981: 421.

Martin Gagne, Eric Thibault. Control of Inclusion Characteristics in Direct Cast Steel Billets [J]. Canada Institute of Mine and Metal, 2000, (8): 312–321.

Sezer, A. Trends in steel cord reinforcement for today's PCR and TBR tyres [J]. International Polymer Science and Technology. 2008. Vol. 35 (1), pp. T1–T6.

Sarychev A. V., Yu. A. Ivin, K. V. Kazyatin, V. V. Pavlov, L. A. Kulichev. Production of cord steel at the Magnitogorsk Metallurgical Combine [J]. Metallurgist. 2007. Vol. 51 (11), pp. 660–662.

Wang Haitao, Hao Ning, Jin Qingfeng. Study on fracture of C72DA [J]. Steel Making, 2005, 21(6): 18–22. (in Chinese).

Wang Xinhua. The study of non-metallic inclusions of heavy rail steel. Beijing [R], University of Science & Technology Beijing, 2001 (in Chinese).

Wang Xinhua, Wang Lifeng, Zhuo Xiaojun, et al. The research of non-metallic inclusions of 70# steel [R]. Beijing, University of Science & Technology Beijing, 2003 (in Chinese).

Zhao Zhongfu, Yu Xinhe, Hong Jun, et al. Study of controlling of inclusions in tire cord steel [J]. Iron and steel, 2009, 44(3): 40–44. (in Chinese).

Architectural, Energy and Information Engineering – Sung & Chen (Eds)
© 2016 Taylor & Francis Group, London, ISBN 978-1-138-02791-6

Porosity characteristics of coal reservoir in the Daqing exploration area

Sh.H. Wang & Y.Z. Wang
Exploration and Development Research Institute, Daqing Oilfield Company Ltd., Daqing, China

ABSTRACT: This paper studies the coal structure in the Daqing exploration area, to discuss the types of pore in different basins. Scanning electron microscopy and low-temperature nitrogen adsorption method are combined to analyse the pore characteristics of coal reservoirs. The results indicate that the types of pore dominated by plant tissue pore are independent of each other from west to east. The main types of pore in tectonic coal are gas pore and emposieu. There are a lot of open fractures accompanied by increasing deformation grade, improving pore connectivity and permeability. There are three types of low-temperature nitrogen adsorption cures in the Daqing exploration area, which represents different types of pore structure. From type I to type III, pores are converted from the open type to the non-breathing pore type whose one end is closed. Finally, they are replaced by the ink bottle pores. The number of ink bottle pores is responsible for the difference in the adsorption capacity of coal.

Keywords: Daqing exploration area; pore characteristics; low-temperature nitrogen adsorption method; pore structure

1 INTRODUCTION

Coal is a complicated porous medium. Adsorbability and permeability of its pore structure for Coalbed Methane (CBM) has drawn the attention of researchers extensively. The porosity of coal enables the coal reservoir to store gas and allows CBM to desorb, diffuse and percolate. For this reason, it is of significance to study the pore structure characteristics for the exploration and development of CBM as well as for the evaluation of its mineability.

The fractal theory deals with a simple and effective method to explain the irregular and complex phenomena in nature. In recent years, the fractal theory has played an important role in the description of reservoir pore structure. Katz and Kroch et al. proved that porous reservoir bodies such as sandstone and carbonatite have the fractal feature. Fu Xuehai et al. deemed that the coal has a better fractal feature in a certain pore range, so that the approximate quantitative information of pore distribution in coal can be obtained by using the fractal theory. By this way, reservoir can be assessed accurately and effectively. Ma Limin and Wen Huijian believed that the fractal dimension is the comprehensive quantitative characterization of the complicated reservoir pore structure by establishing the relationship between parameters of fractal dimension and micro-pore structure; the bigger the fractal dimension is, the stronger the reservoir anisotropy is. Thus, the fractal dimension is an ideal parameter of complicated reservoir pore structure in a quantitative description[15].

The coalbed in the Daqing exploration area, mainly occurring in the early Cretaceous epoch representative of Jixi and Hegang basins and the Huhehu depression of Hailar Basin, has a better CBM exploratory prospect. This paper is intended to explore micro-reservoir space types and pore structure characteristics of coal in different basins based on the evolutionary features of the Daqing exploration area. This will provide the scientific basis for the coal reservoir physical property.

2 GEOLOGICAL SETTING AND COAL SAMPLES

In the three basins of the Daqing exploration area, coal basins include Hailar Basin in the west and Jixi and Higang Basins in the Sanjiang region in the east as representatives (Figure 1). The Huhehu depression is the one where the original texture coal mainly develops (Figure 2a). The wooden structure of indigenous coal can be seen, with clear traces of plant tissues, better integral massive structure, undeveloped cleats, and high handfeel strength. According to Ju Yiwen's deformed coal classifications[16], the coal samples from Hegang and Jixi Basins belong to cataclastic coal and speckle coal in the brittle deformation series. The cataclastic coal (Figure 2b) is complete in the original texture relatively with a visible banded structure. It con-

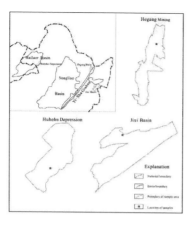

Figure 1. Sketch map of coal samples in the Daqing exploration area.

Figure 2. Types of coal in the aqing exploration area. a. samples of uhehu depression b. samples of Hegang basin c. samples of jixi basin.

tains two groups of cleats: high handfeel strength and some partial fine-grained fragments. In the speckle coal (Figure 2c), the original texture can be seen indistinctly, where groups of cleats and lump structures are developed. The lumps are displaced relatively. Due to the low handfeel strength, they become angular fragments at a size of 1–5 cm when crumbled by hand.

3 PORE CHARACTERISTICS IN SEM

The Scanning Electron Microscope (SEM) can be used for comprehensive analyses on the information of coal samples in undestroyed coal original states based on characteristics such as varied coal structure, strong anisotropy and rich micro-phenomena.

Original texture coal: the coal from the Huhehu depression belongs to lignite. A large amount of intact plant tissue pores can be observed in the SEM. The cell cavities deform to different extents because of the compaction effect (Figure 3a), but they arrange in a uniform direction with similar shapes, which indicates the feature of plant tissues (Figure 3b and c). The fine stratification and fissures can be seen in locality (Figure 3f). A few inter-clast pores and intercrystal pores develop

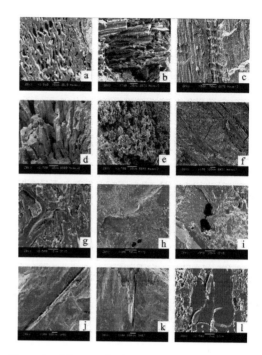

Figure 3. Microscopic characters by SEM of original texture coal and tectonic coal. a. cell of cross-section b. cell of oblique-section c. wall of cell d. inter-clast pores e. intercrystal pores f. micro-beddings g. plant tissue pores filled with minerals h. blowhole i. emposieu j. open fracture k. open fracture on friction surfaces l. shrinkage fracture in vitrinite.

(Figure 3d and e), but their blowholes do not. This type of pore is more independent of poor connectivity. Its fractures occur in interlaminations, but the fractures connecting the pores are rare, which contribute less to the permeability of coal. So, this is to the disadvantage of the migration and deposit of CBM.

Tectonic coal: the coal in Jixi and Hegang Basins develops better between gas coal and coking coal. Although parts of plant tissue pores remain in the coal (Figure 3g), the pores are filled nearly with minerals. A large amount of blowholes (Figure 3h) and emposieu (Figure 3i) exist in the coal with lithification. The number and types of fractures increase in an open state (Figure 3j). Some microstructures such as friction surfaces (Figure 3k) occur with anabatic deformation. At the same time, plenty of shrinkage joints occur in vitrinite (Figure 3l). The original texture is destroyed in the tectonic coal, so that the pore structure becomes complicated. Its specific surface area increases to improve gas adsorbability. A large number of micro-fractures and shrinkage joints form to connect bridges between pores, which improves seepage capability among coal pores to some extent.

4 CHARACTERISTICS OF PORE STRUCTURE

The shape of low-temperature nitrogen adsorption curve represents the type of pore structure to some degree. The pore structures of the coal in the Daqing exploration area are classified into three basic types based on coal sample adsorption and desorption curve characteristics as follows:

Type I: porosity characteristics of coal samples from the Huhehu depression in the Hailar Basin. Adsorption and desorption curves are in parallel without distinct hysteresis loops (Figure 4, I-a), indicating that the coal reservoir has a large number of open breathing holes and a few non-air holes whose one end is closed. The pore size is distributed in twin peaks (Figure 4, I-b): the first peak occurs at 2–3 nm and the second peak appears at 20–30 nm. This indicates that these small pores contribute more to the specific surface area, but the micro-pore contribute less (Figure 4, I-c).

Type II: porosity characteristics of coal samples from the Hegang Basin. It is known from Figure 4, II-a that adsorption and desorption curves have hysteresis loops in the relative pressure at $p/p_0 > 0.5$, indicating that their pores are in composite with multi-pore form and contain ink bottle pores and non-breathing pores whose one end is closed. The pore size distribution curve shows that the micro-pore develops well. These micro-pores have the greatest contribution to the pore volume (Figure 4, II-c), and the peak value exists at 3–4 nm. The micro-pores contribute mainly to the specific surface area, but small pores contribute partly (Figure 4, II-c).

Type III: porosity characteristics of coal samples from the Jixi Basin. This type of adsorption and desorption curves have distinct hysteresis loops

(Figure 4, III-a). The adsorption curve rises up gently all the time, and desorption curve goes down sharply at the pressure of about 0.5. The reason for this phenomenon is the presence of plenty of ink bottle pores. The pore volume is distributed in two peaks, but pores at a size of 3–4 nm have the greatest specific surface area.

The coal pores in the Daqing exploration area have distinct zone characteristics. The average pore diameter becomes small with the deformation intensity of coal from west to east. The Total pore volume decreases at an order of magnitude, but increases at an order of magnitude compared with the specific surface area. It is known from the above phenomena that the pore structure changes and contents and kinds increase with the metamorphic grade of coal. Especially, the occurrence of plenty of ink bottle pores leads to the inflection point of adsorption and desorption curves. On the other hand, the destruction of tectonic activity makes the pore structure further complicated, which shows a significant difference in coal to gas adsorption capacity.

5 CONCLUSIONS

The experiments show that the characteristics of the pore structure of coal reservoir are impacted by tectonic setting in the Daqing exploration area. The tectonic activity becomes increasingly active from west to east, and coal changes from original texture coal into tectonic coal. This leads to a more complicated reservoir pore microstructure and increasing anisotropy.

The original plant tissue pores are gradually replaced by blow holes and emposieu with deformation from the riginal texture coal to tectonic coal, so that the micro-pores prevail. The occurrence of numerous open fractures changes the disadvantages of disconnected pores, and improves the permeability of pores.

Due to the tectonic action, the low-temperature nitrogen adsorption and desorption curves present three types, which prove that the pores are converted from the open type to the non-breathing pore type whose one end is closed. Finally, they are developed as the ink bottle pores. The micro-pore becomes the greatest contributor to the specific surface area. The occurrence of a large number of ink bottle micro-pores is a major reason for the difference in the adsorption capacity of coal.

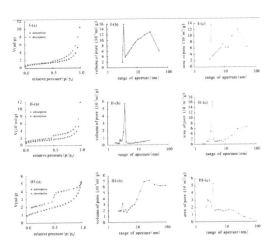

Figure 4. Types of pore structure in the aqing exploration area.

REFERENCES

P. Chen & X. Y, Tang. 2001. The research on adsorption of nitrogen in low temperature and micro-pore properties in coal. Journal of China Coal Society, 26(5): 552–556.

D. Duan, K. Gao & CH. A, Tang, et al. 2009. Study on mechanism of pore pressure in coal and gas outburst process. Safety in Coal Mines, 40(1): 3–6.

Y. B, Hu, Y. SH, Zhao & D, Yang, et al. 2002. Relationship between permeability and fractal dimension of coal mass. Chinese Journal of Rock Mechanics and Engineering, 21(10): 1452–1456.

Y. W, Ju, B, Jiang & Q. L, Hou. 2004. The new structure-genetic classification system in tectonically deformed coals and its geological significance. Journal of China Coal Society, 29(5): 513–517.

A. J, Katz & A. H, Thompson. 1985. Fractal stone pores-implications for conductivity and formation. Phys Rev Lett. 54(3): 1325–1328.

C. E, Kroch. 1988. Sandstone fractal and Euclidean pore volume distributions. Geo Phys Res. 93(B4): 3286–3296.

L. R. Li, Y. Y, Zhao & ZH. X, Li, et al. 2004. Fractal characteristics of micropore structure of porous media and the meaning of fractal coefficient. Journal of the University of Petroleum, China (Edition of Natural Science), 8(3): 105–108.

L. M, Ma, CH. Y, Lin & M. W, Fan. 2012. Quantitative classification and evaluation of reservoirs based on fractal features micropore structure. Journal of Oil Gas Technology, 34(5): 15–20.

R. D, Peng, Y. C, Yang & Y, Jv, et al. 2011. Computation of fractal dimension of rock pores based on gray CT images. Chinese Science Bulletin, 56(26): 2256–2266.

H. J, Wen, L, Yan & F, Jiang, et al. 2007. The fractal characteristics of the pore texture in low porosity and low permeability reservoirs. Journal of Daqing Petroleum Institute, 31(1): 15–18.

T, Zhang & SH. Y, Xu, Yang Ke. Application of fractal dimension of micro-pore structure [J]. Journal of Daqing Petroleum Institute, 2010, 34(3): 43–48.

S. H, Zhang. 2008. Study on coal reservoir physical properties in eastern margin of Ordos Basin. Beijing: China University of Geosciences.

Architectural, Energy and Information Engineering – Sung & Chen (Eds)
© *2016 Taylor & Francis Group, London, ISBN 978-1-138-02791-6*

An empirical study on college English writing instruction based on the *New Horizon College English* online learning system

Z. Ma
Beijing Wuzi University, Beijing, China

ABSTRACT: To improve college students' writing competence, this empirical study focuses on examining the effectiveness of new instruction via the *NHCE online learning system*, and of the traditionally accepted teaching model. For this purpose, subjects were recruited and randomly divided into the Control Group and the Experimental Group. For the whole semester, the subjects took the English writing course in classrooms and language labs, respectively, and received a separate writing training by one teacher. Data were collected from sheets of the Pretest and Posttest of both groups. The statistical analysis shows that students under the innovative writing instruction enlightened by Constructivism, Collaborative learning theory and Computer-Aided Language Learning performed better than their counterparts with regard to the writing ability. In addition, a complementary interview was conducted as qualitative analysis. As a result, the new writing mode based on *the NHCE online learning system* is highly recommended.

Keywords: writing competence; effectiveness; new instruction via *NHCE online learning system*

1 INTRODUCTION

NHCE (New Horizon College English) Online Learning System as a platform is devised by Foreign Language Teaching and Research Press in Beijing, China. The publisher is one of four major foreign language teaching presses throughout Chain. Nearly one-fourth of the present Chinese college students take NHCE as their English learning textbooks. It not only corresponds to *New Horizon College English-Reading and Writing*, and *New Horizon College English-Listening, Viewing and Speaking,* and the two textbooks with related recordings, exercises, and any other useful information as online learning assisting service for teachers and students, but also provides forums for teachers to release teaching syllabus, assignments and notice the same time for students to upload and store anything necessary, so as to share with others, chat and communicate online or by email. The *NHCE online* teaching management platform is comprised of online curriculum and teaching management functions, and could serve for the construction of knowledge, collaborative learning as well as dynamic and process evaluation.

Constructivism is one crucial branch of cognitive learning theory, as well as the development and breakthrough of modern educational theory. On the examination of progress of children's cognition, Piaget (1977) asserted that learning occurs by an active construction of meaning, rather than by passive recipiency. Collaborative learning refers to an instruction method, in which students at various performance levels work together in small groups in or outside the classroom towards a common goal (Gokhale, 1995). Levy (1997) defined CALL (Computer-Aided Language Learning) as "the search for and study of applications of the computer to language teaching and learning". According to Warschauer & Healey (1998), the development of CALL can be divided into three stages: behaviorist CALL, communicative CALL and integrative CALL, with the latter being developed from the former.

Therefore, with testing and comparison as adopted approaches, this study attempts to testify that, as far as writing proficiency is concerned, students with the *NHCE-based* writing instruction can do better than those without. This instruction is set as multiple stages in the whole writing process, namely Prewriting step, Writing step, Post writing step, with foreign and domestic teaching approaches on Blog, BBS, Wiki and Email as backup. The tests were carried out by the research conductor, in both the Experimental Group with CALL (Computer-Aided Language Learning) methodology in language labs, and the Control Group under the traditional teaching mode in classrooms. After the experiment for one term, the quantitative data of the Pretest and Posttest were analyzed by using the t-test, and the qualitative interview was administered to shed light on the results obtained.

2 STUDY PROCEDURES

In order to testify the similarity of writing competence of the students from the CG and the EG, a Pretest was conducted at the beginning of the research. For the reliability of the Pretest, another experienced College English teacher joined with the study conductor to grade the students' writing competence. They referred to the Cet-4 Global Scoring principle, that is, composition is graded by way of raters' general impression on content and language. Before grading, the two teachers practised scoring several sheets to ensure similar standards. As soon as the data of the Pretest was collected, the Independent-Samples t-test was run to study the writing proficiency of the subjects. Based on Table 3, the results indicate that students in both the CG and the EG stand on almost the same level of English writing. With the results of the Pretest, the rest of the research could be conducted. Thus, it can be assumed that any difference and progress in students' writing competence may be due to the writing instruction itself.

The teaching experiment was conducted for 12 weeks, during which the CG followed the traditional writing mode and the EG was instructed with the new writing pedagogical approach based on the *NHCE online system* in language labs where computer, web, earphone and interactive facilities were available. At the end of the experiment, in order to examine the students' achievements and the effectiveness of the computer-based writing instruction, a Posttest was conducted in the 15th week of the final exam, in which subjects were given the same directions on writing format, limits of letters and time, and similar subject and title to those in the Pretest. As compositions were collected, the two raters in the previous test manipulated the identical grading principle and standard to score the papers. Then, various tests were conducted to examine whether there was any significant difference between the CG and the EG with regard to writing proficiency.

3 DATA COLLECTED AND DISCUSSION

In this section, the descriptive statistics of the CG and the EG in both the Pretest and Posttest are presented in Tables 1 and 2. To test whether the CG and the EG bear the same-level writing competence so that the reliability of the experiment can be assured, an Independent t-test on the scores

of the CG and the EG in the Pretest is presented in Table 3. When the innovative and traditional writing instructions were taken, respectively, in the EG and CG, data, as shown in Table 4, from the Posttest were drawn and analyzed to examine whether there was any significant difference in students' grades, and whether their writing ability was improved or not. When checking the mean scores of the Pretest and Posttest in both groups, the researcher observed that grades from the Posttest were lower than those from the Pretest; therefore, the Paired Samples Statistics of the EG and CG in Tables 5 and 6 were used to examine the significant difference. At the same time, relevant discussions, analysis and conclusion were put forward.

The descriptive statistics of the EG and CG are presented in Tables 1 and 2, with the Minimum Score, Maximum Score, Mean (Average) Score and Std. (Standard) Deviation being listed for the Pretest and Posttest, respectively. In the Pretest, the Mean Score of the CG (89.38) was a little higher than that of the EG (89.23), and Independent Samples t-test for the Pretest was used to examine whether there was a significant difference.

For the scores in the Pretest between the CG and EG, as shown in Table 3, equal variances assumed $F = .056$, Sig. $= .814$, greater than .05, indicating the equal variance of the CG and EG, so the corresponding P-value of the 2-tailed t-test was .935, greater than .05, therefore there lies no significant difference in the scores of CG and EG, hence the

Table 1. Descriptive statistics of the EG in the pretest and posttest.

		Descriptive statistics			
EG	N	Mini-mum	Maxi-mum	Mean	Std. Deviation
Pretest score	26	77	98	89.23	6.371
Posttest score	26	73	92	84.35	5.238
Valid N (listwise)	26				

Table 2. Descriptive statistics of the CG in the pretest and posttest.

		Descriptive statistics			
CG	N	Mini-mum	Maxi-mum	Mean	Std. Deviation
Pretest score	26	74	99	89.38	7.072
Posttest score	26	68	90	79.65	5.138
Valid N (listwise)	26				

validity of the study is guaranteed and the experiment can be carried out as scheduled.

After the treatment, the scores of the Posttest were compared in the Independent-Samples t-test, as shown in Table 4. The mean score of the EG was 84.35, higher than that of the CG (79.65). Significant difference will be accounted for by the results of Spss. Statistics. Here, equal variances assumed F = .246, Sig. = .622, greater than .05, which shows that the variance is equal. Moreover, the corresponding P-value of the 2-tailed t-test was 0.002, less than .05, therefore there is asignificant difference between the scores of CG and EG in the Posttest. The results statistically proved that

with *the NHCE* online writing teaching technique, compared with the conventional writing approach, adopted on the CG and EG subjects' writing competence is enhanced.

In Table 5, the scores of both the Pretest and Posttest of EG are compared. Since Sig. (2-tailed) = 0.001, less than 0.05, the score of the Pretest and that of the Posttest are significantly different. The comparison of the scores of the Pretest and Posttest for the CG is provided in Table 6.

For the CG also, Sig. (2-tailed) = 0.000, less than 0.05, there is a significant difference between the scores of the Pretest and Posttest. However, for the EG, the mean score of Posttest is 84.35, lower than

Table 3. Independent-samples t-test for the pretest.

	Levene's test for equality of variances		t-test for equality of means				
	F	Sig.	t	df	Sig. (2-tailed)	Mean difference	Std. error difference
Equal variances assumed	.056	.814	−.082	50	.935	−.154	1.867
Equal variances not assumed			−.082	49.465	.935	−.154	1.867

Table 4. Independent-samples t-test for the posttest.

	Levene's test for equality of variances		t-test for equality of means				
	F	Sig.	t	df	Sig. (2-tailed)	Mean difference	Std. error difference
Equal variances assumed	.246	.622	3.261	50	.002	4.692	1.439
Equal variances not assumed			3.261	49.981	.002	4.692	1.439

Table 5. Paired-samples t-test for the EG in the pretest and posttest.

Paired samples test

		Paired differences							
					95% confidence interval of the difference				Sig. (2-tailed)
		Mean	Std. deviation	Std. error mean	Lower	Upper	t	f	
EG	Pretest score – Posttest score	4.885	6.581	1.291	2.227	7.543	3.785	25	.001

Table 6. Paired-samples t-test for the CG in the pretest and posttest.

Paired samples test

					95% confidence interval of the difference				
		Mean	Std. deviation	Std. error mean	Lower	Upper	t	F	Sig. (2-tailed)
CG	Pretest score – Posttest score	9.731	6.453	1.266	7.124	12.337	7.689	25	.000

that of the Pretest (89.23). For the CG, the mean score of the Posttest is 79.65, far lower than that of the Pretest (89.38). On retrospection and investigation of all experimental processes, the researcher found that since the Posttest was held at the final exam, though similar subjects, directions, limits of time and words, grading principle, raters' performance were identical with those in the Pretest, students' in the EG and CG might feel nervous and worried in the exam itself, leading to difficulty of writing, thus the psychological factor in both groups may account for backwardness of results in the Posttest. However, even if the students' performance in the Posttest was influenced, the mean score was apparently higher for the EG (84.35) than for the CG (79.65) in the Posttest, so the above statistics may further prove that the *NHCE* online system-based writing instruction has significant effectiveness on writing competence.

4 CONCLUSIONS

Based on the analyzed statistical results, the solutions to the research questions raised were collected: the significant difference did lie between the performances of students' under *NHCE* online system-based writing instruction and those within the traditional teaching mode. In other words, learners adopting the new teaching methodology via CALL can achieve more than those following the old unchanged teaching strategy, as far as their writing proficiency is concerned. Yet, from the comparison of mean scores of the Pretest and Posttest for both CG and EG, it was found that the mean scores of the Posttest for two groups are relatively lower than the previous; the underlying reasons were investigated by way of interviews with students. The interviewees in the EG all welcomed the computer-based writing measures,

spoke highly of the inspirable brainstorming and discussion steps, and expressed sincere gratitude towards those seriously editing and revising their drafts. They told that once in the language lab with the computer and net available, they will be enthusiastic enough to participate for any writing task, since writing was no longer an individual activity, in which one inevitably encountered boredom and tension. Now, they took each writing assignment as a fun-game, a good chance to communicate with teachers and fellows, and most importantly, they found that their storage of vocabularies and phrases was enriched, the application of language in compositions was easy and free, and more thoughts and ideas occurred whenever a topic was issued. However, meanwhile, not only subjects in the EG but also those in the CG complained about the Posttest since it was taken at the final exam, during which students wrote under certain stress, and poor performance may be attributed to inevitable nervousness. Even under the same condition, it still could be concluded that students undergoing computer-aided writing instruction could write more proficiently than learners in traditional classrooms.

REFERENCES

Gokhale, A. 1995. Collaborative learning enhances critical thinking. *Journal of Technology Education*, 7 (1): 22–30.

Levy, M. 1997. *Computer-assisted Language Learning: Context and Conceptualization*. New York: Oxford University Press.

Piaget, J. 1977. *The Development of Thought: Equilibration of cognitive structures*. New York: The Viking Press.

Warsehauer, M. & Healey, D. 1998. Computers and Language Learning: An Overview. *Language Teaching*. 31: 57–7.

Architectural, Energy and Information Engineering – Sung & Chen (Eds)
© *2016 Taylor & Francis Group, London, ISBN 978-1-138-02791-6*

A review of studies on carbon-motivated border tax adjustments in the United States

Yi Nan Ding
Graduate School, International Business School, Beijing Foreign Studies University, Beijing, China

Long Peng
International Business School, Beijing Foreign Studies University, Beijing, China

Jie Tang
School of English and International Studies, Beijing Foreign Studies University, Beijing, China

ABSTRACT: This paper aims to review a heated scholarly debate around carbon-motivated border tax adjustments (BTAs). BTAs can be simply understood as tariffs imposed on energy-intensive products from foreign countries that are not actively limiting greenhouse gas emissions. Supporters of BTAs point to the benefits of increased economic competiveness, carbon leakage prevention and decreased free riders. By contrast, protesters warn that BTAs may raise problems such as GATT-rules consistency, computational complexity, and, most importantly for developing countries, equity issues. In conclusion, this paper points out the necessity for further empirical research on the real effects of BTAs.

Keywords: border tax adjustments; economic competiveness; carbon leakage; WTO rules

1 INTRODUCTION

In the past three decades, China has achieved economic takeoff, becoming the second largest economy in the world. However, sustaining high rate economic growth requires compatible energy supply. Since the 1990s, China has actively established and furthered energy relations with the Middle East, African, Latin American and Asian countries to solve its energy shortage caused by inadequate indigenous supply and continuing rising demand. In contrast to the Western countries that are market-driven economies in energy, China strives to ensure its energy security through the promotion of its National Oil Companies (NOCs). From the perspective of Western scholars, it is to "control" energy supply instead of relying on market. Besides, as the second largest economy and the largest energy consumer in the world, China's energy policy now has a great impact not only on itself, but also on the world economy. Against this backdrop, Western scholars, researchers and pundits dedicate themselves to investigating the strategic nature of China's quest for foreign energy and its potential implications, so as to offer sound policy recommendations to their respective officials. Three schools of thoughts have, therefore, emerged from this debate: Mercantilism, Liberalism and the Middle Ground school. This article aims to review this

heated scholarly debate with an attempt to see how Western scholars and pundits view China's quest for foreign energy or China's energy diplomacy.

2 CHINA'S QUEST FOR FOREIGN ENERGY

2.1 *China's energy security*

Since the beginning of the reform and opening up, China's economy has expanded at a spectacular average annual growth rate, and so has its energy consumption. Gradually, its domestic energy supply was no longer sufficient to meet the demand. In 1993, the oil demand exceeded supply. The discrepancy between energy demand and domestic supply for the first time in history emerged as a problem for the nation to solve. However, during the 1990s, the problem imposed little threat to China's energy security because it imported comparatively small quantities of oil and the global oil market during that time was abundant in oil supply with a relatively low price (Christie 2009).

However, things began to change from the very beginning of the 21st century. The 9/11 terrorist attack, the Iraq War and the Afghanistan War rendered the Middle East, the most important area of oil importing for China, insecure. At the same time, domestic demand for energy increased dramatically due to its rapid economic development and urbanization. Thus, energy shortage threatens to disrupt economic growth and social stability. In such situation, the Chinese government has taken three steps to meet its growing demand for energy to power its economic growth and social activities: 1) expanding foreign oil from the Middle East; 2): diversifying energy sources by reaching out to Africa, Russia, Central Asia and the Americas; and 3) securing oil transport routes (Lai 2007).

2.2 *China's energy diplomacy*

The early stage of China's quest for foreign energy started in 1993 when it became a net oil importer. As a new player in the international oil market, China had to buy oil on spot market. During this period, the Middle East was the major oil origin, which has accounted for the largest share of China's imported oil since then. Major exporting countries in this area were Sudan, Iran and Iraq. Since the early 21st century, motivated by growing domestic energy demand and volatile international environment, the Chinese government has adopted a more comprehensive energy diplomacy. First, China has been diversifying its sources for oil with an emphasis on the Middle East, Africa, Russia, Central Asia and Latin America. Second, China has been diversifying its modes to acquire oil from the international market, such as long-term supply contracts and equity investments. Third, China has been offering package deals to oil suppliers for their energy, for example, oil-for-infrastructure and oil-for-loan. Fourth, in order to reduce transit risks, China has been diversifying and securing its energy transport routes, for example, constructing gas and oil pipelines linking Myanmar and China. Fifth, to improve its NOCs' competitiveness, China has been providing them with financial and political support (Christie 2009).

These measures indicate China's strong resolvability and momentum to secure its energy supply, which has invited academic discussions about the target or strategic nature of China's energy diplomacy. The viewpoints of these scholars or pundits fall into three categories: the Mercantilism school, the Liberalism School and the Middle Ground school.

3 THE STRATEGIC NATURE OF CHINA'S QUEST FOR FOREIGN ENERGY

3.1 *The mercantilism school*

Mercantilism refers to the kind of economic practice that advocates government control of national economy with an aim to enhance state strength at the cost of competing states. Another remarkable feature of mercantilism is its promotion of governmental pro-

tection of national industries through various means, tariffs and subsidies, for instance (Goldstein 2003). The Mercantilism School in this debate believes that China's energy diplomacy is mercantilist in nature because the main Chinese actors in oil market are state-owned companies, not private enterprises, and the Chinese government supports them economically and politically to "hunt and nationalize" foreign oil (Holslag 2006; Leverett 2007; Herberg 2011). According to this school, China's mercantilist approach to energy supply has generated a negative impact on the international oil market and energy exporting countries, and has imposed threats to the interests of the US-led Western countries.

To begin with, the Mercantilism School argues that China's mercantilist approach helps it to "lock up" oil resources by formulating unilateral relations with energy exporting countries, which might choose to trade off their oil reserves for loans or infrastructure. The supply of oil from these countries will thus contract and then cause a rise in oil price in the international oil market (Leverett 2007). Second, China's growing interest in foreign oil makes the energy exporting countries increasingly dependent on oil as the main commodity, which renders them more vulnerable to negative price fluctuations and precludes them from diversifying their economies. In other words, the "neo-colonial" feature of China's energy policy will bring about a negative influence in the long term that could override the short-term economic profits (Taylor 2007). Third, China's quest for foreign energy is believed to have imposed threats to the US-led Western countries in the following ways. In terms of energy issue, China's search for foreign energy in the Middle East, Africa, Latin America and Russia has made it a major competitor to the United States for oil reserves. Politically, China has maintained energy relations with states that the US lists as "pariah states" or "rouge states", such as Sudan, Iran and Iraq, which hinders the Western efforts to address the issue of proliferation of nuclear weapons and human rights violation. With regard to security matters, in recent years, China has been improving its naval capacities to secure its major oil transport route, the Strait of Malacca, which, according to this school, will challenge the US dominance in this area (Leverett 2007). Last but not the least, growing competition among China, Japan, India and Korea to promote their NOCs and control foreign energy supplies is weakening the public confidence in fair access to energy supply in the future and deepening strategic distrust (Giljum 2009; Collins 2011).

3.2 *The liberalism school*

Liberal theories of international relations propose that deepening economic interdependence among states will raise the prospects for cooperation and moderate those of conflict. Given the crucial significance of energy to the economic and national security and the desire for a peaceful environment, China's energy dependence will encourage it to secure its energy supply through diplomacy, international market and multilateral institutions rather than through military tools that result in conflict and confrontation. As demonstrated in China's energy activities, China has been actively cooperating with exporting countries, as well as other importing countries through various forums and international institutions, such as the WTO, the International Energy Agency (IEA) and the Shanghai Cooperation Organization (SCO) (Ziegler 2006). Instead of competing in a "zero-sum" game depicted by the Mercantilism School, China is working cooperatively in a "win-win" situation. Another misinterpretation from the Mercantilism school, according to the Liberalism School, is that the Chinese government plays a major role in its NOCs' oversea activities. For the last two decades, China's energy sector has been greatly liberalized and decentralized. It is the NOCs' stakeholders (shareholders and banks), not the government, that play the major role in pursuing profits (Downs 2007).

As for the impact of China's quest for foreign energy, the Liberalism still holds different even conflicting opinions. First, China does not "lock up" oil reserves for itself. Two-thirds of oil bought by China's NOCs in 2006 were sold on the international market. China increases rather than contracts oil supply to the global market, which can help reduce price fluctuations and supply disruptions caused by unexpected incidents (Downs 2007; Leung 2010). Therefore, China's energy policy strengthens international energy security rather than harms it. Second, China's NOCs help energy exporting countries construct infrastructure, such as railroads, schools, hospitals, or offer them loan in return for their energy, which, without any doubt, will help them develop and thrive (Hanauer & Morris 2014). Third, China's energy activities do not fundamentally threaten US economic and political interests. On the contrary, they are, to some extent, helpful. Infrastructure constructed by China in Africa, for instance, helps reduce transaction costs and expand regional oil markets for possible American investors (Yergin 2006; Hanauer & Morris 2014).

3.3 The middle ground school and policy recommendations

The Middle Ground School, as its designation suggests, stands in between. This school contends that China's energy diplomacy reflects both mercantilism and liberalism under certain conditions. In energy countries where China has nurtured unilateral agreements such as Sudan and Iran, policies or strategies lean on mercantilism. One the other hand, when it comes to securing the maritime transport routes, China relies on other countries, such as the United States. In such circumstances, China looks for multilateral cooperation or free riding. According to the Middle Ground School, China has been quite pragmatic and flexible in employing energy policies to fit certain situations and maximize its energy interests (Lieberthal & Herberg 2006; Christie 2009; Lee 2010; McCarthy 2013).

Although these schools hold different opinions towards the strategic nature of China's energy diplomacy, they converge on what their governments should do about China, which is to cooperate with, rather than confront against it. Let us take the United States for an example. As the first two largest economies and energy consumers in the world, energy policy adopted by these two countries will have tremendous impact on each other and more importantly on the world. Therefore, the US and China have vital common interests on energy issues. Both countries share an interest in maintaining stability in energy exporting regions, securing energy transits, avoiding global oil price fluctuations and supply disruptions as well as accelerating the research and development of clean energy. It is undeniable that strategic distrust between the US and China still exists and will do in the years to come. However, the common interests will impel the two to work out some solutions.

4 CONCLUSIONS

This article has thus presented the main arguments of each school on the Western debate about China's quest for foreign energy, respectively—whether China's energy diplomacy reflects mercantilism or liberalism or both. Apparently, these three schools interpret Chinese energy activities form different perspectives. The Mercantilism School, to a great extent, follows the rationale of realism that emphasizes states' instinct and resolve to pursue national interests and power at the expense of other rivals, which usually leads to conflicts. The liberalism School, on the other hand, is in favor of the power of economic interdependence and multilateral institutions to facilitate cooperation among states. The Middle Ground School synthesizes the theoretical rationales of the former two schools and analyses China's search for foreign energy on a case-by-case basis.

Mercantilism has gradually become a word used to describe China's energy activities that go against the US, or the Western interests. Some use it to criticize Chinese government for directing its national oil companies to acquire foreign energy assets, especially oil. However, it should not be neglected that mercantilism was the dominant economic theory and practice in the West between the 16th and 18th centuries, although they now have replaced it with reliance on global oil market. China is not the only one that subordinates its foreign policy to energy interests. The United States, for example, has done some political intervention and even fought wars to serve its oil interests (Iran and Iraq). Both China and the Western countries have employed certain strategies to secure energy supplies. As Mr Zweig once commented, it is the responsibility of Chinese leaders to secure the nation's energy supply, and China has the right to do that through market strategies, which the West countries should recognize (Zweig 2006).

"Energy is a politicized commodity." Without electricity and gasoline, life goes dark and slow in a modern society. Without energy, economy dies. The vital importance of energy to economic and national security has made it preponderant over other commodities in foreign trade. Neither can China, nor the west countries, afford to leave energy to market forces alone.

ACKNOWLEDGMENTS

This work was supported by the Studies of Chinese folk Financial Development motives and Governance, and the Project of National Natural Science Foundation of China (No.71373029).

REFERENCES

Christie, E.H. (Ed.) 2009. China's foreign oil policy: genesis, deployment and selected effects. FIW Research Reports (03): 1–83.

Collins, G. 2011. Asia's rising energy and resource nationalism Implications for the United States, China, and the Asia-Pacific Region. The National Bureau of Asian Research. (31): 1–80.

Downs, Erica S. 2007. The fact and fiction of Sino-African energy relations. China Security 3(3): 42–68.

Downs, Erica S. 2007. China's quest for overseas oil. Far Eastern Economic Review. 52–56.

Giljum, J.P. 2009. The future of China's energy security. The Journal of International Policy Solutions. 11: 12–24.

Goldstein, J.S. 2003. International Relations. New York: Longman.

Hanauer, Larry & Lyle J. Morris. 2014. Chinese engagement in Africa -drivers, reactions and implications for US policy. RAND Corporation Reports. 1–173.

Herberg, Mikkal. 2011. China's energy rise and the future of US-China energy relations. New America Foundation.1–16.

Holslag, Jonathan. 2006. China's new mercantilism in Central Africa. African and Asian Studies 5(2): 134–169.

Lai, H.H. 2007. China's oil diplomacy: is it a global security threat? Third World Quarterly 28(3): 519–537.

Lee, M.S.L. 2010. China's energy security: the grand "hedging" strategy. United States Army Command and General Staff College. 1–61.

Lieberthal, Kenneth & Mikkal Herberg. 2006. China's search for energy security: implications for US policy. The National Bureau of Asian Research. 17(1): 1–52.

Leung, G.C.K. 2011. China's energy security-perception and reality. Energy Policy 39: 1330–1337.

Leverett, Flynt. 2007. The geopolitics of oil and America's international standing. Committee on Energy and Natural Resources. 1–9.

McCarthy, Joseph. 2013. Crude "oil mercantilism"? Chinese oil engagement in Kazakhstan. Pacific Affairs 86(2): 257–278.

Taylor, Ian. 2007. Unpacking China's resource diplomacy in Africa. Center on China's Transnational Relations. (19): 1–34.

Yergin, Daniel. 2006. Ensuring energy security. Foreign affairs 85(2): 69–82.

Ziegler, Charles. 2006. The energy factor in China's foreign policy. Journal of Chinese Political Science.1–26.

Architectural, Energy and Information Engineering – Sung & Chen (Eds)
© 2016 Taylor & Francis Group, London, ISBN 978-1-138-02791-6

Face detection based on multiple skin color spaces

M. Li
*Department of Computer Application, Shanghai Technical Institute of Electronics and Information, Shanghai,
China*

Q.H. Liu
School of Mechatronic Engineering and Automation, Shanghai University, Shanghai, China

ABSTRACT: Focusing on face detection problem in color images under complex background, this paper proposes a face detection method based on the multiple color spaces. Considering the contribution of each color component to the skin color, this method combines with the ellipse skin color model of Cb-Cr color space, the circular skin color model of Cg-Cb color space and the parallelogram skin color model of Cg-Cr color space for skin detection and face detection. The experimental results show that the method can effectively separate skin color and background in color images. Appling with a few morphological processing, non-face regions can be wiped off and face region is detected successfully.

Keywords: YCbCr color space; YCgCb color space; YCgCr color space; clustering; face detection

1 INTRODUCTION

In the early 1960s, H. Chan et al. proposed the concept of face recognition, and had carried on the preliminary research. Along with the rapid development of machine vision, artificial intelligence, pattern recognition, data mining and artificial neural networks and other techniques, face recognition as an emerging biometric technology has been rapid development. Face detection is the first and crucial step of face recognition, and its purpose is to detect whether there is someone in the given image or video, so as to determine the scope of face and extract some information of face such as location, size, and pose. Face detection can provide many bases for facial feature point location and facial matching recognition. Because of its accurate, convenient, concealment and good expandability, face recognition has been widely used in many fields, such as entrance guard system, video surveillance, security authentication, video conference, the public security system, and human-computer interaction. Therefore, as the important part of face recognition, face detection has a strong application space and use value.

Among many features of human body, skin color is an important characteristic of human body surface, especially facial skin color. Because skin color cannot change with different person's facial details, expression and direction, skin color as the important clue of face detection is feasible.

Comparing with other face detection methods, such as neural networks, machine learning, Support Vector Machine (SVM), and template matching, the face detection method based on skin color has the advantages of a small amount of calculation and fast processing speed, which has become one of the hotspots in the research field of face recognition at present. However, skin color is easily influenced by light, background and other factors outside; the expression of skin color model is different in a different color space. Most of the face detections are in a single color space and the detection effect remains to be improved. Selection of color space and skin model is the key of the face detection method based on skin color, which is closely related to the detection results.

2 STEP OF MULTIPLE COLOR SPACE FACE DETECTION

2.1 Color space selection

So far, many color spaces can be used for skin detection; the common color space is RGB, rgb, YUV, YCbCr, HIS, YIQ, and Lab [1]. Some studies have suggested that chrominance have a less effect on human skin color and the appearance differences are mainly caused by brightness. Comparing several color spaces, three color components in RGB are related to the brightness.

It is very difficult to remove the influence of the brightness on each color component. The correlation between the three color components is also strong. So, for face detection, good results would not be obtained. While the normalized RGB removes the relative brightness of RGB, each component has brightness information, so it cannot achieve the ideal effect too. YCbCr [2] is derived from YUV color space, just like HIS and HSV, which has the characteristic of luminance and chrominance separation. YCbCr has become one of the widely used color spaces in skin detection in recent years. Similar to YCbCr space, De Dios proposed YCgCr [3] space. Wang [4] verified that the space has better skin color clustering than YCbCr color space. The skin color characteristics can be expressed effectively in YCgCr. Based on YCbCr and YCgCr, another color space named YCgCb can be concluded. Reference [5] also verified YCgCb space had good color clustering. However, some researchers have shown that red is the largest component in skin color and green is lowest. Considering all of the color components have a certain proportion in skin color, the contribution of each color component should not be discarded. Therefore, this paper combines YCbCr, YCgCr and YCgCb for skin and face detection. The color space distribution discipline is explored and the skin color model is established in YCbCr, YCgCr and YCgCb color space. Synthesizing the skin color detection results in multiple color space, human face could be detected to distinguish the other skin regions.

YCbCr, YCgCr and YCgCb color spaces have the characteristic of luminance and chrominance separation, where Y represents brightness, Cb represents blue, Cr represents red, and Cg represents green, respectively.

Extracting skin samples from plenty of color images and obtaining each component's distribution range, skin color three-dimensional distribution diagrams in different color spaces is shown in Figure 1. All of skin color points are composed of discrete space and concentrated in a small continuous area. It is very convenient to cluster and express. As can be seen from Figure 1, the clustering effect of YCbCr, YCgCr, and YCgCb is obviously better than RGB's.

(a) RGB space   (b) Cb-Cr space   (c) Cg-Cb space   (d) Cr-Cg space

Figure 1.   Three-dimensional space distribution of skin color.

2.2 *Establishment of the skin color model*

For skin detection, it is very important to establish a suitable skin color model to express the color distribution discipline. The mathematical model is the best presentation for the color distribution in color space. In the inspected images, all of the pixels that are conformed to the skin color model are determined as skin, therefore the skin color model is the precondition of skin color detection. In [6], the scope for the Cb and Cr in YCbCr color space is set. If the Cr and Cb value of the pixels meets the Cr = [133 173] and Cb = [77 127], the pixels that belong to a rectangle area are skin color. In [7], skin color is detected in Cb-Cr color space with ellipse fitting. In [8], the skin color model is, respectively, set up in Cg-Cr and Cg-Cb color space to get the merged skin color detection results.

In order to find out the skin color distribution discipline of YCbCr, YCgCr and YCgCb color spaces to detect skin color precisely, the skin color model is established on the basis of [7] and [8]. Removing the influence of Y component and only considering chrominance components, three-dimensional skin color distribution is reduced to two-dimensional plane. Projecting from YCbCr to Cb-Cr subspace, projective shape is similar to an ellipse. Projecting from YCgCb to Cg-Cb subspace, projective shape is similar to a circle. Projecting from YCgCr to Cg-Cr subspace, projective shape is similar to a parallelogram. The results are shown in Figure 2 (a), Figure 4 (a) and Figure 6 (a).

To accurately statistic clustering characteristic of skin pixels in each color space, at the same time, to obtain the best skin color model fitting boundary, 150 images are chosen. All of the images contain face under the conditions of different backgrounds. Some of them come from digital camera taking; some of them come from the Internet downloading. People in the images are of different gender and age. Overall, 200 pieces of skin images whose sizes are different are manually clipped, including the face skin and the other areas of body. The total skin color pixels are 15,992,880, all of them make up skin color training samples.

1. Establishment of the skin color model in Cb-Cr subspace

According to (1), RGB of all the pixels in skin color training samples are converted to YCbCr, and then projected to Cb-Cr color space. Cb-Cr color space distribution is shown in Figure 2 (a). After observation, the graphic that is formed from all of skin color pixels clustering is similar to an ellipse. An ellipse is fitted to the graphic boundary. To the greatest extent, skin color pixels are included inside the boundary. A small amount of isolated points outside the boundary are discarded. The red dot in the figure is the skin color clustering point, and the black line is the skin color fitting boundary model, as shown in Figure 2 (b).

$$\begin{bmatrix} Y \\ Cb \\ Cr \end{bmatrix} = \begin{bmatrix} 16 \\ 128 \\ 128 \end{bmatrix} + \begin{bmatrix} 0.2568 & 0.5041 & 0.0979 \\ -0.1482 & -0.2910 & 0.4392 \\ 0.4392 & -0.3677 & -0.0714 \end{bmatrix} \begin{bmatrix} R \\ G \\ B \end{bmatrix} \quad (1)$$

After many times' fitting and calculation, an ellipse model with translation and rotation is established in Cb-Cr color space. Elliptic mathematical expression is given in (2). The elliptic center locates at (110,150). The long axis radius is 26, the short axis radius is 16. θ, and the rotation angle of the ellipse is $3\pi/4$. If the pixels' values of the inspected images meet (2), that is, the pixels are falling in the elliptic region, they are judged to be skin color pixels, otherwise will be ruled out, as the non-skin color pixels:

$$\frac{(x-110)^2}{26^2}\cos\theta + \frac{(y-155)^2}{16^2}\sin\theta \le 1$$
$$\begin{cases} x = 26\cos\alpha\cos\theta - 16\sin\alpha\sin\theta \\ y = 26\cos\alpha\sin\theta + 16\sin\alpha\cos\theta \end{cases} \quad (2)$$

2. Establishment of the skin color model in Cg-Cb subspace

According to (3), RGB of all the pixels in skin color training samples are converted to YCgCb, and then projected to Cg-Cb color space. Cg-Cb color space distribution is shown in Figure 3 (a). After observation, the graphic that is formed

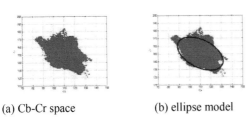

(a) Cb-Cr space   (b) ellipse model

Figure 2.   Cb-Cr space skin color distribution fitting model.

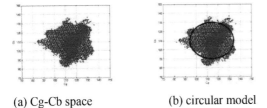

(a) Cg-Cb space   (b) circular model

Figure 3.   Cg -Cb space skin color distribution fitting model.

from all of skin color pixels clustering is similar to a circle. The circle is fitted to the graphic boundary. The blue dot in the figure is the skin color clustering point, and the black line is the skin color fitting boundary model, as shown in Figure 3(b).

$$\begin{bmatrix} Y \\ Cb \\ Cr \end{bmatrix} = \begin{bmatrix} 16 \\ 128 \\ 128 \end{bmatrix} + \begin{bmatrix} 0.2568 & 0.5041 & 0.0979 \\ -0.3180 & -0.4392 & -0.1212 \\ -0.1482 & -0.2910 & -0.4392 \end{bmatrix} \begin{bmatrix} R \\ G \\ B \end{bmatrix} \quad (3)$$

After many experiments, when the center of the skin color fitting boundary circle is located at (112,110), 18 for radius size, the boundary fitting effect is best. The expression of the circular model in Cg-Cb color space is given in (4). When the values of the pixels in the inspected images satisfy (4), these pixels are skin color pixels. Otherwise, the possibility of skin color pixels will be ruled out, as the non-skin color pixels:

$$\frac{(x-112)^2 + (y-110)^2}{20^2} \le 1$$
$$\begin{cases} x = 20\cos\theta + 112 \\ y = 20\sin\theta + 110 \end{cases} \quad (4)$$

3. Establishment of the skin color model in Cg-Cr subspace

According to (5), RGB of all the pixels in skin color training samples are converted to YCgCr, and then projected to Cg-Cr color space. Cg-Cr color space distribution is shown in Figure 4 (a). After observation, the graphic that is formed from all of skin color pixels clustering is similar to a quadrilateral. The parallelogram is fitted to the graphic boundary. The green dot in the figure is the skin color clustering point, and the black line is the skin color fitting boundary model, as shown in Figure 4 (b).

$$\begin{bmatrix} Y \\ Cb \\ Cr \end{bmatrix} = \begin{bmatrix} 16 \\ 128 \\ 128 \end{bmatrix} + \begin{bmatrix} 0.2568 & 0.5041 & 0.0979 \\ -0.3180 & 0.4392 & -0.1212 \\ 0.4392 & -0.3677 & -0.0714 \end{bmatrix} \begin{bmatrix} R \\ G \\ B \end{bmatrix} \quad (5)$$

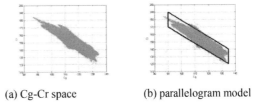

(a) Cg-Cr space   (b) parallelogram model

Figure 4.   Cg-Cr space skin color distribution fitting model.

After many times' calculation, the expression of a parallelogram is determined. The skin color fitting boundary is 4 lines whose equations are given in (6). In the inspected images, if the values of pixels are in the scope of expression, namely $Cg \in [90, 135]$, $Cr \in \left[-\frac{10}{9}Cg + 270, -\frac{10}{9}Cg + 290 \right]$, these pixels are skin color pixels. Otherwise they are excluded, as non-skin color pixels.

$$\begin{cases} Cg = 90 \\ Cg = 135 \\ Cr = -\dfrac{10}{9}Cg + 270 \\ Cr = -\dfrac{10}{9}Cg + 270 \end{cases} \quad (6)$$

2.3 Multiple color space of skin detection

Using skin color models in YCbCr, YCgCb and YCgCr color spaces, the same image is processed in each color space, respectively. Then, three binary images are obtained. Through "and" operations of three binary images, the intersection part of them are extracted. Not common place is wiped off in these binary images. To the maximum extent, skin regions are ensured to be preserved and things of background are removed. Figures 5 and 6 show the result of the combination of three segmentation images. There are single-player and multiplayer skin detection results under some complex background. The results show that skin color detection based on multiple color space can eliminate most of the non-skin regions effectively, including the background objects and the non-skin regions of human body, such as clothes and hair.

2.4 Removal of non-face regions

Although the vast majority of non-skin regions have been removed, there are still some noise to deal with. In order to locate face accurately, first, the residual small noise in the background is removed by the median filter. Then, for the binary image, mathematical morphology processing is used. Because there are some low gray level areas on the face such as eyebrows, eyes, and mouth, they easily tend to form holes of the face. So, a series of morphology processing such as open and close operation, region filling, expansion and corrosion must be handled. These processing regions are beneficial for separating face regions and non-face regions better.

After effective processing of the binary image, there have been some candidate regions of face. Some place of human's body exposed to the outside such as hands, arms, legs and other parts meets the requirements of the skin color model. Some regions existing in the background of images

(a) The original image   (b) Detection result

Figure 5. Single skin color detection under a complex background.

(a) The original image   (b) Detection result

Figure 6. Multiplayer skin color detection under a complex background.

are close to skin color or the same as skin color. All of them are likely to be mistaken for face region. Therefore, further screening is needed. According to some prior knowledge, the obvious non-face regions could be eliminated. Mainly by means of shape, size, density and other geometric properties of the candidate regions for discrimination, the following rules should be considered:

a. *Height-to-width ratio.* The aspect ratio of face is usually between 0.8 and 1.2. Considering that the facial angle and face connects with neck, the aspect ratio is appropriately relaxed. The preferred ratio is between 0.8 and 2.4.
b. *Size.* Small candidate regions can be filtered according to the side length of the shape. If a region's width is less than 20 or height is less than 30, it can be ruled out and deleted.
c. *Area ratio.* If the area of a candidate region is less than 64, or the area ratio of the candidate region to all candidate regions' average area is less than 0.6, this region is a non-face region.

3 EXPERIMENTAL RESULTS AND DISCUSSION

In order to verify whether the face detection method based on multiple color spaces has the versatility and effectiveness, 218 color images from digital camera taking and the Internet downloading are selected with a single-player or multiplayers. The size and background complexity of the images are different. There are various facial expressions and postures of characters. The total number of face in all of the images is 389. The experimental results are summarized in Table 1 and shown in Figures 7 and 8. The detection ratio = detected face number/ total number of face, the false detection ratio = error number/total number of face.

From Table 1, it can be seen that the detection ratio is high under the complex background, but there still exists false detection that is mainly caused by the following factors:

a. *Shelter on the face.* In the color image, there are various shelters such as glasses, beard, and hair on someone's face. These shelters may make face segment into several blocks. It is very difficult to discriminate such face.

b. *Something similar to skin color in the background.* In the background of the image, sometimes there are some objects such as stone, sand, and cement road. When these objects connect with face, separating among them becomes very difficult.

Further work will focus on how to solve the above problem to achieve a better detection ratio.

Table 1. The results of face detection(%).

Type of photos	Total number of photos	Detection ratio	False detection ratio
Single	132	91.6	6.0
Mutiple	86	90.8	7.2

(a) Original image(b) Detection result of skin(c) Detection result of face

Figure 7. Single face detection under a complex background.

(a) Original image(b) Detection result of skin(c) Detection result of face

Figure 8. Multiple face detection under a complex background.

4 CONCLUSIONS

In this paper, a face detection method based on multiple color spaces is proposed. After analyzing and comparing the distribution discipline and the clustering of multiple color spaces in detail, YCbCr, YCgCb and YCgCr color spaces are selected at the same time for skin color detection. Removing the influence of luminance component, different skin color models are established in Cb-Cr, Cg-Cb and Cg-Cr color spaces. Three binary images are obtained on the basis of three skin color models. By intersecting all of the binary images, skin color and background can be separated effectively. Furthermore, some face candidate regions are formed by mathematical morphology processing in the intersection image. Non-face regions are filtered out using prior knowledge. At last, face region is screened and marked. After verification of large amounts of data and experiments, the face detection method based on multiple color spaces can be applied to all kinds of color images under a complex background. The efficiency of this method is high and the computational complexity of this method is low. This method can provide a good premise for the follow-up development of face recognition.

REFERENCES

[1] L. Liang, H. Ai, G. Xu, and B. Zhang. "A Survey of Human Face Detection," Journal of Computers, vol. 25, May. 2002, pp. 449–458.

[2] S.L. Phung, A. Bouzerdoum, and D. Chai. "A Novel Skin Color Model YCbCr Color Space and Its Application to Human Face Detection," Proc. IEEE Symp. Image Processing, IEEE Press, Jan. 2002, pp. 289–292, doi: 10.1109/ICIP.2002.1038016.

[3] J.J. De Dios, and N. Garcia. "Face Detection Based on a New Color Space YCgCr," Proc. IEEE Symp. Image Processing, IEEE Press, Feb. 2003, pp. 909–912, doi: 10.1109/ICIP.2003.1247393.

[4] J. Wang,Y. Lin, and J. Yang. "First Face-region Location Based on a Novel Color Space YCgCr," Computer Science, vol.34, May. 2007, pp. 228–233.

[5] Z. Zhang, and Y. Shi. "Skin Color Detecting Based on Clustering in YCgCb Color Space under Complicated Background," Proc. IEEE Symp. Information Technology and Computer Science, IEEE Press, Feb. 2009, pp. 410–413, doi: 10.1109/ITCS.2009.222.

[6] D. Chai, and K.N. Ngan. "Locating Facial Region of a Head-and-shoulders Color Image," Proc. IEEE Symp. Automatic Face and Gesture Recognition, IEEE Press, 1998, pp. 124–129, doi: 10.1109/AFGR.1998.670936.

[7] R.L. Hsu, M. Abdel-Mottaleb, and A.K. Jain. "Face Detection in Color Images," IEEE Transaction on Pattern Analysis and Machine Intelligence, vol. 24, May. 2002, pp. 696–706.

[8] Z. Zhang, and Y. Shi. "Skin Color Detection YCgCr and YCgCb Color Space," Computer Engineering and Applications, vol. 46, Jun. 2010, pp. 167–170.

Architectural, Energy and Information Engineering – Sung & Chen (Eds)
© *2016 Taylor & Francis Group, London, ISBN 978-1-138-02791-6*

A monitoring data correlation and storage approach for Health Condition Evaluation of Hydro Turbine Generator Sets

F.W. Chu, Z.H. Li, Y.C. Wu & X.L. Cui
Huazhong University of Science and Technology, Wuhan, China

ABSTRACT: Extracting the health condition information about Hydro Turbine Generator Set (HTGS) needs to analyze a large amount of states monitoring data. A reasonable data management approach is of great importance for Health Condition Evaluation (HCE). This paper analyzed the data requirements of health condition evaluation and the characteristics of monitoring data from the Hydropower Plant Optimal Maintenance Information System (HOMIS). Techniques such as Time Synchronization (TS), Operation Condition Synchronization (OCS) and Event Synchronization (ES) were proposed to ensure the effective correlation of monitoring data. Thus, an intelligent software storage strategy based on correlation rules and a hierarchical distributed hardware storage structure were adopted to control the scale of historical data. This approach improves its analysis efficiency and application value, and has been applied successfully in the Gezhouba Power Plant. The concrete application shows its good prospects.

Keywords: Hydro Turbine Generator Sets; Health Condition Evaluation; Integrated Monitoring System; Synchronization; data correlation; intelligent storage

1 INTRODUCTION

With the birth of strict technical regulatory standards and increasing market competition, the power generation enterprises have made the Health Condition Evaluation (HCE) as the key business. At present, as the continuous improvements of sensor technology and monitoring technology, a variety of special monitoring systems (Han et al. 2003, Liu et al. 2010 and Stone et al. 2008) and control systems provide large amounts of operating states data for HCE of Hydro Turbine Generator Set (HTGS).

However, data from specialized monitoring systems and control systems are managed by different information platforms, and data are not transparent to each other and lack organization, which makes it hard for sharing and exchanging data and providing data services for staff. As each system is dispersed in LAN, lacking cooperation, and its states data storage is independent (Xie et al. 2013), there is no correlation between the data from different systems. Moreover, as online monitoring systems are continuously acquiring and analyzing data, all available data will be stored in accordance with the time, and it is bound to form a mass of historical monitoring data, which will make it difficult to quickly and efficiently locate data on important operation conditions and evaluate the health condition of HTGS.

In this study, a monitoring data correlation and storage approach for HTGS health condition evaluation is presented in the Hydropower plant Optimal Maintenance Information System (HOMIS) (Li et al. 2007). Based on the analysis of equipment HCE's demand for states monitoring data and the states monitoring data's characteristics of HOMIS, techniques such as Operation Condition Synchronization (OCS) and Event Synchronization (ES) are used to enhance the correlation of states monitoring data. Thus, an intelligent software storage strategy based on association rules and a hierarchical distributed hardware storage structure are used to control the scale of historical data and improve the efficiency of its analysis and the value of application to meet the needs of HCE.

2 DATA REQUIREMENTS OF HCE

With the complexity of HTGS's structure and the uncertainty of the operating environment, less objectivity when using a single information to reflect the operating status, and the intricate relationship between the fault symptom and the causes of the malfunction, there are clear data requirements for HCE.

2.1 *HCE needs comprehensive states monitoring data*

The HTGS is a complex system consisting of hydraulic, mechanical, electric, control and auxiliary equipments. The health condition of the sub-device within HTGS is often affected by the health condition of the associated equipment. It is really hard to satisfy the requirements of HCE and fault diagnosis only by monitoring part of HTGS (Xie et al. 2013). HCE of faulty device involves all aspects of HTGS. It requires comprehensive states monitoring data to make an accurate judgment (Yang et al. 2009 and Kezunovic et al. 2012).

2.2 *HCE needs states monitoring data with a strong correlation*

As the HTGS has the characteristics of quick start-stop and response, it often undertakes the task of regulating load and frequency in the grid, and its operation condition changes frequently. The operating states values of equipments are different under different operation conditions (Yang et al. 2009), such as the increase in the amplitude and frequency of the generator internal discharge caused by operating overvoltage occurring when the generator circuit breaker inputs during the synchronization condition, and the poor stability of HTGS caused by unstable operation conditions occurring during the process of load adjustment. Simultaneously, the operating state changes significantly with the appearance of relevant operating and the fluctuation of vital operating parameters. The conclusions of HCE, obtained from the analysis of one aspect of decentralized independent states data, are often unscientific or even wrong. HEC needs states monitoring data with a strong correlation.

2.3 *HCE needs states monitoring data with a high application value*

According to the aforementioned information, HTGS has a complex constitution. A large number of measuring points and states monitoring data are usually stored based on time series, which will lead to the geometric growth of monitoring historical data scale and bring massive data, in most cases, which has a poor analysis and use value, that makes it difficult to store, retrieve and use data (Hou et al. 2007), and it is not conducive to HEC. Data scale needs to be controlled effectively, and data need to be stored selectively in order to save space and improve its application value.

3 DATA CHARACTERISTICS OF HOMIS

3.1 *Framework of CMAS in HOMIS*

HOMIS consists of an online Condition Monitoring Analysis System (CMAS) (Li et al. 2007) and enterprise WAN Remote Monitoring and Diagnosis Information System (RMDIS). CMAS, as shown in Figure 1, integrates control systems in the control domain and independent specialized monitoring systems in the maintenance domain together as a unified system with industrial real-time field bus and Ethernet (see Figure 1). The control domain contains the Locate Control Unit (LCU), Automatic Voltage Regulator (AVR) and Governor (GOV), which are directly involved in the power generation control process and have complete data collection of a large number of states of HTGS. These data play an important role in carrying out a comprehensive analysis and diagnosis of HTGS. The maintenance domain consists of the Integrated Monitoring and Analysis Unit (IMAU), Control Monitoring and Analysis Unit (CMAU), Machinery Monitoring and Analysis Unit (MMAU), Electricity Monitoring and Analysis Unit (EMAU) and Fault Recorder (FR). With the use of standardized CAN field bus, each monitoring system can access conveniently in this network that helps to establish a multilateral relation between those systems that are dispersed in maintenance network.

3.2 *Data characteristics of HOMIS*

As shown in Figure 1, HOMIS contains not only a plurality of specialized monitoring and analysis units from maintenance domain, but also many control systems of the control domain, and different monitoring objects have different change rates of operating states that make it have its own data characteristics.

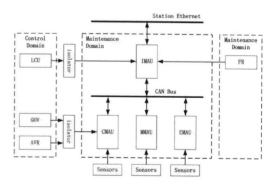

Figure 1. Framework of CMAS in HOMIS.

3.2.1 Characteristic of multiple data sources

On the premise of cost savings, reducing maintenance work and ensuring reliability, HOMIS, on the one hand, makes full use of states data acquired by the existing systems, including LCU, AVR, GOV and FR, on the other hand, installs sensors and acquires states data of HTGS, which are not existing in control systems to meet the requirements of HCE, that makes the data of HOMIS to have multiple sources.

3.2.2 Characteristic of data isomerism

As discussed above, HOMIS has multiple data sources. Because of different manufacturers and hardware, each data source may have different data description dimensions and storing forms (e.g. binary text files, data file, table files or pictures files). The result data obtained (e.g. stability analysis results of HTGS may contain time domain analysis result, frequency domain analysis result and axis orbit analysis result) are different due to different analysis methods. Meanwhile, there are also different data types, including analog type and switch type. All the above-mentioned factorsmeans that data of HOMIS are isomeric.

3.2.3 Characteristic of time multi-scale

Different monitoring objects have different operating states and different operating states means different changing frequencies, such as partial discharge monitoring data of nanosecond level, vibration monitoring data of second level, and transformer's oil analysis data of hour level. Data of HOMIS show the characteristic of time multi-scale.

3.2.4 Characteristic of massive volume

Because of the data characteristic of multiple sources and time multi-scale, and the continuous online data acquisition, massive historical monitoring data will be generated with the accumulation of time. For example, the vibration and swing states' sampling rate usually is 1 K, when calculating data space based on 24 sampling channels, and the space of raw data goes to 3.96 GB each day. For the partial discharge signal, the space of raw data can be up to 506.25 GB each day, which causes great difficulty in the data analysis and storage.

4 MONITORING DATA CORRELATION AND STORAGE

4.1 Key techniques of monitoring data correlation

The integration between various specialized monitoring systems and control systems is not a simply connection with each other to share data, which is just a "collection". To realize the true sense of integration, ensure the effective monitoring data correlation, and meet the requirements of HCE, a real-time interaction and behavior synergy between different subsystems is needed (Xie et al. 2013). Therefore, some key techniques are used in the process of data correlation.

4.1.1 Technique of time synchronization

It is well known that monitoring data having an inconsistent time stamp is meaningless for HCE or other analysis work. In order to realize the correlation of monitoring data between different systems, it must ensure forced time synchronization in every system. According to different data sources, the process of time synchronization, as shown in Figure 2, mainly includes the following:

a. *Time synchronization of set layer*. It includes synchronization of LCU, GOV and AVR in the control network and synchronization of IMAU and FR in maintenance network. Because those systems directly connect with the GPS time server through station Ethernet, they can use time synchronization service component of operating system or GPS synchronization software to obtain time information from GPS time server every 10 minutes and update the time to the local system to implement forced time synchronization with the GPS time server.

b. *Time synchronization of device layer*. It includes synchronization of CMAU, MMAU and EMAU in maintenance network, which do not directly connect with the GPS time server. IMAU broadcasts time information to all other units by CAN bus every 10 seconds and those systems receive it from CAN bus and update the time to the local system to implement forced time synchronization with IMAU .

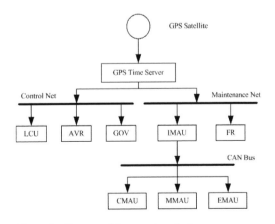

Figure 2.   Process of time synchronization.

By the technique of time synchronization, time of each system in HOMIS is consistent with the GPS time server, ensuring time synchronization and correlation of state data from each data source, which is essential for HCE.

4.1.2 *Technique of operation condition synchronization*

Operating states and health condition of HTGS are closely associated with the set operation condition. The same operating state's value may be totally different under different operation conditions. For example, the peak values of swing and vibration of HTGS under the load rejection condition are much larger than the load stable condition. Therefore, operation conditions are very important to HCE. The OCS process of HOMIS is shown in Figure 3.

a. CMAU obtains data of operation condition states (including water head, active power, reactive power, frequency, stator voltage, stator current, guide vane opening, and wheel opening) by the way of acquisition and communication.

b. Then, CMAU uses the Petri Net model and states data to distinguished operation conditions of HTGS in real time (Li et al. 2008) and then broadcasts the Operation Condition Information (OCI, including code, name, start time and end time of the current operation condition) to all other specialized monitoring and analysis units through CAN bus every 50 milliseconds.

c. IMAU stores the OCI as a record in the database and other units upload operating states data to IMAU of this condition process based on the specific operation condition, which are stored in IMAU.

Through the above steps, each unit in HOMIS can know the current operation condition of HTGS in real time, and the OCS can be realized so that operating states data are associated with the operation condition.

4.1.3 *Technique of event synchronization*

Various types of operations and abnormal alarms are collectively referred to as an event in HOMIS. During the event process, operating states data are of great importance to HEC. The ES process of HOMIS is shown in Figure 4:

a. Each unit distinguishes the state of switch quantities, and judges whether the operation events and alarm events happened or not in real time. At the same time, each unit detects whether there is an abnormality in monitoring objects using the threshold detection method based on the state data and performance data. Then, it generates an alarm event.

b. Once an event is detected, each unit broadcasts the event information (including code, name, start time and end time of the current event) to all other units through CAN bus.

c. Similar to the OCS, IMAU stores the event information as a record in the database and other units upload operating states data to the IMAU of this event process based on the specific event that is stored in the IMAU.

Through the above steps, each unit in the HOMIS can know the current event of HTGS in real time, and ES can be realized so that operating states data is associated with the events.

4.2 *Storage of state monitoring data*

After the integration of different systems, the HOMIS holds a large amount of state data if all of them are completely stored without any distinction, which will bring a severe challenge to the storage and reduced Signal-to-Noise Ratio (SNR) of historical data, seriously affect data query efficiency, and make it difficult to quickly locate the

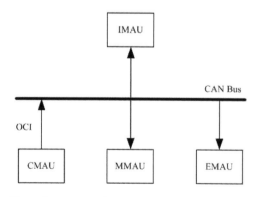

Figure 3. Process of operation condition synchronization.

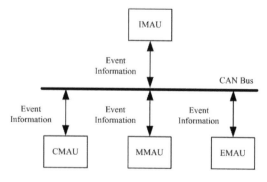

Figure 4. Process of event synchronization.

important characteristic process and find causes for the problem in time. The safety storage of massive data is also a crucial issue that needs a reasonable design of data storage structure. A selective storage software mechanism based on correlation rules and a hierarchical distributed storage hardware structure is employed to realize intelligent and safe storage of massive data.

4.2.1 The selective storage software mechanism based on correlation rules

Mostly HTGSs are in the steady operation state. Station staff usually pay more attention to the equipment health condition and operating states of the dynamic process. It is not necessary to completely store a large amount of operating data generated under the steady condition. Because of the complexity and coupling characteristics of HTGS, there is a strong correlation between states data. Based on certain correlation rules, states data with a typical characteristic are associated and stored effectively in the HOMIS to facilitate comprehensive analysis and provide much clues for HCE and fault diagnosis. The correlation rules used in the HOMIS are as follows:

a. *Operation condition correlation rule*. Based on the aforementioned information, one of the CMAU's core functions is to distinguish the current operation condition of HTGS in real time and then broadcast the OCI to all other units through CAN bus. After receiving the current OCI, they sample and store states data during the operation condition process automatically, and then those feature data will be sent to the IMAU and stored by the IMAU. While the HTGS is in stable operating condition, states data are stored partially with a certain interval time, for example 3 minutes state data will be stored every 30 minutes under the stable operating condition.

b. *Event correlation rule*. The HTGS will periodically carry out some routine activities (defined as events), such as pumps start and stop, valves open and close, and circuit breaks on and off. When specialized monitoring units detect the occurrence of an event in their own domain, they broadcast the event information through CAN bus to the other units. Then, all of them sample and store states data of a certain time scale (such as 1 minute before and after the occurrence of time event) and send them to the IMAU to store.

c. *Abnormality correlation rule*. Abnormality (including fault) usually occurs randomly, such as some state data exceeds the threshold value, state variation rate exceeds the threshold value, performance index drops, load rejection occurs, lighting impulses, and switch misoperation

occurs. Similar to the event correlation rule, once those monitoring units detect the occurrence of abnormality, they broadcast the abnormality information through CAN bus to the other units. Then, all of them sample and store states data of a certain time scale (such as 3 minutes before and after the occurrence of time event) and send them to the IMAU to store.

4.2.2 The hierarchical distributed storage hardware structure

Based on the characteristics of the HOMIS network structure, safety consideration, and the data requirements of HCE, the HOMIS adopts a hierarchical distributed storage hardware structure (see Figure 5) for data storage, which consists of three storage layers as follows:

a. *Device storage layer*. It refers to the states monitoring data obtained by each unit that is stored locally. Based on the consideration of cost and function demand, the device storage layer usually stores only a recent period data, such as raw data of current with 72 hour will be stored for the reason of providing enough necessary original sampling data for further analysis and failure accident recalling.

b. *Set storage layer*. It refers to the states monitoring data of HTGS uploaded by each unit that is stored in the IMAU. Data are stored according to the correlation rules mentioned above and the IMAU is formed as the set layer data center. Similarly, the IMAU of each HTGS also retains only a recent period data (e.g. the past six months) for further analysis and failure accident recalling.

c. *Station storage layer*. In the station layer, the maintenance data server, as the maintenance data warehouse of HOMIS, fully backs up historical data from the IMAU of each HTGS, so as to provide enough monitoring historical data with various functional levels and high application values for HEC.

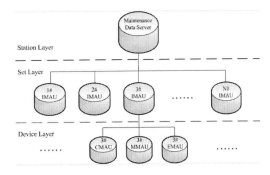

Figure 5. Hierarchical distributed storage hardware structure.

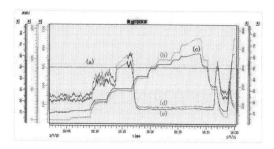

Figure 6. Curve of the upper guide swing when the HTGS changes its active power output, (a) frequency (Hz), (b) active power (MW), (c) guide vane opening (%), (d) X upper guide swing (μm), (e) Y upper guide swing (μm), and x-axis indicates time (M/D/Y H: M: S).

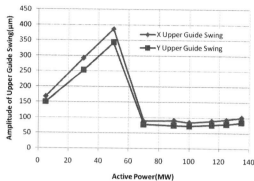

Figure 7. Trend curve of maximum amplitude of the upper guide swing with active power.

5 APPLICATION NOTES

Currently, the proposed approach is applied for data management in the HOMIS, which has been employed by Gezhouba Power Plant to evaluate the health condition of all 22 HTGSs, and achieved good effects.

Figure 6 shows the trend curve of the upper guide swing of HTGS when it changes its active power output. The machine is a 13.8 kV, 146 MVA HTGS. The amplitude of the upper guide swing becomes large when the active power changes from 0 to 50 MW and achieves the peak value when the active power is 50 MW. Then, the amplitude of the upper guide swing becomes small with the increase in active power and achieves a small stable value.

Table 1 shows the maximum amplitude of the upper guide swing in each change range of active power. According to the HTGS operating procedures of Gezhouba Power Plant, the alarm threshold of the upper guide swing is 400 μm. When the active power value ranges between 30 and 50 MW, the amplitude of the upper guide swing is close to the alarm threshold, which means the set is working under a risky condition.

The trend curve of maximum amplitude of the upper guide swing with active power is shown in Figure 7 based on the data in Table 1. From Figure 7, we can know that the stability of this set deteriorates when its active power changes between 30 and 60 MW, and this range is called the unstable operation region, during which it is usually advised to forbid the operation because the set will be damaged seriously when operating under this condition.

Table 1. Maximum amplitude of the upper guide swing in each active power range.

Range of active power (MW)	Maximum amplitude of the X upper guide swing (μm)	Maximum amplitude of the Y upper guide swing (μm)
(0,5)	168	151
(5,30)	292	253
(30,50)	387	343
(50,70)	89	78
(70,90)	91	75
(90,100)	85	73
(100,115)	90	77
(115,125)	94	79
(125,135)	101	85

6 CONCLUSIONS

The HCE of HTGS is very important for a power plant, and it cannot be done well without a reasonable data organization and management. Based on the analysis of data requirements of HCE and the data characteristics of HOMIS, a data management approach was employed in the HOMIS and the realization of the approach was introduced in detail. With the successful implementation of this approach, the states monitoring data can be organized effectively based on the correlation rules, and the scale of data can be controlled that help staff to be liberated from the analysis of massive data and improve the efficiency of evaluation and diagnosis. The application note of HCE based on this approach in Gezhouba Power Plant has shown good prospects, and further efforts will be made on the automated HCE of HTGS.

REFERENCES

Han, Y. & Song, Y. 2002. Condition Monitoring Techniques for Electrical Equipment—A Literature Survey. IEEE Transactions on Power Delivery, vol. 18: 4–13.

Hou, X. 2007. Research and Implementation on Query Optimization based on Data Partition in Massive Data Management, M.D. dissertation, National University of Defense Technology, Changsha, P.R. China.

Kezunovic, M. & Dong, Y. 2012. Information exchange needs for new fault location applications in T&D systems. 16th Electrotechnical Conference, IEEE, Mediterranean: 536–539.

Li, Z., Ai, Y. & Shi, H. 2007. Optimal Maintenance Information System of Gezhouba Hydro Power Plant. Proceedings of 2007 IEEE PES General Meeting, Tampa, FL, United States: 315–320.

Li, Z., Yang, X. & Bi, Y. 2008. Digitization of Hydroturbine Generator Sets and its Engineering Applications. Automation of Electric Power Systems. vol. 32: 76–80.

Liu, M., Han, D. & Li, Z. 2010. Online UHF PD Monitoring for Transformer under the Integrated Framework of HOMIS. High Voltage Engineering, vol. 8: 1975–1980.

Stone, G.C., Lloyd, B. & Sasic, M. 2008. Experience with continuous on-line partial discharge monitoring of generators and motors. Proceedings of IEEE Conference on Condition Monitoring and Diagnosis, Beijing, China: 212–216.

Xie, G. & Li Z. 2013. Integrated Monitoring of Hydroelectric Generators Based on Community Intelligence. Electric Power Automation Equipment. vol. 33: 153–159.

Yang, Z., Tang, W.H. & Shintemirov, A. 2009. Association Rule Mining-Based Dissolved Gas Analysis for Fault Diagnosis of Power Transformers. IEEE Transactions on systems, man, and, cybernetics, vol. 39: 597–610.

Architectural, Energy and Information Engineering – Sung & Chen (Eds)
© 2016 Taylor & Francis Group, London, ISBN 978-1-138-02791-6

The key technology and experimental research of electrochemical grinding

Ying Guan
Shenyang Polytechnic College, Shenyang, China

ABSTRACT: Based on the study of the electrochemical principle, fully considering the influence of various experimental parameters of electrochemical grinding, we set the basic experiment data. We analyzed the surface quality of the processed parts under the different parameters and the stability and reliability of the experimental data obtained by using Minitab software. It is ascertained that there is a changing trend in the effect of electrochemical grinding surface quality state parameters. Thus, we obtained the adjustment range of the key parameters of electrochemical grinding and processing law, and laid the foundation for a better and more efficient practice to improve the efficiency of electrochemical machining.

Keywords: electrochemistry; grinding; experimental data; surface quality

1 INTRODUCTION

With the rapid development of manufacturing industry, some special processing technology is gradually replacing the traditional processing technology, and constantly reflects the advantages in some fields. As a kind of special machining, electrochemical machining involves the use of non-contact type processing, in which high-speed local anodic dissolution actualizes the electrochemical reaction under the electric field. Compared with traditional machining, electrochemical machining can be used to complex surface and cavity hole, and has a good processing effect on the material of high hardness and high tenacity. It not only contains the advantages of traditional processing, and more improved processing stability and consistency of the product, but also has been receiving increased attention in various fields and widely used in parts of processing field such as engine blade, artillery grooves, automobile forging die, steam turbine of integral impeller wheel, special spline and large special-shaped hole. However, the factors affecting electrochemical machining are more complex, and the change in the relationship between various parameters is subtle. Therefore, in order to understand the electrochemical machining technology better, we collected and analyzed the experimental data of the key parameters in the process of electrochemical grinding, and then improved the parts' quality of electrochemical grinding.

2 ELECTROCHEMICAL GRINDING PRINCIPLE

The electrochemical grinding system is shown in Figure 1. The system consists of pulse power supply 1, the negative pole 2, grinding wheel 3, workpiece 4, the positive pole 5, and electrochemical grinding fluid 6. By connecting the grinding wheel 3 with the negative pole and continuously bubbling into the electrochemical grinding fluid between the clearance of the grinding wheel and the bronze, an electrochemical reaction occurs between the workpiece and the negative pole. The metal bond of the grinding wheel surface is the dissolution of iron ions that adhere to the surface of the grinding wheel, and form the oxide films that have the function of insulation at this time. So, the electrical conductivity of electrolyte in the electrochemical machining process will be directly affected by the thickness of the film. When it becomes balanced, the wheel grinding surface will keep a certain amount of abrasive removal. The oxide film is nonconductive films in this process, so it makes the current reduce faster, which effectively prevents the electrochemical reaction. With the development of electrochemical grinding, the film will gradually be ground on the casting and become thinner. When the resistance is reduced to a certain value, the electrochemical reaction will take place again. If the grinding wheel and the spindle are not in a straight line in the process of grinding wheel, or in the state of eccentricity, it will reduce the life of the grinding wheel and the quality of the processed parts in the process of grinding.

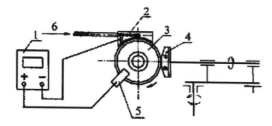

Figure 1. Electrochemical grinding principle diagram.

Table 1. Sample roughness comparison before and after the experiment.

Number	1#	2#	3#	4#	5#	6#	7#	8#
Before the experiment Ra (μm)	1.281	1.824	1.950	1.832	0.688	1.746	1.911	1.789
After the experiment Ra (μm)	0.32	0.30	0.28	0.31	0.30	0.29	0.31	0.32

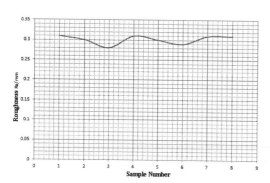

Figure 2. Roughness variation diagram.

3 EXPERIMENTAL RESEARCH OF ELECTROCHEMICAL GRINDING

3.1 Analyzing surface roughness of electrochemical grinding

Setting the condition of 24 v working voltage, initial machining gap of 0.03 mm, 200 g/L as the concentration of NaCl electrolyte, and 0.9 mm/min as the electrolytic speed, all the experimental samples numbered in advance have been electrochemically grinded and tracked. Roughness instrument was used to test all samples for roughness before and after testing. The experimental data are provided in Table 1.

Compared with the experimental data, the level of electrochemical grinding parts' surface roughness is obviously higher than that of machining parts, and the fluctuation value of roughness was in the range of 0.04, as shown in Figure 2. Under

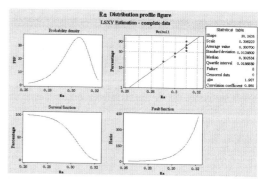

Figure 3. The stability of processing quality and consistency analysis diagram.

a certain condition of electrochemical machining, it can achieve a stable state, in order to ensure the consistency of the product.

We used the Minitab software to analyze that the probability distribution of the surface roughness area, which was less than 90%, electrochemical surface roughness was normally distributed, and the risk priority number reached up to 1.957, whose ability is very strong in the manufacturing industry. The analysis data showed that its standard value was 0.3007 and the standard deviation was 0.012, as shown in Figure 3. It proved that electrochemical grinding was reliable at the same time.

3.2 The gap of electrochemical grinding

We changed the electrochemical machining gap to evaluate the parameter changes in current and removal height. There are 4 grades in power supply voltage, and we changed the electrochemical grinding voltage and gap in accordance with the requirements of the experiment, and measured the average current and removal height, as shown in Table 2. From the experimental data, it was found that with gradual increasing voltage, the average current increased gradually, and removal height also increased gradually in the same gap.

The above experimental data could not prove that the product is stable or reliable, so we used the range and deviation method again to sort the data, as shown in Table 3.

In the process of voltage variation, while machining gap was different, the average current also changed, as shown in Figure 4. However, when the machining gap was 0.07 mm, the characteristic was relatively flat and stable. Thus, it can be initially determined that the machining gap will directly affect the change in average current. and with the increase in machining gap, the deviation of the average current will decrease.

Table 2. Electrochemical grinding voltage and gap change data.

Voltage (V)		6	12	18	24
Gap $\Delta = 0.03$ mm	Average current I1(A)	18	23	34	68
	Removal height H_1 (mm)	0.018	0.032	0.073	0.132
Gap $\Delta = 0.05$ mm	Average current I2(A)	13	19	26	58
	Removal height H_1 (mm)	0.013	0.023	0.064	0.108
Gap $\Delta = 0.07$ mm	Average current I3(A)	9	15	19	50
	Removal height H_3 (mm)	0.008	0.019	0.048	0.088

Table 3. Electrochemical grinding voltage and gap deviation data.

Voltage (V)		6	6	6	18
Gap $\Delta = 0.03$ mm	Current deviation I1(A)	5	11	34	50
	High deviation H_1 (mm)	0.014	0.041	0.059	0.114
Gap $\Delta = 0.05$ mm	Current deviation I2(A)	6	7	28	45
	High deviation H_2 (mm)	0.01	0.041	0.044	0.095
Gap $\Delta = 0.07$ mm	Current deviation I3(A)	6	4	31	41
	High deviation H_3 (mm)	0.011	0.029	0.040	0.080

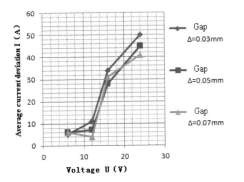

Figure 4. Current under a different voltage deviation and poor value.

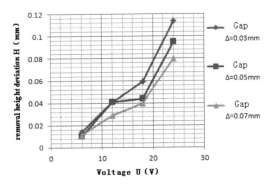

Figure 5. Under a different voltage removal height deviation and range value.

Removal height is closely related to the voltage and the machining gap, as shown in Figure 5. At a low voltage, the deviation of the removal height is small under a different machining gap. However, with the gradual increase in the voltage, the deviation of the removal height under a different machining gap became increasingly large, and the removal height will be shown as the product machining efficiency. Therefore, when we select the machining voltage and the machining gap, we will fully consider the effect on the removal height.

Considering the data in Figures 4 and 5, with the increase in the voltage, the machining gap is smaller, the data obtained more stability. That is to say, when the machining gap reached a steady state, the machining efficiency and accuracy will achieve the best value, which is the most ideal machining state of electrochemical grinding. In fact, the stability of machining gap is also correlated with the removal height. Therefore, we must comprehensively consider the relationship between various parameters in the electrochemical machining process. We can-not improve the quality of electrochemical machining through the adjustment of a single parameter.

3.3 Removal speed of electrochemical grinding

In the electrochemical machining process, we use different electrolytes to measure the relationship between the machining gap and the removal speed, as shown in Table 4.

Figure 6 shows the relationship between the electrochemical machining gap and the solution speed. In the NaCl electrolyte, the machining gap is a hyperbola relationship with the solution speed. With the increasing machining gap, the electrolyte ohmic pressure drop increased, the current density reduced, and the solution rate decreased. The only difference is that the machining gap increases to a certain value, and the dissolved speed is zero.

Table 4. Electrochemical machining gap characteristic curve.

Solution speed mm/min		0.5	1.0	1.5	2.0	2.5	3	3.5
NaCl	Machining gap (mm)	0.05	0.06	0.10	0.15	0.20	0.25	0.30
	Removal speed (mm)	3.30	2.20	1.35	0.90	0.70	0.54	0.50
NaClO₃	Machining gap (mm)	0.025	0.05	0.08	0.10	0.15	0.20	0.23
	Removal speed (mm)	2.80	1.28	0.82	0.65	0.40	0.15	0

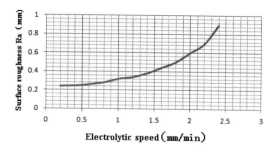

Figure 7. Electrochemical machining surface roughness and electrolytic rate.

The relationship between electrochemical electrolytic speed and surface roughness is shown in Figure 7. When we adjust the electrolytic speed, the machining speed of parts is inversely proportional to it. That is to say, when the electrolytic speed increases gradually, the surface roughness of parts decreases. When the electrolytic speed reaches a certain value, the surface roughness of parts will reach a stable value.

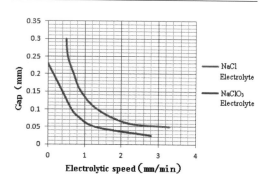

Figure 6. Electrochemical machining gap and electrolytic speed.

Table 5. Electrochemical electrolytic speed and surface roughness data.

Number	1#	2#	3#	4#	5#	6#	7#	8#
Electrolytic speed (mm/min)	0.2	0.5	0.7	0.9	1.2	1.6	2.0	2.4
Surface roughness Ra (mm)	0.24	0.25	0.27	0.30	0.34	0.44	0.63	0.92

We called the solution speed of zero gap "cutting off gap". "Cutting off gap" is unique to the blunt electrolyte. In the active NaCl electrolyte, even if the relative distance between the workpiece and the cathode is very long, the solution rate will not be equal to zero, thus there is no "cutting off gap".

3.4 The relationship between electrochemical electrolytic speed and surface roughness

We adjusted the electrolytic speed in the machining process and recorded the surface roughness data of the samples under different electrolytic speeds, as shown in Table 5.

4 CONCLUSIONS

In this paper, through the comparison of the experimental data, we analyzed the level of electrochemical grinding parts' surface roughness and found it to be significantly higher than that of machining parts, which ensured the consistency of the product. With gradually increasing voltage, the average current increases gradually, and removal height also increases gradually in the same gap. With the machining gap increasing, the electrolyte ohmic pressure drop increases, the current density reduces, and the solution rate decreases. When the machining gap reachs a steady state, the machining efficiency and accuracy will achieve the best value, which is the most ideal machining state of electrochemical grinding. When the electrolytic speed increases gradually, the surface roughness of parts decreases. When the electrolytic speed reaches a certain value, the surface roughness of parts will reach a stable value.

REFERENCES

Asit Baran Puri, Simul Banerjee. Multiple-response optimisation of electrochemical grinding characteristics through response surface methodology [J]. The International Journal of Advanced Manufacturing Technology, 2013, 645.

Fritz Klocke, Andreas Klink, Udo Schneider. Electrochemical oxidation analysis for dressing bronze-bonded diamond grinding wheels [J]. Production Engineering, 2007, 12.

R.N. Goswami, S. Mitra, S. Sarkar. Experimental investigation on electrochemical grinding (ECG) of alumina-aluminum interpenetrating phase composite [J]. The International Journal of Advanced Manufacturing Technology, 2009, 407.

K.Z. Molla, Alakesh Manna. Optimization of Electrochemical Grinding Parameters for Effective Finishing of HybridAl/(Al2O3+ZrO2) MMC [J]. International Journal of Surface Engineering and Interdisciplinary Materials Science (IJSEIMS), 2013, 12.

Pedro Tartaj, Jose M. Amarilla. Iron oxide porous nanorods with different textural properties and surface composition: Preparation, characterization and electrochemical lithium storage capabilities [J]. Journal of Power Sources, 2010.

Ricardo Alcántara, Gregorio F. Ortiz, Pedro Lavela, José L. Tirado, Wolfram Jaegermann, Andreas Thißen. Rotor blade grinding and re-annealing of LiCoO2: SEM, XPS, EIS and electrochemical study [J]. Journal of Electro analytical Chemistry, 2005, 5842.

Suvadeep Roy, Ardhendu Bhattacharyya, Simul Banerjee. Analysis of effect of voltage on surface texture in electrochemical grinding by autocorrelation function [J]. Tribology International, 2007, 40.

I. Sandu, T. Brousse, D.M. Schleich, M. Danot. The chemical changes occurring upon cycling of a SnO2 negative electrode for lithium ion cell: In situ Mössbauer investigation [J]. Journal of Solid State Chemistry, 2005, 1792.

Architectural, Energy and Information Engineering – Sung & Chen (Eds)
© 2016 Taylor & Francis Group, London, ISBN 978-1-138-02791-6

Experiment on the regeneration performance of a new partitions filler

Li Ning Zhou, Hai Zhu Zhou & Jing Jing Yan
China Academy of Building Research Tianjin Institute, Tianjin, China

Zhi Jia Huang
Anhui University of Technology, Ma'anshan, Anhui, China

ABSTRACT: The filler is the important heat and mass transfer components of the dehumidifier and regenerator in the liquid desiccant system. The traditional gas-liquid direct contact filler meets an unfavorable factor, which is gas with liquid. In this paper, a new indirect gas-liquid contact partitions filler, which has a specific surface area of 286 m^2/m^3 and a porosity of 0.86, is proposed. Both regeneration performance of the partitions filler and 5090 wet curtain have been tested in a cross-flow regeneration module laboratory. LiBr solution is used as the desiccant, and the regeneration effect is described by parameters such as regeneration rate, renewable efficiency, average mass transfer coefficient, and volumetric mass transfer coefficient. The influence of the regeneration performance of the system is analyzed with the solution inlet temperature. A comparison of the regeneration performance between the new filler and the 5090 wet curtain is made. The result shows that the volumetric mass transfer coefficient of the filler is 7 to 36 percent higher than the 5090 wet curtain one when the solution temperature is between 43 and 60°C, and the quality of air with liquid in the new filler is 58 to 87 percent lower than the 5090 wet curtain one when the face velocity is between 0.387 and 0.645 m/s.

Keywords: partitions filler; 5090 wet curtain; cross-flow; regeneration experiment

1 INTRODUCTION

Liquid desiccant has many advantages such as low-grade solar energy or waste heat driven, and environmental-friendly cycle working fluid. In recent years, the temperature- and humidity-independent control technology has been well developed in a number of public and civil buildings[1–4]. However, the liquid desiccant has some problems such as gas with liquid and caustic property of liquid[5–6]. Because of the gas with liquid phenomenon, during its application in the civilian workplace, long-term inhalation of LiBr can cause skin rash and central nervous system disorders[7–8]. On the other hand, because the liquid has a caustic property, when it is used in industrial sites, it will affect the quality of products. Thus, these above problems limit this technology. To solve the problem of gas with liquid, the operation of the demister should be improved[9–11], but installing the defogging device cannot fundamentally solve the issue of gas with liquid. In the traditional filler, the air and the solution contact directly, which results in the gas with liquid phenomenon. Thus, a new partitions filler is designed. In the new filler, the air and the solution contact indirectly, which can solve the problem of

gas with liquid from the fundamental solution. Although many studies have been carried out on the regeneration performance of the regenerator at home and abroad[12–16], there are only a few studies on the regeneration performance of the partitions filler. In this paper, experimental tests were carried out to understand the performance of the filler, and a comparison of the regeneration performance between the new filler and the 5090 wet curtain was also made.

2 EXPERIMENTAL SUBJECT

The filler is important heat and mass transfer components in the liquid desiccant dehumidification/regeneration process, and its performance is directly related to the performance of the system. In this paper, the regeneration performance of both the partitions filler and 5090 wet curtain was tested in a cross-flow regeneration module laboratory. The air and the solution contact directly in the 5090 wet curtain, but the contact is indirect in the partitions filler. The flow channel of the 5090 wet curtain is shown in Figure 1(a), and the flow channel of the new partitions filler is shown in Figure 1(b).

Figure 1a.   The flow channel of the 5090 wet curtain.

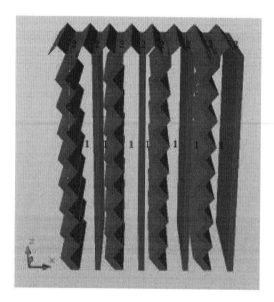

Figure 1b.   The flow channel of the partitions filler.

1. The flow channel of the air. 2. The flow channel of the solution.

As shown in Figure 1, the air is represented in the y-axis in channel 1 and the liquid represented in the negative z-axis in channel 2 in both the 5090 wet curtain and new partitions filler. However, in the 5090 wet curtain, the air and the solution

Table 1.

	Size (mm)	Porosity	Specific surface area (m²/m³)
5090 wet curtain	800 × 400 × 350	0.95	500
new filler	800 × 400 × 350	0.86	286

shared one flow channel, and had a direct contact, so the phenomenon of gas with liquid occurs easily. In addition, as shown in Figure 1(b), in the new partitions filler, each layer of the filler is comprised of two pieces, the solution flows into the two pieces of the filler, and the air flows into the two layers of the filler. The solution and the air flow in a different flow channel, thus achieving the indirect contact of the solution and the air, which solves the problem of gas with liquid effectively. The performance comparison of the two fillers is presented in Table 1.

As shown in Table 1, with the same size, the performance of the 5090 wet curtain is superior than that of the new filler, but both fillers has air contact with solution in different ways. The new filler has a direct contact between the air and the solution, which can effectively avoid the problem of gas with liquid. In order to compare the regeneration performance of the two fillers, both the partitions filler and the 5090 wet curtain were tested in a cross-flow regeneration module laboratory.

3 EXPERIMENTAL STUDY

3.1 Experimental facility

The experimental system consists of five major components: regeneration module, dehumidification module, heat source, cold source, and heat exchanger. The regeneration module uses the form of a cross-flow; the dehumidification module uses the countercurrent form; the heat source uses electric heaters instead of waste heat; the cold source uses cooling tower; and the heat exchanger uses the plate heat exchanger. Both the 5090 wet curtain and new partitions filler were used in the regeneration module, whose dimensions are 780×400×350 mm. The unit flow chart is shown in Figure 2.

Unit processes: in the dehumidifier, the desiccant has a strong solution contact with air within the dehumidifying, then the dilute solution which after dehumidification through the dehumidification outer circulation pump exchanging heat with the hot strong solution in the economizer, and then heated through the heater into the regenerator for regeneration; through a small change in the regeneration solution concentration, part of the solution

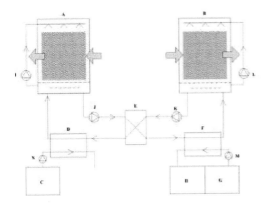

Figure 2. Unit flow chart: (A) dehumidifier, (B) regenerator, (C) cooling water tank, (D) cooler, (E) economizer, (F) heater, (G) heating water tank, (H) water storage tank, (I) dehumidification internal circulation pump, (J) dehumidification outer circulation pump, (K) regeneration outer circulation pump, (L) regeneration internal circulation pump, (M) heating water pump, and (N) cooling water pump.

was then passed to the regeneration internal circulation pump regeneration, another portion of the solution through the regeneration internal circulation pump exchanging heat with the cold dilute solution in the economizer, and then through the cooler for cooling, entering the dehumidifier and dehumidified, through the dehumidifier where the concentration of the solution is changed to a small degree, the portion of the solution can then be dehumidified through the dehumidification internal circulation pump, so as to complete a cycle.

3.2 Experimental condition

Through the fixing solution inlet flow (flow rate 1.60 m³/h), to change the heater heating temperature (heating temperature range 45 to 75°C), the face velocity is changed (face velocity range of 0.387 m/s~0.645 m/s). The experiments were conducted on 18 sets to study the regeneration performance of the two fillers.

The parameters measured in the experiment include: solution flow rate, inlet and outlet temperature, density of the solution import and export, and wet and dry bulb temperature of the air import and export.

4 RESULTS AND ANALYSIS

4.1 Evaluation indicators of regeneration performance

In the present study, some important parameters are used for evaluating the performance of the

regeneration including: m_{reg} (regeneration rate), η_{reg} (regeneration efficiency), k_a (average mass transfer coefficient) and k_v (volumetric mass transfer coefficient).

m_{reg} is defined as the rate at which the moisture is removed from the solution (kg/s). It can be calculated as:

$$m_{reg} = m_a(d_{a,out} - d_{a,in}) \qquad (1)$$

where $d_{a,in}$ and $d_{a,out}$ are the humidity ratio of process air at the inlet and exit from the regeneration, respectively, in kgv/kg d_a.

k_a is defined as the rate of moisture flux passing through a unit area (kg/m² s). It can be obtained from the measured data as follows:

$$k_a = \frac{m_{reg}}{A(d_{av} - d_{eq})} \qquad (2)$$

where A is the interfacial area of contact between the liquid desiccant and the air inside the regeneration. If this area is assumed to be fully wetted by the solution, then $A = a_p V$, where V is the volume of the filler in m³; a_p is the filler density in m²/m³; $d_{av} = (d_{a,in} + d_{a,out})/2$ is the average process air humidity ratio across the regeneration; and d_{eq} is the humidity ratio of air in equilibrium with LiBr solution at the interface.

Because of the constraints of the filler wetting and surface solution residence time, the filler mass transfer surface area A is often less than the filler actual surface area. The size of A is not only related to the geometric characteristics of the filler, but also to the gas-liquid two-phase flow and its physical characteristics; Thus, the mass transfer area A is difficult to be measured directly. Therefore, to define the volumetric mass transfer coefficient to reflect the mass transfer of the unit volume, the volumetric mass transfer coefficient is defined as the mass transfer unit volume (kg/m³ s). It is determined by the following formula:

$$k_v = \frac{m_{reg}}{V \cdot \Delta X} \qquad (3)$$

where ΔX is the mass transfer driving potential.

η_{reg} is defined as the actual humidity ratio drop of the process air to the maximum possible drop. It is calculated as follows:

$$\eta_{reg} = \frac{d_{a,in} - d_{a,out}}{d_{a,in} - d_{eq}} \times 100\% \qquad (4)$$

The quality of air with liquid defined as the quality of solution carried by per volume air:

$$m_s = \frac{m_2 - m_1}{m_a t} \qquad (5)$$

where m_s is the gas with liquid rate;

m_1 and m_2 are the filter paper weight before and after the experiment, respectively, clicked in the regenerator outlet;

t is the experiment time.

4.2 Effect of solution temperature on regeneration performance

The effects of the solution temperature on the rate of regeneration, volumetric mass transfer coefficient, average mass, transfer coefficient, and regeneration efficiency are shown in Figures 3–6.

As shown in Figure 3, under the same conditions of the regenerator in the same cross-section wind speed and the same solution flows, the regeneration rate increases as the solution temperature increases. This is because as the solution temperature increases, the water vapor partial pressure of the solution is increased, and the mass transfer driving potential between the air and the solution increases, so the regeneration rate increases.

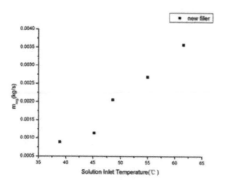

Figure 3.   Effect of the solution temperature on m_{reg}.

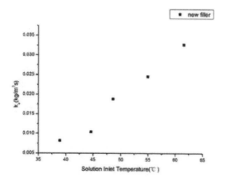

Figure 4.   Effect of the solution temperature on kv.

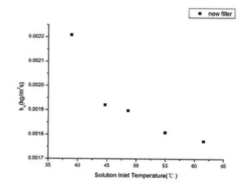

Figure 5.   Effect of the solution temperature on ka.

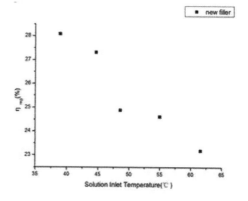

Figure 6.   Effect of the solution temperature on η reg.

As shown in Figure 4, under the same experimental conditions, the volumetric regenerator mass transfer coefficient increases as the solution temperature increases. From the above analysis, it can be understood that the regeneration rate increases as the solution temperature increases, so as the unit volume mass transfer increases, the volumetric mass transfer coefficient reflects per unit volume mass transfer of the packed, so the volumetric mass transfer coefficient increases with the rise in the temperature of the solution.

From Figure 5 under the conditions of the same cross-section wind speed and the same solution flows, the average mass transfer coefficient of the regenerator decreases rapidly with increasing temperature of the solution. This is because as the solution inlet temperature increases, its equivalent moisture decreases, and the degree of reduction is greater than the degree of increase in the regeneration rate. This results in the decline of the average mass transfer coefficient.

As shown in Figure 6, under the same experimental conditions, the regeneration efficiency decreases as the solution temperature increases. This is because the surface water vapor partial pressure of the solution and the moisture content of the outlet air increases as the solution temperature increases, and the equivalent moisture content of the solution that contact with the air increases, but the range of the latter is greater than the former, as shown in equation (1.4), and the regeneration efficiency decreases with increasing temperature of the solution.

4.3 Comparison of the regeneration performance of the two fillers

Figures 7–9 show the comparison of the regeneration performance of the two fillers under the same experimental conditions. Figure 10 shows the quality of air with liquid along with the change of face velocity in both the new partitions filler and 5090 wet curtain.

As shown in Figure 7, under the same experimental conditions, although the average mass transfer coefficient of the two fillers is reduced as the inlet solution temperature increases, the average mass transfer coefficient of the new partitions filler is always higher than that of the 5090 wet curtain.

As shown in Figure 8 under the same experimental conditions, the volumetric mass transfer

coefficient of the two fillers increases with increasing temperature, and the new partitions filler's volumetric mass transfer coefficient is higher than the 5090 wet curtain's and has the same trend with the regeneration rate.

As shown in Figure 9, under the same experimental conditions, the regeneration rate of the new partitions filler is higher than that of the 5090 wet curtain, and the regeneration rate of both the fillers increases as the temperature increases, but as the solution temperature increases, the new partitions filler's regeneration rate is more higher than the 5090 wet curtain's. This is because the indirect contact of the solution and the air in the partitioned fillers affects the heat transfer, which could make the solution to maintain at a higher temperature, resulting in a higher water vapor partial pressure and mass transfer driving potential, so the higher the temperature, the more obvious the advantages, and as the temperature rises, the regeneration rate of the new filler is greater than that of the 5090 wet curtain.

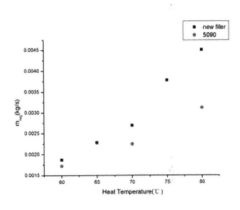

Figure 9.   Effect of the heat temperature on mreg.

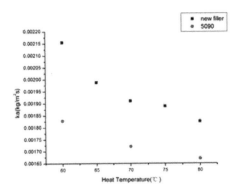

Figure 7.   Effect of the heat temperature on ka.

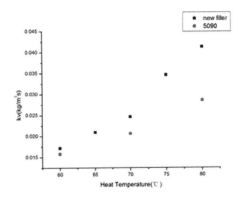

Figure 8.   Effect of the heat temperature on kv.

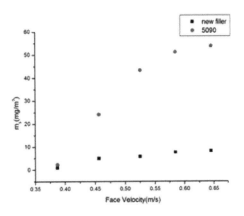

Figure 10.   Effect of the face velocity on ms.

As shown in Figure 10, in the 5090 wet curtain, the quality of air with liquid increases as the face velocity increases. This is because with the increase in the face velocity, the ability of gas with liquid increases. However, in the new partitions filler, the quality of air with liquid changes to a small extent with the increasing face velocity, and quality of air with liquid can be reduced by 58%–87%. This is mainly because the indirect contact between the air and the solution in the new partitions filler reduces the chance of the air carrying the solution, so the quality of air with liquid in the new partitions filler can be significantly reduced compared with the 5090 wet curtain.

5 CONCLUSIONS

1. Due to the fact that the traditional filler meets an unfavorable factor, which is gas with liquid, in this paper, a new partitions filler is designed, which has a specific surface area of 286 m^2/m^3 and a porosity of 0.86. The experimental study found that the regeneration rate of the new filler and the volumetric mass transfer coefficient increased with the rise in the temperature of the solution, and that the average mass transfer coefficient and the regeneration efficiency decreased with increasing temperature of the solution.

2. Since the filler mass transfer area is difficult to measured directly, it often makes the measured average mass transfer coefficient inaccurate. Moreover, the regeneration rate and the volumetric mass transfer coefficient have the same trend, and can reflect the regeneration performance of the regenerator at the macroscopic level, with the use of the volumetric mass transfer coefficient to evaluate the regeneration performance of the regenerator being more scientific.

3. Because the indirect contact between the solution and the air in the new partitions filler affects its heat transfer, the solution was maintained at a higher temperature, and the greater the temperature the more obvious the advantages; therefore, as the temperature increases, the volumetric mass transfer coefficient of the new filler is higher than that of the 5090 wet curtain. When the temperature of the solution being heated is at 60~80°C, the volumetric mass transfer coefficient of the new partitions filler is 7% to 36% higher than that of the 5090 wet curtain.

4. The use of the new partitions filler can significantly reduce the quality of air with liquid, and the quality of air with liquid of the new filler is 58 to 87 percent lower than that of the 5090 wet curtain when the face velocity is between 0.387 and 0.645 m/s. In addition, the quality of air with liquid changes to a small extent with the increase in the face velocity in the new partitions filler.

REFERENCES

An Shouchao, Wang Jin, Liu Inhaul, et al. Liquid desiccant packed tower liquid entrainment and pressure drop problem [J]. HV&AC, 2007, 37(4): 109–112.

M.M. Bassuoni. An experimental study of structured packing dehumidifier/regenerator operating with liquid desiccant [J]. Energy, 2011, 36: 2628–2638.

Chen Xiaoyang, Cong Lin, Zhang Ting, et al. liquid desiccant air handling technology research progress [J]. HV&AC, 2011, 41 (1): 21–27.

Chen Yujian, Pei Qingqing, Xu Guiquan, et al. Liquid desiccant air conditioning air with liquid ion detection method and its application [J]. Building Energy & Environment, 2009, 28(3): 37–40.

Gu Zhongxuan, Liu Yunjie, Huang Douyu, et al. Temperature and humidity independent control air-conditioning systems in large public buildings in the Beijing area applications [J]. HV&AC, 2011, 41(1): 53–54.

Jiang Fuguang, Cai Zhangli, Lu Zhu. BTA on the corrosion of carbon steel in lithium bromide solution [J]. The refrigerating, 1999, 18(4): 29–31.

Liu Shuanqiang, Liu Xiaohua, Jiang Yi, et al. Nanhai E # 3 office temperature and humidity measurement and analysis of individually controlled air conditioning system [J]. HV&AC, 2011, 41 (1): 55–59.

Li Yan, Hu Rong, Feng Tingting. Temperature and humidity independent control air-conditioning system in application of the Xiangxi courtyard two villa in Qingdao [J]. HV&AC, 2011, 41(1): 42–47.

Luo Lei, Zhang Xiaosong, Yin Yonggao. The experimental study of heat and mass transfer coefficients of Forks flow packing regenerator [J]. Journal of Engineering Thermophysics, 2008, 29(7): 1215–1217.

Luo Xulu, Li Wenwu, Cao Zhixi. The refrigerant rust change lithium bromide solution corrosive research [J]. Light Industry Machinery, 2011, 29(1): 104–107.

Ritunesh Kumar, P.L. Dhar, Sanjeev Jain. Development of new wire mesh packings for improving the performance of zero carryover spray tower [J]. Energy, 2011, 36: 1362–1374.

Sun Jian, Shi Mingheng, Zhao Yun. Experimental study of the performance of liquid desiccant air conditioning regeneration [J]. Engineering Thermo Physics, 2003, 24 (5): 867–869.

Wang Qiang, Wang Gang. Theoretical research and experimental analysis of the solution regeneration of liquid desiccant air conditioning system performance [J]. Refrigeration and Air Conditioning, 2010, 24 (4): 124–128.

Wang Shunli. Preliminary study of liquid desiccant air conditioning indoor air quality [D]. Guangzhou: Guangzhou University, 2007.

Yin Yonggao, Li Shiqiang, Zhang Xiaosong. Adiabatic and heat-regenerative process thermal performance comparison [J]. Chemical Industry and Engineering, 2010, 61 (S2): 157–162.

Zhang Liang, Liu Jianhua, Zhang Haijiang, et al. Liquid desiccant screen corrugated packing application characteristics research [J]. Fluid Machinery, 2011, 39 (6): 48–52.

Architectural, Energy and Information Engineering – Sung & Chen (Eds)
© 2016 Taylor & Francis Group, London, ISBN 978-1-138-02791-6

Summary of the beam-column joints connection in a precast concrete structure

Y.L. Liu & T. Qin
College of Civil and Architectural Engineering, Hebei United University, Tangshan, China

ABSTRACT: With the development of construction in China, the prefabricated concrete frame structure has been rapidly developed. This article compares the merits and drawbacks of precast concrete and cast concrete structures. It summarizes the types of beam-column joints connections of precast concrete structures, analyzes the review of all kinds of beam-column joints connections. Finally, it discusses the key points that need to be further studied and the development direction in the future.

Keywords: precast fabricated concrete; beam-column joints; connection

1 INTRODUCTION

In China, the construction industry has become a large energy consumer. However, it has many defects, such as traditional operation mode, low level of industrialization, energy consumption and high pollution. So, the state proposed the national development strategy of building energy conservation and the housing industry, attempting to eliminate the constraints of building industrialization process bottlenecks.

The so-called building industrialization refers to the use of intensive, industrial pipeline operations to complete the prefabrication of components, parts and equipment, which constitute most of the construction process. Then, they are shipped to the construction site for assembling an integrated whole. This is the method adopted in the modern construction. Since the beginning of the 1940s, the process of industrialization has been gradually propelled in Europe and other countries. Especially in Japan, its architectural industrialization rate has exceeded 70%, with the systematic, advanced and mature technology support.

Precast concrete structures have a short construction period, high product quality, low energy consumption, little environmental pollution and other advantages. It is the inevitable choice to achieve sustainable development in the construction industry.

2 CONNECTION TYPES OF BEAM-COLUMN JOINTS

Currently, cast concrete has many advantages and thus plays an important role in the most construction. However, it is undeniable that it

Table 1. The comparison between precast concrete and cast concrete.

	Cast concrete	Precast concrete
Period	Long production cycle	Production cycle greatly reduced
Economy	Relatively high cost	Relatively low cost
Quality	Has some quality defects	Has high average quality
Environment	Not conducive to environmental protection	Very environmental friendly

exposed flaws in the project. For example, a long construction period, and the difficulty to guarantee the quality of the project and the control of the cost [1]. Relative to these issues, precast concrete shows certain advantages. It is different from cast concrete. It pours not at the construction site, but in other places. It will be cast into individual components, and later transported to the construction site to be assembled into structures. At present, precast concrete is widely used. For example, the precast prestressed concrete pedestrian suspension bridge in the Honduras city of San Pedro Sula [2] and the North Caro Linna IJL Financial Center in the US city of Charlotte [3]. A comparison between precast concrete and cast concrete is presented in Table 1.

3 CONNECTION TYPES OF BEAM-COLUMN JOINTS

The beam-column connection of frame structure, also known as beam-column joints or frame joints, mainly refers to the frame beam and frame column's intersect node core area and near the core area of the beam ends and column ends [4]. According to the construction of fabricated concrete structures, it can be divided into dry node connections. Wet connections is grouting or pouring cast concrete between two connecting members. Dry connections is not needed for cast concrete. It is connected by embedded steel or other steel parts in the inner member, welded or bolted.

4 DRY CONNECTIONS OF BEAM-COLUMN JOINTS

4.1 *Corbel connections*

With high carrying capacity and a more reliable delivering vertical force, the corbel connections are quite common in dry connections. For single or multi-plant and other large spaces, their carrying capacity has higher requirements for building corbels. So, an obvious corbel with a higher carrying capacity is used. For building residential or commercial buildings of high demand, inconspicuous corbels are mostly used, which will not affect the appearance [5]. There a variety of inconspicuous corbels, such as steel inconspicuous corbels and concrete inconspicuous corbels.

(1) Obvious corbels.

In the multi-plant of prefabricated reinforced concrete, obvious corbels are widely used, as shown in Figure 1. Obvious corbels have a large bearing capacity, security force, good node rigidity, convenience of construction and installation. There are two types of obvious corbels connection: rigid connection and hinged connection. Their structural details are not the same. However, due to the characteristics of the obvious corbels, the main consideration is the performance of bearing capacity while ignoring the aesthetics and relatively taking up space. So, obvious corbels should be

used only for less demanding aesthetic buildings, such as plants and crane beam support.

(2) Inconspicuous corbels.

When buildings for residential or commercial buildings with high aesthetic requirements, people tend to make use of inconspicuous corbels, which will not affect the aesthetics and the space, as shown in Figure 2. Adaptation to changes in the appearance of the structure and properties simultaneously brings disadvantages, especially the static and dynamic performance of the design is very unfavorable. If the half height of the beam can withstand shear, the other half height of the beam makes the column corbel and the corbel does not exceed the requirements of the edge of the beam, so that the end of the beam and the corbel's reinforcement are relatively complex. Therefore, not all the nodes connection modes are suitable.

(3) Inconspicuous steel corbels.

When the shear is large and the half beam cannot afford the shear of the beam, corbels can be made of steel corbels. In this way, the height of the inconspicuous corbel can be reduced, to increase the height of the gap at the end of the beam and the gap shear capacity [6]. If wrapping the steel up in the concrete in the production process, then the corbel with the ordinary cast reinforcement corbel looks similar. If making the steel sticking out or making the steel into the beam end, the connection is not visible at the side seams. However, in order to prevent fire and corrosion, caution must be taken to pour after installation, as shown in Figure 3.

4.2 *Steel hanger connections*

Steel hanger connections use less amounts of steel. Its significant advantage is the ability to develop simple column templates. Steel hanger connection can prevent eccentric load caused by occasional distortion of the beam. A pin that is located below the steel hanger can resist the torque of the beam, as shown in Figure 4. In the same way, a rigid con-

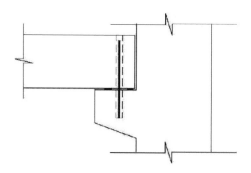

Figure 1. Obvious corbel node.

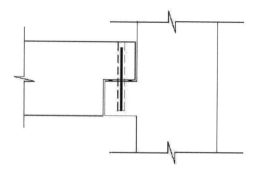

Figure 2. Inconspicuous corbel node.

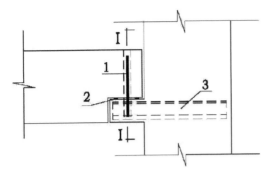

Figure 3. Inconspicuous steel corbel (hinged connection). 1. Grouting pin after adding; 2. Neoprene sheet; 3. Steel.

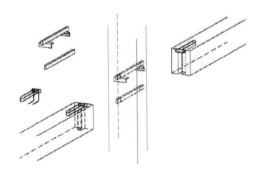

Figure 4. Steel hanger connection (hinged connection).

nection can be made between the beam and the column. The drawbacks of the connection are relatively complex structure, quality of construction and installation of high accuracy requirements, connecting components in a large member, and relatively weak affordability. It is not suitable to withstand large loads [7].

4.3 *Welded connections*

One method of overseas dry-type connections is welded connections, as shown in Figure 5. The connection has no significant plastic hinge and welding seams under repeated earthquake loads prone to brittle failure. So, the seismic performance of the connection is less ideal[8]. However, welded connections eliminate the step of pouring on site and the need to carry out the necessary conservation of concrete, which saves the time. As for the weld joint that set a good plastic hinge, its superiority is still quite obvious. So, the current direction of dry connections includes developing welded connections of good structure deformation properties. In the construction of the welding process, components should be fully arranged, so that it can reduce welding residual stress effectively.

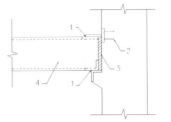

1.Welded connection; 2.Stud; 3.Fine aggregate concrete; 4.Precast beam.

Figure 5. Welded connection (rigid connection). 1. Welded connection; 2. Stud; 3. Fine aggregate concrete; 4. Precast beam.

4.4 *Bolted connections*

Bolted connections have some advantages: joint's installation is quick and neat, but the drawbacks are that the location at the time of pre-bolts would have a little bias and in order to avoid the bends, thread damage and contamination, one must be extremely careful to protect them during transportation and installation. If a bolt threaded or bolt holes have been damaged, their repair or replacement is more complicated. In this connection, the connection structure is generally more complex and there are more connecting components[9]. This connection is the application of the corbel and the bolted connection, as shown in Figure 6. Figure 6 (a) shows the case for obvious corbel and precast beams in the bolted connection. Figure 6 (b) shows the case for inconspicuous corbel. Connections in Figure 6 (a) and Figure 6 (b) can resist a small moment and torque of the beam end, which are hinged connections.

5 WET CONNECTIONS OF BEAM-COLUMN JOINTS [10]

Wet connections mean that prefabricated or prefabricated with cast-in-site components connected together to form the frame structure in cast concrete. There are several common wet connections, as shown in Figure 7. In Figure 7 (a), the beam is precast concrete beam, and precast slab is on the prefabricated beam. On the beam and floor surfaces, there are the core area of beam-column joints and column layout steel, and then pouring concrete. Precast beam and the surface of cast concrete work together to form a composite beam by reinforced ligation. In this way, the workload of site templates can be reduced, but the overhanging longitudinal steel bars of the beam bottom's construction is more difficult in the core part of the anchor node. So, the column size is larger than

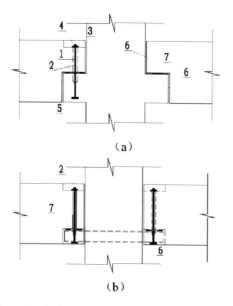

(a)

(b)

Figure 6. Corbels and precast beams of bolted connections (hinged connections). 1. Bolt; 2. Grout; 3. Plate; 4. Nut; 5. Screw and nut of pouring; 6. Grout; 7. Adjustable abutment.

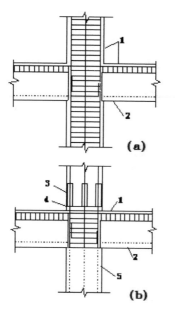

(a)

(b)

Figure 7. Common wet connections. 1. Cast concrete; 2. Precast beam; 3. Steel sleeve; 4. Mortar; 5. Precast column.

the beam size. Furthermore, the core area of the beam-column joints' reinforced is dense, so this part is not easy to pour concrete densely.

The differences between Figure 7(b) and Figure 7(a) are that beams and columns are precast. Concrete is poured at the core area of precast beams

and beam-column joints plate surface, the column longitudinal reinforcement is through metallic sleeve. Then, they can fill together by grouting. In this way, pouring work is less on site, but it cannot avoid the issues of pouring concrete in the beam bottom longitudinal reinforcement anchorage and joint core.

6 SUMMARY

Precast concrete structures are an important building architecture with many advantages, including: rapid pace of construction, the production of precision, low level of construction difficulties, reduce or avoid wet work, and favorable environment. Many countries regard it as an important and main structure.

There are a wide variety of connection types of beam-column joints of precast concrete structures. There is not a complete summary nowadays. The different specifications also lead to different forms of connection, which is not conducive to an in-depth study in beam-column connection nodes in the future.

Precast concrete beam-column joints connecting research is still in its infancy at home and abroad. The rapid development of the housing industry urgently requires us to advance with the time and to keep up with the pace of the time. This research must be continued for further improvement.

REFERENCES

[1] Huiying Wang. Research on Industrialized House System of Precast Concrete Structure [D]. Guangzhou: Guangzhou University, 2007(5). (in Chinese).
[2] Li Zhen-qiang, Rigoberto R C. Precast prestressed cable 2 stayed pedestrian bridge for buffalo industrial park [J]. PCI Journal, 2000(3): 22–33.
[3] Stewart H, Hamva S M, Gleich H A. Curved precast facade adds elegance to IJL financial center and parking structure [J]. PCI Journal, 2000(3): 34–35.
[4] Qinjian Jiang, Zhiqiang Zhong. China precast concrete industry overview in 2011 [J]. China Concrete, 2012, 01: 94–99. (in Chinese).
[5] Jiuru Tang. Seismic performance of reinforced concrete frame joints [M]. Publishing house of Dongnan University, 1989. (in Chinese).
[6] Xinqi Fu, et al. Prefabricated buildings of reinforced concrete [M]. Beijing: Publishing house of China building industry, 1985. (in Chinese).
[7] Haas A M. Precast Concrete: Design and Applications [M]. New York: Applied Science Publishers, 1983.
[8] Robert A. Hartland. Design of Precast Concrete [M]. Great Britain: Surrey University Press, 1975.
[9] Englekitk R. Seismic design considerations for percast concrete multistory buildings [J]. Journal of precast/ prestressed Concrete Institute. 1990, 35(3): 40–54.
[10] Wenjing Ju. The joint connections Summary of Precast concrete frame [J]. Sichuan Building Materials, 2011(6): 38–41. (in Chinese).

Architectural, Energy and Information Engineering – Sung & Chen (Eds)
© 2016 Taylor & Francis Group, London, ISBN 978-1-138-02791-6

A new space shift keying system on difference equation

Peng Cheng Guo
Institute of Communication Engineering PLA, University of Science and Technology, Nanjing, China

Fu Qiang Yao, Sheng Yong Guan & Yong Li
Nanjing Telecommunication Technology Institute, Nanjing, China

ABSTRACT: A new Space Shift Keying system is put forward on Difference Equation, and two kinds of antenna mapping structure are given. First of all, we mainly introduce how to apply the Difference Equation to the antenna mapping. Secondly, the algorithm used by receiver and the complexity of decoding are analyzed. On the system's transmission performance, we compare the difference between the proposed scheme and the traditional technology of Space Shift Keying. At last, through the comparison of the two kinds of antenna mapping structures with uniformity and randomness, we reach a conclusion that the system of convolutional antenna mapping has a better performance.

Keywords: difference equation; space shift keying

1 INTRODUCTION

Multiple-antennas have been developed rapidly in the wireless communication systems in recent years. As the key technique of 4 G mobile communication, the MIMO technique can improve the system performance anti-noise and anti-fading, such as the Vertical Bell Laboratories Layered Space-time (V-BLAST) architecture. However, simultaneous transmission on the same frequency from multiple transmitting antennas causes high Inter-Channel-Interference (ICI). Based on MIMO technique, R. Mesleh and H. Haas have put forward the Spatial Modulation technology in [1]. SM [2] can avoid ICI and the need of accurate time synchronization among antennas through adopting only one antenna to transfer information at any instant. The antenna index takes along additional source of information, which increases the spectral efficiency.

Space Shift Keying (SSK) [3] has a similar transmitting model with SM. The information bits is mapped into space constellation to encode the transmit antenna number. Compared with SM, SSK cancels the traditional digital modulation and has lower complexity of system.

Based on Difference Equation, this paper proposes a new Space Shift Keying system, aiming to explore a way of anti-jamming transmission in air—space. As we all known, the Differential Frequency Hopping [4] is a new frequency hopping spread spectrum system, mainly used for short wave communication. The correlation is set up between the adjacent hopping frequencies through the data. This paper applies the correlation to the antennas, exploring to achieve a new Space Shift Keying system, which can be called Differential Space Shift Keying (DSSK) for convenience. In this paper we will employ two kinds of T function structure: Structure

One is a simple linear T function, Structure Two is the convolutional T function combined with TCM. Also, the comparison of performance will be given.

2 THE BASIC PRINCIPLE OF DSSK

In the Differential Frequency Hopping (DFH), the basic principle of sending information may be expressed as: the current frequency value f_n is decided by the last frequency value f_{n-1} and current information symbol X_n. A mathematical expression is $f_n = G(f_{n-1}, X_n)$. In this formula, $G(\cdot)$ is a mapping function from sending information to sending frequency, and express a special kind of modulation. In the DSSK, we will apply the T function used for the relationship between the antennas just like the G function in the DFH, as $y_n = T(y_{n-1}, X_n)$, which can be supposed to a kind of differential implicit function in the form, the current antenna index y_n is decided by the last antenna index and current input information.

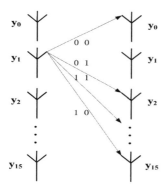

Figure 1. The direction graph of T function.

2.1 T function structure

T function can be regarded as a directed graph, in which each node represents a transmitting antenna. Considering an antenna group with sixteen antennas as an example, as shown in Figure 1, the directed graph has sixteen nodes and each node has four components, which contains two bits information, as (00, 01, 10, 11). It can be shown as $BPH = 2$, in which BPH (Bits Per Hop) expresses the bits number when the antennas switch every time.

Structure One shows a simple linear T function: the current antenna index y_n relates to current information sequence X_n in addition, just to last antenna index y_{n-1}. Also, the relationship between the data and antenna index is linear. When $BPH = 1$, the function expression is given by

$$y_n = G(y_{n-1}, X_n) = y_{n-1} + x_n + 1 \qquad (1)$$

With eight antennas as an example, a set of relationship between the data and antenna index can be shown in Table 1, setting the starting antenna as $y_i = 3$.

Structure Two is the convolutional T function combined with TCM. A set of structure with eight antennas is shown in Figure 2, in which the input information conducts convolutional coding and the output information points to the current antenna position, with $BPH = 1$.

In the Figure 2, input data X_n enters the convolutional encoder firstly, and then operates with the data in the register X_n and M_1. There will be a three bits data as output, which can select the antenna index through number switching. The input information digits mean the BPH, and the output information digits M_2 means the transmitting antenna index$s$, the corresponding relation to N_t.

The convolutional T function may not reflect the relationship between the current antenna and

Table 1. Relationship between the data and antenna index.

yi	y3			
data	0	1	1	0
y_n	y_4	y_6	y_0	y_1

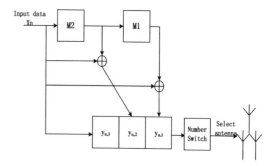

Figure 2. Convolutional T function structure.

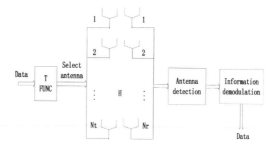

Figure 3. Transmission principle diagram of DSSK.

the last antenna intuitively, However, the relationship implicit in the convolution of the input data, which can be proved in [5].

2.2 Transmission model of DSSK

The basic transmission principle diagram is shown in Figure 3. The input data $N_t = 2^s$ enters into the DSSK system and selects the corresponding antenna index through T function mapping. Also the initial value should be preset in the T function. The antenna index needs to be detected to demodulate data in the receiving end, after channel x transmission. When the current antenna position y_n and last antenna position H are notarized, the effective information can be demodulated by inverse transformation of T function, as

$$X_n = T^{-1}(y_{n-1}, y_n) \qquad (2)$$

$T^{-1}(\bullet)$ means the inverse transformation of T function in (2), then the T function should be reversible.

2.3 Receiving algorithm

The receiving end needs firstly to estimate the antenna position after receiving information. We will apply the Maximum Likelihood Detection [3] to detect antenna index in this paper. Setting transmission symbol $N_t = 2^s$ and the predicted antenna index ℓ and then

$$\hat{\ell} = \arg\max_{\ell} \rho_Y(y \mid x, H) = \arg\min_{\ell} \left\| y - \sqrt{\rho} h_\ell \right\|_F^2$$
$$= \arg\max_{\ell} \mathrm{Re}\left\{ \left(y - \frac{\sqrt{\rho}}{2} h_\ell \right)^H h_\ell \right\}, \tag{3}$$

Among (3), $1 \le \ell \le N_t$,

$$\rho_Y(y \mid x, H) = \frac{\exp\left(-\left\| y - \sqrt{\rho} H x_\ell \right\|_F^2 \right)}{\pi^{N_t}} \tag{4}$$

The receiving end may get the antenna index through ML detection. In the Structure One, the data can be demodulated by inverse transformation of T function. However, In the Structure Two, Viterbi decoding can be employed because of the convolutional coding of the transmission antenna. At this time, Viterbi decoding could be regarded as the path search to antenna position, and also correct the errors of detected antenna index.

2.4 Complexity

In this part, we compare DSSK's complexity to that of SSK. Complexity depends on the workload, corresponding to the implementation of basic operation. So we should begin from ML detection to demodulating sending information in the process of algorithm complexity analysis of antenna demodulation.

In our presentation, we quantify complexity by the number of additions required in the detection process. The number of multiplications could be shown to have a similar value for both detectors, which is given by

$$\delta_{DSSK} = \delta_{SSK} = N_r M \tag{5}$$

M means the total size of the modulation constellation, and is equal to the transmitting antenna numbers: $M = N_t$. The number of additions of SSK is given by

$$\zeta_{SSK} = 2N_r M - M \tag{6}$$

In DSSK, the computation of additions of

$$\zeta_{DSSK} = 2N_r M - M + 1 \tag{7}$$

The number of additions in the Structure Two reduces to

$$\zeta_{DSSK'} = 2N_r M - M + 2L \tag{8}$$

where L represents the state numbers, as a result, the algorithm complexity of DSSK is higher than that of SSK.

3 PERFORMANCE ANALYSIS

The simulation conditions 1: AWGN channel, the simulation points with *num* = 131072, the receiving antenna numbers with $N_r = 8$, $BPH = 2$, different transmitting antenna numbers. The performance simulation of DSSK in the Structure One and the performance comparison to SSK can be shown in Figure 4.

In Figure 4, the performance of DSSK gets better with the less numbers of transmitting antenna under the condition of same receiving antenna numbers. The double numbers of transmitting antenna may bring almost same SNR growth with 2 dB in $BER = 10^{-3}$. When $N_t = 8$, the performance of DSSK approach by comparing the curves of $BPH = 1$ to that of $BPH = 2$. We observe that BPH has a little effect on the performance of DSSK. Meanwhile, the performance of DSSK is slightly worse that of SSK. The number of SNR reduces about 1 dB when $BER = 10^{-3}$. The performance could be improved through optimizing structure and receiving algorithm.

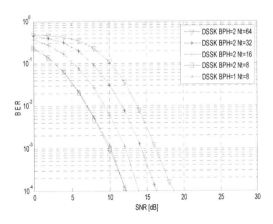

Figure 4. Performance of DSSK with different BPH.

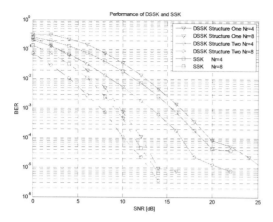

Figure 5. Performance of DSSK and SSK.

The simulation conditions 2: AWGN channel, the simulation points with *num* = 131072, the receiving antenna numbers with $N_r = 8$ and $N_r = 4$, the transmitting antenna numbers with $N_t = 8$. The performance comparison of DSSK with two structure and SSK is given in Figure 5.

In Figure 5, the performance of DSSK in Structure Two is better than that in Structure One and SSK with the same resource of antennas. In particular, the performance of DSSK in Structure Two is about 1.5 dB better than that of SSK and 2 dB better than that of DSSK in Structure One in $BER = 10^{-3}$ when $N_r = 8$. The performance of DSSK in Structure Two is about 2 dB better than that of SSK and 3 dB better than that of DSSK in Structure One in $BER = 10^{-3}$ when $N_r = 4$. In addition, the performance of DSSK and SSK in $N_r = 8$ is better than that in $N_r = 4$ with the same transmitting antenna numbers. It can be seen that the relative numbers of transmitting and receiving antenna play an important influence on the system performance.

The system performance of DSSK in Structure Two has been increased greatly by antenna convolutional coding and Viterbi decoding. Nevertheless, this structure exist a certain limit. For example, the numbers of transmitting antenna lie on information convolution output and BPH numbers depend on information convolution input. Also, the change of either of both means the change of convolutional structure.

4 PERFORMANCE TEST OF T FUNCTION

As mentioned above, T function has the function of data modulation and demodulation, and also produces antenna hopping sequence. Then we will adopt uniformity and randomness experiment to the antenna hopping sequence as DFH.

4.1 Uniformity experiment

We may divide the uniformity experiment to one—dimensional distribution test and two-dimensional continuity test and apply χ^2 guidelines [6] to the experiment.

The experiment conditions: the input data with $s = 131072$, $BPH = 1$, the transmitting and receiving antenna numbers with $N_t = N_r = 8$. The result of one-dimensional distribution test and two—dimensional continuity test of DSSK in two kinds of structure is given in Table 2.

In the one-dimensional distribution test the calculated value of χ^2 in the Table 2 is less than the theoretical value $\chi^2_{0.05}(N-1)$ under the specified level. We can see that the one-dimensional distribution of two kinds of structure is fine. Also, in the two-dimensional continuity test the calculated value of χ^2 in the Table 2 and Table 3 are greater than the theoretical value $\chi^2_{0.05}(N-1)$ under the specified level. We can find that the two-dimensional continuity of two kinds of structure is bad. In comparison, the Structure One is slightly better than the Structure Two in the one-dimensional distribution and the two structures approach in the two-dimensional continuity. We can promote the two-dimensional continuity of antenna hopping sequence through optimizing T function design.

4.2 Randomness experiment

Randomness experiment is mainly estimated through the power spectrum of the antenna hopping sequence. The flatter the power spectrum is, the better the randomness of system is. On the other hand, the randomness is worse. In this part we will estimate the power spectrum of DSSK with two antenna mapping structures through Bartlett method.

The experiment conditions: the input data with $s = 131072$, $BPH = 1$, the transmitting and receiving antenna numbers with $N_t = N_r = 8$, rectangular window, the length of window $L = 1024$, without overlapping of every piece of data.

Table 2. Test result of DSSK in the two Structures.

	One-dimensional test	
Test index	Theoretical χ^2	Calculated χ^2
Structure one	13.7841	2.3921
Structure two	13.7841	6.2281
	Two-dimensional test	
Test index	Theoretical χ^2	Calculated χ^2
Structure one	82.2447	393250
Structure two	82.2447	393270

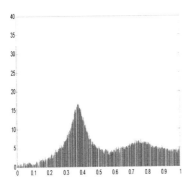

Figure 6. Curve of structure one.

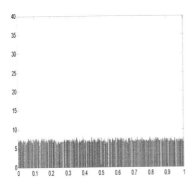

Figure 7. Curve of structure two.

The curve of power spectrum is given in Figure 6 and 7 with the normalized frequency.

We can find that the curve of power spectrum in Structure One is not flat in Figure 6, which reflects the worse randomness of T function in the simple linear structure as (1). Meanwhile, the curve of power spectrum in Structure Two is very flat in Figure 7, which reflects the better randomness of T function in the convolutional structure.

In DSSK, the randomness of antenna sequence produced by convolutional structure enhanced significantly. The convolutional structure transformed the simple linear operation in T function to convolutional operation and got the better randomness, which can increase the system performance of resistance to decipher and intercept.

5 PERFORMANCE ANALYSIS OF ANTI-JAMMING

DSSK applies this idea of DFH to the antenna mapping and could be an exploration to the anti-jamming methods in airspace.

The anti-jamming ability of DFH communication mainly involves anti-resistant jamming ability,

anti-tracking jamming ability and resistance to multipath jamming ability. Among these, anti-resistant jamming ability mainly relies on larger band width of Frequency Hopping and forces the resistant jammer to pay larger price on the bandwidth and power. Likewise, DSSK has the similar advantage. The sender could make the antenna signals be distributed in the broad space, which make the jammer difficultly carry out signal suppression cover the whole space or pay larger price on the signal jamming.

DFH has a benefit anti-tracking jamming ability relying on high jumping speed and random Frequency Hopping pattern. For DSSK, the receiver could receive information from different space channels relying on antenna hopping pattern. The antenna hopping pattern should be controlled by random data and won't repeat. Thus, the jammer will difficultly grasp the path of antenna jumping and apply jamming to the changing space channels.

6 CONCLUSION

In this paper, we proposed a new Space Shift Keying technology (DSSK) based on the idea of Different Frequency Hopping (DFH). The correlation between frequencies will be applied to antenna mapping, aiming to explore a way of anti-jamming transmission in airspace. Also, two kinds of structures of antenna mapping had been put forward in DSSK and compared with the traditional SSK. The simulation results showed that the performance of DSSK with convolutional structure is better than SSK. Meanwhile, we can find that DSSK has a unique advantage on the jamming in airspace through the performance analysis of anti—jamming on DSSK.

REFERENCES

[1] R. Mesleh & H. Haas, C.W. Ahn, et al. *Spatial Modulation—OFDM* [A]. 11th International OFDM-Workshop 2006 (InOWo'06), 2006. 288–292.

[2] R. Mesleh, H. Haas& S. Sinanović, et al. *Spatial Modulation* [J]. IEEE Trans. Veh. Technol, 2008, 255 57(4): 2228–2241.

[3] J. Jeganathan, A. Ghrayeb & L. Szczecinski, et al. *Space Shift Keying Modulation for MIMO Channels* [J]. IEEE Trans. Wireless Commun. 2009, 8(7): 3692–3703.

[4] Yao Fuqiang. *Communication Anti-jamming Engineering and Practice*, Second Edition [M]. Beijing: Publishing House of Electronics Industry, 2012.

[5] Yang Baofeng, Shen Yuehong. *Study Method of Equivalent Convolutional Code Structure for Differential Frequency Hopping* [J]. Journal of Jilin University. 2006, 24(5): 495–500.

[6] John G. Proakis & Masoud Salehi. *Digital Communications, Fifth Edition* [M]. Beijing: Publishing House of Electronics Industry, 2011.

Architectural, Energy and Information Engineering – Sung & Chen (Eds)
© 2016 Taylor & Francis Group, London, ISBN 978-1-138-02791-6

LCC risk management of grid assets based on FAHP

J.S. Qin, B.J. Li & D.X. Niu
School of Business and Management, North China Electric Power University, Beijing, China

ABSTRACT: This paper firstly introduced the LCC (Life Cycle Cost) structure of grid assets, then conducted risk identification of each part of the cost, finally established the LCC risk assessment model based on FAHP (Fuzzy Analytic Hierarchy Process), and did a case analysis of the target project. Those not only provided an effective decision-making help to the LCC risk management of the grid assets, but also put forward suggestions to optimize the cost of the grid assets plan.

Keywords: FAHP; Grid Assets; LCC; Risk Management

1 INTRODUCTION

Improving grid assets management is an effective protection of national strategic energy security, guarantee of economic lifelines smooth, basis of power grid enterprises to improve the quality of services and fulfill the social responsibility, and important means to improve the state-owned assets earnings and to achieve the goal of scientific development[1]. Only by strengthening the LCC asset management can the reliability of the power grid equipment be ensured, while extending equipment life and reduce LCC[2].

This paper firstly introduced the LCC structure of grid assets, then conducted risk identification of each part of the cost, finally established the LCC risk assessment model based on FAHP, and did a case analysis of the target project.

2 RISK IDENTIFICATION OF LCC GRID ASSETS

Stages involved in the LCC grid assets management include feasibility studies and decision-making, procurement, installation, operation, maintenance and scrap recycling process. With the extension of time, the entire LCC is also increasing. According to the characteristics of the entire grid assets investment process, the LCC grid assets are divided into five parts, namely, the initial investment costs, production and operation costs, maintenance costs, the costs of failure and scrap recycling costs. Risks of these five parts of cost are identified in Table 1.

Table 1. Risk identification of these five parts of costs.

Costs	Risks
The initial investment costs	Electrical equipment procurement, forecasting, equipment location, quality of personnel
Production and operation costs	Operating risk, electricity load control risk, legal risk, management risk
Maintenance costs	Maintenance planning uncertain risk, operational risk, risk management overhaul
The costs of failure	The number of failures in the equipment life cycle prediction, failure load loss forecasting, skill quality of repair personnel
Scrap recycling costs	Retirement application and approval risk, waste materials distribution risk, the risk of waste disposal recycling materials, unused supplies statistical risk, the risk of idle materials disposal

3 RISK ASSESSMENT MODEL OF LCC BASED ON FAHP

3.1 Calculation of the index weight coefficient based on AHP

Through the pair wise comparison of the importance of the elements of a low level relative to its previous level, construct a judgment matrix. According to the expressed hierarchy in Table 1, establish the top-down judgment matrix one by one.

Then find the normalized feature vectors of the judgment matrix corresponding to the maximum eigenvalue, and then use the square root method to find the feature vectors and roots of matrices. Each component of the corresponding feature vectors of judgment matrix meeting the consistency is the weights of index layer. By calculation, the weight set of index U_i of the main factor layer can be obtained. Weights of each layer of LCC are shown in Table 2[3].

3.2 Risk assessment model of LCC based on method of fuzzy

In order to facilitate risk management and operations, the paper will classify the risks, and the principle of classification is shown in Table 3.

Single factor evaluation. Considering the possibility of risk occurrence and the severity of the consequences caused, conduct a single factor evaluation of five factors of U_1, U_2, U_3, U_4, U_5 focused on the secondary index factors. The evaluation form of initial cost risk U_1 is as Table 4, and the single factor results of other factors are no longer listed.

Comprehensive Risk Evaluation of LCC. Compose the weight vector corresponding U_i and the evaluation matrix R_i, and get the evaluated vector B_i by normalization as shown below.

$$B_1 = \omega_1 R_1$$
$$= (0.0671, 0.1390, 0.4314, 0.1491, 0.1663, 0.0472)$$
$$\times \begin{bmatrix} 0.1 & 0.4 & 0.2 & 0.2 & 0.1 \\ 0.1 & 0.2 & 0.4 & 0.2 & 0.1 \\ 0.1 & 0.5 & 0.2 & 0.1 & 0.1 \\ 0.2 & 0.5 & 0.2 & 0.1 & 0 \\ 0.4 & 0.4 & 0.1 & 0.1 & 0.1 \\ 0.1 & 0.4 & 0.3 & 0.2 & 0 \end{bmatrix}$$
$$= (0.1648, 0.4303, 0.2159, 0.1253, 0.0804)$$

(1)

Similarly, you can get the total evaluation vector B

$$B = \begin{bmatrix} 0.1648 & 0.4303 & 0.2159 & 0.1253 & 0.0804 \\ 0.1237 & 0.4047 & 0.2256 & 0.1697 & 0.0763 \\ 0.1000 & 0.4048 & 0.2212 & 0.1739 & 0.1000 \\ 0.1000 & 0.4067 & 0.2176 & 0.1757 & 0.1000 \\ 0.1565 & 0.4673 & 0.2163 & 0.1164 & 0.0435 \end{bmatrix}$$

(2)

Based on these above, conduct a comprehensive evaluation of the LCC risk, and the synthesized system of the weight vector $\omega = (0.1803, 0.0870, 0.1945, 0.4955, 0.0427)$ corresponding $(U_1, U_2, U_3, U_4, U_5)$ and B.

Table 2. Weights of each layer.

The second level index weights	The third level index weights
0.1803	T_{11}=0.0671
	T_{12}=0.1390
	T_{13}=0.4314
	T_{14}=0.1491
	T_{15}=0.1663
	T_{16}=0.0472
0.0870	T_{21}=0.5688
	T_{22}=0.1280
	T_{23}=0.0659
	T_{24}=0.2372
0.1945	T_{31}=0.6333
	T_{32}=0.1062
	T_{33}=0.2605
0.4955	T_{41}=0.6687
	T_{42}=0.0882
	T_{43}=0.2431
0.0427	T_{51}=0.0825
	T_{52}=0.0814
	T_{53}=0.2715
	T_{54}=0.5646

Table 3. Risk levels classification principle.

Risk level	I	II	III	IV	V
Risk degree	Very low	Low	Medium	High	Very high

Table 4. The evaluation table of initial cost risk U_1.

R1	Risk evaluated values				
	Very	Low	Medium	High	Very high
T11	0.1	0.4	0.2	0.2	0.1
T12	0.1	0.2	0.4	0.2	0.1
T13	0.1	0.5	0.2	0.1	0.1
T14	0.2	0.5	0.2	0.1	0
T15	0.4	0.4	0.1	0.1	0.1
T16	0.1	0.4	0.3	0.2	0

$$S = \omega \times B$$
$$= (0.1803, 0.0870, 0.1945, 0.4955, 0.0427)$$
$$\times \begin{bmatrix} 0.1648 & 0.4303 & 0.2159 & 0.1253 & 0.0804 \\ 0.1237 & 0.4047 & 0.2256 & 0.1697 & 0.0763 \\ 0.1000 & 0.4048 & 0.2212 & 0.1739 & 0.1000 \\ 0.1000 & 0.4067 & 0.2176 & 0.1757 & 0.1000 \\ 0.1565 & 0.4673 & 0.2163 & 0.1164 & 0.0435 \end{bmatrix}$$
$$= (0.1162, 0.4130, 0.2187, 0.1632, 0.0920)$$

(3)

4 PREFERENCES, SYMBOLS AND UNITS? LCC GRID ASSETS RISK MANAGEMENT AND CONTROL

(1) Risk management and control of the initial investment cost

To avoid the risk of initial investment costs, on the one hand, the quality of site personnel should be improved, on the other hand advanced technology and methods should also be actively introduced, fully taking into account the policy risks, management risks, market risks, technology risks and force majeure risks.

(2) Risk management and control of operating costs of production

In the process of a production, firstly, strictly enforce regular inspection and maintenance tasks to avoid the risks and losses of the occurrence of accidents. Secondly, ensure the skilled workers and technicians who run the production, to make them have plenty of experience and safety awareness.

(3) Risk management and control of maintenance costs

Currently, there is a certain degree of lack of maintenance plan and uncertainty on the grid asset maintenance planning system. Reasonable arrangements for maintenance of electrical equipment, saving maintenance costs, reducing maintenance costs and improving reliability of the system is an effective way to reduce maintenance costs.

(4) Risk management and control of the costs of failure

For the risk brought by the failure uncertainty frequency can be predicted by increasing the quality of personnel, improving the efficiency of fault diagnosis method, inputting advanced cable fault tester and other means to circumvent.

(5) Risk management and control of scrap recycling costs

In the scrap recycling process, the application for approval of the system should be improved to reduce the losses caused by this. For the distribution risk of waste materials, the logistics management should be improved. For retirement or disposal of unused material must be paid by the special management department to ensure standardized and scientific processing.

5 CONCLUSIONS

This paper firstly introduced the LCC structure of grid assets, and then analyzed factors affected each risk, and conducted risk identification of each part of the cost. Followed by case studies, according to the maximum principle [4] that the level of risk is the maximum level of the comprehensive evaluation values, membership (0.4130) of low risk is larger than other risk membership from the calculation results, so the system LCC is low risk level.

ACKNOWLEDGMENTS

This work was partially supported by the Natural Science Foundation of China (71071052) and (71471059).

REFERENCES

[1] Shen Jingjing. Exploration of assets life cycle management for grid enterprises [J]. East China Electric Power, 2012,12: 122–123, In Chinese.
[2] Zhou Jiangxin, Guan Jun, Su Weihua, Li Qianyu. Life Cycle Cost Power Grid Planning Considering Risk Decision [J]. East China Electric Power, 2013, 04: 68–72, In Chinese.
[3] Yu Tian, Ye Qing. CIM risk assessment model of construction project life cycle cost [J]. Journal of Huaqiao University (Natural Science), 2013, 05: 29–33, In Chinese.
[4] Wu Guowei, Zhou Hui, Pan Weiwei, Wang Chong, Hou Yunhe. Risk Assessment of the Transformer Based on Life Cycle cost [J]. East China Electric Power, 2013, 03: 32–41, In Chinese.

Architectural, Energy and Information Engineering – Sung & Chen (Eds)
© *2016 Taylor & Francis Group, London, ISBN 978-1-138-02791-6*

A simple wind turbine power output simulation method based on speed control

Xu Li & Yan Feng Meng
IEECAS, Beijing, China
UCAS, Beijing, China

Shu Ju Hu & Hong Hua Xu
IEECAS, Beijing, China

ABSTRACT: This paper studied the modeling and analysis of wind turbine power output simulation. Based on the similarity principle, a new simple kind of wind turbine power output simulation method using a speed control is proposed. An experimental platform of 17 kW direct drive wind turbines is set up. The experimental results of power output simulation of wind turbines show that this method can be simple and accurate for the simulation of wind turbine power output.

Keywords: wind power; WTS; power output simulation

1 INTRODUCTION

With the development of wind power integration capacity and the proportion of grid connected, studies of characteristics of wind power have become increasingly important. The problem includes low-voltage ride-through. Therefore, a model that can simulate the wind turbine power output characteristics within the laboratory, to study wind power technology, has important significance.

There are three main aspects of the power output simulation of wind turbines, including wind turbine aerodynamic characteristic, mechanical characteristics and power grid characteristic. In this paper, all the three aspects are modeled and analyzed. An experimental platform to simulate a 17 kW direct drive wind turbine is set up, and the experiment of power output characteristic is carried out.

2 WIND TURBINE CHARACTERISTIC

2.1 *Aerodynamic characteristic*

Power absorbed by the wind turbine can be calculated according to Eq. (1) [1–6]:

$$P_m = \frac{1}{2} C_p(\lambda, \beta) \rho \pi R^2 v^3 \qquad (1)$$

where P_m stands for the power absorbed; C_p is the wind power coefficient related to λ, β, with λ being

the tip speed ratio and β being the pitch angle; ρ is the air density; R is the radius of the wind turbine; and v is the wind speed.

The wind power coefficient is calculated according to Eq. (2):

$$C_p(\lambda, \beta) = 0.22 \left(\frac{116}{\lambda_i} - 0.4\beta - 5 \right) \cdot e^{-\frac{12.5}{\lambda_i}} \qquad (2)$$

where λ_i is defined in Eq. (3):

$$\frac{1}{\lambda_i} = \frac{1}{\lambda + 0.08 \cdot \beta} - \frac{0.035}{\beta^3 + 1} \qquad (3)$$

Wind turbine power output under different wind speeds is shown in Figure 1.

Table 1. Wind turbine working points.

Wind speed (m/s)	Turbine speed (p.u.)	power (p.u.)
14	1.0	1.0
12	1.0	1.0
11	0.917	0.77
10	0.83	0.579
9	0.75	0.42
8	0.667	0.296
7	0.583	0.198

2.2 Aerodynamic characteristic simulation

For each wind speed, there is a maximum power that the turbine could absorb under a certain turbine speed. Usually, the wind turbine is designed to work under that speed. Therefore, for each wind speed, there is a working point for the turbine.

Based on Eq. (1), the working point of the wind turbine is presented in Table.1

All the data in Table 1 are measured in p.u. Using the data in Table 1, the MPPT (maximum power point tracking) curve is constructed, as shown in Figure 2.

Using data from Figure 2 as a lookup-table, the power output of the wind turbine could be calculated instantly.

2.3 Mechanical characteristic

Normally, the drive train could be simply modeled as given in Eq. (4) [6]:

$$J_T \frac{d\Omega}{dt} = T_T - T_{GT} - B_T.\Omega \quad (4)$$

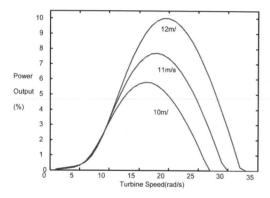

Figure 1. Wind turbine power output.

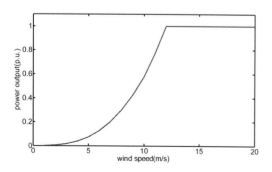

Figure 2. MPPT curve.

where J_T is the drive train inertia; T_T is the turbine torque; T_{GT} is the generator torque; and B_T is the friction factor.

Therefore, the speed of the drive train is determined by two main factors: turbine torque and generator torque. Both factors are related to the wind speed either directly or indirectly.

The dynamic characteristic of the drive train can be simplified as a one-order inertial link.

2.4 Power grid characteristic

The direct-drive wind turbine is connected to the power grid by a full-power convertor. The convertor is composed of two parts: generator side and grid side.

The PMSM could be described as given in Eq. (5):

$$\begin{cases} u_d = -R_s i_d - L_d \dfrac{di_d}{dt} + \omega_r L_q i_q \\ u_q = -R_s i_q - L_q \dfrac{di_q}{dt} - \omega_r L_d i_d - \omega_r \psi_f \end{cases} \quad (5)$$

Normally, the $i_d = 0$ method is used in the generator side to achieve maximum stator current utilization. Under this circumstance, the control block diagram of the generator is shown in Figure 3.

The iq_{ref} is determined by the control system directly.

Similar to the generator side, the current control loop is almost the same. In addition, the current reference is determined by the voltage outer loop, as shown in Figure 4.

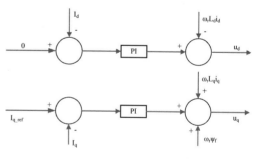

Figure 3. Control block diagram of the generator side.

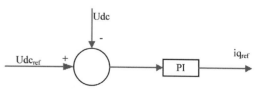

Figure 4. Voltage loop of the grid side.

The active power current iq is determined by the voltage loop. The reactive power current is determined by the control system directly.

3 EXPERIMENTS

3.1 Experimental platform

Based on the above discussion, a 17 kW direct-drive wind power platform is set up. The structure of the experimental platform is shown in Figure 5.

As shown in Figure 5, for each given wind speed produced by the control system, a motor speed is given by a lookup-table. The speed reference is transmitted to the motor driven by an inverter under the speed-control method. The PMSM generator is driven by the motor connected by a gear box. The PMSM is connected to the power grid by a full power AC-DC-AC convertor. The system is controlled by a control system based on a SIEMENS plc.

The motor and the generator is shown in Figure 6.

3.2 Experimental parameters

The platform is set up to simulate a 10 kW wind turbine. The rated speed of the generator is 234 RPM. The rated wind speed is 12 m/s. The grid voltage is set to 300 V. The Dc voltage is set to 500 V.

There are three working conditions that are tested using the experimental platform: constant wind speed situation, step up wind speed situation and random wind speed situation.

A pc monitor interface is programmed to collect importations from the motor, the generator and the convertor using the C# language.

3.3 Experimental results

The experimental results of a constant wind speed (12 m/s) are shown in Figure 7.

Channel 1 is the grid voltage (200 v/div); channel 2 is the convertor DC voltage (500 v/div); channel 3 is the convertor gird current (20 A/div); and channel 4 is the rotor speed (234RPM/div).

The experimental results of the step-up wind speed (from 1 to 12 m/s, 1 m/2 each step) are shown in Figure 8.

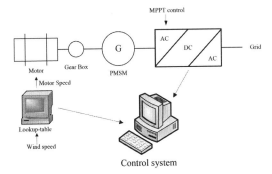

Figure 5. Structure of the experimental platform.

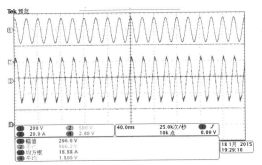

Figure 7. Experiment result of wind speed at 12 m/s.

Figure 6. Photo of a motor and a generator.

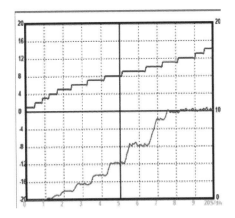

Figure 8. Wind speed and active power output.

473

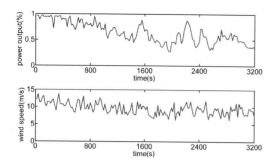

Figure 9. Random wind speed result.

In Figure 8, the blue curve indicates the wind speed and the purple curve indicates the active power output. The vertical axis of the blue curve is on the left side, and the vertical axis of the purple curve is on the right side.

As shown in Figure 8, the power output of the wind turbine is simulated by the platform properly with a slight delay due to the inertial link of the drive train.

The experimental result of the random wind speed is shown in Figure 9.

Figure 9 shows the recording of the working points every 20 second.

As shown in Figure 9, under the random wind speed situation, the platform could represent the power output characteristic properly.

4 CONCLUSIONS

In this paper, a simple method to simulate direct-drive wind turbine power output characteristic was studied. A 17 kW direct-drive experimental platform was set up. The experiments of three typical conditions were carried out. The result showed that this method could represent the power output characteristic of the wind turbine. This method is easy to achieve and carry out. Based on this method, the experiments of grid connected wind power such as power flow simulation could be performed easily.

REFERENCES

Liu Qihui, He Yikang, and Zhao Rende, "Imitation of the Characteristic of Wind Motor", Proceedings of the CSEE, Vol. 26, No. 7, pp. 134–139, 2006.

Li Xu, Meng Yanfeng, Hu Shuju, Xu Honghua. Analysis and simulation of wind turbine mechanic characteristics simulator [J]. Renewable Energy Resources, 2014, 08: pp. 1150–1154.

W. Li, D. Xu, W. Zhang, H. Ma, Research on wind turbine emulation based on DC motor, in 2nd IEEE Conference Industrial Electronics and Applications, 2007, pp. 2589–2593.

L. Lu, Z. Xie, X. Zhang, S. Yang, R. Cao, A Dynamic Wind Turbine Simulator of the wind turbine generator system, in Second International Conference Intelligent System Design and Engineering Application, 2012, pp. 967–970.

W. Ru, Y. Wang, X. Song, Z. Wang, Development of wind turbine simulator for wind energy conversion systems based on permanent magnet synchronous motor, International Conference Electrical Machines and Systems, 2008, pp. 2322–2326.

Weiwei Li, Dianquo Xu, Wei Zhang, and Hongqi Ma, "Research on Wind Turbine Emulation based on DC Motor", Industrial Electronics and Applications, 2007. ICIEA 2007.

Architectural, Energy and Information Engineering – Sung & Chen (Eds)
© 2016 Taylor & Francis Group, London, ISBN 978-1-138-02791-6

Using TRMM satellite precipitation over the data-sparse basins for hydrological application

Xing Liu, Xiao Dong Li, X.L. Li, S.X. Cai & T.Q. Ao
State Key Laboratory of Hydraulics and Mountain River Engineering, College of Water Resource and Hydropower, Sichuan University, Chengdu, China

Fa Ming Liu
China Railway Eryuan Engineering Group Co. Ltd., Chengdu, China

ABSTRACT: In order to evaluate the accuracy of the TRMM precipitation data in the Laotian basin, TRMM 3B42 V7 precipitation data were used to derive a distributed hydrological model to simulate daily streamflow from year 2000 to 2004. Four stations were used, namely the Ban Pak Bak station, the Ban Mout station, the Kasi station, and the Hin Heup station. The former two are located in the Nam Khan river basin, while the latter two are located in the Nam Like river basin. The results show that the daily streamflow simulation fed with TRMM precipitation data could basically reflect the daily streamflow processes at the four stations and are better at determining rain occurrence and the time to peak. Furthermore, the calibrated parameters in the Nam Khan river basin are more suitable than those in the Nam Like basin. The results indicate that the satellite precipitation has a promising prospect in the study of hydrologic processes and water resources management.

Keywords: TRMM satellite precipitation; distributed hydrological model; data-sparse basins; hydrological simulation

1 INTRODUCTION

Precipitation is one of the most important factors in the process of hydrological cycle. As a basic input for the hydrological model, the precision of precipitation data has a tremendous influence on the simulation results [1, 2]. Moreover, due to the spatial and temporal variability, a non-uniform distribution feature has often been observed [3]. Precipitation remains one of the hydrographical factors, which is difficult to observe and estimate precisely [4]. Currently, precipitation estimation is mainly derived from two sources, i.e., rain gauge station observations and ground radar measurements. Rain gauge, though as a direct precipitation measuring instrument, is technologically mature and widely used. Its measurements cannot reflect the spatial variation of rainfall effectively because the distribution of rainfall stations is sparse and the effective radius of point measurements is very limited [5, 6]. In comparison with rain gauge, the ground radar system can provide an instantaneous spatial distribution of precipitation over the basin indirectly, thus helping to offset the bias of rain gauge observations partly. However, because of its limited detecting area, high costs

of establishing and maintaining infrastructure, and technological reasons, the area covered by the ground radar system is limited and there is no perfect radar network for many regions [7]. It still cannot meet the requirements of the study carried out on the large-scale basin. These drawbacks of conventionally obtained rainfall data impose a remarkable restriction on the application of the distributed hydrological model in rainfall-runoff simulation and hydrological forecast. Recently, satellite remote sensing data such as TRMM [8], CMORPH [9], and PERSIANN [10] have emerged as a preferable alternative or supplement to conventional precipitation observations [11, 12] due to their high spatial-temporal resolution and availability over vast ungauged regions. Hence, they improved the study and application of distributed hydrological model in many fields immensely.

Hydrological study in data-sparse and ungauged basins has been an active and difficult issue in the field of hydrology and water resources research. Restricted by disadvantageous strategic status, climate and traffic problems in regional economic development and water security system, numerous data-sparse and ungauged basins exist in many remote parts of the world, particularly in developing countries. Aiming at achieving major advances in the capacity to make reliable predictions in "ungauged" basins, the International Association of Hydrological Sciences (IAHS) recently launched an initiative called the Decade on Predictions in Ungauged Basins (PUB). Sivapalan and Takeuchi [13] presented three methods for the predictions of data-sparse basin responses: 1) interpolation or extrapolation of responses information from gauged to ungauged basins, 2) measurements by remote sensing (e.g radar, satellites), 3) application of process based hydrological models to reduce the dependency on specific precipitation inputs. Thus, the distributed hydrological model, which is suitable for the hydrological study of the data-sparse basin and makes full use of hydro-meteorological data, becomes a feasible approach to solve the PUB problem. Obviously, for data-sparse and ungauged basins, using satellite rainfall products to derive the distributed hydrological model is an ideal choice to tackle the PUB problem. In this paper, the Nam Khan river basin and the Nam Like river basin, two data-sparse basins located in Laos which are in tropic zone, were chosen as the study area. So, it is suitable to utilize TRMM satellite rainfall products for streamflow process simulation. The TRMM 3B42V7 rainfall products were adopted to simulate the rainfall-runoff process based on the physical-based distributed hydrological model BTOPMC.

2 STUDY REGION, DATA AND METHODOLOGY

2.1 Study region

Nam Khan river (Figure 1) is an important tributary of Mekong river. It flows through the northern mountainous part of Laos with both sloping banks covered by forests and crops, drains into Mekong river at Luang Prabang. The basin is located within 101°56′E–103°42′E and 19°22′N–21°1′N, with an area of about 7,620 km². There are two gauge stations, namely the Ban Pak Bak station and the Ban Mout station, with a drainage area of 5800 km² and 6100 km², respectively.

Nam Like river (Figure 1), as a branch of Nam Ngum river which is one of the tributaries on the left bank of Mekong river, originates from Phou Khoun mountain located in Luang Prabang Province and flows through Vientiane Province to Hin Heup in a north-south ward direction on the whole, then joins Nam Xong river, an anabranch of Nam Like river on its left bank, and finally flows into Nam Ngum river downstream the reservoir Nam Ngum. The Nam Like river basin is located within 101°54′E–102°31′E and 18°29′N–19°27′N. Two gauge stations are

available in this basin with a drainage area of 374 km² (Kasi station) and 5115 km² (Hin Heup station) respectively. The upstream of Nam Like river, which is also the upper catchment of the Kasi station, is a plateau with elevations ranging from 1000 m to 1500 m, and the altitude decreases to 300 m–800 m as it flows into the hilly region, the middle or lower reaches of Nam Like river. The width of Nam Like river is normally around 100 m, with the widest part being about 200 m and the narrowest part being 40 m–50 m. The annual mean rainfall of the Nam Like river basin is around 1745 mm, about 90% of which falls in the rainy season (between May and October), with the rest 10% falling during the dry period (the rest of the year).

2.2 TRMM satellite products

The Tropical Rainfall Measuring Mission, a joint US–Japan satellite mission, was launched in 1997 to monitor tropical and subtropical precipitation and to estimate its associated latent heating covering the latitude band 50°N–50°S. The TRMM satellite carries five rain measuring instruments: Microwave Imager (TMI), Visible Infrared Scanner (VIRS), Lightning Imaging Sensor (LIS), Clouds and Earth's Radiant Energy System (CERES), and Precipitation Radar (PR), which is the first space-borne precipitation radar. Among them, the CERES was out of service due to mechanical failure. So far, the TRMM rainfall products have updated up to version 7. Merged with other data from multiple satellites, products version 7 present many new features such as larger coverage, higher spatial-temporal resolution, longer time series and being able to update real-timely, which leads to a new wave of research about the application of TRMM rainfall products version 7 in meteorology, climatology, hydrology and other fields. TRMM rainfall products include several types, i.e. 3B40, 3B41, and 3B42, among which type 3B42 data has the highest resolution in space (0.25° × 0.25°) and time (3h). In this regard, we used TRMM-3B42V7 in this study for streamflow stimulation of the two study basins.

2.3 BTOPMC model description

The physically based distributed hydrological model BTOPMC [14–16] was developed from the semi-distributed hydrological model TOPMODEL [17]. It has the characteristics of parsimonious calibrated parameters, relatively low requirement on inputs, simple to operate, and able to take advantage of satellite remote-sensing data. The SCE-UA algorithm (Duan, Sorooshian [18]) was adopted to calibrate five essential parameters of BTOPMC, which include: saturated soil transmissivity T0, decay factor m, maximum storage capacity of the

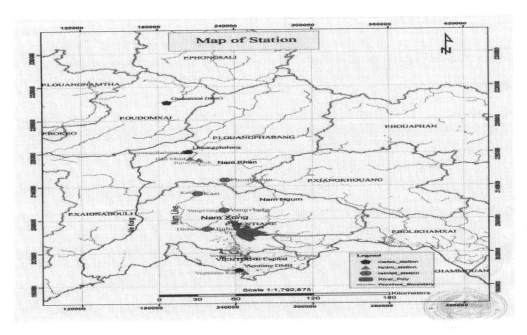

Figure 1.   Location of the study area, river network and distribution of the stations.

root zone Srmax, saturated local saturation deficit Sbar0(k) of sub-basin k, and Manning roughness coefficient n. The former four are for runoff generation and the last one for flow routing. All these parameters have explicit physical meanings. BTOPMC divides the entire basin into sub-basins (lumped) that consist of a number of grid cells (distributed). Then, Runoff generation sub-model and its parameters Runoff calculation are carried out for each grid cell by applying the assumptions and concepts of TOPMODEL to each sub-basin, and finally the Muskingum—Cunge (M—C) method [19] was adopted to calculate flow routing.

The performance of the model is mainly evaluated by the volume ratio (Vr) and the Nash—Sutcliffe Coefficient (NSCE; Nash and Sutcliffe [20]). The volume ratio Vr is calculated as follows:

$$V_r = \frac{TV_{sim}}{TV_{obs}} \tag{1}$$

where TVsim is the total volume of simulated discharges and TVobs is the volume of observed discharges.

2.4   Data preparation

The available data for this study are: (i) a Digital Elevation Model (DEM) provided by the Computer Network Information Center, Chinese Academy of Sciences (http://www.gscloud.cn/), with the spatial resolution of 30 m; (ii) land use/ vegetation coverage data with a resolution of 1 km that were constructed from digitalizing and interpreting remote-sensing images provided by the International Geosphere Biosphere Programme, and the soil data with a resolution of 1 km provided by the Food and Agriculture Organization of the United Nations (FAO); (iii) meteorological data of the Luang Prabang meteorological station, the VangViang rain gauge station, the Ban Pak Bak hydrological station, the Ban Mout hydrological station, the Kasi hydrological station, and the Hin Heup hydrological station (Fig. 1) obtained from the Laos Meteorological Administration. (iv) TRMM-3B42V7 rainfall data with a high spatial resolution of 0.25° × 0.25° and time resolution of 3 h were downloaded from http: //trmm.gsfc.nasa. gov/. The DEM data, land use/ vegetation coverage data and soil data are resampled to the resolution of 300 m to match the model calculation. The three-hourly TRMM-3B42V7 data were aggregated to daily resolution to derive the BTOPMC model.

3   RESULTS AND DISCUSSIONS

The TRMM 3B42 V7 daily precipitation data were used to derive the distributed hydrological model to perform daily streamflow simulations at the the Ban Pak Bak station, the Ban Mout station located in the Nam Khan river basin and the Kasi station, the Hin Heup station located in the Nam Like river basin during the 5-year period from 2000 to 2004, so as to assess the feasibility of the TRMM precipitation on streamflow process simulation in Laos data-sparse basins. The Nam Khan river basin and the Nam Like river basin were both calibrated for the years from 2000 to 2002, but validated for the period from January 2003 to August 2004 (Nam Khan river basin) and from January 2003 to November 2004 (Nam Like river basin). The volume ratio (Vr) and NSCE served as the objective functions for parameter calibration, which was conducted by using the effective and efficient SCE-UA global optimization algorithm. Besides, the runoff depth and the peak time were also adopted as statistic indicators to validate the effectiveness of streamflow simulation based on the BTOPMC model.

Figure 2 shows simulated and observed streamflow forced by TRMM rainfall products at the Ban Pak Bak, Ban Mout, Kasi, and Hin Heup stations. The results indicate that the daily streamflow process can be represented by large and good agreement existing between the simulated daily streamflow process and TRMM daily precipitation series, except significant biases for both high-flow simulation (especially for peaks) and low-flow simulation.

The volume ratio (Vr) values vary from 0.54 to 1.80. Vr values at the Kasi station for the calibration and validation period are 0.54 and 0.57,

respectively, which suggest a negative bias of stimulation. Vr values at the Ban Pak Bak and Ban Mout stations for the calibration period are 1.64 and 1.80, respectively, which suggest a positive bias of stimulation. On the one hand, TRMM rainfall products are derived from near-ground precipitation information, which contains evaporation loss during the precipitation process, and thus might be the reason for the overestimation of total simulated streamflow volume. On the other hand, since the study basins are located in the mountainous region of north Laos that is characterized by convective rainfall and orographic rainfall, which are usually underestimated by satellite rainfall products, they might be the cause of underestimation of simulated streamflow peaks. Beyond these, while the spatial resolution of TRMM rainfall products is $0.25° \times 0.25°$, the drainage area of the Kasi station is only 374 km^2, which leads to few TRMM grids over the Kasi river basin.

NSCE values vary between 25.65% and 83.07% during both the calibration period and the validation period, which is not very reasonable. Although the NSCE value for the Hin Heup station reaches the highest 83.07% for the calibration period, it decreases to 44.47% for the validation period (Table 1). An intercomparison of the NSCE values obtained from both calibrating and validating periods manifests that the NSCE value for the Nam Khan river basin is better than that for the Nam Like river basin, which indicates that the calibrated

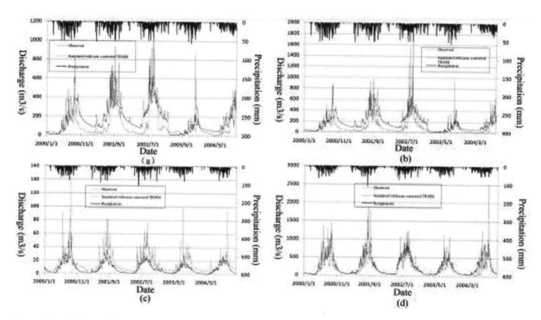

Figure 2. Hydrographs of daily stream flow with TRMM precipitation for model calibration and validation (2000–2004) at (a) Ban Pak Bak station, (b) Ban Mout station, (c) Kasi station, (d) Hin Heup station.

Table 1. The statistical indicators of daily streamflow simulations at the four stations.

Indicator			Stations			
			Ban Pak Bak	Ban Mout	Kasi	Hin Heup
NSCE (%)		calibration	39.77	25.65	39.75	83.07
		validation	32.7	51.2	26.92	44.47
Depth of total runoff (m)	Simulated	calibration	3.57	3.58	6.04	5.95
		validation	0.86	0.86	2.94	3.02
	Observed	calibration	2.19	1.99	11.1	5.54
		validation	0.71	0.68	5.17	2.48
V_1		calibration	1.64	1.8	0.54	1.08
		validation	1.22	1.26	0.57	1.22
Mean precipitation (m)		calibration	5.34	5.28	7.17	6.73
		validation	2.67	2.67	3.86	4.03
Peak time (date)	Simulated	calibration	2001/9/3	2001/9/3	2000/8/17	2001/8/10
		validation	2004/8/2	2004/8/2	2004/9/11	2004/9/12
	Observed	calibration	2002/8/17	2002/8/18	2000/9/1	2001/8/17
		validation	2003/8/26	2004/8/31	2004/9/11	2004/9/11

parameters perform better in the Nam Khan river basin than those in the Nam Like river basin.

For peak representation simulation, we employed the ratio value R to evaluate its performance. The ratio value R can be expressed as follows:

$$R = \frac{N}{M} \quad (2)$$

where R is the ratio value; N is the number of correctly simulated peaks above a given peak flow Q (m³/s); and M is the total number of observed peak flows above a given peak flow Q (m³/s).

For the Nam Khan river basin, when Q is 200 m³/s, R for the Ban Pak Bak and Ban Mout stations are 87.94% and 86.21%, respectively; when Q is 400 m³/s, R for the Ban Pak Bak and Ban Mout stations are 84.44% and 87.18%, respectively. For the Nam Like river basin, when Q is 20 m³/s or 40 m³/s, R for the Kasi station is 40.33% or 4.44%; when Q is 500 m³/s or 700 m³/s, R for the Hin Heup station is 86.78% or 59.00%. These statistical summaries suggest that peak representation simulation performs well except at the Kasi station due to the low peak flow.

4 CONCLUSIONS

The simulated streamflow can represent daily flow mostly at the Ban Pak Bak, Ban Mout, Kasi and Hin Heup stations, except for the underestimation of high flow and low flow. The NSCE and Vr values indicate that the calibrated parameters made the model perform better over the Nam Khan river basin than over the Nam Like river basin. Overall, this study demonstrates the feasibility of TRMM

rainfall data in streamflow stimulation over the data-sparse area of Laos, although the daily-flow simulation results do not match very well, they are still acceptable. Furthermore, TRMM precipitation data are more precise and representative for the tropical area than other satellite precipitation data. In conclusion, it can be suggested that the satellite-based rainfall, e.g. TRMM data, has a good potential for useful application to hydrological simulation and water balance calculations at monthly or seasonal time steps, which is a useful merit for regions where rain gauge observations are sparse or of bad quality. However, several shortcomings, such as the TRMM overestimates the rainfall in some years and areas and underestimates in other years and areas, and failed to detect the extreme rainfall, reduced the accuracy of stream flow simulation at short time steps and other applications including drought monitoring and flood forecasting.

REFERENCES

[1] Meng, J., et al., Suitability of TRMM satellite rainfall in driving a distributed hydrological model in the source region of Yellow River. Journal of Hydrology, 2014. 509: p. 320–332.

[2] Sorooshian, S., et al., Water and energy cycles: Investigating the links. World Meteorological Organization Bulletin, 2005. 54(2): p. 58–64.

[3] LIU, Y., Q. FU, and P. SONG, Satellite retrieval of precipitation: An overview. Advances in Earth Science, 2011. 26(11): p. 1161–1172.

[4] Yong, B., et al., Hydrologic evaluation of Multisatellite Precipitation Analysis standard precipitation products in basins beyond its inclined latitude

band: A case study in Laohahe basin, China. Water Resources Research, 2010. 46(7).

[5] Collischonn, B., W. Collischonn, and C.E.M. Tucci, Daily hydrological modeling in the Amazon basin using TRMM rainfall estimates. Journal of Hydrology, 2008. 360(1-4): p. 207–216.

[6] Jia, S., et al., A statistical spatial downscaling algorithm of TRMM precipitation based on NDVI and DEM in the Qaidam Basin of China. Remote sensing of Environment, 2011. 115(12): p. 3069–3079.

[7] Gu, H., et al., Hydrological assessment of TRMM rainfall data over Yangtze River Basin. Water Science and Engineering, 2010. 3(4): p. 418–430.

[8] Huffman, G. J., et al., The TRMM multisatellite precipitation analysis (TMPA): Quasi-global, multiyear, combined-sensor precipitation estimates at fine scales. Journal of Hydrometeorology, 2007. 8(1): p. 38–55.

[9] Joyce, R.J., et al., CMORPH: A method that produces global precipitation estimates from passive microwave and infrared data at high spatial and temporal resolution. Journal of Hydrometeorology, 2004. 5(3): p. 487–503.

[10] Sorooshian, S., et al., Evaluation of PERSIANN system satellite-based estimates of tropical rainfall. Bulletin of the American Meteorological Society, 2000. 81(9): p. 2035–2046.

[11] Sawunyama, T. and D. Hughes, Application of satellite-derived rainfall estimates to extend water resource simulation modelling in South Africa. Water SA, 2008. 34(1): p. 1–9.

[12] Yong, B., et al., Assessment of evolving TRMM-based multisatellite real-time precipitation esti-

mation methods and their impacts on hydrologic prediction in a high latitude basin. Journal of Geophysical Research: Atmospheres (1984–2012), 2012. 117(D9).

[13] Sivapalan, M., et al., IAHS Decade on Predictions in Ungauged Basins (PUB), 2003–2012: Shaping an exciting future for the hydrological sciences. Hydrological Sciences Journal, 2003. 48(6): p. 857–880.

[14] Ao, T., Development of a distributed hydrological model for large river basins and its application to Southeast Asian rivers, 2001, Ph. D. thesis, University of Yamanashi, Kofu, Japan.

[15] Ao, T., et al., Relating BTOPMC model parameters to physical features of MOPEX basins. Journal of Hydrology, 2006. 320(1): p. 84–102.

[16] Takeuchi, K., T. Ao, and H. Ishidaira, Introduction of block-wise use of TOPMODEL and Muskingum-Cunge method for the hydroenvironmental simulation of a large ungauged basin. Hydrological Sciences Journal, 1999. 44(4): p. 633–646.

[17] Beven, K., et al., Topmodel. Computer models of watershed hydrology., 1995: p. 627–668.

[18] Duan, Q., S. Sorooshian, and V.K. Gupta, Optimal use of the SCE-UA global optimization method for calibrating watershed models. Journal of hydrology, 1994. 158(3): p. 265–284.

[19] Fread, D., Flow routing. Handbook of hydrology, 1993: p. 10.1–10.36.

[20] Nash, J. and J. Sutcliffe, River flow forecasting through conceptual models part I—A discussion of principles. Journal of hydrology, 1970. 10(3): p. 282–290.

Architectural, Energy and Information Engineering – Sung & Chen (Eds)
© 2016 Taylor & Francis Group, London, ISBN 978-1-138-02791-6

A method for the pulse detection of the IR-UWB PPM modulation

Lin Zhu, He Zheng & Hao Zhang
Chongqing Communication Institute, Chongqing, China

ABSTRACT: It is difficult to choose the best decision threshold because of the time-varying wireless communication channel parameters. Based on this, we put forward a detection method which does not need a decision threshold. This method samples the signals on the positions which the pulse takes the '1' and '0' information, then demodulate the signal by comparing the maximum values of the samples. The simulation result shows that we can use this method to demodulate the signal without a decision threshold.

Keywords: IR-UWB; pulse detection; PPM

1 INTRODUCTION

IR-UWB (Impulse Radio-Ultra Wideband) is a kind of new communication technology that developed rapidly in recent years. Its information carrier are narrow pulse sequences instead of sine waves, and the duration of the pulse is very short. The main features of the IR-UWB include high multipath resolution [1,2], high processing gain [3], low detection or interception probability of the signals [4], and high penetration ability. IR-UWB has broad application prospects in many fields such as wireless communication, radar imaging, and accurate positioning [5].

The detection of the pulse is one of the core technologies of the IR-UWB. But because the short duration and Low duty cycle of the pulse, it is difficult to detect the IR-UWB pulse. At present the pulse correlation detection [6,7], the non-correlation detection based on the power of pulse [8], and the edge threshold detection method [9]. These methods all need a decision threshold. The time-varying wireless communication channel parameters increase the difficulty of threshold optimization choice, and have a great influence on pulse detection effect. So it makes sense to look for another detection method which doesn't depend on a decision threshold.

Here we proposed a new detection method for the PPM, which doesn't need a decision threshold, and there for avoid the threshold selection difficulties when the channel parameters vary. The simulation result shows verify the correctness of this method.

2 PULSE COMPARISON DECISION DETECTION

We design this method for the 2PPM signal, can combine TH (Time Hopping) DS (Direct Spread), form TH-PPM and DS-PPM, also can be applied to high order PPM modulation.

2.1 The signal model

DS-2PPM modulation is a typical UWB wireless communication modulation mode, so we take this type of signal to be the received signal to describe the pulse comparison decision method. Equation 1 is the waveform of DS-2PPM signal [10].

$$S^{(k)}(t) = \sum_{j=-\infty}^{\infty} \sum_{n=0}^{N_s-1} a_{j,1}^{(k)} g_n^{(k)} p\left(t - jT_f - \left(n + \frac{a_{j,0}^{(k)}+1}{2} \right) T_c \right) \quad (1)$$

Among them: $p(t)$ is pulse wave function, $g_n^{(k)} \in \{-1,+1\}$ is the pseudo random sequence of the k-th user. T_c is the pulse repetition period, T_f is information period, $a_{j,0}^{(k)}, a_{j,1}^{(k)} \in \{-1,+1\}$ are two independent binary sequences of the k-th user. $N_s = T_f/T_c$, N_s pulses represent an information symbol.

Figure 1 shows one cycle of the DS-2PPM signal when the DS code is 7 bits barker-code ([1110010]).

2.2 *Detection process*

This article describes the basic principles of comparison decision detection based on a 2 PPM modulation. The pulse may appear in one of the two positions in each pulse cycle of the 2 PPM modulation. As the figure 2 shows, the position of solid line pulse means information is '1', and the position of the dashed line pulse means '0'. Whenever the pulse is synchronized on the receiver, the start position of the pulse cycle is obtained, which is the position of t_0 in figure 2. Also we can calculate the positions of the pulse carried the information '1' should appear, which is t_1, and the position of the pulse carried the information '0' should appear, which is t_2.

After the pulse synchronization, we enter the comparison decision process, as the figure 3 shows. The received signal will be sampled in t_1 and t_2, then we compare the two sampling values on the two position, if the value on the t_1 position is greater than the value on the t_2 position, we determine the received information is '1', otherwise we determine the received information is '0'. We determine the information of every cycle according to this principle.

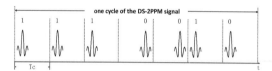

Figure 1.   One cycle of the DS-2PPM signal.

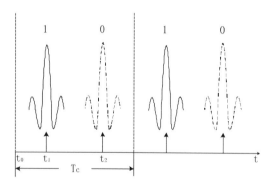

Figure 2.   The time domain waveform of the 2 PPM pulse.

Figure 3.   The process of pulse receive.

3 SIMULATION

3.1 *Experiment model*

The simulation parameters are given as follows:

1. Pulse modulation: DS-2 PPM;
2. The length of pseudo-random codes: Gold codes with a length of 31;
3. Sampling frequency: 250 MHz;
4. Pulse duration: 100 ns;
5. Pulse repetition period: 500 ns;
6. SNR: –5 dB.

3.2 *Simulation result*

We simulated the algorithm with System Generator and Simulink. Figure 4 shows the time domain waveform after the synchronization and filtering of the received pulse sequence. As the figure shows, we can determine the position of pulse by comparing the sampling values of the t_1 and t_2 time, then determine the information.

Figure 4 clearly shows that when the SNR is –5 dB, the sample value of the pulse position is greater than the value of another position, and the difference of the two values is obvious. So it is easy to decide the information bit is '1' or '0'.

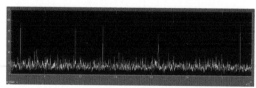

Figure 4.   The time domain waveform of received pulse sequence.

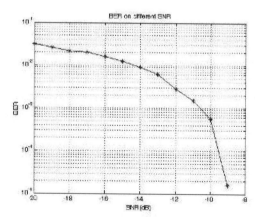

Figure 5.   BER curve of the comparison decision method on different SNR.

Figure 5 is the simulation BER curve on different SNR. When the SNR is greater than −9 dB, there is no error.

Figure 5 clearly shows that we can effectively demodulate the 2PPM signal without a decision threshold by using the pulse comparison decision method we described here.

4 CONCLUSIONS

In this article we described a pulse detection method which does not need a decision threshold. This method samples the signals on the positions which the pulse takes the '1' and '0' bit information, then compares the maximum values of the samples to decide the information. On the simulation environment, the BER is lower than 10^{-3} when the SNR is −10 dB. That means this method is capable to complete the pulse detection without decision threshold, and therefor avoid the threshold selection difficulties when the channel parameters vary.

REFERENCES

[1] Guangrong Yue, Lijia Ge, Shaoqian Li. 2003. Performance of UWB time-hopping spread-spectrum impulse radio in multipath environments. *The 57th IEEE semiannual Vehicular Technology Conference.* April 22–25. Korea.

[2] Win M Z, Ramirez-Mireles F, Scholtz R A, et al. 1997. Ultra-wide bandwidth signal propagation for outdoor wireless communications. *IEEE VTC conference.* Phoenix Arizona.

[3] Withington P. 2000. Time modulated ultra-wideband for wireless applications [DB/OL]. *http://www. time-domain.com/files/downloads/techpapers/Pulsonoverview701.pdf.2000-12-31.*

[4] Zhenyu Zhan, Fanxin Zheng, Lijia Ge. 2003. Ultra Wide Band Impulse Radio Modulated with Time-Hopping Code and its Applications to Communication countermeasures. *Electronic Warfare.* V18, N3, May 2003, 32–36.

[5] Robert J. Fontana, Steven J. Gunderson. 2002. Ultra-Wideb and precision asset location system. *2002 IEEE Conference on Ultra Wideband Systems and Technologies.*

[6] M.Z. Win R.A. Scholtz. 2000. Ultra-wide bandwidth time-hopping spread-spectrum impulse radio for wireless multiple-access communication. *IEEE Trans. On Comm.* 2000. COM-48(4): 679–691.

[7] J.D. Choi W.E. Stark. 2002. Performance of autocorrelation receivers for ultra-wideband communication with PPM in multipath channels. *2002 IEEE Conference on ultra wideband system and technologies.* Digest of papers. May 2002.

[8] Architecture Proposal for LDR-LT Verification Platform. Itergrated Project PULSERS. Contract No 506897, http://www.pulsers,info.

[9] Larry W. Fullerton. 1987. Spread Spectrum Radio Transmission System [P]. *United States Patent*: 4, 641, 317, 1987-02-03.

[10] Lijia Ge, Fanxin Zheng, Yu Li Liu and Guang Rong Yue. 2005. *Ultra Wide Band Wireless Communications.* Chapter 4. National Defense Industry Press.

Architectural, Energy and Information Engineering – Sung & Chen (Eds)
© 2016 Taylor & Francis Group, London, ISBN 978-1-138-02791-6

Square internal and external flow section recharge well with filter layer

Na Li & Wang Lin Li
University of Jinan, Jinan, China
Shandong Provincial Engineering Technology Research Center for Groundwater
Numerical Simulation and Contamination Control, Jinan, China

Yi Cheng Yang
Shandong Survey and Design Institute of Water Conservancy, Jinan, China

Zhi Yuan Yin & Fang Xu
University of Jinan, Jinan, China
Shandong Provincial Engineering Technology Research Center for Groundwater
Numerical Simulation and Contamination Control, Jinan, China

ABSTRACT: Recharge well with filter layer is a kind of well that has the function of filtering granular impurities existing in water. This kind of recharge well with filter layer is often set in pollution-free or slightly polluted seasonal rivers. However, there are many problems on the existing recharge well with filter layer, such as low efficiency, clogging and easy to wash. On the basis of analysis of these issues, considering the recharge process of recharge well with filter layer and aiming at the key factors of the influence and restriction from seepage recharge well recharge quantity, a new square internal and external flow section recharge well with filter layer is designed. The recharge pool of square internal and external flow section recharge well with filter layer is a kind of hollow square cylinder wellhead so that water can flow from the top, bottom, or internal and external side into the well. It can increase the single-well recharge volume and improve anti-silt and anti-rush capability, and it can also save materials such as sand-gravel. In this paper, the theoretical calculation formula of single-well recharge volume is provided. A 1:200 scale model is constructed and tested to study the indoor steady flow recharge, and analyze its single-well recharge volume compared with the existing recharge well with filter layer. The analysis results indicated that: (1) comprehensive reduction factor of square internal and external flow section recharge well with filter layer in confined aquifer is relatively smaller than the ordinary recharge well with filter layer; (2) the single-well recharge volume of square internal and external flow section recharge well with filter layer is far greater than the ordinary recharge well with filter layer, so the square internal and external flow section recharge well with filter layer has the stronger recharge ability.

Keywords: recharge well with filter layer; square; internal and external flow section; recharge pool; the optimization design

1 INTRODUCTION

To make full use of rain-flood resources, recharge well is used, which has the function of filtering granular impurities existing in water. This kind of recharge well is often set in pollution-free or slightly polluted seasonal rivers of Shandong peninsula. The recharge well is also called the recharge well with filter layer, automatic in-filtering recharging well or mechanical infiltration well.

Nowadays, a recharge well with filter layer generally consists of a recharge pool and an ordinary recharge well. The recharge pool is used to collect the river water and transport river water into the recharge well. It is usually buried underground with the shape of upside-down square or cylinder, and sands and gravel are backfilled to the recharge pool as a two-stage filter material. Recharge well transfers water into the aquifer and often lies in the middle of recharge pool bottom. Its wellhead is covered with a perforated concrete well[1]. Based on the ordinary recharge well with filter layer, Wanglin Li improved the structure and put forward a kind of multiple flow section recharge well with filter layer that could increase the single-well recharge volume[2].

Main problems of the existing recharge well with filter layer are low efficiency, easy to be blocked and flushed. In addition, it uses too much building material and the cost is relatively high. The above problems seriously influence the promotion of the recharge well with filter layer.

Aiming at the key factors for restricting recharge volume, this paper designed a new square internal and external flow section recharge well with filter layer with a larger recharge volume.

2 THE DESIGN OF SQUARE INTERNAL AND EXTERNAL FLOW SECTION RECHARGE WELL WITH FILTER LAYER

Focusing on the improvement of the recharge pool, structure, expanding the cross-sectional area and saving the occupied area of the recharge pool, a new square internal and external flow section recharge well with filter layer, which can increase the single-well recharge volume, was designed.

The key points of designing the square internal and external flow section recharge well with filter layer are as follows:

1. In order to prolong service lives, the recharge pool is set above the earth of the river or canal bottom;
2. To improve the single-well recharge volume to a great extent, the top surface, outer side, inner side and bottom of the recharge pool are all taken as cross-sectional;
3. Aiming to improve the resistance of being rushed and ensure the stability of the recharge pool, reinforced concrete and masonry are used as the material;
4. Each surface of the discharge flow section is covered with a layer of geotechnical fabric. Geotechnical fabric is directly fixed to the bottom of the recharge pool to replace the sand gravel filter layer;
5. The part of the recharge pool adjacent to the ground is made into an impermeable body, which can prevent sewage recharge at low water table;
6. The top of the recharge pool is made into a slope along the flow direction and its slope ratio is 1%–10%, which can help flow scour sediment at the surface of the recharge pool, and the recharge pool would have a better ability to prevent silt.

Figure 1 shows the design of the square internal and external flow section recharge well with filter layer.

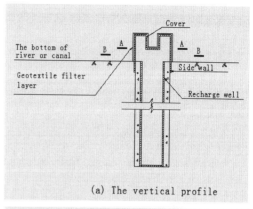

(a) The vertical profile

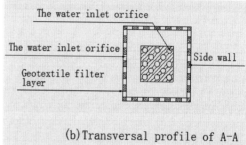

(b) Transversal profile of A-A

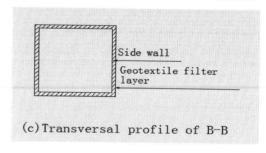

(c) Transversal profile of B-B

Figure 1. Structure diagram of square internal and external flow section recharge well with filter layer.

3 THE CALCULATION OF RECHARGE VOLUME FOR THE SQUARE INTERNAL AND EXTERNAL FLOW SECTION RECHARGE WELL WITH FILTER LAYER

3.1 Calculation formula of recharge volume for square internal and external flow section recharge well with filter layer

Taking fully penetrating well in a confined aquifer as an example, the calculation method of recharge volume for the square internal and external flow section recharge well with filter layer is similar to the ordinary recharge well with filter layer[3-4], and the single-well recharge volume is controlled by the seepage of the square internal and external flow section recharge pool and ordinary recharge well flow. The steady flow model is shown in Equations (1)–(5), and Equation (1) represents the square internal and external flow section recharge pool, Equation (4) represents the hypothetical relationship between the water head Hf after flowing through the square internal and external flow section recharge pool and the effective recharge head Hn outside the recharge well with filter layer.

In the solution process, first, water head Hf was calculated in Formula (4), then Hf was substituted into formula (3), and then the effective irrigation water height of the recharge well Hn could be obtained. Then, Hf and Hn were substituted into Formulas (1) and (2), and the corresponding single well recharge volume could be obtained.

$$Q_f = K_f A_f \frac{H_w - H_f}{m_f} \tag{1}$$

where Q_f = recharge volume of the recharge pool (m³/s); K_f = permeability coefficient of the square internal and external flow section recharge pool (m/s); A_f = cross-sectional area of the square internal and external flow section recharge pool, namely the sum of inflow area including the top surface, the outer surface, the inner side surface and the bottom surface (m²); H_w = outer surface water height of the recharge pool, namely the recharge designed water level of the river or canal (m); H_f = water head when the water flow through the filter layer of the recharge pool (m); and m_f = thickness of the filter layer (m):

$$Q_c = \frac{2\pi k_o M (H_n - H_o)}{\ln \frac{R_o}{r_w}} \tag{2}$$

where Q_c = amount of the recharge water (m³/s); K_o = permeability coefficient of the underground water-bearing sand (m/s); M = thickness of the underground aquifer (m); H_o = groundwater level before recharge (m); R_0 = radius of influence (m); and r_w = filter pipe radius (m):

$$Q_c = Q_f \tag{3}$$

$$H_n = \beta_c H_f \tag{4}$$

where H_n = effective irrigation water height of the recharge well, namely the actual outer surface water height of the recharge pool (m); β_c = synthetical reduction coefficient of the square internal and external flow section recharge well with filter layer:

$$H_f = \frac{H_w + \alpha_c H_o}{1 + \alpha_c \beta_c} \tag{5}$$

where α_c = steady flow coefficient of the square internal and external flow section recharge well with filter layer:

$$\alpha_c = \frac{2\pi k_o M m_f}{k_f A_f \ln \frac{R_o}{r_w}} \tag{6}$$

3.2 The result of indoor steady flow recharge experiment

A 1:200 scale model was constructed and tested to study the indoor steady flow recharge. The recharge device was a cuboid (1.8 m × 0.7 m × and 1.3 m), it was composed of the recharge system and the measurement system. The recharge system included water resource, water diversion system, recharge pool, recharge well, aquifer and drainage system. The measurement system could monitor intake water level, groundwater level and water volume[5].

The indoor experiment of the square internal and external flow section recharge well with filter layer was conducted. The height of the recharge pool well head and the recharge water head were 0.11 m and 0.925 m, respectively. The results of experiment indicated that the single-well recharge volume of the square internal and external flow section recharge well with filter layer was 4.34 m3/min.

3.3 The theoretical calculation of single-well recharge volume

Based on the experimental results of the square internal and external flow section recharge well with filter layer, the following parameters were chosen: K_f = 0.000269 m/s, K_o = 0.00023 m/s and β_c = 0.76. The single-well recharge volume of the square internal and external flow section recharge well with filter layer could be calculated from the theoretical equation, which was 4.28 m³/min.

4 COMPARATIVE ANALYSIS OF THE SINGLE-WELL RECHARGE VOLUME BETWEEN THE SQUARE INTERNAL AND EXTERNAL FLOW SECTION RECHARGE WELL WITH FILTER LAYER AND THE ORDINARY RECHARGE WELL WITH FILTER LAYER

The indoor experiment of the ordinary flow section recharge well with filter layer was conducted. The height of the recharge pool well head and the recharge water head were 0.110 m and 0.925 m, respectively. The experimental results indicated that the single-well recharge volume of the ordinary flow section recharge well with filter layer was 0.97 m³/min. Therefore, under the same conditions, the single-well recharge volume of the square internal and external flow section recharge well with filter layer was about 4.47 times than the ordinary recharge well. It was far greater than the ordinary recharge well with filter layer.

5 CONCLUSIONS

On the basis of the existing recharge well with filter layer, this paper analyzed its current situation and deficiency. On the improvement and optimization of the existing recharge well with filter layer, a new square internal and external flow section recharge well with filter layer was designed. It can increase the single-well recharge volume and improve anti-silt and anti-rush capability, it also saves sand-gravel materials. Moreover, by analyzing the theoretical calculation and experimental result of the recharge volume of the fully penetrating square internal and external flow section recharge well with filter layer in confined aquifer, it was verified that the square internal and external flow section recharge well with filter layer has the stronger recharge ability.

It can be used in the underground reservoir recharge project.

ACKNOWLEDGMENTS

This work was financially supported by the Shandong Province Science and Technology Development Plants (2013GSF11606) and the Shandong Province Natural Foundation (ZR2014EFM023). Li Wanglin is the corresponding author.

REFERENCES

[1] LI Wang-lin. Theory and Method of Structure Design for Recharge Well with Filter, Groundwater. Vol. 31, No. 1 (2009), 126–129. (in Chinese).
[2] LI Wang-lin. Multiple Flow Water of Recharge Well with Filter, China Rural Water and Hydropower, Vol. 12 (2012), 112–114. (in Chinese).
[3] MAO Yong-xi. Seepage Computation Analysis & Control [M]. Beijing: China Water Power Press, 2003.
[4] XUE Yu-qun, ZHU Xue-yu. Groundwater dynamics [M]. Beijing: Geological Publishing House, 2001. (in Chinese).
[5] MA Chao-qun, SHU Long-cang, LI Wei, WANG Bin-bin, ZHANG Jian, WANG Xiao-hui. Experimental Simulation on Recharge Efficiency of Infiltration Pond [J]. International Journal Hydroelectric Energy, 2011, 29(3), 60–63. (in Chinese).

Architectural, Energy and Information Engineering – Sung & Chen (Eds)
© *2016 Taylor & Francis Group, London, ISBN 978-1-138-02791-6*

Detection and energy efficiency evaluation of a domestic solar water heating system

Han Fang Liu & Fu Sheng Dong
Kunming Construction Engineering Quality Test Center, Kunming, China

ABSTRACT: A domestic solar water heating system, which was commonly used at residential district at present, was tested. Based on the analysis and calculation, it was found that the annual solar assurance rate could reach 70%, and that a single system could replace conventional energy about 0.296 tons of standard coal. This finding can provide some basis for the energy efficiency evaluation of the real estate project that used this kind of solar water heating system. Thus, the domestic solar water heating system has high economic and environmental benefits, as revealed by the evaluation.

Keywords: renewable energy; domestic solar water heating system; energy efficiency evaluation

1 INTRODUCTION

In recent years, the real estate industry has been developing rapidly and water use in building has been constantly increasing with the development of society, and the combination of solar energy heat utilization with buildings has become close, promoting the use of solar energy that plays a very important role in conventional energy conservation. Furthermore, energy and environment is one of the main problems that restricts the growth of the national economy and steady improvement of people's life quality. Therefore, the prospects of the development and utilization of solar energy [1–2] are self-evident. For the efficiency evaluation of solar energy, there has been a relatively accurate computation method.

At present, the solar water heating system used in building includes three types: Household heat collection—household heat accumulation system, Centralized heat collection-household heat accumulation system, and Centralized heat collection—centralized heat supply system. The three systems used for different buildings have their own characteristics. The Centralized heat collection-household heat accumulation system is widely used in high-rise (≥15 F) buildings, and the Centralized heat collection—centralized heat supply system is frequently used in small high-rise (10~15 F) buildings. The two systems are more complicated in design and installation than the household heat collection—household heat accumulation system. The Household heat collection—household heat accumulation system is mainly used in multistoried building, which includes wall mounted and home

solar system, while the residents who lived in the bottom of wall mounted system have limited use of solar system for the serious shade, so if conditions permit, the Vacuum tube solar water heating system is generally used in buildings.

At present, residential areas in China are mainly multilayer building, so the mount of domestic solar water heating system used in building is conceivable. According to the standard, per person per family 44 L/d, three people to calculate, each family only needs to configure a 132 L hot water system. Systems are mainly used in marketplace, which have 20 vacuum tube, 24 vacuum tube, and 30 vacuum tube, which means that the system having 20 vacuum tubes is enough for a family whose theoretical volume is 150 L and collecting area is 2.5 m^2. In this paper, a test research is conducted to evaluate the energy efficiency of the proposed water heater.

2 THE PERFORMANCE TESTING OF THE SYSTEM

For testing the system, the static test method was chosen, water supply was stopped during the test, hot water was replaced with cold water before the commencement of the experiment, and 24 h was used as the test cycle. The main test data included ambient wind velocity, ambient temperature, solar radiation intensity, and the water tank temperature. Three temperature probes were used (up, middle, and down) for more accurate measurements. The temperature of the tank was determined by the average value, and temperature recorder made data storage every minute.

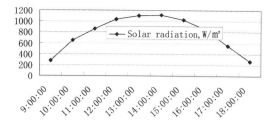

Figure 1. The solar radiation on the test day.

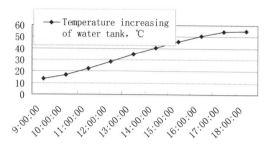

Figure 2. Temperature increasing of the water tank.

A total of 20 vacuum tubes of a solar water system were chosen for the test towards the south. The system had a 152 L water tank and the collecting area was 2.49 m^2, with an inclination of 30° and no sunscreen during the test. The solar radiation on the test day is shown in Figure 1. Solar radiation was in good condition, cumulative radiation was 22 MJ, and the average wind speed was 2.5 m/s, which met the test conditions.

The equipments used in the test were completely measured and checked.

In order to more clearly observe the rise in temperature, the average temperature rise of several nodes were measured, as shown in Figure 2. We found that the temperature rise of the system was in good condition, and the average water temperature rise was 41°C in the testing period.

3 EFFICIENCY EVALUATION OF THE SYSTEM

Several parameters were examined for the evaluation of system efficiency: the annual solar guarantee rate, the annual amount of conventional energy substitution and carbon dioxide emission reduction, sulfur dioxide emission reduction, soot emission reduction, annual cost saving, and the static payback period.

(1) The annual solar guarantee rate calculation

① Solar guarantee rate $f = \dfrac{Q_c}{Q_T}$

Here, f—solar guarantee rate, dimensionless; Q_c—heat of system collecting, MJ; and Q_T—the total heat demand, MJ, design temperature: 55°C, cold water: 15°C.

Also, $Q_c = Q_G + c_{pw}\rho_w V_w (t_i - t_f)/1000000$

Here, Q_c—heat of the system, MJ; Q_G—the use of heat, MJ, the static test is 0; c_{pw}—the heat capacity of water, J/(kg·K); ρ_w—the density of water, kg/m³; V_w—the volume of the the water tank, m³; t_i—end temperature of the water tank, °C; t_f—start temperature of the water tank,°C.

②The number of days is X_1 when local solar radiation is less than 8 MJ/m^2, the measured solar guarantee rate is f_1; the number of days is X_2 when local solar radiation is in the range of 8 MJ/m^2~13 MJ/m^2, the measured solar guarantee rate is f_2; the number of days is X_3 when local solar radiation is in the range of 13 MJ/m^2~18MJ/m^2, the measured solar guarantee rate is f_3; the number of days is X_4 when local solar radiation is more than 18 MJ/m^2, and the measured solar guarantee rate is f_4.

So, the annual solar guarantee rate calculation f_Q is given by

$$f_Q = \frac{x_1 f_1 + x_2 f_2 + x_3 f_3 + x_4 f_4}{x_1 + x_2 + x_3 + x_4}$$

(2) The annual amount of conventional energy substitution

Through the test, the heat of the system obtained is Q_1 when local solar radiation is less than 8 MJ/m^2; heat of the system obtained is Q_2 when local solar radiation is in the range of 8 MJ/m^2~13 MJ/m^2; heat of the system got is Q_3 when local solar radiation is in the range of 13 MJ/m^2~18 MJ/m^2; and heat of the system obtained is Q_4 when local solar radiation is more than 18 MJ/m^2;

So, the annual amount of conventional energy substitution Q_{bm} is given by

$$Q_{bm} = \frac{(x_1 Q_1 + x_2 Q_2 + x_3 Q_3 + x_4 Q_4)}{29309 \times 65\%}$$

The results are summarized in Table 1. We found that the annual solar guarantee rate of the system was up to 70%, which showed good economic benefits. The amount of annual conventional energy substitution of the system was 0.296 tons standard coal, which showed highly good environmental benefits. To make environmental benefit assessment for the real estate that installed this system, system sets are multiplied to obtain the project results.

(3) Environmental benefit assessment

①Calculation of carbon dioxide emission reduction

$$Q_{CO2} = 2.47 Q_{bm}$$

Table 1. The result of annual solar guarantee rate calculation and amount of annual conventional energy substitution.

No.	Parameters	The radiation on testing day (MJ/m^2)			
		J < 8	8 ≤ J < 13	13 ≤ J < 18	J ≥ 18
1	The measured radiation on testing day, MJ/m^2	7	12	17	22
2	The measured heat of system (Q_1, Q_2, Q_3, Q_4), MJ	6.81	11.35	15.81	20.45
3	Days (X_1, X_2, X_3, X_4)	67	65	71	162
4	The measured solar guarantee rate on testing day (f_1, f_2, f_3, f_4)	30.7%	51.1%	71.3%	92.2%
5	The annual solar guarantee rate f_0	69.5%			
6	The annual amount of conventional energy substitution A (tons standard coal)	0.296			
Remark	X_1, X_2, X_3, X_4 calculated by professional programming software				

Table 2. The result of Environmental benefit assessment.

Parameters	The annual amount of conventional energy substitution (*ton/year*)	Carbon dioxide emission reduction (*ton/year*)	Sulfur dioxide emission reduction (*ton/year*)	Soot emission reduction (*ton/year*)	Static payback period (*year*)
Results	0.296	0.730	0.006	0.003	2.99

Here, Q_{CO2}—carbon dioxide emission reduction, unit: ton/year; Q_{bm}—annual amount of standard coal, unit: ton/year; and 2.47—carbon dioxide emission factor of standard coal, dimensionless.

②Calculation of sulfur dioxide emission reduction

$$Q_{SO2} = 0.02Q_{bm}$$

Here, Q_{SO2}—sulfur dioxide emission reduction, unit: ton/year; Q_{bm}—annual amount of standard coal, unit: ton/year; and 0.02—sulfur dioxide emission factor of standard coal, dimensionless.

③Calculation of soot emission reduction

$$Q_{FC} = 0.01Q_{bm}$$

Here, Q_{FC}—soot emission reduction, unit: ton/year; Q_{bm}—annual amount of conventional energy substitution, unit: ton/year; and 0.01—standard coal sulfur soot factor, dimensionless.

④Calculation of annual cost saving, using the price of electricity 0.5 *Yuan/kW·h*

$$C = \frac{(x_1Q_1 + x_2Q_2 + x_3Q_3 + x_4Q_4)}{3.6 \times 0.9} \times 0.5$$

Calculation of the static payback period

$$y = \frac{N}{C}$$

Here, N—price of the solar water system, 2600 *Yuan*.

The results are summarized in Table 2. We can clearly observe the annual environmental benefit parameters of the system. To make environmental benefit assessment for real estate that installed this type system, system sets are multiplied to obtain the project results. It only needs three years to recover the cost, calculated based on 10 years design lifetime, which shows good economic benefits.

4 CONCLUSIONS

This paper studied on a commonly used domestic solar water heating system. Based on the calculation and analysis, it was found that the annual solar guarantee rate of the system was up to 70%, the amount of energy saving of the system was 0.296 tons of standard coal, and the result of environmental benefit assessment was obtained, which can provide a basis for the project such as environmental benefit assessment that installed this type solar water system. It has high economic and environmental benefits, and only needs three years to recover the cost, calculated based on 10 years design lifetime, and has good economic benefits.

REFERENCES

[1] Lv Lan, Tang Xiao, Ma Xiaohong, The analysis among the testing results of different solar water systems, Water & Wastewater Engineering, 2003, 24(4): 41–44.

[2] Wang Hongyu. Design of the Solar Energy Water Heating System for Multi-Layer Buildings [J]. Journal of Jiansu of Science and Technology (Natural Science), 2001, 22(6): 6–9.

[3] Wu Xiaochun, Yang Tian. Discussion on the application of solar water system in high rise building [J]. J Water & Wastewater Engineering, 2012, 38(12): 77–81.

[4] GB20095-2006, Assessment code for performance of solar water heating systems [S].

[5] GB/T50801-2013, Evaluation standard for application of renewable energy in buildings [S].

Architectural, Energy and Information Engineering – Sung & Chen (Eds)
© 2016 Taylor & Francis Group, London, ISBN 978-1-138-02791-6

Research on three-dimensional parallel fast-acquisition method of GPS signals in CUDA

H.J. Li, X.B. Wu & W. Zhang
School of Aerospace Engineering, Beijing Institute of Technology, Beijing, China

ABSTRACT: A hot issue in the field of satellite navigation currently is how to shorten the acquisition time of receivers. To complete the fast-acquisition of the C/A code of GPS software receivers, a method of three-dimensional parallel fast-acquisition of GPS signals in CUDA is put forward, with which all the correlation integral amplitudes of 32 satellites, 41 frequencies and 2046 code phases can be calculated by running the CUDA program only once, therefore shortening the acquisition time of the receivers. For the present, the three-dimensional parallel fast-acquisition module can acquire signals in as fast as 66.2 ms. Such high efficiency will lay a solid foundation for shortening the TTFF and improving the overall performance of the receivers.

Keywords: software receiver; fast-acquisition; CUDA; GPU

1 INTRODUCTION

Previous receivers use many kinds of digital signal processors, including ASIC, FPGA/DSP/ARM and CPU[1]. The rapid development of Graphical Processing Units (GPUs) and the proposal of Compute Unified Device Architecture (CUDA) programming model in recent years make GPUs a new choice[2]. The traditional application of GPUs is confined to processing computing tasks of rendering images. With the improvement of their programmability, the research on using GPUs to complete general computing becomes more and more active. The use of GPUs in other fields except rendering images is called GPGPU (General-Purpose computing on Graphics Processing Unit). GPGPU usually adopts a unified model of CPUs and GPUs with the former responsible for complex logic and physical management, and the latter parallel computing. Current GPUs already have a strong ability of parallel computing, and its float-point performance can be achieved more than ten times than that of CPUs. This is why the CUDA programming model, introduced by NVIDIA in 2007, is designed to support joint CPU/GPU execution of an application. Programmers no longer need to use the graphics API to access the GPU parallel computing capability[3]. Correlators are the key to and the most important part of a receiver. The acquisition speed mainly depends on the quantities and efficiency of correlators. Latest GPUs have nearly 5000 cores that are developed by NVIDIA. Compared with a traditional digital signal processor, those provide a large number of correlators and high-speed operation. The study of those provides a new hardware and software environment at the same time.

Fast-acquisition is the key to completing real-time software receivers, currently, GPS software receivers generally adopt the code phase parallel acquisition method based on FFT[4], the acquisition time of which is very long, hence its low-efficiency. This paper designs a three-dimensional parallel fast-acquisition method in CUDA, which performs three-dimensional acquisition at the same time in PRN, carrier frequency and code phase, greatly shortening the acquisition time. It is fully used that GPUs have a large number of parallel computing units and a high operation speed. The method of exhaustion is adopted with space instead of time. The optimized three-dimensional parallel fast-acquisition module can acquire signals in as fast as 66.2 ms. Nowadays, the fastest capture algorithm takes above 1 s, but three-dimensional parallel fast-acquisition module takes much less than 1 s. It shortens the time of capture algorithm and improves the efficiency greatly.

2 THE THREE-DIMENSIONAL PARALLEL FAST-ACQUISITION BASED ON CUDA

2.1 Test platform

The GPU real-time workstation of Concurrent Computer Corporation is used to test the fast-acquisition method. Table 1 shows the software

Table 1. Configuration of the GPU workstation.

Configuration	CPU	GPU
Cores	Intel core i7 3370k	NVIDIA Tesla C2075
Basic frequency	3.5 GHz	1.15 GHz
The number of cores	4	448
Compiler	GCC	NVCC 4.0
Memory	8 G	6 G
Operation system	RedHawk Linux 6.3	

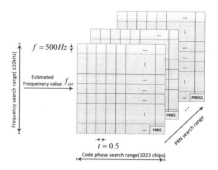

Figure 1. The range of three-dimensional search.

and hardware configurations of the GPU workstation.

The GPU workstation has installed a real-time operating system named RedHawk Linux 6 which optimizes CUDA. The Concurrent and RedHawk Linux guarantee the real-time of the fast-acquisition.

The Intermediate Frequency (IF) digital signal is produced by a sampler named NS 210 from OLinkStar CO., Ltd. IF sampling frequency is 16.367667 MHz. Digital IF frequency is 4.123968 MHz. Quantization bit number is 2 bits.

2.2 The three-dimensional parallel fast-acquisition module

The frequency search range for the 32 GPS satellites of a classical receiver is ±10 kHz and the code phase search range is 1023 chips. The step size in the frequency and code phase search is 500 Hz and 0.5 chip respectively[5]. The total number of the three-dimensional search units is

$$N_{cell} = 32 \times \left(\frac{2 \times 10000}{500} + 1 \right) \times \frac{1023}{0.5} = 2684352 \quad (1)$$

The number of frequency and the code phase of each satellite are 41 and 2046 respectively. Figure 1 is the range of three-dimensional search.

The CUDA program, called the three-dimensional parallel fast-acquisition module in this paper, can configure 2684352 correlators on the GPU at the same time. Running the CUDA program once can calculate the correlation integral amplitudes of 32 satellites, 41 frequencies and 2046 code phases.

The size of the IF digital signals is 16368. Its time is 1 ms. The structure of threads is that one grid is divided into 32 * 41 blocks, each block searching code phases in one satellite and one frequency. Each block is made of 1024 threads, with each thread calculating the correlation integral amplitudes of the two code phases. For example, the k thread calculates the k and k + 0.5 code phases. Figure 2 is the algorithm flow chart.

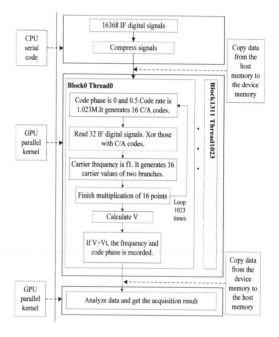

Figure 2. Algorithm flow chart.

The algorithm includes four stages.

The first stage is the initialization of GPUs. The C/A code table of 32 satellites and 41 frequencies are copied to the constant memory on the GPU.

The second stage processes data in a serial way on the CPU. The main tasks are shown as follows.

1. The CUDA program reads the IF digital signals of 1 ms length.
2. Compress the IF digital signals from 16 KB to 4 KB.

The third stage is to process data on the GPU. The main tasks are shown as follows.

1. The compressed IF digital signals are copied to the global memory on the GPU from the

internal storage, and then to the shared memory on the GPU.

2. Each thread calculates the two code phases it is supposed to according to the number of threads.
3. Obtain the carrier frequencies based on the number of blocks by looking up the saved table.
4. Generate 16 C/A codes, and save them into an unsigned integer.
5. Multiply the 16 C/A codes by the 16 IF digital signals. That is, xor one unsigned integer with the other unsigned integer.
6. Generate 16 carrier signals in I branch and Q branch, multiply them by the stripped IF digital signals, and then make an accumulation.
7. Repeat steps 4–6 for 1023 times and calculate the correlation integral amplitudes Ip1, Qp1, Ip2, and Qp2.

$$Ip = \sum_{n=1}^{1023} ip, \quad Qp = \sum_{n=1}^{1023} qp \qquad (2)$$

Calculate the correlation integral amplitudes V1 and V2 of the two code phases.

$$V = \sqrt{Ip^2 + Qp^2}. \qquad (3)$$

8. Judge the result according to V1, V2 and the threshold value. If the correlation integral amplitude is greater than the threshold value, indexes of the frequency and the code phase will be saved. At the same time, correlation integral amplitudes are also saved for later analysis.

The main tasks in the final stage are shown as follows.

1. Data is copied to the internal storage.
2. Indexes are analyzed and the result of the acquisition obtained.

The three-dimensional parallel fast-acquisition module can complete the calculation of the correlation integral amplitudes of all the search units by only one calculation. Figure 3 shows the correlation amplitude of PRN 3. The numbers from 0 to 2046 on the code phase axis are on behalf of 2046 code phases. The numbers from 0 to 40 on the frequency axis are on behalf of the 41 frequencies values searched by the acquisition module. In Figure 3, correlation amplitudes do not have an obvious peak and any correlation amplitudes are greater than the threshold value, which means PRN 3 is not acquired.

Figure 4 shows the correlation amplitudes of PRN 14.

An obvious peak of the correlation amplitudes can be found in Figure 4, which is higher than any other correlation amplitudes and greater than the threshold value. This means PRN 14 is com-

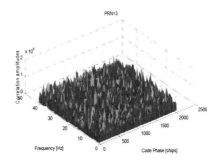

Figure 3.　Correlation amplitudes of PRN 3.

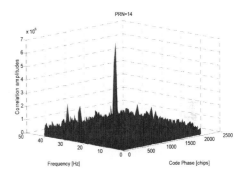

Figure 4.　Correlation amplitudes of PRN 14.

pletely acquired. The roughly estimated value of its Doppler shift and code phase is 2500 Hz and the 540.0th code phase, respectively.

In order to prove the correctness of the acquisition by the three-dimensional parallel fast-acquisition module, an acquisition algorithm is implemented in MATLAB in the meantime. After that, we compare these two algorithms. The acquisition algorithm in MATLAB first uses parallel code phase acquisition algorithm to acquire the accurate code phase and the inaccurate frequency. Then, it uses parallel frequency acquisition algorithm to obtain the accurate frequency. Table 2 shows the results of the acquisition. The 2nd and 3rd columns are the results of acquisition in MATLAB. The 4th and 5th columns are the results of the three-dimensional parallel fast-acquisition module. To make code phase and Doppler shift more accurate, a method is adopted. Two maximum correlation amplitudes are V_1 and V_2, respectively, on the Doppler shift section. Code phases are C_1 and C_2. C is more accurate code phase.

$$C = \frac{C_1 V_1 + C_2 V_2}{V_1 + V_2}. \qquad (4)$$

In the similar way, two maximum correlation amplitudes are V_1 and V_2, respectively, on the code phase section. Doppler shifts are f_1 and f_2. f is more accurate Doppler shift.

$$f = \frac{f_1 V_1 + f_2 V_2}{V_1 + V_2}. \tag{5}$$

The above results prove the correctness of the three-dimensional parallel fast-acquisition module.

2.3 Optimization of the three-dimensional parallel fast-acquisition module

After analyzing the three-dimensional parallel fast-acquisition module, we find that some calculations can be merged. The first algorithm is shown as follows. When searching different PRNs, the same frequency and different code phases, correlators generate the same carrier signals, different PRNs and different C/A signals, which means the multiplication of the IF digital signals by the carrier signals can be merged. In the second algorithm, when searching the same PRN, the same code phase and different frequencies, correlators generate the same code phase signals and different carrier signals. Therefore, the multiplication of the IF digital signals by the C/A signals can also be merged.

The second algorithm will be tested here. Analysis of the kernel of the three-dimensional parallel fast-acquisition module shows that the calculation of different blocks is the same before generating the carrier signals. The kernel can combine some blocks. After that, one block will calculate more than one frequency. Although the task of the single thread calculation becomes heavier; the overall efficiency is higher. Table 3 shows the computing time of the acquisition in different thread structures after the combination.

As is shown in Table 3, with the increase of the frequency number of each block, the acquisition time decreases. The time decrease is caused by the merging of the multiplication of the IF digital signals by the C/A signals. Under current conditions, the three-dimensional parallel fast-acquisition

Table 3. The computing time of the acquisition in different thread structures.

Thread structures and blocks number	Threads number of each block	Frequencies number of each block	Acquisition time/ms
32 × 41	1024	1	981.4
32 × 21	1024	2	749.6
32 × 14	1024	3	714.7
32 × 10	1024	4	513.9
32 × 8	1024	5	424.3
32 × 7	1024	6	357.4
32 × 6	1024	7	312.9
32 × 5	1024	8	264.5
32 × 4	1024	10	220.3
32 × 3	1024	14	154.3
32 × 2	1024	20	110.2
32 × 1	1024	41	66.2

module can acquire signals within as fast as 66.2 ms, which is very efficient.

3 CONCLUSIONS

The acquisition time of traditional GPS software receivers is very long, and not efficient enough. This paper proposes a new method of three-dimensional parallel fast-acquisition of GPS signals in CUDA, which can simultaneously acquire signals in three-dimensional space by using GPUs with many parallel processing units and an excellent CUDA architecture. Optimized three-dimensional parallel fast-acquisition module can acquire signals within as fast as 66.2 ms, which is faster and more efficient than any other existing acquisition algorithms. It will lay a solid foundation for shortening the TTFF and improving the overall performance of receivers.

REFERENCES

[1] Haak, Ulrich, Büsing, et al, 2012. Performance Analysis of GPU Based GNSS Signal Processing. *Proceedings of the 25th International Technical Meeting of The Satellite Division of the Institute of Navigation (ION GNSS 2012)*: 2371–2377. Nashville: ION.

[2] Yang Jing & Liu Yifei. 2012, Parallel-acquisition of GPS signal based on graphic processing unit. *Journal of Chinese Inertial Technology*: (04) 430–434.

[3] Kirk D B, Wen-mei W H. 2012, Elsevier, *Programming massively parallel processors: a hands-on approach*: 2–34.

[4] Van Nee D J R, Coenen A. 1991. New fast GPS code-acquisition technique using FFT. *Electronics Letters*: 27(2): 158–160.

[5] Xie Gang. 2009. *Principles of GPS and receiver design*. Beijing: Publishing House of Electronics Industry.

Table 2. Results of the acquisition.

PRN	Doppler shift Hz	Code phase chip	Doppler shift Hz	Code phase chip
14	2600	540.39	2716.5	540.20
15	−1850	194.88	−1749.5	194.68
17	−800	140.19	−734.4	140.17
21	−2150	748.02	−2161.8	747.80
22	2800	589.07	2810.7	588.81
24	−950	519.20	−802.5	519.15
25	4250	767.70	4286.9	767.37

Architectural, Energy and Information Engineering – Sung & Chen (Eds)
© 2016 Taylor & Francis Group, London, ISBN 978-1-138-02791-6

Equivalent circuit model for a sodium-nickel chloride battery

C.Y. Guo, X.X. Wu & H. Xu
Naval University of Engineering, Wuhan, China

ABSTRACT: To improve the use of sodium-nickel chloride battery in the applications such as electric vehicle, Uninterruptable Power Systems (UPS) and telecom backup systems, an Equivalent Circuit Model (ECM) of the sodium-nickel chloride battery is proposed based on the investigations of the traditional models. And the model parameters are identified based on the experimental data published by Centro Elettrotecnico Sperimentale Italiano (CESI). Finally a simulation model was set up based on Matlab/Simulink, the comparison between simulation and test data is carried out. The comparison shows that ECM is also suitable for the modeling of sodium-nickel chloride battery.

Keywords: Sodium-nickel chloride battery; equivalent circuit model; DP Model

1 INTRODUCTION

The sodium-nickel chloride battery, a high-temperature secondary battery system, has been explored for use in electric vehicle applications because of its high specific energy, power density, long cyclic life, zero maintenance and with a remote diagnostic capability. Vehicles powered by sodium-nickel chloride battery such as the Mercedes A Class car, Vito van4 and the Th!nk City compact car have logged millions of kilometers [1]. And these advantages can also make it particularly suitable for applications such as Uninterruptable Power Systems (UPS) and telecom backup systems [2].

In recent years, a large development effort has been dedicated to the model of power batteries, especially lithium-ion batteries [3–4]. But a little research work has been done on the sodium-nickel chloride battery [5]. And these models of sodium-nickel chloride battery mostly based on the Modeling of reaction kinetics and transport processes in the cell, are not robust so as to predict accurate battery voltage, current, as well as the estimation of batteries' State of Charge (SOC) and State of Health (SOH) used in the battery management system for these applications described above.

To improve the use of sodium-nickel chloride battery in these applications, an equivalent circuit model of the sodium-nickel chloride battery is proposed based on the investigations of the traditional models. And the remainder of this paper is organized as follows. In Section 2, the equivalent circuit battery models are described. The DP Model for a sodium-nickel chloride battery is introduced in Section 3. The battery simulation model and the comparison between simulation and test data are discussed in Section 4, followed by conclusions presented in Section 5.

2 EQUIVALENT CIRCUIT MODELS OF BATTERY

Equivalent circuit models such as the Rint model, the Thevenin model, or the DP model are now widely used because these models can be used to successfully simulate battery performance for various chemistries, including Valve-Regulated Lead-Acid (VRLA), Nickel Metal Hydride (Ni-MH), and lithium-ion batteries [6].

The Rint model, as shown in Figure 1 and Equation (1), is the simplest battery model which consisting of an ideal voltage source (V_{oc}) and a constant equivalent internal series resistance (R_0)

$$V_{terminal} = V_{OC} - I_{battery}R_0 \qquad (1)$$

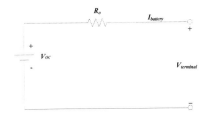

Figure 1. Schematic diagram of the Rint model.

497

The Thevenin model, as shown in Figure 2 and Equations (2), connects a parallel RC network in series based on the Rint model, describing the dynamic characteristics of the battery. It is mainly composed of open-circuit voltage (V_{oc}), internal resistances (the ohmic resistance R_0, the polarization resistance R_1) and equivalent capacitances (C_1).

$$\begin{cases} V_1 = -\dot{V}_1 R_1 C_1 + /I_{battery} / C_1 \\ V_{terminal} = V_{oc} - V_1 - I_{battery} R_0 \end{cases} \quad (2)$$

The DP Model, as shown in Figure 3 and Equations (3), connects a parallel RC network in series based on the Thevenin model to refine the description of polarization characteristics and simulate the concentration polarization and the electrochemical polarization separately. It is mainly composed of open-circuit voltage (V_{oc}), internal resistances (the ohmic resistance R_0, the polarization resistance R_1 and R_2) and the effective capacitances (C_1 and C_2).

$$\begin{cases} V_1 = -\dot{V}_1 / R_1 C_1 + I_{battery} / C_1 \\ V_2 = -\dot{V}_2 / R_2 C_2 + I_{battery} / C_2 \\ V_{terminal} = V_{oc} - V_1 - V_2 - I_{battery} R_0 \end{cases} \quad (3)$$

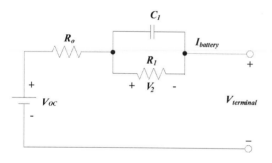

Figure 2. Schematic diagram of the Thevenin model.

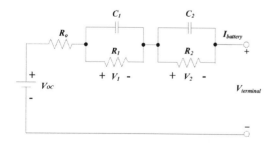

Figure 3. Schematic diagram of the DP model.

3 THE MODEL OF SODIUM-NICKEL CHLORIDE BATTERY

3.1 The DP model of sodium-nickel chloride battery

In order to descript the polarization characteristics of a sodium-nickel chloride battery, The DP Model is chosen to model the sodium-nickel chloride battery herein. And the schematic of DP Model used in this paper is shown in Figure 4.

As shown in Figure 4, the DP Model only uses a series resistance and two RC circuits to model the internal resistance and transient behavior of the sodium-nickel chloride battery. And the electrical behavior of DP Model can be expressed by Equation (4):

$$\begin{cases} \begin{bmatrix} \dot{U}_1 \\ \dot{U}_2 \end{bmatrix} = \begin{bmatrix} -\dfrac{1}{R_1 C_1} & 0 \\ 0 & -\dfrac{1}{R_2 C_2} \end{bmatrix} \begin{bmatrix} U_1 \\ U_2 \end{bmatrix} + \begin{bmatrix} \dfrac{1}{C_1} \\ \dfrac{1}{C_2} \end{bmatrix} I_{battery} \\ V_{terminal} = V_{OC} - U_1 - U_2 - R_0 I_{battery} \end{cases} \quad (4)$$

where R_0, the ohmic resistance of the battery;

R_1 and R_2, the polarization resistances of the battery;

C_1 and C_2, the polarization capacitances of the battery;

U_1 and U_2, the voltages across C_1 and C_2 respectively;

V_{OC}, the open circuit voltage of the battery;

$V_{terminal}$, the terminal voltage of the battery;

$I_{battery}$, the discharge or charge current of the battery.

3.2 The parameters of DP model

These parameters in the DP Model of sodium-nickel chloride battery include three resistors (R_0, R_1, R_2), two capacitors (C_1, C_2), and open circuit voltage of the battery (V_{OC}). And all these parameters as functions of the current State-of-Charge

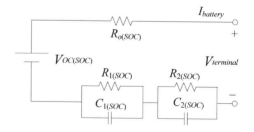

Figure 4. DP model of sodium-nickel chloride battery.

Table 1. The parameters of DP model.

1C discharge rate

SOC(%)	V_{oc}(V)	R_0(Ω)	R_1(Ω)	R_2(Ω)	C_1(F)	C_2(F)
99.41	285.68	0.6243	/	/	/	/
80.35	/	0.5211	0.06245	0.05616	518.4	30634.2
79.94	279.21	0.5175	/	/	/	/
60.54	/	0.5767	0.07141	0.07165	865.5	19545.0
60.32	278.40	0.5287	/	/	/	/
42.65	/	0.5248	0.1579	0.1088	1270.4	14841.3
42.35	277.38	0.5329	/	/	/	/
24.61	/	0.5254	0.2498	0.1337	1657.1	11657.8
24.23	275.56	0.5271	/	/	/	/
12.06	267.34	0.5186	0.1847	0.1229	2465.9	10413.5

0.5C discharge rate

SOC(%)	V_{oc}(V)	R_0(Ω)	R_1(Ω)	R_2(Ω)	C_1(F)	C_2(F)
99.56	285.94	0.7893	/	/	/	/
79.74	/	0.5166	0.06432	0.05435	476.1	29406.2
79.52	278.12	0.5729	/	/	/	/
59.86	/	0.6134	0.07628	0.07383	865.5	18857.8
59.52	277.49	0.5817	/	/	/	/
42.22	/	0.6103	0.1852	0.1171	1226.4	13571.9
41.97	276.71	0.6154	/	/	/	/
24.74	/	0.5224	0.3663	0.1701	1697.8	10767.1
24.56	274.65	0.4039	/	/	/	/
8.97	262.23	0.4934	0.1636	0.1054	2634.5	11236.3

(SOC) are showed in Table 1 which derived from the experimental data published by CESI [7].

Based the dates in Table 1, these parameters as functions of SOC are derived by utilizing Matlab Curve fitting box as show:

$$V_{OC} = 320.6 * (SOC - 0.566)^5 \\ - 13.56 * (SOC - 0.79)^2 \\ + 0.03279 * e^{126.8*(SOC-0.96)} + 278.792$$

$$R_0 = (2.106 * 10^{-9} * e^{18.28*SOC} + 0.5392) \\ \times (1 - 0.005 * (T_{batt} - 305))$$

$$R_{1(SOC,T)} = (0.2447 * e^{-\left(\frac{SOC - 0.2648}{0.1765}\right)^2} \\ + 0.06543) * (1 - 0.005 * (T_{batt} - 305))$$

$$R_{2(SOC,T)} = (0.09479 * e^{-\left(\frac{SOC - 0.2669}{0.234}\right)^2} \\ + 0.05675) * (1 - 0.005 * (T_{batt} - 305))$$

$$C_{1(SOC)} = 3134 * e^{-2.476*SOC} + 108.7$$

$$C_{2(SOC)} = 931 * e^{3.889*SOC} + 9270$$

4 THE SIMULATION OF SODIUM-NICKEL CHLORIDE BATTERY

To evaluate the validity of the DP model of sodium-nickel chloride batter, a Matlab simulation program using the battery model is carried out

and the results are compared with the laboratory experiment. These equations are implemented in Matlab/Simulink as shown in Figure 5.

The battery equivalent circuit model and the accuracy of its parameters identification were verified with constant power stage-discharge experiment published by CESI[7]. The experiment was made up with ten discharge stages which included 20 kW and 5 kW stage-discharge in turn. The discharge power and current curves are shown in Figure 6.

Figure 7 shows that the simulated voltage variation tendency is approximately the same as the

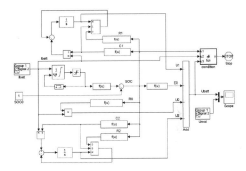

Figure 5. Simulink model of sodium-nickel chloride battery in Matlab/Simulink.

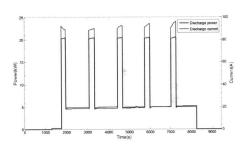

Figure 6. The load current and power profiles of CESI testing cycles.

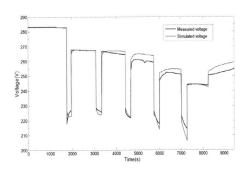

Figure 7. The terminal voltage profiles of the simulation and experiment.

measured and the error is amongst ±4%. It is demonstrated that the equivalent circuit model is highly accurate and exactly simulates the dynamic performance.

5 CONCLUSIONS

This paper investigates the equivalent circuit model of a sodium-nickel chloride battery for using in the battery management system. The comparison of simulation and experimental results shows the DP model can be used to express complicated sodium-nickel chloride battery performance via computer simulation.

REFERENCES

[1] R. Bull, A. Tilley. Development of New Types of Zebra Batteries for Various Vehicle Applications [J]. EVS-18, Berlin German. 2001.10.

[2] Silvio Restello, Nicola Zanon, Emiliano Paolin. Sodium Nickel Batteries for Telecom Hybrid Power Systems [J]. 2013 INTELEC (r), Hamburg. 2013.10.

[3] Hongwen He, Rui Xiong, Jinxin Fan. Evaluation of Lithium-Ion Battery Equivalent Circuit Models for State of Charge Estimation by an Experimental Approach [J]. Energies, 2011(4): 582–598.

[4] Jonathan Brand, Zheming Zhang, Ramesh K. Agarwal. Extraction of battery parameters of the equivalent circuit model using a multi-objective genetic algorithm [J]. Journal of Power Sources, 2014 (247): 729–737.

[5] Damla Eroglu, Alan C. Modeling of reaction kinetics and transport in the positive porous electrode in a sodium-iron chloride battery [J]. Journal of Power Sources, 2012 (203): 211–221.

[6] Xiaosong Hua, Shengbo Li, Huei Peng. A comparative study of equivalent circuit models for Li-ion batteries [J]. Journal of Power Sources, 2012 (198): 359–367.

[7] C. Bossi, A. Buonarota, E. Micolano. Risultati delle prove di laboratorio condotte su accumulatori avanzati (Results of the laboratory tests on advanced storage batteries) [C]. RSE (former CESI), Milano, 2005.03.

Architectural, Energy and Information Engineering – Sung & Chen (Eds)
© 2016 Taylor & Francis Group, London, ISBN 978-1-138-02791-6

Hybrid communication method for data gathering in wireless sensor networks

H.Y. Yuan & D.G. Dai
Department of Computer Science, Shaoguan University, Shaoguan, China

ABSTRACT: When cluster heads transmit their data to the sink via multi-hop communication, the cluster heads closer to the sink are burdened with heavy relay traffic and tend to die early. On the contrary, if all cluster heads transmit datas to the sink via single-hop communication, the cluster heads further from the sink will die much more quickly than those closer to the sink. In this paper, we first develop an analytical model to derive the optimal cluster radius. Then we propose a hybrid communication method where the nodes can transmit data to the sink in either single-hop or multi-hop. Finally, we conduct extensive experiments and show that our method outperforms LEACH and HEED in terms of network lifetime by balancing energy consumption.

Keywords: Wireless sensor network; Data gathering; Energy consumption balance; Network lifetime

1 INTRODUCTION

Wireless sensor networks have attracted much research attention in recent years and can be used in many different applications, including battlefield surveillance, machine failure diagnosis, biological detection, inventory tracking, home security, smart spaces, environmental monitoring, and so on (I. Akyildiz et al. 2002). A wireless sensor network consists of a large number of tiny, low-power, cheap sensor nodes. The sensor nodes in a wireless sensor network are usually deployed randomly inside the region of interest or close to it. The sink give commands to all the sensor nodes and gather data from the sensor nodes.

Being different from the traditional wireless networks, the sensor nodes are limited in energy, so the method for data gathering and routing must be energy efficient in order to prolong the lifetime of the entire network (K. Dasgupta et al. 2003). In order to prolong network lifetime, the sensor nodes can be organized hierarchically by grouping them into clusters. A better clustering algorithm may average the workload on each sensor node. The important factor that impacts the energy consumption is the cluster radius (H. Chen & S. Megerian 2006). If the difference in the cluster radius is big among all the generated clusters, the cluster-head having large cluster radius will consume a larger amount of energy than the others.

In clustered wireless sensor networks, the sensor nodes do not transmit their collected data to the sink, but to designated cluster heads which aggregate the data packets and send them to the sink via single-hop or multi-hop communication mode. For single-hop mode, the cluster heads furthest away from the sink are the most critical nodes, while in multi-hop mode, the cluster heads closest to the sink are burdened with a heavy relay traffic load and die first (S. Olariu & I. Stojmenovic 2006). In case of sensor nodes failure or malfunctioning around the sink, the network connectivity and coverage may not be guaranteed. No matter how many remaining sensor nodes are still active, none of them can communicate with the sink. As a result, the system lifetime becomes short.

In this paper, we develop an analytical model to estimate optimal cluster size and propose an optimal communication method for periodical data gathering applications in sensor networks. In our model, the cluster heads are allowed to communicate with the sink with hybrid communication method instead of only multi-hop or only single-hop.

2 RELATED WORK

Many protocols have been proposed for data gathering or communication in wireless sensor networks. The LEACH (W.R. Heinzelman et al. 2002) is a routing protocol for forming clusters in a selforganized homogeneous sensor network when the sink is located far from the sensor nodes. In LEACH, some nodes are elected as cluster heads while the other nodes communicate with the sink through cluster heads. This protocol randomly rotates the job of cluster heads based on the node's remaining energy in order to uniformly balance the energy consumption throughout the network.

The LEACH allows only single-hop clusters to be constructed. On the other hand, H. Su & X. Zhang (2006) proposed the similar clustering protocol where sensors communicate with their cluster-heads in multi-hop communication mode. The HEED (O. Younis & S. Fahmy 2002) extended LEACH by incorporating communication range limits and cost information. In HEED, the initial probability for each sensor to become a cluster head is dependent on its residual energy. Later on, sensors that are not covered by any cluster heads double their probability of becoming a cluster head. This procedure iterates until all sensor nodes are covered by at least one cluster head.

Y. Yu et al. (2004) studied the problem of constructing a data gathering tree over a wireless sensor network in order to minimize the total energy for compressing and transporting information from a set of source nodes to the sink. They investigated a tunable data compression technique that effective tradeoffs between computation and communication costs. They derive the optimal compression strategy for a given data gathering tree and investigate the performance of different tree structures for networks deployed on a grid topology.

J. Yang & D.Y. Zhang (2010) proposed a cluster-based data gathering and transmission protocol (CDAT) for wireless sensor networks. The CDAT achieves a good performance in terms of lifetime by a clustering method of balancing energy consumption and data prediction transmission strategy. The initial probability of sensor for cluster head election is derived from mathematical relation between application's seamless coverage ratio and numbers of required cluster heads.

C. Tharini & P. Vanaja Ranjan (2010) proposed a scheme for data gathering in wireless sensor networks. This scheme exploits temporal correlation in sensor data and uses this characteristic to efficiently gather at the sink using joint compression and prediction algorithm.

Our work differs from the above works since we focus on how to prolong the lifetime of sensor networks using hybrid communication method. We first derive the optimal cluster radius and then propose a hybrid communication method for data gathering in wireless sensor network.

3 SYSTEM MODEL

3.1 *Energy consumption model*

In this paper, we assume there is an energy-efficient MAC protocol in the underlying MAC layer,

energy will be consumed only when performing sensing task, processing raw data, and transmitting and receiving data for itself and other sensor nodes. We also assume all sensor nodes have power control and can use the minimum required energy to send information to the recipients.

We use a simplified radio model shown in (W.R. Heinzelman et al. 2002) for the radio hardware energy dissipation. Both the free space and the multi-path fading channel models are used in the model, depending on the distance. The energy spent for transmission of a 1-bit packet over distance d is between the transmitter and receiver.

$$E_{Tx}(l,d) = \begin{cases} l \times E_{elec} + l \times efs \times d^2, & d \leq d_0 \\ l \times E_{elec} + l \times emp \times d^4, & d > d_0 \end{cases} \quad (1)$$

and to receive this message, the radio expends energy is:

$$E_{Rx} = l \times E_{elec} \quad (2)$$

We assume that the radio channel is symmetric, which means the cost of transmitting a message from A to B is the same as the cost of transmitting a message from B to A.

3.2 *System assumptions*

In this paper, we assume that N sensors are randomly located in the square W×W observation region, and are stationary that they do not change their locations once deployed, the sink is located outside the region at position $(W/2, W+D)$. The sensor nodes are grouped into clusters. With clustering, sensor nodes transmit their sensed data to their cluster heads. Cluster heads aggregate the received data and forwards it to the sink. Figure 1 depicts an application where sensors periodically transmit their data to the sink. The figure illustrates that cluster heads transmit the aggregated data in single-hop and multi-hop communication mode.

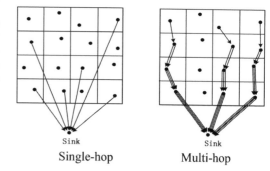

Figure 1. Sensor network model.

We assume that data aggregation technique is available, and data collected from sensor nodes in a cluster is packed into one packet. Furthermore, the sensor nodes are location-unaware and have the same characteristics. It's also assumed that the sensed data is highly correlated, thus the cluster heads can always aggregate the data gathered from its members.

3.3 Hybrid communication method

Notice that in the multi-hop mode, the smaller distance between cluster heads and the sink, the more energy consumes. On the other hand, in single-hop mode, cluster heads which are farther from the sink dissipate more energy than those closer to the sink.

To balance the energy consumption, our network model employs the hybrid communication method which consist of the single-hop and multi-hop mode. The cluster heads use single-hop mode in some rounds but multi-hop mode in the other rounds. We use parameter λ to measure how often the single-hop mode is used. Suppose T is the total rounds that the sensor network can perform, T_s is the number of rounds that single-hop mode is used and T_m is the number of rounds that multi-hop mode is employed. Let $\lambda = T_s / T$ be the frequency with which the single-hop mode is used.

3.4 Problem statement

Our objective is to find the optimal cluster radius r and the parameter λ for the hybrid communication method to maximize the network lifetime. In order to extend the lifetime of the sensor network, we need to maximize the lifetime of the node who dies first. The problem of maximizing the lifetime can be written as

$$
\begin{aligned}
Objective: \quad & \max\{T\} \\
Subject\ to \quad & 0 \le \lambda \le 1 \\
& T_s \times E_{single\_hop} + T_m \\
& \quad \times E_{multi\_hop} \le E_{init}
\end{aligned}
\tag{3}
$$

where we denote E_{single_hop} and E_{multi_hop} as the average energy consumption of single-hop and multi-hop mode in a round, E_{init} as the sensor initial energy. The constraints indicate that we can use hybrid communication method.

4 SOLUTIONS FOR THE PROBLEM

4.1 The optimal cluster radius

Mhatre and Rosenberg (2004) defined characteristic distance as that distance which when used as the inter-sensor node distance, minimizes the energy consumption for sending a data packet from a source sensor to a destination sensor. This characteristic distance is:

$$
d_{char} = \sqrt{2E_{elec} / efs}
\tag{4}
$$

Hence, we set the width of each ring is d_{char}. Only width d_{char} can the energy consumption rate be minimized.

4.2 Solutions for the problem

In this section, we first study the energy consumption. Let us consider a single cluster with radius r. We assume that N_k denotes the number of sensors in set P_k. The average number of sensors in a cluster is In this section, we first study the energy consumption. Let us consider a single cluster with radius r. The average number of sensor nodes in the cluster is:

$$
N_{clu} = \frac{N}{W \times W} \times \left(\sqrt{2}r\right)^2 = \frac{2Nr^2}{W \times W}
\tag{5}
$$

To keep the total energy dissipation within the cluster as small as possible, the cluster head should be positioned at the centroid of the cluster area S. In this case, the square of distance between cluster members and the cluster head is given as:

$$
d_{clu}^2 = \iint (x^2 + y^2)dxdy = r^2 / 2
\tag{6}
$$

The transmitter energy consumption of cluster members is given by as:

$$
E_1 = (N_{clu} - 1) \times len \times (2E_{elec} + efs \times d_{clu}^2)
\tag{7}
$$

where we denote len as the length of the data packet.

Each cluster head receives the data from all member nodes in the cluster, aggregates and transmits to the sink. The sink is located at the coordinate $(x0, y0)$, the distance between the sink and a cluster head with the coordinate (xi, yi) can be expressed as:

$$
d_{to\_sink} = \sqrt{(x_i - x_0)^2 + (y_i - y_0)^2}
\tag{8}
$$

The energy required for the transmission of aggregation data is:

$$
E_2 = len \times (E_{elec} + emp \times d_{to\_sink}^4)
\tag{9}
$$

The number of clusters can be given by:

$$N_{cluster} = \left\lceil W \times W / (2r^2) \right\rceil \qquad (10)$$

Thus, the average energy consumption of single-hop mode in a round is:

$$E_{single\_hop} = (E_1 + E_2) \times N / N_{cluster} \qquad (11)$$

Let us consider the average energy consumption of multi-hop transmission. The total transmitter energy consumption of cluster members is:

$$E_3 = (N - N_{cluster}) \times len \times (2E_{elec} + efs \times d_{clu}^2) \qquad (12)$$

The number of transmission data among cluster heads is given by is:

$$N_c = \sum_{i=1}^{\lceil W/(\sqrt{2}r) \rceil - 1} i \times \left\lceil W / (\sqrt{2}r) \right\rceil \qquad (13)$$

The energy consumption of transmission data among cluster heads can be expressed as:

$$E_4 = N_c \times len \times (2E_{elec} + efs \times (2r)^2) \qquad (14)$$

The energy required for the transmission of the data from cluster heads to the sink is:

$$E_5 = len \times (E_{elec} + emp \times D^4) \qquad (15)$$

where we denote D as the distance between the sink and observation region border.

Thus, the average energy consumption of multi-hop in a round is:

$$E_{multi\_hop} = (E_3 + E_4 + E_5) / N \qquad (16)$$

Hence, the original problem (3) is reformulated as

$Objective:$ $\max\{T\}$
$Subject\ to$ $0 \le \lambda \le 1, T_s = \lambda \times T, T_m = T - T_s$
$T_s \times E_{single\_hop} + T_m \times E_{multi\_hop} \le E_{init}$
$constraints\ (11), (16)$

$$\qquad (17)$$

Problem (17) is a linear programming problem, which is easier to solve than the problem (3).

4.3 Clustering algorithm

Clustering a sensor network means partitioning its sensors into clusters, each one with a cluster head and some sensors as its members.

Our algorithm is a distributed cluster heads competitive algorithm, here cluster head selection is primarily based on the residual energy and position of each node. Every sensor become a tentative cluster head with the same probability p which is a predefined threshold. The algorithm pseudocode for an arbitrary sensor node is given in Figure 2.

5 SIMULATION AND RESULTS

We compared the performance of communication method with LEACH and HEED in this section. We used the same energy consumption model as [5]. The parameters of simulations are listed in TABEL I. Every result shown below is the average of 50 independent experiments.

We first use the number of alive sensor nodes to compare the performance of the three methods. Figure 3 shows the comparison results, we assume $D = 100\ m$. It is shown that the first sensor node halts for the starvation of battery in LEACH and

```
state←candidate;
broadcast Residual_Energy_Msg
receive Residual_Msg;
update neighborhood table NT[];
t←the broadcast delay time for competing a cluster head;
while (the timer1 (T) for cluster head election is not expired)
{    if (CurrentTime<t)
    {    if (a Head_Msg is overheard from a neighbor NT[i])
        {    state←plain;    NT[i].state=head;    }
        else   continue;
    }
    else   if (state=candidate)
    {    state←head;    broadcast Head_Msg;    }
}
while (the timer2 for cluster join is not expired)
{    if (state=plain && have not sended Join_Msg)
        Send (Join_Msg to the nearest cluster head);
    else   receive (Join_Msg from its neighbor plain nodes);
}
```

Figure 2. Cluster head selection pseudocode.

Table 1. Simulations parameters.

Parameter	Value
N	100
W	$50\ m$
Sensor initial energy	$0.5\ J$
E_{elec}	$50\ nJ/bit$
efs	$10\ pJ/bit/m^2$
emp	$0.0013\ pJ/bit/m^4$
d_0	$80\ m$
Data packet size len	$2048\ bits$

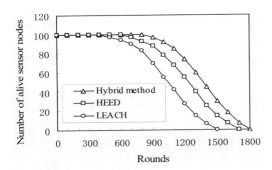

Figure 3.   Number of alive sensor nodes comparison.

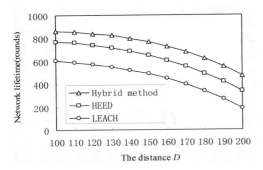

Figure 4.   Network lifetime comparison.

HEED is earlier than in our method. As shown in Figure 3, our method clearly improves the lifetime of sensor nodes.

We then compare the network lifetime to evaluate the performance of the three methods. We define the network lifetime to be the number of rounds until the first sensor node is drained of its energy. Figure 4 shows the network lifetimes of the three methods when the distance D changes from 100 m to 200 m. However, our method outperforms LEACH and HEED in terms of network lifetime.

6   CONCLUSION AND FUTURE WORK

In this paper, we present a hybrid communication method where the cluster heads can transmit data in either single-hop or multi-hop communication. Simulation experiment results show that the hybrid communication method can efficiently balance the energy consumption and prolong the network lifetime.

In the future, we plan to investigate modifications to the method that would allow sensor nodes to be added to the network. We also plan to study the data gathering problem with delay constraints for individual sensor nodes, in order to attain desired tradeoffs between the delay experienced by the sensors and the lifetime achieved by the system.

ACKNOWLEDGEMENTS

This research has been supported by the science and technology project of Shaoguan (Project No. 2014CX/K252, 2012CX/K91) and Science Foundation of Hunan Province (Project No. 11JJ3074).

REFERENCES

C. Tharini, & P. V. Ranjan. 2010. An efficient data gathering scheme for wireless sensor networks. *European Journal of Scientific Research* 43(1): 148–155.

H. Chen, & S. Megerian. 2006. Cluster sizing and head selection for efficient data aggregation and routing in sensor Networks. *In proceedings of WCNC 2006*, 2318–2323.

H. Su, & X. Zhang. 2006. Energy-efficient clustering system model and reconfiguration schemes for wireless sensor networks. *In proceedings of CISS 2006*.

I. F. Akyildiz, W. Su, Y. Sankarasu, & E. Cayirci. 2002. Wireless sensor networks: a survey. *Comput. Netw* 393–422.

J. Yang, & D. Y. Zhang. 2010. Cluster-based data aggregation and transmission protocol for wireless sensor networks. *Journal of Software* 21(5): 1127–1137.

K. Dasgupta, K. Kalpakis, & P. Namjoshi. 2003. An efficient clustering-based heuristic for data gathering and aggregation in sensor networks. *In proceedings of WCNC* 1948–1953.

O. Younis, & S. Fahmy. 2004. A hybrid, energy-efficient, distributed clustering approach for ad hoc sensor networks. *IEEE Transactions on mobile computing*, 3(4). 336–379.

S. Olariu, & I. Stojmenovic. 2006. Design guidelines for maximizing lifetime and avoiding energy holes in sensor networks with uniform distribution and uniform reporting. *In proceedings of INFOCOM 2006*, 1–12.

W. Heinzelman, & A. Chandrakasan. 2002. An application-specific protocol architecture for wireless microsensor networks. *IEEE Transactions on Wireless Communications* 1(4): 660–669.

V. Mhatre, & C. Rosenberg. 2004. Design guidelines for wireless sensor networks: communication, clustering and aggregation. *Ad Hoc Networks* 2(1): 45–63.

Y. Yu, B. Krishnamachari & V. K. Prasanna. 2008. Data Gathering with Tunable Compression in Sensor Networks. *In Proceedings of IEEE Trans. Parallel Distrib.* 276–287.

Architectural, Energy and Information Engineering – Sung & Chen (Eds)
© 2016 Taylor & Francis Group, London, ISBN 978-1-138-02791-6

A review on formation mechanism, hazards and existence of Environmental Persistent Free Radicals (EPFRs)

Z.Q. Cheng, T. Wang, J. Yang, F. Yang & H.Y. Guo
Faculty of Environmental Science and Engineering, Kunming University of Science and Technology, Kunming, China

H. Zhang
Yunnan Institute of Food Safety, Kunming University of Science and Technology, Kunming, China

ABSTRACT: Compared with the traditional short-life free radicals, Environmental Persistent Free Radicals (EPFRs) are generated from the reaction between organic matters and transient metals. EPFRs are resonance stabilized and can exist for tens of minutes to several hours. Besides, they can induce oxidative stress response of biological systems. Researches indicated that EPFRs exist widely in environment, including in soil, atmosphere and surface water. And their powerful environmental persistence and toxicity cannot be overlooked. They trigger lung and cardiovascular disease more easily than separated particles or original organic compounds do. In this paper, the mechanism of EPFRs formation, their hazards and existence will be summarized. And some frontier research trends will be also introduced. In recent years, much attention has been paid to the new type pollutant, EPFRs. Further research on them will help to have a good understanding of their environmental behavior and risk.

Keywords: environmental persistent free radicals; transient metal; hazards; soil; airborne particulate matters; biochars

1 INTRODUCTION

Free radicals are molecules or molecular fragments containing a single unpaired electron. They may be created in a number of ways, including synthesis with very dilute or rarefied reagents, reactions at very low temperatures, or breakup of larger molecules. The first organic free radical identified was triphenylmethyl radical. This species was discovered by Moses Gomberg in 1900 at the University of Michigan USA. In 1961 Leighton demonstrated that the polluted air could produce free radicals. Until the late 1960s (Hanst, Wong et al. 1982), people confirmed the existence of free radicals in an experiment about the mechanism of photochemical smog formation. In general, free radicals are reactive chemically, some (e.g. HO•) being extremely reactive. HO• has the greatest oxidability among all the known chemical substances and can oxidize almost all the organic matters, and the reaction rate is very fast. However, the lifetime of these free radicals is very short, and they are unstable. Small organic radicals, such as, methyl vinyl, or phenyl, are somewhat less reactive but are also less stable. Free radicals in human body function on the immune system. They are able to kill virus and bacteria. While, the free radicals coming from surroundings, such as the ones produced by smoking and pollutants, can induce the injury and stress response of cells, thus causing biological damage. Some researchers found that there were resonance stabilized radicals, such as phenoxyl, propargyl and cyclopentadienyl that are not highly reactive with molecular species and undergo radical—radical recombination reactions to form PAH and possibly soot at moderately high temperatures in the post-flame and cool-zone region of combustion and thermal processes (Mulholland, Lu et al. 2000). Compared with the traditional free radicals, persistent radicals are those whose longevity is due to steric crowding around the radical center, which makes it physically difficult for the radical to react with another molecule (Griller & Ingold 1976). Persistent radicals are generated in great quantity during combustion, and are supposed to be responsible for the oxidative stress resulting in cardiopulmonary disease and cancer that has been attributed to exposure to airborne fine particles (Lomnicki, Truong et al. 2008). In this paper, we focus on a new type of free radicals that have long lifetime and great persistence—EPFRs. Their formation mechanism, hazards and existence will be introduced as follow.

2 MECHANISM OF EPFRs FORMATION

The term of EPFRs is proposed under the comparison with traditionally concerned free radicals that have a short life time (Dellinger, Lomnicki et al. 2007, Lomnicki, Truong et al. 2008, Truong, Lomnicki et al. 2010, Vejerano, Lomnicki et al. 2010). They can exist for tens of minutes to several hours, are resonance stabilized, and generate more organic matters derived from oxidative stress response of biological systems, such as semiquinone, phenoxyl, cyclopentadienyl (Lomnicki, Truong et al. 2008). Formation of EPFRs is attributed to the interaction between organic matters and solid particles containing transition metal elements, such as Cu, Fe and Mn (Truong, Lomnicki et al. 2010, Vejerano, Lomnicki et al. 2010). These radicals are protected by the particles, so that their lifetime increase from a few milliseconds to several hours or days. It is believed that EPFRs have a high reactivity and could be generated in low-temperature region of combustion and thermal processes or in incineration of hazardous wastes and engine combustion.

By taking a common organic pollutant, catechol as an example, the brief mechanism of EPFRs formation is as shown in Figure 1: Firstly, catechol physisorbs on the surface containing transient metal elements through hydrogen bonding. Secondly, strong chemical bonds form via one or two molecules H_2O elimination (viz. a & c). Lastly, in order to stabilize their lone pair electrons, the derivative chemicals obtain electrons from the metal ions to generate EPFRs (viz. b & d), and then the metal ions are reduced.

The above process occurs at temperatures from 150 to 400°C, the metal ion could be Fe, Cu or other transient mental. But EPFRs formed on different mental ion have significant different persistence. For instance, the half-lives of EPFRs on iron range from 24 to 111 h, compared to the half-lives on copper of 27 to 74 min (Vejerano, Lomnicki et al. 2010).

3 HAZARDS OF EPFRs

EPFRs are strongly harmful to biological organisms. In the catalytic cycle formed by EPFR-metal complex, EPFRs trigger lung and cardiovascular disease more easily than separated particles or original organic compounds do (Balakrishna, Lomnicki et al. 2009, Fahmy, Ding et al. 2010). Meanwhile, EPFRs show prolonged environmental and biological lifetime. Shrilatha Balakrishna et al. exposed human bronchial epithelial cells (BEAS-2B) to 2-monochlorophenol (MCP230, one kind of EPFR) or the CuO/silica substrate, demonstrating that EPFR-containing particles have enhanced cytotoxicity compared with the substrate species (Balakrishna, Lomnicki et al. 2009). And there was evidence that its greater cytotoxicity was connected with its ability to generate more cellular oxidative stress and reduce the antioxidant defenses of the epithelial cells concurrently (Balakrishna, Lomnicki et al. 2009). The particles containing EPFRs pose a threat to our heath by generating DNA damage and decreasing pulmonary function (Knaapen, Shi et al. 2002, Simoneit 2002, Maskos, Khachatryan et al. 2005). Maud Walsh et al. found that when human body inhales EPFRs inflammatory responses in the lung will occur, which leads to chronic lung diseases, such as asthma or Chronic Obstructive Pulmonary Disease (COPD) in humans (Walsh, Cormier et al. 2010). A considerable amount of research proposed that formation of ROS that can initiate cancer is massively likely to be catalyzed or mediated by semiquinone-type radicals, though its mechanisms are poorly understood. Currently in China, few research studies have carried out on toxicology of EPFRs.

4 EPFRs IN DIFFERENT MEDIA

4.1 EPFRs in soil

Stable free radicals exist in solid soil humic acid, approximately on the order of 10^{18} radicals/g. Some researchers found that there are two kinds of stable free radicals in humic acid through the analysis of the EPR spectrum: a semiquinone of a catechol-resorcinol-type copolymer and a quinhydrone-type radical (Steelink & Tollin 1962). In tropical and subtropical soil, transition metal elements are rich. Meanwhile, sunshine time is long and ultraviolet radiation is very strong. So it is easy to produce

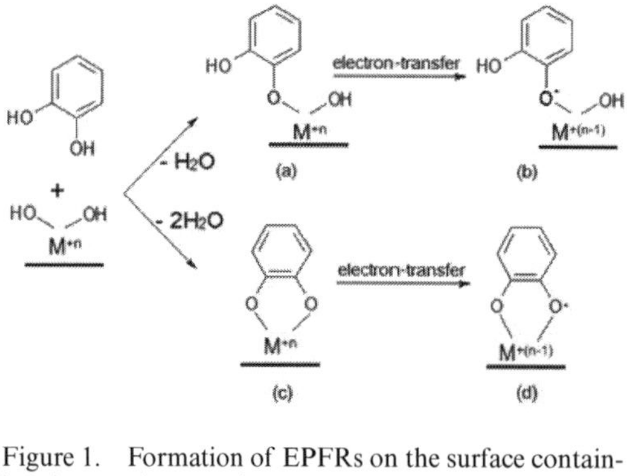

Figure 1. Formation of EPFRs on the surface containing the transient metal.

EPFRs. Albert Leo N. dela Cruz et al. discovered EPFRs in soils contaminated with PCP from Georgia and Montana, and sediments contaminated with PAHs from Washington (dela Cruz, Gehling et al. 2011). Chemisorption and electron transfer from PCP or PAHs to transition metals and other electron sinks in soil is responsible for EPFRs formation. Similarly, some researchers also detected EPFRs in soils and sediments at Superfund sites that was a former wood treating facility containing PCP as a major contaminant. To simulate EPFRs formation in natural soil, Hao Li et al. carried out an experiment about catechol degradation on hematite-silica surface under UV irradiation (Li, Pan et al. 2014). They found that stabilized semiquinone or quinine and phenol radicals formed, and due to that catechol degradation under UV irradiation decreased in hematite-coated particles. All the research results show that under UV light, interaction between soil minerals and organic pollutants can produce EPFRs.

4.2 EPFRs in airborne particulate matters

With publication of epidemiological data that demonstrate a clear correlation between exposure to Particulate Matter (PM), the impact of exposure to PM on human health has become a major environmental issue. Airborne PM is generated by a variety of sources including industrial processes and combustion of biomass and fossil fuels (Zheng, Cass et al. 2002). PM can provide places that hazardous organic matters can exist long and favourable conditions for formation of EPFRs. In 1996, XY Li etc al. confirmed the free radical activities of PM_{10} in vitro by its ability to deplete supercoiled plasmid DNA and provided evidence that PM_{10} causes lung inflammation and epithelial injury due to its free radical activity (Li, Gilmour et al. 1996). Giuseppe L. Squadrito et al. found $PM_{2.5}$ contained abundant semiquinone radicals, typically 10^{16} to 10^{17} unpaired spin s/gram, and that these radicals were stable for several months (Squadrito, Cueto et al. 2001). In combustion processes, phenoxyl- and semiquinone-type radicals that are stabilized and resistant to oxidation when associated with metal oxide containing particles (i.e. EPFR), can be formed in high temperatures (Lomnicki, Truong et al. 2008). Some researchers studied the lifetime of EPFRs in $PM_{2.5}$, and revealed three types of radical decay—a fast decay, slow decay, and no decay (Gehling & Dellinger 2013).

4.3 Persistent free radicals in biochars

Many literatures have suggested that free radicals are generally formed during pyrolysis or by heat treatment of organic chemicals. But, without transition metals, the pyrolysis of organic materials can also generate free radicals, like the pyrolysis of coal samples. Since biochars generally come from biomass charring, so there is a problem that whether or not do biochars have free radicals or even persistent free radicals. It is worth noting that Shaohua Liao et al. observed stable free radical signals in biochars produced from pine and chitin (Liao, Pan et al. 2014). Guodong Fang et al. investigated the activation of hydrogen peroxide (H_2O_2) by biochars for 2-chlorobiphenyl (2-CB) degradation and found that H_2O_2 could be activated by biochar that produces hydroxyl radical (•OH) to degrade 2-CB (Fang, Gao et al. 2014). Meanwhile, they discovered that biochar contained persistent free radicals of typically $\sim 10^{18}$ unpaired spins/gram. All these conclusions remind us to pay attention to the potential risk and harm of relatively persistent free radicals in biochars.

4.4 EPFRs in other media

Similar to semiquinone-type radicals previously observed in combustion generated particulate and contaminated soils, two types of organic radicals that are resonance stabilized were founded in crude oil: one is an asphaltene radical species (g = 2.0035), and the other is a new type of radical (g = 2.0041-47) (Kiruri, Dellinger et al. 2013). Besides, it is well recognized that EPFRs can also exist in aqueous media. Since EPFRs can induce oxidative stress in living cells. So the idea that EPFRs can promote Reactive Oxygen Species (ROS) formation in aqueous media instead, will be considered to be reasonable. Lavrent Khachatryan et al. succeeded to experimentally demonstrate the formation of ROS in the presence of EPFRs by using a 5,5-dimethyl-1-pyrrolineN-oxide (DMPO) spin-trapping agent in conjuction with EPRs pectroscopy (Khachatryan, Vejerano et al. 2011). This work provides us a new way to study EPFRs.

5 CONCLUSIONS

EPFRs is a new type pollutant that can exist almost everywhere including in soil, atmosphere and surface water. So their environmental hazards cannot be overlooked. And their existence and transformation will bring uncertainty to determination of organic pollutants. Besides, EPFRs have a long lifetime and strong environmentally persistence and toxicity. So they should have special characteristics of environmental fate. We should study further to recognize their formation and transport mechanism at the micro level. The research group lead by Dellinger has laid the foundations by revealing the mechanism for EPFRs formation, but their work currently concentrate upon EPFRs on mineral particles or simply in aqueous solutions. Formation mechanism and behavior in natural environment is an unworked area and should be paid more attention.

REFERENCES

Balakrishna, S., S. Lomnicki, et al. 2009. Environmentally persistent free radicals amplify ultrafine particle mediated cellular oxidative stress and cytotoxicity. Part Fibre Toxicol 6: 11.

dela Cruz, A. N., W. Gehling, et al. 2011. Detection of environmentally persistent free radicals at a superfund wood treating site. Environmental science & technology 45(15): 6356–6365.

Dellinger, B., S. Lomnicki, et al. 2007. Formation and stabilization of persistent free radicals. Proceedings of the Combustion Institute 31(1): 521–528.

Fahmy, B., L. Ding, et al. 2010. In vitro and in vivo assessment of pulmonary risk associated with exposure to combustion generated fine particles. Environmental toxicology and pharmacology 29(2): 173–182.

Fang, G., J. Gao, et al. 2014. Key Role of Persistent Free Radicals in Hydrogen Peroxide Activation by Biochar: Implications to Organic Contaminant Degradation. Environmental science & technology 48(3): 1902–1910.

Gehling, W. & B. Dellinger 2013. Environmentally persistent free radicals and their lifetimes in PM2.5. Environ Sci Technol 47(15): 8172–8178.

Griller, D. & K. U. Ingold 1976. Persistent carbon-centered radicals. Accounts of Chemical Research 9(1): 13–19.

Hanst, P. L., N. W. Wong, et al. 1982. A long-path infrared study of Los Angeles smog. Atmospheric Environment (1967) 16(5): 969–981.

Khachatryan, L., E. Vejerano, et al. 2011. Environmentally persistent free radicals (EPFRs). 1. Generation of reactive oxygen species in aqueous solutions. Environ Sci Technol 45(19): 8559–8566.

Kiruri, L. W., B. Dellinger, et al. 2013. Tar balls from deep water horizon oil spill: Environmentally persistent free radicals (EPFR) formation during crude weathering. Environmental science & technology 47(9): 4220–4226.

Knaapen, A. M., T. Shi, et al. (2002). Soluble metals as well as the insoluble particle fraction are involved in cellular DNA damage induced by particulate matter. Oxygen/Nitrogen Radicals: Cell Injury and Disease, Springer: 317–326.

Li, H., B. Pan, et al. 2014. Formation of environmentally persistent free radicals as the mechanism for reduced catechol degradation on hematite-silica surface under UV irradiation. Environmental Pollution 188: 153–158.

Li, X., P. Gilmour, et al. 1996. Free radical activity and pro-inflammatory effects of particulate air pollution (PM10) in vivo and in vitro. Thorax 51(12): 1216–1222.

Liao, S., B. Pan, et al. 2014. Detecting free radicals in biochars and determining their ability to inhibit the germination and growth of corn, wheat and rice seedlings. Environ Sci Technol 48(15): 8581–8587.

Lomnicki, S., H. Truong, et al. 2008. Mechanisms of product formation from the pyrolytic thermal degradation of catechol. Chemosphere 73(4): 629–633.

Lomnicki, S., H. Truong, et al. 2008. Copper oxide-based model of persistent free radical formation on combustion-derived particulate matter. Environmental science & technology 42(13): 4982–4988.

Maskos, Z., L. Khachatryan, et al. 2005. Precursors of radicals in tobacco smoke and the role of particulate matter in forming and stabilizing radicals. Energy & fuels 19(6): 2466–2473.

Mulholland, J. A., M. Lu, et al. 2000. Pyrolytic growth of polycyclic aromatic hydrocarbons by cyclopentadienyl moieties. Proceedings of the Combustion Institute 28(2): 2593–2599.

Simoneit, B. R. 2002. Biomass burning—a review of organic tracers for smoke from incomplete combustion. Applied Geochemistry 17(3): 129–162.

Squadrito, G. L., R. Cueto, et al. 2001. Quinoid redox cycling as a mechanism for sustained free radical generation by inhaled airborne particulate matter. Free Radical Biology and Medicine 31(9): 1132–1138.

Steelink, C. & G. Tollin 1962. Stable free radicals in soil humic acid. Biochimica et biophysica acta 59(1): 25–34.

Truong, H., S. Lomnicki, et al. 2010. Potential for misidentification of environmentally persistent free radicals as molecular pollutants in particulate matter. Environmental science & technology 44(6): 1933–1939.

Vejerano, E., S. Lomnicki, et al. 2010. Formation and stabilization of combustion-generated environmentally persistent free radicals on an Fe (III) 2O3/silica surface. Environmental science & technology 45(2): 589–594.

Walsh, M., S. Cormier, et al. 2010. By-products of the Thermal Treatment of Hazardous Waste: Formation and Health Effects. EM (Pittsburgh, Pa.): 26.

Zheng, M., G. R. Cass, et al. 2002. Source apportionment of PM2. 5 in the southeastern United States using solvent-extractable organic compounds as tracers. Environmental Science & Technology 36(11): 2361–2371.

Architectural, Energy and Information Engineering – Sung & Chen (Eds)
© 2016 Taylor & Francis Group, London, ISBN 978-1-138-02791-6

A state-testing device for power system operators

Bin Zhou, Jia Liu & Wen Ling Fan
Maintenance Branch Company, Chongqing Electric Power Company, Chongqing, China

ABSTRACT: Human error has become one of the important factors that affect the safety operation of the power system. However, personnel state-testing devices for power system operators before operation have not been applied. A state-testing device for power system operators is proposed in this paper, which can be used to test the personnel state of power system operators with state data saved. This state-testing device has convenient operation and perfect functions, suitable for testing states of power system operators before operation, to ensure the safe and stable operation of the power system.

Keywords: human error; power system; state-testing device

1 INTRODUCTION

1.1 Power system operation

Safe operation is the premise for power system to provide continuous, stable electric energy. At present, the application of an intelligent electronic device, unattended and remote program control technologies further ensure the reliable operation of equipment. Meanwhile, substation failure caused by human error has been highlighted.

Power system operation steps have an order of complexity, with a high level of risk of error. It requires the operators to be not only familiar with the principle structure, connection mode, protection configuration and mechanical structure, but also well known about three-phase current, voltage, load, power factors, and principle and measurement methods of current's positive, negative and zero sequence. Along with the aging of experienced operators, the training for new staff cannot catch up with the speed of power system development, which greatly increases the potential risks. On the other hand, the whole power system accident statistics show that the number of accidents decreased year by year, but the analysis report still exposes many problems in the organization and management, security guarantee, personnel quality and safety consciousness.

1.2 Personnel state-testing devices

Personnel state-testing devices for power system operators before operation have not been applied at present. Therefore, in order to improve the reliability of substation operation as well as reduce the malignant disoperation and personal injury accidents, there is an urgent need for research on the working state of power system operators and a state-testing device ,which can test the work state of operators before operation to ensure the safe and stable operation of power system.

In order to overcome the defects of the existing technology, this paper presents a state-testing device with convenient operation and perfect functions based on temperature and alcohol detection, which can be used to test the personnel working state of power system operators with state data saved. This test device comprises a data acquisition unit, MCU, extended memory chips, LED display, a data transmission unit and a power supply, which has convenient operation and perfect functions, particularly suitable for testing power operator state.

2 DEVICE STRUCTURE

The structure of the state measurement power operator proposed in this paper is shown in Figure 1, and the different units are given in Table I. This measurement device contains a data collection unit, single chip microcomputer (16 bit), Extended memory chip, LED display, data transmission unit and power source. Data collection unit contains keyboard, body temperature detection circuit and alcohol detection circuit. Keyboard, body temperature detection circuit and alcohol detection circuit are connected to the single chip microcomputer, respectively, and the single chip microcomputer is connected to the LED screen. Extended memory chip is connected to the data transmission unit,

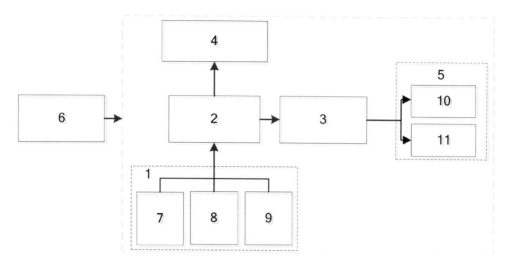

Figure 1. Structure of the device. 1-Data collection unit, 2-Single chip microcomputer, 3-Extended memory chip, 4-LED display, 5-Data transmission unit, 6-Power source, 7-Keyboard, 8-Body temperature detection circuit, and 9-Alcohol detection circuit.

Table 1. Units of the device.

Units	Sub-element
1. Data collection unit	Keyboard
	Body temperature detection circuit
	Alcohol detection circuit
2. Single chip microcomputer	
3. Extended memory chip	
4. LED display	
5. Data transmission unit	USB bus
	WLAN card
6. Power source	

with power source providing electricity energy to drive the whole device.

There are ten digital input keys, one enter key, one return key, and one indicator light. Body temperature detection circuit contains temperature sensor, operational amplifier and A/D convertor. The infrared thermopile temperature sensor ZTP 135 is used in the temperature sensor, and OP07 is applied in the operational amplifier. Alcohol detection circuit contains alcohol sensor, operational amplifier and A/D convertor. MQ3 gas sensor is used in the alcohol sensor. Data transmission unit contains USB bus general interface chip and WLAN card, connecting the expand memory chip and the power source. The output voltage of power source is direction current with 6 V (4 AA batteries in PRC).

3 APPLICATION

First, users should plug in 6V-DC power or put in four AA batteries and turn on the switch of the device. Then, the ID of operator should be input though the keyboard of the data collection unit. With setup and configuration completed, the operator can execute the body temperature test and the sobriety test against the requirements to ensure that workers are really suitable for the work. The data collection unit collects all the data from the body temperature detection circuit and the alcohol detection circuit, and then the data will be compared with the standard values by the microcomputer and output an evaluation result. Moreover, the values of body temperature, blood-alcohol concentration as well as the evaluation result are displayed in the LED screen. All the data and result will be saved in the extended memory chip and transported to the servers though the USB interface or wireless fidelity.

4 ADVANTAGES

Compared with the existing technologies, the device proposed in this paper has the following advantages:

1. This device has the advantage of convenience with the realization of body temperature detection and alcohol density detection in one device. Using the device proposed in this paper, we can have the body temperature test and the sobriety

test by the body temperature detection circuit and the alcohol detection circuit and get the test results, respectively, to see whether the operators are sick or under the influence. It greatly spares us from the troubles of using a thermometer and the alcohol concentration detector separately.

2. This device has the function of cloud data store with the data interface. After the body temperature test and the sobriety test, the device will upload all the test data to a corresponding computer recording system or database, in order to facilitate the management of operations and operators in the power system for the supervisors. There are USB bus general interface chip and WLAN card installed on the device; users can choose one of the transmission ways to transport all the data conveniently.

5 CONCLUSIONS

As an important factor in the risk evaluation of power grid, human error has a significant impact on the power system reliability and safety. With the development of science and technology, it is necessary to develop a personnel state-testing device for power system operators to improve the human reliability.

In this paper, a state-testing device for power system operators is proposed, which can be used to test the personnel state of power system operators with state data saved to the cloud data store. This state-testing device has convenient operation and perfect functions, suitable for testing states of power system operators before operation, to ensure the safe and stable operation of the power system. Moreover, this equipment is believed to be helpful to control the risk of the operation task in advance and is a beneficial attempt to lay a foundation for the human error control explorations.

REFERENCES

Bogner, Marilyn Sue Ed. Human error in medicine. Lawrence Erlbaum Associates, Inc, 1994.

Fujita Y., Hollnagel E. Failures without errors: quantification of context in HRA, Reliability Engineering & System Safety. 83 (2004): 145–51.

Guo Y.J. Reliability of power systems and power equipment, Automation of Electric Power Systems. 25 (2001): 53–56.

Hollnagel E. Cognitive Reliability and Error Analysis Method. Elsevier, Oxford, 1998.

John M.O., James C.G, Joel K. Advanced information systems design: technical basis and human factors review guidance. Washington, DC: USA NRC, NUREG/CR-6633, 2002.

Kim M.C., Seong P.H., Hollnagel E. A probabilistic approach for determining the control mode in CREAM, Reliability Engineering & System Safety. 91 (2006): 191–199.

Lee Y.H., Lee J.W., Park J.C., et al. Development of An Extremely Low Frequency Human Error Assessment Technology for Digital Devices and Its Applications to the Nuclear Fields. Korea Atomic Energy Research Institute, Daejeon (Korea, Republic of), 2013.

Noroozi A., Khan F., MacKinnon S., et al. Determination of human error probabilities in maintenance procedures of a pump. Process Safety and Environmental Protection, 92 (2014): 131–141.

Paulo V.R.C., Isaac L.S., et al. Human factors approach for evaluation and redesign of human-system interfaces of a nuclear power plant simulator. 29 (2008): 273–284.

Reason J. Human error. Cambridge University Press, UK. Cambridge, 1990.

Shaw J.A. Distributed control system: cause or cure of operator errors. Reliability Engineering and System Safety, 39 (1993): 263–271.

Swain A.D., Henry E.G. Handbook of human-reliability analysis with emphasis on nuclear power plant applications. Final report. No. NUREG/CR-1278; SAND-80-0200. Sandia National Labs. Albuquerque, NM (USA), 1983.

Williams J.C. A data-based method for assessing and reducing human error to improve operational performance. Human Factors and Power Plants, Conference Record for 1988 IEEE Fourth Conference on. IEEE, (1988): 436–450.

Zhang J.J., Ding M., Li S.H. Impact analysis of human error on protection system reliability, Automation of Electric Power Systems. 36 (2012): 1–5.

Architectural, Energy and Information Engineering – Sung & Chen (Eds)
© *2016 Taylor & Francis Group, London, ISBN 978-1-138-02791-6*

Evaluation of substation supply reliability considering human errors

Y.K. Bao, Z. Li, D.S. Wen & C.X. Guo
College of Electrical Engineering, Zhejiang University, Hangzhou, Zhejiang Province, China

L. Zhang & S.H. Pang
Maintenance Company of State Grid Henan Electric Power Company, Zhengzhou, Henan Province, China

ABSTRACT: Conventional studies related to substation reliability only consider equipment failures, neglecting the effects of human errors which might occur in substation switching operation and protection system. This paper proposes a more realistic substation reliability evaluation approach which incorporates the impact of human errors. The main contribution of this paper includes two aspects: 1) Making an introduction of human errors and human reliability assessment in power systems; 2) Proposing an approach to assess the substation reliability considering human errors. A typical 4/3 circuit substation is studied as an example to demonstrate the influence of human errors on substation supply reliability.

Keywords: Substation supply reliability; failure model; minimal cut sets; protection system failures; human errors

1 INTRODUCTION

Substation supply reliability plays an important role in power system, since substation originated failures can have significant impact on both bulk transmission systems and distribution systems. It might lead to load shedding or incur cascading failures, and much worse, it might incur a wider range of blackout [1]. So it is very important to guarantee substation connectivity and supply reliability. However, continuous supply of electricity is not possible owing to random system failures and human errors.

Many reliability related research was conducted in order to improve substation supply reliability. Lots of publications have described how to assess the connectivity using minimal cut sets method [2–3]. Fuzzy fault tree approach [4] and Monte Carlo approach [5] are proved to be efficient methods to evaluate substation reliability. A "generalized n+2 state system model" is proposed to modify the conventional three-state model [6]. Since equipment failure rate is not consistent forever, it might be influenced by many factors, such as maintenance measures, external environment and internal aging conditions. Paper [7] proposed a novel failure-rate model taking these influencing factors into consideration. In [8], the reliability of high-voltage substations subject to protection system failures is analyzed using event tree analysis. Another important factor that affects substation reliability is human errors [9], which is usually

neglected. Human errors are defined as failures on the part of human to perform a prescribed act. Sometimes human errors might result in damage to equipment or whole system. On one side, substation cannot operate without human intervention. On the other side, human behavior has a high degree of uncertainty, and human errors cannot be eliminated. So it is necessary to take human errors into consideration when evaluating substation reliability.

This paper is organized as follow. First, aspects regarding human errors and human reliability assessment in power system are introduced in section II. Then in section III, the influences of human errors on substation supply reliability is analyzed from two aspects: substation switching operation and protection system failures. Section IV illustrates the procedure of substation reliability evaluation with a flow chart. Then the typical 4/3 circuit substation is studied as an example to demonstrate the influence of human errors on substation supply reliability.

2 HUMAN ERRORS AND HUMAN RELIABILITY ASSESSMENT

2.1 *Definition and classification of human errors*

It is recognized that system function would be affected by human factors positively or negatively. Human error can be simply defined as human actions with negative influence on system, which exceed the scope of the system acceptability. With the development of science and technology, the reliability of hardware and software has improved a lot. We gradually recognize that human errors become a great threat to system safety, because they could not be totally eliminated. Through statistic analysis, we can see that human error contribution to industry accident has increased from around 20% to 80% over the years [10].

The reason why people make mistakes is complex and the type of human errors is various. Human error classification is he base for further study of human behaviors in order to improve the system reliability and reduce the occurrence of accidents. There are many different ways of classifying human errors according to different standards, such as error reasons, error sources and error contents. Various classification standers are adopted for different research purposes. Research [11] introduces a kind of classification according to error contents, including design error, operator error, fabrication error, maintenance error, contributory error and inspection error.

2.2 *Human errors in power system*

As power system is becoming more complex, an increasing number of people are involved in. It has been noted that human operators are not entirely reliable and human performance might be affected by various factors. As a fundamental model of human performance, human information processing model, shown in Figure 1, is widely accepted. These processes could be divided into 3 steps, namely perception of information generated from the system, processing information and making a decision, and action implement. Human errors might occur in any phase of this process.

The underlying causes of errors are what disturb the above process. Many researches try to figure out the Human Error Inducing Factors (HEIF) aiming at reducing human errors [12]. HEIF usually represents one aspect of operation characteristics. Based on research, several common factors that might cause human errors in power system are listed in table 1.

2.3 *Human Reliability Assessment*

Human Reliability Assessment (HRA), part of Probabilistic Safety Assessment (PSA), is usually used to assess human performance under certain working conditions. Generally, HRA consists of both qualitative analysis and quantitative analysis. Through HRA, we can figure out how likely the operator would finish the task successfully, and quantify the probability of human errors during operation [13].

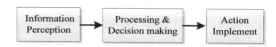

Figure 1.   General human performance model.

Table 1.   Factors that contribute to human errors.

Factors	Descriptions and explanations
Knowledge & experience	Professional knowledge learned from training, and experience acquired through long-term work.
Team factors	Quality of team communication and cooperation.
Procedural factors	Correct and realistic work procedures guiding human operators what to do.
Mental and physical factors	Attention to the task and stress level; mental and physical fitness to present operation.
Environmental factors	Working conditions, like light, noise, space, etc. Social factors regarding family pressures and organizational pressures such as work relations.
Motivation & awareness	Individual aspects like job satisfaction and leadership style; role awareness of his duty.
Situation factors	Relevant factors about working situations, like task complexity, task urgency, task and time load.

3   IMPACT OF HUMAN ERRORS ON SUBSTATION RELIABILITY

3.1   *Human errors in substation switching operation*

Substation switching operations may be required for a number of purposes, such as performing maintenance or replacement of substation equipment, isolation of substation equipment and making temporary or permanent reconfiguration of the substation. The procedure for performing the substation switching operation will be provided in the form of a switching instruction.

Another function of substation switching operation is to enhance substation supply reliability. As we know, standby system often finds applications in substation. When running components fail to work due to failure, it is supposed to be replaced by the backup through switching operations. Human operators play an important role in this process, and they should meet the demand, both emotionally and physically.

When taking switching operations, operators must have enough knowledge and skill to accomplish operation tasks successfully. As is stated in

Section II, human operation could not be completely reliable, and human errors may occur owing to various reasons. So substation switching operation might fail because of human errors, and substation reliability decreases as a result.

3.2 Protection system failures

Protection system is an important part of power substation to protect the power system from incurring a widespread fault. Usually when the protection system senses a fault in its protection zone, it will send a trip signal to corresponding circuit breakers. Consequently the tripping circuit will isolate the faulted component from the healthy part of the system. However, the protection system itself can also fail, which might lead to a disaster.

Two main protection system failure models are refuse operation and maloperation. The former failure will lead to a widespread fault and more components will be disconnected. While the latter one might incur unexpected outage. There are several common reasons leading to these two failures models, such as inappropriate setting, wrong measurements, signal failures, telecommunication channel failures, hardware failures and software failures.

3.3 Effect of human errors and protection system failures on substation equipment

Similar to protection system own failures, failures caused by human errors could be classified into refuse operation and maloperation. In the proposed model, two main kinds of human failures are considered.

- Human incorrect actions that cause protection system inadvertent operation. For example, dispatchers or operators on duty put the protection equipment into operation mistakenly.
- The protection facilities are not completely repaired during maintenance, which might lead the system operating with hidden faults. For example, the parameters are not reset according to the operation model.

Figure 2 shows an 8-state protection system Markov model incorporating protection system failures and effects of human errors [14]. Definitions of the abbreviations and parameters are listed in Table 2. In Figure 2, c_3 represents the proportion of maloperation in the total amount of maloperation and refuse operation. c_5 and c_6 represent the rates failing to detect the unannounced state of maloperation and refuse operation respectively. Their values can be determined by (1) and (2):

$$c_5 = c_3(1-c_1)\lambda_p \tag{1}$$

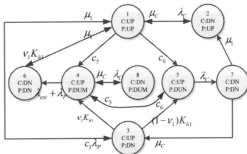

C: Component   P: Protection   UP: Normal State   DN: Failure State
DUM: Inadvertent Operation   DUN: Hidden Failure leading Rejection

Figure 2.   Reliability model of protection system.

Table 2.   Symbols notation.

Symbols	Notation
λ_c	Failure rate of protected component
μ_c	Repair rate of protected component
λ_p	Failure rate of protection system
λ_{ext}	Rate that component is isolated due to extent failure
μ_1	Repair rate of protection system
υ_1	Percentage of human error causing maloperation
Kh_1	Failure rate of protection system caused by human errors
c_1	Efficiency of self-checking and routine test

$$c_6 = (1-c_3)(1-c_1)\lambda_p \tag{2}$$

State 1 is the normal operation state with both the component and protection system functioning well. When a component fault occurs and protection system responds correctly, the state will move from state 1 to state 2; if component is repaired, it will return to 1. If a protection system failure is detected, it will enter state 3 from state 1. If the latent failure of refuse operation is not diagnosed, it will move to state 5. Then if the component fails in this case, the system enters state 7. The transition from state 7 to state 2 or to state 3 might occur in relation to repair rate. If the latent failure of maloperation is not diagnosed, it will move to state 4 from 1. Latent failure of maloperation might be triggered by internal or external factors, then the component will be isolated, namely state 6. When a component failure occurs, the state will move to 8 from 4. There also exists bidirectional transition between state 4 and state 5. Let $P_1 = p_1, p_2, ..., p_8$ donate steady state probability, then probability of each state can be obtained using the following equations:

$$P_1A = 0 \qquad (3)$$

$$\sum_{i=1}^{8} p_i = 1 \qquad (4)$$

where p_i is steady state probability of state i, and a_{ij} is the transition rate from state i to state j. Incorporating human errors and protection system failures, the substation equipment reliability indices can be obtained using Markov process analysis shown in Figure 4. The failure frequency (λ) and unavailability (U) are:

$$\lambda = p_1 a_{12} + p_1 a_{13} + p_1 a_{16} + p_4 a_{46} + p_4 a_{48} \qquad (5)$$

$$U = p_2 + p_3 + p_6 + p_8 \qquad (6)$$

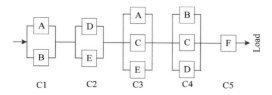

Figure 3. Minimal cut set of example system.

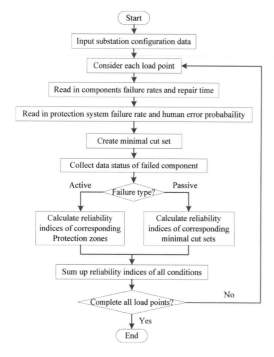

Figure 4. Flow chart for the evaluation of electrical substations.

The rate of protection system reject-action could be expressed as:

$$\lambda_{re} = p_5 a_{57} \qquad (7)$$

4 SUBSTATION RELIABILITY ASSESSMENT PROCEDURE

It is assumed that continuity of service is the criterion of success for any outgoing feeder. Furthermore, it is assumed that any available incoming feeder can supply the total need of the station. The modes of component failures can be classified by the failure condition of component. There are two common types of failure modes as follows.

1. Passive Failure: Passive failure is the failure which does not impact the operation of protective devices. The failure component must be taken out and repaired before being re-connected to the system and carrying power flow.
2. Active Failure: Active failure is defined as the failures that cause the operation of protective devices in primary protection zone. After tripped by these protective devices, such as circuit breakers, the failure component can be safely removed.

Minimal Cut Set (MCS) is the smallest group of the components which might cause system failure. Minimal cut set analysis is widely used to judge the connectivity of substation configuration. An example system is shown in Figure 3, alphabets A-F denote substation components. It includes five minimal cut sets, namely C1, C2, C3, C4 and C5.

From C1 in Figure 3, if only one of the components A and B fails, the system will not fail. However, if such component A and B fail simultaneously, the system will fail. Another example as C5, the solely failure of component F can cause the system failure. F is defined as first order minimal cur set, and A, B are defined as second order minimal cur set.

The flow chart for reliability evaluation of substations is shown in Figure 4. Each load points is analyzed. When all conditions are considered, the evaluation of substation supply reliability is completed.

The reliability indices subject to active failure can be obtained from Figure 4 and the failure frequency and unavailability could be obtained using equation (5) and (6). If the relay rejects to react, other related components might be isolated, and the failure frequency could be calculated using (7). The reliability indices subject to passive failure are calculated using minimal cut set method, and the failure rate can be obtained using (8).

518

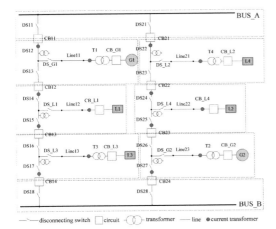

Figure 5. 4/3 circuit breakers substation configuration.

$$\lambda_{pa} = \sum_{i=1}^{n} \lambda_i \qquad (8)$$

where λ_i is the failure rate of MCS i and n is the total number of MCS of the substation.

5 CASE STUDY

The substation configuration of a 4/3 circuit breakers substation is wildly used. The circles named G represent the generators, while the blocks named L refer to load points. For the 4/3 circuit breakers substation, on each branch there are three fields. Therefore, two branches are needed to satisfy the assumption of "2 generators and 4 loads connected to the substation" [8]. The 4/3 circuit breakers substation is shown in Figure 5.

The substation is divided into several protection zones represented by dotted boxes in Figure 5. Which of these protection systems reacts to the component fault mainly depends on the fault location. For the components that are within one protection zone, the effects on the protection will be the same. Therefore, the whole substation is divided into several fault zones. Taking fault zone G1 for example, this zone is constructed by seven components: two disconnecting switches, two current transformers, one voltage transformer, one transformer and one cable. When the initiating fault occurs in the G1 zone, such as a short circuit or an explosion, the protection system should react and G1 will be isolate. However, if the protection system fails to respond, it will lead to a widespread fault and G1&L1 will be isolated as a result.

To analyze the passive failure, minimal cut set is determined from an adjacent matrix based on

Table 3. Component data for system in figure 5.

Component	Acive failure rate	Repair time/ hour	Passive failure rate	Switch time/ hour
Busbar	0.024	8	0.024	4
Line	0.06	4	0.031	0.8
Circuit breaker	0.03	2	0.03	1.5
Disconnector	0.003	8	0.002	1
Transformer	0.05	24	0.025	20

Table 4. Evaluation results of case study.

Load point	Part 1 Failure frequency (per year)	Part 2 Failure frequency (per year)
L1	0.4222	0.5051
L2	0.4532	0.5359
L3	0.4722	0.5568
L4	0.4692	0.5538

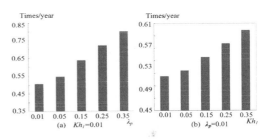

Figure 6. Impact of failure rate of protection system/ human error probability on the failure frequency of L1.

the depth first search theory. The reliability indices of failure frequency and unavailability can be obtained. The results are illustrated in table 4. In part 1, results are calculated with protection system failures and human errors neglected, but considered in part 2.

It is obvious to see that, the failure frequency is higher when taking protection system failure and human errors into consideration. To make a further explanation of this effect, we adjust human error probability (Kh_1) and the failure rate of protection system (λ_p), and the corresponding curves are shown in Figure 6 (a) and (b) respectively.

In Figure 6 (a), the failure frequency of L1 increases from 0.5051 to 0.8009 when λ_p varies from 0.01 to 0.35; In Figure 6 (b), the failure frequency of L1 increases from 0.5051 to 0.5789 when Kh_1 varies from 0.01 to 0.35. It can be concluded that, human errors and protection system failures have an obvious influence on substation reliability.

6 CONCLUSIONS

This paper proposes a more realistic approach to assess the substation reliability incorporating various failure models, protection system failures and human errors. With the use of Markov model and minimal cut set method, we could evaluate substation supply reliability. Analysis of 4/3 circuits substation reliability illustrates that protection system failures and human errors have a great effect on substation reliability. So, in order to maintain the power system reliable, the protection system should be checked to eliminate hidden faults. On the other hand, some measures should be taken to manage human errors and enhance human reliability.

REFERENCES

[1] R. E. Brown, "Electric Power Distribution Reliability", New York: Marcel Dekker, 2002.

[2] G. B. Jasmon and O. S. Kai, "A new technique in minimal path and cut-set evaluation", *IEEE Trans. Reliab.*, vol. 34, no. 2, pp. 136–143, Jun. 1985.

[3] R. N. Allan., R. Billinton and M. F. De Oliveira, "An Efficient Algorithm for Deducing the Minimal Cuts and Reliability Indices of a General Network Configuration. Reliability", *IEEE Trans. Reliab.*, Vol. 25, no. 4, pp. 226–233, Oct. 1976.

[4] M. Verma, C. Rukmini and S. Goyal, "Electrical substations reliability evaluation using fuzzy fault tree approach". *in Power, Control and Embedded Systems (ICPCES), 2012 2nd International Conference on,* Allahabad, 2012.

[5] R. Billinton and G. Lian, "Station reliability evaluation using a Monte Carlo approach". *IEEE Trans. Power Del.*, vol. 8, no. 3, pp. 1239–1245, Jul. 1993.

[6] R. Billinton, H. Chen and J. Zhou, "Generalized n+2 state system Markov model for station-oriented reliability evaluation", *IEEE Trans. Power Syst.*, Vol. 12, no. 4, pp. 1511–1517, Nov. 1997.

[7] D. L. Duan, X. Y. Wu and H. Z. Deng, "Reliability Evaluation in Substations Considering Operating Conditions and Failure Modes", *IEEE Trans. Power Del.*, vol. 27, no. 1, pp. 309–316, Jan. 2012.

[8] F. L. Wang., W. Bart, M. Gibescu, et al, "Reliability evaluation of substations subject to protection system failures", *in Power Tech (POWERTECH), 2013 IEEE Grenoble*, Grenoble 2013.

[9] D. O. Koval and H. Landis Floyd, "Human element factors affecting reliability and safety. Industry Applications", *IEEE Trans. Ind. Appl.*, vol. 34, no. 2, pp. 406–414, Mar./Apr. 1998.

[10] E. Hollnagel, "The reliability of cognition: Foundations of human reliability analysis", London, UK: Academic Press, 1993.

[11] W. Rankin, et al., "Development and evaluation of the Maintenance Error Decision Aid (MEDA) process", *International Journal of Industrial Ergonomics*, vol. 26, no. 2, pp. 261–276, 2000.

[12] F. Badenhorst and J. Tonder, "Determining the factors causing human error deficiencies at a public utility company", *J Hum Resour Manage*, vol. 2, no. 3, pp. 62–69, 2004.

[13] P. Pyy, "An approach for assessing human decision reliability", *Reliab Eng Sys Saf.*, vol. 68, no. 1, pp. 17–28, 2000.

[14] J. J. Zhang, M. Ding and S. H. Li, Impact Analysis of Human Error on Protection System Reliability, *Automation of Electric Power System*, vol. 36, no. 8, pp. 1–5, 2012.

Architectural, Energy and Information Engineering – Sung & Chen (Eds)

Adsorption behavior of m-cresol in coal gasification wastewater by magnetic functional resin

Y.B. Wang, X.W. Hu, S.W. Jin, Y.L. Zhang & L.J. Xia
Faculty of Environmental Science and Engineering, Kunming University of Science and Technology, Kunming, China

ABSTRACT: Magnetic functional resin was employed to adsorb m-cresol, and the effects of operational conditions including pH, salts, initial concentration on treatment effect and the desorption efficiency and regeneration times were investigated. The results showed that the Freundlich model was more appropriate to describe the adsorption pattern of m-cresol in solution. When the pH was 8, the resin had the highest adsorption efficiency. The adsorption of m-cresol on the resins was inhibited in the presence of Na_2SO_4. When the concentration of Na_2SO_4 was < 2 g/L, the removal efficiency of m-cresol declined. When the concentration of Na_2SO_4 was > 2 g/L, the removal efficiency of m-cresol elevated. With the increasing initial concentration, the adsorption capacity of m-cresol on the resins increased. The desorption efficiencies measured > 90% and the regeneration times > 10.

Keywords: magnetic functional resin; m-cresol; adsorption; coal gasification wastewater

1 INTRODUCTION

Along with the development and expansion of gas industry in China, large amounts of wastewater were produced by the gasification process, which contains a large number of m-cresol. M-cresol is harmful to people's health. The traditional bio-chemistry treatment method cannot effectively remove m-cresol. So, adsorption removal of m-cresol was adopted.

Adsorption is simple process, in which no new chemical substances are introduced, without affecting subsequent treatment processes, to carry out a series of research at home and abroad. Magnetic resin in the adsorption of target substances in water shows a better adsorption effect and efficiency. In recent years, a growing number of scholars have dedicated to research and development in this area. However, most studies have been designed for ion exchange resins to remove macromolecular organic compounds such as NOM and dye in water. In addition, Mergen's studies suggested that it also has an effect on magnetic resin adsorption of small molecule organic matter. So, it has a great significance to master the law and mechanism of magnetic resin adsorption of small molecular organic matter such as m-cresol. It is also important to improve the effect of magnetic resin on the treatment of coal gasification wastewater. Furthermore, it can solve the environmental problems by the development of the coal-to-oil industry.

This study used a new type of magnetic anion exchange resin (NDMP) to adsorb m-cresol.

It examines the influence of magnetic functional resin on the adsorption of m-cresol by pH, temperature, salinity, initial concentration. It also explores the mechanism of magnetic functional resin to adsorb m-cresol. Thus, it lays a foundation for the resourcezation of coal-gasification wastewater.

2 EXPERIMENTAL

2.1 Chemicals

The resin NDMP was purchased from Jiangsu Jinkai Resin Chemical Co. Ltd, as well as M-cresol, sodium chloride, sodium sulfate, and sodium hydroxide were used for analysis.

2.2 Adsorption

Briefly, three conical flasks containing 100 mL of 200 mg/L m-cresol solution were shaken at different temperatures (278, 293, and 308 K) with 0.100 g of resin at 130 rpm for 24 h. The amounts of m-cresol in the solutions at different temperatures were obtained by the measurement of the concentration of m-cresol. To explore the effect of pH on adsorption, 100 mL of 200 mg/L m-cresol solution and 0.100 g resin were introduced into a series of 250-mL conical flasks with the pH of the solutions varying from 6 to 10, which was adjusted by using 1 mol/L HCl or 1 mol/L NaOH aqueous solution. To determine the effect of Na_2SO_4 on adsorption, the amounts of m-cresol solution and resin were identical with those of the experiments for the pH series. The concentrations of Na_2SO_4 varied from 0 to 6 g/L. To investigate the effect of the initial concentration on adsorption, 100 mL of different concentrations of m-cresol solution and 0.100 g resin were introduced into a series of 250-mL conical flasks. The concentrations of m-cresol varied from 75 to 200 mg/L, respectively, and the effect of solution chemistry on adsorption was conducted at 293 K for 1 h. All experiments were repeated three times to obtain the average values. The adsorption capacity of m-cresol (qt, mg/g) was calculated using Eq. (1):

$$q_t = \frac{(C_0 - C_t)V}{W} \quad (1)$$

where V is the volume of the solution (given in liters), and W is the weight of the dry resin (given in grams). C_0(mg/L) and C_t(mg/L) represent the initial concentration and concentration at time t (min), respectively. The variables qt and C_t are replaced, respectively, by q_e and C_e to represent the adsorption capacity and m-cresol concentration at adsorption equilibrium, respectively.

2.3 Regeneration

A series of parallel experiments investigating m-cresol adsorption were conducted at 293 K with 100 mL of 200 mg/L m-cresol solution and 0.100 g of resin in conical flasks. The resin was filtered after adsorption equilibrium. NaOH aqueous solution and NaCl aqueous solution were used for regeneration.

3 RESULTS AND DISCUSSION

3.1 Adsorption isotherms

Figure 1 shows the adsorption isotherm of m-cresol onto NDMP at three different temperatures. The adsorption capacities of the m-cresol decreased for NDMP as the temperature was increased. The Langmuir model and the Freundlich model are the most frequently used equations for fitting the experimental data of an isotherm and can be expressed as follows:

$$Langmuir \quad q_e = \frac{Q_m K_L C_e}{1 + K_L C_e} \quad (2)$$

$$Freundlich \quad q_e = K_F C_e^{1/n} \quad (3)$$

where q_e (mmol/g) is the equilibrium amount of m-cresol adsorption; Q_m(mmol/g) and K_L(L/mmol) represent the maximum adsorption capacity and the affinity of binding sites, respectively; and K_F (L/g) and n are the Freundlich constants, which

indicate the adsorption capacity and intensity, respectively.

The fitting results from the isotherms of the adsorption of m-cresol on NDMP are listed in Table 1. According to the values of correlation coefficients (R^2), the Freundlich equation described the data better than the Langmuir equation, suggesting a heterogeneous adsorption of m-cresol on NDMP. The heterogeneous adsorption was possibly due to other interactions except ion exchange, such as backbone adsorption.

3.2 Effects of pH solution

As shown in Figure 4, at pH <8.0, The adsorption capacity for NDMP was enhanced with increasing pH. The adsorption capacity of NDMP decreased at pH >8.0. The maximum adsorption capacity was 49.9 mg/g at pH = 8.0.

This is due to the fact that m-cresol is a weak polar monobasic acid. In the acidic solution, the main form is $CH_3C_6H_4OH$, and the primarily reaction is physical adsorption; in the alkaline solution, the main form is $CH_3C_6H_4O-$, and the primarily reaction is the ion exchange. This resin is strongly alkaline ion exchange resin, so the removal rate in the alkaline state is higher than that in the acidic state. Moreover, the primary reaction on this resin is ion exchange. With further increasing the pH,

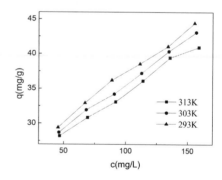

Figure 1. Adsorption isotherms of m-cresol.

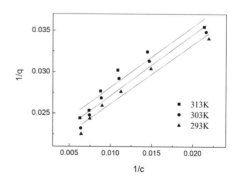

Figure 2. Langmuir adsorption isotherm of m-cresol.

522

resin surface functional groups take negative electricity under alkaline conditions, and generate electrostatic repulsion with the functional groups such as -OH and -OR which take negative electricity in wastewater. The removal efficiency of organic matter was declined. So, when the pH is over 8, the removal efficiency of m-cresol was decreased.

3.3 *Effects of salts*

In the coal gasification process, the pH of gasification wastewater was alkaline. In the subsequent wastewater treatment, it is commonly used for pH adjustment. So, there was a large number of SO_4^{2-} in wastewater. Figure 5 shows that Na_2SO_4 has an inhibitory effect. The concentration of Na_2SO_4 was in the range of 0 ~ 6 g/L, the trend of m-cresol removal was decreased and then increased, when the concentration of SO_4^{2-} was 2 g/L, the adsorption quantity of m-cresol was minimum, i.e. only 11.66 mg/g.

This is due to the fact that this resin is strongly alkaline anion exchange resin. When the salt concentration is low, it forms competition between SO_4^{2-} and the solute, and the electrostatic force of SO_4^{2-} with the functional groups on resin is greater. The adsorption capacity is decreased by increasing salt concentration. However, due to the less amount of functional groups on its surface, the decreasing adsorption capacity ends soon. When the ion exchange was severely weakened, removal of solutes in water by resin primarily relies on the π-π effect. According to the theory of the interactions between the ionic and adsorbate molecules, the dielectric constant of m-cresol is much lower than water, resulting in the the salting-out effect. It plays a catalytic role for adsorption. Therefore, the adsorption capacity increases slowly with increasing salt concentration.

3.4 *Effects of initial concentration*

As shown in Figure 6, in the range of 75–200 mg/L, the adsorption quantity increases gradually with the increase in initial concentration. When the initial concentration is greater than 125 mg/L, the trend of increase in the adsorption quantity changes slowly.

This phenomenon is caused by the higher concentration, causing the liquid-solid concentration difference greater between the two phases, prompting that m-cresol molecules can more easily overcome the limitations between the two phases to adsorb on the resin surface. So, with the increasing initial concentration, the adsorption capacity increases. As the initial concentration increases, the trend of adsorption capacity becomes slow, which suggested that when the initial concentration is at a low concentration range, the adsorbent can provide enough adsorption sites. When the concentrations rises, the number of m-cresol in aqueous solution will be more than the number of adsorption sites on resin gradually, which

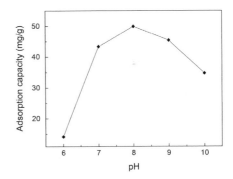

Figure 4.   Effect of pH on adsorbing m-cresol.

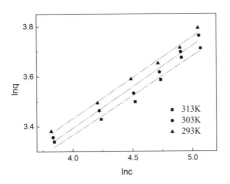

Figure 3.   Freundlich adsorption isotherm of m-cresol.

Table 1.   Adsorption isotherm parameters of m-cresol under different temperatures.

Temperature (K)	Langmuir			Freundlich		
	$Q_m/(mg \cdot g^{-1})$	b	R^2	K	$1/n$	R^2
293	52.632	0.267	0.969	8.356	0.327	0.993
303	52.632	0.258	0.945	7.957	0.329	0.983
313	50.000	0.274	0.933	8.199	0.315	0.978

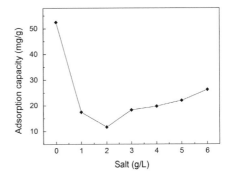

Figure 5.   Effect of salts on adsorbing m-cresol.

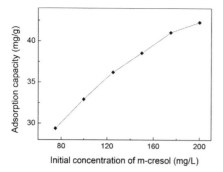

Figure 6. Effect of initial concentration on adsorbing m-cresol.

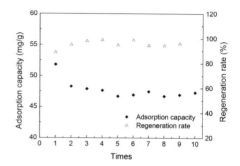

Figure 7. Regeneration and stability of NDMP.

creates competitive adsorption between adsorbates, so the adsorption increases slowly at higher concentrations.

3.5 *Regeneration and stability*

As shown in Figure 7, both the adsorption and regeneration efficiencies of resin were relatively stable. The average adsorption capacity of m-cresol on resin was 47.77 mg/g. The average regeneration efficiency of resin was 95.37%. During the repetition of the adsorption/desorption process 10 times, the regeneration efficiencies of resin were all over 90%, so the resin regeneration was at least 10 times. This indicates that the resin has a better reusability, and has a good value for industrial use.

4 CONCLUSIONS

1. The Freundlich equation described the data better than the Langmuir equation, suggesting a heterogeneous adsorption of m-cresol on NDMP.
2. The best pH for adsorption m-cresol is 8.0; Na_2SO_4 has an inhibitory effect on resin to adsorb m-cresol. When the concentration of SO_4^{2-} was 2 g/L, the adsorption quantity of m-cresol was minimum; the adsorption quantity increases gradually with the increase in initial concentration.
3. When the adsorption/desorption process was repeated 10 times, the average adsorption capacity of m-cresol on resin was 47.77 mg/g. The average regeneration efficiency of resin was 95.37%.

REFERENCES

Abdelkader, N.B., Bentouami, A., Derriche, Z., Bettahar, N., de Me´norval, L.C., 2011. Synthesis and characterization of Mg-Fe layer double hydroxides and its application on adsorption of Orange G from aqueous solution. Chemical Engineering Journal, 169 (1–3), 231–238.

Boyer, T.H., Singer, P.C. 2005. Bench-scale testing of a magnetic ion exchange resin for removal of disinfection by-product precursors. Water Research, 39 (7), 1265–1276.

Dabrowski A, Podkoscielny P, Hubicki Z, et al. 2005. Adsorption of phenolic compounds by activated carbon-acritical review. Chemosphere, 58: 1049–1070.

Dursunm G, Cecek H, Dursun A Y. 2005. Adsorption of Phenol from Aqueous Solution by Using Carbonized Beet Pulp. Journal of Hazardous Materials, 125(1/3): 175–182.

Humbert H, Gallard H, Suty H, et al. 2005. Performance of selected anion exchange resins for the treatment of a high DOC content surface water. Water Research, 39(9): 1699–1708.

Kaewsuk J, Seo GT. 2011. Verification of NOM removal in MIEX-NF system for advanced water treatment. Separation and Purification Technology, 80(1): 11–19.

Lu Z C, Qiu Y F, Meng G H, et al. 2012. Application of resin in advanced treatment of coking wastewater. Ion Exchange and Adsorption, 28(5): 423–431.

Pan B C, Xiong Y, Li A M, et al. 2002. Adsorption of Aromatic Acids on an Aminated Hypercrosslinked Macroporous Polymer. React. Funct. Polym., 53(2/3): 63–72.

Senturk H B, Ozdes D, Gundogdu A, et al. 2009. Removal of phenol from aqueous solutions by adsorption onto organomodified Tirebolu bentonite: Equilibrium, kinetic and thermodynamic study. Journal of Hazardous Materials, 172(1): 353–362.

Shuang C.D., Li P.H., Li A.M., et al. 2012. Quaternized magnetic microspheres for the efficient removal of reactive dyes [J]. Water Research, 46(14): 4417–4426.

Shuang C D, Pan F, Zhou Q, et al. 2012. Magnetic Polyacrylic Anion Exchange Resin: Preparation, Characterization and Adsorption Behavior of Humic Acid. Industrial & Engineering Chemistry Research, 51: 4380–4387.

Susan H, Singer P C. 2010. Removal of bromide and natural organic matter by anion exchange. Water Research, 44(7): 2133–2140.

Zhang W.H., Wei C.H, Yan B. 2012. Composition characterization of dissolved organic matters in coking wastewater. Environmental Chemistry, 31(5): 702–707.

Architectural, Energy and Information Engineering – Sung & Chen (Eds)
© 2016 Taylor & Francis Group, London, ISBN 978-1-138-02791-6

Raptor codes for P2P VOD with dynamic sizes

W.C. Duan, P. Zhai, B. Zhou & C.J. Chen
Information Engineering College of Zhengzhou University, Zhengzhou, Henan, China

ABSTRACT: In this paper, we propose using Raptor codes with dynamic block sizes and symbol sizes for P2P VOD instead of LT codes to solve the problem of data loss in P2P VOD and video interruption. Raptor codes can reduce the complexity of scheduling algorithms and improve the throughput of encoding and decoding, and the dynamic sizes can reduce overhead and interrupt latency. Moreover, feedback information can reduce unnecessary encoded symbol generation. At last, the experiment proved that the idea is effective and improve the quality of VOD service.

Keywords: raptor codes; dynamic sizes; P2P VOD; interrupt latency

1 INTRODUCTION

Currently the representative of the most popular P2P streaming media is PPLive, PPStream, Sop-Cast, and so on. P2P streaming data block transfer has been distributed from the traditional use of the tree to the current Mesh distribution which combines pull-push data distribution and protocol of Gossip[1]. In order to achieve high throughput in P2P networks, researchers have proposed a variety of P2P streaming media-based data distribution algorithm environment. And through the use of a specific forward error correction code-fountain code[2] which can improve network utilization so that the distribution of data can achieve the best performance. Fountain codes are rateless so that they can generate an unlimited number of independent and unique encoded data[3]. The receiver can successfully restore the original data if it can get a little more encoded data than the amount of the original data from the peers. Fountain codes are a supplement to unreliable but lightweight transport protocol like UDP[4], it not only provides reliable transmission but also greatly increases data transfer performance of P2P networks.

A few years ago, B. Li and C. Wu who came from the University of Toronto proposed the use of LT codes for P2P streaming media. They proposed that LT codes can solve the problem of coding block schedule in P2P streaming media. They compared four different coding schemes in a P2P streaming media, and evaluated the feasibility of LT codes[5]. Then Suh used LT codes to build an analytical model for P2P streaming system[6]. Oh used online codes to propose the prototype of P2P streaming recently, and suggested to use the property of rateless fountain code to

dynamically adjust symbol size to avoid excessive number of start-up delay and interrupt latency[7]. Raptor codes are the latest research in the field of fountain codes, Raptor codes combined with LDPC and HALF on the basis of the LT, so that decoding complexity reduce from $O(nlogn)$ to $O(n)$[8]. Raptor codes have a great advantage in the codec complexity and overhead rates than LT codes and online codes. Therefore, this paper proposes to use Raptor codes in P2P VOD system on the basis of this study, it can improve network throughput of the P2P streaming media and reduce the encoding and decoding time. At the same time due to the uncertainty of the network, fixed symbol size and block size may cause interrupt latency, therefore, we propose to use dynamically adjusting the sizes of symbols and blocks to adapt to changing network environments to reduce interrupt latency.

2 RAPTOR CODES

Raptor codes have very small overhead of decoding and codec time $(O(Klog(1/\varepsilon)))$. The encoding consists of two phases: pre-code and LT coding[10]. This procedure of encoding is coarsely illustrated in Figure 1. in the pre-coding phase, the K original symbols will be precoded into K' intermediate symbols by the LDPC code. In the LT coding phase, the encoder chooses randomly a degree d between 1 and K from a specifically designed degree distribution and selects also randomly d intermediate symbols. Each encoded symbol is then generated by XORing the set of chosen intermediate symbols. Repeating the second phase, we can continuously generate encoded symbols.

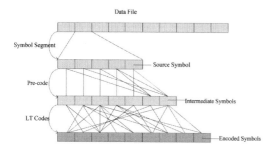

Figure 1. Encoding of raptor codes.

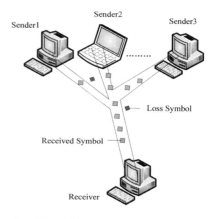

Figure 2. The working diagram of raptor codes in P2P VOD.

Although encoded symbols may be lost during transmission, receiver can restore the original symbols if it can receive enough encoded symbols. In general, the number of encoded symbols required for successful fountain decoding is calculated by $K(1+\varepsilon)$ [12]. ε is the symbol overhead with a very small real number. Upon reception of a given threshold $K(1+\varepsilon)$ of encoded symbols, the receiver starts the decoding process. At first, the decoder use iterative to restore intermediate symbols and afterwards, it use Gaussian elimination to restore original symbols.

In summary, Raptor codes have good performance, small overhead and are very close to the ideal of fountain codes. It is suitable for application in P2P VOD.

3 DYNAMIC ADJUSTMENT OF RAPTOR CODES IN P2P VOD

When the client requests a video, a request is needed to send to the management peer. Then the peer sends a list of information of peers to the client, receiver chooses the peers for sending according to the selection mechanism of peers, then receiver sends a message with the block sizes M bytes and the symbol size L bytes to the peers. The senders begin to work after they receive the message. The working diagram of Raptor codes in P2P VOD is shown in Figure 2.

3.1 Sender

1. According to the size of the block size and the symbol size, the sender intercept M bytes with symbol size L bytes, and K original symbols will be obtained. If the last symbol of a block would not consist of l bytes, it is padded with 0 s to l.
2. The encoder uses the LDPC codes and Half codes to generate intermediate symbols. The intermediate symbols of all the senders are the

same. These ensure that the receiver can decode the encoded symbols.
3. Each encoded symbol has a random seed value[13]. According to the seed value, the encoder chooses randomly a degree d between 1 and K based on a designed degree distribution, and selects also randomly d intermediate symbols, Each encoded symbol is then generated by XORing the set of chosen intermediate symbols.

The header of each encoded symbol includes the BlockID, the SymbolID and degree d. This information is necessary for receiver to match the symbol to the correct position in the block. Each encoder must use a unique seed value, or the encoder will generate the same encoded symbol. These redundant encoded symbols will increase the overhead of decoding. Then the Packet of encoded symbols according to Adaptive Packet Distribution Algorithm[11] will be carried to make the Packet that receiver receive meeting the buffer size of receiver[9].

The workflow of Video data transmission is shown in Figure 3 below.

3.2 Receiver

Upon reception of a given threshold $K(1+\varepsilon)$ of encoded symbols, the receiver starts the decoding process. The degree d can be found from the encoded symbols, and then original symbols will be restored by Gaussian elimination[8].

1. If the current block is successfully restored, the receiver sends an ACK message to all senders to tell them to stop transmission of the block and tell them the block size and the symbol size of the next block Depend on dynamic adjustment mechanism, send the information to the sending node.

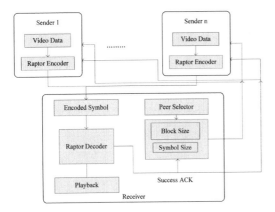

Figure 3.   The workflow of video data transmission.

2. If the decoding is unsuccessful, the receiver will continues to receive encoded symbols to decode again.

4   ANALYZE OF DYNAMIC OPTIMIZATION OF PARAMETERS

Symbol size determines the encoding and decoding overhead of a single block, the block size determines the throughput of data, the number of repair symbols affect the overhead of encoding and decoding. Therefore, in order to reduce the encoding and decoding overhead, increase data throughput and reduce the interruption, this paper analysis symbol size, block size and the number of repair symbols. At the same time due to the uncertainty of the network, if the symbol size and block size are fixed, interruption may occur, so this paper propose to dynamically adjust the block size and the symbol size.

4.1   Symbol size

The symbol size affects the encoding and decoding time. Theoretically, Raptor codes have a decoding time $O(K\log(1/\varepsilon))$, therefore, by choosing a symbol size and block size as large as possible, the total amount of encoding and decoding time could be decreased in theory. However, decoding relies heavily on the matrix inversion algorithm, which Occupies 92% of the computation time, the theoretical optimum cannot be reached with large block size. In addition, due to limitations of the MTU of Ethernet, so symbol size should be set less than 1450 bytes, and if symbol size is larger than 1450 bytes, sender need to fragment the symbols, receiver need to reassemble the symbols, which will increase additional overhead.

4.2   Block size

Block size determines the entire data throughput. If the block is selected is very large, then the receiver will need to wait longer time to receive a sufficient number of encoded symbols for decoding. But if the block is selected is very small, although a single block encoding and decoding time overhead is small and the recipient does not need a lot of delays for decoding, but the overhead of transmission will increase.

4.3   Number of repair symbols

Fountain code is a non-systematic code. Ideally, decoder will begin to work when it has received K encoded symbols, but actually it is unable to achieve. When encoded symbols lose or damage during transmission, receiver needs additional encoded symbols to ensure the successful decoding. The additional encoded symbols are called repair symbols. So decoder need to receive K $(1+\varepsilon)$ encoded symbols for successful decoding. ε is overhead rate, and the smaller ε is, the higher the encoder and decoder performance is, and the larger the effective throughput of the network is.

Before i_{th} blocks put into video buffer, the data of video buffer has been played out, interruption will be generated. In order to improve the user's experience, this paper proposes to use dynamic symbol size and block size to avoid interruption.

4.4   Dynamic optimization of the size of block and symbol

Before i_{th} locks put into video buffer, the data of video buffer has been played out, interruption will be generated. In order to improve the user's experience, this paper proposes to use dynamic symbol size and block size to avoid interruption.

$$T_i \leq T_i^{remain} \tag{1}$$

where T_i is the time from requesting i_{th} block to put it into video buffer, T_i^{remain} is the time that video buffer can still play when receiver request ith block.

$$T_i = T_i^{en/decoding} + T_i^{trans} \tag{2}$$

$$M_i = L_i K_i \tag{3}$$

$$M_i = V_{i-1} T_i^{trans} - \Delta M_i \tag{4}$$

$$T^{en/decoding} = aK_i \tag{5}$$

527

$$T_i = \left(\frac{a}{L_i} + \frac{1}{V_{i-1}} \right) M_i \qquad (6)$$

where M_i is i_{th} block size, L_i is i_{th} symbol size, K_i is the number of symbol, V_{i-1} is the average network speed of transmission of data i_{th} block, ΔM_i is the overhead of header of packet, a determined by ε is a constant.

Since the current speed is unpredictable, we use the speed of $(i-1)_{th}$ block instead of the speed of i_{th} block.

$$\left(\frac{a}{L_i} + \frac{1}{V_{i-1}} \right) M_i + \Delta \le T_i^{remain} \qquad (7)$$

$$\left(a + \frac{L_i}{V_{i-1}} \right) K_i + \Delta \le T_i^{remain} \qquad (8)$$

When T_i^{remain} and V_{i-1} is determined, we can dynamically adjust the block size and the symbol size to avoid interruption of video playback.

5 EXPERIMENTAL RESULTS

Experimental code is implemented by C on an Intel Pentium i7 CPU and Linux as operating system. The experiment consists of two parts. The first part is the comparison of LT codes and Raptor codes. The second part is the influence of dynamical the block size and the symbol size to the interruption. The experiment in this paper is implemented by single-thread Without Parallel Algorithms.

In our experiments we investigated a range of block sizes with k = 32, 64, 128, 256, 512, 1024 and a symbol size l = 64, 128, 256, 512, 1024 bytes. For every combination of block and symbol size, randomly chosen video data was encoded and decoded on the test machine. The unit of block size is the number of symbol(K) the result of throughput with the block size is 32 and the symbol size is 32,64,128,256,512,1024 byte system is shown in Table 1.

As one can observe the influence of the symbol size on the encoding and decoding speed is minimal yet, larger symbol sizes increase the speed slightly.

To study the influence of block size on throughput, we investigated a range of block sizes with k = 32, 64, 128, 256, 512, 1024 and a symbol size L = 1024 bytes, the result of Raptor codes and LT codes shown in Figure 4:

As can be seen from Figure 4, Raptor codes has advantage than LT codes in encoding and decoding, and the performance is twice than the performance of LT codes. Due to the encoding and

Table 1. Throughput with different symbol size and K = 32.

Symbol size (Byte)	32	64	128	256	512	1024
Throughput Mbit/s	23.8	24	23.9	24.1	24	24.1

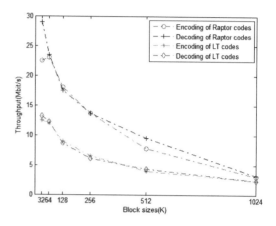

Figure 4. Throughput of encoding and decoding.

decoding speed of LT codes is lower than the network transmission bandwidth, the speed is the bottleneck restricting P2P VOD system. But the max speed of Raptor codes is 25–30 Mbit/s, the bottleneck will not exist. And Figure 4 show that the larger the block size is, the smaller the throughput is, because in the decoding process of Raptor codes, the bigger the block is, the higher the complexity of the matrix inversion is. So in order to improve the speed of encoding and decoding, the small block is better. However, block sizes chosen too small (k ≤ 32) have an adverse effect on the encoding performance, as the content witching overhead for the CPU increases.

To decode successfully, receiver needs to receive K (1 + ε) encoded symbols. ε is overhead rate, and the smaller ε is, the larger the effective throughput of the network is. Because the channel of the packet loss rate is different, ε is different, and the number of repair symbols is different. To test out the relationship between K and ε, assuming the packet loss rate of channel is 0.05. that is to say, the encoded symbols generated by Raptor codes is randomly deleted, and the probability is 0.05, K = 10, 100, 1000, 2000, L = 1024 bytes. The experimental result is shown in Figure 5.

When the receiver receives 9 repair symbols and K is between 10 and 2000, the probability of decoding success is 99.9 %. Different values of K can get approximate curve. That is to say, the consistent of decoding of Raptor codes is very good.

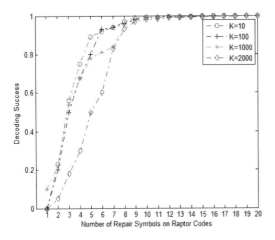

Figure 5. Decoding success of Raptor codes depending on the number of repair symbols.

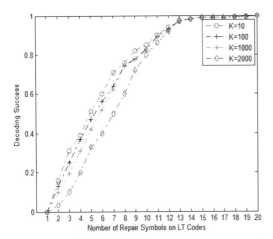

Figure 6. Decoding success of LT codes depending on the number of repair symbols.

So in the design of the upper application, N = K + 9 can be set as a generic threshold.

Figure 6 is the overhead of decoding success rate of LT codes, it need 13 repair symbols to successful decoding, that is bigger than Raptor codes. So in the P2P network with the same packet loss rate, Raptor codes can significantly increase the effective bandwidth utilization.

In the second part of experiment, the network environment is based on a benchmarking speed, and randomly fluctuates. The speed changes all the time. Because the influence of the symbol size on the throughput of encoding and decoding is minimal and the influence of K is relatively large, then according to the limit of K and L, K and L are dynamically adjusted for transmission. The two different video used in the experiment are the same

Table 2. Performance comparison between the fixed size and the dynamic size.

File	Network speed	Type	Interrupt latency
Data1 176717K B2626s	80 ± 10 KB/s	Dynamic Size	0 s
		Fixed Size K = 1024, L = 1024byte	16.4 s
Data2 385164K B2626s	150 ± 10 KB/s	Dynamic Size	3.2 s
		Fixed Size K = 1024, L = 1024byte	21.3

except resolution. The experimental results are shown in table 2.

As depicted in Table 2, we could observe that dynamic size generate less the time of interrupt than fixed size. So dynamic size proposed by this paper is better.

6 CONCLUSIONS

In this paper, we propose to use raptor codes in the P2P VOD instead of LT codes, raptor codes have an advantage in throughput and overhead. Through the analysis of the parameters, this paper propose to use dynamic sizes to reduce the interrupt latency in P2P VOD. The Raptor codes haven't introduced parallel operation, if Raptor codes with parallel operation is used in P2P VOD. Raptor codes may have better performance. This will be the next work direction for the author.

REFERENCES

[1] Zimmermann, R. & Liu, L.S. 2008. Peer-to-Peer Streaming. Encyclopedia of Multimedia. Springer US, 2008: 708–714.
[2] Thomos, N. & Frossard, P. 2008. Collaborative video streaming with Raptor network coding. Multimedia and Expo, 2008 IEEE International Conference on. IEEE, 2008: 497–500.
[3] Shokrollahi, A. 2006. Raptor codes. Information Theory, IEEE Transactions on, 52(6): 2551–2567.
[4] MacKay D.J.C. 2005. Fountain codes. IEE Proceedings-Communications, 152(6): 1062–1068.
[5] Ahmed, E. & Wagner, A.B. 2011. Optimal Delay-Reconstruction Tradeoffs in Peer-to-Peer Networks. Selected Areas in Communications, IEEE Journal on, 29(5): 1055–1063.
[6] Wu, W. & Lui, J. 2012. Exploring the optimal replication strategy in P2P-VoD systems: Characterization and evaluation. Parallel and Distributed Systems, IEEE Transactions on, 23(8): 1492–1503.

[7] Oh, H.R., Wu, D.O. & Song, H. 2011. An effective mesh-pull-based p2p video streaming system using fountain codes with variable symbol sizes. Computer Networks, 55(12): 2746–2759.

[8] Zhu, M., Qu, Y. & Zhang, K. 2014. An improved ensemble of variable-rate LDPC codes with precoding. Information Theory (ISIT), 2014 IEEE International Symposium on. IEEE, 2177–2181.

[9] Huang, S., Sanna, M. & Izquierdo, E. 2014. Optimized scalable video transmission over P2P network with hierarchical network coding. Image Processing (ICIP), 2014 IEEE International Conference on. IEEE, 3993–3997.

[10] Marfia, G., Roccetti, M. & Cattaneo, A. 2011. Digital fountains+ P2P for future IPTV platforms: a test-bed evaluation. New Technologies, Mobility and Security (NTMS), 2011 4th IFIP International Conference on. IEEE, 1–5.

[11] Park, G.S. & Song, H. 2013. A fountain codes-based hybrid P2P and cloud storage system. ICT Convergence (ICTC), 2013 International Conference on. IEEE, 78–79.

[12] Zhu, K.Y., Wang, H.Y. & Sun, W.Z. 2014. A Distributed Fountain Code for Cooperative Communications. Chinese Journal of Electronics, 42(7): 1249–1255.

[13] Mu, J.J., Jiao, X.P. & Cao, X.Z. 2009. A Survey of Digital Fountain Codes and Its Application. Chinese Journal of Electronics, 37(7): 1571–1577.

Architectural, Energy and Information Engineering – Sung & Chen (Eds)
© *2016 Taylor & Francis Group, London, ISBN 978-1-138-02791-6*

Improved dynamic software watermarking algorithm based on R-tree

L. He & J.F. Xu

School of Information Engineering, Zhengzhou, Henan, China

ABSTRACT: In order to solve the low data embedding rate and the poor tamper-proof capability of the R-tree software watermarking algorithm, we propose a sharing software watermarking method based on the m-n variable carrying rule to pretreat software watermarking. While embedding watermarking, the algorithm combined the original sequence of entries in the R-tree nodes with the original watermarking to form a new embedded watermarking. Theoretical analysis and experimental results show that the scheme can effectively increase the data embedding rate of watermarking, and improve the robustness and the tamper-proof capability.

Keywords: dynamic software watermark; R-tree; m-n variable carrying rule; tamper-proof

1   INTRODUCTION

Software watermarking is an important branch of digital watermarking technology[1,2], which effectively protects the interests of consumers and software developers by embedding some information that identifies the author, publisher, owner, and user information of software.

Dynamic software watermarking technology based on data structures uses the topological properties or node redundant information of specific data structures (e.g. lists, graphs, trees) to hide the watermark[3–5]. The literature[6] has proposed a dynamic software watermarking algorithm based on the m-n variable carrying rule to pretreat watermarking, combined with a perfect hash function, and used permutation graph to encode. This method can effectively control watermarking-sharing granularity and its data embedding rate is higher, but its ability to resist the attack of Split Classes is weak[6]. In the literature[7], Ibrahim Kamel and Qutaiba Albluwi proposed the R-tree dynamic software watermarking technology, in which the watermark was embedded on the redundancy in the order of entries inside the R-tree nodes and encoded by the Variable-Base Factorial Number System (VF-NS). The method relies on the R-tree data structure that was built on the program and has strong robustness, but its data embedding rate is lower and tamper-proof capability is weak[7].

Based on the above analysis, this paper proposes a new dynamic software watermarking scheme based on the R-tree structure. The scheme is based on the m-n variable carrying rule to pretreat software watermarking, and then uses the Variable-Base Factorial Number System to encode watermarking, which can improve the data embedding rate. At the same time, the program will combine the original sequence of entries in the R-tree nodes with the original watermarking to form a new embedded watermarking to improve the tamper-proof capability.

2   THEORETICAL OVERVIEW

2.1   *Watermarking-sharing based on m-n variable carrying rule*

Software watermarking-sharing[8] is a pretreatment process. This paper uses the m-n variable carrying rule[8] to decompose the watermark into a plurality of sub-watermarks.

Theorem 1: Let P_n^m be the permutation of m elements from the series $\{1, 2, \ldots, n\}$, then *any integer* $0 - P_n^m - 1$ can be represented in the form

$$\sum_{i-1}^{m} a_i P_{n-m+i-1}^{i-1}, 0 \leq a_i \leq n-m+i-1, 1 \leq i \leq m$$

The sequence $(a_m, a_{m-1}, \ldots, a_1)$ is called the m-n variable carrying number, which is a bit i that meets every $0 \leq a_i \leq n - m + i - 1$ and $n - m + i$ into one of the number systems.

When we convert W to the m-n variable carrying value $(a_m, a_{m-1}, \ldots, a_1)$, the $(a_m, a_{m-1}, \ldots, a_1)$ will be saved as the sharing watermark $(W_m, W_{m-1}, \ldots, W_1)$.

For instance, let m = 3 and n = 6, then $P_6^3 = 120$. If the watermark W = 106, then $W \in [0,\ P_6^3 - 1]$.

Then, $(106)_{10} = 2* P_3^0 + 1* P_4^1 + 5 * P_5^2$.

Finally, W = 106 will be saved as the sharking watermark $(W_3, W_2, W_1) = (5,1,2)$.

2.2 Zero coding

Let $(W_m, W_{m-1}, ..., W_i, ..., W_1)$ be the embedded watermark, the sequence $(E_1, E_2, ..., E_q)$ is the permutation of entries inside the node, and then save $W_i (m \geq i \geq 1)$ to the ordered entries as follows:

Let $i \% q = j$, then E_j is coded as E_0, and other location subscript sequentially encoded as the natural number sequence $\{1, 2, ..., n-1\}$.

This encoding form is called the "zero coding."

2.3 Variable-Base Factorial Number System (VF-NS)

Theorem 2[7]: Let $n \in 0 \sim m! - 1$, then any integer n can be represented in the form:

$$\sum_{k=1}^{m-1}(a_k * k!)\, 0 \leq a_k \leq k, k = \{1, 2, ..., m-1\} \tag{1}$$

If the natural number $n = \sum_{k=1}^{l}(a_k * k!)$, $0 \leq a_k \leq k, k = \{1, 2, ..., 1\}$, $a_l \neq 0$, then $(a_l, a_{l-1}, ..., a_1)$ is called a variable-base factorial number, which is VF-NS encoding.

For instance, the integer 769 in decimal (base-10) can be represented as:

$$(769)_{10} = 1 * 1! + 0 * 2! + 0 * 3! + 2 * 4! + 0 * 5! + 1 * 6!,$$

which is equivalent to 102001 in the factorial system. Namely,

$$(769)_{10} = (102001)_{VF-NS}.$$

In order to realize the uniqueness of the watermark, we create a one-to-one mapping between all permutations of R-tree node entries and all possible watermark values W. The VF-NS provides us a suitable solution for this one-to-one mapping.

2.4 R-tree related features

R-trees are extension of the B+-tree for multidimensional objects[9]. R-tree uses the object boundary defining technology, a geometric object is represented by its Minimum Bounding Rectangle (MBR), and the nodes are divided into two types:

1. Non-leaf nodes contain entries of the form (ptr, R), where ptr is a pointer to a child node in the R-tree and R is the MBR that covers all rectangles in the child nodes. As depicted in Figure 1, A, B and C are the non-leaf nodes, where A contains MBR of its child nodes D, E, F.
2. Leaf nodes contain entries of the form (obj-id, R), where obj-id is a pointer to the object description and R is the MBR of the object. As

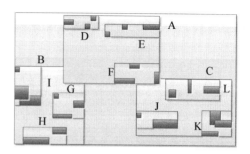

Figure 1.   Example of R-tree MBR.

shown in Figure 1, D, E, F, G, H, I, J, K and L are the leaf nodes.

Because the structure of the leaf nodes is similar to that of the non-leaf nodes, the proposed watermarking scheme does not differentiate between the leaf and non-leaf nodes.

Compared with the B-tree, the biggest advantage of R-tree is not put under any conditions on the order of entries inside the node. The R-tree software watermarking algorithm takes advantage of this feature, and hides the watermark by changing the order of entries in the tree.

3 IMPROVED R-TREE SOFTWARE WATERMARKING ALGORITHM

3.1 Overview of the algorithm

In the literature[7], the R-tree software watermarking algorithm has strong robustness, but the data embedding rate is low and tamper-proof capability is weak.

To overcome this problem, we plan to improve the original algorithm by providing the following R-tree software watermarking algorithm.

First, we use the m-n variable carrying rule to divide W into $W' = (W_m, W_{m-1}, ..., W_1)$, then we combine W' with the sequence E_R of the original entries to form a new sharing watermark $W_R = (W'_m, W'_{m-1}, ..., W'_1)$, then we use VF-NS to encode W_R to the form $W_f = (W'_m, W'_{m-1}, ..., W'_1)$, then embed the watermark Wf to the nodes in turn. Figure 2 shows the whole framework of the algorithm.

3.2 R-tree watermark embedding

The R-tree watermark embedding algorithm is divided into the following six steps:

1. Node number: R-tree grows bottom-up[9], and we use this structure to hide the watermark in one node, multiple nodes or all the nodes. We

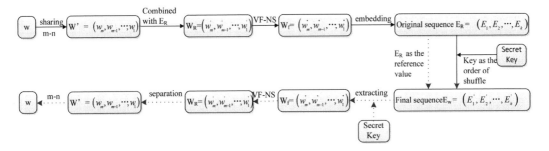

Figure 2.   Improved R-tree software watermark embedding and extraction process.

use the method of breadth-first traversal to traverse R-tree[10], while every node number is $N_i(i = 1, 2, ..., N)$.

2. Sharing watermark: we convert W to m-n Variable Carrying number $W' = (W_m, W_{m-1}, ..., W_1)$ and the software owner needs to record the number of sub-watermarks m.

3. Watermark embedding: according step (1), we embed W_m in N_N, embed W_{m-1} in N_{N-1} ... ...; if the last layer node is used up, we can embed watermarks $W_i(m \leq i \leq 1)$ in the former layer nodes, until the watermarks are embedded in full. Meanwhile, we should do the following steps from (4) to (6).

4. Combining the entry sequence with sharing watermarks: according to "zero coding" and step (2), we combine the sharing watermarks $W_i(1 \leq i \leq m)$ with the sequence of the first k entries $E_R(k \leq R \leq K)$ (k is the minimum value of the entries inside the node and K is the maximum value of the entries, and $k = K / 2$) inside the R-tree nodes $N_i = (i = 1, 2, ..., N)$, so $E_{i\%k}$ is encoded as E_0, other entry subscripts R is encoded as the natural number. So, the entry number is $(E_1, E_2, ..., E_0, ..., E_{k-1})$. Then, we add W_i $(1 \leq i \leq m)$ and the sum $k(k-1)/2$ of entry subscripts together, and save the summation as new watermarks $W_i'(1 \leq i \leq m)$.

5. VF-NS Coding: we use VF-NS to encode $W_i(1 \leq i \leq m)$ to a new watermark W_i' $(1 \leq i \leq m)$.

6. Shuffle operation: according to a secret instruction that is arranged in advance with only the software owners knowing this directive. The secret instruction equals to "Secret Key", the directive here as "Rotate Left". We then shuffle the entries $(E_1, E_2, ..., E_0, ..., E_{k-1})$ in the following steps, where each value of $W_i'(1 \leq i \leq m) = $ corresponds to the number of every cycle times.

Let us assume a watermark value $W_i(1 \leq i \leq m) = $ "311" and $E_R = (E_1, E_0, E_2, E_3)$, as shown in Figure 3. The first digit in W_i' is "3" and then the entries (E_1, E_0, E_2, E_3) will be shifted three positions, resulting in the new sequence (E_3, E_1, E_0, E_2).

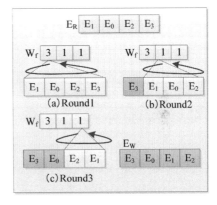

Figure 3.   Watermark embedding process.

The first entry (E_3) will be fixed and the entries (E_1, E_0, E_2) will be shifted only once because the second digit is "1", resulting in a new intermediate order (E_3, E_0, E_2, E_1). The remaining two entries (E_2) and (E_1) will be shifted once since the third digit is "1".

The final (watermarked) order of the entries would be (E_3, E_0, E_1, E_2). We denote the result node as $E_W = (E_3, E_0, E_1, E_2)$.

3.3   Watermark extraction process

The watermark extraction algorithm can be divided into the following steps:

1. Extracting the watermark $W_f = (W_m', W_{m-1}', W_1')$: according to the "Secret Key", we compare E_R with E_W, to match each entry in the nodes. The number of circular shift operations that are needed to align an element from E_R with the corresponding element from E_W is used to construct the watermark $W_f = (W_m', W_{m-1}', W_1')$.

As shown in Figure 3, the original entry sequence $(E_R = (E_1, E_0, E_2, E_3)$, after embedding the watermarks, becomes $E_W = (E_3, E_0, E_1, E_2)$, thus we can get the embedded watermark

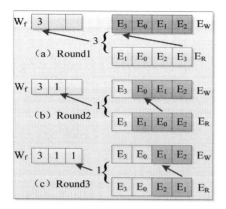

Figure 4.   The process of extracting the watermark.

$W_i'' =$ "311". Figure 4 shows the watermark extraction procedure.

2. Separation and recovery watermark: Separate the watermark $W_f = (W_m', W_{m-1}', W_1')$ one by one, each watermark value $W_i''(1 \le i \le m)$ minus $k(k-1)/2$; then convert W_f to decimal format number based on VF-NS, which results in the value $W' = (W_m, W_{m-1}, ..., W_1)$, and then according to the m-n variable carrying rule, revert W' to the original watermark W.

4   PERFORMANCE ANALYSIS

4.1   *Anti-attack capability*

Attacks against software watermarks can be divided into four main categories: subtractive attacks, distortive attacks, additive attacks and collusion attacks[11].

Because the watermark is encoded in the relative order of the entries inside the R-tree nodes, subtractive attacks cannot be invoked against the proposed watermarking technique. If the attacker eliminates the watermarks, he needs to remove half of the entries in the node. In this case, the performance of the code will be affected significantly or it might not work properly.

In distortive attacks, once the owner has found the attacks, when inserting new nodes into R-tree, the watermark embedding algorithm will be operated and creates the watermark again. So, the software owners can still prove their ownership. In collusion attacks, the final result is to eliminate the watermarks, which will also affect the normal operations of the program. Thus, collusion attacks cannot be invoked against the proposed watermarking algorithm.

4.2   *Tamper-proof capability*

We propose to combine the original sequence of entries inside R-tree nodes with the original watermark to form a new embedded watermark. Then, even without adding new nodes, when the watermark is extracted, we still need to separate the embedded watermark. Therefore, if the attacker does not know the coding sequence of the original entries, it is difficult to separate the watermark and tampering attacks will become invalid.

As analyzed in Table 1, the proposed watermarking algorithm can effectively resist the common software watermarking attacks, and can effectively improve the tamper-proof capability.

4.3   *Data embedding rate*

In the literature[7], the watermark is embedded in the first K/2 entries inside all the nodes of R-tree, and the embedded watermarks in each node are the same. According to VF-NS, suppose we have k items, the permutation of k items = $P_k^k = k!$, excluding the initial state k!-1. For embedding the watermark in the k positions, the maximum value of the watermark can represented as (k + 1)!-1.

In this paper, the proposed scheme stores the watermark in multiple nodes or all the nodes, and each watermark value in each node is not the same. Compared with the original algorithm, the improved algorithm only needs to select the appropriate m and n, when the embedded watermark in k entries, the watermark value will be far greater than (k+1)!-1.

4.4   *Overload performance analysis*

There are two major types of the impact on its performance, space overload and time overload.

In the physical space, R-tree software watermarking algorithm does not occupy additional disk space, but hide the watermark by changing the order of the original entries inside the node.

Table 1.   Comparison of robustness and tamper-proof capability.

Algorithm	Additive attacks	Subtractive and collusion attacks	Distortive attacks	Tamper-proof capability
Original algorithm	Weak	Strong	Strong	Weak
Improved algorithm	Strong	Strong	Strong	Strong

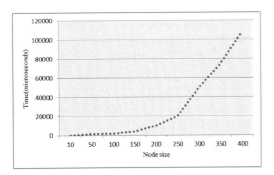

Figure 5. Cost of embedding watermark.

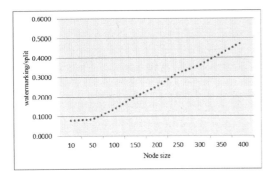

Figure 6. The ratio between the embedding time and a node split time.

Therefore, embedding watermark cannot add additional space and not affect the size of the R-tree.

In the time performance, as shown in Figure 5, the watermark embedding time is increasing quadratic with the number of entries in the node. For example, for the R-tree with node size 200, the watermark insertion cost is close to 20 ms, which is equivalent to one disk access. However, this cost is incurred at the split operation only, thus the overhead is acceptable.

In Figure 6, the y-axis shows the ratio between the embedding time and a node split time; the x-axis shows the number of entries inside the R-tree node. It should be noted that the ratio of the watermark insertion to the split cost is below 0.25 for node sizes below 200, while for node sizes 200–400, this ratio is between 0.25 and 0.5. In practice, most R-trees have a node size below 400 and the split operation is the least frequent operation, and thus an increase in its time with a reasonable fraction (0.2–0.5) is acceptable.

5 CONCLUSION

Based on the analysis and research of the R-tree software watermarking, we use the m-n variable carrying rule to share the watermark with sub-watermarks, and combine the internal sequence of entry inside R-tree nodes with the original water-mark. After extracting the watermark, we need to separate the embedded watermark to obtain the original watermark. Compared with the original algorithm, the improved algorithm does not increase the complexity of the space and time, and effectively improves the data embedding rate and the tamper-proof capability. In the future, we will study to combine the software features with the watermark, to improve the elusiveness and availability of the software watermarking algorithm.

REFERENCES

[1] Zhu, W.F. 2007. *Concepts and techniques in software watermarking and obfuscation.* Auckland: The University of Auckland New Zealand.
[2] Liu, M. 2011. *The Research of Dynamic Graph Software Watermarking Algorithm Based on Tamper-proof Technology.* Guangzhou: Jiangxi University of Science and Technology.
[3] Zhang, L.H. & Yi, X. 2003. A Survey on Software Watermarking. *Journal of Software* 14(2): 268–277.
[4] Wang, Y.S. & Xu, J.F. 2012. Improved PPCT Hybrid Coding Scheme. *Computer Engineering and Applications* 48(34): 107–111.
[5] Collberg, C. & Thomborson, C. 1999. Software watermarking: Models and dynamic embeddings. *Proceedings of the 26th ACM SIGPLAN-SIGACT symposium on Principles of programming languages*: 311–324.
[6] Li, S.Z. & Wang, X.M. 2012. Dynamic Graph Software Watermarking Algorithm Based on m-n Variable Carrying Rule. *Computer Engineering* 38(21): 17–21.
[7] Kamel, I. & Albluwi, Q. 2009. A robust software watermarking for copyright protection. *Computers & Security* 28(6): 395–409.
[8] Wang, X.M. 2013: *The Research of Software Watermarking Sharing and Encoding Algorithm Based on The Tamper-proof Technology.* Ganzhou: Jiangxi University of Science and Technology.
[9] Guttman, A. 1984: R-trees: a dynamic index structure for spatial searching. In New York, *ACM:* 47–57.
[10] Zhang, M.B. & Lu, F. 2005: The Evolvement and Progress of R-tree Family. *Chinese Journal of Computers* 28(03): 289–300.
[11] Li, L. 2008: A Suvery on Attacks on Software Watermarking. In Qingdao, *The 15th at the Chinese academy of electronic information theory academic annual meeting and proceedings of the first national network coding academic conference(I):* 525–530.

Architectural, Energy and Information Engineering – Sung & Chen (Eds)
© *2016 Taylor & Francis Group, London, ISBN 978-1-138-02791-6*

Chloride binding capability of fly ash concrete exposure to the marine environment

Bo Da, Hong Fa Yu & Hai Yan Ma
College of Aerospace Engineering, Nanjing University of Aeronautic and Astronautic, Nanjing, China

ABSTRACT: The total chloride (C_t) and free chloride (C_f) were determined by calculating the natural diffusion and the Chloride Binding Capability (CBC) of Ordinary Portland Cement Concrete (OPC) and Fly Ash Concrete (FAC). In addition, the influences of exposure time, curing age and content of Fly Ash (FA) on CBC were discussed in this paper. It shows that whether it is OPC or FAC, the CBC of concrete was independent of the exposure time in the marine environment. However, with the increasing curing age, CBC tended to stabilize gradually. The CBC of FAC increased first, and then it reduced with the increase in the FA content. When the FA content was up to 30%, the CBC reached to maximum. Therefore, in terms of concrete structures in the actual marine environment, in order to ensure the structural strength, it is better to prolong the wet curing age to 180 d and make the FA content not to be lower than 30%, which will enhance the service life of these structures.

Keywords: marine environment; fly ash concrete; exposure time; curing age; chloride binding capability

1 INTRODUCTION

In the marine environment, due to the erosion of sea water, corrosion of the steel bar and the concrete occurs, which leads to the early damage of structures and loss of durability, and has become an important problem in engineering practice. The marine environment is typical of the chloride environment. Therefore, there is a need to focus on the study of the diffusion process with chloride ions into the concrete in the marine environment, to solve the structure damage problem caused by the galvanic corrosion of reinforced bar in the chloride environment. Therefore, the study on the durability of concrete structures in the marine environment is very urgent and necessary[1–3], with the Chloride Binding Capability (CBC) of concrete being one of the important parameters of the diffusion of chloride ions in the equation.

The method of chloride ion diffusion has achieved many breakthroughs and innovations after many years of study. The diffusion theory and life assessment model of chloride ions has been further developed, combined with a preliminary study on the factors influencing the chloride binding capability of concrete, which has also made some progress and began to guide the engineering practice[4–5]. Collepardi[6–7] proposed to use the Fick law to describe the chloride ions in concrete apparent diffusion behaviors, but Verbeck[8] pointed out that the diffusion experiment of concrete, regardless of

CBC, and the direct application of the Fick law will lead to erroneous results. Hooton[9] pointed out that when the concrete is exposed to the chloride environment, chloride ions enter into the transfer mechanism within the concrete, which are of at least 6 kinds: adsorption, diffusion, binding, penetration, capillary effect and dispersion. At present, domestic and foreign scholars have studied the chloride ion adsorption isotherm and obtained the CBC of concrete by using many methods, such as the balance method[10], the natural diffusion method[11] and the migration method[12].

In this paper, the natural diffusion method[13] is used to determine the total chloride (C_t) and free chloride (C_f) of different curing agse of Ordinary Portland Cement Concrete (OPC), and to calculate the CBC of OPC. Based on this study, the influences of Fly Ash (FA) content on the CBC of Fly Ash Concrete (FAC) are discussed.

2 EXPERIMENTAL PROCEDURE

2.1 Raw materials

The cement uses P·II 42.5 Portland Cement, whose chemical compositions are given in Table 1 and physical properties are given in Table 2. The FA is from Zhenjiang, whose chemical compositions are also given in Table 1. The fine aggregate is the river

Table 1. Chemical compositions of main cementitious materials %.

Material	SiO$_2$	Al$_2$O$_3$	MgO	CaO	Na$_2$O	Fe$_2$O$_3$	SO$_3$	MnO$_2$	TiO$_2$	I.L	K$_2$O
P•II	21.53	4.60	0.96	64.09	0.12	3.37	2.09	—	—	1.84	0.62
FA	52.37	32.13	2.16	0.47	0.33	4.13	0.25	—	—	0.61	1.30

Table 2. Basic physical properties of portland cement.

Cement	a/m^2•kg^{-1}	b/%	c/%	Setting time/h: min		Compressive strength/MPa		Flexural strength/MPa	
				Initial	Final	3d	28d	3d	28d
P•II	341	0.3	26.0	1:43	2:39	35.4	65.2	6.4	9.2

a: Specific surface area; b: Residue on 80 μm sieve; c: Water requirement for normal consistence.

sand, whose apparent density is 2605 kg/m^3, mud content is 1.0%, and fineness modulus is 2.74. The coarse aggregate is basalt, whose maximum grain is 12 mm, with a continuous grade of 5~16 mm. The high range water reducer is naphthalene type super plasticizer powder, whose water-reducing percent is above 20%, in which the Cl⁻ is below 0.01% and Na$_2$SO$_4$ content is below 2%.

2.2 Mixture proportions

In order to meet the needs of the research, this paper designs 5 groups of concretes. The mixture proportion, slumps and the compressive strength (100 mm × 100 mm × 100 mm) of the concretes in standard curing 28d are listed in Table 3. Among these, the basic of concrete (F0) mixture proportion is that cement is 500 kg/ m^3 and w/b is 0.34. A series of FAC (F10–F40) are designed, which are incorporated with FA, whose proportions are 10%, 20%, 30% and 40%, respectively.

2.3 Experimental methods

2.3.1 Sample preparation
Cement, stone, sand, additive and fly ash were first mixed for 1 minute in a blender, and then water was added and stirring was continued for another 3 minutes. The slumps were measured before casting, and concrete prisms with dimensions of 40 mm × 40 mm × 160 mm were cast and cured in a sealed condition for 1 day. They were demolded and moved into a saturated limewater with (20 ± 3)°C. The curing age were 7 d, 28 d, 56 d, 180 d, 365 d, 730 d and 1095 d. The exposure times were 7 d, 28 d, 90 d, 180 d and 730 d.

According to the ASTM D1141–2003, artificial Seawater was prepared using the chemical reagents NaCl, MgCl · 6H$_2$O, Na$_2$SO$_4$, CaCl$_2$ and KCl. The chemical compositions of the artificial seawater are given in Table 4.

2.3.2 Sampling and analysis
After collecting the sample from the Artificial Seawater, the powder was collected by using the drilling method in the middle of the 1/3 of two sides. The drill with a 6 mm diameter and distance between two holes was 10~20 mm. At least 16~24 holes were drilled for each sample to obtain 5 g powder for the analysis. Powder collected was sieved using a 0.15 mm sieve to remove the coarse aggregate. According to the Chinese Standard JTJ270–98, this paper uses acid soluble chloride measures C_t and water soluble chloride measures C_f. So, the bound chlorides (Cb) can be obtained by the following equation:

$$C_b = C_t - C_f \tag{1}$$

2.4 Data processing

According to the measured C_t and C_f values at depths of 0~5 mm, 5~10 mm, 10~15 mm and 15~20 mm, the relationship between C_t and C_f can be fitted as follows:

$$C_t = KC_f \tag{2}$$

where K is the coefficient, in accordance with the relationship between the C_t and C_f of OPC exposure to Japan seawater for 10~50 years[14].

Based on the study of Nilsson et al.[15], the CBC of concrete R is given by

$$R = \frac{\partial C_b}{\partial C_f} \tag{3}$$

So, the chloride CBC of concrete is given by

$$R = \frac{\partial C_b}{\partial C_f} = K - 1 \tag{4}$$

Table 3. Mixture proportions and slump of concrete.

| No. | W/B | Material content/kg·m⁻³ | | | | | Slum p/mm | f_{cu}/MPa |
		Cement	FA	Sand	Stone	Water		
F0	0.34	500	–	657	1168	168.5	195	71.2
F10	0.34	450	50	657	1168	168.5	190	91.4
F20	0.34	400	100	657	1168	168.5	195	72.8
F30	0.34	350	150	657	1168	168.5	155	76.3
F40	0.34	300	200	657	1168	168.5	208	83.1

Table 4. Chemical compositions of the artificial seawater (kg/m³).

NaCl	Na₂SO₄	MgCl·6H₂O	KCl	CaCl₂
24.5	4.1	11.1	0.7	1.2

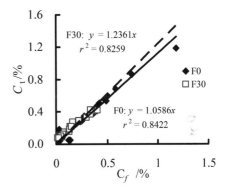

Figure 1. Relationship between C_t and C_f at the curing age 28 d of concrete.

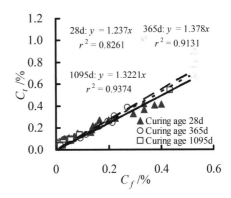

Curing age 28d, 365d and 1095d

Figure 2. Relationship between C_t and C_f at different curing ages of FAC (F30).

3 RESULTS AND DISCUSSION

3.1 *Effect of the exposure time*

Figure 1 shows the relationship between C_t and C_f at the curing age 28 d of concrete (F0, F30). Figure 2 shows the relationship between C_t and C_f at different curing ages of FAC (F30). The results show that the data separately fit a straight line for each type of exposure time of F0~F40. This means that the CBC of concrete does not have a relationship with the exposure time. Therefore, in the later CBC of OPC and FAC do not need to consider the effect of the exposure time.

Curing age 28 d, 365 d and 1095 d.

3.2 *Effect of the curing age*

3.2.1 *Effect of the OPC curing age*

Figure 3 shows the CBC at the curing age 7~1095 d of OPC (F0). As shown in Figure 3, obviously, as the curing age is prolonged, the CBC of OPC increases significantly. At the curing age of 730 d, the CBC of OPC is maximum; at the curing age of 7 d, it increased by 5.53 times; at the curing age of 28 d, it increased by 4.51 times. When the curing age extends further to 1095 d, the CBC decreases slightly, but does not change significantly, and finally it tended to stabilize gradually.

3.2.2 *Effect of the FAC curing age*

Figure 3 shows the CBC at the curing age 7~1095 d of different FAC contents (F0~F40). As shown in Figure 3, obviously, as the curing age is prolonged, the CBC of FAC increases significantly. At the curing age of 1095 d, the CBC of F10 is maxi-

mum; at the curing age of 7 d, it increases by 2.70 times; at the curing age of 28 d, it increases by 1.54 times; at the curing age of 180 d, the CBC of F20 is maximum; at the curing age of 7 d, itincreases by 2.62 times; at the curing age of 28 d, it increases by 1.41 times; at the curing age of 730 d, the CBC of F30 is maximum; at the curing age of 7d, it increases by 2.16 times; at the curing age of 28 d, it increases by 1.63 times; at the curing age of 365 d, the CBC of F40 is maximum; at the curing age of 7 d, it increases by 2.01 times; and at the curing age of 28 d, it increases by 1.74 times. When the curing time extends further to 1095 d, the CBC decreases slightly, but does not change significantly, and finally it tends to stabilize gradually.

This article uses the logarithmic function to fit the relationship between the CBC and the curing age, as shown in Figure 3. The results obtained are given in Formula (5):

$$R = a\ln(T) + b \tag{5}$$

where R is the CBC of concrete; T is the curing age; and a, b are the fitting parameters. Under the condition of different fittings, the fitting parameters are different. The relevant parameters are given in Table 5.

3.3 Effect of the FA content

From Figure 4, we can find that the different contents of FAC (F0, F30), C_t, at the standard curing

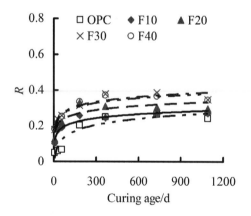

Figure 3. Chloride binding capability at different curing ages of concrete (OPC, F10~F40).

Table 5. Relationship between CBC and the curing age.

| Sample | Relevant parameters | | |
	a	b	r
F0	0.0505	−0.0831	0.933
F10	0.0327	0.0563	0.961
F20	0.0343	0.0922	0.881
F30	0.0404	0.1032	0.955
F40	0.0395	0.1028	0.948

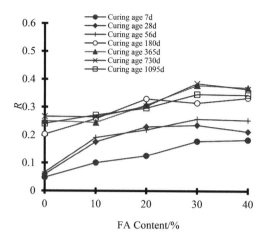

Figure 4. Chloride binding capability at different curing ages of FAC (F10–F40).

age of 28d showed a good linear relationship with C_f in the seawater environment. Using Excel to fit the different contents of C_t and C_f, we found that the slope of the line is different and the fly ash content of 30% concrete slope is maximum.

As shown in Figure 4, the different contents FAC (F0~F40) exhibit the effect of the CBC in the marine environment. In addition, obviously the CBC of FAC increases first, and then reduces with the increase in the FA content. When the FA content was up to 30%, the CBC reached to maximum. Therefore, in terms of concrete structures under the actual marine environment, in order to ensure the structural strength, it is better to prolong the wet curing age to 180 d and make the FA content not to be lower than 30%, which will enhance the service life of these structures.

4 CONCLUSIONS

1. In the marine environment, whether it is OPC or FAC, the Chloride Binding Capability (CBC) of concrete was independent of the exposure time. However, with the increasing curing age, the CBC tended to stabilize gradually.
2. In the marine environment, for the different curing ages of FAC, the CBC of FAC increased first, and then reduced with the increase in the FA content. When the FA content was up to 30%, the CBC reached to maximum.
3. Therefore, in terms of concrete structures under the actual marine environment, in order to ensure the structural strength, it is better to prolong the wet curing age to 180 d and make the FA content not to be lower than 30%, which will enhance the service life of these structures.

ACKNOWLEDGMENTS

The authors gratefully acknowledge the financial supports from the National Key Basic Research Development Plan of China (973 Plan) under Grant No. 2009CB623203 and the National Natural Science Foundation of China under Grant No. 21276264.

REFERENCES

[1] Chalee W, Jaturapitakkul C. Effect of W/B ratios and fly ash fineness on chloride diffusion coefficient of concrete in marine environment. Materials and Structures, 2009, 42: 502–14.

[2] Cheewaket T, Jaturapitakkul C, Chalee W. Long term performance of chloride binding capacity in fly ash concrete in a marine environment. Construction and Building Materials, 2010, 24: 1352–1357.

[3] Fajardo G, Valdez P, Pacheco J. Corrosion of steel rebar embedded in natural pozzolan based mortars exposed to chlorides. Construction and Building Materials, 2009, 23: 768–74.

[4] Ann Ki Yong, Song Ha-Won. Chloride threshold level for corrosion of steel in concrete. Corrosion Science, 2007, 49: 4123–4133.

[5] Jensen OM, Korzen MSH, Jakobsen HJ, Skibsted J. Influence of cement constitution and temperature on chloride binding in cement paste. Advances Cement Research, 2000, 12: 57–64.

[6] Collepardi M, Marcialis A, Turrizzani R. The kinetics of penetration of chloride ions into the concrete [J]. Cement, 1970, 4: 157–164.

[7] Collepardj M, Marcialis A, Turrizzani R. Penetration of chloride ions into cement pastes and concretes [J]. American Ceramic Society, 1972, 55: 534–535.

[8] Verbeck G. Mechanisms of corrosion of steel in concrete. Corrosion of metals in concrete [R]. Construction and Building Materials, 1987, 49: 211–219.

[9] Hong K, Hooton R. Effects of cyclic chloride exposure on penetration of concrete cover [J]. Cement and Concrete Research, 1999, 29(9): 1379–1386.

[10] Tritthart J. Chloride binding in cement II. The influence of the hydroxide concentration in the pore solution of hardened cement paste on chloride binding [J]. Cement and Concrete Research, 1989, 19(5): 683–91.

[11] Glass GK, Wang Y, Buenfeld NR. An investigation of experimental methods used to determine free and total chloride contents [J]. Cement and Concrete Research. 1996, 26(9): 1443–1449.

[12] Whiting, D. Rapid Measurement of the Chloride Permeability of Concrete [J]. Public Roads, 1996, 45(12): 1831–1842.

[13] LENG Faguang. Study on concrete penetration and diffusion properties of Chloride ion [J]. Shanxi Construction Magazine (in Chinese), 1999, 15(4): 9–10.

[14] Mohammed T.U, Hamada H. Relationship between free chloride and total chloride contents in concrete [J]. Cement and Concrete Research, 2003, 33(9): 1487–1490.

[15] Nilsson L O, Massat M, Tang L. The effect of nonlinear chloride binding on the prediction of chloride penetration into concrete structure [A]//Malhotra Y M ed. Durability of Concrete [C]. Detroit American Concrete Institute, 1994. 469–486.

Architectural, Energy and Information Engineering – Sung & Chen (Eds)
© 2016 Taylor & Francis Group, London, ISBN 978-1-138-02791-6

Investigation on enrichment behaviour of trace elements in coal-fired fine particulate matters

D.B. Li & Q.S. Xu
Electric Power Research Institute of Guangdong Power Grid Co. Ltd., Guangzhou, P.R. China

X.D. Ren, X. Cao & J.Y. Lv
School of Environmental Science and Engineering, North China Electric Power University, Baoding, Hebei, P.R. China

ABSTRACT: Certain constituents in coal, even with very low contents, could be environmentally toxic during the combustion of coal, such as trace elements. Special attention has been paid to these elements due to their abundant amount, various emission forms and potential hazards in the coal combustion process. This paper provides an investigation on several aspects relating to trace elements and their enrichment behaviour in coal-fired fine particulate matters, such as their modes of occurrence in coal, their partitioning and enrichment in fine particles during coal combustion, their enrichment mechanism in different emission streams, and a series of emission control strategies on trace elements. Also a schematic diagram of the enrichment mechanism is drawn, based on the investigation and our previous work.

Keywords: trace elements; fine PM; enrichment; coal combustion

1 INTRODUCTION

Besides SO_X, NO_X and CO_2, the release of fine Particulate Matter (PM) from daily combustion processes has also been a significant toxicological and environmental issue over the past decades. Abundant studies have identified the relationship between acute respiratory or cardiovascular distress and the exposure to ambient fine PM (Davidson et al. 2005, Schwartz et al. 1996), since these fine particles are capable of being inhaled into human respiratory system more deeply, compared with the coarse ones (Pope et al. 2006). Not only having potential negative impact on human health, fine PM can also cause a series of environmental problems, such as deterioration of the atmospheric visibility. According to the results of numerous studies, combustion of fossil fuels, especially pulverized coal in power plants, is believed to be a major anthropogenic source of PM in the urban atmosphere of the globe (Lighty et al. 2000).

Although sulfates, nitrates and major elements are the main constituents of combustion-generated particles, another kind of hazardous substance, trace elements, for example, Hg, As, Se, Cr, Co, Cd, Ni, Pb, Zn, are found to be enriched abundantly in fine PM after combustion. With high volatility, many of these elements will undergo the process of vaporization at high temperatures in the combustion zone. Further, due to the high surface-to-volume ratios of fine particles, especially PM2.5, gaseous elements tend to deposit on their surface rather than the coarse ones at low temperatures downstream, resulting in the enrichment of trace elements on these fine PM (Quann et al. 1982).

Because of its relatively high reserves as energy source, coal will still contribute a large part to fuel combustion process in power generation industry in the future. However, fine PM with submicron diameters (<1 μm) cannot be captured efficiently by traditional Particle Emission Control Devices (PECDs) in power plants such as ESPs or fabric filters as coarse particles (McElroy et al. 1982), meaning that a large amount of PM enriched of trace elements will still be emitted into the ambient air annually, if no more effective emission control measures be developed and taken.

2 STUDY METHODS

Increasing concern about the environmental impact of certain trace elements has led to the implementation of some related emission standards all over the world since the promulgation of the U.S. Clean Air Act (CAA), in which they are listed as potential Hazardous Air Pollutants (HAPs). For taking targeted and efficient control strategies to trace elements from coal combustion process, a fundamental understanding of their enrichment behaviour in fine PM is an essential step. This paper starts from the investigation of modes of occurrence of some trace elements in original coal, discusses their partitioning and enrichment during coal combustion process, concludes their enrichment mechanism in coal-fired fine particulate matter, and finally gives some information about currently available emission control strategies on trace elements in coal-fired power plants.

3 RESULTS AND DISCUSSION

3.1 Modes of occurrence of trace elements in coal

One important influencing factor, which greatly determines the transformation and partitioning of trace elements in combustion process and even their ultimate enrichment behaviour in the vapor and particulate matter, is the modes of occurrence of these elements in the original coal. Generally they exist in these three modes: associated with organic matter, included forms in which the elements are surrounded by mineral, and excluded forms in which they are in discrete state (James et al. 2014). Even if under the same environment of high temperatures, these various modes will result in differences in the vaporization degree of trace elements, since their difficulty levels of release from the surrounding environment are diverse.

Using specific analytical techniques like Inductively Coupled argon Plasma Atomic Emission Spectroscopy (ICP-AES), Inductively Coupled argon Plasma Mass Spectroscopy (ICP-MS), or Neutron Activation Analysis (NAA), different elements in coal and their modes of occurrence could be determined. Goodarzi (Goodarzi et al. 2006) have concluded that the combustion of coals with high content of sulfur in Canadian power plants usually results in high contents of As, Cd, Hg, Mo, Ni and Pb in fly ash, since these above elements in coal are mostly combined with sulfide minerals. Senior et al. (Senior et al. 2000) have detected subbituminous coals from the western U.S. that are widely used by U.S. utilities, and the results show a close association of trace elements with minerals. In these coals, a large portion of Cr is combined with silicates, part of As exists in the form of arsenates, and Zn is mainly in sphalerite, whereas approximately half of these elements are associated with organic matter of the coal. In a study conducted by Wagner et al. (Wagner et al. 2005) which investigate the trace elements content in South Africa coals, the Zn is believed to associate with pyrite, and these obtained values are generally lower compared with that of other areas in the world. Querol et al. (Querol et al. 1995) have done a detailed investigation on Spanish subbituminous coals, and have drawn some general conclusions

Table 1. Classification of elements according to their affinities.

Affinity		
Organic	S, C, N, B, Be, Ge, V, W, Zr	
	Mineral category	-
	Clay minerals	Al, Ba, Bi, Cr, Cs, Cu, Ga, K, Li, Mg, Na, Ni, P, Pb, Rb, Sn, Sr, Ta, Th, Ti, U, V, Y
Inorganic	Iron sulfides	As, Cd, Co, Cu, Fe, Hg, Mo, Ni, Pb, S, Sb, Se, Ti, W, Zn
	Carbonates	C, Ca, Co, Mn, W
	Sulfates	Ba, Ca, Fe, S
	Sulfides	Co, Cu, Ni, Pb, S, W
	Tourmaline	B

about the affinities of major and trace elements with coal, as listed in Table 1.

It is worth noting that various modes of occurrence of trace elements in coal are inevitable results from extensive researches all over the world, since the original plants formed into coals, the coalification process and the geological environment of coal fields at different countries or regions are not quite the same.

3.2 Partitioning and enrichment of trace elements during coal combustion

When feeding pulverized coal into a furnace, trace elements in coal particles, no matter they are associated with organic matters, in included forms or in excluded forms, will all experience a high temperature environment in the combustion system. After passing through the combustion zone, there are several emission streams in which these trace elements could partition, including bottom ash, fly ash and flue gas.

Many studies, including our own works, have investigated the partitioning and enrichment behaviour of various trace elements after combustion, and the results suggest that most of them are combined with the fly ash and later are captured by conventional Particle Emission Control Devices (PECDs) like ESPs or fabric filters. But for the part of trace elements in vapor phase, it is much harder to capture them by conventional PECDs. For the purpose of efficiently reducing the emission amount of trace elements and mitigating their adverse impact on human health and the environment, having a quite clear understanding of their partitioning and enrichment in the particulate and vapor phase is of great importance.

As mentioned earlier, fine particles have a larger tendency than the coarse ones to enrich trace elements due to their high surface-to-volume ratios. Since these fine particles are in the size range of low collection efficiency of conventional PECDs in coal-fired power plants, and can remain suspended in the ambient air for longer periods of time after they are emitted, trace elements enriched in fine particles have more negative impacts than the coarse ones.

For certain elements with high vapor pressure like Hg, As, Se, they will volatilize completely under a high temperature environment in the furnace. According to numerous experimental and modeling studies, As and Se are partially emitted into the ambient air in vapor phase, but for Hg, whose vapor pressure is extremely high, almost fully be emitted as vapor in flue gas.

In order to clearly describe the different volatility, partitioning and enrichment behaviour of trace elements, several classification methods of them have been put forward during the past decades. According to their enrichment behaviour based on volatility (e.g. boiling point), Clarke (Clarke 1993) has classified trace elements into three basic groups, as shown in Figure 1.

Class I: Elements which are approximately equally distributed between the bottom ash and fly ash, or show no significant enrichment or depletion in the fly ash, such as Mn, Rb, Sm, and Zr.

Class II: Elements which are enriched in the fly ash and depleted in the bottom ash, or show increasing enrichment with decreasing fly ash particle size, such as Cd, Ge, Pb, Sb, Sn, Ti and Zn.

Class III: Elements which are volatized, but are not enriched on the fly ash. They are emitted fully in the vapor phase, such as Hg, Br and Cl.

But in many cases, the boundary of the classification of these trace elements are not so evident, since some elements often show enrichment behaviour both in the vapor phase and fly ash, such as B, Se and I, or both in the fly ash and bottom ash, such as Ba, Cr and Ni.

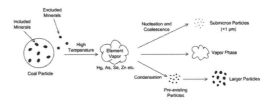

Figure 2. Enrichment mechanism of trace elements in fine particles during coal combustion.

3.3 Enrichment mechanism of trace elements in coal-fired particulate matter

Based on abundant studies about the formation of fine particles involving trace elements in coal (Linak et al. 1994) and our previous work (Lu et al. 2008), we conclude the mechanism by which the trace elements are enriched in fine particles during coal combustion process, and draw a schematic diagram of this process as shown in Figure 2.

In the combustion zone of the furnace, due to the thermal effect of high temperatures, various trace elements which are associated with organic matter or included in minerals in coal will vaporize in different degrees. Subsequently, as entering into the cooler part downstream the combustion system, some of these element vapors will homogeneously nucleate and coalescence to form submicron particles of only a few nanometers in a vapor-to-particulate growth process (Seames 2003), while some of them will heterogeneously condense or react with other chemical species on the pre-existing fine particles surface, to larger their diameters at the same time. In addition, mass accumulation of trace elements on fly ash particles could also be achieved by physical or chemical adsorption. For certain element with relatively high vapor pressure, such as Hg, it almost totally remains in the vapor phase even at the exit of the boiler.

3.4 Available emission control strategies on trace elements

After much effort has been paid for several years, pollution control technologies of SO_X, NO_X and CO_2 in coal-fired power plants have been well developed and are relatively mature. Nevertheless, almost no targeted emission control strategy of trace elements that of high efficiency has been put forward and widely adopted in practical combustion utilities.

As mentioned earlier, a large portion of trace elements emitted from the boiler exit are enriched in fly ash particles, especially in these of submicron sizes. Currently, prevalent PECDs in a coal-fired power plant mainly include ESPs, fabric filters, and wet scrubbers. For coarse fly ash particles in

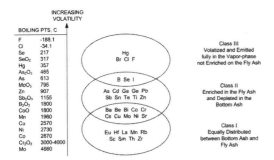

Figure 1. Categorization of trace elements based on volatility behaviour.

Table 2. Control measures of trace elements in fine particles.

Categories	Control strategies
Agglomeration approaches	Electrostatic agglomeration Magnetic aggregation Wet agglomeration
Modification of conventional PECDs	Hybrid ESP/baghouse Wet ESP

flue gas, these devices usually could achieve satisfying collecting effects. But because of their special capturing mechanisms, the ability to collect fine PM of submicron sizes of these devices are relatively low, which means most trace elements could not be removed efficiently by conventional PECDs.

With a series of physical or chemical agglomeration approaches, such as electrostatic agglomeration, magnetic aggregation and wet agglomeration, the ability of control devices to aggregate fine PM into larger particles can be improved, which is obviously favorable for the latter removal of coarse particles. In addition, the collection efficiency of trace elements enriched in the fine PM could also be maximized through modifying conventional PECDs, for example, the hybrid ESP/baghouse and wet ESP. And these above approaches have been discussed in detail in our previous work (Lu et al. 2014), as listed in Table 2.

For the part of trace elements of vapor phase in the flue gas, many studies pay attention to the gas cooling process, and accordingly several control methods have been applied into practical processes, such as spray dryer absorption systems and condensing wet scrubbers. In these apparatus, trace elements vapor could be cooled and then condense to form larger particles, which will be captured by following pollution control devices. But in recent decades, sorbents including minerals, fly ash and activated carbon are through to be good alternatives of trace elements control (Uberoi et al. 1991).

4 CONCLUSIONS

A general introduction about the relationship between trace elements in fine particulate matter and the atmospheric environment is given at the initial of this study. Then some problems of vital importance of trace elements are then investigated and discussed: their modes of occurrence in original coal, their partitioning and enrichment during coal combustion process, the mechanism governing their enrichment behaviour in particulate and vapor phase, especially in fine particles, and currently available emission control strategies on trace elements that are used in coal-fired power plants. But in a word, all the studies focusing on trace elements characteristics during coal combustion process are finally targeted to control their emission and mitigate their adverse impacts on human beings.

REFERENCES

Clarke, L.B. 1993. The fate of trace elements during coal combustion and gasification: an overview. *Fuel* 72(6): 731–736.

Davidson, C.I. & Phalen, R.F. & Solomon, P.A. 2005. Airborne particulate matter and human health: A review. *Aerosol Sci. Technol.* 39(8): 737–749.

Goodarzi, F. 2006. Characteristics and composition of fly ash from canadian coal-fired power plants. *Fuel* 85: 1418–1427.

James, D.W. & Krishnamoorthy, G. & Benson, S.A. & Seames, W.S. 2014. Modeling trace element partitioning during coal combustion. *Fuel Process. Technol.* 126: 284–297.

Lighty, J.S. & Veranth, J.M. & Sarofim, A.F. 2000. Combustion aerosols: Factors governing their size and composition and implications to human health. *J. Air Waste Manage. Assoc.* 50(9): 1565–1618.

Linak, W.P. & Wendt, J.O.L. 1994. Trace metal transformation mechanisms during coal combustion. *Fuel Process. Technol.* 39 (2): 173–198.

Lu, J. & Li, D. 2008. *J. Combust. Sci. Technol. (in Chinese)* 14: 55–60.

Lu, J. & Ren, X. 2014. Analysis and discussion on formation and control of primary particulate matter generated from coal-fired power plants. *J. Air Waste Manage. Assoc.* 64(12): 1342–1351.

McElroy, M.W. & Carr, R.C. & Ensor, D.S. & Markowski, G.R. 1982. Size distribution of fine particles from coal combustion. *Science* 215: 13–19.

Pope, C.A. III & Dockery, D.W. 2006. Health effects of fine particulate air pollution: lines that connect. *J. Air Waste Manage. Assoc.* 56(6): 709–742.

Quann, R.J. & Neville, M. & Janghorbani, M. & Mims, C.A. & Sarofim, A.F. 1982. Mineral matter and trace-element vaporization in laboratory-pulverized coal combustion system. *Environ. Sci. Technol.* 16: 776–781.

Querol, X. & Fernandez, T.J.L. & Lopez, S.A. 1995. Trace elements in coal and their behavior during combustion in a large station. *Fuel* 74 (3): 331–343.

Schwartz, J. & Dockery, D.W. & Neas, L.M. 1996. Is daily mortality associated specifically with fine particles? *J. Air Waste Manage. Assoc.* 46(10): 927–939.

Seames, W.S. 2003. An initial study of the fine fragmentation fly ash particle mode generated during pulverized coal combustion. *Fuel Process. Technol.* 81(2): 109–125.

Senior, C.L. & Lawrence E. Bool III & Srinivasachar, S. & Pease, B.R. & Porle, K. 2000. Pilot scale study of trace element vaporization and condensation during combustion of a pulverized sub-bituminous coal. *Fuel Process. Technol.* 63: 149–165.

Uberoi, M. & Shadman, F. 1991. Simultaneous condensation and reaction of metal compound vapors in porous solids. *Ind. Eng. Chem. Res.* 30: 624–624.

Wagner, N.J. & Hlatshwayo, B. 2005. The occurrence of potentially hazardous trace elements in five Highveld coals, South Africa. *Int. J. Coal Geol.* 63(3–4): 228–246.

Architectural, Energy and Information Engineering – Sung & Chen (Eds)
© 2016 Taylor & Francis Group, London, ISBN 978-1-138-02791-6

Research on the melting-ice sequence based on risk analysis

X.Y. Zhou & H.Q. Li
Intelligent Electric Power Grid Key Laboratory of Sichuan Province, Chengdu, Sichuan, China
School of Electrical Engineering and Information, Sichuan University, Chengdu, Sichuan, China

X. Wang
Sichuan Economic Research Institute, Chengdu, Sichuan, China

ABSTRACT: This paper builds an exponential failure probability model of the icing line according to the metal deformation theory, and proposes comprehensive assessment indices of the severity to measure the consequence after the fault of the icing line. Based on the risk theory, this paper combines the failure probability and severity of the icing line to build the risk index, which can reflect the security level of the grid, and then makes a specific melting-ice sequence strategy by considering the icing rate and risk index of the icing line. Taking the IEEE 30-bus system as an example, the analysis of simulation results proves the effectiveness and practicality of the proposed method.

Keywords: power system; ice disaster; risk analysis; security assessment; melting-ice strategy

1 INTRODUCTION

Snow disasters make great damage on the grid, and a large number of transmission lines malfunction because of over icing (Zheng, et al, 2009). It endangers the safe operation of the power system seriously, and even causes huge economic losses. To reduce the damage caused by ice storm, a direct and effective way is used to measure the melting-ice (Liqiang, et al, 2008). However, limited by the configuration level of ice-melting devices and safe operating conditions of the power system, it is unable to start the work of melting-ice on all icing lines at the same time. Therefore, when a wide range of ice storm occurs, scientific and rational decision-making of melting-ice sequence becomes very important for carrying out the work of melting-ice in time and effectively, improving the efficiency of melting-ice and reducing the safe operation risk of the grid.

The study of ice disaster in the power system receives more attention, mainly in the aspects of icing mechanism, melting-ice technology and prevention strategies. These studies provide a strong theoretical support to the defence work of ice storm; however, the current research on how to make the melting-ice sequence program is still not perfect. It becomes a problem to make a specific, comprehensive and practical model of melting-ice sequence, which needs to be solved urgently.

This paper adopts the icing rate of the line to characterize the severity of icing, and puts forward the failure probability function of the icing line in an exponential form in accordance with the metal deformation theory; next, the paper combines electric betweenness, which can reflect the topological structure with the operating state, and builds comprehensive assessment indices of the severity from three aspects, namely low voltage, circuit overload and load losses, to measure the impact on the system caused by the fault of the icing line. Then, the paper combines the failure probability and severity of the icing line to build the risk index, and proposes the method of making a specific melting-ice sequence further. Finally, the results of the IEEE 30-bus system simulation can verify the rationality and feasibility of the method, and it can provide effective reference information for the arrangements of melting-ice work.

2 THE FAILURE PROBABILITY MODEL OF THE LINE IN ICE STORM

The literature (Jones, et al, 1998) has considered the effect of wind speed and rainfall on icing growth, and calculates the total amount ice into which the lines capture water droplets to freeze, so the ice thickness growth rate in hours is

$$\Delta H = \frac{1}{\rho_i \pi} [(\rho_w Hg)^2 + (3.6VW)^2]^{\frac{1}{2}} \qquad (1)$$

where $W = 0.067Hg^{0.846}$ is the liquid water content in air; ρ_w, the density of water; ρ_i, the density of ice; V, the wind speed; and Hg, the precipitation.

Considering the ice thickness H_t of line at time t, the ice thickness of the line is forecasted after Δt hours:

$$H(t + \Delta t) = H_t + \Delta H \Delta t \qquad (2)$$

where $H(t+\Delta t)$ is the forecasting ice thickness of the line at time $(t+\Delta t)$.

The line model is different in the grid, so that each line that withstands maximum stress is not the same. Therefore, it is not comprehensive to only use the ice thickness to measure the icing severity of lines. This paper adopts the icing rate to measure the icing severity of lines, which can describe the icing situation of lines quantitatively. It is defined as the ratio of the ice thickness H and the design ice thickness H_n of lines. The expression is as follows:

$$\eta = \frac{H}{H_n} \qquad (3)$$

Based on the metal deformation theory, when the strain of towers or lines (with an obvious yield phenomenon) is small, the stress increases with increasing strain. When the stress reaches the limit, with increasing strain, the stress will decrease in an exponential form rapidly, as shown in Figure 1.

After the stress reaches the limit, the failure probability of the line is inversely proportional to the stress, and the failure probability of the line will rise exponentially. The failure probability function of the line has following expression (Yunyun, et al, 2013):

$$P_f = \begin{cases} \exp\left(\dfrac{H - H_n}{H_n}\right) & H \leq 2H_n \\ 1 & H > 2H_n \end{cases} \qquad (4)$$

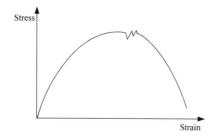

Figure 1. Stress and strain analysis of the metal.

The failure probability model based on the ice thickness is only related to the icing rate of the line. This model is based on the physical stress analysis of transmission lines, and therefore it is more accurate.

3 COMPREHENSIVE ASSESSMENT INDICES OF THE SEVERITY

3.1 The severity of low voltage

The severity of low voltage risk reflects the severity degree of bus voltage drop after the accident. Its function is shown in Figure 2.

The corresponding severity function of low voltage is expressed as follows:

$$S(u_i) = \begin{cases} 0 & u_i \geq U_N \\ \dfrac{U_N - u_i}{U_N - U_{\lim}} & u_i < U_N \end{cases} \qquad (5)$$

where u_i is the voltage of node I; U_N, the rated voltage of the node; and $U_{\lim}$, the low voltage limit of the node, which is generally taken as 90% of U_N.

Due to different nodes and lines in the system having different importance, this article introduces electric betweenness as weighting factors to reflect the topological structure importance of grid elements, and combines it with the severity of the operating state (Yang, et al, 2013). Considering electric betweenness as weights, the severity index of low voltage in the system is defined as follows:

$$S_{ev}(U) = \sum_{i=1}^{N} B_e(n)S(u_i) \qquad (6)$$

where $B_e(n)$ is the electric betweenness of node n and N is the number of nodes in the system.

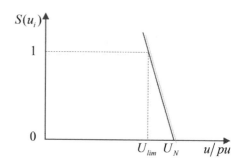

Figure 2. Severity function of low voltage risk.

3.2 The severity of circuit overload

The severity of circuit overload risk reflects the severity degree of transmission line overload after the accident. Its function is shown in Figure 3.

The corresponding severity function of circuit overload is expressed as follows:

$$S(p_l) = \begin{cases} 0 & p_l \le P_w \\ \dfrac{p_l - P_w}{P_{\lim} - P_w} & p_l > P_w \end{cases} \tag{7}$$

where p_l is the active power of the circuit; $P_{\lim}$, the limit transmission power of the circuit; and P_w, the power risk threshold of the circuit, which is generally taken as 90% of $P_{\lim}$.

As with the severity of low voltage, considering electric betweenness as weights, the severity index of the circuit overload in the system is defined as follows:

$$S_{ev}(P) = \sum_{l=1}^{Y} B_e(l)S(p_l) \tag{8}$$

where $B_e(l)$ is the electric betweenness of circuit 1 and Y is the number of circuits in the system.

3.3 The severity of load loss

The ice storm accident often leads to the loss of load, even causing huge economic losses to some users that have important load, so paying a close attention to the load risk is very important. Due to different types of load, this article introduces the load economic factor as weights to reflect differences of impact that different load losses have on the system. The load loss ratio is defined as follows:

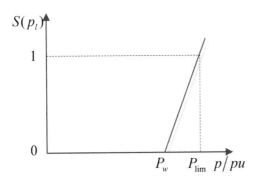

Figure 3. Severity function of low voltage risk.

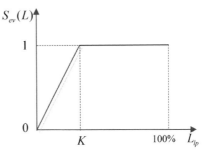

Figure 4. Severity function of load loss risk.

$$L_{lp} = \frac{\sum_{i=1}^{W} w_i L_i}{L_0} \times 100\% \tag{9}$$

where w_i is the economic factor of the load node; W, the set of the nodes with load losses; L_i, the amount of load losses by node i; and L, the total amount of the load before accident.

The severity function of load loss is shown in Figure 4.

The corresponding severity function of load loss is expressed as follows:

$$S_{ev}(L) = \begin{cases} \dfrac{L_{lp}}{K} & L_{lp} \le K \\ 1 & L_{lp} > K \end{cases} \tag{10}$$

where K is the risk threshold of load loss ratio, which is taken as 30% in this paper.

4 THE SCHEME OF MELTING-ICE SEQUENCE DECISION-MAKING

4.1 Risk index

There are two kinds of factor that affect the stable operation of the power grid in ice disaster situations: one is the icing conditions of lines, which can directly affect the reliable operation of the lines; the other is the severity of the line fault, because some lines playing an important role in the power grid will cause a serious harm to the safe operation of the system once failure occurs.

Risk is expressed as the product of the probability and severity of the accident. This article introduces the icing situation to the failure probability model, and uses the comprehensive severity index to reflect the severity of line failure, and then takes the combination of both as the power system risk index under the icing line failure, reflecting the impact of the icing line failure on system security risks, measuring the dangerous level of icing lines in the ice storm. Its expression is as follows:

$$R_{risk} = P_f \times [S_{ev}(U) + S_{ev}(P) + S_{ev}(L)] \tag{11}$$

4.2 The method of making the melting-ice sequence

The ice thickness of the line increases over time by the effect of ice storm weather, thus the security risk of the system continues to rise. This paper makes the melting-ice sequence decision-making scheme by taking into account the icing condition of the line and the operation risk of the grid. The division of the melting ice lines according to the ice rate of lines, and the specific rules of melting-icing sequence decision-making are given below:

1. If the icing rate of lines is greater than λ_1, these lines enter into the icing warning state. Thus, the melting-ice line set is formatted.
2. In the melting-ice line set, the lines with an icing rate greater than λ_2 are in the serious icing situation and prone to the line disconnection fault. It is necessary to take timely measures to give priority to these lines for melting ice. The severity degree of covering ice has become a major factor affecting the melting-ice sequence. The greater the icing rate is, the more the melting-ice order of the line is;
3. In the melting-ice line set, for the lines with an icing rate less than λ_2, the safety risk level of the grid should be the main consideration when melting ice. The melting-ice sequence is determined by the risk index of melting-ice lines; therefore, the larger the risk index is, the more the melting-ice order of the line is.

5 TEST SCENARIO

This paper takes the IEEE 30-bus system as an example, assuming that the system is affected by the ice storm. In the process of simulation, the parameters are set as follows: $\lambda_1 = 0.4$, $\lambda_2 = 0.7$, $H_n = 10$ mm, the diameter of the line is determined by the type, and the time of melting-ice for one line is two hours.

Assuming that there will be a freezing weather disaster lasting 12 hours from 10:00 to 22:00 one day by the weather forecast, to facilitate analysis, this paper takes the mean value of meteorological factors in this period of time. The icing load situation at 10:00 and the growth rate of lines are given in Table 1.

The melting-ice sequence is determined in accordance with the proposed method in this

Table 1. The icing load situation and the growth rate of lines at 10:00.

Line	Ice thickness (mm)	Icing rate	Icing growth rate (mm/h)
L1	3	0.3	0.6284
L2	5	0.5	0.4548
L4	3	0.3	0.4970
L8	4	0.4	0.2845
L10	8	0.8	0.4675
L20	4	0.4	0.3162

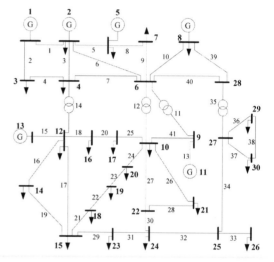

Figure 5. Electrical connection diagram of the IEEE 30-bus system.

paper. First, the lines with the icing rate greater than 0.4 are screened out, so we can get the melting-ice line set, that is {L2, L4, L10, L20}. The icing rate of L10 is 0.8, which is more than 0.7, so we choose L10 as the first melting-icing line to avoid the line disconnection fault by the serious icing load. Then, we can find the ice thickness of lines at different times according to the growth rate forecast, and then obtain the icing rate of lines as well as the corresponding risk index. The result of the melting-ice line sequence from 12:00 to 16:00 is given in Table 2.

As shown in Table 2, L1 will enter into the melting-ice line set after two hours of melting-ice for L10. The icing rate of these lines in the set is less than 0.7, so this paper sorts them in accordance with the value of risk. At 12:00, the corresponding fault probability of L2 is great, because the icing situation of L2 is more serious than the other three

Table 2. Melting-ice sequence of lines from 12:00 to 14:00.

Time	Line	Ice thickness (mm)	Icing rate	Risk index	Prior line of melting-ice
12: 00	L1	4.2568	0.4257	1.6252	L2
	L2	5.9096	0.5909	2.0325	
	L8	4.5690	0.4569	1.4282	
	L20	4.6324	0.4632	0.7038	
14: 00	L1	5.5136	0.5514	1.8433	L1
	L4	4.9880	0.4988	1.4485	
	L8	5.1380	0.5138	1.5118	
	L20	5.2648	0.5265	0.7496	
	L24	4.6088	0.4609	0.5435	

lines. In addition, the comprehensive assessment index of severity becomes higher after the fault of L2. It is indicated that L2 plays an important role in the system. Once L2 malfunctions, it will lead to a serious consequence. From this, the dangerous level of L2 with the top risk index becomes higher. It is shown that L2 has a great influence on the system after fault, and the safe operation risk degree of the system will reduce if the icing load of L2 melts. We choose L2 as the prior melting-ice line, which can enhance the ability of the grid to withstand the ice storm. By this analogy, we can get the melting-ice sequence, that is L10-L2-L1-L8-L4-L20, according to the proposed method in this paper.

6 CONCLUSIONS

In the case of ice storm, the stable operation of the power system is under serious threat. Based on the risk analysis, this paper combines the failure probability and severity of the icing line to calculate the risk index, which can help operators grasp the operation level of the grid fully. Furthermore, a specific method of making the melting-ice sequence is proposed, to reduce security risk and ensure stable operation of the grid in the process of melting-ice further. This method can provide a reasonable reference for the arrangements of melting-ice and the development of relevant programs.

REFERENCES

Jones KF. A Simple Model for Freezing Rain Ice Loads [J]. Atmospheric research, 1998, 46(1): 87–97.

Li Zheng, Yang Jingbo, Han Junke. Analysis on Transmission Tower Toppling Caused by Icing Disaster in 2008 [J]. Power System Technology, 2009, 33(2): 31–35.

Pan Liqiang, Zhang Wenlei, Tang Jihong. Overview of The Extraordinarily Serious Ice Calamity to Hunan Power Grid in 2008 [J]. Power System Technology, 2008, 32(26): 20–25.

Xie Yunyun, Xue Yusheng, Wen Fushuang. Space-time Evaluation for Impact of Ice Disaster on Transmission Line Fault Probability [J]. Automation of Electric Power Systems, 2013, 37(18): 32–41.

Zhao Yang, Li Huaqiang, Wang Yimiao. A Complex Network Theory and Conditional Probability Based Risk Assessment Method for Disastrous Accidents [J]. Power System Technology, 2013, 37(11): 3190–3196.

Architectural, Energy and Information Engineering – Sung & Chen (Eds)
© 2016 Taylor & Francis Group, London, ISBN 978-1-138-02791-6

Cascading failures forecasting research considering brittle relevance

X.X. Qi, Y.J. Yu & D.G. Liu
Electric Power Research Institute, Xinjiang Uygur Autonomous Region, China

X.Y. Lv & H.Q. Li
Intelligent Electric Power Grid Key Laboratory of Sichuan Province, Chengdu, Sichuan, China
School of Electrical Engineering and Information, Sichuan University, Chengdu, Sichuan, China

ABSTRACT: From the perspective of systematics, a new cascading failures forecasting method is proposed in this paper. First, the electrical structure relevance degree is established based on the electrical structure theory to reflect the influence of fault components. Second, the brittleness theory is introduced to further analyze the occurrence mechanism of cascading failures, and the comprehensive brittle relevance index is proposed to identify the most likely occurred cascading failure sequence, by considering the combination of the structure and state of the power grid. Lastly, compared with the traditional method by using the IEEE 30 bus system, the feasibility and validity of the model are verified.

Keywords: electrical structure; brittleness theory; comprehensive brittle relevance degree; cascading failures forecasting

1 INTRODUCTION

The fault of a component in the system can easily lead to cascading failures in the power grid, and finally lead to the blackout (Libao, et al, 2010). Therefore, the depth study on the mechanism of cascading failures from the viewpoint of the overall operation of the complex grid and the study of looking for the cascading failures sequence have become the biggest problem of system operators.

So far, domestic and foreign researchers have built multiple analysis methods and models of cascading failures, including the pattern search method and the model analysis method. The method used by Xiaodan et al. (2006) shows a forecasting approach, reflecting the influence degree of the occurred fault on the subsequent fault. An approach to explain the relevance between fault components and normal components has been proposed by Yingying et al. (2012). However, the above references forecast subsequent failures only by considering the state of the grid, and ignoring the change in the network topology during the progress of cascading failures, which leads to one-sidedness of characterizing the overall operation of the complex power grid. A distance index has been proposed by Yimiao et al. (2012), characterizing the margin between the steady state and the critical state. Meanwhile, a comprehensive forecasting index with a combination of the distance index and electric betweenness has beenproposed by Yimiao et al. (2012). Although the state

and structure of the power grid are considered as well, the inter-stage faults are in close contact. The above index cannot reflect the influence of normal components that are affected by fault components.

To solve the above problems, from the perspective of systematics, this paper proposes a cascading failures model by considering the combination of the structure relevance and the state relevance of the grid. First, the electrical topology is proposed to substitute the pure topology according to Kirchhoff Laws, which captures the electrical properties of the grid. Second, this paper introduces the brittleness theory to further analyze the occurrence mechanism of cascading failures, and shows that the comprehensive brittle relevance degree plays a decisive role in the development of cascading failures. Then, a comprehensive brittle relevance index with a combination of the state and structure of the grid is proposed, which can forecast subsequent failures effectively. Lastly, this model is compared with the traditional method by using the IEEE 30 bus system, and the feasibility and validity of the model are verified.

2 ELECTRICAL STRUCTURE MODEL

Multiple structure models are proposed on the premise of ignoring the electrical properties of the components in the network (Yijia, et al, 2007), and there are differences between these methods and the actual power grid (Yudong, et al, 2014). In Koc et al. (2014), an electrical structure is proposed based on electrical science. Figure 1 shows the electrical topology of the IEEE 30 power system.

In Figure 1, the electrical distance is expressed by the equivalent impedance between two nodes, and the magnitude of equivalent impedance is expressed by the thickness of connecting lines. The equivalent impedance $Z_{eq,ij}$ is equal to the voltage U_{ij} when a unit current is added between node i and node j, which is given as follows:

$$Z_{eq,ij} = \frac{U_{ij}}{I_{ij}} = U_{ij} \qquad (1)$$

The equivalent impedance $Z_{eq,ij}$ is computed as follows:

$$Z_{eq,ij} = (Z_{ii} - Z_{ij}) - (Z_{ij} - Z_{jj}) \qquad (2)$$

where Z is the node impedance matrix; the smaller the value of $Z_{eq,ij}$, the more parallel the branches of the area share the transfer flow. A higher number of parallel branches imply a more robust network against cascading failures, which can effectively resist disturbance.

To quantify the electrical connection of the electrical structure, the electrical connection degree of the power grid is defined as the sum of all branches' equivalent impedance for a network G:

$$Z_G = \sum_{i=1}^{N} \sum_{j=i+1}^{N} Z_{eq,ij} \qquad (3)$$

The electrical connection degree Z_G reflects the degree of the electrical structure. The smaller value of Z_G reflects that there are more parallel branches. Therefore, a relatively more homogeneous distribution of the power flow over these parallel paths

Figure 1. Electrical topology of the IEEE 30 power system.

increases the robustness of the power grid against cascading failures.

From the perspective of systematics, considering the operation condition of the whole power grid, the electrical structure importance degree of the line l is defined as follows:

$$\Delta Z_G^l = \frac{Z_{G-l} - Z_G}{Z_G} \qquad (4)$$

where the equivalent impedance Z_{G-l} is the electrical connection degree of the grid by removing l; the value of ΔZ_G^l is higher, and the probability of causing the next level failure is greater. The electrical structure importance degree can evaluate important lines from the aspect of the electrical structure, which forms the basis for the study of the change in the electrical structure.

3 COMPREHENSIVE BRITTLE RELEVANCE MODEL BASED ON THE BRITTLENESS THEORY

The brittleness of a complex system is defined as follows: due to a combination of the internal and external interference factors of the complex system, the subsystem collapses. The collapse of the subsystem affects the rest of the subsystems either directly or indirectly, and finally the system cascading collapses. The attribute of the complex system is brittle.

3.1 *Electrical structure brittle relevance index*

In the electrical structure of the power grid, the structure relevance can be reflected by the change in the electrical structure importance degree of branches. The importance degrees of the rest of the branches change when the fault lines break down, and it is possible that the relatively unimportant branch changes to a relatively important branch in the new electrical structure. This change in the process embodies the electrical structure changes after the failure at the top level. Under the normal operation of the power grid, in the event of the $p-1$ level failure, the electrical structure importance degree of branch l is $\Delta Z_G^l(p-1)$ from Eq. (4). In the event of the p level failure, when the line m breaks down, the electrical structure importance degree of branch l is computed as follows:

$$\Delta Z_G^{l-m}(p) = \frac{Z_{G-l-m}(p) - Z_{G-m}(p)}{Z_{G-m}(p)} \qquad (5)$$

where $Z_{G-m}(p)$ is the electrical connection degree after the p level failure, and $Z_{G-l-m}(p)$ is the new electrical connection degree by removing l after the p

level failure. A new importance degree of branch l is reflected in Eq. (5).

After the p level failure, the structure brittle relevance degree $Z_{ml}(p)$ of line l affected by the fault line m is defined as follows:

$$Z_{ml}(p) = \Delta Z_G^{l-m}(p) - \Delta Z_G^l(p-1) \qquad (6)$$

where $\Delta Z_G^l(p-1)$ is the structure importance degree and $\Delta Z_G^{l-m}(p)$ is the structure importance degree of line l after the p level failure.

$Z_{ml}(p)$ is measured by the change in the structure importance degree of branch l. From Eq. (6), it can be found that the higher the value of $Z_{ml}(p)$, the greater the influence of branch m, and it is more likely to evolve into a new brittle source, excite the system's brittleness, and finally lead to the blackout.

3.2 State brittle relevance index

Considering the actual operation of the power system, the transfer of the flow is a significant factor that causes the brittle collapse of the power grid. When the brittle resource is excited, the power flow redistribution caused by load fluctuation will lead to certain components arriving at the branch thermal limit, and then the protectors take action immediately.

After the brittle source failure, the change inpower is computed as follows:

$$\Delta P_l = P_{l-m} - P_l \qquad (7)$$

where P_l is the power flow of line l in the normal operation of the power grid, and P_{l-m} is the power flow in the new operation of the power grid when the line m breaks down.

ΔP_l reflects the power change in branches and the relevance degree between each branch.

After the brittle source m failure, the degree of the approximation to the power limit μ_l of branch l is computed as follows:

$$\mu_l = (P_{l\max} - P_{l-m}) \Big/ P_{l\max} \qquad (8)$$

where $P_{l\max}$ is the active power thermal limit of branch l.

While the change in the flow and the degree of the approximation to the power limit are comprehensively considered, the state brittle relevance degree under the p level failure is defined as follows:

$$P_{ml}(p) = \Delta P_l(p) P_m(p-1) - \Delta P_l(p) P_m(p-1) \mu_l(p) \qquad (9)$$

where $\Delta P_l(p)$ is the change flow of line l caused by the failure of line m; $P_m(p-1)$ is the active power of the brittle source m under the $p-1$ level failure; and $\mu_l(p)$ is the degree of the approximation to the power limit of line l. The higher the $P_{ml}(p)$ values, the more the influence of fault lines on the normal lines. The $P_{ml}(p)$ can assess the most vulnerable branch under the current operation of the power grid completely.

3.3 Comprehensive brittle relevance index

The cascading failures are the results of the iteration among all branches relevance, which is called thebrittleness process. The relevance among all branches is closely related to the state and structure of the power grid.

The comprehensive brittle relevance degree of a branch is defined as follows:

$$S_{ml}(p) = \frac{1}{2}(Z_{ml}(p) + P_{ml}(p)) \qquad (10)$$

where $Z_{ml}(p)$ and $P_{ml}(p)$ are the brittle relevance index by normalizing. The higher the $S_{ml}(p)$ values, the larger the influence of fault lines on the normal lines. The branch with the highest value of $S_{ml}(p)$ will become the most vulnerable line in the power grid, which is selected as the next level failure.

4 CASCADING FAILURES FORECASTING MODEL

4.1 Cascading failures forecasting index

Taking the cumulative effect into consideration, when the p level failure occurs, the comprehensive prediction indicator of the $p+1$ level failure can be obtained as follows:

$$\alpha_{ml}(p+1) = S_{ml}(p) + \frac{S_{ml}(p) - S_{ml}(p-1)}{S_{ml}(p-1)}(p=1,2,...) \qquad (11)$$

where the first part is the comprehensive forecasting index after the p-level failure; the second part represents the rate of change in the comprehensive forecasting index based on the $p-1$ level failure after the p level failure; $\alpha ml(p+1)$ is the forecasting index to comprehensively reflect the current and historical relevance status of the system as well as the electrical structural relevance after the p level failure, which can be used to predict the serious failure modes of the system.

4.2 Prediction procedure of cascading failures

Figure 2 shows the flowchart of the cascading failure comprehensive forecasting model.

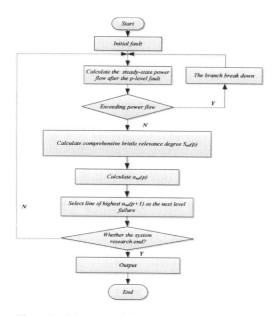

Figure 2. Flow chart of the cascading failures forecasting model.

5 TEST SCENARIO

In this paper, only 37 branches are considered (without considering 4 transformer branches) in the IEEE 30 buse system. Figure 3 shows the IEEE 30 buse system.

Several lines are chosen as the brittle sources to cascade the sub-sequences. To exemplify, the comprehensive brittle relevance forecasting index after L10 faults is shown in Figure 4.

As shown in Figure 4, branch L35 with the highest value is selected as the next level fault, whereas the branches with the lowest value have a low correlation with the fault branch L10.

The brittle processes are obtained by using the proposed method in this paper, asshowed in Table 1.

Analyzing the above-mentioned results available, in the brittle processes 1 and 4, the generator changes the power transmission path that provides to the load after the L4 and L8 failure. L3 and L6 have the biggest power flow change, and both are strongly connected with the fault branches. So, L3 and L6 are more likely to break down to aggravate the transferring flow around generators, and lead to the breaking down of L6 and L7.

In the brittle process 2, sfter L10 breaks down, L36 undertakes the heavier power flow transmission task than L35, and has a close connection with L10. Therefore, L35 is influenced significantly by L10. After both L36 and L10 break down, the generator 8 is connected to the power grid through

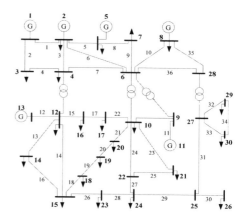

Figure 3. Electrical connection diagram of the IEEE 30 bus system.

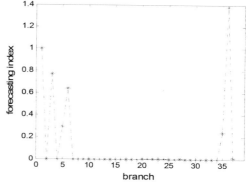

Figure 4. Brittle relevance forecasting index after L10 failures.

Table 1. Part of brittleness processes and cascading failures forecasting indices.

Brittleness procedure	Brittle source	$a_{ml}(2)$	Second failure	$a_{ml}(3)$	Third failure
1	L4	2.00000	L3	1.98705	L6
2	L10	1.38348	L36	1.04316	L27
3	L13	1.03378	L14	1.54237	L21
4	L8	1.58593	L6	2.00000	L7
5	L26	1.01063	L27	2.00000	L36

L35, L31, L29, transferring the power flow to the nodes 29, 3 and 26 through L10, L36 before L35, L10 break down change their transmission path to L24, L27, and cause the power flow transfer in the range of the power grid. At this time, the important degree of L27 becomes bigger in the electrical structure, and may break down easily to worsen the cascading failures. Similarly, this is true for the brittle processes 3 and 5. As the initial faults, both L13 and L26 undertake the heavier power transmission task. The branches around L13, L26 are influenced by the fault branches to break down, and then cause the power flow transfer in the range of the power grid, and finally worsen the cascading failures.

In comparison with the sequence L8-L35-L28 of the traditional forecasting method proposed by Yimiao et al. (2012), the sequence L8-L6-L7 forecasted in this paper is more consistence with the actual one. L8, as the initial fault, undertakes the power flow transmission task of the generator 5. After L8 breaks down, L6 will undertake more transmission power flow as the substitution of L8, and it is also heavily influenced by L8 in the electrical structure. Obviously, L6 is more likely to be the next level fault than L35. The superiority and the validity is reflected by using the forecasting method proposed in this paper.

6 CONCLUSIONS

A cascading failure forecasting model based on the comprehensive brittle relevance degree of branches is proposed in this paper. The conclusion is summarized as follows:

1. Considering the structure brittle relevance degree and the state relevance degree comprehensively can perfect the study of the cascading failures, making up for the consideration of only the structure.
2. The higher the comprehensive brittle relevance degree values, the more vulnerable the branch.

This index has a higher identification to the next level failure, and the effective sequence is obtained by using the proposed model.

REFERENCES

Cao Yijia, Chen Xiaogang, Sun Ke. Identification of vulnerable lines in power grid based on complex network theory [J]. Electric Power Automation Equipment. 2007, 26(12): 1–5. (in Chinese).

Koc Y, Warnier M, Koojj R, et al. Structural vulnerability assessment of electric power grids [C]//Networking, Sensing and Control (ICNSC), 2014 IEEE 11th International Conference on. IEEE, 2014: 386–391.

Shi Libao, Shi Zhongying, Yao Liangzhong, et al. A review of mechanism of large cascading failure blackouts of modern power system [J]. Power System Technology, 2010, 34(3): 48–53. (in Chinese).

Tan Yudong, Li Xinran, Cai Ye, et al. Critical node identification for complex power grid based on electrical distance [J]. Proceedings of the Chinese Society for Electrical Engineering, 2014, 34(1): 146–152. (in Chinese).

Wang Yimiao, Li Huaqiang, Xiao Xianyong, et al. Cascading failures forecasting based on running state and structure [J]. Power System Protection and Control, 2012, 40(20): 1–5. (in Chinese).

Wang Yingying, Luo yi, Tu Guangyu, et al. Correlation model of cascading failures in power system [J]. Transactions of China Electrotechnical Society, 2012, 27(2): 204–209. (in Chinese).

Yu Xiaodan, Jia Hongjie, Chen Jianhua. A Preliminary Research on Power System Cascading Contingency Forecasting [J]. Power System Technology, 2006, 30(13): 20–25. (in Chinese).

Architectural, Energy and Information Engineering – Sung & Chen (Eds)
© 2016 Taylor & Francis Group, London, ISBN 978-1-138-02791-6

Sensitive equipment risk assessment of voltage sag fault levels

D.G. Liu, Y.J. Yu & X.X. Qi
Electric Power Research Institute, Xinjiang Uygur Autonomous Region, China

B.J. Liu & H.Q. Li
Intelligent Electric Power Grid Key Laboratory of Sichuan Province, Chengdu, Sichuan, China
School of Electrical Engineering and Information, Sichuan University, Chengdu, Sichuan, China

ABSTRACT: The defects in the assessment methods for traditional equipment sensitivity to voltage sag are analyzed. The necessity of conducting the risk assessment of the failure level of sensitive equipment caused by voltage sag in the electricity market environment is expounded. This paper established one kind of probability model for the failure level of the sensitive equipment caused by voltage sag on the basis of the voltage sag severity index. With the economic index characterizing the consequences of an event, an approach to assess the risk of the sensitive equipment caused by voltage sag is proposed, which can be used directly by decision makers and designers. The assessment results for the IEEE 9-bus system shows that the proposed approach is feasible and rational.

Keywords: voltage sag; sensitive equipment; severity index; failure level; risk assessment; economic index

1 INTRODUCTION

In many power quality problems, the sensitive equipment failure event caused by voltage sag is most frequent and also the most serious problem. The existing research on the impact of voltage sag on equipment mathematical description, the model building method and the applicability of the conclusion is very immature (Djokic, et al, 2005). The assessment on the failure level of the sensitive equipment caused by voltage sag is of great significance for designers' reasonable planning, for electric power supply enterprises to develop power solutions, improve the efficiency of industrial production and reduce wear and tear (Wagner, et al, 1990).

The existing sensitive equipment voltage sag assessment method is roughly divided into the measured statistics method and the uncertainty assessment method. These methods consider the uncertainty of sensitive equipment failure characteristics from different angles and different levels, but the results that are to a large extent influenced by the model, based on much experience and assumptions, has a very big subjectivity (Gupta, et al, 2006).

The risk assessment methodevaluates the possibility of failure and severity. Possibility expresses the accident and severity describes the consequences (Milanovic, et al, 2006). This paper considers the failure probability sensitive equipment affected by voltage sag and the consequences of equipment failure by using the basic principle of risk assessment to set up two aspects, namely the sensitive equipment failure probability model and the equipment failure consequences model, ending up with the sensitive equipment voltage sag fault level of the risk model (Chan J Y, et al, 2007). The application of the risk model can be related to the design and economic decisions.

2 THE PROBABILITY OF EQUIPMENT FAILURE CAUSED BY VOLTAGE SAG

2.1 *Voltage sag probability model*

There are many causes of voltage sag in the system, which mainly include: system fault, start of the motor and transformer cut. In this paper, the risk assessment of voltage sag only consider the voltage sag caused by the system failure, to facilitate the discussion and calculation.

In this paper, the causes of the system failure can be divided into two categories, namely equipment failure and external causes. All of these factors are random, which are described by using stochastic models. The ordinary dimensional stochastic model cannot very well simulate the failure factors; therefore, the study by Khanh et al. (2008) proposed a binary random model to better simu-

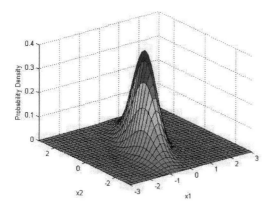

Figure 1. Example of the bivariate normal distribution.

late the above situation. The ordinary binary distribution model is shown in Figure 1.

We assume that the equipment failure rate model conforms to the uniform distribution. So, the failure rate in the position i can be calculated by the following formula:

$$P_{eq-trans(i)} = \frac{\alpha_{eq} \cdot N_{trans}}{m_t} \qquad (1)$$

where

N_{trans}: the number of transformer faults in a test system;

m_t: the total number of the distribution transformer; and

α_{eq}: equipment factors that lead to the failure percentage contribution.

The line fault is generally used to express the number per unit length line fault every year. A section of the line failure rate is calculated by the following formula:

$$P_{eq-line(i)} = \alpha_{eq} \cdot N_{line} \cdot \frac{l_i}{\sum\limits_{k=1}^{m_b} l_k} \qquad (2)$$

where

N_{line}: the number of the line fault in a test system;

m_b: the total number of lines; and

l_i: the length of the line i.

The fault probability distribution caused by external factors is associated with the location of the fault, and follow the binary random model. As a result, each location of the failure rate is calculated by the following formula:

For the transformer i,

$$P_{ex-trans(i)} = \alpha_{ex} \cdot N_{trans} \cdot W_{trans(i)} \qquad (3)$$

For the line i,

$$P_{ex-line(i)} = \alpha_{ex} \cdot N_{line} \cdot \frac{l_i \cdot W_{line(i)}}{\sum\limits_{k=1}^{m_b} l_k \cdot W_{line(k)}} \qquad (4)$$

where

α_{ex}: external factors that lead to the failure percentage contribution;

$\alpha_{eq} + \alpha_{ex} = 100\%$; and

$W_{trans(i)}$, $W_{line(i)}$: weighted factors of the failure rate for the transformer i and line i under the bivariate normal distribution model considering the fault location.

The bivariate normal distribution model of the joint probability density function is given by

$$f(x,y) = \frac{1}{2\pi \cdot \sigma_x \sigma_y \cdot \sqrt{1-\rho^2}} \cdot \exp\left[-\frac{z}{2(1-\rho^2)}\right]$$
$$= \Phi(\mu_x, \mu_y, \sigma_x, \sigma_y, \rho) \qquad (5)$$

where

$$z = \left(\frac{x-\mu_x}{\sigma_x}\right)^2 - \frac{2\rho \cdot (x-\mu_x) \cdot (y-\mu_y)}{\sigma_x \sigma_y} + \left(\frac{y-\mu_y}{\sigma_y}\right)^2$$
$$(6)$$

μ_x, μ_y, σ_x, σ_y: the mean and standard deviation of the two variables x and y, and

ρ: correlation coefficient. If fault co-ordinates are independent variables, then $\rho = 0$.

Within the scope of the $\Delta s_i = \Delta x_i \Delta y_i$, the probability of the failure of location (x_i, y_i) can be calculated using the following formula:

$$F(\Delta x_i, \Delta y_i) = \iint\limits_{\Delta s_i} f(x,y)dxdy = \frac{f(x_i, y_i) \cdot \Delta s_i}{\sum\limits_{k=1}^{\infty} f(x_k, y_k) \cdot \Delta s_k}$$
$$(7)$$

If $\Delta s_i = \Delta s_0 = const(\forall i = 1, m)$ and m is large enough, the above equation can be normalized as follows:

$$F(\Delta x_i, \Delta y_i) \approx \frac{f(x_i, y_i) \cdot \Delta s_0}{\sum\limits_{k=1}^{m} f(x_k, y_k) \cdot \Delta s_0} = \frac{f(x_i, y_i)}{\sum\limits_{k=1}^{m} f(x_k, y_k)} \qquad (8)$$

In the distribution network system, if the network node position has a uniform distribution, then the approximation can be made at the co-ordinate (x_i, y_i) fault location as follows:

Weight failure rate for transformer i

$$W_{trans(i)} = \frac{f(x_i, y_i)}{\sum_{k=1}^{m_t} f(x_k, y_k)} \quad (9)$$

Weight failure rate for line i

$$W_{line(i)} = \frac{f(x_i, y_i)}{\sum_{k=1}^{m_b} f(x_k, y_k)} \quad (10)$$

2.2 Voltage sag on the distribution of eigenvalues

To obtain the distribution of voltage sag corresponding to the characteristic value of voltage sag, the estimate of voltage sag can be made. Using the actual data grid in Chan J Y et al. (2009), 183 voltage sags can be monitored within 13 years. The characteristic value of voltage sag, i.e. the amplitude and duration, meets a certain distribution curve. This paper, by using the data, obtains the fitting curve to predict a specific amplitude and duration of voltage sag occurrence probability.

Based on the results shown in Figures 2 and 3, we can get a certain MDSI sag occurrence probability distribution.

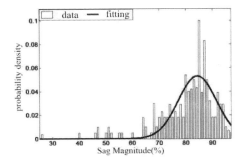

Figure 2. Sag magnitude.

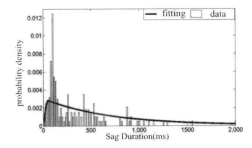

Figure 3. Sag duration.

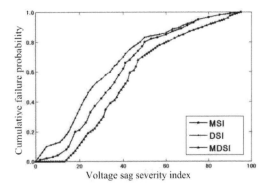

Figure 4. Cumulative failure probability of PC.

$$P_{db} = F_D(x) \quad (11)$$

where

P_{db}: occurrence probability of a specific MDSI voltage sag, the independent variable x is for MDSI values, and

$F_{D(x)}$: distribution function of a specific MDSI sag.

2.3 Sensitive equipment response caused by voltage sag at the fault level

Based on the paper by Chan J Y et al. (2011) that provides PC test information collection and analysis, 38 curves of the voltage sag resistance is used to establish the model. In conclusion, the corresponding MSI, DSI, and MDSI curves of PC cumulative failure are shown in Figure 4.

Based on the results shown in Figure 4, we can extract the fault probability distribution of the equipment, in particular the MDSI voltage sag, as follows:

$$P_f = F_f(x) \quad (12)$$

where

P_f: the failure probability of the equipment in a specific MDSI voltage sag, the independent variable x is for MDSI values, and

$F_{f(x)}$: equipment failure probability distribution function when a specific MDSI sag occurs.

Based on the above comprehensive description, we can obtain the probability of the sensitive equipment's failure affected by the voltage sag, $P(D|X_0)$, as follows:

$$P(D|X_0) = P_{trans}(P_{line}) \times \int_0^{100} F_D(x) F_f(x) dx \quad (13)$$

561

3 THE RISK MODEL OF EQUIPMENT FAILURE CAUSED BY VOLTAGE SAG

3.1 The consequence of equipment failure caused by voltage sag

The response of the sensitive equipment for voltage sag at the fault level is mainly manifested in the economic consequences of the failure.

For individual equipment, the consequences of its fault can be divided into three parts: equipment maintenance cost I_{mt}, update costs I_{ch} and production loss caused by the fault I_l:

$$E[I(D)] = E[I_{mt}(D)] + E[I_{ch}(D)] + E[I_l(D)] \quad (14)$$

Equipment maintenance cost is expressed as K; we considered that the equipment should be updated by several times of failure, update equipment costs in terms of K_c, and β for the number of the fault before being eliminated:

$$E[I_{mt}(D)] = K \quad (15)$$

$$E[I_{ch}(D)] = \frac{K_c}{\beta} \quad (16)$$

The production loss caused by a single device failure related to the position of the equipment in the production line and its importance, introducing an important degree coefficient λ, the largest economic loss (coefficient $\lambda = 1$) caused by equipment failure is K_r:

$$E[I_l(D)] = K_r \times \lambda \quad (17)$$

Assuming that the objective (a transformer secondary distribution network or a single line) of the study contains a sensitive equipment number N, the consequences of the sensitive equipment voltage sag at the fault level is given by

$$E[I(D)] = (K + K_c/\beta + K_r \times \lambda) \times N \quad (18)$$

In this paper, the specific risk theory was applied to the power system voltage sag of risk assessment, and the voltage sag sensitive device failure risk is defined as the product of the probability of the sensitive equipment failure after voltage sag and the consequences of sensitive equipment failure:

$$R(X_0) = P(D \mid X_0) \times E[I(D)] \quad (19)$$

where
X_0: the currently running state of the system;
$R(X_0)$: the risk indicators of the system voltage sag that lead to the equipment failure;

$P(D|X_0)$: the probability of equipment failure when the system is in the current running status under the voltage sag; and
$E[I(D)]$: the expectations of equipment failure consequences when the voltage sag occurs in the system.

4 ANALYSIS OF EXAMPLES

We perform the risk assessment risk assessment for the failure level of the sensitive equipment caused by voltage sag using the IEEE3 machine 9 nodes system, as shown in Figure 5. In this paper, as an example of the PC, ignoring the motor start may produce the transformer voltage sag, and only consider the voltage sag caused by the transformer or line fault.

We assume $\alpha_{eq} = \alpha_{ex} = 50\%$; at the same time, we set up the coordinate system for the transformer and each branch in the system, respectively: T1(1,0), T2(0,0), T3(0,1), L5-8(0,0), L5-7(0,1), L7-4(0,2), L8-6(1,0), and L6-9(1,1), L9-4(1,2). We then calculate the PC fault risk within 1 h under the current state of voltage sag. We consider the accident for the transformer T1~T3 fault and 6 break lines, and assume the failure probability caused by equipment factors and external factors as $P_{eq-trans(i)}$/ $P_{eq-line(i)}$ and $P'_{ex-trans(i)}/P'_{ex-line(i)}$, shown in Table 1.
Here,

$P_{ex-trans(i)}$: the transformer fault probability in the position i caused by external factors considering the weight of the position, and

$P_{trans(i)}$: transformer comprehensive failure probability in position i.

We calculate the voltage sag PC failure probability caused by the failures of the transformer and line according to Formulas (11)~(13), in which Formulas (11)~(12) are distribution functions obtained by curve fitting using the measured data. By treating different systems, we can perform different tests to obtain a more accurate distribution.

Voltage sag may result in the sensitive equipment failure or stoppage, leading to the corre-

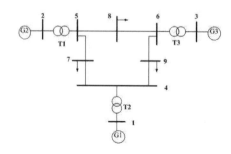

Figure 5. Single-line diagram of the IEEE 9-bus system.

Table 1. The equipment failure rate indices.

Device/line	$P_{eq\text{-}trans(i)}/$ $P_{eq\text{-}line(i)}$	$P'_{ex\text{-}trans(i)}/$ $P'_{ex\text{-}line(i)}$	$W_{trans(i)}/$ $W_{line(i)}$	$P_{ex\text{-}trans(i)}/$ $P_{ex\text{-}line(i)}$	$P_{trans(i)}/$ $P_{line(i)}$
T1	1.67×10^{-5}	2.49×10^{-5}	0.5142	1.2804×10^{-5}	2.9504×10^{-5}
T2	1.89×10^{-5}	3.64×10^{-5}	0.2429	0.8842×10^{-5}	2.7742×10^{-5}
T3	2.06×10^{-5}	2.65×10^{-5}	0.2429	0.6437×10^{-5}	2.7037×10^{-5}
L5-8	3.76×10^{-5}	7.22×10^{-5}	0.1086	0.7841×10^{-5}	4.5441×10^{-5}
L5-7	5.33×10^{-5}	6.66×10^{-5}	0.2298	1.5305×10^{-5}	6.8605×10^{-5}
L7-4	3.22×10^{-5}	5.12×10^{-5}	0.1086	0.5560×10^{-5}	3.7760×10^{-5}
L8-6	2.56×10^{-5}	7.15×10^{-5}	0.0845	0.6042×10^{-5}	3.1642×10^{-5}
L6-9	2.44×10^{-5}	6.16×10^{-5}	0.3788	2.3334×10^{-5}	4.7734×10^{-5}
L9-4	3.96×10^{-5}	8.59×10^{-5}	0.0845	0.7259×10^{-5}	4.6859×10^{-5}

Table 2. The equipment risk value.

Device/line	Probability	PC	Consequence	Risk-value
T1	2.9504×10^{-5}	45000	45000	13276.8
T2	2.7742×10^{-5}	65000	65000	18032.3
T3	2.7037×10^{-5}	45000	45000	12166.7
L5-8	4.5441×10^{-5}	30000	30000	13632.3
L5-7	6.8605×10^{-5}	15000	15000	10290.8
L7-4	3.7760×10^{-5}	35000	35000	13216.0
L8-6	3.1642×10^{-5}	25000	25000	7910.5
L6-9	4.7734×10^{-5}	20000	20000	9546.8
L9-4	4.6859×10^{-5}	30000	30000	14057.7

sponding economic losses. We assume that each line contains PC sets, as shown in Table 2. Based on Formulas (14)~(18), assuming that malfunction lasts for 6 h, to resume production after the 12 h, each PC of failure will lead to the loss of 10000 yuan, thus we can calculate the consequences of each transformer or each route failure of voltage sag. Then, we obtain the risk for the failure level of PC caused by voltage sag.

5 CONCLUSIONS

1. Based on the voltage sag severity index, this paper established a sensitive equipment level of the voltage sag failure probability model, based on the measured data. There is no artificial assumption, and the results are acceptable.
2. From the perspective of economic indicators, this paper established the consequences of the sensitive equipment level of the voltage sag fault model, using economic indicators, and risk models can be directly used in economic analysis and decision-making.
3. Based on the above two points, we can obtain the sensitive equipment level of the voltage sag

failure risk model. This model makes up for the previous voltage sag sensitivity equipment evaluation that only considered the disadvantages of probability, and decision makers and designers can directly make use of the results.

4. Finally, we simulated the model in the IEEE3 machine 9 nodes system, and calculated the risk of the sensitive equipment in the transformer and line voltage sag at the fault level, providing the basis for the actual operation. The result of the example shows the feasibility of the model.

REFERENCES

Chan J Y, Milanovic JV, 2007. Severity indices for assessment of equipment sensitivity to voltage sags and short interruptions [C]. *Power Engineering Society General Meeting, Tampa.*

Chan J Y, Milanovic J V, Delahunty A, 2011. Risk-based assessment of financial losses due to voltage sag [J]. *Power Delivery, IEEE Transactions on*, 26(2): 492–500.

Chan J Y, Milanovic J V, Delahunty A, 2009. Generic failure-risk assessment of industrial processes due to voltage sags[J]. *Power Delivery, IEEE Transactions on*, 24(4): 2405–241.

Djokic S Z, Desmet J, Vanalrne G, et al, 2005. Sensitivity of personal computers to voltage sags and short interruptions [J]. *IEEE Transactions on Power Delivery*, 20(1): 375–383.

Gupta C P, Milanovic J V, 2006. Probabilistic assessment of equipment trips due to voltage sags [J]. *IEEE Transactions on Power Delivery*, 21(2): 711–718.

Khanh B Q, Won D J, Moon S I, 2008. Fault distribution modeling using stochastic bivariate models for prediction of voltage sag in distribution systems [J]. *Power Delivery, IEEE Transactions on*, 23(1): 347–354.

Milanovic J V, Gupta C P, 2006. Probabilistic assessment of financial losses due to interruptions and voltage sags, part II: practical implementation [J]. *IEEE Transactions on Power Delivery*, 21(2): 925–932.

Wagner V E, Andreshak A A, Staniak J P, 1990. Power quality and factory automation [J]. *IEEE Transactions on industry applications*, 26(4): 620–626.

Architectural, Energy and Information Engineering – Sung & Chen (Eds)
© 2016 Taylor & Francis Group, London, ISBN 978-1-138-02791-6

Study on the preliminary mining scheme of the deep orebody of Haikou phosphate of Yunnan Phosphate Chemical Group Co., Ltd

Yao Ji Li & Xiao Shuang Li
National Engineering Research Center of Phosphate Resources Development and Utilization, Kunming, China
Yunnan Phosphate Chemical Group Co. Ltd., Kunming, China
China Sino-Coal International Engineering Group, Shenyang Design and Research Institute, Shenyang, China

Meng Lai Wang
Yunnan Phosphate Chemical Group Co. Ltd., Kunming, China

ABSTRACT: Taking the Haikou phosphate rock of the largest open-air phosphate mining company-Yunnan Phosphate Chemical Group Co., Ltd (YPC) as the engineering background, a study was conducted on the preliminary scheme mining of the deep ore body of gently inclined and thin to thick phosphate deposits with a soft interlayer, according to the actual technical and economic conditions of the phosphate mine, on the basis of the comprehensive research method of site investigation, theory analysis and engineering analogy. Related research could provide technical and theoretical guidance for the underground mining engineering construction of Yunnan Phosphate Chemical Group Co., Ltd and a large number of similar occurrence conditions of phosphate mines, surrounding the area of Yunnan Dianchi Lake.

Keywords: Phosphate mine; preliminary scheme mining of deep ore body; gently inclined and thin to thick phosphate deposits with soft interlayer

1 INTRODUCTION

Phosphate resources in China are distributed mainly over Yunnan, Guizhou, Hubei, Hunan and Sichuan provinces, whose total resource is 16.7 billion tons, and rank second in the world. However, the average grade of the phosphate in China is only 16.95%, ore grade is greater than 30%, the rich ore is only 5.8% of the total, and the low and medium grade phosphate rocks that cannot be directly used is as high as 90% of the total. At the same time, China is the world's first big country that has a rich source of phosphate rock, whose annual consumption is more than 60 million tons of phosphate rock, annual production of rich ore is about 30 million tons, and the remaining rich ore can only maintain mining for more than ten years [1–6].

Yunnan is a big province of phosphorus resources in China, and its phosphate rock resources reserve is 3.959 billion tons, accounting for 23.59% of the total amount in China.

At the same time, Yunnan province in China is a powerful province containing phosphorus and phosphorus chemical industry, and its annual output of the phosphate rock is nearly 30% of the total. The high concentration of phosphate fertilizer production capacity accounts for 45% of the total, and the yellow phosphorus production capacity accounts for 43% of the total. The efficient use of the phosphate rock resource has an important role to play in the construction of the state-level base of phosphate and compound fertilizer and implements the strategy for the development of the West [7–10].

2 THE GENERAL SITUATION OF HAIKOU PHOSPHATE MINE

2.1 *The current situation of mine*

The mining time of Haikou phosphate mine is relatively late. The initial design of Haikou phosphate mine is divided into 4 mining areas: the first mining period is mainly in No.1 and No.2 mining area, the No.3 mining area is the recovery of high grade ore for ore blending, and the second mining period is mainly in No.3 and No.4 mining area. The No.3 mining area is the largest scale in production over all the four mining areas in Haikou phosphate mine. Its reserves accounts for about half of the mining area, and the average grade of I and II phosphate rock is 29.26%. The mining levels of No.1, No.2 and No.3 mining areas are +2340 m,

+2310 m and +2310 m, respectively, and the No.3 mining area is the main mining area. The polish and back nest burial site of the No.4 mining area have been depleted, and a village in the mining area has affected the next stoping. As the village relocation costs are very high, more than 10 million tons phosphate resources in the No.4 mining area are currently not considered for mining. The production capacity of Haikou phosphate mine at present is 2.5 million tons per year, as the scrub mine is 1 million t every year and the flotation mine is 1.5 million t every year.

2.2 Ore body characteristics

Haikou phosphorus deposit has two layers of ore, stuffed with a layer of sandy dolomite, and the seam occurrence is conditioned by tilting anticline. Due to the backplane silicon-based dolomite limestone dissolution, under the effect of gravity, the seam direction traction cave in the hillside, so the small fold and small fracture develop well, and appears to be of microwave type. The angle of inclination of phosphorus deposit edge part sometimes becomes steeper, and the thickness of its phosphate rock layer becomes larger. In general, the thickness of the upper seam is larger than the lower seam, and stable. The seam inclinations of the No.1 and No.2, No.3 and No.4 mining areas are 4° to 7°, 5° to 7°, and 8° to 10°, respectively.

The seam thickness of the upper seam is commonly 8 m to 10 m, the No.3 mining area is 5 m to 8 m, and the average layer thickness of the upper seam of Haikou phosphate mine is 7.24 m. There is a layer of relatively stable granular phosphorites in the top of the No.3 mining area, and it gradually becomes thin to pinch near the No.1 mining area. The grade of the phosphate ore of the lower seam is commonly 25% to 30%, and the highest grade can be up to 36%. With the increasing depth of the ore body, the grade of the phosphate ore gradually reduces to 15% to 20%.

For the lower seam of Haikou phosphate mine, except the top bioclastic phosphate rock with a stable horizon outside, the other type of phosphate rock is staggered development block, and it is divided into the same seam with the upper seam. The seam thickness of the lower seam is commonly 5 m to 8 m, the seam thickness in the No.1 and No.4 mining areas is 4 m to 5 m, and the average layer thickness of the lower seam of Haikou phosphate mine is 4.37 m. The seam thickness in No.1, No.1 and No.4 mining areas is not stable. Many sections of the No.1, No.1 and No.4 mining areas are out of mining sections. The No.4 mining areas, though having a large area, cannot be mined, and in the table and outside orebody, mutual wear is removed very powerfully. Except the individual

lots with a high grade of phosphate ore, the grade of phosphate ore in most areas is 15% to 20%.

2.3 Engineering geological condition

Mining rock strata occurrence is gentle, and basically contains no water, thus being conducive to the stability of the slope. The ore body mining step height is 10 m, the slope angle of the soft rock and weathering fissures is 35° to 40°, the slope angle of semi-hard strata is 40° to 50°, and the slope angle of hard strata is 50° to 60°.

The ore hardness coefficient is 3 to 15, commonly ranging between 5 and 10. The coefficient of volumetric expansion is 1.40, and its soil natural angle is 36° to 38°. The direct roof is siliceous dolomite containing phosphorus, the direct floor is dolomitic limestone, and the interlayer stratum is sandy dolomite containing phosphorus. The thickness of the interlayer stratum is 1.76 m to 22.46 m, and its average thickness is 10 m. The hardness coefficient of the surrounding rock is 6 to 12, the coefficient of volumetric expansion is 1.46, and its soil natural angle is 30° to 36°.

3 THE PRELIMINARY MINING SCHEME OF THE DEEP UNDERGROUND ORE BODY

3.1 The selection of the mining method

The filling mining method has many advantages, such as high recovery rate, low depletion rate, strong ability to adapt to the ore body, environmental protection, safety, and reliability. However, the cost of the filling method is very expensive, especially for the low grade, low value of the metal and nonmetal mines. As the grade of the phosphate ore of Haikou phosphate mine is low, its value is cheap, and it cannot be used for the filling method in mining. According to the condition of mining technology of the ore body, the selection of mining methods can have a comprehensive method, room and pillar method, the sublevel open stope method and the slicing and caving method.

A comprehensive method is basically similar to the room and pillar mining method, which is fully open section along the tendency of the ore body, moves forward along the strike of the ore body, and leaves the irregular points column in the mining process. It is fit for a thin ore body with a thickness thinner than 3 m to 4 m, and medium thick ore body mining can be layered, but less applied. The large angle of the ore body can use comprehensive shrinkage for exploitation.

The room and pillar method is used to divide the ore body block in stope along the strike, and leave the rule pillar, mining from bottom to top propulsion, and each ore block can push the bench at the same time. The room and pillar method is generally used in a thin ore body, and medium thick ore body mining can also be used. With the development of mine equipment and the improvement of the drilling equipment, there are more cases of application of the room and pillar method in the medium thick ore body, and the adaptation of the dip angle of the ore body is also very large, and the angle of the ore body is commonly smaller than 30°.

Based on the characteristics of the comprehensive method and the room and pillar method, combined with the specific ore body occurrence Haikou phosphate mine, the room and pillar method is appropriate, and a comprehensive method should to be excluded.

The sublevel open stoping method is mainly used in inclined and above the ore body. It is also used in a slowly inclined ore body, such as the Kaiyang phosphate mine. The ore body inclination of Haikou phosphate changes greatly, thus the sublevel open stoping method is not suitable for application.

Although layered sublevel caving stoping has the following advantages, such as process flexible, adaptable to changes in the ore body, and simple operation, it has the following shortcomings, such as the low degree of mechanization, the big intensity of labor, the low production efficiency, the small capacity, quantities waterproof and drainage, the big stope support workload and big wood consumption, and poor ventilation condition, so the layered sublevel caving stoping method is not suitable for application.

On the basis of a comprehensive analysis and comparison, the room and pillar method is recommended for the underground mining method for Haikou phosphate rock.

3.2 Room and pillar method

3.2.1 The stope structure parameters
Mining area is along the ore to layout, and its length is 50 m to 80 m. It is divided into 4 to 6 chambers, the width of the room is 10 m to 12.5 m, and the width of the continuous column ranges between 4 m and 5 m. Ore pillar is reserved in the room, the diameter is 3 m to 4 m, and the distance between two pillars is 8 m to 10 m. According to the ore body dip, the height of the phase height is 40 m to 60 m, which is divided into two sections, whose section height is 20 m to 30 m. The stope length ranges from 50 m to 60 m, and from 4 m to 5 m thick at the bottom of the column is reserved as the sublevel or needle of the next phase or segment.

3.2.2 Quasi and cutting
Loading position is arranged along the vein ore body footwall, and the haulage roadway is arranged along the vein in the ore body footwall. The centerline of each chamber layout drawing of chute, at the bottom of the ore room decorates scraper cavern tiberium spike, the district rise layout in the ore room floor and along the center line of the room. Each room is contacted by the roadway, which is diverged in the ore body, and the cutting roadway is drilled along the bottom of the ore room boundaries.

3.2.3 Recovery
When the thickness of the ore body is less than 3 m, it should use the whole layer recovery, and when the thickness of the ore body ranges between 3 m and 8 m, it can use pull bottom pick mining top way for recovery. The stoping sequence is the first pull and then pick bottom top, with each height ranging between 2 m and 3 m. When the thickness of the ore body is more than 8 m, it should use the slicing or deep hole mining for recovery, and the working surface uses the inverse tilting stage for propulsion.

3.2.4 Rock caving
Shallow hole drilling machine is the YT-27 rock drill, which makes horizontal shallow holes and lateral caving. It makes upper and horizontal drilling when pulling bottom or slicing. The diameter of the shallow hole is 40 mm, and the depth of the shallow hole is mostly 3 m, with no more than 5 m. Hole spacing ranges from 0.6 m to 0.8 m and row spacing from 0.8 m to 1.0 m. It has an alternative arrangement.

Long and medium hole drilling machine used is the YG90 rock drill, which makes up or down parallel hole. The diameter of the shallow hole is 55 mm, and the depth of the shallow hole is mostly 6 m to 15 m. Hole spacing ranges from 1.1 m to 1.3 m and row spacing from 1.8 m to 2.0 m. It has a staggered charge package. Cartridge diameter is 45 mm, bulk explosive, and its way of initiation is detonating cord detonator.

3.2.5 Ore removal
The caving ore is slipped into a well by the Electric LHD, whose volume ranges between 2 m³ and 4 m³, and the loading is done in the haulage roadway below the slip.

The location for a scraper is inconvenient, the winch is applied, and the scraper's volume is 0.4 m³. It is taken as the auxiliary carrier tools for the ore body.

3.2.6 Roof control
Although the overall firmness of the ore body roof rock is good, timely pumice processing and necessary support should be done after blasting, so as

to ensure the safety of the mine. For the poor stability area of the roof, bolt metal mesh roof support should be used after the roof cutting, which can increase the sprayed concrete support in the special area.

3.2.7 *Goaf treatment*
Mined out area should be filled by waste rock or filling materials as far as possible, and the roadway that communicates with the stope should be closed in time.

4 CONCLUSIONS

The gently inclined thin and medium thick phosphate ore body mining problem still exits at present, such as the matching of the mechanical equipment, the mining method and the mining technology.

Based on the analysis from the perspective of safety, economy and technology, the feasibility mining scheme of deep ore body of Haikou phosphate mine is studied. Combined with the specific geological and mining conditions, the room and pillar mining method for the underground mining scheme is determined, and the mining method of stope structure parameters is presented.

ACKNOWLEDGMENTS

This work was financially supported by the National Science and Technology Support Program during the Twelfth Five-Year Plan Period of China (No. 2011BAB08B01, No. 2013BAB05B06) and the Talents of Science and Technology plan of China (No. 2014HAB014).

REFERENCES

[1] Yang Li, Xinming Wang, Jianwen Zhao. Optimal selection of mining method of transition from open-pit to underground mining in Shirengou [J]. Metal Mine, 2011, 46(7): 19–23. (in Chinese).

[2] Xiaoshuang Li. Study on Stability of the open pit slope and deformation and failure rule of the underground stope and overlying rock strata after transformation from open-pit to underground Mining [D]. Chong Qing: Chongqing University, 2010. (in Chinese).

[3] Shengqing Xie. Research of the Huangmailing phosphorite safe and smooth transfer technology from open pit to underground [D]. Chang Sha: Central South University, 2011. (in Chinese).

[4] Clough A. K. Variable factor of safety in slopes stability analysis by limit equilibrium method [J]. 1st Eng., 1988, 69(3): 149–155.

[5] Yang Li, Wang Xinming, Zhao Jianwen. Production steady and slope stability analysis during the transition period of open pit to underground in Miaogou Iron Mine [J]. Metal Mine, 2013, 48(1): 7–10. (in Chinese).

[6] Xiaoshuang Li, Yaoji Li. Phosphate rock resources exploitation status and prospect in China [C] //China's mining technology conference corpus in 2012. Ma Anshan: Metal mine magazine, 2012: 20–23. (in Chinese).

[7] Darrah, P.R. The Rhizosphere and Plant Nutrition—a Quantitative Approach [J]. Plant and Soil, 1993, 115: 1–20.

[8] Yamagami T., Ueta Y. Search for noncircular slip surfaces by the Morgenstern-Price method [J]. Proc., the 6th Int. Conf. Number. Methods in Geomech., 1988, 23(6): 1219–1223.

[9] Yaoji Li, Xiaoshuang Li, Gun Huang Research on the large state-owned phosphate mines transform from open pit to underground mining [M]. Chongqing: Chongqing University Press, 2011 (in Chinese).

[10] Yaoji Li, Xiaoshuang Li, Dongming Zhang. Deep underground mining technology of phosphate mine [M]. Bei Jing: China Metallurgical Industry Press, 2013 (in Chinese).

Architectural, Energy and Information Engineering – Sung & Chen (Eds)
© 2016 Taylor & Francis Group, London, ISBN 978-1-138-02791-6

Design of sensor based on laser trigonometry

Zhong Hu Yuan, Xue Lian Zhang, Sa Liu & Ying Kui Du
College of Information and Engineering, Shenyang University, Shenyang, China

ABSTRACT: This paper introduces optical sensor design based on laser triangulation principle. We construct the principle formula of optical triangulation method for displacement measurement by using direct laser triangulation principle. We proposed a new automatic calibration identification algorithm based on the theory of image corrosion and minimum convex hull and adopted Zhang Zhengyou classical calibration algorithm to design and achieve automatic camera calibration system based on checkerboard areas automatic recognition and feature point extraction. The operation is very convenient and efficient. The experimental results show that the system has very good calibration accuracy and reliability under the condition of complicated illumination and background.

Keywords: Laser triangulation method; Image corrosion; Minimum convex hull; Zhang Zhengyou uncertainty perspective algorithm

1 INTRODUCTION

Laser triangulation method is a non-contact 3D measurement method. It combines with modern electronic technology and can be widely used in industrial measurement, medical diagnosis, computer vision and other industries. The method has the advantages of fast measurement speed, high precision, non-contact [1].

Laser triangulation measurement object morphology principle is already proposed. At present, The object shape measurement methods based on the principle have: scanning method, the fringe projection method, encoding method, phase measurement [2]. These methods needed to measure the system parameters (the relative positions light source and cameras, light source and the measured object, camera and the measured object). This paper presents a geometric shape measurement method based on laser triangulation principle. establishing the mathematical model of sensor. We need simply put observed quantities into the relationship and get measured quantities. We apply Zhang Zhengyou uncertain perspective method to calibrate the internal and external parameters of camera. The calibration method is simple and has high precision.

2 DESIGN OF SENSOR

Direct laser triangulation measurement principle [3,4] is shown in Fig. 1. The laser incident beam is perpendicular to the test surface normal of the object. Incident point moves along the optical axis when object moves or surface changes. Position sensitive detector CCD receives the scattered light from the incident light spot of the lens, forming image.

It can be seen from the figure that the angle between the laser beam and normal direction is 0 degree. The angle of the reflected beam AA′ and the normal direction is called as α. The angle of the photosensitive unit and reflected beam AA′ is called as β. The object distance AO of incident light spot and receiving lens optical center is l1. The image distance OA′ of forming image distance in photosensitive unit about light through the lens is l2. With the movement distance y of the reference plane, spot movement distance is x in photosensitive

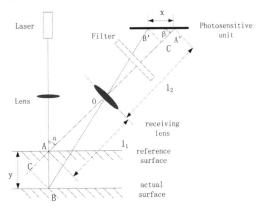

Figure 1. Direct laser triangulation measurement principle.

unit. Lens focal length is f. We respectively make vertical lines of A′A, AA′ extension lines through B, B′ and pedals respectively are C, D. We get the follow formula by the triangle similarity.

$$\frac{\overline{B'D}}{\overline{BC}} = \frac{\overline{OD}}{\overline{OC}} = \frac{l_2 - \overline{DA'}}{l_1 + \overline{AC}} \qquad (1)$$

where,

$$\overline{B'D} = x\sin\beta, \overline{BC} = \overline{BA}\sin\alpha, \overline{DA'}$$
$$= x\cos\beta, \overline{AC} = \overline{BA}\cos\alpha, \overline{BA} = y$$

It can be launched:

$$y = \frac{xl_1\sin\beta}{l_2\sin\alpha \mp x\sin(\alpha+\beta)} \qquad (2)$$

We take "−" when the actual surface is under the reference plane. Conversely, we take "+". l1, l2, α, β, f are known after light path of system is determined. As long as the displacement X of photosensitive unit is calculated, object movement distance y can be determined.

3 CALIBRATION OF SENSOR

3.1 *Camera parameters calibration*

The calibration process of sensor parameters includes calibration of camera parameters and structure light plane parameters [5, 6]. We make sensor calibration parameters based on Zhang Zhengyou uncertainty perspective method [7]. In the calibration calculation process, We assume that one point of the image coordinate system is $m_i = (\mu_i, \nu_i)^T$. Its corresponding 3D space coordinate point is $p_i = (x_i, y_i, z_i)^T$. Then, We have follow equation [8].

$$s\tilde{m}_i = M_i\tilde{p}_i \qquad (3)$$

where, s is a non—zero scale factor. M_i is camera calibration matrix. We make $M_i = K[R \ t]$. The rotation matrix R of 3×3 and translation vector t of 3×1 are external parameters of the camera. K is internal parameters of the camera. We make K as [9]: $K = \begin{bmatrix} f_x & 0 & \mu_0 & 0 \\ 0 & f_y & v_0 & 0 \\ 0 & 0 & 1 & 0 \end{bmatrix}$, where, f_x and f_y are effective focal length of camera. Their value respectively are projection in direction x and y axis. It makes pixels as company. (μ_0, v_0) is main point coordinate. It shows intersection image coordinate about main axis and image plane.

$$s\begin{bmatrix} \mu_i \\ v_i \\ 1 \end{bmatrix} = \begin{bmatrix} M_{11} & M_{12} & M_{13} & M_{14} \\ M_{21} & M_{22} & M_{23} & M_{24} \\ M_{31} & M_{32} & M_{33} & M_{34} \end{bmatrix}\begin{bmatrix} x_i \\ y_i \\ z_i \\ 1 \end{bmatrix} \qquad (4)$$

That is, the formula contains three equations:

$$s\mu_i = M_{11}x_i + M_{12}y_i + M_{13}z_i + M_{14} \qquad ①$$

$$s\nu_i = M_{21}x_i + M_{22}y_i + M_{23}z_i + M_{24} \qquad ②$$

$$s = M_{31}x_i + M_{32}y_i + M_{33}z_i + M_{34} \qquad ③$$

①/②③/③ Then:

$$x_iM_{11} + y_iM_{12} + z_iM_{13} + M_{14} - \mu_ix_iM_{31} \\ - \mu_iy_iM_{32} - \mu_iz_iM_{33} = \mu_iM_{34}$$

$$x_iM_{21} + y_iM_{22} + z_iM_{23} + M_{24} - \nu_ix_iM_{31} \\ - \nu_iy_iM_{32} - \nu_iz_iM_{33} = \nu_iM_{34}.$$

We get n known points in calibration block. Assume that they are space coordinate (x_i, y_i, z_i) $(i = 1, ..., n)$, image coordinate $(\mu_i, \nu_i)(i = 1, ..., n)$. Then we have 2n linear equations:

$$\begin{bmatrix} x_1\, y_1\, z_1\, 1\, 0\, 0\, 0\, 0 & -\mu_1x_1\, -\mu_1y_1\, -\mu_1z_1 \\ 0\, 0\, 0\, 0\, x_1\, y_1\, z_1\, 1 & -\nu_1x_1\, -\nu_1y_1\, -\nu_1z_1 \\ ...... \\ x_n\, y_n\, z_n\, 1\, 0\, 0\, 0\, 0 & -\mu_nx_n\, -\mu_ny_n\, -\mu_nz_n \\ 0\, 0\, 0\, 0\, x_n\, y_n\, z_n\, 1 & -\nu_nx_n\, -\nu_ny_n\, -\nu_nz_n \end{bmatrix} \cdot \begin{bmatrix} M_{11} \\ M_{12} \\ M_{13} \\ M_{14} \\ M_{21} \\ M_{22} \\ M_{23} \\ M_{24} \\ M_{31} \\ M_{32} \\ M_{33} \end{bmatrix} = \begin{bmatrix} \mu_1M_{34} \\ \nu_1M_{34} \\ ... \\ ... \\ ... \\ ... \\ ... \\ ... \\ ... \\ \mu_nM_{34} \\ \nu_nM_{34} \end{bmatrix} \qquad (5)$$

Then, We have $P \cdot M = U$. making $M_{34} = 1$. At last, we get $M = P^{-1}U$.

The relationship between M matrix and intrinsic and extrinsic parameters of camera is:

$$M_{34}\begin{bmatrix} M_1^T & M_{14} \\ M_2^T & M_{24} \\ M_3^T & 1 \end{bmatrix} = \begin{bmatrix} f_x & 0 & \mu_0 & 0 \\ 0 & f_y & v_0 & 0 \\ 0 & 0 & 1 & 0 \end{bmatrix} \cdot \begin{bmatrix} r_1^T & t_x \\ r_2^T & t_y \\ r_3^T & t_z \\ 0^T & 1 \end{bmatrix} \qquad (6)$$

where, M_i^T $(i = 1\text{~}3)$ is row vector consisting of the front three elements of the i row in M matrix.

$M_{i4}(i=1\sim3)$ is the forth column elements of the i row in M matrix. $r_i^T(i=1\sim3)$ is the irow in rotation matrix R. t_x, t_y, t_z is respectively three components of translation vector t.

It deformed as:

$$M_{34}\begin{bmatrix} M_1^T & M_{14} \\ M_2^T & M_{24} \\ M_3^T & 1 \end{bmatrix} = \begin{bmatrix} f_x r_1^T + \mu_0 r_3^T & f_x t_x + \mu_0 t_z \\ f_y r_2^T + \nu_0 r_3^T & f_y t_y + \nu_0 t_z \\ r_3^T & t_z \end{bmatrix} \quad (7)$$

We get: $M_{34} \cdot M_3 = r_3$ and $|r_3| = 1$ (orthonormal matrix). It can be launched: $M_{34} = \frac{1}{|M_3|}$. The following formula can be obtained:

$$r_3 = M_{34} \cdot M_3$$

$$\mu_0 = (f_x r_1^T + \mu_0 r_3^T)r_3 = M_{34}^2 M_1^T M_3$$

$$\nu_0 = (f_y r_2^T + \nu_0 r_3^T)r_3 = M_{34}^T M_2^T M_3$$

$$f_x = M_{34}^2 \cdot |M_1 \times M_3|$$

$$f_y = M_{34}^2 \cdot |M_2 \times M_3|$$

We can furtherly obtain:

$$r_1 = \frac{M_{34}}{f_x}(M_1 - \mu_0 M_3)$$

$$r_2 = \frac{M_{34}}{f_y}(M_2 - \nu_0 M_3)$$

$$t_z = M_{34} -$$

$$t_x = \frac{M_{34}}{f_x}(M_{14} - \mu_0)$$

$$t_y = \frac{M_{34}}{f_y}(M_{24} - \nu_0)$$

The camera intrinsic and extrinsic parameters are calibrated. We complete the calibration work of sensor.

3.2 Checkerboard corners sequence

We adopted minimum convex hull method to determine the four poles of checkboard. The four corners of checkboard are four pole points. Determination of four pole points is important for checkerboard corners sequence. As shown in Fig. 2, We selects the minimum ordinate point P_0 as the starting point in step (1); we connects P_0 and its surrounding points, calculating the counterclockwise rotation corner of ov shaft and each connections and marking minimum rotation corner θ_{1min} as corresponding point P_1 in step (2); we repeat step (2), determine the points P_2, P_3, ..., P_n until points are identified by $(\theta_{1min}, \theta_{2min}, ..., \theta_{kmin}, ..., \theta_{Nmin})$ back

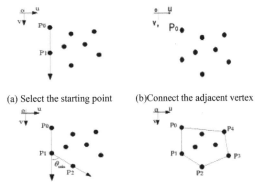

(a) Select the starting point   (b)Connect the adjacent vertex

(c)Calculate the minimum angle (d)Generate the minimum

Figure 2. The generation process of minimum convex hull.

Figure 3. Minimum con- Figure 4. Checkerboard vex hull sequence. four poles.

Figure 5. The 16 position calibration images.

toP_0. Finally we get the minimum convex hull vertices $(P_0, P_1, ..., P_k, ..., P_{N-1})$.

The minimum convex hull vertices sequence generated by checkerboard corners is shown in Fig. 3; we connect fixed-points sequence $(P_0, P_1, ..., P_k, ..., P_{N-1})$ and calculate two adjacent convex hull edge angle $(\phi_0, \phi_1, ..., \phi_k, ..., \phi_{N-1})$ in step (4); we calculate the minimum four corners in $(\phi_0, \phi_1, ..., \phi_k, ..., \phi_{N-1})$. The four angle corresponding to the vertex is the four candidate pole points in step (5); we use the center of gravity method to verify four candidate poles in step (6), The results are shown in Fig. 4.

571

4 EXPERIMENT AND RESULT ANALYSIS

The experiment uses the Basler A312fc camera of PENTAX 16 mm lens. Computer operating system is Windows7 X64, i7–2600 3.4GHz CPU, and 4G RAM. The shot image size is 640 × 480, We collect several images and adopt suggests in reference [10] to select 16 images (see Fig. 5) of moderate position and angle as calibration images. Corner extraction algorithm of this paper combined Zhang Zhengyou classic calibration algorithm and select Visual C++2010 based on OpenCV2.4.2. Experimental results contrast is shown Table 1 with classic matlab Calibration Toolbox in reference [10].

We can obtain from Table 1 that the proposed algorithm has smaller error and greatly improves the efficiency of calibration.

Experiments show that this method is widely applicable. Fig. 6 is shot under dark background and complex environment. The proposed algorithm also can quickly and accurately locate checkerboard corner.

5 CONCLUSIONS

This paper established the mathematical model of sensor by using direct laser triangulation. We take

Table 1. Experimental results comparison.

Parameter type	Reference [11] results	The paper results	Error
Focal length (fx)	1875.74888	1877.40085	−1.65197
Focal length (fy)	1878.22324	1878.08795	0.13529
Main point (u0)	370.32896	372.18348	−1.85452
Main point (v0)	219.68863	221.31043	−1.62180
Distortion (k1)	−0.32854	−0.28251	0.61105
Distortion (k2)	2.62790	1.09013	1.53777
Distortion (p1)	−0.00634	−0.00682	0.00048
Distortion (p2)	−0.00665	−0.00427	−0.00238
Time consuming (s)	327	4.7	

Figure 6. Corner extraction under the complex background.

automatic, fast and accurate extraction of checkerboard corners as breakthrough, combining Zhang Zhengyou classical calibration algorithm. Finally we realized the automatic calibration of camera. Experimental results show that the calibration system proposed by the paper has automaticity, accuracy and rapidity under in different scenarios and light conditions.

ACKNOWLEDGEMENTS

It is a project supported by Science and Technology Foundation of Liaoning, No. LT2013024. The corresponding author is Yingkui Du.

REFERENCES

[1] Howlett, Eric M. "High-resolution inserts in wide-angle head-mounted stereoscopic displays." *SPIE/ IS&T 1992 Symposium on Electronic Imaging: Science and Technology*. International Society for Optics and Photonics, 1992.

[2] Shenghua, Ye, et al. "Visual Inspection Technology and its Application." *Engineering Science*. 1999, 1(1): 49–52.

[3] Wang, Shaoqing, et al, "Rebuilding Principle Formula of Optical Triangulation Method of Displacement Measurement in View of the Lambert Theory." *ACTA OPTICA SINICA*, 2009.

[4] Wu, Jian-feng, Wen Wang, and Zi-chen Chen. "Study on the Analysis for Error in Triangular Laser Measurement and the Method of Improving Accuracy." *Mechanical & Electrical Engineering Magazine*. 2013, 20(5): 89–91.

[5] Tsai, Roger Y. "A versatile camera calibration technique for high-accuracy 3D machine vision metrology using off-the-shelf TV cameras and lenses." *Robotics and Automation, IEEE Journal of* 1987, 3(4): 323–344.

[6] Wang Peng."Study on Key Techniques for Automatic 3D Structured-light Scanning System." Tianjin University, 2008.

[7] Zhang, Zhengyou. "A flexible new technique for camera calibration." *Pattern Analysis and Machine Intelligence, IEEE Transactions on* 2000, 22(11): 1330–1334.

[8] Dong, Xiaoxiao, Limei Song, and Feng Dong. "A 3D coordinate location method based on single camera." *Control Conference (CCC), 2014 33rd Chinese. IEEE*, 2014.

[9] He Juan, "Research on fast corner extraction algorithm in camera calibration." *National University of Defense Technology*, 2011.

[10] Bouguet, Jean-Yves. "Camera calibration toolbox for Matlab." 2004.

[11] Barreto, Joao, et al. "Automatic camera calibration applied to medical endoscopy." *BMVC 2009-20th British Machine Vision Conference. The British Machine Vision Association (BMVA)*, 2009.

Architectural, Energy and Information Engineering – Sung & Chen (Eds)
© 2016 Taylor & Francis Group, London, ISBN 978-1-138-02791-6

Experimental study on steel frame structure considering the initial geometric imperfection

Lu Jin, Man Zhou & Dan Liu
School of Civil Engineering, Shenyang Jianzhu University, Shenyang, China

ABSTRACT: The scaled test of two span double-layer steel frame and braced frame structure is carried out. The influence of initial geometric imperfections on the deformation property is taken as the key research. The distribution of initial defect frame including beam and column before the test is used to the finite element model considering the random initial defect. The results show that the test and the finite element analysis are in good agreement with the consideration of contrast. At last, the correctness of the finite element model of random initial imperfections is proved.

Keywords: Initial geometric imperfection; finite element analysis; deformation property; random initial imperfections; experimental research; advanced analysis

1 INTRODUCTION

Recently, the study on advanced analysis and design method has not been involved in the effect of initial geometrical imperfections on the deformation property yet. To get the deformation properties of the steel frame structure with the initial geometric imperfections and validate reliability of analyzing this kind of problem by using the finite element method ANSYS [1–3], experimental study of single span and double steel frame structure is carried out in this paper.

2 TEST SURVEY

The frame is designed considering the section and beam size with the ratio of 1: 3. The final height of the pure steel frame is 1500 mm, the span height is 1500 mm, and the overall frame height is 3000 mm. According to the specification [4], Q235 hot-rolled h-beam with the feature of smaller section is selected as the material of test model of beam and column, and the parameters of beam-column section are all H100 × 100 × 6 × 8. The entire frame is rigid connection.

In the experiment, two concentrated loads of the vertical and horizontal are applied at the same time. The vertical concentrated load is applied on the distributive girder at the top of the frame by a 30 tons of hydraulic push-pull jack, and the horizontal load is applied on the center section of the top frame beam by a horizontal 25 tons of MTS. During the experiment, the change of horizontal load by load factor is only considered.

Table 1. Mean results of coupon tests.

Standard parts	$E_s(Mpa)$	$f_y(Mpa)$	$f_u(Mpa)$
Steel bar	1.91E+05	309	450
Web plate	2.14E+05	265	387
Flange	1.81E+05	308	408

Table 2. Initial imperfections of column specimens (mm).

Layer		Column		Beam
		Left	Right	
First	δ	1.13	−1.34	1.12
	Δ	1.34	0.56	
Second	δ	1.63	−1.15	1.36
	Δ	2.87	3.54	

Average performance indicators of materials quality testing results are shown in table 1.

Before the test, frame is needed to measuring the initial geometric imperfection. To fully consider the true state of the specimens when simulating. Initial defects of steel frame model before loading is presented in table 2.

3 EXPERIMENTAL PHENOMENA

When the load is about 50 KN, frame tends gradually yield state, then load increases modestly and pillars of lateral growth obviously. When the load increases to 52.14 kN, displacement adds up to

Figure 1. Deformation patterns of instability for model I.

268 mm, the results show that the frame achieves ultimate bearing capacity at this time. Comprehensive experiment phenomenon is found that joint cracking cause the bearing capacity of the frame can't continue to increase, finally frame bear load decreases, and displacement under the action of the MTS continue to increase to 300 mm.

Figure 1 shows the deformation of the model I when unstable failure at last.

4 FINITE ELEMENT ANALYSIS

Steel frame structure is simulated by finite element program ANSYS12.0, considering the initial geometric imperfection according to the measured model specimens [5,10]. The nonlinear analysis of structure is used by arc-length method, and finally some parameters can be got such as the ultimate bearing capacity of the steel frame structure, deformation of the beam, instability mode and internal forces by Supported [6–8]. Contrast test and finite element results are analyzed.

In finite element model, the model of beam and column is simulated by Beam188, and the connection of the beam and column is used with rigid joint. Bar unit can only carry tensile stress. Using translational displacement (UX, UY&UZ) and rotational displacement (ROX, ROY & ROZ) of constraint column foot nodes both within and outside the plane, the rigid joint of column is simulated.

In the finite element model, several parameters such as the initial bending of pillars, initial vertical degree of pillars and initial bending of beam are applied according to the measured values in table 2. In the model analysis, steel material properties are chosen two directivity equivalent strengthen stress-strain curve, then steel is taken 1% of the elastic modulus at strengthening stage. In the simulation, the real material properties such as elastic modulus E and yield stress can be got according to the coupon tests in table 1.

5 COMPARED WITH THE RESULTS OF FINITE ELEMENT ANALYSIS AND THE TEST

Nonlinear factors of structural deformation are studied by the finite element software ANSYS, especially the influence of the steel frame real defects on the deformation properties should be considered based on the test and deformation performance of the pure frame [9]. In steel frame research, it can be seen that the yield strength of sway size and test relative lateral displacement results are well matched, and errors are within 5%. The finite element analysis and the test compared with the load displacement curve is very close to the trend, as shown in Figure 2. But finite element analysis in the frame work of the yield has been increased gradually after load, no drop test., it was caused by constitutive model of the definition of the bilinear isotropic hardening stress-strain curve itself lacks the descending segment and limiting the use of Newton Rap son method.

Based on the key research and analysis of steel frame structure of deformation and stress, observation of a careful analysis of the model of stress nephogram and buckling phenomenon in the actual test and post stress data. It follows that the

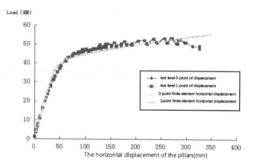

(a) Load-displacement curves of point 0 and 1

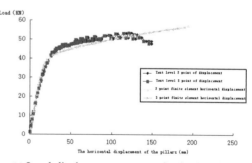

(b) Load-displacement curves of point 2 and 3

Figure 2. Load-displacement curves of steel structure model.

model should be test phenomena stress nephogram and test results show a good agreement, as shown in Figure 3.

To observe the stress distribution and size with increasing of load change, it follows that the lower flange and pull column side beam should stress on the flange force and compression column side beam in almost at the same time in the growth. Two tension side heel stress are together, simulation of the distribution of stress and test record of the stress numerical trend to maintain good, moreover, the overall framework at failure location such as the beam flange weld, the situation of buckling heel condition with finite element simulation coincide, it's proved that the stress state of the true test of frame structure and the results of finite element analysis are fit well.

If the individual is defined in the yield top side load near shift or horizontal load, the finite element software ANSYS deformation size analysis results in load and are larger than the test value, but not more than 10%, as shown in table 3.

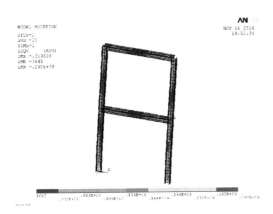

Figure 3. The stress nephogram of model.

Table 3. Experimental and finite element analysis of the data.

Form	Data	Test	Finite element	Error
Bare frame	Yield displacement (mm)	65.02	59.87	7.90%
	Yield load (KN)	43.75	41.23	5.76%
	Ultimate displacement (mm)	268.78	—	—
	Ultimate load (KN)	52.14	54.03	3.62%

6 CONCLUSIONS

In this paper, experimental study was carried out on 1:3 scale of 1 groups of single span two-storey steel structure of the pure frame structure, measurement of install rear beam frame model and the true initial geometric imperfection column, and the ultimate bearing capacity of frame, the distribution of stress and strain control section can be got and the relationship between load and deformation level framework is experiment. Analysis and comparison on a practical framework for the simulation of finite element software and the test results, through test and study the following conclusions can be obtained:

1. Specification of sway should not exceed is 6 mm, however, the structure researches the limit bearing capacity of the frame deformation have been far more than the regulation limits. Fully considering the random distribution of the whole structure and the single member initial imperfection in this experiment, The results show that the structure of the actual bearing capacity and serviceability limit state is directly related to the framework of deformation constraints, and it is necessary to probe deeply into the ultimate bearing and deformation performance of structure under the limit condition.

2. Adding the measured steel frame with initial geometrical imperfections in the finite element model, compare the experimental data of steel frame model with finite element simulation software considering initial defects results framework, the final deformation performance of frame structure can be found, the yield load structure fit will with ultimate bearing capacity, the test samples reveal that by using the finite element software to simulate structure model considering the initial geometric imperfections is correct.

ACKNOWLEDGEMENTS

Natural Science Foundation Project(grant number: 51208314) and Liaoning Province Education Department Project(grant number: L2013234).

CORRESPONDING AUTHOR

Name: Zhouman, Email: 383976439@qq.com, Mobile phone,18040201221.

REFERENCES

[1] Kim, S.E., Chen, W.F. 1996. Practical Advanced Analysis for Braced Steel Frame Design [J]. *Journal of Structural Engineering*, 122(11): 1266–1274.

[2] Zhang, Y.C., Jin, L., Shao, Y.S. 2011. Practical Advanced Design Considering Random Distribution of Initial Geometric Imperfections [J]. *Advances in Structural Engineering*, 14(3): 387–397.

[3] Jin, L., Zhang, Y.C., Zhao, J.Y., Shao, Y.S. 2010. Consideration of initial geometric imperfections in advanced analysis for steel frame, *Progress in Steel Building Structures*, Vol. 12, No. 3, pp. 19–26. (in Chinese).

[4] GB50017–2003 (2003). Code for Design of Steel Structure, China Planning Press, China (in Chinese).

[5] Kim S E, Chen W F. 1999. Design guide for steel frames using advanced analysis program [J]. *Engineering structures*, 21(4): 352–364.

[6] Kim S E, Lee D H. 2002. Second-order distributed plasticity analysis of space steel frames [J]. *Engineering Structures*, 24(6): 735–744.

[7] Liew J Y R, White D W, Chen W F. 1993. Second-order refined plastic-hinge analysis for frame design. Part I [J]. *Journal of Structural Engineering*, 119(11): 3196–3216.

[8] Zhang, Y.C., Jin, L., Shao, Y.S. 2010. Study on deformation performances of steel frames with imperfections considering random distribution [J]. *Journal of Building Structures*, 31(Suppl 1): 1–6. (in Chinese).

[9] Li, G.Q., Liu, Y.S., Zhao, X. 2006. Advanced analysis for steel frames and design of system reliability [M]. *Beijing: China Building Industry Press*: 244–249. (in Chinese).

[10] Jin, L., Zhang, Z.N., Wang, C.G., Zhang, Y.C. 2012. Nonte Carlo-Based analysis method considering random initial imperfections [J]. *Engineering Structures*, 39(Suppl. II): 93–96.

Architectural, Energy and Information Engineering – Sung & Chen (Eds)
© 2016 Taylor & Francis Group, London, ISBN 978-1-138-02791-6

A probability based MAC channel congestion control mechanism for vehicular networks

Yu Wang

Bengbu Navy Petty Officer Academy, Bengbu, China

ABSTRACT: In recent years, IEEE 802.11p/WAVE (Wireless Access in Vehicular Environments) is an emerging family of standards intended to support wireless access in Vehicular Networks. Broadcasting data and control packets is expected to be crucial in this environment. Both safety-related and non-safety-related applications rely on the broadcasting for the exchange of data packets or states. Most of the broadcasting packet is designed to be delivered on a given frequency during the Control Channel (CCH) interval set by the IEEE 1609.4 draft standard. However, the situation of the collision during the process of the channel switch can happen. Therefore, the PCC (Probability based Congestion Control) is proposed in the paper. The mechanism can estimate the numbers of neighboring vehicles by utilizing the broadcasting Hello messages with each vehicle, and then calculating the expected displacement interval values through the probabilistic models and the valued should be added before the original MAC contention window in advance. The mechanism can reduce the collision situations and increase the transmission throughput. For example, a given number of vehicles is 60, the simulation results show that about 45%~60% reduction in packet loss probability could be achieved by using the PCC mechanism.

Keywords: vehicular networks; probabilistic model; congestion control; media access control; MAC

1 INTRODUCTION

The safety-related messages can be transmitted with the method of the periodic and the event triggering among the vehicles under the situation of the Vehicular Communication Networks, such as the J2735 Basic Safety Message defined by the Society of Automotive Engineers, the Cooperative Awareness Message and the Decentralized Environmental Notification Message. The transmission of the above messages can cause the channel congestion under the situation of the driving environment in which there are many numbers of the vehicles and make the urgent safety-message cannot be correctly transmitted and received. The American try finding out the optimal congestion control mechanism with the CAMP program conducted by the eight big car factories, while the EU specify the Decentralized Congestion Control protocol from the access layer, the network and the transmitting layer, the equipment layer and the managing layer, in which the standard of the ETSI TS 102 687 [2] specifies the whole framework of the Decentralized Congestion Control, utilizing the transmit power control, the transmit rate control and the transmit data rate control to solve the condition of the channel congestion. The protocol belongs to the control protocol in the access layer. The standards of the control protocol in other layers (such as the Media Access Control) is continually defined from 2012 to 2013. The paper discusses the problem of the channel congestion based on the viewpoint of the Media Access Control for ensuring the right transmission of the safety-related messages and providing the a more safe driving environment.

The ITS of the Department of Transportation in America planned to design the communication standard of the IEEE 1609 from 2009. The IEEE 1609 standard uses the IEEE 802.11p as the bottom communication technology. The channels in the IEEE 802.11p can be divided into 7 sub channels with 10 MHz, which is composed of a Control Channel (used for transmitting the broadcasting messages and building up the on-line) and six Service Channels (used for transmitting the provided service messages.) The CCH is mainly applied in the communicating application in the Safety aspect, and the SCH is mainly applied in the communicating application in the Non-safety aspect. The function of the multi-channel operation in the IEEE 1609.4 [3] is the operation of coordinating the CCH and the SCH.

At present, the signal antenna in the IEEE 1609.4 standard specification is switched between the CCH and SCH within the 50 ms, while many numbers of the WAVE may suffer the condition of the channel congestion during the process of the channel switch, as shown in the figure 1. If the WAVE Vehicle A wants to transmit a MPDU (MAC Protocol Data Unit) at the bottom of the CCH in the CCH interval, such as vehicle safety message and the Vehicle is still within the SCH interval, the MPDU must be buffered in he queue of the MAC layer. When many MPDU need to be transmitted at the bottom of the CCH in the queuing belonging to many neighboring WAVE Vehicle (the Vehicle A and the Vehicle B, as shown in the figure 1), the same random back-off slot value (such as transmitting the voice message, and its CWmin value is 3) should be simultaneously selected at the bottom of the CCH interval after the Vehicle finishes within the Guard interval (the interval is regarded as the interval needed by the channel switch and the synchronization), as shown in the table 1 [4]. In this way, many WAVE vehicles can obtain the CCH and can transmit each MPDU at the same time so that the MPDU collision among the vehicles can happen. Therefore, the paper will study how to design a mechanism in the MAC under the decentralized environment and cooperate with the neighboring vehicles to dynamically adjust to the value of the original CWmin

and find out the optimal CWmin. The use of the mechanism can effectively reduce the probability of the packet collision and improve the efficiency of the transmission, especially transmitting the broadcasting packets under the environment of the channel switching (without the acknowledgment), such as vehicle safety message.

At present, many researches about the improving mechanism of the channel utilization and the channel switching are discussed in many papers ([5]–[9]), while the patents related to the IEEE 1609.4 Multi-Channel are being applied just by the IEEE 802.11p/1609 standards for the Kapsch [10]–[11]. There are no researches of reducing the channel congestion and the technologies of the well-combined factory channel switching. Therefore, the paper adopts it as the entry point to realize the operating mechanism of the IEEE 1609.4 Multi-channel channel switching. The framework of the paper is as follows: the second chapter is the researches of the related papers, the third chapter proposes the MAC Channel Congestion Control Mechanism based on the probability, the fourth chapter is the results and the discussions of the simulations, and the last chapter is the conclusion.

2 RELATED PAPER RESEARCHES

Both the American standards [12]–[14] and the EU standards adopt to the multi channel operating framework with the 5.9 GHz frequency, and how to effectively adopt these channels to improve the whole efficiency of the network within the tolerable delay time is one of an important issue. At present, the America adopts the IEEE 1609.4 standards to specify the multi channel operating framework, while the EU will do the formulation about the related standards in 2012. The specifications of the IEEE 1609.4 standards divide the channel access time within each second regarding the Sync Interval as the unit, as shown in the figure 2. Each Sync Interval is composed of a CCH Interval and its neighboring SCH Interval, and the preset length of each channel interval is 50 millisecond, in which its initiation is the Guard Interval and includes the time needed by the channel switching and the synchronization. The Control Channel Interval is adopted to control the transmitting & receiving Vendor Specific Action in the channels, the Timing Advertisement and other management frames, while the Internet communicating Protocol data frame is specified in the transmission and the reception of the service channels within the service channel interval. The four kinds of the channel access methods of the IEEE 1609.4 standards specifications in the access parts of the channel meet the requirements of different applications,

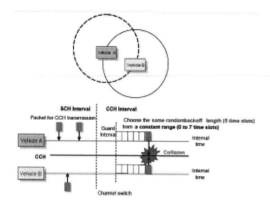

Figure 1. The diagram of the channel congestion caused in the channel switching among the vehicles.

Table 1. The parameters used by the EDCA.

Traffic type	Message type	AIFSN	CWmin	CWmax
Voice	Accident	2	3	7
Video	Possibility of accident	3	3	7
Best effort	Warning	6	7	15
Background	General	9	15	1023

as shown in the figure 2. The continuous access is allowed to continuously stay in the signal channel, the alternating access must comply with the specifications of the synchronized intervals and be periodically switched between the control channel and the service channel, the immediate access is allowed to switch to the service channel before the control channel intervals are not finished and the extended access is allowed to delay the staying time in the service channel so that the control service intervals are not needed to switch to the control channel.

The problem of the broadcast storm caused by the broadcast packets in the VANETs has been widely known and discussed [15]. Therefore, the main purpose of many researching methods are to solve the problem of the channel contestant and the packet collision under the situation of the VANETs. The back off based on the MAC layer and the re-transmitting mechanism is proposed to increase the throughput of the workshop communication broadcasting packets by Alapati and other authors. Although the proposed framework is similar to the framework proposed by the paper, its method does not consider the reaching-ability of the broadcasting information. A kind of the analyzed model proposed by Campolo and other authors is adopted to estimate the broadcasting efficiency of the CCH belonging to the IEEE 802.11p/WAVE. The model is required to compute the size of the contestant windows and know how the number of the vehicles can influence the success probability of the packet transmission during the process of the WAVE channel switch. The three broadcast suppression technologies based on the probability and the timer are proposed by Tonguz and other authors, that is, the weighted p-persistence, the slotted 1-persistence and the slotted p-persistence. In the meantime, the information received from the GPS and the received signal power values is are regarded as the technological designing basis. The mechanism proposed in the paper regards the one-dimensional highway with the double-way signal way and the general two-dimensional square topology as the verified environment. The restricted flooding mechanism

is proposed by the Reumerman and other authors. The strength of the signals received from the neighboring nodes judges whether the neighboring nodes can be selected as the transmitting packet nodes. Although the method can be applied in the sensing networks, the improvement of the throughout can not be considered. The mechanism proposed by the Mylonas and other authors regards the local information adding to the probability and the vehicle speed as the estimation basis for the rebroadcasting packets, while the method can not rely on the GPS. Suriyapaibon wattana and other authors adopt the GPS to obtain the information of The Last One in the accident sites and then the information should be broadcast again by the car. A kind of the congestion supervision and control mechanism is proposed by Fallah and other authors. Their researches mainly observe the channel occupancy, the feedback measure, the controlled network parameters (such as the ratio of the broadcasting information, the size of the contestant windows and the range of the broadcasting information) and the relationships among the network efficiency.

3 THE MAC CHANNEL CONGESTION CONTROL MECHANISM BASED ON THE PROBABILITY

3.1 *Text and indenting*

The paper is based on the probability and designs a mechanism in the MAC layer for reducing the packet (especially the broadcasting packets) collision, which is called as the Probability based Congestion Control (PCC) approach (the mechanism is combined with the random back-off mechanism formulated by the IEEE 802.11 and provides the compatible ability.) The PCC mechanism includes two functions, the Contestant Estimation Function and the Expected Offset Window Calculation Function. The Offset Window Calculation Function adopts the historic data to compute the number of the broadcast Hello packets delivered by the WAVE vehicle within a certain interval so that the number of WAVE vehicles can be obtained. After the number of the WAVE vehicles in the CCH is obtained, the Offset Window Calculation Function can conduct the computed offset window length (the length depends on the number of the WAVE vehicles) before the random back-off mechanism formulated by the original IEEE 802.11 is conducted. Then a time slot value should be randomly selected from the window length. After the value is selected, it should be combined with the random back-off mechanism formulated in the IEEE 802.11. At this time, a new time slot value should be randomly selected for reducing probability of

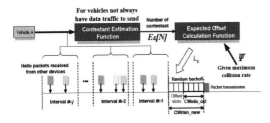

Figure 2. PCC mechanism combines the contestant estimation function with the offset window calculation function.

obtaining the channels and transmitting the packets simultaneously by the WAVE vehicles. The following operations are as shown in the figure 2.

3.2 The contestant estimation function

The probability of the packet collisions is from the following behaviors, many vehicles compete the same medium and transmit the MPDU packets after the channel switching. When the speed caused by the MPDU packets is larger than the transmitting speed, the caused MPDU packets must be put in the queue. In addition to the data rate, the transmission power, the number of the competing vehicles is also one of the variables for influencing the collision. Therefore, the Contestant Estimation Function is adopted to estimate the possible number of the neighboring vehicles. Each vehicle periodically transmits the Hello message packet within the CCH interval (we set one second to transmit once referring to the actual suggested vales.) The size of the Hello information packets is 300 bytes, and its content includes the MAC_id timestamp of the transmitters and the position information obtained from the GPS. It transmits once each second with the robust PHY (adopting to the BPSK 1/2) and the 33 dbm (2 watt) transmitting power. Each vehicle receives the Hello information packets from the neighboring vehicles and calculates the data of all neighboring vehicles through the formula (1). Each vehicle records the information received from the Hello information packets through the table (such as the MAC_id of transmitting the vehicles, the position information of the GPS and the Quiet period). The Quiet period is the differential value between the measuring finished interval and the final received Hello information packets timestamp. Each vehicle after receiving the Hello information packets can adopt the received MAC_id to judge whether the table information should be updated (such as the position information and the Quiet period.) When the Quiet period in the table is larger than the pre-set threshold (If the simulation is set as two seconds, that is, having twice repeating time of the Hello information packets, a Hello information packet is allowed to be lost in the situation.) At this time, the MAC_id can be removed from the table (it represents that the vehicle has left.) Estimating the number of the neighboring possible vehicles can calculate the number of the different MAC_id in the k intervals, as shown in the following formula:

$$E_k[N] = N * \left[\frac{\sum_{kth} \text{Traffic time}_{kth} * a + (1-a) * \text{Traffic } (N-1)}{\text{Total inverval time}} \right]$$

(1)

in which N represents the number of the vehicles, $E_k[N]$ is the average traffic time by integrating all k interval traffic number and the (k−1) interval (as shown in left part of the figure 2). The formula (1) effectively estimates the neighboring vehicle number with the traffic transmission, in which α is the constant weighting factor(the value satisfies with the equation: $0 < \alpha < 1$.)

3.3 The expected offset calculation function

The function mainly estimates the number of the neighboring vehicles with the traffic transmission for the contestants on the basis of the formula (1). The increased displacement length before the initial CWmin can be calculated through the formula (2). The significance of the formula (2) is to give the estimated $E_k[N]$ and the Maximum collision rate. Later, the minimum slot time length L_k can be obtained through the Probabilistic model. (its unit is slots.)

$$L_k = \arg\min_N \left[1 - \frac{\prod_{i=0}^{E_k-1}(N+1-i)}{(N+1)^{Ek}} \leq \psi \right]$$

(2)

The changing situation of the new CW_{min} can be verified through the simulation, as shown in the equation (3). The equation (3) is value by pulsing the displacement length L_k of the increasing neighboring vehicle numbers and the original CW_{min} in the MAC.

$$CW_{min\_new} = L_k + CW_{min\_old}$$

(3)

The PCC mechanism is divided into six steps:

1. The Vehicle utilizes the Expected Offset Calculation Function for calculating the offset window length in the k-1 interval.
2. The Vehicle randomly selects a value from the offset window length and combines with the slot time selected from the traditional 802.11 random backoff mechanism. Later, the value becomes the waited backoff time finally transmitting the packet.
3. When the PCC mechanism finishes, it is back to the Vehicle whether there is the state of having the broadcast packets.
   1. The vehicle is in the SCH state and judges whether there are the broadcast packets to be transmitted. If no, the Vehicle should enter the L is $CW_{min\_new} = L_k + CW_{min\_old}$ ten mode. Otherwise, it should enter the Receiving mode.
   2. The vehicle receives the transmitted broadcast Hello packets from other Vehicles.

(calculating the historic traffic information, that is the number of the Vehicle.)

3. The Vehicle utilizes the Contestant Estimation Function for estimating the contenders in the k interval.

4 SIMULATION VERIFICATION AND ITS DISCUSSION

The simulators can have the IEEE 802.11p and the IEEE 1609. In order to observe and study the influence of the road topology and the car mobility patterns to the PCC mechanism. Therefore, aiming to the road topology, the straight linear highway topology is adopted, as shown in the figure 3. As to the condition of the car mobility, the highway speed should be considered, and the speed of the vehicles in the highway is 90 km/hr or above it. In addition, mobility model developed by the NCTUns6.0 simulators regards the situations of the speeding up, the speeding down and the swerving of the cars as the moving basis.

The figure 3 is the topology and the situation of the highway (10 km) with four lines, and the transmitting range of each vehicle is 500 bytes (the overhead between the MAC and the PHY, including the path history, GPS correction and security information.) The simulating times of each data have 10 times and the simulating time of each data is 500 seconds. In addition, the situation assumes that each vehicle has the data transmitted in the same time. The table 2 is other parameters simulated by the NCTUns.

The figure 4 compares the results of the back off number between the PCC mechanism and the fixed CWmin value. The figure 4 shows that the low congestion happens, its fixed CWmin mechanism experiences over 300 back off, the failure of each transmission causes the size of the contention window becoming twice. When the situation of the congestion become serious, the fixed CWmin mechanism can produce a larger back off value, while the PCC mechanism can have the maximum limitation of the Collision rate and adopt to the random back off mechanism twice. Therefore, the performance of the back off number is better and its value is lower than the value in the fixed CWmin mechanism.

Comparing the PCC mechanism with the fixed CWmin mechanism, the figure 5 observes that the changing of the vehicle numbers and the probability of the packet collision. With the increasing of the neighboring vehicle numbers, a lower CWmin value (CWmin = 32) can cause a higher probability of the packet collision (there are 60 vehicles, and its collision probability is 0.6.) The proposed PCC mechanism can dynamically adjust to the displacement length no matter how many vehicles have and

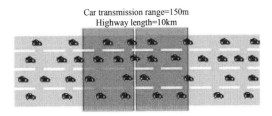

Figure 3. The topology and the situation of the highway (10 km) with four lines.

Table 2. Other parameters simulated by the NCTUns.

Environment parameter	Value
MAC type	802.11p
Data rate	3 Mbps
Transmission range	150 meters
Mobility model	Constant velocity

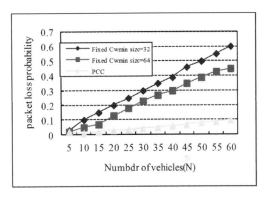

Figure 4. The comparison of the back off number between the PCC mechanism and the fixed CWmin value (from 32 to 64.).

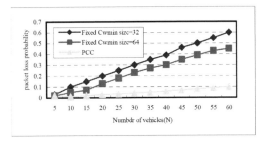

Figure 5. The changing of the vehicle numbers, the observation of the PCC mechanism, and the result of the packet loss ratio.

reach to the minimum probability of the packet collision (its value is about 0.1). The above simulated results show that the designed PCC mechanism has a lower collision probability, its mechanism is easier to be practical. The firmware in the IEEE 802.11p MAC is just needed to be modified.

5 CONCLUSIONS

The PCC mechanism proposed in the paper can not violate the specification of the present IEEE 1609 and can obtain the compatibility of the 802.11p MAC for breaking through the concept of the channel congestion during the process of designing the channel switching. The broadcast Hello packet number transmitted by the WAVE vehicle can be calculated within a certain interval through the historical information. According to the number of the WAVE vehicle taking part in the CCH, the proper offset window length can be calculated. In addition, the PCC mechanism proposed in the paper is compared with the fixed CWmin mechanism, its probability of the packet loss can be reduced 45% to 60% when the number of the vehicles are 60. In this way, the bandwidth resources can be effectively utilized and the transmitting efficiency can be improved. When the mechanism can be applied in the IEEE 1609.4 protocol, the transmitting technology should be applied in the Netcom manufacturers which produce the WAVE/DSRC related equipment. The equipment of the manufacturers can offer more service abilities for the service suppliers through the method.

REFERENCES

Chen, J.H. & Sheu, S.T. & Yang, C.A. 2003, Proc. IEEE Proceedings on Personal, Indoor and Mobile Radio Communications (PIMRC), *A new multichannel access protocol for IEEE 802.11 ad hoc wireless LANs.*

Mcnew, J.P. & Moring, J.T. & Dessouky, K.I. Mar. 26, 2009, U.S. Patent 20090081958, *METHOD AND SYSTEM FOR BROADCAST MESSAGE RATE ADAPTATION IN MOBILE SYSTEMS.*

Mcnew, J.P. & Moring, J.T. & Dessouky, K.I. Sept. 25, 2008, U.S. Patent 20080232433, *SYSTEM AND METHOD FOR SHORT RANGE COMMUNICATION USING ADAPTIVE CHANNEL INTERVALS.*

Mehta, S. & Kwak, K.S. 2010, EURASIP Journal on Wireless Communications and Networking, *Performance Analysis of Binary Exponential Backoff and Improved Backoff for WPAN.*

Ni, S.Y. et al., 1999, Proc. ACM MOBICOM, *The Broadcast Storm Problem in a Mobile Ad Hoc Network.*

So, H.S. & Walrand, J. & Mo, J. 2007. Proc. IEEE Wireless Communications and Networking Conference (WCNC), *McMAC: A Parallel Rendezvous Multi-Channel MAC Protocol.*

So, H.S. & Walrand, J. 2005. Univ. of California, Berkeley, CA, Tech. report, *McMAC: A Multi-Channel MAC Proposal for Ad-Hoc Wireless Networks.*

Suthaputchakun, C. & Ganz, A. 2007. Proc. IEEE Vehicular Technology Conference (VTC-Spring), *Priority Based Inter-Vehicle Communication in Vehicular Ad-Hoc Networks using IEEE 802.11.*

Wang, M.F. Ci, L.L. & Zhan, P. & Xu, Y.J. 2008, Proc. International Colloquium on Computing, Communication, Control, and Management (CCCM), *Multi-Channel MAC Protocols in Wireless Ad Hoc and Sensor Networks..*

2010, IEEE Std. 1609.3, *IEEE Standard for Wireless Access in Vehicular Environments (WAVE) -Networking Service.*

2011, IEEE Draft Std. 1609.1, *Draft Standard for Wireless Access in Vehicular Environments (WAVE)-Remote Management Services.*

2011, IEEE Draft Std. 1609.2, S, *Draft Standard for Wireless Access in Vehicular Environments (WAVE)–Security Services for Applications and Management Messages.*

Architectural, Energy and Information Engineering – Sung & Chen (Eds)
© 2016 Taylor & Francis Group, London, ISBN 978-1-138-02791-6

Study on the technology of Multi-wedge Cross Wedge Rolling (MCWR) forming automobile semi-axis shafts

Chuan Liu, Xue Dao Shu & Ji Dong Ma
Faculty of Mechanical Engineering and Mechanics, Ningbo University, Ningbo, China
Zhejiang Provincial Key Laboratory of Part Rolling Technology, Ningbo, China

ABSTRACT: With the rapid development of automobile industry, the demand for automobile semi-axis has increased generally. As an effective tool, Cross Wedge Rolling (CWR) gives us an opportunity to form automobile semi-axis professionally and in a mass production. However, the cost of dies is high when we usually form automobile semi-axis by single-wedge cross wedge rolling. In addition, MCWR could reduce the cost of dies definitely. Therefore, the study on the deformation of MCWR forming automobile semi-axis is of great significance in the long run. In this paper, the software PRO/E is adopted for setting up the three-dimensional models, and the advanced explicit dynamic finite element ANSYS/LS-DYNA and DEFORM are also used, and then the simulation of rolling automobile semi-axis with MCWR was analyzed systematically. The influence that the rolling parameters has on rolling force was achievedby using a practical computed method. It showed that side wedge forming angles and side wedge transition angles have a little influence on the force, while the coefficient friction has a critical influence on the force. In addition, during the process of rolling forming, the axial force generated is almost balanced from a general analysis. These research results have provided a theoretical basis for realizing professional and mass production of the MCWR automobile semi-axis.

Keywords: MCWR; automobile semi-axis; force parameter

1 INTRODUCTION

Automobile semi-axis shafts are drive shafts of driving wheels, which are power transmission shafts, as shown in Figure 1. Two driving wheels are at the different speeds of rotation in the process of driving; therefore, the two driving wheels cannot be connected by one driving shaft but are driven by two semi-axis shafts that are connected with a differential mechanism. Automobile semi-axis shafts not only transmit torque from the engine, which are an important force transmission part of the car, but also withstand the vertical force and lateral force generated from the wheel, as well as the traction force and the longitudinal force generated from the braking force.

Therefore, the automobile semi-axis shafts are an important carrier transmission system, and one of the vulnerable parts of the automobile currently. The popular methods of producing automobile semi-axis shafts are forging and single wedge cross wedge rolling forming at home and abroad. Single wedge cross wedge rolling forming has lots of advantages such as high production efficiency, material utilization, high quality and low cost compared with the forging method. However, the investment of mold is large by single wedge cross wedge rolling; hence, it will further reduce the costs and weight of the mold and equipment by applying the MCWR technology[1]-[2].

MCWR is one kind of plastic forming technology, which conducts radial pression and axial extension on the raw shafts simultaneously by couples of wedges called the main wedge and side wedges. In addition, MCWR is an advanced precision long shaft parts near net shaping technology, which has many advantages such as saving roll surface, reducing weight of the equipment, high efficiency, saving materials and low costs compared with the

Figure 1. Automobile semi-axis shafts.

single wedge CWR, and also this technology is one of the most effective technologies to produce long shafts professionally and economically[3][4].

Therefore, studying a new effective and energy saving technology on producing large-scale long shafts meets the urgent requirement of the development of the society and the market.

At present, the research study of automobile semi-axis shafts by MCWR is almost vacant worldwide. Therefore, this paper established three dimensional rigid-plastic finite models of automobile semi-axis shafts by MCWR based on the Deform-3D software and ANSYS/LS-DYNA. On the basis of the model construction, the influence rolling parameters on the rolling force were achieved by a practical computed method. It was shown that side wedge forming angles and side wedge transition angles have a little influence on the force, and the coefficient friction has a critical influence on the force. The above research results provided a theoretical basis for realizing professional and mass production of the MCWR automobile semi-axis[5][6].

In conclusion, the study of automobile axle MCWR on automobile semi-axis shafts has critical significance.

2 DESIGNING THE MOLD AND ESTABLISHING THE FINITE ELEMENT MODEL

2.1 Design the mold

There are three ideas to design the mold for automobile semi-axis shafts by MCWR. One is rolling simultaneously the side short parts and the middle long parts, and the advantage of this idea is that the surface of the mold is shorter than other designation, while the disadvantage is that the force is not symmetric in the process of rolling. Another idea is rolling the side short parts of shafts after having rolled the middle long parts of shafts; therefore, the mold could be designed symmetrical, while the length of the die surface is too long. The last idea is to lengthen one of the side wedges to rolling the one side short part of the shaft on the basis of the second idea.

One can design a mold called No. 1 for automobile semi-axis shaft, as shown in Figure 2(a) from the first idea, and this mold's length is 2313.16 mm, which could be worked on an H800 rolling machine.

One can design a mold called No. 2 for automobile semi-axis shaft, as shown in Figure 2(b) from the second idea, and this mold's length is 2538.81 mm, which could be worked on an H1000 rolling machine. Then, one can design a mold called No. 3

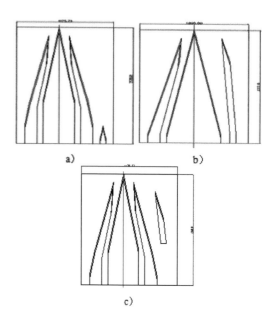

a)   b)

c)

Figure 2.   Benz rear axis mold.

for automobile semi-axis shaft, as shown in Figure 2(c) from the second idea, and this mold's length is 2320.00 mm, which could be worked on an H800 rolling machine.

In conclusion, we selected the No. 1 mold to perform the simulation analysis.

2.2 Establishing the finite element model

To simplify the problem, the following assumptions are made as follows: 1) since rolling is carried out at a high temperature, the elastic deformation of shafts in rolling is much smaller than the plastic deformation, so the flattening deformation of the mold and the elastic deformation of the die could be ignored to some extent, and the die could be deemed as rigid, which will reduce the amount of calculation in the analysis, with the elastic modules E = 210 GPa. 2) The weight of the shafts is ignored and deemed as plastic, with the plastic modules E = 90 GPa. 3) Since the whole rolling process is completed within 3 to 4 seconds, and the time of heat transfer between the die and the shaft with air is intensely short, the rolling temperature could be deemed as constant during the process. 4) To simplify, the friction between the die and the shaft is coulomb friction, assuming the friction between the die and the shaft is the same in all the contact portions. 5) The shaft is not contacted with rolling guides in the process of rolling, and the rolling condition of rolling dies is perfectly symmetrical. The final finite element model is shown in Figure 3.

Figure 3. Finite element simulation mold by multi-wedge cross wedge rolling.

Table 1. Process parameters by mult-wedge cross wedge rolling.

Diameter of dies(mm)	800	Coefficient of friction	0.3
			0.4
Side forming angles(°)	35		0.5
	30		
	25		
Side transition angles(°)	25	Rolling temperature(°C)	1050
	35		
	45		

3 THE RULES OF TECHNOLOGICAL PARAMETERS ON FORCE PARAMETERS

3.1 Process parameters by mult-wedge cross wedge rolling

Using the finite element model, which has been established to perform finite element simulation under the condition of typical process parameters such as side wedge forming angles, the side wedge transition angles and coefficient of friction are given in Table 1.

3.2 Influence rule of side forming angles on the force

Under the condition of side transition $\alpha_z = 25°$ and friction coefficient $\mu = 0.5$, the variation of force parameters affected by different sides wedge angles in the process of rolling is shown in Figure 4. As can be seen from Figure 4, as the side wedge forming angles increase, the force will increase, but to a small extent.

3.3 Influence rule of side transition angles on the force

In the condition of side wedge forming angles $\alpha_i = 30°$, and the coefficient friction $\mu = 0.5$, the variation of force parameters affected by different side wedge transition angles in the process of rolling is shown in Figure 5. As can be seen from Figure 5,

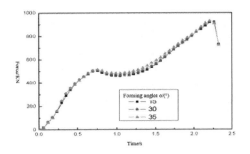

Figure 4. Influence rule of side forming angles on the force.

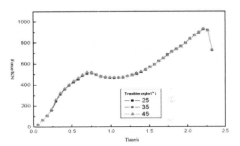

Figure 5. Influence rule of side transition angles on the force.

the increase in side wedge transition angles does not have an obvious effect on the force.

Therefore, changing the transition angles has not obvious effects on the variation of force.

However, the transition angles will influence the surface quality of shafts in the rolling transition section. In conclusion, in terms of surface quality of the transition section, we should choose small transition angles.

3.4 Influence rule of coefficient friction on the force

Under the condition of side transition $\alpha_z = 25°$ and side wedge forming angles $\alpha_i = 30°$, the variation of force parameters affected by different coefficient frictions in the process of rolling is shown in Figure 6. As can be seen from Figure 6, as the coefficient friction increases, the force will increase, and the extent is large. Among the force parameters, it is practically necessary to consider the coefficient friction.

4 INFLUENCE RULES OF THE AXIAL FORCE

The influence rules of the axial force are shown in Figure 7. From the figure, it can be seen that different side forming angles, different side transition

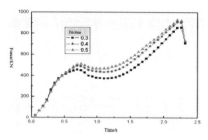

Figure 6.   Influence rule of friction on the force.

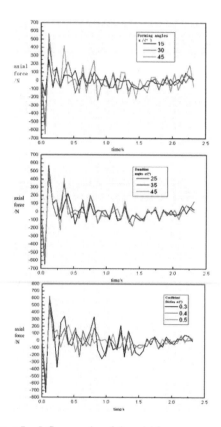

Figure 7.   Influence rules of the axial force.

angles and different coefficient frictions have the same effects on the axial force to a certain extent. Considering that the axial force is extremely smaller than the radial force and tangential force, the axial force could almost be negligible. Besides, during the process of rolling forming, the axial force generated is almost balanced from a general analysis; in other words, it means that the mold is balanced from the axial direction. In addition, asymmetric rolling just generate an imbalance that

can be resolved by the plastic deformation of the shafts in a little moment.

5   CONCLUSIONS

Based on the finite element simulation, the law of automobile semi-axis shaft's force influenced by technological parameters is studied, from which the following conclusions can be drawn:

1. Side wedge forming angles and side wedge transition angles have a little influence on the force.
2. The coefficient friction has a critical influence on the force.
3. Different side transition angles and different coefficient frictions have the same effects on the axial force to a certain extent. During the process of rolling forming, the axial force generated is almost balanced from a general analysis.
4. The above results will provide a theoretical basis for realizing professional and mass production of the MCWR automobile semi-axis.

ACKNOWLEDGMENTS

This paper was supported by the National Natural Science Foundation of China (Grant No. 51475247).

REFERENCES

[1] SHU xuedao, Valery Ya. Schukin, G. Kozhevnikova, et al. Forming technologies and theories of CWR [M], Beijing: China science & technology press, 2014, 2–10.
[2] PENG wenfei, SHEN fa, SHU xuedao, et al. Research on influences of axial movement for cross wedge rolled asymmetric shafts [J]. China Mechanical Engineering, 2014(2): 311–314.
[3] YU penghui, SHU xuedao, PENG wenfei, et al. Influence of Process Parameters on Grain Size of Hollow Axle in Multi-wedge Cross Wedge Rolling [J]. Hot Working Technology 2014(43): 96–99.
[4] YANG Cui-ping, ZHANG Kang-sheng, HU Zheng-huan. Numerical simulation study on the cause of ellipse generation in two-roll cross wedge rolling the hollow parts with uniform inner diameter [J]. Journal of University of Science and Technology Beijing, 2012(34): 1426–1431.
[5] ZHANG ting, SHU xuedao. Study on forming mechanism Rule of Hollow Railway Axle in Multi-wedge Synchrostep Cross Wedge Rolling [D]. Ningbo University, 2013, 35.
[6] JIANG Yang WANG Bao-yu HU Zheng-huan, et al. The effect of process parameter on non- circularity of thick-walled hollow axle during cross wedge rolling [J]. Journal of Plasticity Engineering, 2012(1): 19–24.s

Architectural, Energy and Information Engineering – Sung & Chen (Eds)
© 2016 Taylor & Francis Group, London, ISBN 978-1-138-02791-6

Image color transfer method based on clustering segmentation and particle swarm optimization correction

Zhi Zheng
School of Mathematics and Computer Science, Fujian Normal University, Fuzhou, China

Jian Huang
Fujian Preschool Education College, Fuzhou, China

ABSTRACT: To improve the bad performance in the image color transferring, we presented an image color transfer method based on clustering segmentation and correction. First, we transmit the target and source images into $L\alpha\beta$ color space, using affinity propagation clustering to partition the target and source image, respectively. Second, we search the best match block from all blocks by the mean value and variance of luminance, and use the higher moments and lower moments information to transfer. Then, we filter out the wrong color in the color image. Finally, we use the Multi-swarm Cooperative PSO (MCPSO) to estimate the missing color and complete the whole process. The simulation results show that the transfer effect of this method is better than the traditional one, and can be applied in many fields.

Keywords: affinity propagation clustering; segmentation; color transfer; multi-swarm cooperative PSO; correction

1 INTRODUCTION

Image color transfer has been the hot field of machine vision and image processing in recent years. It means that the color can be obtained from the source image A, without changing the target image texture B and the content, the source image color information can be accessed automatically, and a new visual effect of target image B can be obtained appropriately. This technology includes color image to color transfer, color image to gray-scale transfer, and grayscale to color image transfer. It has the widespread application prospect, which can be used to color the old photographs and films, color the medical image and the night view image, also the creation of advertising and art.

The algorithm was first proposed in [1], by using the statistical information to achieve the same distribution of every color channel of two color images, so as to change the color information. It is the global transfer, but not suitable for a rich color image. Based on [1], the literature [2] matched the luminance and texture information to achieve the color image to grayscale transfer. Due to the low accuracy of lower moments information matching, the pixel with closed color, because the luminance texture change obviously, will occur with the transmission error. In recent years, many scholars have made a lot of improvements in the transfer effect

and efficiency [3–7], but those methods are mainly focused on improving the matching accuracy in the transfer process.

In this paper, for color image transmission errors to grayscale, we proposed a color transmission method based on clustering segmentation and correction. First, we use the affinity propagation clustering (AP) algorithm to divide the source image and target image into several blocks, and then calculate the luminance mean and variance of each block, so that each block can be corresponded, respectively, to the similarity. Furthermore, we match and transfer the color of the corresponding block according the higher moment and lower moment information of pixels in each block; next, by the correlation of adjacent pixels in initial formed color image, we remove the possible wrong color of the pixel point. Through the MCPSO, we then estimate the lost color by fitness function, and finally complete the color of the target image. The experiments show that this method can effectively reduce the transmission error rate and make the image more smooth, natural and achieve the good effect in transfer.

2 AP ALGORITHM

The traditional method searches the best matching point of the target image in the whole source image. For one image with rich color but with a little change in luminance, it is easy to lead to transmission error in the regions with a similar brightness because the matching precision is not enough. In this paper, we divide the image according to the visual content first. We then divide the source image and the target image into several blocks, corresponding, respectively, to the similarity. In this paper, we use the affinity propagation clustering (AP) algorithm for image segmentation. The AP algorithm is presented at the "SCIENCE" in 2007 [10]. It can avoid the dependence on the initial cluster center selection of traditional clustering algorithms [8,9], so it has a wide range of applications, which is suitable for large-scale data, and quickly achieves a better clustering effect [9,11].

The main parameters of the AP algorithm are: similarity matrix s, $s(i,k)$ as the similarity between the sample x_k and x_i, which represents the suitable degree of x_k as the representative point of x_i. The bigger the parameter $s(i,k)$ is, the more suitable the x_k as the representation of x_i will be. Among the responsibility matrix $R[r(i,k)]_{m*n}$ and availability degree matrix $A[a(i,k)]_{m*n}$, $r(i,k)$ means x_i points to candidate point x_k, and represents whether x_k is appropriate as the class representative point of x_i. $a(i,k)$ means the candidate class representative point x_k points to x_i, and represents whether x_i, to choose the x_k as its class representative point. For each sample point x_i, according to the iterative update by Formulas (1)–(3), we get x_k as its class representative point that satisfies $\text{argmax}\,(r(i,k)+a(i,k))$. The damping factor $\lambda \in [0,1)$ is used to eliminate vibration and accelerate the speed of convergence:

$$r^{(t)}(i,k) = (1-\lambda)*(s(i,k)$$
$$- \max_{k's.t.k'\neq k}\{a^{(t)}(i,k')+s(i,k')\}) + \lambda * r^{(t-1)}(i,k) \quad (1)$$

$$a^{(t)}(i,k) = (1-\lambda)*(\min\{0, r^{(t-1)}(k,k)$$
$$+ \sum_{i's.t.i'\notin\{i,k\}} \max\{0, r^{(t)}(i',k)\}\}) + \lambda * a^{(t-1)}(i,k) \quad (2)$$

$$a^{(t)}(k,k) = (1-\lambda)*(\sum_{i's.t.i'\notin\{i,k\}} \max\{0, r^{(t)}(i',k)\})$$
$$+ \lambda * a^{(t-1)}(i,k) \quad (3)$$

The segmentation procedures are as follows.

Step 1. Because the target image is grayscale and has no color information, we choose the distance dc from the center point of the image and each pixel's luminance l, neighbourhood luminance mean value lm, neighbourhood luminance variance value lsd, skewness value ls and kurtosis value lk in the target image as the properties of sample point, and the neighbourhood size is 3*3; select the luminance of each pixel and the channel α, β in $L\alpha\beta$ color space as the properties of the sample point in the source image.

Step 2. Initialize the parameters of the AP algorithm in the source image and the target image, respectively. Then, set $R[r(i,k)]_{m*n}$ and $A[a(i,k)]_{m*n}$ for 0, calculate $s(i,k) = -\|x_k - x_i\|^2$, and assume the probability of each sample point as the class representative point is the same, so making $s(k,k) = p$, the value of p influences the final number of clusters. Here, we let p equal to the average of all the elements in s, and set λ and the number of iterations.

Step 3. According to (1)–(3), iterative and update the summary until they satisfy the number of iterations, output n representative points and assign the other pixels to the corresponding classes.

3 BLOCK MATCHING AND TRANSFER

After dividing the source image and the target image into several blocks, respectively, we correspond the blocks of the target image and the source image. Finally, we transfer the color between the corresponding blocks preliminarily. Block transfer can reduce the transmission error due to the closed luminance information, but different visual contents before undividing greatly. In addition, we improve the transmission effect, and reduce the workload for the subsequent color correction. Because the source image has the color information but the target image only has luminance information, we choose the luminance means and variance as the matching criteria.

The steps are as follows:

Step 1. Calculate the luminance means u and variance σ of each block in the source image and the target image, and use the $d = \sqrt{(\sigma_i - \sigma_j)^2 + (u_i - u_j)^2}$ to calculate the similarity between each block.

Step 2. Transfer the color space from RGB to $La\beta$, and select l, lm, lsd, ls and lk as the pixel properties between the corresponding blocks in the source image and the target image. According to (4), transfer the best color channel of the pixel point in each block of the source image to the corresponding block of the target image, and finally back to the RGB space performance.

$$d = \sqrt{\begin{array}{l} k_1(l_i - l_j)^2 + k_2(lm_i - lm_j)^2 + k_3(lsd_i - lsd_j)^2 \\ + k_4(ls_i - ls_j)^2 + k_5(lk_i - lk_j)^2 \end{array}} \quad (4)$$

Among them, $\sum_{i=1}^{5} k_i = 1$, i, j represent the pixels of the source image and the target image, respectively.

4 DELETING THE POSSIBLE WRONG COLOR

Although through the block transfer we can avoid the error due to the similar luminance but a different color to a large extent, there are still some transmission error colors, which need to be removed and then to be corrected. Because the adjacent pixel luminance and color are closed, in the neighbourhood of the point, if the difference between pixel adjacent colors' mean value and the mean value include the point that exceeds a threshold T, the color of the point is considered as an error, which can be expressed as follows:

$$\sqrt{\left(U_{\alpha 1} - U_{\alpha 2}\right)^2 + \left(U_{\beta 1} - U_{\beta 2}\right)^2} > T \tag{5}$$

where $U_{\alpha 1}$, $U_{\alpha 2}$ represent the α channel mean value of the adjacent pixel and the α channel value of all pixels including the pixel in the neighbourhood, respectively. Similarly, $U_{\beta 1}$, $U_{\beta 2}$ represent the mean value of the β channel. Though the experiment, too big neighbourhood area will increase the possibility of the error point, and easily regard the correct information as an error. So, we set neighbourhood size as 3*3 and T=0.58 to get a better effect, and use (5) to judge all pixels in the initial colored target image. If it exceeds the threshold, the color is wrong, and then we keep the luminance of the pixel and remove the color information. At the same time, we record the number of wrong points in the target image, and use the error rate *Erate* to measure the effect of transfer as follows:

$$Erate = \frac{N_{error}}{N} \tag{6}$$

where N_{error} represents the number of wrong points and N represents the number of all pixels.

5 CORRECTION BASED ON MCPSO

After deleting the wrong color, there are some lost colors in the target image, and we need to estimate the missing color. We use MCPSO [12] to estimate the missing color. MCPSO is based on the PSO, which use the biological symbiosis in nature. Using the relationship of the master-slave population, we simulate the symbiotic community, and through the exchange of information between symbiotic groups to improve the evolution capability. First, the master-slave population update and evolve independently. After each update, we pass the best current position from the slave group to the master group. Through the exchange of information between master-slave populations, the master group guide the update of status according to the slave group or the best position of individual. Compared with a single population, it is better able to find the global optimum, and prevent the local optimum. MCPSO is updated as follows:

$$v_i(t+1)^M = w * v_i(t)^M$$
$$+ c_1 * rand * (pbest^M - x_i(t)^M)$$
$$+ c_2 * rand * (gbest^M - x_i(t)^M)$$
$$+ \phi * c_3 * rand * (gbest^Q - x(t)^M) \tag{7}$$

$$x_i(t+1) = x_i(t) + v_i(t+1) \tag{8}$$

where w is the inertia factor, which reflects the ability of search for the solution space; c_1, c_2 are the acceleration factors *rand* is the random number between (0, 1); *pbest* is the best position particles themselves have experienced; *gbest* is the best location entire population have experienced. M represents the master group, Q represents the slave group, and f represents the migration factor, determined by the following formula:

$$\varnothing = \begin{cases} 0 & gbest^Q < gbest^M \\ 0.5 & gbest^Q = gbest^M \\ 1 & gbest^Q > gbest^M \end{cases} \tag{9}$$

Suppose that there are k pixels that need to estimate the missing color information, the color information of pixel r_i that needs to be estimated are a_i and β_i. For the adjacent pixels having the similar luminance and color, the sum of weighted squared difference of all estimated pixels' color information and that of neighbourhood pixels should be minimized, and the fitness function is set as follows:

$$\min\left(\sum_{i=1}^{k} \sqrt{\sum_{n \in Neighbor(i)} e^{\frac{l_{n\max} - l_{n\min}}{|l_i - l_n|}(\alpha_i - \alpha_n)^2}}\right) \tag{10}$$

s.t $-125 \le \alpha i \le 125$

where a_n, l_n, $l_{n\max}$ and $l_{n\max}$ represent the a channel of N pixels in the r_i neighbourhood, luminance value, neighbourhood maximum brightness, and neighbourhood minimum luminance, respectively.

Color correction steps:

Step 1. Initialize the parameters of the master and slave groups in MCPSO. Then, set the number of iterations, the number of particles M, the solution of space dimension k, w, c_1, c_2, $x(0)$ and $v(0)$. In order to increase the diversity of individuals in the slave group, add a random number between (0,1) into the learning factor and inertia weight when each status is updated.

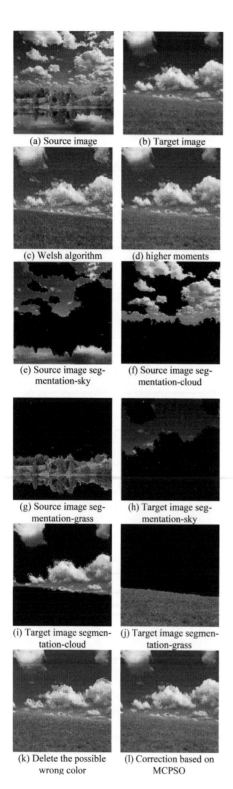

(a) Source image   (b) Target image

(c) Welsh algorithm   (d) higher moments

(e) Source image seg-   (f) Source image seg-
mentation-sky   mentation-cloud

(g) Source image seg-   (h) Target image seg-
mentation-grass   mentation-sky

(i) Target image segmen-   (j) Target image segmen-
tation-cloud   tation-grass

(k) Delete the possible   (l) Correction based on
wrong color   MCPSO

Figure 1. Experimental results.

Step 2. The state of the particle is updated according to (7) and (8) until reaching the number of iterations, and then output the optimal solution.

Step 3. Similarly, make the same estimation for the β channel.

Step 4. Assign the optimal solution of α and β channels to the pixels whose color information is missing, respectively. Then, convert the entire target image into RGB space to display.

6 ANALYSIS OF THE EXPERIMENTAL RESULTS

We select a group of source image and target image, transfer the color in accordance with the method in this paper, and compare the result with the traditional method. As shown in Figure 1, obviously, the traditional method proposed in [2] only uses the lower moments information of luminance, in a similar luminance region, emerge a large number of transfer wrong color; the method proposed in [3] uses the higher moments information, which can improve the matching accuracy, and obtain the good transfer effect, but there are also a few mistakes. The method proposed in this paper, although it reduces the error to a certain extent, there is still some mistake after segmenting the block and delivering it in the corresponding blocks. We remove the possible wrong color and correct, the final effect can be further improved, and the image seems to be more natural.

7 CONCLUSIONS

In this paper, for the possible transmission errors in color image transfer to grayscale, we put forward the color method that is based on AP clustering segmentation and MCPSO correction. First, we divide the source image and the target image into several blocks according to AP clustering, and then correspond each block by similarity, and transfer the color in the corresponding blocks; next, we remove the possible wrong pixel's color according to the correlation of adjacent pixels in the initially formed color image, and then use MCPSO to estimate the lost color, and calculate the optimal solutions of the channel; finally, we realize the color of the target image. Though the observation and the error rate verification, the proposed method can effectively reduce the transmission error rate, and makes the image more smooth and natural, achieving a good transmission effect. However, through experiments we found that the more the error colors need to be corrected, the worse the result is, so we will improve the method in the future work.

ACKNOWLEDGMENTS

This research was supported by the Education Department Project of Fujian Province under Grants JA10064 and the Science & Technology Department Project of Fujian Province under Grants JK2011007.

REFERENCES

[1] Reinhard, E., Adhikhmin, M., Gooch, B., & Shirley, P. 2001. Color transfer between images. *Computer Graphics and Applications, IEEE* 21(5): 34–41.

[2] Welsh, T., Ashikhmin, M., & Mueller, K. 2002. Transferring color to greyscale images. *SIGGRAPH '02 Proceedings of the 29th annual conference on Computer graphics and interactive technique* 21(3): 277–280.

[3] G. Zhao, S. Xiang, & H. Li. 2004. Application of higher moments in color transfer between images. *Journal of Computer-Aided Design & Computer Graphics* 16(1): 62–66.

[4] X. Qian, L. Xiao, & H. Wu. 2006. Application of fuzzy color cluster in color transfer between images. *Journal of Computer-Aided Design & Computer Graphics* 18(9): 1332–1336.

[5] L. Zhu, S. Sun, X. Gu, R. Xia, & M. Ye. 2010. Image colorization based on color transfer and propagation. *Journal of Image and Graphics* 15(2): 200–205.

[6] D. Kong, X. Xiao, Z. Xu, & J. Guo. 2009. A new algorithm for colorizing grayscale image based on pixel correlation. *Journal of Beijing University of Technology* 35(5): 708–714.

[7] S. Shi, L. Wang, W. Jin, & Y. Zhao. 2010. Color night vision research based on multi-resolution color transfer. *Acta Photonica Sinica* 39(3): 553–558.

[8] X. Xu, Z. Lu, G. Zhang, C. Li, & Q. Zhang. 2012. Color image segmentation based on improved affinity progagation clusterin. *Journal of Computer-Aided Design & Computer Graphics* 24(4): 514–519.

[9] S. Zhou, Z. Xu, & X. Tang. 2011. Method for determining optimal number of clusters based on affinity progagation clustering. *Control and Decision* 26(8): 1147–1152.

[10] BJ, F. 2007. Clustering by passing messages between data points.. *Science (New York, N.Y.)* 315(5814): 972–976.

[11] Y. Xiao, & J. YU. 2008. Semi-supervised clustering based on affinity progapation algorithm. *Journal of Software* 19(11): 2803–2813.

[12] B. Niu, L. Li, & X. Chu. 2009. Novel multi-swarm cooperative particle swarm optimization. *Computer Engineering and Applications* 45(3): 28–34.

Architectural, Energy and Information Engineering – Sung & Chen (Eds)
© 2016 Taylor & Francis Group, London, ISBN 978-1-138-02791-6

Design of a landslide disaster remote monitoring and forecasting system

G.L. Sun, G.L. Zhu & Z.G. Tao

State Key Laboratory for GeoMechanics and Deep Underground Engineering,
China University of Mining and Technology, Beijing, China
School of Mechanics and Civil Engineering, China University of Mining and Technology (Beijing),
Beijing China

ABSTRACT: In order to reduce landslide disaster losses, the landslide monitoring and forecasting system should be developed urgently. Based on the mechanics relationship between the sliding force and the shear strength of the landslide, by using the effective theory, practical technology and innovative materials, a landslide disaster remote monitoring and forecasting system is successfully developed. As for energy consumption, we use the low power design methodology in both hardware circuit design and software design. To solve the signal transmission instability problem in the landslide monitoring field, the Zigbee wireless network technology is used to realize the effective transmission of information, and the redundancy network topology structure is also designed. Then, the information is transmitted to the analysis and processing center through the Beidou satellite. The center will give an early warning decision according to the treatment results. The warning system is successfully applied in many places in China. The application results show that this system can realize the early warning function of landslide hazard. It also can provide sufficient time for field personnel and units to take effective preventive measures. Good engineering application results are achieved.

Keywords: landslide disasters; remote monitoring system; wireless sensor network

1 INTRODUCTION

The landslide is a kind of frequent global geological disaster of a large magnitude and wide distribution. It is not only a direct disaster, but also can cause many secondary disasters. For forecasting this geological disaster, many studies have been conducted (Hungr et al., 2014; Ramli et al., 2010). The most effective one is a new type of remote slope stability intelligent monitoring and warning system developed by professor M.C. He and his research team, who are from the China University of Mining and Technology (Beijing) (He et al., 2009).

To ensure the safety, stable and efficient operation of the system, on the basis of the analysis of field application, a wireless way is used to form a network. This wireless sensor network can perform real-time monitoring, sensing and transmission of various monitoring objects' integrated information, and then the information is sent out via the wireless way (Anastasi et al., 2009; Scaioni et al., 2012).

There are many advantages when the wireless sensor network is applied in slide monitoring, which are as follows. (1) Fault tolerance: the redundant nodes reduce the monitor blind spot, so the system has a very strong fault tolerance. (2) Equality from network: all nodes are equal in the wireless sensor network. The nodes coordinate their behaviors by the distributed algorithm without human intervention and any other preset network facilities. They can be placed quickly and automatically at any time. (3) Dynamic network topology: switching between sensor nodes, conversion between two working states, mutual interference between the wireless channels, comprehensive factors such as environmental changes, all requests the sensor network system to adapt such changes and to reconfigure for a dynamic performance.

Due to the difficulty in designing power circuits in the landslide monitoring field, the system is usually powered by batteries. The battery capacity is limited, but the environmental condition is not favorable for charging. Therefore, during the design of the hardware system and the software system, measures should be taken to reduce their power consumption and to extend the system's working time and save energy.

2 LANDSLIDE DISASTERS REMOTE MONITORING AND FORECASTING SYSTEM

2.1 Mechanical principle

Landslide occurrence depends on the balanced relationship between the sliding force and the ski-resistance. Hence, a complex mechanical system is obtained when the mechanical sensing equipment traverses the sliding surface. The functional relationship between the measurable mechanical quantity (perturbation) and the non-measurable mechanical quantity (sliding force) can be derived from the mechanical sensor's prestress P. The mechanical model of the landslide disaster remote monitoring and forecasting system is shown in Figure 1.

Based on the mechanical model of the system, the sliding friction can be written as follows:

$$F_\varphi = (P_n + G_n)\tan\overline{\varphi} + cl \tag{1}$$

In the limit equilibrium state,

$$G_t = P_t + F_\varphi \tag{2}$$

Advanced sliding force function is given by

$$G_t = k_1 P + k_2 \tag{3}$$

$$k_1 = \cos(\alpha + \theta) + \sin(\alpha + \theta)\tan\overline{\varphi} \tag{4}$$

$$k_2 = G\cos\alpha\tan\overline{\varphi} + cl \tag{5}$$

where F_φ is the sliding friction resistance on the sliding surface (kN); G_t is the sliding force on the sliding surface (kN); P is the artificial disturbing force (kN); G is the gravity of landslide (kN); α is the angle sliding surface and the horizontal plane (°); θ is the monitoring anchor's incidence angle (°); φ is the weighted average soil internal friction angle of landslide (°); C is the sliding surface soil

cohesion (kPa); and l is the length of the sliding surface (m).

The above analysis shows that the unmeasurable mechanical quantity G_t of the natural mechanical system in the process of slope development can be calculated through functional relationships with the help of the artificial mechanical system's measurable mechanical quantity P. Therefore, the slope internal dynamic information can be obtained in real time, which provides the basis for the effective judgment of landslide.

2.2 System structure and principle

Landslide disasters remote monitoring and forecasting system includes the data acquisition transmission system and the indoor data processing system. The monitor anchor cable disturbing force can be gathered by the mechanical sensing equipment. Then, the information is transmitted by a wireless network. Indoor remote terminal receiving device receives monitoring data automatically via the satellite, and the mechanical information is sent to the data analysis and processing center. Then, the dynamic monitoring curve of information can be obtained and released to the Internet. Thus, users can access the network to master the field monitoring slope's steady state and release early warning information for the unstable slope if necessary. Landslide disasters remote monitoring and forecasting system's working flowchart is shown in Figure 2.

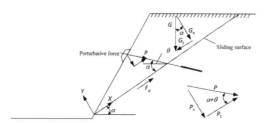

Figure 1. The mechanical model of landslide.

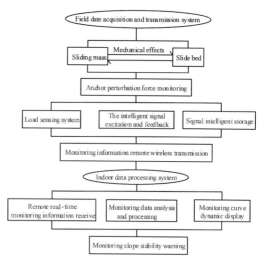

Figure 2. Working flowchart of the system.

3 WIRELESS NETWORK AND ITS TOPOLOGY

In the field, a wireless communication technology is needed to choose to connect each monitoring point. Meanwhile, the topology of the wireless network has an important impact on the information transmission.

3.1 Wireless communication technology choice

In order to save costs and reduce power consumption, convenient for engineering site wiring, this research connects the independent monitoring points as a panel-shaped network in a small region by wireless communication. Then, these monitoring points' data are long-distance range centralism transmitted by a Beidou user machine.

Currently, the common short distance wireless communication technologies mainly includes ZigBee, Bluetooth, WiFi, infrared, and radio frequency technology. Their performance parameters are given in Table 1.

Because the field devices usually have features of small data amount, low transmission rate and powered by battery, the transmission equipment must have characteristics of low cost and low power consumption. The comparative analysis indicated that Zigbee technology has unique advantages in the field of slope monitoring data transmission. Consequently, this research finally selects the ZigBee wireless technology to transmit the sensor's gathering information.

3.2 Wireless communication network topology

In a wireless sensor network, the massive wireless independent nodes cooperate with the divisional labor to complete the data acquisition and the transmission function mutually. In view of the complex slope environment and the system structure, to reduce the system power consumption, a suitable redundant topology of wireless sensor networks for the landslide monitoring warning system is designed. It is a hierarchical network structure composed of three parts: sensor nodes, gateway nodes and remote management center. The bottom layer is terminal wireless sensor nodes deployed in the monitoring of the environment. In the monitoring area, a number of wireless sensor nodes are artificially arranged to acquire the slope sliding dynamic force real-time data. After the data are processed by a microprocessor built in the sensor nodes, the sensor nodes form a self-organizing network via ZigBee communication, and cooperate with group nodes to send to the central station with a multi-hop way. Then, the data are sent to the monitoring center in Beijing by the Beidou satellite. The server in the monitoring and control center completes the work of data receiving, storage, analysis and display.

In this redundancy structure, instead of communicating to the central station directly, statuses of all nodes are equal and each of them will automatically search for a closer node for data forwarding. If any node does not work properly, other nodes' normal data transmission will not be affected. If one node does not work, the related nodes automatically adjust the network routing and the data are transmitted via other nodes.

4 HARDWARE DESIGN OF THE WIRELESS SENSOR NODE

Wireless sensor network is a kind of self-organizing network formed by sensor nodes. On the basis of the network, the sensor network node mainly consists of four parts: sensor module, microprocessor module, wireless communication module and power supply and management module.

4.1 Sensor and exciting circuit

The sensor module is responsible for monitoring regional information collection and data conversion.

Monitoring point uses the steel string type load pressure sensor, as shown in Figure 3. There are three steel strings distributed with 120-degree angular on the sensor ring. While the sensor is operating, the strong pulse stimulation signal is acted on the coil with an external cable. Vibrating string near the coil vibrates to cut the magnetic field under the effect of the change in the magnetic

Table 1. Comparison of different short-range wireless transmission technologies.

Parameter	ZigBee	Blue-tooth	WiFi	Radio frequency
Working Frequency/GHz	2.4	2.4	2.4	5.8
Coverage area/m	50~300	10	<30	1~10
Transmission rate /bps	250 K	1 M	<108M	212k
Number of connected devices	65000	7	255	2
Access time	30 ms	10 s	3 s	3 ms
Maximum power consumption/mW	1~3	<100	100	0
Use cost	Low	Low	General	Low

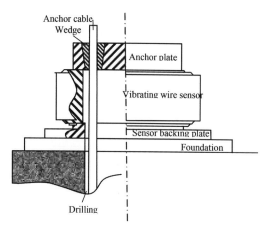

Figure 3. Steel string pressure sensor.

field generated by the variation of the current in the induction coil produced. Thus, the induced current is produced in the coil. The microcontroller measuring system gets the induction current feedback through the external cable to determine the vibration frequency of a vibrating string, and the value of the current pressure can be obtained by calculating the vibration frequency.

4.2 Microcontroller and measuring circuit

The microprocessor module is responsible for controlling the sensor nodes and data processing.

The system uses the microcontroller LPC2103 that is produced by PHILIPS Company. It is a real-time simulation of ARM7TDMI-S CPU, and has 8 kB and 32 kB embedded high-speed flash memory. A 32 bit code can operate at a maximum clock rate ensured by a 128 bit memory interface and the unique accelerating structure. It enables the performance of the interrupt service program and the DSP algorithm improved by about 30% compared with the Thumb mode.

Microcontroller is suitable for the requirements of this system because it has a small size and low power consumption. The feedback signal triggered by the sensor is shaped into the microcontroller, and the signal frequency can be measured.

4.3 Power management circuit

The power supply module ensures the entire system's power supply. The power management module uses the 5 V lithium battery LTC3455 of Linear Company. It has a USB power management function and double DC-DC rechargeable function. It also can realize external power supply with seamless switching and battery management.

4.4 Communication module interface circuit

The wireless communication module is responsible for communication with other nodes. As the interface circuit of the microcontroller and the CC2420, its role is to unify the microcontroller with the communication module interface to realize communication.

5 DESIGN OF WIRELESS SENSOR NODE SOFTWARE

The wireless sensor network node is a resource-constrained embedded system and requires a higher operating system. Some existing embedded operating systems cannot be applied to the sensor network node. The concurrency of the sensor node's arrival message requires the operating system to perform frequently occurring operations in a short period of time. The tinyOS operating system which is based on module programming language nesC compilation has solved problems of few sensor node resources with strong concurrency. Therefore, the system uses the TinyOS operating system.

TinyOS is much simpler than the general thread lightweight technology and a two-layer scheduling method. Task preemption is not allowed between each other. To ensure the hardware interrupt response, the interrupt processing thread can preempt the user's task and low priority interrupt processing thread.

Because the nodes of the wireless sensor network are powered by batteries, the energy is limited. TinyOS energy management can be realized by three interrelated parts. First, use the Std-Control.stop command to stop the device and then switch it to the low-power state. Second, the TinyOS can identify the current hardware status by HPLPowerManagement components and changes the processor into the low-power mode. Third, the timer of TinyOS can work in the power saving mode with very low power consumption.

TinyOS adopts an active communication mechanism, which not only improves the efficiency of the CPU, but also reduces the energy consumption.

6 CONCLUSIONS

1. Hardware and software systems of wireless sensor nodes have provided the important material base and the technical support for the point's information transmission. Based on the actual situation, the topology structure of the landslide monitoring and early warning system of wireless sensor network ensures each monitor-

ing point's information transmission safely and effectively.

2. After the system is developed successfully, its application scope is already spread more than 10 areas of the nation and builds more than 160 monitoring points in total. According to the users' feedback information, the system application effect is well and many landslides have realized the success early warning. For example, on February 20, 2008, West-East gas pipelines in Yan'an landslide monitors, a landslide surface cracks warning was issued. On March 15, 2008, surface cracks warning was issued again successfully. For the evacuation of production facilities and personnel, there was sufficient time in the monitoring area, thus avoiding personnel casualty and economic loss. On August 2012, there were landslides and cracks disaster caused by rainfall and excavation in the monitoring area in Benxi Iron and Steel (Group) Nanfen open-pit iron. It was successful to achieve early warning to avoid rolling stones and sliding body to personnel and equipment damage. In addition, safety control management and safety mining slope excavation were realized.

REFERENCES

Anastasi G, Conti M, Di Francesco M, Passarella A, 2009. Energy conservation in wireless sensor networks: A survey. Ad Hoc Networks, 7(3): 537–568.

He M-c, Tao Z-g, Zhang B, 2009. Application of remote monitoring technology in landslides in the Luoshan mining area. Mining Science and Technology (China), 19(5): 609–614.

Hungr O, Leroueil S, Picarelli L, 2014. The Varnes classification of landslide types, an update. Landslides, 11(2): 167–194.

Ramli MF, Yusof N, Yusoff MK, Juahir H, Shafri HZM, 2010. Lineament mapping and its application in landslide hazard assessment: a review. Bulletin of Engineering Geology and the Environment, 69(2): 215–233.

Scaioni M et al. Wireless Sensor Network Based Monitoring on a Landslide Simulation Platform. In: Wireless Communications, Networking and Mobile Computing (WiCOM), 2012 8th International Conference on, 21–23 Sept. 2012 2012. pp 1–4. doi: 10.1109/WiCOM.2012.6478743.

Architectural, Energy and Information Engineering – Sung & Chen (Eds)
© 2016 Taylor & Francis Group, London, ISBN 978-1-138-02791-6

Research on development and sorts of solar cell

Hai Kuang
Jiangxi Science and Technology Normal University, Nanchang, Jiangxi, China

Shi An He
Jiangxi Changyun Company, Nanchang, Jiangxi, China

Ying Luo
Jiangxi Science and Technology Normal University, Nanchang, Jiangxi, China

ABSTRACT: The development of solar cell is reviewed, and the sorts of solar cell are introduced, including Monocrystalline Silicon cell, polycrystalline silicon solar cell, thin film solar cell and CdTe solar cell, and so on. And the merits and demerits of these solar cells are discussed. It is pointed out that the polycrystalline silicon solar cell has been the dominant producers in the solar cell industry because of its low cost, little pollution, and good stability. The application of solar cell can offer renewable clean energy for human society . It is pointed out the industry of polycrystalline silicon should be carried out on a large scale and so as to improve the solar cell.

Keywords: solar cell; development; sorts; polycrystalline silicon

1 INTRODUCTION

With the development of world economy, energy and environmental issues become increasingly important, which is directly related to the sustainable development of society and economy. The limitations, renewability of conventional energy sources and bad environment pollution make people to realize that clean renewable energy has become a very important way. Solar cell generating electricity with no moving parts is clean and widespread. So, the development of solar cell meets the requirement of environmental protection and sustainable development. According to forecasts, the photovoltaic power generation will be one of humanity's basic energy by mid-century [1].

2 DEVELOPMENT OF SOLAR CELL

Human understanding of photovoltaic effect can be traced in 1839, in which French scientist E. Becquerel discovered the photovoltaic effect of liquid, i.e., photovoltaic phenomenon. It has been 160 years [2]. Then three scientists in the Bell Labs successfully developed the monocrystalline silicon solar cell, which plays a decisive role in the practical application of solar cell and acts as a milestone in the history of the development of solar cell. So far, the conversion efficiency has been greatly improved, and the process has also been optimized. Table 1 shows the development of solar cell. But the basic structure and mechanism of a solar cell has not changed and it can change solar radiation directly into electricity by the photovoltaic effect.

3 THE SORTS OF SOLAR CELL

Currently, the sorts of solar cell on the market are the following: monocrystalline silicon cell, polycrystalline silicon cell, amorphous silicon thin film cell and CdTe cell, CIS cell. In addition to the above commodities solar cell, GaAs cell is mainly used in space technology due to its high price, as well as nanometer TiO2 dye-sensitized solar cell, polycrystalline silicon thin-film cell and organic solar cell, and so on.

3.1 *Silicon solar cell*

The main raw materials of silicon cell are silicon slices made from a high-purity silicon scrap in the electronics industry in semiconductor processing. The band gap of silicon material is 1.12 eV, so to be more fully absorbed solar radiation required around 100 μm thick silicon material. Currently, monocrystalline cell is the highest efficiency solar cell other than GaAs solar cell. Moreover, it has achieved the industrialization. The representative

Table 1. Development of solar cell.

Time	Main Researcher	The main contribution
1954	D.M. Chapin	They reported monocrystalline silicon solar cell with 4.5% efficiency
1957	Hoffman	They inverted monocrystalline silicon solar cell with 8% efficiency
1976	RCA lab	They reported amorphous silicon cell whose the efficiency is 1 to 2%
1991	Grätzel	He developed Nano TiO2 solar cell with efficiency of 7%
1999	M.A. Contreras	They developed Copper indium tin (CIS) cell and its efficiency reached 18.8%
2001	Dhere R.G	They studied CdTe solar cell whose efficiency is 16.4%
2004	Martin Green	His Monocrystalline silicon cell efficiency reached 24.7%
2004	Schultz	He researched Polysilicon cell and its efficiency reached 20.4%

manufacturers are Shell solar in the Netherlands, Isofoton in Spain, Microsol in India, and so on.

Because of high-purity silicon is expensive, monocrystalline silicon cell is developed toward high efficiency and thin film. German researchers have confirmed efficiency of solar cell with 40 μm thick silicon can reach very high[3]. Possibility, ultra-thin silicon solar cell can achieve industrial production by improving the production process, and may reach the efficiency obtained in the laboratory. Hitachi Company in Japan made a double-sided junction monocrystalline silicon solar cell with SiO_2 as anti-reflection and surface passivation, whose front and back efficiency is 15% and 10.5%, respectively[4]. By optimizing the design and production technology, the efficiency of silicon solar cell is doubled in the past 20 years.

3.2 Polycrystalline silicon solar cell

Difference between polycrystalline silicon cell and monocrystalline cell is polycrystalline silicon includes a number of different sizes and different orientations of the grains. The cost of polycrystalline silicon cell is reduced by eliminating the production of monocrystalline and saving the silicon material. Its disadvantage is the grain boundaries, dislocations, vacancies and impurities that have a certain impact on the conversion efficiency.

Currently, polysilicon production technologies are mainly used in reduction methods, thermal decomposition of silane[5] and the purification of metallurgical grade silicon using the zone melting method. Like monocrystalline silicon cell, polycrystalline silicon cell is developed toward ultra thin in order to save the expensive high-purity silicon material. Some manufacturers made the cell with 180 μm silicon or even thinner. Because of different grain orientations, polycrystalline silicon cell can be presented in different colors in the sunlight. Therefore, polycrystalline silicon cell can be used as a good decoration.

3.3 Amorphous silicon thin film solar cell

Amorphous silicon (a-Si: H) is an alloy of silicon and hydrogen. 1 μm of amorphous silicon cell can absorb most of sunlight. Currently, it is the most successful thin film solar cell. Its structure is n-i-p type and mainly made by SnO2. Now there is the amorphous silicon cell with two or three p-n junctions. In 1990, conversion efficiency of P/i (α-Si)/n-c-Si (HIT) cell in Japan's Sanyo Company reached 15.8%[6]. In 1994, Japan produced Back Surface Field (BSF) structure cell by PECVD and its conversion efficiency is18.9% [7].

Amorphous silicon cell cannot be used in large scale due to light-induced metastable effects. Also, this makes amorphous silicon from 1996's 12% share in 2005 down to 4.7%. The research work focused on improving the efficiency and stability at present.

3.4 CdTe cell

CdTe is a compound of II–VI and its bandgap is 1.45 eV, well matched with the solar spectrum and can be a large-scale production. The performance of CdTe cell is very stable. The market share is slow due to the toxicity of cadmium.

3.5 (CIS/CIGS) thin film solar cell

CIS is a ternary compound semiconductor whose bandgap is 1.04eV. P-type or N-type semiconductor can be obtained by adjusting chemical composition of CuInSe2 appropriately. Its performance of anti-jamming and anti-radiation is good due to without the aid of additional impurities. So service life of PV crystal made of it is up to 30 years[8]. Since the 1980s, the research work in CIS of ARCO Solar Company has gradually become advanced. It works with the Solar Energy Research Institute of Chinese, which promotes the rapid development of CIS thin-film batteries. CIGS production methods are mainly vacuum evaporation or sputtering. With a three-step Co-evaporation method, CIGS solar cell has been made by Ramanathan and others, and its efficiency reached 18.8% in 1999[9]. But because of the production process is difficult to control and the poor uniformity and repeatability of a large area, the industrialization process of CIS cell is slow.

3.6 Dye-sensitized TiO$_2$ solar cell

Dye-sensitized TiO$_2$ solar cell is actually a photo-electrochemical cell. Grätzel in Switzerland added dye-sensitized matters to this type of battery and efficiency of the cell is 7.1% in 1991[10]. Since then it has become one of the hot research. However, the instability of the liquid electrolyte is problem. Therefore, the stability problem is an important direction of this cell research.

But which is the current leading product?

In fact, since Bell Labs successfully developed crystalline silicon solar cell, crystalline silicon cell has become the main products and will not change in the next 20 years due to its high efficiency and stability, low environmental impact, moderate band gap, and enduring performance. The costs of polysilicon are lower than monocrystalline material. Moreover, polysilicon can be directly prepared as a large-scale product with simple equipment and it has the advantages of the simple manufacturing process, energy-saving, and so on. So monocrystalline cell has a greater potential for cost reduction.

4 CONCLUSIONS

In recent years, the solar cell is increasingly emphasized. Currently, silicon is the main material used to make the cell. Now, polysilicon successfully become the most important material instead of monocrystalline silicon. We should take the advantage of our resources and improve our solar cell.

REFERENCES

[1] Zhang Yao-ming. Sunlight economic and energy revolution[J]. Jiangsu Science and Technology Information, 2004 (4): 1–4.
[2] He Yun-Ping, Wang Duo. Industrial Silicon Technology Advances. Beijing: Metallurgical Industry Press, 2003.
[3] Annual Report 2003 of Fraunhofer Insitirur Solar Energiesysteme, 36.
[4] Shen Hui, Zeng Zu-qin. Solar photovoltaic technology. Chemical Industry Press, Beijing. 2005, 44.
[5] Xi Zhen qiang, Yang De-ren, Chen Jun. The development of casting polysilicon [J]. Materials Review, 2001 (02): 67–69.
[6] Yoshihiro Hamaleawa. Recent advances in solar photovoltaic technologies in Japan, Solar Energy Materials, 1991, 23: 139.
[7] K. Fujimoto Y, Sogawa K, Shima K, et al. High efficiency silicon solar cell by plasma-CVD method, Solar Energy Materials and Solar cell, 1994, 34: 193.
[8] MVYakushev, AVMudryi, VFGremenok, EPZretskaya. Influence of growth conditions on the structural quality of Cu (InGa) Se2 and CuInSe thin films Thin Solid Films, 2004, 452: 133–136.
[9] Ramanathan, K. Contreras Properties of 19.2% efficiency ZnO/CdS/CuInGaSe2 [J] Thin film solar cell. 2003, 11: 225–230.
[10] O'Regan B, Grätzel MA. Low-cost, high-efficiency solar cell based in dye-sensitized colloidal TiO2 films. Nature, 1991, 353: 737–739.

Architectural, Energy and Information Engineering – Sung & Chen (Eds)
© 2016 Taylor & Francis Group, London, ISBN 978-1-138-02791-6

Multi-layer soil humidity and temperature monitoring system for sugarcane farmland based on WSN

Xiu Zeng Yang & Hai Sheng Li
The Department of Physics and Electronic Engineering, Guangxi Normal University for Nationalities, Guangxi, China

ABSTRACT: A multi-layer soil humidity and temperature monitoring system for sugarcane farmland based on the wireless sensor network is designed. The whole monitoring system consists of a coordinator device, multiple router devices and collector devices, a host PC with user interface software. A multi-layer humidity and temperature sensor is applied in the system to get different- depth humidity and temperature in soil. The testing result shows that the system is stable, user-friendly and easy to use. The system, due to the advantages above, can apply in monitoring humidity and temperature in sugarcane field.

Keywords: sugarcane farmland; humidity and temperature; WSN; CC2430; ZStack2007

1 INTRODUCTION

With the development of wireless communication technology, wireless sensor network, which features its short distance, low cost, low power and low rate, has been widely applied in the agricultural environment monitoring [1–3], for example, many researchers in academy and industry have devoted to study and design different Zigbee-based wireless monitoring systems to monitor some important plant growth environment parameters, such as light, carbon dioxide, moisture and temperature, and so on.

It is well known that the growth of plant is much related to the soil moisture and temperature in the farmland. Some study results of sugarcane growth have revealed that, under condition of proper temperature, when the soil relative humidity is in the range of 80–90%, the growth speed of sugarcane is fastest, and when the soil relative humidity is lower than 40%, the growth speed of sugarcane becomes slowest, and even led to sugarcane plant death. It is obvious that how to collect and control the soil humidity and temperature in sugarcane field is very important in the period of sugarcane growth.

In applying Zigbee-based electric devices into sugarcane farmland for monitoring soil humidity and temperature, it is very difficult to install and deploy these devices in sugarcane farmland because of the number of Zigbee-based nodes is great. Furthermore, how to power these amounts of electric devices is also become more difficult, this is because that sugarcane farmland is away from city, which is out range of power supply capability.

From the above analysis, the Wireless Sensor Network (WSN) is the perfect solution for designing sugarcane soil humidity and temperature monitoring system. In order to accurately monitor the soil humidity and temperature of sugarcane farmland, a multi-layer soil humidity and temperature monitoring system, which in the technology of wireless sensor network is employed, is studied and implemented. Aimed at the nature of the soil of sugarcane field, the multi-layer soil humidity and temperature sensor is designed. The multi-layer sensor, because of embedding a lot of humidity and temperature sensors with different position, it can easily collect the humidity and temperature of different layer.

2 THE OVERVIEW OF ZIGBEE

Zigbee is one of the new international standard for wireless network communication, which features low power, low rate, and short distance communication. Zigbee standard based on IEEE 802.15.4, which is sponsored by Zigbee Alliance, is very suited to the wireless sensor network (WSN) with long device battery lifetime and low latency. The Zigbee standard includes Physical Layer (PHY), Media Access Control Layer (MAC), Network Layer (NWK), Application Layer (APL), and Security Service. As shown in Figure 1, the Application Layer is divided into Application Support

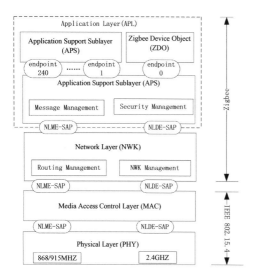

Figure 1. Zigbee layer.

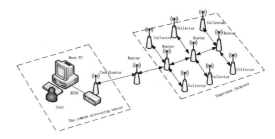

Figure 2. Architecture of humidity and temperature monitoring system.

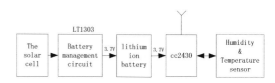

Figure 3. Block diagram of collecting device.

Sub layer (APS), Application Framework (AF), and Zigbee Device Object (ZDO).

Currently, the latest Zigbee standard is ZStack 2007. The Zigbee standard defines three logical devices types, which are called as Zigbee coordinator, Zigbee router and Zigbee end device. When these devices are deployed in an area, if these devices are power up, these devices can self-form a multi-hop wireless network. The network is consisting of a Zigbee Coordinator, multiple Zigbee routers and Zigbee end devices. The Zigbee standard supports star, tree, and mesh topologies. The Zigbee network involves two types of address: the 64-bit IEEE address and the 16-bit network address. The former is the MAC address, which is unique in the world, and is assigned by the manufacturer during installation. The latter is the address, which is only unique in a certain wireless network, is assigned when the device joins a network. The Zigbee network adopts a distributed addressing scheme for assigning the network address to ensure network address is unique.

3 SYSTEM ARCHITECTURE

Figure 2 shows the architecture of design system that is applied to monitor the humidity and temperature of sugarcane farmland soil. The monitoring system is consisting of a coordinator device, multiple router devices and collector devices, a host PC with user interface software. The coordinator device is the first apparatus in the network. When the device is powered up, it can automati-

cally choose a channel and a network identifier to form a network and periodically broadcasts beacon to inform other nodes joining the network. When these router and collector devices detect existence of the network, they automatically join the network. The routers mainly perform function of multi-hop routing, communication with its child device and permission other device joining network. Collector device is an end device in the wireless sensor network, whose main function is to collect information. Due to equipped with multilayer humidity and temperature sensor, the end device can automatically acquire different humidity and temperature information of sugarcane soil. These humidity and temperature data, which is relayed by end-device's father device (Router node), are transmitted to coordinator (sinking node). Finally, these data are sent to host PC, and gets stored in hard disk.

4 THE HARDWARE DESIGN OF COLLECTING DEVICE

The block diagram of collecting device is shown in Figure 3. It makes up of solar cell, battery operation system, lithium ion battery, CC2430 and humidity and temperature sensor. Because this collecting device is deployed in the wild, it is necessary to supply this device power with solar cell. Due to the nature of solar cell, a battery management circuit, between solar cell and lithium ion battery, is employed to enhance the performance of solar power. Apparently, the function of battery management circuit, which the IC of LT1303 is applied, is regulating outputs of 3.7V. The LT1303 is a micro-power step-up DC/DC converter ideal for use in small, low voltage, battery-operated system. The CC2430 is a true system-on-chip solution for system where a low power consumption is required. The CC2430 combines the excellent performance of the leading CC2420 RF transceiver with an industry-standard enhanced 8051 MCU, 32/64/128 KB flash memory, 8 KB RAM and many other powerful features. Combined with the industry leading ZigBee protocol stack (Z-Stack) from Figure 8 Wireless, the CC2430 provides the market's most competitive ZigBee solution.

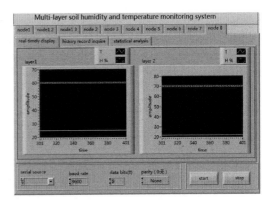

Figure 5.   The graphical user monitoring interface on host PC.

5   THE ANALYSIS PROCESS OF COORDINATOR DEVICE STARTING

The analysis process of coordinator starting and configuring wireless sensor network is shown in Figure 4. The staring process mainly involves Sapi layer, ZDO layer and NWK layer in the Zstack 2007. After the coordinator is powered, Sapi layer calls the function of Sapi Init(), in which the ZB_ENTRY_EVENT event is set. After the function of zb_StartRequest() is called, the function of ZDOInitDevice() in ZDO layer is called, in which the ZDO network_INIT event is set. After the function of ZDOinitDevice() is called, the function of FormationRequest() in the NWK layer is called to request formatting a WSN. If the WSN

formation is successful, the function of Formation Comfirm CB() is called for confirming the Request from ZDO layer, and ZDO_NETWORK_START event is set. After the ZDO_NETWORK_START event is responded by ZDO layer, the function of ZDApp_NetworkStarEvt() is called, in which ZDO_STATE_CHANGE_EVT event is set. After the ZDO_STATE_CHANGE event is responded in Sapi layer, the function of SAPI_SartConfirm() is called to inform OS the WSN starting is successful.

6   THE DESIGN OF HOST PC MONITORING INTERFACE

The host PC monitoring interface facilitates the user to display, store and analyze data collected by the system. The graphical user monitoring interface of system is shown in Figure 5. After logging in the system successfully, user can click the tabs in the top of interface to inquire different node's soil humidity and temperature data in the sugarcane field. If the tab of real-timely display is clicked, the two layers soil information of humidity and temperature are both displayed in the waveform charts. If user want to inquire history information, clicks the tap of history record to inquire.

7   CONCLUSIONS

The system is designed for monitoring multi-layers' soil humidity and temperature of the sugarcane field. The testing result shows that the system is stable, user-friendly and easy to use. The system, due to the advantages above, can apply in monitoring humidity and temperature in sugarcane field.

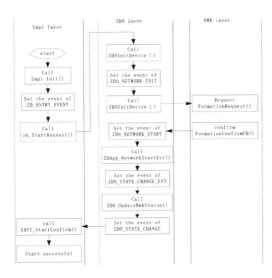

Figure 4.   The starting flow chart of coordinator forming WSN.

ACKNOWLEDGEMENTS

This article is supported by University scientific research project funding projects of Guangxi (2013YB266, KY2015LX541).

REFERENCES

[1] Kim Y, Evans R G, Iversen W M, Pierce F J. Instrumentation and control for wireless sensor net-work for automated irrigation [C]. 2006 ASABE Annual International Meeting, Portland, 2006. ASABE Paper No. 061–105.

[2] Raul Morais, Valente A, Serôdio C. A wireless sensor network for smart irrigation and environmental monitoring [C]. EFITA/WCCA Vila Real Portugal, 2005: 845–850.

[3] Qiao Xiao-jun, Zhang Xin, Wang Cheng, et al. Application of the wireless sensor networks in agriculture [J]. Transactions of The Chinese Society of Agricultural Engineering, 2005, 21(2): 232–234.

Architectural, Energy and Information Engineering – Sung & Chen (Eds)
© 2016 Taylor & Francis Group, London, ISBN 978-1-138-02791-6

Investigation on combination structures of solid particles in ER polishing fluid

Y.W. Zhao, X.M. Liu & D.X. Geng
Engineering Training Center, Beihua University, Jilin Province, P.R. China

ABSTRACT: The rheological properties of Electrorheological (ER) polishing fluid are changed enormously as the complex microstructure formed by the polarized particles. The microstructure formed by polarized particles perpendicular to the electrodes is observed by using CCD camera. On the basis of dielectric polarization model and the interacting force between particles in ER polishing fluid, the combination structures of ER polishing fluid mixed with abrasive particles of different grain size are acquired under applied external electric field, and the influence on the interaction force between the ER particles and different grain sizes of abrasive particles to the combination structures of ER polishing fluid is analyzed. The combination microstructure of ER polishing fluid has been analyzed with the influencing factors such as the interaction force between the particles and the grain size of abrasive particle, and so on.

Keywords: electrorheological fluid; combination structures; electrical field strength

1 INTRODUCTION

ER fluid-assisted polishing is a navel polishing process employing the ultra-fine abrasive particles mixed into ER fluid to complete the material removal by the ER effect. The shear yield stress and the viscosity of ER polishing fluid have undergone a tremendous change as the ER particle and the abrasive particles will polarize to aggregate into complex microstructure of columns when the electric field is applied [1]. Kuriyagawa has observed the polishing behavior of ER particles and the abrasive particles by using CCD camera in the ER fluid-assisted polishing process [2]. W.B. Kim et al. have investigated the microstructure of ER polishing fluid under electric field and found that the motion of abrasive particles is consistent with the ER particles [3].

The solid particles of the ER polishing fluid will attract each other to form a chain-like structure along the field direction under applied external electric field. Simulation, the abrasive particles or ER particles will be combined into particle chains and distributed among particles to form a similar unit cell of Body-Centered Cubic (BCC) structure [4]. The interaction force of the abrasive particles in contact with the workpiece surface is investigated by the combination structures of solid particles in ER polishing fluid to analyze the surface material removal mechanism in ER fluid-assisted polishing processes.

2 MICROSTRUCTURE OF ER POLISHING FLUID

ER fluid, which consisted of silicone oil and starch particles, is used as the medium. The ER polishing fluid between the electrode and the workpiece is mixed from ER fluid and ultra-fine abrasives with a certain percentage.

The ER particle and the addition of ultra-fine abrasive particles into ER fluids are dielectrically polarized with an electric field. The ER particles surrounding the abrasive particles form a

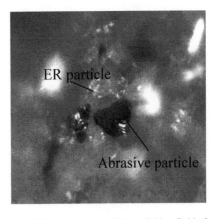

Figure 1. Microstructure of ER polishing fluid after an electric field is applied.

chain-like structure and give bonding strength to the embedded abrasive particles. The microstructure formed by particles perpendicular to the electrodes is observed by using CCD when the electrical field is 0.3 KV/m in Figure 1. It can be seen in Figure 1, the starch particle and SiC particle are the white and black particles respectively in the picture. The abrasive particles are attracted by ER particle into particle chain to distribute among ER particles. So, the polarized ER particles and abrasive particles can form a specific combination structure after an electric field is applied.

3 INTERACTING FORCE BETWEEN SOLID PARTICLES IN ER POLISHING FLUID

The solid particles are made up of ER particle and abrasive particle in ER polishing fluid, so the interacting forces exerted on the ith particle include the action of all polarized particles (j_1th ER particle and j_2th abrasive particle) except the ith particle. The interacting forces exerted on the ith particle in ER polishing fluid is given by

$$F_i^{el}(\{R_j\}) = \sum_{i \neq j_1} F_{epij_1}^{el}(R_{ij_1}, \theta_{ij_1}) + \sum_{i \neq j_2} F_{apij_2}^{el}(R_{ij_2}, \theta_{ij_2}) \quad (1)$$

where $F_{epij_1}^{el}$ and $F_{apij_2}^{el}$ are the interacting forces exerted on the ith particle by j_1th ER particle and j_2th abrasive particle, R_{ij} is the distance between polarized particles, and θ_{ij} is the angle of the electric field direction and the joint line of the two dipoles.

The interaction of particles due to electric polarization is shown in Figure 2.

As shown in Figure 2, the interacting forces exerted on the ith particle by jth particle using the following equation [5]:

$$F_{ij}^{el}(R_{ij}, \theta_{ij}) = F_0 \left(\frac{d_{eo}}{R_{ij}}\right)^4 [(3\cos^2\theta_{ij} - 1)e_r + \sin 2\theta_{ij}e_\theta] \quad (2)$$

where F_0 is given as

$$F_0 = \frac{3p_ip_j}{4\pi\varepsilon_0\varepsilon_f d_{ep}^4} \quad (3)$$

where p_i and p_j are the dipole moments of two interacting particles, respectively, ε_0 is the permittivity of free space, ε_f is the relative permittivity of dielectric fluid, R_{ij} is the distance between two interacting particles, and d_{ep} is the diameter of ER particles.

The expressions for p_{ep} and p_{ap} are given in the following equations.

$$p_{ap} = \frac{\pi\varepsilon_0\varepsilon_f\beta_{ep}d_{ap}^3 E(1 + \beta_{ep}d_{ep}^3 / 4R_{ij}^3)}{2[1 - (\beta_{ap}d_{ap}^3 / 4R_{ij}^3)(\beta_{ep}d_{ep}^3 / 4R_{ij}^3)]} \quad (4)$$

$$p_{ep} = \frac{\pi\varepsilon_0\varepsilon_f\beta_{ep}d_{ep}^3 E(1 + \beta_{ap}d_{ap}^3 / 4R_{ij}^3)}{2[1 - (\beta_{ep}d_{ep}^3 / 4R_{ij}^3)(\beta_{ap}d_{ap}^3 / 4R_{ij}^3)]} \quad (5)$$

Assuming that all solid particles are approximated to be spherical in shape of the same size and the ER–abrasive particles are nearly touching in this paper. The simulation conditions as shown in Table 1.

The interacting forces of pair particles under the non-uniform electric field are depicted in Figure 1. The attraction force between ER particles is highest, secondary between ER particles and abrasive particles, the electrophoretic forces on ER particles and abrasive particles are negligible.

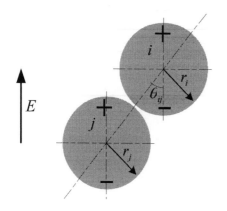

Figure 2. The interaction of particles due to electric polarization.

Table 1. Parameters on particles of ER polishing fluid.

Parameters	Value
Relative permittivity of Al$_2$O$_3$ particles	5.7
Relative permittivity of starch particles	20
Relative permittivity of silicone oil	2.7
Permittivity of vacuum	8.854×10^{-12} F/m
Diameter of diamond particles	10 μm
Diameter of starch particles	10 μm

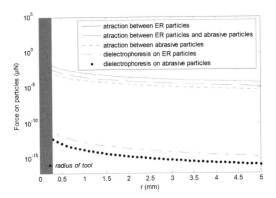

Figure 3. The interacting forces between polarized particle.

4 INFLUENCE FACTORS OF COMBINATION STRUCTURES OF SOLID PARTICLES

The combination structures of the polarized particles have relations to the composition of the polishing fluid, electric field strength, the interaction force among particles, the volume fraction of abrasive particles, grain size of abrasive particles and granularity as well as polarization capability of the particles. The microstructure of ER polishing fluid mainly depends on the interaction force between the pair particles, the grain size of abrasive particle, the volume fraction of abrasive particles, and so on.

4.1 The influence of the interaction force

It can be seen from the Equation (2) that the factors to affect the interaction force between the particles are mainly the relative dielectric constant, grain size and gap of the attractive particles as well as the electric field strength. It is proportional to the electric field strength and the relative dielectric constant, and it varies inversely with the gap of the attractive particles. As the interaction force increases with the increase of the electric field strength, the attractive particles combine more closely. Simultaneously, the polarized particle that has the greater relative dielectric constant will strongly attract the other particles. That is to say, the closer of the particles, the larger interaction force between polarized particles, and the particles gathered into a chain as the interaction force is greater than Brown force.

4.2 The influence of the volume fraction of particles

The volume fraction of particles determines the distribution particles and the distance between the particles. And the interaction force is inversely proportional to the distance between the particles.

The far distance between particles, the weaker of the interaction force the closer of the particles, the larger force between them. Therefore, the particles with less distant gathered into a chain preferentially.

4.3 The influence of the size of the abrasive particles

As the size of ER particles is constant in ER polishing fluid and the grain size and species of abrasive particle will vary with different materials of the workpiece. The grain size of abrasive particle has an immediate effect on electric dipole moment of particles, the distance between the particles and the angle of the electric field direction and the joint line of the two dipoles.

5 COMBINATION STRUCTURES OF SOLID PARTICLES IN ER POLISHING FLUID

The polarized particles will form multiplex combination structures in ER polishing fluid mixed with abrasive particles of different grain size under applied external electric field. The distribution of polarized particles is depended on the angle of the electric field direction and the joint line of the two dipoles. When the angle is greater than zero ($0 \le \theta_{ij} \le 55°$), the pair of particles attracts each other, on the contrary, that repels each other. The abrasive particle and ER particle will be combined into particle chains with the mode of single particle or micromass and distributed among particles linked chain-like structure to form a similar unit cell of Body-Centered Cubic (BCC) structure, as shown in Figure 4.

As the relations between particles shown in Figure 4, assuming that ith particle and jth particle

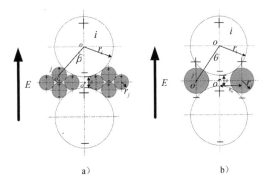

Figure 4. Combination structures of solid particles.

are nearly touching, the small particle will aggregate into micromass to distribute among particles linked chain-like structure, the angle of the electric field direction and the joint line of the pair of particles β can be given as follows:

$$\beta = \arccos\left(r_i - r_j + \frac{\delta}{2}\right) / (r_i + r_j) \qquad (6)$$

where r_i and r_j are the diameter of ith particle and jth particle, respectively. δ is the gap between two interacting particles.

When the attracted particle is a single particle, the angle of the electric field direction and the joint line of the pair of particles is given by

$$\theta = \arccos\left(r_i + \frac{\delta}{2}\right) / (r_i + r_j) \qquad (7)$$

As the attractive force between polarized particles is greater than Brownian force, the combination structure of solid particles with abrasive particles of different grain size can be obtained including six different forms by Equations (5) and (6) as shown in Figure 5.

Figure 5(a) and 5(b) is the combination structure of the abrasive particles of different grain size within 0~3 µm. The small abrasive particle will disperse on the surface of ER particle or aggregate into micromass to distribute among particles. When the grain size is in the range of 4~14 µm, the abrasive particle will be combined into particle chains or distributed among particles linked chain-like structure with single particle (Figure 5c and 5d). When the grain size is the range of 15~37 µm, the ER particle will play adhesive to attract the abrasive particle to form the particle chain (Figure 5e). When the grain size is greater than 15~38 µm, the ER particle will aggregate into micromass firstly as the small size relatively to distributed on the surface of abrasive particle (Figure 5f).

6 CONCLUSION

The paper investigates combination structures of solid particles in ER polishing fluid mixed with abrasive particles of different grain size under applied external electric field. The attracting force between polarized particles plays a key role on combination structures and the distribution of polarized particles is acquired by the interacting force equation between particles. According to the abrasive particles of different grain size, the combination structures of solid particles are built including six different forms. The polarized abrasive particles are attracted by near ER particle with the mode of single particle or micromass and combined into particle chains to form a similar unit cell of Body-Centered Cubic (BCC) structure when the electric field is applied.

ACKNOWLEDGEMENTS

The authors would like to acknowledge the financial support for this investigation from program for the Dr. Scientific Research Starting Foundation of Beihua University (199500004), PR China. Corresponding author. Tel.18604498852. email: jlulxm@163.com(X. M. Liu)

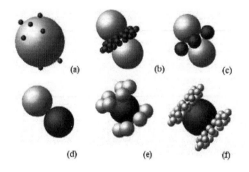

Figure 5. Combination structures with abrasive particles of different grain size.

REFERENCES

[1] Zhang, L. et al. 2010. An integrated tool for five-axis electrorheological fluid-assisted polishing, *International Journal of Machine Tools and Manufacture* 50: 737–740.

[2] Kuriyagawa, T. & Saeki, M. K. Syoji. 2002. Electrorheological fluid-assisted ultra-precision polishing for small three-dimensional parts, *Journal of the International Societies for Precision Engineering and Nano technology* 26: 370–380.

[3] Kim, W. B. et al. 2004. Development of a padless ultraprecision polishing method using electrorheological fluid, *Journal of Materials Processing Technology* 155–156: 1293–1299.

[4] Wen, W. & Sheng, P. 2003. Two- and three-dimensional ordered structures formed by electromagnetorheological colloids, *Physica B*, 338: 343–346.

[5] Kim, W. B. et al. 2003. The electro- mechanical principle of electrorheological fluid-assisted Polishing, *International Journal of Machine Tools and Manufacture* 43: 81–88.

Architectural, Energy and Information Engineering – Sung & Chen (Eds)
© 2016 Taylor & Francis Group, London, ISBN 978-1-138-02791-6

Experimental study on pneumatic space bending flexible joint

Hong Bo Liu
College of Mechanical Engineering, Beihua University, Jilin Province, P.R. China

De Xu Geng, Xiao Min Liu & Yun Wei Zhao
Engineering Training Center, Beihua University, Jilin Province, P.R. China

ABSTRACT: In this paper, we develop a new type of pneumatic space bending flexible joint, which is mainly composed of elastic rubber tube and constrain components. The joint is equivalent to three artificial muscles connected in parallel, which has three degrees of freedom. The joint can elongate in the axial direction and bend arbitrarily in space when the air pressure is supplied. The axial elongation and space bending of the joint have been done by the experimental device, and the relationship deformation of the joint and air pressure is required. This study laid the foundation for further application of the joint.

Keywords: flexible joint; axis elongation; space bending

1 INTRODUCTION

The robots have been widely applied in military, medical, service and entertainment, industrial and other fields, and have become an indispensable part of people's daily life and work[1-2]. The core of all kinds of robot and motion of mechanical parts consists of active joints, the driving mode and structural properties, which largely determine the performance of the robot. The driving modes mainly include motor driving, fluid driving, and functional material driving. Pneumatic flexible joint as a new type of pneumatic actuators, compared with the traditional actuator, has many advantages such as good flexibility, clean, and simple structure. Its particular advantage has attracted the attention of scholars at home and abroad, which has good prospects for development[3-4].

Now, the existing pneumatic artificial joint has not fully met the special needs in the field of bionic and other special robots[5-6]. The pneumatic space bending flexible joint proposed in this paper can realize axial elongation and space bending movement. From the perspective of bionics, the joint has the structural and functional characteristics of the finger joint.

2 THE STRUCTURE AND WORKING PRINCIPLE OF THE JOINT

The structure of the pneumatic space bending flexible joint is shown in Figure 1. The inside of the joint is elastic rubber tubes and springs, and its outside is a set of closely constrain components. The top and bottom of the pneumatic space bending flexible joint are the covers. The down cover and the air interface are fixedly connected into a whole. Elastic rubber tube, constrain component and the plug form a good seal cavity called the artificial muscle. The pneumatic space bending flexible joint is equivalent to three artificial muscles in parallel, leading to a joint with three degrees of freedom, so that it can elongate in the axial direction or bend arbitrarily in space when air pressure is supplied.

Monomer artificial muscle only performs axis elongation when the compressed air is supplied; the elongation will increase with the increase in the air pressure. The joint will restore to the original state with the help of its own elastic force when the pressure is discharged.

The joint elongates along the axial direction when the same air pressure is supplied to each

Figure 1. 3-dimensional model of the joint.

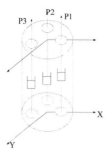

P2
P3 · P1

X
Y

Figure 2. Force analysis of the joint.

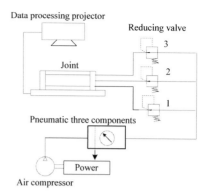

Data processing projector

Reducing valve
3

Joint
2

1

Pneumatic three components

Power

Air compressor

Figure 3. The principle of the experiment.

artificial muscle, that is $Pi = P$, and i (i = 1,2,3) is the serial number of the artificial muscles (Figure 2). The joint elongates and bends in the required direction simultaneously when different air pressures are supplied to each artificial muscle. When the air pressure is supplied only to $P1$, the joint will bend in the direction of $P2, P3$. When the same air pressure is supplied to $P2, P3$, the joint will bend in the direction of $P1$. It is obvious that we can control the elongation, bending direction and bending angle of the joint by adjusting the value of Pi.

3 THE PRINCIPLE OF THE EXPERIMENT AND DATA ANALYSIS

The experimental system of the joint is shown in Figure 3, and the experimental devices are mainly composed of air compressor, data processing projector, three pneumatic components and precision reducing valve. The parameters of the joint are listed in Table 1.

3.1 Experimental study on the axial elongation of the joint

We ensure that the air pressures supplied to each artificial muscle are the same, that is $Pi = P$, and i (i = 1,2,3). With an increase of 5% of the air pressure each time, we obtain the data of the axial elongation of the joint. We then reduce the air pressure when it reaches 0.35 Mpa, and obtain the axial elongation of the joint corresponding to the air pressure. We can obtain the curves of the joint under the two different states (Figure 4).

As shown in Figure 4, the elongation of the joint increases with the increase in the air pressure, and the curing deformation is nonlinear. The elongation reaches up to 31.8% of its own length when the air pressure reaches up to 0.35 Mpa. The curves of the joint under the two different states do not agree well with each other. Deformation of rubber tube materials has certain hysteretic characteristics, which results in the pressurized curve

Table 1. The parameters of the joint.

Total length of joint	52 mm
Active length of deformation	42 mm
The thickness of constraints ring	2 mm
Outer diameter of the rubber	9 mm
Inner diameter of the rubber	6 mm
Wire diameter of the spring	0.6 mm
The number of active coils	76
The pitch diameter of the spring	4.1 mm
The elastic shear modulus of spring	79 Gpa
The elastic shear modulus of rubber tube	1.367 Mpa

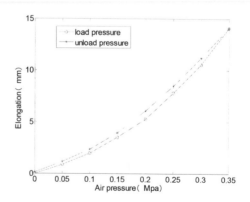

Figure 4. Axial deformation curve of the joint.

being lightly lower than the depressurized curve. By flexible joint statics experimental data, we obtain the experience formula of the axial elongation of deformation joints as follows:

$$\Delta l = 87.74P^2 + 8.884P + 0.1013 \quad (1)$$

612

3.2 Experimental study on bending deformation of the joint

Controlling the air pressure supplied only to P1, and then adjusting the pressure P from 0 to 0.35 MPa, we obtain the bending angle of the joint under certain conditions of pressure (figure 6). As shown in Figure 6, the bending angle of the joint increases with the increase in the air pressure and the curing deformation is nonlinear. The joint bending angle empirical formula for using a single muscle is as follows:

Figure 5.   The picture of the elongation (P = 0.35 Mpa).

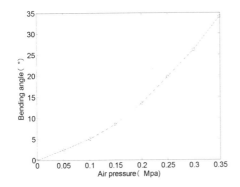

Figure 6.   Bending deformation curve of the joint.

Figure 7.   The picture of the deformation (P1 = 0.35 Mpa, P2 = P3 = 0).

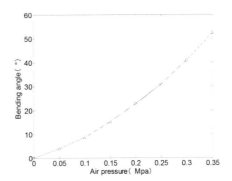

Figure 8.   Bending deformation curve of the joint (P1 = 0, P2 = P3 = 0.35 MPa).

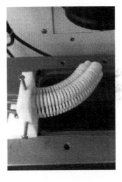

Figure 9.   The picture of the elongation.

$$\theta = 196.5P^2 + 28.01P + 0.1817 \qquad (2)$$

Controlling the air pressure supplied to P2 and P3, and making sure the air pressure P1 is 0, and then adjusting the air pressure P (P2 = P3) from 0 to 0.35 MPa, we obtain the bending angle of the joint under certain conditions of pressure (Figure 8). The bending angle reaches up to 52° when the air pressure is 0.35 MPa. Joint The bending angle empirical formula for using two muscles is as follows:

$$\theta = 239.6P^2 + 63.91P + 0.2721 \qquad (3)$$

4   CONCLUSIONS

This paper develops a new type of pneumatic space bending flexible joint and conducts some experiments. The experimental results indicate that the the elongation and bending of the joint have some degree of nonlinearity. Based on the results of data analysis, the elongation of the joint reached up to 31.8% of its own length and the bending angle of using two muscles can reach up to 52° when the air pressure is 0.35MPa. The joint with three degrees of freedom can elongate in the axial direction or bend arbitrarily in space when the air pressure is supplied, so that it has a good flexibility and practicality.

ACKNOWLEDGMENTS

The project was supported by the National Natural Science Foundation (51275004), the Key Science & Technology Brainstorm Project of Jilin Provincial Science & Technology Department (20130206026GX), the Science and Technology Innovation Development Plan of Jilin City Science and Technology Bureau (201464043), the Youth Science Foundation Project of Jilin Provincial Science & Technology Department (20150520109JH), and the program for the Dr Scientific Research Starting Foundation of Beihua University (199500005), PR China.

REFERENCES

[1] Yuanshen Zhang, Mingchun Liu, et al. 2008. *Hydraulic and Pneumatic*. 7: 13–15.
[2] Yingfei Sun & Aihua Luo. 2012: *Science Technology and Engineering*. 12: 2912–2918.
[3] Pengcheng Huang, Qinghua Yang, Guanjun Bao, et al. 2013. *Robotics*. 35: 67–72.
[4] Wei Fan, Guangzheng Peng, et al. 2003. *Machine and hydraulic*. 4: 32–36.
[5] Geng Dexu, Zhao Ji, Zhang Lei & Zhao Yunwei. 2011. *Applied Mechanics and Materials*. 46: 2883–2887.
[6] Shanghui Li, Qinghui Yang, Guanjun Bao, et al. 2009. *Journal of Zhejiang University of Technology* 37(6): 662–666.

Architectural, Energy and Information Engineering – Sung & Chen (Eds)
© 2016 Taylor & Francis Group, London, ISBN 978-1-138-02791-6

Study on the anti-lateral mechanical properties of the new reinforced ceiling

Z.N. Zhang, J.S. Bai, G.C. Li & Z.D. Tan
Civil Engineering of Shenyang Jianzhu University, Shenyang, China

ABSTRACT: Anti-lateral mechanical properties can be greatly improved by using reinforced ceiling. In this paper, a new reinforced form was designed, and a comparison was made between the reinforced ceiling and the ordinary one using the finite element software. The results indicate that the lateral mechanical properties of the reinforced ceiling are superior to those of the ordinary one. The hysteretic curve of the reinforced ceiling is smooth, and its seismic behavior is better.

Keywords: reinforced ceiling; anti-lateral mechanical properties; finite element analysis; hysteretic curven; Seismic properties

1 INTRODUCTION

Light steel keel ceiling has been widely used in the building as a popular decoration design and fitment[1]. Considerable research[2-3] on the mechanical properties of suspended ceiling has been done in overseas, especially in Japan[4]. With the wide application of the suspended ceiling, the scientific and economic value of the strengthening technology has become increasingly pronounced though the rare research at the domestic level. The seismic and anti-lateral mechanical properties of the reinforced ceiling are superior to those of the common ceiling owing to their better stiffness, strength and stability. In view of the above advantages, a new form of reinforced ceiling was designed to strengthen the boundary of the keel, and a comparison was made on the anti-lateral mechanical properties between the reinforced ceiling and the ordinary one using the finite element software in this paper.

2 NEW REINFORCED CEILING

According to the requirement of the document "Code for quality acceptance of decoration and fitment engineering", reverse support[5] should be given when the length of the suspender is more than 1.5 m. The length of the suspender used in the reinforced ceiling is 1.5 m, and the distance of the joint between the slant rod and the rail is 20 mm. The details are shown in Figure 1. In order to simplify the simulation, the end of the suspender was considered as a whole, while it should be connected by connectors in the real situation. The end of the reinforcing rod is connected to a horizontal bar and a longitudinal rod, respectively, and each of them has a corresponding groove. The connectors are shown in Figure 2.

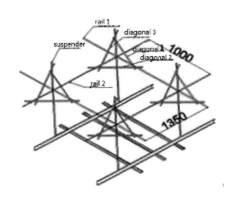

Figure 1. The structure of the new reinforced ceiling.

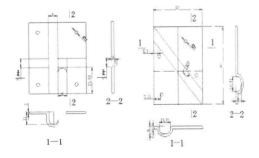

Figure 2. The firmware of the reinforcing bar.

3 FINITE ELEMENT ANALYSIS

3.1 *Modeling*

3.1.1 *Constitutive relation*

A finite element analysis of ceiling is carried out by using ABAQUS[6-7]. As shown in Figure 3, the constitutive relation of the steel is ideal elastic-plastic. The elastic modulus of the material is 2.06×10^5 N/mm^2, and the yield stress is 235 MPa[8].

3.1.2 *Element type and mesh generation*

All the material of the ceiling used was the solid element in order not to consider the warping of the light-gage steel. The division of the grid is shown in Figure 4.

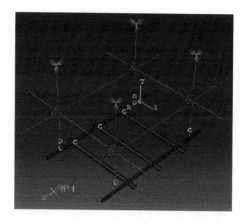

Figure 5. The boundary conditions of the reinforced ceiling.

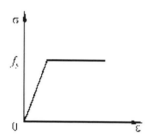

Figure 3. The ideal elastic-plastic constitutive relation curve of the steel.

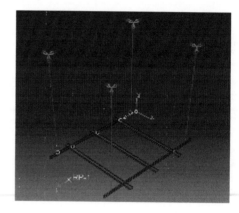

Figure 6. The boundary conditions of the normal ceiling.

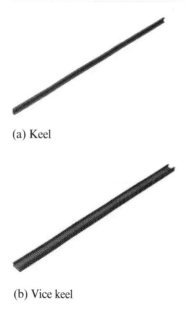

(a) Keel

(b) Vice keel

Figure 4. The division of the grid in the model.

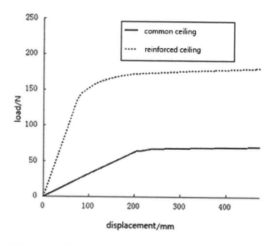

Figure 7. The comparison of the load-displacement curves.

3.1.3 Boundary conditions

The boundary conditions of the ceiling are shown in Figure 5 and Figure 6. The displacement load is given at point RP-1, with a value of 100 mm.

3.2 The results of the finite element simulation

As shown in Figure 7, the lateral displacement of the reinforced ceiling is smaller than that of the common ceiling in the elastic stage under the same force, and the reinforced ceiling can resist the lateral force from the subject effectively, which means the stiffness of the reinforced ceiling is stronger than that of the ordinary one.

4 EQUIVALENT STATIC ANALYSIS

4.1 Modeling

The seismic behavior of the reinforced ceiling and common ceiling is simulated using the finite element software. The boundary conditions of the reinforced ceiling and the common ceiling are shown in Figure 5 and Figure 6, respectively. The displacement-load curve is given at point RP-1, and the displacement is 500 mm. The details of the loading system are shown in Figure 8.

Figure 9 and Figure 10 show the hysteretic curves of the reinforced ceiling and the common ceiling under the circulating force, which reflected the deformation features, stiffness degradation and the energy consumption of the ceiling under the circulating force, which is a basis for the restoring force model and nonlinear seismic response analysis[9–10]. The hysteretic curve of the reinforced ceiling is smooth, as can be seen from the figure, and the curve of the bearing capacity is symmetric, which means a high ductility and a better energy dissipation ability.

Figure 11 and Figure 12 show the skeleton curve of the reinforced ceiling and common ceiling, which present a linear state at the beginning of the loading. It indicated that the reinforced ceiling in the elastic stage, the suspender and the

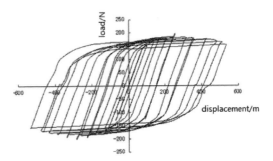

Figure 9. Load-displacement curve of the reinforced ceiling.

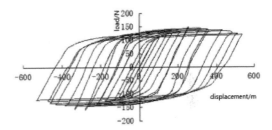

Figure 10. Load-displacement curve of the common ceiling.

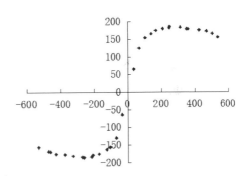

Figure 11. The skeleton curves of the reinforced ceiling.

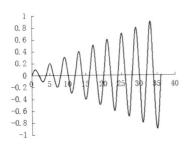

Figure 8. The loading system.

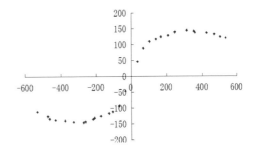

Figure 12. The skeleton curves of the common ceiling.

617

stiffener are not yielded. The slope of the tangent line decreases gradually when the displacement reaches 100 mm, which showed that part of the member is yielded. With the increasing load, the skeleton curve reaches the peak when the displacement reaches 250 mm, and then the curve declines, showing a better ductility of the reinforced ceiling.

5 CONCLUSIONS

A new reinforced technology is proposed to improve the anti-lateral mechanical properties of the suspended ceiling. The anti-lateral mechanical properties of the reinforced ceiling is superior to those of the common ceiling, and the lateral displacement of the reinforced ceiling is smaller than that of the common ceiling under the same lateral force based on the finite element simulation. The seismic properties of the reinforced ceiling are better according to the hysteretic curve.

ACKNOWLEDGMENTS

This project was supported by the National Science and Technology Supporting Plan (serial no. 2012BAJ13B05). The authors gratefully acknowledge their support.

REFERENCES

[1] Architectural Institute of Japan. JASS 26 Construction Standard Specification. Japan: Skill at hall, 2010.
[2] Baoxin Tian, Yujun Liu. The leakage and prevention of light steel ceiling. Shanxi Architecture, 2000, (3): 69.
[3] Futatsugi Shuya, Minewaki Shigeo, Okamoto Hajime, Takahashi Hiromu, Yamamoto Masato, Tomioka Hirokazu, Kamoshita Naoto, Aoi Atsushi. A study on falling down of suspended ceilings without seismic bracing [J]. AIJ Journal of Technology and Design, 2014, 20(44): 149–152.
[4] GB50017-2003 Code for Design of Steel Structures. Beijing: China Planning Press, 2003 (in Chinese).
[5] GB 50210–2001 Code for Quality Acceptance of Decoration and Fitment Engineering. Beijing: China Architecture & Building Press, 2001.
[6] Germain Adriaan. Abaqus. Brev Publishing, 2012.
[7] Li Fan, Zhang N. Analysis of hysteretic curve of md of structures in seismic test. Industrial Construction, 2007, (s1): 283–286.
[8] Qingzhi Liu, Zuozhou Zhao, Xinzheng Lu, Jiaru Qian. Simulation methods for hysteretic curve of steel braces. Building Structure, 2011, 41(8): 63–67.
[9] Yao G C. Seismic performance of direct hung suspended ceiling systems. Journal of Architectural Engineering, 2000, 6(1): 6–11.
[10] Yiping Shi, Yurong Zhou. Detailed examples of ABAQUS finite element analysis. Beijing: Chine Machine Press, 2006 (in Chinese).

Architectural, Energy and Information Engineering – Sung & Chen (Eds)
© 2016 Taylor & Francis Group, London, ISBN 978-1-138-02791-6

Development of automatic testing systems of PXI bus developed flight parameter recorder

W.J. Li & W.H. Wang
State Grid Sichuan Technical Training Center, Chengdu, China

S.Q. Li
Southwest Jiaotong University, Chengdu, China

L. Sun
Avic Chengdu Aircraft Industry(Group), Chengdu, China

N. Liu
New Oriental Corporation, Chengdu, China

Z. Huang
State Grid Sichuan Maintenance Company, Chengdu, China

L. Hao
Air China, Chengdu, China

Z.L. Huang
Ocean University of China, Qingdao, China

ABSTRACT: This paper introduces the Flight Data Recorder Automatic Test System. The system can test the reliability of the Flight Data Recorder to ensure that the FDR records the flight data accurately before air crash, to help analyze the reasons for the crash. The working principle of the Automatic Test System is that the virtual instruments produce signals and the PXI bus transmits signals at a high speed. Then, the signal conditioning board amplifies and shapes the signal to input more than 100 key parameters with a high precision and high speed. These parameters simulate the plane's signal input, and the software system reads the return value. Finally, the system can judge these values automatically and locate the malfunction. We develop the software system in the Visual C# in .net development environment. This system can expand the test coverage greatly, reduce troubleshooting time and improve the efficiency of maintenance. It has many advantages such as small volume, high precision and abundant source.

Keywords: flight data recorder; automatic test system; virtual instruments; PXI bus; Visual C

1 DESCRIPTION OF THE TEST OBJECT THE INTRODUCTION OF THE TEST OBJECT

The purpose of the flight Data Recorder is to collect and record the information of flight parameters, which can save critical flight information in the case of accidents. It has a great effect on fault detection and incident traceability. The Aircraft Flight Data Recorder in the structure mainly consists of three parts, namely flight information collection devices, controllers and memory (commonly known as the black box).

1. Flight information gathering device is a core component of the flight parameter recorder, which is mainly used to receive imitation information, the frequency of the signal and digital signal, and then convert these signals into a digital serial code, and also to provide a stable power supply and voltage for the primary potentiometer sensor.
2. The controller is part of the parameter recorder on board, mainly for the input and display auxiliary information. It can put through the memory on board via a manual remote control and inspect each device in the machine parameter recorder.

3. The parameter memory records dual-code information from the collecting device on a magnetic carrier, and can guarantee to preserve the complete record when it has a malfunction. It consists of the matching components, reel mechanism, control devices, motor speed controller and commutation institutions function modules in the structure.

The Flight Data Recorder is shown in Figure 1.

1.1 *Measured signal*

Flight parameter recording system sensor sends 10 kinds of signals, as shown in Figure 1. It can form a complete and detailed analysis of the test signal about the complete and precise extent of the following flight parameter record. According to the printed test report, the system provides fault data to experts as a support.

The following is a group of 10 signal sets:

0~6.3 V direct current voltage
0~33 V direct current voltage
0~5 V a frequency of 400 Hz AC voltage
0~45 V a frequency of 400 Hz AC voltage frequency of 7 to 100 Hz sine wave
2.4~4.5 V frequency of 8.2 to 10 kHz square wave
0~360° synchronic signals
0~360° cosine signal of the transformer

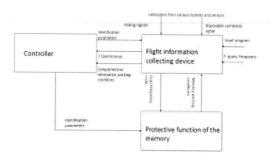

Figure 1. Flight data recorder chart.

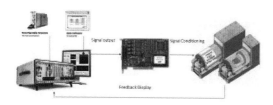

Figure 2. Flight data recorder system hardware components.

2 THE STRUCTURE OF THE AUTOMATIC TEST SYSTEM

2.1 *Hardware section*

1. The case with PXI slotting from National Instruments company, equipped with two PXI boards, namely imitation input and output, one digital multimeter PXI board card, one digital signal output PXI board, one PXI switch control board, equipped with its own Windows operating system, and NI Measurement Studio 7.1 tool library.
2. Signal conditioning board is used for shaping and amplifying a signal conditioning of the output signal. Additionally, it decodes the input signal and converse A/O.
3. Connecting cable is used for the docking flight information collection device and memory.

2.2 *Software section*

Development Language: Microsoft Visual C#

1. Automatic Test: the operators simply configure the test before the test tasks via the computer-controlled automated test to the product. Then, the testing system will automatically run the test item, noting the failure point and automatically entering into the flight reference database when the test is completed.
2. Manual Test: this test module helps the operator manually to conduct a detailed test on individual items and to automatically enter into the flight reference database based on the testing result.
3. Expert Systems: the reference database can provide experiential learning and inquiry.
4. Test Reports: it can generate reports for record test results, and the format of the report can be maintained.
5. System Maintenance: it has several functions such as system connection test, user rights management, system metering.
6. System Assist: it helps in documentation toguide the user's actions.

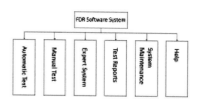

Figure 3. Software system.

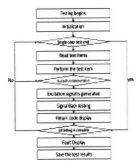

Figure 4. Automatic testing system operation.

3 TEST RESULT

It takes 6 minutes and 30 seconds to completemore than 100 test items. It shows "the testing process failed" and aborts the test process if it is beyond the set range of error or has no signal response section. If we check the test after the continuing option, we can skip the failed items and continue until all the testing is completed. Then, we can export the test data and print the reports.

4 CONCLUSIONS

The development of the automatic test system is to compensate the lack of factory routine testing methods. In the system test software, the key technologies used in the design process are as follows:

1. The application of the virtual instrument technology makes an effective comprehensive test switchboard to replace several different test equipment. Also, this technology is more accurate and operates at high speed in switching between sources. It uses the VISA (Virtual Instrument Software Architecture) technology to ensure the compatibility between the different test interfaces and interchangeabilities.
2. It adopts the software development environment, which is consistent with the VPP standard to ensure the portability between different platforms and operating systems.
3. It uses modular software design methods to improve the flexibility of the system software, portability and maintainability, thereby reducing the complexity of the system.

Currently, one Flight Data Recorder automatic test system can replace the workload of five workers in the two days of testing, which is completely unattended. We believe that continuing efforts to optimize the sound will allow the test system to play a greater role efficiently on the basis of the stage work.

BIOGRAPHY

Li Wujin received the B.S. degree from USTC Chengdu, China. Currently, he is a lecturer at State Grid Sichuan Technical Training Center, Chengdu, China. His research interest is substation maintenance in power system.

REFERENCES

[1] C.L. Chen (China Machine Press 2008): Automatic Testing and Interface Technology, p. 239–264.
[2] S.L. Xie (Tsinghua University Press 2004): Visual C#. NET 2003 Development and Skills, p. 57–71.
[3] M. Shi (Peking University Press 2001): SQL Server 2000 Programming Guide, p. 98–120.
[4] M. J. Ma (Tsinghua University Press 2007): PCI, PCI-X and PCI Express Principles and Architecture, p. 134–145.

Architectural, Energy and Information Engineering – Sung & Chen (Eds)
© 2016 Taylor & Francis Group, London, ISBN 978-1-138-02791-6

Meter reading acquisition design and implementation of a wireless communication network

S.Z. Hou, Y. Yang & G.Q. Yang
North China Electric Power University, Baoding City, Hebei Province, China

ABSTRACT: The artificial intelligence wireless meter reading system is based on the micro-power short-range wireless communication technology, using the combination of digital signal single-chip RF transceiver chip, chip technology with a micro-controller and a small amount of peripheral devices constituting a special or general purpose wireless communication module. The current meter reading method brings a lot of trouble to users and staff, whereas the use of wireless meter reading can not only save resources, but also improve the accuracy, timeliness, and efficiency. Using the si4438 wireless communication chip as a wireless transceiver wireless communication module and using the wireless mesh network, the appropriate routing protocols can be selected. The system has a good communication quality, low cost, fast networking, high stability, reliable data, facilitative monitoring, and good scalability.

Keywords: wireless meter reading; networking; routing protocols; Si4438 chip

1 INTRODUCTION

With the progress of science and technology, especially the rapid development of computer technology, communication technology, VLSI technology, and measurement technology, people's demand for the intelligent level of living and home conditions has been increasing. The indoor user's measuring instruments data sent to the CCMS center wirelessly has become the goal. Water, electricity, gas, heat and other public utilities administration also hope that the new technology can solve long-term problems of their meter reading difficulty and charge difficulty, in order to achieve the purpose of savings in manpower, user-friendly and improve management. In recent years, with the gradual realization of the national power grids and commercial electric power operations, "A table one, meter reading to home" policy and the "sub-users, subtime pricing" tariff system have been progressively implemented and enforced. The National power sector has also established a "centralized meter reading, banking network," the long-term planning, which will undoubtedly increase the meter reading work enormously. However, the existing meter reading model is clearly not fully adapted to the modern life, and cannot meet the development needs of our society and economy[1]. Changing the existing meter reading mode, research and promotion of the "automatic meter reading technology" has become the increasingly urgent demands and an inevitable trend in the future development.

Because the wireless meter reading system involves thousands of families, its cost has a direct impact on its popularity and the user's acceptance of the system, so the cost has become an important factor to measure the success of the system. The user's data on the meter, which the wireless meter reading system transfers at the power control centers' command, will be accurate. The wireless network must be self-healing function, so a high reliability is also an important requirement for the wireless meter reading system, including data collection reliability, reliability of data transmission and signals interference. In the event of a failure, it can inform the cause of the error to the user in time via devices such as LCD and alarm system.

2 WIRELESS METER READING SYSTEM

2.1 *The overall structure of the wireless meter reading system*

The overall block diagram of a wireless meter reading system is shown in Figure 1.

The overall structure of the wireless meter reading system includes four parts: center management station, concentrators, data acquisition and collection terminal. The management of the central station constitutes the higher computer system, concentrators, and data acquisition, and collection terminals constitute the lower computer systems[2]. Center management station mainly sets up the command sent, and

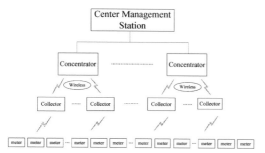

Figure 1. Overall block diagram of a wireless meter reading system.

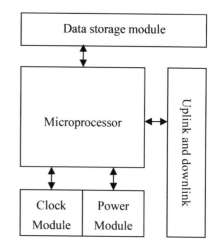

Figure 2. Concentrator chart.

then the system detects and displays the operating status and data communication with the meter concentrator for the meter reading system. Concentrator is a communication hub for the automatic meter reading system, which is responsible for receiving the instruction of the central management station, collecting the users' electricity data and statistical calculation data, and is the communication tie between the central management station and the meter.

2.2 Concentrator system components

Concentrator plays a very important role in the whole meter reading system, which not only controls the governed meter according to the commands sent by the master, but also collects the meter data and uploads it to the master, so the concentrator must be able to communicate with the master and the meter simultaneously. Concentrator with the master station is carried out via the GPRS communication, so the concentrator should have the GPRS modules. Concentrator communicates with the meter using a wireless way, so the concentrator should have a wireless communication module[3]. The centralized structure is shown in Figure 2.

A description of each module is given as follows:

1. Power management module: it is responsible for the management of the power of the concentrator, and provides different power supply interfaces.
2. Data storage module: it stores the meter reading data and concentrator parameter information of the electric meter, including real-time data, historical data and event data, and history data save two months' data generally.
3. The microprocessor: it is the management center of the concentrator, which is responsible for managing the various functions of the task, mainly the meter reading tasks, communication tasks, the data management tasks, and protocol processing tasks.
4. The clock module: it provides the clock information to the system.

3 CHIP SELECTION

Using the SILICON LABS, highly integrated wireless communication chip Si4438, Si4438 wireless transceiver is designed for smart meters in the China market. It has the characteristics of high performance, low power consumption and high integration, and can extend the wireless transmission distance and battery life, with smart meters having 30% lower material costs at the same time. All parts offer an outstanding sensitivity of –124 dBm while achieving extremely low active and standby current consumption. The 58 dB adjacent channel selectivity with 12.5 kHz channel spacing ensures robust receiving operation under harsh RF conditions. The Si4438 offers exceptional output power of up to +20 dBm with an outstanding TX efficiency. The high output power and sensitivity results in an industry-leading link budget of 144 dB, allowing extended ranges and highly robust communication links.

MCU selection AT89C51. AT89C51 is a cost-effective 8-bit microprocessor with advanced features, which has been widely used in many embedded systems, especially in 4 K bytes of internal EPROM for the user to provide a great convenience, when the application is small and does not add the external expansion of the CPU ROM. Idle mode and power-down protection mode ensures that the CPU is operating in a low-power state, and that the internal data RAM is not lost.

4 THE WIRELESS COMMUNICATION UNIT

Wireless communication module is shown in Figure 3. It enables to complete the terminal

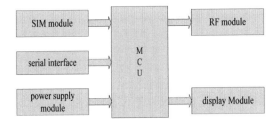

Figure 3. Wireless communication module.

energy data acquisition and receiving, but also to have the routing and coordination functions, and can be connected to a computer to complete the task of the network control and data management. The core circuit consists of the MCU, RF module and transmitting antenna, display module, power supply module, SIM modules and serial interface.

5 PROTOCOL SELECTION

The protocol follows DL/T645-2007 agreements and national grid "power user information collection system communication protocol."

The 645 agreement between meters and acquisition device uses the half duplex communication mode. Gathering station is given priority to equipment and meters from the station. In the system, each meter has a unique address code[4]. The address domain can fill high AAH, used as wildcard bytes for low abbreviated addressing data to read the meters, electric meter response when actual address. Low 5 is used to represent the function of the frame, including reading, writing data, reading and writing the address, data clearing, freezing and other data. Data length field represents the total number of bytes of data domain, L represents the length of the data field of the total number of bytes of the data field, when reading the data field length cannot be greater than 200 and writing the data field length cannot be greater than 50. Data frame before sending the data in bytes as the 33 h operation, when receiving the data in bytes as minus 33 h operation. Transmission when the sender adds 33 h processing in bytes, the receiver to reduce 33 h operation in bytes. Between data frames in the check code exactly 256 bytes of data and to check code.

6 CONCLUSION

This paper designs a wireless meter reading system based on the Si4438 wireless chip and the DL/T645-2007 communication protocol. Using the wireless communication chip Si4438 as the wireless transceiver to design the wireless communication module, and the wireless mesh network for networking, realized with efficient, fast, and reliable electricity power system customer information collection.

7 SOFTWARE NETWORK ALGORITHM

Referring the ZIGBEE Union standards, IEEE802.15.4 standard is the basis for ZigBee, WirelessHART and other norms, describing the low-rate wireless personal area network physical layer and media access control protocol, belonging to the IEEE 802.15 working group. On the 868/915M, 2.4 GHz ISM band, the data transfer rate is up to 250 kbps. The advantages of its low power consumption and lowcost make it to be used widely in many areas.

Wireless ad hoc network is a kind of more jump frequency of temporary autonomous system of mobile network, and there is no infrastructure. It is made up of a bunch of mobile terminals node with a wireless radio frequency transceiver, with the characteristics of self-organization and self-healing, and there are effective mechanisms for managing and tracking mobile users. It is not only a multi-hop broadband wireless network, but also a high-capacity, high-speed distributed network. It can be seen as the convergence of WLAN and MANET, which has convenient network construction and low cost, and also has the incomparable advantages of traditional wired and wireless network system. It can provide wireless broadband access services to LAN, campus network or MAN. Due to self-organizing and self-healing capabilities, network administrators do not have to manually configure the WMN network. In the event of a node, link failure, it can be automatically adjusted to complete the network self-healing. Most of this network node is basically stationary, not with the battery as a power source, the topology changes are small, high reliability, high network coverage, easy to maintain, it is a kind of wireless meter reading self-organizing networks, which is appropriate in the residential area, allows multiple networks coexist, automatically distinguishes between different network topology dynamic structure variables and dynamic routing.

Traditional wireless mesh routing protocols are single channel routing protocols, and depending on the route discovery process, they can be divided into two categories: table driven routing protocol and on-demand routing protocol. Table driven routing protocol is also called the first type routing protocol, in which each node needs to maintain one or more tables to record the routing node to other nodes. In order to maintain the consistency of the network topology and routing information, all nodes are for regular or trigger to update the routing table. On-demand routing protocol is also called the reactive routing protocol. It mainly includes the DSDV, CGSR and WRP typical protocol. The main difference is that each node maintains a number of different tables, and update the information, when the network topology changes with different ways of transmission in the network. In the on-demand routing protocol, each node does not need to maintain the latest routing information and routing can be created only when needed. When the source node needs to send the data to the destination node, the source node starts the routing discovery mechanisms to find the best path to reach the destination node[5]. AODV, DSR and TORA are three typical protocols. To make full use of the wireless resources in order to improve the network capacity, a new routing protocol is proposed for the use of multi-channels. It mainly includes the hybrid routing protocol (HwMP), Multi-Channel Routing Protocol (MCRP) and Radio Link Quality Source Routing Protocol (MR-LOSR).

REFERENCES

[1] Sizu Hou, Hui Wu, Xiaoxu Zhang. Embedded Linux transplant on the platform of meter reading concentrator based on AT91SAM9260 [C]. AMSMT, 2013, pp. 2118–2122.
[2] Li Li. Design of new architecture of AMR system in Smart Grid [J]. Industrial Electronics and Applications (ICIEA), 2011 6th IEEE Conference 21–23 June 2011.
[3] Parikh, P.P., Kanabar, M.G., Sidhu, T.S., Opportunities and Challenges of Wireless Communication Technologies for Smart Grid Applications [J]. Power and Energy Society General Meeting, 2010IEEE, Page(s): 1–7.
[4] Sizu Hou, Xiaoxu Zhang, Hui Wu. Meter reading concentrator software design and implementation based on linux [C], CSETA2013, 2013(766): 894–894.
[5] Guilin Zheng, Yichuan Gao, Lijuan Wang. Realization of Automatic System Based on ZigBee with Improved RoutingProtocol [C]. Power and Energy Engineering Conference, 2010: 1–6.

Architectural, Energy and Information Engineering – Sung & Chen (Eds)
© 2016 Taylor & Francis Group, London, ISBN 978-1-138-02791-6

The research on electric power communication network risk assessment based on the performance parameters

S.Z. Hou, G.Q. Yang & Y. Yang
North China Electric Power University, Baoding City, Hebei Province, China

ABSTRACT: The electric power communication network is an important part of the power system. Its security is of great significant for making power system safe and stable. This article applies the theory of risk analysis to the electric power communication network, and realizes the collection of performance data in the network by the CORBA north interface technology. Electric power communication network parameters were classified comprehensively, the weight calculation method was introduced, and the scoring criteria were given in view of the performance parameters. Finally, the article calculates the value of all risks. The result of the analysis indicates that the electric power communication network security risk assessment can provide a quantitative basis for the operation and dispatch personnel, which also contributes to finding the weak link of the electric power communication network in time, avoiding serious accidents and improving the level of safe operation.

Keywords: electric power communication network; risk assessment; indicator system

1 INTRODUCTION

With the development of smart grid and modern communication technology, electric power communication system has bore the core business of the grid such as security and stability control. Power communications and power grid security are closely linked, and the importance is also more highlighted in the grid security. Due to the communication network in the process of planning, construction and operation maintenance, the original concept, technology and standard cannot adapt to the needs of the rapid development of the modern power grid. In addition, the original network space truss structure, system capacity, performance, reliability requirements have shortcomings, which poses a great threat to the safety of power grids. Therefore, strengthening the safety of the electric power communication system and ensuring the normal operation of equipment is the first task of communication management. To discover the network vulnerability and potential safety hazard that may arise, research and analysis on the electric power communication network risk assessment is necessary.

2 BASIC DATA ACQUISITION

At present, much of the SDH transmission network in the electric power communication network is in a different manufacturer equipment environment.

Figure 1. The adapter schematic diagram.

There are differences in aspects such as architectures, information management model and network management protocol, especially the difference between the network Element Layer (EML) and the Network Layer (NML) is more outstanding. In this paper, a common north interface adapter model has been adopted, which realizes the communication between the common interface and the private interface based on the CORBA (Common Object Request Broker Architecture) technology and can successfully access all kinds of performance data from the network management. The adapter schematic diagram is shown in Figure 1.

3 THE RISK ASSESSMENT PROCESS

3.1 *The establishment of the index system*

The establishment of the electric power communication transmission network risk assessment index system should satisfy some important principles that include the integrity, practicability, comparability, maneuverability, and independence.

a. The integrity is that the index system should fully cover all the factors that influence the risk of the electric power communication transmission network; none of the factors should be left out.

b. The practicality refers to the indices chosen, which should possess specific meaning, evaluation basis, and also the realistic foundation.

c. The comparability indicates that the rules should be consistent with the current ones as far as possible when designing the index names. The truth is that the stronger the comparability of indicators is, the higher the reliability of evaluation results eventually is.

d. The operability expresses that the selected indicators are collected from the existing statistical data in principle, such as the network performance data; in addition, it is important to pay attention to the practicality and convenience when few indicators need to be confirmed.

e. The independence requests to reduce the meaning overlap between the indicators as far as possible, for example, the indicators that are at the same level cannot contain each other, which can guarantee that the indicators reflect the actual situation of the equipment from different aspects.

Electric power SDH transmission network is composed of different equipment, at the different levels of the operation environment and the management maintenance. A combination of various factors makes the whole network in the different risk conditions. This paper uses the general hierarchical index system, which could eliminate the differences in the electric power communication network, is beneficial to the objectivity of risk analysis and comprehensive, and can reflect the risk point closest to the real situation of risk, as shown in Figure 2.

3.2 Calculation of the index weight

This article applies the AHP method for the electric power communication network risk assessment.

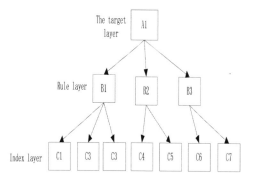

Figure 2. Hierarchical indicator model.

Based on the hierarchical structure model of the risk evaluation system about the power communication network, we construct the judgment matrix $A = (a_{ij})$. Among them, scale value a_{ij} represents the ratio of the relative importance about the lower index of Y:

$$A = \begin{bmatrix} a_{11} & a_{12} & a_{13} & a_{14} \\ a_{21} & a_{22} & a_{23} & a_{24} \\ . & ... & . & . \\ a_{n1} & a_{n2} & a_{n3} & a_{n4} \end{bmatrix} \quad (1)$$

The assignment is obtained by consulting the expert and research, which is given by the 1–9 scale method. The digital scale of a_{ij} and its significance is presented in Table 1.

The elements of the matrix $A = (a_{ij})$ need to be satisfied as follows:

$$\begin{cases} a_{ij} > 0 & (i, j = 1,2,...n) \\ a_{ii} = 1 & (i = 1,2,...n) \\ a_{ij} = 1/a_{ji} & (i, j = 1,2,...n) \end{cases} \quad (2)$$

We can put any of a matrix using the formula for iteration:

$$b_{ij} = \sqrt[n]{\prod_{k=1}^{n} a_{ik} \bullet a_{kj}} \quad (3)$$

Table 1. Scale value and definition of a_{ij}.

Scale numerical	Meaning
1	Two elements have the same importance
3	Comparing two elements, an element is more important than another element slightly
5	Comparing two elements, an element is more important than another element obviously
7	Comparing two elements, an element is more important than another element strongly
9	Comparing two elements, an element is extremely more important than the other element
2,4,6,8	Paired differences between things, take values in the middle of the adjacent judgment
Reciprocal	If the elements I and j importance for a_{ij}, j and the ratio of the importance I to $a_{ji} = 1/a_{ij}$

Finally, the weight is calculated as follows:

$$w_j = \frac{c_j}{\sum_{k=1}^{n} c_k} \ (j = 1, 2, \ldots n) \qquad (4)$$

Among them,

$$c_j = \sqrt[n]{\prod_{k=1}^{n} b_{jk}} \ (j = 1, 2, \ldots n) \qquad (5)$$

3.3 Risk value calculation

The change in each parameter will make the running status of the corresponding equipment change. The possibility of the electric power communication network risk can be summed up as the scope of different equipment unit parameters deviating from the standard. Taking 10 group performance parameter values of different periods of time when the equipment operates normally, the average of the 10 sets of data is considered as the standard values of the performance parameters. The threshold scores are given in Table 2.

The table can be divided into six threshold segments, with each threshold value corresponding to the fluctuating range of parameter deviating from

Table 2. The threshold value score.

Parameter	Unit	Threshold 1 (0)	...	Thresold 5 (10)
PMP_TPL	dBm	±1.5	...	±12
PMP_RPL	dBm	±2	...	±13
PMP_OPT_LTEMP	°C	±5	...	±14
PMP_OPT_LBIASN	mA	±3	...	±15
PMP_PJE	T	±20	...	±180
PMP_ES	T	±18	...	±40

Table 3. The risk level.

Risk Level	Value at Risk	Processing Instructions
No risk	0~3	Do not need to deal with
Slight risk	3~5	To prevent risk increase
Moderate risk	5~6	Arrange time to deal with risks
Significant risk	6~8	As soon as possible to deal with risks
Catastrophic risk	8~10	Immediately to deal with risks

the standard value, different threshold segments assigned by the value representing the size of the possibility of a risk, and the resulting score values devices and optic cable. Considering the index system comprehensively, we obtain the risk value of the risk points of the electric power communication network. The specific calculation method is as follows:

$$R = \sum_{i}^{N} \sum_{j}^{M} S_i \cdot W_j \qquad (6)$$

Here, R is the risk value of the risk point; W_j is the index weight; S is the score of performance parameters; and N, M represents the different indicators and business number.

3.4 The risk level

According to the risk values of the risk point, judging the risk level of the equipment, the risk level settings as given in Table 3.

4 THE RESULTS OF THE RESEARCH

The whole process of the electric power communication network risk assessment is mainly composed of data acquisition module, calculation module, and risk display module. The risk calculation module is the core part of the risk assessment process. The risk display module is through a chart or form showing the evaluation results, comprehensively divided into the following categories, as shown in the following figure.

a. Show the risk status according to the equipment.
b. Show the risk statistics according to the risk level.
c. Show the changes in risk by time.

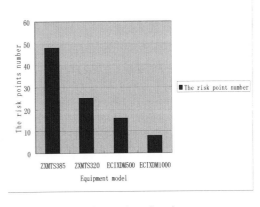

Figure 3. Risk point number of equipment.

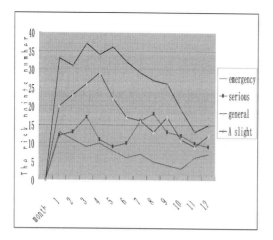

Figure 4.　Risk profile statistics.

5 CONCLUSIONS

This paper introduces the basic principle of the electric power communication network risk assessment and the result of the research. The whole assessment process using real-time performance parameters as the basis for risk analysis fully considers the subjective and objective reasons, and may conclude the risk. Constructing a hierarchical index system and the analysis result is scientific and reasonable. At present, the research results have been successfully applied in some regional power network risk assessment, and the results are the same as the actual risk status. The research greatly increases the security of the electric

power communication network and, at the same time, saves a large number of human resources, providing an important guarantee for the safe and efficient operation of the power grid.

REFERENCES

[1] Li Li-li. The dangerous point analysis and precontrol management in electric power communication system [J]. Electric power security technology. (2011).13(12).
[2] Wei Ming-hai. Electric power communication system analysis of production safety [J]. Shaanxi electric power. Vol. 6. (2008). P. 36–38.
[3] Zhao Zhen-dong, Lou Yun-yong, Zhang Ya-dong, et al. Structure of reliability evaluation model for electric power communication network [J]. Electric Power Technology. Vol. 9 (2010). P. 74–77.
[4] Xiao Long, Qi Yong, Li Qian-mu. Evaluation of information security risk assessment based on AHP and fuzzy comprehensive [J]. Computer engineering and application. Vol. 11-9 (2009). P. 82–85.
[5] Zhao Zi-yan, Liu Jian-ming. A new service risk balancing based method to evaluate reliability of electric power communication network [J]. Power System Technology. Vol. 10 (2011). P. 209–213.
[6] Cui kai, Fang Dazhong. Study on Probabilistic Assessment Method for Power System Transient Stability. Power System Technology. Vol. 10 (2013). P. 44–49.
[7] Han Xiao-tao, Yin Xiang-gen. Application of Fault Tree Analysis method in Reliability Analysis of Substation Communication System [J]. Power System Technology. Vol. 1–2 (2004). P. 50–59.
[8] Satty, T.L., The Analytic Hierarchy Process. Wiley, New York. (1980).
[9] Zhao Xia, Zhao Cheng-yong. Fuzzy Synthetic Evaluation of Power Quality. Power System Technology. (2005). P. 11–16.

Architectural, Energy and Information Engineering – Sung & Chen (Eds)
© *2016 Taylor & Francis Group, London, ISBN 978-1-138-02791-6*

Research of ultra-low power intelligent flow totalizer based on the online computing method of gas compression factor Z

Xun Liu
Chengdu Qianjia Technology Co. Ltd., Institute of Technology, Chengdu, China

ABSTRACT: In the paper, the system design of the ultra-low-power intelligent flow totalizer was described. For this purpose, three computing methods of gas compression factor Z were introduced, namely the NX-19[1,2] algorithm, AGA8-92DC algorithm[3,4] and SGERG-88[4,5] algorithm. With the host computer (PC computer) and the totalizer, using the method of curve fitting and interpolation, the low-power high-precision online calculations of gas compression factor Z were implemented based on the AGA8-92DC algorithm and the SGERG-88 algorithm. The simulation results show that the best orders of curve fitting of the AGA8-92DC algorithm and SGERG-88 algorithm were three. In the AGA8-92DC algorithm, the absolute maximum relative error of factor Z was 0.0513%, and in the SGERG-88 algorithm, the absolute maximum relative error of factor Z was 0.0516%. Based on these methods, the ultra-low power intelligent flow totalizer was developed with a high precision grade.

Keywords: ultra-low power intelligent flow totalizer; gas compressibility factor Z; AGA8-92DC algorithm; SGERG-88 algorithm; online computing method

1 INTRODUCTION

Traditional Roots gas flow meter and gas turbine flow meter with the word wheel readings could given the traffic condition (in current temperature and pressure case) volume of gas. However, with the development of natural gas application, the traffic volume of gas could not meet the users' needs. In order to facilitate trade settlement, traffic volume must be transformed into the standard volume (20 degrees Celsius, a standard atmospheric pressure). The intelligent gas flow totalizer could solve this problem. In order to calculate the real-time high-precision standard volume of gas, it is necessary to calculate the real-time high-precision compressibility factor Z of gas. For the calculation of the compression factor Z of gas, the NX-19 algorithm was widely used. Although the NX-19 algorithm was computationally simple, there are limitations. It requires that the real relative density of the gas be less than 0.75, the mole fraction of carbon dioxide of the gas be less than 0.15, and the mole fraction of nitrogen of the gas be less than 0.15; meanwhile, the calculation precision of compression factor Z could not meet the requirements of high-grade (A grade and B grade) measuring instruments[6]. In the actual use, the gas properties may exceed these limits. In order to develop measuring instruments with a high computational

precision, other algorithms must be selected. The AGA8-92DC algorithm and SGERG-88 algorithm can be selected. However, these algorithms are quite complex, require high resources (computing power and memory resources), and could not achieve real-time high-precision online calculation. In this paper, with the host computer (PC computer) and the totalizer, using the methods of curve fitting and interpolation, the low-power high-precision calculation methods of gas compression factor Z based on the AGA8-92DC algorithm and the SGERG-88 algorithm were implemented. Thus, using the ultra low-power system design, the ultra low-power intelligent flow totalizer was ultimately developed based on the computing method of gas compression factor Z. Finally, the totalizer test data and user interface were generated.

2 SYSTEM DESIGN OF THE INTELLIGENT FLOW TOTALIZER

The system design of the intelligent flow totalizer mainly includes the main circuit board, the turbine sensor pre-amplifier circuit, roots sensor pre-amplifier circuit, communication module, the current module, memory modules, LCD circuit, and circuit interface board. The system design diagram is shown in Figure 1.

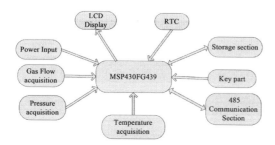

Figure 1. System design diagram.

Figure 2. Components of the computing system.

The key to the system design is designed for low power consumption, and the most important is the choice of MCU. In this design, the hardware system chooses the low-power chip MSP430FG439.

In addition, the temperature measurement section, pressure measurement section, and flow measurement section (using the new magnetic sensor, which directly outputs square wave pulse with a very low power consumption) were also designed with low power thinking.

In power management section, 485 communications section and LCD display part also carried out a low-power design.

3 ONLINE HIGH PRECISION COMPUTING OF COMPRESSION FACTOR Z

3.1 *Components of the computing system*

Figure 2 shows the computing system. The system consists of PC and totalizer, and the communication between them employs the RS485 bus.

3.2 *Computing implementation*

Step1: based on the AGA8-92DC algorithm and the SGERG-88 algorithm, the precise gas compression factors Z in a given molar fraction of natural gas, the pressure range and temperature range (adjusted by software settings according to the actual situation) were calculated, using the software in PC developed by the author. Compression factors Z obtained based on the AGA8-92DC algorithm and the SGERG-88 algorithm are two two-dimensional arrays of values. Selecting a few typical pressure values as references, for example 0 Mpa, 1.75 Mpa, 3.50 Mpa, 5.25 Mpa, 7.0 Mpa, etc, curve fittings of compression factor values were performed. The order of curve fitting is quite important. If the order of curve fitting is smaller, although the calculation of totalizer is simpler, the final error of compression factor Z is larger.

If the order of curve fitting is larger, although the error of compression factor Z is smaller, the amount of calculation of the totalizer is larger. When the order reaches a certain value, any increase of the order could not reduce the errors obviously.

Step 2: the compression factors Z within the ranges of temperature and pressure were estimated using the interpolation method. Based on the comparison of the estimated value and the actual value of the compression factors, the effect of this method was evaluated. The specific interpolation algorithm is as follows:

In the given pressure and temperature ranges, arbitrarily select a temperature (t) and a pressure (P), and find the nearest from the pressure value P two fitting curves. The pressure (denoted by P_1) of a fitting curve is greater than the pressure value P, and the pressure (denoted by P_2) of another fitting curve is less than the pressure value P. Calculate the second compression factor values (denoted by Z_1 and Z_2) at the temperature t, respectively.

The formula is as follows:

$$Z_1 = a_n t^n + a_{n-1} t^{n-1} + \cdots + a_1 t + a_0 \tag{1}$$

where $a_n, a_{n-1}, \ldots a_1, a_0$ are the corresponding parameters of n-order curve fitting at pressure P_1

$$Z_2 = b_n t^n + b_{n-1} t^{n-1} + \cdots + b_1 t + b_0 \tag{2}$$

where $b_n, b_{n-1}, \ldots b_1, b_0$ are the corresponding parameters of n-order curve fitting at pressure P_2.

After linear interpolation, obtain the third compression factor value (Z_t) at pressure P and temperature t, which can be calculated as follows:

$$Z_t = Z_1 + (Z_2 - Z_1)/(P_2 - P_1) * (P - P_1) \tag{3}$$

According Z_t and Z, the corresponding relative error ΔZ of compression factor Z can be obtained as. follows:

$$\Delta Z = (Z_t - Z)/Z * 100\% \tag{4}$$

Figure 3 (a) shows the exact value of the gas compression factor based on the AGA8-92DC algorithm. Figure 3 (b) shows the exact value of

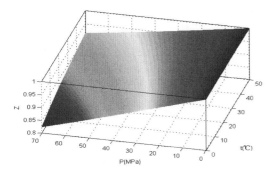

Figure 3(a). The exact value of the gas compression factor Z based on the AGA8-92DC algorithm.

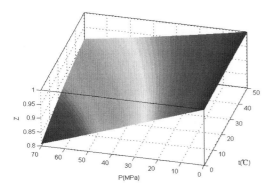

Figure 3(b). The exact value of the gas compression factor Z based on the SGERG-88 algorithm.

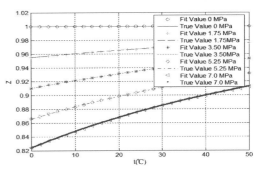

Figure 4(a). The fitting curves of the AGA8-92DC algorithm under the second-order curve fitting.

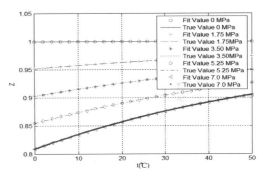

Figure 4(b). The fitting curves of the SGERG-88 algorithm under the second-order curve fitting.

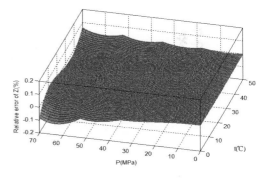

Figure 5(a). The relative error of the AGA8-92DC algorithm under the second-order curve fitting.

the gas compression factor based on the SGERG-88 algorithm.

Figure 4 (a) shows the fitting curves of the compression factor based on the AGA8-92DC algorithm under the second-order curve fitting. Figure 4 (b) shows the fitting curves of the compression factor based on the SGERG-88 algorithm under the second-order curve fitting.

Figure 5 (a) shows the relative error of the AGA8-92DC algorithm under the second-order curve fitting. Figure 5 (b) shows the relative error of the SGERG-88 algorithm under the second-order curve fitting.

The maximum absolute value $|\Delta Z|$ of the relative error of the AGA8-92DC algorithm under the second-order curve fitting is 0.1155%. The maximum absolute value $|\Delta Z|$ of the relative error of the SGERG-88 algorithm under the second-order curve fitting is 0.1162%.

Thus, using the second-order curve fitting and interpolation algorithm, the relative errors of the compression factor based on the two algorithms

are found to be larger[6], and thus the errors do not meet the requirement of the totalizer.

Figure 6 (a) shows the fitting curves of the compression factor based on the AGA8-92DC algorithm under the third-order curve fitting. Figure 6 (b) shows the fitting curves of the compression factor based on the SGERG-88 algorithm under the third-order curve fitting.

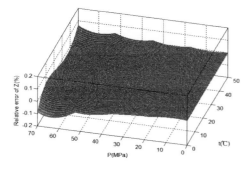

Figure 5(b). The relative error of the SGERG-88 algorithm under the second-order curve fitting.

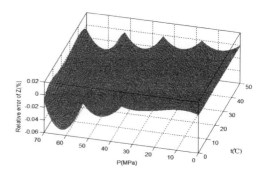

Figure 7(a). The relative error of the AGA8-92DC algorithm under the third-order curve fitting.

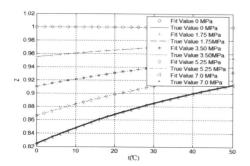

Figure 6(a). The fitting curves of the AGA8-92DC algorithm under the third-order curve fitting.

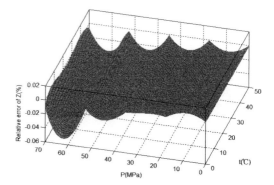

Figure 7(b). The relative error of the SGERG-88 algorithm under the third-order curve fitting.

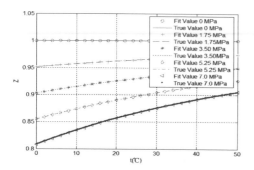

Figure 6(b). The fitting curves of the SGERG-88 algorithm under the third-order curve fitting.

Figure 7 (a) shows the relative error of the AGA8-92DC algorithm under the third-order curve fitting. Figure 7 (b) shows the relative error of the SGERG-88 algorithm under the third-order curve fitting.

The maximum absolute value $|\Delta Z|$ of the relative error of the AGA8-92DC algorithm under the third-order curve fitting is 0.0513%. The maximum absolute value $|\Delta Z|$ of the relative error of the SGERG-88 algorithm under the third-order

curve fitting is 0.0516%. Hence, the errors meet the requirement of the totalizer[6]. Thus, according to the simulation results, the third-order of the curve fitting was selected.

Step 3: Record the parameter of each curve fitting, i.e., the third-order curve fitting (a,b,c,d) and its corresponding pressure value (P).

According to Figure 1, through RS485 or RS232 communication, the parameters are passed to the totalizer, and then these parameters are stored in the non-volatile storage devices of the totalizer, such as Flash or EEPROM, or ferroelectric.

Step 4: Flow totalizer (with pressure and temperature sensors) based on real measured gas temperature (t) and pressure (P), find the nearest from pressure P two fitting curves. The pressure (denoted by P_1) of a fitting curve is greater than the pressure value P, the pressure (denoted by P_2) of another fitting curve is less than the pressure value P. Calculate the second compression factor values (denoted by Z_1 and Z_2) at the real measured temperature t, respectively.

The formula is as follows:

$$Z_1 = a_1t^3 + b_1t^2 + c_1t + d_1 \qquad (5)$$

where a_1, b_1, c_1, d_1 are the corresponding parameters of the third-order curve fitting at pressure P_1:

$$Z_2 = a_2t^3 + b_2t^2 + c_2t + d_2 \qquad (6)$$

where a_2, b_2, c_2, d_2 are the corresponding parameters of the third-order curve fitting at pressure P_2.

After linear interpolation, obtain the third compression factor value (Z_t) at pressure P and temperature t, which can be calculated as follows:

$$Z_t = Z_1 + (Z_2 - Z_1)/(P_2 - P_1)*(P - P_1) \qquad (7)$$

According to the above steps, the third compressibility factor Z_t is obtained. It fully meets the high-grade (A grade and B grade) precision requirement of the standard EN1776[6].

Step 5: using the above method, the compression factor Z_n of gas under the standard conditions (a standard atmospheric pressure and 20 degrees Celsius) is calculated.

According to the following formula, the accurate volume of gas at the standard conditions is calculated as follows:

$$V_n = V_t(P_t/101.325)(293.15/(T_t + 273.15))(Z_n/Z_t) \qquad (8)$$

where V_t is the measured gas volume at the actual gas temperature T_t and pressure P_t; P_t is the actual measured gas pressure (absolute pressure, in Kpa); and T_t is actually the measured gas temperature (in degrees Celsius).

4 TEST DATA AND USER INTERFACE FIGURE

System power measurement result: in the normal operation condition, the current of the overall system is maintained in less than 180 µA. A system with the 16 Ah lithium battery, in the normal operation, can be used for more than 9 years. User interfaces of the totalizer are shown in Figure 8.

In the main interface, the information about the standard condition instantaneous flow, cumulative standard conditions flow, temperature, pressure, time, and the flow percentage of full scale can be displayed.

5 CONCLUSIONS

In this paper, the system design of the ultra-low-power totalizer was described. Then, according to the combined method of PC and totalizer, using the algorithms of curve fitting and linear interpolation, the online high-precision calculations of the gas compressibility factor Z based on the AGA8-92DC algorithm and the SGERG-88 algorithm were realized in the totalizer. The simulation results show that the third-order curve fitting is the best choice. While using the low-power system design, the ultra-low-power intelligent flow totalizer was developed based on the online calculation method of the gas compressibility factor Z. The totalizer has been in batch production and use, and achieves a good economic benefit.

REFERENCES

[1] ISO 5167-1:2003. Measurement of fluid flow by means of pressure differential devices inserted in circular cross-section conduits running full—Part 1: General principles and requirements.
[2] ISO 5167-2:2003. Measurement of fluid flow by means of pressure differential devices inserted in circular cross-section conduits running full—Part 2: Orifice plates.
[3] ISO 12213-1:2006. Natural gas-Calculation of compression factor—Part 1: Introduction and guidelines.
[4] ISO 12213-2:2006. Natural gas-Calculation of compression factor—Part 2: Calculation using molar-composition analysis.
[5] ISO 12213-3:2006. Natural gas-Calculation of compression factor—Part 3: Calculation using physical properties.
[6] EN1776:1998. Gas Supply systems-Natural Gas measuring Stations-Functional requirement.

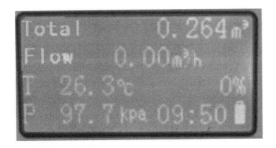

Figure 8. The main interface of the totalizer.

Architectural, Energy and Information Engineering – Sung & Chen (Eds)
© *2016 Taylor & Francis Group, London, ISBN 978-1-138-02791-6*

An Energy-Efficient Clustering Routing strategy for the mobile sensor network

D.Y. Zhang & Z.C. Da
University Science Park of IoT, Nanjing University of Posts and Telecommunications, Nanjing, China

ABSTRACT: Traditional wireless sensor network routing protocol has not well adapted to the mobile environment. This paper proposes an Energy Efficient Clustering Routing (EECR) strategy for mobile sensor networks. In this proposed strategy, it adds the residual energy and mobility as cluster head selection parameters in the threshold formula, and uses the dynamic slot allocation to improve the data transfer phase. So, the nodes with more residual energy and less mobility can be selected with a higher probability. The simulation results show that the proposed EECR routing strategy has prolonged the network life, reduced the average energy consumption and decreased the control overhead in the network.

Keywords: mobile sensor network; cluster; energy conservation

1 INTRODUCTION

Based on its different methods of data aggregation, WSN can be divided into static wireless sensor networks and mobile sensor networks[1]. In the static wireless sensor network, sensor nodes are relatively static and in close proximity to each other. It transmits directly by wireless link, and completes data fusion and forwarding by multi-hop transmission. For many applications, such as sea exploration, wildlife protection, and traffic congestion control, mobile sensor nodes need to be deployed in a network that would cause a frequent topology change and thus result in high packet loss, especially the increase in the node energy consumption.

Currently, to reduce energy consumption and prolong the network lifetime as much as possible, network clustering is an effective way. Its main feature is that the sensor nodes can be divided into several clusters according to the specific rules. Each cluster selects a cluster head node and forwards the collected sensor data to the base station through the cluster heads[2]. However, with the enhancement of node mobility, most of the existing routing algorithms will face the problems of reduced reliability of data transmission and higher overhead cost[3].

In the current research on the clustering routing protocol, LEACH is one of the most representative self-organizing clustering routing protocols[4]. It forms a sufficient number of clusters in a self-organizing manner. In order to avoid the premature death of the cluster head, LEACH take turns choosing the cluster head node, balancing energy consumption between the nodes, so that the network life cycle is extended.

Each round of the LEACH algorithm can be divided into two stages: cluster head selection and data transmission. At the beginning of each round in the network, each node randomly generates a value between 0 and 1, and compares it with the threshold formula $T(n)$. If a node generates a random number that is less than $T(n)$, then the node is elected as the cluster head and broadcasts the message to other nodes in the network. The threshold formula $T(n)$ is calculated as follows:

$$T(n) = \begin{cases} \dfrac{p}{1 - p\left[r \bmod \left(\frac{1}{p}\right)\right]} & n \in G \\ 0 & otherwise \end{cases} \quad (1)$$

where p is the proportion of cluster nodes; r is the index of the current round; and G is the set of nodes that have not been cluster heads in the most recent residue $1/p$ rounds.

The LEACH protocol has an equal probability for the selection of each cluster head. However, there are also some disadvantages in the LEACH algorithm. The most prominent problem is that it does not take into account the problem of the remaining energy of candidate nodes in selecting a cluster head. It will lead to the elected cluster head with less residual energy. LEACH was designed based on the static wireless sensor network. Although it can also support part of mobile nodes, its performance in the mobile sensor network is poor.

Kim D[5] proposed a LEACH-M protocol that supports the mobility of a sensor node by adding its membership declaration to the LEACH protocol. The sensor node, which does not receive a request from its cluster head during two consecutive frames, can also recognize that it has moved out of the cluster, and thus it will broadcast a cluster joint request message in order to join in a new cluster and avoid losing more packets. Therefore, the LEACH-M protocol increases the successful packet delivery rate at the cost of increased control overhead.

Through the above analysis, the existing routing algorithms have the advantages of a simple cluster structure and less packet loss. However, the generated cluster structure is not stable and requires a frequent re-clustering of the network. It increases the control overhead, increases the network energy consumption and reduces the network lifetime.

To address these problems, this paper proposes an Energy Efficient Clustering Routing (EECR) strategy for mobile sensor networks. It adds the residual energy and mobility as cluster head selection parameters in the clustering stage. So, the nodes with more residual energy and less mobility can be selected with a higher probability, thus saving the node energy.

2 NETWORK MODEL AND ENERGY MODEL

2.1 Network model

To simplify the network model, we adopt a few reasonable assumptions as follows:

1. Each sensor node is the same type as uniform physical characteristics such as energy consumption and antenna gain. Each node has an Identity (ID) with equal initial energy.
2. The node has enough capacity in data fusion.
3. All sensor nodes have a random initial position, but the position can be learned by GPS or other positioning monitoring algorithms.
4. The BS is stationary after deployment and its location is in the center of the monitoring area.
5. All the sensor nodes in the network are in the movable state, but the moving speed is slow, and does not move out of the area range. Each node moves by the RWP model[6].

2.2 Energy model

This paper focuses on the energy consumption of the wireless communication module, and uses the same energy dissipation model as in[7] as the energy model.

Figure 1. Wireless communication energy consumption model.

As shown in Figure 1, the energy spent for the transmission of a k-bit packet over the distance d is as follows:

$$E_{Tx}(k,d) = E_{Tx-elc}(k) + E_{Tx-mp}(k,d)$$

$$= \begin{cases} kE_{elec} + k\varepsilon_{fs}d^2, d < d_0 \\ kE_{elec} + k\varepsilon_{mp}d^4, d \geq d_0 \end{cases} \quad (2)$$

$$E_{Rx}(k,d) = E_{Rx-elec}(k) = kE_{elec} \quad (3)$$

Here, E_{elec} is the transmission circuit energy consumption, and the energy coefficients of transmission amplification circuit are ε_{fs} = 10 pJ/bit/m^2 and ε_{mp} = 0.0013 pJ/bit/m^4. According to the above energy model, the energy consumption of each sensor node can be obtained.

3 THE EECR STRATEGY

3.1 Cluster head selection

In the LEACH-M protocol, the selection of the cluster head adopts the same method as that of the LEACH protocol. Each node randomly generates a value between 0 and 1 and compares it with the threshold formula $T(n)$. If a node generates a random number that is less than $T(n)$, then the node is elected as the cluster head. This clustering approach allows each node to have equal probability to be elected as the cluster head with balanced network energy consumption. However, in the mobile sensor network, node mobility cannot be ignored. In order to reduce the influence of mobile nodes in the network, mobility should be considered as a parameter, taking into account in the threshold formula the cluster head selection time. Therefore, this paper first proposes a mobile factor as a parameter in the threshold formula.

In the network model proposed in this paper, each node moves by the RWP model. We also assume that the nodes set their speed and direction of the next step before each cluster head selection. At this moment, the node has its own speed information. Choosing the node with a relatively small movement speed to be the cluster head is conducive to maintain the stability of the clusters. To meet these requirements, we add the mobile factor in the threshold formula as follows:

$$Fv = \frac{V_{max} - V_{current}}{V_{max}} \tag{4}$$

where the $V_{current}$ is the current speed and V_{max} is the maximum speed of the node. In the RWP model, the nodes moves at a constant speed $V_{current}$ at every movement. When the speed $V_{current}$ increases, the value Fv decreases, and reduces the probability of cluster head election. As a result, the nodes with lower mobility have high probability to be the cluster head.

We also consider the residual energy of nodes when the cluster head is selected. The node with the largest residual energy, which is selected as a cluster head, will contribute to increasing the lifetime of a network. Through the above analysis, each node saves two energy information: current node residual energy $E_{n-current}$ and the initial energy E_{init}. We add the energy factor $E_{n-current} / E_{init}$ as one of the parameters into the threshold formula. When the difference between $E_{n-current}$ and E_{init} is larger, it indicates that the node's residual energy is less and it is unsuitable to be the cluster head.

Taking the node mobility and residual energy as parameters to the original formula, and we get the improved $T(n)'$ as follows:

$$T(n)' = \begin{cases} \dfrac{p}{1 - p[r \bmod(1/p)]} \cdot \left(\dfrac{E_{n-current}}{E_{n-init}} \cdot \dfrac{V_{max} - V_{n-current}}{V_{max}} \right) & n \in G \\ 0 & n \notin G \end{cases} \tag{5}$$

where P is the ratio of the cluster head in the network, and we will determine the optimal ratio through simulation experiments. The range of the improvement threshold value is between 0 and 1, and it conforms to the number range that the node randomly generates. The more residual energy makes sure that the cluster heads are capable of prolonged extra work, and the lower speed copes with the situation of the cluster head going out of reach because of high speed.

3.2 Data transmission

In the data transmission phase, the cluster head broadcasts an advertisement message as well as its location to the sensor nodes within its transmission range. The nodes select the closest cluster head to join and send the adding information. Cluster heads allocate the transmit data time slot for each member node.

To avoid the situation that mobile nodes move outside of the scope of the original area and cannot transmit data, this paper improves the data transfer phase: adding dynamic slot allocation and allowing the nodes that moved out of the original cluster quickly add new cluster. This can effectively avoid the frequently network topology changes due to the sensor nodes' movement.

In the EECR, at each start of every TDMA time slot, cluster heads send the data request message Date_Req to the member nodes. If the member node receives the message, on the one hand, this means it is still within the cluster structure and transfers the collected data. On the other hand, it means the nodes may have moved out of the scope of the original cluster and mark it as freedom node.

At this moment, the node broadcasts Join_Req to join the nearby cluster structure. The cluster head that received Join_Req will send the ACK message containing the location, whether there are free slots and other information to that freedom node. Then, the freedom node chooses the nearest cluster head with idle slots. This process is shown in Figure 2.

In the EECR, the cluster head dynamically updates the slot allocation scheme. When it does not receive the answer before the end, it is considered that the sensor node has moved out of the range and the node is marked as mobile nodes. If it still has not received the answer at the next slot, the original node is considered to have moved outside and marked the time slot as free slots. Then, it assigns the free slots to the new added nodes and avoids frequent network topology changes.

In summary, the EECR strategy can be described as follows:

Step 1: Based on the small-scale and low velocity moving mobile sensor networks, we choose the RWP model as the moving model of nodes and the free space model as the energy model.

Step 2: Network initializes and the base station advertises its location to all the nodes in the network. Nodes calculate the residual energy according to the energy model and initial energy of nodes.

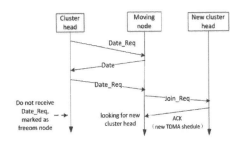

Figure 2.   The process in ACK response in the EECR.

Step 3: Add the residual energy and mobility as cluster head selection parameters to the threshold formula and get the improving threshold formula $T(n)'$.

Step 4: Cluster heads generate the slot allocation scheme according to the received member nodes message and allocates the time slot after cluster formation.

Step 5: Data transmission begins. Nodes transmit the sensor data to the cluster head and then forward to the base station.

4 SIMULATION AND ANALYSIS

The simulations were performed in MATLAB and in the area of a square $100\ m \times 100\ m$ area, randomly distributed 100 sensor nodes with the 0.5 J initial energy. The base station is located at (50, 50). We choose the RWP model as the moving model of nodes and the free space model as the energy model[8]. The maximum moving speed is 3 m/s. The detailed simulation parameters are listed in Table 1.

As shown in Figure 3, the average energy consumption, which mainly results from the transmission between the cluster head and the member node, varies with the increase in the rate between the number of cluster heads and total sensor nodes. In order to improve the working efficiency and satisfy the network coverage, we use 5% in the latter simulation, namely, $p = 0.05$.

Table 1. Simulation parameters.

Parameter	Value
Network coverage	100 m × 100 m
Initial energy	0.5 J
E_{elec}	50 nJ/bit
Data packet size	500 bits
ε_{fs}	10 pJ/bit/m²
Control packet size	10 bits

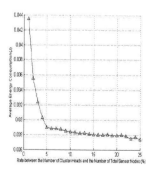

Figure 3. Percentage of cluster heads.

4.1 Comparison of the network lifetime

Figure 4 shows the percentage of surviving nodes in the LEACH, LEACH-M and EECR protocols. We can see that the percentage of surviving nodes in the EECR is significantly higher than that in the LEACH and LEACH-M. This is due to the fact that the EECR takes energy and mobility factors into account during cluster head selection, and EECR can well reduce the energy consumption. Compared with the LEACH-M and LEACH, it still has 20% nodes surviving until the 900 rounds in the network and effectively prolongs the network lifetime.

4.2 Comparison of the average energy consumption and control overhead

Figure 5 shows the average energy consumption for successfully receiving a packet, respectively, with the EECR, LEACH and LEACH-M. The EECR algorithm has better performance and reaches the target of energy conservation. This is because it takes into account the residual energy and node mobility at cluster heads selection. In addition, a more stable link also leads to fewer membership changes and less control overhead. Therefore, the EECR consumes less power compared with both LEACH and LEACH-M.

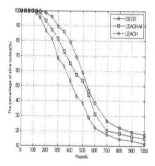

Figure 4. Network lifetime.

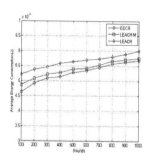

Figure 5. Energy consumption.

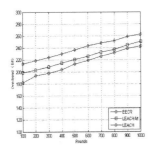

Figure 6.   Average overhead.

Figure 6 shows the control overhead for successfully receiving a data packet, respectively. With the operation rounds of network, nodes have a significantly lower average control overhead in the EECR algorithm. This is due to the consideration of mobility at cluster head selection and the clusters are more stable. Therefore, these avoid the frequent control message transmission.

5   CONCLUSIONS

In this paper, we proposed an energy efficient clustering routing strategy for mobile sensor networks (EECR). The proposed clustering protocol allows a sensor node to elect itself as a cluster-head based on its residual energy and mobility, and to balance the network energy consumption better. The simulation analysis shows that the proposed EECR routing strategy has a prolonged the network life, reduced the average energy consumption and decreased the control overhead. It obviously improved the energy saving performance of the mobile sensor network.

ACKNOWLEDGMENTS

This work was supported by the National Natural Science Foundations of P. China (NSFC) under Grant No. 61071093, the National 863 Program No. 2010AA701202, the Returned Overseas Project, Jiangsu Province Major Technology Support Program No. BE2012849, and the Graduate Research and Innovation project No. KYLX_0812.

REFERENCES

[1] Sara G S, Sridharan D. Routing in mobile wireless sensor network: a survey [J]. *Telecommunication Systems*, 2014, 57(1): 51–79.

[2] Lee S, Choe H, Park B, et al. LUCA: An energy-efficient unequal clustering algorithm using location information for wireless sensor networks [J]. *Wireless personal Communications*, 2011. 5(6), pp. 715–731.

[3] Olascuaga-Cabrera J G, Lopez-Mellado E, Mendez Vazquez A, et al. A self-organization algorithm for robust networking of wireless devices [J]. *IEEE Sensors Journal*, 2011, 11(3): pp. 771–780.

[4] Heinzelman W R, Chandrakasan A, Balakrishnan H. Energy-efficient communication protocol for wireless microsensor networks [C]. //System Sciences, 2000. Proceedings of the 33rd Annual Hawaii International Conference on. IEEE, 2000.

[5] Kim D, Chung Y. Self-Organization Routing Protocol Supporting Mobile Nodes For Wireless Sensor Network [J]. *Computer and Computational Sciences, 2006. IMSCCS '06. First International Multi-Symposiums on*, 2006: 622–626.

[6] Hyytiä E, Virtamo J. Random Waypoint Mobility Model in Cellular Networks [J]. *Wireless Networks*, 2007, 13(2): 177–188.

[7] Heinzelman W R, Chandrakasan A P, Balakrishnan H. An application specific protocol architecture for wireless micro sensor networks [J]. *IEEE Trans on Wireless Communications*, 2002, 1(4): 660–670.

[8] Sha K, Gehlot J, Greve R. Multipath Routing Techniques in Wireless Sensor Networks: A Survey [J]. *Wireless Personal Communications*, 2013, 70(2): 807–829.

Architectural, Energy and Information Engineering – Sung & Chen (Eds)
© 2016 Taylor & Francis Group, London, ISBN 978-1-138-02791-6

Construction methods of 16-QAM sequences with zero-correlation zone for AS-CDMA communication systems

F.X. Zeng & L.J. Qian
Chongqing Key Laboratory of Emergency Communication, Chongqing Communication Institute, Chongqing, China

Z.Y. Zhang
College of Communication Engineering, Chongqing University, Chongqing, China
Chongqing Key Laboratory of Emergency Communication, Chongqing Communication Institute, Chongqing, China

ABSTRACT: This paper presents two constructions producing Zero-Correlation-Zone (ZCZ) sequences over 16-Quadrature Amplitude Modulation (QAM) constellation. The proposed QAM ZCZ sequences are based on the existing quaternary ZCZ sequences. Under some conditions, new sequences are optimal on the theoretical upper bound of ZCZ sequences. These proposed sequences can be applied to Approximately Synchronized (AS) Code-Division Multiple-Access (CDMA) communication systems so as to fully suppress Multiple Access Interfere (MAI) and Multiple Path Interfere (MPI), and to improve those systems' synchronization.

Keywords: information theory; communication; AS-CDMA, ZCZ sequences; QAM constellation

1 INTRODUCTION

In 1994, a new Code-Division Multiple-Access (CDMA) communication system, called Approximately Synchronized (AS) CDMA, was introduced [Suehiro 1994]. In comparison with conventional CDMA communication system, new system's specialization lies at spreading sequences employed. Here, new system uses sequences with zero correlation zone as its spreading spectrum signals instead of optimal sequences with minimal magnitude of correlation in conventional system. The advantages of new system are remarkable, including no Multiple Access Interfere (MAI), no Multiple Path Interfere (MPI), and excellent synchronization.

Inspired by new system's benefit, ZCZ sequences are quickly developed, and a large number of construction methods are proposed, such as in [Fan a, b, 2000], [Krone, 1984], [Matsufuji, 1999], and [Takatsukasa, 2004]. However, no ZCZ sequences over Quadrature Amplitude Modulation (QAM) constellation have been known apart from the ones given by the authors up to today. It is well-known that employment of QAM signals in a communication system can result in higher data transmission rate [Anand, 2008]. It is natural to combining with AS-CDMA communication system and QAM signals so as to greatly improve communication

performance. As a consequence, design of QAM ZCZ sequences is one of all-important issues in a QAM AS-CDMA system.

The rest of this paper is organized as follows. In Section 2, we briefly recall some concepts referred to in this paper. In Section 3, construction methods of 16-QAM ZCZ sequences are presented. An example appears at Section 4. Finally, we conclude this paper in Section 5.

2 PRELIMINARIES

In this section, we will recall the definitions of periodic correlation, ZCZ sequences, and 16-QAM constellation, and their relevant results.

2.1 *16-QAM constellation*

The 16-QAM constellation is in fact the set:

$$\{a + bj \mid a, b \in \{\pm 1, \pm 3\}\} \tag{1}$$

where $j_2 = -1$.

The 16-QAM constellation can be produced by two independent quaternary variables. In mathematical terms, the 16-QAM constellation is given by

$$\{(1+j)(j^{a_0}+2j^{a_1})\,|\,a_0,a_1\in Z_4=\{0,1,2,3\}\} \qquad (2)$$

and

$$\{(1+j)(j^{a_0}-2j^{a_1})\,|\,a_0,a_1\in Z_4=\{0,1,2,3\}\} \qquad (3)$$

respectively.

2.2 ZCZ sequences

Let $\underline{u}_r=(u_r(0),u_r(1),\ldots,u_r(N-1))$ be a complex sequence of period N, where $1\le r\le M$. We have the following definitions and their relevant conclusions.

Definition 1. For $\forall r,s\in[1,M]$, we refer to

$$R_{u_r,u_s}(\tau)=\sum_{t=0}^{N-1}u_r(t)\overline{u_s(t+\tau)} \qquad (4)$$

as a periodic correlation function. In particular, when $r=s$, $R_{u_r,u_r}(\tau)$ is said to be an autocorrelation function, or else to be a cross-correlation function, where $\overline{x}$ denotes complex-conjugate of x and the addition $t+\tau$ is performed modulo N.

Definition 2. For given non-negative integer T, for $\forall r,s\in[1,M]$ if we have

$$R_{u_r,u_s}(\tau)=\begin{cases}N & \tau=0 \text{ and } r=s\\ 0 & 0<|\tau|\le T \text{ and } r=s\\ 0 & |\tau|\le T \text{ and } r\ne s,\end{cases} \qquad (5)$$

we say that the sequence set $\{\underline{u}_r\,|\,1\le r\le M\}$ is a ZCZ sequence set, denoted by $ZCZ_4(N,M,T)$ and $ZCZ_{16\text{-QAM}}(N,M,T)$ for quadriphase and 16-QAM sequences, respectively.

A ZCZ sequence set must satisfy the following lower bound [Tang, 2000].

Lemma 1. For $ZCZ(N,M,T)$, we have

$$M(T+1)\le N. \qquad (6)$$

Definition 3. For two sequences $\underline{u}_r$ and $\underline{u}_s$ of period $\underline{N}$, we define a new sequence of period $2N$ by

$$(u_r(0),u_s(0),u_r(1),u_s(1),\ldots,u_r(N-1),u_s(N-1)). \qquad (7)$$

We refer to this new sequence as interleaved sequence denoted by $I[\underline{u}_r,\underline{u}_s]$.

3 CONSTRUCTION OF 16-QAM ZCZ SEQUENCES

In this section, we will give construction methods of 16-QAM ZCZ sequences. The presented sequences come from transformation of the known quadriphase ZCZ sequences.

Throughout this section, we employ the following symbols.

- N: The period of quadriphase ZCZ sequences employed.
- M: The number of quadriphase ZCZ sequences employed.
- T: The width of ZCZ of quadriphase ZCZ sequences employed.
- $\Omega_4=\{\underline{u}_r\,|\,1\le r\le M\}=ZCZ_4(N,M,T)$.
- M': The number of the resultant 16-QAM ZCZ sequences.
- k: The times that iteration happens.
- ψ_k: Quadriphase ZCZ sequences survived when the k-iteration happens.

Construction 1:

By following the steps below, we will complete the constructions of the proposed sequences.

1. ***Step 1:*** Initialization. $k\leftarrow 1$, $\psi_k\leftarrow\Omega_4$.
2. ***Step 2:*** Arbitrarily choose two sequences in ψ_k, say, $\underline{u}_r$ and $\underline{u}_s$. Set $\Pi_k=\{\underline{u}_r,\underline{u}_s\}$
3. ***Step 3:*** Construct two 16-QAM sequences of period N below.

$$v_k^{(+)}(t)=(1+j)[j^{u_r(t)}+2j^{u_s(t)}] \qquad (8)$$

and

$$v_k^{(-)}(t)=(1-j)[j^{u_s(t)}-2j^{u_r(t)}]. \qquad (9)$$

4. ***Step 4:*** Construct the interleaved sequence from Eqs. (8) and (9) below.

$$\underline{w}_k=I[\underline{v}_k^{(+)},\underline{v}_k^{(-)}]. \qquad (10)$$

5. ***Step 5:*** $\psi_{k+1}\leftarrow\psi_k-\Pi_k$.
6. ***Step 6:*** Turn Step 2 with $k\leftarrow k+1$ when $|\psi_{k+1}|\ge 2$, or else turn Step 7.
7. ***Step 7:*** End.

Theorem 2. In Construction 1, the 16-QAM sequence set $\{\underline{v}_k^{(+)},\underline{v}_k^{(-)}\,|\,1\le k\le M'\}$ forms $ZCZ_{16\text{-QAM}}(N,2M',T)$, and $\{\underline{w}_k\,|\,1\le k\le M'\}$ is $ZCZ_{16\text{-QAM}}(2N,2M',T)$, where $M'=\left[\frac{M}{2}\right]$ which is the maximal positive integer not excelling $M/2$.

Further, the following theorem reflects the performance of Construction 1.

Theorem 3. If Ω_4 satisfies Lemma 1 and M is an even number, $\{\underline{v}_k^{(+)},\underline{v}_k^{(-)}\,|\,1\le k\le M'\}$ satisfies Lemma 1 as well.

Construction 2:

By following the steps below, we will construct the required sequences.

1. ***Step 1:*** Initialization. $k\leftarrow 1$, $\psi_k\leftarrow\Omega_4$.
2. ***Step 2:*** Arbitrarily choose a sequence in ψ_k, say, $\underline{u}_r$. Set $\Pi_k=\{\underline{u}_r\}$

3. **Step 3:** Construct two 16-QAM sequences of period N below, in accordance with T odd or even, respectively.

When T is odd, that is, $T = 2T' + 1$, we construct

$$v_k^{(+)}(t) = (1+j)\left[j^{u_r(t)} + 2j^{u_r(t+T'+1)}\right] \qquad (11)$$

and

$$v_k^{(-)}(t) = (1-j)[j^{u_r(t+T'+1)} - 2j^{u_r(t)}]. \qquad (12)$$

When T is even, that is, $T = 2T'$, we construct

$$v_k^{(+)}(t) = (1+j)[j^{u_r(t)} + 2j^{u_r(t+T')}] \qquad (13)$$

and

$$v_k^{(-)}(t) = (1-j)[j^{u_r(t+T')} - 2j^{u_r(t)}]. \qquad (14)$$

4. **Step 4:** Construct the interleaved sequence from Eqs. (11) and (12) or Eqs. (13) and (14) below, depending on T odd or even, respectively.

$$\underline{w}_k = I[\underline{v}_k^{(+)}, \underline{v}_k^{(-)}]. \qquad (15)$$

5. **Step 5:** $\psi_{k+1} \leftarrow \psi_k - \Pi_k$.
6. **Step 6:** Turn Step 2 with $k \leftarrow k+1$ when $|\psi_{k+1}| \geq 1$, or else turn Step 7.
7. **Step 7:** End.

Theorem 4. In Construction 2, the 16-QAM sequence set $\{\underline{v}_k^{(+)}, \underline{v}_k^{(-)} | 1 \leq k \leq M\}$ forms $ZCZ_{16\text{-}QAM}(N, 2M, T')$ or $ZCZ_{16\text{-}QAM}(N, 2M, T'-1)$ depending on T odd or even, respectively, and $\{\underline{w}_k | 1 \leq k \leq M\}$ is $ZCZ_{16\text{-}QAM}(2N, M, T'-1)$ or $ZCZ_{16\text{-}QAM}(2N, M, T'-2)$, depending on T odd or even, respectively.

Similarly, the following theorem represents the performance of Construction 2.

Theorem 5. If Ω_4 satisfies Lemma 1 and T is odd, $\{\underline{v}_k^{(+)}, \underline{v}_k^{(-)} | 1 \leq k \leq M'\}$ also satisfies Lemma 1.

For the reader who wants to know the proofs of Constructions 1 and 2, please refer to [Zeng, 2011].

4 AN EXAMPLE

We will give an example to help the reader understand our constructions. Consider the quadriphase ZCZ sequences $ZCZ_{16\text{-}QAM}(64,4,14)$ [Torii, 2001] as follows.

$\underline{u}_1 = $ (00000123020203210000123020203210000002 3010202213000003012202010 32),

$\underline{u}_2 = $ (01230202032100000123131321033333301232 0200321222201233131210 31111),

and

$\underline{u}_4 = $ (0321000001230202032111112301313103212 2220123202003213333 23011313).

By using Construction 1, we set that $\Pi_1 = \{\underline{u}_1, \underline{u}_2\}$ and $\Pi_2 = \{\underline{u}_3, \underline{u}_4\}$. Thus, Eqs. (8) and (9) produce the 16-QAM ZCZ sequences below.

$\underline{v}_1^{(+)} = $ (3+3j, −1+3j, −1−j, 3−j, 3+3j, −3−j, 1+j, −1−3j, 3+3j, 1−3j, −1−j, −3+ j, 3+3j, 3+j, 1+j, 1+3j, 3+3j, −1+3j, −1−j, 3−j, −3+3j, 1−3j, −3−3j, −1+3j, 1+j, 3−j, 3−3j, 1−3j, 1−j, 3−j, 3+3j, −1+3j, −1−j, 3−j, −3−3j, 3+j, −1−j, 1+3j, 3+3j, 1−3j, −1−j, −3+j, −3−3j, −3−j, −1−j, −1−3j, 3+3j, −1+3j, −1−j, 3−j, 3−3j, −1+3j, 1−j, −3+j, −3−3j, −1+3j, 1+ j, 3−j, −3+3j, −1+3j, −1+ j, −3+ j)

$\underline{v}_1^{(-)} = $ (−1+j, −1+3j, −3+3j, −3+j, −1+j, −3−j, 3−3j, 1+3j, −1+j, 1−3j, −3+ 3j, 3−j, −1+j, 3+j, 3−3j, −1−3j, −1+j, −1+3j, 3+3j, −3+j, −1−j, 1−3j, 3+3j, −3+j, 1−j, −1+3j, 3−3j, −3+j, 1+j, 1−3j, −3−3j, −3+j, −1+j, −1+3j, −3+3j, −3+j, 1−j, 3+j, −3+3j, −1−3j, −1+j, 1−3j, −3+3j, 3−j, 1−j, −3−j, −3+3j, 1+3j, −1+j, −1+3j, 3+3j, −3+j, 1+j, −1+3j, −3−3j, 3−j, 1−j, −1+3j, 3−3j, −3+j, −1−j, −1+ 3j, 3+3j, 3−j)

$\underline{v}_2^{(+)} = $ (3+3j, 1−3j, −1−j, −3+j, 3+3j, 3+j, 1+j, 1+3j, 3+3j, −1+3j, −1−j, 3−j, 3+3j, −3−j, 1+j, −1−3j, 3+3j, 1−3j, −1−j, −3+j, −3+3j, −1+3j, −1+j, −3+j, 3−3j, 1−3j, 1+j, −3+j, 3−3j, −1+3j, 1−j, −3+j, 3+3j, 1−3j, −1−j, −3+j, −3−3j, −3−j, −1−j, −1−3j, 3+3j, −1+3j, −1−j, 3−j, −3−3j, 3+j, −1−j, 1+3j, 3+3j, 1−3j, −1 −j, −3+j, 3−3j, 1−3j, 1+j, 3−j, 3−3j, −3−3j, 1−3j, 1+j, −3+j, −3+3j, 1−3j, −1+j, 3−j)

and

$\underline{v}_2^{(-)} = $ (−1+j, 1−3j, −3+3j, 3−j, −1+j, 3+j, 3−3j, −1−3j, −1+j, −1+ 3j, −3+3j, −3+j, −1+j, −3−j, 3−3j, 1+3j, −1+j, 1−3j, −3+3j, 3−j, −1−j, −1+3j, 3+3j, 3−j, 1−j, 1−3j, 3−3j, 3−j, 1+j, −1+3j, 3−3j, 3−j, −1+j, 1−3j, −3−3j, 3−j, −1−j, −1+3j, 3−3j, −3−j, −3+3j, 1+3j, −1+j, −1+3j, −3+3j, 3−j, −3+j, 1−j, 3+j, −3+3j, −1−3j, −1+j, 1−3j, −3+3j, 3−j, 1−j, 1−3j, 3−3j, 3−j, −1−j, 1−3j, 3+3j, −3+ j).

Apparently, the resultant 16-QAM ZCZ sequences satisfy Lemma 1, which means that the proposed sequences in this example are optimal. Figure 1 gives the autocorrelation function of the 16-QAM sequence $\underline{v}_1^{(+)}$. Figure 2 shows the cross-correlation function between the 16-QAM sequences $\underline{v}_1^{(+)}$ and $\underline{v}_1^{(-)}$. Figure 3 gives the auto-correlation function of the 16-QAM sequence $\underline{v}_2^{(+)}$. Figure 4 shows the cross-correlation function between the 16-QAM sequences $\underline{v}_1^{(+)}$ and $\underline{v}_2^{(-)}$. Finally, Figure 5 shows the cross-correlation function between the 16-QAM sequences $\underline{v}_1^{(-)}$ and $\underline{v}_2^{(-)}$. Apparently, there exists a zero correlation zone with width 14 in all the figures.

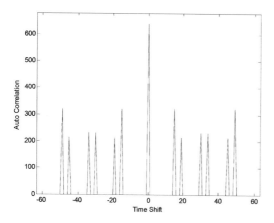

Figure 1. Autocorrelation function of the proposed $\underline{v}_1^{(+)}$.

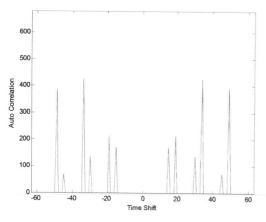

Figure 2. Cross-correlation function of the proposed $\underline{v}_1^{(+)}$ and $\underline{v}_1^{(-)}$.

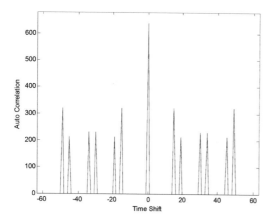

Figure 3. Autocorrelation function of the proposed $\underline{v}_2^{(+)}$.

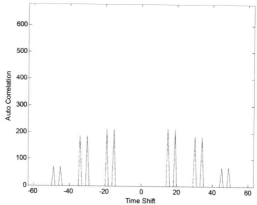

Figure 4. Cross-correlation function of the proposed $\underline{v}_1^{(+)}$ and $\underline{v}_2^{(-)}$.

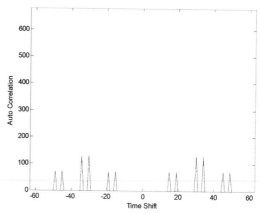

Figure 5. Cross-correlation function of the proposed $\underline{v}_1^{(-)}$ and $\underline{v}_2^{(-)}$.

5 CONCLUSION

This paper discusses the construction methods of ZCZ sequences over 16-QAM constellation, and the proposed sequences are optimal under some conditions. An example given verifies that the proposed methods are feasible.

ACKNOWLEDGEMENTS

This work was supported by Chongqing Science and Technology Committee (cstc2014pt-sy40003), and National Natural Science Foundation of China (NSFC) under Grants 60872164, 61002034, 61271003, and 61471336, China Postdoctoral Science Foundation Grant

2014M552318, Natural Science Foundation Project of CQ Grant cstc2014jcyj A40050, and Chongqing Postdoctoral Science Special Foundation Grant Xm2014031.

REFERENCES

Anand, M. et al. 2008. Low-correlation sequences over the QAM constellation. *IEEE trans. on inf. theory*, 54(2): pp. 791–810.

Fan, P.Z a. et al. 2000. Generalized orthogonal sequences and their applications in synchronous CDMA systems. *IEICE trans. on fundamentals*, E83-A(11): 2054–2069.

Fan, P.Z b. et al. 2000. A class of binary sequences with zero correlation zone. *Electr. lett.*, 35(10): 777–779.

Krone, S.M. et al. 1984. Quadriphase sequences for spread spectrum multiple-access communication. *IEEE trans. inf. theory*, IT-30(3): 520–529.

Matsufuji, et al. 1999. Spreading sequence sets for approximately synchronized CDMA system with no co-channel interference and high data capacity. *Proc. of 2nd int. symp. wireless pers. multimedia commun.*, 333–339, Sept. 1999. Netherlands: Amsterdam.

Suehiro, N. 1994. A signal design without co-channel interference for approximately synchronized CDMA systems. *IEEE j. sel. areas commun.*, SAC-12(5): 837–841.

Takatsukasa, K. et al. 2004. Formlization of binary sequence sets with zero correlation zone. *IEICE trans. on Fundamentals*, E87-A(4): 887–891.

Tang, X.H. et al. 2000. Lower bounds on the maximum correlation of sequence set with low or zero correlation zone. *Electron. lett.*, 36(6): 551–552.

Torii, H. et al. 2001. New method for constructing polyphase ZCZ sequence sets. *Proc. of 2nd WSEAS multiconf. on applied and theoretical mathematics*, 100–104, Dec. 2001, Australia: Cairns.

Zeng, F.X. et al. 2011. 16-QAM sequences with zero correlation zone from the known quadriphase ZCZ sequences. *IEICE trans. on fundamentals*, E94-A(3): 1023–1028.

Architectural, Energy and Information Engineering – Sung & Chen (Eds)
© *2016 Taylor & Francis Group, London, ISBN 978-1-138-02791-6*

Per capita carbon footprint characteristics based on household consumption in Beijing, China

L.N. Liu
Research School of Arid Environment and Climate Change, MOE Key Laboratory of Western China's Environmental Systems, Lanzhou University, Lanzhou, China

J.S. Qu & J.J. Zeng
Research School of Arid Environment and Climate Change, MOE Key Laboratory of Western China's Environmental Systems, Lanzhou University, Lanzhou, China
Information Center for Global Change Studies, Chinese Academy of Sciences, Lanzhou Library, Lanzhou, China

S.D. Zhang
Department of Geography, University of Hong Kong, Hong Kong

ABSTRACT: This paper analyzes the carbon footprint from household consumption based on the IPCC's reference approach and Input-Output Analysis (IOA) in Beijing, China, from 1995 to 2012. The main purpose of this study emphasizes the characteristics of carbon footprint based on three levels: rural and urban Household CO_2 Emissions (HCEs) and per capita HCEs, the direct and indirect per capita HCEs, and per capita HCEs caused by residents' lifestyle (housing, transportation, food, goods, and services) in Beijing. The results show that (1) rural, urban and Beijing's total HCEs and per capita HCEs all exhibit an increasing trend; both total HCEs and per capita HCEs in urban Beijing are extremely higher than that in rural Beijing; (2) Beijing's per capita indirect HCEs are obviously higher than per capita direct HCEs; (3) per capita HCEs caused by the residents' housing, services, transport and goods show a significant positive trend, while that caused by the residents' food shows a slow growth trend. In consideration of insignificant per capita carbon emissions from these household sectors, this work suggests the prioritization of the substantial contribution of household consumption to carbon emissions while making energy conservation, emission reduction and adaptation policies.

Keywords: carbon footprint; household consumption; Beijing; China

1 INTRODUCTION

Recently, a large number of studies focusing on carbon emissions have turned to the consumption level analyses based on household units (Qu et al., 2013, Liu et al., 2013). Some researchers calculated CO_2 emissions from household consumption in a variety of countries including China (Qu et al., 2013; Liu et al., 2013), Denmark (Munksgaard et al., 2000), USA (Bin et al., 2005), Japan (Nansai et al., 2008) and Spain (Sánchez-Chóliz et al., 2007). These studies have shown that the household consumption from different countries is one of the most important sectors that resulted in CO_2 emissions. Particularly, per capita CO_2 emissions have been the most intractable problems to be considered in meeting the international agreements on global climate change (Clarke-Sather et al., 2011). Based on previous studies, a large number of arti-

cles have reported on the study of CO_2 emissions at a large level, but there is a lack of research at the Chinese city level.

Over the past few decades, global warming has become an increasing concern to a great number of researchers and policy makers (Ang, 2009). It is particularly important to cope with global warming and to analyze its causes. Many studies have concluded that greenhouse gas emissions are the main contributors to the greenhouse effect and climate change, while CO_2 is the largest contributor among the six main kinds of GHGs (IEA, 1996). China put forward a target of 40–45% reduction of CO_2 emissions intensity for climate change mitigation by 2020; meanwhile, Chinese government is making an array of measures to promote a low carbon economy and publish a China's pathway to reduce CO_2 emissions including short-term and long-term reduction goals (Zhou et al., 2013).

Beijing is the cultural, economic, educational and political center in China, which is one of the 20 largest cities in the world with a heavily polluted environment (UNEP, 2007).

This paper selects Beijing city as the research object to analyze its carbon footprint from household consumption by calculating HCEs. The rest of this paper is organized as follows. Section 2 discusses the data source and analytical methods. Section 3 discusses household carbon emissions in Beijing and also the carbon footprint characteristics. Finally, Section 4 makes some conclusions.

2 DATA SOURCE AND RESEARCH METHOD

2.1 Data source

In this study, total household CO_2 emissions are associated with direct and indirect emissions from urban and rural household consumption (Munksgaard et al., 2000; Qu et al., 2013). Also, according to Christopher (2011), the total household carbon footprint mainly comes from the following five residents' lifestyles: housing, services, food, transportation, and goods.

Annual data for direct household energy consumption are obtained from the China Energy Statistical Yearbook (1996–2012), and annual data for indirect household consumption data are obtained from the China Statistical Yearbook (1996–2013). In addition, urban and rural population data are obtained from the China Population Statistical Yearbook (1996–2013).

2.2 Estimation of household CO_2 emissions

In this work, the direct household carbon emissions are calculated based on the reference approach recommended by IPCC (2006), which are expressed in Eq. (1) as follows:

$$E_d = \sum_{i=1}^{i=n} (f_i \times e_i \times c_i \times o_i) \times 44/12 \qquad (1)$$

where "E_d" is the amount of total direct household carbon emission (t CO_2); "i" is the number of fuel types; "n" represents the total numbers of fuel types; "f_i" denotes the fuel consumption of the average household (10^4 t); "e_i" is the Net Calorific Value (NCV) of the fuel "f_i" (TJ/10^4 t); "c_i" is the Carbon Emission Factor (CEF) of the fuel "f_i" (t C/TJ); "o_i" expresses the fraction of carbon oxidized (COF) for the fuel "f_i"; and 44/12 is the ratio of the molecular weight of CO_2/C.

Eq. (2) gives the estimate approach for the emissions (t CO_2) from the electricity consumption as follows:

$$E_e = e_{ei} \times c_e \times 10^{-3} \qquad (2)$$

where "E_e" (t CO_2) is the emissions from the electric power consumption; "e_{ei}" (kWh) is the electric power consumption in the ith household; and "c_e" (t CO_2/MWh) is the CO_2 emission factor of the electricity sector in the study area based on the 2012 Baseline Emission Factor for regional power grids in China[1].

Indirect household carbon emissions are calculated by using an input-output method following Qu et al. (2013). Applying the China's input-output tables of 2007 and the data from the China Energy Statistical Yearbook and China Statistical Yearbook, we calculate the CO_2 emissions factors for different types of goods and services.

The indirect CO_2 emissions from household consumption are estimated in Eq. (3) as follows:

$$E_j = \sum_{j=1}^{j=n} (I_j \times C_j \times 10^{-3}) \qquad (3)$$

where "E_j" is the total indirect HCEs (t CO_2); "j" is the number of household consumption items; "n" represents the total number of household consumption items; "I_j" (RMB) expresses the consumption of household goods and services; and "C_j" denotes the CO_2 emissions (kg CO_2/RMB) from the consumption of goods and services.

3 RESULTS AND DISCUSSIONS

3.1 Per capita HCEs in Beijing

Over the period 1995–2012, household carbon emissions in Beijing, urban Beijing, and rural Beijing show a gradually increasing trend (see Figure 1(a)), although HCEs in rural Beijing show a declining trend from 2011 to 2012. Furthermore, there is an apparent growth on the trends of HCEs in Beijing and urban Beijing, but there is a slow growth in rural Beijing. Meanwhile, HCEs in urban Beijing city are greatly higher than the average level of HCEs in Beijing.

Per capita HCEs in Beijing, urban Beijing, and rural Beijing also show a gradually increasing trend from 0.96, 1.12, 0.68 t CO_2/person to 3.63, 3.93, 2.40 t CO_2/person, respectively (see Figure 1(b)), from 1995 to 2012, while per capita HCEs in rural Beijing city shows a declining trend in the period 2011–2012. Per capita HCEs in urban Beijing are obviously higher than per capita HCEs in rural Beijing from the period 1995–2012, due to the more

1 The National Development and Reform Commission of China, 2012. Baseline Emission Factor for Regional Power Grids in China, 2012. Download at http://www.doc88.com/p-997213839413.html.

urban household consumption expenditure in spite of the less urban population. Meanwhile, per capita HCEs in urban Beijing are also higher than the average level of per capita HCEs in Beijing.

3.2 Direct and indirect HCEs

In this study, per capita HCEs in Beijing are calculated in terms of per capita direct and indirect household consumption. It is clear that over the past eighteen years, per capita indirect HCEs are greater than per capita direct HCEs in Beijing (see Figure 2). Figure 2 also shows that Beijing's per capita indirect HCEs exhibit a sharp increase from 0.77 tons in 1995 to 3.43 tons in 2012, with an annual growth rate of 9.27%. Most of the growth in per capita indirect HCEs can be explained by the increasing level of private consumption, especially due to an increase in the demand for housing, transportation and service (see Figure 4), over the whole period 1995–2012. However, Beijing's per capita direct HCEs remain a rather stable trend in the period 1995–2012, mainly because of using clean energy products instead of traditional energy products (e.g. using LPG and natural gas rather than coal).

3.3 Per capita household sector's carbon emissions

In the period 1995–2012, per capita HCEs caused by the residents' housing, services, transport and goods show a significant positive trends, while per capita HCEs caused by the residents' food show a slow growth trend (see Figure 3). Figure 3 also shows that per capita HCEs from residents' housing are larger than that from other residents' lifestyles. Following the housing, per capita HCEs from food and service also exhibit a large part from period 1995–2007, while per capita HCEs caused by transportation surpass per capita HCEs caused by food from 2008 to 2012.

In Beijing, per capita HCEs by housing consumption account for the highest part of per capita HCEs (see Figure 4); however, there is a slight descend trend for the ratio in total per capita HCEs from 1995 to 2012. From Figure 4, we can also find that the ratio of per capita HCEs from residents' service and transportation in total per capita HCEs exhibits an increasingtendency. In addition, the ratio of per capita HCEs caused by residents' food displays a slow downward trend. Meanwhile, the ratio of per capita HCEs from goods exhibits a downward tendency from the period 1995–2002, and then shows an upward trend in the period 2003–2012.

From the above analysis of per capita HCEs from the resident's lifestyle, we can conclude that household CO_2 emissions from housing are larger than those caused by other residents' lifestyles. On the other hand, household CO_2 emissions caused by service and transportation display a gradually rising trend. We should consider the environmental impact of household consumption and the residents' lifestyle when formulating some measures to reduce emissions or making energy conservation policies in Beijing.

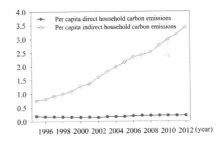

Figure 2. Per capita direct HCEs and indirect HCEs in Beijing.

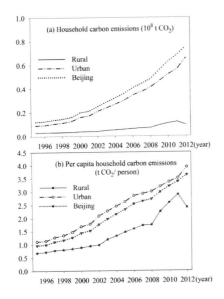

Figure 1. Household carbon emissions (a) and per capita household carbon emissions (b) in Beijing.

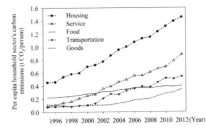

Figure 3. Per capita HCEs caused by the residents' lifestyle.

651

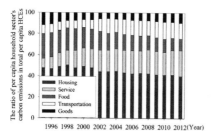

Figure 4. The ratio of per capita HCEs caused by the residents' lifestyle in total per capita HCEs.

4 CONCLUSIONS

This paper makes a first step toward comparing urban Beijing, rural Beijing and Beijing's total HCEs and per capita HCEs. From the above analysis on HCEs and per capita HCEs, we can clearly find that both total PCEs and per capita HCEs in urban Beijing, rural Beijing and Beijing all exhibit an increasing trend. We also obtain the result that the urban Beijing has greater total HCEs and per capita HCEs compared with rural Beijing. In addition, the growth trend of total HCEs and per capita HCEs in urban Beijing is very obvious, while in rural Beijing, it is very slow.

Second, we analyze the direct and indirect per capita HCEs in Beijing. Based on the results, we conclude that per capita indirect HCEs in Beijing have a significant increase from 1995 to 2012. On the other hand, per capita direct HCEs are nearly stable from 1995 to 2012.

Third, the results of this paper quantify that the residents' lifestyle can have a significant impact on household CO_2 emission.

Lastly, we should consider the prioritization of the substantial contribution of household consumption to carbon emissions when making energy conservation, emission reduction and adaptation policies. Based on the above analysis, we could further analyze the optimal household consumption patterns optimized to minimize environmental burdens and the effective policy tools to shift the consumption patterns. Moreover, further research on the influencing factors of HCEs and the environment in Beijing is required.

ACKNOWLEDGMENTS

This paper was funded by the National Natural Sciences Foundation of China (NSFC) Project-Household Carbon Emission and its Developing Path in China: A Scenario Research based on the Dynamic Emission Baseline (Grant No: 41371537) and the Strategic Priority Research Program of Chinese Academy of Sciences-Climate Change: Carbon Budget and Related Issues (Grant No: XDA05140100). We also thank the financial support from CAS.

REFERENCES

Ang, J.B. 2009. CO_2 emissions, research and technology transfer in China. *Ecological Economics* 68 (10): 2658–2665.

Bin, S., Dowlatabadi, H. 2005. Consumer lifestyle approach to US energy use and the related CO_2 emissions. *Energy Policy* 33 (2): 197–208.

China Energy Statistical Yearbook (1996–2012). *Energy Statistics Division of National Bureau of Statistics.* Beijing: China Statistics Press, 1997–2013.

China Statistical Yearbook (1996–2012). *The People's Republic of China National Greenhouse Gas Inventory.* Beijing: China Environmental Science Press, 1997–2013.

China Population Statistical Yearbook (1996–2005), China Population & Employment Statistical Yearbook (2006–2012). National Bureau of Statistic of China.

Christopher, M. J., Daniel, M. K., 2011. Quantifying carbon footprint reduction opportunities for U.S. households and communities. Environmental Science & Technology 45: 4088–4095.

Clarke-Sather, A., Qu, J.S., Wang, Q., Zeng, J.J., Li, Y. 2011. Carbon inequality at the sub-national scale: A case study of provincial-level inequality in CO_2 emissions in China 1997–2007. *Energy Policy* 39: 5420–5428.

IEA (International Energy Agency). 1996. World Energy Outlook 1996. Paris, France: *Organization for Economic Cooperation and Development.*

IPCC (Intergovernmental Panel on Climate Change). 2006. *IPCC Guidelines for National Greenhouse Gas Inventories.* <http: //www.ipcc-nggip.iges.or.jp/public/2006 gl/>. (Accessed on the16th, October, 2014).

Liu, L.N., Qu, J.S., Zeng, J.J., Wang, Q.H., Wang, L. 2013. Analysis the influence factors of China's household carbon intensity. *Environment, Energy and Sustainable Development.* CRC Press, 443–448.

Munksgaard, J., Pedersen, K.A., Wier, M. 2000. Impact of household consumption on CO_2 emissions. *Energy Economics* 22: 423–440.

Nansai, K., Inaba, R., Kagawa, S., Moriguchi, Y. 2008. Identifying common features among household consumption patterns optimized to minimize specific environmental burdens. *Journal of Cleaner Production* 16(4): 538–548.

Qu, J.S., Zeng J.J., Li Y., Wang Q., Maraseni T., Zhang L., Zhang Z.Q., Clarke-Sather A. 2013. Household carbon dioxide emissions from peasants and herdsmen in northwestern arid-alpine regions, China. *Energy Policy* 57: 133–140.

Sánchez-Chóliz, J., Duarte, R., Mainar, A. 2007. Environmental impact of household activity in Spain. *Ecological Economics* 62(2): 308–318.

UNEP (United Nations Environment Programme). 2007. 'Beijing 2008 Olympic Games: An Environmental review'.

Zhou, X.Y., Zhang, J., Li, J.P. 2013. Industrial structural transformation and carbon dioxide emissions in China. *Energy Policy* 57: 43–51.

Architectural, Energy and Information Engineering – Sung & Chen (Eds)
© 2016 Taylor & Francis Group, London, ISBN 978-1-138-02791-6

Load monitoring for highly sub-metered buildings using a multi-agent system

J.L. Chu, Y. Fang, Y. Yuan & P. Chen
Shanghai Municipal Engineering Design Institute (Group) Co. Ltd., Shanghai, China

ABSTRACT: Energy conservation is becoming critical with respect to the increasing electricity demand. In order to perform further analysis of energy saving strategies and techniques, the ability to measure electricity consumption accurately is necessary. Various types of devices and technologies are developed for monitoring the electricity consumption of buildings. In this paper, the proposed load monitoring method uses a Multi-Agent System (MAS) with the average-consensus algorithm. The system architecture and computational results are presented. In addition, industry experience is considered. Data collected from computer simulations is used for validation of the proposed method.

Keywords: load monitoring; sub-metering; smart buildings; multi-agent system; energy conservation

1 INTRODUCTION

In recent years, electricity consumption has increased significantly with respect to the rapid economic growth in China. However, the power infrastructure has not expanded as quickly as the increasing demands of electricity markets. The power grid, therefore, is forced to operate at lower reserve capacity and stability margins. Energy conservation is becoming a critical problem. Thus, the ability to measure energy consumption accurately is important.

Buildings are among the largest consumers of electricity. The purpose of this research is to develop a method that measures electricity consumption of highly sub-metered buildings for further analysis, e.g., cost-benefit analysis of an equipment upgrade, behavioral conservation (Parker et al. 2006), investigation of energy efficiency. To this end, a solution that uses a Multi-Agent System (MAS) is adopted to handle the disadvantages of conventional hardware-based load monitoring methods. In addition, computer simulations are developed for validation.

There are two classes of approaches to electricity load monitoring of appliances: Intrusive Load Monitoring (ILM) (Ridi et al. 2013, 2014) and Non-Intrusive Load Monitoring (NILM) (Bergman et al. 2011, Parson et al. 2012, Egarter et al. 2013). In general, NILM measures the electricity consumption using a smart meter at the main breaker level. In contrast, ILM uses low-end metering devices. Surveys of ILM and NILM approaches are reported in (Ridi et al. 2014) and (Zeifman & Roth 2011, Zoha et al. 2012), respectively. In (Berges et al. 2008), it is indicated that much efforts have been made on improving NILM techniques. However,

much further research remains to be conducted on overcoming the disadvantages of conversional hardware-based ILM methods.

The hardware-based sub-metering method is able to separate different components of the electricity consumption accurately. However, the networking and information collection of a large number of sub-meters is challenging. Note that the conventional centralized architecture is required to collect all meter readings to a central node. Thus, a multi-agent based distributed solution is proposed using the average-consensus method for information sharing among agents. The contributions of this research include: 1) a distributed architecture is designed for the network of sub-meters, 2) a MAS is applied for information sharing and data fusion among sub-meters, 3) computational results are presented for validation.

The remaining of this paper is organized as follows: agents and the MAS are described in Section 2. The proposed multi-agent based load monitoring method is presented in Section 3. Section 4 provides the computational results. The conclusion is stated in Section 5.

2 AGENTS AND MAS

A MAS is a system that comprises two or more agents or intelligent agents that cooperate with each other through achieving local goals (Weiss 1999). This study adopts the definition of agents as proposed by Wooldridge (Wooldridge 2008): an agent is a computer system that is situated in some environment and is capable of autonomous actions in this environments in order to meet its design objectives. According to this definition, the agent comprises not only software but also hardware.

For comparison, the centralized and multi-agent based system frameworks are shown in Figure 1 and Figure 2, respectively. F1 and F10 denote the 1st and 10th floor. In the centralized scheme, all sub-meters in each floor are connected to an information center for collecting measurements. However, the multi-agent based distributed scheme requires that each agent communicates with adjacent agents only by two-party messages. For example, the agent in each floor only communicates with agents that are in its adjacent floors. All sub-meters in the same floor can be modeled as the sensors of the agent on this floor. The agent framework is shown in Figure 3. As a result, the

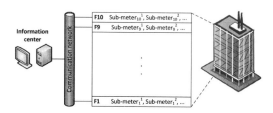

Figure 1. System framework of the centralized load monitoring.

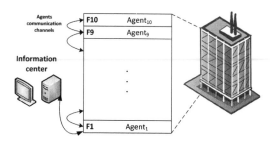

Figure 2. System framework of the multi-agent based load monitoring.

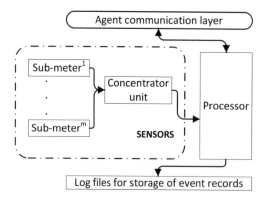

Figure 3. Agent framework.

commination network and electrical wiring can be simplified.

3 MULTI-AGENT BASED LOAD MONITORING

The average-consensus protocol without communication time-delay is presented in (Olfati-Saber & Murray 2004). The Continuous-Time (CT) model is defined as follows:

$$\dot{x}_i(t) = u_i(t) \tag{1}$$

$$u_i(t) = \sum_{j=1}^{n} a_{ij}(x_j(t) - x_i(t)) \tag{2}$$

where $x = (x_1, ..., x_n)^T$ is a vector that includes values of agents. x_i is the value of the i th agent. Note that x_i can be a single value or a row vector with respect to the number of sensors of the agent. n denotes the number of agents. Thus, the communication network of n agents is modeled as a directed graph, denoted by $G = (V, E, A)$. $V = \{1, 2, ..., n\}$ is the set of agents. The interconnection topology of this network is specified by a weighted nonnegative $n \times n$ adjacency matrix $A = [a_{ij}]$. If an active communication link exists from agent i to agent j, a_{ij} is a positive value, otherwise $a_{ij} = 0$. In addition, it is assumed that $a_{ii} = 0$ for all $i \in V$. $N_i = \{i \in V | a_{ij} \neq 0\}$ and $J_i = N_i \cup \{i\}$ are the sets of neighbors and inclusive neighbors of agent i, respectively. $e_{ij} = (v_i, v_j)$ denotes an edge of G and $E = [e_{ij}]$. This edge starts from agent v_i and points to agent v_j. If $e_{ij} \in E$, $a_{ij} > 0$.

The network dynamics of agents can be defined as follows:

$$\Delta = (\Delta_{ij}) \tag{3}$$

$$\Delta_{ij} = \begin{cases} deg_{out}(v_i) = \sum_{j=1}^{n} a_{ij}, & if \ i = j \\ 0, & if \ i \neq j \end{cases} \tag{4}$$

$$L = L(G) = \Delta - A \tag{5}$$

$$\dot{x}_i(t) = \sum_{j=1}^{n} a_{ij}(x_j(t) - x_i(t)) \tag{6}$$

$$\dot{x}(t) = (A - \Delta)x(t) = -Lx(t) \tag{7}$$

where L is called the graph Laplacian matrix. $deg_{out}(\cdot)$ denotes the operation of calculating the out-degree of an agent.

4 COMPUTATIONAL RESULTS

For simplicity, a building with five floors is used in this case. In addition, the network topology of agents is set as a line, i.e., communication links exist between Agent 1 and Agent 2, Agent 2 and Agent 3, Agent 3 and Agent 4, and Agent 4 and Agent 5. In order to enhance the convergence speed of the average-consensus algorithm, the adjacency matrix A is optimized by maximizing the second smallest eigenvalue of the Laplacian matrix L using a genetic algorithm. Details of the genetic algorithm is reported in (Russell et al. 2009). The value of matrix A used in this computer simulation is shown as follows:

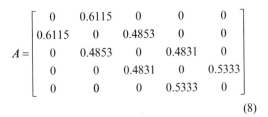

$$
A = \begin{bmatrix}
0 & 0.6115 & 0 & 0 & 0 \\
0.6115 & 0 & 0.4853 & 0 & 0 \\
0 & 0.4853 & 0 & 0.4831 & 0 \\
0 & 0 & 0.4831 & 0 & 0.5333 \\
0 & 0 & 0 & 0.5333 & 0
\end{bmatrix}
$$

(8)

Three sub-meters are installed in each floor and belong to an agent. Therefore, the electricity consumption of heating, illumination, and other appliances are measured separately. Note that the MAS can be expanded flexibly with more sub-meters that measure other types of electricity consumption.

The information sharing processing of five agents is shown in Figure 4 from the perspective of Agent 1. Before sharing information, each Agent has an initial vector, in which each element represents the locally measured electricity consumption in kWh by each sub-meter. The disagreement vector vanishes exponentially. See Figure 5. After 22 iterations, agents reach interactive consistency and the vector of Agent 1 is {158.1, 116.9, 137.3}. A total of five agents participate in the information sharing process. Thus, the total amount of electricity consumption is $5 \times \{158.1, 116.9, 137.3\} = \{790.5, 584.5, 686.5\}$(kWh).

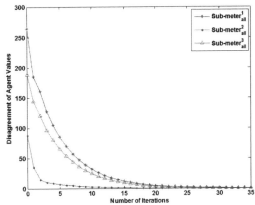

Figure 5. Disagreement of agent values.

5 CONCLUSION

A load monitoring algorithm is proposed to measure electricity consumption of highly sub-metered buildings using a MAS. While the conventional ILM scheme is a mature technology, the performance and flexibility may be enhanced by the proposed multi-agent scheme. Further enhancements can be achieved: 1) modeling of the communication network of sub-meters, 2) enhancing the system resilience with respect to the agent failure, and 3) optimizing the convergent speed of the information sharing process. As the structure of multi-agent based system can be reorganized and extended flexibly, this hierarchical scheme can be expanded by modeling more intelligent electronic devices as agents in a lower-lever layer. As a result, distributed artificial intelligence technologies can be applied to achieve better performance and robustness.

REFERENCES

Berges, M., Goldman, E., Matthews, H.S., & Soibelman, L. 2008. Training load monitoring algorithms on highly sub-metered home electricity consumption data. *Tsinghua Science and Technology* 13(S1): 406–411.

Bergman, D.C., Jin, D., Juen, J.P., Tanaka, N., Gunter, C.A., & Wright, A.K. Distributed non-intrusive load monitoring. *Innov. Smart Grid Technol. (ISGT)*, Hilton Anaheim, CA, Jan. 2011.

Egarter, D., Sobe, A., & Elmenreich, W. 2013. Evolving non-intrusive load monitoring. *Applications of Evolutionary Computation*: 182–191. Berlin: Springer.

Olfati-Saber, R. & Murray, R.M. 2004. Consensus problems in networks of agents with switching topology and time-delays. *IEEE Transactions on Automatic Control* 49(9): 1520–1533.

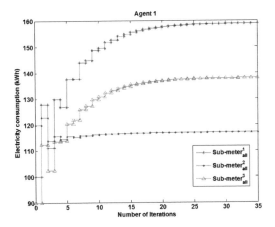

Figure 4. Information sharing process of five agents.

Parker, D., Hoak, D., Meier, A., & Brown, R. How much energy are we using? Potential of residential energy demand feedback devices. *Proc. 2006 Summer Study Energy Effic. Build.*, Asilomar, CA, Aug. 2006.

Parson, O., Ghosh, S., Weal, M., & Rogers, A. Non-intrusive load monitoring using prior models of general appliance types. *Proc. 26th Conf. Artif. Intell. (AAAI)*, Toronto, CA, Jul. 2012.

Ridi, A., Gisler, C., & Hennebert, J. 2013. Unseen appliances identification. *Progress in Pattern Recognition, Image Analysis, Computer Vision, and Applications*: 75–82. Springer Berlin Heidelberg.

Ridi, A., Gisler, C., & Hennebert, J. A survey on intrusive load monitoring for appliance recognition. *Proc. 22nd Int. Conf. Pattern Recognit. (ICPR)*, Aug. 2014.

Ridi, A., Gisler, C., & Hennebert, J. Appliance and state recognition using hidden Markov models. *Int. Conf.*

Data Sci. Adv. Anal. (DSAA), Shanghai, China, Oct. 2014.

Russell, S.J., Norvig, P., Canny, J.F., Malik, J.M., & Edwards, D.D. 2009. *Artificial Intelligence: A Modern Approach*. New Jersey: Prentice Hall.

Weiss, G. 1999. *Multiagent Systems: A Modern Approach to Distributed Artificial Intelligence*. Cambridge: The MIT Press.

Wooldridge, M. 2008. *An Introduction to Multiagent Systems*. New Jersey: John Wiley & Sons.

Zeifman, M. & Roth, K. 2011. Nonintrusive appliance load monitoring: Review and outlook. *IEEE Transactions on Consumer Electronics* 57(1): 76–84.

Zoha, A., Gluhak, A., Imran, M.A., & Rajasegarar, S. 2012. Non-intrusive load monitoring approaches for disaggregated energy sensing: A survey. *Sensors* 12(12): 16838–16866.

Architectural, Energy and Information Engineering – Sung & Chen (Eds)
© 2016 Taylor & Francis Group, London, ISBN 978-1-138-02791-6

Design of a spam short messages semantic recognition system based on cloud computing

X.F. Pan

Information Technology Department, Hainan Vocational College of Political Science and Law, Haikou, China

ABSTRACT: With the growth of mobile phone users, spam short messages are flooding. The traditional short message spam filtering system has a higher rate of false positives, so a new spam short messages semantic recognition filtering system based on cloud computing is proposed. The system uses the Naive Bayes algorithm to identify spam short messages semantic, and chooses cloud corpus as the training set. The analysis shows that the system has a high recall rate and a higher precision rate. This design provides a new idea for spam short message filtering.

Keywords: cloud computing; mapReduce; semantic recognition; spam short message; SMS

1 INTRODUCTION

With the popularity of mobile applications, Short Message Service (SMS) has become a high-growth services, with 98.1% users communicating and exchanging with other people through SMS [ISC 2014]. However, with the development of SMS, there is an explosive growth of SMS spam. The so-called spam short messages refer to the messages that are sent without the recipient's consent, which contain contents that are illegal or advertising in nature, and against the legitimate rights and interests of the recipient. The Statistics of ISC (2014) show that only 0.1% of users did not receive spam short messages weekly. Spam short messages have an impact on the normal life of people. Therefore, identifying and filtering spam short messages is very important.

2 EXISTING SPAM SHORT MESSAGES RECOGNITION SYSTEM

The spam short messages filtering technology is primarily through two ways [Yu, Y. & Chen, Y.C. 2013]. One way is that SMS center operators join the short messages filtering module that filters spam messages in real time, but in this method, the SMS load is too large. Another way is to filter the mobile terminal, which allows the user to choose the personalized shielded content, but this way will consume more mobile phone resources, and due to the limited phone resources, real-time filtering is not sufficient.

SMS center filtering module is modeling by the pattern of the user's messages sending, builds black and white lists, and restricts spam short messages being sent. For instance, Yang, N.N. & Wang, Q. (2009) designed a filter module based on the Struts framework; Zhong, Y.E. et al. (2009) proposed an algorithm based on a sample for filtering; Zhu, W.H. & Wang, M.Q. (2012) used a spam phone number filtering method based on the SMS submission pattern; Yu, Y. & Chen, Y.C. (2013) analyzed and applied social behavior in the offline spam message filter. However, because of the huge number of short messages, SMS center processing efficiency is not high. Also, the short messages could be shielded because of improper classification, and with the false positives of the user's spam short messages, short messages could not pass to the users. So, the short message service center module is difficult to achieve a comprehensive and accurate filtering of spam short messages.

The mobile phone terminal filtering technology is borrowed from the already quite mature spam filtering technology. However, short messages have three characteristics different from e-mail: short message only contains text and number, it does not contain other information such as attachments and hyperlinks; short message contents are colloquial, forms are not standardized; short message length limit is 140 English characters or 70 Chinese characters. Currently, the spam short messages recognition method in mobile phones mainly uses characteristic filtering and content filtering. For example, Jin, Z. et al. (2008) achieved a self-adaptive filtering system based on the Naive Bayes

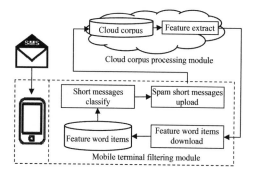

Figure 1. Spam short messages filtering system model diagram.

and support vector machine. Wang, S.Q. & Zhang, Y.L. (2011) proposed a new filter method based on CAPTCHA and Winnow. Zhao, Y.G. & Gong, L. (2011) researched spam short messages filtering based on behavior recognition and SVM. These characteristic filtering methods introduce smart technology for filtering, and can automatically filter spam short messages. Due to few message contents, and a huge number of same features, characteristic filtering of spam short messages is more likely to misjudge. In terms of content filtering, some scholars have proposed semantic recognition and word terms [Zhang, Y.J. & Liu, J.L. 2013, Xu, Y.H. & Liu, M.Y. 2013]. These methods can distinguish spam short messages from the message content, and have better classification results. To prevent some keywords of spam short messages from processing, Liu, J.L. et al. (2012) applied semantic recognition based on lexical chain for filtering. Because of colloquial text semantic, semantic text is difficult to judge. Also, due to limited resources, the mobile client, SMS spam corpus is not big enough, and content filtering has a higher rate of false positives.

3 SPAM SHORT MESSAGES SEMANTIC RECOGNITION SYSTEM BASED ON CLOUD COMPUTING

Based on the above SMS spam filtering system, this paper presents a spam short messages semantic recognition system based on cloud computing. The system's corpus of spam short messages is stored in the cloud, and calculated for each phone user's keywords, thus achieving a personalized filter spam system. The mobile phone anti-obsessed system structure is shown in Figure 1.

The system consists of two parts: mobile terminal filtering module and cloud corpus processing module. When a phone receives a short message,

in accordance with feature word items, the mobile terminal filtering module could classify this message as a normal short message or a spam short message. At the same time, the module will send the message to the cloud corpus if the user finds classification errors. Through the accumulation of cloud corpus, the cloud corpus processing module extracts each user's personalized feature word items. The mobile client downloads these features to update local feature word items, constantly revising the short message classification. The Semantic Recognition system through cloud feedback can improve the success rate of spam messages classification.

4 SHORT MESSAGE SEMANTIC CLASSIFICATION

The short message semantic classification uses machine learning techniques for filtering message, by analyzing the content. The message can be divided into two types: normal short message and spam short message. Commonly used algorithms have artificial neural network algorithms, KNN, SVM, and Naive Bayes algorithm [Yu, Y. & Chen, Y.C. 2013]. The Naive Bayesian classification algorithm has a high accuracy, simplicity, fast speed and other characteristics, and it is widely used. This paper also uses the semantic classification based on the Naive Bayes algorithm.

4.1 Short message text representation

The content of short message is composed of text and punctuation. Currently, the most common content represents the vector space model (VSM) [Zhang, Y.J. & Liu, J.L. 2013, Xu, Y.H. & Liu, M.Y. 2013, Yu, Y. & Chen, Y.C. 2013]. It is expressed as feature item collections, such as $T = \{t_1, t_2, ..., t_n\}$, where each item can be words or phrases, and each item has a feature weight and is used to indicate the level of importance in the message. For the Chinese text, because there is no delimiter between words, the text should be carried out word segmentation. Through the training message sets (including normal corpus and spam corpus) segmentation, the system can statistic words frequency as the feature items weight [Xu, Y.H. & Liu, M.Y. 2013]. Word segmentation program is developed by the Institute of Computing Technology of Chinese Academy science (ICTCLAS) [Zhang, H.P. & L, Q. 2012].

4.2 Naive Bayes semantic classification

Short message filtering is actually a binary classification problem, judging whether a given message is spam or normal short messages. Here, each message is represented as an n-dimensional feature vector T. The Bayesian probability formula is given in Equation (1). This equation can be used to calculate the probability that T, respectively, belongs to classes m_1 and m_2. Then, it will classify the messages to the class which T has the greatest probability. Here, m_1 represents normal short messages classification and m_2 represents spam short messages classification:

$$p(m_i \mid T) = \frac{p(T \mid m_i)p(m_i)}{p(T)} \quad i = 1, 2 \qquad (1)$$

where $p(T)$ is constant for all short message classifications; therefore, it can be ignored when comparing the maximum probability, and just simply calculate the values of $p(T \mid m_i)\,p(m_i)$. $p(m_i)$ represents the probability of message class m_i, and can be expressed by the total ratio of m_i class training corpus. $p(T \mid m_i)$ represents the probability that T appears under the conditions of m_i class, and can be calculated by the joint probability of feature items under the conditions of m_i class, which is given in Equation (2):

$$p(T \mid m_i) = \prod_{k=1}^{n} p(t_k \mid m_i) \quad i = 1, 2 \qquad (2)$$

In practice, spam short messages have more obvious features, while normal short messages content is rich and varied, and features are not obvious. Features extraction results in missing some features of normal short messages. In other words, the prior probabilities of these feature items appear toe b 0%. So, when using the Naive Bayes classification algorithm, results are instable. For this particular short message classification application, using quadrature sums instead of Equation (2), avoiding a direct impact on the classification results, we get:

$$p(T \mid m_i) = \sum_{k=1}^{n} p(t_k \mid m_i) \quad i = 1, 2 \qquad (3)$$

For classification, the algorithm not simply selects $max\{p(T \mid m_1),\ p(T \mid m_2)\}$ as a basis for classification, but also improves the weight value of the probability of spam short messages, which can reduce the false positive rate of normal short messages. The weight coefficient selection is 2.8. When $p(T \mid m_1) > 2.8\ p(T \mid m_2)$, T is determined to be spam short messages, otherwise it is determined to be normal short messages [Xu, Y.H. & Liu, M.Y. 2013].

4.3 Short message classification evaluation index

Content-based short message filtering classification problems are essentially within the text classification. This paper uses a common evaluation index to assess the filtering system, which includes the recall rate and the precision rate.

Recall rate refers to the ratio of correct filtering spam short messages accounting for actual spam short messages. It reflects the system's ability to find spam short messages. The larger the recall rate, the fewer the omission of spam short messages.

Precision rate refers to the ratio of correct filtering spam short messages accounting for filtering spam short messages (which contain error filtering). It reflects the system's ability to find spam short messages. The larger the precision rate, the fewer the false positive rates of spam short messages.

5 CLOUD CORPUS IMPLEMENTATION

Through the training corpus, the short message filtering technology based on the semantic classification can update and extract the latest classification feature items, and can filter the spam short messages according to different mobile phone users' personalized understanding. Due to limited mobile phone resources, the system stores the corpus in the cloud, and implements the corpus algorithm based on cloud computing.

5.1 Calculating feature items of cloud corpus

After word segmentation, short messages are represented by feature vector. Due to the unstructured feature of the Chinese text, the vector of text representation usually reaches a million or even more dimensions. In order to improve the efficiency and accuracy of classification, the system needs to reduce the dimension of feature vectors. The commonly used feature dimension reduction method is based on the following: word Document Frequency (DF), Information Gains (IG), x^2 statistics (CHI), and mutual information (MI) [Xu, Y.H. & Liu, M.Y. 2013]. This paper selects information gain as a method of feature dimension reduction. Information gain method uses the amount of information of t_k feature items to measure the importance of this feature. The information gain calculation of feature t_k is defined in Equation (4):

$$
\begin{aligned}
IG(t_k) = & -\sum_{i=1}^{2} p(m_i) \lg p(m_i) \\
& + p(t_k) \sum_{i=1}^{2} p(m_i \mid t_k) \lg p(m_i \mid t_k) \\
& + p(\bar{t}_k) \sum_{i=1}^{2} p(m_i \mid \bar{t}_k) \lg p(m_i \mid \bar{t}_k) \quad (4)
\end{aligned}
$$

$p(m_i)$ represents the probability of category mi appearing in the training corpus; $p(t_k)$ represents the probability of feature t_k appearing in the training corpus; $p(m_i|t_k)$ denotes the probability that t_k appears in the training corpus, the feature belongs to m_i; and $p(m_i|\overline{t_k})$ denotes the probability that t_k does not appear in the training corpus, the feature belongs to m_i.

By calculating the information gain of cloud corpus for each feature, the system sorts information gain in descending order, and ultimately chooses 1000 highest gain words as feature items and downloads to the phone.

5.2 Feature extraction algorithm based on cloud computing

Since 2004, Google Company has proposed Google File System [Ghemawat, S. et al. 2003], BigTable [Chang, F. et al. 2008.] and MapReduce [Dean, J. & Ghemawat S. 2004] technologies. With the introduction of these technologies, cloud computing has emerged, as a new storage to manage and analyze the pattern of mass data, and has been widely used by many of the industry's companies. MapReduce is a programming model of cloud computing. MapReduce will divide the user's original data into blocks, and hand them over to different Map task deals, and then the Map task will output pairs of Key/Value from input data parsing. The user-defined Map function acts on these Key/Value pairs and gets the corresponding intermediate results. These results are written to the local hard disk. The Reduce task reads data from the hard disk, sorts key values, and organizes the same key values together. Finally, the user-defined Reduce function acts on these key values and output the final result. Google has a variety of opensource MapReduce implementations; the most widely used version is Hadoop MapReduce. The feature extract algorithm of this paper is achieved by Hadoop MapReduce.

The feature extract algorithm of cloud computing is described as follows:

```
mapper (username, normal-corpus, spam-corpus):
    for each t_k in normal-corpus:
        Word segmentation by ICTCLAS
        emit (t_k, 1)
    for each t_k in spam-corpus:
        Word segmentation by ICTCLAS
        emit (t_k, 1)
reducer (word, normal-values, spam-values):
    s_1 = 0
    for each value in normal-values:
        s_1 = s_1+value
    s_2 = 0
    for each value in spam-values:
        s_2 = s_2+value
```

Table 1. Spam short messages filtering results.

	Recall rate	Precision rate
NB	92.3%	90.6%
CAP	94.8%	99.1%
BR	96.7%	97.1%
SR	97.85%	99.8%
SRCC	99.5%	100%

c_1 = normal-values.count; c_2 = spam-values.count;

$p(m_i) = s_i/(s_1+s_2)$;
$p(t_k) = (s_1+s_2)/(c_1+ c_2)$; $p(\overline{t_k}) = 1 - p(t_k)$;
$p(t_k|m_i) = s_i/c_i$; $p(\overline{t_k}|m_i) = 1 - p(t_k|m_i)$;
$p(m_i|t_k) = p(t_k|m_i) \, p(m_i)/p(t_k)$;
$p(m_i|\overline{t_k}) = p(\overline{t_k}|m_i) \, p(m_i)/p(\overline{t_k})$;
Calculate $IG(t_k)$ by Equation (4)
emit $(t_k, IG(t_k))$

6 EXPERIMENTAL RESULTS AND ANALYSIS

The experiment corpus of normal short messages published from the School of Computing of the National University of Singapore (NUS SMS Corpus) [Chen, T. & Kan, M.Y. 2012], and the experiment corpus of spam short messages are obtained from phone user feedback. These Corpuses contain 31465 normal short messages and 1392 spam short messages, which are divided into two groups: training corpus and testing corpus. The semantic recognition algorithm based on cloud computing (SRCC) is compared with the NB [Jin, Z. et al. 2008], CAP [Wang, S.Q. & Zhang, Y.L. 2011], BR [Zhao, Y.G. & Gong, L. 2011] and SR [Xu, Y.H. & Liu, M.Y. 2013] algorithms. The results are presented in Table 1.

The precision rate of the SRCC algorithm reaches up to 100%, and the recall rate is higher than that of the other algorithms. Compared with the common spam messages filtering method, the SRCC algorithm is very effective. If the training corpus contains more content, the algorithm has the more accurate spam classification.

7 CONCLUSIONS

This paper designs a spam short messages semantic recognition system, which is based on cloud computing. Users can personalize filter spam short messages by this system. The analysis shows that under the premise of a high recall rate, the system also maintains a higher precision rate. However, the design of the system has also a shortcoming. The system is not introduced for mechanical self-learning, but has to rely manually to maintain the cloud corpus. Future research will improve this shortcoming.

ACKNOWLEDGMENTS

This paper was supported by the Hainan Provincial Natural Science Foundation of China (No. 20156244).

REFERENCES

Chang, F., Dean, J., Ghemawat, S., Hsieh, W.C., Wallach, D.A., Burrows, M., Chandra, T., Fikes, A. & Gruber, R.E. 2008. Bigtable: A distributed storage system for structured data. *ACM Trans. on Computer Systems* 26(2): 1–26.

Chen, T. & Kan, M.Y. 2012. Creating a Live, Public Short Message Service Corpus: The NUS SMS Corpus. *Language Resources and Evaluation* 47(2): 299–335.

Dean, J. & Ghemawat S. 2004. MapReduce: Simplified data processing on large clusters. *Communications of the ACM* 51(1): 107–113.

Ghemawat, S., Gobioff, H. & Leung, S.T. 2003. The google file system. *In: Proc. of the 19th ACM Symp. on Operating Systems Principles. New York: ACM:* 29–43.

ISC 2014. *SMS survey report in the first half 2014.* Beijing: Internet Society of China.

Jin, Z., Fan, J. & Chen, F. 2008. Spam message self-adaptive filtering system based on Native Bayes and support vector machine. *Computer Applications* 28(3): 714–718.

Liu, J.L., Feng, W.L. & Gao L. 2012. Semantic recognition of altered Chinese junk short messages based on lexical chain. *Computer Engineering and Applications* 48(19): 135–139.

Wang, S.Q. & Zhang, Y.L. 2011. Research on SMS spam filtering based on CAPTCHA and Winnow. *Computer Engineering and Design* 32(1): 313–315.

Xu, Y.H. & Liu, M.Y. 2013. Content-based junk short message filtering for mobile phone Content-based junk short message filtering for mobile phone. *Journal of Beijing Information Science and Technology University* 28(1): 51–55.

Yang, N.N. & Wang, Q. 2009. Design of Junk Short Message Filter Module Based on Struts Framework. *Computer Engineering* 35(1): 283–285.

Yu, Y. & Chen, Y.C. 2013. Analysis and Application of Social Behavior in Offline Spam Message Filter. *Journal of Chinese Computer Systems* 34(8): 1877–1818.

Zhang, H.P. & L, Q. 2012. Chinese Named Entity Recognition Using Role Model. *http://ictclas.nlpir.org/newsdownloads?DocId=389.*

Zhang, Y.J. & Liu, J.L. 2013. Spam short message classifier model based on word terms. *Journal of Computer Applications* 33(5): 1334–1337.

Zhao, Y.G. & Gong, L. 2011. Research of Filtering Method for Short Message Based on Behavior Recognition and SVM. *Microcomputer Information* 28(1): 176–177.

Zhong, Y.H., Fu, Y., Chen, A.L. & Guan, N. 2009. Filtering algorithm of junk SMS based on sample. *Application Research of Computers* 26(3): 933–935.

Zhu, W.H. & Wang, M.Q. 2012. Spam phone number filtering method based on SMS submission pattern. *Journal of Computer Applications* 32(12): 3565–3568.

Architectural, Energy and Information Engineering – Sung & Chen (Eds)
© 2016 Taylor & Francis Group, London, ISBN 978-1-138-02791-6

Existence of solutions for a coupled system to nonlinear fractional boundary value problems

Yun Hong Li

College of Sciences, Hebei University of Science and Technology, Shijiazhuang, Hebei, China

ABSTRACT: In this paper, we consider a coupled system with integral boundary conditions of fractional boundary value problem. Using the contraction mapping principle, Schauder fixed-point theorem, the existence of solutions is obtained. Two examples are given to illustrate our results.

Keywords: solutions; fractional differential equations; fixed point theorem

1 INTRODUCTION

In recent years, differential equations have gained importance due to their various applications in science and engineering such as heat conduction, control theory, chemical physics, etc. The study of fractional boundary value problems was initiated by many authors, one may see [1–10] and references therein. However, as far as we know, few results can be found in the literature concerning a coupled system of nonlinear fractional differential equations with integral boundary value problems. As a result, this paper is concerned with the existence of solutions for the following coupled system with integral boundary conditions

$$D_{0+}^{\alpha_1} u(t) = f_1(t, u(t), v(t), K_1 u(t), H_1 v(t)), 0 < t < 1,$$
$$D_{0+}^{\alpha_2} v(t) = f_2(t, u(t), v(t), K_2 u(t), H_2 v(t)), 0 < t < 1,$$

$$u(0) = u'(0) = 0, u(1) = \lambda_1 \int_0^1 u(s)ds,$$

$$v(0) = v'(0) = 0, v(1) = \lambda_2 \int_0^1 v(s)ds.$$

$$(1)$$

where $2 < \alpha_i \le 3, 0 < \lambda_i < \alpha_i$, $D_{0+}^{\alpha_1}, D_{0+}^{\alpha_2}$ are the Rieman-Liouville fractional derivative, and

$$k_i(t,s):[0,1]\times[0,t] \to R, \quad h_i(t,s):[0,1]\times[0,t] \to R,$$

$$(K_i x)(t) = \int_0^t k_i(t,s)x(s)ds,$$

$$(H_i x)(t) = \int_0^t h_i(t,s)x(s)ds,$$

be continuous functions, and

$$l_i = \sup_{t\in[0,1]} \int_0^t k_i(t,s)ds < \infty,$$

$$m_i = \sup_{t\in[0,1]} \int_0^t h_i(t,s)ds < \infty, \quad i=1,2. \quad (2)$$

2 PRELIMINARIES

In this section, we introduce definitions and preliminary facts which are used throughout this paper. These definitions can be found in works [11, 12].

Definition 2.1 The fractional integral of order $\alpha > 0$ of a function $y:(0,\infty) \to R$ is given by

$$I_{0+}^\alpha y(t) = \frac{1}{\Gamma(\alpha)} \int_0^t (t-s)^{\alpha-1} y(s)ds,$$

provided the right side is pointwise defined on $(0,\infty)$.

Definition 2.2 For a continuous function $y:(0,\infty) \to R$, the Rieman-Liouville derivative of fractional Order $\alpha > 0$ is defined as

$$D_{0+}^\alpha y(t) = \frac{1}{\Gamma(n-\alpha)} \left(\frac{d}{dt}\right)^{(n)} \int_0^t (t-s)^{n-\alpha-1} y(s)ds,$$

where $n = [\alpha] + 1$, provided the right side is pointwise defined on $(0,\infty)$.

$(u(t), v(t))$ is a solution of the coupled system (1) if and only if $(u(t), v(t))$ is a solution of the system of integral equations

$$u(t) = \int_0^1 G_1(t,s)f_1(s,u(s),v(s),K_1u(s),H_1v(s))ds,$$

$$v(t) = \int_0^1 G_2(t,s)f_2(s,u(s),v(s),K_2u(s),H_2v(s))ds,$$

where $G_i(t,s)$ $(i = 1, 2)$ is the Green's function [2] defined as follows:

$$G_i(t,s) = \begin{cases} \dfrac{t^{\alpha_i-1}(1-s)^{\alpha_i-1}(\alpha_i - \lambda_i + \lambda_i s) - (\alpha_i - \lambda_i)(t-s)^{\alpha_i-1}}{(\alpha_i - \lambda_i)\Gamma(\alpha_i)}, & 0 \le s \le t \le 1, \\[4mm] \dfrac{t^{\alpha_i-1}(1-s)^{\alpha_i-1}(\alpha_i - \lambda_i + \lambda_i s)}{(\alpha_i - \lambda_i)\Gamma(\alpha_i)}, & 0 \le t \le s \le 1. \end{cases}$$

Lemma 2.1 Fix $2 < \alpha_i \le 3$, $0 < \lambda_i < \alpha_i$, the Green's function $G_i(t,s) \ge 0$, for all $t,s \in [0,1]$, and satisfied the following inequalities:

$$t^{\alpha_i-1}G_i(1,s) \le G_i(t,s) \le \frac{\alpha_i}{\lambda_i}G_i(1,s), \forall t,s \in (0,1),$$

where

$$G_i(1,s) = \frac{\lambda_i s(1-s)^{\alpha_i-1}}{(\alpha_i - \lambda_i)\Gamma(\alpha_i)}, \quad i = 1,2.$$

3 MAIN RESULTS

Theorem 3.1 Let $f_1, f_2 : [0,1] \times R^4 \to R$ be continuous functions. Assume there exist positive constants $L_{i1}, L_{i2}, L_{i3}, L_{i4}$, such that

$$\begin{aligned} &|f_i(t,u(t),v(t),K_iu(t),H_iv(t)) \\ &-f_i(t,x(t),y(t),K_ix(t),H_iy(t))| \\ &\le L_{i1}|u-x| + L_{i2}|v-y| \\ &+ L_{i3}|K_iu - K_ix| + L_{i4}|H_iv - H_iy| \end{aligned}$$

Further

$$(L_{i1} + L_{i2} + l_iL_{i3} + m_iL_{i4})\frac{1}{(\alpha_i - \lambda_i)\Gamma(\alpha_i)} < 1.$$

where l_i, m_i is defined in (2), $i = 1,2$. Then the coupled system (1) has a unique solution.

Proof. let $X = C[0,1]$ endowed with the norm

$$|u| = \sup_{t \in [0,1]} |u(t)|.$$

Define $E = X \times X$ endowed with the norm

$$|u,v|_E = \max\{|u|, |v|\}.$$

Obviously, E is a Banach space.
Let operator $T : E \to E$,

$$T(u,v)(t) = (T_1(u(t),v(t)), T_2(u(t),v(t))),$$

where

$$T_1(u(t),v(t)) = \int_0^1 G_1(t,s)f_1(s,u(s),v(s),K_1u(s),H_1v(s))ds,$$

$$T_2(u(t),v(t)) = \int_0^1 G_2(t,s)f_2(s,u(s),v(s),K_2u(s),H_2v(s))ds.$$

For all (u,v), $(x,y) \in E$, we have

$$\begin{aligned} &|T(u,v) - T(x,y)|_E \\ &= |(T_1(u,v),T_2(u,v)) - (T_1(x,y),T_2(x,y))|_E \\ &= |(T_1(u,v) - T_1(x,y), \ T_2(u,v) - T_2(x,y))|_E. \end{aligned}$$

Using the condition of theorem 3.1 and Lemma 2.1, we get

$$\begin{aligned} &|(T_1(u,v) - T_1(x,y)| \\ &= |\int_0^1 G_1(t,s)f_1(s,u(s),v(s),K_1u(s),H_1v(s))ds \\ &- \int_0^1 G_1(t,s)f_1(s,x(s),y(s),K_1x(s),H_1y(s))ds| \\ &\le (L_{11}|u-x| + L_{12}|v-y| + L_{13}|K_1u - K_1x| \\ &+ L_{14}|H_1v - H_1y|)\int_0^1 G_1(t,s)ds \\ &\le (L_{11}|u-x| + L_{12}|v-y| + l_1L_{13}|u-x| \\ &+ m_1L_{14}|v-y|)\frac{\alpha_1}{\lambda_1}\int_0^1 G_1(1,s)ds \\ &\le (L_{11} + L_{12} + l_1L_{13} + m_1L_{14})\frac{|u-x,v-y|_E}{(\alpha_1 - \lambda_1)\Gamma(\alpha_1)} \\ &= \mu_1|u-x,v-y|_E \\ &= \mu_1|(u,v) - (x,y)|_E \end{aligned}$$

664

Similarly, we can get

$$|(T_2(u,v) - T_2(x,y)| \leq \mu_2 \left\| (u,v) - (x,y) \right\|_E,$$

Let $\mu = \max\{\mu_1, \mu_2\}$, obviously, $\mu < 1$. Hence, we have

$$\left\| T(u,v) - T(x,y) \right\|_E \leq \mu \left\| (u,v) - (x,y) \right\|_E,$$

therefore T is a contraction. Thus, the conclusion of the theorem 3.1 holds by the contraction mapping principle. The proof is completed.

Theorem 3.2 Let $f_1, f_2 : [0,1] \times R^4 \to R$ be continuous functions. Suppose that one of the following conditions is satisfied.

(H1) There exist nonnegative functions $a_i(t) \in L[0,1]$, and constants $b_i, c_i, d_i, e_i, \rho_i, \theta_i, \sigma_i, \delta_i$, such that

$$|f_i(t,x,y,z,w)| \leq a_i(t) + b_i |x|^{\rho_i} + c_i |y|^{\theta_i} + d_i |z|^{\sigma_i} + e_i |w|^{\delta_i}$$

where $b_i, c_i, d_i, e_i > 0, \rho_i, \theta_i, \sigma_i, \delta_i > 1, i = 1,2$.

(H2) There exist constants $b_i, c_i, d_i, e_i, \rho_i, \theta_i, \sigma_i, \delta_i$, such that

$$|f_i(t,x,y,z,w)| \leq b_i |x|^{\rho_i} + c_i |y|^{\theta_i} + d_i |z|^{\sigma_i} + e_i |w|^{\delta_i}$$

where $b_i, c_i, d_i, e_i > 0, \rho_i, \theta_i, \sigma_i, \delta_i > 1, i = 1,2$.

Proof. We shall prove this result by using the Schauder fixed-point theorem. The Banach space E and operator T are defined in theorem 3.1, then we define $U = \{(u,v) | (u,v) \in E, \left\| (u,v) \right\|_E \leq r, t \in [0,1]\}$.

If the condition (H1) holds, then

$$r \geq \max\left\{ \frac{5\alpha_i}{\lambda_i} \int_0^1 G_i(1,s) a_i(s)ds, \right.$$

$$\left(\frac{5b_i}{(\alpha_i - \lambda_i)\Gamma(\alpha_i)} \right)^{\frac{1}{1-\rho_i}}, \left(\frac{5c_i}{(\alpha_i - \lambda_i)\Gamma(\alpha_i)} \right)^{\frac{1}{1-\theta_i}},$$

$$\left. \left(\frac{5d_i l_i^{\sigma_i}}{(\alpha_i - \lambda_i)\Gamma(\alpha_i)} \right)^{\frac{1}{1-\sigma_i}}, \left(\frac{5e_i m_i^{\delta_i}}{(\alpha_i - \lambda_i)\Gamma(\alpha_i)} \right)^{\frac{1}{1-\delta_i}} \right\}$$

where $i = 1,2$.

Obviously, U is the ball in the Banach space E.

For any $(u,v) \in U$, applying the condition (H1) and Lemma 2.1, we have

$$|T_1(u(t),v(t))|$$

$$= \| \int_0^1 G_1(t,s) f_1(s,u(s),v(s),K_1u(s),H_1v(s))ds$$

$$\leq \int_0^1 G_1(t,s)(a_1(s) + b_1 |u(s)|^{\rho_1} + c_1 |v(s)|^{\theta_1}$$

$$+ d_1 |K_1 u(s)|^{\sigma_1} + e_1 |H_1 v(s)|^{\delta_1})ds$$

$$\leq \int_0^1 G_1(t,s) a_1(s)ds$$

$$+ (b_1 r^{\rho_1} + c_1 r^{\theta_1} + d_1 l_1^{\sigma_1} r^{\sigma_1} + e_1 m_1^{\delta_1} r^{\delta_1}) \int_0^1 G_1(t,s)ds$$

$$\leq \frac{\alpha_1}{\lambda_1} \int_0^1 G_1(1,s) a_1(s)ds$$

$$+ (b_1 r^{\rho_1} + c_1 r^{\theta_1} + d_1 l_1^{\sigma_1} r^{\sigma_1} + e_1 m_1^{\delta_1} r^{\delta_1}) \frac{\alpha_1}{\lambda_1} \int_0^1 G_1(1,s)ds$$

$$\leq \frac{r}{5} + \frac{b_1 r^{\rho_1} + c_1 r^{\theta_1} + d_1 l_1^{\sigma_1} r^{\sigma_1} + e_1 m_1^{\delta_1} r^{\delta_1}}{(\alpha_1 - \lambda_1)\Gamma(\alpha_1)} \leq r.$$

Similarly, one has $|T_2(u(t),v(t))| \leq r$, That is, we get $\left\| T(u,v) \right\|_E \leq r$, thus, $T : U \to U$ is obtained.

Now we are in the position to let (H2) be satisfied. Choose $r > 0$ and

$$r \leq \min\left\{ \left(\frac{(\alpha_i - \lambda_i)\Gamma(\alpha_i)}{4b_i} \right)^{\frac{1}{\rho_i - 1}}, \left(\frac{(\alpha_i - \lambda_i)\Gamma(\alpha_i)}{4c_i} \right)^{\frac{1}{\theta_i - 1}}, \right.$$

$$\left. \left(\frac{(\alpha_i - \lambda_i)\Gamma(\alpha_i)}{4d_i l_i^{\sigma_i}} \right)^{\frac{1}{\sigma_i - 1}}, \left(\frac{(\alpha_i - \lambda_i)\Gamma(\alpha_i)}{4e_i m_i^{\delta_i}} \right)^{\frac{1}{\delta_i - 1}} \right\}.$$

where $i = 1,2$.

For any $(u,v) \in U$, applying the condition (H2) and Lemma 2.1, we have

$$|T_1(u(t),v(t))|$$

$$= | \int_0^1 G_1(t,s) f_1(s,u(s),v(s),K_1u(s),H_1v(s))ds |$$

$$\leq \int_0^1 G_1(t,s)(b_1 |u(s)|^{\rho_1} + c_1 |v(s)|^{\theta_1} + d_1 |K_1 u(s)|^{\sigma_1}$$

$$+ e_1 |H_1 v(s)|^{\delta_1})ds$$

$$\leq (b_1 r^{\rho_1} + c_1 r^{\theta_1} + d_1 l_1^{\sigma_1} r^{\sigma_1} + e_1 m_1^{\delta_1} r^{\delta_1}) \frac{\alpha_1}{\lambda_1} \int_0^1 G_1(1,s)ds$$

$$\leq \frac{b_1 r^{\rho_1} + c_1 r^{\theta_1} + d_1 l_1^{\sigma_1} r^{\sigma_1} + e_1 m_1^{\delta_1} r^{\delta_1}}{(\alpha_1 - \lambda_1)\Gamma(\alpha_1)} \leq r$$

Similarly, one has $|T_2(u(t),v(t))| \leq r$, That is, we get $\left\| T(u,v) \right\|_E \leq r$, thus, $T : U \to U$ is obtained. In view of the continuity of G_1, G_2, f_1, f_2, the operator T is continuous.

Let $\Omega \subset U$ be bounded, which is to say there exist two positive constant $L > 0$ such that $\left\| (u,v) \right\|_E \leq L$ for all $(u,v) \in \Omega$. Define now

$$B_i = \max_{0 \leq t \leq 1, 0 \leq u \leq L, 0 \leq v \leq L} |f_i| + 1,$$

Then, for all $(u,v) \in \Omega$, it is satisfied that

$$T_i(u,v) = \leq \frac{B_i \alpha_i}{\lambda_i} \int_0^1 G_i(t,s) f_i ds \int_0^1 G_i(1,s)ds \leq \frac{B_i}{(\alpha_i - \lambda_i)\Gamma(\alpha_i)}$$

that is, the set $T(\Omega)$ is bounded in E.

For each $(u,v) \in \Omega$, we have

$$\left| \frac{dT_i(u,v)}{dt} \right|$$

$$= \left| \int_0^1 \frac{t^{\alpha_i-2}(1-s)^{\alpha_i-1}(\alpha_i - \lambda_i + \lambda_i s)}{(\alpha_i - \lambda_i)\Gamma(\alpha_i - 1)} f_i ds \right.$$

$$\left. - \int_0^t \frac{(t-s)^{\alpha_i-2}}{\Gamma(\alpha_i - 1)} f_i ds \right|$$

$$\leq B_i \left(\frac{1}{(\alpha_i - \lambda_i)\Gamma(\alpha_i - 1)} + \frac{1}{\Gamma(\alpha_i)} \right)$$

$$\triangleq N_i, \quad i = 1,2.$$

As consequence, for all $t_1, t_2 \in [0,1]$, $t_1 < t_2$, we have

$$|T_i(u(t_2), v(t_2)) - T_i(u(t_1), v(t_1))|$$

$$= \int_{t_1}^{t_2} |(T_i(u,v))'(t)| \, dt \leq N_i(t_2 - t_1),$$

and the set $T_i(\Omega), i = 1,2$ is equicontinuous. As a consequence, together with Ascoli-Arzela theorem, we can get that T is a completely continuous operator. Therefore, the Schauder fixed point theorem implies that the coupled system (1) has at least one solution in U. The proof is completed.

4 EXAMPLE

Example 4.1 Let us consider the following coupled system

$$D_{0+}^{2.5}u(t) = f_1(t,u(t),v(t),K_1u(t),H_1v(t)), \quad 0 < t < 1,$$

$$D_{0+}^{2.6}v(t) = f_2(t,u(t),v(t),K_2u(t),H_2v(t)), \quad 0 < t < 1,$$

$$u(0) = u'(0) = 0, u(1) = 1.5\int_0^1 u(s)ds,$$

$$v(0) = v'(0) = 0, v(1) = 1.6\int_0^1 v(s)ds. \quad (3)$$

where

$$K_1u(t) = \int_0^t \sin(t-s)u(s)ds,$$

$$K_2u(t) = \int_0^t \cos(t-s)u(s)ds,$$

$$H_1v(t) = \int_0^t e^{(t-s)}v(s)ds,$$

$$H_2v(t) = \int_0^t e^{0.1(t-s)}v(s)ds,$$

and

$$f_1 = \frac{|u(t)|}{(t+4)^2(|u(t)|+1)} + \frac{|v(t)|}{(\sin t + 5)(|v(t)|+2)}$$

$$+ \frac{K_1u(t)}{\ln(t+16)} + \frac{H_1v(t)}{\ln(t+81)},$$

$$f_2 = \frac{|u(t)|}{(t+3)^2(|u(t)|+2)} + \frac{|v(t)|}{(\cos t + 2)(|v(t)|+3)}$$

$$+ \frac{K_1u(t)}{\ln(t+25)} + \frac{H_1v(t)}{\ln(t+64)},$$

Obviously,

$$|f_1(t,u(t),v(t),K_1u(t),H_1v(t))$$

$$- f_1(t,x(t),y(t),K_1x(t),H_1y(t))|$$

$$\leq \frac{1}{16}|u-x| + \frac{1}{10}|v-y| + \frac{1}{4}\|K_1u - K_1x\|$$

$$+ \frac{1}{9}\|H_1v - H_1y\|.$$

$$|f_2(t,u(t),v(t),K_2u(t),H_2v(t))$$

$$- f_2(t,x(t),y(t),K_2x(t),H_2y(t))|$$

$$\leq \frac{1}{18}|u-x| + \frac{1}{6}|v-y| + \frac{1}{5}\|K_2u - K_2x\|$$

$$+ \frac{1}{8}\|H_2v - H_2y\|.$$

Further $\mu_1, \mu_2 < 1$. By the condition in theorem 3.1, we know that the coupled system (3) has a unique solution.

Example 4.2 In the coupled system (3), let

$$f_1 = (t+1)^2 + \frac{u^{\frac{1}{4}}(t)}{2t+1} + \frac{v^{\frac{1}{2}}(t)}{\cos 3t + 1}$$

$$+ (K_1u(t))^{\frac{1}{3}} + (H_1v(t))^{\frac{1}{4}},$$

$$f_2 = (t+2)^3 + \frac{u^{\frac{1}{5}}(t)}{t+1} + \frac{v^{\frac{1}{8}}(t)}{\sin t + 1}$$

$$+ (K_2u(t))^{\frac{1}{6}} + (H_2v(t))^{\frac{1}{7}}.$$

$K_iu(t), H_iv(t), i = 1,2$ is defined in example 4.1, by the condition (H1) and theorem 3.2, we know that the coupled system (3) has at least one solution.

THANKS

The project is supported by the Science Foundation of Hebei University of Science and Technology (XL201144).

REFERENCES

[1] X. Zhang, L. Liu, Y. Wu, Multiple positive solutions of a singular fractional differential equation with negatively perturbed term, Math. Comput. Model. 55 (2012) 1263–1274.

[2] A. Cabada, Z. Hamdi, Nonlinear fractional differential equations with integral boundary value conditions, Appl. Math. Comput. 228 (2014) 251–257.

[3] C. Li, X. Luo, Y. Zhou, Existence of positive solutions of the boundary value problem for nonlinear fractional differential equations, Comput. Math. Appl. 59(2010) 1363–1375.

[4] Y. Sun, M. Zhao, Positive solutions for a class of fractional differential equations with integral boundary conditions, Appl. Math. Lett. 34 (2014) 17–21.

[5] C. Zhai, L. Xu, Properties of positive solutions to a class of four-point boundary value problem of Caputo fractional differential equations with a parameter, Commu Nonlinear Sci Numer Simulat. 19 (2014) 2820–2827.

[6] X. Yang, Z. Wei, W. Dong, Existence of positive solutions for the boundary value problem of nonlinear fractional differential equations, Commu Nonlinear Sci Numer Simulat. 17 (2012) 85–92.

[7] X. Xu, D. Jiang, C. Yuan, Multiple positive solutions for the boundary value problem of a nonlinear fractional differential equation, Nonlinear Anal. 71(2009) 4676–4688.

[8] L. Zhang, B. Ahmad, G. Wang, R. Agarwal, Nonlinear fractional integro-differential equations on unbounded domains in a Banach space, J. Comput. Appl. Math. 240 (2013) 51–56.

[9] Cabada, G. Wang, Positive solutions of nonlinear fractional differential equations with integral boundary value conditions, J. Math. Anal. Appl. 389 (2012) 403–411.

[10] T. Jankowski, Positive solutions to fractional differential equations involving Stieltjes integral conditions, Appl. Math. Comput. 241 (2014) 200–213.

[11] I. Podlubny, Fractional differential equations, mathematics in science and engineering, Academic Press, New York, 1999.

[12] K.S. Miller, B. Ross, An introduction to the fractional calculus and fractional differential equations, Wiley, New York, 1993.

Architectural, Energy and Information Engineering – Sung & Chen (Eds)
© 2016 Taylor & Francis Group, London, ISBN 978-1-138-02791-6

Prediction method for the monolayer production contribution of the multilayer gas well based on the BP artificial neural network

J.H. Li, P. Hu & X. Xiang
*A.A. Key Laboratory of Exploration Technologies for Oil and Gas Resources, Yangtze University,
Wuhan, Hubei, China*
Institute of Petroleum Engineering, Yangtze University, Wuhan, Hubei, China

ABSTRACT: Tight oil and gas reservoirs are mainly exploited by the method of multilayer commingled production. However, due to the physical difference of each layer and the different fluid flow rule of each layer, multilayer reservoir exploitation will show some different features with single layer reservoir exploitation. Thus, it is hard to evaluate the contribution of each layer. By using the BP neural network technique of nonlinear mapping ability, and based on the actual parameters of a tight gas reservoir, a three-layer geological model without crossflow is established. Then, we identify 5 different parameters, namely porosity, thickness H, gas saturation S_g, permeability K and the fracture length L, which are related to the production Q. We use any 20 groups for training and the method of merging and splitting to improve accuracy. Finally, we select other 5 groups for testing. The result shows that the reservoir physical property of parameters can be used to predict the yield contribution of each layer of commingled gas reservoir by using this kind of nonlinear mapping technology.

Keywords: multilayer commingled production; productivity contribution rate; BP artificial neural network; method of merging and splitting

1 INTRODUCTION

Monolayer yield of a low permeability tight gas reservoir is generally low, and cannot reach the industrial capacity, while the use of multilayer commingled production can not only increase the single well productivity, but also improve the development benefit of gas field. However, multilayer commingling production is not conducive to clearly grasp the contribution of each monolayer production, but this issue decides the success or failure of the development of the reservoir. So, the determination of each layer's contribution is significant to oil and gas field development[1-3]. Currently, people find the production of monolayer by oil testing, production logging, geological analysis and geochemical method. However, all these methods have their own defects. This paper will use the BP artificial neural network[4-5] and its nonlinear mapping ability to find out the relationship between the physical property of the reservoir and the contribution of each layer's production to achieve the predictive effect.

2 BP NEURAL NETWORK

Artificial Neural Networks (ANNs) were developed during the 2nd half of the 20th century, inspired from the biological nervous system. Artificial neural networks are nonlinear mathematical models relating inputs to output in complex nonlinear multi-variable systems. They have been successfully used for function approximation, data processing and pattern recognition in different scientific fields. Artificial neural networks are, in fact, made up of several simple processing units connected together and organized according to the network type. Because of being inspired from biological neurous, processing units are called neurous.

The application of the neural network is in great demand. Artificial neural networks include three basic processing elements: unit, network topology, and training rules. The ANN processing unit is the basic unit of operation. It simulates the functions of neurons. A processing unit possesses multiple input and output paths. The role of information transfer is to enter the ports, while outputs carry the information from one processing unit to the next. As shown in Figure 1, x is the neuron input, w is the adjustable input weight, and θ is the offset signal used for modeling the neurons excitation threshold. u(·) and f(·) are the basic function and the activation function, respectively.

Network topology determines the route and approach of information transfer between each processing unit and layer. At present, dozens of neural network topologies are available for application. During the period from 1986 to 1988,

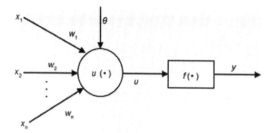

Figure 1. General neuron model.

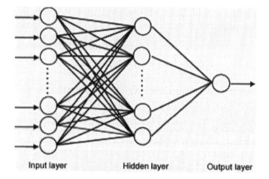

Input layer   **Hidden layer**   **Output layer**

Figure 2. The neural network model.

psychologists Mcclelland and Rumelhart proposed the famous multi-network Back Propagation (BP) algorithm. A three-layer back propagation neural network model, as shown in Figure 2, can simulate any continuous function. This algorithm is still one of the most widely used neural network learning algorithms.

Training is an essential feature of a neural network. Required accuracy is achieved using repeated training and adjustment. Weights and sums of data typically use the transfer function f(x), as does training of a network system for pattern recognition to manage the weighted value. The output value is then obtained using the conversion function.

Similar to the mechanisms of the brain, a neural network process can be divided into two primary phases to accomplish the task. The first stage is the study period in which self-improvement occurs. During this stage, a network modifies synaptic weights under certain learning rules, and works to make the measure function reach a minimum. Moreover, the state of the calculating unit remains unchanged, and all the weights can be modified by training. The second stage is the implementation period, where a neural network processes input data and generates the corresponding output data. At this point, the connection weights are fixed and the calculation unit reaches a steady-state condition.

3 SPECIFIC METHODS

3.1 *Analysis of how the parameters influence the output of each layer*

This paper is based on the model of three layers without crossflow between the layers, and using the numerical simulation method, it studies how the reservoir's physical properties affect the production of each layer.

According to the physical property of reservoir sensitivity analysis, the flow rate is related to the pressure gradient, permeability, viscosity, and cross section.

We retain the common parameters of each layer and find out the difference. We consider the parameters width, pressure gradient and cross section as the same. We identify 5 different parameters: porosity, thickness H, gas saturation S_g, permeability K and the fracture length L.

3.2 *Analysis and research on the value of the input*

After knowing the relevant parameters, it is necessary to find how to input them simply and also with high accuracy.

We select 25 sets of data from numerical simulation software for the training and prediction of the network. The information of the 25 parameters can be represented by a 4×1 matrix. Specific transformation is as follows:

$$P_i = \frac{\phi_i h_i s_{gi} k_i}{\sum \phi_i h_i s_{gi} k_i}$$

where p_i is called the storage coefficient. We use p_i and fracture length L as input parameters and the production of each layer q_i as output parameters. Then, the input layer has four neurons, intermediate hidden layer 20 and output 3, and use traingdx as the training function. Finally, we can find the mapping relationship between the input layer and the output layer, and then predict the yield percentage.

As can be seen from Figure 2, the error of group 4, 5, 15, 20, 23 and 25 is more than 20%, and the overall error is relatively large. After studying the results, we can conclude that the percentage of the yield of these groups has a smaller value, some even less than one percent. In the prediction process, the error is inevitable, and the cardinal number is small, so the results of the relative error is large, which affects the accuracy. We can also find that the disparity of the yield percentage of each layer in these groups is large, some layer between the yield ratio reaches more than 2 orders of magnitude, which makes it hard to achieve the accuracy in training. To improve this situation, we cannot eliminate the error, but we can consider how to eliminate the difference between the percentage of production. This paper will use the method of merging and splitting to reduce the difference.

3.3 *The method of merging and splitting*

The yield contribution of each layer is q_1, q_2, q_3 and the maximum value is q_{max}. Specific transformation is as follows:

$$q_1' = q_{max}/(q_{max} + q_u)$$

$$q_2' = q_{max}/(q_{max} + q_d)$$

where q_u, q_d, respectively, represent the former and the latter of q_i after removing the maximum q_{max}. For example, if q_{max} is q_2, then q_u is equal to q_1, and q_d is equal to q_3.

In the same way, p_1, p_2, p_3 can be converted into p_1', p_2' training the network by using the input values P_i 'and the output values q_i'. After obtaining the predicted value q, we can use the following formula to calculate the yield percentage of the original layers as follows:

$$q_{max}/(q_{max} + q_u) = q_1'$$

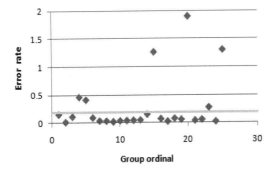

Figure 3. The network prediction error rate.

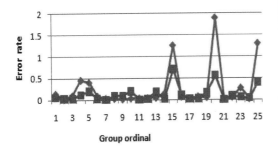

Figure 4. Error rate comparison between the previous method and the improved method.

$$q_{max}/(q_{max} + q_u) = q_2'$$

$$q_1 + q_2 + q_3 = 1$$

It is easy to calculate q_1, q_2, q_3. The following table is a error rate comparison between the previous method and the improved method.

It can be seen from Figure 3, the red line is lower than the blue line. In the 25 group forecast, only 3 groups exceeded twenty percent, and the overall error rate is low, indicating that the improved method has a high accuracy.

4 APPLICATION EXAMPLE

Based on the actual parameters of a tight gas reservoir, we establish a three-layer geological model without crossflow, and select 25 sets of

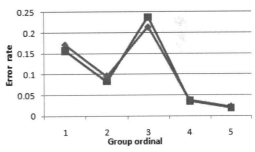

Figure 5. Comparison between the predicted and actual values.

Table 1. Parameter information.

plan	L	K	h	Sg	q
Case1	60	0.1	1.2/11.4/11.4	0.59	0.055/0.47/0.474
Case2	80	0.1	1.2/11.4/11.4	0.59	0.126/0.651/0.222
Case3	100	0.1	1.2/11.4/11.4	0.59	0.150/0.714/0.135
Case4	120	0.1	1.2/11.4/11.4	0.59	0.166/0.755/0.078
Case5	140	0.1	1.2/11.4/11.4	0.59	0.182/0.795/0.022
Case6	80	0.1	2.2/10.9/10.9	0.59	0.098/0.448/0.452
Case7	100	0.1	2.2/10.9/10.9	0.59	0.211/0.587/0.200
Case8	120	0.1	2.2/10.9/10.9	0.59	0.247/0.632/0.119
Case9	140	0.1	2.2/10.9/10.9	0.59	0.271/0.661/0.068
Case10	60	0.1	2.2/10.9/10.9	0.59	0.293/0.687/0.019
Case11	100	0.1	4.8/9.6/9.6	0.59	0.205/0.396/0.398
Case12	120	0.1	4.8/9.6/9.6	0.59	0.390/0.455/0.154
Case13	140	0.1	4.8/9.6/9.6	0.59	0.440/0.471/0.088
Case14	60	0.1	4.8/9.6/9.6	0.59	0.470/0.480/0.049
Case15	80	0.1	4.8/9.6/9.6	0.59	0.497/0.489/0.013
Case16	120	0.1	7/8.5/8.5	0.59	0.293/0.352/0.353
Case17	140	0.1	7/8.5/8.5	0.59	0.507/0.367/0.124
Case18	60	0.1	7/8.5/8.5	0.59	0.559/0.371/0.069
Case19	80	0.1	7/8.5/8.5	0.59	0.588/0.375/0.037
Case20	100	0.1	7/8.5/8.5	0.59	0.615/0.375/0.01
Case21	140	0.1	8.0/8.0/8.0	0.59	0.333/0.333/0.333
Case22	60	0.1	8.0/8.0/8.0	0.59	0.554/0.333/0.112
Case23	80	0.1	8.0/8.0/8.0	0.59	0.604/0.333/0.062
Case24	100	0.1	8.0/8.0/8.0	0.59	0.632/0.333/0.033
Case25	120	0.1	8.0/8.0/8.0	0.59	0.657/0.334/0.008

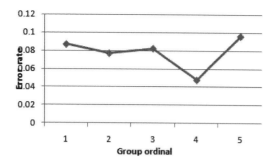

Figure 6. The network prediction error rate.

data, using 20 sets of data for training and five for prediction.

As shown below, the maximum error rate is less than 14%, and the value is mainly from the layer that has the minimum output.

5 CONCLUSIONS

1. Storage coefficient is used to simplify the input value (porosity φ, thickness h, gas saturation sg, and permeability k) of the BP neural network. The input layer has four neurons, intermediate hidden layer 20 and output 3, and uses traingdx as the training function. Finally, we can find the mapping relationship between the input layer and the output layer, and then predict the yield percentage.
2. By using the method of merging and splitting, the disparity of the yield percentage of each layer is reduced, and, finally, the calculation accuracy of the BP neural network is improved.
3. The results show that the method can correctly predict the yield contribution of each layer, and has a high accuracy, which adds a new idea for calculating the yield contribution of each layer of a multilayer gas well.

ACKNOWLEDGMENTS

The authors thank the support provided by the Scientific Research and Technological Development Projects of China National Petroleum Corporation.

REFERENCES

[1] CHAI Rui, LI Zhenduo Commingled gas production technology of Changqing gas field with composite gas reservoir [J]. Natural Gas Industry, 2002, 22(2): 104–1061.
[2] XU Xianzhi, KUANG Guohua, CHEN Fenglei, et al A method of pressure analyzing for multi layered commingled reservoir [J]. Acta Petrolei Sinica, 1999, 20(5): 43–471.
[3] WANG Xiaodong, LIU Ciqun Productivity analysison commingled production wells [J]. Oil Drilling & Production Technology, 1999, 21(2): 56–611.
[4] WU Yan, WANG Shou-jue. A new algorithm of improving the learning performance of neural network by feedback [J]. Journal of Computer Research and Development, 2004, 41(9): 1488–1492.
[5] WU Wei, WANG Jian, CHENG Ming-song, et al. Convergence analysis of online gradient method for BP neural networks [J]. Neural Networks, 2011(24): 91–98.

Architectural, Energy and Information Engineering – Sung & Chen (Eds)
© 2016 Taylor & Francis Group, London, ISBN 978-1-138-02791-6

The approximate realization of mini-max algorithm for measurement matrices in compressed sensing

Bing Jie Li, Xu Wei Li & Long Yan
School of Science, Air Force Engineering University, Xi'an, China

ABSTRACT: For fixed orthogonal basis, we propose a new approximate algorithm for the design of measurement matrices in compressed sensing based on the incoherence rule between the measurement matrices and sparse basis. According to the definition of incoherence, the mini-max model of obtaining the measurement matrix, together with the approximate algorithm of solving this model, is established in this paper. The obtained measurement matrix is incoherent with the orthogonal basis. The minor the coherence, the less the required measurement dimensions in the process of compressed sampling, the more information contained in the original signal, and the higher the probability of re-construction. This method is of important application values in signal processing. Finally, considering the discrete cosine basis as the fixed orthogonal basis, a numerical example is presented to verify the effectiveness of the method.

Keywords: compressed sensing; sparse basis; measurement matrix; mini-max method

1   INTRODUCTION

Constructing the stable measurement mode of decreasing dimension being incoherent with transformation, which guarantees that the information is not destroyed in the compressed process, is the core issues among the three problems in the compressed sensing theory. In the literature [4][9], we know that the designing of measurement matrices is the core problem in the compressed sampling theory, which determines directly whether it is realized successfully. Presently, the measurement matrices used in the compressed sensing theory are mostly the following cases: Gaussian random matrices in [6][7], Bernoulli matrices in [1], Fourier random matrices in [1], uniformly spherical matrices in [1], and so on. However, all the above measurement matrices only ensure a high probability to restore the signal. Since the signal reconstruction problem is an optimal problem based on given underdetermined system of equations as a restriction condition, the Restricted Isometry Property (RIP) and incoherence criterion with sparse basis are very important which can be seen in [1][2]. In compressed sensing theory, from the literature [3] [5], we can obtain that the constraint of measurement matrices is loose; therefore, Donoho presents three necessary conditions of measurement. In the literature 8, the author has studied the design algorithm of optimal measurement matrices by using the incoherence criterion, and presented mini-max algorithm base on incoherence criterion for the

case of K-sparse or compressible signals. However, the mini-max algorithm only for low-dimensional case is feasible. This paper presents a measurement matrix design methodology from low to high expansion, which stays a little coherence between the high-dimensional measurement matrix and any fixed orthogonal.

2   THE MINI-MAX METHOD OF DESIGN OF MEASUREMENT MATRICES BASED ON INCOHERENCE CRITERION

Let $x \in R^N$ be an original date, and an orthogonal basis satisfying $x = \Psi\Theta$ be $\Psi \in R^{N \times N}$, where Θ is the sparse projection satisfying $x = \Psi\Theta$. The matrices A^{CS} is called information operator if the matrices Φ is the measurement matrix and y is the measurement satisfying $y = \Phi x \in R^M$ or $y = \Phi\Psi\Theta = A^{CS}\Theta$, where $M \ll N$. In the compressed sensing theory, sampling rate is not determined by a signal wideband but the two fundamental criterions of sparsity and incoherence to some extent.

Definition 1 The coherence of orthogonal basis pair (Φ, Ψ) is defined as follows

$$\mu(\Phi, \Psi) = \sqrt{N} \max_{1 \le k, j \le N} |\langle \Phi_k, \Psi_j \rangle| \qquad (1)$$

if (Φ, Ψ) is a pair of $N \times N$ orthogonal basis satisfying the following conditions

(1) $\Phi^T \Phi = I$; (2) $\Psi^T \Psi = I$, where $\tilde{A}^{CS} = \Phi\Psi$,

$$\mu(\tilde{A}^{cs}) = \sqrt{N} \max_{1 \le k, j \le N} | \tilde{A}^{cs}_{k,j} |.$$

Theorem 1 (see the literature [8][9]). Suppose that a pair of $N \times N$ orthogonal basis is (Φ, Ψ), then $\mu(\Phi, \Psi) \ge 1$ or $\mu(\tilde{A}^{cs}) = \sqrt{N} \max_{1 \le k, j \le N} | \tilde{A}^{cs}_{k,j} | \ge 1$

By this theorem, the norm of every row and column of $\tilde{A}^{CS}$ is 1.

In ideal case, when the elements of per row of $\tilde{A}^{cs}$ is all $\pm 1/\sqrt{N}$, we can obtain that $\mu(\Phi, \Psi) = 1$ meaning that incoherence between Ψ and Φ is the largest, and they are not denoted sparsely each other, which helps signal reconstruction and compressed sampling in high efficiency. Oppositely, when there is only one component of per row vector is 1 and the others are 0, we have $\mu(\Phi, \Psi) = \sqrt{N}$ meaning that the elements of per row highly agglomerate and the probability of original signal is 0 by compressed observation and precise reconstruction, hence, we obtain that $1 \le \mu(\Phi, \Psi) \le \sqrt{N}$.

Theorem 2 (see the literature [9]). Suppose that original date x is k-spars in Ψ filed, if we select M vectors as measurement $\tilde{\Phi}$, and assumed that

$$M \ge C \cdot \mu^2(\Phi, \Psi) \cdot K \cdot \log N \qquad (2)$$

where C is some fixed positive constant, then the problem of compressed sensing reconstruction can be transformed into the following optimal problem with constraints

$$\min_{\Theta \in R^N} | A^{cs}\Theta |_1 \quad s.t \ A^{cs}\Theta = y \qquad (3)$$

whose solution Θ^* is equal to the exact solution Θ of the following problem in high probability

$$y = \tilde{\Phi}x, x = \Psi\Theta$$

Particularly, suppose that $\mu(\Phi, \Psi) = 1$, we can precisely restructure original signal in high probability by just only $O(K \cdot \log N)$ observations. The theorem 2 is just suitable for the k-sparse case, but in compressible situation, it is also suitable for (Φ, Ψ) when the incoherence orthogonal basis is composed of random orthogonal basis Φ and arbitrary fixed orthogonal basis Ψ. The less the coherence $\mu(\Phi, \Psi)$ of Ψ and Φ is, the smaller about the measurement required by the compressed sampling, which means the more information of x included by compressed measurement y. By searching the orthogonal basis Φ which is highly inherent with the fixed orthogonal basis Ψ, we can establish the following problem

$$\min \max_{1 \le k, j \le N} \sqrt{N} | \langle \Phi_k, \Psi_j \rangle | \qquad (4)$$

where $\Phi_k = (\phi_{1k}, \phi_{2k}, \cdots, \phi_{Nk})^T$, $k = 1, 2, \cdots, N$, and $\Phi = (\Phi_1, \Phi_2, \cdots, \Phi_N)$ satisfying $\Phi\Phi^T = I$.

Clearly, the problem (4) is a mini-max problem.

Theorem 3 (see the literature [9]). The mini-max problem (4) that searches the orthogonal basis Φ highly inherent with the fixed $N \times N$ orthogonal basis Ψ is equivalent to the following problem by introducing a positive real number S and regarding S, $\phi_{1k}, \phi_{2k}, \cdots, \phi_{Nk}$, ($k = 1, 2, \cdots, N$) as variables

$$\begin{cases} \min S \\ S.t \ \sqrt{N} |\langle \Phi_k, \Psi_j \rangle| \le S, \quad 1 \le k, j \le N \\ \Phi\Phi^T = I \\ S \ge 1 \end{cases} \qquad (5)$$

By the theorem 3, we can obtain the orthogonal basis Φ which is highly incoherent with fixed orthogonal basis Ψ, and also attain the value of optimal coherence S by solving the problem (5). In Φ field, we stochastically and uniformly choose M vectors satisfying (2) as measurement, which accomplishes the highest possible probability by using the measurements as little as possible.

The author of this paper has validated the theoretical feasibility of the mini-max method of design of measurement matrices based on incoherence criterion by the numerical example in which the fixed orthogonal basis Ψ is discrete cosine basis in low dimension case in reference 6, then gained the orthogonal basis Φ which is highly incoherent with Ψ, when $N = 9$, we can gain $S^* = 1.1940$. However, we must emphasize the method introduced in this paper is just only a feasible method which has been different with actual application yet. When the original dates is in high dimension, the problem (5) is a large-scale problem, for example, if $N = 1000$, the problem (5) totally contains 106 unknown variables and 2500501 constraint conditions; if N = 10000, it contains more unknown variables and constraint conditions, which must construct the correspondingly heuristic algorithm to solve the problem. Since the constraint condition is nonlinear, it is difficult to construct the heuristic algorithm.

3 APPROXIMATION ALGORITHM

The mini-max problem (4) that searches the orthogonal basis Φ highly inherent with the fixed orthogonal basis Ψ is minimizing the maximum of the information operator $\tilde{A}^{CS} = \Phi\Psi$ by searching an orthogonal basis Φ. $\tilde{A}^{CS}$ is also orthogonal matrices, let $\tilde{A}^{cs} = (a_{ij})_{N \times N}$, if we find a orthogonal matrices $\tilde{A}^{CS}$ which make $\min \max_{1 \le i, j \le N} | a_{ij} |$, then the orthogonal basis $\Phi = \tilde{A}^{CS}\Psi^T$ which is highly inherent with the fixed orthogonal basis Ψ. The process searching a orthogonal matrices $\tilde{A}^{CS}$ which make $\min \max_{1\le i,j \le N} |a_{ij}|$ has nothing to do with orthogonal basis Ψ. By searching the $\tilde{A}^{CS}$, we can establish the following problem

$$\begin{cases} \min s \\ S.t \, |a_{ij}| \le s, \quad 1 \le i,j \le N \\ \sum_{j=1}^{N} a_{ij}a_{kj} = 0, \quad 1 \le i,k \le N, i \ne k \\ \sum_{j=1}^{N} a_{ij}a_{kj} = 1, \quad 1 \le i,k \le N, i = k \end{cases} \qquad (6)$$

When the dimension N is small, we can get orthogonal matrices $\tilde{A}^{CS}$ by solving the problem (6), then can get $\Phi = \tilde{A}^{CS}\Psi^T$ and select M vectors as measurement matrices further. However, when the dimension N is big, solving the problem (6) will be very complicated and hard to realize, so can we expand the dimension from low to high? That is getting orthogonal matrices $\tilde{A}^{CS}$ in low dimension by solving the problem (6) frist, then expanding the dimension from low to high.

Theorem 4 If $N \times N$ matrices $2N \times 2N$ is orthogonal matrices, then $\tilde{A}^{cs}_{2N\times 2N} = \frac{1}{\sqrt{2}}\begin{pmatrix} \tilde{A}^{cs} & \tilde{A}^{cs} \\ \tilde{A}^{cs} & -\tilde{A}^{cs} \end{pmatrix}$ block matrices is also orthogonal

Proof. Let $\tilde{A}^{cs}_{2N\times 2N} = (\bar{a}_{ij})_{2N\times 2N}$, Then

$$\tilde{A}^{cs}_{2N\times 2N} \tilde{A}^{cs}_{2N\times 2N}{}^T = \frac{1}{\sqrt{2}}\begin{pmatrix} \tilde{A}^{cs} & \tilde{A}^{cs} \\ \tilde{A}^{cs} & -\tilde{A}^{cs} \end{pmatrix} \cdot \frac{1}{\sqrt{2}}\begin{pmatrix} \tilde{A}^{csT} & \tilde{A}^{csT} \\ \tilde{A}^{csT} & -\tilde{A}^{csT} \end{pmatrix}$$

$$= \frac{1}{2}\begin{pmatrix} 2\tilde{A}^{cs}\tilde{A}^{csT} & 0 \\ 0 & 2\tilde{A}^{cs}\tilde{A}^{csT} \end{pmatrix} = \begin{pmatrix} I_{N\times N} & 0 \\ 0 & I_{N\times N} \end{pmatrix} = I_{2N\times 2N}$$

By the theorem 4, if we find the orthogonal matrices A^{CS} in low dimension which satisfy $\min \sqrt{N} \max_{1\le i,j \le N} |a_{ij}|$, we have

$$\sqrt{2N} \max_{1\le i,j \le 2N} |\bar{a}_{ij}| = \min \sqrt{N} \max_{1\le i,j \le N} |a_{ij}|$$

to the expanded $2N \times 2N$ orthogonal matrices. For any fixed orthogonal basis $\Psi_{N\times N}$(known matrices), let $\Phi_{N\times N} = \tilde{A}^{cs}_{N\times N}\Psi^T_{N\times N}$. For any fixed orthogonal basis $\Psi_{2N\times 2N}$, let $\Phi_{2N\times 2N} = \tilde{A}^{cs}_{2N\times 2N}\Psi^T_{2N\times 2N}$, because the expanded orthogonal matrices are only the feasible answer of the problem 6. The coherence is as follow

$$\mu(\Phi_{2N\times 2N}, \Psi_{2N\times 2N}) = \mu(\Phi_{N\times N}, \Psi_{N\times N})$$

$\Phi_{N\times N}$ and $\Psi_{N\times N}$ is least coherence, but $\Phi_{2N\times 2N}$ and $\Psi_{2N\times 2N}$ is not always best incoherence.

Considering $N = 2$, $\tilde{A}^{cs}_{2\times 2} = \frac{1}{\sqrt{2}}\begin{pmatrix} 1 & 1 \\ 1 & -1 \end{pmatrix}$ can be get from the problem 6. For any fixed 2×2 orthogonal basis $\Psi_{4\times 4}$ (known), let $\Phi_{2\times 2} = \tilde{A}^{cs}_{2\times 2}\Psi^T_{2\times 2}$, then $\mu(\Phi_{2\times 2}, \Psi_{2\times 2}) = 1$, let

$$\tilde{A}^{cs}_{4\times 4} = \frac{1}{\sqrt{2}}\begin{pmatrix} \tilde{A}^{cs}_{2\times 2} & \tilde{A}^{cs}_{2\times 2} \\ \tilde{A}^{cs}_{2\times 2} & -\tilde{A}^{cs}_{2\times 2} \end{pmatrix} = \frac{1}{2}\begin{pmatrix} 1 & 1 & 1 & 1 \\ 1 & -1 & 1 & -1 \\ 1 & 1 & -1 & -1 \\ 1 & -1 & -1 & 1 \end{pmatrix}$$

For any fixed 4×4 orthogonal basis $\Psi_{4\times 4}$, let $\Phi_{4\times 4} = \tilde{A}^{cs}_{4\times 4}\Psi^T_{4\times 4}$, then $\mu(\Phi_{4\times 4}, \Psi_{4\times 4}) = 1$.

4 NUMERICAL EXAMPLE

Set fixed orthogonal basis Ψ is discrete cosine basis (see the literature[8] [9]). $\Psi = (\psi_{mn})_{N\times N}$, where

$$\psi_{mn} = w(m)\cos\left(\frac{\pi(2n-1)(m-1)}{2N}\right) \qquad (7)$$

$$w(m) = \begin{cases} \sqrt{\dfrac{1}{N}}, & m = 1 \\ \sqrt{\dfrac{2}{N}}, & 2 \le m \le N \end{cases} \qquad (8)$$

When $N = 4$, then

$$\Psi_{4\times 4} = \begin{pmatrix} 0.5000 & 0.5000 & 0.5000 & 0.5000 \\ 0.6533 & 0.2706 & -0.2706 & -0.6533 \\ 0.5000 & -0.5000 & -0.5000 & 0.5000 \\ 0.2706 & -0.6533 & 0.6533 & -0.2706 \end{pmatrix}$$

$$\Phi_{4\times 4} = \tilde{A}^{cs}_{4\times 4}\Psi^T_{4\times 4}$$

$$= \begin{pmatrix} 1.0000 & 0.0000 & 0.0000 & 0.0000 \\ 0.0000 & 0.3827 & 0.0000 & 0.9239 \\ 0.0000 & 0.9239 & 0.0000 & -0.3827 \\ 0.0000 & 0.0000 & 1.0000 & 0.0000 \end{pmatrix}$$

By that analogy, for any fixed $2^n \times 2^n$ orthogonal basis Ψ (known matrices, n is a positive integer), we always start from $\tilde{A}^{cs}_{2\times 2}$, then obtain the corresponding $2^n \times 2^n$ orthogonal matrices $\tilde{A}^{cs}$ according to the expansion formula of Theorem 4. Let $\Phi = \tilde{A}^{cs}\Psi^T$, then $\mu(\Phi, \Psi) = 1$. On the part of Compressed Sensing theory, when original sampled number is 2^N, we can often find $2^n \times 2^n$ orthogonal matrices $\tilde{A}^{cs}$. Selecting M vectors as the measurement matrices in $\Phi = \tilde{A}^{cs}\Psi^T$, then Ψ and Φ have the largest incoherence, meaning that they are not denoted sparsely each other which helps signal reconstruction and compressed sampling in high efficiency. When original sampled number is not 2^n, we not always find measurement matrices Φ that make $\mu(\Phi, \Psi) = 1$, but can obtain the better incoherence measurement matrices by the idea of expansion of dimension from low to high.

Considering $N = 3$, $\tilde{A}_{3\times3}^{cs} = \dfrac{1}{3}\begin{pmatrix} 2 & -1 & 2 \\ 2 & 2 & -1 \\ 1 & -2 & -2 \end{pmatrix}$ can be get from the problem 6. For any fixed 3×3 orthogonal basis $\Psi_{3\times3}$, then $\mu(\Phi_{3\times3}, \Psi_{3\times3}) \approx 1.14$.

$\tilde{A}_{3\times3}^{cs}$ can expand to $\tilde{A}_{6\times6}^{cs}$, $\tilde{A}_{12\times12}^{cs}$, $\tilde{A}_{24\times24}^{cs}$, $\tilde{A}_{48\times48}^{cs}$, $\tilde{A}_{96\times96}^{cs}$ and more higher dimensions by Theorem 4, such that coherence between the corresponding sparse basis and measurement matrices is about 1.14.

Similarly, we have $\mu(\Phi_{6\times6}, \Psi_{6\times6}) \approx 1.095 < 1.14$.

So in the expansion to high dimensions, the measurement matrices from higher low dimensions are better than from lower low dimensions. However, not coherence of the higher low dimensions is always better than the lower low dimensions.

Considering $N = 5$, we can get orthogonal matrices $\tilde{A}_{5\times5}^{cs}$ by solving the problem 6, here

$$\tilde{A}_{5\times5}^{cs} = \frac{1}{11}\begin{pmatrix} -6 & 3 & 6 & 6 & -2 \\ 6 & 6 & 3 & -2 & -6 \\ 6 & -2 & 6 & 3 & 6 \\ 2 & 6 & -6 & 6 & 3 \\ -3 & 6 & 2 & -6 & 6 \end{pmatrix}$$

The discrete cosine basis $\Psi_{5\times5}$ is the matrix

$$\begin{pmatrix} 0.4472 & 0.4472 & 0.4472 & 0.4472 & 0.4472 \\ 0.6015 & 0.3717 & 0.0000 & -0.3717 & -0.6015 \\ 0.5117 & -0.1954 & -0.6325 & -0.1954 & 0.5117 \\ 0.3717 & -0.6015 & -0.0000 & 0.6015 & -0.3717 \\ 0.1954 & -0.5117 & 0.6325 & -0.5117 & 0.1954 \end{pmatrix}$$

$\Phi_{5\times5} = \tilde{A}_{5\times5}^{cs}\Psi_{5\times5}^{T}$ is the matrix

$$\begin{pmatrix} 0.2846 & -0.3201 & -0.8770 & 0.0289 & -0.2158 \\ 0.2846 & 0.9265 & -0.2436 & -0.0319 & -0.0136 \\ 0.7725 & -0.1690 & 0.1954 & 0.2734 & 0.5117 \\ 0.4472 & -0.0547 & 0.3643 & -0.0338 & -0.8143 \\ 0.2033 & -0.0866 & 0.0246 & -0.9603 & 0.1683 \end{pmatrix}$$

$\mu(\Phi_{5\times5}, \Psi_{5\times5}) = \sqrt{5} \cdot \dfrac{6}{11} = 1.2197 > 1.095.$

In the same way, considering $N = 7$, then $\mu(\Phi_{7\times7}, \Psi_{7\times7}) = \sqrt{7} \cdot 0.4438 = 1.1742$, the value is larger than $\mu(\Phi_{6\times6}, \Psi_{6\times6}) = 1.095$, but smaller than $\mu(\Phi_{5\times5}, \Psi_{5\times5}) = 1.2197$. Finding the statistical

regularities, we find, the coherence of even-order is smaller than of odd-order. So generally speaking, in the compressed sensing, selecting even dimension sampling is superior to odd dimension sampling.

5 CONCLUSIONS

In this paper, the author presents the mini-max method of design of measurement matrices based on incoherence criterion in the literature [8], but this method needs complex calculation, and is too hard to complete for higher dimensions. This paper gives the approximate realize method, which if only to get the lower dimension measurement matrices expand to high dimension for the low dimension expand to high dimension, it is easy to obtain the low dimension measurement matrices. For example, for known fixed sparse orthogonal basis, it is easy to get the 7 dimensions measurement matrices, whose optimal coherence is 1.1742. For this sparse orthogonal basis, if the sampling number is 2^n times the number 7, then the coherence stays 1.1742 meanwhile the measurement matrices expand to the corresponding high dimension. It is worth noting that the dimension is 2^n times the number 7, the optimal coherence is superior to 1.1742, which is hard to achieve. Even so, for the much high dimension, the coherence of the designed measurement matrices in this paper is much smaller than that of the common method (between 1.4 and 3.0).

REFERENCES

[1] Blanchard, J.D., Cartis, C. & Tanner, J. 2009. Decay properties of restricted isometry constants. *IEEE Signal Processing Letters* 16(7): 572–575.

[2] Candès, E. & Romberg, J. 2007. Sparsity and incoherence in compressive sampling. *Inverse Problems* 23(3): 969–985.

[3] Candès, E. & Tao, T. 2006. Near optimal signal recovery from random projections: Universal encoding strategies. *IEEE Trans. Info. Theory* 52(12): 5406–5425.

[4] Donoho, D.L. 2006. Compressed sensing. *IEEE Trans. Information Theory* 52(4): 1289–1306.

[5] Donoho, D.L., Tsaig, Y. & Johnson, H.L. 2006. Extensions of compressed sensing. *Signal Processing* 86(3): 533–548.

[6] Eldar. Y. & Mishali, M. 2009. Robust recovery of signals from a structured union of subspaces. *IEEE Trans. Inf. Theory* 55(11): 5302–5316.

[7] Guo, Y. & Yang, Z. 2009. The compressed sensing of speech signal based on approximate KLT field. *Electron & Information transaction.* 2(31): 2948–2952.

[8] Li, B.J., Lv, Y., Ye, M. & Li, G.F. 2011. The Mini-max method of design of measurement matrices for compressed sensing based on incoherence criterion. *Journal of Force Engineering University* 12(5): 81–84.

[9] Sun, L.H. & Yan, Z. 2010. The distributed speech compression and reconstruction based on compressed sensing. *Signal Processing* 26(6): 824–829.

Architectural, Energy and Information Engineering – Sung & Chen (Eds)
© 2016 Taylor & Francis Group, London, ISBN 978-1-138-02791-6

On-site PD test technology of converter transformer in UHVDC

Y.Q. Li, J.H. Han, W. Wang, H.B. Zheng, X.G. Li & Y.B. Shao
State Grid Henan Electric Power Research Institute, Zhengzhou, Henan Province, China

ABSTRACT: The four types of converter transformers used in the ±800 kV UHVDC project of Hami-Zhengzhou are introduced in this paper. First, this study presents the technical parameters of converter transformers and field test requirements of partial discharge for the converter transformer-based regulations and standards. Second, the test procedures of partial discharge for converter transformers in UHVDC are given in detail in this study. Then, the rated voltage and rated capacity of the test equipment are verified to meet the test conditions by theoretical analysis. Finally, the test results verify the theoretical analysis. The test data and field test experience in this study are very helpful for future partial discharge of converter transformers in theUHVDC project.

Keywords: ultra high voltage DC; converter transformer; partial charge; assessment

1 INTRODUCTION

The largest DC transmission project of the world is the Hami-Zhengzhou UHVDC, whose maximum transmission power is 8000 MW, rated voltage is ±800 kV, and rated current is 5000 A.

Insulation of converter transformer is the key character, bearing the AC and, especially, the DC voltage. It is important to perform insulation experiments to ensure transformer safety, and the ACLD and PD test is the most important experiment in hand-over experiments of converter transformer.

Partial discharge is the discharge that presented in poles though not run-through that because of the weakness existing inside the insulation or production process, and dielectric was poor and even broken down when PD existed for a long time. The test of PD is effective on evaluating the insulation of the transformer or preventing problems occurring during production and installation.

2 PARAMETERS AND REQUIREMENTS

2.1 Parameters of converter transformers

In the UHVDC transmission project, there are 24 transformers divided into four types based on different DC voltages of valve halls with each capacity being 376.6 MVA. Y/Y-800, Y/Δ-600, Y/Y-400, and Y/Δ-200 were different from each other in relation to the ratio, no-load loss and entrance capacitance, as presented in Table 1.

2.2 Requirements of the ACLD and PD test

The procedure for the partial discharge test on-line side windings is shown in Figure 1. It refers to DL/T274–2012 and GB 1094.3–2003.

Here,

As $U_1 = 1.1 U_m / \sqrt{3}, U_2 = 1.3 U_m / \sqrt{3}, U_3 = 1.5 U_m / \sqrt{3}$,

and $Um = 550$ kV.

The voltages of different transformer valve windings are presented in Table 2.

Table 1. Parameters of converter transformer.

Parameters	Y/Y-800	Y/Δ-600	Y/Y-400	Y/Δ-200
Type	ZZDFPZ-376600/500–800	ZZDFPZ-376600/500-600	ZZDFPZ-376600/500-400	ZZDFPZ-376600/500-200
Rated voltage	$\dfrac{525}{\sqrt{3}} / \dfrac{159.8}{\sqrt{3}}$	$\dfrac{525}{\sqrt{3}} / 159.8$	$\dfrac{525}{\sqrt{3}} / \dfrac{159.8}{\sqrt{3}}$	$\dfrac{525}{\sqrt{3}} / 159.8$
No-load loss	173.61	155	173.61	155

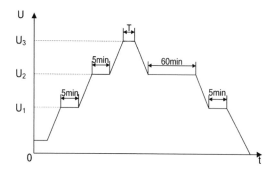

Figure 1. Voltage increasing procedure for the partial discharge test of the UHV converter transformer.

Table 2. Test voltages of different converter transformers.

Type	Test stage	Voltage of line winding (KV)	Voltage of valve windings (KV)
Y/Y	U1	476	145
	U2	413	126
	U3	349	106
Y/Δ	U1	476	251
	U2	413	218
	U3	349	184

3 SCHEME OF THE ON-SITE EXPERIMENT

The following two methods for boast were chosen, which refers to the difference in valve winds: unilateral boast and symmetrical boast.

3.1 Unilateral boast

The Y-Y converter transformer is boasted directly because of the low test voltage on valve windings, as shown in Table 2. It has a little difference with the normal power transformer PD test, except the interference that results from coronal generated by the high voltage; meanwhile, the capacity of test equipments is considered. The voltage applying mode for the PD test in the Y-Y converter transformer is shown in Figure 2.

3.2 Symmetrical boast

Because of the difference in the ratio, voltage boasted on Y-Δ valve windings is very high, effectively resulting from the coronal generated by the test line and equipments are increased, and the symmetrical boast is chosen to avoid these factors, as shown in Figure 3.

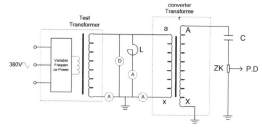

Figure 2. Voltage applying mode for the partial discharge test in the Y-Y converter transformer.

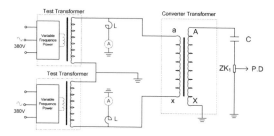

Figure 3. Voltage applying mode for the partial discharge test in the Y-Δ converter transformer.

4 EQUIPMENTS AND PARAMETERS

4.1 Variable frequency power

The power of the variable frequency cabinet is 450 kW, and the range of frequency is 30 Hz~300 Hz. It is beneficial to decrease the capacity of compensation as the frequency regulates continuously when the whole test circuit resonates.

4.2 Test transformer

The choice of test transformer rated voltage refers to the biggest voltage in the experiment proceed as the unilateral boast and the symmetrical boast. The capacity of the test transformer is 450 kvar and the rated voltage of high winding is 200 kV. There are three rated voltages of low winding, namely 300 V, 350 V, and 400 V, and the range of frequency is 100 Hz~300 Hz.

4.3 Compensation reactor

Entrance capacitance was calculated from the data of the delivery test, and then the value of the compensation reactor could be determined as the variable frequency power working at the range of frequency.

Referring to those factors, the capacity of the fixed compensation reactor is 2000 kVar, the rated voltage is 200 kV, the rated current is 10 A, and the inductance is 20 H.

Considering that the entrance capacitances of valve winding are different when the symmetrical boast, adjustable reactor is needed in order to keep the two terminals of valve winding in full compensation, the circulating current is avoided because of the different compensations between the terminals of valve winding. The capacity of the adjustable reactor is 500 kvar, the rated voltage is 50 kV, the rated current is 10 A, and the range of the inductance is 1~5 H.

5 ANALYSIS OF THE TEST DATA

5.1 Verification of test equipments

The entrance capacitance of the transformer is 62300 pF calculated from the delivery test data. This is the foundation data to check the equipments whether or not they are suitable for test under $1.3 U_m / \sqrt{3}$.

5.1.1 Frequency in test
Considering that the compensation reactor $L = 20$ H, the frequency f when completely compensated is as follows:

$$f = \frac{1}{2\pi\sqrt{LC}}$$
$$= \frac{1}{2 \times 3.14 \times \sqrt{20 \times 62300 \times 10^{-12}}} = 142.5 Hz \quad (1)$$

5.1.2 Power in test
1. Active power

The active power of the test transformer under the test voltage is as follows:

$P \Delta (K \times f_N/f_S)^{1.9} \times (f_S/f_N)^{1.6} \times P_0$
K——multiple of voltage
f_N——rated frequency, Hz
f_S——test frequency, Hz
P_0——no-load loss, kW
When f = 142.5 Hz, K = 1.3 × 550/525 = 1.362:

$$P = (K \times f_N / f_S)^{1.9} \times (f_S / f_N)^{1.6} \times P_0$$
$$= (1.362 \times \frac{50}{142.5})^{1.9} \times (\frac{142.5}{50})^{1.6} \times 173.61 = 235.2 kW$$
$$(2)$$

The capacitive reactive power of the converter transformer valve side is as follows:

$$Q_C = U^2 \times 2\pi f C$$
$$= 126000^2 \times 2 \times 3.14 \times 142.5 \times 62300 \times 10^{-12}$$
$$= 885 k \text{ var} \quad (3)$$

The inductive reactive power of the compensation reactor is as follows:

$$Q_L = U^2 / 2\pi f L = 885 k \text{ var} \quad (4)$$

3. Current in test

The capacitive current of the converter transformer valve side under U_2 is as follows:

$$I_C = U \times 2\pi f C$$
$$= 126000 \times 2 \times 3.14 \times 142.5 \times 62300 \times 10^{-12}$$
$$= 7.02 A \quad (5)$$

The resistive current of the converter transformer valve side under U_2 is as follows:

$$I_R = P / U = 235200 / 126000 = 1.87 A \quad (6)$$

The total current that inflow in the converter transformer valve side under U_2 is as follows:

$$I = \sqrt{I_R^2 + I_C^2} = \sqrt{1.87^2 + 7.02^2} = 7.26 A \quad (7)$$

The compensation current of the reactor is as follows:

$$I_L = U / 2\pi f L = \frac{126000}{2 \times 3.14 \times 142.5 \times 20} = 7.04 A \quad (8)$$

The current out of the variable frequency power is $I_R^{\boxtimes} = I_R \times 200000 / 500 = 748 A$, as the ratio of the test transformer is 400. All the parameters of equipments are suitable for ACLD of converter transformer Y/Y-800.

5.2 Comparison of the experimental result

Each of the experimental data was recorded when ACLD of Y/Y-800 at the U_2 stage, as shown in Table 3.

Entrance capacitive is calculated from the formulas mentioned above. The output voltage of variable frequency power is bit lower than the result of the theoretical calculation due to the capacitance-heightened voltage. The difference between the current and the frequency is inexact because of the estimation on entrance capacitance.

Table 3. Test voltages of different converter transformers.

Valve voltage (kV)	Voltage (V)	Current (A)	Frequency (Hz)	Entrance capacitive (pF)
126	301	730	143.6	63577

Table 4. Test data of the Y/Δ-600 converter transformer.

Valve voltage (kV)	Voltage (V)	Current (A)	Frequency (Hz)	Entrance capacitance (pF)
218	262	610	177.5	40726

Table 5. Test data of the Y/Y-400 converter transformer.

Valve voltage (kV)	Voltage (V)	Current (A)	Frequency (Hz)	Entrance capacitance (pF)
126	302	537	153.3	55088

Table 6. Test data of the Y/Δ-400 converter transformer.

Valve voltage (kV)	Voltage (V)	Current (A)	Frequency (Hz)	Entrance capacitance (pF)
218	261	536	162.1	48625

5.3 Experimental data of all transformers

The accuracy of the theoretical calculation is proofed in the on-site experiment as the example of Y/Y-800, and the experimental data of other types converter transformers are presented in Tables 4–6.

6 CONCLUSIONS

The ACLD and partial discharge test of the converter transformer in the UHVDC project is researched in this paper. Two methods of boast are presented for different types of the converter transformer. The adaptability of equipments was checked by taking the Y/Y-800 as an example, and the accuracy was proofed in the on-site test, and the entrance capacitance of four types of transformer was calculated based on the on-site test data. Thus, it is the best reference for future study on the converter transformer in the project.

REFERENCES

[1] GB 1094.3 Power transformer, part3: insulation level, insulation test and external insulation air gap [S], 2003.
[2] DL/T 274–2012 Hand-over test of ±800 kV HVDC electric equipments.
[3] SUN You-liang, WANG Qing-pu, LI Wen-ping. R&D of converter transformer for ±800 kV DC transmission project [J]. Power Equipment, 2006, 7(16): 17–20.
[4] LI Wen-ping, CHEN Zhiwei, SONG Xiusheng. ±800 kV HVDC converter transformers engineering structure of the main insulation [J]. Electrical Equipment, 2007, 8(3): 1–7.
[5] IEC 60076-3, Power transformers-Part 3: Insulation levels, dielectric tests and external clearances in air [S], 2000.
[6] IEC 60076-5, Power transformers-Part 5: Ability to withstand short circuit [S], 2006.
[7] LIU Fan, ZHANG Jun, YAO Xiao, et al. Recognition of PD mode based on KNN algorithm for converter transformer [J]. Electric Power Automation Equipment, 2013, 33(5): 89–93.
[8] CHEN Xiao lin, CAVALLINI A, MONTANARI G C. Improving high voltage transformer reliability through recognition of PD in paper/oil systems [C]. 2008 International Conference on High Voltage Engineering and Application. China: [s.n.]. 2008: 9–13.
[9] Ostrenko M, Andriienko B, Ryzhyi V. Finite element method application for HVDC electrical insulation strength problems solution [C]. IEEE International Conference on Computation-al Technologies in Electrical and Electronics Engineering. 2010.

Architectural, Energy and Information Engineering – Sung & Chen (Eds)
© 2016 Taylor & Francis Group, London, ISBN 978-1-138-02791-6

Study on ACSD for a neutral-point reactor of ultra-high voltage AC

J.H. Han, W. Wang, Y.Q. Li, H.B. Zheng, X.G. Li & Y.B. Shao
State Grid Henan Electric Power Research Institute, Zhengzhou, Henan Province, China

ABSTRACT: The insulation characteristics of the neutral-point reactor in the Ultra-High Voltage AC (UHVAC) pilot project are studied in this paper. First, based on power frequency overvoltage analysis and the on-site test data in the project, this study presents the potential distribution and wave process of the reactor under impulse voltage. Then, the feasibility and applicability of the on-site short-duration AC withstand voltage test (ACSD) for neutral-point reactors are proofed theoretically. Finally, the test method proposed in this paper for the insulation test of a neutral-point reactor is verified on site, and the data obtained in the test are available for future on-site hand-over test for neutral-point reactors.

Keywords: UHVAC; neutral-point reactors; wave process; ACSD; hand-over test

1 INTRODUCTION

The 1000 kV Jindongnan-Nanyang-Jingmen UHVAC pilot project started at Changzhi substation in southeastern Shanxi Province, passed through Nanyang substation in Henan Province, and ended at Jingmen substation in Hubei Province. The single circuit stretches 640 kilometers, crossing the Yellow River and the Han River. It has a transforming capacity of 6000000 kVA, with a nominal voltage of 1000 kV and the highest operation voltage of 1100 kV.

Due to its long-distance power transmission, shunt reactors are used at each station to compensate for the capacitive reactive power, which suppresses the power frequency overvoltage.

According to the simulation analysis of a single-phase short-circuit reclosing accident using the electromagnetic transient ATP-EMTP procedure, the highest voltage that the neutral-point of a high voltage shunt reactor can withstand is 170.6 kV. During the trial operation of the UHVAC pilot project, the insulation breakdown of a neutral-point reactor occurred. The analysis found that it was mainly caused by the winding turn-to-turn insulation damage [1–4].

Currently, the withstand voltage tests mainly assess the main insulation of the reactor, while there is no effective assessment of the longitudinal insulation itself. In view of this, this paper focuses on the assessment method of the winding turn-to-turn insulation, checks the calculation of its internal electric field distribution and the dynamic thermal stability on field tests, and proposes the method of the on-site short-duration AC withstand voltage test for the neutral-point reactor.

2 PARAMETERS AND INSULATION LEVEL OF THE UHVAC REACTOR

In the 1000 kV Jindongnan-Nanyang-Jingmen UHVAC pilot project, Changzhi substation installed a set of high-voltage shunt reactors with a capacity of 960000 kvar; Nanyang substation has two sets of high-voltage shunt reactors, installed in Changnan I and Nanjing I separately, with a capacity of 720000 kvar for each; Jinmeng substation installed a set of high voltage shunt reactors with a capacity of 600000 kvar.

When the fault transmission line clears from both sides, the capacitive and inductive couplings between the non-fault and open-circuit lines will continually provide the secondary arc current to the fault line. If the arc current is too large, it will lead to the reclosing failure. Usually, a set of delta reactors are installed to compensate for the capacitive coupling, thereby inhibiting the horizontal component part of the arc current; a set of star connected reactors are installed at both ends of the line to offset the vertical component of the arc current.

There is one neutral-point reactor installed in each set of reactors in the UHVAC pilot project. In this study, the neutral-point reactor of high voltage shunt reactors for Changnan I in Nanyang substation will be researched as an example. The technical parameters of its installed neutral-point reactor are outlined in Table 1.

Table 1. Parameters of a neutral-point reactor in the UHVAC project.

Unit type	Nominal capacity (kvar)	Rated current (A)	Insulation level (kV)	
XKD-361/154	329	30	Line terminal LI/AC750 /325	Neutral termina LI/ AC200/85

3 TEST METHOD

3.1 Scheme of the ACSD test

According to [5], the head of the reactor should be pressurized and the tail end should be grounded when performing the on-site ACSD test of the neutral-point reactor. Because of the large capacity of the reactor itself, a lot of inductive reactive power should be compensated. Therefore, capacitor banks are employed to compensate the inductive power in this study, and by increasing the test frequency, the test power capacity can be lowered. As shown in Figure 1, serial-parallel compensation is used to perform the ACSD test.

3.2 Calculation of test parameters

The impedance value of the neutral-point reactor is 412.49 Ω at the frequency of 50 Hz in Changnan I. In order to reduce the size of the compensation capacitor, variable-frequency power supply is used to increase the test frequency. Then, the impedance of the reactor will correspondingly increase and the inductive reactive power will reduce. Considering the on-site condition and test capability, six capacitors with 50 nF are used to compensate the reactive power by a series and parallel combination. Five of them are parallel to the neutral-point reactor to reduce the inductive current and then connected with one capacitor in series to produce a high-voltage resonance. In theory, on the fully resonant state, the circuit frequency can be calculated as $f = 253$ Hz.

The test voltage is $U = 260$ kV according to [5], so the current flowing through the reactor is as follows:

$$I_L = \frac{U}{j\omega L} = -j\frac{U}{2\pi fL} \qquad (1)$$

Then, the current flowing through the parallel compensation capacitor can be expressed as follows:

$$I_C = \frac{U}{1/j\omega C} = j2\pi fUC \qquad (2)$$

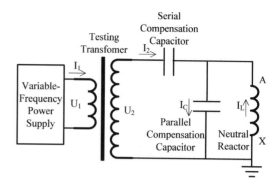

Figure 1. ACSD test for the neutral-point reactor with a serial-parallel compensation scheme.

The current flowing through the series compensation capacitor and the secondary side of the test transformer is as follows:

$$I_2 = I_C - I_L = j(2\pi fUC - \frac{U}{2\pi fL}) \qquad (3)$$

On the basis of formulas (1) to (3), the current flowing through the neutral-point reactor is 124 A in the first 60 s of the ACSD test, while the current flowing through the series capacitor and test transformer is only around 20 A.

Considering that the rated current of the neutral-point reactor is 30 A, an equivalent and thermal stability analysis to determine whether the reactor could withstand the impact of this large current is carried out in detail.

4 THEORETICAL ANALYSIS

4.1 Equivalence analysis

The normal operating frequency of the reactor is 50 Hz. In order to meet the test requirements of reactive power compensation, the frequency of the test voltage needs to be increased. So, the assessment of insulation equivalence should be demonstrated under the voltage of the increased frequency. According to the electromagnetic field theory,

$$\lambda = cT = \frac{c}{f} \qquad (4)$$

where λ is the wavelength (in meters); c is the speed of light (in meters per second); T is the cycle (in seconds); and f is frequency (in hertz).

Electromagnetic wave spreads at the speed of light in the wire. When the frequency of sine wave is very low, the wavelength is correspondingly very long. Also, when the wavelength is much longer than the total length of the wire, it will be considered as an electrostatic field. When $f = 50$ Hz, the electromagnetic wavelength in the free space is about 6000 km. Under the test frequency (approximately 250 Hz), the electromagnetic wavelength is around 1200 km. Therefore, considering the electric and magnetic fields within the range of tens of kilometers near the excitation source, the analysis and calculation can be done following the criterion of electrostatic field and constant magnetic field. In other words, under power and test frequency conditions, the reactor can be analyzed and calculated as in the electrostatic field.

Therefore, the insulation assessment of the reactor is equivalent to the actual operating power frequency under the improved power frequency.

4.2 Thermal stability assessment under the impact of large current

According to [6], the average thermal stability of the windings is as follows:

$$\theta_1 = \theta_0 + \frac{2 \times (\theta_0 + 235)}{\frac{106000}{J^2 \times t} - 1} \tag{5}$$

where θ_1 is the average temperature of the windings under the impact of large current (in Celsius degree); θ_0 is the winding starting temperature (in Celsius degree); J is the current density (in A/mm²); and t is the duration time (in seconds).

Based on a large shock current with 124A and duration of 60s, the calculation results of thermal stability are provided in Table 2.

The average temperature of the winding in thermal stability time does not exceed 250°C, so the thermal stability of the neutral-point reactor winding meets the criterion.

4.3 Dynamic stability assessment under the impact of large current

The axial dynamic model of the reactor winding is shown in Figure 2. Each pancake coil can be sim-

Table 2. Thermal stability assessment.

Items	Winding
θ_0 (°C)	105
J (A/mm²)	1.74
T (s)	60
θ_1 (°C)	106.2

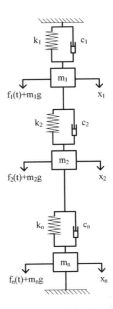

Figure 2. Axial dynamic model of a reactor.

plified to a quality point; the block between pancakes is simplified to the spring.

The assessment results of axial dynamic stability are summarized in Table 3. From the safety factor value listed in Table 3, we can draw the conclusion that the axial dynamic stability meets the requirement.

For the radial dynamic stability assessment, first, the maximum leakage flux density can be calculated as follows:

$$B_m = \frac{\mu_0 \cdot N \cdot I_m}{H_w} = \frac{1.256 N \cdot I_m}{H_w} \times 10^{-3} \tag{6}$$

where $\mu_0 = 1.256 \times 10^{-6}$ is the magnetic permeability of vacuum (in henries per meter, H/m); N is the number of the coil turns; and Hw represents the coil geometry height (in millimeters).

Therefore, the radial forces per unit length of each coil turn can be calculated as follows:

$$F_u = \frac{1}{2} B_m \cdot I_m \tag{7}$$

Then, the forces of the whole inner winding, in newtons, can be expressed as (8):

$$F_r = \pi D_m \cdot N \cdot F_u = \frac{1.97 (I_m N)^2 D_m}{H_w} \times 10^{-6} \tag{8}$$

where D_m is the average diameter of the winding coil, in millimeters.

683

Table 3. Axial dynamic stability assessment.

Axial force (kg/mm)	Axial strength (kg/mm)	Displacement (mm)	Maximum allowable displacement (mm)	Safety factor
18.30	666.35	0.38	25.00	36.41

Table 4. Radial dynamic stability assessment.

Radial force (kg/mm)		Safety factor
1.62	16.83	10.39

So, the radial force acting on each pancake coil can be expressed as follows:

$$F_C = \frac{F_r}{\pi D_m \cdot M_1} \tag{9}$$

In (9), M_1 represents the number of the pancake coils in the reactor winding.

The calculation results of radial short-circuit forces and safety factors are summarized in Table 4. So, the radial dynamic stability meets the requirement.

5 FIELD TEST ANALYSIS

According to Section 3, when U_2 reaches the test voltage, the neutral-point reactor does not have insulation breakdown. It is considered that it withstands the assessment made by the ACSD test. The actual test voltage of the neutral-point reactor reaches 258.5 kV, and the current flowing through is 122A.

During the test, an infrared tester is used to closely monitor the reactor body temperature rise under the test voltage, and the tester shows no abnormal temperature rising.

We also use the spectrum response method to perform the winding deformation test on the neutral-point reactor after the impact of large current. The before and after comparison result of the winding deformation patterns shows a good agreement, which indicates that the reactor winding does not deform after this large current test. Finally, we use the oil sample to carry out the chromatographic analysis, and the result shows no changes in the oil gas before and after the test.

6 CONCLUSIONS

The ACSD test of neutral-point reactors installed in the high-voltage shunt reactors of the UHVAC pilot project is studied in this paper. First, we analyzed the insulation breakdown of a neutral-point reactor during the trial operation of the UHVAC pilot project, and the analysis shows that the accident is mainly caused by the winding turn-to-turn insulation damage. Then, the feasibility and applicability of the on-site short-duration AC withstand voltage test for the neutral-point reactor are proofed theoretically. Finally, the feasibility of this assessment method is verified in the field test, which achieves the assessment criterion for neutral-point reactors.

REFERENCES

[1] H.M. Ahn, J.Y. Lee, J.K. Kim, et al, *Finite-element analysis of short-circuit electromagnetic force in power transformer* [J]. IEEE Transactions on Industry Applications, 2011, 47(3): 1267–1271.

[2] E. Rahimpour, J. Christian, K. Feser, et al, *Transfer function method to diagnose axial displacement and radial deformation of transformer windings* [J]. IEEE Transactions on Power Delivery, 2003, 18(2): 493–505.

[3] S. Gopalakrishna, K. Kumar, B. George, and V. Jayashankar, *Design margin for short circuit withstand capability in large power transformers* [C]. Int. Conf. on Power Engineering, Singapore, Dec. 3–6, 2007.

[4] J. Faiz, B.M. Ebrahimi, and W. Abu-Elhaija, *Computation of static and dynamic axial and radial forces on power transformer windings due to inrush and short circuit currents* [C]. Int. Conf. on Applied Electrical Engineering and Computing Technologies, Amman, Jordan, Dec. 6–8, 2011.

[5] IEC 60076-3, *Power transformers-Part 3: Insulation levels, dielectric tests and external clearances in air* [S], 2000.

[6] IEC 60076-5, *Power transformers-Part 5: Ability to withstand short circuit* [S], 2006.

Architectural, Energy and Information Engineering – Sung & Chen (Eds)
© *2016 Taylor & Francis Group, London, ISBN 978-1-138-02791-6*

Self-adaptive discrete differential evolution algorithm for assembly line balancing problem-II

H.J. Zhang, Q. Yan, G.H. Zhang & Y.P. Liu
Zhengzhou Institute of Aeronautical Industry Management, Zhengzhou, Henan, China

ABSTRACT: To solve the assembly line balancing problem-II, a Self-adaptive Discrete Differential Evolution algorithm (SaDDE) was proposed. The SaDDE employed the method of indirect coding based on priority and introduced a gradually approaching procedure. In the mutation operation, the self-adaptive double mutation strategy was developed. The experimental results showed that the SaDDE converged faster than that of the original discrete differential evolution algorithm.

Keywords: assembly line balancing; differential evolution algorithm; self-adaptive; discrete

1 INTRODUCTION

Assembly line is a most type of production organization, which helps to improve the production efficiency and the utilization of manufacturing resources. However, the line unbalance will lead to blocking and starvation, too much work in process, which ultimately results in decreasing productivity and seriously affects production efficiency and manufacturing resource utilization. This paper considers only ALBP-II. ALBP-II falls into the NP-hard class of combinatorial optimization problems, whose complexity will increase in a geometrical ratio along with increase in the number of tasks[1]. Numerous research efforts have been directed towards the development of computer efficient approximation algorithms or heuristics, e.g. genetic algorithm[2], ant colony algorithm[3], particle swarm optimization algorithm[4], Tabu search algorithm algorithm[5]. However, there are few reports about the Differential Evolution (DE) algorithm for solving ALBP-II. The reason may be that the original DE involves the floating-point vectors encoding over continuous spaces, but ALBP-II is a typical discrete optimization problem.

The paper is structured as follows: firstly, the mathematical model of ALBP-II is built, and the method of encoding and decoding oriented on stations is developed; secondly, the feasible population sequence is obtained based on priority; then a kind of self-adaptive double mutation strategy is proved to improve the diversity of population and the speed of convergence; finally, the experimental results show that the proposed method is effective and superior.

2 DESCRIPTION AND MODELING OF ALBP-II

2.1 ALBP-II description

ALBP-II consists of assigning tasks to a given number of stations such that the precedence relations among the tasks are satisfied and the cycle time is minimized. These elements can be summarized by a *precedence diagram*. It contains a node for each task, node weights for the task times, and arcs for the direct precedence constraints. A precedence diagram is a directed graph $G = (C, E)$, where V is a set of tasks and E is a set of edges, each joining two tasks. The tasks are denoted by numbers 1, 2 ... I and the edge between vertex i and vertex j by $[i, j]$. The edges have direction, thus $[i, j] \neq [j, i]$. A feasible ALBP must satisfy the precedence relation among the tasks, that is the sequence of tasks assigned to stations should be a *Feasible Task Sequence* (FTS). Fig. 1 shows a precedence diagram with n=9 tasks, and one of FTS is $1 \rightarrow 2 \rightarrow 3 \rightarrow 4 \rightarrow 5 \rightarrow 6 \rightarrow 7 \rightarrow 8 \rightarrow 9$.

2.2 ALBP-II modeling

Model Assumptions: (1) the number of stations is given. The line layout is only serial, not paral-

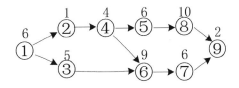

Figure 1. Precedence diagram.

lel. (2) the task times are known deterministically and independent of each other. A task can be performed at any station. (3) there is no assignment restrictions besides the precedence constraints. (4) the task is the smallest unit of manufacturing process. (5) the total task time of each station can't exceed the limit on cycle time.

The mathematical model of ALBP-II is as follows:

$$\min CT$$

$$s.t. \quad S_i \cap S_j = \phi, i \neq j, \forall i, j \in E;$$

$$\bigcup_{k=1}^{K} S_k = E;$$

$$0 < T(S_k) \leq CT, \forall k \in \{1, 2, \cdots, K\};$$

$$a \leq b, \forall v_{i,a} = 1, v_{j,b} = 1, pro_{i,j} = 1.$$

3 SADDE ALGORITHM

3.1 Individual encoding

The paper employs the indirect encoding method based on priority (PZTS) [4]. Each individual is a gene with the length of I. Each task is indexed by an unique number, and the value of gene locus is the priority value of the corresponding task $P_i \in E$. The greater the priority value, the greater the possibility of choosing the task with zero indegree. The zero indegree of vertex with the greatest priority value should be assigned firstly, and then all the arcs whose head endpoint is the assigned vertex are deleted, until all vertexes in G are assigned. Therefore, according to the Fig. 2, FTS is $1 \rightarrow 2 \rightarrow 3 \rightarrow 4 \rightarrow 5 \rightarrow 6 \rightarrow 7 \rightarrow 8 \rightarrow 9$.

3.2 Individual decoding

Suppose instead that $I \geq K$, the concrete steps are as follows:

Step 1: Initialize the index of current station $k = 1$, and the cycle time CT is given.

Step 2: If $k < K$, calculate the idle time of k station, $IdleT_k = CT - T(S_k)$, and to step 3; If $k = K$, it means that the current station is the last station without tasks. In the situation, all unassigned tasks should be performed in the station. Then the optimal solution is gotten, and update $CT = \min\{CT, \max\{T(S_k)|_1^K\}\}$.

Step 3: Calculate the set of unassigned tasks which have zero indegree: $\Theta = \{j \mid j \in E, Idg_j = 0, \sum_{k=1}^{K} v_{j,k} = 0\}$. The number of tasks $N_I = |\Theta|$ and the number of unused stations $NK = K - k$. If $NI = NK$, FTS would be got according to PZTS and update $CT = \min\{CT, \max\{T(S_k)|_1^K\}\}$; if $N_I > N_K$ then go to Step 4.

Figure 2.   An example priorith-based chromosome.

Step 4: Calculate the set of tasks whose completion time is less than $IdleT_k$: $\Theta_s = \{j \mid j \in \Theta, t_j \leq IdleT_k\}$. If $\Theta_s \neq \phi$, assign the greatest priority of task $i \in \Theta_s$ to the current station, that is $S_k = S_k \cup \{i\}$, update $pro_{i,j} = 0$ $(\forall j \in E)$, $v_{i,k} = 1$ and $T(S_k)$, then turn to Step 3; if $\Theta_s = \phi$ then go to Step 5.

Step 5: Close the current station and open a new station. Update $k = k + 1$, and then go to Step 2.

3.3 Mutation operation

The paper employs the self-adaptive double mutation strategy, in order to improve the diversity of population and the speed of convergence. Maintaining population diversity is critical for avoiding the premature convergence. Therefore, the paper defines the *diversity of individual position* (DIP):

$$DIP^{(g)} = \frac{1}{NP} \cdot \frac{1}{K-1} \cdot \sum_{i=1}^{NP} \sqrt{\frac{1}{I} \sum_{j=1}^{I} (x_{i,j}^{(g)} - x_{best,j})^2}$$

If $DIP^{(g)} \leq \varepsilon \in (0,1)$ (ε: the threshold of DIP), SDDE would employ the DE/rand/1/bin strategy, otherwise the DE/best/2/bin strategy.

$$V_i^{(g)} = \begin{cases} X_{r1}^{(g)} + F \cdot (X_{r2}^{(g)} - X_{r3}^{(g)}) & DIP^{(g)} \leq \varepsilon, \\ X_{best} + F \cdot [(X_{r1}^{(g)} - X_{r2}^{(g)}) \\ \quad + (X_{r3}^{(g)} - X_{r4}^{(g)})] & DIP^{(g)} > \varepsilon. \end{cases}$$

3.4 Discrete operation

The set of same values is denoted by E_s; the set of values of gene locus is denoted by E_g. The detailed rules of discrete operation are as follows:

Calculate the set $E - E_g + E_s$, and then the elements of the set are randomly assigned to each task of the same priority. In this way, the same values of priority are updated.

3.5 SaDDE process

Step 1: Setting parameters NP, F, CR and ε. The cost function is $f(X) = CT = \max\{T(S_k)|_1^K\}$.

Step 2: Population initialization: according to Section 2.1, the NP target vectors $X_i^{(1)}$ are randomly generated; then according to Section 2.2, the target vectors are decoded and the global optimal individual $X_{best} = arg\,min\{f(X_i^{(1)})|_1^{NP}\}$ is gotten.

Step 3: Mutation operation: the mutation individual $V_i^{(g)} = \{v_{i,j}^{(g)} \mid j \in E\}$ is gotten, according to Section 2.3.

Step 4: Crossover operation: the same to that of the simple DE without any alteration.

Step 5: Discrete operation: the trial vector $U_i^{(g)}$ is gotten after the discrete operation, according to Section 2.4.

Step 6: Selection operation: the same to that of the simple DE without any alteration.

Step 7: Updating optimal individual: calculate the optimal individual X_{best}^{g+1} in the new generation, $X_{best}^{(g+1)} = arg\,min\{f(X_i^{(g+1)})\,|_i^{NP}\}$, and then compared with the global optimal individual in the last generation X_{best}, update the global optimal individual: $X_{best} = arg\,min\{f(X_{best}^{(g+1)}), f(X_{best})\}$.

Step 8: Termination judgment: the above steps are repeated generation after generation until some specific stopping criteria are satisfied.

4 CASE STUDY OF SADDE

4.1 Cases and parameters settings

The performance of the proposed SaDDE algorithm was evaluated on a set of Scholl test problems[6]. We used the Matlab 7.0 to implement the algorithm and the system configurations are listed as follows: Intel Pentium 4 CPU 3.00 Ghz, 3 GB of memory, WIN 7.

4.2 Experimental results

For the example of Buxey problem, the number of stations k is set from 7 to 14. For each number of stations, the algorithm is run 10 times independently. The algorithm could find the optimal solution in 10 seconds (see Table 1). The convergence map of the SaDDE algorithm at $K = 14$ is plotted in Fig. 3. The original DDE algorithm trapped in the local optimum at $G = 21\sim64$, while SDDE had gotten the global optimum ($CT = 25$) at $G = 24$. From the results it can be concluded that the SaDDE algorithm is usually able to find

Table 1. The result of buxey problem.

Number of stations/K	Theoretical optimum CT/s	CT by SaDDE /s	Average operation time/s
7	47	47	9.39
8	41	41	9.39
9	37	37	9.21
10	34	34	9.20
11	32	32	9.22
12	28	28	9.16
13	27	27	9.15
14	25	25	9.11

Table 2. The result of Arcus2 problem.

Number of stations/K	Theoretical optimum CT/s	CT by SaDDE/s	Average operation time/s
5	30080	30080	876.56
10	15040	15040	898.65
13	11570	11570	886.21
15	10035	10035	892.80
17	8855	8855	879.22

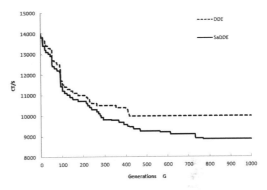

Figure 3. The convergent tendency of SaDDE compared with DDE (Buxey).

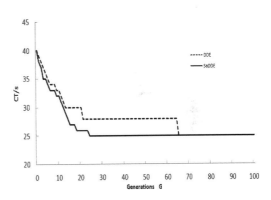

Figure 4. The convergent tendency of SaDDE compared with DDE (Arcus2).

optimum solutions for ABLP-II with a less number of generations than the original DDE algorithm.

For the example of Arcus2 problem, the number of stations K is set at 5, 10, 13, 15 and 17. For each number of stations, the algorithm is run 100 times independently. The algorithm could find the optimal solution in 1000 seconds (see Table 2). The convergence map of the SaDDE algorithm at $K = 17$ is plotted in Fig. 4. The original DDE algorithm is trapped in the local optimum at $G = 416$, while SDDE had gotten the global optimum ($CT = 8855$) at $G = 770$. From the results it can be concluded that the SaDDE algorithm is usually able to find optimum solution for ABLP-II with a less number of generations than the original DDE algorithm.

5 CONCLUSIONS

In this paper, a self-adaptive discrete differential evolution algorithm (SaDDE) has been introduced for ABLP-II. The proposed SaDDE incorporates self-adaptive double mutation strategy, which can improve the diversity of population and simultaneously keep convergence speed. Based on the PZTS, the paper designs a method of individual decoding that can assign all tasks to stations effectively. Utilizing the set of School problems for test, the results show that SaDDE can solve the differential scaling ABLP-II effectively.

ACKNOWLEDGEMENTS

The paper is supported by the Innovation Scientists and Technicians Troop Construction Project of Henan Province (134200510024), the Soft Science Research Project of Henan Province (132400410782), the General Research Project of Zhengzhou City (20140583) and the Higher Education Key Research Project of Henan Province (15A630050).

REFERENCES

[1] E. Erel, I. Sabuncuoglu, H. Sekerci. Stochastic assembly line balancing using beam search [J]. International Journal of Production Research, 2005, 43(7): 1411–1426.

[2] Özcan Mutlu, Olcay Polat, Aliye Ayca Supciller. An iterative genetic algorithm for the assembly line worker assignment and balancing problem of type-II [J]. Computers & Operations Research. 2013, 40(1): 418–426.

[3] Y.D Li, J. Lu. Improved ant colony optimization for assembly line balancing-II problem [J]. Computer Integrated Manufacturing Systems, 2012. 18(4): 754–760. (In Chinese).

[4] J.P Dou, C. Su, J. Li. Discrete particle swarm optimization algorithms for assembly line balancing problems of type [J]. Computer Integrated Manufacturing Systems, 2012, 8(5): 1021–1030. (In Chinese).

[5] R. Pastor, C. Andres, A. Duran. Tabu search algorithms for an industrial multi-product and multi-objective assembly line balancing problem with reduction of the task dispersion [J]. Journal of the Operational Research Society, 2002, 53(12): 1317–1323.

[6] A. Scroll. Balancing and sequencing of assembly lines [M]. Heidelberg, Germany: Physica-Verlag, 1995.

Architectural, Energy and Information Engineering – Sung & Chen (Eds)
© 2016 Taylor & Francis Group, London, ISBN 978-1-138-02791-6

The application and prospect of wet electrostatic precipitate in large coal-fired power plants

Pei Di Wang, Xue Tao Wang, Yu Liu & Jian Xing Ren
College of Energy and Mechanical Engineering, Shanghai University of Electric Power, Shanghai, China

ABSTRACT: With the improvement of national environmental protection requirements, wet electrostatic precipitators, which have been widely concerned to meet the latest requirement of "Emission standard of air pollutants for thermal power plants", are an efficient and reliable dust removal technology. In this paper, the principle, classification and technical characteristics of WESP were described. And the development and application prospect of WESP in large coal-fired power plants was proposed. In addition, the application of WESP in a 1000 MW coal-fired unit was introduced.

Keywords: WESP; application; prospect, large coal-fired power plants

1 INTRODUCTION

With the rapid development of economy in our country, the atmospheric pollutant emissions also increased dramatically. Especially the growing mass haze weather of our country has brought great harm to people's healthy life in recent years, which has caused high attention of the people. According to statistics, the highest amount of soot emissions comes from coal-fired power plants in various industries in China.[1]

Under the circumstances of green environmental protection, it is dictated in the latest *Emission Standard of Air Pollutants for Thermal Power Plants* (GB 13223-2011) that the dust emission concentration of coal-fired power plants is required to be lower 30 mg/m³ and the dust emission in key areas mustn't be higher 20 mg/m³. More strict requirements of the dust emissions from coal-fired power plants in China were put forward to meeting the improvement of standards.[2]

It's difficult for dry electrostatic precipitator to meet the latest emissions standards. In particular, the effect of dry electrostatic precipitator to remove PM2.5 and PM10, which have attracted widespread attention, can not meet the requirements. According to the advanced technologies and experiences in foreign countries, the Wet Electrostatic Precipitator (WESP) has a prominent removal effect of the sulfuric acid, toxic heavy metals and the fine particulate matter, especially the PM10 and PM2.5. WESP also has the advantages to eliminating the phenomenons like gypsum rain and acid rain. The WESP, which has been listed in national 863 programs and international science

and technology cooperation projects, has attached great importance and achieved vigorously support. Therefore, wet electrostatic precipitator is deemed as a kind of efficient dust removal technology in the thermal power plant of our country, which can also achieve the combination to reduce the emissions of a variety of pollutants and has a broad development prospect.[3]

2 PRINCIPLE AND CLASSIFICATION OF WESP

2.1 *Principle of WESP*

As shown in Figure 1, the principle of the wet electrostatic precipitator has three processes, roughly similar with the principle of dry ESP approximation. Particles are charged under a high voltage power supply when flue gas passing though WESP. Particles with negative electric charge move to the collecting electrode, and are adsorbed to the surface of collecting plates or lines. Water mist, which sprays from a sprinkler, forms a continuous water film on the surface of collecting electrode. The captured dust flows into the hopper with the flushing fluid, so as to make the flue gas purification.[4]

Wet ESP is completely different from dry ESP because of different mode of deashing. Dry ESP usually removes dust through mechanical shaking cleaning dust or sonic soot cleaning. However, Wet electrostatic precipitator makes use of the liquid to flush off the dust deposited on the surface of the collecting electrodes and the dust is removed with washing fluid.

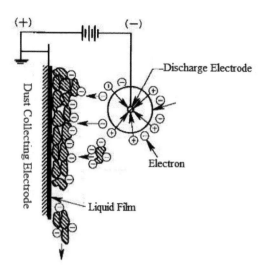

Figure 1. Principle of WESP.

2.2 Classification of WESP

The wet electrostatic precipitator has been studied abroad for a history of nearly 30 years. There are more species of along with the continuous development. As the change of the standard, the categories of WESP also becomes different[5].

Specific classifications are shown in Table 1.

3 TECHNICAL FEATURES OF WESP

When it is compared with dry ESP, WESP has characteristics of high dust removal efficiency, low gas pressure loss, less energy consumption, low operation cost and simple maintenance. etc. WESP can work under the flue gas acid dew point temperature. It can be combined with other flue gas treatment equipment and has diversification design forms because of the compact structure.

The specific technical characteristics are as follows.[6]

Table 1. Classification of WESP.

Classification criteria	Category	Introduction
According to the structure	Tubular WESP	The collecting electrodes of Tubular WESP are several extremely multiple parallel round or polygonal metal pipe. And discharge electrodes are evenly distributed between the plates. It can only be used for vertical-flowing flue gas. Tubular WESP has higher dust removal efficiency and takes up less space.
	Plate WESP	The collecting electrodes of Plate WESP are several parallel plates. It has good liquid film characteristics and can be used for both vertical-flowing flue gas and horizontal-flowing flue gas.
According to the material	WESP with metallic electrodes	WESP with metallic electrode has the roughly similar structure with dry ESP. It can generate a stable HV electric field and has a high removal efficiency of aerosol and fine particles like PM2.5.
	WESP with FRP-electric electrodes	WESP with FRP-electric electrode, structure of which is pipe-type, generally is used in metallurgical and chemical industry in our country, and has a stable HV electric field, less water consumption and a high removal efficiency of aerosol and fine particles like PM2.5.
	WESP with flexible electrodes	Material of WESP with flexible electrodes is organic synthetic fiber. The soaked fiber electrodes have the electrical conductivity to collect droplets and dusts in the flue gas. This kind of material provides low mechanical strength and lacks stability of electric field, No water film flushing and the low removal efficiency of aerosol and fine particles like PM2.5.

3.1 Advantages of WESP

1. No accumulation of dust on electrode and no dust re-entrainment so that the phenomenon of back corona and corona obstruction can be eliminated and the dust removal efficiency can be promoted.
2. No moving parts, minimum consumable parts. And maintenance expenses can be greatly cut down.
3. No influence of the dust property, such as resistivity.
4. As a result, WESP has an effective coordinated control of multiple pollutants. It's able to effectively remove heavy metals (like mercury, arsenic and selenium), fine particulate matter (PM10 and SO_3 dust aerosols), etc. According to a study[7], removal efficiency of submicrometer aerosol particles (like HCl, H_2SO_4) can even reach 99% when flue gas is passing through double-field of WESP, and the dust emissions can be below 10 mg/m^3, even below 5 mg/m^3. Therefore WESP can efficiently remove various pollutants in the flue gas.

3.2 Disadvantages of WESP

1. WESP can be applied only in low gas temperature as it uses washing water. Otherwise, dust particles will be dry and the removal effect will be reduced.
2. Large water consumption. So recirculating water treatment system and sewage treatment sys-tem are necessary.
3. Not suitable for flue gas of high dust load or high SO$_x$ concentration[8,9].

4 PROSPECTS AND APPLICATION OF WESP

4.1 Development and prospects in large coal-fired plants

WESP was firstly applied in the sulfuric acid industry and metallurgical industry in 1907 and began to be used in coal-fired power plants in 1986. According to incomplete statistics, more than 50 sets of WESP systems have been applied in coal-fired plants to control fine particle, reduce the turbidity of flue gas and remove acid mist in the United States, Europe and Japan[10].

At present WESPs are mainly used in small and medium chemical and metallurgical industries and small thermal power plants. The engineering application of WESP in large coal-fired plants is roughly blank. And WESP could be applied with other pollution control equipment as a group to

Table 2. Removal results of some pollutants of WESP (Standard conditions).

Unit load	Removal rate of SO_3	Exit concentration of PM2.5 (mg/m^3)	Outlet concentration of droplet (mg/m^3).
100%	71.5%	0.32	7.6
75%	67.6%	0.28	3.4

achieve the goal of ultra-low emission. This kind of composite applications is efficient to reduce phenomenon of gypsum rain and remove fine particles (like PM10 and PM2.5), sulfate aerosols, toxic heavy metals and other flue gas pollutants. So WESP and its composite applications enjoyed broad space for development in thermal power plants in China, especially in the large coal-fired power plants.

4.2 Application of WESP in a 1000 MW coal-fired units

A 1000 MW unit of Jiahua power plant was equipped with WESP, which is arranged behind the system of ESP, WFGD and MGGH.

The technical requirements are as follows:

1. Removal rate of dust: ≮70%
2. Removal rate of PM2.5: ≥70%
3. Removal rate of fog drop: ≥70%
4. Removal rate of SO$_3$: ≥20%
5. Exit concentration of dust: ≯5 mg/Nm^3
6. Pressure drop: ≯200 Pa
7. Air leak rate: ≯1.0%
8. Circulation area: 2 × 168 m^2

The WESP was put into the operation in June 2014. According to the test, the missions of dust is as low as 2 mg/m^3. In addition, removal results of some other pollutants of WESP are shown in Table 2.

5 CONCLUSIONS

1. At present the fine dust pollution of large coal-fired power plants has received widely attention. As an efficient dust-removing equipment, WESP can meet the latest Emission Standard which request the exit concentration of dust must be lower than 10 mg/m^3.
2. According to introduction of the application of WESP in a 1000 MW coal-fired unit, WESP is able to effectively remove fine particles, sulfate aerosols and droplet, etc.

3. At present WESPs are mainly used in small and medium chemical and metallurgical industries and small thermal power plants in China. So WESP and its composite applications enjoyed broad space for development in thermal power plants, especially in the large coal-fired power plants.

REFERENCES

[1] Zhang Yanqi, Tan Qing, Wang Weimin, etc. The Application of The WESP For The Flue Gas Dedusting of Power Plant. *Shanghai Energy Conservation*, 2013, (12): 34–37.

[2] Chen Jun, Lin Zuhan, Yan Xun. Application of new type WESP to cleaning units on tail gas desulfuration and denitrification processes. *Electric Power Technology And Environmental Protection*, 2014, 30(3): 46–48.

[3] Shi Chaoslin, Pan Weiguo, Guo Ruitang.etc. Development Situation of the WESP in Themal Power Plant. *Power & Energy*, 2013, 34(5): 493–499.

[4] Chen Zhaomei, Gao Zhifeng, Lv Mingyu. Application of WESP in Coal-fired Power Plants for 'Ultra-clean Emission'. *Power System Engineering*, 2014, 30(6): 18–20.

[5] Liu, He-zhong, Tao, Tiu-gen. Exploration Application of Wet Electric Dust Catcher to Engineering. *Electric Power Survey & Design*, 2012, (6): 43–47.

[6] Zhao Qinxia, Chen Zhaomei, Zhou Chaojiong, Yin deshi. Discussion on wet ESP technology and its application prospect in coal-fired power plants. *Electric Power Technology And Environmental Protection*, 2012, 28(4): 24–26.

[7] Bologa A, Paur HR. Novel wet electrostatic precipitator for collection of fine aerosol. *Journal of Electrostatics*, 2009, 67(2–3): 150–153.

[8] Yatavelli L N R. *Capture of Soluble Mercury Using Membrane -based Wet Electrostatic Precipitator as a Function of Temperature.* Ohio University, 2005.

[9] Chikayuki Nagata, Kazuaki Miyake, Shiri Suzuki. Industrial and Advanced Applications of Wet type Electrostatic Precipitator Technology. *Proceeding of the 11th Conference of ESP*, 2005: 505–517.

[10] Tian Wen-hua, Li Jing-sheng. Wet ESP Technology Application and Installation Quality Control of a Thermal Power Plant. *Power Generation & Air Condition*, 2014, 35(158): 21–23.

Architectural, Energy and Information Engineering – Sung & Chen (Eds)
© 2016 Taylor & Francis Group, London, ISBN 978-1-138-02791-6

The control of PM2.5 for coal-fired power plants WESP

Huan Liu, Fang Qin Li, Ji Fa Zhang, Ji Yong Liu & Hai Gang Ji
Shanghai University of Electric Power, Shanghai, China

ABSTRACT: The reduction of fine particles from power plant flue gas is an important way to reduce PM2.5. Therefore, a stricter emission standard of air pollutants of power plants isintroduced. In this review, the dust sample was taken from a power plant. By studying the particle size distribution of the electrostatic precipitator, a higher particle removal efficiency of 10 μm~45 μm was found. For the particle with the diameter below 1 μm, the removal efficiency reduced rapidly. In addition, it is difficult to remove the fine particle with the diameter below 0.5 μm. Dry electrostatic precipitator obtains low efficiency on super fine particles; moreover, it is hard to meet the demand of the new standards. Thus, the Wet Electrostatic Precipitator (WESP) has a beneficial effect on the control of PM2.5.

Keywords: PM2.5; particle size distribution; Wet Electrostatic Precipitator (WESP)

1 INTRODUCTION

The atmospheric PM2.5 is a major factor that can trigger acid rain and photochemical smog and thus reduce the visibility. When people inhaled PM2.5 from the air, it could cause a variety of diseases. Coal-fired flue gas contains a large number of fine particles, which include about 40% PM10, and there are 40~70% PM2.5 in PM10[1]. Therefore, reducing the particulate emission of coal-fired power plant flue gas is particularly important. In 2012, it was reported thatthe Ambient Air Quality Standard (AAQS) will reach all-sided implementation by 2016 (the daily average allowed emission concentration is less than or equal to 75 μg/m³ and the annual average concentration is 35 μg/m³)[2,3]. After studying the fine ash composition before and after the coal-fired power plant electrical precipitator, we found that most large particles of dust existed before the precipitator, and the ash percentage of PM10 and PM2.5 accounted for 39.35% and 2.42%, and as high as 92.47% and 92.47% after the precipitator. That is to say, the ordinary Electrostatic Precipitator (ESP) obtains low efficiency on the removal of PM2.5[4].

In recent years, a large number of new technologies have been applied to control fine particle emission. The collection efficiency of fine dust is shown in Figure 1.

As shown in Figure 1, as the particulate diameter reduced from 10 μm to 1.0 μm, the corresponding dust collecting efficiency of each technology reduced rapidly, except the wet electrostatic technology. The dust collecting efficiency of this technology is infinitesimally affected by

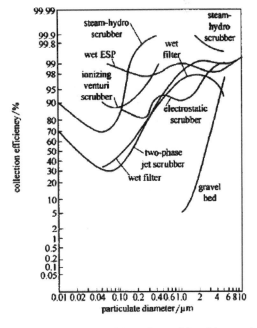

Figure 1. Collection efficiency for particles with control technologies[5].

fine particle diameter, although it obtains a high collection efficiency in the size range of 0.06~10 μm. Based on the above research background, in view of the coal-fired power plant flue gas characteristics and the perspective of the removal of PM2.5, it is significant to study on wet electrostatic precipitation.

2 THE WORKING PRINCIPLE OF WESP

At present, the main comprehensive treatment of coal-fired boiler flue gas is the electrostatic precipitator (bag dust remover, electrostatic fabric filter) and the wet limestone—gypsum method (ammonia method, magnesium oxide method and double alkali method) desulfurization[6]. After desulfurization, the flue gas contains dust and a large amount of desulfurization products and aerosol. The dust emission concentration is difficult to meet the new national standards. This part of the dust particle size is about 0.05~2 μm. In order to remove fine particles (PM2.5) in flue gas and make the dust emission concentration satisfy the new national emission standards, we need to increase the dust removal device. Because after the wet desulfurization process, the flue gas is almost wet and saturated, and the best dust removal equipment is not the bag filter but the WESP.

The WESP worked by metal wire with the action of high DC voltage, and the flue gas was ionized by the electriferous wire. Dust charged in the electric field under the electric field force and moved to the collecting plate. When the dust moved to the surface of a dust collecting plate, it would be removed with the liquid film flow. The three operation stages of the WESP are the same as the dry ESP, which include charge, collecting and cleaning. The collecting electrodes of the WESP are cleansed by water flushing rather than rapping. The dust particles are altered as mud and then drain off [7].

3 EXPERIMENTAL APPROACH

In this experiment, we collected the samples (the dust in collector outlet) from a domestic coal-fired power plant. Its 600 MW units burned the same coal, and the #4 unit installed dry ESP and the #8 unit installed WESP, respectively. The dust samples were measured by a laser particle size distribution analyzer. The particle size distribution data of the collector outlet is shown in the following chart.

Dust charge is the key step of the electrostatic precipitator, and the corona electric field has two charged mechanisms[1]. One is the electrostatic force that makes the ions do directional movement and charged by particle collision ions, which is known as the electric field charge. Another is particle charging by the ion diffusion phenomenon, which is called the diffusion charge; this process depends on the thermal energy of the ion.

Based on the chart, in the removal particulate matters of the dry electrostatic precipitator, the most abundant particle size was 10 μm~45 μm and the least content was the particles whose diameter was below 0.5 μm. The content of the fine particles

whose diameter was less than 1μm decreased rapidly. This can be explained by the major charge process due to the particle size[8]. More than 10 μm large particulate matter was given priority to with electric field charge, and the electrostatic force was great, so that they could be removed effectively. The particles whose size is about 1 μm is in the electric field and diffusion charge mixed zone, and its charged ability is poor, so the removal efficiency dropped significantly. The particles whose size is less than 0.5 μm are mainly charged by the diffusion of charged particles, so that it is difficult to remove them.

Because the dust removal mechanism is the same, the particles removed by the WESP are similar to those removed by the dry ESP. The most abundant particle size was 10 μm~45 μm, and the least content was the particles whose diameter was below 0.5 μm. The content of the fine particles whose diameter was less than 1μm decreased rapidly. However, compared with the dry ESP, the WESP can remove more super fine particles. This is because there is no secondary blowing dust phenomenon in the WESP. Moreover, in the high temperature environment, the electric field existed large amounts of charged droplets. This increased the probability of collision charged of the submicron particles greatly, made the speed of the micron particles approach to the collector sharply, and made the WESP to capture more particles under a high gas velocity.

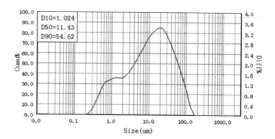

Figure 2. The particle size distribution of removed particles in flue gas (ESP).

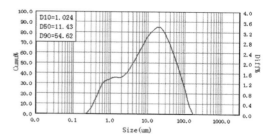

Figure 3. The particle size distribution of removed particles in flue gas (WESP).

Table 1. All levels of particle size percentage and cumulative percentage of removed particles (WESP).

No.	Particle size/μm	Interval/%	Accumulation/%
1	0.212–0.271	0.11	0.11
2	0.271–0.345	0.5	0.61
3	0.345–0.440	1.12	1.73
4	0.440–0.561	1.85	3.58
5	0.561–0.715	2.42	6
6	0.715–0.911	2.66	8.66
7	0.911–1.161	2.81	11.47
8	1.161–1.479	2.9	14.37
9	1.479–1.885	2.87	17.24
10	1.885–2.403	2.94	20.18
11	2.403–3.062	3.22	23.4
12	3.062–3.902	3.65	27.05
13	3.902–4.972	4.2	31.25
14	4.972–6.336	4.77	36.02
15	6.336–8.074	5.34	41.36
16	8.074–10.28	5.92	47.28
17	10.28–13.11	6.35	53.63
18	13.11–16.70	6.66	60.29
19	16.70–21.28	6.84	67.13
20	21.28–27.12	6.73	73.86
21	27.12–34.56	6.26	80.12
22	34.56–44.04	5.56	85.68
23	44.04–56.13	4.77	90.45
24	56.13–71.52	3.89	94.34
25	71.52–91.14	2.93	97.27
26	91.14–116.1	1.87	99.14
27	116.1–147.9	0.76	99.9
28	147.9–188.5	0.1	100
29	188.5–240.3	0	100
30	240.3–306.2	0	100

Table 2. All levels of particles size percentage and cumulative percentage of removed particles (WESP).

No.	Particle size/μm	Interval/%	Accumulation/%
1	0.212–0.271	0	0
2	0.271–0.345	0.21	0.21
3	0.345–0.440	0.93	1.14
4	0.440–0.561	1.9	3.04
5	0.561–0.715	2.72	5.76
6	0.715–0.911	3.16	8.92
7	0.911–1.161	3.33	12.25
8	1.161–1.479	3.42	15.67
9	1.479–1.885	3.34	19.01
10	1.885–2.403	3.22	22.23
11	2.403–3.062	3.31	25.54
12	3.062–3.902	3.64	29.18
13	3.902–4.972	4.06	33.24
14	4.972–6.336	4.47	37.71
15	6.336–8.074	4.86	42.57
16	8.074–10.28	5.29	47.86
17	10.28–13.11	5.51	53.37
18	13.11–16.70	5.51	58.88
19	16.70–21.28	5.55	64.43
20	21.28–27.12	5.66	70.09
21	27.12–34.56	5.63	75.72
22	34.56–44.04	5.47	81.19
23	44.04–56.13	5.19	86.38
24	56.13–71.52	4.61	90.99
25	71.52–91.14	3.93	94.92
26	91.14–116.1	3.01	97.93
27	116.1–147.9	1.63	99.56
28	147.9–188.5	0.41	99.97
29	188.5–240.3	0.03	100
30	240.3–306.2	0	100

4 CONCLUSIONS

As the particles' size is different, their charged ways are diverse, and the electrostatic force is also different. As a result, different sizes of particles obtain different removal efficiencies. The removal efficiency of 10 μm~45 μm particle is great. The particle whose diameter is below 1 μm, the removal efficiency reduced rapidly. The dry ESP obtains low efficiency on super fine particles. Ine th WESP, the electric field of discharge electrode existed large amounts of charged droplets. This increased the probability of collision charged of the submicron particles greatly, made the speed of the micron particles approach to the collector sharply, and made the WESP to capture more fine particles (PM2.5)

REFERENCES

[1] Han Jingjing, Wang Liping, Li Jie. Selection and Application of Technology of Highly Efficient Removal for Flue Gas of Coal-Fired Power Plant [J]. *Environmental Science and Management*. 2011, 36(1): 86–89.

[2] McKenna JD, Turner JH, McKenna JP. Fine Particle (2.5 Microns) Emissions: Regulation, Measurement, and Control [M]. Hoboken, New Jersey: John Wiley & Sons, Inc. 2008: 1–247.

[3] Ehrlich C, Noll G, Kalkoff WD, et al. PM10, PM2.5 and PM1.0 Emissions from industrial plants Results from measurement programs in Germany [J]. *Atmospheric Environment*. 2007, 41: 6236–6254.

[4] Haiyan Fan. Studies on characteristics of Formation and Distribution of superfine Particulates from Coal-fired Boiler and Heavy Metal Contents in Them [D]. Hang Zhou: Zhejiang University, 2002.

[5] Green DW, Peny R H. Peny's chemical engineer's handbook [M]. New York: McGraw-Hill Co., Inc. 2008: 22–55.

[6] Zhang Lei, LI Yantao, AI Hua, et al. The Control of Fine Particulates for Coal-fired Boiler [J]. *Industrial Safety and Environmental Protection*. 2012, 38(6): 28–30.

[7] Chen Zhaomei, Zhao Qinxia, Yang Yanxia, et al. Wet ESP Technology and Its Application Prospect in Power Plant [C]. Proceedings of 14th Conference of ESP. 2011.

[8] Wang Fengming. Study on motion of charged particles in electric field of Electrostatic precipitator [D]. Hebei University, 2006.

Architectural, Energy and Information Engineering – Sung & Chen (Eds)
© 2016 Taylor & Francis Group, London, ISBN 978-1-138-02791-6

Iterative carrier synchronization for SCCPM systems based on EM algorithm

R. Zhou & K.G. Pan
PLA University of Science and Technology, Nanjing, China

H.Y. Liu
The Company of Panda Handa Technology, Nanjing, China

ABSTRACT: Regarding the carrier synchronization difficulty of serially concatenated CPM systems at low SNR, a code-aided carrier synchronization algorithm based on EM algorithm is proposed. The soft decoding information is utilized to coarse synchronization and fine synchronization based on EM algorithm and Maximum Likelihood (ML) criterion, the effective synchronization is achieved by using the joint iteration of synchronizer and decoder, simulations show that. At the low SNR, not only the efficient carrier synchronization can be achieved, but also the almost ideal performance is obtained greatly.

Keywords: CPM; synchronization; SCCPM

1   INTRODUCTION

Continuous phase modulation is a class of high efficient modulation technique [1]. CPM signals have the continuous carrier phase, the concentrated energy of main lobe spectrum and fast attenuation, small adjacent channel interference and high spectrum efficiency, which is especially suitable for the application in the situation of limited spectrum resources. Then, CPM signals have constant envelope characteristics, and also are not sensitive to the nonlinear characteristics of the high power amplifier which can be worked in nonlinear amplification area, thus the efficiency of the system can be improved greatly. Compared with the Phase Shift Keying modulation, the CPM signals have more efficient bandwidth and power efficiency. Meanwhile, it was found that convolution and CPM signal can be combined to Serially Concatenated Code CPM system (SCCPM), which has a similar structure with turbo codes, and then by using the iterative decoding detection method, the superior system performance will be obtained. But the synchronization of SCCPM system is also difficult at low SNR for its high detection complexity.

In recent years, the synchronization method based on Coding Assistance (CA) in iterative detection system has become a hot topic [2–3], which is based on the maximum likelihood criterion, and makes full use of soft information provided by

the CPM soft demodulator, the synchronizer and demodulator are iterated jointly to obtain a superior system performance. Reference [4] proposes a CA synchronization algorithm based on maximum likelihood criterion and simplified pilot, which has large estimation accuracy and range, but band efficient has been down for being data-aided. In [5] a called a priori information (APPA) synchronization algorithm is proposed by utilizing extrinsic information of Turbo decoder as synchronization prior information. In [6] and [7], the posteriori information and extrinsic information of decoder is used for iteration of phase respectively, on the premise that the synchronization of carrier frequency and timing are precise, the better performance of the system, the better the synchronization. A non-data-aided synchronization method based Laurent decomposition is proposed in [8], which is suitable to most CPM signals. Reference [9] proposes a joint timing and phase estimation algorithm based on Laurent decomposition.

CA synchronization method and common synchronization algorithm of CPM are researched in above documents, but few documents are researched for the synchronization algorithm for SCCPM system. It is a nice synchronization strategy that using CA method at the low SNR, therefore, a CA iteration carrier synchronization based EM algorithm is proposed for SCCPM system by using the soft information provided by CPM decoder.

2 THE CPM MODULATION AND SYSTEM MODEL

2.1 CPM modulation

CPM signals can be described by

$$s(t) = \exp\{j\,\varphi(t,\alpha)\} \tag{1}$$

Where $\varphi(t,\alpha)$ is the information-bearing phase:

$$
\begin{aligned}
\varphi(t;\alpha) &= 2\pi h \sum_{i=0}^{k-L} \alpha_i q(t - iT) \\
&= \eta(t, C_k, \alpha_k) \\
&\quad + \Phi_k, \Phi k, \quad kT \le t \le (k+1)T
\end{aligned} \tag{2}
$$

Where h represents modulation index, α_k is the data symbols, $\alpha_k \in \{\pm 1\}$, $q(t) = \int_0^t f(\tau)d\tau$ is the phase response pulse, which satisfies $q(t) = \begin{cases} 0 & t \le 0 \\ 1/2 & t \ge LT \end{cases}$, $\eta(t, C_k, \alpha_k)$ is the phase branch, Φ_k is the phase state, which can be defined as follows:

$$\eta(t, C_k, \alpha_k) = 2\pi \sum_{i=k-L+1}^{k} \alpha_i q(t - iT) \tag{3}$$

$$\Phi_k = \pi h \sum_{i=0}^{k-L} \alpha_i \mod 2\pi \tag{4}$$

The vector $C_k = (\alpha_{k-L+1}, \ldots, \alpha_{k-2}, \alpha_{k-1})$ is the transmitting symbols.

2.2 System model

The system model of the algorithm is shown in Fig. 1. The received signal after through the transmission channel is:

$$r = s(t)e^{j(2\pi\Delta ft + \theta)} + n(t) \tag{5}$$

Where Δf is carrier frequency, θ is carrier phase, channel is zero-mean complex Gaussian white noise channel, whose double-sided power spectral density is $N_0/2$.

3 CODE-AIDED CARRIER SYNCHRONIZATION ALGORITHM

3.1 The EM algorithm

Dempster and Rubin have proposed the concept of EM algorithm, which is used to compute iteratively the ML estimation of a parameter [10]. Replace $p(r|b)$ in formula (6) with $\ln p(r|b)$, then $\hat{b}_{ML}$ will be

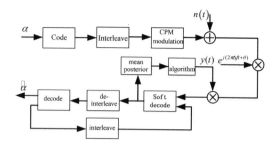

Figure 1. CPM system model.

$$\hat{b}_{ML} = arg_b max \ln p(r|b) \tag{6}$$

The two main steps in EM algorithm are called the Expectation and the Maximization, they are constructed as follows:

E-step

$$Q(b|\hat{b}^{(ln)}) = \int_\alpha p(\alpha|r, \hat{b}^{(n-1)}) \ln p(z|b)d\alpha \tag{7}$$

M-step

$$\hat{b}^{(n)} = \underset{b}{argmax}\, Q(b|\hat{b}^{(n-1)}) \tag{8}$$

Where b is a vector of unknown parameters, while $\hat{b}^{(n-1)}$ and $\hat{b}^{(n)}$ is the last and the current estimation respectively, r is the data received, which is related to the complete data z, defined as $z = [z : z \triangleq (r, \alpha)]$.

3.2 Carrier parameters estimation

The log-likelihood function about parameter vector $b = [\Delta f\, \theta]$ in AWGN channel is defined as

$$
\begin{aligned}
\ln p(r|\alpha, b) &\propto \frac{1}{N_0} \text{Re} \left\{ \int_{kT}^{(k+1)T} r(t)s^*(t) \right\} \\
&= \frac{1}{N_0}\sqrt{\frac{2E_s}{T}} \sum_k \text{Re}\{Z_k(C_k, \alpha_k)e^{-j\Phi_k}e^{-j(2\pi k\Delta f + \theta)}\}
\end{aligned} \tag{9}
$$

Where

$$Z_k(C_k, \alpha_k) = dt \int_{kT}^{(k+1)T} r(t)h^l(t - kT) \tag{10}$$

In formula (10), l is the number of the required matched filters, whose pulse response can be expressed as

$$h^1(t) = \begin{cases} e - j\eta(kT - t, C_k, \alpha_k) & kT \le t \le (k+1)T \\ 0 & else \end{cases} \tag{11}$$

698

Where $\eta(kT-t,C_k,\alpha_k)$ is the phase branch, representing a math in CPM state transition trellis.

The change of phase in CPM signals can be described by CPM state transition trellis. Therefore, the posterior probability $p(\alpha|r,\hat{b}^{(n-1)})$ in Q function is

$$
\begin{aligned}
p(\alpha|r,\hat{b}^{(n-1)}) &= p(\Phi_k,\alpha_{k-L+1},\ldots,\alpha_{k-1},\alpha_k|r,\hat{b}^{(n-1)}) \\
&= p(\Phi_k,C_k,\alpha_k|r,\hat{b}^{(n-1)}) \\
&= p(\Delta_k|r,\hat{b}^{(n-1)})
\end{aligned} \tag{12}
$$

Where $\Delta_k = \sigma_k \xrightarrow{\alpha_k} \sigma_{k+1}$, corresponds to a state transition in the CPM trellis.

Substitution formula (12) and (9) into (7) yields:

$$
\begin{aligned}
Q(b|\hat{b}^{(n)}) &= \sum_\alpha p(\alpha|r,\hat{b}^{(n-1)}) \ln p(r|\alpha,b) \\
&= \frac{1}{N_0}\sqrt{\frac{2E_s}{T}}\sum_k \mathrm{Re}\Big\{\sum_{\Delta k} p(\Delta_k|r,\hat{b}^{(n-1)}) \\
&\quad \cdot Z_k(C_k,\alpha_k)e^{-j\phi_k}e^{-j(2\pi k\Delta f + \theta)}\Big\}
\end{aligned} \tag{13}
$$

Where $Z_k(C_k,\alpha_k)e^{-j\Phi_k}$ can be got from matched filter of receiver, $p(\Delta_k|r,\hat{b}^{(n-1)})$ comes from soft demodulator of CPM signals.

But it is hard to compute parameter b directly for there is no closed expressions in $\hat{b}^{(n)} = argmax_b Q(b|\hat{b}^{(n-1)})$.

In the paper, we sample the parameters discretely and search the value satisfied the formula (8) with specific steps. For example, sampling carrier frequency Δ_f and yielding discrete value Δf_l

$$
\Delta f_l = l\Delta f_{\min}, \quad l\in\{-W,-W+1,\ldots,W-1,W\} \tag{14}
$$

Where $\Delta f_{\min}$ represents the minimum steps of searching, W is the size of the window, so $W\Delta f_{\min}$ represents the estimated carrier frequency.

So the normalized carrier frequency and phase will be

$$
\begin{aligned}
\widehat{\Delta f}^{(n)} &= \arg\max_{\Delta f} \frac{1}{N_0}\sqrt{\frac{2E_s}{T}}\sum_k \mathrm{Re}\Big\{\sum_{\Delta_k} p(\Delta_k|r,\hat{b}^{(n-1)}) \\
&\quad \cdot Z_k(C_k,\alpha_k)e^{-j\phi_k}e^{-j(2\pi k\Delta f_l + \theta)}\Big\}
\end{aligned} \tag{15}
$$

$$
\begin{aligned}
\hat{\theta}^{(n)} &= \arg\Big\{\frac{1}{N_0}\sqrt{\frac{2E_s}{T}}\sum_k \mathrm{Re}\Big\{\sum_{\Delta_k} p(\Delta_k|r,\hat{b}^{(n-1)}) \\
&\quad \cdot Z_k(C_k,\alpha_k)e^{-j\phi_k}e^{-j2\pi k\widehat{\Delta f}^{(n-1)}}\Big\}\Big\}
\end{aligned} \tag{16}
$$

In the paper, this method is called EM algorithm.

4 SIMULATION RESULTS AND ANALYSIS

The simulation results are as follows: the convolution code with generation matrix (7,5) and code rate 1/2, the parameters of CPM are 2RC, M = 2, h = 1/2, the length of code is 1024 bits with random interweave, the demodulator and decoder are all with five iterations, the iteration times of carrier synchronization is 15, and the normalized carrier frequency is $\Delta f = 10^{-4}$, the phase offset is $\theta = \pi/10$.

Figure 2 and Fig. 3 show the relationship between normalized carrier frequency and iterative number and the relationship between normalized carrier phase and iterative number respectively. It can be found that carrier frequency and carrier phase are almost closed to the real value respectively, after the number of the iteration is 10, carrier frequency is closed to $\Delta f = 10^{-4}$, carrier phase is closed to $\theta = \pi/10$, which are the value we set at first. Meanwhile, the bigger the SNR, the faster the speed of the system synchronization parameters close to the real value. As shown in Fig. 2 and Fig. 3, the system synchronization parameters are closed to the real value after the iterative number is 5 in $Eb/N_0 = 3$ dB, while the number of iteration is 10 in $Eb/N_0 = 1$ dB.

Fig. 4 and Fig. 5 show MSE of normalized carrier frequency and phase estimation respectively. Compared with the ideal estimation performance based on date-aided, the curve of DA is given. It is found that the proposed method can estimate the system parameters perfectly at the low SNR. The estimation performance of the proposed method is similar to the DA, and closes to MCRB when $Eb/N_0 \geq 1.5$ dB. When $Eb/N_0 \leq 1.5$ dB, because of the high BER of system, the compute of the path posterior probability is not accurate, so the

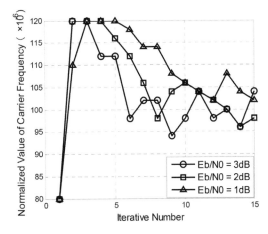

Figure 2. The relationship between normalized carrier frequency and iterative number.

699

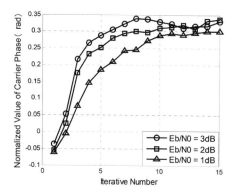

Figure 3. The relationship between normalized carrier phase and iterative number.

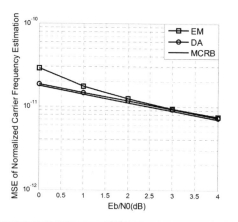

Figure 4. MSE of normalized carrier frequency estimation.

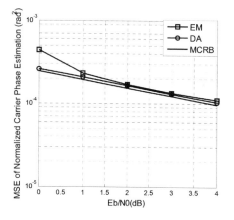

Figure 5. MSE of normalized carrier phase estimation.

effect of the proposed method is influenced, then the BER of the system is influenced, reciprocating alternately, and the process goes into a vicious circle.

5 CONCLUSIONS

This paper proposed a code-aided iteration carrier synchronization method based on EM to resolve the difficulty in accurately synchronization under low SNR for CPM iteration system, and simulated for the proposed scheme, and showed that the proposed scheme closes to the MCRB, and reached the ideal effective. Besides, there is just a slight system performance degradation When $Eb/N_0 \leq 1.5$ dB, system performance closes to MCRB when $Eb/N_0 \geq 1.5$ dB.

REFERENCES

[1] Anderson J B, Aulin T, Sundberg C E, Digital Phase Modulation [M]. New York: Plenum Press, 1986: 256–263.
[2] Herzet C, Noels N, Lottici V. Code-aided turbo synchronization [J]. Proceedings of the IEEE, 2007, 95(6): 1255–1271.
[3] Zhang Zhong-pei, Gao Zhong-jie, Xu Jun-hui. Code-aided synchronization algorithm for high-order QAM [J]. Journal of University of Electronic Science and Technology of China, 2011, 40(6): 825–828.
[4] Shi Z P, Tang F J, Yan H. Joint assisted carrier synchronization algorithm by pilot and code at extremely low SNR [J]. Journal of Electronics and Information Technology, 2011, 33(10): 2506–2510.
[5] Zhang Li, Alister B. Iterative carrier phase recovery suited to Turbo coded systems [J]. IEEE Trans. on Wireless Communications, 2004, 3(6): 2267–2276.
[6] Lottici V, Luise M. Embedding Carrier Phase Recovery Into Iterative Decoding of Turbo-Coded Linear Modulations [J]. IEEE Transactions on Communications, 2004, 52(4): 661–669.
[7] Li Zhang, Burr G, Alister. Iterative Carrier Phase Recovery Suited to Turbo-Coded Systems [J]. IEEE Transactions on Wireless Communications, 2004, 3(6): 2267–2276.
[8] A.N.D'Andrea, A. Ginesi, and U. Mengali, "Frequency detectors for CPM signals," IEEE Trans. Commun., vol. 43, pp. 1828–1837, Feb./Mar./Apr. 1995.
[9] Liu Xiao-ming, Liao Cong. Joint Timing and Phase Estimation Algorithm for CPM Signals [J]. Computer Engineering, 2012, 38(21): 103–107.
[10] A. P. Dempster, N. M. Laird, and D. B. Rubin. "Maximum-likelihood from incomplete data via the EM algorithm". J. Roy. Stat. Soc., 39(1): 1–38, January 1977.

Architectural, Energy and Information Engineering – Sung & Chen (Eds)
© 2016 Taylor & Francis Group, London, ISBN 978-1-138-02791-6

The study on the construction techniques of Jiangxi traditional dwellings

Yi Ping Tang

School of Tourism and Urban Management, Jiangxi University of Finance and Economics, Nanchang, China

ABSTRACT: The traditional dwelling in Jiangxi Province has its unique original ecology and integrity. This paper is organized around the main theme of construction techniques. It analyzes the spatial structure and the form of traditional dwellings. Similar to the building technology of wooden structure, it analyzes the construction technology of the wall, foundation treatment and construction of traditional dwellings. It also summarizes the construction techniques systematically. Finally, it provides the suggestions of protection and inheritance.

Keywords: Jiangxi traditional dwelling; Construction Techniques; Carpentry work

1 THE PREVIOUS PREPARATION OF TRADITIONAL DWELLINGS

1.1 *Selecting the site of construction*

Traditional dwellings in Jiangxi Province all reflect the unique folk customs and local cultural characteristics: from site selection to construction, people will ask the geomancer to practice geomancy before building. The selection of site mainly considers the following aspects:

First, the front part of the house must be open and wide, with no block. In addition, the center of the door must aim at the exacatin of the mountain. This means gathering wealth.

Then, there must be clean and high-quality water around the house. So, it can guarantee the basic needs of life.

Finally, the foundation of the house must be flat and wide. The foundation of Jiangxi traditional dwelling is underground. So, it needs artificial rammed earth above to cover with concrete made by lime, cement and sand, in order to prevent moisture.

1.2 *Choosing material*

In the building of traditional dwellings, wood, bamboo, mud and stone have always been the main building materials. These materials come from nature. Because of different regions, there are big differences in the morphology and properties, and the regional constructing techniques are often based on the local natural materials, as well as on a deep understanding and flexible use of the materials. Using widely available local materials

for building is convenient, and it can also reduce the cost of residential construction to a maximum extent. The pursuit of residential building is not the best, but it is most suitable. The choice and use of materials and tools reflects the craftsman wisdom in construction everywhere.

1.2.1 *The choice and use of wood*

One must be more careful in choosing the lumber before construction. Especially, the materials used for the main room and the main house should be selected carefully. The wood chosen is different because of different components. The craftsman has a specific method for identifying whether the wood is good or not.

Identifying the decay degree of the lumber has several aspects as follows:

First, it is by observing. When the color is pure and uniform, it is a good wood. When it is black and not uniform, it is a bad wood.

Second, it is by breaking off or pinching the wood with hands deeply. If the wood is broken and layered easily, it is proved that it may have the signs of rotting.

Finally, it is by knocking the wood heavily, and listening to the sound. If the sound is dull, it indicates that the central of the wood is empty, and there may be a moth inside.

The wood that is to be cut needs to be air dried. The place where there is no light best lined and avoiding overlapped is chosen. So, it is beneficial to the moisture volatilizing. After several days of moisture volatilizing, it is moved to a sunny place. The drying days is determined by the different seasons. When there is a tiny crack on the surface, the wood can be classified, and stored, and prepared for processing.

1.2.2 *The selection and application of soil*

The craftsman's mastery of the material is mainly manifested in the understanding of the wood. However, the soil and the stone are very important building materials in Jiangxi traditional dwelling construction. The soil of Jiangxi can be roughly divided into sand soil and cohesive soil.

The soil used for making the rammed earth wall is usually made of local materials. It is used in the time of near Spring Festival, and the water is moderate, as well as in the time of slack seasons. However, not all the soil can be used to make the wall. The choice of the soil property is very important.

The soil used for making the wall is good when it can be found with crushed stone or crushed brick as the aggregate. In order to enhance the rigidity of the wall, the shrinkage and deformation of the wall is reduced. When constructing the wall, it is better to use old wall soil because its property is relatively stable, it has no big deformation, and it is difficult to crack. It should be adjusted according to the specific characteristics of the soil. The most crucial thing is to control soil moisture.

1.2.3 *The selection and application of stone materials*

There are two kinds of building stone in a local place: one is the Mihua stone and the other is the sandstone. This is a general classification. Local people have made a more detailed classification over a long period of time using the two kinds of stone material. With respect to stone texture, shape, color and the degree of difficulty in processing, the stone color can be roughly divided into red sandstone, white sandstone and yellow sandstone stone. Red sandstone is the hardest stone. Its core particles are relatively close, smooth, and not bibulous. Yellow sandstone has hardness that is lower than the general sandstone. It is very brittle and easy to weathering after absorbing water. Therefore, it is used less in places such as the eaves wall.

1.3 *Production and application of tools*

Not only the application of big woodwork has many kinds, but also its range is quite different. It contains two categories: measurement of line drawing and making. The measurement and line drawing tool is mainly used for measuring component specifications, drawing lines and marks. Production tools are cutting, sawing, digging, and drilling. They are used to process components after drawing lines.

1.3.1 *Ruler, ink and tenon plate*

This is the most common tools used for measuring and drawing lines in the area of Jiangxi Province during the frame construction of residential house. The most commonly used feet by the craftsman is square and live. Carpenter square is also called the Right angle ruler or Norma. It is mainly used to draw vertical lines and parallel lines, and to measure whether the surface component is straight, and to test whether two adjacent surfaces and after assembly is vertical. The square is usually made by the craftsman. It can measure the angle and oblique Slash drawled arbitrary angle. It can also be used to detect whether the component is correct or not. The method used is similar to square.

One of the master carpenter tools is ink fountain, which is also essential. It is a special tool. It has an elastic line of light and flexible with diverse forms. The end is like a leaf on the head of ancient wooden buildings. There is a central set line wheel, with a tail cut ink well and a small line from the rear through the ink silk wound on the reel. The end of the thread is often tied to hitch a heavy bomb. On the one hand, the painting line is easy to grasp; on the other hand, it is taken as a temporary error correction for wire drawing a line. The ink fountain has horns and perform better.

Bamboo pen is commonly used by cutting the bamboo. It is the ink attachment. It is mainly used to draw the short term, and depends on the ruler. It can be used to make a straight line. Bamboo pen is made of ductile bamboo or horn. The tip can be slightly cut into the arc in order to facilitate the marking pen rotation angle.

A mortise and tenon plate is used for tenon shape lofting, and the tool is also used as a painting component according to the actual need. Before its first use, the wood must be cooked with sesame oil. Then, a line is draw above and set off. Its two ends are dovetail shaped. It can be easily used as a template.

1.3.2 *Adzes and plans*

In wooden processing, the adze is used to cut blocks roughly flat. Thus, it greatly speeds up the progress of work. It is not very easy to process the blocks with adzes in the beginning. This must be completed by skilled craftsmen. At the same time, craftsmen must be familiar with all kinds of wood's characteristic, texture direction, and scar knot. Thus, it will be easy to split, cut or cut flat.

Nowadays, the plane is the most commonly used wood tools. Dado plane is used for fine processing of wood surface roughness. It has a certain flatness and a degree of finishing. In addition to dado planes, there are some special forms of the plane, such as groove planes, rabbet plans, and mounding planes. They are used for processing the components' edge roughness and shape.

1.3.3 Axe, saw and chisel

Axes are widely used not only for wood cutting, chopping, and peeling, but also for knocking at the chisel hole and assembling wooden parts. Axes has double and single edge points. A single blade axe is partial to just one side. The other side is not completely flat. However, slightly concaving to its body is generally about 3 millimeters. The single blade axe is oriented. It can cut the material smoothly. So, it is only suitable for cutting and peeling, but not for splitting. The double axe has both knife grinding inclined sides. The axes are in the middle. It is used flexibly. Also, it can be used for cutting and splitting. However, it is not suitable for peeling. In addition to the professional woodworking, in China, the double axe is widely used in the northern region and single blade axes are preferred in southern region.

Saw is one of the important tools for a woodworker's manual operation. Its main function is to cut the wood followed by the preparation of shovel. According to the different purposes, it can be divided into vertical cutting (cutting along) and cross-cut saw. Longitudinal cutting saw used for parallel to grain direction is anatomy. Cross cutting saw vertical grain direction is disruption. According to the different structures and shapes, it can be divided into frame-saw, inadvertently, cross cut saw and wire saw.

The commonly used chisel is chisel and gouge. Chisel is usually used with axe and hammer. The chisel blade of gouge is thinner than the body, but the body is wider. It mainly relies on the wrist for cutting and blowing components on the part of the plane that cannot be planned. Its edge is thin and sharp. Chisel tools always are single blade.

2 THE CONSTRUCTION SYSTEM OF TRADITIONAL DWELLING STRUCTURE

The brick and wood are the main materials to build houses in Jiangxi traditional dwellings. In these houses, some perfectly carry the weight by woods, and some are mixed with woods and bricks.

2.1 The independence of the bearing system and the maintenance system

One of the important features in the timberwork building is the separation of the bearing system and the maintenance system. When the house falls, the wall will not collapse. This vividly demonstrates the feature of its frame structure.

Timberwork building is the most widely used and distributed types of houses in Jiangxi Province. According to the combination of components, it can be divided into beam-lifted timber system, column-and-tie timber system and combination timber system.

2.1.1 Beam-lifted timber system

Beam-lifted timber system uses the beam to support the purlin, which belongs to the beam-column system. The transfer mode of the roof load is from the rafter to the purlin then to the beam, short column, pillar and then to the ground. Its component uses lots of materials. In addition, it is stable for the dispersion and transmission of the load, so it can be used to build multi-storey pavilion and tower, or a variety of buildings with a special flat.

2.1.2 Column-and-tie timber system

Column-and-tie timber system uses the pillar to bear the purlin, which belongs to the purlin-column system. The transfer mode of the roof load is from the rafter to the purlin, and then to the column, and finally to the ground. Because the tie-beam is the only contacts component, the roof load is directly borne by the columns. As they are closed, it is difficult to get a big space. So, it is commonly used for every column landing. It truncates the column that is not landing on the ground, and puts the landing on the tie-beam.

2.1.3 The combination timber system

The combination timber system has the characteristics of the beam-lifted timber system and the column-and-tie timber system.

It always uses the column to bear the purlin. Sometimes it also uses the beam and column to bear the purlin at the same time. This method combines the structural model of north and south, but also the result of adaptation to local conditions.

2.2 The combination of the supporting system and the maintenance system

There is still a large number of traditional dwellings made by brick and wood or stone and wood that jointly support the load. Also, the system of the vertical and horizontal wall supports the load. The hybrid supporting system is economic and practical, and is widely used in Jiangxi. In the traditional dwellings, timber frames are used to support the load. There are many houses that use the gable as part of the supporting system.

2.2.1 Hybrid supporting system

In brick or stone and wood hybrid bearing housing, a common practice is to place the purlin on the partition wall or gable directly without the beam, which is named the flush gable roof putting on purlin. It is named the flush gable roof putting on girder if the method does not use the column, which is used in the eave wall and columns where it cannot be seen clearly. In residential house, it is common to using the purlin type. When the gable is made into fire-sealing gable, there is a method of putting the purlin into the gable.

2.2.2 Vertical and horizontal wall supporting system

If the internal space is divided into small parts, it is restricted by the load-bearing wall. The vertical and horizontal bearing wall system is mainly used for smaller single building. Generally, there are one to two bays. Also, there is no wall or a transverse wall. It bears the weight directly on the upper part of the wall. This kind of the system is made with less wood, whose roof covers on the wall directly. This is a poor binding with exterior-protected construction. Interior space is not flexible and free as the timber frame bearing system, which is commonly distributed in the countryside or villages.

The links between the elements of the wood frame bearing system are closed overall. The node can simplified as hinged. Moreover, it has cushioning properties. It can gradually decentralize energy caused by external conditions. It has a better seismic performance. However, the seismic performance of the hybrid supporting system is worse because of the heavy load from the roof and the wall will collapse inward during earthquake. However, the structure of wooden frame houses is more complex. Nevertheless, the hybrid supporting system is simpler, economical, and more durable.

3 LARGE WOODEN PARTS CONSTRUCTION TECHNOLOGY OF TRADITIONAL DWELLINGS

In the construction process of big wood components in traditional dwellings, the column is the most important component in the construction process. It needs the craftsmen's advice to take the residue of the column and the tie-beam together, which is installed by the craftsman himself, stitching fixed beams using tenon connections. Beam-column connections tenon is slightly different in different locations.

Drawing processing is not required before processing building components. Design and construction of the whole house are formulated in the heart of craftsmen. They direct the dimensions on the timber, and then the workers processed as required. Generally, during the processing of components, timber for the central room and cube must be a good choice. For circle, it is round; for

square, it is square; For the tie-beam, it is purlins and columns, or where there is no guidance, it can be processed into less round and not party.

3.1 Stress component

The roof trusses of traditional local-style dwelling houses bear the load from the roofing and loads from the frame itself. In the Jiangxi area, an ordinary three-room wooden architecture probably needs more than 8000 tiles, and more than 120 wooden rafters. If coupled with mud, it used in the tile, which is undoubtedly a great weight. These loads to the wooden frame is transferred to the foundation. Therefore, it is very important to take the wooden structure components of stress into consideration.

Timberwork traditional local-style dwelling houses of Jiangxi Province are built with a kind of lifting beam and Chuan-dou-shi structure as the main characteristics of timberwork residence. Pillars in the whole system play a very important role. The wooden frame's four main components to the Liang Tie-beam are associated with the poles. All members are subject to the force will eventually be transferred to the foundation through the poles. Therefore, the design of the column tenon mouth has become the key to the whole wooden structure force, to decide whether it is reasonable or not. The craftsmen considered members of the force and ultimately made the association between the various components of the wooden tenon. The major ones are the following principles:

Mortising on wood should not be too big to avoid as much as possible.

Tenon member should be as long as possible. It can penetrate the best poles. If it is not able to do through the joints, tenon is best to wear to the middle column.

If both horizontal and vertical components intersect at the same location, the small force members avoid large force members, that is, to the main longitudinal and transverse supplement. The main force component must be allowed to wear at least to the tenon middle column.

3.2 The component connections

Tenon connection is characterized by the use of traditional architecture. Ancient craftsmen created many types of tenon connections. Traditional wooden structure houses have been known for seismic capability, which is the credit of each tenon member operably linked to a large extent.

Beam fixing also uses the mortise and tenon joint link way by joining together the different positions of mortise and tenon joint link beams, which are slightly different.

3.2.1 Dovetail

Dovetail tenon connections are the most common kind. Multi-use cartesian plane is used for stitching. Stitching beams use this link way as much as possible. Multi-connection single slot is a multi-slot. When both ends of the member is large, based on the actual situation, the slot number increases.

3.2.2 Open tenon and dowel

Mortise and tenon bright dark residential use in creating process is similar. Besides, it is more for the four corners of the house or at the door stairs and other parts. The open tenon joints are visible. The dark tenon joints are invisible and more beautiful from a visual point of view. Dowels have a variety of forms.

3.2.3 Set tenon

This connection set tenon residential buildings in the process of creating a rare. Under normal circumstances, such as when used for splicing, it occurs due to terrain and other factors that have led to the length of the column deficit.

These flexible tenon structures are used to effectively strengthen the integrity of the wood frame. It is summed up by the local folk craftsmen effective long-term way to withstand earthquakes. Folk craftsmen consider the connections between components. The focus is on between the various components in what tenon is more appropriate.

3.2.4 Tiebeam

Strictly speaking, this is not a tenon construction, but a way of connecting members. It is the role of each of the vertical frame members townhouse mutual string from the formation of a whole. Due to the tie-beam to wear all the pillars of collusion the body of the tongue and mouth column can only wear, it is not like hanging tongue firmly fixed as the pillar which allows the column within a certain range to be loosed but not off the tongue, which in effect under earthquake loading. Timber frame is often deformed, but do not fall apart, which plays a certain role. Meanwhile, due to the tie-beam to frame through each of row houses all columns, wall the columns are fixed in the same plane, thereby reducing the frame distortion. There is deformation of the probability, so that the roof load can be transmitted to the foundation by a reasonable means.

3.3 Component installation

The building of traditional dwelling is a kind of construction way that involves stitching and assembling of prefabricated building elements. Characterizing with the construction of traditional residence is the integration of design and construction activities. At the beginning of the component production, the great carpenter must take into account that how each component should be installed and to think about how to make from the angle of the installation. In this kind of thinking, there lies the he main problem of the scape overlapping each other. When there are several components, the main consideration should be how to install. Thinking is the core of artifacts between should be how to this kind of thinking is often embodied in the component joint on the handling of is to hang on: use straight mortise; consider from the main components and sub-components; and it must make the tendons of bearing into hang down. Instead of making it straight mortise in order to guarantee the rationality of the component force and feasible installation, some special structures of mortise and tenon joint are also considered for the convenient installation and specialized production. For example, cutting the square off from a joint edge is to facilitate member joining together during installation.

4 THE CONSTRUCTION TECHNOLOGY OF TRADITIONAL LOCAL-STYLE DWELLING HOUSES WALL

Wall is a building palisade structure. It is an important part of the residence and not a single house is without the wall. Jiangxi folk common brick hybrid structure houses walls are part of the bearing system and the building technology is an important part of civil construction technology.

4.1 Brick masonry

The Jiangxi folk blue bricks are adopted as a basic for clay bricks. It is hard and its color is soft. It has a very good adornment effect. Therefore, most houses do water the wall. That is to say, they do not paint the wall. However, the blue bricks are general residential use local brick kiln fire. Because there was no strict regulation, its size has different specifications due to regional differences.

Traditional architecture of the wall has a variety of building by laying bricks or stones. Bricks have the specifications of the size. Wall structure choice led to different methods of masonry. In the long-term practice, according to the local geographical location and climate conditions, Jiangxi folk craftsmen chose the suitable lining technology. It is generally divided into two kinds: solid roll build by laying bricks or stones and build by laying bricks or stones. The stress that the wall and metope beared is different. So, folk craftsmen often combined two ways in practice. As a result, they produced the flower roll build by laying bricks or stones that makes the stone roller and bucket method of combination of build by lying bricks or stones.

4.1.1 Solid stone build by laying bricks or stones

Solid stone build by laying bricks or stones. The brick and tile stick and build by laying bricks or stones brick joints between mortar filling. In order to ensure the bearing capacity of the wall, it must be built by laying bricks or stones. And stagger the long distance is at least 1/4 brick. Due to reasons such as the blue brick flies in the specification is inconsistent. By using the solid build by laying bricks or stones method in residential side wall are often mixed with several kinds of masonry in the method, or there are different types and specifications of one side wall brick, such as adding some steep bricklaying in the blue brick wall. It can not only combine the advantages of a variety of build by laying bricks or stones method. But not be afraid that the craftsman goes wrong in the process of building, improving the efficiency of the work.

4.1.2 A bucket build by laying bricks or stones

Open build by laying bricks or stones to fights the wall of build by laying bricks or stones method. From the side of the ordinary brick after laid flat and build by laying bricks or stones formed by one named hollow. The empty bucket is empty bucket wall; Fill in the mud and rubble in the bucket for irrigation pipe wall. Jiangxi Province is located in the middle and lower reaches of the Yangtze River. It is on the hot summer and cold winter weather. The roof load is not. Use an empty bucket wall not only save labor and material. In wall insulation also has really cannot match advantage in general. Filling the bucket wall stability and wind performance better than the cavity wall thick masonry walls can adapt to the external walls of masonry. The light weight thin-walled cavity wall. Not the collision. For interior masonry.

4.2 Earth and stone masonry

Adobe wall is constructed by adobe block which is filled with clay mixed with water and mashed into a mold made of dry plucked out after forming. The general size is bigger than grey brick. It has the advantage of heat preservation and heat insulation performance. Warm in winter and cool in summer to live in this house. Although it's not firing. Better fire performance. But it is afraid of the rain scour erosion so the base of general wall body should use brick wall or stone wall. Adobe masonry blocks generally mixing clay and lime. Commonly basic same as the brick masonry method. But compared with green walls is not easy to guarantee smooth

walls masonry should pay attention to the mud spread evenly when building daub.

The stone wall was built with stone. Jiangxi folk stone wall which contains a stone wall and rough wall two commonly. Stonewall bar is generally used polished. More square shape stone. Rough wall used mostly stone shape is not very structured. So, there are gaps between the stones. But these are not in the mortar to fill the gap. But in pieces. Thin stone to fill.

5 TRADITIONAL LOCAL-STYLE DWELLING HOUSES FOUNDATION TREATMENT AND CONSTRUCTION TECHNOLOGY

The foundation load-bearing part of the bottom of the building in contact with the ground and its role is to load the upper part of the building passed to the foundation. Building a solid foundation is whether or not the key. Half timber house after hundreds of years of wind and rain test can be preserved. With its solid foundation inseparable. Jiangxi common foundation generally constitutes two basic materials: brick and rubble.

5.1 The processing of brick foundation

No matter which kind of foundation, the start of the work is digging groove not only to the slots in the base building foundation, but also to consolidate the base tank at the bottom of the soil. In order to make a more solid foundation, it can be divided into three layers of soil compaction. So that solid foundation and smooth underside. Jiangxi folk craftsmen in brick masonry foundation typically begins at the bottom of the base slot to make concrete block gravel cushion. The height of the cushion needs to be determined according to the height of the building cushion thickness of ordinary residence at about 30 cm to 50 cm, with the pouring of the mortar mix concrete. The traditional lime-based mortar. Add rice or other ash material. Now has more than using cement mortar. To enhance the bearing capacity of the foundation, while saving material brick, masonry foundation tends to make admission. Make the bottom and top surface of the wide and narrow patterns. Masonry should pay attention on both sides of the contract must be consistent. This cannot affect the axis of the vertical wall.

5.2 The processing of rubble foundation

Traditional local-style dwelling houses building construction is generally adjusted to measure the local conditions of local materials. Near the natural rocks more places, in order to save the cost of brick. On the premise of not affecting the quality, the rubble was used as a construction material because of its low cost. The quality of the material is solid. Rubble masonry based in Jiangxi dwellings is common. Rubble can be divided according to the size and shape of the chaotic rubble and peace rubble. Rubble angular flat plane the construction easier to operate. Rubble and chaos of irregular shape. Construction is difficult and must be performed by an experienced chef operation. There are certain tips when choosing a stone. For example, when laying the first layer of rubble, we should choose large flat stones. And we must make great face in order to ensure carrying capacity. Rubble foundation has as filler between the stones gravel with a sheet. But also the use of mortar pouring ensure dense void-free basis to ensure that the upper part of the building will not loose and tilt base.

6 THE INHERITANCE OF TRADITIONAL DWELLING BUILDING SKILLS

Traditional local-style dwelling houses in Jiangxi Province face problems nowadays, so we should take both residential living environment and cultural environment into consideration. On the one hand, the modern material civilization has changed the traditional way of life and living style. And traditional houses are facing the impact of new material, and thus new technology and new energy. Architectural form has also gradually lost its original technical frame. On the other hand, the craftsmen hold pessimistic attitude on the inheritance of the traditional houses, making the building skills in the inheritance and protection of traditional folk house at a disadvantage position.

Jiangxi traditional residence construction technique has been inherited by the form of oral body. There is no system of writing and standardized processes. The inheritance way will lead to the loss of residence construction techniques. The more complete preservation of the houses was investigated on the spot measurement by relevant experts and scholars, and drew the construction drawings completely and made the residential building structure form. Construction methods and materials. The characteristics are completely reflected. It requires a combination of terrain rendering spatial distribution and residential buildings to conduct a comprehensive and detailed analysis, so as to make the whole process of creating skills and systematic standardized. It turnsg the construction craft from communication and language of storage into a complete, and detailed construction drawings and symbols into words. All of these factors will create an effective inheritance of traditional residential technology.

REFERENCES

Min Kang. 2010. Analysis of Static Performance of Mortise and Tenon Joint in Traditional Timber Structure. Beijing: Beijing Jiaotong University.
Wei Zhang. 2014. Xiang Xi Tujia Traditional Houses Build Skills and Heritage Studies. Wuhan: Hubei University of Technology.
Xin Yi. 2010. Rural Renewal Planning and Related Planning Law and Regulations in Germany. *Urban Planning International:* 25(2): 11–16.
Yi Xiong. 2011. Research on Traditional Construction Techniques and Inheritance of Northern Hainan. Wuhan: Huazhong University of Science and Technology.
Zhuo Liu. 2010. A Study on the Construction Techniques of Hubei Traditional Dwelling. Wuhan: Wuhan University of Technology.

Architectural, Energy and Information Engineering – Sung & Chen (Eds)
© 2016 Taylor & Francis Group, London, ISBN 978-1-138-02791-6

Simulation of flow field in mid-pressure butterfly valve of nuclear steam turbine

W. Rao
China Nuclear Power Engineering Co. Ltd., Shenzhen City, Guangdong Province, China

X.R. Zhang, Y.F. Liu, J.Q. Gao & Y.H. Huo
School of Energy, Power and Mechanical Engineering, North China Electric Power University, Baoding City, Hebei Province, China

ABSTRACT: In this paper, the use of CFD software FLUENT is aimed at numerical simulations for the mid-pressure butterfly valve of nuclear power steam turbine. Calculation model uses the incompressible flow Reynolds equations when turbulence model adopts the standard mode. Discrete pressure equations and the coupled equations adopt the semi-implicit method. The numerical simulations of the flow characteristics are obtained when main valve is fully open and control valve opens under three working conditions. Clear velocity distribution in valve under different operating conditions is represented, which as well as shows the formation and development of the vortexes.

Keywords: FLUENT; numerical simulation; flow field; butterfly valve

1   INTRODUCTION

Nuclear power whose development in the world accords with the trend of energy utilization is an effective way to meet the power demand and reduce the environmental pollution. The starting mode of mid-pressure cylinder in the steam turbine startup mode has a lot of advantages, for example, the uniform heating reduce the waste of life and the shortening of start-up time reduce the fuel costs of boiler startup and so on. Numerical simulation of flow field in the mid-pressure butterfly valve of nuclear power steam turbine is studied in this paper. Butterfly valves in mid-pressure cylinder have the effect of regulating and truncating medium flow. It's necessary to have a precise understanding of hydraulic characteristics for the design of the large diameter butterfly.

Valve analysis used experiment methods in the past were performed, which required a number of equipments, a lot of time, fund and so on. With the development of the Computational Fluid Dynamics (CFD), the approach for using the technique of computational fluid dynamics has been substantially appreciated in mainstream scientific research and industrial engineering communities. There have been also many reports on valves used Computational Fluid Dynamics analysis. Amirante[1] maked a comparison on the experimental study and numerical simulation of regulators, confirming the reliability of the CFD method[2]. Some scholars analyzed cavitations' generation and development of butterfly valve[3-4]. Liu jian & Li futang maked flow field simulation for simple butterfly valve[5]. Huang guoquan & Cao zhongwu analyzed the numerical simulation for centric type butterfly valve[6]. Cui baoling et al. simulated the flow field in butterfly valve whose structure is improved by CFD[7].

The numerical simulations which have been studied for internal flow characteristics of butterfly valve are simple with a single butterfly disk, The author, using CFD software FLUENT, makes numerical simulations for the butterfly valve with a main valve and a control valve on the flow field under different working conditions. Keep the main valve fully opened and take it as the basis to analyze the flow field in different control valve's openings, and then make numerical simulations for the internal flow field so as to reveal the velocity distribution in the valve, and it as well as could provide the basis for the improvements of butterfly valve's design.

2   NUMERICAL SIMULATION

2.1   *Model building*

Take the inlet valve of the steam turbine's mid-pressure cylinder about a 1000 MW nuclear power plant as the research object, where the pressure and temperature are relatively low but it is well worth making researches based on what have been talked. The inlet valve studied in this paper is butterfly valve that is mainly composed of valve body, valve shaft and valve disk. The most important thing in the process of simulation is to establish and simplify the model. In order to get the flow field inside butterfly valve, a lot of work needs to be prepared. Firstly, professional CAD software SolidWorks is used to create three-dimension geometric models for the butterfly valve which consists of a main valve and a control valve. As shown in figure 1, the left one is the main valve and on the contrary the control valve is on the right.

The valve's opening, shutdown and regulating fluid flow in the channel is through reciprocating rotary of valve shaft connected to the valve disk and the actuator outside valve body. The valve's structure is shown in figure 2.

2.2 *Flow model and mesh division*

The model of figure 1 above is actual valve's size, but there is a question that the upstream length before the main valve disk and downstream length behind control valve disk are too short to get the steady flow if we take it as the research object. The upstream length L_1 and downstream length L_2 should be several times of the pipe diameter in order to get enough entry length and a fully devel-

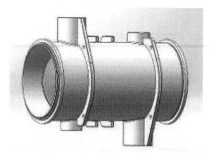

Figure 1. The entity figure of butterfly valve in the mid-pressure cylinder.

Figure 2. The 3-d model of butterfly disk.

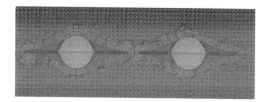

Figure 3. Mesh division near valve disk of y = 0 cross section.

oped flow downstream (valve diameter D is 1.2 m). We will take the upstream length L_1 five times of the pipe diameter and L_2 ten times of the pipe diameter as new studied model.

Because of butterfly disk's irregular structure, using ANSYS ICEM CFD software, the butterfly disk area is used with unstructured grids while areas in front of main valve and at back of control valve are adopted the structured grids for mesh division.

Flow analysis around the valve disks is the important part of the research, thus boundary layer is added to the main valve's and control valve's surface (Fig 3), eventually the whole grid total reaching 1.25 million.

2.3 *Control equation and calculation method*

The flow inside butterfly valve is three-dimension flow and the fluid medium is saturated steam. The method discreteness the equations in a collocated grid node with second order upwind scheme to ensure the accuracy of the simulation results. FLUENT, which uses the Semi-Implicit Method for Pressure-Linked Equations (SIMPLE) algorithm with an iterative line-by-line matrix solver and a multigrid acceleration, provides the velocity and pressure as solutions for the butterfly valve flow. The flow is stationary and the maximum residual error for convergence criteria of flow calculation is less than 10^{-4}. The inlet boundary is set by velocity import and pressure export is set as export boundary.

3 THE RESULTS ANALYSIS OF FLOW FIELD

Main valve that ensures safety for start-up and operation of steam turbine is one of the important valves to control the steam channel. Main valve of the steam turbine has usually only two kinds of situations: fully opened and closed. During normal operation of the steam turbine, main valve is kept fully opened, whose function is to quickly shut down under emergency, cutting off the steam into the steam turbine passage and preventing the happening of the accident. Control valve of steam turbine is used to adjust the steam flow rate and pressure of the valve. When the butterfly disk rotates to different angles, the flow area in pipe immediately changes, regulating the amount of admission and adapting to the need in different working conditions.

Keep the main valve fully opened, using FLUENT software solver, and evaluate the internal flow fields for comparative analysis with control valve's three openings for 100%, 50% and 10%. With the operation condition, steam parameters in different openings of control valve are listed in the table below.

3.1 Control valve at full opening

To make a study for the flow field inside the valve, the flow of y = 0 cross section is selected to analyze the velocity distribution in this paper. The main valve is on the left side and the control valve is on the right side (the same below). Keep the main valve and control valve both fully open at this moment in figure 4.

When the main valve and control valve are both fully open, the flow area of flow channel is relatively the largest and the velocity symmetry in valve is better than other situations at this moment. Because of the large transition area for flow rate, the velocity gradient is relatively small and the entire flow is smoother. The flow area decreases, leading to the increase of flow velocity, due to the block of valve disks, which gives rise to a certain local hydraulic loss. The velocity distributions around the main valve and control valve are in conformity with the real situation.

3.2 Control valve at half opening

In velocity nephogram 5, when the main valve is fully open, around which velocity gradient is small, velocity is well-distributed just because of the main valve's minimum block. When the control valve is in the half opening, the flow area around control valve decreases and high velocity is distributed between valve body and control valve disk, which has an impulsive force on pipe wall.

Because of the shape and structure of the butterfly valve disk, the flow upper through the control valve disk is close proximity to tube wall. The flow below control valve disk moves to upper area (Fig. 6) due to the low pressure area behind control valve disk, developing small vortexes at the same time (Fig. 7). Compared to first simulation result, the flow field within valve is no longer stable as the former.

3.3 Control valve at 10% opening

We take the control valve disk as the research boundary and flow condition in front of and at back of the control valve disk is totally different. At the entrance section, steam parameters remain roughly unchanged. At this time the control valve's nearly shutting down, so the flow area is very small. After the steam flows through the passage between butterfly disk and valve body, the volume of steam expands sharply and the flow rate of steam increases rapidly in such a short trip, appearing complex three-dimension flow phenomenon (Fig. 8). The flow through control valve's bottom moves upward, appearing clear vortexes behind control valve disk (Fig. 9).

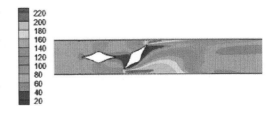

Figure 5. Velocity nephogram for control valve's half opening.

Table 1. Steam parameters in different openings of control valve.

Opening	Steam flow kg/s	Pressure bar	Density kg/m³	Inlet velocity m/s
100%	268	9.35	3.87	61.3
50%	128	4.58	1.86	61
10%	35	1.26	0.51	61.9

Figure 4. Velocity nephogram for valves' full openings.

Figure 6. Velocity vector diagram for control valve's half opening.

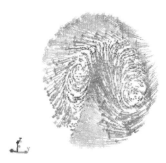

Figure 7. Velocity vector diagram of a cross section behind control valve.

Figure 8. Velocity nephogram for control valve's 10% opening.

Figure 9. Velocity vector diagram for control valve's 10% opening.

4 CONCLUSIONS

FLUENT has been applied to investigate three dimensional fluid phenomena of butterfly valve flows. The computational studies have shown that the flow field through the butterfly valve is complicated and what the simulation result is depends on the openings of valve disk. With the FLUENT used, it is convenient to observe clear velocity distribution in valve under different operating conditions.

Based on the simulation results of three working conditions above, it could be concluded that: (1) when main valve and control valve are fully opened, the flow area is large, so the overall velocity gradient is relatively small. The impulsive force that valves and pipe wall are suffering from is small and the flow inside valve is relatively not complex and stable. (2) When the control valve is half opened, the decrease of flow area leads to flow velocity's change. Flow velocity increases obviously especially when the flow goes through control valve disk, behind which vortexes appear to generate. (3) With the continuous decrease of the opening for the control valve, the increase of velocity gradient around butterfly disk is extremely obvious and the range of vortexes behind control valve disk becomes larger than the former.

REFERENCES

[1] Amirante, G. Del Vescovo & A. Lippolis. 2006a. Evaluation of the flow forces on an open centre directional control valve by means of a computational fluid dynamic analysis. *Energy Convers Manage* (47): 1748–1760.

[2] R. Amirante, G. Del Vescovo & A. Lippolis. 2006b. A flow forces analysis of an open centre hydraulic directional control valve sliding spool. *Energy Convers Manage* (47): 114–131.

[3] Baran, G et al. 2007. On cavitation and cavitational damage at butterfly valves [C] //*Proceeding of the 2nd IAHR International Meeting of Workgroup on Cavitation and Dynamic Machinery and systems.* Timisoara, Romania: [s.n.].

[4] Baran, G et al. 2010. Control the cavitation phenomenon of evolution on a butterfly valve. *IOP Conference Serious: Earth and Environmental Science* 12(1): 1–9.

[5] Liu, J et al. 2008. Numerical simulation and Analysis on 3D flow in large diameter butterfly valve. *Fluid Machinery* 36: 30–33.

[6] Huang, G.Q et al. 2011. The research on numerical simulation of centric type butterfly valve flow field. *Machinery Design and Manufacture*: 186–188.

[7] Cui, B.L et al. 2013. Butterfly disc structure improved design and numerical analysis based on CFD. *Journal of Drainage and Irrigation Machinery Engineering* 31(6): 13–17.

Research on the tendency of indoor interface design

Qi Zou & Yong Wen Yang
School of Architecture, South China University of Technology, Guangzhou, Guangdong, China

ABSTRACT: Indoor space is an important part of architectural space, and its interface design is the deepening and continuing of its architectural design. This paper summarizes the main techniques of contemporary interior space's interface design in order to help people understand the interior design and the architecture design better through elaborating several concepts of architectural interior space's interface and studying several typical contemporary Western interior design cases.

Keywords: interior design; architecture design; space; interface

1 INTRODUCTION

The enclosure of interface and space should be described as 'the interface reveals the space' instead of 'the interface encloses the space'. The nature of space is originated from the enclosed interface. The enclosure of interior space and interface includes floors, ceilings and walls.

1.1 *Floors*

Floors function as a supporting indoor space both materially and visually, and carry indoor human activities. There are two main approaches in floor design. The first one forms plentiful interior space changes through setting the floor different elevations. This approach is very commonly used in large public space in order to divide the whole space into regional spaces with different functions. The second one is the pattern change of the floor, using different materials and colors to make an interesting collage of indoor floor space aiming to an active indoor atmosphere, for example, stone mosaic in lobbies, printed carpets in ballrooms and colorful plastic collages in children's activity room.

1.2 *Ceilings*

Ceilings are a very important part of indoor space's interface. In many cases, walls and floors are often obscured by a large number of furniture, making it difficult for people to form an overall visual experience. As a result, the features of the space are mainly reflected by the ceilings that unify heterogeneous scenes and organize the spaces orderly. Meanwhile, in several large open spaces, such as banquet halls

and lobbies in office buildings, using the changes in ceilings is the main approach to organize spaces with small furniture and simple walls. Besides, indoor ceiling lighting can influence the atmosphere greatly.

1.3 *Walls*

Walls have a crucial impact on human vision because the visual experience of the height of the space is mainly from the walls. The horizontal and vertical partition of walls and the arrangement of windows and doors are related to the blending of the space, and the artistic style of the indoor space is mainly reflected by the design of the walls. What's more, partitions and big furniture, which function as splitting the space, often appear as walls.

2 MAIN TRENDS

2.1 *Transparent interfaces*

The use of glass and steel has reached a very high level in the buildings designed by the modern architect Mies Van der Rohe; contemporarily, the use of glass and other transparent materials is still a wide and important method of designing.

The lack of indoor space's interface often can change the features of the space, and contributes to the formation of different degrees of open space. Transparent surfaces have the dual characteristics of both open and closed surfaces, making the outdoor space and the indoor space integrated and forming mentally spatial continuity. People can feel the depth and expression of the space through transparent interfaces. In the contemporary architecture design, many large public spaces, such as libraries, office buildings and shopping malls, are with a large scale of glass curtain wall to create a transparent visual effect and form a perfect transition between the outdoor space and the indoor space.

2.2 *Cutting indoor interfaces*

From Richard Meier's architectural work, we can see that he is a real master in cutting indoor interfaces. In Grotta House, New Jersey, Meier made the structure independently with walls, separated perpendicular walls with ceilings so as to create several independent surfaces and show the original characteristics of the interfaces: a sense of direction and ductility. Meier used windows to connect ceilings and walls so as to make the indoor space loose and relaxed and fused with each other. In this case, the indoor space is no longer a closed entity's inner visual experience, but it is the combination of both indoor space and outdoor space while people can enjoy the landscape in the building as well as enjoy the indoor scenes from outside.

2.3 *Multi-folding*

Monomer space usually acquires very jagged interfaces and forms a strong dynamic visual impact through folding the interfaces multi-angle, multidirection and repeatedly. In this way, the collapsible nature of the interface can be fully exploited.

Hotel Puerta America, in Madrid, Spain, has 12 floors, and every floor is designed by one of the world's top architects or interior designers. The fourth floor, designed by Plasma Institute, forms an image of anti-traditional hotel by using a lot of stainless steel with a radiation effect on the public space, broken triangle blocks to make the walls and ceilings look uneven, fully taking advantage of increasing the surface area of reflection to create a space filled with blurred streamer atmosphere.

In the rooms, the Plasma also uses a large number of broken mirror, geometrical stainless steel and linear lighting to create a dynamic visual impact, and strengthen the spatial experience of advanced fashion.

2.4 *Fragments*

Deconstruction doctrine appeared to promote the disintegration of the traditional architectural form, and many classic architectural styles and forms are often collaged disassembled and disorderly in one building to make people focus on the deep description of time and space experience, which is what the building wants to express, instead of focus on subjective aesthetic form of the building.

Interior interface design techniques, often by colliding architectural form, the body's own burst and the dismemberment of the surface to form fragments effect so as to create a new order and organizational logic to form a strong visual impact in indoor space. The spatial experience brought by the fragments also adds the concept of time into indoor space. Zaha Hadid started using the form of fragments when she designed in 1982 in The competition program The Peak Club in Hong Kong. Later, in Hadid's works such as the Vitra Fire Station, Weil Gallery and Moonsoon Hotel, people can see the influence of fragments.

2.5 *Curly interfaces*

Curl interfaces are one of the important techniques of contemporary architecture and interior design, in which the space can be sporty and organized by curling. Curly interfaces make vertical facade elements disappear, so as to make the plane and the facade of space and even the ceilings as a whole while forming a natural continuation of indoor space and making the space flow between the interfaces to maintain the unity of the whole. Interfaces function as props separating space, and have become one of the goals of space aesthetics because of their beautiful colors and flowing lines. Now, a new kind of space constructing logic has been established, and the traditional dull experience of space has been broken and replaced, of which is the kind of spiritual freedom. Curly interfaces are used by many architects such as Eyebeam Studio, FOA, and Rem Koolhaas.

2.6 *Collages*

Space is inseparable from its material properties because absolute, homogeneous and non-feature space cannot exist in real life. Architectural materials are an important factor in determining the spatial characteristics and temperament, and the physical beauty also determines the quality of the space. Interfaces' material properties include the interfaces' materials and forms, and many other factors. Interface is the condition to create a space objectively and the material is the method to construct the interface. The use of the material can be adjusted to different interfaces and spaces, such as shaping the directions of the interfaces and the spaces. With the development of architecture, people pay more attention to the culture and ecology, and the materials of interfaces have been valued more.

3 SUMMARY

This article focuses on the main characteristics and tendencies of contemporary indoor space, analyzes the reasons for the formation of a wide range of spatial forms, and also analyzes the contemporary indoor space's difference from the past.

Multiple changes in the development of contemporary interior space is the root cause of interface design to produce a change. Approach is the way to create a sense of space. The results are not pursued, not one-sided pursuit of design techniques in isolation, but they should be placed in the context of the development of the interior space to further understand the relationship between the practices and the development of space, to the future of architectural design interior design practice and practice to provide guidance and inspiration.

REFERENCES

Cuito & Aurora. 2002. *Rem Koolhaas/ OMA*. Krefeld: Te Neues Publishing Group.

Jean Louis Cohen. 2001. *Frank Gehry, architect.* New York: Guggenheim Museum Publications.

Jeremy Meyerson & Philip Ross. 2003. *21st Century office.* New York: Rizzoli International Publications.

Philip Jodidio. 1999. *Building a New Millennium—Architecture Today and Tomorrow*. Cologne: Benedikt Taschen Verlag.

Philip Jodidio. 1996. *Richard Meier*. Cologne: Benedikt Taschen Verlag.

Architectural, Energy and Information Engineering – Sung & Chen (Eds)
© 2016 Taylor & Francis Group, London, ISBN 978-1-138-02791-6

Experimental study on high energy level dynamic compaction of collapsible loess with large thickness and low water content

C.L. Wang & H.H. Geng
No. 12 of Dewai street, Beijing, China

E.X. Song
Tsinghua University, Beijing, China

ABSTRACT: Dynamic compaction tests are carried out to study the high energy level dynamic compaction parameters and their effects on the large thickness collapsible loess foundation with low water content in Lanzhou region. Analyses are performed on the regularity of the main physical and mechanical indexes of soil in each test area before and after dynamic compaction. According to the parameters for dynamic compaction, such as coefficient of consolidation, coefficient of collapsibility, the treatment effect of dynamic compaction is analyzed and the effective influence depths of dynamic compaction under the energy levels of 8000 kN · m, 10000 kN · m or 12000 kN · m for the collapsible loess with large thickness and low water content are calculated. The test results can provide a reference for the design and construction of engineering projects of the same kind.

Keywords: collapsible loess; high energy level dynamic compaction; coefficient of consolidation; coefficient of collapsibility

1 INTRODUCTION

Loess is widely distributed in China, especially concentrated in the middle and western regions of China[1]. The distribution area of loess is approximately 4.4×10^5 km². The maximum thickness of loess can reach 400 meters deep. In recent years, with the implementation of western development strategy, more and more engineering projects are constructed in the loess area. For the study of how to improve the bad traits of loess foundation and how to satisfy the requirements of upper structure, the engineering profession has made great achievements and accumulated rich experience[2]. However, with the decreasing construction land, construction projects are developed from low terrace toward high terrace where the depth of loess varies from shallow to deep. As the thickness of loess layer increases, theoretical studies and engineering experiences for the the treatment of large thickness of collapsible loess are limited[3]. Therefore, it needs further investigations to study the properties of large thickness of collapsible loess.

Dynamic compaction is an effective approach to deal with the collapsible loess. It has the features of economic, efficient and better treatment effect, and thus has been widely applied in many engineering projects[4]. However, the dynamic compac-

tion method also has constraints, for instance in order to obtain good reinforcement effect, it needs certain requirements for water content of the loess. In case the water content is too small (less than 10%), the soil should be humidified. Humidifying treatment usually requires longer time, whereas some large-scale projects have great demands for water, and the Loess Area is often short of water resources[5].

The main objective of this study is to evaluate whether high energy level dynamic compaction can be directly used for the collapsible loess foundation with low water content in some projects, ignoring the humidifying process to shorten the duration and reduce the requirement of water. Furthermore, this paper investigate the effective influence depth and reinforcement effect for more than 8500 kN·m compaction level and provide reference for the high energy level dynamic compaction applied in large thickness collapsible loess foundation treatment in future.

2 TESTING AREA

The testing ground, which located in Lanzhou City, Gansu Province, is the testing section of a large foundation project. The study site belongs to a loess hilly region at the edge of the basin, and almost no engineering experience exists for this area. The project site is a collapsible loess ground with large thickness and the collapsible level is four. The nature characteristics of the soil in the testing area are given in Table 1.

During site investigation, only a small amount of groundwater outcrops are found. The water depth for these outcrops is 62.3 m, being judged as local distribution of stagnant water. However, groundwater outcrops are not observed in most of the field area. Therefore, it doesn't need to consider the influence of groundwater in this study.

3 EXPERIMENTAL SCHEME

Three typical areas with the size of 30 m × 20 m are selected as the sites for dynamic compaction test. Before the dynamic compaction test, humidification is not carried out within the study areas. Dynamic compaction tests are performed with the energy levels of 8000 kN · m, 10000 kN · m and 12000 kN · m. All the tamping points are arranged in a square grids. Then, full tamping with the energy level of 1000 kN · m is carried out after the tamping points bulldozed. The full tamping hammer marks are overlapped by 1/4. The parameters of dynamic compaction test for each testing area are shown in Table 2.

After the dynamic compactions, three exploratory wells are excavated in each testing area. One of these wells should be located at the soil of tamping point. Samples from exploratory wells are taken to perform laboratory tests, such as physical and mechanical tests, compaction coefficient test and coefficient of collapsibility test.

4 TEST RESULTS AND ANALYSIS

4.1 Dynamic compaction results for the testing area

During the dynamic compaction tests, the impact number of single point is controlled according to the average settlement of the last two hits, the soil uplift surrounding the tamping pit, etc. The specific parameters are shown in Table 3.

The tamping tests show that the soil around the pit is compacted gradually and no large bulge deformation occurs, as the impact number increasing. In the testing process, when the energy level for a single dynamic compaction is given, larger settlement developed during the initial six hammer tamping, then the settlement became smaller and gradually converged during the later six hammer tamping. In the period of first pass tamping, as the soil becomes compacted, subsequent impact number of single point decreases gradually. The test result analysis shows that the greater the energy level is applied, the more impact numbers of single point is needed and the greater cumulative settlement will be developed. With the increasing tamping energy levels, in order to ensure the effectiveness of the dynamic compaction, the space between tamping points should increase and the impact number should also increase.

In general, the soil below the surface by more than 0.1 m–0.3 m could reach dense state after the dynamic compaction. Low water content within the loess does not significantly affect the dynamic compaction with high energy level.

4.2 Variations of physical and mechanical indexes of the soil

When the dynamic compaction finished, exploratory sampling in the testing area is carried out to identify the physical and mechanical parameters of the soil by laboratory tests. Three exploration wells are arranged in each testing area. Two of them are in the range between tamping points and one is

Table 1. Natural characteristics of the soil in testing areas.

Soil	Natural water content (%)	Natural density (g/cm³)	Natural void ratio	Plasticity index
② Loess-like silt	8.095	1.461	1.005	9.499
③ Malan loess	7.535	1.507	0.943	10.078
④ Silty clay	18.99	1.88	0.717	11.27

Table 2. Dynamic compaction parameters for testing area.

Partition number	Click level (kN · m)	Hammer weight (T)	Hammer diameter (m)	Fall distance (m)	Impact number	Tamping points distance (m)	Arrangement of tamping points
Q1	8000	50.1	2.5	16	2	4.0	Square
Q2	10000	48.0	2.7	21	3	4.0	Square
Q3	12000	54.5	2.6	22.1	4	4.0	Square

Table 3. Dynamic compaction process parameters for the testing area.

| Partition number | Click level (kN · m) | Impact number | | | | Average settlement of the last two hits (mm) | Cumulative settlement (mm) |
		First time	Second time	Third time	Fourth time		
Q1	8000	10	6			40–90	730–2010
Q2	10000	12	10	6		20–80	1100–2450
Q3	12000	15	12	8	6	25–90	1780–2850

Table 4. Compression modulus before and after the dynamic compaction.

| Sampling depth (m) | Undisturbed soil | The energy level of 8000 kN · m | | The energy level of 10000 kN · m | | The energy level of 12000 kN · m | |
		Soil between tamping points	Soil of tamping point	Soil between tamping points	Soil of tamping point	Soil between tamping points	Soil of tamping point
1	17.2	20.4	23	19.8	25	20.2	18.7
2	18.2	19.4	22.2	19.6	19.2	24.4	19.8
3	15.2	27.8	23.8	21.3	17.9	16.3	32.8
4	17.1	22.2	20.2	13.9	16.7	22.7	28.6
5	12.4	20.4	20.2	22.5	19.8	23.8	27.0
6	17.4	20.2	21.5	20.8	17.9	33.3	26.8
7	15.3	20.8	14.0	20.8	20.2	29.9	31.0
8	17.0	17.1	16.8	22.5	20.2	20.0	20.3
9	11.6	11.8	11.7	22.2	10.7	29.4	21.5
10	9.0	9.8	9.1	18.3	9.3	12.5	10.3
11	7.4	7.3	7.3	6.3	7.8	12.7	15.7
12	6.5	6.8	5.6	5.7	6.6	9.9	12.2
13	8.3	8.0	8.2	8.3	8.8	10.8	9.2
14	7.6	7.8	7.5	7.7	7.9	7.9	7.5
15	9.1	9.9	9.0	9.0	9.7	9.0	9.3

located on tamping point. The depth of exploratory well is 15 m and the sampling interval is 1.0 m. Compression modulus values after the dynamic compaction are shown in Table 4.

The physical and mechanical indexes of the soil are enhanced and improved after the dynamic compaction. Some indexes vary greatly in the shallow layers, but vary small in deep depth, such as the soil density, the void ratio and the compression modulus. Dynamic compaction with high energy level can effectively reinforce the collapsible loess foundation within a certain range. Under the energy levels of 8000 kN · m, 10000 kN · m and 12000 kN · m, the effective reinforced depths of dynamic compaction are 7 m, 9 m and 11 m respectively.

4.3 Compaction coefficient after dynamic consolidation

The compaction coefficients after dynamic compaction are shown in Fig. 1. For the testing areas under the energy levels of 8000 kN · m, 10000 kN · m and 12000 kN · m, the compaction

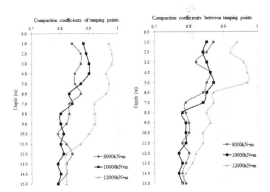

Figure 1. Variations of compaction coefficients for testing area.

coefficients of the soil at the tamping points, in the range of 1.0–15 m, vary between 0.79–0.87, 0.79–0.90 and 0.82–0.98, respectively. The compaction coefficients of the soil between the tamping points, in the range of 1.0–15 m, vary between 0.78–0.83

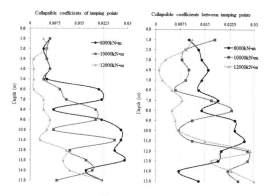

Figure 2. Variations of collapsible coefficient for testing area.

0.77–0.85 and 0.78–0.97, respectively. The value 0.84 is often seen as the required minimum compaction coefficient. Thus it can be concluded that when single-point tamping with the energy levels of 8000 kN · m, 10000 kN · m and 12000 kN · m applied, the effective influence depths of dynamic compaction are about 6 m, 7 m and 11 m, respectively.

4.4 Coefficient of collapsible after dynamic consolidation

The collapsibility coefficients after the dynamic compaction are shown in Fig. 2. When the dynamic compaction is completed, for the testing areas under the energy levels of 8000 kN · m, 10000 kN · m and 12000 kN · m, collapsibility coefficients of the soil at the tamping points, in the range of 1.0–15 m, vary between 0.004–0.029, 0.004–0.026 and 0.001–0.022 respectively. The collapsibility coefficients of the soil between tamping points, in the range of 1.0–15 m, are 0.007–0.027, 0.007–0.028 and 0.001–0.029 respectively. After the treatment by single-point tamping with the energy level of 8000 kN · m, 10000 kN · m and 12000 kN · m, the loess collapsibility in the treated ranges of 6 m, 7 m and 11 m respectively will be eliminated. The test results are consistent with those from previous researches, and the correlation between collapsibility coefficient and the compaction coefficient is clear.

5 CONCLUSIONS

1. For the collapsible loess foundation with low water content, direct using of high energy level dynamic compaction without humidification can play an effective improvement effect. Thus in order to avoid the defects of humidification, such as long duration, high water consumption, etc., this method can improve the economics of the project and shorten the construction period. Therefore, it is applicable for the engineering projects which have high requirements.

2. For the collapsible loess of large thickness and low water content in Lanzhou region, the effective treated depths by dynamic compaction are respectively 6 m, 7 m and 11 m under the energy levels of 8000 kN · m, 10000kN · m or 12000 kN · m. When carrying out high energy level dynamic compaction for the collapsible loess of large thickness and low water content, the values of correction coefficient is recommended between 0.21–0.31.

3. The collapsible loess foundation of large thickness and low water content has complex features. The test results of this project still have some discreteness in practical applications. In order to achieve better treatment effect, targeted treatment parameters should be developed according to the actual situation of the project.

ACKNOWLEDGEMENTS

This study was partly supported by the National 973 Program of China (No. 2014CB047003).

REFERENCES

He W.M., J. Fan. Evaluation of valuation of collapsible loess subgrade treated by dynamic compaction [J]. Chinese Journal of Rock Mechanic sand Engineering, 2007, 26(S2): 4095–4101.

Huang X.F., Z.H. Chen, X.W. Fang, et al. Study on foundation treatment thickness and treatment method for collapse loess with large thickness [J]. Chinese Journal of Rock Mechanics and Engineering, 2007, 26(Supp. 2): 4332–4338.

Lü X.J., X.N. Gong, J.G. Li. Research on parameters of construction with dynamic compaction method [J]. Rock and Soil Mechanics, 2006, 27(9): 1628–1633.

Nian T.K., H.J. Li, Q. Yang, et al. Improvement effect of high energy dynamic compaction under complicated geological conditions [J]. Chinese Journal of Geotechnical Engineering, 2009, 31(1): 139–142.

Zhan J.L., W.H. Shui. Application of high energy level dynamic compaction to ground improvement of collapsible loess for petrochemical project [J]. Rock and Soil Mechanics, 2009, 30(S2): 469–472.

Architectural, Energy and Information Engineering – Sung & Chen (Eds)
© 2016 Taylor & Francis Group, London, ISBN 978-1-138-02791-6

Traffic organization planning design of Nanyang old town renewal

Zheng Tan
Wuhan University of Technology, Wuhan, Hubei, China
Nanyang Institute of Technology, Nanyang, Henan, China

Hang He
Nanyang Institute of Technology, Nanyang, Henan, China

ABSTRACT: City historic district is different from the ancient town of historical relics and ancient villages, has undertaken important functions of the city, and is a witness to the history of the city. The city is still an important part of the development. However, with the city's economic development, urban motorized transports speed up the process, the rapid growth of vehicle ownership, historical district of the former only to meet the needs of pedestrians, horse-drawn carriage and a small car, while planning and construction of the road transport system has not been adapted to the need of modern cities to develop rapid motorization. Traffic problem is becoming the bottleneck of city's historic district, restricting the sustained and healthy development. Also, a large number of cultural relics and historic streets serve as the cured limit of the expansion of space historic district transportation facilities. Above all, how to break the historical limitations of the city's historic district road traffic to road traffic organization as a means of optimizing the development of transport organization mode for the historic district is the historic district city's sustainable and healthy development process need to explore and solve problems.

Keywords: old town renewal; traffic organization; planning design

1 INTRODUCTION

Nanyang, located in the southwest region of Henan Province, has a history of more than three thousand years. There are many ancient artifacts, including the famous tomb of Zhang Heng Division St., Zhang Zhong jing medical shrine, the temple of Marquis Wu Liang and national level architectural heritage Nanyang town hall. The renovation of the old city is located in the center of Nanyang City, which is from the south of Plains Road to the north of Jianshe Road. The Old Town is the main body of the ancient city of Nanyang City, with a total area of 168.86 hectares of land planning. With an existing total population of 34,687, Nanyang town hall, any Courtyard and some provincial and municipal listed building are concentrated in the Old City area. The Nanyang old town pattern in the Qing Dynasty is shown in Figure 1.

With the gradual formation of market economy, the environment, economic development speed greatly accelerated expansion of Nanyang City District to expand the unprecedented scale and speed, the gap between the New and Old constantly pulled a large urban renewal is imperative.

However, it is found that the Nanyang town Renewal is very difficult in the research process. Recently, high population and building density, but poor construction quality, heritages damaged in serious condition, deterioration of urban living environment, and the government have not put a large capital investment. The renovation of the mechanism of how to build a virtuous circle has become Nanyang town Renewal key, and the protection of the road is particularly a critical update.

In the context of rapid motorization, the essence of historical and cultural district road traffic problems is the main mode of transportation modern car dealers, high density motor vehicle road traffic flow characteristics, instead of walking past the traditional traffic and slow traffic characteristics. That is the original traffic patterns and motorized transport gap between demand and contradictions. The problem is by no means a simple road design and construction traffic will be able to solve, need transportation system measures and strategies to address these issues. The traditional lack of human traffic planning, traffic is organized in conjunction with the regional strategy update, existing planning methods are mostly focused on the test vehicle traffic is smooth, such planning methods in many

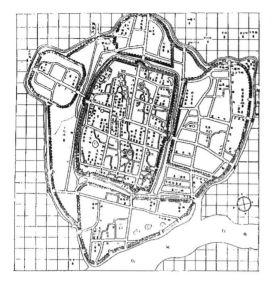

Figure 1. Nanyang old town pattern.

aspects of the protection of historical and cultural district with the conflict. For example, road widening will inevitably lead to changes in historical and cultural district street, surrounded by the destruction of the bonding interface, and even pose a threat to historic buildings, pedestrians and public transport interests that do not receive the attention they deserve. Car ownership rates and increased usage often lead to a conflict between pedestrian and vehicle parking and narrow streets in between. Increased historical and cultural districts and the surrounding real estate development activities within the historical and cultural blocks exacerbated these contradictions.

2 TRAFFIC SITUATION

2.1 Current state of roads

Nanyang Old Town retains the basic pattern of the original of old town in the Qing Dynasty. Its most important feature is the large number of alleyways. Currently, in addition to the people, the way around the Old Town, the Zhongjing Road, Jianshe Road, Qiyi Road, Riverside Road, outside, inside there are 91 alleyways, with a total length of 32,459 meters. Existing public transport routes to focus on the periphery of major cities in the old town roads, internal accessibility is dependent on a small tricycle. Old Town has an existing public car park, located near People Xinhua Road intersection on the north side, an area of about 1600 square meters, and another former Jin Hanfeng, in front of the Crown Building, People's Theater,

Chenzhou Parishes parking spots for commercial entertainment configuration.

2.2 Current problems

The extant Old Town Road Traffic has the following questions:
Poor road accessibility, including some roads are narrow; poor road quality, surviving many unpaved roads; Inadequate roads, poor drainage; Broken road, T-shaped intersection more other factors, it cannot meet the traffic requirements, thus inducing a mixed picture of many transport uses, so that the traffic situation is more complicated and difficult to Shuli, seriously affecting the quality of life.

Section in the form of road unreasonable main road too wide sidewalk so that small business operators have an opportunity to warn that serious street, only damaged the environment, but also affected the normal traffic; Share is serious, not only damaging the environment, but also affecting the normal traffic.

Static lack of transport facilities, the entire old town is now a public car park just Xinhua intersection, as well as several small parking spots configured for business and entertainment, all the rest of the street, sidewalk parking, disorder, affecting the city and traffic.

3 PLANNING IDEAS

3.1 Sort of road capabilities

The fundamental goal of the planning and construction of the road is to ease traffic flow, but the functionality of the old city street is mixed, so when planning and adjustment, the first step should be positioned in the regional road network in the city, clear the road network structure, hierarchy function, form the overall coordination of road network, reasonable traffic diversion, to neighborhood protection purposes. Do not break down regional inherent tranquil atmosphere of the historic district of the original premise of the full assurance of its transportation accessibility, and become fully enhance the quality of life and protect the residents of the development of tourism in the Old Town. When neighborhood while maintaining the original structure, strengthen the internal branch of beautiful and comfortable streets and travel space distribution function.

3.2 Overall protection

Not only is the city urban road traffic carrier, it is reflected in the urban landscape and architecture of the window. When Nanyang Rebuilding roads inside the old city, a beautiful part of the landscape may be preferred, strong historical context, urgent transportation needs, and is typical of the streets focused, selective protection and construction, highlighting its cultural and historical atmosphere, ease block traffic. It is required to achieve the readability of the city's historical context, the degree of reversibility and minimum intervention, and the feasibility of urban transport, sustainability and fully tap the potential maximum traffic capacity of the street.

3.3 Implement transportation demand management

Located in the old town area of Nanyang City, the bustling commercial area, and a large number of residents gathered inside, business, sports, concentration, leading to strong traffic to the heart. In the process, on the one hand, it is to protect the evacuation of the old part of town residents, strengthening traffic demand management in the Old Town, and strictly control access right of the vehicle. On the other hand, it is to raise the level of public service transport, public transport services will radiate throughout the protected areas. Protected areas should pay attention to the traffic access facilities, which will greatly optimize the regional road resources, and create a pleasant and comfortable walking space and original neighborhood environment.

4 ROAD NETWORK PLANNING MEASURES

4.1 Adjust the plan as a whole road network

Road Planning General: Road planning should not undermine the overall style of the neighborhood, and improve the appearance of the municipal road infrastructure pleasant degree of modernization and cross-section, following the neighborhood context of space and spatial scales. To protect the overall style of the historic district, both inside and outside the transportation needs of neighborhoods, and meet the needs of motorized travel, without destroying the historic district is relatively comfortable and peaceful environment, you need to block road traffic system to sort out, promote internal traffic and external traffic, transit traffic and isolate human traffic and control the development of motorized transport, tourism promotion organizations to optimize the traffic environment.

4.2 Selection and design of road sections

For different levels of roads, it requires a different section design road, to extend features imagery

of the city. Urban trunk function is to allow the passage of vehicles to meet the needs of urban transport, but in the space on the roads should be designed to be pleasant, especially within the region is a strong business climate, should provide loose and comfortable walking space, while city roads need to provide entry into the neighborhood, these nodes need to focus on landscape coordinate and guide the transition to modern life historical and cultural district. City roads and regional branches to ease traffic flow-based, cross-sectional sides should give full consideration to the style of the historic district. Affected relics of ancient buildings, regional roads are narrow line width, so the road section of the design process, in addition to the need prevailing outside the motor vehicle lanes on both sides of the pedestrian and non-motorized vehicles set up in the same section, so that non-motorized vehicles complementary sidewalk use of road space saving resources. The entire road section can be combined with the construction of advance and retreat to form flexible and vivid visual space. Part of the road section taking into account the status quo of existing trees to be retained in the trees on one side can be set individually non-motorized vehicles and pavements, and the two sides of the non-motorized vehicles and pedestrian walkways set to the same section.

4.3 Parking facilities

Parking facilities are mainly for public parking, some residents supporting the parking and non-motorized parking. Parking is arranged in a relatively concentrated "Nanyang Old Town" periphery, avoiding the introduction of a motor vehicle causing damage style of protected areas. Public parking and ancillary parking is a combination of the four directions, East and West sub-set of the population. Convenient traffic in favor of parking from different directions, different functions parking services based primarily on the area of land to determine the range of features and services. Parking services are set in the newly built building ground floor underground parking garage, a large area of open space to avoid causing disharmony to the space style area. Coach to be set in the ground segment generally, bus parking ground combined Minyuan Stadium renovation plans to build parking.

Non-motorized vehicles mainly with the walls inside the back part of the alleyways, related green space, corner zone and public parking, taxi stops set up temporary parking space for non-motorized vehicles, try to set up in every neighborhood roads mouth as a non-motor vehicle parking area residents and visitors access point.

Taxi stops should be close to the passenger distribution point, the use of green belt isolated non-motorized vehicles. At the same time, the construction and management of parking along the road should be strengthened, and there should be an appropriate increase in parking fees and reduced parking demand. Sites are set to the middle part of the protection zone around the road, so that passengers can conveniently walk through protected areas.

5 CONCLUSIONS

With the rapid development of urban construction, after the founding of the streets and buildings are built and aging. With the increase in population density and motor vehicles, the old city layout widespread confusion, narrow streets, low road density, overcrowding, traffic congestion, environmental pollution, municipal and public facilities such as a shortage of many problems, restricting further the Old Town development, but do not meet the needs of modern life. The purpose of this study historic district is committed to the integration of modern city life. By adjusting the existing transportation system, it is ensured that the original historic district pleasant interior dimensions while meeting the requirements of modern traffic travel, historic district to achieve full integration of sustainable development and green transportation system. The priority is to meet the needs of the majority of the traffic to rail transportation as the main public transport system and improve the overall level of transit service, a substantial increase in the proportion of public transport passenger share; at the same time, efforts to guide and control the car use; positive for bicycle traffic and pedestrian traffic to create a good environment, to create a low-carbon, green transportation demonstration area.

REFERENCES

Ashihara, Yoshinobu. 2006. *The Aesthetic Townscape.* Japan: Bai Hua Publishing.
Elizabeth, H. J. 2007. *Space of belonging-home, culture and identity in 20th century French autobiography.* Amsterdam: Rodopi B.
Frith, S. 1991. Knowing one's place: the culture of cultural industry. *Cultural study from Birmingham.* 1: 135–155.
Ge, Y. H. 2010. *A new history preservation method in recent China: a case study of Nanjing 1912.* Cardiff: Cardiff University.
Markus, H. R. and Kitayama, S. 1991. Culture and the self: implications for cognition, emotion, and motivation. *Psychological Review* 98(2), 224–253.
Peter Booth and Robin Boyle. 1993. See Glasgow, see culture. In Bianchini, F. and Parkinso. (ed) *Cultural Policy and Urban Regeneration: The West European Experience.* New York: Martin's Press, 35.
Wisker, G. 2001. *The postgraduate research handbook: succeed with your MA, MPhil, EdD and PhD.* Basingstoke: Palgrave Macmillan.
Zukin, S. 1995. *The Cultures of Cities.* Cambridge: Blackwell Publishers.

Architectural, Energy and Information Engineering – Sung & Chen (Eds)
© 2016 Taylor & Francis Group, London, ISBN 978-1-138-02791-6

An efficient communication method of CAN bus

Jian Qun Wang, Jing Xuan Chen, Ning Cao, Xiao Qing Xue & Rui Chai
Beijing Institute of Technology, Beijing, China

ABSTRACT: A dynamic autonomous synchronization method in CAN bus communication based on frame identity is proposed to relieve the collisions of communication, in which a queue of frame identities is stored in each CPU, and then locates the frame that current CPU is about to send in the queue. If and only if the frame is the next frame to send in the queue, the very CPU sends the data, so that the frame crash is avoided and the reliability and real-time ability of information collection are guaranteed. The method has been tested in an information collection system and achieved great results.

Keywords: CAN communication; frame queue; conflict arbitration; dynamic autonomous synchronization

1 INTRODUCTION

The production of field bus has brought significant convenience to the design, installation, operation and maintenance of industrial control system due to its simple structure, good real-time ability, excellent openness and wonderful reliability. In the mean time, Controller Area Network (CAN) has been widely employed in various fields such as automotive industry, process control, numerical control machine and so forth because of its high communication speed, strong reliability, easy to connect and great cost-effective[1]. But with the boom of technical industry, the amount of nodes and data transmitted is increasing sharply, the real-time ability and reliability of conventional CAN are hard to guarantee. Thus, it is of great realistic and practical value to optimize the conventional CAN in order to improve the real-time ability and avoid the conflicts of frames so that the requirements of multiple nodes and data can be met.

CAN protocol is an event triggered protocol, in which, each frame transmits according to its priority. Taking data frame as an example (see Fig. 1), there is a 11-bits long identity sign (29 bits in 2.0B) used to mark the priority of each datum. CAN employs the Carrier Sense Multiple Access with Collision Detection (CSMA/CD) mechanism in order to arbitrate the access to bus while two or more nodes wish to transmit. In the arbitration, the frame with a higher priority wins and goes on transmission in bus while the one with a lower priority stops transmitting and begins to monitor the bus level. Note that this scheme means that no bandwidth is wasted during the arbitration process and thus the real-time ability of communication is ensured. Besides, CAN protocol provides various inspections such as ACK inspection, bit stuffing inspection to ensure the reliability of communication. Reference[2] illustrates the principles and advantages of such arbitration in detail.

When the load rate of network is relatively low in CAN communication, the frames in the system are hard to crash so that the real-time ability of transmission is comparably wonderful. But, with the increase of amount of data transmitted and frequency of communication, the collisions of frames are inevitably violent because each CPU sends the data randomly and some send several frames at a time without intervals. According to the arbitration mechanism, once conflict occurs, the node with a lower priority will exit the bus competition to guarantee the transmission of the one with a higher priority. Suppose an extreme case, where the load of network is extremely heavy, the node with a low priority was mistaken as a broken-down node and excluded from the communication system by bus because of repeatedly faulty sending, which, of course, will lead to the loss of multiple frames, the failure of transmission and the decrease of real-time ability of communication. At the same time, because the "fault silence" is hard to achieve in the field bus, if a CPU with a high priority breaks down and keeps sending data to bus, the bandwidth of bus is wasted and the security and reliability of network are hard to ensure. Then the so-called "Babbling idiot" fault happens. Reference[3] discusses this problem in detail and puts forward a method named FlexCAN with the capability of tolerating such problem by using a simple bus guardian in the architecture of CAN, which will be discussed in the second part of this paper.

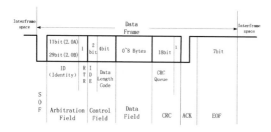

Figure 1. Data frame in CAN.

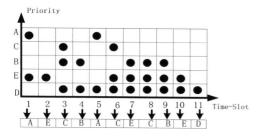

Figure 2. Transmission delay in conventional CAN arbitration.

Fig. 2 gives an example to illustrate the influence on reliability and real-time ability of communication under the circumstance of mass nodes and data[4].

The unify transmission delay model for field bus put forward by Tong and his colleagues in Harbin Institute of Technology[5] and the solution of M/D/1/FIFO model declared by Liao in Wuhan Institute of Technology all show that when the nodes and load in network are enormous, the collisions caused by competing bus resource will be extremely serious, that is, the frame with a low priority can hardly sent to bus and the real-time ability of communication is awful.

2 CURRENT REFORMS TO CONVENTIONAL CAN

Nowadays, there are lots of papers that discuss the CAN protocol and propose some reform plans from communication reliability, real-time ability and security. In order to solve the problems above, current reforms are as follows.

2.1 Dynamic priority algorithm

Many researchers have taken adjusting priority of each frame dynamically into account to avoid such circumstance. Though these methods vary in adjustment conditions and scopes, the fundamental ideology of dynamic priority algorithm is that when collision occurs, the priority of the node that fails in this arbitration will be increase. And if it fails again in the next arbitration, its priority will be further enhanced. Thus, the node suffering from repeated failures will have a pretty superb priority, which is to said, it would have large opportunity to win the access to bus and send its data successfully[4-5].

This method performs well in solving Babbling idiot and enhancing the real-time ability of communication. But, this method has to return the priority to the original state after transmitting the information, which raises the cost of maintenance and management. And in the process of raising the priority, the priority of two frames may be the same and causes the collision. The problem of frame collision has not been solved yet.

2.2 Reforms on CAN protocol: TTCAN, FlexCAN

TTCAN protocol is a reform to CAN protocol. Different from the event-trigger mechanism of CAN, TTCAN employs the time-trigger mechanism and divides the communication of CAN into different basic periods by main timer. The main problem of this method is that TTCAN can not be used with CAN in a specific system because TTCAN requires to take own window. If we plan to use TTCAN in practical projects, we have to change the protocol of all the ECUs in the system into TTCAN, which costs numerous work and budget and has no realistic significance. Apart from that, TTCAN is also born with the drawbacks such as too many error frames under sever environment, the high expense caused by preserved error frame, the low use ratio of time due to different usages of Slot, the difficulty in realizing arbitration window and so on[6].

2.3 A load dispersion method

Ding in Electronic technology[7] puts forward a method to reduce the times of collisions through dispersing the load in bus. The method introduces a concept of initial period offset that gives multiple frames with multiple identities different offsets, and disperses the time that frames with different identities send their own information. This method avoids the unordered situation that CPUs sends information at any time to some degree and has positive effects in practical projects. However, when the number of nodes in a system changes, this method can hardly fit in the dynamic change and unordered circumstance would continue. What is more, this method does not fully take system's fault tolerance into considerations.

3 A DYNAMIC AUTONOMOUS SYNCHRONIZATION METHOD BASED ON IDS OF FRAMES

3.1 *The method proposed*

Most of current methods solve the problems above in terms of reforming conventional CAN protocol and have gained practical meaningful achievements. But these methods usually can not go well with current CAN protocol, solve the collisions of frames caused by unordered sending situation and, moreover, most approaches have not adequately taken the variation of nodes and fault tolerance of system into account.

In recent years, microelectronic technology is growing by leaps and bounds, and meanwhile, the processing speed and store memory of CPU, especially the interrupt handing ability, are sharply increasing. Utilizing software to avoid bus collision, relieve frame-crash, and enhance the reliability and real-time ability of transmission may be an efficient method.

As is mentioned in the context, every node in CAN system has the right to send information to the bus and monitors the arbitration by bus when collides due to the broadcast characteristic of CAN, which leads to severe frame crash, especially when there are multiple nodes in the system and numerous data being sending. Therefore, a dynamic autonomous synchronization method based on frame identity queue is introduced in this paper. Each CPU stores a queue arranging the priority of each frame from the highest to the lowest according to the priority (i.e. the value of ID) of the data frame to be sent. When the system works, every CPU receives all the data on the bus (i.e. switching the acceptance filter to bypass mode) and adjusts the ID queue dynamically. When a CPU is about to send a frame (named A, for example), the software will judge that whether all the frames possessing the higher priority than A in the queue, have been received. If so, A will be sent immediately. That is to say, every frame of CPU can find the time slot to send itself. As long as every node communicates strictly obeying this rule, the system can make sure that in each time slot, there is only one node to send data and the collision of nodes and crashes of frames are eliminated naturally, and the amount of data in CAN communication increases.

In order to meet the requirements of dynamic changes of nodes, this method also involves the approaches to delete or add IDs of the frame. At the same time, the method can satisfy various communication speeds and the need of efficient communication by endowing one CPU different multiple IDs.

3.2 *The contents of the method*

Figure 3 displays the principle of this method. The data of CAN bus connect with the bus driver of each node firstly, and then the driver transforms the differential signal to the single end signal that is connected with controller of CAN. CPU judges whether the data are received or sent completely through query flags or interruptible identification and then reads or writes the data frames of CAN.

The moment the system starts, every CPU stores a default queue of IDs as the initial queue in its memory which is programmed at the stage of system design. After CPU finishing the initialization of CAN in system, CPU enters the software shown in the red box of Fig. 3 by means of query or interruptible reception. Once there is information being transmitted in the bus, CPU will receive the frame, extract the ID of the frame and then store it in an array (which is so-called queue of frames' identities). According to the value of the IDs, the one that is smaller in number (i.e. higher in priority) stands before the one that is bigger in number (i.e. lower in priority), and then inserts the very frame that this CPU is going to send, and there, a complete queue forms. After forming a complete queue, CPU takes two points into considerations to judge whether to send the frame or not: the first is whether the frame that stands before the to-send frame has been sent; the second is whether the time interval between the current time of the system and the time when this CPU sends the latest frame successfully is greater than the CPU's sending period.

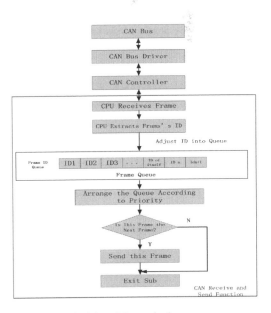

Figure 3. Principles of the method.

If all these two requirements are met, CPU sends the frame immediately.

Otherwise, as is shown in the left of Fig. 4, if CPU finds that the frame before the frame that it is to send has not been received, CPU will wait a time period, after which the frame has still not been received, and this CPU will regards the former node has broken down and delete the ID of the former node from the queue. The time interval between two adjacent frames T_{th} in the figure means under the circumstance of normal communication, the longest time interval that completing a transmission of a frame with maximum length. The minimum of T_{th} equals to the bits of the frame multiplies the bit time. For instance, the current baud rate of CAN communication is 250 Kbps; the maximum length of data field is 8 Bytes; CAN 2.0B communication mode is adopted; the total length of the frame contains the 32 bits of ID and 64 bits data ups to 96 bits. And therefore, the minimum of T_{th} is 96 multiplies 4 equaling to 384 microsecond. Taking the delays of reacting, receiving and sending into account, the value may be increased appropriately to 500 microsecond. When CPU waits over this time interval, it regards the node has been broken down and delete it from the queue so that the normal communication is guaranteed and the efficiency of communication is improved to the utmost.

The right of Fig. 4 shows the approach to add a new ID. While a new node is connected in the system and sends data to bus or the old node needs to add new IDs due to the increase of the demand of communicational efficiency or the node that has been deleted intends to join in the system again, as soon as the new frame with new ID is sent to bus, every CPU will arrange the position of the new comer in the queue using the method mentioned before in its memory and give the right of using the bus to the new comer in next communication period in order to avoid the collision.

3.3 The advantages of the method

The predominant advantage of the method is building a queue of IDs of frames to guarantee that one CPU sends data to bus at each time and collisions are voided. This method can support the reliability of network and prevent the "cyber storm" from happening, especially under the circumstance of multiple nodes and heavy load of network. Besides, the method predominates in following aspects:

- As is discussed before, this method can adjust the queue in each CPU dynamically. The queue in CPU is real-time updating according to the situation of the system, which accounts great convenience to add or delete the IDs and fits well in nodes-changing situations.
- As is discussed before, this method can judge whether the node is working regularly or not according to the parameter: T_{th}. The nodes that do not work well can be deleted from the system easily so that the communication efficiency is guaranteed in a large scale.
- As is discussed before, this method can grant one CPU multiple IDs so as to raise the communication efficiency of the CPU and meet the demands of different elements in the system.
- As is discussed before, this method moves the frame backward an interval unit as soon as the frame was not sent correctly and sends the frame in next time slot to avoid the loss of data and improve the fault tolerance of system.

4 TESTS AND RESULTS

Figure 5 is the working principle figure of 8 channels pressure signal collection card: TGWH3028. And the material object is shown in Figure 6. As is shown in Figure 5, U1 is the ARM chip of 32 bits: STM32F103, which is low price, excellent performance and possesses abundant I/O ports and built-in CAN controller to control the collection chip HX711 with 8 channels and 24 bits and finally realizes collection function. The collection information travels to host computer through CAN and then goes to personal computer to achieve signal acquisition. The JP1 in Fig. 5 functions setting the site number of CAN. There can be 16 CPUs in the bus at the same time, which is convenient for the expansion of functions in system.

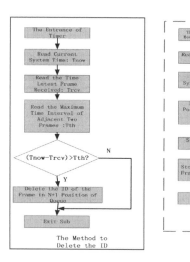
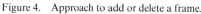

Figure 4. Approach to add or delete a frame.

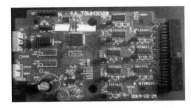

Figure 5. The principle of collection and CAN communication.

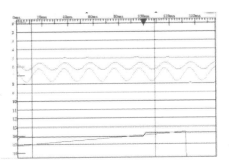

Figure 6. The material object of TGWH3028.

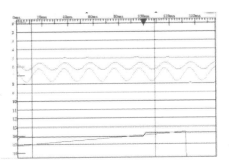

Figure 7. The waveforms of signals collected.

Each TGWH3028 is set different site number of CAN by JP1. The IDs of CAN are granted according to the state of JP by CPU automatically. The method proposed in Part 3 based on frame-identity-queue is employed to carry on the communication of CAN and achieves the real-time pressure acquisition function.

Figure 7 shows the waveforms of pressure collection signal acquired by TGWH3028. In the figure, channel 6 and 7 represent the pressure signals of test bed. As can be seen, the curves of pressure signals are smooth and continuous, which shows that there is no data loss on process of communication, and that is to say, the continuity and real-time ability of signals are guaranteed, and thus, the effectiveness of the proposed method is justified.

5 CONCLUSIONS

In this paper, a dynamic autonomous synchronization method for CAN communication based on the queue of IDs of frames is proposed to realize a quasi synchronous communication result. The purpose of the method is setting up a quasi synchronous principle, in which every CPU sends information according to an agreed consequence so that the possibility of collisions is reduced and the situation on which CPUs send data in no order is avoided and thus the flow of effective data in communication and real-time ability of data in network are heighten, and by all means, the problem of low communication efficiency and bad real-time ability in CAN communication is solved.

One of the drawbacks of this method is that CPU receives all information in bus and thus increases the burden of CPU. But, on one hand, a intrinsic defect of CAN that it is necessary that propagation delays on the shared medium be negligible with respect to the bit time in order for the arbitration mechanism to behave correctly limits the communication speed of CAN in a large scale[8]. Generally speaking, the maximum communication speed of CAN merely ups to 1 Mbps at the distance of 40 meters. On the other hand, with the development of microrelectronic technology, the word length of most normal CPUs has reached 64 bits and domain frequency has already exceeded 2 GHz. As a result, in terms of normal network communication, especially in-vehicle cyber fields, the process ability of CPU nowadays is capable of this task and the influence on the real-time ability of transmission caused by arranging the queue of IDs dynamically is neglectable.

Along with the development of industrial technology, the engineering system is becoming much more complicated, and therefore the applications of CAN communication in electronic automobile control and engineering filed will be even more common. The real-time ability of CAN communication plays an important part in maintaining the security of vehicles. The proposed method has great effects on improving the security of automotive and there is no need to change the existing hardware by this method, and therefore the method deserves an extensive prospect of applications.

REFERENCES

[1] Hanxing Chen, Research on the controller area network, 2009. International Conference on Networking and Digital Society (ICNDS) 2005: 251–4.
[2] Farsi, M. Ratcliff, K. Barbosa, An overview of controller area network. Computing & Control Engineering Journal, 1999, 10(3): 113–20.

[3] Tong Weiming, Gao Hongwei, Chen Peiyou, A study on transmission laitance characteristics of CAN. Chinese Journal of Scientific Instrument, 2007 (28): 295–297.

[4] G. Buja, R. Pimentel, and A. Zuccollo, Overcoming Babbling idiot Failures in CAN Networks: A Simple and Effective Bus Guardian Solution for the Flex-CAN Arbitration, IEEE Trans on Industrial Informations, 2007, 3(3).

[5] A. Farahani and G. Latif Shabgahi, A Novel Method for Softening Some of the Drawbacks of CAN, Proc. of the Int. Corif. Signal Processing Systems, Page 771–775, Singapore, 2009.

[6] Ferreira J, Oliveira A, Fonseca P. et al. An experiment to assess bit error rate in CAN [C]. RTN 2004–3rd Int. Workshop on Real-Time Networks satellite held in conjunction with the 16th Euro micro Intl Conference on Real-Time Systems, June 2004.

[7] Ding Xuyang. A method to disperse the load of communication network bus in automotive controller: Chinese patent. 201010528846.4[P]. 2011-02-16 (in Chinese).

[8] G. Cena and A. Valenzano, Fastcan: a High-Performance Enhanced CAN-Like Network, IEEE Trans on Industrial Electronics, 2000, 47(4).

Architectural, Energy and Information Engineering – Sung & Chen (Eds)
© *2016 Taylor & Francis Group, London, ISBN 978-1-138-02791-6*

Economic comparison of pure electric logistics vehicles and traditional fuel logistics vehicles

W.Z. Tao & D.J. Liu

Department of System Engineering, School of Traffic and Transportation, Beijing Jiaotong University, Beijing, China

ABSTRACT: As a transshipment station of commodities, logistics industry undertakes considerable scale of vehicle operating business. The PELV (Pure Electric Logistics Vehicle) is put into use by the government as a newly developed and eco-friendly means of transportation. According to both the characteristics of logistics industry and the specificity of PELVs, the vehicle's life cycle was divided into two parts: one to four years and five to eight years. By building mathematic model, one can get and compare the annual costs of the PELVs and the TFLVs (Traditional Fuel Logistics Vehicles) of those eight years. Then, it is possible to find which one of the vehicles is cheaper to operate and at what time among those years the economy is the best. Finally, through sensitivity analysis, the main factors affecting the economy of pure electric cars were analyzed. These will provide a basic idea for reducing the cost of vehicles for logistics companies.

Keywords: pure electric logistics vehicles; traditional logistics vehicles; mathematic model; economy

1 INTRODUCTION

Electric vehicles are becoming increasingly important as not only do they reduce noise and pollution. Electric vehicles can also reduce carbon emissions, because reaching zero release of carbon dioxide requires using the energy produced from non-fossil-fuel sources such as nuclear and alternative energy [1-2].

About 24% of the world's energy consumption is the vehicle energy consumption. As an alternative to traditional fuel vehicle, electric vehicles have great advantages in reducing energy consumption and environmental pollution. Compared to the ordinary engine driving vehicles, EV has the advantages of high level of efficiency, low level of environmental pollution, low rate of noise, multiple energy resources available and energy feedback[3].

Through the comparison between the economy of pure electric vehicles and of traditional logistics vehicles, the paper fully demonstrates the various economic factors of two kinds of vehicles in their whole life cycle, and finally based on the project of Beijing "pure electric logistics vehicle demonstration operations," the economy of the two types of vehicles is analyzed with an example, which provides a reference for the government to make policies and for the enterprises to replace traditional fuel vehicles.

2 THE COSTS STRUCTURE OF PELVS AND TFLVS

In this paper, the author considers the full life-cycle costs of the two vehicles, when calculating the life-cycle costs of the vehicles which include vehicle purchase costs, fuel costs, labor costs, vehicle maintenance costs and the cost of battery (which the fuel vehicles do not have)[4]. The costs included are as follows:

2.1 The costs of vehicles

Each of the pure electric vehicles would be given 60,000 yuan by the financial subsidies[5]. At the same time, the government gives 60,000 yuan as the local financial subsidies.

Vehicle purchase costs a one-time purchase tax in addition to the payment of the vehicle, so the total costs of the car are the sum of the cost of vehicles and vehicles purchase tax.

The cost of PELVs are PC_{11}, so the costs of traditional vehicles are PC_1 as follows:

$$PC_1 = 1.1PC_{11} - 12 \qquad (1)$$

The same type of fuel vehicle's price is PC_2, and the TFLV does not have related subsidies. It is

$$PC_2 = 1.1PC_{21} \qquad (2)$$

2.2 Tax-insurance costs

As for the travel tax, according to *the People's Republic of China Travel Tax Law Implementation Regulations*, the tax of commercial trucks is calculated as 96 yuan per ton curb weight, setting the vehicle's curb weight to be t ton and the useful life to be n years, so the traditional fuel vehicle's travel tax is:

$$Ins_{22} = 96 \times t \times n \qquad (3)$$

So the tax-insurance costs of PELVs and TFLVs are as follows:

$$Ins_1 = 10\% \times PC_{11} + Ins_{13} \qquad (4)$$

$$Ins_2 = 10\% \times PC_{21} + 96 \times t \times n + Ins_{23} \qquad (5)$$

Ins_{13} and Ins_{23} are the insurance of PELVs and TFLVs respectively.

2.3 Labor costs

The two vehicles are of the same type, so they have the same amount of laden, and thus they have the same labor costs. The labor costs are

$$Sal_1 = Sal_2 \qquad (6)$$

2.4 Vehicle maintenance costs

Because the PELVs do not need a series of mechanical transmission equipment including engine, clutch, gearbox and so on, the vehicle's failure rate is reduced, and the maintenance costs can be reduced by more than 70%[6]. So the two vehicles' relationship of maintenance costs is:

$$MR_1 = 0.3MR_2 \qquad (7)$$

2.5 Fuel costs

Assuming that under the same year distribution mileage, providing cycling the annual distribution of mileage for L km, PELVs are α kWh consumption per hundred kilometers, the charging efficiency of pile is δ, and the electricity charge is A_1 yuan/kWh, then the fuel costs of PELVs are calculated as follows:

$$FC_1 = n \cdot L \cdot \frac{\alpha}{100\delta} \cdot A_1 \qquad (8)$$

Assuming that the fuel consumption of the ordinary fuel vehicle driven for a hundred kilometers is βL, the charges of fuel are A_2 yuan/L, so the fuel costs of TFLVs are calculated as follows:

$$FC_2 = n \cdot L \cdot \frac{\beta}{100} \cdot A_2 \qquad (9)$$

2.6 The cost of battery replacement

To the PELVs, during the period of useful life of 8 years, the battery needs to be replaced once, assuming that the cost of battery replacement of PELVs is B yuan.

3 ECONOMIC COMPARISON OF THE TWO VEHICLES

With the comparison of the two vehicles and the use of the mathematical model, the total costs are calculated. The paper followed year time node and, as time n grows, constantly calculate the two vehicles' costs, at last getting the relationship of PELVs and TFLVs and the economic comparison of the two vehicles in different stage.

3.1 Before PELVs replacing the battery

When the time of two vehicles is less than 4 years, the battery don't need to be replaced for the PELVs, so their costs are:

$$C_1 = 1.1PC_{11} - 120000 + Ins_{13} + MR_1 + n \cdot L \cdot \frac{\alpha}{100\delta} \cdot A_1 \qquad (10)$$

$$C_2 = 1.1PC_{21} + 96t \cdot n + Ins_{23} + MR_2 + n \cdot L \cdot \frac{\beta}{100} \cdot A_2 \qquad (11)$$

Through the comparison of C_1 and C_2, we can get the result that in which year the two vehicles' economy is the same. If $n > 4$, it means before the PELVs replacing its battery, the economy of TFLVs is better than that of the electric logistics vehicle. If $n < 4$, it means when the using time is n, the economy of electric logistics vehicle is beyond that of traditional fuel vehicle.

3.2 After PELVs replacing the battery

When the using time of two vehicles is more than 4 years, the costs of battery replacement need to be considered, and as the increase of maintenance costs, at the same time, the maintenance costs of TFLVs will also increase. Because of the subsidies of the electric vehicle, we can also consider the replacement of PELVs (we cannot consider the worthy of the abandonment of vehicles), the costs of battery replacement of PELVs, and replacement of PELVs and TFLVs are as follows:

$$C_1 = 1.1PC_{11} - 120000 + Ins_{13} + MR_1'$$
$$+ n \cdot L \cdot \frac{\alpha}{100\delta} \cdot A_1 + B \qquad (12)$$

$$C_2 = 1.1PC_{21} + 96t \cdot n + Ins_{23} + MR_2'$$
$$+ n \cdot L \cdot \frac{\beta}{100} \cdot A_2 \qquad (13)$$

$$C_3 = (1.1PC_{11} - 120000) \times 2 + Ins_{13} + MR_1$$
$$+ n \cdot L \cdot \frac{\alpha}{100\delta} \cdot A_1 \qquad (14)$$

Through the comparison of the economy's three options, we can get the economy of each option of each year during those 5–8 years, therefore we can select the best solution at each time.

4 CASE STUDY

The paper relies on "PELVs demonstration operation", compared Beijing Foton Omar 3T PELVs[7] with Omar 3T TFLVs. The price of PELVs which contains purchase tax is 270,000 yuan, and the price of TFLVs is 80,000 yuan. The purchase tax is 10% of the vehicle price, so the price of TFLVs which contains purchase tax is 88,000 yuan. For the PELVs could get the extra 120,000 yuan financial subsidies, the purchase costs of electric vehicle and fuel vehicle are 150,000 and 88,000 yuan respectively. Through the project we can get the data as follows: the wage of drivers who drive the two vehicles are same, which is 2,500 yuan; the maximum daily driving mileage of pure electric logistics vehicle is 100 km, so daily vehicle mileage is calculated in accordance with 100 km, and it also meets the need of driving in urban; the using time of vehicles is in accordance with 360 days a year; the energy rate of pure electric logistics vehicle and of traditional fuel logistics vehicle are 23 yuan/km and 80 yuan/km respectively; the maintenance cost of traditional logistics vehicle is 3,000 yuan, and the failure rate of pure electric logistics vehicle is 30% of the rate of the traditional fuel vehicle, so the costs of maintenance of pure electric logistics vehicle can be calculated based on that. Their economies can be compared through the data above.

Table 1. The economic indicator of the two vehicles.

Costs	Purchase vehicle yuan	Tax insurance yuan	Labor yuan	Fuel yuan	Maintenance yuan
PELV	150,000	-	$30,000n$	$8100n$	$900n$
TFLV	88,000	$288n$	$30,000n$	$28,800n$	$3000n$

Table 2. The comparison of the two vehicles when used 1–4 years.

Time	One year yuan	Two year yuan	Three year yuan	Four year yuan
PELV	189,000	228,000	267,000	306,000
TFLV	150,088	212,176	274,264	336,352
Difference	38,912	15,824	−7264	−30,352

4.1 Before PELVs replacing the battery

Through the data above, we can build the economic indicator of PELVs and TFLVs in the using time $n(n < 4)$, the expression shown in table 1.

The data in table 1 can be solved by formula (1–13), then can obtain the costs and relationship of PELVs and TFLVs in the 1–4 years. It shows in table 2.

It can be seen by the above table, when the two vehicles are used for more than 2 years, the economy of pure electric logistics vehicle is better than traditional fuel vehicle. By further calculation, the result shows the time when, at the durable years n = 2.685 ≈ 2.7 years, the two vehicles' economy are equal. And when the time of two vehicles is four years, the total costs of pure electric logistics vehicle are 30,352 yuan less than traditional fuel vehicle, so it can be seen, as the increase of time, the costs of PELVs are far less than TFLVs before PELVs was replaced the battery.

4.2 After PELVs replacing the battery

When the durable years of two vehicles are more than 4 years, PELVs need to be considered replacing the battery. So we can adopt two options: the first one is replacing the battery and the second one is replacing the new PELVs. For the program of replacing the battery, people need to consider the increase of maintenance costs because of the increase of the using time, and the costs will be twice than that of the first four years, the maintenance costs of TFLVs will be twice than that of the original vehicles. When the two vehicles are used for more than 4 years, the costs of the programs are shown in table 3.

Table 3. The costs of pure electric logistics vehicle and traditional fuel vehicle when used beyond 4 years.

Costs	Purchase vehicle yuan	Tax insurance yuan	Labor yuan	Fuel yuan	Maintenance yuan	Replace battery yuan
Replace battery	150,000	-	30,000n	8100n	3600 + 1800 $(n-4)$	100,000
TFLV	88,000	288n	30,000n	28,800n	12000 + 6000 $(n-4)$	-
Replace PELV	150,000 × 2	-	30,000n	8100n	900n	-

Table 4. The comparison of pure electric logistics vehicle replaces the battery and traditional fuel vehicle costs.

Time	One year yuan	Two year yuan	Three year yuan	Four year yuan
Replace battery	445,900	485,800	525,700	565,600
TFLV	401,400	466,528	531,616	596,704
Replace PELV	495,000	534,000	573,000	612,000

Table 5. The influence of various factors change to PELCs whole life costs rate of change.

Uncertainty factors	The rate of uncertainty factors					β
	−10	−5	0	5	10	
The vehicle price	2.65	1.33	0	−1.33	−2.65	26.52
Fuel costs	1.15	0.57	0	−0.57	−1.55	11.46
Battery prices	1.77	0.89	0	−0.89	−1.77	17.68

The data above can be put into the formula (1–11) and formula (14), (15) to obtain the costs of PELVs and TFLVs during the beginning of the fifth year and the annual costs of their difference. They are shown in table 4 and table 5 above.

It can be seen from the table 4, due to the large cost of battery, when pure electric logistics vehicle is replaced the battery in the fifth year, the costs of PELVs are more than TFLVs, when the two vehicles used beyond 6 years, the economy of PELVs exceeds traditional fuel vehicle. When the two vehicles used 8 years, the total costs of PELVs are less than TFLVs by 31,104 yuan.

By the table 5, it can be seen that the economy of replacing pure electric logistics vehicle is less than TFLVs. So according to the comparison of the three options, the economy of PELVs with replaced battery is the best, the second one is TFLVs and replacing the PELVs is the last one.

5 PELVs SENSITIVITY ANALYSIS

Parameter sensitivity of automobile fuel economy refers to the degree of influence where a certain

parameter change will be imposed on the whole life cost, which is the rate of the whole life cost caused by the rate of parameter change. Obviously, the parameter sensitivity is a dimensionless quantity.

The paper has made a single factor sensitivity analysis on the whole life cost with several parameters such as fuel costs, battery prices and vehicle prices, which are also the main factors in affecting the economy of both.

By setting factors changing in the range of −10% to 10%, we can calculate the change rate of the whole life cost under varied cases and the sensitivity coefficients $\beta = \Delta A/\Delta F$ of each factors, Among which, ΔA is the corresponding change rate (%) of evaluation index in the case that the uncertainty F has a change of ΔF (or 20%).

Influence that the change of various factors has imposed on the rate of whole life cost is shown in Table 5.

Through horizontal comparison, it shows that the order of PELVs sensitivity factors is followed by vehicle prices, battery prices and fuel costs. Therefore, by increasing the investment of PELVs R&D to reduce the vehicle price, one can make pure electric cars become more of the economical vehicles.

6 CONCLUSION AND OUTLOOK

Due to the characteristics of the vehicle itself, pure electric logistics vehicles and the traditional fuel logistics vehicles cannot complete a contrast through the simple sum of the costs. Based on the characteristics of the pure electric logistics vehicles, the paper makes economical comparison time apart from the fourth year, fully considers the characteristics of vehicles, and uses different formulas in the two periods. The paper ends up with two conclusions: 1) when the operating time of two cars is less than four years and while the time of pure electric logistics vehicle is more than 2 years, its economy will take advantages over traditional fuel cars. 2) when the time of two cars is 5–8 years, we need to additionally consider the cost of replacing the battery. At present, there exist two ways which are direct replacement of pure electric vehicles and replacement of the electric car batteries, through calculating, the paper makes a conclusion that replacing the battery is better in economy, when the operation of two kinds of vehicles is more than six years, the economy of pure electric logistics exceeds traditional fuel logistics vehicles.

REFERENCES

[1] John Lowery. Electric Vehicle Technology Explained [M]. 2nd Edition, Weley, Somerest, NJ, USA, 2012/06: 29.

[2] Li Xing-hu. Energy Issue of Pure Electric Vehicle in China [J]. World Electric Vehicle Journal. 2011(1): 91–97.

[3] Zhang Gui-rong, Zhang Heng-hai, Li Hou-yu. The Driving Control of Pure Electric Vehicle [C]. Procedia Environmental Sciences, 2011 (10): 433–438.

[4] Yang Feng, Fu Jun. Economic Comparison and Analysis of Pure Electric Vehicles [J]. Transaction of Wuhan University of Technology, 2009, 31(2): 286–288.

[5] Energy-saving and new energy vehicle demonstration and extension of financial assistance fund management Interim measures of "Ten city thousand vehicles" project.

[6] HU Shu-hua, Yang Wei. China's Electric Vehicle Industry Strategy Analysis [J]. Beijing Automotive, 2004(3): 20–25.

[7] "Energy saving and new energy vehicle demonstration to promote fiscal subsidy funds management interim measures" (Finance Building [2009] No. 6)— BJ5036XXYEV-1 Pure Electric Cargo Truck.

Architectural, Energy and Information Engineering – Sung & Chen (Eds)
© 2016 Taylor & Francis Group, London, ISBN 978-1-138-02791-6

Argumentation visualization techniques and application systems

W. Zhang, S.S. Guo, S.J. Gao & L. Cao
School of Information Science and Technology, Qingdao University of Science and Technology, Qingdao, China

ABSTRACT: Argumentation visualization techniques and argumentation software systems implement argument diagramming techniques by computer. It is very useful for constructing, organizing, explaining, and evaluating arguments in an efficient and convenient way. Argumentation visualization techniques are effectively used in education, informal logic, complex legal reasoning, business, and scientific reasoning. We compare different argumentation visualization techniques and single-user and collaborative-user systems. Finally, we conclude and present future problems.

Keywords: argumentation; argument visualization; argument system

1 INTRODUCTION

The real-life argumentation is very complex, and the research of argumentation visualization techniques is very useful for organization, explanation and evaluation of arguments. In the legal and other argumentation fields, the first problem is processing large-scale arguments. It is also a difficult problem to remember, retrieve and explain voluminous arguments. It has been shown in a psychology study when people use a meaningful method to organize a great deal of information, it can enhance the ability to remember, retrieve and explain information.

As early as 1913, Wigmore proposed the use of argument diagramming techniques to promote rational thinking. Modern computer software can construct a useful argument diagram and aid the reasoning of amounts of arguments. Argumentation visualization techniques are a powerful method to evaluate and analyze arguments. Using graph analysis argumentation is rooted in the education of argument and critical thinking[1]. Argument diagramming is a general arguments explaining technique in the argumentation theory, fallacy study, and particularly introductory logic and informal logic[2]. Argument evaluation depends on the argument structure shown in the argument diagram.

2 ARGUMENTATION VISUALIZATION TECHNIQUES

Argumentation visualization techniques theory roots in argument diagramming. Argument diagramming is a basic method of informal logic. There are two essential elements: (1) nodes—they represent proposition, and are promises and conclusion in the argument graph; (2) arcs—i.e. the arrows between nodes, representing the inference. A group of nodes connected by arrows represent a series of arguments. In an argument diagramming, there are amounts of arguments; there is a finial conclusion that represents the proposition. In addition, the conclusion is supported by a series of arguments; each middle conclusion can be a promise of the next argument. A group of premises can be combined together to support the conclusion in various ways. There are two special argument structures: joint argument and convergence argument. In the joint structure, every premise depends on each other and supports the same conclusion jointly. If a premise is deleted, the other premise's support strength is discounted compared with the joint support of the conclusion. In the convergence structure, every premise independently supports the conclusion. Even a premise is deleted, and the other premise's support strength is undamaged.

The contemporary argument visualization method arises from the 19th century 50 ears Whately's argument diagramming, which is connected by premises and conclusions. In the early 20th century, Wigmore using the argument diagramming described legal cases in detail. About 40 years later, two methods appeared and consolidated the formal basis. The first is the Beardsley model[3]. Beardsley explicitly put forward the basic model and composition of the argument structure. He also concluded the argument diagramming. The second is the Toulmin model[4]. Toulmin put forward the argument structure, which composed of six components to comprehend a domain-independent reasoning. Later, the method is becoming very popular in the teaching of critical thinking.

The Wigmore model, Toulmin model and Beardsley model are three main traditional argument diagramming models; the research of computer argument visualization is also along with the three main branches. Because of too detailed arguments and reasoning type, the Wigmore model is difficult to be correctly used by practitioners. Anderson, Schumm and Twining using the graph method simplified and updated the Wigmore model to make it well structured and easily organized[5].

Because of pedagogical needs, Scriven extended the Beardsley model and started the development of informal logic. Another foresighted, productive, interdisciplinary, and outstanding scholar is Walton, who extended informal logic to deal with reasoning, for example fallacies and conventional reasoning model. Among these, the most famous is the argument scheme[6].

3 ARGUMENTATION DIAGRAMMING MODEL

There are three famous traditional argumentation models: Wigmore model, Toulmin model and Beardsley model. They are the three main branches in the research area of computer science.

3.1 Wigmore model

The Wigmore model includes 30 kinds of different components and appointments[7]. It is very complicated and comprehensively describes the argumentation procedures, but it is difficult to be implemented in computer.

The Wigmore model using the graph represents the argumentation structure and procedures. In the court, there are two kinds of arguments: prosecution and defense. Node 1 is the evidence that will be certified by prosecution. Lined arguments are defense arguments, for example 9, 16–19, and the other arguments are prosecution arguments. The model includes Box and Arrow, which connect boxes. Box represents argument, which can be divided into two classes: conclusive evidence supporting other arguments and negative evidence denying other arguments. Arrow represents reasoning. In accordance with the actual court debate, Wigmore classified the arguments and reasoning, nodes including (1) "□"—represents witness arguments; (2) "○"—represents indirect evidence or circumstantial evidence; (3) "□"—represents supporting evidence; (4) ">"—represents explanation evidence. Reasoning including: (1) "→"—represents direct support; (2) "↑"—represents powerful support; (3) "⊁"—represents indirect support.

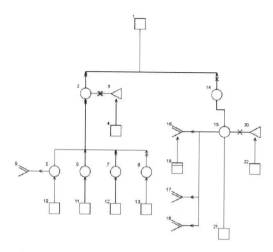

Figure 1.   An example of the Wigmore model.

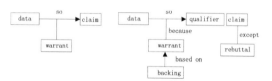

Figure 2.   Toulmin model: basic scheme and complete model.

3.2 Toulmin model

Philosopher Toulmin put forward a general argument scheme[4] to analyze argumentation. Recently, the Toulmin argument scheme became influential in argumentation computation. The Toulmin model is depicted in Figure 2. The left is a basic scheme; there are three elements in an argument: Claim, Data and Warrant. The right is a complete model, which includes Backing, Qualifiers and Rebuttal.

There are 6 elements in the Toulmin complete model:

(1) Claim C, i.e. someone tries to testify that is the valid conclusion; (2) data D, i.e. the facts as an argumentation basis (evidence, living example or statistical figures); (3) warrant W, it is the bridge between data and claim to ensure that the claim is legally based on the data; (4) backing B, it is the argument of warrant and by answering the query of warrant to support it; (5) qualifier Q, the constraint condition from warrant to conclusion (the conclusion is obtained affirmatively or possibly, for example possibly, hypothetically, certainly); and (6) rebuttal R, preventing from data to claim.

Table 1. Argumentation visualization software and their properties.

Types & names		Application domain	Features	Argument model
A	Araucaria	law, education, news etc.	interdisciplinary, transform of different diagramming model	multiple model
	Rational	philosophy teaching, business	using visualization method quickly construct, modify and evaluate argument	close to Beardsley model
	Anthena	education	compute the value of every node in accordance with weighted sum of acceptability value and correlation value of child-node	Beardsley model
	ArguMed	law	Based on basic logic	similar to Toulmin model
	ConvinceMe	scientific reasoning	Join probability component	variant of Beardsley model
B	Belvedere	scientific reasoning, evidence reasoning	based on prove and assumption, it is fully tested in the classes	interactive Beardsley model
	Compendium	education and science	based on IBIS	Toulmin model
	Truthmapping	education, media, politics, science, law, philosophy, social problem	online unitedly construct argument	Beardsley model

Type A: single user argumentation systems; Type B: collaborative argumentation systems

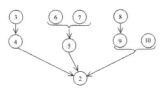

Figure 3. Beardsley model.

3.3 *Beardsley model*

The Beardsley model uses an inverted tree to represent the argument graph. As depicted in Figure 3, the root node ② is under the tree and represents conclusion; leaf nodes ③, ⑥, ⑦, ⑧ and ⑩ represent premises; intermediate node is the intermediate conclusion from premise to conclusion. For example, nodes ⑥ and ⑦ infer ⑤, and it is convergence argument. Nodes ④, ⑤ and ⑨ or ⑩ jointly infer node ③, and it is a joint argument. Nowadays, the method appears in many formal logic textbooks.

4 ARGUMENTATION VISUALIZATION SYSTEMS

Argument diagramming has been used in education and legal domain for years. Now, the Argumentation-support Visualization System has also been used to solve problematic issues in both academic and business communities. Argumentation visualization system can be divided into two kinds: single user argumentation system and collaborative argumentation system. Table 1 presents the argumentation visualization software and their properties.

4.1 *Type A: single-user argumentation system*

1. Araucaria
Araucaria[8] uses the argumentation visualization method to support the analysis text argument structure. Araucaria has many users; on average, 2000 users download the software from different IP every year.

Most of the Araucaria users are tutors and their students. They are tightly connected with graduated students in some universities, such as philosophy major in Winnipeg, legal theory major in Groningen, especially in the course of critical thinking of South American. Tutors use Araucaria to automatically match argument graph in the class; compared with the traditional method, Araucaria offers convenience to the tutors.

The Ontario Court of Justice evaluated Araucaria in 2004. The judges proved that the argumentation support system is very useful when dealing with large-scale law cases.

Every day there used to be 60–70 cases related to traffic accidents in the Ontario Court of Justice, and they were handled by the fixed model earlier. When they used Araucaria, these small-scale cases were dealt with quickly than before by setting small argumentation schemes.

2. Reason!Able/Rational

Reason!Able[9] is an education software supporting education reasoning technique. It aims at constructing, analyzing arguments and using visualization method to quickly and conveniently structure, modify and evaluate arguments. Reason!Able argument tree includes claim, reasoning, and counter advice. By the year 2007, Reason!Able updated to Rational and used in business.

3. Athena

Athena[10] marks the relevant value to show the acceptability and relation of arguments. The fill degree shows the acceptability value of an argument. The percentage on the edges shows the relation value between two arguments. Athena computes the value of each node according to the weighted sum of the acceptability value and the relation value. The nodes whose values were lower than the threshold will be deleted. The left is the undeleted argument graph, and the right deletes some lower value arguments. Above all, the mechanism is helpful to pay close attention to primary arguments.

4. Argue!/ArguMed

ArguMed[11] is updated from Argue!. It aims at dealing with law arguments. The visualization feature is based on the basic logic and similar to the Tulmin model.

5. ConvinceMe

ConvienceMe[12] is used in science reasoning education by generating cause-effect net to structure and analysis argument. It is based on Thagard's Theory of Explanatory Coherence. In ConvienceMe, the nodes are assumption or evidence, and the undirected arcs show interpretation relation or conflict relation. The user conclusion is generated from the cause-effect net.

4.2 Type B:collaborative argumentation system

1. Belvedere

Belvedere[13] may be the most famous argumentation system. It is a multi-user and argument graph based argumentation tool. Belvedere includes argument represent, ontology, visualization, analysis and feedback. Initially Belvedere is used in complicated scientific argumentation for middle school students. The newest version turned from complicated scientific argumentation into simple argumentation between data and assumption. Belvedere can establish query map and argumentation model. In Belvedere, there are data nodes, hypothesis nodes and unknown nodes. Undirected arcs represent support, oppose or unknown relation between nodes.

2. Compendium

Compendium[14] is a free system which is updated from Questmap, then it is used for science argumentation. Questmap originally aimed to mediate dispute by establishing visualization information graph and support collaboration argumentation in law education. It is rotted in Issue Based Information System. Compendium allow multi-user discuss issue related problem and collaboratively recognize and solve problem then reach agreement. To a certain degree Compendium can construct argument. The focus issue can be divided into some sub-issues. Questmap supply some additional nodes, including issue, claim, reason, support and data nodes. Arguments are structured by these nodes.

3. Truthmapping

Truthmapping (http://truthmapping.com) is a free system to solve the most common problem in argumentation. It overcomes the defect of message board, e-mail, even conversation and widely used in education, media, politics, science, law, philosophy, and social problem. Truthmapping interface is visualized, users easily participate in a topic by finding out an argument, counter-example, weaken an argument to proceed argumentation and use a simplified structure to construct these nodes.

5 CONCLUSION AND FUTURE PROBLEMS

By argument visualization techniques, users can deal with large-scale arguments, remember and retrieve them. These argumentation visualization systems can well describe the relation between arguments, and improve the efficiency of reasoning. The development of computer makes argument visualization possible. The argumentation visualization system can be used in scientific reasoning, teaching of critical thinking, intelligent analysis or corporation problem solving, and business and even military affairs. In addition, current artificial intelligence research offers precise evidence reasoning, specific graphics with semantic and computing method.

However, so far, these software technologies rarely supply or do not support structuring information processing. Available argumentation systems permit the users to store argument data according to event, object, figure and their relation, but do not permit to represent the argument how to support or undermine the hypothesis of the facts.

Nowadays, there are many problems existing in the argument visualization technique, which are listed as follows:

1. At present, how many newest factual reasoning and argument visualization methods are available?
2. Semantics of graphical annotation: what is the basic theory of evidence reasoning? Does it includ the legal, philosophical, rhetorical, logical and mathematical theory?
3. What argumentation software are available or on development to reason about facts or represent argument diagramming? How to validate their effectiveness using experimental methods?
4. What are the application environments (such as investigation of crime, intelligent analysis and legal education)?
5. Do the argument visualization techniques support automatically evaluating hypothesis?
6. In the current human-computer interaction, how to enhance the effectiveness of argumentation visualization software? How to manage a large number of argument visualization?
7. How to investigate and test argument visualization research method?

REFERENCES

[1] Walton, D. Fundamentals of Critical Argumentation [M]. Cambridge: Cambridge University Press, 2006.

[2] Hongzhi Wu, Jian Tang. An introduction to informal logic [M]. BeiJing: People's Publishing House. 2009:422.

[3] Reed, C., Walton, D., & Maclagno, F. Argument diagramming in logic, law and artificial intelligence [J]. The Knowledge Engineering Review, 2007, 22, 87–109.

[4] Stephen E. Toulmin. The Uses of Argument [M]. Cambridge: Cambridge University Press, 1958.

[5] T. Anderson, D. Schum, and W. Twining. Analysis of Evidence [M]. Cambridge: Cambridge University Press, 2nd edition, 2006.

[6] Douglas Walton, Chris Reed, and Fabrizio Macagno, Argumentation Schemes [M], Cambridge University Press, 2008, 443.

[7] Wigmore, John Henry, The Principles of Judicial Proof [M]. Boston: Little, Brown and Company. 2nd ed. 1931.

[8] Chris A. Reed and Glenn W.A. Rowe. Araucaria: Software for argument analysis, diagramming and representation [J]. International Journal on Artificial Intelligence Tools, 14(3–4), 2004: 961–980.

[9] Tim J. van Gelder. Argument mapping with Reason!Able [J]. The American Philosophical Association Newsletter on Philosophy and Computers, 2002:85–90.

[10] Bertil Rolf and Charlotte Magnusson. Developing the art of argumentation: A software approach [C]. Proceedings of the Fifth Conference of the International Society for the Study of Argumentation, 2002.

[11] Bart Verheij, Artificial argument assistants for defeasible argumentation[J]. Artificial Intelligence, 150(1-2), 2003:291–324.

[12] Patricia Schank and Michael Ranney. Improved reasoning with Convince Me[C]. CHI '95: Conference Companion on Human Factors in computing Systems, New York: ACM Press, 1995: 276–277,.

[13] Daniel Suthers, Arlene Weiner, John Connelly, and Massimo Paolucci. Belvedere: Engaging students in critical discussion of science and public policy issues[C]. The 7th World Conference on Artificial Intelligence in Education, 1995:266–273.

[14] Albert Selvin, Simon Buckingham Shum and Maarten Sierhuis et al. Compendium: Making meetings into knowledge events[C]. Proceedings Knowledge Technologies, 2001.

Architectural, Energy and Information Engineering – Sung & Chen (Eds)
© 2016 Taylor & Francis Group, London, ISBN 978-1-138-02791-6

Gamification, interactive C-programming online platform

W. Zhang, G.Z. Liu, G. Tong & S.S. Guo
School of Information Science and Technology, Qingdao University of Science and Technology, Qingdao, China

ABSTRACT: C-programming language is the first basic computer language for information science and technology students. We need to transfer learning and education from the classroom to e-space, from text books to PC, from concrete buildings to digitals. To address major problems in C-programming learning, we implement a gamification, interactive online platform that combines the advantages of lecture and online platform. The key idea is to activate the students' learning interest by gamification and peer learning, and make them allocate half of the class time to students' online learning and the other half to the teacher's presentation, so that they can gain self-pacing and instant feedback on their courses. The platform and the new blending learning method can improve the students' learning effects and make them to learn actively.

Keywords: C-programming, gamification, interactive, Online learning system

1 INTRODUCTION

C-programming is the first programming language course offered to the computer science, information science, integration circuits and software engineering major students in the university. C-programming language lays a solid foundation to the subsequent course of computer language programming. The Given education has been calcified for 500 years. Traditional education is concentrated on teaching in the classroom, and teachers are the protagonists in the classroom. Students are passive learners. To overcome these disadvantages of traditional education, we establish a gamification, interactive online learning platform.

With the development of computer and Internet, students can learn by using network resources anywhere anytime. However, these resources loosely exist on the Internet with weak pertinence, so they do not meet the needs of teaching in the university. We design and implement an online learning platform. First, the introduction of the gamification concept can activate students to learn with interest. Second, we propose a blended teaching method "classroom teaching+Internet learning", which can meet the needs of university teaching. Third, by forum, online exercise and exam or interactive discussion among students and teachers, it can transfer exam-oriented education to application practice education. In our college, we establish the unified C-programming learning platform to 6 different information majors using unified syllabus, unified lesson preparation, unified exercise, unified exam and unified checking.

2 KEY IDEAS AND SYSTEM GOAL

We propose some key ideas of the online learning platform.

(1) Active Learning

Before every class, students can prepare on our online platform, they can read lecture notes of the teacher, watch videos and according to the content of videos, make exercise. In 1972, Craik and Lockhart proposed that learning and retention is related to the depth of mental processing. If these resources are combined with classroom teaching, students can learn better than ever before[1].

(2) Gamification

We combine C-programming language knowledge points with games[2]; students must master the knowledge points, then they can pass the corresponding stage of the games. In this way, we can initiate the students' impetus and enthusiasm. We set 10 (20 or 30) C-programming choices, filling or true or false questions, and 2–4 players in a game. From the beginning of the game, the student players race to be the first to answer a question, if the answer is right, then he wins some gold coin. The players' degree is decided by the amount of gold coins. The more thegold coins he has, the more the privilege he will have, such as to download some learning resources or to get the corresponding scores.

(3) Self-pacing

In the classroom, the students may take notes or discuss in private, then they cannot pace with the teacher even for five minutes[3]. On the contrary, our online platform supplies videos and interactive participation to the students, which can make up the disadvantage of classroom teaching. The students can pause and rewind and even can mute the video. Self-pacing is helpful for computer language learning.

(4) Instant feedback

Traditionally, the ordinary paper homework feed backs to the students in one or two weeks and at the moment, the students may forget all the answers of the questions. The traditional method is of low efficiency and the effect is poor. In our platform, by online exercises, the students can fill the answers again and again until the answer is correct. The whole process is attractive, and the students gain instant feedback. This method can instantly give the learning achievement and enhance the students' self-confidence.

(5) Peer learning

We set up a learning forum such as QQ group or facebook-like, not for amusement but for real learning[4]. The students can discuss, ask questions or answer the other students' questions. The teachers can monitor the forum and answer questions[5]. This method can enhance peer learning as well as student-teacher learning to make full use of the forum, and make the students fully master the knowledge points as the syllabus requires.

3 ONLINE PLATFORM FUNCTION MODULES

The gamification and interactive C-programming online platform includes five function modules, as depicted in Figure 1:

(1) Course introduction

This module includes two parts: curricular introduction and teacher introduction. The curricular introduction is composed of syllabus, curriculum content, corresponding requirement and curriculum provision. The teacher introduction includes teaching staff, teaching achievement and teachers' resume.

(2) Learning resource

This function module includes five parts:
Course wares download (free)
Teaching video download (free)
Exercise and Simulated Tests download (free)
Technical e-books download (free)
Extracurricular small routine download (using gold coins)

(3) Online exercise (gamification)

According to the whole requirement of the Information Science and Technology College in QUST, the platform adopts a step-by-step system. We set

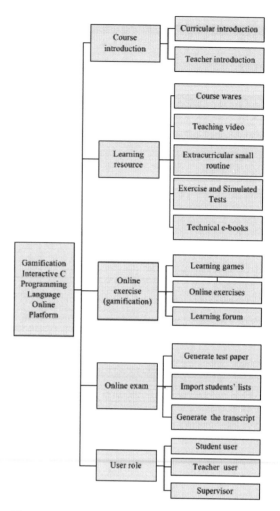

Figure 1. The platform function modules.

three terms to pointedly accomplish the ultimate goal of the C-programming course. In the first term, students learn the basic knowledge and the corresponding training to master the basic knowledge points of the course. In the second term, students develop a special project to skillfully use the C-programming language. In the third term, students develop a comprehensive project to gain proficiency in the use of the C-programming language.

This module is the kernel module of the platform. It includes learning games, online exercises, and learning forum.

Knowledge points are embedded in games. Only the player who masters the knowledge point can pass the corresponding level of the game. The winner will win certain gold coins. After the game, the correct answer is published to the players.

(I) Learning games

This module combines C-programming knowledge points with games. C-programming knowledge.

(II) Online exercises

This module integrate the teacher's prelection schedule into homework exercises. After one lesson, the students randomly extract exercises from the question bank. The students must submit the answers in a specific time, and then they can gain regular grades. Others who cannot accomplish the exercise will lose some regular grades. Close to the end of the term, students can make practice test and evaluate by themselves. Thus, they can work out the review plan for the next stage and get good grades.

(III) Learning forum

Students can post in the learning forum. The student user and teacher user can answer the questions. They also can discuss some problems and alternate concepts. If a student user gives the correct answer, he will win rewards, such as gold coins.

(IV) Online exam

This module implements online exam. The teachers can automatically or manually select a test paper from the question bank. Before the test, the students' list is automatically loaded in the system. The students log into the system by student ID. After the test, the transcript is automatically generated.

(V) User role

There are three user roles: student user, teacher user and supervisor. Different users have different authorities.

(I) Student user

There are two modules of student user. The first is personal center; he can see his own learning schedule, the amount of gold coins, post and regular grades. The second is question uploading; he can upload the challenge question or a good topic. If the question is checked by the teacher, then it can be enrolled in the question bank, and the student will be rewarded. The method can enhance the initiative of the students.

(II) Teacher user

There are four modules of teacher user. The first module is personal center. Teacher user can examine and modify the basic information. The second module is uploading such as courseware and teaching video. The third module is grading the students.

The fourth module is managing the students. At the beginning of the term, the teacher imports the students' lists, and create the student user lists for every students.

(III) Supervisor

The supervisors manage all the users in the platform including student users and teacher users. He can increase, modify, and delete the student users and teacher users.

4 CONCLUSIONS

The gamification and interactive online C-programming platform is designed from the point of view of learners. The features highlight the introduction of game and the interaction of the user. It is different from traditional online platforms.

According to the step-by-step system concept, we set three terms to learn and master the C-programming language. This method combines the advantages of lecture and online platform. This platform can satisfy the needs of teaching and also can initiate the students. It is also a tool that transfers the C-programming course from exam-oriented to practical application.

The interactivity is reflected in learning forum, learning game, online test and online exercise. The gamification is reflected in game integrating knowledge points. Only the students master the basic grammar and usage of C-programming, and they can pass the game level. By three stages, incentive mechanism students can master the kernel knowledge of C-programming, and skillfully use them. The platform can create a blending education model "classroom teaching+internet learning", which can improve the study efficiency, make up the disadvantage of classroom teaching, and meet the needs of university education.

REFERENCES

[1] Wei Zhang, Shaosheng Guo. Merging Daily Scene and Project Development into C-Programming. Applied Science, Material Science and Information Technologies in Industry. 2014(2): 2174–2177.

[2] Gamification Learning, http://baike.baidu.com/, 2014/12/30.

[3] Wei Zhang. The Research and Construction of C++ Programming Digital Education Platform. Journal of QingDao University of Science and Technology. 2010, 6, 26: 79–80.

[4] Shaosheng Guo, Wei Zhang. The Construction and Research of Object-Oriented Programming On-line Learning Platform. Lecture Notes in Management. Science. 2013(11): 168–171.

[5] ZHANG John Xue-xin. PAD Class: A New Attempt in University Teaching Reform. Fudan Education Forum 2014. 12(5): 5–10.

Architectural, Energy and Information Engineering – Sung & Chen (Eds)
© 2016 Taylor & Francis Group, London, ISBN 978-1-138-02791-6

Big data manufacturing for aviation complicated products

Hai Jun Zhang, Qiong Yan, Hui Yin Li & Shi Jie Dang
Zhengzhou Institute of Aeronautical Industry Management, Zhengzhou, Henan, China

ABSTRACT: While aviation manufacturing enterprises carry out many information systems, a large amount of data is obtained. It is important to know how to use these data for more value and decision support. In the paper, first, the big data manufacturing is defined; second, the process of the MBD-based aviation product data is analyzed; finally, a novel architecture based on MBD for aviation complicated products is proposed, which is based on the definition of BDMfg and the analysis of MBD technology. The architecture is a standardized and open architecture, which can support the rapid and flexible deployment of application to maximize the value of data.

Keywords: big data; manufacturing; aviation complicated products; model-based definition

1 INTRODUCTION

Aviation complicated products have the characteristics of long development cycle, large amount of technology information, and multi-disciplinary professionals[1,2]. The development trend of aviation industry collaboration tools is from single tools (e.g. CAx, EAM, LES) to large information systems (e.g. MES, ERP, CRM, PLM); that of design drawings is from 2D to 3D design model, and then Model-Based Definition (MBD). The amount of information technologies makes great improvement in the new product development speed and the design efficiency. The real-time sensing, acquisition, monitoring of production process promote data sharing, which is helpful to the digitization of the product's life cycle. While the information systems are run in aviation manufacturing enterprises, a large amount of data is obtained. There are demands to collect, manage and analyze the structured and unstructured data. What is important is how to use these data for more value and decision support.

2 BIG DATA MANUFACTURING

According to the definition of big data[3] and the characteristics of digital manufacturing[4], the Big Data Manufacturing (BDMfg) is summarized as a revolutionary new manufacturing enterprises operating mode, which serves all product data as core assets of potential value, adopts big data technology and method to collect, process, analyze and integrate them, and then help manufacturing enterprises operators forecast the market direction in time, guide product design and pricing, precise marketing, and refine the internal management process. The implementation of BDMfg has four stages, as shown in Figure 1:

1. BDMfg Acquisition: the massive data is the foundation of the BDMfg model. The data may be structured, semi-structured or unstructured, which is acquired from the way of the Internet of things technology (e.g. embedded system, RFID, sensor), GPS and all kinds of digital manufacturing information systems.
2. BDMfg Pretreatment: it analyzes, extracts and cleans the data, in order to filter "noise" (not care or wrong data) for high-quality data.

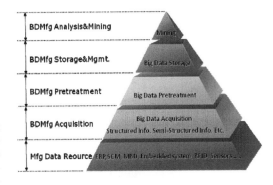

Figure 1. Stages of BDMfg implementation.

3. BDMfg Storage and Management: the collected data should be stored and the database should be set to manage and call the data.
4. BDMfg Analysis and Mining: the large, incomplete, noisy, fuzzy and random application data should be refined for the hidden value through the big data mining, fusion, machine learning techniques.

3 MODEL-BASED DEFINITION

MBD highly integrates Product Manufacturing Information (PMI) such as product design, product process, and product manufacturing. MBD has powerful expression and can present the characteristics of the product more vividly, and express the design intention more accurately. Therefore, it is also easier to abstract engineering information and mine data, and is beneficial to BDMfg. According to previous research[5], the process of MBD-based aviation complicated products data is as follows, as shown in Figure 2.

3.1 MBD-based product design

The product engineers design the three-dimensional (3D) CAD models, then define the product MBD models, and finally create the Engineering Bill of Material (EBOM) and Assembly Require-

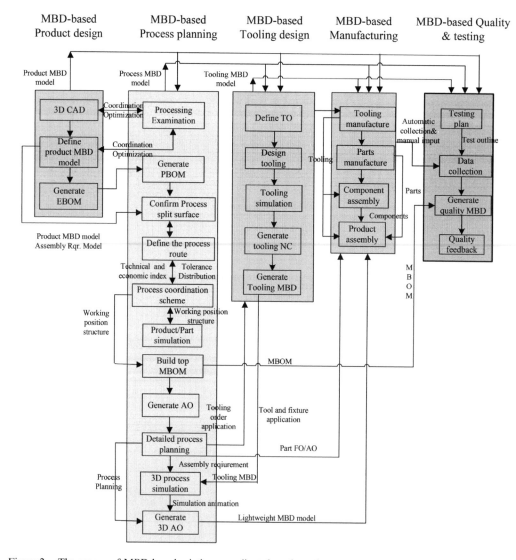

Figure 2. The process of MBD-based aviation complicated products data.

ment Model (ARM), according to the product functional requirements.

3.2 MBD-based process design

The PBOM (Product Bill of Material) is created by the process department according to the 3D-CAD models and the official released EBOM. The PBOM can be automatically associated with components (e.g. batch sorties, process route) and 3D CAD models. The whole machine assembly process is simulated, and then the top-level structure file (manufacturing bill of material, MBOM) is created. After the component geometric and the part material properties are simulated, the Assembly Outlines (AO) and Fabrication Outline (FO) will be prepared. The process is designed in detail, and its process MBD model is built.

3.3 MBD-based tooling design

According to the tooling ordering application, the process MBD model, the product process program and product MBD model, the tooling structure are designed, and the structure and form of the clamping and positioning of tooling are determined, and finally the tooling MBD model is designed, which will be simulated in DELMIA DPM. The tooling NC program is prepared.

3.4 MBD-based manufacturing

The production sequence is divided into tooling manufacturing, parts manufacturing, component assembly and product assembly. The workers can view the 3D process instruction information from the Manufacturing Execution System (MES) terminals, and then start the site processing, the assembling operation and quality data collection. Production equipments and testing equipments process and measure, respectively, according to the NC program and measurement program, and in strict accordance with the process MBD model defined for processing and measurement, which realizes a real-time control of flexible production lines.

3.5 MBD-based quality and testing

The inspection plan is established in the quality management department, according to the product MBD model, process MBD model and tooling MBD model. Through the automatic input of device or manual collection by workers, the quality data is fed back to the quality MBD model, and then a single product quality vehicles report is formed, upon which relies the quality assessment and statistical analysis.

4 BDMFG ARCHITECTURE FOR AVIATION COMPLICATED PRODUCTS

Based on cloud computing[6], a big data manufacturing architecture for aviation complicated products is built, which can be applied to store and analyze the big data in the development process of aviation complicated products. The architecture is a hybrid architecture, which can integrate the RDB and NoSQL database, as shown in Figure 3:

1. **The infrastructure layer** contains server devices, network devices, and storage devices, for example, PC servers, blade servers, and multi-array cluster. The layer is the underlying physical devices to keep big data manufacturing systems running.
2. **The virtual resource layer** contains computing resource pools, network resource pools and database resource pools (RDB and NoSQL) in order to virtualize the underlying physical devices, for example, VMWare and Hyper-V.
3. **The middleware layer** is the core layer of BDMfg architecture that contains SLA management, account management, charge management, mapping and reduction, resources optimization scheduling, distributed cache, disaster recovery support, interface management, monitoring management and safety management.
4. **The data manipulation layer** contains Hadoop tools, ELT tools, OLAP analysis, stream analysis, distributed text analysis, multimedia analysis, data mining and data modeling.
5. **The application layer** contains aviation complicated products collaborative design, engineering analysis (HPC), production optimization, predictive maintenance, business intelligence and virtual desktop.

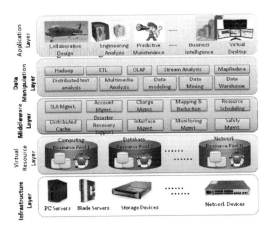

Figure 3. BDMfg architecture.

5 CONCLUSIONS

Data of digital manufacturing increases quickly. The data analysis and processing is important to the aviation manufacturing enterprises in this age of big data. Therefore, this paper proposes a BDMfg architecture for aviation complicated products, which is based on the definition of BDMfg and the analysis of MBD technology. The architecture is standardized and open, which can support the rapid and flexible deployment of application. Combined with the MBD technology, the cloud computing technology and big data technology synthetically, the potential value of enterprise data can be developed.

ACKNOWLEDGMENTS

This paper was supported by the Innovation Scientists and Technicians Troop Construction Project of Henan Province (134200510024) and the Soft Science Research Project of Henan Province (132400410782).

REFERENCES

[1] W.H. Fan, B.Y. Liu. Develop of Digital Collaborative Design Technology for Complicated Products. Aeronautical Manufacturing Technology, 2013, (3): 44–46.

[2] L.T. Zhang. Research on Information Integrating and Sharing Technology of Collaborative Design for Complex Product. Machinery Design & Manufacture, 2014, 2(2): 259–261.

[3] Viktor Mayer-Schönberger, Kenneth Cukier. Big Data: A Revolution That Will Transform How We Live, Work, and Think, Eamon Dolan/Houghton Mifflin Harcourt, 2013.

[4] D.C. Liu, L. Zheng, Z.Z. Li etc. Advanced Manufacturing Technology and Digital manufacturing. Machinery, 2001, 39(7): 7–10.

[5] H.J. Zhang, S. Zhang, Q. Yan. Study on the Archives Management System of Aviation Products based on MBD. Applied Mechanics and Materials, 2013, 321–324: 2396–2399.

[6] X.L. Jia. Thinking about MBD Technology for the Aviation Manufacturing Enterprises. Aeronautical Manufacturing Technology, 2013, 3: 50–54.

Architectural, Energy and Information Engineering – Sung & Chen (Eds)
© 2016 Taylor & Francis Group, London, ISBN 978-1-138-02791-6

Analysis and implementation of an RFID tags secure access protocol based on EPC C1G2

Zhong Wen Li, Qian Li & Ling Tir
College of Information Science and Technology, Chengdu University, Chengdu, China

ABSTRACT: Radio Frequency Identification (RFID) technique is one of the main methods for data collection in the Internet of Things (IOT). Unfortunately, the weak security capability of RFID systems has greatly hindered the development of RFID applications in the IOT. Electronic Product Code Class-1 Generation-2 (EPC C1G2) has many security risks, such as plaintext transmission of tag information, easy compromised passwords, and weak tag authentication. Some studies on designing security protocols for RFID tags either are incompatible with EPC C1G2 or present security flaws. In this paper, a bi-authentication method between the tag and reader in the RFID tag secure access protocol based on EPC C1G2 is proposed by using the Burrows-Abadi-Needham (BAN) method, and implemented. Using Eclipse 3.7, we developed a simulated reader myAlien on the Rifidi platform. The experimental results show that this protocol can be successfully implemented on myAlien, and its bi-authentication is effective.

Keywords: EPC C1G2; tags; authentication; protocol; reader; BAN logic

1 INTRODUCTION

The protocol given in the EPC C1G2 standard is not secure, the exposed label may destroy the user's privacy, and the attacker can easily track and clone the tag. In 2008, Cai et al [1] proposed an RFID mutual authentication protocol. However, it could not resist the attack of fake tags. Protocols presented by Duc et al. [2] and Li Fang [3] are two-way authentication in design. However, their protocols are based on the same assumptions, namely, that the backend authentication database has unlimited computing power. Later in 2009, Choi et al. [4] proposed a security protocol compliant with the EPCGen2 standard, and this protocol is vulnerable to reader impersonation attack [5]. Sun et al. [6] proposed an authentication protocol Gen2+ based on the EPCGen2 standard. At the end of each authentication phase, this protocol needs to update the Key pool value. Therefore, as the update is not synchronized, it might be vulnerable to DOS attack. Both [7] and [8] designed a RFID mutual authentication protocol based on the EPC C1G2 standard. It can effectively prevent disclosure, replay, camouflage, location tracking, DoS and other common attacks. Unfortunately, it missed the treatment on password disclosure. In 2012, we proposed a RFID tags security access protocol based on EPC C1G2 in [9]. It not only is effective against information disclosure, replay, camouflage,

location tracking, DoS attacks and other common attacks, but also ensures that the password is not disclosed. In addition, the analysis results show that this protocol meets the security requirement of RFID systems based on EPC C1G2. This paper further discusses the formal validation and application of this protocol.

2 PROPOSED RFID TAGS SECURE ACCESS PROTOCOL WITH BAN LOGIC

2.1 Symbol table

Table 1 outlines the notations used in the proposed protocol.

2.2 The RFID tags secure the access control protocol

Each tag in the factory randomly selects a random 32-bit access password PWA, a 32-bit kill password PWK and a 32-bit key K_i. All these values as well as EPC_i and PC_i are stored in the tag and the backend database. This protocol was proposed in [9], and its flow is as follows:

1. Readerain $_i$: the reader generates a 32-bit random number R_{rl} and sends the query request and R_{rl} to the tag$_i$.

Table 1. The notations.

Notation	Meanings	Notation	Meanings
Rri/Rti	the ith 32bits random number generated by reader/tag, i = 1,2,...,n.	EPCi	EPC code of tag i
PWA$_i$	access password of tag i	TIDi	32-bit ID information of tag i
CRC($\cdot$)	CRC checking function	P $\lhd$ X	*P sees X*
PWKi	the kill password of tag i, i = 1,2,...,n	P$\vert$es	*P believes X*
A$\Vert$B	*A* and *B*	P$\vert \Rightarrow$ X	*P controls X*
y	variable	P$\vert \sim$ X	*P said X*
A$\oplus$B	*A* XOR *B*	# (X)	*fresh (X)*
K$_i$	the 32bits-length key of tag i	$< X >_y$	*X* combined with the formula *y*
M$_i$	the ith message passed between tag, reader and database	$P \overset{K}{\leftrightarrow} Q$	*P* and *Q* may use the shared key *K* to communicate.
PC$_i$	protocol control bit of tag i	$P \overset{y}{\Leftrightarrow} Q$	The formula y is a secret known only to P and Q

2. Tag$_i$→agthe: after receiving the query message, tag$_i$ generates a 32-bit random number R_{t1}, and calculates M1 = CRC $(EPC\Vert R_{t1}\Vert R_{r1})$ $\oplus K_i$, and then it sends M1 to the reader.

3. Reader. Tag$_i$: after receiving M1, the reader generates a 32-bit random R_{r2}, calculates M2 = M1$\oplus R_{r2}$, then it sends {ACK (M2), R_{r2}} to tag$_i$.

4. Tag$_i$→ago: after receiving {ACK (M2), R_{r2}}, tag$_i$ calculates y = M2$\oplus R_{r2}$, and judges whether y is identical to M1. If these two values are the same, then M3 = $[CRC(EPC_i\Vert R_{t1} \Vert R_{r1}] \oplus K_i$ is calculated and {M3, PC_i, $R_{t1}\oplus K_i$} is sent to the reader. Otherwise, the operation ends without any action.

5. Readerthe opera: after receiving {M3, PC_i, $R_{t1}\oplus K_i$}, the reader sends {M3, PC_i, $R_{t1}\oplus K_i$, R_{r1}} to the database.

6. DataBaseaBasese: after receiving the message from the reader, the database traverses all the tag data according to the reader's reading and writing permissions, and then finds out whether there exists tag$_j$ that makes $[CRC(EPC_j\Vert R_{t1} \Vert R_{r1})] \oplus K_j$ equal to M3. If tag$_j$ is found, the database will send {$PWAi\oplus R_{r1}$ and $PWKi\oplus R_{r1}$} to the reader, and then turn to step 7. Otherwise, it ends the operation without doing anything.

7. Readering.$_i$: once the reader receives {$PWAi\oplus R_{r1}$ 和 $PWKi\oplus R_{r1}$}, it generates a 32-bit random number R_{r2}, and then sends {$ReqRN$(M2), R_{r2}} to tag$_i$.

8. Tag$_i$→ago n: after tag$_i$ receives {$ReqRN$(M2) R_{r2}}, it then calculates y = M2$\oplus R_{r1}$, and checks whether y is the same as M1. If the two values are the same, tag$_i$ will generate a 32-bit random number R_{t2}, calculates y = $R_{t2}\oplus R_{r1}$, then sends the handle and y to the reader, and goes to step

9. Otherwise, it ends the operation without doing anything.

9. Readering.$_i$: after the reader receives the handle information, it calculates M5 = $PWAi\oplus R_{r2}$, M6 = $PWKi\oplus R_{r2}$, and sends {M5, M6} to tag$_i$.

The next step is the same as Step 8 of the EPC C1G2 standard to start the normal communication between the tag and the reader. Figure 1 shows the timing diagram of this protocol.

2.3 Bi-authenticational analysis with BAN logic

Using the similar method in [10], we verify the bi-authentication of this protocol.

Lemma 1. The reader can authenticate the Tag$_i$. Proof: We idealized the protocol by transferring the generic form into the following idealized form:

1. Reader → Tag$_i$: Rr1
2. Tag$_i$ → Reader: $< R_{t1} >_{K_i}$

The following assumptions were used to analyze the proposed protocol:

A1: Reader$\vert$1: Read$\overset{K_i}{\leftrightarrow}Tag_i$
A2: Tag$_i\vert$ag Read$\overset{K_i}{\leftrightarrow}Tag_i$
A3: Reader $\vert$3:R_{t1})
A4: Reader$\vert$ 4Tag$_i\vert$ R_{t1}

The main steps of the proof are as follows:
B1: Reader $\lhd < R_{t1} >_{K_i}$ (Using (1) and the Receiving rule)
B2: Reader $\vert$2Tag$_i\vert \sim R_{t1}$ (Using A2, B1, and the Message-meaning rule)
B3: Reader $\vert$3Tag$_i\vert$aR_{t1} (Using A3, B2, and the Nonce-verification rule)
B4: Reader $\vert$4:R_{t1} (Using A4, B3, and the Jurisdiction rule)

Therefore, the reader can authenticate the Tag_i.

Lemma 2. The Tag_i can authenticate the reader.

Proof: We idealized the protocol by transferring the generic form into the following idealized form:

1. $Tag_i \rightarrow Reader$: M_1
2. $Reader \rightarrow Tag_i$: $<R_{r2}>_{M_1}$

3 THE IMPLEMENTATION OF THE PROPOSED PROTOCOL ON MYALIEN

The following assumptions were used to analyze the proposed protocol:

C1: $Tag_i | ag foll \overset{M_1}{\leftrightarrow} Tag_i$

C2: $Reader | lowing \overset{K_i}{\leftrightarrow} Tag_i$

C3: $Tag_i | ag R_{r2})$

C4: $Tag_i |\equiv Reader |\Rrightarrow R_{r2}$

The main steps of the proof are as follows:

D1: $Tag_i \lhd <R_{r2}>_{M_1}$ (Using (4) and the Receiving rule)

D2: $Tag_i |\equiv Reader |\sim R_{r2}$ (Using C2, D1, and the Message-meaning rule)

D3: $Tag_i | ag)der|)R_{r2}$ (Using C3, D2, and the Nonce-verification rule)

D4: $Tag_i | ag R_{r2}$ (Using C4, D3, and the Jurisdiction rule)

Therefore, the Tag_i can authenticate the Reader.

Lemma 3. The reader believes $EPCi$ of the Tag_i.

Proof: We idealized the protocol by transferring the generic form into the following idealized form:

1. $Reader \rightarrow Tag_i$: $Rr1$
2. $Tag_i \rightarrow Reader$: $<EPCi>_{K_i}$

The following assumptions were used to analyze the proposed protocol.

E1: $Reader | Reader \overset{K_i}{\leftrightarrow} Tag_i$

E2: $Tag_i | agader \overset{K_i}{\leftrightarrow} Tag_i$

E3: $Reader |3: EPC_i)$

E4: $Reader |4 Tag_i |\Rrightarrow EPC_i |$

The main steps of the proof are as follows:

F1: $Tag_i \lhd <EPC_i>_{K_i}$ (Using (6) and the Receiving rule)

F2: $Tag_i | \equiv Reader| \sim EPC_i$ (Using E2, F1, and the Message-meaning rule)

F3: $Tag_i | ag)der|)EPC_i$ (Using E3, F2, and the Nonce-verification rule)

F4: $Tag_i | ag EPC_i$ (Using E4, F3, and the Jurisdiction rule)

The emulation platform—Rifidi Emulator was considered and the build environment—Eclipse 3.7 [11] to improve the simulated RFID reader

myAlien designed in [12]. After creating a number of new tags, we manually drag the tags into the myAlien's antenna to allow the reader to recognize these tags. The bi-authentication process of tag and reader is shown in Figure 1.

In Figure 1, the data listed below every step is the key data sent to the reader/tag by tag/reader, and each row represents one data. As shown in Figure 1, information is exchanged between the tag and the reader seven times to complete the bi-authentication. As mentioned above, the real databased is not used; instead, we use java collection object to act as a database. The information exchanged between reader and database is not described in Figure 1 during the bi-authentication process.

Taking Tag_i as an example, the meaning of each step in Figure 1 is described as follows:

1. step 1 (Reader-> Tag_i): the reader generates a 32-bit random number R_{r1} and sends it to the tag_i, where $R_{r1} = 740A0D03$.
2. step 2 (Tag_i->Reader): the tag_i receives R_{r1} and generates 32-bit random numbers R_{t1} itself, then calculates M1 = $CRC(EPC_i\|R_{r1}\|R_{t1})\oplus K_i$, and sends M1 to the reader, where M1 = 5634853B.
3. step 3 (Reader-> Tag_i): after the reader receives M1, it generates 32-bit random number R_{r2}, and calculates M2 = M1$\oplus R_{r2}$, then sends {R_{r2}, ACK(M2)} to Tag_i. Here, the message {R_{r2}, M2} is designed to be transmitted directly, where R_{r2} = 1639968677 and M2 = 931883678.
4. step 4 (Tag_i->Reader): after the tag_i receives {R_{r2}, M2}, it calculates y = M2$\oplus R_{r2}$ and generates 32-bit random M3. Then, it sends {M3, CRC-16, PC_i, $R_{r1}\oplus K_i$} to the reader, where M3 = 1446282555, CRC-16 = 4D18, PC_i = 44D7 and $R_{r1}\oplus K_i$ = 570649846.

```
Reader->Tag:      74 0A 0D 03
Tag->Reader:      56 34 85 3B
Reader->Tag:      1639968677
                  931883678
Tag->Reader:      1446282555
                  4D 18
                  44 D7
                  570649846
Reader->Tag:      1646356121
                  931883678
Tag->Reader:      453701178
                  1
Reader->Tag:      213619522
                  800265696
                  1

This tag is not cloned!
```

Figure 1. Bi-authentication process.

Here, after the reader receives {M3, PC_i, $R_{r1} \oplus K_i$}, it sends {M3, PC_i, $R_{r1} \oplus K_i$, R_{r1}} to the back-end database. After receiving these messages sent by the reader, the back-end database traverses all the tag data according to the reader's reading and writing permissions. If there exists tag_i that makes $(EPC_i \| R_{t1} \| R_{r1})] \oplus K_i$ equal to M3, then {$PWA_i \oplus R_{r1}, PWK_i \oplus R_{r1}$} is sent to the reader. This whole process is handled in the background and is not shown in Figure 1.

1. step 5 (Reader->Tag_i): after the reader receives {$PWA_i \oplus R_{r1}, PWK_i \oplus R_{r1}$}, it generates 32-bit random number R_{r3}, and sends R_{r3} and M2 to Tag_i, where $R_{r3} = 1646356121$ and M2 $= 931883678$.
2. step 6 (Tag_i->Reader): after the tag receives R_{r3} and M2, it calculates $y = M2 \oplus R_{r1}$ and compares y with M1. If these two values are the same, then tag_i generates one 32-bit random R_{t2}, calculates $y = R_{t2} \oplus R_{r1}$, and sends y and handle to the reader, where $y = 453701178$ and handle = 1.
3. step 7 (Reader-> Tag_i): after the reader receives the handle and y, it calculates $y \oplus R_{r1}$ to get Rt2, and then calculates M5 $= PWA_i \oplus R_{t2}$, M6 $= PWK_i \oplus R_{t2}$, and finally sends {M5, M6, handle} to the Tag_i, where M5 = 213619522, M6 = 800265696 and handle = 1.

By this time, the bi-authentication process is over, and the tag and the reader can communicate based on the EPC C1G2 standard.

4 SUMMARY

RFID is one of the main methods for data collection in the Internet of Things (IOT). However, the weak security capability of RFID systems has greatly hindered the applications development in the IOT. Some studies on designing secure protocols for RFID tags either do not conform to EPC C1G2 or suffer from security flaws. We proposed a RFID tags secure access protocol that is based on the EPC C1G2 standard with a higher security performance. In this paper, after verifying its bi-authentication by the formal method BAN, we implement this protocol on our newly simulated RFID reader myAlien, which is deployed on the emulation platform—Rifidi Emulator. Finally, the experimental results show that this RFID tags secure access protocol and myAlien can work well together. In the future, from the practical point of view, we will further evaluate myAlien's efficiency and improve our RFID tags secure access protocol.

ACKNOWLEDGMENTS

This project was supported by the Sichuan Prov. Science and Technology Foundation (2014SZ0107), the Sichuan Prov. Education Department Foundation (13ZA0296), and the Universities cooperation research project (20804). The corresponding author is Qian Li.

REFERENCES

[1] Cai Q L, Zhan Y J, Wang Y H. A minimalist mutual authentication protocol or RFID system & BAN logic analysis [C], ISECS International Colloquium Computing, Communication, Control, and Management. Guangzhou: IEEE Press, 2008: 449–453.
[2] Duc D N, Park J, Lee H, et al. Enhancing security of EPC global gen-2 RFID tag against traceability and cloning [C] // Proceedings of the IEEE Int Conf Symposium on Cryptography and Information Security Communications, 2006: 269–277.
[3] Li Fang, Huang Shengye. A protocol based on dynamic secret key for RFID in accordance with EPC C1G2 standard [C], Proceedings of the Anti-Counterfeiting, Security and Identification, 2008: 459–462.
[4] Choi E Y, Lee D H, Lin J. Anti2cloning protocol suitable to EPC global Class21 Generation22 RFID systems [J]. Computer Standards & Interfaces, 2009, 31 (6): 1124–1130.
[5] Deng Miaolei, Huang Zhaohe, Lu Zhibo. Secure RFID Authentication Protocol for EPCGen2, Computer Science, 2010, 37(7): 115–117.
[6] Sun H M, Ting W C. A Gen2 based RFID authentication protocol for security and privacy. IEEE Transactions on Mobile Computing, 2009, 8 (1): 1–11.
[7] Nie Peng. Efficient mutual authentication protocol on EPC global Class 1 Gen 2 standard, Computer Engineering and Applications, 2011, 47(10): 92–94.
[8] Wang Yi-wei, Zhao Yue-hua, Li Xiao-cong. RFID Authentication Scheme Based on EPC-C1G2 Standard, Computer Engineering, 2010, 36(18): 153–157.
[9] Zhongwen li, Chengbin Wu. A secure access control protocol of RFID tags based on EPC C1G1, Proceedings of 2012 International conference on machine learning and cybernetics, 2012: 532–537.
[10] M. Burrows, M. Abadi and R. Needham, A logic of authentication, *ACM Transactions on Computer Systems*, vol. 8, no. 1, pp. 18–36, 90.
[11] The IBM development community, Bundle application based on OSGi should be developed with Eclipse [EB/OL]. http://www.ibm.com/developerworks/cn/opensource/os-ecl-osgibdev/index.html#main.[2006-07-17].
[12] Weidao Zhang. EPC global secure architecture based on CL-PKC [D]. Fujian: Xiamen Univ., (2011).

Architectural, Energy and Information Engineering – Sung & Chen (Eds)
© 2016 Taylor & Francis Group, London, ISBN 978-1-138-02791-6

Engineering electronic archives several double bearing problems

Lu Ji Zhang, Ru Liu, Qiong Wu & Xiao Qing Dong
Beijing Municipal Institute of Science and Technology Information, Beijing, China

ABSTRACT: In 1995, in the "five-year plan" of the fifth plenary session of the fourteenth of "speed up the process of national economy informatization strategic task" is a peak gradually implement informatization engineering in China. Meanwhile, electronic file has also gradually become the hottest topic. The archives field around the relative merits of traditional archives and electronic archives disputes has become the focus of academic circles at that time. Most scholars believe that the transition from the traditional archives to the electronic archives is a kind of inevitable trend, but the shift cannot occur overnight, but can only change gradually and needs a longer period. So, it should meet the challenge of the era of electronic archives, and pragmatic traditional archives work, seek development on the basis of inheritance, a dual-track double bearing has important significance. With the development of technology, the passage of time, the fact that double bearing is fit to become today's archives management mode is worthy of question. This project was funded by the Beijing financial fund.

Keywords: Engineering electronic archives; control; dual-track

1 A DUAL-TRACK, DOCUMENTS, THE ORIGIN OF THE DOUBLE BEARING

Archives dual-track double bearing has a long history in China. As early as in 1972, premier Zhou Enlai in his report on a clear instruction requests to strengthen the application of the computer. According to the instructions, the relevant government departments began to take action actively, introduce to deal with or take advantage of home computer statistical data and some necessary and important scientific computing data. So, in the mid-70s, first in the national bureau of statistics and statistical system of other departments began to use, in electricity, seismic, geological and meteorological department began to use a computer to deal with some data, do scientific computing, highly structured, or other data are in progress. If this is the starting point of China's informationization, electronic archives also arise.

The emergence of electronic archives, not only changed the way to store, has also changed the file management paradigm. However, when the electronic file has many limitations, country and industry a new electronic file management of the relevant national standards, industry standards, regulations for the file, save for a long time value to electronic and paper form two sets of preservation. In general, the archives of double-track double bearing refer to methods for the management of the archives to take paper and electronic.

2 SECOND, UNDER THE DOUBLE BEARING ON SEVERAL ISSUES OF FILE MANAGEMENT

With the further development of information technology, increasing the coverage of information systems of all kinds of institutions, information technology has almost permeated all aspects of social activities, as the technology foundation of the modern social activities, is gradually changing the operation mode of social activities. File as a product of human social activities and social activities must change the way of operation in relation to the formation, transmission, storage and utilization of their own on the motion law of the nature of change. This also means that the file processing, running process of a tendency on monorail system are applicable.

Archives in people's thinking, but, still there is a psychological dependence of paper files, full of not sure about the electronic file. At present, the cause of archives dual-track double bearing common thinking is fundamentally archives still stays in the era of paper documents, electronic files and technical environment also is not familiar with, not sure, the electronic file can last the preservation and the long-term availability of lack of cognition. Second, it is the national archives administration lack documents paperless strategic planning, system design and system specification, makes the dependency on paper document a lack of institutional choice.

Although our country promulgated the "electronic signature law", by forms of law, confirm the digital information in the legal position in the social and economic activities and the conditions for the existence of credentials, electronic data can be used as legal evidence was determined by the law, made his electronic documents and paper documents as evidence. However, the linking problem and related files regulation were not solved, and the specification depends on archives management institutions to solve.

As "electronic signature law" did not solve the linking problem and related files regulation and people's psychological dependence to paper files, all kinds of archives, organs and institutions at all levels in China has not formed in the true sense of electronic archives. Is formed by the electronic file, are the domestic departments, institutions at various levels and of archives, in order to make the archives information resources can provide network application of this unit, the process finally formed by electronic documents printed in paper document and endorsement form paper files, a large number of paper files to digital form. Indefinitely repeatable archives digitization and cause huge resources and human waste.

As shown in the above, archives dual-track double bearing a knife traditional paper files and electronic files into two of the world. Method of the serial number of two system, transfer process are not the same, and makes the file are not included in the normal business operation process, the archive of chaos caused by increasing the difficulty of finishing.

So, it does not solve the problem of double bearing, and is unable to promote the development of archives information. The informationization in the process of informationization in the developed countries emphasizes "paperless" as a strategic objective to actively promote informatization, visible on the significance of the informatization development.

3 THREE, ARCHIVES MANAGEMENT MODE CHOICE

Archives management mode choice, double bearing the double solution, should not simply rely on the technology to solve, also should be institutionalized scheme to cooperate.

To this end, both at home and abroad have made a lot of explorations and trials. Countries are pouring strength and complete set system, and widely applicable to all kinds of information system of electronic file management requirements specification and file system standard, makes every effort to solve the question in the formation of the

electronic file stage, achieve the goal of permanent preservation of electronic documents, adopt the way of our country are archive department standing on the ground of long-term preservation, by the constraints of the filing, guarantee the reliability of the electronic document, such as CAD electronic document CD storage, archive and file management requirements (GB/T17678.1-2-1999), "electronic filing and management norms" (GB/T18894-2002), provinces and cities and industry competent authorities also formulated the relevant specification or standard, such as the ministry of construction of the urban construction of electronic filing and electronic archives management.

Now, solve the authenticity of electronic file, available for a long time, save for a long time use ways mainly have the following categories:

Obtaining the probative value of electronic documents and legal effect. How to ensure the authenticity of electronic records, integrity, availability and security, is the core issue of electronic archives document value. Through legal means to ensure the authenticity of electronic file, such as China's "electronic signature law" electronic data was determined by the law can serve as legal evidence. Aiming at this problem, the Beijing municipal science and technology commission joint university of posts and telecommunications has designed a kind of applicable to electronic archives of electronic file number chapter, through the digital signature, implements the electronic file of forgery prevention, tamper-proof and prevent denial, guarantee the authenticity of the electronic archives, integrity, availability, and security, so that the electronic file obtains the value the same credentials as paper files.

Denominated in electronic files are available for a long time. Digital information complex coding format, reading and writing methods and carrier materials, are difficult to electronic files are available for a long time. Through to the electronic file exists some relevant factors of system specification design and process control, including the background data and specification, control system to record information circulation links, to ensure the long-term availability of electronic documents. For example, in 2009, the national archives of the electronic document metadata schemes, would be the form of electronic documents, exchange, archive, handed over to the design, process, storage, utilization of metadata capture, description of the general requirements. The system effectively records the content of the electronic document feature, form feature, background information and management process, the electronic government affairs, office automation, provides the reference for the design of documents, and archives the management system.

An electronic document preservation and utilization for a long time. Electronic file with the carrier's dependence, electronic file organization is done by encoding, processing, transmission, storage and electronic files, to use to deal with the coding information technology equipment indirectly. In the aspect of saving for a long time, the construction of digital archives guides to determine the long-term preservation of electronic documents and storage architecture strategy choice. In terms of use, the electronic file depends on the equipment, and cannot exist independently. According to the ministry released the 2011 statistics bulletin of the electronic information industry, China's electronic information industry has become the first big country. Chinese computer that year production is 320 million sets, mobile phones is 320 million units, China mobile phone penetration rate of 73.6 per one hundred people, Internet penetration rate reached 38.3%, mobile Internet users account for 69.4% of the total number of Chinese Internet users. It provides convenient conditions for the use of electronic archives.

Above all, archives dual-track double bearing is not fit to be archives management mode today, although electronic archives in China still exist some problems, such as has nothing to do with the traditional file operation process of e-mail management, the textual data archiving, web page type of file management to archives management institutions to solve, but the advantage of the electronic file has been far beyond the traditional paper files. To informationization, this need archives workers actively looking for an appropriate safety management method, the research archives this whole thing on the basis of information technology in the new law, explore and establish the new idea, the new model based on information management and the new method, build for archive management mode in China.

REFERENCES

[1] Wang Zhijin & Liu Bing, "Enterprises Competitors Tracking Base on Dynamic Environment", Competitive Intelligence, 2008.
[2] GIA White Paper. "Building Strategic Intelligence Capabilities through Scenario Planning", 2005.
[3] GIA White Paper. "MI Trends 2015—The Future of Market Intelligence", 2010.

Architectural, Energy and Information Engineering – Sung & Chen (Eds)
© 2016 Taylor & Francis Group, London, ISBN 978-1-138-02791-6

Research of three-dimensional machining in-process model forward generation technology

Fu Jun Tian, Hong Qi Zhang, Xing Yu Chen, Xiang Xiang Zhang & Wu Si Cheng
No. 38 Research Institute of CETC, Hefei, Anhui, China

ABSTRACT: To solve the problems of data redundancy and feature invalid in the reverse generation of the In-Process Model (IPM), a three-dimensional (3D) machining IPM forward generation method was proposed. The IPM forward generation principle was analyzed, and a part IPM tree was established, which the process information can store in the model by attribute. The mapping relationship between the typical machining feature and the modeling feature, including turning, milling, and drilling, and a 3D process planning procedure was put forward, which takes IPM built as the core. By projecting or offsetting the geometry element of the design model to get the sketch feature, the IPM can change when the design changes. Finally, a 3D machining process planning system was developed, and a part was taken to illustrate the validity of the method.

Keywords: In-Process Model (IPM), three-dimensional process planning, Model-Based Definition (MBD), computer-aided process planning

1 INTRODUCTION

With the deepening application of the Model-Based Definition (MBD) [1–2] technology, more enterprises have abandoned two-dimensional (2D) Engineering Drawing, and adopted a three-dimensional (3D) model to compile machining process planning, which achieves some effects. Corresponding to process drawing, the In-Process Model (IPM) reflects the result of the operation, which is the basis of the machining, and has important significance. However, IPM generation is a difficulty and bottleneck of 3D process planning. To assist the process designer generate IPM conveniently and rapidly, extensive research has been conducted by the Software Company and researchers at home and abroad, for examples, UG WAVE [3] technology and DELMIA machining process planning can realize the generation of IPM. Wan et al. [4] proposed a 3D machining procedure model generation method, constructed the mapping relationship between process ontology and modeling ontology, and presented the positive and negative generation method of the procedure model. Tang et al. [5] proposed an In-Process modeling method by combining blend feature simplification with boundary extraction of machining feature, Boolean subtraction between the former IPM and the processing volume characteristics was operated to obtain IPM. The above research has made beneficial exploration to IPM generation.

Part designer usually uses increased volume for modeling in the process of part design, such as extrude and revolve, but the part machining process is subtract volume from the blank. So, it is difficult to generate IPM from the design model, and directly modify the design model to generate IPM always, which generates a lot of redundant and invalid features, which lead to feature tree disorder. To solve this problem, this paper proposed a 3D machining IPM forward generation method, and studied the key technology.

2 IN-PROCESS MODEL GENERATION PRINCIPLE

From the perspectives of manufacturing, part design model can be considered to be formed by a series of machining activities to the blank model. So, the blank machining process can be expressed as:

$$Design = Blank - \sum_{i=1}^{n}\sum_{j=1}^{m_i} F_{ij} \qquad (1)$$

where *Design* indicates the part design model; *Blank* indicates the blank model; n indicates the number of operations; m_i indicates the feature number of the ith operation; and F_{ij} indicates the jth feature of the ith operation.

Suppose IPM_i is the ith IPM, then the IPM can be expressed as:

$$IPM_1 = Blank - \sum_{j=1}^{m_1} Volume_{1j},$$

$$IPM_i = IPM_{i-1} - \sum_{j=1}^{m_i} F_{ij}, \quad i = 2, 3, \ldots, n \quad (2)$$

That is, every IPM is formed by cutting the machining features of the operation from the blank model or the previous IPM.

3 PART PROCESS MBD MODELING

All information should be definite in the model, which is required by the model-based definition technology. Meanwhile, directly modifying the design model to generate IPM always generates a lot of redundant and invalid features, which leads to feature tree disorder. To realize the model-based process information definition and avoid redundancy features, this paper proposed a part process MBD model.

Part process MBD model taking operation as a basic unit to organization features, which includes machining feature and annotation, reflecting the machining process and feature machining order. To conveniently represent and store the part process MBD model, this paper describes the model tree, which is described in the graph theory, and named the part process MBD model tree, as shown in Figure 1.

Suppose that n represents the number of the operation, the blank model includes M_0 features, including machining feature and annotation, the ith operation including M_i features. Then, the part process MBD model tree can be defined as follow:

Part process MBD tree is an ordered tree, can be denoted by IPM, which is a finite set composed of N ($N = n + \sum_{i=0}^{n} M_i + 1$) nodes, which can

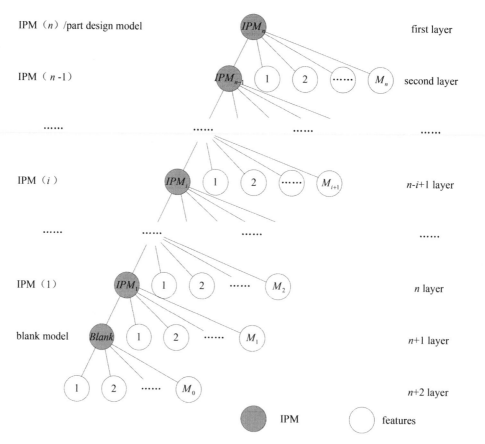

Figure 1. Part process MBD model tree.

be represented as D, and the relationships of D, which can be represented as R. The root node of IPM is IPM_n, $IPM_n \in D$, and IPM_n only including one branch node, which is named IPM_{n-1} and M_n leaf nodes. The branch node IPM_{n-1} is the first node, and the branch node is also a part process MBD model tree. The leaf nodes are ordered by the machining sequence.

The definition of the part process MBD model tree is a recursive definition, for its branch node is also the part process MBD model tree. Besides, the general characteristics of the tree and the part process MBD model possess the following characteristics:

1. Part process MBD model tree is an order tree.
2. A part process MBD model tree is composed of N nodes.
3. Suppose the operation number is n, the depth of the tree is $n + 2$.
4. Except the lowermost layer, every layer only has one branch node.
5. The degree of the ith node is $M_i + 1$;
6. The ith operation corresponds with the $n - i + 2$th layer of the MBD tree, and the node number of the layer is $M_i + 1$.
7. Beside the first layer and lowermost layer, the node number of the ith layer is $M_{n-i+2} + 1$.

Part process MBD model tree taking IPM as mainline, reflecting the machining procedure. From the lowermost of the part process MBD model tree, cutting some machining feature and annotation it, then formed a new IPM. Based on IPM, cutting some machining feature and annotation to form a new IPM, until to the last operation.

Part process MBD model tree can reflect the part feature machining sequence. The first one is the machining sequence of different layers, with the lower layer prior to the higher layer. The second one is the same layer, with the left node feature prior to the right node feature.

After the MBD model is built, process information can be saved in the model by attribute. For example, operation information can be saved in the IPM node, including operation number, operation name, equipment, and fixtures.

4 IN-PROCESS MODEL AIDED GENERATION METHOD

The mapping is between the machining feature and the modeling feature. The modeling feature in the commercial Computer Aided Design (CAD) system is oriented part design, which including increase volume and subtract volume, but the traditional machining process usually subtracts the volume. In addition, in the process of IPM generation, process designer usually oriented the machining method rather than the geometry modelling method. To solve this problem, this paper refers to the classification of machining feature based on the STEP AP224 standard, and builds the mapping relationship between the machining feature, machining method and modeling feature. Table 1 gives the mapping relationship of turning feature, milling feature and drilling feature.

From Table 1, we can conclude that to the same machining feature if its machining method different, the modeling feature is different too. The modeling feature of the machining feature was related to the machining method; when the machining method is turning, the modeling feature usually revolves; when the machining method is milling, the modeling featuring usually extrudes and sweeps.

The Process of IPM Forward Aided Generation. In the traditional 3D process planning system, first, it usually processes planning, and then generates 3D IPM, so process planning and IPM generation are two independent processes. In the MBD condition, the core of the process planning is IPM generation, the process planning is integrated into IPM generation, and process information is saved in IPM directly. The process of IPM forward aided generation is shown in Figure 2.

Table 1. Mapping relationship between the machining feature and the modeling feature.

machining feature type		machining method	modeling feature	sample
turning	cylindrical	turning, grinding	revolve	
	end face	turning, grinding, milling	revolve (turning, milling), extrude (grinding)	
	slot	turning	revolve	
milling	planar face	milling, shaping, grinding	extrude	
	pocket	milling	extrude	
	slot	milling	extrude	
	step	milling, shaping	extrude	
	tool path	milling	sweep	
drilling	hole	drill, milling, boring, grinding	hole, revolve	

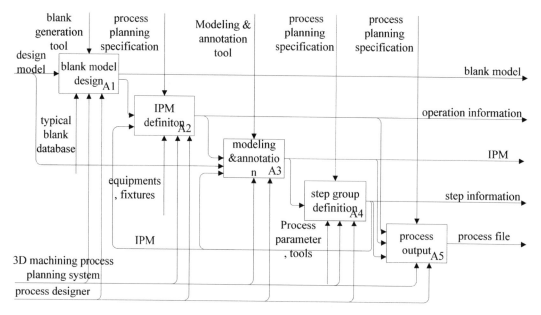

Figure 2. The process of IPM forward aided generation.

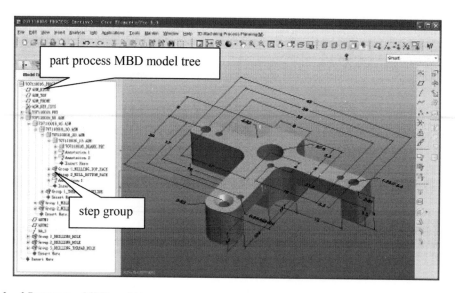

Figure 3. 3 Part process MBD model tree.

1. Blank model design

    For the simple part, the blank model can generate by getting the maximum dimension; for the complex part, the blank model can be directly modified from the design model.

2. IPM definition

    Define an empty IPM node, and complete operation attribute, including operation number, operation name, equipment, and tools, and then assembly blank model or previous IPM to the node.

3. Modeling and annotation

    Define the machining feature which the operation machined in the IPM node, and then annotation the process dimension to the IPM.

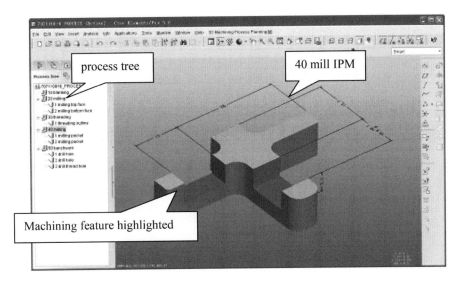

Figure 4. 4 Process information output.

4. Step group definition
   Define the step, and group the machining feature and annotation.
5. Process release
   Save the process result in product data management system.

5 APPLICATION EXAMPLE

Based on the proposed method, using Visual Studio 2005 as the development tool, by secondary development of Pro/E, a 3D machining process planning system was developed. The system included five modules: blank model design, IPM definition, modeling and annotation, step group definition, and process release. Figure 3 shows a part design model, the left region is the part process MBD model tree, the later IPM including the former IPM, also with the machining features which machined in the operation and the annotations, for example, 40 milling operation including 30 threading operation, the machining features which 40 operation machined, and the annotations. All process information is defined in the IPM node by attribute. When all IPM definition finished, it generated the structuring process, as shown in Figure 4. When process designer click an operation node in the left region, the corresponding IPM will visualization in the right region, and the machining feature which machined in the operation will be highlighted.

6 CONCLUSIONS

3D process planning is a research focus in the field of manufacturing, and IPM generation is a difficulty and bottleneck. This paper proposed a 3D machining IPM forward generation method, and studied the key technology. First, analyzing the IPM generation principle, it built the part process MBD model, which can reflect machining sequence, and realize model-based process definition. It then studied the mapping relationship between the machining feature and the modeling feature, and pointed out that modeling feature correlated with the machining method. Finally, it proposed the process of IPM aided forward generation, and provided a sample to validate the proposed method.

ACKNOWLEDGMENTS

This research was financially supported by the National Defense Basic Research Program of China (Grant No. A1120131044), the National Defense Basic technology Research Program of China (Grant No. Z312012B001, B3120131100), and the National Key Technology R&D Program of China (Grant No. 2012BAF12B03).

REFERENCES

Alemanni M, Destefanis F, Vezzetti E. Model-based definition design in the product lifecycle management scenario. In: International Journal of Advanced Manufacturing Technology, 2011, 52: 1–14.

Quintana V, Rivest L, Pellerin R, et al. Will modelbased definition replace engineering drawings throughout the product lifecycle? A global perspective from aerospace industry. In: Computer in Industry, 2010, 61: 497–508.

Tang J J, Tian X T, Geng J H. In-Process modeling method of applying blend feature simplification (in Chinese). In: Computer Integrated Manufacturing Systems, 2013, 19: 1984–1989.

Tian F J, Tian X T, Geng J H, et al. Process planning method driven by process model (in Chinese). In: Computer Integrated Manufacturing Systems, 2012, 17: 1128–1134.

Wan N, Chang Z Y, Mo R. Three-dimensional new mode of machining process planning (in Chinese). In: Computer Integrated Manufacturing Systems, 2011, 17: 1873–1779.

Architectural, Energy and Information Engineering – Sung & Chen (Eds)
© 2016 Taylor & Francis Group, London, ISBN 978-1-138-02791-6

The water-salt system of Rare-Earth Element

Larisa Grigorieva & Angela Orlova
Technical Sciences, The Department of Composite Materials Technology and Applied Chemistry,
Moscow State University of Civil Engineering (MGSU), Moscow, Russian Federation

ABSTRACT: The water-salt system of lanthanum and neodymium trichloroacetates, $La(CCl_3COO)_3$–$Nd(CCl_3COO)_3$–H_2O, has been investigated by the isothermal solubility method at 298 K. Some thermodynamic characteristics are measured for the solid solutions. The effect of the anion on the behavior of salts in the co-crystallization is analyzed.

Keywords: nanomaterials; Rare-Earth Element (REE); solid solutions; trichloroacetate

1 INTRODUCTION

In recent years, rapidly developing new scientific direction has been associated with obtaining and studying nanomaterials. Nanomaterials include nanopowders with a size less than 100 nm, and glass-crystalline materials are distributed in the volume elements with nanoscale structures, and also films and fibers have a nanoscale thickness [1].

Among the thin film and disperse systems, promising materials have been obtained on the basis of simple and complex oxides of elements III-V group. One of the promising new classes of materials is coordination compounds of Rare-Earth Elements (REE) and, based on them, multifunctional nanoscale materials are obtained.

The global consumption of rare-earth elements is estimated to be 115–120000 Tons. The most important application areas of REE are petrochemicals, electronics, glass, ceramic industry, and construction industry.

The use of rare-earth metals has been observed in a number of areas related to the use of undivided REE. However, preferably, the individual compound requires oxysulfides, for example, in the ceramic industry—yttrium, high abrasives—cerium. Unique optical and magnetic properties of nanoparticles based on rare-earth elements have been analyzed by the Research and Development Center [2]. For the successful application of new functional materials and the development of manufacturing technology, it is necessary to establish the relationship between technology and the desired properties, composition, structure and the conditions of their receipt.

The study of solubility and distribution of components between the phases of water-salt systems of rare earth elements has theoretical and practi-
cal significance. These data supplement reference materials on the solubility and are the basis for the development of separation technology and produce REE in the individual state in the form of salts. Since the lanthanide atoms are very similar in the electronic structure and short range, they form a substitutional solid solution, making it difficult to separate them. The study of such systems helps to identify the factors influencing these processes and to establish the concentration area formation of solid solutions. In addition to the processes solvent often is water, so this work is devoted to the study of water—salt systems.

2 THEORY AND EXPERIMENT

The literature data are available for the study of systems of chlorides, nitrates, iodide, sulfate, selenate, bromate ethylsulfates REE, which allows us to compare not only the behavior of salts of the same anion in a series of REE, but also the influence of the anion on the behavior of salts with same cations [3,4]. We studied the system trichloroacetate REE. Trichloroacetic acid anion has a greater weight and size compared with that studied previously. The trichloroacetate ion radius is 3,4Å, and its shortened intermolecular contact Cl ... Cl is equal to 3,647. Heavy and large-trichloroacetate ion smooths the relative difference in lattice components, this reduces the costs of energy on deformation of the crystal lattice.

In this work, we used neodymium and lanthanum trichloroacetates, which were synthesized according to [5]. The salts were obtained as trihydrates. Their composition was established by chemical analysis. The rare-earth element was determined by complexometric titration with

Xylenol Orange [6]. The chlorine was determined via burning a sample in a microbomb with metal sodium and subsequent titration of the chloride ion [7]. The number of water molecules was found by difference and by differential thermal analysis. The water-salt system of lanthanum and neodymium trichloroacetates, $La(CCl_3COO)_3$–$Nd(CCl_3COO)_3$–H_2O, has been investigated by the isothermal solubility method at 298 K. Equilibrium was established in the system for 7 days by continuous stirring with a glass striker followed by the separation and analysis of the equilibrium phases. The sum of the rare-earth elements was determined by complexometric titration [6], and the neodymium concentration was determined by spectrophotometry [8].

The system was found to be eutonic. The eutonic point has the following composition, in wt.%: $La(CCl_3COO)_3$–21,7 ± 1,9; $Nd(CCl_3COO)_3$–40,7 ± 1,9; H_2O–37,6 ± 1,8.

In the system $LaAn_3$—$NdAn_3$—H_2O, the difference in radii crystallizing component is 5.8%. Ethylsulfate, bromate lanthanum and neodymium nonahydrate were isostructural. Structure trichloroacetate (nitrate) lanthanum differs from the structure of trichloroacetate (nitrate) neodymium. Isotherm distribution systems are diverse (Figure 1). Solid solutions of nitrate, bromate ethylsulfates have a field "pseudo- ideal" solid solutions.

To determine the influence of the anion on the formation of solid solutions, we calculated the free energies (ΔG^M) and excess free energy (ΔG^E) formation of solid solutions (Table 1). The calculation was performed using the formula:

$$\Delta G^M = RT(x_1 ln a_1 + x_2 ln a_2) \qquad (1)$$

$$\Delta G^E = RT(x_1 ln \gamma_1 + x_2 ln \gamma_2) \qquad (2)$$

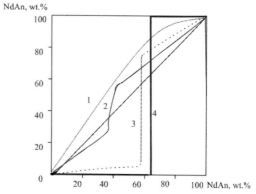

1. $La(C_2H_5SO_4)_3 - Nd(C_2H_5SO_4)_3 - H_2O$
2. $La(BrO_3)_3 - Nd(BrO_3)_3 - H_2O$
3. $La(NO_3)_3 - Nd(NO_3)_3 - H_2O$
4. $La(CCl_3COO)_3 - Nd(CCl_3COO)_3 - H_2O$

Figure 1.   Diagram of distribution systems.

Table 1.   The values of the free energy and the excess free energy of formation of solid solutions in the system $LaAn_3$–$NdAn_3$–H_2O.

Composition of the solid solutions mole fraction $La(NO_3)_3$	ΔG^M, J/mol	ΔG^E, J/mol	Composition of the solid solutions mole fraction $La(BrO_3)_3$	ΔG^M, J/mol	ΔG^E, J/mol	Composition of the solid solutions mole fraction $La(C_2H_5SO_4)_3$	ΔG^M, J/mol	ΔG^E, J/mol
$La(NO_3)_3 - Nd(NO_3)_3 - H_2O$			$La(BrO_3)_3 - Nd(BrO_3)_3 - H_2O$			$La(C_2H_5SO_4)_3 - Nd(C_2H_5SO_4)_3 - H_2O$		
0,987	–49	26	0,96	–183	13	0,070	–523	109
0,977	–72	46	0,93	–275	23	0,155	–862	238
0,956	–106	88	0,90	–351	33	0,230	–992	339
0,916	–142	168	0,86	–430	48	0,320	–1105	456
0,844	–165	301	0,80	–731	154	0,396	–1112	556
0,812	–166	354	0,49	–709	13	0,435	–1109	602
0,790	–166	387	0,41	–673	123	0,483	–1075	644
0,122	–144	255	0,39	–661	129	0,485	–1079	653
0,110	–143	230	0,34	–620	136	0,585	–946	724
0,081	–133	170	0,31	–590	141	0,706	–761	724
0,026	–75	54	0,22	–469	154	0,753	–665	695
			0,20	–433	156	0,788	–611	665
			0,09	–186	170	0,815	–732	615
						0,928	–332	310
						0,968	–188	146
						0,990	–88	50

where R—universal gas constant
- a—the activity of the components
- x—mole fraction of components in the solid phase
- γ—coefficients of the active ingredient.

3 CONCLUSIONS

The highest thermodynamic stability has solid solutions of lanthanum-neodymium ethylsulfates. The breaking system for confluency increases lanthanum-neodymium nitrate, respectively, and increasing the excess free energy of solid solutions of lanthanum nitrate, neodymium. The different behavior of salts in the co-crystallization is probably due to the influence of the anion, its mass, size, structure, nature of the relationship and the structural features of the original components.

ACKNOWLEDGMENTS

We acknowledge the financial support from the Russian Ministry of Education (Contract No. 7.2200.2014/K).

REFERENCES

[1] I.V. Melihov: submitted to Bulletin of the Russian Academy of Sciences (2002).

[2] M.A. Green: *High Efficiency Silicon Solar Cells* (Trans Tech Publications, Switzerland 1987). 2. Chunhui Huang: *Rare earth coordination chemistry: fundamentals and application* (Singapore: Wiley, 2010).

[3] Y. Mishing, in: *Diffusion Processes in Advanced Technological Materials*, edtied by D. Gupta Noyes Publications/William Andrew Publishing, Norwich, NY (2004), in press. V.V. Serebrennikov, Batyreva V.A. and L.F. Serzhenko: submitted to Journal Inorg. Chem. (1981).

[4] G. Henkelman, G. Johannesson and H. Jynsson, in: Theoretical Methods in Condensed Phase Chemistry, edited by S.D. Schwartz, volume 5 of Progress in Theoretical Chemistry and Physics, chapter, 10, Kluwer Academic Publishers (2000). V.V. Serebrennikov and T.N. Tsybukova: submitted to Journal Inorg. Chem. (1982).

[5] G. P. Tilley and I. Roberts: submitted to Journal Inorg. Chem. (1963).

[6] G. Schwarzenbach and H. Flaschka: *Die komplexometrische Titration* (Ferdinand Enke, Stuttgart, 1965; Khimiya, Moscow, 1970).

[7] V. A. Klimova: *Basic Techniques of Organic Microanalysis* (Khimiya, Moscow, 1975). [in Russian].

[8] N. S. Poluektov and L. I. Kononenko: *Spectrophotometric Determination of Individual Rare-Earth Elements* (Naukova Dumka, Kiev, 1968) [in Russian].

Architectural, Energy and Information Engineering – Sung & Chen (Eds)
© *2016 Taylor & Francis Group, London, ISBN 978-1-138-02791-6*

Image research of high-speed train seats

K. Sing Fernandez & C.Q. Xue
School of Mechanical Engineering, Southeast University, Nanjing, China

H.Y. Wang
Science and Technology on Electro-Optic Control Laboratory, Southeast University, Nanjing, China

ABSTRACT: The main objective of this study is to describe the high-speed train seats design associated with the Kansei Engineering approach. The first part involves the expression of the user feelings in elements of the product design, providing an intuitive design basis that improves the efficiency and quality of work of the designer. The second part deals with the product images that are needed in order to build a database. The database design requires obtaining a lot of emotional analysis. Some surveys and analysis of the product design are used in the perceptual engineering approach. Simple applications may be used in the research and analysis of design processes, while complex applications require the help of extensive mathematical analysis and statistical methods. However, for the intermediate level of application complexity, a relatively simple mathematical calculation may be applied.

Keywords: seats trains; Kansei Engineering; emotional design

1 INTRODUCTION

Product design in the modern market has become more complex, because the product has a growing diversity of features and needs to meet the growing demand. Product design should be seen as a medium of communication between people, able to communicate to users its advantages and, mode of use in a proper way.

The Japanese Nagamachi began to study the earliest emotional engineering practices, and later the discipline that is named Kansei Engineering. In Europe, there is also a new field of study, called the "emotional design." Through the use of a cognitive theory, the designers of product design have orientation in the user's perspective and cultural background. In the high-speed train seat design process, there is a need to consider semantic imagery and psychological factors. These are important factors in the success of the seat design. Designers should consider the user's experience and feelings as the fundamental starting point of the design.

2 KANSEI ENGINEERING

Kansei Engineering (KE) is a method of engineering whereby users experience in physical and psychological ways the product images. The result will be quantitative elements that will be converted into elements of the physical design. Kansei Engineering is essentially a method of mathematical analysis using social science research. The prime object of the research is the "human being", because the individual needs of people on Kansei Engineering are important.

The image is very important in Kansei Engineering analysis. There are two types of images: one is direct and the other is indirect. Direct imagery is the first accurate image of a product, and a precise observation, while indirect imagery is a psychological image, a feeling that we get when we see a product. In engineering terms, emotional imagery of the world is very interesting, which includes psychology, ergonomics, semiotics and emotional design.

3 KANSEI ENGINEERING IN THE SEMANTIC DIFFERENTIAL METHOD

The American psychologist Oswald Packard (Charles E. Osgood) in 1942 developed a Semantic Differential method (SD). It is an experimental method for studying measured psychological imagery. In order to use perceptual differential language for the building of engineering systems, the specific process is as follows:

a. From the market, enterprise or magazine, collect the customer's words at perception; usually hundreds of first time words are collected into an emotional vocabulary, and then a number of the most relevant words are chosen for analysis.

b. Through investigation or trial, test the selected words and the design elements of relevance.

c. Using advanced computer technology establish a systematic framework for perceptual engineering. Through the use of artificial intelligence, neural networks and fuzzy logic geometrical methods, establish the relevant databases and computer reasoning systems.

3.1 Semantic differential method

Semantic differential method is a method of psychological measurement, which aims through interviews with subjects or questionnaires to the target consumer group, to test the product sensibility expressed by adjectives. The selection number of the evaluation scale for a group of semantic adjectives must be odd, and is often chosen to be 5 or 7.

SD method (Semantic Difference Method) requires the seat sample assessment scale. The evaluation will typically use a scale of 0 in the center. Starting from the center to the sides, the smaller value should be more representative of the left adjective; the greater value will be more representative of the right side adjective; and 0 is neutral.

The Semantic differential uses positive and negative adjectives; between an adjective and its antonym, there is about a 5 to 11 scale range, as shown in Table 1. Human perception can be reflected in adjectives that can make people think about the product, and the one that is chosen is the one having the most intense reaction between these different words.

The product is evaluated using adjectives. This evaluation composition has three aspects: a total number of 5 to 30 adjectives is more appropriate, and the evaluation level chosen points must be odd (3,5, 7,9, or 11). Generally, the most commonly used adjectives are 5 and 7, and the elements of sample should be in the range of 20 to 50.

4 EXTRACTION OF THE SEAT FEATURE

Considering the appearance, structure and other characteristics of the seat design, we achieved 50 samples in a side view, a front view and isometric view of the main seat image data. After analysis of each of these samples, 15 samples were selected as the most representative (12 samples of the second-class seat and 3 samples of the first-class seat), as shown in Figure 1.

The specific characteristic extraction process of the seat sample is as follows:

a. Get a seat graphics and three-dimensional map;
b. Draw the abstract structure of the shape of the seat, retaining the main elements;

c. Use the Corel Draw graphics software to draw the characteristics of the seat samples;
d. The main features obtained are: a headrest, the seat body, and an armrest, as shown in Figure 2.

After extraction of the sample characteristics, another additional analysis with 44 samples obtained different characteristics, which comprised: a group 16 headrest samples, 12 seat body samples, and 16 handrail samples, as shown in Figures 3–5 for the seat map sample characteristics.

The seat designer can see the sample characteristics in the figure of the seat. Figures 3–5 show us the increasing complexity of the seat design characteristics. The gradual increase in the use requires a curve in the seat, and there are more complex factors to be considered. For passengers, a simple

Table 1. Semantic differential scale (Osgood semantic differential inventory).

Low level	−2−1012	High level
Simple	−2−1012	Complex
Rigid	−2−1012	Soft
Warm	−2−1012	Cool
Heavy	−2−1012	Lightweight
Dangerous	−2−1012	Security

Figure 1. The main seat of samples.

Figure 2. Component of the seats.

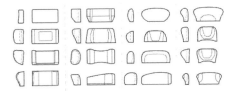

Figure 3.   Headrests.

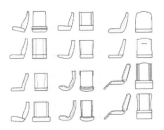

Figure 4.   Main part of the seat.

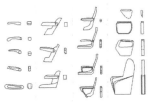

Figure 5.   Armrest.

design is supposed to meet the basic needs only, but comfort and appearance require additional factors to be considered. With a more humane design, seat design will be more prone to consider the needs of passenger comfort, taking environmental requirements and other special needs into consideration and leading to an increased difficulty of seat design.

5   CHARACTERISTICS OF THE EMOTIONAL IMAGERY EVALUATION

The vocabulary of feelings expresses that all kinds of characteristics of the seat are prepared according to the concepts related to Kansei Engineering and passenger demand, as a starting point. First, the list has around 100 words, then the designer can select from the group, a smaller vocabulary that will be able to accurately express the feeling of the respondents. This study used 16 imagery vocabularies as a final conclusion, which are: safety, modern, comfortable, flow lines, technology, clean, atmosphere, durable, luxurious, sophisticated, stable, practical, professional, high-level,

coordination and concise. It should be noted that 70 of the respondents have studied design, and 42 have no background in design.

The variables of the imagery vocabulary of each seat sample were loaded into Excel, and then using the hierarchical analysis software module, the relationship between the study sample data was calculated. The main objective of this first analysis is to classify and select the most representative semantic vocabulary. The lexical semantics include: high level, safe, simple, soft, warm and stable. Six phrases composed of semantic meaning with the opposite phrase were then selected as follows: low-level—high-level, dangerous—security, complex—simple, rigid—soft, cool—warm, lightweight—heavy, resulting in three pairs of semantic vocabulary to analyze in more detail.

The number of selected persons was 112, all of whom have multiple experiences with train seats. 15 samples were evaluated with a perceptual evaluation using the SD. The number of valid questionnaires was 77. Subsequently, the average of the last three digits after the decimal point was taken to get the semantics evaluation of the high-speed train seats sample.

5.1   Results of the experimental analysis

The results of the questionnaire data through use of the statistical analysis software allowed the qualitative and quantitative analysis of the data in the graph. Three pairs of adjectives from six perceptual adjectives were singled out. In Figure 6, "rigid—soft" and "warm—cool" were selected as the first pair of adjectives. In Figure 7, "lightweight—heavy" and "dangerous—security" were selected as the second pair of adjectives. In Figure 8, "high level—low level" and "simple—complex" were selected as the third pair of adjectives.

The figures in this uneven distribution of sets can be analyzed by seat material and color of the backrest. The material gives a warm or cool feeling. The sample 4,13,5,10,6 uses warmer colors (red, purple, orange, yellow) and the leather, textile and wood materials can give the passengers a soft and warm feeling. The relevant parts of the seat shape, as shown in the handrails, were arc-shaped and the rounded shape can also give a soft and warm seat feeling. The material and color in sample 1,3,7,14,12 and 2 express a relatively cold feeling, and although sample 2 has an arc-shaped handrail, the overall shape is triangular, which also gives a cool feeling.

The sets of data about the security and stability showed that, with the exception of sample 3, all other sample are in the first quadrant of the graph. This analysis also shows the importance of stability and security in the design on the seat.

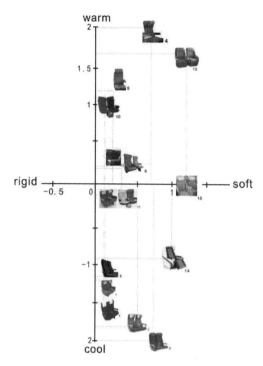

Figure 6. Images of the spatial distribution of the semantic analysis (1).

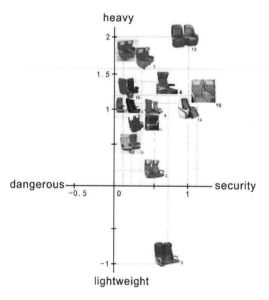

Figure 7. Images of the spatial distribution of the semantic analysis (2).

The paper also analyzes the differences between sexes. The above sample design features a comprehensive analysis, and found that men are in favor of straight lines and tough geometric shapes as

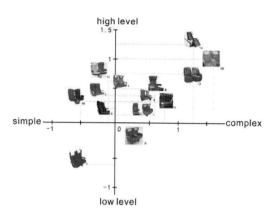

Figure 8. Images of the spatial distribution of the semantic analysis (3).

well as in favor of cool colors; but women prefer the curves, rounded shape and warm colors.

This application of the theoretical framework of Kansei Engineering system permits to take a combination of qualitative and quantitative methods for the seat design, in order to draw some correspondence between the mental and emotional imagery of the consumer product morphology.

6 CONCLUSIONS

This article based on the theory of Kansei Engineering for high-speed rail seats has the following attributes: it uses the semantic differential method for comprehensive evaluation of high-speed train seat; it analyzes the emotional characteristics of passengers; it proposes seat features based on the appearance of structural features and design; it analyzes head restraints, seat body and the armrests, showing in total 54 samples; and it uses three components of the semantic differential method, with the use of vocabulary groups "rigid—soft" and "complex—simple," "dangerous—security" as the space axis x, and with "cool—warm", "low-level—high-level" and "lightweight—stable" as a spatial axis y, establishing a two-dimensional image of the spatial distribution, as well as the analysis and evaluation of the passenger emotional attitude.

ACKNOWLEDGMENTS

We are grateful to Mrs Amalia Fernandez Hechevarria (psychologist), the Engineer Tsachi and Mr Parker Brown for their helpful suggestions. This work was supported by the National Natural Science Foudation of China (Grant No.

71271053), the Aeronautical Science Foundation (No. 20135169016), and the Scientific Innovation Research of College Graduates in Jiangsu Province (No. CXLX13_082).

REFERENCES

Liu G.Y & Shen J. 2001. The basis product forms of the design. Beijing: China Light Industry Press.

Huang G. S. 2001. Color and Design. Beijing: China Textile Press.

M.M. Verver, R. de Lange, & J. van Hoof. 2005. "Aspects of seat modeling for seating comfort analysis", Applied Ergonomics.

Peng D. L & Zhang will be hidden. 2004. Cognitive psychology [M]. Hangzhou: Zhejiang Education Press.

Xu J, Sun Sh. & Q, Park Development. 2007. Optimal design of products based on genetic algorithm modeling imagery [J]. Mechanical Engineering.

Tong W. 2012. EMU seat design aesthetics and comfort [D]: [Master's thesis]. Nanjing: Southeast University.

Gyouhyung K, Maury A. N & Kari B-R. 2008. "Driver sitting comfort and discomfort (part I): Use of subjective ratings in discriminating car seats and correspondence among ratings", International Journal of Industrial Ergonomics.

A. Siefert & S. Pankoke. 2008. "Virtual optimization of car passenger seats: Simulation of static and dynamic effects on drivers' seating comfort", International Journal of Industrial Ergonomics.

Architectural, Energy and Information Engineering – Sung & Chen (Eds)
© *2016 Taylor & Francis Group, London, ISBN 978-1-138-02791-6*

An improved WSVM prediction algorithm in Cognitive Radio

Hao Chen, Jin Chen & Zhan Gao
College of Communication Engineering, PLA University of Science and Technology, Nanjing, China

ABSTRACT: Support Vector Machine (SVM) is one of the most useful methods of spectrum prediction in Cognitive Radio (CR) owning to its well-known validity and solidity. In this paper, in order to improve the accuracy of SVM in spectrum prediction of CR, a new spectrum prediction algorithm based on Weighted-SVM (WSVM) is proposed. The algorithm changes the weight of an important parameter in SVM according to both sampling time and receiving SNR of each sample. The simulation result shows that the proposed algorithm provides a more ideal result than traditional SVM with the same samples.

Keywords: cognitive radio; spectrum prediction; support vector machine; receiving SNR; sampling time

1 INTRODUCTION

Nowadays, wireless spectrum has been commonly recognized as a vital important resource. The study by the Federal Communications Commission (FCC) has indicated that over all of the frequency bands, some are very busy. On the contrary, others are believed to indicate low utilization, especially in the 3–6 MHz bands [1]. This view is supported by studies of the FCC's Spectrum Policy Task Force, which has reported vast temporal and geographic variations in the usage of allocated spectrum, with utilization ranging from 15% to 85% [2]. In order to improve spectrum utilization, CR has been through a rapid development due to its advantage of sensing spectrum holes independently and access to them to make both authorized users and cognitive users communicate properly.

CR was originally proposed by Dr Joseph Mitola [3]. The basic idea is as follows: under the precondition that does not interrupt authorized users who own the frequency band, cognitive users dynamically perceive the spectrum hole to communicate. Spectrum prediction is a key step in CR, because when authorized users begin to use the spectrum, cognitive users must give it up and make the prediction to find another idle spectrum to communicate swiftly [4].

There are already some methods in spectrum prediction at the present time. The Hidden Markov Model (HMM) was used in [5] for spectrum prediction; In [6], a multi-layer sensor Artificial Neural Network (ANN) was proposed; In [7], Least-Squares SVM (LS-SVM) was introduced to make spectrum prediction.

The remainder of this paper is organized as follows. Section 2 gives a brief introduction of tradi-tional SVM. The proposed algorithm is described in Section 3 in detail. In section 4, simulation results between the proposed algorithm and the traditional SVM are summarized. Finally, Section 5 concludes the paper and provides a glimpse of future work.

2 TRADITIONAL SVM

The main idea of SVM can be described as follows: first, mapping the samples from low-dimensional space to high-dimensional space by defining a certain nonlinear kernel function; then construct the optimal separating hyper-plane by finding the Support Vectors (SV) that nearest to and parallel to the optimal interface in high-dimensional space.

In Figure 1, two different shapes represent two groups of samples. H is the optimal separating hyper-plane. H_1 and H_2 pass through two types of samples that are nearest to H. H_1 and H_2 are paral-

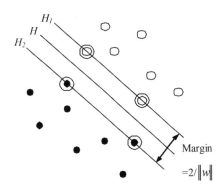

Figure 1. The optimal separating hyper-plane.

lel to H. The samples on H_1 and H_2 are Support Vectors, which contribute to the optimal separating hyper-plane [4].

SVM has two main applications: Support Vector Classification (SVC) and Support Vector Regression (SVR). SVR handles with data fitting and Image reconstruction recovery and time series prediction. In this case, this paper mainly discusses about SVR.

In SVR, first, we need to set up a continuous function:

$$y = r(x) + \delta \tag{1}$$

where $r(x)$ is the target function and δ is the zero-expect noise.

Then, mapping the samples to n-dimensional space in order to build concerned linear model, it can be described by the function as follows:

$$f(x, \omega) = \omega \cdot \Phi(x) + b \tag{2}$$

where x is the sample vector; $f(x, \omega)$ is the estimation function; ω is the normal vector of the hyperplane; $\Phi(x)$ is a mapping function; and b is the threshold.

ε (insensitive loss function) was proposed by Vapnik, and represents allowable training loss, which is used to realize linear regression in high-dimensional space. The formula can be written as follows:

$$L_\varepsilon(y, f(x, \omega)) = \begin{cases} 0, & |y - f(x, \omega)| \le \varepsilon \\ |y - f(x, \omega)| - \varepsilon, & \text{otherwise} \end{cases} \tag{3}$$

The optimal interface theory requires that H can not only separate two types of samples correctly, but also maximize the separate margin. The process of right regression corresponds to the minimization of empirical risk and the process of maximizing the separate margin corresponds to the minimization of confidence risk. The optimized function is given by

$$\min \frac{1}{2} \| \omega \|^2 + C \sum_{i=1}^{l} (\xi_i + \xi_i^*) \tag{4}$$

Subject to:

$$\begin{cases} y_i - f(x_i) \le \varepsilon + \xi_i^* \\ f(x_i) - y_i \le \varepsilon + \xi_i \\ \xi_i, \xi_i^* \ge 0 \quad i = 1, \dots, n \end{cases} \tag{5}$$

$f(x_i) = \omega \cdot \Phi(x) + b$ is the estimation function, and C is the penalty factor balancing the tradeoff between the maximization of the margin and the minimization of the empirical risk; and $\xi_i + \xi_i^*$ are the slack variables that represent the training error.

3 NEW ALGORITHM BASED ON WSVM

In SVR, the penalty factor C of each sample is the same, which means the same requirement of precision and penalty to deviation of every sample. However, in practical, temporal prediction for instance, the recent data is more important than the early data; or those sample points of greater density will be more helpful to the accuracy of the estimation result. That is to say, the training error of the important sample points should be more accurate, and for those sample points of less importance, requirement of accuracy of training error is not that high. So, if we put a different penalty factor C on each sample point when using SVM, then the estimation result could be more precise. This kind of SVM is called the Weighted-SVM (WSVM) [8].

The existing WSVMs often consider the sampling time or the sample density, but a few take the receiving SNR in consideration. Because when one sample point has a higher receiving SNR, it is more reliable and its weight should be increased; on the other hand, when this sample point has a lower receiving SNR, we think it is less reliable and its weight should be decreased. At the same time, the recent samples are more important than earlier samples, which means they demand a lower training error and their weights should be increased; the earlier samples are less important and their weights should be decreased.

Therefore, this paper proposes an improved WSVM, taking both sampling time and receiving SNR into account. The basic idea is a different penalty factor C according to the related weight based on both sampling time and receiving SNR.

As shown in Figure 2, t_i is the sampling time of each sample point and T is the length of the sampling time. R_i is the receiving SNR of each sample point. Because the receiving SNR R is different from each other, we need to normalize it in order to normalize W_i:

$$R_{norm} = \frac{R_i - R_{min}}{R_{max} - R_{min}} \tag{6}$$

Figure 2. Sampling time and receiving SNR.

After all the above steps, the weight of the penalty factor C can be described as follows:

$$W_i = \frac{1}{2}\left(\frac{t_i}{T} + R_{norm}\right) \qquad (7)$$

Then, each penalty factor of each sample point can be written as follows:

$$C_i = W_i * C \qquad (8)$$

Based on all the above considerations, we propose the following improved algorithm:

Algorithm : the proposed improved WSVM

1. Divide the sample set into training set and test set, extract the sampling time and receiving SNR of each sample point in training set.
2. Use grid search, try every possible pair of (C, ξ), then use cross validation, pick the best pair of (C, ξ).
3. Normalize the receiving SNR and compute the weight of penalty factor C_i of each sample point using function (6) and (7).
4. Train the SVM using the weighted penalty factor C_i in function (8) with training set and obtain a prediction model.
5. Predict the test set using the SVM prediction model concluded in step 4.
6. Analyze the results.

4 SIMULATION RESULTS AND ANALYSIS

Simulation has been made between the proposed algorithm and the traditional SVM.

Figure 3 shows the performance between the proposed improved algorithm and the traditional SVM when the number of samples ranges from 40 to 95 using the same sample set. It could be eas-

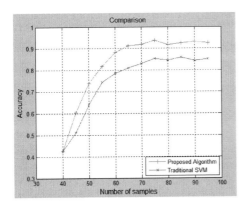

Figure 3.   Comparisin between the proposed improved algorithm and the traditional SVM.

ily found that the prediction accuracy of the proposed improved algorithm is higher than that of the traditional SVM. Furthermore, because the proposed improved algorithm adds a few addition and multiplication, the calculation is a little more than the traditional SVM. So, the proposed improved algorithm will improve the precision of prediction at the cost of a little calculation.

5 CONCLUSIONS

In this paper, spectrum prediction and traditional SVM are briefly introduced, and an improved algorithm based on the sampling time and the receiving SNR is proposed. The simulation result shows that the proposed improved algorithm performs better than the traditional SVM in relation to prediction accuracy, but the calculation is a little more than the traditional SVM. The future work includes: 1) the weight of the penalty factor C of the earliest sample points is too small because their t_i are almost 0, optimization needs to be done; 2) the proportion between the sampling time and the receiving SNR is still not optimal, and we will find the best proportion in future study.

REFERENCES

[1] Cabric, D., Mishra, S.M., Brodersen, R.W., 2004. "Implementation issues in spectrum sensing for cognitive radios," Signals, Systems and Computers, 2004. Conference Record of the Thirty-Eighth Asilomar Conference on, vol. 1, pp. 772, 776 Vol. 1, 7–10 Nov.
[2] FCC, 2002. "Spectrum Policy Task Force Report", ET Docket No. 02-155, Nov 02.
[3] Mitola, J. & Maguire, G.Q., Jr., 1999. "Cognitive radio: making software radios more personal," Personal Communications, IEEE, vol. 6, no. 4, pp. 13, 18, Aug.
[4] Yuqing Huang, Hong Jiang, Hong Hu & Yuan Cheng Yao, 2009. "Design of Learning Engine Based on Support Vector Machine in Cognitive Radio," Computational Intelligence and Software Engineering, 2009. CiSE 2009. International Conference on, pp. 1, 4, 11–13 Dec.
[5] Akbar, I.A. & Tranter 2007, W.H., "Dynamic spectrum allocation in cognitive radio using hidden Markov models: Poisson distributed case," SoutheastCon, 2007. Proceedings. IEEE, pp. 196, 201, 22–25 March.
[6] V. K. Tumuluru, P. Wang & D. Niyato, 2010. A neural network based spectrum prediction scheme for cognitive radio [C] 2010 IEEE International Conference on Communications. Cape Town: IEEE press: 1-5.
[7] Wu, C., Yu, Q. & Yi, K. 2012, "Least-squares support vector machine-based learning and decision making in cognitive radios," Communications, IET, vol. 6, no. 17, pp. 2855, 2863, November 27.
[8] Shu-xin Du & Sheng-Tan Chen 2005, "Weighted support vector machine for classification," Systems, Man and Cybernetics, 2005 IEEE International Conference on, vol. 4, pp. 3866, 3871 Vol. 4, 10–12 Oct.

Architectural, Energy and Information Engineering – Sung & Chen (Eds)
© 2016 Taylor & Francis Group, London, ISBN 978-1-138-02791-6

Research on strengthening methods of copper-based material

Hai Kuang
Jiangxi Science and Technology Normal University, Nanchang, Jiangxi, China

Shi An He
Jiangxi Changyun Company, Nanchang, Jiangxi, China

Ying Luo
Jiangxi Science and Technology Normal University, Nanchang, Jiangxi, China

ABSTRACT: High strength and good electrical conductivity possessed by a copper-based alloy at the same time is difficult. So, it is limited in the application. The strengthening methods, alloying method and compositing method are introduced in this paper. The comparative analysis shows that the requirement of both high strength and high conductivity is difficult to achieve by the copper-based alloy due to the limitations, while higher strength and better design property exist in the composite copper-based material. Meanwhile, the performance of material at high temperature is improved by the compositing method. Therefore, the compositing method is the best way for achieving high strength and good conductivity of copper-based material.

Keywords: copper-based materials; strengthening; alloying method; compositing method

1 INTRODUCTION

Much valuable physical and chemical property exists in copper, such as high thermal conductivity, high electrical conductivity, good corrosion resistance and fine ductility. However, its mechanical property is not very good. For example, its strength value is only 230 MPa~290 MPa, and will soon be lost in the tempering process[1,2]. So, it is limited in application. On the other hand, the increasingly high demand for material requires the rapid development of science and technology. Its strength and electrical conductivity properties at high temperature of copper and copper alloys is difficult to balance, which cannot fully meet the requirement of high-tech development. For example, in accordance with the development of microelectronics technology, LSI lead frame material required indicators: tensile strength $\geqq 600$ MPa and conductivity $\geqq 80\%$ IACS[3~6]. So, it has become the focus of current research topics to improve the strength of copper.

Currently, There are two main ways for strengthening copper[7,8]. First, the introduction of alloying elements to the copper matrix to form an alloy, called the alloying method, which is the traditional method. Second, the introduction of the second phase to the formation of composites, i.e. composites method, is a new strengthening method.

2 ALLOYING METHOD[9–12]

Alloying method includes solid solution strengthening, precipitation strengthening, grain refinement, cold deformation and failure to strengthen.

2.1 Solid solution strengthening method

An appropriate amount of alloying elements is added to the copper to form a solid solution. The strength of the alloy will be improved in general, which is called the solid-solution strengthening method, so that the main mechanism is lattice distortion caused by introducing atoms into the copper matrix, which impede dislocation, and then the material to be strengthened.

However, after entering the copper crystal alloy elements, the different size of an atom will cause lattice distortion, and this will have a strong scattering effect on the movement of electrons, which leads to the decrease in electrical conductivity. Therefore, although the solid solution strengthening of Cu can improve the strength, it greatly reduces the conductivity of the copper matrix.

2.2 Precipitation strengthening method

Precipitation strengthening is aging heat treatment on precipitation. Alloy elements should have the following two conditions: first, the solid solubility of copper at a high and low temperature difference is large, so that aging can produce enough strengthening phase; second, little solubility in copper at room temperature, to ensure high conductivity of the matrix.

2.3 Grain refinement method

Dislocation motion is effectively impeded by the Grain boundary, which is the characteristic of the grain refinement method. Atomic arrangement at grain boundaries is non-uniform, and there are many impurity defects to impede dislocation from one grain to another grain, thereby improving the strength of the alloy. However, grain refinement for improving the strength of the material is limited, and it is only suitable for copper moderate strength requirements.

2.4 Cold deformation with aging strengthening

Based on the metal deformation theory, the alloy can further improve its strength by cold working. However, this application is limited. So, this method cannot be generally used alone, but usually with other intensive methods, for example, the use of solid-solution strengthening with cold deformation and aging strengthening. The reason is that the solubility of some alloying elements in the copper is sharply reduced with decreasing temperature, resulting in precipitation, which will hinder dislocation motion and thus plays a strengthening effect.

3 COMPOSITING METHOD

Composite material refers to a combination of two or more materials made up of a multiphase material. Typically, there is a continuous phase, which known as the matrix, and the other is the dispersed phase, called the enhanced phase. Depending on the form of reinforcement, copper-based composite materials can be divided into particles-reinforced material and fiber-reinforced material. Particles are dispersed copper matrix composites reinforced composite material.

Here we take Al_2O_3/Cu composite materials as an example to introduce compositing method.

3.1 Powder metallurgy

The preparation of Al_2O_3/Cu composite material by Powder Metallurgy (PM) is a mature technology with short production cycles and low costs. The main process is as follows: first, a certain proportion of Cu powder and Al_2O_3 particles

are mixed, then pressed, and finally sinter at high temperature.

This method is restricting to improve the performance of Al_2O_3/Cu composite materials. It is difficult to achieve good-refined and well-distribution of Al_2O_3 particles[13].

3.2 Composite electro-deposition method

Composite electro-deposition method is a new way of producing metal-based composites[14]. The ceramic, mineral or resin particles and the matrix metal or alloy co-deposited on the cathode surface to form composite coating, thereby significantly improving the performance of the material. Copper sulfate and copper fluoroborate are usually used for Al_2O_3/Cu composite electro-deposition.

Composite electro-deposition method does not require high-temperature conditions, and the preparation process is simple, also its cost is low, but the particles evenly distributed in the bath is difficult to control.

3.3 Mechanical alloying

Mechanical Alloying (MA) is a high-tech non-equilibrium alloy powder preparation which successfully developed by American scientists Benjami in 1970. The Cu powder with fine Al_2O_3 particles is mixed and milled in a high-energy ball for a long time, so that to achieve the atomic level of the powder, while the hard phase of Al_2O_3 particles are uniformly embedded in the Cu particles, and eventually form a solid solution alloy. The composite powder is then molded, sintered, and machined.

The equipment of Mechanical Alloying method is simple, but a longer production cycle, and the larger grain size, so performance of composite materials is greatly affected.

3.4 Stirring and casting method

Stirring and casting method is a simple and common way. Its process is adding solid Al_2O_3 reinforcing particles into Cu solution under mechanical stirring gradually. Ultrasonic agitation may be employed in the melting process in order to strengthen the stirring effect. And the reinforcing particles and copper were mixed uniformly, then be directly cast molding.

The process and equipment of this method is relatively simple, but the biggest drawback is difficult to control the content of the reinforcing particles Al_2O_3, and the second is hard to deal with the distribution due to poor wettability and a large proportion of the difference between Al_2O_3 particles and Cu matrix. Furthermore, the particles and the matrix are also prone to interfacial reactions[15], resulting in poor properties of Al$_2$O$_3$/Cu composite material.

3.5 Internal oxidation method

This method is the use of principle of the oxidation-reduction reaction: adding unstable compound powder into the alloy powder to produce the alloy of the component, and more stable ceramic particles will be formed, and then it becomes composite materials after mixing and sintering[16]. This method usually used to produce Al$_2$O$_3$/Cu matrix composites. Oxidation is a comprehensive concept within thermodynamics and kinetics. Generally speaking, the conditions for the Cu-Al alloy oxidation are: preferential oxidation and $C_O D_O \gg C_B D_B$, where the C_O, D_O, C_B and C_B represents concentration and diffusion coefficient of O and Al in Cu substrate, respectively. Preferential oxidation is a thermodynamic condition and the latter is a dynamic condition. From the thermodynamic point of view, the first internal oxidation of Al is preferentially oxidized.

Compared with other conventional methods such as mixing casting method or powder metallurgy method, the Al$_2$O$_3$/Cu composites made by the oxidation method have finer particles, disperse more evenly and the densification is better. Furthermore, the connection of the particles and the substrate is good due to the interface is not contaminated. So the material properties are superior organization. However, the oxidation also as a conventional powder metallurgy method, it is difficult to obtain fully dense materials. In order to further enhance the composite density and uniformity of the structure, plastic deformation must be carried out in an appropriate manner, such as extrusion, rolling system, and so on. Also oxidation process is more complicated and the cost is high. You can not produce goods in large size and complex shape.

3.6 Reactive spray deposition method

Reactive spray deposition method is a novel rapid solidification process, which combines the advantages of stir casting method and powder metallurgy method, and it overcomes some serious shortcomings such as the interfacial reactions. The reaction injection method composites the process of metal liquefied, ceramic particle reinforced synthetic reaction and rapid solidification process together, i.e., it injects melted Cu, Al and oxygenates liquid metal into the reaction vessel, then gets the enhanced particles.

This method can make the base metal Cu fine grain size, reinforcing phase Al$_2$O$_3$ particles uniformly distributed, but also to ensure a solid combination of enhanced particle Al$_2$O$_3$ and the matrix Cu, and thus prepared Al$_2$O$_3$/Cu composite

high performance. But the equipment is expensive and not suitable for mass production.

4 CONCLUSION

Both high strength and good conductivity in copper alloys are difficult to get. Due to their own limitations, the copper-based alloy is difficult to achieve high strength and high conductivity requirements. The composite material simultaneously exerts a synergistic effect both of reinforcing elements and the matrix material, and also it has a considerable design freedom. Meanwhile, due to the electron scattering caused lattice is stronger than that caused by the second phase, so compositing process will not significantly reduce the conductivity of the copper substrate. In addition, the property of material at room temperature and high temperature is improved. Therefore, the development of compositing method for high strength and high electrical conductivity copper-base alloy is the main direction.

ACKNOWLEDGMENT

This project was supported by the Technology Research Program of Jiangxi Science and Technology Normal University (2013XJYB007). The corresponding author is Kuanghai.

REFERENCES

[1] Liu DeBao, Cui Chunxiang. Review and prospect for fabricating techniques of high strength and electric conductivity copper-matrix composites. Journal of Tianjin Institute of Technology, 19 (4), 2003: 29–33.

[2] Min Guanghui, Song Li, Yu Huashun. High Strength and Electric Conductivity Copper-based Composites Functional Materials, 28 (4), 1997: 342–345.

[3] Chen Yisheng, Han Baojun. Progress of high strength and high conductivity copper alloy. Journal of Southern institute of metallurgy, 25 (2), 2004: 17–21.

[4] Clyne T. W, Withers P J. An Introduction to Metal Matrix Composites Cambridge University Press, Cambridge, 1993: 373–392.

[5] Ding Yutian, Li LaiJun, Xu Guangji, et al. The status-quo of contact wire and popular topics of its research. Electric wire and cable, 2004 (2): 3–9.

[6] Nie Cunzhu, Zhao Naiqin. Review of Meta-l Matrix Composite Materials for Electronic Packaging. Metal heat treatment, 28 (6), 2003: 1–5.

[7] Korb G, Korab J, Groboth G. Thermal expansion behaviour of unidirectional carbon-fiber-reinforced copper-matrix composites [J]. Composites, Part A,1998, 29: 1563.

[8] Lee breeze, Zhao, Zhang Jianqing situ metal matrix composites Materials Science and Engineering, 20 (3), 2002: 453–457.

[9] Zhao Dongmei, Dong Qiming, Liu Ping, Jin Zhihao, Huang Jinliang try to explore the best high strength and high conductivity copper alloy composition of functional materials, 32 (6), 2001: 609–611.

[10] Korab J. Thermal conductivity of unidirectional copper matrix carbon fiber composites [J]. Composites Part A, 2002, 33: 577–581.

[11] Ibrahim LA et al. Particulate Reinforced Metal Matrix Composites-a Review. J Mater Sci 26, 1991: 1137–1156.

[12] Joanna Groza Heat-Resistant Dispersion-Strengthened Copper Alloys. Journal of Materials Engineering and Performance, 1 (1), 1992: 113–121.

[13] Zhang Yun, Wu Jianjun, Guobin, et al. Internal oxidation of copper aluminum Materials. Material Science and Technology [J] 1999, 7 (2): 91–95.

[14] CBiselli, Morris D G. Microstructure and Strength of Cu-Fe in situ Composite obtained from Pre-alloyed Cu-Fe Powders [J]. Aeta Metall,1994, 42 (1): 163–176.

[15] S E Broyles, K R Anderson, J R Groza. Creep Deformation of Dispersion Strengthening Copper. Metal Trans A, 1996, 27A (5): 1217–1227.

[16] Shen Yutian, Cui Chunxiang, Meng Fanbin, et al. Thermodynamic analysis of internal Oxidation of Cu-Al alloy and oxidation behavior of Cu. Powder metallurgy technology [J]. 2001, 19 (3): 28–32.

Architectural, Energy and Information Engineering – Sung & Chen (Eds)
© 2016 Taylor & Francis Group, London, ISBN 978-1-138-02791-6

Research on project cost control theory of construction enterprise based on fuzzy earned value method

Li Ge
Construction Management Office, Huazhong University of Science and Technology, Wuhan, China

Yang Pan
School of Management, Huazhong University of Science and Technology, Wuhan, China

ABSTRACT: Currently, construction projects are always uncertainties, it is unable to achieve accurate and effective cost schedule integration control. For the above problem, we plan to use three key intermediate variables (the net investment, the actual investment, and the completed investment), two variable index difference analysis (cost variance and schedule variance) and two indicator variables (cost deviation rate and schedule variance rate) to analyze the cost of the construction projects and schedule performance. We also use the level of risk factor α cut set the quantitative performance evaluation and the prediction index, where α is determined by the Gray theory model, it allows us to obtain the accurate project cost schedule control information in an uncertain environment. Finally, we empirically verified the feasibility of this method analysis.

Keywords: cost schedule control; fuzzy earned value method; gray theory model; progress monitoring

1 INTRODUCTION

In recent years, as the scale and complexity of the construction projects is growing, the following two issues have become a hot topic for many scholars: how to control the cost and the schedule in the construction business management and project supervision effectively; how to reduce the risk of large scale projects implemented scheduling scheme. For the actual construction project, there is a complex relationship between its cost and schedule. With the progress of the project, the progress of the lag will result in increased costs. However, the budget problems also need to make appropriate adjustments to progress. An increase in investment of resources and personnel processes rationing may make shorter duration, but it can also lead to a corresponding increase in costs, and simultaneous occurrence of force majeure will have some impact on the project cost and schedule. Babu [1] has introduced the project quality objectives to the optimization model, DiXundi [2] has expressed the duration and cost as a non-linear function, and algorithms to solve the model with NSGAII. Qingfu [3] has established a multi-objective optimization algorithm based on decomposition to optimize the multi-objective model. These studies did not consider the feasibility of including scheduling scheme, resulting in large-scale projects is difficult to complete in the implementation process as planned due to various risk factors. This problem will also lead the contractor to cut corners in order to reduce costs, and causes an adverse impact on the cost and quality of the project.

This study is an integrated management research for the progress of the construction cost of the project under an uncertain environment, which is based on the Fuzzy earned value method. Morteza has used fuzzy number as an Earned Value, the fuzzy theory is combined with the earned value method, thereby the completion time and the cost of the project can be predicted accurately. [4] Leila has created Fuzzy earned value method and improved Earned Value Index System. [5] In this paper, we improve the fuzzy earned value in construction projects under the conditions of the duration of the process is uncertain or unknown. We introduce a trapezoidal fuzzy membership function to describe the project cost schedule performance situation, using α cut set of fuzzy numbers sorting method to quantify the fuzzy data. And we determine the level of risk factor α with gray method, proposing an improved Fuzzy earned value method.

2 THE BASE MODEL

Project Management Institute definition of earned value method in 2000 "Project Management Body of Knowledge" is: earned value management, a technology used to measure and report project performance from initiation to closeout. That is, the earned value method is based on earned value completion budget, using three key intermediate variable values to measure the schedule and the cost of the project. Earned value method is an overall technical method to measure and reflect the progress of project management. We should master the "three, two, two" principle to use earned value method: three key intermediate variables, two variable index difference analysis, and two variable index.

3 THREE KEY INTERMEDIATE VARIABLE VALUES

Three key intermediate variable values include BCWS, ACWP, and BCWP. Among them, Budgeted Cost of Work Scheduled (BCWS planned investment) represents a point in time required to complete the work should be invested or spent money accumulated value of the cost. It is equal to the product of the plan quantities and budget price, is a benchmark or baseline measure project progress and cost of the project. The formula is expressed as follows:

$$BCWS_t = \sum_{i=1}^{n}[Rb_i(t) \times Qs_i(t)]$$

Actual cost of work performed (ACWP actual investment) represents the total amount of a point in time has been spent to complete the actual cost of the work. It is equal to the product of the amount of the completed project and the actual price. The formula is expressed as follows:

$$ACWP_t = \sum_{i=1}^{n}[Rc_i(t) \times Qp_i(t)]$$

In the above formula, n represents the number of budget items, Rc_i represents the i-item's actual price, QP_i represents the i-item's the amount of the completed projects, t represents a point in time. ACWP reflects the actual cost of the project.

Budgeted cost of work performed (BCWP completed investment) is an earned value: a point in time required to complete the work already accumulated value of funds invested. It is equal to the product of the budget price and the quantity of the completed project, and it reflects the actual project

progress and job performance that meets the quality standards. It reflects the conversion of amount of investment to the outcome of the project. The formula is expressed as follows:

$$BCWP_t = \sum_{i=1}^{n}[Rb_i(t) \times Qp_i(t)]$$

In the above formula, n represents the number of budget items, Rb_i represents the i-item's budget price, QP_i represents the i-item's the amount of the completed projects, t represents a point in time. BCWP reflects the actual cost of the project and the actual value of the project. By calculating the amount of the actual project completion and budget price, BCWP established the statement of accounts for the investment budget of a certain time period.

4 TWO VARIABLE INDEX DIFFERENCE DATA

Two variable index difference data are cost variance and schedule variance. Cost variance is represented by CV: CV = BCWP − ACWP. In the formula, CV > 0 means project costs within budget, CV = 0 means project costs in line with budget, and CV < 0 means project costs exceed budget.

Schedule variance is represented by SV: SV = BCWP − BCWS. In the formula, SV > 0 means progress ahead of schedule, SV = 0 means progress in line with plan, and SV < 0 means progress behind schedule.

5 TWO VARIABLE INDEXES

Two variable indexes are cost deviation rate and schedule variance rate. The cost deviation rate is represented by CPI, and CPI = BCWP/ACWP. In the formula, CPI > 1.0 means project costs within budget, CPI = 1.0 means project costs in line with budget, and CPI < 1.0 means project costs exceed budget.

The schedule variance rate is represented by SPI, and SPI = BCWP/BCWS. In the formula, SPI > 1.0 means progress ahead of schedule, SPI = 1.0 means progress in line with plan, and SPI < 1.0 means progress behind schedule.

6 FUZZY EARNED VALUE MODEL

Technical engineer evaluates the progress of the project by project and experiences. Construction project's fuzzy generally represented by trapezoidal fuzzy numbers and triangular fuzzy numbers.

Table 1. Construction project risk index.

First index	First index risk weight	Second index	Second index risk weight
Natural risks U1	0.103	Earthquakes, typhoons and other natural disasters U11	0.268
		Adverse geological hydrological conditions U12	0.256
		Adverse weather conditions U13	0.476
Technical risk U2	0.219	Survey design flaws U21	0.368
		Technical solutions unreasonable U22	0.311
		Construction design unreasonable U23	0.321
Economic risk U3	0.128	Rising interest rates U31	0.459
		Inflation U32	0.211
		Rising Price U33	0.33
Manage risk U4	0.414	Control and coordination difficult U41	0.149
		Management system is imperfect U42	0.231
		Contract Dispute U43	0.441
		Unskilled construction workers U44	0.179
Social Risk U5	0.136	The introduction of policies and regulations U51	0.456
		Processions and strikes U52	0.544

Because of the triangular fuzzy number can be expressed as [a, b, b, d] or [a, c, c, d], this study describes the project progress of trapezoidal fuzzy numbers. We use α cut set fuzzy numbers classic sort to blur the schedule and cost performance evaluation and the prediction index. For trapezoidal fuzzy number $\hat{A} = [a,b,c,d]$, and

$$\hat{A}_\alpha = [a + \alpha(b-a), d - \alpha(d-c)](0 \le a \le 1) .$$

7 CASE STUDY

The building entrusted to carry out the design and construction of a construction worker, the project involves a variety of professional, payment for projects to be paid as a percentage of the construction price construction contract. The project has a long duration and tight construction schedule, and it has high cost and schedule control requirements. So, it is more suitable for us to use the earned value method to control project's cost and schedule. Construction project risk index is shown in Table 1.

The construction projects are scored by expert, scoring results are shown in Table 2.

After calculate, we get the gray evaluation weight matrix:

$$R = \begin{pmatrix} 0.294 & 0.211 & 0.275 & 0.183 & 0.115 \\ 0.651 & 0.317 & 0.109 & 0.291 & 0.123 \\ 0.284 & 0.221 & 0.175 & 0.302 & 0.382 \\ 0.401 & 0.423 & 0.187 & 0.331 & 0.284 \\ 0.218 & 0.591 & 0.281 & 0.109 & 0.118 \end{pmatrix}$$

Table 2. Scoring results.

Expert 1 score D1	Expert 2 score D2	Expert 3 score D3	Expert 4 score D4	Expert 5 score D5
1	0.5	1	0.5	1
1.5	1	1	2	1
4.5	3	4	4.5	3.5
2	2.5	2	1	2
2.5	2	3	1.5	2
2.5	1	2	1.5	2
1	1	1.5	0.5	0.5
1	1.5	1	1	1
1.5	2	2	1.5	1
4	3.5	3	3	2.5
2.5	2	1	2	2.5
3.5	3	2	1.5	2
3	2	2.5	3	2
1	1	0.5	1	0.5
1	1	0.5	0.5	1

By transforming the R matrix, we get the comprehensive evaluation vector:

$$B = W \times R = (0.234, 0.112, 0.345, 0.221, 0.083)$$

Risk value $Z = B \times C$ is 1.2342, it belongs to the general risks and corresponding to the level of risk coefficient of 0.3.

REFERENCES

[1] Babu A J G, Suresh N. Project Management with Time, Cost, and Quality Considerations [J]. European Journal of Operational Research, 1996, (02): 320–327.

[2] Diao Xundi, Li Heng, Zeng Saixing. A Pareto Multi-objective Optimization Approach for Solving Time-costquality Tradeoff Problems [J]. Technological and Economic Development of Economy, 2011, (01): 22–41.

[3] Zhang Qingfu, Li Hui. MOEA/D: A Multi-objective Evolutionary Heydari. Algorithm Based on Decomposition [J]. IEEE Transactions on Evolutionary Computation, 2007, (06): 712–731.

[4] Morteza Bagherpour, Abalfazl Zareei, Siamak Noori, Mehdi. Designing a control mechanism using earned value analysis: an application to production environment. The International Journal of Advanced Manufacturing Technology, 2010, 49(5–8): 419–429.

[5] Moslemi Naeni, Leila. Evaluating fuzzy earned value indices and estimates by applying alpha cuts. Expert Systems with Applications, 2011, 38(7): 8193–8198.

Architectural, Energy and Information Engineering – Sung & Chen (Eds)
© 2016 Taylor & Francis Group, London, ISBN 978-1-138-02791-6

Study on analysis model of industrial international competitiveness

Li Ping Wang & Ming Hao Liu
School of Economics and Management, Henan Polytechnic University, Jiaozuo, China

ABSTRACT: Industrial international competitiveness has become a hot topic in recent years. Relevant studies have more and more appeared on the books, journals, and the internet. Different scholars established different analysis model for the different countries in order to explain the situation more exactly. So each analysis model has its own advantage and limitation. How to apply the special model for the special country becomes an important question. This paper systematically reviewed the major analysis models of industrial international competitiveness, such as Porter's, Cho's, Jinbei's, and so on, which lay the foundation for the further study.

Keywords: analysis model; diamond model; industrial international competitiveness

1 INTRODUCTION

The research of industrial international competitiveness is an important branch of the field of international competitiveness research. The research on the industrial international competitiveness not only helps us to understand and grasp the strength and potential of industries in international competition, and then guide industry to participate in international economic and technology exchanges and the international division of labor, but also to deeply understand the international gap and competition status of industry development, and then promptly adjust the orientation and emphasis of industrial development and provide a scientific decision-making basis for the government's macroeconomic policies, industrial policy, trade policy and foreign policy.

However, the research on the industrial international competitiveness started relatively late, so both theoretical research and the practice of application need to be supplied and improved. The academic controversy on the concept of the industrial international competitiveness is still in continuous. Some representative definitions are as following:

1. Porter's definition of the industrial international competitiveness has the greatest influence in the world. He pointed out that the industrial international competitiveness is the ability that specific industries of a country, under the international free trade conditions (on the conditions of non-tariff trade barriers), provide consumers (including the productive consumption) more products with higher productivity in the international market than other countries', and the ability that continues to make more profit.

2. The definition of industrial international competitiveness presented by Jin Bei is also very popular in the country. He considered that the industrial international competitiveness is the productivity that specific industry of one country reflected by selling their products in the international market. From this definition, we can see that the essence of industrial international competitiveness is the comparative productivity of a country's specific industry relative to foreign competitors, and it reflects the scale of the international market share and profitability of the industry's products. Although the expression of this definition is different from Porter's, there are still some similar characteristics. For example, both of them stressed that the level of productivity of the industry has an important influence on the industrial international competitiveness.

In addition, there are many scholars brought forward their definition of the industrial international competitiveness. They generally think that the industrial international competitiveness is a relative productivity, but their different perspectives on the industry caused the different interpretations and expression on the industrial international competitiveness.

2 THE ANALYSIS MODEL OF COMPETITIVENESS STRENGTH BY PORTER

The representative industrial international competitiveness analysis model must be the competitive advantage six essential analysis model by Pro.

Porter and this model also can be called Diamond model or National Rhombus chart [1]. Porter believed that there are six issues that can determine the competitive advantage of industrial development: factors of production, demand conditions, the state of the related and auxiliary industries, corporate strategy, structure and competitors, government and opportunities, the relationship of these six interrelated factors that influence each other can be showed by Fig. 1.

Porter made the competitive dynamic and evolvement as the precondition of theory analysis, and divided the process of industries participating in the international competitiveness into four phases:

Firstly, the factor-driven stage. The industrial international competitive strength which belongs to this stage profited from some basal factors of production, such as the natural resources or abundant and low-cost workforce.

Secondly, the investment-driven stage. Compared with the factor-driven stage, this upgrade of industrial international competitive strength in this stage was driven by investment not by demand, invest to these areas that could promote the production factors toward more advanced direction, and enhanced the modern basic establishment's construction at the same time. Also the industrial international competitiveness intensified. Besides, whether the government can carry out the proper policy was also important.

Thirdly, the innovation-driven stage. In this stage, the contribution rate to the industrial international competitiveness by the natural gift of production factors decrease more and more. On the contrary, the disadvantages of production factors

stimulate the enterprises to introduce and apply advanced international technology, and constantly make improvement and innovation to these technologies. What must be emphasized is that the enterprise's ability of digestion and absorption of advanced foreign technology is the key for a country to achieve industry innovation-driven stage and also the fundamental difference between innovation-driven and investment-driven.

Fourthly, the wealth-driven stage. As the name suggests, at this stage the driving force is the acquired economic wealth, rather than investing more natural factors of production or human capital, also not through a positive innovation to enhance the international competitiveness of the industry. A more obvious feature that the transition to the wealth-driven stage is the wide range of enterprises mergers and takeovers inside and outside of industries, reflecting that the enterprises hoping to strengthen the stability by reducing competition, but this is not able to fundamentally enhance their competitive advantage and upgrade the international competitiveness of industry. On the contrary, the enterprises will lose further pioneering and innovative spirit by weakening of the competitive degree. Therefore, the stage is the recession of international competitiveness of industry.

3 THE NINE FACTORS MODEL BY CHO

Porter's "Diamond model" can be properly used for explaining the international competitiveness of the industry in developed country, but it can not be proper in developing or less developed country. It's mainly because the factors that influence the international competitiveness of the industry are very different in countries which have different economic level. For this reason, many scholars made revision and expansion on the "Diamond model", of which, Cho composed a "nine factors model" that is more appropriate to developing countries (Fig. 2).

The nine factors model emphasizes the people and material's influence on the international competitiveness of the industry, which form the inherent influence of competitiveness. The people's factor mainly refers to the politicians and government officials, entrepreneurs, professional managers and professional technical master, as well as workers. The material factor mainly refers to the gift resources, domestic demand, and related industries and business environment. The external factor mainly refers to the opportunity. He had the same view with Porter that the opportunities' impact that can not be ignored on the international competitiveness of industry. At the same time, he also emphasized that in the different stages of

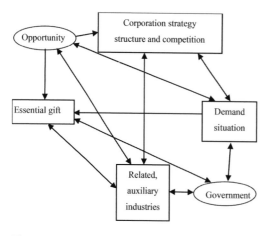

Figure 1. The analysis model of industrial international competitiveness.

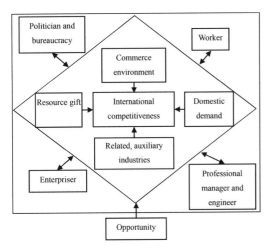

Figure 2.   The nine factors model by Cho.

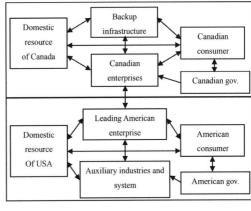

Figure 3.   The "double diamond model" by Krugman.

industries development, different factors' impact on the different industries will also be different, and these factors interact with each other, and form a dynamic model that can affect the international competitiveness of industries in developing countries [2].

In addition, Denning believes that it needs to make the "transnational business activities" as the third exogenous variables paratactic with government and opportunities. Krugman took the Canadian industry as an example to study, and he found that Canadian companies can take full advantage of the transnational business opportunities by the Canada-US's Free Trade Agreement, borrowed the market capacity of the United States to expand its economy scale and avoid the restrictions by the small national economies to some industries international competitive strength. Then it would change the "diamond model" into the "double diamond model" (Fig. 3). Meng etc, have made further change about the "double diamond model" and improved it into the "generalized double diamond model" in order to apply to all small national economies, and they also took the South Korea and Singapore as example and made empirical analysis. Steven etc believed that it should consider government as the fifth deciding factor in the diamond model. Posey etc appealed that we should carefully study the impact of the national culture on the competitive advantage sources. Naruila thought that it should take the "cumulative technology" as the exogenous variables in Diamond model. Some academics have also introduced the "system competitiveness" notion and believed that the competitiveness depends on the macro, meso and micro-level of coordinate system [3–10].

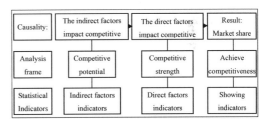

Figure 4.   The analysis frame of the international competitiveness of the industrial products.

4   THE ANALYSIS MODEL OF INDUSTRIAL INTERNATIONAL COMPETITIVENESS BY JINBEI

Jinbei used the national competitiveness model by Porter to analyze the international competitiveness of China's industries. In his view: the research vision of international competitiveness of the industry should focus on the areas that economic analysis is easier to grasp and the causal relationship that is relatively clear. Therefore, we must first started the study from the international competitiveness of industrial products, from the market share of domestic industrial products and profit status and its direct and indirect determinant factors and established an economic analysis model that suitable for the actual situation of China's industrial development and easy to conduct more in-depth international comparative study. As Fig. 4, this model thought that the international competitiveness of the industrial strength in a certain country or region could be analyzed from the reasons and results [11].

5 OTHERS ANALYSIS MODEL OF THE INDUSTRIAL INTERNATIONAL COMPETITIVENESS

Based on the Diamond model by Porter and nine factors model by Cho, Zhu Chunkui established a hierarchical structure model about the competitive sources and divided it into direct sources, indirect sources and the ultimate source, a total of three levels and 16 factors [12]. The international competitiveness of the industry is divided into three forms: absolute competitiveness, comparative competitiveness and different competitiveness and established a triangular model of the industrial international competitiveness on the base of different assumptions about the analysis of the industrial international competitiveness by Zhao hongbin [13]. In addition, there are a lot of analysis model which are not enumerated here.

6 CONCLUSIONS

In short, there are many factors affecting the industrial international competitiveness, both domestic and international factors, both macro and micro factors, both within the industry and out of the industry, furthermore, a variety of factors are often intertwined, mutual influenced and interacted to become a complex network system. Therefore, these factors and their interaction mechanism should be in-depth analyzed in order to more profoundly understand the essence and connotation of industrial international competitiveness.

ACKNOWLEDGMENTS

This work is supported by Youth Project of National Social Science Fund (13CJY045); Henan Province Education Department Science and Technology Research Project (14A630027); Henan Province College Humanities and Social Science Research Project (2014-ZD-067); The Key Scientific Research Project of Henan Province (15A790032).

REFERENCES

[1] M. Porter. Competitive advantage. Beijing: Huaxia Press, 1997.
[2] Cho, Dong-Sung. A dynamic approach to international competitiveness: the case of Korea. Journal of Far Eastern Business, 1994, (1): 17–36.
[3] Rugman, A.M., D'Cruz, R., The double diamond model of international competitiveness: the Canadian experience. Management International Review, 1993, (2): 17–39.
[4] Moon, H. Chang, Rugman, A.M., Verbeke, A. A generalized double diamond approach to the global competitiveness of Korea and Singapore. International Business Review, 1998, (7): 135–150.
[5] Brian Leavy. Organization and competitiveness—towards a new perspective. Journal of General Management, 1999, (3): 13–52.
[6] Dunning, J.H., Sarianna M. Lundan. The geographical sources of competitiveness of multinational enterprises: an econometric analysis. International Business Review, 1998, (7): 115–133.
[7] Nicholas J.O'Shaughnessy. Michael Porter's competitive advantage revisited. Management Decision, 1996, (6): 12–20.
[8] Heinz, Weihrich. Analyzing the competitive advantages and disadvantages of Germany with the TOWS Matrix—an alternative to Porter's Model. European Business Review, 1999, (1): 9–22.
[9] Dryden Spring. An international marketer's view of Porter's New Zealand study. Business Quarterly, 1992, (3): 65–69.
[10] Donald N. Thompson. Porter on Canadian competitiveness. Business Quarterly, 1992, (3): 55–58.
[11] J. Bei. Competitive economics. Guangzhou: Guangdong Economic Press, 2003.
[12] Z. Chunkui. The theoretic study on industrial competitiveness. Productivity Study, 2003, (6): 182–183.
[13] Z. Hongbin. Industrial competitiveness—an overview. Modern Finance and Economics, 2004, (12): 67–70.

Architectural, Energy and Information Engineering – Sung & Chen (Eds)
© *2016 Taylor & Francis Group, London, ISBN 978-1-138-02791-6*

Dispersal Paxos

Y.Y. Chen & D.G. Li

School of Electronic and Computer Engineering, Peking University, Beijing, China

ABSTRACT: Paxos is the most widely used consensus algorithm in the distributed system where data replication is provided and replicas multiply the storage cost. In this paper, we propose a new Paxos variant called Dispersal Paxos, which introduces the Information Dispersal Algorithm (IDA) to split data into fragments and stores the fragments instead of replicas in Paxos. We propose several rules of using the IDA in Paxos, present the phases of Dispersal Paxos, and analyze the storage cost. The experiments show that the Dispersal Paxos reduces 2/3 of storage cost and communication traffic.

Keywords: Paxos; Information Dispersal Algorithm; storage cost

1 INTRODUCTION

Data replication is provided to against data lost caused by disk corruption or natural disaster in the distributed system, but data replicas will multiply the storage cost. Another problem of data replication is the replicas may be inconsistent. Paxos is an important asynchronous algorithm which is based on a message delivery model and provides consistency between replicas in the distributed system where servers may slow, crash or restart and messages may lost, delay or duplicated, but not corrupted (non-Byzantine fault). Paxos makes replicas consensus, but the replicas still multiply the storage cost in Paxos. We propose Dispersal Paxos that introduces a dispersal algorithm (Information Dispersal Algorithm, IDA) to Paxos, stores the fragments split by IDA instead of replicas to reduce storage cost and communication traffic at the mean time. In this paper, we present Classical Paxos and IDA briefly, propose Dispersal Paxos and analyze its storage cost.

2 RELATED WORK

Paxos [1,2] becomes the most widely used consensus algorithm after it was presented. Many Paxos variants are proposed to improve Paxos in different aspects. Multi-Paxos sends value stream instead that sends them one by one to reduce the message delays. Byzantine Paxos [3] verifies the values between servers to prevent the Byzantine fault. Fast Paxos [4] sends messages to acceptors directly without through the leader to reduce the message delays too. Cheap Paxos [5] introduces F auxiliary servers to tolerate the F failures of main servers. Egalitarian Paxos [6] decentralizes to optimal the latency.

And in this paper, we introduce IDA [7] to Paxos to reduce the storage cost and communication traffic.

3 CLASSICAL PAXOS

Paxos make data consistency through the phases of proposing data, agreeing data, and storing data. Consensus means that Paxos cannot choose two different data, that is, if data v is chosen, then Paxos can only choose v later. Any data v is chosen, it must has agreed by some majority servers, because any two majority servers have at least one server in common that make sure they cannot choose two different data. When the data v is chosen, to make sure any data agreed by servers must be v, servers only can propose v. So before server want to propose some data, it asks some majority servers what data they have agreed. If v is chosen, that is, some majority servers have agreed v, when server asks some majority servers to propose data, at least one server has agreed v and it will propose v to make sure consensus. So Paxos asks some majority servers what data they have agreed and choose the latest one to propose, send the data to some majority servers. If some majority servers agreed the data, which means that, they choose the data as consensus one, and then it broadcasts the data to servers.

4 INFORMATION DISPERSAL ALGORITHM

An Information Dispersal Algorithm (IDA) splits data of length L into n fragments, the length of each fragment is L/m (m < = n), and every m fragments can reconstruct the data. The sum of all fragments is L*n/m.

Take n = 5 and m = 3 for instance and we call this IDA (3, 5) below. The transform matrix A is a 5 * 3 matrix, every 3 * 3 sub matrix of A is an invertible matrix. Let cut source file (S) into 3 pieces, S1, S2 and S3. Destination file (D) has 5 pieces, d1, d2, d3, d4 and d5, which can split with the equation $A*S = D$.

$$\begin{bmatrix} a_{11} & a_{12} & a_{13} \\ a_{21} & a_{22} & a_{23} \\ a_{31} & a_{32} & a_{33} \\ a_{41} & a_{42} & a_{43} \\ a_{51} & a_{52} & a_{53} \end{bmatrix} * \begin{bmatrix} s_1 \\ s_2 \\ s_3 \end{bmatrix} = \begin{bmatrix} d_1 \\ d_2 \\ d_3 \\ d_4 \\ d_5 \end{bmatrix} \quad (1)$$

We can get 5 pieces of D, d1, d2, d3, d4 and d5 from (1), but now we only look at 3 pieces of them (d1, d2 and d3 for example), and we can know the equation below is still right.

$$\begin{bmatrix} a_{11} & a_{12} & a_{13} \\ a_{21} & a_{22} & a_{23} \\ a_{31} & a_{32} & a_{33} \end{bmatrix} * \begin{bmatrix} s_1 \\ s_2 \\ s_3 \end{bmatrix} = \begin{bmatrix} d_1 \\ d_2 \\ d_3 \end{bmatrix} \quad (2)$$

We can still short (2) to the equation $A*S = D$, but now A is a 3 * 3 invertible matrix and D has just 3 pieces. So we can reconstruct S with the equation $S = A^{-1}*D$, that is,

$$\begin{bmatrix} s_1 \\ s_2 \\ s_3 \end{bmatrix} = A^{-1} * \begin{bmatrix} d_1 \\ d_2 \\ d_3 \end{bmatrix} \quad (3)$$

So we can split the source file of length F into 5 fragments of length F/3, 5F/3 in total, and every 3 fragments can reconstruct the source file.

5 DISPERSAL PAXOS

We use IDA to split the data into fragments and store the fragments instead of replicas to reduce storage cost in Classical Paxos.

5.1 *Introduction of Dispersal Paxos*

We describe the rules when we take IDA to Classical Paxos below.

1. Servers and fragments need to numbered and the fragments only can be transmit to servers with the corresponding number.
   We numbered servers and fragments and only transmit fragments to the corresponding number servers to prevent the same fragment stores in the different servers which may happened in the situation like a fragment transmit to a server in one Paxos round, and the same fragment transmit to another server in the same or followed Paxos round. The two same fragments in two different servers will be considered as one fragment when we use the fragments to reconstruct the data. And we must unicast different fragments to different servers, rather than broadcast same replicas to all servers in Classical Paxos.

2. The transform matrix A in IDA must same when we run the Dispersal Paxos on same data more than once.
   We cannot change the transform matrix when we run Dispersal Paxos more than once, because different transform matrix on the same data will have different fragments, and these different fragments cannot reconstruct the data back. Since the Paxos allow to be run more than once on the same data and still maintain consensus, that is, we can still run Paxos on the same data even it has reached the consensus. For example, Paxos has reached the consensus, but not all servers stored the consensus data, when we access the server without the consensus data, it can run Paxos again to return the consensus data. So, we cannot change the transform matrix on some data, because it may run Paxos again even it has reached the consensus. And we still need the transform matrix to reconstruct the data. For convenience, we can just use the same transform matrix on all data in the system.

3. The fragments should be labeled with which data they belong to.
   In Dispersal Paxos, fragments instead of replicas stored in servers and the fragments are different even they are split from the same data, so we should labeled the fragments with which data they belong to and let servers know which data's fragment they stored to agree data to reach consensus. We also must know which data the fragments belong to and make sure that we won't get fragments from the different data's together to reconstruct the data. We can label the fragments with data digest and transmit the fragments to servers with data digest.

4. Some majority of servers should agree the data to choose it as consensus data because any two majorities have at least one server in common in Classical Paxos. And we need at least $\frac{n+m}{2}$, $2x = n + m$ servers agree the data to choose it as consensus data if we use IDA (m, n) in Dispersal Paxos.
   If we have 5 servers, then a majority of servers, that is 3 servers, should agree the data to choose the consensus data in Classical Paxos, because any 3 servers have at least one in common. But in Dispersal Paxos with 5 servers and IDA (3, 5), we need at least 4 servers to agree the data to choose it as consensus one, not only 3 servers,

because any 4 servers have at least 3 servers in common and 3 fragments in the common servers can reconstruct the data. If we still have 3 servers to agree the data, that is one server in common, and one fragment in that common server cannot reconstruct the data.

Let's extend it to IDA (m, n) with n servers. Then how many servers should agree the data to choose the consensus data? We suppose x servers.

$$\overbrace{\underbrace{S,\cdots,S,\overbrace{S,\cdots S}^{m},S,\cdots S}_{x}}^{n}$$

We have n servers and any two x servers have at least m servers in common, which m fragments in the common servers can reconstruct the data. We want to express x with m and n.

$$\overbrace{\underbrace{S,\cdots,S,\overbrace{S,\cdots S}^{m},S,\cdots S}_{x},\underbrace{S',\cdots S'}_{x}}^{n}\;\overbrace{S,\cdots S,S',\cdots S'}^{m}$$

We assume m virtual servers S', thus any two x servers have at least m servers in common, equals to sum of any two x servers more than sum of n servers and m virtual servers, so $2x = n + m$, that is, $x = \frac{n+m}{2}$. So we need at least $\frac{n+m}{2}$ servers agree the data to choose the consensus data in Dispersal Paxos.

We also can consider Classical Paxos as a special case of Dispersal Paxos which just one fragment, data itself, can reconstruct the data, that is, IDA (1, n). The majority of servers should agree the data in Classical Paxos, which "the majority" means more than $x = \frac{n+1}{2}$ servers.

5.2 Phases of Dispersal Paxos

We describe the phases of Dispersal Paxos with 5 servers and IDA (3, 5) for instance. We need to number the servers before run Dispersal Paxos.

In the preparation phase, client collects fragments with the data digest instead of replicas from servers.

1. Client chooses a new ballot number and sends it to servers to run a new round of Dispersal Paxos.
2. Upon receipt of message with ballot number from client, servers send fragments (v1-v5) with the data digest v' which is agreed latest to client.

    In the agreement phase, server received the fragments with the data digest from client and vote for the corresponding data.
3. After receiving the fragments with the data digest from at least 4 servers. If more than 3 fragments with same data, that is same data

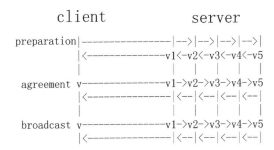

Figure 1. Phases of Dispersal Paxos.

digest, are received, client will reconstruct the data with the fragments, or it will choose a new data, then use IDA (3, 5) to split the data into fragments with numbers and transmit the fragments with the data digest to the corresponding number servers.
4. Upon receipt of fragments with data digest, servers cast their votes and sent the votes back to client.

    In the broadcast phase, client broadcasts the fragments with the data digest to servers.
5. If client receives the votes from at least 4 servers, that is, the data are chosen as the consensus data. Then client broadcasts the fragments with the data digest to servers.
6. Upon receiving the fragments with the data digest of consensus data, servers send acknowledgement messages back to client.

5.3 Analysis of storage cost

Classical Paxos with 5 servers need at least 3 servers store the data to make it consensus, 5 servers at top, that is, storage cost is 3 to 5 times of data. Dispersal Paxos with 5 servers and IDA (3, 5), each fragment's length is 1/3 of data, at least 4 servers store the fragments to make it consensus, 5 servers at top, that is, storage cost is 4/3 to 5/3 times of data. When all servers store the data or fragments, storage cost is 5 times of data in Classical Paxos and 5/3 times of data in Dispersal Paxos, that is, storage cost of Dispersal Paxos is 1/3 of Classical Paxos. When the minimum servers store data or fragments still maintain consistency, storage cost is 3 times of data in Classical Paxos and 4/3 times of data in Dispersal Paxos, that is, storage cost of Dispersal Paxos is 4/9 of Classical Paxos. Under the same conditions, which means that, under the minimum and maximum servers it takes to maintain consistency, storage cost of Dispersal Paxos is 1/3 to 4/9 of Classical Paxos.

Let's extend it to IDA (m, n) with n servers. Classical Paxos needs at least some majority of servers store the data to make it consensus, n servers at top, that is, storage cost is $\frac{n+1}{2}$ to n times of data. Dispersal Paxos with n servers and IDA (m, n), each fragment's length is 1/m of data, at least $\frac{n+m}{2}$ servers store the fragments to make it consensus, n servers at top, that is, storage cost is $\frac{n+m}{2m}$ to $\frac{n}{m}$ times of data. So under the same conditions, storage cost of Dispersal Paxos is $\frac{1}{m}$ to $\frac{m+n}{m(n+1)}$ of Classical Paxos.

6 EXPERIMENTS

We implement Dispersal Paxos presented above on the distributed system of 1client and 5 servers, program above the layer of libpaoxs [8], take IDA (3, 5) as our dispersal algorithm.

We can know from analysis of storage cost above, under the same conditions, storage cost of Dispersal Paxos is 1/3 to 4/9 of Classical Paxos. We can see from Figure 2, storage cost of Classical Paxos and Dispersal Paxos is almost proportional to the data that we store, and storage cost of Dispersal Paxos also is almost proportional to Classical Paxos, the ratio is close to 1/3, less than 4/9. The reason is that when we run Paxos, if first round of Paxos cannot make consistency, we will continue to run Paxos again try to make it consensus instead of waiting the data are accessed and then run Paxos to make it consensus. So, we do not run Paxos once, more servers even all servers store data or fragments, and we can know from analysis of storage cost above, the ratio of Dispersal Paxos to Classical Paxos is close to 1/3 on that situation.

When the size of data is large enough, the communication traffic of Paxos mainly associated with the data, and the data digest, message head, the other commands in the communication of Paxos play negligible or minor roles. We can see from Figure 3, communication traffic of Dispersal Paxos is almost proportional to Classical Paxos, the ratio is close to 1/3. We also can see the ratio

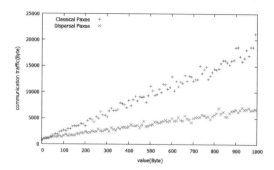

Figure 3. Communication traffic of Classical Paxos and Dispersal Paxos.

of communication traffic is more random than storage cost. Because the storage cost is almost the same no matter how many round of Paxos are run to make it consensus, but running extra round needs extra communication traffic.

7 CONCLUSION

In this paper, we briefly presented Classical Paxos and IDA first, and proposed Dispersal Paxos that introduce IDA to Paxos, store fragments instead of replicas. We proposed several rules of using IDA in Paxos, proposed the phases of Dispersal Paxos, and analyzed the storage cost. The experiments also show that when IDA (3, 5) used in the Dispersal Paxos of 5 servers, storage cost and communication traffic will reduce about 2/3 compared with Classical Paxos.

REFERENCES

[1] Leslie Lamport. The part-time parliament. ACM Transactions on Computer Systems, 16(2): 133–169, 1998.
[2] Leslie Lamport. Paxos made simple. ACM SIGACT News (Distributed Computing Column), 32(4): 18–25, December 2001.
[3] Castro M, Liskov B. Practical Byzantine fault tolerance. OSDI. 99: 173–186, 1999.
[4] Lamport L. Fast paxos. Distributed Computing, 19(2): 79–103, 2006.
[5] Lamport L, Massa M. Cheap paxos. Dependable Systems and Networks, International Conference on IEEE, 307–314, 2004.
[6] Iulian Moraru, David G Andersen, Michael Kaminsky. Egalitarian Paxos. 2012.
[7] Rabin M O. Efficient dispersal of information for security, load balancing, and fault tolerance. Journal of the ACM (JACM), 1989, 36(2): 335–348.
[8] Primi M. Paxos made code. Master thesis, Informatics of the University of Lugano, 2009.

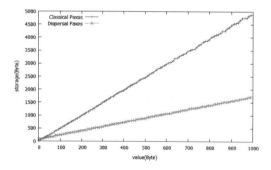

Figure 2. Storage cost of classical paxos and dispersal paxos.

Author index

796